W J Riddick

MODERN
WELDING
TECHNOLOGY

MODERN WELDING TECHNOLOGY

HOWARD B. CARY

Vice President
Welding Systems
Hobart Brothers Company
also
Vice President
American Welding Society

PRENTICE-HALL, INC., Englewood Cliffs, New Jersey 07632

Library of Congress Cataloging in Publication Data
Cary, Howard B.
    Modern welding technology.

    Includes bibliographical references and index.
    1. Electric welding.   2. Welding.
I. Title.
TK4660.C37   671.5′2   78-2966
ISBN 0-13-599290-7

EW-498

Editorial production supervision
and interior design by: James M. Chege

Page layout by: Rita K. Schwartz

Manufacturing buyer: Gordon Osbourne

Printed in the United States of America

10 9 8 7 6 5 4 3

Prentice-Hall International, Inc., *London*
Prentice-Hall of Australia Pty. Limited, *Sydney*
Prentice-Hall of Canada, Ltd., *Toronto*
Prentice-Hall of India Private Limited, *New Delhi*
Prentice-Hall of Japan, Inc., *Tokyo*
Prentice-Hall of Southeast Asia Pte. Ltd., *Singapore*
Whitehall Books Limited, *Wellington, New Zealand*

CONTENTS

PREFACE *ix*

1 WELDING BACKGROUND *1*

1-1 The Importance of Welding   *1*
1-2 Welding Joins All Metals   *3*
1-3 Where Welders Work   *5*
1-4 Historical Development of Welding   *11*
1-5 The Welding Industry and Its Future   *16*
Questions   *20*
References   *20*

**2 THE FUNDAMENTALS
    OF WELDING**   *21*

2-1 Basic Definitions   *21*
2-2 The Welding Processes and Grouping   *25*
     Arc welding   *26*
     Brazing   *30*
     Oxy fuel gas welding   *30*
     Resistance welding   *30*
     Other welding processes   *30*
     Soldering   *31*
     Solid state welding   *31*
2-3 Welders and Welding Operators — The

Method of Applying   *31*
2-4 Welding Procedures   *33*
2-5 Welding Electricity   *35*
2-6 Welding Arcs   *38*
2-7 The Physics and Chemistry of
     Welding   *42*
Questions   *44*
References   *45*

**3 WELDING PERSONNEL SAFETY,
    TRAINING, AND
    QUALIFICATION**   *47*

3-1 The Welding Occupation   *47*
3-2 The Safety and Health of Welders   *49*
     Electrical shock   *50*
     Arc radiation   *52*
     Air contamination   *55*
     Fire and explosion   *58*
     Compressed gases   *62*
     Cleaning and chipping welds and other
     hazards   *63*
     Welder protection   *64*
     Safety precautions for arc
     welding   *64*

CONTENTS

       Safety precautions for oxyacetylene
       welding and cutting   *64*
3-3   Welder Training Programs   *65*
3-4   Training School Facilities   *67*
3-5   Qualifying and Certifying Welders   *69*
3-6   Training and Qualifying the Other
       Welding Personnel   *70*
Questions   *73*
References   *73*

4   THE ARC WELDING PROCESSES—USING CC POWER   *75*

4-1   Welding With Constant Current   *75*
4-2   The Shielded Metal Arc Welding
       Process—SMAW   *79*
4-3   The Gas Tungsten Arc Welding
       Process—GTAW   *95*
4-4   The Plasma Arc Welding Process
       (PAW)   *107*
4-5   Carbon Arc Welding Process
       (CAW)   *115*
4-6   The Stud Arc Welding Process
       (SW)   *119*
4-7   Other Processes Using Constant
       Current Power   *126*
Questions   *129*
References   *130*

5   THE ARC WELDING PROCESSES—USING CV POWER   *131*

5-1   Welding With Constant Voltage   *131*
5-2   Submerged Arc Welding (SAW)   *134*
5-3   Gas Metal Arc Welding (GMAW)   *146*
5-4   Flux-Cored Arc Welding (FCAW)   *160*
5-5   Arc Welding Variables   *169*
5-6   Electroslag Welding Process   *178*
5-7   Selecting the Correct Arc Welding
       Process   *192*
Questions   *196*
References   *196*

6   THE OTHER WELDING PROCESSES   *197*

6-1   Brazing   *197*
6-2   Oxy Fuel Gas Welding   *206*
6-3   Resistance Welding   *213*
6-4   Solid State Welding   *221*
6-5   Soldering   *227*

6-6   Electron Beam Welding   *231*
6-7   Laser Beam Welding   *234*
6-8   Thermit Welding   *237*
6-9   Miscellaneous Welding Processes   *239*
Questions   *241*
References   *241*

7   PROCESSES RELATED TO WELDING   *243*

7-1   Adhesive Bonding   *243*
       Types of adhesive for bonding   *244*
7-2   Arc Cutting   *245*
7-3   Oxygen Cutting   *251*
7-4   Other Thermal Cutting Processes   *258*
7-5   Thermal Spraying   *259*
7-6   Heat Forming and Straightening   *262*
7-7   Pre- and Postheat Treatment   *264*
7-8   Joining of Plastics   *268*
Questions   *269*
References   *270*

8   POWER SOURCES FOR ARC WELDING   *271*

8-1   Types of Welding Power Sources   *271*
8-2   Generator Welding Machines   *276*
8-3   Transformer Welding Machines   *279*
8-4   Rectifier Welding Machines   *283*
8-5   Multiple Operator Systems   *288*
8-6   Selecting and Specifying
       a Power Source   *291*
8-7   Installation and Maintenance
       of Power Sources   *292*
Questions   *297*
References   *298*

9   WELDING EQUIPMENT   *299*

9-1   Electrode Holders and Cables   *299*
9-2   Electrode Wire Feeders   *307*
9-3   Arc Motion and Standardized
       Machines   *313*
9-4   Fixtures and Positioners   *317*
9-5   Custom Fixtures and Automatic
       Welding   *322*
9-6   Automated Welding   *329*
9-7   Auxiliary Equipment   *332*
Questions   *334*
References   *334*

10   FILLER METALS AND MATERIALS FOR WELDING   *335*

10-1   Types of Welding Materials   *335*
10-2   Covered Electrodes   *338*
10-3   Solid Electrode Wires   *345*

10-4 Flux-Cores or
    Tubular Electrodes   *348*
    Mild steel electrodes   *351*
    Stainless steel tubular wires   *352*
10-5 Fluxes for Welding   *352*
10-6 Other Filler Metals   *355*
10-7 Shielding Gases Used
    in Welding   *356*
10-8 Fues Gases for Welding
    and Cutting   *361*
10-9 Gas Containers and Apparatuses   *364*
Questions   *369*
References   *369*

11 WELDABILITY AND PROPERTIES OF METALS   *371*

11-1 Properties of Metals   *371*
11-2 Metal Specifications
    and Groupings   *380*
11-3 Identification of Metals   *381*
11-4 Heat and Welding   *385*
11-5 Weldability of Metals   *401*
Questions   *403*
References   *404*

12 WELDING THE STEELS   *405*

12-1 Steel Classifications   *405*
12-2 Welding Low-Carbon
    and Low-Alloy Steels   *411*
12-3 Welding the Alloy Steels   *414*
12-4 Welding Stainless Steels   *419*
12-5 Welding Ultra-High
    Strength Steels   *426*
Questions   *429*
References   *430*

13 WELDING THE NONFERROUS METALS   *431*

13-1 Aluminum and Aluminum Alloys   *431*
13-2 Copper and Copper-Base Alloys   *442*
13-3 Magnesium-Base Alloys   *447*
13-4 Nickel-Base Alloys   *451*
13-5 Reactive and Refractory Metals   *455*
13-6 The Other Nonferrous Metals   *458*
Questions   *459*
References   *460*

14 WELDING SPECIAL AND DISSIMILAR METALS   *461*

14-1 Cast Iron and Other Irons   *461*
14-2 Tool Steels   *465*
14-3 Reinforcing Bars   *469*
14-4 Coated Steels   *472*
14-5 Other Metals   *474*
14-6 Clad Metals   *477*
14-7 Dissimilar Metals   *481*
Questions   *485*
References   *485*

15 DESIGN FOR WELDING   *487*

15-1 Advantages of Welded
    Construction   *487*
15-2 Weldment Design Factors   *490*
15-3 Welding Positions, Joints, Welds,
    and Weld Joints   *499*
15-4 Weld Joint Design   *502*
15-5 Weldment Design and Influence of
    Specifications   *523*
15-6 Weldment Conversion from Other
    Designs   *527*
15-7 Weldment Design Problems   *533*
15-8 Welding Symbols   *535*
Questions   *540*
References   *540*

16 THE COST OF WELDING   *541*

16-1 The Elements of Weldment Cost   *541*
16-2 Weld Metal Required For Weld Joints   *543*
16-3 Filler Metal and Materials Required   *547*
16-4 Time and Labor Required   *551*
16-5 Power and Overhead Costs   *553*
16-6 Total Weld Cost Formulas
    and Examples   *554*
16-7 Reducing Welding Costs   *557*
Questions   *559*

17 WELDMENT EVALUATION AND QUALITY CONTROL   *561*

17-1 Weld Reliability   *561*
17-2 Welding Codes and Specifications   *564*
17-3 Welding Procedures and Qualifying
    Them   *568*
17-4 Weld Testing—Destructive Testing   *584*
17-5 Visual Inspection   *590*
17-6 Nondestructive Examination   *593*
17-7 Weld Defects—Causes and
    Corrective Actions   *601*
17-8 Workmanship Specimens,
    Standards, and Inspection Symbols   *613*
Questions   *620*
References   *620*

CONTENTS

18 WELDING PROBLEMS AND SOLUTIONS   623

18-1   Arc Blow   623
18-2   Welding Distortion and Warpage   626
18-3   Weld Stresses and Cracking   636
18-4   In-Service Cracking   639
18-5   Weld Failure Analysis   644
18-6   Other Welding Problems   651
Questions   653
References   653

19 WELDING FOR REPAIR AND SURFACING   655

19-1   The Need For Weld Repair and Surfacing   655
19-2   Analyze and Develop Rework Procedure   656
19-3   Repair Welding   659
19-4   Rebuilding and Overlay   663
19-5   Surfacing for Wear Resistance   666
19-6   Surfacing For Corrosion Resistance   671

19-7   Other Surfacing Requirements   673
Questions   674
References   674

20 SPECIAL APPLICATIONS OF WELDING   675

20-1   Pipe Welding   675
20-2   Arc Spot Welding   685
20-3   One-Side Welding   690
20-4   Underwater Welding   694
20-5   Narrow Gap Welding   698
20-6   Gravity Welding   700
20-7   Remote Welding   702
20-8   Sheet Metal Welding   704
Questions   707
References   708

APPENDIX   711

A-1   Glossary of Welding Terms   711
A-2   Organizations Involved With Welding   715
A-3   Conversion Information   718
A-4   Weights and Measures   721

INDEX   723

PREFACE

This book removes the mystery of welding and presents welding as a technology rather than an art. Welding technology is becoming more complex each year. New welding processes and methods are being introduced which provide greater capabilities for joining metals. Welding is still growing and this book will enable the reader to acquire the necessary know-how to grow with it. It provides the necessary practical information for people involved in welding and responsible for welding operations.

The book explains all of the welding processes, and will assist in selecting the optimum process of any metal-joining situation. The major emphasis is on the arc welding processes since they are the most widely used in industry and construction, and relates all the arc welding processes to manual covered electrode welding which is still the most popular, and explains the similarities and differences.

The safety, training, testing and qualification of welding personnel receives much attention. The book explains the different equipment required to use the arc welding processes and assists in selecting and specifying the equipment. Emphasis is on the welding of steel since it is the most widely used metal. However, the weldability of other metals is discussed along with the selection of filler metals and shielding methods for joining them. The development of welding procedures and the use of welding schedules for automatic welding is explained.

The design considerations for welds are covered and cost estimating and calculating is presented in detail. Weld quality and evaluation including nondestructive examination, testing and quality control and weld problems are given broad coverage. Finally, the special application and the use of welding by different industries is included.

The book reflects current practice in North America and utilizes the terminology, standards and specifications established by the American Welding Society. However, standards and practices of other industrial countries are considered since welding is an international technology. In view of this and in the interests of wider understanding, metric references are included.

The book contains many tables on welding procedure which, if followed, would result in quality welds. However, since the expected results and production conditions vary widely, it is highly recommended that the data from the tables be used as a starting point when establishing procedure. A new procedure should be thoroughly tested to determine that it meets established goals before it is accepted.

A systematic program of analyzing and answering the welding questions will evolve and will allow the reader to tackle and solve any welding problem.

I wish to thank the American Welding Society for permission to utilize data from its many standards, codes and specifications. I also want to thank Hobart Brothers Company for permission to use many photographs and technical data. Finally, I wish to acknowledge the cooperation of many people and organizations for their assistance in supplying data, pictures, etc., for the book.

Troy, Ohio **Howard B. Cary**

MODERN
WELDING
TECHNOLOGY

1

WELDING BACKGROUND

1-1 THE IMPORTANCE OF WELDING

1-2 WELDING JOINS ALL METALS

1-3 WHERE WELDERS WORK

1-4 HISTORICAL DEVELOPMENT

1-5 THE WELDING INDUSTRY
     AND ITS FUTURE

1-1 THE IMPORTANCE OF WELDING

Welding is the most important way to join metals. It is the only way to join two or more pieces of metal to make them act as one piece. Welding is widely used to manufacture or repair all products made of metal. Look around, almost everything made of metal is welded; the world's tallest building, moon rocket engines, nuclear reactors, home appliances, and automobiles barely start the list.

The use of welding is still increasing, primarily because it is the most economical and efficient way to join metals. If a joint is welded it is a permanent joint. Obviously, if the joint must be disassembled occasionally it should not be welded. Thus we should change our statement to "welding is the most economical method to permanently join metal parts." To join two members by bolting or riveting requires holes in the parts to accommodate the bolts or rivets. These holes reduce the cross-sectional area of the members to be joined by up to 10%. The joint may also require the use of one or two gusset plates as well as the bolts or rivets thus increasing the amount or weight of material required and the cost. This expense can be eliminated by the use of a weld. The greatest economy of a welded design will be obtained if the cross-sectional area of the entire structural member is reduced by the amount of the bolt holes. This can be done since the entire cross section of a member of a welded design is utilized to carry the load. The amount of material required is reduced as well as its cost, as is shown by Figure 1-1. This same design concept applies to joining plates used to build a ship or a container. In view of this material savings, ships and storage tanks are no longer riveted.

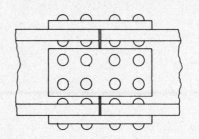
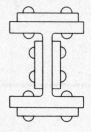

RIVETED SPLICE IN WIDE FLANGE MEMBER

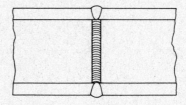

WELDED SPLICE IN WIDE FLANGE MEMBER

FIGURE 1-1 *Comparison of welded and riveted structural joints.*

Pipe joining by welding offers similar economies. The wall thickness of a pipe should be heavy enough to carry the required load. However, if the pipe is joined by screw threads a heavier wall thickness is used to allow for cutting away a portion of the thickness for the threads, as shown by Figure 1-2. The thinner pipe wall thickness is used for the entire welded pipe system. This reduces the amount of metal required and the cost. The inside surface of the welded joint is also smoother. Large diameter pipes are no longer connected together with screw threads and fittings.

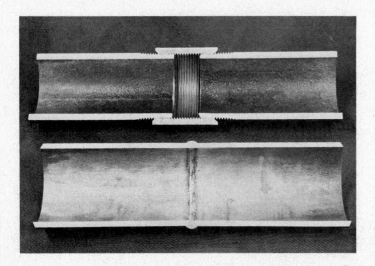

FIGURE 1-2 *Pipe joints welded and threaded.*

Converting castings to weldments allows the designer to reduce weight by reducing metal thickness. In a sense, welding is a design concept which allows freedom and flexibility not possible with cast construction. Heavy plates can be used where strength is required and thin ones can be used where possible. The uniform thickness rule and minimum thickness required for foundry practice are not necessary for weldments. Additionally, high strength materials can be used in specific areas while normal strength materials are used where required.

Welding is the best way to protect and conserve materials by protecting their surface with special metal overlays. Corrosion and wear of metals account for losses running into billions of dollars annually. Together they are responsible for an untold loss of lives. Waste from both of these destructive forces can be greatly reduced by welding. Special alloys are weld-deposited on base metals to provide corrosion-resistant surfaces. Also, hard surfacing overlays can be made by welding

to provide special alloys with wear-resistant surfaces. A typical application is shown by Figure 1-3 in which a cinder crusher roll is resurfaced with hard weld metal. Weld surfacing is used to reduce the costly abrasive and corrosive wear of machinery.

FIGURE 1-3 *Overlayed part—cinder crusher roll.*

There are many ways to make a weld and there are many different kinds of welds. The welder behind the hood making sparks is using one of the more popular welding processes, known as arc welding. Some welding processes do not cause sparks; in some cases electricity is not used, and in some cases there is not even extra heat. Welding has become complex and technical. It requires considerable knowledge to select the proper welding process for critical work. Our job is to learn enough about welding so we can utilize its many advantages. Some of these advantages are:

Welding is the lowest-cost joining method

It affords lighter weight through better utilization of materials

It joins all commercial metals

It can be used anywhere

It provides design flexibility

It is also important to know the limitations of welding. The limitations are:

Much welding depends on the human factor

It often requires internal inspection

It is restricted by some specifications and codes

Most of these limitations can be overcome by means of good controls and normal supervision.

Welding is also an economical manufacturing method of almost any type of operation wherever there is a need to perform a joining function. In the high-volume production industries it is not uncommon to see welding operations intermixed with bending, machining, forming, assembly, etc. Welding is an important manufacturing process taking its place with other metalworking operations to help bring us quality metal products at economical prices.

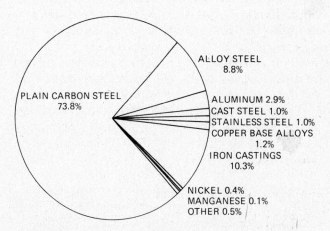

FIGURE 1-4 *Metal production by types in the United States.*

1-2 WELDING JOINS ALL METALS

All metals can be joined by one welding process or another. There is a saying, "If it's metal, weld it," and it is certainly true. However, to state that all metals can be welded should be qualified by stating that all metals commercially used for structural or strength parts are weldable. Some metals are easy to weld, and others are difficult to weld. The metals that are easily weldable can be welded in thickness from the very thinnest, about the thickness of this paper page, to the thickest or heaviest produced. The difficult-to-weld metals require special procedures and techniques that must be developed for specific applications.

Some metals may never be welded or joined. Later chapters will provide some of the properties of metals such as melting temperature, density, thermal conductivity, tensile strength, ductility, etc. These properties provide hints as to whether or not a metal can be welded. For example, mercury is a liquid at room temperature and cannot be welded, while sodium and potassium melt just below the temperature of boiling water and are of no use as a strength member and cannot be welded. In general, metals that have a low melting temperature or low strength could not be welded. Furthermore, some metals are so scarce or expensive that they would not be used where welding would be required.

The physical and mechanical properties, availability, and price all help determine if a metal will be used in applications in which welding is required.

The more abundant and stronger metals are those most frequently welded. Figure 1-4 shows the commercial metals according to their annual production in the United States.[1,2] This chart may be somewhat misleading since many metals used in the U.S.A. are im-

ported. However, the ratio of metals is important and it is believed that these ratios represent usage and are similar in most of the industrialized countries.

Plain carbon steel is by far the most widely used metal. For this reason we provide maximum information for welding ordinary mild steel. Iron castings are shown to be the second largest type of metal produced. Most iron castings are used without welding; however, on occasion iron castings are joined or repaired and so it is necessary to know how the various types of cast iron are welded. Alloy steels make up the next largest group of metals. This group encompasses many special types from the low-alloy high-strength steels, the heat-treated steels, the ultrahigh-strength steels and many other classes. Each type of alloy steel involves different welding procedures which must be closely followed for successful welding.

Aluminum and aluminum alloys represent the next largest group. Aluminum is continually finding wider application, especially when weight is a factor. The different aluminum alloys have different properties and require different procedures for welding.

Copper and copper base alloys such as brasses and bronzes make up the next largest group. These metals are used when electrical conductivity, corrosion resistance, or heat conductivity is important. These metals are often joined and require special attention.

Stainless steel and cast steels are tied for the next position. Cast steels are normally welded like rolled steels of the same composition. There are many different types of stainless steels having specific properties. Welding procedures have been specifically developed for joining each type.

3

Nickel and nickel alloys represent the next largest group. This is a small percentage of the total, but it is very important since certain nickel alloys are the best metal to use for certain types of service.

The magnesium group is the smallest, though important. Magnesium is the lightest structural metal and has many applications when welding is required.

The remaining metals represent a small percentage of the total, but they are used when their particular characteristic is required and these applications often require that they be welded.

Steel is produced in many different forms. Figure 1-5 shows the production ratios of these various forms of steel.[3] Sheet steel represents the greatest volume. This isn't surprising considering that most of the steel used in automobiles, trucks, buses, home appliances, office machinery, and furniture is sheet steel. It is also used for many containers, cans, and so on, and much industrial and electrical machinery. The welding of sheet metal involves many of the arc welding processes and is a very important segment of the welding industry.

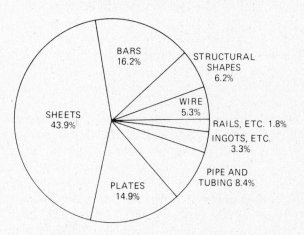

FIGURE 1-5 *Steel production by types in the United States.*

Steel bars represent the second largest product form. Many bars are machined or cut and used, but a sizable portion of bars go with structural shapes, the fourth largest group, and are used in the construction industry for building bridges, and buildings.

Steel plates represent the third largest group and are used to make tanks, boilers, machinery, ships, and other weldments. In such applications, plates are normally welded.

Pipe and tubular products represent the fifth largest group. All large diameter pipe and much of the smaller pipe is joined by welding. The welding of pipe requires special techniques and procedures.

The last remaining groups—wires, rails, and ingots may not involve much welding. However, rails are now being welded for the railroads. Ingots may be rolled into other forms, and wire and rods are usually drawn to small diameter for their end use.

Thus, we can see the broad use of metal, especially the steels, and how welding relates to them. It is important that we know how to successfully weld each metal in any form. To do this it is necessary that we organize the information about each metal and provide welding procedures and schedules.

Information about each metal is provided by specifications. Most specifications are originated by engineering societies, technical groups, and trade associations, which group the metals according to composition, strength, or some other characteristics. Later chapters are devoted to metal groupings, steels, nonferrous metals, and so on, and the specifications are explained.

Table 1-1 is a summary of metals according to composition groupings. They are related to the more common welding processes, and a rating system is given which indicates how they can be welded. This is only an indication, but a chapter number reference is given in which more information is available.

All metals *can* be welded, but some are easier to weld than others. In other words, they possess **weldability**. Weldability is defined as "the capacity of a metal to be welded under the fabrication conditions imposed into a specific, suitably designed structure and to perform satisfactorily in the intended service." All metals cannot be joined by each welding process. Actually some of the newer welding processes were developed to join specific metals. Certain metals are known as "difficult to weld," which means that specific precautions and procedures are required to produce efficient joints. Other factors complicate the making of successful welds. For example, the thickness of the metal section involved is a variable or factor that must be considered. Because of this and the complexities of each process and filler metal a chapter reference is given. Each welding process chapter explains the capabilities of that process. Chapter references are given for each metal grouping with information for welding them.

Another of the major accomplishments of the welding industry has been to develop the proper materials to join these different metals and alloys. These materials, or **filler metals** as they are commonly called, fill in the weld joint and usually provide joints as strong as the metals being joined.

The term filler metals means "the metal to be added in making a welded, brazed, or soldered joint" and

TABLE 1-1   Summary of metals welded by various processes

BASE METALS WELDED / WELDING PROCESSES	Chapter For Metal	Shielded Metal Arc	Gas Tungsten Arc	Plasma Arc	Submerged Arc	Gas Metal Arc	Flux Cored Arc	Electroslag	Braze	Gas
Aluminums	13-1	C	A	A	No	A	No	Exp	B	B
Copper base alloys										
Brasses	13-2	No	C	C	No	C	No	No	A	A
Bronzes	13-2	A	A	B	No	A	No	No	A	B
Copper	13-2	C	A	A	No	A	No	No	A	A
Copper nickel	13-2	B	A	A	No	A	No	No	A	A
Irons										
Cast, malleable, nodular	14-1	A	B	B	No	B	B	No	A	A
Wrought iron	14-1	A	B	B	A	A	A	No	A	A
Lead	13-6	No	B	B	No	No	No	No	No	A
Magnesium	13-3	No	A	B	No	A	No	No	No	No
Nickel base alloys										
Inconel	13-4	A	A	A	No	A	No	No	A	B
Nickel	13-4	A	A	A	C	A	No	No	A	A
Nickel silver	13-4	No	C	C	No	C	No	No	A	B
Monel	13-4	A	A	A	C	A	No	No	A	A
Precious metals	13-6	No	A	A	No	Exp	No	No	A	B
Steels										
Low carbon steel	12-2	A	A	A	A	A	A	A	A	A
Low alloy steel	12-3	A	A	A	A	A	A	A	A	A
High and medium carbon	12-3	A	A	A	B	A	A	A	A	A
Alloys steel	12-3	A	A	A	B	A	A	A	A	A
Stainless steel	12-4	A	A	A	A	A	B	A	A	A
Tool steels	13-4	A	A	A	No	C	No	No	A	A
Titanium	13-5	No	A	B	Exp	A	No	No	No	No
Tungsten	13-5	No	B	B	No	No	No	No	No	No
Zinc	13-6	No	C	C	No	No	No	No	No	C
Chapter For Welding Processes		4-2	4-3	4-4	5-2	5-3	5-4	5-5	6-1	6-2

Metal or Process Rating: A = Recommended or easily weldable, B = Acceptable but not best selection or weldable with precautions, C = Possible-usable but not popular or restricted use or difficult to weld, No = Not recommended or not weldable.

includes electrodes and welding rods. The term can be broadly interpreted to cover shielding methods including fluxes and gases. The filler metals and materials for welding will be explained in Chapter 10.

1-3 WHERE WELDERS WORK

The application of welding is practically unlimited. Welding is used everywhere. Almost everything we use in our daily life is welded, or is made by equipment that is welded. It is used by all industries that use metal and many that do not. The relative amount of welding used by different industries can be estimated based on U.S. government population statistics. The government population figures show the number and percentage of welders and flamecutters employed by industry groups according to their Standard Industrial Classification (SIC). This classification system[4] of the U.S. government is used to provide categories which are groups of establishments primarily engaged in manufacturing the same or similar products. The number of welders and flamecutters employed by these different industry groups is based on 1970 population data[5] provided by the U.S. Department of Labor. This listing, shown by Table 1-2, gives the rank order of the amount of welding done within the different industrial groups. Figure 1-6 is a more graphic presentation showing the number of welders and flamecutters employed by each group and the percentage of the total welders and flamecutters employed in the U.S.A. The ranking of the order of industries shown by the table is based on the assumption that the greatest amount of welding is done by those industrial groups that employ the greatest number of welders. This assumption can be slightly misleading

TABLE 1-2 Number of welders and flamecutters employed by various industry groups.

Rank Order	Industry Grouping and Definition	SIC No.	Welders & Flamecutters %	Number
1	Fabricated Metal Products	34	16.8	90,268
2	Machinery, Except Electrical	35	15.7	84,266
3	Construction	15-17	10.3	55,416
4	Motor Vehicles and Equipment	371	7.7	41,383
5	Aircraft, Ships, Railroad Equipment	372, 3 and 4	7.6	40,405
6	Primary Metal Industries	33	7.2	38,711
7	Electrical Machinery, Equipment and Supplies	36	6.0	32,638
8	Transportation, Communications and Utility Services	40-49	5.9	31,503
9	Repair Services, Auto and Misc.	75 and 76	5.1	27,500
10	Wholesale Trade	50	3.7	19,991
11	Mining	10-14	2.2	12,025

since there are many persons in industry who do welding, but may not be classified as welders but instead are called boilermakers, ironworkers, plumberfitters, and so on. Additionally, the category **flamecutter** involves the oxygen-flamecutting process, but may not involve actual welding. From the surveys available these data are the most reliable.

Each of the industries shown by the different Standard Industrial Classification numbers uses welding extensively—but in different ways under different working conditions on different metals and under different codes and specifications. In spite of this, the following statistics have emerged.

The greatest number of welders are employed in maintenance and repair welding, as determined by combining the ninth rank group, "Repair Services—Auto and Miscellaneous," SIC 75 and 76 with the sixth ranked group, "Primary Metal Industries," SIC 33 and the eighth ranked group, "Transportation, Communication, and Utility Services," SIC 40 through 49.

The repair services group is involved almost entirely with maintenance and repair welding. The welding work done in the primary metal industries is concerned primarily with maintenance and repair. Some of it, however, might be done on the product, that is, the casting that needs to be weld-repaired before it is finished. Also, an analysis of the utility services category, which includes gas and water distribution, electrical distribution, railroads, buslines, truck lines, and so on, reveals that the major use of welding is repair and maintenance. Combining these three groups we find that approximately 18% of the welders are engaged in maintenance and repair welding. Also, some of the welding in all industries involves maintenance and repair work. Since maintenance and repair are such important welding activities Chapter 19 will be completely devoted to it.

The single largest number of welders (16.7%) in one group is employed by the "Fabricated Metal Products" group. This industry group includes manufacturers of

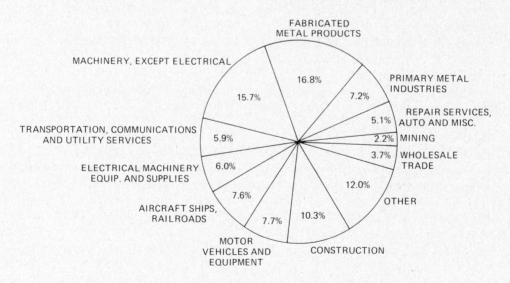

FIGURE 1-6 *Percentages of welders and flamecutting—industry groups.*

FIGURE 1-7 *Fabricated metal—large tank being welded.*

FIGURE 1-8 *Machinery—power shovel, lower frame.*

pressure vessels, nuclear reactors, tanks, heat exchangers, large diameter pipes, plate trusses, weldments, sheet metal work, prefabricated metal buildings, architectural and ornamental metal work, and so on. Most of this work is governed by codes. A typical application is Figure 1-7, which shows a tank being welded. Sheet metal work is different since it uses different processes and procedures, even though it is within this Standard Industrial Classification group.

The second largest number of welders (15.6%) is employed by the group known as "Machinery, Except Electrical." This group includes manufacturers of farm machinery, construction, mining, and materials handling equipment such as power shovels, bulldozers, cranes, and so on, metalworking machinery, such as press brakes, punches, industrial trucks, overhead cranes, food processing machinery, textile and woodworking machinery, paper-making and printing machinery, as well as office machinery. This is a broad group of companies. However, these companies are not governed by codes, but usually by company standards. Most of these products are made of steel using plates, shapes, and sheet metal. Figure 1-8 is typical and shows a heavy machinery weldment.

The third largest number of welders are employed by the "Construction" industry, which employs 10.3% of the welders. This group includes contracting companies that build tunnels, subways, dams, hydroelectric projects, power houses, chemical plants, and structural

7

steel for bridges and buildings, and welding contractors operating at the construction site (SIC 1799). The construction industry uses many welding processes under all types of conditions. Much of their work involves structural welding which is covered by codes. Figure 1-9 shows a typical field erection scene. This Standard Industrial Classification also includes pipeline contractors. However, pipe welding is quite specialized and is governed by the specific application and code involved.

FIGURE 1-9 *Construction — welding on high-rise building.*

Three types of pipe welding will be covered in Chapter 20. A power piping assembly being welded is shown by Figure 1-10.

The fourth largest number of welders (7.7%) is employed by the "Motor Vehicle and Equipment" group which is the major segment of the transportation industry. This industry group includes companies that manufacture automobiles, buses, trucks, trailers, and associated equipment. Traditionally, this industry has used welding to a very great extent. It is not governed by a welding code. In general, it involves light welding, usually on a mass production basis. It is an industry that uses semiautomatic and automatic welding almost entirely. Figure 1-11 shows welding on an auto body.

FIGURE 1-10 *Pipe work — piping subassembly.*

The fifth largest number of welders (7.6%) is employed by the remaining segments of the transportation industry. These include the "Shipbuilding and Boat Building" group, which employs 4.7%, the "Aircraft and Spacecraft" industry which employs 1.6%, and the "Railroad Equipment" group which employs 1.3% of the welders and flamecutters. The welding done by these rather diverse groups is all somewhat different.

FIGURE 1-11 *Automobiles — welding an auto body.*

In general, ship welding involves plate work and is governed by strict codes, both government and commercial. Figure 1-12 shows welding on a ship hull.

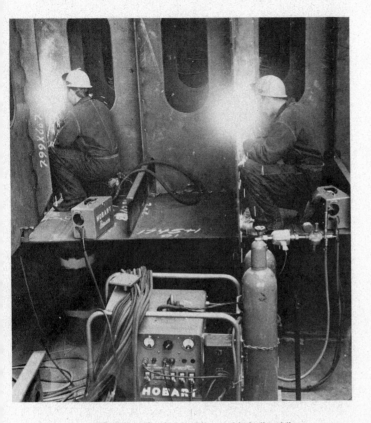

FIGURE 1-12 *Shipbuilding—ship hull welding.*

FIGURE 1-13 *Aircraft—tubular frame.*

Aircraft welding is governed by government and military standards. It is usually performed on thin, nonferrous metals. Figure 1-13 shows welding on an airplane frame. The welding in the railroad industry for rolling stock is regulated by several codes. A welder working on a railroad car is shown by Figure 1-14.

The sixth largest number of welders (7.2%) are employed by the group "Primary Metal Industries." These industries include steel mills, iron and steel foundries, smelting and refining plants. These companies produce plate, structural shapes, pipe, sheet metal, and castings of all commercial metals. Most of the welding in this group is for maintenance and repair work and was mentioned as part of the maintenance and repair group.

The seventh largest number of welders (6%) is employed by the group "Electrical Machinery Equipment and Supplies." This group includes companies which produce electrical generators, transformers, switch gear, electric motors, welding machinery, battery chargers, and household appliances. Welding done in this group runs from the heaviest to some of the lightest. Welding codes are not normally involved.

FIGURE 1-14 *Railroads—box car.*

FIGURE 1-15 *Direct metal sculpture—Tsutakawa fountain.*

The eighth largest number of welders (5.8%) are employed by the group known as "Transportation, Communications, Electric, Gas, and Sanitary Services," in other words, the transportation and utility group. This group includes the railroads, the truck lines, airlines, cross-country pipelines, and telephone companies, gas and electric utilities, and the sanitary services companies. Most of the welding done by these companies is of the repair and maintenance type.

The ninth largest group of welders (5.1%) belongs to the group known as "Automotive Repair, Services, and Garages" and "Miscellaneous Repair Services." Companies in this group would include the automobile repair shops, welding repair shops (SIC 7692), garages, and companies that specialize in the repair of heavy equipment. This group also belongs in the maintenance and repair group mentioned as the largest.

The tenth largest number is somewhat hard to determine since by actual count the "Wholesale Trades" group employs 3.7% of the welders and flamecutters, whereas the next largest group, "Mining," employs 2.2% of the welders and flamecutters. However, an analysis of the wholesale trade group indicates that the bulk of the people involved here are flamecutters working for steel service centers that perform flamecutting operations in conjunction with steel warehousing. Little welding is done. Therefore, the mining group is probably the next largest concentration of welders and here again much of the work done is of a repair and maintenance nature.

The remaining 12% of the welders and flamecutters are employed by the rest of the different industries throughout the country. The type of welding done is similar to that previously mentioned, and undoubtedly, maintenance and repair welding employs the largest number of people. Probably the least number of people that use welding are artists. The welding they do may have as great an impact as some of the larger groups.

(a)

(b)

(c)

(d)

(e)

Direct metal sculpture can be made from different metals and utilizes the more popular welding processes. Figure 1-15 shows a silicon bronze and copper nickel fountain by George Tsutakawa and the welding involved.

The geographical location of welders and flame-cutters is also of interest. The largest number of welders are employed in the Great Lakes States, but Pennsylvania ranks first with 8.1% of the welders, Ohio second with 7.7%, California third with 7.6%, Texas fourth with 7.2%, Michigan fifth with 6.9%, Illinois sixth with 6.4%, New York seventh with 4.8%, Indiana eighth with 4.1%, Wisconsin ninth with 3.2%, and Louisiana tenth with 2.8%. The Northeastern section of the country has the second largest number of welders, the Southeast ranks third, the Southwest fourth, the Western states fifth, and the Central part of the country accounts for the least number. The major metropolitan areas in rank show Detroit first with 3.4% of the welders, Chicago second with 3.3%, Los Angeles third with 3.1%, Philadelphia with 2.3%, followed by Houston 1.9%, New York 1.7%, Pittsburgh 1.7%, St. Louis 1.3%, Cleveland 1.3%, and Milwaukee 1.25%.

The analysis shows where the welders are and in which industries they are working. It also gives an indication of the type of work they do.

1-4 HISTORICAL DEVELOPMENT OF WELDING

Welding, which is one of the newer metalworking trades, can trace its historic development back to ancient times. The earliest example comes from the Bronze Age. Small gold circular boxes were made apparently by pressure welding lap joints together. It is estimated that these boxes were made more than 2,000 years ago and are presently on exhibit at the National Museum in Dublin, Ireland. During the Iron Age it appears that the Egyptians and other people in the Eastern Mediterranean area learned to weld pieces of iron together. Many tools and weapons have been found which were apparently made approximately 1000 B.C. Items of this type are on exhibit in the British Museum in London. Other examples of early welded art are displayed in the museums of Philadelphia and Toronto. Items of iron and bronze that exhibit intricate forging and forge welding operations have been found in the pyramids of Egypt.

During the Middle Ages the art of blacksmithing was developed to a high degree and many items of iron were produced which were welded by hammering. One of the largest welds from this period was the Iron Pillar of Delhi in India, which was erected about the year 310 A.D. It was made from iron billets welded together. It is approximately 25 ft (7.6 m) tall with a diameter of 12 in.

(300 mm) at the top and 16 in. (400 mm) at the bottom. Its total weight is 12,000 lb (5.4 metric tons). Other pillars were erected in India at about this same time and a few other large weldments made by the Romans have been discovered in Europe and in England. Other examples of welded work have been found in Scandinavia and in Germany. However, it was not until the 1800s that welding as we know it today was discovered and used.

Sir Humphrey Davies of England is credited with providing a foundation for modern welding with two of his discoveries. One was the discovery of acetylene and the second was the production of an arc between two carbon electrodes using a battery. In the mid-1800s, the electric generator was invented and arc lighting became popular. The period of 1877 to 1903 provided a great number of discoveries and inventions pertaining to welding. During this period gas welding and cutting were developed. Arc welding with the carbon arc and metal arc was developed and resistance welding, much as it is known today, became a practical joining process. Auguste De Meritens, working in the Cabot Laboratory in France, used the heat of an arc for joining lead plates for storage batteries in the year 1881. It was his pupil, however, the Russian, Nikolai N. Benardos, working in the French laboratory, who was granted a patent for welding. He, with a fellow Russian, Stanislaus Olszewaski, secured a British patent in 1885, and an American patent in 1887. The patents show the forerunner of today's electrode holder and are shown by Figure 1-16.

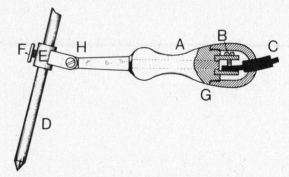

FIGURE 1-16 *First electrode holder.*

This work was the actual beginning of arc welding or at least carbon arc welding. Benardos' efforts were apparently all restricted to carbon arc welding, although he was able to weld iron as well as lead. Carbon arc welding became increasingly popular during the late 1890s and early 1900s.

WELDING BACKGROUND

Apparently Benardos was not successful with a metallic electrode, and in 1892 C. L. Coffin of Detroit was awarded the first U.S. patent for an arc welding process using a metal electrode. This was the first record of the metal melted from the electrode actually carried across the arc to deposit filler metal in the joint to make a weld. At about the same time, N. G. Slavianoff, a Russian, presented the same idea of transferring metal across an arc, but to cast metal in a mold.

In about 1900 Strohmeyer introduced a coated metal electrode in Great Britain. There was a thin wash coating of clay or lime, but it did provide a more stable arc. Oscar Kjellberg of Sweden invented the covered or coated electrode during the period of 1907 to 1914. Figure 1-17 shows his concept of welding with coated stick electrodes. They were produced by dipping short lengths of bare iron wire in thick mixtures of carbonates, silicates, etc., and allowing the coating to dry in air. It is interesting to note that wash-coated electrodes and the dipped type covered electrode are still manufactured in some countries.

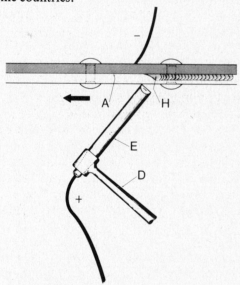

FIGURE 1-17 *Kjellberg covered electrode.*

Meanwhile, the resistance welding processes were developed, including spot welding, seam welding, projection welding, and flash butt welding. Professor Elihu Thompson is given credit for originating resistance welding. His series of patents began in 1885. He originated a company which later bore his name, the Thompson Electric Welding Company and developed the different resistance welding processes between that time and 1900.

Thermit welding was invented by a German named Goldschmidt in 1903 and was used to weld railroad rail.

Gas welding and cutting were also perfected in this same period. The production of oxygen, and later the liquifying of air, along with the introduction in 1887 of a blow pipe or torch, helped the development of both welding and cutting. Before 1900 hydrogen and coal gas were used with oxygen. However, in about 1900 a torch suitable for use with low-pressure acetylene was developed and the oxyacetylene gas welding and cutting processes were launched.

During the period of about 1900 to 1918, the oxyacetylene welding and cutting process plus the carbon arc welding and metal arc welding process with lightly covered electrodes were used primarily for repair and maintenance work.

World War I brought a tremendous demand for metal material production, and welding was pressed into service. Many companies sprang up in America and in Europe to manufacture various types of welding machines and electrodes to meet this requirement. Many innovations were tried and most were successful. The British built the first all-welded ship, H.M.S. Fulagar, and the Dutch started to weld fuselages of fighter planes. Perhaps the most famous incident was the repair work on German ships interned in New York harbor that were sabotaged by their crews. By means of arc welding these ships were quickly repaired and put back into service to deliver material from the U.S. to Europe.

Immediately after the war, in 1919, twenty members of the Wartime Welding Committee of the Emergency Fleet Corporation, under the leadership of Comfort Avery Adams, founded the American Welding Society. It was founded as a nonprofit organization dedicated to the advancement of welding and allied processes.

Alternating current welding was invented in 1919 by C. J. Halslag; however, it did not become popular until the 1930s when the heavy-coated electrode found widespread use.

In 1920, automatic welding was introduced. It utilized bare electrode wire operated on direct current and utilized arc voltage as the basis of regulating the feed rate of the electrode wire. Automatic welding was invented by P. O. Nobel of the General Electric Company. It was used to build up worn motor shafts and worn crane wheels. It was also used by the automobile industry to produce rear axle housings.

During the 1920s various different grades and types of welding electrodes were developed and made available. Mild steel or low carbon steel, usually with a carbon of 0.20% or less, was used for welding practically all grades of rolled steel. Higher carbon electrodes and alloy steel electrodes were also developed. Copper alloy rods were developed for carbon arc welding and brazing.

There was considerable controversy during the 1920s about the advantage of the heavy-coated rods versus light- or wash-coated rods. Heavy-coated rods made by dipping were more expensive. Coating applied by extrusion on the rod was less expensive. The heavy-coated electrodes, which were made by extruding the coating on the bare rod, were developed by Langstroth and Wunder of A. O. Smith Company and were used by the company in 1927. In 1929 Lincoln Electric Company produced extruded electrode rods and these were sold to the public. By 1930 covered electrodes had come into their own and specifications were being written for them. At the same time welding codes appeared which required higher-quality weld metal and this in turn increased the use of covered electrodes.

During the 1920s there was considerable interest in shielding the arc and weld area by externally applied gases. It was realized that the atmosphere of oxygen and nitrogen in contact with the molten weld metal caused brittle and sometimes porous welds. In view of this, research work was done to utilize gas-shielding techniques. Alexandre and Langmuir did considerable work in chambers using hydrogen as a welding atmosphere. They utilized two electrodes starting with carbon electrodes but later changing to tungsten electrodes. The hydrogen was changed to atomic hydrogen in the arc. It was then blown out of the arc forming an intensely hot flame of atomic hydrogen burning to the molecular form and liberating heat. This arc produced half again as much heat as an oxyacetylene flame. This was named the atomic hydrogen welding process. Atomic hydrogen never became extremely popular but was used during the 1930s and 1940s for special applications of welding and later on for welding of tool steels.

During this same period H. M. Hobart and P. K. Devers were doing similar work but using atmospheres of argon and helium. In their patents applied for in 1926, arc welding utilizing gas independently supplied around the arc was a forerunner of the gas tungsten arc welding process. They also showed welding with a concentric nozzle and with the electrode being fed as a wire through the nozzle. This was the forerunner of the gas metal arc welding process. Neither of these processes was developed until later.

The covered electrode became the mainstay of the welding industry, even though it was applied manually. At the same time many efforts were made to improve automatic welding utilizing bare wire. Such efforts included the use of a covering on electrodes which were then coiled and at the welding point the coating was milled away to introduce welding current to the core wire. Another method, the Una-method, utilized a woven screen of fine wire saturated with coating material, which was then wrapped around the electrode wire as it was fed into the arc below the current contact point.

This process had some commercial utilization, but never became too popular. The Fusarc process developed in England was another variation, and here small wires were wrapped around the main large electrode wire and the areas between were filled with coating material. This process became popular in Europe and is still used in Great Britain.

One of the more specialized welding processes was developed in 1930 at the New York Navy Yard. This process, known as stud welding, was developed specifically for attaching wood decking over a metal surface. The process welded studs, screws, etc., to the base metal by means of a special gun, which automatically controlled the arc. Fluxing elements on the end of the stud improved the properties of the weld. Stud welding became popular in the shipbuilding and construction industries and in manufacturing.

The automatic process that became extremely popular was the submerged arc welding process. This "under powder" or smothered arc welding process was developed by the National Tube Company for a new pipe mill at McKeesport, Pennsylvania. It was designed to make the longitudinal seams in the pipe. This process was patented by Robinoff in 1930 and it was later sold to Linde Air Products Company who renamed it "Unionmelt" welding. Submerged arc welding was used during the defense buildup in the late 1930s and early 1940s in both the shipyards and in ordnance factories. It is one of the most productive welding processes and remains popular today.

Gas tungsten arc welding had its beginnings from an idea by C. L. Coffin to weld in a nonoxidizing gas atmosphere, which he patented in 1890. The concept was further refined in the late 1920s by Hobart, when he used helium for shielding, and Devers, who used argon. The threat of World War II indicated the need to weld magnesium to build fighter planes. Engineers of the Northrup Aircraft Company, in conjunction with Dow Chemical Company, began a program to develop a welding process for joining magnesium. The inert gas-shielded process developed by Hobart and Devers was ideal for welding magnesium and also for welding stainless and aluminum. It was perfected in 1941, patented by Meredith, and named "Heliarc" welding, since helium was initially used for shielding. Figure 1-18 shows scenes of early torches developed by Meredith. It was later licensed to Linde Air Products who developed the water-cooled torch. The gas tungsten arc welding process has become one of the most important arc welding processes.

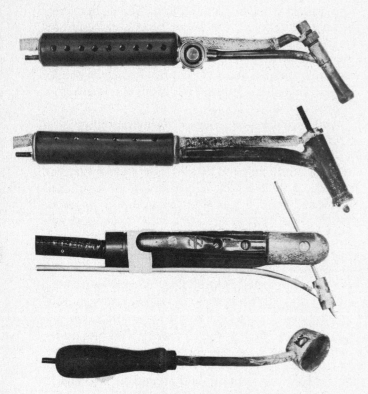

FIGURE 1-18 *Meredith "heliarc" torch.*

inert gas metal arc welding but could now be used for economically welding steels. The CO_2 arc is a hot arc and the larger electrode wires required fairly high currents. Efforts were made to make the process more acceptable for the welder and this in turn led to smaller diameter electrode wires and refined power supplies. The outcome of this development was the short circuiting arc variation which was known as Micro-wire, short arc, and dip transfer welding, all of which appeared late in 1958 and early in 1959. This variation of the process allowed all-position welding on thin materials. It soon became the most popular of the gas metal arc welding process variations.

Another variation was the use of inert gas with small amounts of oxygen which provided the spray-type arc transfer. This variation became popular in the early 1960s for welding agricultural equipment. The latest variation of gas metal arc welding is the use of pulsed current. In this variation the current is switched from a high value to a low value at a rate of once or twice the line frequency.

Soon after the introduction of the CO_2 welding process a variation utilizing a special electrode wire was developed. This wire, described as an inside-outside electrode, was tubular in cross section with the fluxing agents on the inside. These wires could be used in the same equipment as the gas metal arc welding process.

The other concept invented by Hobart and Devers was the gas shielded metal arc welding process successfully developed at Battelle Memorial Institute in 1948 under the sponsorship of the Air Reduction Company. This development utilized the gas-shielded arc similar to the gas tungsten arc, but replaced the tungsten electrode with a continuously fed electrode wire. Figure 1-19 shows the first practical gas metal arc welding gun designed by the author. One of the basic changes that made the process more acceptable was the use of small diameter electrode wires and the constant voltage principle of control and power source. This principle had been patented earlier by H. E. Kennedy. The initial introduction of GMAW was for welding nonferrous metals, particularly heavy aluminum plate. The gas metal arc welding process is faster than gas tungsten arc welding. The high deposition rate led users to try the process on steel. The cost of inert gas was relatively high and the cost savings were not immediately available.

In 1953 Lyubavskii and Novoshilov announced the use of welding with consumable electrodes in an atmosphere of CO_2 gas. The CO_2 welding process immediately gained favor since it utilized equipment developed for

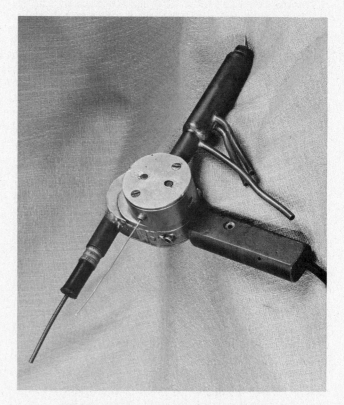

FIGURE 1-19 *First gas metal arc semiautomatic gun.*

The process was called "Dualshield," which indicates that external shielding gas was utilized, as well as the gas produced by the flux in the core of the wire, for arc shielding. This process, invented by Bernard, was announced in 1954 but was patented in 1957 at which time it was reintroduced by the National Cylinder Gas Company.

Soon afterwards in 1959 an inside-outside electrode was produced which did not require external gas shielding. Initially, the weld deposits were of lower quality. The absence of shielding gas gave the process popularity for noncritical work. This process was the self-shielding process later called "Innershield" by Lincoln Electric Company. Both the gas-shielded and self-shielding systems are widely used today and are rapidly growing in popularity.

The electroslag welding process was announced to the Western world by the Russians at the Brussels World's Fair in Belgium in 1958. It had been used in the Soviet Union since 1951 but was based on work done in the U.S.A. by R. J. Hopkins, who was granted patents in 1938 and 1940. The Hopkins process, as it was then called, was never used to a very great degree for joining.

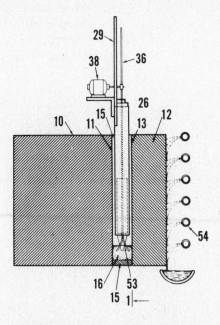

FIGURE 1-20 *Hopkins electro-slag welding.*

Figure 1-20 shows the Hopkins process for joining steel pieces. In Russia, there was a large demand for heavy weldments in order to build large machines and tools. The process was perfected and equipment was developed at the Paton Institute Laboratory in Kiev, Ukraine, U.S.S.R., and also at the Welding Research Laboratory in Bratislava, Czechoslovakia. The first production use in the U.S.A. was at the Electromotive Division of General Motors Corporation in Chicago, where it was called the

Electro-molding Process. It was used in 1959 for the fabrication of welded diesel engine blocks. The process and its variation, using a consumable guide tube, have become popular and are widely used for welding thicker materials. Electroslag uses equipment similar to submerged arc welding but there is no arc after start up.

Another vertical welding method called Electrogas was introduced in 1961 by the Arcos Corporation. It utilized equipment developed for electroslag welding but employed a flux-cored electrode wire and an externally supplied gas shield. It is an open arc process since a slag bath is not involved. It is a variation of flux-cored arc welding. A newer development uses self-shielding electrode wires and a variation uses solid wire but with gas shielding. These methods allow the welding of thinner materials than can be welded with the electroslag process.

The plasma arc welding process, which is very similar to gas tungsten arc welding, was invented by Robert Gage in 1957. Plasma arc welding uses a constricted arc or an arc through an orifice, which creates an arc plasma that has a higher temperature than the tungsten arc. It is also used for metal spraying and for cutting. As a cutting process it became popular for nonferrous metals. It is used for spraying both wires and powders. Plasma arc welding is popular for low-current welding and is becoming increasingly popular in higher-current applications.

The electron beam welding process, which uses a focused beam of electrons as a heat source in a vacuum chamber, was developed in Germany and France. The Ziess Company and the French Atomic Energy Commission are credited with developing the process in the late 1940s. In the last few years the process has gained widespread acceptance for welding. Its popularity is increasing since recent developments have allowed it to be taken out of the vacuum chamber which was its major disadvantage. More and more applications are being found for this process, which should grow in popularity in the future.

Friction welding, which uses high rotational speeds and upset pressure to provide friction heat, was developed in the Soviet Union, but additional work was done in Great Britain and the U.S.A. It is a specialized process and has applications only where a sufficient volume of similar parts are to be welded because of the initial expense of equipment and tooling. This process, also called inertia welding, will find more uses and will become more popular.

CHAPTER 1

WELDING BACKGROUND

The newest of the welding processes is laser welding. The laser originally developed at the Bell Telephone Laboratories was used as a communications device. Because of the tremendous concentration of energy in a small space it proved to be a powerful heat source. It has been used for cutting metals and other materials. The early problems involved short pulses of energy; however, today continuous pulse type equipment is available. The equipment is still extremely expensive and bulky, but in time it should be reduced in cost and size. The laser is now finding welding applications in routine metalworking operations.

There are many other variations of these processes, which are not specifically processes themselves. These will be discussed along with the basic process. Undoubtedly, additional welding processes and methods will be developed and as the need arises they will be adapted to metalworking requirements.

1-5 THE WELDING INDUSTRY AND ITS FUTURE

Welding is now the universally accepted method of permanently joining all metals. In some respects, it might be considered a mature industry but it is still a growing industry. The true impact of welding on the metalworking industry should be measured in the value of the parts produced by welding, the amount of money saved by the use of welding over other metal fabrication

FIGURE 1-21 *Welding industry sales in the United States.*

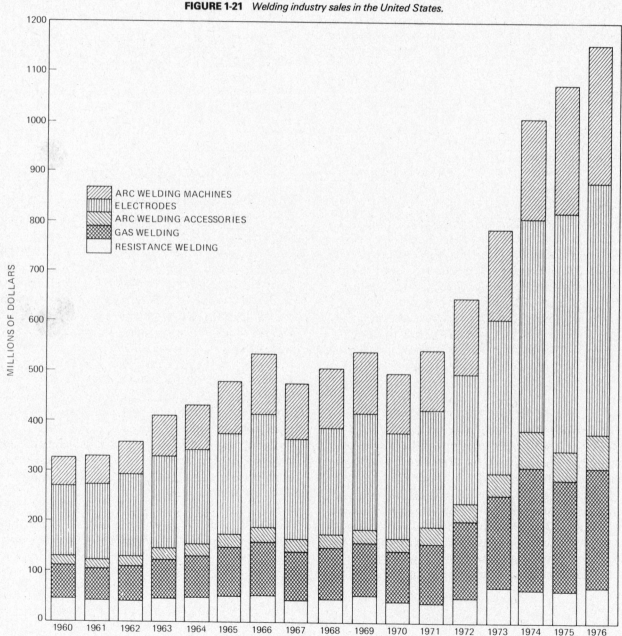

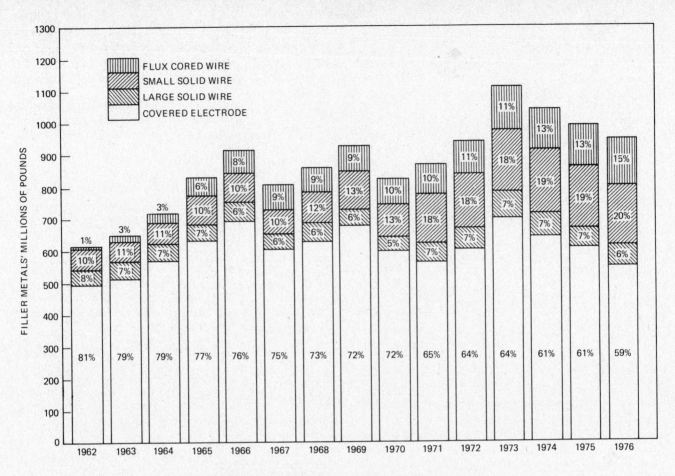

FIGURE 1-22 *Welding filler metal sales in the United States.*

processes, and in the value of products made possible by welding. However, this information is impossible to determine.

Historical data are available that record the growth of the welding equipment and materials industry, which in turn gives an indication of the projected growth for the future. Figure 1-21 is a record of the industry since 1960,[6] and shows growth greater than the gross national product for the country. The arc welding segment of the industry seems to be growing the fastest with recent years showing maximum growth. The growth of the welding industry has been approximately 6% per year and the industry shipments are currently in excess of $800 million. Conventional electric arc welding equipment and filler metals represent over two-thirds of this total. Each segment of the industry and each welding process has its own growth patterns which will be explored later.

In order to make a projection we must determine the past historic growth patterns, determine the present position, and consider those factors that will have an impact on the growth in the future. Future growth of the arc welding processes depends on factors that may have an impact on the industries served by welding.

A convenient starting point for such an analysis is to consider the chart shown by Figure 1-6, which shows industries that use welding to the greatest degree. The future of these industries will largely determine the future of welding. Another clue is the data shown in Figure 1-4, which shows the amount and type of metal produced in the country. These data relate to processes and welding methods. Another indicator is Figure 1-5, which shows the types of steel most widely used throughout the country.

It is possible to estimate the amount of each type of arc welding that is being done in the U.S.A. and how it is being applied. The most suitable data for a starting point is shown by Figure 1-22, the "welding filler metals sold in the U.S.A."[6] from 1962 to the present. This shows the amount and percentage of filler metals sold based in the following categories:

1. Covered electrodes (stick electrodes) all types
2. Submerged arc welding electrode wire [solid steel larger than 1/16 in. (1.6 mm) in diameter]
3. Gas metal arc welding electrode wire [solid steel wire 1/16 in. (1.6 mm) and smaller]
4. Flux-cored arc welding electrode wire

17

At this point we can project into the future. This is done by charting the bar graph information into line charts and extending these lines for five years as shown by Figure 1-23. This shows that based on the percentage of the total;

> Covered electrodes have been decreasing steadily for the last 14 years dropping from 81% to 59% and projected to 45%.
>
> Submerged arc welding has remained constant at about 5% to 7%.
>
> Gas metal arc welding has almost doubled, rising from 10% to 20%, and is projected to double again in the next ten years.
>
> Flux-cored welding is increasing, but at a slower rate.

This information shows that semiautomatic welding will greatly increase, machine and automatic welding will increase modestly, but manual welding is decreasing at least as a percentage of the total.

After analyzing recent trends in welding and manufacturing it becomes evident that the following must be considered with regard to the future of welding:

1. There will be continuing need to reduce manufacturing costs and to improve productivity, since:

   (a) Wage rates for the people in manufacturing industries will continue to increase.

   (b) The cost of metals for producing weldments and filler metals will also continue to be more expensive.

   (c) Energy and fuel costs will increase and shortages may occur.

2. There will be a continuing trend towards the use of higher-strength materials, particularly in the steels and lighter-weight materials.

3. There will be more use of welding by manufacturing industries, probably decreasing the use of castings.

4. There will be a trend towards higher levels of reliability and higher-quality requirements.

5. The trend towards automatic welding and automation in welding will accelerate.

Productivity is considered the amount of welding that can be done by a welder in a day. This is determined by several factors, the most important of which is the operator factor or duty cycle. Operator factor for a welder is the number of minutes per eight-hour period that is spent actually welding. The different methods of welding have different average duty cycles. Manual

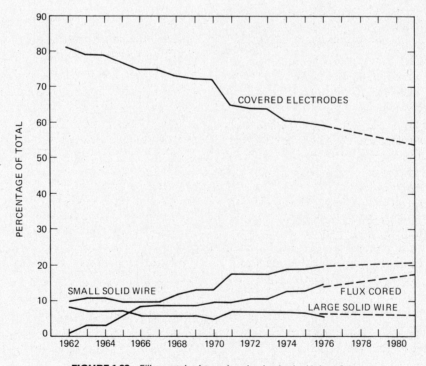

FIGURE 1-23 *Filler metal sales and projection in the United States.*

welding has the lowest operator factor with semiautomatic welding approximately double and machine welding the next highest, with automatic welding approaching 100%. Efforts will be made to utilize those processes that have the highest-duty cycles. The expected trend will be away from manual welding towards semiautomatic welding and to machine or automatic welding when possible.

Another factor affecting productivity of welders relates to the deposition rate of the welding process. The higher current processes have the highest deposition rates, thus the submerged arc welding process and the electroslag welding process will remain important as costs must be reduced.

The next factor deals with increasing material costs. It is imperative to obtain the maximum utilization of filler metals. The cold wire type processes, gas tungsten arc and plasma arc, can actually deposit 100% of the filler metal purchased. Submerged arc welding, when the electrode only is considered approaches 100% as does electroslag welding. Gas metal arc welding will give about 95% utilization. Flux-cored welding is the lowest of the continuous wire processes, normally in the 80% plus range. Covered electrodes have the lowest utilization because of the stub end and coating loss that results in approximately 65% of the weight of the filler metals purchased actually being deposited in the weld joint.

Another factor closely related to filler metal efficiency and operator factor is the total deposit of weld metal to produce a given weldment. If the amount of weld metal can be reduced to make a weld it is an economic savings, thus there is an advantage to methods such as narrow gap welding. The higher penetration characteristics of CO_2 welding gives it an advantage over shielded metal arc welding because fillet weld sizes can be reduced and the same weld strength retained.

In forecasting the arc welding field, we will consider each process separately since each has its own historical development and utilization and will have a different future. However, the arc welding processes will continue to dominate the welding industry.

The shielded metal arc welding process is the oldest of the current arc welding processes but is losing ground in the total arc welding market. This trend will continue and manual electrode welding in the near future may represent only a third of arc welding.

The percentage of filler metal used by submerged arc welding has remained almost constant through the years. It is impossible to differentiate between filler metal used for electroslag welding and submerged arc welding; however, both processes will increase modestly.

Gas metal arc welding will continue to accelerate

since it is being substituted for shielded metal arc, gas welding, brazing, and resistance welding. This process, since it is a continuous wire process with high filler metal utilization, will continue to rise at the highest rate.

The flux-cored arc welding process started from a lower base and has been gaining modestly. This trend will continue; however, lower filler metal utilization and higher filler metal cost will keep it from growing as fast as gas metal arc welding.

Gas tungsten arc welding will grow as fast or faster than the total welding market. There are three reasons; it is adaptable to automation, it is being used on high quality work, and for welding newer thin specialty metals.

Plasma arc welding will grow faster than gas tungsten arc as soon as its capabilities are better known.

Special automated fixtures will become increasingly important. It is expected that fixtures will soon be specified by the type of work and the size of work they are expected to perform, thus we will have automatic machines for seamers, for tank making, for pipe welding, for attaching spuds, for overlaying, and for other special applications. Tape-controlled automatic machines, numerically controlled machines, and computer-controlled machines will become more common place in the years ahead. Every effort will be made to reduce the amount of manual labor involved in making welds.

Some of the newer processes and some which are of a more specialized nature will grow quite rapidly; however, they will never become large segments of the total welding industry. These include electron beam welding, laser beam welding, friction welding, ultrasonic welding, diffusion welding, and cold welding. For example, the electron beam welding equipment market totaled $9.3 million in 1965, $28 million in 1970, and is predicted to reach $50 million by the early 1980s. Laser beam welding is predicted to grow even more rapidly from $4 million in 1970 to $70 million in the early 1980s.

With increased emphasis on welding as a basic manufacturing technology the growth rate in the future will approximate 8% per year and welding equipment shipments are expected to more than double in the next five years. Growth rate is expected to be shared by all of the different welding processes. However, the more conventional arc welding processes may not grow as fast as the more exotic processes primarily because of a larger base.

QUESTIONS

1. Why is welding the most economical way to permanently join metals?

2. Why can thinner wall pipe be used with welded joints?

3. Why do bolt holes reduce the strength of a bolted butt splice joint?

4. What are five advantages of welded construction?

5. What metal is not welded?

6. What metal is most often welded?

7. Are all other metals welded with equal ease?

8. Steel is produced in many forms. What type is the most popular?

9. What type of work is most often performed by welders?

10. Welders work in many industries. What industry type employs the most welders?

11. Where does the construction industry rank as an employer of welders?

12. The largest number of welders are employed in what part of the U.S.?

13. How do artists use welding?

14. When was arc welding, as we know it today, invented?

15. Who invented metal arc welding?

16. Who invented the heavy-covered electrode?

17. When was automatic welding first used?

18. What is the electrode utilization factor for a covered electrode?

19. Why is this utilization factor so low?

20. What arc welding process seems to be growing the fastest?

REFERENCES

1. *Statistical Abstracts of the United States* (U.S. Department of Commerce, Bureau of the Census, Washington D.C., 1973 edition).

2. *Metal Statistics, 1973 American Metal Market* (Fairchild Publications, Inc., New York, New York 10003).

3. *Annual Statistical Report* (American Iron & Steel Institute, Washington, D.C., 1972).

4. *Standard Industrial Classification Manual* (Statistical Policy Division, Office of Management and Budget, U.S. Government Printing Office, Washington, D.C., 20402, 1972).

5. *Where the (Welding) Action Is* November, 1973, p. 96y, provides 1970 Census of Population Data, Welding Design and Fabrication.

6. File Facts issue for years shown, *Welding Engineer*.

2

THE FUNDAMENTALS
OF WELDING

2-1   BASIC DEFINITIONS

2-2   WELDING PROCESS AND GROUPINGS

2-3   WELDERS AND WELDING OPERATORS

2-4   WELDING PROCEDURES

2-5   WELDING ELECTRICITY

2-6   WELDING ARCS

**2-7   THE PHYSICS AND CHEMISTRY
OF WELDING**

2-1   BASIC DEFINITIONS

To understand welding it is necessary to be familiar with some of the basic terms used by the industry. The American Welding Society (AWS) provides the majority of definitions,[1] and many of these are given in this chapter and throughout the book whenever the need arises. The official AWS definitions will be used. There are, however, some slightly obscure definitions and slang terms. These and the AWS terms will be presented in the definition section of the Appendix.

Welding is "a materials joining process used in making welds," and a **weld** is "a localized coalesence of metals or nonmetals produced either by heating the materials to suitable temperature with or without the application of pressure or by the application of pressure alone and with or without the use of a filler material." **Coalesence** means a growing together or a growing into one body and is used in all of the welding process definitions.

A **weldment** is an assembly of component parts joined by welding. A weldment can be made of many or few metal parts. A weldment may contain metals of different compositions and the pieces may be in the form of rolled shapes, sheet, plate, pipe, forgings, or castings. To produce a usable structure or weldment there must be weld joints between the various pieces that make the weldment. The **joint** is "the junction of members or the edges of members which are to be joined or have been joined." There are five basic types of joints for bringing two members together for welding. These joint types or designs are also used by other skilled trades.

The five basic joints shown by Figure 2-1 are:

B, Butt joint:—parts in approximately the same place.

C, Corner joint:—parts at approximately right angles and at the edge of both parts.

E, Edge Joint: an edge of two or more parallel parts.

L, Lap Joint: between overlapping parts.

T, T Joint: parts at approximately right angles, not at the edge of one part.

A **weld** is a localized coalescence of metal at the junction of metal parts in a specific area. In a weld filler metal may or may not be used and heat with or without pressure is used, but the result is a continuity of solid metal parts.

THE FUNDAMENTALS OF WELDING

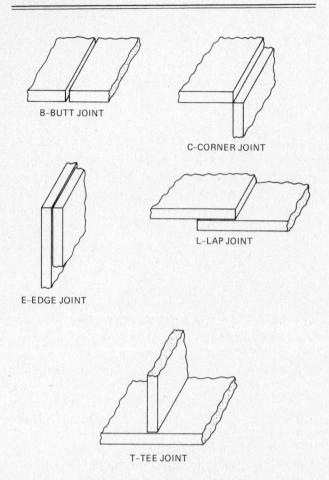

FIGURE 2-1 *The five basic joint designs.*

tions. In all of the arc welding processes, the heart of the welding system is the welding power source. This piece of equipment provides the electrical power to sustain the arc so that it can be used for making welds. There are many types, sizes, and variations. Some generate electricity from rotating energy sources, and are called welding generators. Others take the power available from the lines and change it to power suitable for arc welding. These are known as *transformers* or *transformer-rectifier welding machines.* Both alternating and direct current can be used for some of the arc welding processes. The welding process will determine the type of power source required.

The most important part of the welding system is the welder or welding operator, the human element. There is a difference between welders and welding operators and this is primarily a difference of the manipulative skills involved. The welder must exercise skill and ability to manipulate equipment to produce welds. The welding operator may monitor or operate an automatic welding machine.

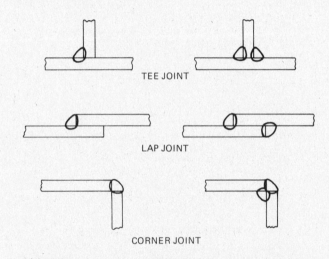

FIGURE 2-2 *Applications of fillet welds—single and double.*

It is important to distinguish between the *joint* and the *weld*—each must be described to completely describe the *weld joint.* There are many different types of welds and they are best described by their shape when shown in cross section. The most popular weld is the *fillet weld* named after its cross-sectional shape. Fillet welds are shown by Figure 2-2. The second most popular is the *groove weld* and there are seven basic types of groove welds. These are shown by Figure 2-3. There are other types of welds: the *flange weld,* the *plug weld,* the *slot weld,* the *seam weld, surfacing weld,* and the *backing weld.* Joints are combined with welds to make weld joints. Examples are shown by Figure 2-4. Welds and joints are completely described in Chapter 14 on design.

There are approximately fifty different distinct welding processes. They are subdivided into seven groups.

The arc welding group of processes are the most popular and widely used for metal joining. There are eight distinct arc welding processes and numerous varia-

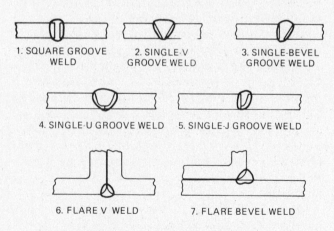

FIGURE 2-3 *The seven basic groove welds.*

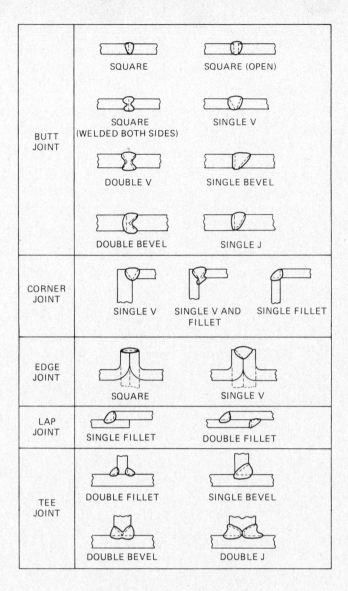

BUTT JOINT	SQUARE	SQUARE (OPEN)	
	SQUARE (WELDED BOTH SIDES)	SINGLE V	
	DOUBLE V	SINGLE BEVEL	
	DOUBLE BEVEL	SINGLE J	
CORNER JOINT	SINGLE V	SINGLE V AND FILLET	SINGLE FILLET
EDGE JOINT	SQUARE	SINGLE V	
LAP JOINT	SINGLE FILLET	DOUBLE FILLET	
TEE JOINT	DOUBLE FILLET	SINGLE BEVEL	
	DOUBLE BEVEL	DOUBLE J	

FIGURE 2-4 *Some typical weld joints.*

Another way of dividing or categorizing welding processes relates to whether filler metal is or is not used. **Filler metal** is "the material to be added in making a welded, brazed, or soldered joint." It becomes the weld fillet or weld metal in a groove weld. In some welding processes, the filler metal is carried across the arc and deposited in the weld. In others filler metal is not carried across the arc but is melted by the heat of the arc and added to the molten puddle. If the weld metal passes through the arc, it is provided by an electrode. If it is melted by the heat of the arc and added to the puddle it is called a welding rod. Welding electrodes and welding rods have special composition requiring detailed specifications to completely describe them. Proper selection of filler metals is important, normally their properties should match the properties of the metal being welded. This metal is called the

base metal which is defined as "the material to be welded, soldered, or cut." This is the preferred term. In some countries the word *parent* metal is used, for thermal spraying the word *substrate* is used.

The type of base metal often dictates the welding process that can be used.

To completely describe the making of a weld, it is necessary to specify the welding process to be employed and to indicate the method of applying the process. It is also necessary to describe the **welding procedure,** which is "the detailed methods and practices including joint design details, materials and method of welding in order to describe how a particular weld or weldment is to be made." It is becoming more and more important to completely describe and document the entire welding procedure.

To insure that the welds conform to demanding specifications, specialized inspection techniques are used. These include destructive and nondestructive testing methods. Nondestructive testing includes visual inspection, magnetic particle inspection, radiographic inspection, liquid penetrant inspection, and ultrasonic inspection. Welding quality control is required by most codes and is a necessary requirement for most manufactured products.

Welding is often done on structures in the position in which they are found. In view of this, techniques have been developed to allow welding in any position. Certain welding processes have "all-position" capabilities while others may be used in only one or two positions. The welding positions are defined by the American Welding Society. There are four basic welding positions. They are shown by Figures 2-5, 2-6, and 2-7 and are described as follows:

Flat: "when welding is performed from the upper side of the joint and the face of the weld is approximately horizontal." *Flat welding* is the preferred term; however, the same definition is sometimes called *downhand.*

Horizontal: "the axis of the weld is approximately horizontal but the definition varies for groove and fillets."

Overhead: "when welding is performed from the underside of the joint."

Vertical: "the axis of the weld is approximately vertical."

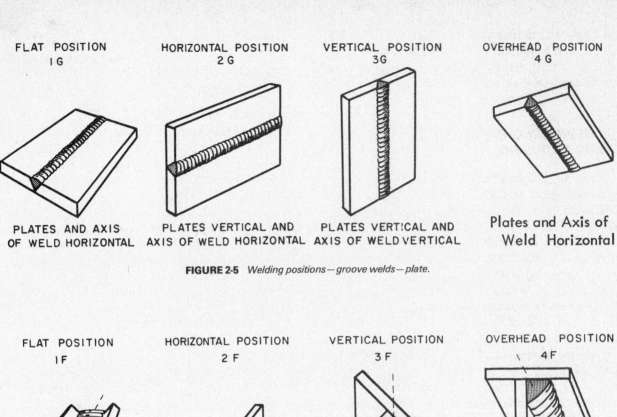

FLAT POSITION
1 G

HORIZONTAL POSITION
2 G

VERTICAL POSITION
3 G

OVERHEAD POSITION
4 G

PLATES AND AXIS
OF WELD HORIZONTAL

PLATES VERTICAL AND
AXIS OF WELD HORIZONTAL

PLATES VERTICAL AND
AXIS OF WELD VERTICAL

Plates and Axis of
Weld Horizontal

FIGURE 2-5 *Welding positions—groove welds—plate.*

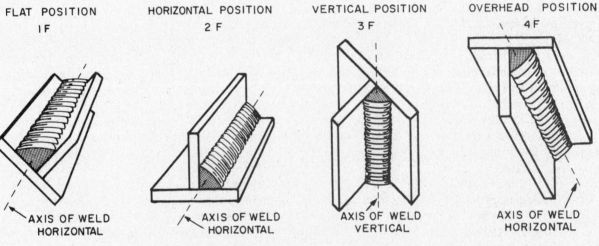

FLAT POSITION
1 F

HORIZONTAL POSITION
2 F

VERTICAL POSITION
3 F

OVERHEAD POSITION
4 F

AXIS OF WELD
HORIZONTAL

AXIS OF WELD
HORIZONTAL

AXIS OF WELD
VERTICAL

AXIS OF WELD
HORIZONTAL

FIGURE 2-6 *Welding positions—fillet welds—plate.*

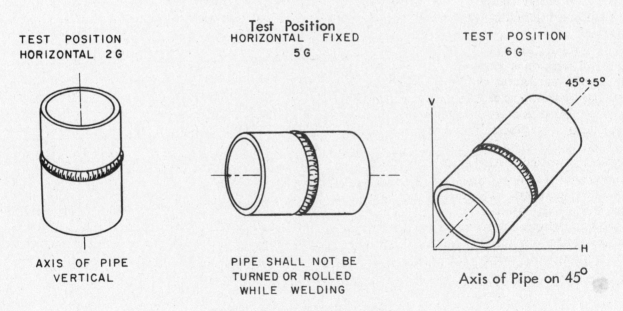

TEST POSITION
HORIZONTAL 2 G

Test Position
HORIZONTAL FIXED
5 G

TEST POSITION
6 G

45°±5°

AXIS OF PIPE
VERTICAL

PIPE SHALL NOT BE
TURNED OR ROLLED
WHILE WELDING

Axis of Pipe on 45°

FIGURE 2-7 *Welding positions—pipe welds.*

More terms and definitions will be presented in later chapters. It is important at the beginning of the book to at least briefly define these terms so that you will better understand their meaning as you read further in the book.

2-2 THE WELDING PROCESSES AND GROUPING

The American Welding Society has made each welding process definition as complete as possible so that it will suffice without reference to another definition. They define a process as "a distinctive progressive action or series of actions involved in the course of producing a basic type of result." The official listing of processes and their grouping is shown by Figure 2-8, the AWS Master Chart of Welding and Allied Processes. The welding society formulated process definitions from the operational instead of the metallurgical point of view. Thus the definitions prescribe the significant elements of operation instead of the significant metallurgical characteristics. The AWS definition for a welding process is "a materials joining process which produces coalescence of materials by heating them to suitable temperatures with or without the application of pressure or by the application of pressure alone and with or without the use of filler material." AWS has grouped the processes together according to the "mode of energy transfer" as the primary consideration. A secondary factor is the "influence of capillary attraction in effecting distribution of filler metal" in the joint. Capillary attraction distinguishes the welding processes grouped under "Brazing" and "Soldering" from "Arc Welding," "Gas Welding," "Resistance Welding," "Solid State Welding," and "Other Processes." The distinguishing feature of the latter groups of welding processes is the "mode of energy transfer." "Adhesive Bonding" is also included since it is being increasingly used to join metals.

The welding society deliberately omitted the designation of *pressure* or *nonpressure* since the factor of pressure is an element of operation of the applicable welding process. The designation "Fusion Welding" is not recognized as a grouping since fusion is involved with many of the processes. Other terms or factors such as the type of current used in arc or resistance welding processes, whether electrodes are *consumable* or *nonconsumable* or *continuous* or *incremental* are not shown in process groupings. These and other items characterize the methods by which the processes are performed. In some countries the welding processes are grouped differently; for example, in the United Kingdom Group I is designated for welding processes using heat with pressure and Group II is for welding processes requiring

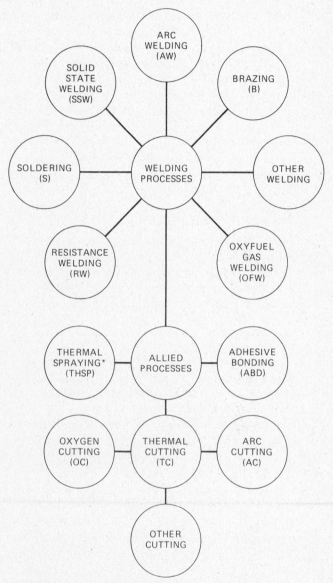

*SOMETIMES A WELDING PROCESS.

FIGURE 2-8 *AWS master chart of welding and allied processes.*

heat alone. In Germany, there is a distinction between pressure welding and fusion welding; the former includes ultrasonic, friction, forge, resistance, stud, and diffusion welding; the latter includes gas welding, electroslag welding, arc welding, plasma welding, electron beam and laser welding. Other countries refer to the type of energy involved, i.e., thermochemical, electrothermic, mechanical energy, or focalized energy.

Welders sometimes distinguish between welds made with the addition of filler metal and those made by fusing only the joint edges together, an autogenous weld. However, it is best to relate to the welding process. Other ways of classifying are indicating the use of a nonconsumable or continuously fed electrode or the heat source, or by considering an exposed puddle process versus nonexposed puddle which relates more to the skill of the welder. The AWS designation will be used throughout the book.

The arc welding processes are defined as "a group of welding processes which produce coalescence of metals by heating them with an arc or arcs with or without the application of pressure and with or without the use of filler metal." *Coalescence* is defined as a "growing together" or "growth into one body" and is regarded as more applicable to all types of welding than the term *consolidation* which might imply the use of external force or could mean bolting, riveting, nailing, etc. Coalescence is used in all welding process definitions.

Arc Welding

The arc welding group includes eight specific popular processes, each separate and different from the others but in many respects similar. The arc welding group of processes is defined in the next few pages.

The *carbon arc welding* (CAW) process is the oldest of all the arc welding processes and is considered to be the beginning of arc welding. The welding society defines carbon arc welding as "an arc welding process which produces coalescence of metals by heating them with an arc between a carbon electrode and the work. No shielding is used. Pressure and filler metal may or may not be used." Figure 2-9 shows the single carbon arc process in use. It has limited applications today, but a variation or twin carbon arc welding is more popular. Another variation uses compressed air for cutting and gouging.

The development of the metal arc welding process soon followed the carbon arc. This developed into the currently popular *shielded metal arc welding* (SMAW) process defined as "an arc welding process which produces coalescence of metals by heating them with an arc between a covered metal electrode and the work. Shielding is obtained from decomposition of the electrode covering. Pressure is not used and filler metal is obtained from the electrode." Figure 2-10 shows this extremely popular process which is widely used today to weld

FIGURE 2-9 *Carbon arc welding.*

FIGURE 2-10 *Shielded metal arc welding.*

steels. It is usually applied manually. It can also be used for metal cutting.

Automatic welding utilizing bare electrode wires was used in the 1920s, but it was the *submerged arc welding* (SAW) process that made automatic welding popular. Submerged arc welding is defined as ''an arc welding process which produces coalescence of metals by heating them with an arc or arcs between a bare metal electrode or electrodes and the work. The arc is shielded by a blanket of granular fusible material on the work. Pressure is not used and filler metal is obtained from the electrode and sometimes from a supplementary welding rod.'' Submerged arc welding is shown by Figure 2-11. It is usually applied by machine or automatic methods; however, it can be applied semiautomatically. It is normally limited to the flat or horizontal position.

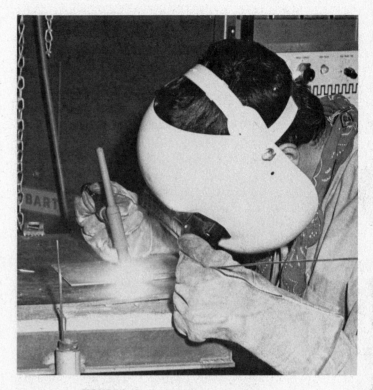

FIGURE 2-12 *Gas tungsten arc welding.*

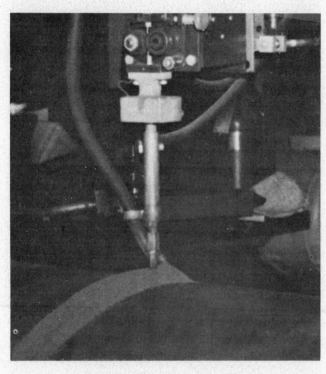

FIGURE 2-11 *Submerged arc welding.*

The need to weld nonferrous metals particularly magnesium and aluminum challenged the industry. A solution was found called *gas tungsten arc welding* (GTAW) and was defined as ''an arc welding process which produces coalescence of metals by heating them with an arc between a tungsten (nonconsumable) electrode and the work. Shielding is obtained from a gas or gas mixture. Pressure may or may not be used and filler metal may or may not be used.'' The process developed in the late 1930s also became known as Heliarc, or TIG welding and was immediately popular in the air-

craft industry where it was used to join ''hard to weld'' metals. GTAW welding is shown by Figure 2-12. Inert gases are used for shielding and it is normally applied manually, although automatic applications are becoming more popular. It has limited use for cutting.

A very close companion process was developed in the mid-1950s, which is similar to gas tungsten arc except the arc is constricted in such a way to produce a plasma. It is called *plasma arc welding* (PAW) and is defined as ''an arc welding process which produces a coalescence of metals by heating them with a constricted arc between an electrode and the work piece (transferred arc) or the electrode and the constricting nozzle (nontransferred arc). Shielding is obtained from the hot ionized gas issuing from the orifice which may be supplemented by an auxiliary source of shielding gas. Shielding gas may be an inert gas or a mixture of gases. Pressure may or may not be used and filler metal may or may not be supplied.'' Plasma arc welding is shown by Figure 2-13. Filler wire when it is used is normally a ''cold'' or nonelectrical rod which is added to the molten puddle of the weld by the welder. Plasma welding has been used for joining some of the thinner materials.

FIGURE 2-13 *Plasma arc welding.*

It is normally manually applied. It is also popular as a cutting process.

Another welding process also related to gas tungsten arc welding is known as *gas metal arc welding* (GMAW). It was developed in the late 1940s for welding aluminum and has become extremely popular. It is defined as "an arc welding process which produces coalescence of metals by heating them with an arc between a continuous filler metal (consumable) electrode and the work. Shielding is obtained entirely from an externally supplied gas or gas mixture." This process sometimes called MIG welding, is shown by Figure 2-14.

The electrode wire for GMAW is continuously fed into the arc and deposited as weld metal. This process has many variations depending on the type of shielding gas, the type of metal transfer, and the type of metal welded. It is capable of welding many different metals and is becoming one of the most popular of the arc welding processes. The semiautomatic method of

FIGURE 2-14 *Gas metal arc welding.*

application is the most popular, but the process is also applied automatically.

A variation of gas metal arc welding has become a distinct welding process and is known as *flux-cored arc welding* (FCAW). It is defined as "an arc welding process which produces coalescence of metals by heating them with an arc between a continuous filler metal (consumable) electrode and the work. Shielding is provided by a flux contained within the tubular electrode. Additional shielding may or may not be obtained from an externally supplied gas or gas mixture." This process is shown by Figure 2-15. It was developed in the mid-1950s. The coating material on the outside of a covered electrode with its normal functions are now included in the core or center of the tubular electrode wire. There are two variations; one utilizes external shielding gas and the other does not. Both are used primarily for welding steels and are normally applied by the semiautomatic method of application.

The final process within the arc welding group of processes is known as *stud arc welding* (SW). This process is defined as "an arc welding process which produces coalescence of metals by heating them with an arc between a metal stud or similar part and the work. When the surfaces to be joined are properly heated they are brought together under pressure. Partial shielding may be obtained by the use of ceramic ferrule surrounding the stud. Shielding gas or flux may or may not be used." This process was developed in the mid-1930s. It is shown by Figure 2-16. There are several

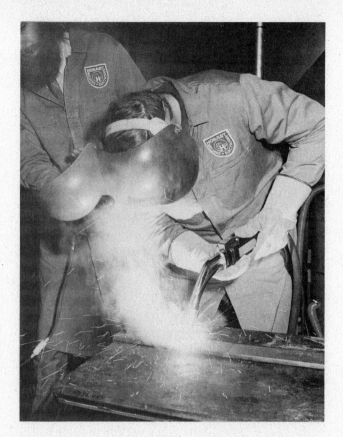

FIGURE 2-15 *Flux-cored arc welding.*

FIGURE 2-16 *Stud welding.*

variations of the process and it is normally applied as an automatic method.

Electroslag welding (EW) a nonarc welding process, borrowed from the "other welding processes" group, is included here since it employs equipment used by the gas metal arc, flux-cored arc, and submerged arc welding and is defined as "a welding process producing coalescence of metals with molten slag which melts the filler metal and the surfaces of the work to be welded.

FIGURE 2-17 *Electroslag welding.*

The molten weld pool is shielded by this slag which moves along the full cross section of the joint as welding progresses. The process is initiated by an arc which melts the slag. The arc is then extinguished and the conductive slag is maintained in a molten condition by its resistance to electric current passing between the electrode and the work." It is thus not an arc welding process although its initiation starts with an arc. This process was invented in the early 1930s in the U.S.A., but became popular when equipment was designed for its use in Russia in the early 1950s. There are two major variations, the upward-moving system and the consumable-guide system. The consumable-guide variation is shown by Figure 2-17. Electroslag welding is used normally to make welds in the vertical position and on steels. It is applied automatically.

CHAPTER 2

THE FUNDAMENTALS OF WELDING

TABLE 2-1 Welding processes and letter designation.

Group	Welding Process	Letter Designation	Chapter
Arc Welding	Carbon Arc	CAW	4-5
	Flux Cored Arc	FCAW	5-4
	Gas Metal Arc	GMAW	5-3
	Gas Tungsten Arc	GTAW	4-3
	Plasma Arc	PAW	4-4
	Shielded Metal Arc	SMAW	4-2
	Stud Arc	SW	4-6
	Submerged Arc	SAW	5-2
Brazing	Diffusion Brazing	DFB	6-1
	Dip Brazing	DB	6-1
	Furnace Brazing	FB	6-1
	Induction Brazing	IB	6-1
	Infrared Brazing	IRB	6-1
	Resistance Brazing	RB	6-1
	Torch Brazing	TB	6-1
Oxyfuel Gas Welding	Oxyacetylene Welding	OAW	6-2
	Oxyhydrogen Welding	OHW	6-2
	Pressure Gas Welding	PGW	6-2
Resistance Welding	Flash Welding	FW	6-3
	High Frequency Resistance	HFRW	6-3
	Percussion Welding	PEW	6-3
	Projection Welding	RPW	6-3
	Resistance-Seam Welding	RSEW	6-3
	Resistance-Spot Welding	RSW	6-3
	Upset Welding	UW	6-3
Solid State Welding	Cold Welding	CW	6-4
	Diffusion Welding	DFW	6-4
	Explosion Welding	EXW	6-4
	Forge Welding	FOW	6-4
	Friction Welding	FRW	6-4
	Hot Pressure Welding	HPW	6-4
	Roll Welding	ROW	6-4
	Ultrasonic Welding	USW	6-4
Soldering	Dip Soldering	DS	6-5
	Furnace Soldering	FS	6-5
	Induction Soldering	IS	6-5
	Infrared Soldering	IRS	6-5
	Iron Soldering	INS	6-5
	Resistance Soldering	RS	6-5
	Torch Soldering	TS	6-5
	Wave Soldering	WS	6-5
Other Welding Processes	Electron Beam	EBW	6-6
	Electroslag	ESW	5-6
	Induction	IW	6-9
	Lasser Beam	LBW	6-7
	Thermit	TW	6-8

Brazing (B)

Brazing is "a group of welding processes which produces coalescence of materials by heating them to a suitable temperature and by using a filler metal, having a liquidus above 450 °C (840 °F) and below the solidus of the base materials. The filler metal is distributed between the closely fitted surfaces of the joint by capillary attraction." A braze is a very special form of weld, the base metal is theoretically not melted. There are seven popular different processes within the brazing group. The source of heat differs among the processes. Braze welding relates to welding processes using brass or bronze filler metal, where the filler metal is not distributed by capillary action.

Oxyfuel Gas Welding (OFW)

Oxyfuel gas welding is "a group of welding processes which produces coalescence by heating materials with an oxy fuel gas flame or flames with or without the application of pressure and with or without the use of filler metal." There are four distinct processes within this group and in the case of two of them, *oxyacetylene welding* and *oxyhydrogen welding,* the classification is based on the fuel gas used. The heat of the flame is created by the chemical reaction or the burning of the gases. In the third process, *air acetylene welding,* air is used instead of oxygen, and in the fourth category, *pressure gas welding,* pressure is applied in addition to the heat from the burning of the gases. This welding process normally utilizes acetylene as the fuel gas. The oxygen thermal cutting processes have much in common with the welding processes.

Resistance Welding (RW)

Resistance welding is "a group of welding processes which produces coalescence of metals with the heat obtained from resistance of the work to electric current in a circuit of which the work is a part, and by the application of pressure." In general, the difference of the resistance welding processes has to do with the design of the weld and the type of machine necessary to produce the weld. In almost all cases the processes are applied automatically since the welding machines incorporate both electrical and mechanical functions.

Other Welding Processes

This group of processes includes those which are not best defined under the other groupings. It consists of the following processes: *electron beam welding, laser beam welding, thermit welding,* and other miscellaneous welding processes in addition to electroslag welding which was mentioned previously.

Soldering (S)

Soldering is "a group of joining processes which produces coalescence of materials by heating them to a suitable temperature and by using a filler metal having a liquidus not exceeding 450°C (840°F) and below the solidus of the base materials. The filler metal is distributed between the closely fitted surfaces of the joint by capillary attraction." There are a number of different soldering processes and methods.

Solid State Welding (SSW)

Solid state welding is "a group of welding processes which produces coalescence at temperatures essentially below the melting point of the base materials being joined without the addition of a brazing filler metal. Pressure may or may not be used." The oldest of all welding processes *forge welding* belongs to this group. Others include *cold welding, diffusion welding, explosion welding, friction welding, hot pressure welding,* and *ultrasonic welding.* These processes are all different and utilize different forms of energy for making welds.

The welding processes, in their official groupings, are shown by Table 2-1. This table also shows the letter designation for each process. The letter designation assigned to the process can be used for identification on drawings, tables, etc., and will be used throughout this book. Also shown by Table 2-1 is the chapter number in the book for complete information about each specific process.

Allied and related processes include adhesive bonding, thermal spraying, and thermal cutting. These processes will be described in Chapter 7.

2-3 WELDERS AND WELDING OPERATORS—THE METHOD OF APPLYING

Manual arc welding requires a high level of manipulative skill on the part of the welder. There is, however, more than one method of applying the different welding processes and some require very little manipulative skills. The title we use for the individual doing the welding indicates the manipulative skill level involved. By definitions: The **welder** is "one who is capable of performing a manual or semiautomatic welding operation." The **welding operator** is "one who operates machine or automatic welding equipment."

The definitions do not indicate the actual level of manipulative skill involved since both of them cover two methods of making welds. This tends to create confusion since a welder trained to do semiautomatic welding using one process may not be able to do manual welding with another process. This is not so important for the

welding operator since the difference in skill for machine welding and automatic welding is not so great. The American Welding Society has established four specific methods of applying the many welding processes. These are based on the following interpretations:

Manual: Done, made, operated by, or used with the hand or hands.

Semiautomatic: Operated partly automatically and partly manually.

Machine: Mechanism serving to transmit and modify force and motion so as to perform some kind of work.

Automatic: Having a self-acting or self-regulating mechanism that performs a required act as a predetermined point in an operation.

The four methods of applying are shown by Figure 2-18 and are defined below.

MA, Manual Welding: Welding wherein the entire welding operation is performed and controlled by hand.

SA, Semiautomatic Welding: Welding with equipment which controls only the filler metal feed. The advance of welding is manually controlled.

ME, Machine Welding: Welding with equipment which performs the welding operation under the constant observation and control of an operator. The equipment may or may not perform the loading and unloading of the work.

AU, Automatic Welding: Welding with equipment which performs the entire welding operation without constant observation and adjustment of the controls by an operator. The equipment may or may not perform the loading and unloading of the work.

This concept of varying degrees of control in the hands of the welder can be better understood when we consider the normal activities involved when making an arc weld manually. These are defined and broken down into the following functions:

1. Maintain the arc—holding and controlling the correct arc length. This also means controlling the arc voltage in shielded metal arc welding.

Semiautomatic

Automatic

Manual

Machine

Method of Application → ↓ Activities to Make Weld	MA Manual	SA Semiautomatic	ME Machine	AU Automatic
Maintain the arc	Individual	Machine	Machine	Machine
Feed filler metal	Individual	Machine	Machine	Machine
Provide joint travel	Individual	Individual	Machine	Machine
Provide joint guidance	Individual	Individual	Individual	Machine
Applied by	The Welder		The Welding Operator	

FIGURE 2-18 *Man-machine relationship re method of applying.*

Welding Process	MA Manual	SA Semiautomatic	ME Machine	AU Automatic
Shielded metal arc*	Most popular	Not used	Not used	Special
Gas tungsten arc*	Most popular	Possible—rare	Used	Used
Plasma arc*	Most popular	Not used	Used	Used
Submerged arc*	Not possible	Little used	Most popular	Popular
Gas metal arc*	Not possible	Most popular	Used	Popular
Flux cored arc*	Not possible	Most popular	Used	Popular
Electroslag	Not possible	Possible—rare	Most popular	Used
Torch brazing*	Most popular	Used	Used	Used
Oxyfuel gas*	Most popular	Not used	Little used	Little used
Thermal cutting*	Most popular		Popular	Popular

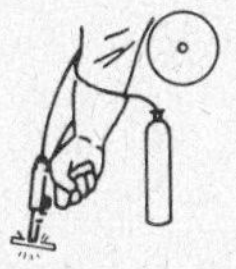

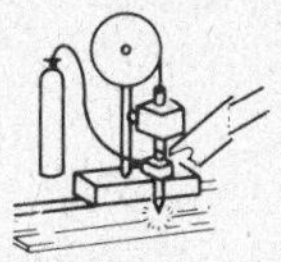

*Manipulative skills required for these processes when applied
manually or semiautomatically.

FIGURE 2-19 *Possible methods of applying the various processes.*

2. Feed filler metal into the joint—feeding the electrode or "cold" filler wire into the joint.

3. Provide travel along the joint—providing relative motion or progression along the weld joint.

4. Provide joint guidance—following the joint and providing uniform fill or size.

The person–machine relationship, shown by Figure 2-18, shows that in manual welding the individual has control over all four of these functions and in automatic or automated welding the same four functions are completely controlled by the machine. The skill required by the individual is greatest when all four functions are under the control of the person and diminishes as the functions are taken over by the machine.

All of the arc welding processes can be analyzed with respect to the "method of applying." AWS uses these methods of applying also for gas welding, brazing, and cutting operations. The method of applying the resistance welding process, solid state welding processes, and most of the others is dictated by the process and the machine. Certain of the welding processes may be applied only as a manual process while others are applied as semiautomatic, machine, or fully automatic. The chart of welding processes—Method of Application, Figure 2-19—shows this relationship for some of the more popular processes. This figure also shows the processes where manipulative skills are normally involved. For these processes aptitude testing is suggested and skill training is required.

The method of application is extremely important when writing procedures or assessing the economic capabilities of a process.

2-4 WELDING PROCEDURES

As welding becomes a modern engineering technology it requires that the various elements involved be identified in a standardized way. This is accomplished by writing a procedure which is simply a "manner of doing" or "the detailed elements (with prescribed values or range of values) of a process or method used to produce a specific result." The AWS definition for a welding procedure is "the detailed methods and practices including all joint welding procedures involved in the production of a weldment." The joint welding procedure mentioned includes "the materials, detailed methods and practices employed in the welding of a particular joint."

A welding procedure is used to make a record of all of the different elements, variables, and factors that are involved in producing a specific weld or weldment. Welding procedures should be written whenever it is necessary to:

Maintain dimensions by controlling distortion.

Reduce residual or locked up stresses.

Minimize detrimental metallurgical changes.

Consistently build a weldment the same way.

Comply with certain specifications and codes.

Welding procedures must be tested or qualified and they must be communicated to those who need to know. This includes the designer, the welding inspector, the welding supervisor, and last but not least, the welder.

When welding codes or high-quality work is involved this can become a *welding procedure specification* known as a "WPS," which lists in detail the various factors or variables involved. Different codes and specifications have somewhat different requirements for a welding procedure, but in general a welding procedure consists of three parts as follows:

1. A detailed written explanation of how the weld is to be made.
2. A drawing or sketch showing the weld joint design and the conditions for making each pass or bead.
3. A record of the test results of the resulting weld.

If the weld meets the requirements of the code or specification and if the written procedure is properly executed and signed it becomes a *qualified welding procedure* known as a "QWP."

The variables involved in most specifications are considered to be essential variables. In some codes the term nonessential variables may also be used. Essential variables are those factors which must be recorded and if they are changed in any way, the procedure must be retested and requalified. Nonessential variables are usually of less importance and may be changed within prescribed limits and the procedure need not be requalified.

Essential variables involved in the procedure usually include the following:

1. The welding process and its variation.
2. The method of applying the process.
3. The base metal type, specification, or composition.
4. The base metal geometry, normally thickness.
5. The base metal need for preheat or postheat.
6. The welding position.
7. The filler metal and other materials consumed in making the weld.
8. The weld joint, that is, the joint type and the weld.
9. Electrical or operational parameters involved.
10. Welding technique.

Some specifications also include nonessential variables and these are usually the following:

1. The travel progression (uphill or downhill).
2. The size of the electrode or filler wire.
3. Certain details of the weld joint design.
4. The use and type of weld backing.
5. The polarity of the welding current.

The procedure write-up must include each of the listed variables and describe in detail how it is to be done. The second portion of the welding procedure is the joint detail sketch and table or schedule of welding conditions.

Tests are performed to determine if the weld made to the procedure specification meets certain standards as established by the code or specification. If the destructive tests meet the minimum requirements the procedure then becomes a *qualified procedure specification* "QPS." The writing, testing, and qualifying procedures become quite involved and are different for different specifications and will be covered in detail in a later chapter.

In certain codes, welding procedures are prequalified. By using data provided in the code individual qualified procedure specifications are not required, for the standard joints on common base materials using the shielded metal arc welding process.

The factors included in a procedure should be considered in approaching any new welding job. By means of knowledge and experience establish the optimum factors or variables in order to make the best and most economical weld on the material to be welded and in the position that must be welded.

Welding procedures take on added significance based on the quality requirements that can be involved. When exact reproducibility and perfect quality are required, the procedures will become much more technical with added requirements, particularly in testing. Tests will become more complex to determine that the weld joint has the necessary properties to withstand the service for which the weld is designed.

Procedures are written to produce the highest-quality weld required for the service involved, but at the least possible cost and to provide weld consistency. It may be necessary to try different processes, different joint details, and so on, to arrive at the lowest-cost weld which will satisfy the service requirements of the weldment.

The contents of a welding procedure are brought out at this early stage to help you realize the importance of defining the factors involved in making a successful weld.

2-5 WELDING ELECTRICITY

The electrical arc welding circuit is the same as any electrical circuit. In the simplest electrical circuits, there are three factors:

(a) Current—flow of electricity.

(b) Pressure—force required to cause the current to flow.

(c) Resistance—force used to regulate the flow of current.

Current is a "rate of flow." Current is measured by the amount of electricity that flows through a wire in one second. The term ampere is the amount of current per second that flows in a circuit. The letter I is used to designate current amperes.

Pressure is the force that causes a current to flow. The measure of electrical pressure is the volt. The voltage between two points in an electrical circuit is called the difference in potential. This force or potential is called electromotive force or EMF. The difference of potential or voltage causes current to flow in an electrical circuit. The letter E is used to designate voltage or EMF.

Resistance is the restriction to current flow in an electrical circuit. Every component in the circuit, including the conductor, has some resistance to current flow. Current flows easier through some conductors than others; that is, the resistance of some conductors is less than others. Resistance depends on the material, the cross-sectional area, and the temperature of the conductor. The unit of electrical resistance is the *ohm*. It is designated by the letter R.

Copper is widely used for conductors since it has the lowest electrical resistivity of common metals. Insulators have a very high resistance and will not conduct current.

A simple electrical circuit is shown by Figure 2-20. This circuit includes two meters for electrical measurement, a voltmeter, and an ammeter. It also shows a symbol for a battery. The longer line of the symbol represents the positive terminal. Outside of a device that sets up the EMF, such as a generator or a battery, the current flows from the negative ($-$) to the positive

($+$). The arrow shows the direction of current flow.

The ammeter is a low resistance meter shown by the round circle and arrow adjacent to the letter I. The pressure or voltage across the battery can be measured by a voltmeter. The voltmeter is a high resistance meter shown by the round circle and arrow adjacent to the letter E.

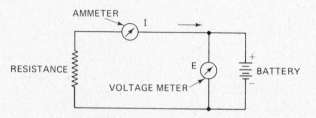

FIGURE 2-20 *Simple electrical circuit.*

The resistance in the circuit is shown by a zigzag symbol. The resistance of a resistor can be measured by an ohmmeter. An ohmmeter is *never* used to measure resistance in a circuit when current is flowing.

The relationship of these three factors is expressed by Ohm's law as follows:

$$\text{Current} = \frac{\text{Pressure}}{\text{Resistance}}$$

or

$$\text{Amperes} = \frac{\text{Volts}}{\text{Ohms}} \quad \text{or} \quad I = \frac{E}{R}$$

where I = current in amperes (flow),

E = pressure in volts (EMF),

R = resistance in ohms.

Ohm's law can also be expressed as:

$$E = IR \quad \text{or} \quad R = \frac{E}{I}$$

By simple arithmetic if two values are known or measured the third value can be determined.

A few changes to the circuit shown by Figure 2-20 can be made to represent an arc welding circuit. Replace the battery with a welding generator since they are both a source of EMF (or voltage) and replace the resistor with a welding arc which is also a resistance to current flow. The arc welding circuit is shown by Figure 2-21. The current will flow from the negative terminal through the resistance of the arc to the positive terminal.

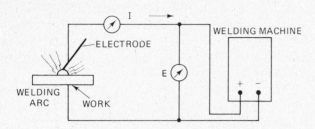

FIGURE 2-21 *Welding electrical circuit.*

In the early days of arc welding when welding was done with bare metal electrodes on steel, it was normal to connect the positive side of the generator to the work and the negative side to the electrode. This provided 65% to 75% of the heat to the work side of the circuit to increase penetration. When welding with the electrode negative, the polarity of the welding current was said to be *straight*. When special conditions (such as welding cast iron or nonferrous metals) made it advisable to minimize the heat in the base metal, the work was made negative and the electrode positive, and the welding current polarity was said to be *reverse*.

In those days, to change the polarity of the welding current it was necessary to remove the cables from the machine terminals and replace them in the reverse position. The early coated electrodes for welding steel gave best results with the electrode positive or reverse polarity; however, bare electrodes were still used. It was necessary to change polarity frequently when using both bare and covered electrodes.

To meet this condition, welding machines were equipped with switches that changed the polarity of the terminals and with dual reading meters. Thus the welder could quickly change the polarity of the welding current. In marking welding machines and polarity switches these old terms were used and indicated the polarity as *straight* when the electrode is negative, and *reverse* when the electrode is positive. In this book, in order to avoid confusion, whenever polarity is discussed the term *electrode negative* (DCEN) will be used instead of straight polarity (DCSP) and *electrode positive* (DCEP) will be used instead of reverse polarity (DCRP).

The ammeter used in a welding circuit is a millivoltmeter calibrated in amperes connected across a high current shunt in the welding circuit. The shunt is a calibrated, very low resistance conductor. The voltmeter shown in the figure will measure the welding machine output and the voltage across the arc which are essentially the same. Before the arc is struck or if the arc is broken, the voltmeter will read the voltage across the machine with no current flowing in the circuit. This is known as the open circuit voltage and is higher than the arc voltage or voltage across the machine when current is flowing.

Another unit in an electrical circuit, and important to welding, is the unit of power. The rate of producing, or of using energy, is called power and is measured in watts. Power in a circuit is the product of the current in amperes times the pressure in volts or:

$$Power = Current \times Pressure$$

or

$$Watts = Amperes \times Volts$$

or

$$P = I \times E$$

where P = Power in watts,

I = Current in amperes,

E = Pressure in volts.

When welding using a 1/8-in. electrode at 100 amperes and an arc voltage of 25 the power would be 2,500 watts. 2,500 watts can be expressed as 2.5 kilowatts. Power is measured by a wattmeter which is a combination of an ammeter and a voltmeter.

In addition to power, it is necessary to know the amount of work involved. Electrical work or energy is the product of power times time and is expressed as watt seconds or joules or kilowatt hours.

$$Work = Power \times Time \quad or \quad W = Pt$$

where W = Work in joules or watt seconds or kilowatt hours,

P = Power in watts or kilowatts,

t = Time in seconds or hours.

Cost of welding calculations involve these work units since the watt hour or kilowatt hour are commercial units of work and are the basis of charges by the electric utility companies.

So far, we have dealt exclusively with direct current electricity, electricity that flows continually through the circuit in the same direction. Alternating current electricity is also important since it is the power furnished by utility companies.

Alternating current is an electrical current which flows back and forth at regular intervals in a circuit. When the current rises from zero to a maximum, returns to zero and increases to a maximum in the opposite direction, and finally returns to zero again, it is said to have completed one cycle. For convenience, a cycle is divided into 360 degrees. Figure 2-22 is a graphical representation of a cycle, and is called a sine wave. It is generated by one revolution of a single loop coil armature in a two-pole alternating current generator. The maximum value in one direction is reached at the 90° point and in the other direction at the 270° point. The number of times this cycle is repeated in one second is called the frequency and is measured in hertz. When a current rises to a maximum in each direction 60 times a second it completes 60 cycles per second or has a frequency of 60 hertz. The frequency of electrical power in North America and other parts of the world is 60 hertz. Fifty hertz is used in Europe and Africa.

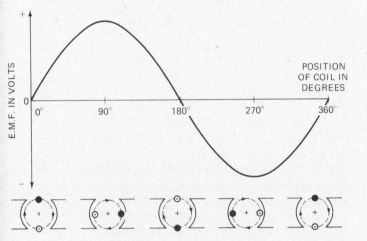

FIGURE 2-22 *Sine wave generation.*

The principle of electrical generation states that "when a conductor moves in a magnetic field so as to cut lines of force an electro-motive force is generated." The lines of force run between the north and south magnetic poles of the generator. The single turn coil rotates within these lines of force or magnetic field and as the conductor cuts the lines of force at right angles the maximum voltage is generated, i.e., at 90° and at 270°. When no lines of force are being cut as at positions 0°, 180°, and 360°, there is no EMF generated. The EMF generated in the one loop coil is taken from the rotating armature by means of slip rings. In welding generators there are usually more than two poles and many hundred loops of wire in the coil.

Alternating current for arc welding normally has the same frequency as the line current. The voltage and current in the AC welding arc follow the sine wave and go through zero twice each cycle. The frequency is so fast that the arc appears continuous and steady to the naked eye. The sine wave is the simplest form of alternating current. It is always assumed that alternating current has a sine wave shape unless otherwise stated.

Alternating current and voltage are measured with AC meters. An AC voltmeter measures the value of both the positive and negative parts of the sine wave. It reads the effective voltage, called the root-mean-square (R.M.S.) voltage. The effective direct current value of an alternating current or voltage is 0.707 times the maximum value.

An alternating current has no unit of its own, but is measured in terms of direct current, the ampere. The ampere is defined as a steady rate of flow, but an alternating current is not a steady current. An alternating current is said to be the equivalent to a direct current when it produces the same average heating effect under exactly similar conditions. This is used since the heating effect of a negative current is the same as that of a positive current. Therefore, an AC ammeter will measure a value called the effective value of an alternating current which is shown in amperes. All AC meters, unless otherwise marked, read effective values of current and voltage.

Ohm's law applies also to alternating current circuits. This is because Ohm's law deals only with voltage, current, and resistance. In alternating current welding circuits there are other factors, and one of the most important is inductance. To understand inductance we must refer to magnetism.

A magnet has a north pole and a south pole, which have identical strength. Between these poles there are lines of force. This effect can be shown by sprinkling iron filings on a sheet of paper and placing it over a magnet. The distinct pattern shows these lines of force running from one pole to the other. Similar lines of force exist around electric conductors that carry direct current. This can be proven by placing a small compass near a current-carrying wire. The needle will deflect when the current is turned off and on. Magnetic lines of force create physical forces between magnets or magnetic fields around current carrying wires. This is the principle of operation of an electric motor. The magnetic properties of a ferromagnetic material such as iron when wrapped with a coil of wire is such that the combination will produce a much stronger magnetic field than the magnetic field produced by the coil alone. The coil of wire around an iron core is a magnetic circuit. Magnetic circuits will have a specific inductance. Inductance expresses the results of a certain arrangement of conductors, iron, and magnetic fields. Inductance involves change since it only functions when magnetic lines of force are cutting across electrical conductors. Inductance is important only in alternating current circuits or in direct current circuits when they are connected or disconnected. When the current is turned off, the magnetic field collapses and the lines of force cut across the wires and induce current in the wires in the same direction as it had been flowing. If the coil is connected to alternating current the lines of force build up to the maximum and then collapse and then build up in the opposite direction to a maximum and collapse each cycle. If another coil is placed on the same iron core and close to the first coil the magnetic lines of force will cut across the second coil and induce an EMF in it. The closer the coils, or the stronger the magnetic lines of force, the greater will be the induced EMF. This is the principle of the transformer and is shown by Figure 2-23. By changing the magnetic coupling of the two coils we can control the output of the second coil (the secondary) and thus the output of the welding transformer. This coupling can be changed by moving the coils closer together or by increasing the strength of the magnetic field between them. The strength of the magnetic field can be changed by putting more or less iron in the area between the coils or by adjusting the availability of the magnetic field in other ways.

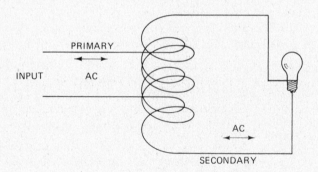

FIGURE 2-23 *Transformer principle.*

The output of a transformer welding machine is alternating current of the same frequency as the input power. A rectifier is a device that conducts current easier in one direction than the other. It has a high resistance to current flowing in one direction and a low resistance to current flowing in the opposite direction. A diode vacuum tube is an efficient rectifier but will not carry sufficient current for welding. Another type, the dry disc rectifier, employs layers of semiconductors such as selenium between plates. Selenium rectifiers are used in small welding machines. The newest and most popular rectifier is the silicon diode. These are made of thin wafers of silicon that have had small

amounts of impurities added to make them semiconductors. The wafers are specially treated and then assembled in holders for mounting in welding machines. The diodes are connected to the output of a welding transformer to produce a rectifier welding machine with D.C. output.

This brief discussion of electricity is presented to help explain the electricity of welding. More about welding electricity will be given in Section 4-1 and 5-1 and Chapter 8 on power source for welding.

2-6 WELDING ARCS

We have used the welding arc for many years without really understanding its complexities. From the beginning of welding, the arc has been utilized as a concentrated source of high temperature heat that can be moved and manipulated to melt the base metal and filler metal to produce welds. As time went on, scientists have investigated the welding arc and learned many of its secrets.

There are two basic types of welding arcs; one uses the nonconsumable electrode and the other the consumable electrode. The nonconsumable electrode does not melt in the arc and filler metal is not carried across the arc stream. The welding processes that use the nonconsumable electrode arc are: carbon arc welding, gas tungsten arc welding, and plasma arc welding. The consumable electrode melts in the arc and is carried across the arc in a stream to become the deposited filler metal. The welding processes that use the consumable electrode arc are: shielded metal arc welding, gas metal arc welding, flux-cored arc welding, and submerged arc welding.

The main function of the arc is to produce heat. At the same time it produces a bright light, noise, and, in a special case, bombardment that removes surface films of the base metal.

A welding arc is a sustained electrical discharge through a high conducting plasma. It produces sufficient thermal energy which is useful for joining metals by fusion.[2] The welding arc is a steady-state condition maintained at the gap between an electrode and an work that can carry current ranging from as low as 5 amperes to as high as 2,000 amperes and a voltage as low as 10 volts to the highest voltages used on large plasma units. The welding arc is somewhat different from other electrical arcs since it has a "point-to-plane" geometric configuration, the point being the arcing end of the electrode and the plane the arcing area of the workpiece. Whether the electrode is positive or negative, the arc is restricted at the electrode and spreads out towards the workpiece.

The length of the arc is proportional to the voltage across the arc. Every welder knows that if the arc length is increased beyond a certain point the arc will suddenly go out. This means that there is a certain current necessary to sustain an arc of different lengths. If a higher current is used a longer arc can be maintained.

The arc column is normally round in cross section and is made of two concentric zones, an inner core or plasma and an outer flame. The plasma carries most of the current. The plasma of a high-current arc can reach a temperature of from 5,000 to 50,000° Kelvin. The outer flame of the arc is much cooler, and tends to keep the plasma in the center. The temperature and the diameter of the central plasma depend on the amount of current passing through the arc, the shielding atmosphere, electrode size and type.

The relationship between current and arc voltage is not in exact accordance with Ohm's law, since the volt-ampere characteristic curve of the welding arc is not a straight line. The curve of an arc, shown by Figure 2-24, takes on a nonlinear form which in one area has a negative slope.[3] The arc voltage increases slightly as the current increases. This is true except for the very low-current arc which has a higher arc voltage. This is because the low-current plasma has a fairly small cross-sectional area, and as the current increases the cross section of the plasma increases and the resistance is reduced. The conductivity of the arc increases at a greater rate than simple proportionality to current.

The arc is maintained when electrons are emitted or evaporated from the surface of the negative pole (cathode) and flow across a region of hot electrically charged gas to the positive pole (anode) where they are absorbed. Cathode and anode are electrical terms for the negative and positive poles.

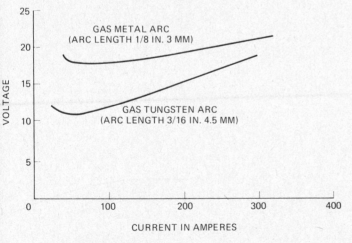

FIGURE 2-24   *Arc characteristic volt amp curve.*

Arc action can be best explained by considering the DC tungsten electrode arc in an inert gas atmosphere as shown by Figure 2-25. On the left the tungsten arc is connected for direct current electrode negative (DCEN).

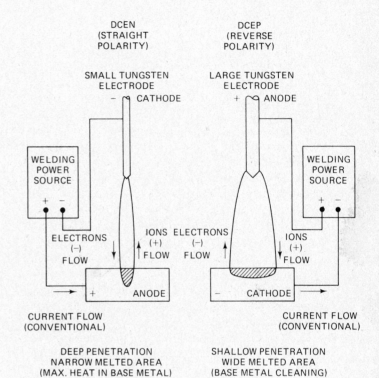

FIGURE 2-25   *The dc tungsten arc.*

When the arc is started the electrode becomes hot and emits electrons. The emitted electrons are attracted to the positive pole, travel through the arc gap and raise the temperature of the argon shielding gas atoms by colliding with them. The collisions of electrons with atoms and molecules produce thermal ionization of some of the atoms of the shielding gas. The positively charged gaseous atoms are attracted to the negative electrode where their kinetic (motion) energy is converted to heat. This heat keeps the tungsten electrode hot enough for electron emission. Emission of electrons from the surface of the tungsten cathode is known as thermionic emission. Positive ions also cross the arc. They travel from the positive pole, the work, to the negative pole, the electrode. Positive ions are much heavier than the electrons, but help carry the current flow of the relatively low voltage welding arc. The largest portion of the current flow, approximately 99%, is via electron flow rather than the flow of positive ions. The continuous feeding of electrons into the welding circuit from the power source accounts for the continuing balance between electrons and ions in the arc. The electrons colliding with the work creates the intense localized heat which provides melting and deep penetration of the base metals.

In the DC tungsten to base metal arc in an inert gas atmosphere, the maximum heat occurs at the positive pole (anode).[4] When the electrode is positive (anode), and the work is negative (cathode), as shown on the right by Figure 2-25, the electrons now flow from the work to the electrode where they create intense heat. The electrode tends to overheat and so a larger electrode with more heat-absorbing capacity is used for DCEP than for DCEN for the same welding current. In addition, since less heat is generated at the work the penetration is not so great. One result of DECP welding is the so-called "cleaning effect" on the base metal adjacent to the arc area. This appears as an etched surface, and is known as cathodic etching and results from positive ion bombardment. This positive ion bombardment also occurs during the reverse polarity half-cycle when using alternating current for welding.

Researchers theorized that if the arc column of the tungsten arc could be reduced in cross section its temperature would increase. If the cross-section area of the conducting column is reduced by constricting it, the only way that the same current would still pass is for its conductivity and its temperature to increase. Constriction occurs in a plasma arc torch by making the arc pass through a small hole in a water-cooled copper nozzle. It is a characteristic of the arc that the more it is cooled the hotter it gets, however, it requires a higher voltage. By flowing additional gas through the small hole, the arc is further constricted and a high velocity, high temperature gas jet or plasma emerges. This plasma is now being used for welding, for cutting, and for metal spraying.

The arc length or gap between the electrode and the work can be divided into three regions, a central region, a region adjacent to the electrode, and a region adjacent to the work. At the end regions the cooling effects of the electrode and the work causes a rapid drop in potential. These two regions are known as the anode and cathode drop, according to the direction of current flow. The length of the central region or arc column represents 99% of the arc length and is linear with respect to arc voltage. Figure 2-26 shows the distribution of heat in the arc which varies in these three regions.[5] In the central region, a circular magnetic field surrounds the arc. This field, produced by the current flow, tends to constrict the plasma and is known as the magnetic pinch

effect. The constriction causes high pressures in the arc plasma and extremely high velocities and this in turn produces a plasma jet. The speed of the plasma jet approaches sonic speed.

The cathode drop is the electrical connection between the arc column and the negative pole (cathode). There is a relatively large temperature and potential drop at this point. This is the point the electrons are emitted by the cathode and given to the arc column. The stability of an arc depends on the smoothness of the flow of electrons at this point. Tungsten and carbon provide thermic emissions since both are good emitters of electrons. They have high melting temperatures, are practically nonconsumable, and are therefore used for welding electrodes. Since tungsten has the highest melting point of any metal, it is preferred.

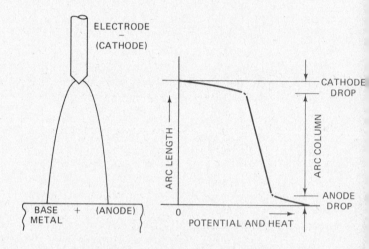

FIGURE 2-26 *Arc length vs voltage and heat.*

The anode drop occurs at the other end of the arc and is the electrical connection between the positive pole (anode) and the arc column. The temperature changes from that of the arc column to that of the anode, which is considerably lower. The reduction in temperature occurs because there are less ions in this region. The heat liberated at the anode and at the cathode is greater than from the arc column.

In the carbon arc, a stable DC arc is obtained when the carbon is negative. In this condition about 1/3 of the heat occurs at the negative pole (cathode), the electrode, and about 2/3 of the heat occurs at the positive pole (anode), the workpiece.

In the consumable electrode welding arc, the electrode is melted and molten metal is carried across the arc. A uniform arc length is maintained between the electrode and the base metal by feeding the electrode into the arc as fast as it melts. The arc atmosphere has a great effect on the polarity of maximum heat. In shielded metal arc welding the arc atmosphere depends on the composition of the coating on the electrode. Usually the maximum heat occurs at the negative pole (cathode). When straight polarity welding with an E6012 electrode, the electrode is the negative pole (DCEN) and the melt-off rate is high and penetration is minimum. When reverse polarity welding with an E6010 electrode (DCEP) the maximum heat still occurs at the negative pole (cathode) but this is now the base metal, which provides deep penetration. This is shown by Figure 2-27. With a bare steel electrode on steel the polarity of maximum heat is the positive pole (anode). This is why bare electrodes were operated on straight polarity (DCEN) so that maximum heat would be at the base metal (anode) to ensure adequate penetration. When coated electrodes are operated on AC the same amount of heat is produced at each polarity of the arc.

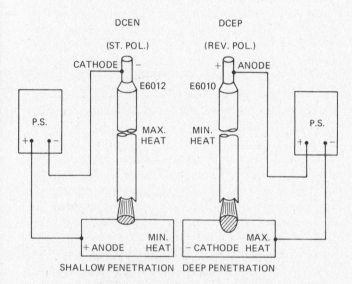

FIGURE 2-27 *The dc shielded metal arc.*

The forces that cause metal to transfer across the arc are similar for all the consumable electrode arc welding processes. The type of metal transfer dictates the usefulness of the welding process. It affects the welding position that can be used, the depth of weld penetration, the stability of the welding pool, the surface contour of the weld, and the amount of spatter loss. The metal being transferred ranges from small droplets, smaller than the diameter of the electrode, to droplets larger in diameter than the electrode. The type of transfer depends on the current density, the polarity of the electrode, the arc atmosphere, the electrode size, the electrode composition, and so on.

The mechanism of transferring liquid metal across an arc has been a mystery to many welders. Scientists have found that several forces are involved. These are:

surface tension, the plasma jet, gravity, at least in flat position welding, and electromagnetic force.

Surface tension of a liquid, causes the surface of the liquid to contract to the smallest possible area. This tension tends to hold the liquid drops on the end of a melting electrode without regard to welding position. This force works against the transfer of metal across the arc. It helps keep molten metal in the weld pool when welding in the overhead position.

The welding arc is constricted at the electrode and spreads or flares out at the workpiece. The current density and the arc temperature are the highest where the arc is the most constricted or at the end of the electrode. An arc operating in a gaseous atmosphere contains a plasma jet which flows along the center of the arc column between the electrode and the base metal. Molten metal drops in the process of detachment from the end of the electrode, or in flight, are given acceleration towards the work piece by the plasma jet.

Earth gravity tends to detach the liquid drop when the electrode is pointed downward and is a restraining force when the electrode is pointing upward. Earth gravity has a noticeable effect only at low currents. The difference between the mass of the molten metal droplet and the mass of the workpiece has a gravitational effect which tends to pull the droplet to the workpiece. An arc between two electrodes will not deposit metal on either.

Electromagnetic force also helps transfer metal across the arc. When the welding current flows through the electrode a magnetic field is set up around it. The electromagnetic force acts on the liquid metal drop when it is about to detach from the electrode. As the metal melts, the cross-sectional area of the electrode changes at the molten tip. The electromagnetic force depends upon whether the cross section is increasing or decreasing. There are two ways in which the electromagnetic force acts to detach a drop at the tip of the electrode. When a drop is larger in diameter than the electrode and it is positive (DCEP) the magnetic force tends to detach the drop. When there is a constriction or necking down which occurs when the drop is about to detach, the magnetic force acts away from the point of constriction in both directions, and the drop that has started to separate will be given a push which increases the rate of separation.[6] Figure 2-28 illustrates these two points. Magnetic force also sets up a pressure within the liquid drop. The maximum pressure is radial to the axis of the electrode and at high currents causes the drop to become elongated. It gives the drop stiffness and causes it to project in line with the electrode regardless of the welding position.

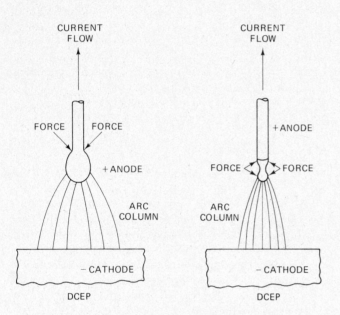

FIGURE 2-28 *The dc consumable electrode metal arc.*

2-7 THE PHYSICS AND CHEMISTRY OF WELDING

Welding follows all of the physical laws of nature and a good understanding of physics and chemistry will help you better understand how welds are made.

Physics deals with energy and motion and is subdivided into such subjects as mechanics, sound, light, friction, magnetism, electricity, heat, and so on. This section briefly describes some of these subjects.

The science of mechanics involves physical laws that relate to forces, motion, and direction. The term **force** is defined as a push or pull, specifically a force is a tendency to produce a change, a change in motion of the body upon which it acts. It is not necessary that the body be in motion; it is only necessary that there is a tendency to produce change. The first law of motion states that a body will remain at rest or in uniform motion, if no force acts upon it. One of the forces that we live with daily is the pull of gravity which acts on all objects on the earth's surface. Another of the laws states that for every action there is an equal and opposite action; that is, forces act in pairs. An opposite action force cannot exist before the action force takes

place. If you push against the wall, the wall pushes back an equal and opposite force equal to your push; otherwise the wall would move. All forces and all opposite forces or reaction forces have direction and also have **magnitude,** which is the amount of force involved. It is possible to graphically represent different forces and magnitudes and direction by means of vector diagrams. Vectors indicate the direction, and the length of the vector indicates the magnitude. A thorough knowledge of this science is necessary in order to design welded structures. It is the basis for establishing the sizes of members and would be the basis for the size of welds to join members together.

The science of sound is important to welding since one welding process and one weld nondestructive examination technique is based on the use of sound. Sound is transmitted through most materials: metals, gases, liquids, etc., but it will not pass through a vacuum. Sound is an alternating type of energy based on vibrations, which are regions of compaction and rarification. A compression wave and rarification wave are alternating pressures or vibrations which allow your eardrums to hear. The hearing of most people is sensitive between 20 and 20,000 vibrations per second. Sound has pitch, loudness, and quality. Pitch is defined as frequency, the higher the frequency, the higher the pitch. Loudness is subjective and is related to intensity, which is the basis of the energy in sound. Quality is the function of wave form based on the frequency and phase of combining vibrations. The use of sound in welding is in the higher-than-audible sound vibrations, above the normal hearing range for people. Ultrasonic vibrations are used to make welds and they are also used to detect voids in metals.

The science of light also involves welding. The laser beam welding process utilizes light energy at very high concentrations to create heat sufficient to cause melting, which can be used for welding or cutting. Light is a by-product of the arc welding processes. Light is given off by the arc and by heated electrodes and base metals. Light is a phenomenon that has never been completely explained to the satisfaction of scientists. Propagation of light is explained by alternating vibrations or the wave theory, while the energy transfer of light is explained by the particle theory. It is sufficient for us to know that light is transmitted through vacuums and gases, but not through all types of materials. Light is of interest to the welder from the point of view of protecting the body from the effects of the light rays. In the electromagnetic spectrum, light ranges from the lower frequencies of infrared through visible light up through ultraviolet and X-rays. The speed of light is independent of the light source, its intensity, or color. Light transmits energy and the higher the frequency the greater the energy; however, light radiates equally in all directions and its strength diminishes by the square of the distance.

The science of friction also involves welding. Here we are interested in dynamic friction, better known as sliding friction. This is the force between two moving bodies and if sufficient force is available heat will be generated. This is the basis for the friction-welding process.

The magnetic theory and its relationship to current flow is explained in the section on welding electricity. One aspect of magnetic fields can be detrimental. Welders call it arc blow which is the deflection of an electric arc from its normal path due to magnetic forces. The leads from the welding machine to the electrode and from the work back carry a heavy current and create a magnetic field. The welding current flowing through the electrode and the base metal, providing it is ferromagnetic, also create magnetic fields. The intensity of the magnetic field is directly proportional to the square of the current flowing. The distribution of a magnetic field in a welding circuit can become quite complex, particularly for nonuniform joint details and also when fixtures are employed. If the distribution of the magnetic field close to the arc is not uniform it may cause the arc to deflect or attract toward the stronger portion, depending on polarity. Arc blow can create difficulties which affect weld quality. One advantage of alternating current is that the arc blow is minimized since the magnetic field is changing at line frequency and does not build up to as great a force.

Chemistry deals with the makeup of all matter. We are most interested in metals in connection with welding. By definition matter is anything that occupies space and has mass or weight. Also by definition elements are those particular kinds of matter which cannot be decomposed or broken down into simpler substances by ordinary means. Pure metals and pure gases are examples of elements. Compounds or mixtures can be broken down into their original elements. The elements are composed of atoms which are identical with each other atom of the same element but are different from atoms of other elements. The molecule is the smallest particle of a substance which has all the properties of that substance. It is a combination of two or more atoms of the same element or of different elements.

Many of the elements have similar properties; for example, some are inert gases, others are noble metals, others are active gases, etc. This allows the elements to be classified and put into groups of families. This classification is called the *periodic table,* which is shown in all chemistry textbooks.

In order to better understand metals we must first consider the atom and its structure. Scientists believe

that each atom is composed of a very small compact nucleus surrounded by empty space in which one or more electrons revolve about the nucleus. It is believed that the nucleus of the atom is the most dense form of matter known. It is made or contains two main types of particles known as protons and neutrons. These particles differ from each other in their charge, but they have about the same weight. The positive particle of matter is called a proton. All atoms have at least one proton in their nuclei. The number of protons in the nucleus is equal to the number of electrons outside the nucleus. Each proton has a charge of plus one, the charge on the nucleus is positive since it contains positively charged protons and no electrons. The neutron in the nucleus is a particle of matter which has a relative weight of 1 but it has no electrical charge. The third item is the electron and it is very light in comparison with the nucleus of the atom. Each electron has a charge of negative 1. Electrons of an atom are located in shells around the nucleus. These shells are more properly called energy levels because electrons in different shells have different amounts of energy. Electrons revolve around the nucleus in these shells in varying distances from the nucleus. Electrons are relatively far from the nucleus so that most of the atom consists of *empty space.* The difference between atoms of different elements is the result of differences in the number of protons and neutrons in the nucleus and the difference in the number and arrangement of the electrons surrounding the nucleus. Electrons in the outer shells have more energy than those in the inner shells. An electron can change from one shell or energy level to another; if it absorbs energy it moves to an outer shell or to a higher energy level. If it gives off energy it drops to a shell closer to the nucleus. Energy emitted when electron drops from a higher to a lower energy level is in the form of electromagnetic radiation, light, or X-rays. The study of the makeup of the atoms is extremely technical and beyond the scope of this book. However, this brief explanation will help you better understand the makeup of metals which will be further explained in the chapter on metallurgy.

Matter can exist in four states. These are known as solids, liquids, gases, and plasmas. Changes from one state to another are brought about by supplying energy in the form of heat. Water in the solid state is ice, by adding heat, the ice changes to water, which is its liquid state and by adding additional heat, will be converted to its gaseous state. The reverse can be done by removing

THE FUNDAMENTALS OF WELDING

heat energy and the gas (steam) turns to the liquid (water) and then to the solid (ice). Most substances can be changed from one physical state to another in the same manner. The temperature of these changes indicates the physical state the element will be at the normal room temperatures.

Each of the elements on the periodic table has its own name and symbol. Most of these elements will combine chemically to form compounds. There is the law of definite composition which states that a chemical compound always contains the same elements with the same ratio of atoms of each. This is not true for mixtures. Most commercial metals are mixtures or alloys in that they are predominantly the element of the pure metal plus additions of other elements but not chemically combined as a compound. Gases also occur as elements, compounds, or mixtures. Air is a mixture of approximately 78% nitrogen and 20% oxygen with small amounts of other elements. Carbon dioxide is a compound and always in the ratio of 1 atom of carbon and 2 of oxygen. Argon is an element and more importantly an inert gas which will not chemically combine with any other element.

Several other chemical definitions relate to welding. One is known as burning or oxidation. This takes place when any substance combines with oxygen usually at high temperatures. An example of this is the combining of acetylene with oxygen. This produces carbon dioxide plus water plus a large amount of heat. We use the heat produced by the burning of acetylene in the flame of the oxyacetylene torch to make welds. In all oxidation reactions heat is given off. Oxidation can occur very slowly as in the case of rusting. If iron is exposed to oxygen at high temperature rapid oxidation or burning will occur with the liberation of more heat. Rapid oxidation or burning does not occur until the kindling temperature of the material is reached. In the case of a liquid this term is called the flash point. Oxidation is very important in welding operations since oxygen of the air is usually present as well as heat.

Another chemical definition is reduction which is the process by which oxygen is taken from another element. The substance used to take the oxygen from the element is called a reducing agent. A reducing agent is an element which adds electrons to another element. Hydrogen is one of the most active reducing agents; however, in the case of the iron and oxygen reaction mentioned above, the iron is the reducing agent. Whenever there is an oxidation reaction there is also a reduction reaction. A common term in oxyacetylene welding is a reducing atmosphere or an oxidizing atmosphere. The flame can be adjusted to provide sufficient oxygen for complete combustion or an excess of acetylene and insufficient oxygen resulting in incomplete combustion. Acetylene is high in hydrogen and carbon. The oxidizing flame would contribute excess oxygen; the reducing flame would contribute hydrogen and carbon.

The last two definitions are common in chemistry but less common in welding. The words are acidic and basic and refer to acid or base substances. A measure of basicity or acidity is by means of the pH scale. A pH of 7 is considered the neutral point. Pure water has a pH of 7 since it has the same number of hydrogen ions (H+) as hydroxide ions (OH−). Acids or acidic substances have an excess of hydrogen ions and have a pH value of less than 7. Bases or basic substances have an excess of hydroxide ions and have a pH value of more than 7. The terms are used in connection with nonmetallic slags used in welding and also used in steelmaking. These slags are related to the coatings of electrodes and have acidic or basic characteristics when heated to steel melting temperatures. Certain types of electrodes are called basic type because their coatings produce basic slags which react with impurities in the weld metal. Basic type coatings are the low hydrogen sodium or potassium containing types which produce basic slags, the basic slags remove appreciable amounts of undesirable phosphorous and sulfur from the molten weld metal. The celleousic-type electrodes are sometimes called the acidic type since an excess amount of hydrogen is present in the arc atmosphere and in the slag.

QUESTIONS

1. What is a weld?
2. What are the five basic joints?
3. What is the most popular weld type?
4. There are approximately how many welding processes?

5. How many arc welding processes are there?
6. How many basic welding positions are there?
7. Define the word *coalescence.* How does it apply to welding?

8. What is the difference between GTAW and GMAW?

9. Can all of the arc welding processes be used in "all positions"?

10. Explain the difference between the welder and the welding operator.

11. Name the four activities performed when making a manual arc weld.

12. Name the four methods of applying an arc weld.

13. Why is a written welding procedure required?

14. What is meant by the "point-to-plane" relationship of a welding arc?

15. Is the voltage across an arc the same for a long arc and a short arc? Explain.

16. What is the difference between a consumable electrode arc and a nonconsumable electrode arc?

17. Where does the deposited metal come from in a nonconsumable electrode welding arc?

18. Name the different branches of physics that relate to welding.

19. Pure metals and pure gases are elements or compounds?

20. Is an oxyacetylene flame reducing or oxidizing or both? Explain.

REFERENCES

1. *Terms and Definitions,* AWS A3.0-76, The American Welding Society, Miami, Florida.

2. Jackson, C. E. "The Science of Arc Welding," *Welding Journal Research Supplement,* June 1960, p. 225s.

3. Lancaster, J. F. "Energy Distribution in Argon-Shielded Welding Arcs," *British Welding Journal,* September 1954, p. 412.

4. Gibbs, E. F. "A Fundamental Study of the Tungsten Arc," *Metal Progress,* July 1960, p. 84.

5. Milner, Salter, and Wilkenson "Arc Characteristics and Their Significance in Welding," *British Welding Journal,* February 1960.

6. Lancaster, J. F., *The Metallurgy of Welding Brazing and Soldering,* New York: American Elsevier, 1965.

7. Hrivnak, Ivan, *The Theory of Mild Steel and Micro Alloy Steel's Weldability,* Bratislava, Czechoslovakia, 1969.

8. Rosenthal, D. "The Theory of Moving Sources of Heat and Its Application to Metal Treatment," *Transactions of the ASME,* November 1946, p. 849.

3

WELDING PERSONNEL SAFETY, TRAINING, AND QUALIFICATION

3-1 THE WELDING OCCUPATION

3-2 THE SAFETY AND HEALTH OF WELDERS

3-3 WELDING TRAINING PROGRAMS

3-4 TRAINING SCHOOL FACILITIES AND AIDS

3-5 QUALIFYING AND CERTIFYING WELDERS

3-6 TRAINING AND QUALIFYING OTHER WELDING PERSONNEL

3-1 THE WELDING OCCUPATION

The increasing use of welding in industry and in construction continues and is creating an increasing demand for trained welding personnel.

Welders can work practically anywhere from the small shop down the street to the largest factories in the major cities. In manufacturing, welders help build everything from automobiles to the largest earth-moving giants. They help build the space vehicles that take astronauts to the moon and millions of other products ranging from oil-drilling rigs to computers. In construction, welders are virtually rebuilding the world, extending subways, building bridges, and so on. They work practically everywhere helping to improve the environment by building and installing pollution control devices, water, and air filters. Welders work in shipyards, since all of the newer ships are welded, and they work on the farm repairing agricultural equipment. There is no lack of variety of the type of welding work that is done throughout the world.

Welding is a challenging career.[1] Work is done under all kinds of conditions both inside and outside. Welding is an exacting work; every weld on a nuclear

power reactor must be perfect. Welding is interesting work. The actual progress and completion of welds happens before the welder's eyes providing a sense of accomplishment of a job well done. Welding is a rewarding career since welders receive high pay. In factories, the pay of the welder is equivalent to that of a skilled machinist. In the construction trade it is equal to that of the skilled craftsman and in the piping trade it is perhaps the highest in the group.

Welding is growing and thus more welders are needed. The future is bright.[2] Even during economic slumps, there is usually a need for welders. During recessions companies rebuild machinery. Maintenance work receives high priority and this creates work for the skilled welder.

The United States Government Census figures[3] show that in 1960 there were 360,000 welders and flame cutters employed in the country. In 1970 there were 535,000 welders and flame cutters employed; 5% were women. This amounts to a 50% gain of welders while the population grew only 13.3%. The government estimates[4] that by 1980 there will be 675,000 welders and flame cutters required. The opportunity and challenges are here for skilled welders.

CHAPTER 3

WELDING PERSONNEL

Before seriously considering a career as a welder, you should review all aspects of the work. Working conditions vary considerably and in the construction trades it can be outside work in all kinds of weather many times at high heights and sometimes below the surface of the earth. Often the welder must be a member of a labor organization for the craft with which welding is associated. This may mean an apprenticeship program which can take up to six years. Sometimes when welders are urgently required, apprenticeship requirements are waived and the new welder can become a journeyman quickly. However, investigate this factor in the particular field of activity you wish to enter. Since welding is considered a tool of the trade in the construction industry, welders may be called boilermakers, plumber, pipe fitters, iron workers, sheet metal workers, etc., even though their principle occupation is welding.

In manufacturing, the variation of work is not as broad. The work is more repetitive unless it happens to be maintenance and repair welding. The need to weld in unusual positions on a wide variety of materials is normally not required.

The welder's job involves physical activities different from most occupations. The welder must hold a consistent arc and work in awkward and sometimes cramped positions for long periods of time. The welder is exposed to heat, hot metal and sparks, fumes, etc., while continually watching the arc. Also, the necessary protective clothing requirement should be considered in making a career decision.

The hours of work are the same as for others in the same factory or on the same project. Pay is similar to other skilled occupations either in construction or in manufacturing. Pay is sometimes regulated by the qualification level of the welder. Qualification tests may be required and are of different levels of severity. The most difficult usually qualify welders to work on the higher paying jobs.

Prospective welders should make field trips to factories and construction sites where welding is being done. You should talk with practicing welders to learn as much as possible about the occupation before making a firm decision to enter it.

Visiting welders at work and reviewing the job of the welder makes you soon realize that the work done by people called welders in different industries and companies, and in different geographical locations varies tremendously. In some production shops, the welder may make the same type of weld on the same part day after day as shown by Figure 3-1. Welders in construction and some industries may find each job totally different. A typical construction job is shown by Figure 3-2 which is on a nuclear piping assembly. Between these extremes there are many variations.

The "Dictionary of Occupational Titles"[5] published by the U.S. Department of Labor lists approximately 22,000 job titles which include over 75 for welding, flame cutting and related work. Job descriptions are written for many of these. An analysis of these has been made to determine the "Work Performed" by welders and "Worker Traits" of welders.[6] The work performed by welders was summarized in the previous paragraphs. The worker traits of welders includes the following:

Training Time
    General Educational Development (GED)
    Special Vocational Program (SVP)
Aptitude
Temperaments
Interests
Physical Demands
Working (Environmental) Conditions

This analysis revealed that training time for General Educational Development for welders was average for industrial and construction craftsmen. The training time for Specific Vocational Preparation—that is, welding training—ranged from a minimum of one day to a maximum of four years. Aptitude for welders was validated against actual job performance. This indicated that Spatial Aptitude, Form Perception, Finger Dexterity, and Manual Dexterity are the most significant for the welder. The temperament of welders was rated by determining a positive preference for ten different factors. This indicated that welders have a preference for activities dealing with things and objects and for activities that are carried on in relation to processes, machines, and techniques.

The physical demands on welders are related to strength, climbing or balancing, stooping, kneeling, crouching, or crawling, reaching, handling, fingering, or feeling, talking and hearing, and seeing. The information determined that welders are involved with heavy work and in some cases medium work. Seeing is involved in all welding except for the helper and the production line welder. This is one factor that is so important to most welders because of the need to continually watch, observe, and see the weld as it is being made.

THE SAFETY AND HEALTH OF WELDERS

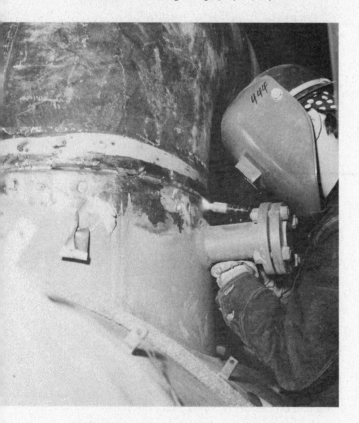

FIGURE 3-1 *Welding on highly repetitive parts.*

FIGURE 3-2 *Welding on nuclear piping—GTAW.*

Finally, the working or environmental conditions were checked. These data show that welders work inside primarily on factory type jobs and both inside and outside for construction type jobs. It shows that all welders are exposed to certain hazards and these are usually in common with other metalworking occupations. Finally, it shows that all of the welder occupational titles surveyed are exposed to fumes with the exception of the spot welder.

Fortunately, in the United States, the State Employment Services, in cooperation with the Department of Labor, provide a trade aptitude test program. Anyone seriously interested in a career in welding should visit a local office of the State Employment Service and request to take the General Aptitude Test Battery.

Scores for Spatial Aptitude, Form Perception, Finger Dexterity, and Manual Dexterity should be well above the minimum. However, it is felt that a strong desire to become a welder may overrule all aptitude tests.

The job of the welder, the flame cutter, and the welding operator represents by far the greatest percentage of jobs in the welding field. There are other opportunities in welding that should be considered. One is the welder-fitter, whose job involves layout or setup work. The applicant must have the skills of the welder, but must also be able to plan and set up work to be welded by the use of fixtures, instructions, and blueprints. Another is the welder who specializes in a particular area such as working on difficult-to-weld materials.

The job of the welder is a worthwhile occupation and a rewarding one. It can be used as a stepping stone to many other careers.

3-2 THE SAFETY AND HEALTH OF WELDERS

The safety and health of industrial and construction workers are receiving increasing attention. All workers engaged in production and construction are continually exposed to potential hazards. There are several potential health and safety problems associated with welding. However, with properly instituted precautionary measures, welding is a safe occupation. Health officials state that welding, as an occupation, is no more hazardous or injurious to the health than other metalworking occupations.[7] Governments are now enacting laws which make recommended safety practices enforceable regulations. In the United States, the ANSI Standard Z49.1, "Safety in Welding and Cutting,"[8] a National Consensus Standard, is enforceable through the provision of the Occupational Safety and Health Act (OSHA).[9]

The more common hazards, which apply to all metalworking occupations, are accidents resulting from falling, from being hit by moving objects, from exposure to excessive noise, from working around moving machinery, from exposure to hot metal, etc.

The hazards which are more or less peculiar to welding are as follows:

1. Electrical Shock
2. Arc Radiation
3. Air Contamination
4. Fire and Explosion
5. Compressed Gases
6. Cleaning and Chipping Welds and Other Hazards

Electrical Shock

The shock hazard is associated with all electrical equipment. This includes extension lights, electric hand tools, and all types of electrically powered machinery. Ordinary household voltage (115 v) is higher than the output voltage of a conventional arc welding machine.

Use only welding machines that meet recognized national standards. Most industrial welding machines meet the National Electrical Manufacturers Association (NEMA) standards for electric welding apparatus.[10] This is mentioned in the manufacturer's literature and is shown on the nameplate of the welding machine. In Canada approval by the Canadian Standards Association is required for certain types of welding machines and this is also indicated on the nameplate.[11] In certain parts of the U.S., and for certain applications, the Underwriters Laboratory approval is required for transformer type welding power sources.[12] The NEMA specification provides classes of welding machines, duty cycle requirements, and no load voltage maximum requirements. In order to comply with the OSHA requirements, manufacturers have recently made changes to improve the safety of the machines. This includes the covering of the output terminals with insulating devices. This is shown by Figure 3-3. They have also

made the ventilating holes smaller so that the welder cannot come in contact with high voltage inside the case. They have changed the cases of the welding machines so that "tools" are required to open the case where high voltage is exposed.

Only insulated type welding electrode holders should be used for shielded metal arc welding. Semiautomatic welding guns for continuous wire processes should utilize low-voltage control switches so that high voltage is not brought into the hands of the welder. In fully automatic equipment higher voltages are permitted, but are inaccessible to the operator during normal operation.

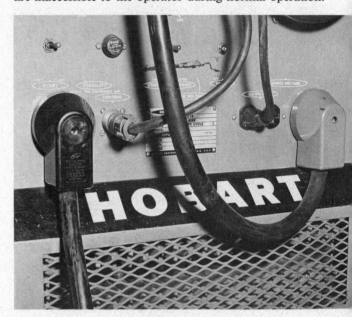

(a)

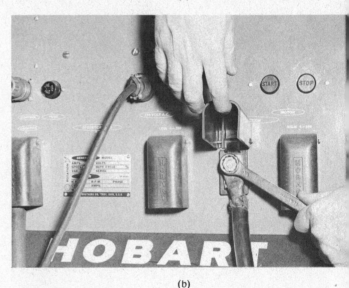

(b)

FIGURE 3-3 *Insulating devices on terminals of a welding machine.*

All electric arc welding machines must be installed in accordance with the National Electrical Code[13] and all local codes. Installation instructions are included in the manufacturer's manual that accompanies the welding machine. The manual also gives the size of power cable that should be used to connect the machine to the main line. Motor generator welding machines feature complete separation of the primary power and the welding circuit since the generator is mechanically connected to the electric motor. However, the metal frames and cases of motor generators must be grounded since the high voltage from the main lines does come into the case. In transformer and transformer rectifier type machines, the primary and secondary transformer windings are electrically isolated from each other by insulation. This insulation may become defective in time if proper maintenance practices are not observed. The metal frame and cases of transformers and transformer rectifier machines must also be grounded to earth. The work terminal of the welding machine should not be grounded to earth.

Disconnect switches should be employed with all power sources so that they can be disconnected from the main lines for maintenance.

It is extremely important when paralleling transformer welding machines that the phases of a three-phase power line be accurately identified. This will insure that the machines will be on the same phase and "in phase" with one another. It is relatively easy to check this by connecting the work leads together and measuring the voltage between the electrode holders of the two machines. This voltage should be practically zero. If it is double the normal open circuit voltage, it means that either the primary or secondary connections are reversed. If the voltage is approximately 1-1/2 times the normal open circuit voltage it means that the machines are connected to different phases of the three phase power line. Corrections must be made before welding begins.

When large weldments, like ships, buildings, or structural parts, are involved it is normal to have the work terminal of many welding machines connected to it. Here again it is extremely important that the machines be connected to the proper phase and have the same polarity. This can be checked by measuring the voltage between the electrode holders of the different machines as mentioned above. The situation can also occur with respect to direct current power sources when they are connected to a common weldment. If one machine is connected for straight polarity and one for reverse polarity, the voltage between the electrode holders will be double the normal open circuit voltage. Precautions should be taken to see that all machines are of the same polarity when connected to a common weldment.

The welding electrode holders must be connected to machines with flexible cables designed for welding application. There shall be no splices in the electrode cable within 10 feet (3 meters) of the electrode holder. Splices, if used in work or electrode leads, must be insulated.

Finally, it is important to locate welding machines where they have adequate ventilation and that ventilating ports are so located that they cannot be obstructed.

Use

Electrode leads and work leads should not be coiled around the welding machines, nor should they ever be coiled around the welder. Electrode holders should not be hung where they can accidentally come in contact with the other side of a circuit. Electrodes should be removed from holders whenever they are not in use. It is absolutely essential that power cables or primary power coming to a welding machine should not be intermixed or come in contact in any way with the welding cables. The welding machine must be kept dry and if it should become wet it should be properly dried by competent electrical maintenance personnel. In addition, the work area must be kept dry. Welders should never be working in water or damp areas since this reduces the resistance to the welder and increases potential electrical hazard.

Welders should not make repairs on welding machines or associated equipment. Welders should be instructed not to open cases of welding machines. They should also be instructed not to perform maintenance on electrode holders, welding cables, welding guns, wire feeders, etc. Instead, they should be advised to notify their supervisors of maintenance problems or potential hazards so that qualified electrical maintenance personnel can make needed repairs.

Maintenance

Welding machines and auxiliary equipment must be periodically inspected and maintained by competent electricians. During maintenance the equipment must be disconnected from main power lines so that there is no possibility of anyone coming in contact with the high input voltage. Maintenance records should be kept on welding power supplies in order to comply with OSHA regulations. Supervisors and maintenance personnel should make routine inspection of welding cables and electrode holders and guns. Welders should report defects or equipment problems to their supervisors. Electrode holders with worn or missing insulators, and worn and frayed cables, should be repaired or replaced. Wire feeding semiautomatic equipment and specialty equipment, designed for gas tungsten arc welding, utilize power contactors. This means that the electrode wire or torch is electrically "cold" except while welding. The trigger on the welding gun or foot switch or programmer closes the contactors which energize the welding circuit. Arc voltage is normally non-hazardous.

Education

Safety rules are required for safe operation of welding equipment. Suggested safety rules are given at the end of this chapter. The management of the welding shop is responsible for the safety of the people working in the shop and therefore has full responsibility to see that the rules are known, practiced, and enforced.

Arc Radiation

The electric arc is a very powerful source of light, visible, ultraviolet, and infrared. It is necessary that welders and others close to the welding arc wear suitable protection from the arc radiation. The brightness and exact spectrum of a welding arc depend on the welding process, the metals in the arc, the arc atmosphere, the length of the arc, and the welding current. The higher the current and arc voltage the more intense the light from the arc. Arc radiation like all light radiation decreases with the square of the distance. Those processes that produce smoke surrounding the arc have a less bright arc since the smoke acts as a filter. The spectrum of the welding arc is similar to that of the sun. Exposure of the skin and eyes to the arc is the same as exposure to the sun. Welders sometimes suspect that there is X-radiation in the arc if they are using a thoriated tungsten electrode for the gas tungsten arc welding process. Radiation is minute and exhaustive tests by the Atomic Energy Commission have proven that such worries are needless.[14]

Heat is radiated from the arc in the form of infrared radiation. The infrared radiation is harmless provided that the proper eye protection and clothing are worn. To minimize light radiation screens should be placed around the welding area so that others are shielded from the arc. Welders should attempt to screen all others

from their arc. Screens and surrounding areas should be painted with special paints which absorb ultraviolet radiation yet do not create high contrast between the bright and dark areas. Light pastel colors of a zinc oxide or titanium dioxide base paint are recommended. Black paint should not be used.

Eye Protection

Welders must use protective welding helmets with special filter plates or filter glasses. The welding helmets should be in good repair since openings or cracks can allow arc light to get through and create discomfort. The curved front welding helmets are preferred over straight front because they reduce the amount of welding fumes that come to the welder's breathing zone. Figure 3-4 shows both types of welding helmets. Fiberglass for light weight is recommended. Welding helmets can be attached to safety hard hats for industrial and construction work. Welding helmets have lense holders for inserting the cover glass and filter glass or plate. The standard size filter plate is 2 × 4-1/4 in. (50 × 108 mm). In some helmets, the lense holders will open or flip upwards. Helmets which accommodate larger-size filter lenses are also available and are used for light duty work. The larger filter glasses are 4-1/2 × 5-1/4 in. (115 × 133 mm) and are more expensive. The filter glasses or plates come in various optical densities to filter out more or less of the light. The shade of the filter glass used is based on the welding process, the type of base metal, and the welding current. Table 3-1 shows the proper filter shades according to American National Standard Z-87.1.[15] A cover plate should be placed on the outside of the filter glass to protect it from weld spatter. Plastic or glass plates are used. Some welders also use magnifier lenses behind the filter plate to provide clearer vision. The filter glass must be tempered so that it will not break if hit by flying spatter, etc. Filter glasses must be marked showing the manufacturer, the shade number, and the letter "H" indicating that it has been treated for impact resistance.

Contact lenses should not be worn while welding or when working around welders. Safety glasses may be required to be worn underneath the welding helmet. These are required since the helmet is usually lifted when slag is chipped or welds are ground. OSHA requirements do not include safety glasses. However, for the welder's own protection it is prudent to observe the use of safety glass when chipping or grinding. Tinted safety glasses with side shields are recommended. People working around welders should also wear tinted safety glasses with side shields.

On occasion welders and others will have their eyes exposed to the arc for a short period of time. This will result in what is known as "arc burn." It is very similar

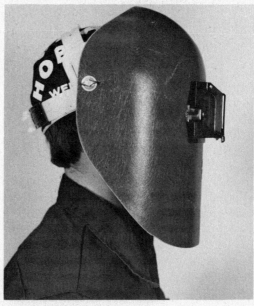

FIGURE 3-4 *Helmets, curved and straight front types.*

TABLE 3-1 Eye protection filter shade selector for welding or cutting (goggles or helmet), from AWS A6.2-73.

Welding or Cutting Operation	Electrode Size Metal Thickness or Welding Current	Filter Shade Number
Torch soldering	–	2
Torch brazing	–	3 or 4
Oxygen cutting		
Light	Under 1 in., 25mm	3 or 4
Medium	1 to 6 in., 25-150mm	4 or 5
Heavy	Over 6 in., 150mm	5 or 6
Gas welding		
Light	Under 1/8 in., 3mm	4 or 5
Medium	1/8 to 1/2 in., 3-12mm	5 or 6
Heavy	Over 1/2 in., 12mm	6 or 8
Shielded metal-arc	Under 5/32 in., 4mm	10
welding (stick)	5/32 to 1/4 in., 4 to 6.4mm	12
electrodes	Over 1/4 in., 6.4mm	14
Gas metal-arc welding (MIG)		
Non-ferrous base metal	All	11
Ferrous base metal	All	12
Gas tungsten arc welding (TIG)	All	12
Atomic hydrogen welding	All	12
Carbon arc welding	All	12
Plasma arc welding	All	12
Carbon arc air gouging		
Light	–	12
Heavy	–	14
Plasma arc cutting		
Light	Under 300 Amp	9
Medium	300 to 400 Amp	12
Heavy	Over 400 Amp	14

to a sunburn of the eye. It is sometimes called an "arc flash" and for a period of approximately 24 hours the welder will have the painful sensation of sand in the eyes. The condition is normally of temporary duration and should not last over 48 hours. The welder who receives an arc flash may not be aware of it at the time. The first indication of an arc burn may come in the middle of the night. Temporary relief can be obtained by using eye drops and eye washes. If the painful sensation lasts beyond one day a doctor should be consulted for treatment.

Skin Protection

Welders should wear work or shop clothes without openings or gaps to prevent the arc rays from contacting the skin. If the arc rays contact the skin for a period of time painful "sunburns" or "arc burns" will result. People working close to arc welding should also wear protective clothing.

For light-duty welding, normally 200 amperes or lower, the level of protection can be reduced. Figure 3-5 shows a welder dressed for light-duty work. Woolen clothing is much more satisfactory than cotton since it will not disintegrate from arc radiation or catch on fire as quickly. Cloth gloves can be used for light-duty work. For heavy-duty work more thorough protective clothing is required. Figure 3-6 shows a welder dressed for heavy-duty welding work wearing leather gauntlet gloves, a leather jacket, leather apron, and spats which also protect against sparks and molten metal. When welding in the vertical and overhead position this type of clothing is required. In all cases a headcap should be used. Flame-retardant clothing should be worn.

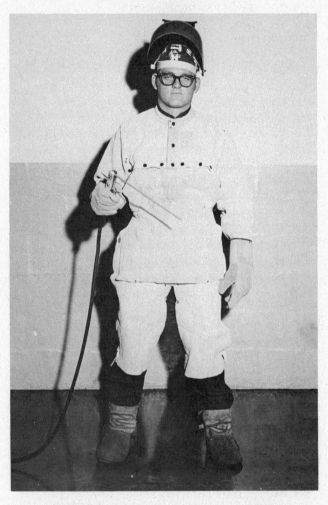

FIGURE 3-6 *Welder dressed for heavy duty welding.*

FIGURE 3-5 *Welder dressed for light duty welding.*

Clothing should always be kept dry and this applies to gloves as well. High top shoes with safety toes are recommended. The leather clothes should be of the chrome tanned type. Leather gloves should not be used to pick up hot items since this will cause the leather to become stiff and to crack.

Other Factors

Welding operations should be isolated from metal-degreasing or solvent-cleaning operations. Chemical-degreasing tanks may use trichloloethylene or other chlorinated hydrocarbons which will decompose to phosgene gas when exposed to arc radiation. Phosgene can build up to dangerous concentrations which will be potentially harmful. Fortunately, the odor of phosgene gas is quickly recognized (it smells like new mown hay) and if it is detected the area should be evacuated and ventilated. Degreasing operations should be at least 200 feet away from welding operations. Care should be taken when welding parts that have been cleaned with these solvents.

Ultraviolet rays from the arc, particularly the high-intensity gas tungsten arc on aluminum, react with the atmosphere and the oxygen in the atmosphere to produce ozone. Ozone is an active form of oxygen which has a sweet smell. It is sometimes evident after a lightening strike or in the generating room of a power house. Ventilation should be used so that ozone concentration will be below the threshold limit values.

Air Contamination

Research work has been performed to investigate the welder's breathing atmosphere, to determine the contamination due to welding, and to overcome problems that may exist. The fumes, odors, dusts, and gases are important only if they present a hazard. There are two basic types of air contamination; particulate matter and gases. Particulate matter is extremely small solids suspended in the air. Smoke from a cigarette is particulate matter in air. Gases are elements or compounds according to their composition. The amount of particulate matter and gases produced by welding depends on the welding process, the metals involved, and the amount of welding current employed. The potential harm from fumes and gases depends upon (1) the chemical composition of the fumes or particulate matter, (2) the concentration at the welder's breathing zone, and (3) the length of time of exposure to the gases and fumes. Major research efforts have been directed toward determination of composition of fumes, their concentration at the welder's breathing areas, and the effect of these fumes on the health of the welder. A Danish study[7] sampled 156 welders and 152 non-welders. The results of their investigation indicated that there is no statistically significant difference between the group of welders and the non-welders. The report concluded that fume concentration should not exceed a specific maximum and that mechanical ventilation is often required. Similar conclusions have come from the American Welding Society's study concerning the welder's environment.[16]

Particulate Matter

The Shielded Metal Arc Welding process releases contaminants into the air. For this reason cartons of electrodes produced in the U.S. state "Welding may produce fumes and gases hazardous to health. Avoid breathing these fumes and gases. Use adequate ventilation, see ANSI Standard Z49.1." The Welding Society study shows the amount of particulate matter produced by different types of welding electrodes. The process that produces the most particulate matter or smoke is the flux-cored arc welding process. This process places the maximum material into the air when used without external shielding gas. This process appears to produce more air contaminants because it is used at higher welding currents than are normally used for covered electrodes.

One source of contaminants is the base metal. Various metals when melted by an arc volatilize and produce air-borne contaminants which may be hazardous. The metals that create hazardous air-borne contaminants are zinc, cadmium, lead, beryllium, and possibly others. These vaporized metals must be kept under control. Welding should not be attempted on any of these unless mechanical ventilation is used or unless the welder is protected in some other way.

Another source of air-borne contaminants occurs from welding on coated materials, not only metal coated, but coated with paint, chemicals, oils, cleaners, etc. Old structural steel may be covered with numerous coats of lead-bearing paints. When this steel work is flame cut or welded, the heat of the flame or arc will cause the coating to volatilize and produce smoke containing lead or other contaminants. Pipe installations may be chemically coated with plastic materials which may produce harmful fumes. Adequate ventilation or protection must be employed.

Gases

Different gases are involved in many of the welding processes. When the coating on covered electrodes disintegrate in the arc, they produce gas to shield the arc from the air. This gas is normally carbon dioxide. Carbon dioxide is also produced by flux-cored electrode wires to help protect the arc area from the atmosphere. In the arc there is the possibility of carbon monoxide, but it readily combines with available oxygen in the atmosphere to produce CO_2 adjacent to the arc. Certain fluxes used for gas welding, for submerged arc welding, for electroslag welding, and for brazing will produce gases when they are heated. Some fluxes, particularly brazing and gas welding fluxes, are high in fluorine compounds and are potentially hazardous. Brazing and gas welding fluxes that contain fluoride are marked on the labels that they do produce potentially harmful gases and that adequate ventilation should be employed.

The gas-shielded welding processes utilize gases in the arc area to protect the weld from the atmosphere. Inert gases are used for gas tungsten arc welding but active gases or mixtures of active and inert gases are used for gas metal arc and flux-cored arc welding. With these processes there is the possibility of the shielding gas displacing the air in the welding area. When welding in small restricted areas, if the weight of the shielding gas is heavier than air, the shielding gas will in time displace the air so that the atmosphere around the welder will become rich in the shielding gas atmosphere. The welder would face a deficiency of oxygen which can be extremely harmful. When welding in enclosed areas extra special precaution should be taken to avoid oxygen deficiency.

Another potential problem is the exhaust gas given off by engine driven welding machines. In enclosed areas, even large rooms, an engine-driven welding machine, if not exhausted to the outside, can produce a buildup of carbon monoxide hazardous for people working in the room. It is similar to running an automobile engine in an enclosed garage. Exhausts of internal-combustion engines should always be channeled to the outside.

A similar type of problem can occur when preheating weldments. Gas-, oil-, charcoal-, or coal-fired furnaces are used for preheating. The burning of these fuels will produce carbon monoxide which must be exhausted. This can be extremely hazardous if the welder and the preheat flame are in an enclosed area.

Ventilation

Ventilation for welding can be accomplished in three different ways: natural ventilation, mechanical ventilation of the welding shop, or local exhaust ventilation of specific welding operations. In general, if the room is sufficiently large or if the welding takes place outdoors, natural ventilation will remove the contaminants from the welder's breathing zone. If the natural ventilation is not sufficient because the room is too small or there are too many welders in the room, or if there is the possibility of welding on contaminating materials, then mechanical ventilation should be provided. It should be provided if the space is less than 1,000 cu ft (284 cu m) per welder, or if the ceiling height

FIGURE 3-7 *Welding booth with mechanical ventilation.*

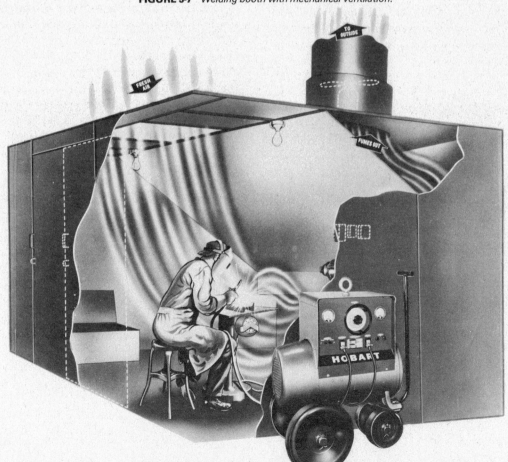

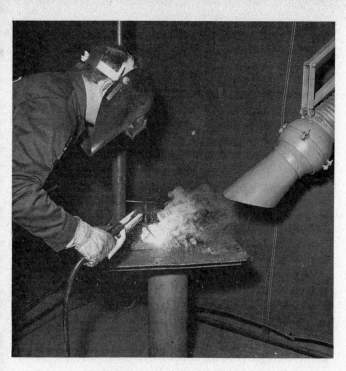

FIGURE 3-8 *Movable hood for local exhaust.*

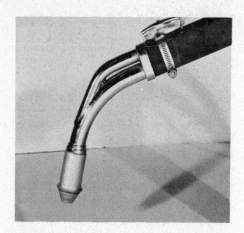

FIGURE 3-9 *Smoke exhaust nozzle on welding gun.*

of the room is less than 16 ft (5 m) or if the welding is done in a confining space which contains partitions, balconies, or other structural barriers. Mechanical ventilation should be at the rate of 200 cu ft (57 cu m) per minute per welder unless local ventilation is supplied. Tabular data, based on welding conditions, are given in "Safety in Welding and Cutting."[8] Figure 3-7 is an illustration of a welding booth equipped with mechanical ventilation sufficient for one welder.

If mechanical ventilation is not suitable then local ventilation may be required. Local exhaust ventilation can be obtained by four different methods: (1) by means of freely movable hoods shown by Figure 3-8 placed as near as possible to the arc, (2) a fixed enclosure with a top and not less than 2 sides surrounding the welder with a sufficient rate of air flow, (3) tables with down draft ventilation of 100 cu ft per minute per sq ft (47 cu m per minute per sq m) of surface area, or (4) a low-volume high-velocity fume exhaust device attached to the welding gun. This device is based on collecting the fumes as close as possible to the point of generation or at the arc.[17] This latter method of fume exhaust has become quite popular for the semiautomatic welding processes, particularly the flux-cored arc welding process. Figure 3-9 shows an apparatus designed for this application, and Figure 3-10 shows the results of two welders, one using the localized exhaust system and one not using it.

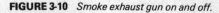

FIGURE 3-10 *Smoke exhaust gun on and off.*

Smoke exhaust systems incorporated in semiautomatic guns provide the most economical exhaust system since they exhaust much less air and eliminate the need for massive air makeup units to provide heated or cooled air to replace the air exhausted. Local ventilation mentioned in (1) or (2) above should have a rate of air flow sufficient to maintain a velocity away from the welder of not less than 100 ft (30 m) per minute. Air velocity is easy to measure using a velometer or air flow meter. Measuring air velocity with a velometer is shown by Figure 3-11. These two systems can be extremely difficult to use when welding on other than small weldments. The down draft welding work tables are popular in Europe but are used to a limited degree in North America. In all cases when local ventilation is used the exhaust air should be filtered.

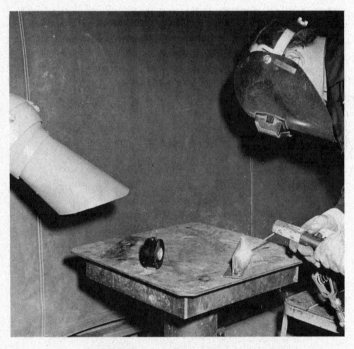

FIGURE 3-11 *Measuring the velocity of arc flow.*

There is one foolproof way to determine if proper ventilating methods are being used. This is done by collecting samples of the atmosphere at the welder's breathing zone under the helmet. The samples are collected by specialized instruments for a specific time period. The samples are then analyzed in calibrated instruments which determine the values of all of the elements that are found in the welder's breathing zone. By comparing the amount of material found and using a factor based on the length of time the sample was taken it can be determined if it meets the limits prescribed. The threshold limit values (TLV) of most hazardous materials are established by OSHA regulations based on numerous tests of government hygienists. The determinations found in the welder's breathing zone should be below these TLV limits. This type of analytical work must be done by highly qualified people who are familiar with the welding operations, testing and sampling techniques, as well as the analytical methods to determine the amount of contaminants found in the air samples from the welder's breathing zone.

All welding and thermal cutting operations carried on in confined spaces must be adequately ventilated to prevent the accumulation of toxic materials, combustible gases, or possible oxygen deficiency. Monitoring instruments should be used to detect harmful atmospheres. Where it is impossible to provide adequate ventilation, air-supplied respirators or hose masks, approved for this purpose, must be used. In these situations, lookouts should be used on the outside of the confined space to insure the safety of those working within it.

Fire and Explosion

Approximately 6% of the fires in industrial plants have been caused by cutting and welding primarily with portable equipment, in areas not specifically designated or approved for such work. The three elements of the fire triangle, fuel, heat, and oxygen are present at all welding operations. The heat is from the torch flame the arc or from hot base metal. The fuel is from the fuel gas employed, or from combustibles in the welding area. The oxygen is present in air but may be enriched by oxygen used with the fuel gas. Many of these industrial fires have been caused by sparks which are globules of molten metal which can travel up to 40 ft (13 meters). Sparks have also fallen through cracks, pipe holes, or other small openings in floors and partitions starting fires in other areas which may go temporarily unnoticed.

Hot pieces of base metal may come in contact with combustible materials and start fires. Fires and explosions have also been caused when this heat is transmitted through walls of containers to flammable atmospheres or to combustibles within containers. Anything that is combustible or flammable is susceptible to ignition by cutting and welding. Welding or cutting on metal which is in contact with urethane foam insulation must be prohibited. All insulating organic foams whether indicated fire retarded or not, should be considered combustible and handled accordingly.

Cutting and welding fires can be prevented by eliminating all combustibles from the welding area. Welding arcs or oxy fuel gas flames rarely cause fires when used in the normal workshop area designed for welding and cutting. Fire and explosion hazards should be considered from two points of view: welding in designated workshop areas and welding with portable equipment in all other areas.

Work Area

A safe work place must be provided for welding and cutting operations.

Floors, walls, ceilings, etc., must be constructed of noncombustible materials. The work area must be kept clean and free of combustible and flammable materials. All fuel gas lines, manifolds, branches, etc., must be installed in accordance with specifications and codes.

Fire-fighting equipment must be installed in the welding workshop areas. The types of extinguishers for the possible different types of fires should be available and identified for the type of fires that might occur.

Fuel Gases

There are a number of different types of fuel gases used for welding and flame cutting. The most familiar is acetylene, but propane, natural gas, methylacetylene-propadiene stabilized, and others are used.

Acetylene is sometimes produced on the premises by means of an acetylene generator. An acetylene generator uses carbide and water to produce acetylene which is then piped through the plant to the welding and cutting departments. Acetylene generators must be installed properly, maintained properly, and operated only by trained and qualified people. The piping systems from acetylene generators to welding and cutting stations must be installed in accordance with National Fire Protection Association Standard No. 51.[18] Carbide must be stored properly and never exposed to moisture or water which creates more acetylene to feed fires.

Acetylene cylinders and other fuel gas cylinders should be stored in a specified well-ventilated area or outdoors and in the vertical position. All cylinders in storage should have their caps on and cylinders either filled or empty should have the valve closed. In a fire situation there are special precautions that should be taken for acetylene cylinders.[19] All acetylene cylinders are equipped with one or more safety relief devices filled with a low melting point metal. This fusible metal melts at about the boiling point of water (212°F or 100°C). If fire occurs on or near an acetylene cylinder the fuse plug will melt. The escaping acetylene may be ignited and will burn with a roaring sound. Immediately

evacuate all people from the area. It is difficult to put out such a fire. The best action is to play water on the cylinder to keep it cool and to keep all other acetylene cylinders in the area cool. Attempt to remove the burning cylinder from close proximity to other acetylene cylinders, from flammable or hazardous materials, or from combustible buildings. It is best to allow the gas to burn rather than to allow acetylene to escape, mix with air, and possibly explode.

If the fire on a cylinder is a small flame around the hose connection, the valve stem, or the fuse plug, try to put it out as quickly as possible. A wet glove, wet heavy cloth, or mud slapped on the flame will frequently extinguish it. Thoroughly wetting the gloves and clothing will help protect the person approaching the cylinder. Avoid getting in line with the fuse plug which might melt at any time.

Apparatus

Gas welding or cutting apparatus must show the approval of an independent testing laboratory. When ordering gas welding or cutting apparatus specify that it must carry the Underwriters Laboratory (UL) or Factory Mutual Engineering Corporation (FM) seal of approval.

Gas apparatus must be properly maintained and repaired only by qualified people. All too often a gas apparatus is allowed to deteriorate before maintenance is performed. Welding gauges, welding regulators, welding torches, welding tips, etc., should all be carefully inspected periodically and maintained at the first sign of deterioration. Oil or grease should never be used on any gas welding or cutting apparatus.

Only approved gas hoses for flame cutting or welding should be used with oxy-fuel gas equipment. Single lines, double vulcanized, or double or multiple stranded lines are available. The double vulcanized or twin hose is preferred. The size of hose should be matched to the connectors, the regulators, and torches. In the U.S. the color green is used for oxygen, red for the acetylene or fuel gas, and black for inert gas and compressed air. The international standard calls for blue for oxygen and orange for the fuel gas. The connections on hoses are right-handed for inert gases and oxygen, and left-handed for fuel gases. The nuts on fuel gas hoses are identified by a groove machined in the center of the nuts. Hoses should be periodically inspected for burns, worn places, or leaks at the connections. They must be kept in good repair and should be no longer than necessary.

PERMIT NO. 10534

For electric and acetylene burning and welding with portable
equipment in all locations outside of shop.

Date _____ _____

Time Started _____ Finished _____

Building _____

Dept. _____ Floor _____

Location on Floor _____

Nature of Job _____

_____

Operator _____

Clock No. _____

All precautions have been taken to avoid any possible fire
hazard, and permission is given for this work.

Signed _____
                                                        Foreman

Signed _____
                                          Safety supervisor or
                                          plant superintendent

_____

PERMIT NO. 10534

Date _____ _____

Bldg _____ Floor _____

Nature of Job _____

Operator _____

_____

PERMIT NO. 10534

INSTRUCTIONS TO OPERATORS

This permit is good only for the location and time shown.
Return the permit when work is completed.

PRECAUTIONS AGAINST FIRE

1. Permits should be signed by the foreman of the welder or
   cutter and by the safety supervisor or plant superintendent.

2. Obtain a written permit before using portable cutting or
   welding equipment anywhere in the plant except in
   permanent safe-guarded locations.

3. Make sure sprinkler system is in service.

4. Before starting, sweep floor clean, wet down wooden floors,
   or cover them with sheet metal or equivalent. In outside
   work, don't let sparks enter doors or windows.

5. Move combustible material 25 feet away. Cover what can't
   be moved with asbestos curtain or sheet metal, carefully
   and completely.

6. Obtain standby fire extinguishers and locate at work site.
   Instruct helper or fire watcher to extinguish small fires.

7. After completion, watch scene of work a half hour for
   smoldering fires, and inspect adjoining rooms and floors
   above and below.

8. Don't use the equipment near flammable liquids, or on
   closed tanks which have held flammable liquids or other
   combustibles. Remove inside deposits before working on
   ducts.

9. Keep cutting and welding equipment in good condition.
   Carefully follow manufacturer's instructions for its use and
   maintenance.

FIGURE 3-12 *Hot work permit—front and back.*

Hot Work Permits

Welding permits or, as they are sometimes called, *hot work permits,* are required by some specifications or companies. These permits must be used when welding or flame cutting is done on items that may involve hazards. One of the more common uses of the work permit is when welding on aircraft. Special precautions such as making out of the welding permit, the stationing of a fire watcher or lookout with proper fire control equipment, and the signing on and off of the welding operation are prescribed. Other types of operations require hot work permits. These are outlined by the instructions given by the National Safety Council Data Sheet 522 (Revision A) entitled "Hot Work Permits (Flame or Spark)."[20] Their sample permit form is shown by Figure 3-12. Instructions and precautions against fire are on the reverse side. Your casualty insurance company may have similar tags. The National Fire Protection Association in their specifications for cutting and welding processes[21] also recommends the use of a hot work permit. A welding permit or hot work permit should be used when portable welding or flame cutting equipment is used for maintenance welding in plants where combustible materials are present, for maintenance on construction equipment in the field, for maintenance on aircraft, for maintenance on ships, and on any other type of potentially hazardous welding or flame-cutting operations. The maintenance department should use a similar permit in your plant. When using portable equipment these are invaluable and can avoid the headlines stating that the fire was caused by "a welder's torch."

Any container or hollow body such as a can, a tank, a hollow compartment in a weldment, or a hollow area in a casting, even though it may contain only air, should be given special attention before welding. The heat from welding the metal will raise the temperature of the enclosed air or gas to a possible dangerously high pressure so that the part or container may explode. Always vent confined air before welding or cutting on a hollow area. Hollow areas can also contain oxygen-enriched air or fuel gases, either of which is extremely dangerous when heated or exposed to an arc or flame.

Explosions and fires may result if welding or cutting is done on containers that are not entirely free of combustible solids, liquids, vapors, dust, and gases. Containers can be made safe for welding or cutting by following prescribed steps. Refer to the welding society's "Safe Practices for Containers that have held combustibles."[22] No container should be considered clean or safe until proved so by tests. Cleaning the container, which is normally made of metal, is necessary in all cases before welding or cutting. Cleaning should be done by experienced personnel familiar with the characteristics of the contents. Cleaning should be done outdoors; if this is impractical, the inside work area should be well ventilated so that flammable vapors will be quickly carried away. Drain all material from the container and remove sludge, sediment, and the like. Dispose of residue before starting to weld or cut. Identify the material that was in the container and match the cleaning method to the material previously contained. If the material is water soluble the container can be cleaned with water. If the material is not readily soluble in water, the container should be cleaned by a hot chemical solution or by steam. Mix the chemicals for cleaning in hot water and pour into the container, fill the container completely with water, and then introduce live steam to heat and agitate the solution. If hot water and steam are not available the cold water method can be used; however, it is less effective and agitation should be handled by means of compressed air. Another way of cleaning the container is to fill the container 25% full with cleaning solution and clean thoroughly. Following this, introduce low-pressure steam into the tank allowing it to vent through openings. Continue to flow steam through the tank for several hours. None of these cleaning methods is perfect, and after cleaning, the tank should be inspected to determine that it is thoroughly cleaned. If it is not, continue the cleaning operation.

After the container is cleaned close all openings and after 15 minutes test a sample of the gas inside the container. Use a combustible gas indicator instrument.

If the concentration of flammable vapors in the sample is not below the limit of flammability repeat the cleaning procedure. When it is determined that the gas or air inside the container is safe the container should be so marked, signed, and dated. Even after tanks have been made safe they should be filled with water, as an added precaution before welding or cutting. Place the container so that it can be kept filled with water to within a few inches of the point where welding and cutting are to be done. Make sure that the space above the water level is vented so that the heated air can escape. See Figure 3-13.

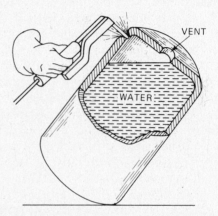

FIGURE 3-13 *Safe way to weld container that held combustibles.*

As an alternate to the water-filling method fill the container with an inert type gas. Flammable gases and vapors will be rendered nonflammable and nonexplosive if mixed with a sufficient amount of inert type gas. Nitrogen or carbon dioxide is normally used. The concentration of flammable gases and vapors must be checked by testing as mentioned previously. The inert gas concentration must be maintained during the entire welding and cutting operation. Hot work or welding permits should be utilized for all welding or cutting operations on containers that have held combustibles.

Hot Tapping

In pipe lining, welders sometimes do **hot tapping.** This is the welding of a special fitting to a line carrying a combustible liquid or gas. Then cutting a hole in the pipe after the fitting has been welded to it. This must be done by experienced people using special equipment with proper precautions. Before attempting such work refer to the American Petroleum Institute (API) publication, "Welding or Hot Tapping on Equipment Containing Flammables."[23]

Compressed Gases

All compressed gas cylinders are potential hazards. The major hazard is the possibility of sudden release of the gas by removal or breaking off of the valve. Escaping gas which is under high pressure will cause the cylinder to act as a rocket smashing into people and property. Escaping fuel gas can also be a fire or explosion hazard.

The Treatment of Gas Cylinders

Gases, used for welding, fuel gases, oxygen, or shielding gases, are normally delivered in cylinders which are manufactured and maintained by the gas supplier in accordance with the regulations of the U.S. Department of Transportation (DOT). In Canada the Board of Transport Commissioners for Canada has this responsibility. In most countries there are laws and regulations concerning the manufacturing, the maintaining, and periodic inspection of portable cylinders for the storage and shipment of compressed gases. All compressed gas cylinders must be legibly marked to identify the gases contained by either the chemical or the trade name of the gas. There is no uniform color coding for identification purposes; however, there is a uniform standard for connection threads at least for North America in accordance with the American-Canadian Standard Compressed Gas Cylinder Valve Outlet and Inlet Connections.[24] Most of the gases are available in portable cylinders of different sizes.

In North America the authorities require that a cylinder be condemned when it leaks or when internal or external corrosion, denting, bulging, or evidence of rough usage exists to the extent that the cylinder is likely to be appreciably weakened. Always inspect cylinders for suspicious areas and report this or any damage done to a cylinder to your gas supplier.

Cylinder Storage

Oxygen cylinders should be stored separately from fuel gas cylinders, and separate from combustible materials. Store cylinders in cool, well-ventilated areas. The temperature of the cylinder should never be allowed to exceed 130°F (54°C). Cylinders should be stored vertically and secured to prevent falling. The valve protection caps must be in place. When cylinders are empty they should be marked empty and the valves must be closed to prohibit contamination from entering. When the gas cylinders are in use a regulator is attached and the cylinder should be secured to prevent falling by means of chains or clamps. Cylinders for portable apparatuses should be securely mounted in specially designed cylinder trucks. Cylinders should be handled with respect. They should not be dropped or struck. They should never be used as rollers. Hammers or wrenches should not be used to open cylinder valves that are fitted with hand wheels. They should never be moved by electromagnetic cranes. They should never be in an electric circuit so that the welding current could pass through them. An arc strike on a cylinder will damage the cylinder causing possible fracture, and requiring the cylinder to be condemned and discarded from service.

Oxygen

Oxygen is one of the most common gases carried in portable high-pressure cylinders. It should always be called oxygen, never air. Combustibles should be kept from oxygen, including the cylinder, valves, regulators, and hose apparatuses. Oxygen cylinders or oxygen apparatuses should not be handled with oily hands or oily gloves. Oxygen does not burn, but will support and accelerate combustion of oil and grease and other hydrocarbon materials, causing them to burn with great intensity. Oil or grease in the presence of oxygen may spontaneously ignite and violently burn or explode. Oxygen should never be used in any air tools and for any of the purposes where compressed air is normally used.

Fuel Gases

There are a number of fuel gases of which acetylene is the most popular. The properties of fuel gases are given in Chapter 10. All are compounds of carbon and hydrogen. All fuel gases are potentially hazardous and should be treated with respect.

When welding or cutting with oxygen and fuel gases the welder should be particularly alert to backfires and flashbacks. A **backfire** is a loud noise associated with the momentary extinguishment and reignition of the flame at the torch tip. This is caused by obstructing the gas flow, by an overheated tip, or a damaged tip. If this occurs the equipment should be shut down immediately and corrective action taken. A **flashback** is the burning back of the flame into the tip or torch or even the hose if an explosive mixture is present in one of the lines. It is sometimes accompanied by a hissing or squealing sound and the characteristic smoky or sharp pointed flame. When this occurs, the equipment should be shut down immediately and corrective action should be taken. Flashback can be caused by improper pressures, distorted or loose tips or damaged seats, kinked hoses, clogged tips, damaged tips, and so on.

In extinguishing the oxygen fuel gas flame the proper sequence is to first close the torch oxygen valve and then the torch acetylene or fuel gas valve. In starting a torch, it is always proper to first open the fuel gas valve followed by the oxygen torch valve.

Shielding Gases

Shielding gases are either inert or active. Inert gases are argon and helium and are stored in high-pressure cylinders. These cylinders must be treated with the same cautions as oxygen cylinders. The active gas normally used for weld shielding is carbon dioxide (CO_2). It is stored as a liquid gas but gasifies upon release. Shielding gases are covered in more detail in Chapter 10.

Cleaning and Chipping Welds and Other Hazards

The slag that often covers the deposited weld metal must be removed. Welds are often chipped and ground. Hand and power tools are employed and the materials removed are propelled through the air to become potential hazards. Safety glasses with side shields should be worn under the welding helmet.

Confined Areas

Confined areas must be given special attention, particularly when welding or flame cutting in them. Normal hazards that can occur to the welder may occur in confined areas and if they are undetected the welder can suffer serious harm before help comes. The buddy system should be used by welders working in these situations. When the situation is considered extra hazardous it is wise to utilize lookout personnel or watchers who will continually watch the welders at work. One lookout may survey a team of welders working in a specific enclosed area. In hazardous cases lifelines with harnesses should be employed. Lifelines should be attached so that workers can be removed through manholes with ease. Some of the reasons for extra special precautions have to do with the contamination of atmosphere in enclosed areas or of oxygen deficiency. Another potential hazard occurs if an oxyacetylene torch is left inside a compartment. Oxygen may leak from the torch and produce an oxygen enriched atmosphere which may support rapid combustion or the fuel gas can leak into the compartment which will create an explosive atmosphere. Gases from preheating might also leak into the compartment. A judgment must be made whether the buddy system or lookout watchers are required as well as other extra special safety precautions.

Radioactive Hot Areas

Welders may be required to work in radioactive "hot" areas. This is due to a repair and maintenance operations necessary in nuclear-fired power plants, etc. In such cases, extra special care and precaution must be made to determine the radiation levels, time of exposure, radiation protection, and all other factors involved. The exposure time may be extremely short and welders may be used to set up automated welding devices and then leave the hot area to operate the devices remotely. Only qualified personnel with knowledge of working in and around radioactive areas should be permitted to make judgments of this type.

Noise

Weld chipping and weld peening produce excessive noise and should be controlled. If weld chipping or peening is required, ear protection must be provided for the welders or others doing this type of work. The carbon arc air gouging system at high currents also produces large amounts of noise and ear protection is required. Plasma arc cutting with high current also creates excessive noise and ear protection is required. Noise measurement instruments are available and should be used to check noise in the work area so that precautionary measures can be taken. Normal arc welding operations do not exceed noise level requirements as specified by OSHA. In combination with other noise producing industrial machinery noise levels may be excessive. It is wise to make noise measurements in welding shops to determine the compliance with OSHA regulations.

Other Hazards

Falling items create hazards. Hard hats should be worn in connection with welding helmets on construction sites and in some plants. Other hazards such as falling from high places, working with heavy objects, and working around heated metals are similar to the harzards encountered by all employees in steel plants, forging shops, structural shops, and so on. Welding electrode stubs, the unused ends gripped in the holder, are usually short and round and act as a roller when stepped on at the wrong angle. Electrode stubs should be placed in containers and not thrown or allowed to remain on the floor or working surface.

Welder Protection

Welders can be protected from all potential hazards by providing them a safe place to work. The shop management has an obligation to provide a safe work place. It is the responsibility of supervisors to enforce safety rules and regulations. The welders and other workers have an obligation to learn safe practices and to obey safety rules and regulations. It is not sufficient that a safe work place be provided, that equipment be approved and properly installed, and that it be maintained in good working order. The use of the equipment and the materials at the work place are the responsibility of the welder, who is expected to work in a safe manner.

Protective Clothing

The types of protective clothing have been mentioned earlier. Protective clothing must be kept in good repair. It must be kept dry. Hard hats should be checked occasionally. Gloves should be clean and not oily. Welding helmets should be checked occasionally for cracks, and filter glasses should be replaced if damaged. Side shields for safety glasses should be used and safety glasses in welding areas are preferably tinted.

If the hazards mentioned in this chapter are properly and adequately handled, the welder is as safe as any other industrial worker. There must be continual vigilance over safety conditions and safety hazards. Safety meetings should be held regularly. The safety rules should be reissued occasionally and they must be completely understood. The following safety rules are recommended for use in every welding shop.

Safety Rules

The following sets of twenty rules, "Safety Precautions for Arc Welding" and "Safety Precautions for Oxyacetylene Welding and Cutting," are suitable for use in the welding shop.

Safety Precautions for Arc Welding

1. Make sure your arc welding equipment is installed properly and grounded and is in good working condition.

2. Always wear protective clothing suitable for the welding to be done.

3. Always wear proper eye protection, when welding, grinding, or cutting.

4. Keep your work area clean and free of hazards. Make sure that no flammable, volitile, or explosive materials are in or near the work area.

5. Handle all compressed gas cylinders with extreme care. Keep caps on when not in use.

6. Make sure that compressed gas cylinders are secured to the wall or to other structural supports.

7. When compressed gas cylinders are empty close the valve and mark the cylinder "empty."

8. Do not weld in a confined space without extra special precautions.

9. Do not weld on containers that have held combustibles without taking extra special precaution.

10. Do not weld on sealed containers or compartments without providing vents and taking special precautions.

11. Use mechanical exhaust at the point of welding when welding lead, cadmium, chromium, manganese, brass, bronze, zinc, or galvanized steel.

12. When it is necessary to weld in a damp or wet area, wear rubber boots and stand on a dry insulated platform.

13. If it is necessary to splice lengths of welding cable together, make sure all electrical connections are tight and insulated. Do not use cables with frayed, cracked, or bare spots in the insulation.

14. When the electrode holder is not in use, hang it on brackets provided. Never let it touch a compressed gas cylinder.

15. Dispose of electrode stubs in proper container since stubs on the floor are a safety hazard.

16. Shield others from the light rays produced by your welding arc.

17. Do not weld near degreasing operations.

18. When working above ground make sure that scaffold, ladder, or work surface is solid.

19. When welding in high places without railings, use safety belt or a lifeline.

20. When using water-cooled equipment, check for water leakage.

Safety Precautions for Oxyacetylene Welding and Cutting

1. Make sure that all of your gas apparatus shows UL or FM approval, is installed properly, and is in good working condition. Make sure that all connections are tight before lighting

the torch. Do not use a flame to inspect for tight joints. Use soap solution to detect leaks.

2. Always wear protective clothing suitable for welding or flame cutting.

3. Keep work area clean and free from hazardous materials. When flame cutting sparks can travel 30–40 feet. Do not allow flame cut sparks to hit hoses, regulators, or cylinders.

4. Handle all compressed gas cylinders with extreme care. Keep cylinder caps on when not in use.

5. Make sure that all compressed gas cylinders are secured to the wall or to other structural supports. Keep acetylene cylinders in the vertical position.

6. Store compressed gas cylinders in a safe place with good ventilation. Acetylene cylinders and oxygen cylinders should be kept apart.

7. When compressed gas cylinders or fuel gas cylinders are empty, close the valve and mark the cylinder "empty."

8. Use oxygen and acetylene or other fuel gases with the appropriate torches and only for the purpose intended.

9. Oxygen should not be used for "AIR" in any way.

10. Never use acetylene at a pressure in excess of 15 pounds per square inch. Higher pressure can cause an explosion.

11. Never use oil, grease or any material on any apparatus or threaded fittings in the oxyacetylene or oxy-fuel gas system. Oil and grease in contact with oxygen may cause spontaneous combustion.

12. Treat regulators with respect. Do not turn valve handle using force.

13. When assembling apparatus, crack gas cylinder valve before attaching regulators (cracking means opening the valve on a cylinder slightly, then closing). This blows out any accumulated foreign material. Make sure that all threaded fittings are clean and tight.

14. Always use this correct sequence and technique for lighting a torch:

(a) Open acetylene cylinder valve.

(b) Open acetylene torch valve 1/4 turn.

(c) Screw in acetylene regulator adjusting valve handle to working pressure.

(d) Turn off acetylene torch valve (you will have purged the acetylene line).

(e) Slowly open oxygen cylinder valve all the way.

(f) Open oxygen torch valve 1/4 turn.

(g) Screw in oxygen regulator screw to working pressure.

(h) Turn off oxygen torch valve (you will have purged the oxygen line).

(i) Open acetylene torch valve 1/4 turn and light with lighter (Use friction type lighter or special provided lighting device only).

(j) Open oxygen torch valve 1/4 turn.

(k) Adjust to neutral flame.

15. Always use this correct sequence and technique of shutting off a torch:

(a) Close acetylene torch valve first, then close oxygen torch valve.

(b) Close cylinder valves—the acetylene valve first, then the oxygen valve.

(c) Open torch acetylene and oxygen valves (to release pressure in the regulator and hose).

(d) Back off regulator adjusting valve handle until no spring tension is felt.

(e) Close torch valves.

16. Use mechanical exhaust at the point of welding when welding or cutting lead, cadmium, chromium, manganese, brass, bronze, zinc, or galvanized steel.

17. If you must weld or flame cut with combustible or volitile materials present, take extra precautions, make out hot work permit, and provide for a lookout, etc.

18. Do not weld or flame cut on containers that have held combustibles without taking extra special precautions.

19. Do not weld or flame cut into sealed container or compartment without providing vents and taking special precautions.

20. Do not weld or cut in a confined space without taking special precautions.

3-3 WELDER TRAINING PROGRAMS

Welding training with emphasis on the manipulative skills is provided by a wide variety of training programs. Welding is taught in high schools, trade, vocational, technical schools, colleges, universities, joint apprentice training schools, in-plant or company schools, armed forces schools, among others.

There are many different types of welding programs available. These range from short adult evening courses designed to provide the minimum skill sufficient to strike an arc and make a simple metal project, to in-depth courses which provide the student with sufficient skill and knowledge to pass difficult qualification tests.

The objective of a training course must be clearly established. A vocational course has the objective to develop entry level skills for a particular type of work. Unfortunately in the field of welding it is difficult to define entry level skills since there are so many different types of jobs requiring different skill levels as previously described. In spite of this, welding training programs can be analyzed from the student's point of view or designed from the instructor's point of view based on the stated objectives of the course and the skill level expected upon completion.

A vocational program to develop entry-level skills for the arc welder should involve from 250 to 300 hours. This course should involve welding with the shielded metal arc welding process in all positions. It should teach the welding of thin and medium thicknesses of steels using single and multiple pass procedures. It should provide training for a variety of electrode types including the cellulosic, the iron powder, and the low-hydrogen types. Both alternating current and direct current welding machines should be used. Related technical topics such as principles of arc welding, welding safety, principles of welding machines, welding electrode identification and selection, weld testing, and inspection would be included. Upon completion of this training program the welder should be able to pass the qualification test necessary for plate or structural welding.

A similar course covering oxyacetylene or oxy-fuel gas welding, brazing, and flame cutting should require from 125 to 150 hours. This course should include the assembly of equipment for gas welding, the proper lighting and adjustment of gas torches of different sizes, welding with the torch on thin materials and medium thickness materials in all positions, and the use of filler rod. In addition it should include brazing using the torch method and flame cutting of steel both manually and with a machine. Related subjects include the principles of combustion of fuel gases with oxygen, the selection of the proper tips for different applications, the adjustment of oxygen and fuel gas for different tips and types of flames, and the selection of filler metals

and welding flames. It should include many of the same items for brazing and the setup of equipment and adjustments for flame cutting and the procedures necessary for quality cuts.

These two courses combined should provide entry-level skills for a combination welder. A combination welder should be able to pass the structural test with shielded metal arc welding, the flame-cutting tests, and the gas welding and brazing test on various materials.

A vocational training program to train welders to use the gas tungsten arc welding process should require from 60 to 75 hours. The gas tungsten arc welding program should include the use of the process for welding nonferrous materials, particularly aluminum—and also stainless steels and carbon steels. This should be done in all positions utilizing thin to medium thickness materials. Related information should include the principles of the gas tungsten arc welding process, the different equipment required, the construction and adjustment of the welding torch, the selection of shielding gas, the use of high frequency current, and the selection of filler metal for welding different base metals. Upon completion of this course the welder should be capable of passing appropriate performance tests.

The vocational welding training course for the gas metal arc welding process should also require from 60 to 75 hours. This should include the fine wire or short circuiting variation, the spray transfer variation, and could include flux-cored arc welding. This program should include welding of thin materials using the fine wire technique, welding thicker materials with the spray transfer method, and the use of the flux-cored process on heavier materials. It should include welding of nonferrous metals if the student has sufficient skill and ability. The related topics include the principles of the process and its variation, the different equipment that is required, including the principle of constant voltage power supplies and wire feeders. It should also cover the selection of filler metal and welding shielding gas. Upon completion of this program the welder should be able to pass appropriate performance tests for thin and medium thickness metals in all positions.

The above preceding programs are vocational in nature, and upon successful completion the welder should have the necessary skill and ability to be employable in the average fabrication plant. These programs can be supplemented with additional programs for specific welding techniques such as pipe welding with uphill or downhill travel, welding thin wall tubing, welding nonferrous materials to aircraft quality standards, welding with the plasma arc process, etc.

In many situations the time available for such thorough training is not available. In view of this, the programs can be abridged to provide training in fewer processes or with types of electrodes or filler metals or on fewer types of base metals in less varied thicknesses, and may be on only one or two positions. Courses can also be reduced if the skill required is restricted to specific applications which are repetitive day after day.

The American Welding Society provides a booklet entitled "Minimum Requirements for Teaching Welding,"[25] which outlines minimum programs similar to those mentioned above. For more exhaustive programs refer to the book "Recommendations for Teaching Welding."[26] Many workbooks and manuals providing specialized welding programs are available. The companion manual to this book covers the four main processes and is recommended. It can be abridged to fit the time available.

No matter what type of welding training program is planned it should be presented in the following four steps:

1. A lecture and discussion—prior to each new phase of instruction. This provides background information. These discussions normally last from 15 to 30 minutes and should include a question and answer period. Audio visuals can be used.

2. An explanation demonstration—the demonstration of the particular new training phase is given. This demonstration may be repeated so that there is no question about the technique employed. Specific points should be brought out to make sure that each student completely understands the new phase and techniques involved. Audio visuals can be used.

3. Supervised individual practice. After the discussion and demonstration the trainees go to their individual booths and practice the techniques demonstrated. Instructors should periodically check the progress of each student and correct any faults that might develop.

4. Periodic practical tests—at the termination of each particular phase, the student is given the test which covers the technique involved. This should be a practical test on which the student makes the prescribed welds. Test pieces should be prepared and tested by fracturing, bending cross sectioning, etc., and evaluated to determine the successful mastery of the phase being tested. Tests such as these are very useful since they help students to overcome the fear of qualification tests that are required many times throughout a welder's career.

The student should also be given training in related subjects and should be periodically tested to determine their understanding of the subjects.

3-4 TRAINING SCHOOL FACILITIES

A well-planned welding training department or school with proper facilities and equipment makes it much easier to provide a successful training program. The well-planned school welding shop must be designed for the safety of the students and to accommodate the number of students that will be enrolled. The following information is presented to assist in the planning of a welding training facility or to assist in remodeling existing facilities.

1. Determine the welding processes that are to be taught now and that may be taught in the foreseeable future.

2. Determine the number of students to be accommodated now and in the foreseeable future.

3. Determine the number of students that will be in the same specific activity at any one time.

4. Plan for welding shop access from the building for students and others. Plan also for access to the welding shop from the outside for ease of delivery of material, equipment, etc.

5. Plan areas for the different activities that will be involved, including a classroom, welding demonstration area, welding booths, test areas, storage areas, metalworking area, instructors' offices and washroom and restroom.

6. The classroom for lectures and discussion should be large enough to accommodate the number of students that will be using it at any one time. It should be possible to darken the room for the use of audio-visual aids. It should include chalkboards, bulletin boards, as well as tables, chairs, etc. Individual learning carrels may be employed and can be a part of the classroom or immediately adjacent to it.

7. The welding demonstration area should be outside the classroom since many welding processes create dust and dirt which is not desirable in the classroom. The welding demonstration area may be elevated so that students can see welds being made. A different demonstration area should be provided for each welding process being taught.

8. Individual welding stations should be provided for each student for each process. Booths for the arc welding processes should provide shield-

ing to prevent arc rays from coming in contact with others. Oxyacetylene or gas welding stations should not be enclosed. Arc welding and the gas welding processes should not be included in the same booth. Students should not be required to work two in a booth.

9. Space should be allocated for metalworking equipment including tools for preparation of practice specimens and preparation for test specimens. This can include flame-cutting equipment, saws, metal shears, iron workers, grinders, etc.

10. Space should be provided for weld testing. Equipment should include a bending press with appropriate dies, a fixture for breaking specimens, for cross sectioning welds, and for tensile testing.

11. Storage space is particularly important for a welding shop. Welding material, projects, working supplies, and materials should have proper storage areas.

12. An office must be provided for the instructors. It should include space for a welding library and for storage of training aids and projection equipment.

13. Sufficient restroom and washroom facilities must be provided. Lockers must be available for students, with a sufficient number provided for the total student enrollment.

In addition to the specific items listed above, there are other general requirements for a welding shop. It should be of fireproof construction throughout and should be planned for ease of maintenance. It should allow for natural light and natural ventilation if at all possible. Proper exhaust ventilation, heating, and air makeup should be provided. If climate dictates cooling should also be provided. The shop must have adequate utilities including sewer, hot and cold water, compressed air, and electric power. The electric power requirements for a welding shop are considerably greater than for other shops. This is calculated based on equipment included. Many shops also have gas distribution systems for fuel gas, oxygen, and, in some cases, shielding gases. Gas distribution systems must be designed by people experienced in the field, and must meet all applicable codes.

The two books mentioned in the previous section "AWS Minimum Code for Teaching Welding"[25] and the "Recommendations for Teaching Welding"[26] should be consulted when planning a new shop or a shop remodeling program.

Special precautions should be taken for fire protection. This includes making everything in the welding shop of noncombustible materials but also providing fire extinguishers and other safety equipment. State and local regulations must be followed. Arc welding booths should be at least 5-ft square and they should include mechanical ventilation.

The welding equipment for a school shop should be representative of the type of machines used in local industry. They should be of industrial quality with 60% duty cycle or higher. Limited input, light-duty machines are not recommended. The welding equipment should be specified for the welding processes to be taught. Both direct current and alternating current machines should be employed for shielded metal arc welding. The equipment should be completely specified including rated output current and type, output voltage, duty cycle, and NEMA class primary input voltage, frequency, and number of phases. The transformer and rectifier type power sources are less noisy than motor generators; however, some motor generators should be employed since this type is still widely used in the construction industry.

The gas-welding and oxygen-cutting equipment should also be of industrial quality and should show appropriate approvals. Sufficient supplies and auxiliary apparatus should be available, including welding tips, hose, regulators, and lighters.

Each student should have safety and protective clothing for the process being learned.

Welding instructors should make arrangements for welding materials including gases for welding, cutting and shielding, filler metals including electrodes, rods and flux, and metal for welding practice. Metals for welding must correspond to the training program. Many companies will cooperate with welding training departments and assist by supplying scrap metal for student use. Maintenance procedures should be established so that all welding equipment is in first-class condition at all times.

The final item required for the welding school shop is furniture or equipment for each work station. It is difficult to find school shop furniture constructed for heavy-duty welding shop use. Many instructors produce their own in the welding shop. Detailed plans and material lists for weld shop equipment are provided in the "Recommendations for Teaching Welding" book.

Welding training aids should be used when they contribute to the program. Training aids include the chalkboard, posters, movies, transparencies, slides and film strips, models, workmanship samples, cutaway equipment, etc. Equipment for using audio and visual aids must be available.

There are many movies available on welding and related subjects. Most are in color with sound. Audio coupled slide and film strip presentations are also available. These should be selected with respect to the subject areas being taught and should fit the objectives of the course.

Models of parts, cutaway parts, and so on, are all of value and should be used when appropriate.

Most welding equipment manufacturers and trade associations in the welding industry and in the metals industry provide much of this material. In addition, training aids and programs are available from many publishers. When analyzing training aids make sure that they use correct terminology, feature safety practices, and that they are technically accurate.

3-5 QUALIFYING AND CERTIFYING WELDERS

Becoming a qualified welder is a very satisfying experience. It means that you have made specified welds under specific conditions and that these welds have been prepared and tested and have passed the requirements of a specific specification. Congratulations! You deserve it. Qualification of welders is an extremely technical subject and carries with it certain legal responsibilities. Definitions of some of the terms involved follow: **Welder qualification** means "the demonstration of a welder's ability to produce welds meeting prescribed standards." **Welder certification** means "certification in writing that a welder has produced welds meeting prescribed standards."

Qualification is defined differently in different specifications, but in general the qualification is the ability of an individual to perform to a required standard. It is the demonstration that the welder has the ability to make a weld that meets the standards of the specification involved. It means taking and passing a prescribed practical welding test.

Certification is a written statement attesting to the fact that the welder has produced welds meeting a prescribed standard. Certification implies that a testing organization, a manufacturer, contractor, owner, or user has witnessed the preparation of the test welds, has conducted the prescribed testing of the welds, and has recorded the successful results of tests in accordance with prescribed acceptance standards.

Registration is the act of filing or registering the welder's written certification to an appropriate or authoritative group or body.

Welders and others sometimes become confused when they encounter the need to weld with a "qualified

welder." This confusion is easy to understand and is due to the great number of different codes, specifications, and government regulations with seemingly different requirements applying to welder qualification.

Before a welder can work on products such as pressure vessels, piping, bridges, public buildings, aircraft, ships, and so on, the welder must be qualified. Requirement for qualification is dictated by the specification that governs the product being welded. In addition to specific product specifications, there are legal requirements such as city or state laws and federal government regulations that require all welding to be done by certified welders.

Certification under one code will not necessarily qualify the welder to weld under a different code, even though the tests are similar. Each industry has its own welding requirements; therefore, it is absolutely essential that the applicable code be used in qualifying the welders.

Welding codes and specifications are written to provide a minimum set of rules for the construction of weldments that will protect the public life and property. Codes or specifications cover welding on the following:

1. Nuclear reactors, pressure vessels, boilers, etc.
2. Piping for high pressures and high temperatures.
3. Cross-country transmission line piping.
4. Buildings, bridges, and structural work.
5. Ships and boats—commercial, Navy and Coast Guard.
6. Ordnance and military engineering material.
7. Aircraft and space vehicles.
8. Shipping containers, tank cars and tank trucks.

Many engineering and technical societies have formulated codes and specifications in order to maintain uniformity in the fields they serve. Foremost among these in the welding field are the American Welding Society and the American Society of Mechanical Engineers.

Many trade associations, in order to establish product uniformity, have established standards. Among those issuing welding standards are the Association of American Railroads and the American Petroleum Institute.

Many of the larger cities have codes that regulate welding on buildings and structures erected within their city.

Most of the states and provinces have specifications covering the welding of highway bridges. In addition, many of them also have industrial commissions that issue codes covering pressure vessels and pressure piping. The bridge codes are similar to the AWS structural code and the pressure vessel and piping codes are similar to those of ASME. State-adopted codes have legal status.

Various federal government departments, branches, and bureaus issue specifications. Some involve welding. The Department of Defense issues the various military (MIL) specifications. Other standards are issued by the Atomic Energy Commission, Department of Transportation, and others. These also have legal status.

Many companies manufacture products not covered by any code or specification. In order to ensure continuing product quality they issue their own codes which are becoming more numerous because of product liability requirements.

The specifications issued by government bodies are enforceable by law and quite often become contractual requirements. Trade association and engineering society specifications are usually involved in purchase agreements as standards of quality acceptance.

Fortunately, most of the welding codes are similar or have a similar requirement for the welding and testing specimens. They also have similar requirements for the welding procedures, etc. However, make sure to consult the specific code under which you are qualifying.

Most of the specifications state that the manufacturer, contractor, owner, or user is responsible for the welding done and that they must conduct tests required to qualify welding procedures used. This responsibility carries the requirement of replacing any work that does not pass inspection or test prescribed by the specification. The contractor, manufacturer, owner, or user is personally unable to observe the work of each and every welder that might be employed. Therefore, they rely upon the fact that each welder has passed qualification tests. Contractors, manufacturers, owners, and users must maintain complete records of procedures and test results of qualified welders and operators. They must also hold and use only as directed various approval stamps.

Welders or welding operators cannot be qualified or certified on their own. Manufacturers, contractors, owners, or users will certify that a welder is qualified based on the successful completion of specific test welds. In some codes, the welder is qualified based on the successful completion of specified tests in that code.

In most codes, however, a welding procedure specification must first be established and qualified before the welders themselves can be qualified. A welding procedure specification is required by most codes and is proof that the procedure will produce satisfactorily welded joints. Following this, each welder must take a performance qualification test.

Qualifying under one code normally will not qualify under another with a few exceptions. One exception involves certain Navy codes, the Coast Guard code, and the American Bureau of Shipping specifications for the welding on ships. Qualifying for one manufacturer under a specification normally will not qualify you for another manufacturer, with a few exceptions. In the piping field procedures are qualified in the name of the National Certified Pipe Welding Bureau or of local contractor associations. Qualified welders may transfer from one member company to another without requalification.

A welding procedure specification requires two signatures, one by the person witnessing welding and testing, and the other by the person employed by the manufacturer, contractor, owner, or user who is responsible. Independent Testing Laboratories do procedure and performance testing for companies and individuals. This can have an advantage since it eliminates possible bias that might occur. The person responsible for welding must also sign procedure and performance test documents.

3-6 TRAINING AND QUALIFYING THE OTHER WELDING PERSONNEL

The "other" personnel in the welding field are in occupations that require a basic understanding of welding. To prepare for these occupations, it is wise to learn to weld and to qualify under one of the more important codes.

The Welding Supervisor

Welding supervisors go under many names but in general they supervise and coordinate the activities of welders. They must aid and assist welders, instruct welders in technique, and judge the work of welders, particularly the quality of welds and weldments. A welding supervisor must have welding experience but need not be the best welder. The supervisor must have knowledge of the processes being used by the people being supervised. Many supervisors are selected from the ranks of welders and are given additional specialized training. This training usually consists of courses on supervision and management, on production control, inspection, cost control and accounting, human relations, etc. Many part-time training programs are available that provide the principles of supervision. These, plus additional courses appropriate to the company or situation involved, are helpful.

In some situations experienced supervisors not involved with welding are given welding training and assigned to manage welding departments. Sufficient welding skill training for the processes that are involved is required plus background technical information.

Some supervisors are required to pass practical and technical tests in welding plus tests related to supervision and management. In Canada, companies doing fabrication and erection of welded steel structures must obtain and maintain certification.[27] A requirement for company certification is that the welding supervisors must have a thorough knowledge of the company welding procedures and specifications, have a good knowledge of the various welding codes and standards involved, and must have knowledge of the welding processes, welding problems, faults, quality control, inspection methods, and the various types of welding equipment. The supervisor must be able to read drawings and blueprints and interpret welding symbols. Furthermore, it is required that the supervisor show proof of experience or take practical tests.

Australia has a code for certifying structural welding supervisors.[28] This code has a strict requirement of experience and training. It requires that supervisors pass written tests covering the aspects of welding and structural steel work.

Many companies have intensive training programs for welding supervisors which include administering performance tests to welders, training welders, checking fit, examining welds being made, verifying that codes and specification requirements are being fulfilled and making certain that welding procedures are correct, etc. It is obvious that the job of welding supervisor is an important and complex job, extremely worthwhile and absolutely necessary.

The Welding Inspector

This is another person associated with welding. The welding inspector inspects and tests welded joints for visible defects, correct dimensions, joint strength, weld bead penetration, and conformance with layout print and work order specifications, applying knowledge of geometry, welding principles and physical properties of the metal. The inspector examines joints to detect flaws such as cracks, spatter, and undercut using a flashlight and magnifying glass, inspects joints for internal defects using nondestructive techniques, tests welds using testing equipment, verifies the alignment and dimensions of product, and sets up machines and fixtures used for work in progress, marks defective pieces and recommends scrapping or methods

for rework, records inspection data, supervises the taking of qualification tests, records data and may certify results. The inspector also varifies that procedures are in order and appropriate to the work and the specifications.

The American Welding Society has established a procedure for certifying welding inspectors.[29] This requires a record of the inspector's experience and training and a test to determine the person's technical knowledge of welding. It is required that the inspector have welding experience; however, a practical test is not required. Upon the successful completion of the test and approval of the application the welding inspector is certified as qualified by the American Welding Society and is registered at their headquarters.

Most states qualify inspectors for pressure vessel and pressure piping work. These are usually done by the State Industrial Commission in conjunction with the National Board of Boiler and Pressure Vessel Inspectors[30] and casualty insurance companies. The National Board provides training and testing to determine that inspectors have the necessary knowledge and ability. Additionally, there are government inspectors for the U.S. Department of Defense military requirements; various cities have inspectors for structural welding on buildings and bridges, and most state highway departments also have welding inspectors. It is impossible to include all the different types of inspectors and the tests and qualifications that are required. For your particular needs you should contact the organization of interest to determine their requirements.

The Welding Technician

Many welding technicians have received specialized training by taking various courses to supplement their knowledge of welding. Most technicians were highly skilled welders who had the desire to take additional training to become a technician. The training should include many subjects available through adult training courses at local schools and colleges, technical society programs, correspondence courses, company-sponsored seminars and courses on subjects that are associated with welding, including drawing and blueprint reading, mathematics, physics, chemistry, metallurgy, business, accounting, English, etc. More and more technicians are graduates of associate degree programs from technical schools, colleges, and universities. An associate degree is obtained with two years of post-high school study or equivalent on-the-job training. In their work,

WELDING PERSONNEL

they apply scientific and engineering knowledge and methods combined with technical skills in support of engineering activities. On the job, they perform semi-professional engineering and scientific functions normally with general supervision by an engineer. Since the technician works closely with the engineers, the engineering technologist must master the language of engineering, mathematics, science, graphics, communications, and specialized subject matter. The technician must be able to apply theory and use capabilities of skilled craftsmen to achieve practical and economical results. The welding technician works with designers to improve the way welding is used on products, machines, structures, and equipment. The technician maintains or improves the properties of metals by the welding processes. The technician must know, understand, and be able to operate processes, procedures, and equipment of the welding industry and must verify the existence of a predetermined quality level.

Upon graduation, the employment opportunities of welding technicians are almost unlimited.

The Welding Instructor

This is another important person in the overall welding field. Welding instructors must have a high degree of welding capabilities. The instructor must be able to make welds in order to instruct others. A welding instructor should have the competence to pass qualification tests in the same field of instruction. The instructor must be able to demonstrate welding in front of students and to work with students to correct their errors of technique. The instructor must also have the ability to communicate to the students, to relate with them, and to impart welding technical knowledge.

In some states vocational instructors are required to have education degrees. In some vocational schools the college degree requirement is eliminated, but practical experience is required. In many cases, specialized training to help instructors teach is required.

There are other types of welding instructors; for example, instructors who teach welding in industrial arts, vocational agriculture, automotive mechanics, and hobby and craft programs. These instructors may only teach a limited area of welding, but must be knowledgeable in this area. It is doubtful if practical tests would be needed to qualify these instructors, but they must be able to communicate, teach, and demonstrate the processes that are offered.

The Welding Engineer

This is one of the newer titles of an engineer and is of great importance to the welding industry. The welding engineer applies scientific principles to the useful arts. Normally the welding engineer is applying several of the sciences simultaneously, and must have knowledge of other engineering fields such as metallurgical, mechanical, electrical, structural, and chemical. The welding engineer is dealing almost exclusively with metals, but needs knowledge of ceramics and chemistry since these greatly influence slags, fluxes, etc. The welding engineer must know the chemistry of flames and of exothermic welding and stress-relieving techniques. The welding engineer must have an understanding of the basic principles of electricity, electronics, and of the complexities of arcs in gases. This engineer must also have a knowledge of physics with respect to metal transfer, heat flow, conductivity, etc., since heat is involved in most welding processes. The concepts of welding and weldment design must be well known since weldments are successful or not depending on their design. The service performance of the weldment is a tribute to the skill or lack of skill of the welding engineer. Welding engineers require a baccalaureate degree from a college or university. Few engineering colleges offer welding engineering programs. There is only one baccalaureate program accredited by the engineering council for professional development. Some universities offer Masters and PhD degrees in other engineering disciplines intimately associated with welding. Welding engineers with Bachelor degrees and with advanced degrees are in high demand.

Registration of welding engineers is not necessary. However, if the welding engineer is involved with public works of any type, registration as a professional engineer under state law is highly recommended. Engineering certification is provided by the Society of Manufacturing Engineers. Certification can be acquired by taking specific tests and maintaining a program of continuing education.

The Welding Sales Representative

This is another important person in the welding industry. Many welding sales representatives learned about welding as welders or technicians. With this practical and technical knowledge of welding they can acquire the necessary product knowledge, sales administration, marketing, accounting, commercial law, and other information necessary to successfully sell welding products. The National Welding Supply Association offers specialized training programs to people wishing to become welding sales representatives. Sales representatives can find employment with national manufacturing organizations or local welding distributors. They can work in almost any part of the country. The remuneration for sales representatives can be very rewarding depending on individual efforts and successes.

The Welding Service Representative

Also important to the welding industry are service representatives who help maintain proper operation of welding equipment. They normally represent manufacturers of distributors. Their primary function is to help users diagnose problems that involve the welding equipment and to correct the problem by servicing it in the customer's plant or by taking it to their own service shop for repair. The welding service representative must know and under-

stand welding because often the difficult problems involve both welding and the intricacies of the equipment. Electrical and mechanical knowledge and experience are necessary as well as the ability to use tools and instruments. The ability to read blueprints and electrical schematic drawings and to analyze problems is extremely important. Welding service technicians are in great demand most of the time.

Remember the welder can be the starting point for any of these different types of work. In every case familiarization with welding is required to successfully function in these related welding jobs.

QUESTIONS

1. Where are welding plants generally located? What products do welders help make?

2. What tests can be taken to determine if a person should become a welder?

3. Name some of the future jobs open to welders who take extra training.

4. Is welding more dangerous or less dangerous than other metalworking jobs?

5. Why shouldn't a welder open up the case of an arc welding machine?

6. Which is more dangerous, welding open-circuit voltage or the voltage in an electric drill?

7. Why is eye and skin protection required against an electric arc?

8. Name the metals that create hazards when welded.

9. What is meant by an oxygen-deficient atmosphere? How is it related to welding?

10. Why is it dangerous to weld on a closed tank or can?

11. Give two reasons for checking an enclosed area before welding in it.

12. What is the purpose of a hot work permit?

13. What should you do if an acetylene cylinder catches on fire?

14. Why is maintenance welding more dangerous than production welding?

15. How do you make a tank that has held a combustible safe for welding?

16. Why should oil and grease be kept away from pure oxygen?

17. What is meant by qualifying a welder?

18. What is meant by certifying a welder?

19. Are the welding tests of all specifications the same? Explain.

20. How can you advance to higher-paying welding-related jobs?

REFERENCES

1. "Opportunities in the Welding Industry," *The American Welding Society,* Miami, Florida, May 1974.

2. *Occupational Outlook Handbook,* Bulletin 1785, U.S. Department of Labor, Bureau of Labor Statistics, Superintendent of Documents, Washington, D.C., 1974–75.

3. "Where are They Now?," *The Welding Engineer,* Morton Grove, Illinois 60053, July 1973, p. 31.

4. *Tomorrow's Manpower Needs,* Bulletin 1737, Vol. IV, Research 1971, Bureau of Labor Statistics, U.S. Department of Labor, Superintendent of Documents, Washington, D.C.

5. *Dictionary of Occupational Titles,* Vols, I and II., plus Supplement 1966 and Supplement 2, 1968, U.S. Dept. of Labor, Superintendent of Documents, Washington, D.C.

6. CARY, H. B., "The Job of the Welder," *The Welding Journal,* Miami, Florida, January 1975.

7. "Report on a Danish Investigation into the Health and Working Environment of Arc Welders," *Welding in the World,* Vol. 10, Nos. 3 and 4, 1972.

8. "Safety in Welding and Cutting," *American Welding Society* ANSI Z49.1, Miami, Florida, 1973. .

9. "Occupational Safety and Health Standards," *Code of Federal Regulations Title* 29—Labor, Part 1910, Supt. of Documents, U.S. Printing Office, Washington, D.C. or *Federal Register,* Vol. 37, No. 202, Part II, October 18, 1972.

10. "Electric Arc Welding Apparatus," *National Electrical Manufacturers Association* ANSI C87.1, New York, 1971.

11. "Construction and Test of Arc-Welding Equipment, Transformer Type," *Canadian Standards Association, (CSA)* C22.2 No. 60, Ottawa, Canada 1959.

12. "Transformer Type Welding Machines," Underwriters Laboratories, *ANSI* C33.2, Chicago, Illinois, 1972.

13. "National Electric Code," National Fire Protection Association (No. 70-1971), *ANSI* C1-1971, Boston, Massachusetts.

14. BRESLIN AND HAMS, "Use of Thoriated Tungsten Electrodes in Inert Gas Shielded Arc Welding," *American Industrial Hygiene Assoc. Quarterly,* 13.4, December, 1952.

15. "Practice for Occupational and Educational Eye and Face Protection." *ANSI* Z87.1, New York, 1968.

16. "The Welding Environment," *American Welding Society,* Miami, Florida, 1973.

17. WIEHE, CARY, AND WILDENTHALER, "Application of a Smoke Extracting System for Continuous Electrode Arc Welding," *The Welding Journal,* June 1974.

18. "Oxygen-Fuel Gas Systems for Welding and Cutting," *National Fire Protection Association, (NFPA)* No. 51, Boston, Massachusetts, 1969.

19. "Handling Acetylene Cylinders in Fire Situations," Compressed Gas Association, *Safety Bulletin* No. SB-4, New York.

20. Hot work permits (flame or sparks) National Safety Council. Data Sheet 522 Revision A. Chicago, Illinois, 1962.

21. "Cutting and Welding Practices," *National Fire Protection Association, (NFPA)* No. 51B Boston, Massachusetts, 1971.

22. "Safe Practices for Welding and Cutting Containers That Have Held Combustibles," *American Welding Society,* A-6.0-65.

23. "Welding or Hot Tapping on Equipment Containing Flammables," *Petroleum Safety Data Sheet* 2201, American Petroleum Institute, New York, November 1963.

24. "Compressed Gas Cylinder Valve Outlet and Inlet Connections," *American Standard* B57.1-65 and *Canadian Standard (CSA)* B96-65.

25. "Minimum Requirements for Training Welders," E3.1-74, *American Welding Society* E3.1-74 Miami, Florida.

26. CARY. H. B., "Recommendations for Teaching Welding," Hobart School of Welding Technology, Troy, Ohio.

27. "Certification of Companies for Fusion Welding of Steel Structures," *Canadian Standards Association, (CSA)* 47.1 Ottawa, Canada.

28. "Certification of Welding Supervisors in Structural Welding," *The Standards Association of Australia,* N.S.W., AS CA60-1970, Sydney, Australia.

29. "Standard for Qualification and Certification of Welding Inspectors," *American Welding Society* QCI-76, Miami, Florida.

30. *National Board Inspection Code,* The National Board of Boiler and Pressure Vessel Inspectors, Columbus, Ohio.

Joint Design

When welding materials thicker than 1/8 in. (3.2 mm) space must be made available to deposit the weld metal. Various weld groove designs are used but the fillet is the most common weld made. Chapter 15, Design for Welding gives complete information on joint design for the shielded metal arc welding process.

Equipment Required to Operate

The welding machine or power source is the heart of the shielded metal arc welding system. Its primary purpose is to provide electric power of the proper current and voltage to maintain a controllable and stable welding arc.

The Welding Circuit

Figure 4-13 is the circuit diagram for shielded metal arc welding. It shows the welding cables used to conduct the welding current from the power source to the arc. The electrode lead forms one side of the circuit and the work lead is the other side of the circuit. They are attached to the respective "work" and "electrode" terminals of the welding machine.

Welding can be accomplished with either alternating current (AC) or direct current electrode negative (DCEN), straight polarity or electrode positive (DCEP) reverse polarity.

The output characteristics of the power source must be of the constant current (CC) type. The normal current range is from 25 amperes to 500 amperes using conventional size electrodes. The arc voltage varies from 15 to 35 volts.

Figure 4-14 shows a typical shielded metal arc welding operation using an engine-driven welding machine.

There are certain items of auxiliary equipment that can be used with the shielded metal arc process. These include low-voltage control circuits which provide for a cutoff of open-circuit voltage until the electrode is touched to the work. Other items include secondary contactors, remote control switches for the contactors, remote control current-adjustment devices, crater-eliminating devices, engine-idling devices for engine-driven machines, etc.

The next important piece of apparatus is the electrode holder which is held by the welder. The holder firmly grips the electrode and carries the welding current to it.

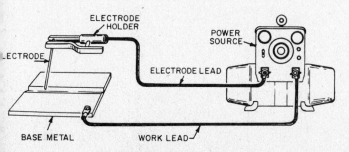

FIGURE 4-13 *Circuit diagram.*

FIGURE 4-14 *Welding operation—engine drive machine.*

Electrode holders are supplied in various types and sizes; two types are shown by Figure 4-15. They are usually designated for their current-carrying capacity. The two more popular types are the pincher type and the collet type.

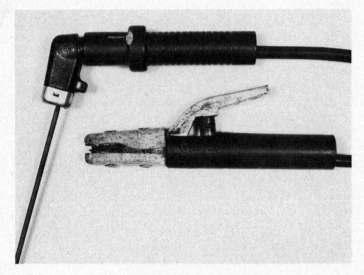

FIGURE 4-15 *Electrode holders.*

A later chapter provides more information on the types, sizes, and selection of electrode holders, cable connectors, work clamps, and other devices in the welding circuit.

Welding Materials

The covered electrode is the only item of material normally required. The selection of the covered electrode for specific work is based on the electrode usability and the composition and properties of the deposited weld metal. In order to properly select an electrode it is well to understand the function of the coating, the basis of specifying, the usability factors, and the deposited weld metal properties.

The coating on the electrode provides (1) gas from the decomposition of certain ingredients of the coating to shield the arc from the atmosphere, (2) the deoxidizers for scavenging and purifying the deposited weld metal, (3) slag formers to protect the deposited weld metal with a slag from atmospheric oxidation, (4) ionizing elements to make the arc more stable and to operate with alternating current, (5) alloying elements to provide special characteristics to the deposited weld metal, and (6) iron powder to improve productivity of the electrode. Before

the late 1920s, bare or lightly-coated electrodes were used for welding. However, since they were more difficult to use and did not produce high-quality weld metal their use has fallen to almost zero.

Welding electrodes have been divided into various categories based on composition. More information about electrodes; manufacturing, specifications, selection, storage, etc., is given in Chapter 10, "Filler Metals and Materials for Welding."

The American Welding Society has established a system for identifying and specifying the different types of electrodes and filler metals. This is shown by Figure 4-16. The mild steel and low-alloy steel-covered electrodes are prefixed by the letter "E," followed by a 4- or 5-digit number. The prefix "E" means electrode.

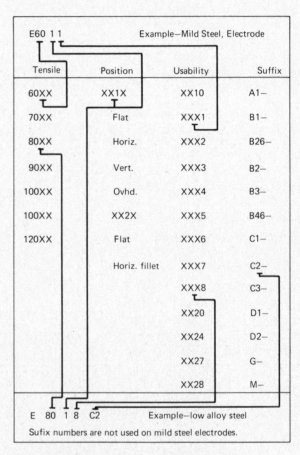

FIGURE 4-16 *Covered electrode identification system.*

The first 2 (or 3) digits indicate the tensile strength in pounds per square inch of the deposited weld metal. The third or fourth digit indicates the position for which the electrode is designed. No. 1 means "all positions," flat, horizontal, vertical, and overhead, and No. 2 means horizontal fillet and flat position only. No. 3 means vertical with downward progression. The fourth or fifth digit is a "usability" rating which

First Two or Three Digits Indicate Tensile Strength and Other Mechanical Properties

Classification	Tensile Strength min. psi	Yield Strength min. psi	Elongation in. 2 in. %, min.
E60XX	62,000	50,000	17
E70XX	70,000	57,000	22
E80XX	80,000	67,000	19
E90XX	90,000	77,000	17
E100XX	100,000	87,000	16
E110XX (1)	110,000	95,000	15
E120XX (1)	120,000	107,000	14

Note 110XX and 120XX low hydrogen type of coating only.

Third (or) Fourth Digit Indicates the Welding Position That Can Be Used

Classification	Flat Position	Horiz. Post.	Vertical Post.	Overhead Position
EXX1X	Yes	Yes	Yes	Yes
EXX2X	Yes	Fillet	No	No

Last Digit Indicates Usability of the Electrode

Classification	Current	Arc	Penetration	Covering & Slag	Iron Powder
F-3 EXX10	DCEP	Digging	Deep	Cellulose-sodium	0-10%
F-3 EXXX1	AC&DCEP	Digging	Deep	Cellulose-potassium	0
F-2 EXXX2	AC&DCEN	Medium	Medium	Rutile-sodium	0-10%
F-2 EXXX3	AC&DC	Light	Light	Rutile-potassium	0-10%
F-2 EXXX4	AC&DC	Light	Light	Rutile-Iron powder	25-40%
F-4 EXXX5	DCEP	Medium	Medium	Low hyd.-sodium	0
F-4 EXXX6	AC or DCEP	Medium	Medium	Low hyd.-potassium	0
F-4 EXXX8	AC or DCEP	Medium	Medium	Low hyd.-iron powder	25-40%
F-1 EXX20	AC or DC	Medium	Medium	Iron oxide-sodium	0
F-1 EXX24	AC or DC	Light	Light	Rutile-iron powder	50%
F-1 EXX27	AC or DC	Medium	Medium	Iron oxide-iron powder	50%
F-1 EXX28	AC or DCEP	Medium	Medium	Low hyd.-iron powder	50%

Note: Iron powder percentage based on weight of the covering.

FIGURE 4-17 *Details of the classification system for mild and low alloy steel electrodes.*

indicates the type of coating, which in turn indicates the type of welding current that is to be used. The exact meaning of each code number is given by Figure 4-17. Note that when the fourth or fifth digit is zero the type of coating and current to use are determined by the third digit. For example, E-6010 indicates a cellulose sodium coating and operates on DC electrode positive while an E-7018 has an iron powder low-hydrogen coating and operates on DC electrode positive or on AC. In order to identify electrodes, each one is type marked or printed with the identifying number as shown by Figure 4-18. Color coding was formerly used for identification but is no longer popular. The color coding is still used for surfacing electrodes.

FIGURE 4-18 *Printed type marking on electrodes.*

THE ARC WELDING PROCESSES—USING CV POWER

The mechanical properties of the deposited weld metal must equal or exceed those of the base metal. Weld metal must also have approximately the same composition and physical properties.

Base metal mechanical properties and composition—the base metal must be identified so that its mechanical properties and composition are known. If the base metal is mild steel, select any E60XX electrode because its deposited metal will overmatch mechanical properties.

The abridged specifications for carbon steel covered arc welding electrodes (A5.1) is shown by Figure 4-19. This provides the AWS classification, the radiographic standard, the mechanical properties, and also the minimum V-notch impact requirements for the E-60XX and E-70XX class of electrodes. The radiographic standard is related to the allowable porosity limits as shown by the specification. The mechanical properties are the minimum tensile strength in pounds per square inch, the minimum yield point in pounds per square inch and the minimum elongation in two inches in percent. The V-notch impact requirements are a minimum at two temperatures, either 0°F (17.8°C) or −20°F (−29°C). There are other requirements in this specification and for this reason the specification itself must be referred to for details.

Welding Position: Electrodes are designed to be used in specific positions. The third (or fourth) digit of the electrode classification indicates the welding position that can be used. Match the electrode to the welding position to be encountered.

Welding Current: Some electrodes are designed to operate best with direct current (DC), and others operate best with alternating current (AC). Some will operate on either. The last digit indicates welding current usability. Select the electrode to match the type of power source to be used.

Joint Design and Fitup: Welding electrodes are designed with a digging, medium, or soft arc for deep, medium, or light penetration. The last digit of the classification also indicates this factor. Deep penetrating electrodes with a digging arc should be used when edges are not beveled or fitup is tight, but light penetrating electrodes with a soft arc are required when welding on thin material or when root openings are too wide.

Service Conditions or Specifications: For weldments subject to severe service conditions such as low temperature, high temperature, or shock loading, select the electrode that matches base metal composition, ductility and impact resistance properties. The low-hydrogen types have the better properties.

Production Efficiency and Job Conditions: Some electrodes are designed for high deposition rates, but may be used only under specific position requirements. If they can be used, select the high-iron powder types—EXX24, 27, or 28. Other conditions may be present which will require experimentation to determine the most efficient electrode.

For the usability of covered electrodes the mild steel electrodes are classified into four general groups.

F-1 High Deposition Group–E-6020, E-6024, E-6027, E-6028

F-2 Mild Penetration Group–E-6012, E-6013, E-7014

AWS Classification	Tensile Strength min. psi	Yield Point min. psi	Elongation in 2 inch %	Radiographic Standard AWS Grade	V Notch Impact min. ft-lb
E6010	62,000	50,000	22	2	20 @ −20° F
E6011	62,000	50,000	22	2	20 @ −20° F
E6012	67,000	55,000	17	Not Rgd.	Not Rgd.
E6013	67,000	55,000	17	2	Not Rgd.
E6020	62,000	50,000	25	1	Not Rgd.
E6027	62,000	50,000	25	2	20 @ −20° F
E7014	72,000	60,000	17	2	Not Rgd.
E7015	72,000	60,000	22	1	20 @ −20° F
E7016	72,000	60,000	22	1	20 @ −20° F
E7018	72,000	60,000	22	1	20 @ −20° F
E7024	72,000	60,000	17	2	Not Rgd.
E7028	72,000	60,000	22	2	20 @ 0° F

FIGURE 4-19 *Abridged specification for mild steel covered electrodes per AWS A5.1-69.*

Welding Position / Weld Type / Material Thickness	Fillet Welds — Inside or Outside				Groove Welds — Square		Groove Welds — Vee (Open Root)		Groove Welds — U	
	Very Thin	Thin	Medium	Thick	Very Thin	Thin	Medium	Thick	Medium	Thick
1 FLAT	F-2	F-2	F-1	F-1 F-4	F-2	F-3	F-3	F-3 F-4	F-4	F-4
1A HORIZ FILLET	F-2	F-3	F-1	F-1 F-4	—	—	—	—	—	—
2 HORIZ	F-2	F-3	F-3 F-4	F-3 F-4	F-2	F-2 or F-3	F-3 and F-4	F-3 F-4	F-4	F-4
3 VERT UP	F-2	F-3	F-4	F-4	F-2	F-2 or F-3	F-3 and F-4	F-3 and F-4	F-4	F-4
3A VERT DOWN	F-2	F-3	—	—	F-2	F-2 or F-3	F-3	F-3	F-3	F-3
4 OVERHEAD	F-2	F-3	F-3 F-4	F-3 F-4	F-2	F-2 or F-3	F-3 and F-4	F-3 and F-4	F-4	F-4
5 PIPE FIXED DOWNHILL	—	—	—	—	F-2	F-2	F-3	F-3	F-3	F-3
5A PIPE FIXED UPHILL	—	—	—	—	F-3	F-3	F-3 and F-4	F-3 and F-4	F-4	F-4

Material Thickness: Very thin = .005″ to .063″ (.125–1.6 mm),
Thin = $\frac{1}{16}$″ – $\frac{1}{4}$″ (1.6–6.3 mm), Medium = $\frac{1}{4}$″ – $\frac{3}{4}$″ (6.3–19 mm),
Thickness = $\frac{3}{4}$″-up (19 mm–up)

FIGURE 4-20 *Usability rating guide for selecting MS and LA electrodes.*

F-3 Deep Penetration Group–E-6010, E-6011

F-4 Low Hydrogen Group–E-6015, E-6016, E-7018

Figure 4-20 is a guide to aid in selecting the covered electrode for specific welding jobs based on the welding position, metal thickness, and weld type.

Electrodes in the same grouping operate and are run the same way. These F numbers correspond to the classification system used in Section IX of the ASME Pressure Vessel Boiler Code.

Deposition Rate

The melting rate of the electrode is related to the welding current. A portion of the arc energy is used to melt the surface of the base metal and part to melt the electrode. The electrode coating also affects deposition rates. The iron oxide types and iron powder types have higher deposition rates.

The melting rate to current is a fairly direct relationship as shown by Figure 4-21. With higher current, the current density in the electrode increases and this increases the melting rate which increases the deposition rate. Electrode size is determined by the job, the welding position, the joint detail, and the skill of the welder.

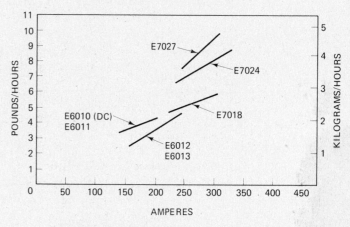

FIGURE 4-21 *Deposition rates — various electrodes.*

Quality of Welds

The quality of the weld depends on the design of the joint, the electrode, the technique, and the skill of the welder. If joint details are varied greatly from established design details, lower quality can result. The fitup of joints must match the design. Some electrodes deposit higher-quality weld metal than others, based on their specifications.

Welding Schedules

Welding schedules are tables of operating parameters that will provide high-quality welds under normal conditions. Strict welding schedules are not as important for manual shielded metal arc welding as for semiautomatic and automatically applied arc welding processes for several reasons. First, in manual shielded metal arc welding the welder controls welding conditions more by the manipulation of the arc than in any of the other arc welding processes. The welder directly controls the arc voltage and travel speed and indirectly the welding

current. Second, in shielded metal arc welding, meter readings are rarely used for the duplication of jobs. It is felt that recommended welding conditions for the different types of electrodes are sufficient for most operations. These are given by Figure 4-22. However, when more complete information is needed see the data provided by the tables of Figure 4-23. The settings given in these tables provide a good starting point when first welding on a new application. They are not necessarily the only welding settings that can be used under every condition. For example, for high-production work, the current settings could be increased considerably over those shown. Such factors as weld appearance, welder skill, and quality level will allow variations from the settings. As the requirements of a new application become better known the settings can be adjusted to obtain optimum welding conditions. Trials are suggested prior to the establishment of a firm procedure for specific applications.

These tables are based on welding low carbon mild steels under normal conditions and show the suggested electrode types for different weld types. Other electrode types and joints may be used. They are made in a consistent manner and are based on the type of welds and the position of welding, and are made by a welder of normal skill.

Type of Electrode	Size		Direct Current		Alternating Current	
	mm	inch	amps	volts	amps	volts
E-6010 F-3	2.38	3/32 × 14	40-80	24-26		
All position	3.18	1/8 × 14	70-130	24-26		
Deep penetration	3.97	5/32 × 14	110-165	21-23	—	—
DC EP	4.76	3/16 × 14	140-225	20-26		
	5.56	7/32 × 18	160-300	20-32		
	6.35	1/4 × 18	200-400	24-34		
E-6011 F-3	2.38	3/32 × 14	50-70	21-26	50-70	25-29
All position	3.18	1/8 × 14	75-130	20-30	75-130	25-28
Deep penetration	3.97	5/32 × 14	120-160	16-26	120-160	24-29
AC or DC EP	4.76	3/16 × 14	150-190	20-28	150-190	24-33
	5.56	7/32 × 18	180-250		180-250	
	6.35	1/4 × 18	200-300	22-28	200-300	30-40
E-6012 F-2	2.38	3/32 × 14	50-90	16-26	80-120	
All position	3.18	1/8 × 14	75-135	16-22	80-120	17-23
Medium penetration	3.97	5/32 × 14	120-205	16-24	120-190	17-25
AC or DC EN	4.76	3/16 × 14	140-255	17-26	140-240	17-28
	5.56	7/32 × 18	220-335	12-28		
	6.35	1/4 × 18	200-400	16-25		

FIGURE 4-22 *Recommended welding conditions for covered electrodes (mild and low alloy steel).*

Type of Electrode	Size mm	Size inch	Direct Current amps	Direct Current volts	Alternating Current amps	Alternating Current volts
E-6013 F-2	2.38	3/32 × 14	50-100	21-24	50-80	20-27
All position	3.18	1/8 × 14	80-140	18-20	80-120	22-24
Light penetration	3.97	5/32 × 14	120-190	20-25	120-170	22-26
AC or DC	4.76	3/16 × 14	160-220	25-28	190-220	20-22
	5.56	7/32 × 18	240-270			
	6.35	1/4 × 18	270-350	18-21	270-350	20-22
	7.94	5/16 × 18	320-420	23-26	320-420	24-30
E-6020 F-1	2.38	1/8 × 14	120-145	20-22	120-145	20-22
Flat and HF position	3.18	5/32 × 14	150-175	24-26	150-175	24-26
Medium penetration	3.97	3/16 × 18	210-240	23-27	210-240	23-27
High deposit rate	4.76	7/32 × 18	240-275	24-28	240-275	24-28
AC or DC	5.56	1/4 × 18	290-320	29-30	290-320	29-30
E-7010			Same as E6010 above			
E-7014 F-2	2.38	3/32 × 14	70-90	20-24	70-90	20-24
All position	3.18	1/8 × 14	120-145	14-17	120-145	14-17
Light penetration	3.97	5/32 × 14	140-250	17-19	140-210	17-19
AC or DC	4.76	3/16 × 18	180-280	20-24	180-280	20-24
	5.56	7/32 × 18	250-375	28-35	250-375	28-35
	6.35	1/4 × 18	300-420	26-31	300-420	26-31
	7.94	5/16 × 18	375-500	28-33	375-500	28-33
E-7016 F-4	2.38	3/32 × 14	70-100	17-21	70-100	20-25
All position	3.18	1/8 × 14	80-130	17-22	40-130	20-22
Medium penetration	3.97	5/32 × 14	120-170	18-19	120-170	18-20
AC or DC EP	4.76	3/16 × 14-18	170-250	17-22	170-250	17-23
	5.56	7/32 × 18	250-325			
	6.35	1/4 × 18	300-350	21-25	300-350	19-22
	7.94	5/16 × 18				
E-7018 F-4	2.38	3/32 × 14	80-110	20-22	80-110	24-26
All position	3.18	1/8 × 14	90-150	20-21	90-150	19-23
Medium penetration	3.97	5/32 × 14	110-230	20-22	110-230	20-25
AC or DC EP	4.76	3/16 × 14	150-300	20-22	150-300	21-28
	5.56	7/32 × 18	250-350	20-24	250-350	20-29
	6.35	1/4 × 18	300-400	20-24	300-400	24-30
	7.94	5/16 × 18				
E-7024 F-1	2.38	3/32 × 12	100-140	27-32	100-140	27-32
Flat and HF position	3.18	1/8 × 14	130-180	27-30	130-180	27-30
Light penetration	3.97	5/32 × 14	180-240	30-33	180-240	30-33
High deposition rate	4.76	3/16 × 14	200-280	22-28	200-280	22-28
AC or DC	5.56	7/32 × 18	250-375	29-32	250-375	29-32
	6.35	1/4 × 18	300-420	28-34	300-420	28-34
	7.94	5/16 × 18	425-500	29-35	425-500	29-35
E-7028 F-1	2.38	1/8 × 14	—	—	—	—
Flat and HF position	3.18	5/32 × 14	—	—	—	—
Medium penetration	3.97	3/16 ×	240-300	31-33	260-320	30-33
High deposition rate	4.76	7/32 × 18	300-400	36-40	320-400	30-35
AC or DC EP	5.56	1/4 × 18	350-450	37-41	370-470	31-37
	7.94	5/16 × 18				
E-8018			Same as E7018 above			

Note: Voltage is based on a normal arc length measured at the arc.

FIGURE 4-22   *(Continued) Recommended welding conditions for covered electrodes.*

FILLET SIZE	WELDING POSITION			
	FLAT 1F	HORIZONTAL 1F	VERTICAL UP 3F (U)	OVERHEAD 4F
1/4	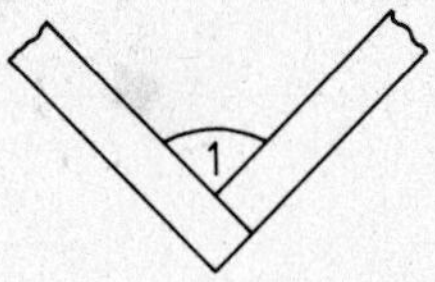	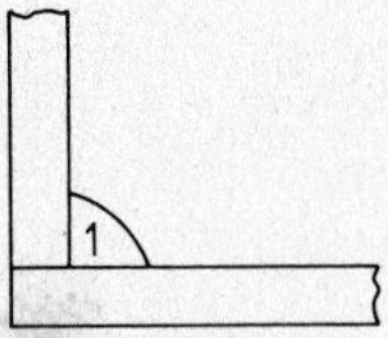	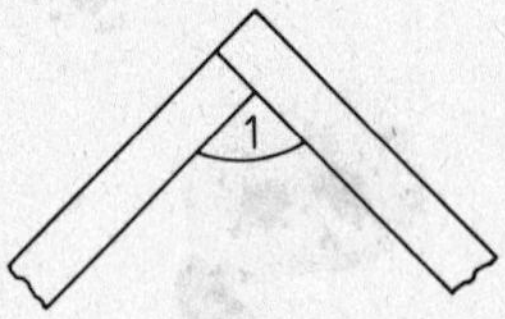	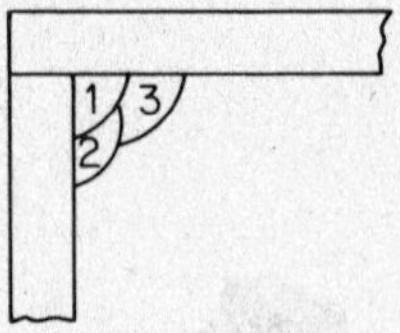
1/2	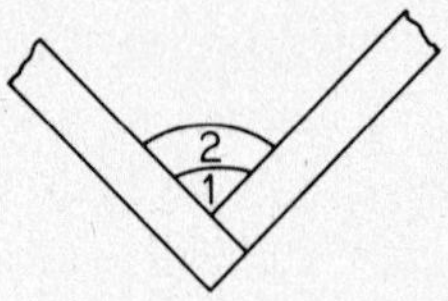	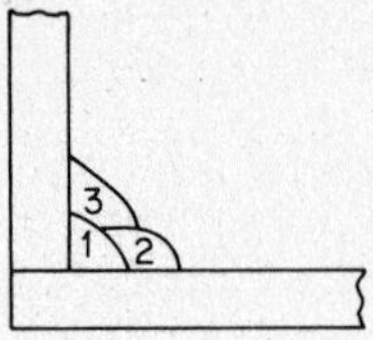	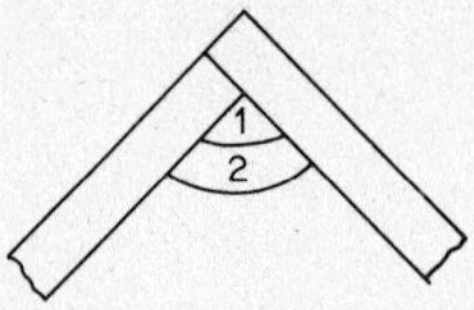	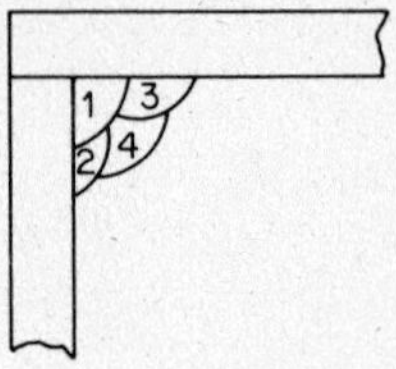
3/4	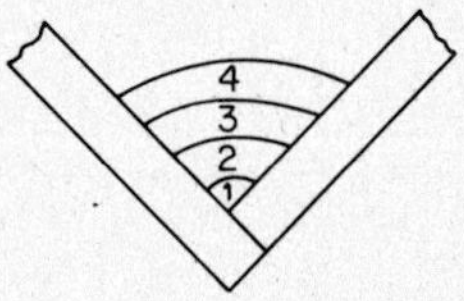	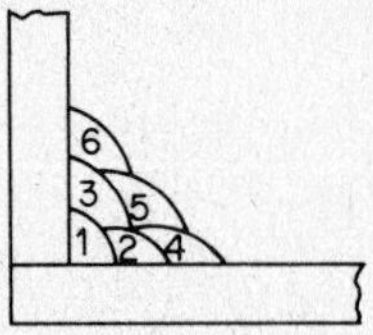	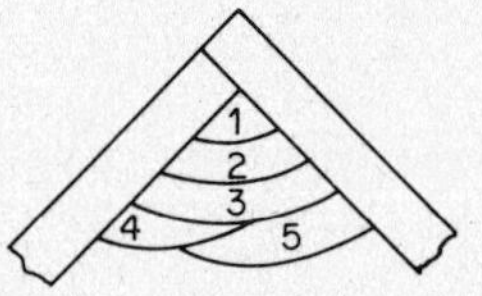	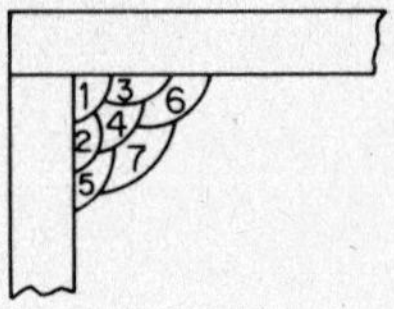

WELDING POSITION	SUGGESTED ELECTRODE TYPE	ELECTRODE DIAMETER FOR MATERIAL THICKNESS INCH		
		1/4	1/2	3/4
1F, 2F	E7024	1/4	1/4	1/4
3F (U)	E7018	5/32	5/32	5/32
4F	E6010	3/16	3/16	3/16
	E7018	5/32	5/32	5/32

MATERIAL THICKNESS (IN.)	WELDING POSITION
1/8	ALL POSITIONS
3/16	
1/4	

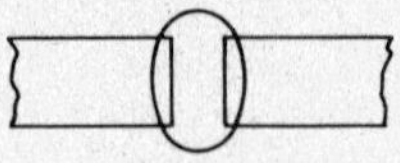

WELDING POSITION	SUGGESTED ELECTRODE TYPE	ELECTRODE DIAMETER FOR MATERIAL THICKNESS (IN.)		
		1/8	3/16	1/4
1G	E6010	3/32	1/8	5/32
2G, 3G(U)	E6010, E6012			
3G(D), 4G	E6014, E6013	3/32	1/8	5/32

FIGURE 4-23 *Welding procedure schedule—various welds.*

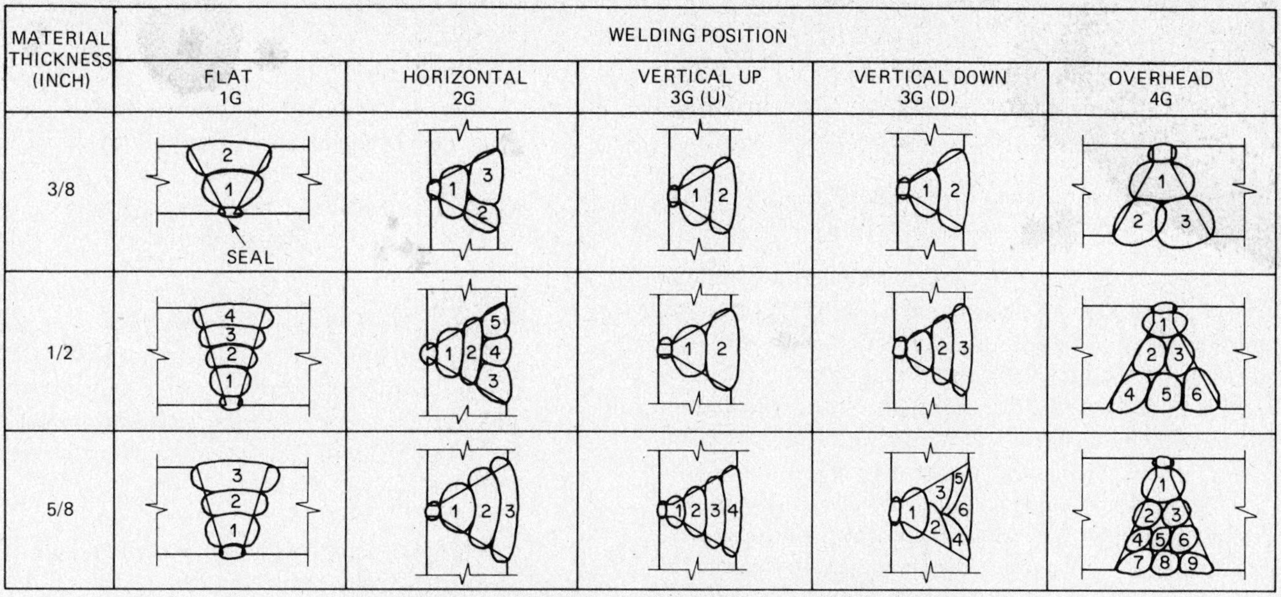

MATERIAL THICKNESS (INCH)	WELDING POSITION				
	FLAT 1G	HORIZONTAL 2G	VERTICAL UP 3G (U)	VERTICAL DOWN 3G (D)	OVERHEAD 4G
3/8					
1/2					
5/8					

WELDING POSITION	SUGGESTED ELECTRODE TYPE	ELECTRODE DIAMETER FOR MATERIAL THICKNESS (IN.)		
		3/8	1/2	5/8
1G	E6010	3/16	3/16	3/16
2G	E6010	3/16	3/16	3/16
	E7018	5/32	5/32	5/32
3G (U)	E6010	5/32	5/32	5/32
3G (D), 4G	E7018	5/32	5/32	5/32

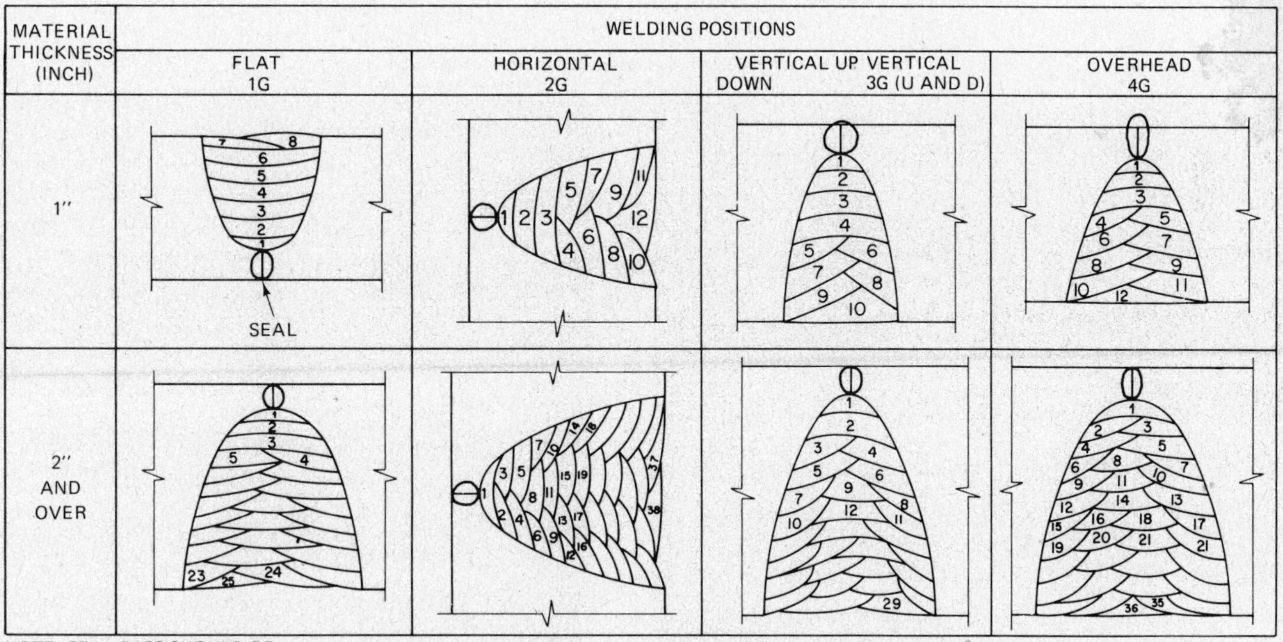

MATERIAL THICKNESS (INCH)	WELDING POSITIONS			
	FLAT 1G	HORIZONTAL 2G	VERTICAL UP VERTICAL DOWN 3G (U AND D)	OVERHEAD 4G
1"				
2" AND OVER				

NOTE: SEAL PASS SHOULD BE E6010, 5/32 OR 3/16" DIAMETER.

WELDING POSITION	SUGGESTED ELECTRODE TYPE	ELECTRODE DIAMETER FOR MATERIAL THICKNESS (IN.)	
		1	2
1G	E7018	1/4	1/4
2G	E7018	5/32	5/32
3G (U)	E6010	3/16	3/16
3G (D), 4G	E7018	5/32	5/32

FIGURE 4-23 (Continued)

CHAPTER 4

THE ARC WELDING PROCESSES—USING CC POWER

*Tips for Welding with the
Shielded Metal Arc Process.*

There is a definite relationship between the welding current, the size of the welding electrodes, and the welding position. These must be selected so that the welder has the molten weld metal puddle under complete control at all times. If the puddle becomes too large, it becomes unmanageable and molten metal may run out of the puddle, particularly in out-of-position welding.

The welder should maintain the steady frying and crackling sound that comes with correct procedures. The shape of the molten pool and the movement of the metal at the rear of the pool serve as a guide in checking weld quality. The ripples produced on the bead should be uniform and the bead should be smooth with no overlap or undercut. The following seven factors are essential for maintaining high quality welding.

1. Correct Electrode Type: It is important to select the proper electrode for each job. This should be based on information presented in the earlier part of this section.

2. Correct Electrode Size: Electrode size choice involves consideration of the type of electrode, welding position, joint preparation, weld size, welding current, the thickness or mass of the base metal, and the skill of the welder.

3. Correct Current: If the current is too high, the electrode melts too fast and the molten pool is large and irregular and hard to control. If the current is too low there is not enough heat to melt the base metal and the molten pool will be too small and will pile up and be irregular. This is shown by Figure 4-24A.

4. Correct Arc Length: If the arc is too long, the metal melts off the electrode in large globules which wobble from side to side giving a wide, spattered, and irregular bead with poor fusion to the base metal. It may also result in porosity, especially with low hydrogen electrodes. If the arc is too short, there is not sufficient heat in the arc at the start to melt the base metal sufficiently and the electrode often sticks to the work.

FIGURE 4-24 *Welding variables travel speed and current.*

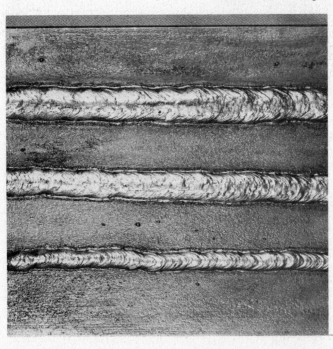

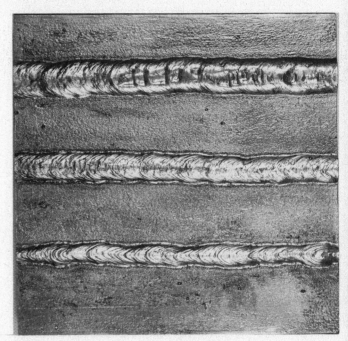

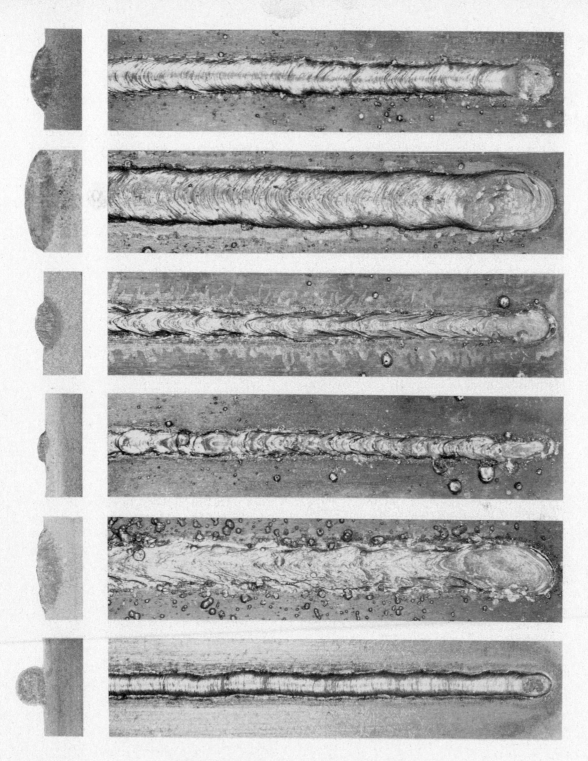

FIGURE 4-25 *Weld examples relating to malpractices.*

5. Correct Travel Speed: When the speed is too fast the weld pool freezes too quickly. Impurities and gases are not allowed to be released. The bead is narrow and the ripples pointed. When the speed is too slow the metal piles up, the bead is too high and wide with a rather straight ripple. This is shown by Figure 4-24B.

The factors: correct current, correct arc length (or voltage) and correct travel speed all relate to heat input. An experienced welder inherently adjusts these factors for the optimum weld under every possible condition. By referring to Figure 4-25 the welder in training can learn to adjust these factors for the best possible weld.

6. **Correct Electrode Angle:** The electrode angle is important, particularly in fillet welding and in deep groove welding. Generally, when making a fillet the electrode should be held so that it bisects the angle between the plates and is perpendicular to the line of the weld. When undercut occurs in the vertical member, lower the angle and direct the arc towards the vertical member.

7. **Correct Manipulation Pattern:** Different manipulation patterns are used for different types of electrodes, different weld designs, and different welding positions. Knowledge of the different patterns is learned in a good welding training program.

Breaking the Arc: In breaking the arc it is important to know whether it will be immediately re-established with the next electrode and continued or whether it is the end of a weld. If the weld is to be continued, the crater should remain, and the arc should be quickly broken. If it is the end of the weld, the arc should not be broken until travel has stopped momentarily to allow the crater to fill.

When weaving is used, the width of the weave and the pause at the end of the weave and other movements become important. The welder must pause at each end of the weave to allow for complete fusion into the side. The welder should quickly move across the center of the weld since heating is more concentrated in the center than at the edges. The width of the weave for low-hydrogen electrodes should not exceed two and one-half times the core wire diameter. For other electrode types, this can be doubled.

Safety Considerations

Safety factors and potential hazards involved with the shielded metal arc process are covered in detail in Chapter 3, "Safety and Health of Welders."

Limitation of the Process

One of the major limitations to the shielded metal arc process is the "built-in break." Whenever an electrode is consumed to within two inches (50 mm) of its original length, the welder must stop. Welding cannot continue since the bare portion of the electrode in the electrode holder should not be used. The welder must stop, chip slag, remove the electrode stub, and place a new electrode in the holder. This occurs many times during the work day and is controlled by the size and length of the electrode. This prohibits the welder from attaining an operator factor or duty cycle much greater than 25%.

Another limitation is the filler metal utilization. The electrode stub loss and the coating loss allows for a total utilization of covered electrode of approximately 65%.

Variations of the Process

There are several variations of the shielded metal arc welding process; among them are:

Gravity Welding.

Fire Cracker Welding.

Massive Electrode Welding.

Arc Spot Welding.

Gravity Welding

Gravity welding has been available for many years, but it did not become significant until simple gravity type feeders were developed in Japan for use in shipyards. Figure 4-26 is an indication of why gravity welding is popular. There is a definite economic advantage if one welding operator can tend three or four or more of the gravity feeders. This is possible in shipbuilding work since the closeness of the welds makes it possible for the welding operator to quickly move from one holder to another and to reload them, start them, and let them operate unattended. This process variation is more completely covered in Chapter 20.

Firecracker Welding

Firecracker welding is a method of welding developed by George Hafergut[3] of the Elin Company of Austria, and for this reason the method is often called the Elin-Hafergut method. It was developed in the late 1930s and was used for building bridges in Europe and was also used for making the longitudinal seams in small tanks and for making welds on railroad cars and bus bodies. It can be used for making square groove butt welds in material from 0.030 in. (0.8 mm) to 0.120 in. (3 mm) for making full fillet lap welds in material of similar thickness and for making fillet welds from 3/16 in. (5 mm) and heavier. Firecracker welding is a method of automatically making welds using a long electrode with a heavy coating that is not electrically conductive. The AWS Type E-6024 and E-7028 can be used.

To make a firecracker fillet weld the work is positioned so the weld is in the flat position. The welding

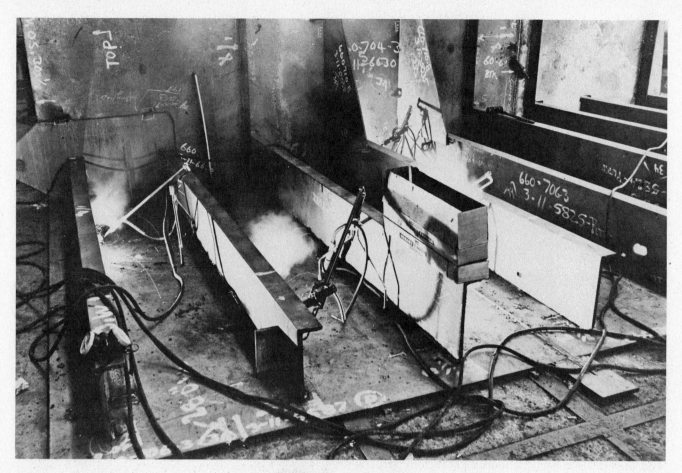

FIGURE 4-26 *Gravity welding.*

electrode is placed in the joint and a retaining bar is placed over it. The arc is started by shorting the end of the electrode to the work. The arc length depends on the thickness of the coating. As the arc travels along the electrode, the electrode melts and makes a deposit on the material immediately underneath it. Once the arc is started, the process proceeds to completion automatically. Figure 4-27 shows the method being used for making butt welds, lap welds, and fillet welds. Electrodes up to 39″ long (1000 mm) and with a 5-, 6-, 7-, or 8-millimeter core diameter have been used. Both alternating current and direct current have been used and there seems to be a preference for alternating current because of excessive arc blow with direct current.

The quality of the weld metal deposited by this method is equal to that produced by manual means. One operator can make several firecracker welds simultaneously.

This method of shielded metal arc welding has been used very little in North America. The popularity of semiautomatic welding is probably responsible. However, for short repetitive welds firecracker welding should be considered.

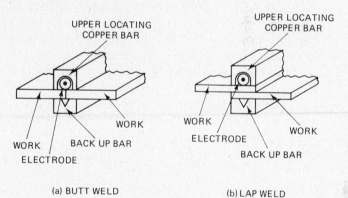

(a) BUTT WELD

(b) LAP WELD

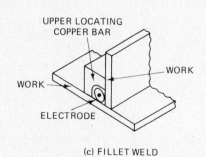

(c) FILLET WELD

FIGURE 4-27 *Firecracker welding.*

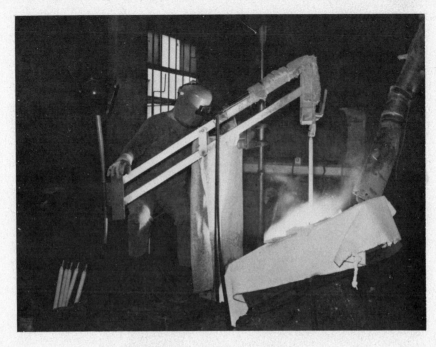

FIGURE 4-28 Using massive electrodes.

Massive Electrodes

There is another variation that employs extremely large diameter and long electrodes. These electrodes are made especially for repairing castings and are used at extremely high-current levels. These electrodes are so large and heavy that they require a manipulator to hold them and to feed them into the arc. Figure 4-28 is an example of these large-size electrodes being used for repairing a faulty casting. Size range and welding current required are shown by Figure 4-29. These electrodes are for very special applications but for this particular requirement it is an excellent welding method.

Arc Spot Welding

The shielded metal arc welding process can be used for arc spot welding. Special spring loaded feeders are used with small diameter electrodes for arc spot welding

thin sheet metal. This method has its major use in the automotive body repair operations. Arc spot welding as a special welding application will be covered completely in Chapter 20.

Typical Applications and Industries Using

Typical applications of the shielded metal arc welding process are as varied and widespread as arc welding processes in general. Shielded metal arc welding will probably always be the mainstay for maintenance and repair welding; because welding is required at remote locations, the jobs are relatively small and each and every one is different.

Shielded metal arc welding will also remain popular in the small production shops where limited capital is available and where the amount of welding is relatively minor to other manufacturing operations.

Electrode Size in inches	Amperage Range DCEP	Approx. Puddle Size in Square inches	Weld Metal Dep. lb/hr
5/16 × 24	300- 500	2- 6	10
3/8 × 24	400- 600	6-10	15
1/2 × 30	600- 950	10-20	25
5/8 × 30	600-1500	20-36	45
3/4 × 36	1200-2100	36-60	60

FIGURE 4-29 *Deposition rate chart—massive electrodes.*

Gas tungsten arc welding is an arc welding process which produces coalescence of metals by heating them with an arc between a tungsten (nonconsumable) electrode and the work. Shielding is obtained from a gas or gas mixture. Both pressure and filler metal may or may not be used. This process is sometimes called TIG welding which indicates tungsten inert gas welding. In Europe it is called WIG welding, using Wolfgram, the German word for tungsten.

Principles of Operation

The gas tungsten arc welding process shown by Figure 4-30 utilizes the heat of an arc between a nonconsumable tungsten electrode and the base metal. The arc is initiated various ways which will be explained later. The arc develops intense heat which melts the surface of the base metal to form a molten pool. Filler metal is not added when thinner materials, edge joints, flange joints, are welded. For all but the thinner materials an externally fed or "cold" filler rod is generally used.

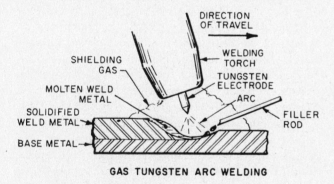

GAS TUNGSTEN ARC WELDING

FIGURE 4-30 *Process diagram.*

The filler metal is not transferred across the arc but melted by it. The arc area is protected from the atmosphere by the inert shielding gas which flows from the nozzle of the torch. The shielding gas displaces the air so that the oxygen and the nitrogen of the air does not come in contact with the molten metal or the hot tungsten electrode. As the molten metal cools coalescence occurs and the parts are joined. There is little or no spatter and little or no smoke. The resulting weld is smooth and uniform and requires minimum finishing.

Advantage and Major Uses

The outstanding features of the gas tungsten arc welding process are:

1. It will make high quality welds in almost all metals and alloys.

2. Very little, if any, postweld cleaning is required.

3. The arc and weld pool are clearly visible to the welder.

4. There is no filler metal carried across the arc and so there is little or no spatter.

5. Welding can be performed in all positions.

6. There is no slag produced that might be trapped in the weld.

This process allows the welder extreme control for precision work. Heat can be controlled very closely and the arc can be accurately directed. GTAW is used in many welding manufacturing operations, primarily on thinner materials and where metal finishing is not desired. It is very useful for maintenance and repair work and for welding die castings and unusual metals. Gas tungsen arc welding is widely used for joining thin wall tubing and for making root passes in pipe joints. The gas tungsten arc welds are usually of extremely high quality.

The manual method of applying is used for the greatest majority of work. However, both machine and automatic methods are increasingly used. The semiautomatic method is rarely used. Torches equipped with filler wire guides and filler metal wire feed systems are available for semiautomatic welding, but they have limited application. This information is summarized by Figure 4-31.

Method of Applying	Rating
Manual (MA)	A
Semiautomatic (SA)	C
Machine (ME)	A
Automatic (AU)	A

FIGURE 4-31 *Method of applying.*

The gas tungsten arc welding process is an all-position welding process as shown by Figure 4-32. Welding in other-than-flat positions depends on the base metal, the welding current, and the skill of the welder.

This process was originally developed for the "hard-to-weld" metals. It can be used to weld more different kinds of metals than any other arc welding process, as shown by Figure 4-33.

Coated base metals, such as steel coated with low melting temperature metals present welding problems. Steel coated with lead, tin, zinc, cadmium, or aluminum

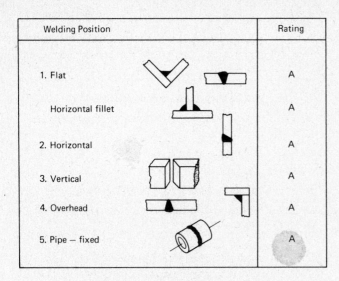

Welding Position	Rating
1. Flat	A
Horizontal fillet	A
2. Horizontal	A
3. Vertical	A
4. Overhead	A
5. Pipe — fixed	A

FIGURE 4-32 *Welding position capabilities.*

Base Metal	Weldability	Chapter
Aluminums	Weldable	13-1
Bronzes	Weldable	13-3
Copper	Weldable	13-3
Copper nickel	Weldable	13-3
Cast, malleable, nodular	Possible but not popular	13-1
Wrought iron	Possible but not popular	14-3
Lead	Possible but not popular	13-4
Magnesium	Weldable	13-5
Inconel	Weldable	13-6
Nickel	Weldable	13-6
Monel	Weldable	13-6
Precious metals	Weldable	13-7
Low carbon steel	Weldable	12-2
Low alloy steel	Weldable	12-3
High and medium carbon	Weldable	12-4
Alloys steel	Weldable	12-5
Stainless steel	Weldable	12-6
Tool steels	Weldable	14-4
Titanium	Weldable	12-9
Tungsten	Possible but not popular	12-8

FIGURE 4-33 *Metals weldable.*

can be welded with GTAW, but the coating is destroyed in the vicinity of the weld and the weld may have substandard qualities.

This process can weld extremely thin metals normally by the automatic method and without the addition of filler metal. Above 0.125 in. (3.2 mm), a joint preparation is usually required; however, this depends on the base metal type and welding position. Also, above this thickness multipass technique is usually required; this is graphically shown by Figure 4-34.

Joint Design

The joint designs used for gas tungsten arc welding are essentially the same as those used for shielded metal arc welding and for gas welding. Some changes are made, but these are usually involved with different metals or for welding piping materials in the fixed position. Joint detail variations for different metals will be covered in Chapter 13 and 14 and special joints for pipe welding will be covered in Chapter 20.

Welding Circuit and Current

Welding circuit for gas tungsten arc welding is shown by Figure 4-35. This circuit diagram shows several optional items. One is the "cold" filler rod mentioned previously, the second is the foot pedal which can be used to regulate the current while welding, and the third is cooling water used for the water-cooled welding torch recommended when welding at high-current levels. Constant current (CC) is used and it may be AC or DC.

Both alternating and direct current can be used for welding with the GTA and direct current can be used with either polarity depending on the job requirements. The polarity of the welding current has an important bearing on the resulting gas tungsten arc welds. Figure 4-36 shows the tungsten arc and aids in explaining the effect of welding polarity. The direct-current electrode-negative DCEN (straight polarity) arc produces the maximum heat in the base metal and the minimum heat

FIGURE 4-34 *Base metal thickness range.*

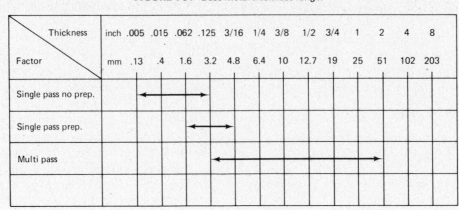

Thickness	inch	.005	.015	.062	.125	3/16	1/4	3/8	1/2	3/4	1	2	4	8
Factor	mm	.13	.4	1.6	3.2	4.8	6.4	10	12.7	19	25	51	102	203
Single pass no prep.			←		→									
Single pass prep.					←	→								
Multi pass							←					→		

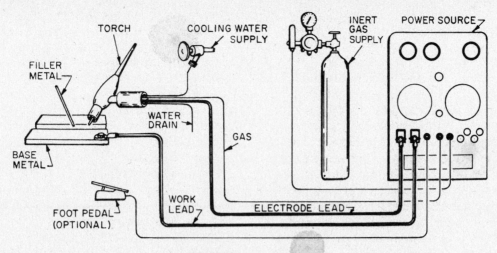

FIGURE 4-35 *Circuit diagram.*

in the electrode. This allows the use of a smaller electrode and obtains deeper penetration in the base metal. When the welding circuit is reversed the direct-current electrode-positive DCEP (reverse polarity) arc produces the maximum heat in the electrode and the minimum heat in the base metal. This means that a larger than normal electrode is required and that the depth of base metal penetration is relatively shallow. With direct current electrode positive (DCEP), ionic bombardment of the base metal occurs which provides the cleaning action necessary for welding aluminum and magnesium.

When welding with alternating current, the voltage goes through zero each half-cycle and the arc must be reignited to provide current flow for that half-cycle. During one half-cycle the base metal, which is negative, is the emitter of electrons (the cathode). Since the relatively cool base metal is not so good an emitter as the hot tungsten electrode the arc might not be reignited. Thus when the electrode is positive, reignition may not occur and no current flows for that half-cycle. This, in effect, is rectification in the arc, since more current flows in one direction than the other. This rectification in the arc can be troublesome. Delayed ignition may occur and complete half-cycles may be skipped but the maximum amount of current flowing in the circuit will be alternating even though a sizable direct current component may be measured. A solution was temporarily obtained by using storage batteries to buck the DC component or to use extra high open circuit voltage or series capacitors to stop the DC component. Modern AC power supplies designed for the gas tungsten arc process have overcome this problem.

Usually superimposed high-frequency current is allowed to operate continuously when welding with alternating current. The high-frequency current will maintain an ionized path for the welding current to follow

across the arc gap during the time the voltage goes through zero. When using alternating current the cleaning action occurs every other half-cycle which is sufficient and good weld penetration is obtained.

When superimposed high-frequency is used with AC gas tungsten arc welding certain precautions are required. These are necessary since welding power sources equipped with high-frequency spark gap oscillators inherently radiate power at frequencies which may interfere with

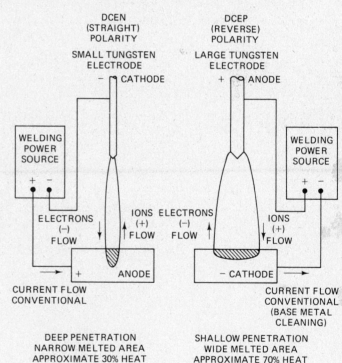

FIGURE 4-36 *The tungsten arc.*

THE ARC WELDING PROCESSES—USING CC POWER

radio communications and television transmission. In view of this, their operation in the United States is subject to control by the Federal Communication Commission. Most countries have similar regulations.

Welding machines containing high-frequency stabilizers or separate high-frequency stabilizers must be installed with special attention to provide earth grounding and special shielding. Manufacturers provide special installation instructions that limit high-frequency radiation. These instructions require that all metal conductors in the area of the machine must be earth grounded. If these instructions are followed the user can post a certificate stating that the high-frequency stabilizer may reasonably be expected to meet FCC regulations.

Equipment Required to Operate

The heart of the welding circuit is the welding machine or power source. The constant current (CC) power source is used for gas tungsten arc welding. The steeper the volt ampere characteristic curve, the more satisfactory the power source. This is because variations in arc length will cause changes in arc voltage but will result in only very small changes in welding current.

This is very important for manual welding. Conventional or constant current welding machines used for shielded metal arc welding can be employed for gas tungsten arc welding. Conventional AC welding machines must be derated 25% due to rectification in the arc. Machines designed for GTAW should be used. These machines include special features, such as high frequency stability, gas, and water valves, etc., making them more suitable for this process. A typical GTAW welding machine operates with a range of 3 to 200 amperes or 5 to 300 amperes with a range of 10 to 35 volts at a 60% duty cycle. The newer gas tungsten arc welding machines offer other advantages such as closed-loop control, programability, remote current control, and current pulsing. Pulsing will be covered later in this chapter. More details of these power sources will be given in Chapter 8.

The *torches* used for gas tungsten arc welding are designed and used only for this process. Torches with handles are used with the manual method of applying.

Automatic torches are similar in design but without a handle. They are designed to be clamped in a holding bracket. The basic construction is the same. Both air-cooled torches, designed for light-duty welding up to approximately 150 amperes, and water-cooled torches, designed for heavy-duty welding up to 600 amperes, are available. Figure 4-37 shows the manual type torches, both the air-cooled and the water-cooled types. A cross-sectional view of a water-cooled manual torch is shown

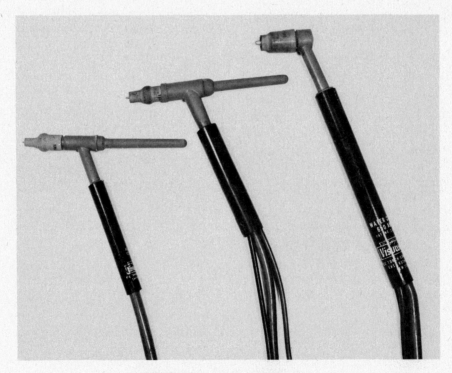

FIGURE 4-37 *Manual GTAW torches.*

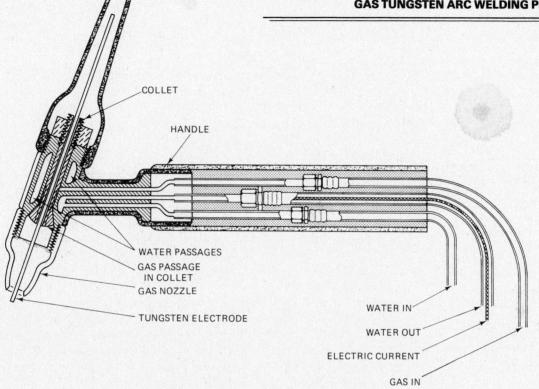

FIGURE 4-38 *Cross section view of a water cooled torch.*

by Figure 4-38. The different types and sizes of welding torches and their capacities are summarized by the table in Figure 4-39. The head angle is measured by the angle between the centerline of the handle and the centerline of the tungsten from the arcing end. Guns are rated by the current carrying capacity normally at a 100% duty cycle. The inside diameter of the nozzle is given by inches or by numbers that represent eighths of an inch. Welding torch nozzles are available in different materials; normally a ceramic is used, however metal nozzles and glass nozzles for visibility, are available.

Special nozzles are available that provide trailing gas shielding for welding titanium and other easily oxidized metals. Devices are also available for providing lamellar shielding gas flow. These are used for welding in deep grooves, on outside corners, or where drafts will disturb the gas shielding. They are not necessary for most welding applications since the torch nozzle must be held close to the work to maintain the necessary arc length. The torches are designed to firmly hold the tungsten electrode, to transmit the welding current to the electrode in the center of the nozzle, and to introduce the inert

Current Capacity 100% Duty Cycle	AC	DC EN	Head Angle	Cooling Method	Maximum Nozzle Inside Dia.	Tungsten Dia. in.	mm	Tungsten Length in.	mm	Torch Weight oz	g
50 @ 60%		x	135°	Air	1/4	0.040-3/32	1.0-2.4	1	25.4	3	85
75	x	x	120°	Air	5/16	0.020-1/16	0.5-1.6	3-7	76.2-177.8	4	113.4
100	x	x	120° or 135°	Air	5/16	0.020-5/32	0.5-4.0	2-3-7	50.8-76.2-177.8	4	113.4
150	x	x	120°	Air	5/16	0.020-5/32	0.5-4.0	3-7	76.2-177.8	5	141.75
200	x	x	120° or 135°	Water	5/16	0.020-3/32	0.5-2.4	2-3-7	50.8-76.2-177.8	5	141.75
250	x	x	120° or 135°	Water	3/8	0.020-1/8	0.5-3.2	2-3-7	50.8-76.2-177.8	5	141.75
300	x	x	120°	Water	3/8	0.020-5/32	0.5-4.0	3-7	76.2-177.8	6	170
350	x	x	120°	Water	11/16	0.010-5/32	0.3-4.0	3-6	76.2-152.4	6	170
500	x	x	120°	Water	11/16	0.010-3/16	0.3-4.8	3-7	76.2-177.8	6	170
650	x	x	135°	Water	5/8	1/8-5/16	3.2-7.9	2-1/2	63.5	7	198.45

FIGURE 4-39 *Gas tungsten arc welding torches—size and capacity.*

THE ARC WELDING PROCESSES—USING CC POWER

shielding gas around the tungsten electrode and to direct it to the arc area. Water-cooled torches are more complicated since they provide water passages in the head of the torch in addition to providing the shielding gas and the welding current. In the water-cooled torch the power cable is enclosed in the return hose for the cooling water. In this way, smaller cables can be used since they are cooled during the operation of the torch. Gas tungsten arc torches come equipped with the necessary hoses and leads attached.

Materials Used

Materials used in gas tungsten arc welding are the filler metal, the shielding gas, and, to a lesser degree, the tungsten electrode. Filler metal is not used, when welding extremely thin metals; however, for most applications filler metal is added to the puddle. The composition of the filler metal should match that of the base metal. The size of the filler metal rod depends on the thickness of the base metal which usually dictates the welding current. Filler metal is normally added to the puddle manually, but automatic feed is sometimes used.

The *electrode* material for gas tungsten arc welding should have the following properties: high melting point, low electrical resistance, good heat conductivity, and the ability to easily emit electrons. Thus tungsten and tungsten alloys are used since tungsten has the highest melting point of any metal (6,170°F) (3,410°C). Four classes of tungsten electrodes have been standardized by the American Welding Society in their specification "Tungsten Arc Welding Electrodes."[4] These four types are shown in Figure 4-40 which also shows the AWS classification and the color code for the tip of the electrode. Tungsten electrodes are available in two finishes: standard or ground. The ground finish provides an extremely smooth and perfectly round elec-

AWS Classification	Type	Tip Color
EWP	Pure Tungsten	Green
EWTh1	1% Thorium added	Yellow
EWTh2	2% Thorium added	Red
EW Zr	1/2% Zirconium added	Brown

Diameter - 0.020 to 0.250 inch   (0.5 to 6.4 mm)

Lengths - 3 to 24 inch   (76 to 610 mm)

FIGURE 4-40 *Tungsten electrodes type, sizes and classification.*

trode which is able to conduct heat from the electrode to the collet of the welding torch. Tungsten electrodes are available with diameters ranging from 0.020–0.250 in. (1–6.4 mm) and in lengths of 3–24 in. (75–610 mm).

Of the four types of electrodes, the EWP class is pure tungsten. The other three have alloy added. The EWTh1 has 1% thorium; the EWTh2 has 2% thorium added; and the EWZr has 1/2 of 1% zirconium added. The addition of thorium and zirconium makes the tungsten alloy more able to emit electrons when hot. It also provides for increased current-carrying capacity of the electrode and provides for a more stable arc and better arc starting.

When installing a new tungsten electrode in the torch the color tip should be at the back end of the torch so that it is not destroyed by the arc. The collet should be checked and it must be the proper size for the tungsten being used. The entire assembly must be tight to the tungsten and the collet tight to the torch so that the heat of the arc is transmitted to the torch body where it can be carried away. The EWP or pure tungsten electrodes are the least expensive and should be used for the less critical operations or for general purpose work on different metals. The EWTh1 or 1% thoriated tungsten provides for easier arc starting, gives a more stable arc, and can be operated at slightly higher temperatures. The EWTh2 or 2% thoriated tungsten is even better for arc starting, is more stable, and has a higher current-carrying capacity. EWZr or zirconated tungsten electrodes also provide for longer life and more stable operation. The table shown by Figure 4-41 gives the continuous welding current range for each of the different types of electrodes related to the electrode size. This table is based on the current-carrying ability of the different electrode types. Under the column, "AC with High Frequency," it is evident that the alloy electrodes will carry higher current than the pure tungsten electrode.

Electrode size must always be selected with care and must be related to the type of current, the type of work, and the amount of current that will be used. The amount of welding current required is found in the welding procedure schedules for welding the particular metal in question. The data provided in these tables is the beginning; for example, if a welding current of 100 amps is required and alternating current is used, it would indicate that a 3/32 in. (2.4 mm) EWP or pure tungsten or a 1/8-in. (3.2 mm) pure tungsten electrode could be used, or if an alloyed electrode were to be used the 1/16-in. (1.6 mm) or the 3/32-in. (2.4 mm) electrode could be used. If the procedure schedule calls for direct current electrode negative (DCEN) or direct current electrode positive (DCEP) different electrode sizes would be needed. Welder preference may also enter into the

| Tungsten Electrode Diameter | | CONTINUOUS WELDING CURRENT—AMPERES | | | |
| | | AC & HF | | D.C.E.N. | D.C.E.P. |
in.	mm	EWP	EWTh1 or 2 or EW Zr	EWTh1 or EWTh2	EWTh1 or 2 or EW Zr
0.020	.5	—	—	5-35	—
0.040	1.0	10-40	15-60	30-100	—
1/16	1.6	30-70	60-100	70-150	10-20
3/32	2.4	70-100	100-160	150-225	15-30
1/8	3.2	100-150	140-220	200-275	25-40
5/32	4	150-225	200-275	250-350	40-55
3/16	4.8	200-300	250-400	300-500	55-90
1/4	6.4	275-400	300-500	400-650	80-125

FIGURE 4-41 *Current ranges for tungsten electrodes.*

size selection. However, experience is also important; for example, if the welder is using pure tungsten electrode and it tends to become overheated or appears to have a "wet surface" the current is too high for the size of the electrode. When the tungsten has this "wet surface" appearance it becomes more susceptible to picking up contamination from the base metal. A larger electrode size of the same type should be selected or an alloyed type electrode of the same size could be used. Too much current or an electrode too small will cause excessive tungsten erosion. Tungsten particles may become deposited in the weld metal.

If the current is too low, or the tungsten electrode is too large in diameter, the arc will wander erratically over the end of the electrode. Grinding the electrode to a point will reduce this problem. It will also help to direct the arc. In general, choose the size of electrode that will be working as close to its maximum current-carrying capacity as possible. The electrode should remain shiny after use and should never be allowed to touch the molten metal. If this happens it will become contaminated and must be reprepared. The electrode should show a balled end and the balled end should not exceed 1-1/2 times the diameter of the electrode. This is shown by Figure 4-42. The angle of pointing the electrode should be related to the welding current and

HEMISPHERICAL END

NOTE CLEAN CONDITION OF ELECTRODES PROTECTED BY SHIELDING GAS.

BALLED END SHOULD NOT EXCEED 1-1/2 TIMES DIAMETER OF ROD.

FIGURE 4-42 *Tungsten electrode arc end condition.*

the thickness of the metal being welded. It usually ranges from 30° to 120°; 60° is the most common angle.

The *shielding gas* used for gas tungsten arc welding is usually an inert gas. **Inert** means that they do not form compounds with other elements. They are monatomic gases and they do not form molecules of two combined atoms like many of the active gases. In the case of welding, only argon and helium are used since the other inert gases are much too expensive for this type of use. Gas selection is based on the metals to be welded. It is necessary to consult the procedure schedule for the particular metal that is to be welded. These tables show the type of gas recommended and the gas flow rate. More information concerning shielding gas will be found in Chapter 10. Argon and helium are both available in compressed cylinders of the high purity required for welding. Argon is more commonly used. It is more readily available and is heavier than helium and slighly heavier than air which provides for a more efficient arc shielding at lower flow rates. Additionally, argon is usually better for arc starting and operates at a lower arc voltage. Helium is much lighter than argon or air and thus tends to float away from the weld zone, and higher flow rates are required. Helium provides for a higher voltage and thus provides more heat in the arc than argon. It is also possible to weld at a higher speed with helium than with argon. There are some cases where helium and argon are mixed together for the optimum shielding gas for a particular metal or weld schedule.

Hydrogen is sometimes added in amounts up to 25% to the inert gas. Hydrogen increases the arc voltage and produces more heat than the pure inert gas. A higher arc voltage is desirable when welding thick materials and metals that have high heat conductivity. It is also an advantage in high-speed automatic welding. Hydrogen, however, cannot be added and used with all metals. The addition of hydrogen may injure some metals and alloys including all aluminums, copper, and magnesium base alloy. It can be used successfully on certain stainless steels and nickel alloys.

For certain applications, very very small amounts of oxygen may be added to the shielding gas. The amount of oxygen added is less than 1%. This tends to help stiffen the arc and is used for direct current electrode negative welding (DCEN) of aluminum.

Nitrogen is not an inert gas, but is sometimes used for welding copper. Nitrogen provides an extra high temperature arc. This is useful in overcoming the high thermal conductivity of copper. Electrode erosion is high when welding with nitrogen. Nitrogen is rarely used for the gas tungsten arc welding process in North America. It is sometimes used where inert gases are not readily available.

Deposition Rates

The gas tungsten arc is used as a source of heat to allow the welder to melt sufficient base metal and filler metal to provide a manageable weld puddle. In this sense, it is similar to the oxyacetylene welding process and the carbon arc welding process. Thus the gas tungsten arc welding process cannot be considered a high-production or high-deposition rate welding process.

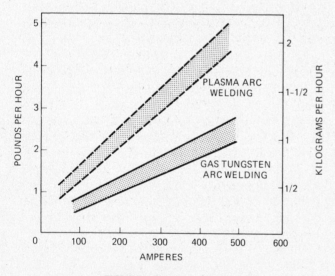

FIGURE 4-43 *Deposition rate.*

The graph shown by Figure 4-43 is a deposition rate curve relating to welding current. This deposition rate is relatively slow and this is one reason why gas tungsten arc welding is not too popular for welding heavy material. For mechanized installations where the filler

wire can be fed in electrically "hot," a higher deposition rate is possible. The deposition rate depends on the diameter of the filler wire and the welding current.

Quality of Welds

The quality of gas tungsten arc welds ranks as high or higher than the quality of any of the arc welding processes. This high level of quality can be obtained with gas tungsten arc welding provided that all of the necessary precautions are taken. Since much of the work done by the gas tungsten arc welding process is on non-ferrous metals it is absolutely essential that cleanliness be considered every step of the way. The work center or work area should be extremely clean. Work tables, fixtures, etc., should be clean. The gloves used by welders should be clean. Filler metal should be clean, the gas must be welding grade, and the apparatus must be in excellent condition. If these conditions are followed and if the welder has sufficient skill, high-quality welds will result.

Heat input and welder technique has much to do with weld quality. Figure 4-44 shows these factors when welding on aluminum. The same situation occurs here as with other arc welding processes. When the heat input is too low which can occur from too low welding current or welding speed too fast, the high small bead is evident and penetration is minor. When the welding current is too low the bead is too high, there is poor penetration and possibility of overlapping at the edges. When welding speed is too fast the bead is too small and the penetration is minimal. When the heat input is too great, which can occur from too high a welding current or too low welding speed, the bead becomes extremely large, usually wide and flat. There is too much penetration and there may be spatter. When the torch is too far from the work, a long arc occurs, the efficiency of the gas shielding is reduced and poor weld appearance will result, especially in welding aluminum.

The following is a brief discussion of some of the common quality problems and the recommended action to overcome them.

Weld Metal Porosity: Porosity is usually caused by oily, wet, dirty base metal, insufficient inert gas coverage, or dirt and heavy oxide coating on the filler rod. In groove welds there should be backing or purging gas and cleaning of the underside of the groove adjacent to the weld. In the case of aluminum, a stainless steel wire brush or chemical cleaning should be used. Inefficient gas shielding can be caused by side drafts of air which disturb the shielding gas envelope. It may also be caused by leaks in the gas system coming from the cylinder to the torch nozzle or it can be caused by impure shielding gas. Equipment should be checked frequently to make sure that the gas system is tight. It is also

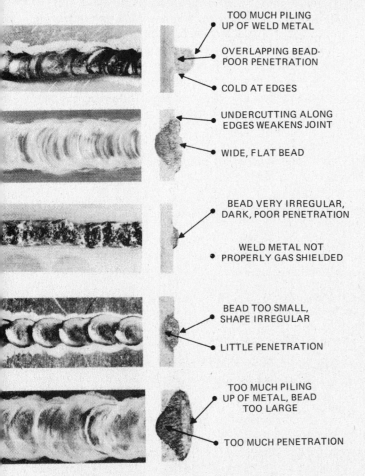

TOO MUCH PILING
UP OF WELD METAL

OVERLAPPING BEAD-
POOR PENETRATION

COLD AT EDGES

UNDERCUTTING ALONG
EDGES WEAKENS JOINT

WIDE, FLAT BEAD

BEAD VERY IRREGULAR,
DARK, POOR PENETRATION

WELD METAL NOT
PROPERLY GAS SHIELDED

BEAD TOO SMALL,
SHAPE IRREGULAR

LITTLE PENETRATION

TOO MUCH PILING
UP OF METAL, BEAD
TOO LARGE

TOO MUCH PENETRATION

FIGURE 4-44 *Quality control factors of GTAW of aluminum.*

important that no moisture from cooling water get inside the gas supply hose.

Dirty welds, particularly on aluminum, are another frequent problem. This can result from a problem in the shielding gas supply. There can be a leak in the hose connection, poor-quality shielding gas, an oversize gas nozzle, the nozzle held too far from the work, insufficient gas flow, or anything else that contributes to poor shielding of the arc area.

Poor penetration is primarily a heat input problem and is related to travel speed and welding current to the base metal thickness and conductivity and joint type. Too much amperage will make the bead too flat and rough and may cause cracking. Insufficient amperage will produce an uneven high crown bead. Travel speed also affects penetration. If the speed is too fast, the bead will have a high crown, be rough, and have insufficient penetration. The welding speed may need to be varied or changed, particularly on small weldments where heat buildup occurs. When the base metal is cold a lower speed is required; as the work piece absorbs heat and rises in temperature, the speed should be increased.

A common quality problem is tungsten inclusions in the weld deposit, which can be detected by radiography. This is sometimes called *tungsten spitting* and is based on using too much current for the size or type of the tungsten electrode. The tungsten type may have to be changed or a larger tungsten electrode employed.

Another serious problem is an unstable arc which is normally a result of a contaminated or dirty electrode. The electrode will become oxidized if the inert gas is not continually surrounding it while it is hot. This is the reason for postflow controls on most gas tungsten arc welding machines. The gas flow should continue after the welding is stopped to keep the tungsten surrounded by inert gas while it is still at an elevated temperature. Another problem can be that the tungsten electrode protrudes too far beyond the end of the gas nozzle. Normally it should not protrude more than 1/16-in. (1.6 mm). Postflow should be adjusted for a longer period of time if the tungsten remains dirty. The other problem with tungsten electrodes is the contamination which occurs if the electrode is allowed to touch the molten metal. This will immediately cause the electrode and the arc to become unstable and at that time welding should be stopped and the electrode redressed.

Dirty filler metal or filler rod with an excessive oxide coating can also create dirty welds. Water vapor or moisture can occur in this heavy oxide coating. The filler rod should be cleaned with sandpaper.

Welding on dirty, oil-impregnated material, or attempting to repair cracks in machinery parts requires the removal of defective material, thorough cleaning, and preheating to help eliminate any absorbed oil, grease, moisture, etc.

Water leaks in the torch can usually be detected by coloring of the weld surface. Condensation can occur on the inside of gas hoses and the water vapor in the arc will cause the tungsten to become contaminated.

In summary, it is necessary that all conditions be correct, that materials used be of the correct specification and cleanliness, that the apparatus be in good working order, and that the proper welding technique be employed.

Weld Schedules

Welding procedure schedules for manually applied gas tungsten arc welding are not as important as schedules for the continuous wire processes. For gas tungsten arc welding it is important to establish welding currents based on the needs of different materials and this is related to the type of metal and the weld joint detail. Welding schedule tables are provided in the chapter for the metal being welded. Refer to Figure 4-33 for this information. Normal practice is to consult these procedure tables and to establish the welding conditions in accordance with them. Once the welding current level is known, the type of welding current to be used and the type of tungsten electrode recommended, it is then possible to establish tungsten electrode size.

The welding procedure tables shown for the different metals also provide weld travel speed which must be used in determining heat input and welding costs. Under mechanized conditions the weld speeds can normally be increased over manual application. The feed rate for filler rod is not given since this is a matter of technique for manual welding.

The weld schedules are related to the weld puddle that must be carried by the welder. This relates to welding position, type of metal, weld joint detail, etc. More experienced welders can carry larger puddles and make welds at a higher speed.

Welding Variables

Gas tungsten arc welding involves a number of variables. Each variable has a specific effect on the weld and there is an interrelationship among variables that affects the final weld. The preselected variables include: tungsten type, tungsten size, nozzle size and gas type. These must be established and are usually part of a welding procedure. The primary variables are welding current, arc voltage, travel speed, pulsing when used, taper, upslope and downslope, when used, for programmed welding. The secondary variables include rod feed speed when used, torch angles and possibly tungsten angles. There are other factors that affect the weld quality and these include clamping, fixturing, heat sinks, heat buildup, backing, purging gas, and high frequency for arc starting.

In general, the factors that are of interest in a weld are penetration of the weld in the base metal, bead width of the deposit, and weld reinforcement or height. It is assumed that weld surface appearance is acceptable and weld metal deposit is of the required quality. It is therefore important to recognize the interrelationship of the primary variables to provide penetration, bead width, and reinforcement. These factors are all influenced by

heat input. The effect of travel speed and current was previously mentioned; however, it is important to consider the conductivity of the metal and the heat sink effect of any fixturing that might be employed.

A major reason for developing pulsed gas tungsten arc welding techniques was to provide deep penetration, which is obtained from the high current while reducing the total heat input to avoid too much molten weld metal. Programming is also useful, particularly on small welds where heat buildup can become a factor.

Tips For Using The Process

Factors involved to produce quality gas tungsten welds are similar to those outlined for shielded metal arc welding. However, there are some extra factors:

1. The arc length and electrode condition.
2. The welding current.
3. The travel speed.
4. The angle of the torch.
5. The shielding gas coverage.

Arc length relates to arc voltage. Normally the arc length should be just slightly longer than the diameter of the tungsten electrode being used. If the arc length is too long the bead will be flat, will usually have poor penetration, and will show evidence of oxidation. If the arc length is too short, the tungsten electrode may occasionally touch the molten puddle, which will cause contamination of the electrode.

The point of the electrode and its condition affect the welding quality and technique. This factor is somewhat different for different types of electrodes and for welding with different currents on different metals.

If the *welding current* is too high, the bead will be flat and rough, and cracking may occur. If the current is too low an uneven high and round bead will result.

Travel speed should be based on watching the weld bead form. Travel speed must be changed as the weldment absorbs heat. A steady forward motion is recommended. If the bead is high and rough and penetration is minimum the speed is too fast. If travel is too slow, there will be a flat bead with excessive penetration. A torch should point towards the direction of the travel with a travel angle of 20–30 degrees. The filler rod should be introduced ahead of the tungsten and should come in at about 20 degrees from the base metal.

The *shielding gas coverage* is a critical factor. It depends on the purity of the gas, the gas flow rate, the integrity of the gas system, and the absence of wind or breezes that might blow the shielding gas away.

The safety factors involved with gas tungsten arc welding are very similar to those involved with the other arc welding processes. There are these exceptions.

The gas tungsten arc seems brighter at the same current than the arc of shielded metal arc welding. This is because the smoke generated in the shielded metal arc is not present with gas tungsten arc welding. The smoke is a shielding mechanism that reduces the brightness of the arc.

The brightness of the gas tungsten arc tends to cause air to break down and form ozone. Thus, adequate ventilation should be provided. The bright arc rays cause fumes from hydrochlorinated cleaning materials or degreasing agents to break down and form phosgene gas. Cleaning operations using these materials should be shielded from the arc rays of the gas tungsten arc.

The final hazard, which is common with other gas-shielded welding processes, is the possibility of displacing the air when welding in enclosed areas such as tanks. Ventilation and other precautions for welding in enclosed areas should be followed. All other factors are similar to other processes and were covered in Chapter 3.

Limitations of the Process

The major limitation of gas tungsten arc welding is its low productivity. A comparison of the deposition rate of gas tungsten arc welding with gas metal arc welding makes this obvious. Another possible limitation of the process is its higher initial cost. The welding power source is more expensive than that used for shielded metal arc welding, and the torch is more expensive than the electrode holder. The justification for this is the ability of the process to weld so many metals in thicknesses and positions not possible by shielded metal arc welding.

Variations of the Process

There are a number of variations of the gas tungsten arc welding process. The more popular of these are:

Pulsed current GTAW.

Manual programmed GTAW.

Hot wire GTAW.

Gas tungsten arc brazing.

Gas tungsten arc cutting.

Gas tungsten arc spot welding.

These variations are described in the remaining portion of this section, except for GTA spot welding, which will be covered in Chapter 20.

The most popular variation of this process is known as pulsed-current gas tungsten arc welding. The pulsed-current mode of welding is primarily a way to control heat input. It offers a number of advantages over conventional or steady-current welding as follows:

1. Control puddle—size and fluidity (especially out of position).
2. Increased penetration.
3. Oscillation travel and dwell control.
4. Travel speed control.
5. Better consistent quality.

In conventional GTA welding the amount of welding current at the arc or heat input is essentially the same except when it is adjusted by the welding machine rheostat or by a foot-controlled rheostat. In some cases more current might be required for deeper penetration and in other cases less current is desired to reduce the size of the molten puddle. The pulsed-current mode provides a system in which the welding current continuously changes between two levels. See Figure 4-45.

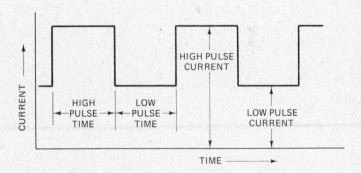

FIGURE 4-45 *Pulsed current time and current relationship.*

During the periods of high-pulsed current heating and fusion takes place and during the low-pulse current periods, cooling and solidification take place. It is as if the foot rheostat were moved up and down to increase and decrease the welding current on a regular basis. Newly designed GTAW power sources equipped with special programmers provide for these high- and low-current periods or pulse current. The machine automatically switches to high current then to low current and will hold each value for a specific time. The pulsed gas tungsten arc makes a weld seam of overlapping arc spot welds. Each arc spot type weld is produced during the high-current pulse time period which provides for maximum penetration and heat. The current then decreases to the "low-pulse current" or background current, which allows the weld to partially cool and solidify, while maintaining a low-current consistent arc. The torch is then moved to the next point along the weld joint and held motionless during the next high-current pulse. This sequence of events continues and the weld that is produced is shown by Figure 4-46.

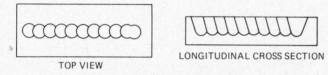

TOP VIEW   LONGITUDINAL CROSS SECTION

FIGURE 4-46 *Pulsed current weld section view.*

There are four factors that must be considered and controlled in order to successfully weld with the pulsed-current mode. These four factors are defined as follows:

1. *High-pulse current* or *pulse current*—the welding current during the high-pulse time period.

2. *Low-pulse current* or *background current*—the current during the low-pulse time period.

3. *High-pulse time*—the time period or duration of the high-current pulse.

4. *Low-pulse time*—the time period or duration of the low-current pulse.

Power sources designed for pulsed current welding are equipped with controls to adjust the high- and low-pulse current and time. In general, the high-current pulse should be adjusted to twice or at least 1-1/2 times the normal steady state current that would be used for the same job. The low-pulse current is usually only sufficient to maintain a stable arc during the low-current pulse period. This is normally set at 25% of the high-pulse current, or less. For example, if the high-pulse current is set at 200 amps and a low-pulse current of 50 amps is desired, the percent of weld current should be 25%.

The time period for each pulse is extremely important. The "high-pulse time" is set to allow the formation of the molten puddle which has the penetration that is desired. This may vary from a 0.20 second to 1.0 second. The low pulse time is set sufficiently long to allow the molten puddle to partially freeze. With proper adjust-

ments it is possible to control the weld puddle, the size of the weld bead, and the penetration depth in any position on any weldable metal. Initial welding conditions are usually a starting point and the welder will be able to adjust the settings for optimum welding based on the material, its thickness, the welding position, the joint detail, etc. For example, if the molten puddle becomes too large the high-pulse time should be reduced. If penetration is too deep the high current should be reduced.

The use of pulse-current welding will allow the welder to develop a rhythm of movement. Normal technique is to hesitate during the high-pulse time and move or oscillate with the torch during the low-pulse time. Filler metal is usually added during the high-pulse period. With additional experience using pulsed current GTA welding, the welder will make the proper adjustments and vary travel speed for consistent high-quality welds. The welder will find that it is easier to do a better welding job with pulsed current than with conventional steady current.

Distortion and warpage are reduced with pulsed current GTA welding especially on lighter or thinner materials. This is due to the lower heat input of the process. Misalignment of joints and the welding of light to heavy sections are made easier with pulsed-current welding. The pulsed current can be adjusted to the point that adequate penetration can be obtained on the heavier section with the high pulse while the low pulse provides the control needed to avoid excessive heating of the thinner member. Pulsed-current welding can be done manually or automatically with or without filler wire. Figure 4-47 shows a weld made in copper using pulsed current.

FIGURE 4-47 *Pulsed current weld on thin material.*

The programming of weld current is often used in automatic welding. It is also used for manual application and can be accomplished by the use of a foot-controlled rheostat for starting and finishing welds. For certain applications additional changes in current are required during the cycle for making a particular weld. To do this programmers are used where the welding current can be made to rise or fall at specific rates to specific values at the command of a switch. A finger switch mounted on the torch will start the preselected program to be used to produce a particular weld. The torch switch can be used to stop the program or to make it repeat. This is known as manual programmed gas tungsten arc welding and is popular for welding tubing and root pass welding of pipe. More information on programmed welding will be covered in Chapter 9.

The ''hot'' wire TIG welding variation uses electrical power on the filler metal. The filler rod that is fed into the weld puddle is ''electrically'' hot compared to the normal filler rod addition which is electrically ''cold.'' The electrical hot wire carries a low-voltage current that preheats the filler rod. It enters the weld pool at an elevated temperature and melts quicker, thus increasing the deposition rate. One of the major applications for the ''hot'' wire GTAW variation is for weld surfacing, particularly overlaying of stainless steel on low-carbon steel. It is used in the machine and automatic method of applying since the hot wire must always be in contact with the molten puddle in order to conduct the preheating current through the filler rod.

The gas tungsten arc welding torch can be used for brazing. It is a good controllable heat source that can be used to heat base metals for brazing.

The gas tungsten arc welding torch can be used for cutting. Cutting metal is effected by melting the base metal with the heat of the arc. This thermal cutting process will be discussed in detail in Chapter 7.

Industries Using and Typical Applications

The use of gas tungsten arc welding is almost as widespread as the use of shielded metal arc welding. It is being used to a great degree for welding nonferrous metals. The aircraft industry is one of the principle users of gas tungsten arc welding for joining aluminum, magnesium, titanium, and stainless steel. Space vehicles are almost all entirely fabricated by the gas tungsten arc welding process. This includes the shells, structures, the various tanks that are required, and the thousands of feet of tubing involved in rocket engines.

Small-diameter thin-wall tubing is almost exclusively welded by this process. Tubes are also welded to tube sheets for heat exchangers with programmed gas tungsten arc welding. Another important use of gas tungsten arc welding is the making of root pass welds in piping—thin and heavy wall, large and small diameter for the process and power industries where high-quality welding is required. Virtually every industry uses GTAW for welding thin materials especially the nonferrous metals.

The repair and maintenance industry is a major user of the process. Gas tungsten arc welding is used for repairing tools and dies, for repairing cast aluminum and magnesium parts, and for repairing highly critical items. Figure 4-48 is an example of reclaiming a cast aluminum housing.

Gas tungsten arc welding is used everywhere. Its use is continuing to increase and this trend will continue.

FIGURE 4-48 *Reclaiming cast aluminum housing.*

4-4 THE PLASMA ARC WELDING PROCESS (PAW)

Plasma arc welding is a process in which coalescence is produced by heating with a constricted arc between an electrode and the work piece (transfer arc) or the electrode and the constricting nozzle (nontransfer arc). Shielding is obtained from the hot ionized gas issuing from the orifice which may be supplemented by an auxiliary source of shielding gas. Shielding gas may be an inert gas or a mixture of gases, pressure may or may not be used, and filler metal may or may not be supplied. It is shown by Figure 4-49.

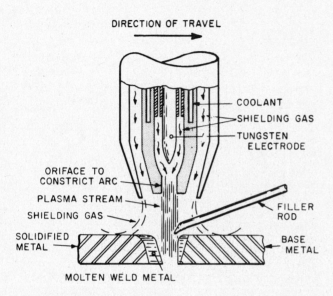

FIGURE 4-49 Process diagram—keyhole mode—PAW.

Principle of Operation

The plasma arc welding process is normally compared to the gas tungsten arc process because of the many similarities. If an electric arc between a tungsten electrode and the work is constricted or reduced in cross-sectional area, its temperature goes up since it carries the same amount of current. This constricted arc is called a plasma and plasma is sometimes called the fourth state of matter. There are two modes of operation, the nontransferred arc and the transferred arc. The **nontransferred mode** means that the current flow is from the electrode inside the torch to the nozzle containing

the orifice and back to the power supply. The **nontransferred mode** is normally used for plasma spraying or for generating heat in nonmetals. The transferred arc means that the current is transferred from the tungsten electrode inside the welding torch through the orifice to the work piece and back to the power supply. The difference between these two modes of operation is shown by Figure 4-50. The transferred arc mode is used for welding metals and will be the subject for the remainder of this chapter.

The plasma is generated by constricting the electric arc passing through the orifice of the nozzle and the hot ionized gases that are also forced through this opening. The plasma has a stiff columnar form and is fairly parallel sided so that it does not flare out in the same manner as the gas tungsten arc. The high-temperature stiff plasma arc, when directed toward the work, will melt the base metal surface and the filler metal that may be added to make the weld. In this way, the plasma acts as an extremely high-temperature heat source to form a molten weld puddle in the same manner as the gas tungsten arc. The higher-temperature plasma, however, causes this to happen faster. When the plasma is used in this way it is known as the "melt in" mode of operation. High temperature of the plasma or constricted arc and the high-velocity plasma jet provide an increased heat transfer rate over gas tungsten arc welding when using the same current. This results in faster welding speeds and deeper weld penetration. This method of operation is used for welding extremely thin material and for welding multipass groove welds and fillet welds. Another method of welding with plasma, is known as "keyhole" welding. In this method the plasma jet penetrates through the work piece and forms a hole known as a *keyhole*. Surface tension forces the molten base metal to flow around the keyhole to form the weld. The keyhole method can be used only for joints where the plasma can pass through the joint. It is used for base metals 1/16 inch (1.6 mm) to 1/2 in. (12 mm) in thick-

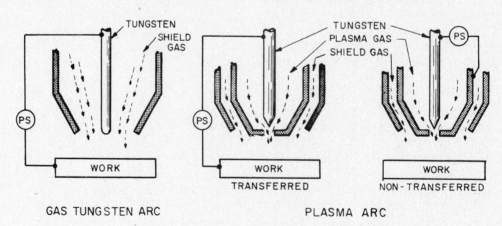

FIGURE 4-50 *Modes of operating the plasma arc.*

ness and is affected by the base metal composition and the welding gases. The keyhole method provides for full-penetration single-pass welding which may be applied either manually or automatically in all positions.

Advantages and Major Use

Advantages of plasma arc welding when compared to gas tungsten arc welding stem from the fact that it has a higher energy concentration. Its higher temperature, its constricted cross-sectional area, and the velocity of the plasma jet create a higher heat content. The other advantage is based on the stiff columnar type arc or form of the plasma which doesn't flare like the gas tungsten arc. These two factors provide the following advantages:

The torch-to-work distance from the plasma arc is less critical than for gas tungsten arc welding due to the columnar form of the plasma. This is particularly important for manual operation since it gives the welder more freedom to observe and control the weld.

High temperature and high heat concentration of the plasma allow for the keyhole effect, which provides complete-penetration single-pass welding of many joints. In this operation, the heat-affected zone and the form of the weld are more desirable. The heat-affected zone is smaller than with gas tungsten arc, and the weld tends to have more parallel sides, which reduces angular distortion.

The higher heat concentration and the plasma jet allow for higher travel speeds. From the welder's point of view, the plasma arc is more stable and is not as easily deflected to the closest point of base metal. Greater variation in joint alignment is possible with plasma than with gas tungsten arc welding. This is particularly important when making root pass welds on pipe and other one-side weld joints. Plasma weld has deeper penetration capabilities and produces a narrower weld. This means that the depth-to-width ratio is more advantageous.

Some of the major uses of plasma arc are its application for the manufacture of tubing. Higher production rates based on faster travel speeds result from plasma over gas tungsten arc welding. Tubing made of stainless steel, titanium, and other metals is being produced with the plasma process at higher production rates than previously with gas tungsten arc welding.

Most applications of plasma are in the low-current range from 100 amperes or less. The plasma can be operated at extremely low currents to allow the welding of foil thickness material.

Plamsa arc welding is also used for making small welds on weldments for instrument manufacturing and other small components made of thin metal. It is being used for making root pass welds on pipe joints and is

used for making butt joints of thin wall tubing. The plasma arc welding process has also been used to do work similar to that done by electron beam welding in the open with a much lower equipment cost.

Plasma arc welding is normally applied as a manual welding process, but is also used in automatic and machine applications. Manual application is the most popular. It is doubtful if semiautomatic methods of application would be useful. Figure 4-51 shows the normal methods of applying plasma arc welding.

Method of Applying	Rating
Manual (MA)	A
Semiautomatic (SA)	No
Machine (ME)	A
Automatic (AU)	A

FIGURE 4-51 *Method of applying.*

The plasma arc welding process is an all-position welding process. Figure 4-52 shows the welding position capabilities.

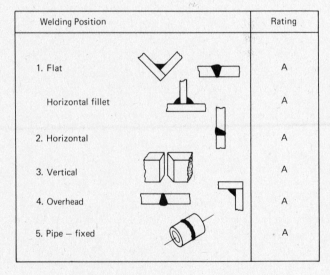

Welding Position		Rating
1. Flat		A
Horizontal fillet		A
2. Horizontal		A
3. Vertical		A
4. Overhead		A
5. Pipe — fixed		A

FIGURE 4-52 *Welding position capabilities.*

The plasma arc welding process is able to join practically all of the commercially available metals. It may not be the best selection or the most economical process for welding some metals. The plasma arc welding process will join all of the metals that the gas tungsten arc welding process will weld. This is shown by the table in Figure 4-53.

109

Base Metal	Weldability	Chapter
Aluminums	Weldable	12-1
Bronzes	Possible but not popular	12-3
Copper	Weldable	12-3
Copper nickel	Weldable	12-3
Cast, malleable, nodular	Possible but not popular	12-1
Wrought iron	Possible but not popular	13-3
Lead	Possible but not popular	12-4
Magnesium	Possible but not popular	12-5
Inconel	Weldable	12-6
Nickel	Weldable	12-6
Monel	Weldable	12-6
Precious metals	Weldable	12-7
Low carbon steel	Weldable	11-2
Low alloy steel	Weldable	11-3
High and medium carbon	Weldable	11-4
Alloys steel	Weldable	11-5
Stainless steel	Weldable	11-6
Tool steels	Weldable	13-4
Titanium	Weldable	12-9
Tungsten	Weldable	12-8

FIGURE 4-53 *Base metals weldable by the plasma arc process.*

Joint Design

Joint design is based on the thicknesses of the metal to be welded and by the two methods of operation. For the keyhole method of operation, the joint design is restricted to full-penetration types. The preferred joint design is the square groove, with no minimum root opening. For root pass work, particularly on heavy wall pipe, the U groove design is used. The root face should be 1/8 in. (3.2 mm) to allow for full keyhole penetration.

For the melt-in method of operation for welding thin gauge, 0.020 in. (0.5 mm), to 0.100 in. (2.5 mm), metals the square groove weld should be utilized. For welding foil thickness, 0.005 in. (0.13 mm) to 0.020 in. (0.5 mm), the edge flange joint should be used. The flanges are melted to provide filler metal for making the weld.

When using the melt-in mode of operation for thick materials the same general joint detail as used for shielded metal arc welding and gas tungsten arc welding can be

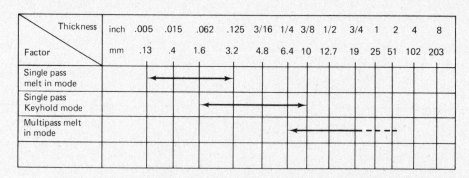

FIGURE 4-54 *Base metal thickness range.*

Regarding the range of thickness welded by the plasma process, consider first the keyhole mode of operation which can be used only where the plasma jet can penetrate the joint. In this mode the process can be used for welding material from 1/16 in. (1.6 mm) through 1/4 inch (12 mm). Thickness ranges vary somewhat with different metals. The melt in mode is used to weld material as thin as 0.002 in. (0.05 mm) up through 1/8 in. (3.2 mm). On the other hand, using multipass techniques, we can weld up to an unlimited thickness of metal. Note that filler rod is used for making welds in thicker material. This information is shown by Figure 4-54.

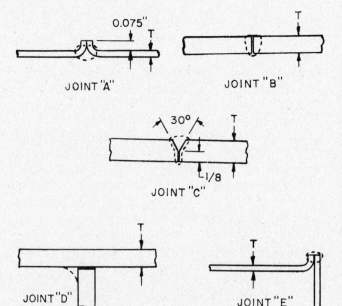

FIGURE 4-55 *Various joints for plasma arc.*

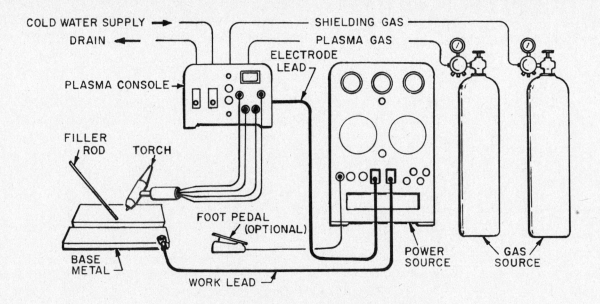

EQUIPMENT ILLUSTRATION

FIGURE 4-56 *Circuit diagram—PAW.*

employed. It can be used for fillets, flange welds, all types of groove welds, etc. It can also be used for lap joints using arc spot welds and arc seam welds. Figure 4-55 shows various joint designs that can be welded by the plasma arc process.

Welding Circuit and Current

The welding circuit for plasma arc welding is somewhat more complex than for gas tungsten arc welding. An extra component is required. It is the control circuit necessary to aid in starting and stopping the plasma arc. The same power source is normally used. There are two gas systems, one to supply the plasma gas and the second for the shielding gas. The welding circuit for plasma arc welding is shown by Figure 4-56.

Direct current of a constant current (CC) type is used. Alternating current is used for only a few applications.

Equipment Required to Operate

Power Source: A constant current drooping characteristic power source supplying the DC welding current is recommended; however, AC/DC type power source can be used. Power source should have an open circuit voltage of 80 volts and should have a duty cycle of 60%. It is very desirable for the power source to have a built-in contactor and provisions for remote control current adjustment. For welding very thin metals it should have a minimum amperage of 2 amps. A maximum of 300 is adequate for most plasma welding applications.

The *welding torch* for plasma arc welding is similar in appearance to a gas tungsten arc torch, but it is more complex. Figure 4-57 shows a typical plasma torch.

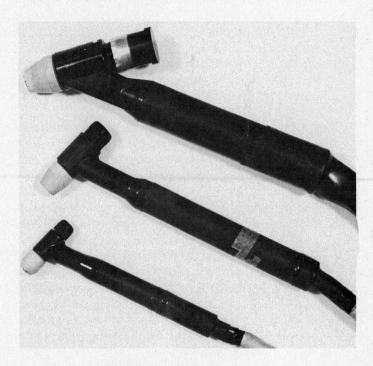

FIGURE 4-57 *Plasma arc torch.*

All plasma torches are water cooled, even the lowest-current range torch. This is because the arc is contained inside a chamber in the torch where it generates considerable heat. If water flow is interrupted briefly the

111

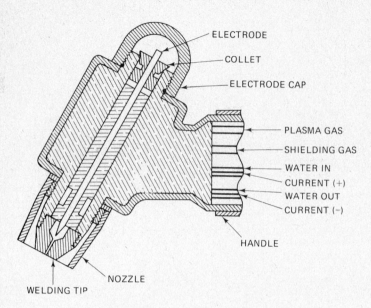

FIGURE 4-58 *Cross section of plasma arc torch head.*

nozzle may melt. A cross section of a plasma arc torch head is shown by Figure 4-58. During the nontransferred period, the arc will be struck between the nozzle or tip with the orifice and the tungsten electrode. Manual plasma arc torches are made in various sizes starting with 100 amps up through 300 amperes. Automatic torches for machine operation of the same basic ratings are also available.

The torch utilizes the 2% thoriated *tungsten electrode* similar to that used for gas tungsten welding. Since the tungsten electrode is located inside the torch, it is almost impossible to contaminate it with base metal.

A *control console* is required for plasma arc welding. The plasma arc torches are designed to connect to the control console rather than to the power source. The control console includes a power source for the pilot arc, delay timing systems for transferring from the pilot arc to the transferred arc, and water and gas valves and separate flow meters for the plasma gas and the shielding gas. Usually the console is connected to the power source and may operate the contactor. The control console will also contain a high-frequency arc starting unit, a nontransferred pilot arc power supply, torch protection circuit, and an ammeter. The high-frequency generator is used to initiate the pilot arc. Torch protective devices include water and plasma gas pressure switches which interlock with the contactor. A *wire feeder* may be used for machine or automatic welding and must be the constant speed type. The wire feeder must have a

speed adjustment covering the range of from 10 in. per minute (254 mm per minute) to 125 in. per minute (3.18 m per minute) feed speed.

Materials Used

Filler metal is normally used except when welding the thinnest of metal. The composition of the filler metal should match that of the base metal. The size of the filler metal rod depends on the thickness of the base metal being welded and the welding current. The filler metal is usually added to the puddle manually, but it can be added automatically.

Plasma and Shielding Gas: An inert gas, either argon helium, or a mixture is used for shielding the arc area from the atmosphere. Argon is more commonly used because it is heavier and provides better shielding at lower flow rates.

For flat and vertical welding, a shielding gas flow of 15 to 30 cu ft per hour (7-14 liters per minute) is usually sufficient. Overhead position welding requires a slightly higher flow rate. Argon is usually used for the plasma gas at a flow rate of 1 cu ft per hour (0.5 liters per minute) up to 5 cu ft per hour (2.4 liters per minute) for welding, depending on torch size and the application. Active gases are not recommended for the plasma gas. In addition to the plasma and shielding gases, cooling water is required.

Quality, Deposition Rates, and Variables

The quality of the plasma arc welds are extremely high and usually higher than gas tungsten arc welds because there is little or no possibility of tungsten inclu-

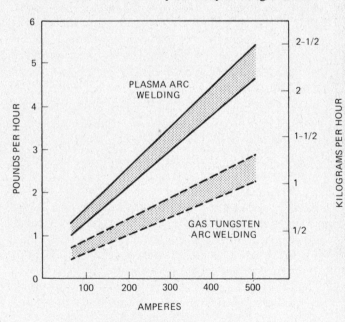

FIGURE 4-59 *Deposition rates.*

Material	Material Thickness in.	Type of Weld	Orifice Dia. in.	Filler Dia. in.	Shield Gas at 20 CFH	Plasma Gas Flow CFH Argon	Weld Current - Amps	No. of Passes	Travel Speed ipm
Stainless steel (1)	0.008	Edge butt	0.093	—	A	0.5	12 DCEN	1	7
	0.008	Edge butt	0.093	—	A-5H$_2$	0.5	10 DCEN	1	13
	0.020	Square groove	0.046	—	A-5H$_2$	0.5	12 DCEN	1	21
	0.030	Square groove	0.046	—	A-5H$_2$	0.5	34 DCEN	1	17
	0.062	Square groove	0.081	—	A-5H$_2$	0.7	65 DCEN	1	14
	0.093	Square groove	0.081	—	A	2.0	85 DCEN	1	12
	0.093	Square groove	0.081	—	A-5H$_2$	2.0	85 DCEN	1	16
	0.125	Square groove	0.081	—	A	2.5	100 DCEN	1	10
	0.125	Square groove	0.081	—	A-5H$_2$	2.5	100 DCEN	1	16
	0.187	Square groove	0.081	—	A-5H$_2$	3.5	100 DCEN	1	7
	0.250	V-groove	0.081	—	A-5H$_2$	3.0	100 DCEN	First	5
	0.250	V-groove	0.081	3/32	A-5H$_2$	1.4	100 DCEN	Second	2
Cop- Mild steel per (1)	0.030	Square groove	0.081	—	A	0.5	45 DCEN	1	26
	0.080	Square groove	0.081	—	A	1.0	55 DCEN	1	17
Aluminum	0.016	Edge butt	0.093	—	He	0.5	18 DCEN	1	24
	0.036	Square groove	0.081	1/16	He	0.05	47 DCEP	1	24
	0.050	Edge joint	0.081	—	He	0.5	48 DCEP	1	22
	0.090	Fillet	0.081	3/32	He	1.4	34 DCEP	1	4

(1) Backing gas 5 to 10 CFH argon.

FIGURE 4-60 *Weld procedure schedule—plasma arc welding—manual application.*

sions in the welds. The skill of the welder is a major factor with respect to the quality of welds. A welder will find the plasma arc welding process easier to use than the gas tungsten arc process which helps insure weld quality.

	SUNKEN BEAD, UNDERCUT TOO MUCH PENETRATION
	WELDING CURRENT IS TOO HIGH OR TRAVEL SPEED IS TOO SLOW
	BEAD TOO SMALL, IRREGULAR LITTLE PENETRATION
	WELDING CURRENT IS TOO LOW OR PLASMA GAS FLOW IS TOO LOW OR TRAVEL IS TOO FAST
	UNDERCUT AND IRREGULAR EDGES
	THE PLASMA GAS FLOW IS TOO HIGH
	PROPER SIZE BEAD EVEN RIPPLE AND GOOD PENETRATION
	CORRECT CURRENT, EVEN TORCH MOVEMENT, PROPER ARC VOLTAGE AND PLASMA GAS FLOW

FIGURE 4-61 *Quality and common faults.*

Deposition rates for plasma arc welding are somewhat higher than for gas tungsten arc welding and are shown by the curve in Figure 4-59.

Weld schedules for the plasma arc process are shown by the data in Figure 4-60.

The process variables for plasma arc welding are shown by Figure 4-61. Most of the variables shown for plasma arc are similar to the other arc welding processes. There are two exceptions: the plasma gas flow and the orifice diameter in the nozzle. These are unique to plasma welding. The major variables exert considerable control in the process. The minor variables are generally fixed at optimum conditions for the given application. All variables should appear in the welding procedure. Variables such as the angle of the tungsten electrode, the setback of the electrode, and electrode type are considered fixed for the application. The plasma arc process does respond differently to these variables than does the gas tungsten arc process. The standoff or torch-to-work distance, is less sensitive with plasma but the torch angle when welding parts of unequal thicknesses is more important than with gas tungsten arc.

Tips for Using the Process

The most important tip for using plasma arc welding is to properly maintain the welding torch. The tungsten electrode must be precisely centered and located with respect to the orifice in the nozzle. The

THE ARC WELDING PROCESSES—USING CC POWER

pilot arc current must be kept sufficiently low, just high enough to maintain a stable pilot arc.

When welding extremely thin materials in the foil range the pilot arc may be all that is necessary.

When filler metal is used, it is added in the same manner as gas tungsten arc welding. However, with the torch-to-work distance a little greater there is more freedom for adding filler metal. Equipment must be properly adjusted so that the shielding gas and plasma gas are in the right proportions. Proper gases must also be used.

Heat input is important for quality welds. Plasma gas flow also has an important effect. These factors are shown by Figure 4-61, "Weld Quality and Common Faults."

It has been found by many production welders that the plasma process is more flexible and easier to use for some applications than gas tungsten arc welding.

The safety considerations for plasma arc welding are the same as for gas tungsten arc welding.

Limitations of the Process

The major limitations of the process have to do more with the equipment and apparatus. The torch is more delicate and complex than a gas tungsten arc torch. Even the lowest rated torches must be water cooled. The tip of the tungsten and the alignment of the orifice in the nozzle is extremely important and must be maintained within very close limits. The current level of the torch cannot be exceeded without damaging the tip. The water-cooling passages in the torch are relatively small and for this reason water filters and deionized water are recommended for the lower current or smaller torches. The control console adds another piece of equipment to the system. This extra equipment makes the system more expensive and may require a higher level of maintainance.

Variations of the Process

The welding current may be pulsed to gain the same advantages as pulsing provides for gas tungsten arc welding. A high current pulse is used for maximum penetration but is not on full time—to allow for metal solidification. This gives a more easily controlled puddle for out-of-position work. Pulsing can be accomplished by the same apparatus as is used for gas tungsten arc welding.

Programmed welding can also be employed for plasma arc welding in the same manner as it is used for gas tungsten arc welding. The same power source with programming abilities is used and offers advantages for certain types of work. The complexity of the programming depends on the needs of the specific application. In addition to programming the welding current it is oftentimes necessary to program the plasma gas flow. This is particularly important when closing a keyhole which is required to make the root pass of a weld joining two pieces of pipe.

The method of feeding the filler wire with plasma is essentially the same as for gas tungsten arc welding. The "hot wire" concept can be used. This means that low-voltage current is applied to the filler wire to preheat it prior to going into the weld puddle.

The plasma arc process is one of the newer of the arc welding processes and has not yet attained wide use. Part of the problem has been the complexity and the delicate nature of the plasma torch. More and more users are finding advantages of plasma welding and it is expected that it will become increasingly important in the next few years. The low current (below 100 amps)

FIGURE 4-62 *Picture of process in use.*

has found many uses for precision welding of extremely thin pieces. Higher-current plasma welding applications are now beginning and it is expected that this process will find more applications in the near future for more of the routine types of welding and joining requirements. Figure 4-62 shows plasma arc welding being used. The aircraft industries, jet engine industry, piping, tubing, and precision instrument industries are users of the plasma arc welding process.

4-5 CARBON ARC WELDING PROCESS (CAW)

Carbon arc welding is a process in which coalescence is produced by heating with an arc between a carbon electrode and the work, and in which no shielding is used. Both pressure and filler metal may or may not be used.

Principle of Operation

Carbon arc welding shown by Figure 4-63 uses a single electrode with the arc between it and the base metal. It is the oldest arc process, is included for historical purposes, but is not popular today. There are two variations that are important, however.

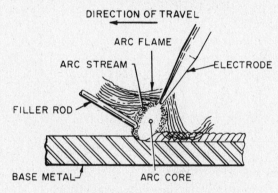

FIGURE 4-63 *Process diagram—CAW.*

In carbon arc welding, the heat of the arc between the carbon electrode and the work melts the base metal and, when required, also melts a filler rod. As the molten metal solidifies a weld is produced. The carbon graphite electrode, considered to be nonconsumable, errodes away fairly rapidly and in its disintegration produces a shielding atmosphere of carbon monoxide and carbon dioxide gas. These gases partially displace air from the arc atmosphere and prohibit the oxygen and nitrogen from coming in contact with the molten metal. Filler metal when employed is normally the same composition as the base metal. Bronze filler metal can be used for brazing and braze welding.

Advantage and Major Use

The single electrode carbon arc welding process is no longer widely used. It is used for welding copper since it can be used at high currents to develop the high heat usually required. It is also used for making bronze repairs on cast iron parts. When welding thinner materials the process is used for making autogenous welds or welds without adding filler metal. Carbon arc welding is also used for joining galvanized steel. In this case the bronze filler rod is added by placing it between the arc and the base metal.

The carbon arc welding process has been used almost entirely by the manual method of applying. In its early use it was mechanized and electromagnetic coils were added around the electrode to direct the arc to the weld joint. Later a rope-type material was fed into the arc and was consumed to create a protective arc atmosphere. This coupled with the magnetic directing coils provided a machine welding system that had fairly high production capacities. This system is no longer of industrial significance. The normal method of applying is shown by Figure 4-64.

Method of Applying	Rating
Manual (MA)	A
Semiautomatic (SA)	No
Machine (ME)	B
Automatic (AU)	No

FIGURE 4-64 *Method of applying.*

The manual carbon arc process is an all-position welding process. It is considered primarily as a heat source and is used to generate the weld puddle which can be carried in any position. Figure 4-65 shows the welding position capabilities.

Weldable Metals

Since the carbon arc is used primarily as a heat source to generate a welding puddle it can be used on metals that are not affected by carbon pickup or by the carbon monoxide or carbon dioxide arc atmosphere. Mild and low carbon steels are most widely welded with the carbon arc process, followed by its use for welding copper. The carbon arc has been used for welding other nonferrous metals. However, it has been supplanted

THE ARC WELDING PROCESSES—USING CC POWER

Welding Position		Rating
1. Flat		A
Horizontal fillet		A
2. Horizontal		A
3. Vertical		A
4. Overhead		A
5. Pipe—fixed		—

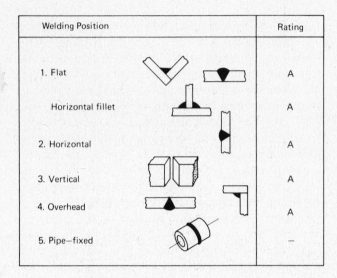

FIGURE 4-65 *Welding position capabilities.*

by the gas tungsten arc and gas metal arc welding processes.

The greatest use of carbon arc welding is its use as a heat source for brazing, and to deposit materials for wear-resistant surfaces. Various filler metals are used. The carbon arc is used for weld repairing iron castings. The filler metal can be case iron or bronze.

The carbon arc is also used in repair of steel castings. In this case, the carbon arc is used primarily as a puddling method for filling casting defects or holes. The high-current capacity is utilized to heat the defective area to a molten stage and feed in filler metal that matches the composition of the casting.

The base metal thickness range and the joint design used are very similar to those of shielded metal arc welding.

Welding Circuit and Welding Current

The welding circuit for carbon arc welding is the same as for shielded metal arc welding. The difference in the apparatus is a special type of electrode holder used only for holding carbon electrodes shown by Figure 4-66. This type holder is used because the carbon electrode becomes extremely hot in use and the conventional electrode holder will not efficiently hold and transmit current to the carbon electrode. The power source is the conventional or constant current type with drooping volt-amp characteristics. Normally a 60% duty cycle power source is utilized. The power source

should have a voltage rating of 50 volts since this voltage is used when welding copper with the carbon arc.

Single electrode carbon arc welding is always used with direct current electrode negative DCEN (straight polarity). In the carbon to steel arc the positive pole (anode) is the pole of maximum heat. If the electrode were positive, the carbon electrode would erode very rapidly because of the higher heat, and would cause black carbon smoke and excess carbon which could be absorbed by the weld metal. Alternating current is not recommended for single-electrode carbon arc welding. The electrode should be adjusted often to compensate for the erosion of the carbon. From 3 to 5 in. of the carbon electrode should protrude through the holder towards the arc.

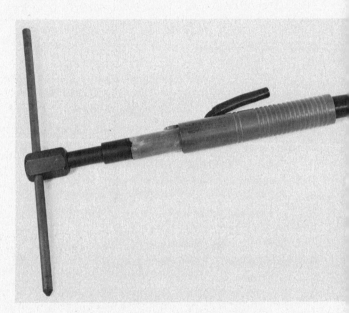

FIGURE 4-66 *Carbon arc electrode holder.*

Carbon Electrode

There are two types of electrodes used for carbon arc welding. One is made of pure graphite and the other of baked carbon. The pure graphite electrode does not erode away as quickly as the carbon electrode. It is more expensive and more fragile. Electrodes are available in diameters ranging from 5/32 in. (4 mm) up through 3/4 in. (19 mm) diameter with a length of either 12 or 17 in. (310 or 460 mm).

Welding Schedules

Since the carbon arc welding process is not widely used, the only schedules that will be provided are those that may be of interest today. The welding schedule for carbon arc welding galvanized iron using silicon bronze filler metal is given by Figure 4-67. A short arc must be

| Material Thickness | | | Electrode Size | | Filler Rod Size | | Welding Current | Arc Voltage |
Gage	in.	mm	in.	mm	in.	mm	amps DC	Electrode Neg.
24	0.024	0.6	3/16	5	3/32	2.4	25-30	13-15
22	0.020	0.7	3/16	5	3/32	2.4	25-30	13-15
20	0.036	0.9	3/16	5	3/32	2.4	30-35	14-16
18	0.048	1.2	1/4	6.4	1/8	3.2	30-35	14-16
16	0.060	1.5	1/4	6.4	1/8	3.2	30-35	14-16
14	0.075	1.9	1/4	6.4	1/8	3.2	30-35	14-16
12	0.105	2.7	1/4	6.4	1/8	3.2	35-40	15-17

FIGURE 4-67 *Welding procedure schedule—galvanized steel—braze welding.*

| THICKNESS OF COPPER | | | DIAMETER OF ELECTRODE AND FILLER ROD | | | | | Welding | Voltage |
| Decimal Inches | Fraction Inches on US Gage | | Electrode Carbon | | Filler Rod | | | Current | Electrode |
	in.	mm	in.	mm	in.	mm		DC Amps	Negative
0.05	18		3/16	4.8	3/32			80	
0.0563	17				3/32	2.4		90	35
0.0625	1/16	1.6	3/16	4.8	1/8			90	
0.07	15				1/8	3.2		100	40
0.078	5/64	1.9			5/32			120	
0.094	3/32	2.4	1/4	6.4	5/32			135	
0.109	7/64	2.8			5/32	11.9		140	40
0.125	1/8	3.2			3/16			150	
0.141	9/64	3.6			3/16			160	
0.156	5/32	3.9	1/4	6.4	3/16			165	
0.172	11/64	4.4			3/16			170	
0.1875	3/16	4.8			3/16	4.8		185	45
0.203	13/64	5.2			1/4			200	
0.219	7/32	5.6			1/4			200	
0.234	15/64	5.9	5/16	7.9	1/4			205	
0.25	1/4	6.4			1/4			215	
0.266	17/64	6.7			1/4	6.4		225	45
0.281	9/32	7.1			5/16			250	
0.3125	5/16	7.9			5/16			250	
0.344	11/32	8.7	5/16	7.9	5/16			255	
0.375	3/8	9.5			5/16			270	
0.406	13/32	10.3			5/16	7.9		290	50
0.4375	7/16	11.1			3/8			300	
0.4688	15/32	11.9	3/8	9.5	3/8			310	
0.5	1/2	12.7			3/8	9.5		325	50

FIGURE 4-68 *Welding procedure schedule for carbon arc welding copper.*

| Electrode Diameter | | Carbon Electrodes | Graphite Electrodes |
in.	mm	DCEN Amps	DCEN Amps
1/8	3.2	15-30	15-35
3/16	4.8	25-55	25-60
1/4	6.4	50-85	50-90
5/16	7.9	75-115	80-125
3/8	9.5	100-165	110-165
7/16	11.1	125-185	140-210
1/2	12.7	150-225	170-260
5/8	15.9	200-310	230-370
3/4	19.0	250-400	290-490
7/8	22.2	300-500	400-750

FIGURE 4-69 *Welding current for carbon electrode types.*

used to avoid damaging the galvanizing. The arc must be directed on the filler wire which will melt and flow on to the joint.

For the welding copper use a high arc voltage and follow the schedule given by Figure 4-68.

Figure 4-69 shows the welding current to be used for each size of the two types of carbon electrodes.

Variations of the Process

There are two important variations of carbon arc welding. One is twin carbon arc welding and the other is carbon arc cutting and gouging. The latter will be completely described in Chapter 7. Twin carbon arc welding is "an arc welding process wherein coalescence

THE ARC WELDING PROCESSES—USING CC POWER

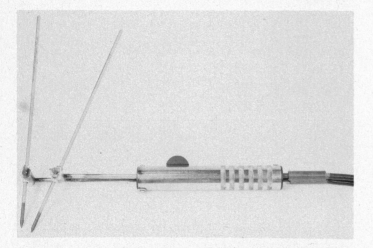

FIGURE 4-70 *Twin carbon electrode holder.*

is produced by heating with an electric arc maintained between two carbon electrodes. Filler metal may or may not be used.'' The twin carbon arc can also be used for brazing.

A special electrode holder is used for the twin electrodes. It is shown by Figure 4-70.

The twin carbon electrode holder is designed so that one electrode is moveable and can be touched against the other to initiate the arc. The carbon electrodes are held in the holder by means of set screws and are adjusted so they protrude equally from the clamping jaws. When the two carbon electrodes are brought together the arc is struck and established between them. The angle of the electrodes provides an arc that forms in front of the apex angle and fans out as a soft source of concentrated heat or arc flame. It is softer than that of the single carbon arc. The temperature of this arc flame is between 8,000° and 9,000°F (4,426° and 4,982°C).

Alternating current is used for the twin carbon welding arc. With alternating current the electrodes will burn off or disintegrate at equal rates. Direct current power can be used but when it is the electrode connected to the positive terminal should be one size larger than the electrode connected to the negative terminal. This will ensure an even burning or disintegration of the carbon electrodes since the positive electrode disintegrates at the higher rate. The arc gap or spacing between the two electrodes is adjustable during welding and must be adjusted more or less continuously to provide the fan shape arc.

The twin carbon arc is a very useful source of heat that can be used for many applications in addition to welding, brazing, and soldering. It can be used as a heat source to bend or form metal. The welding current settings or schedules for different size of electrodes is shown by Figure 4-71.

The twin carbon electrode method is used by the hobbyist and for maintenance work in the home, in the small shop, and on the farm. It is used with the low-duty cycle single-phase limited-input AC transformer welding machines. It can be used in any position and on any materials where the heat is required. It is relatively slow

| Carbon Electrode Diameter | | Welding Current | | Base Metal Thickness | |
in.	mm	Amperes AC	Arc Voltage	in.	mm
1/4	6.4	55	35-40	1/16	1.6
5/16	7.9	75	35-40	1/8	3.2
3/8	9.5	95	35-40	1/4	6.4
3/8	9.5	120	35-40	over 1/4	over 6.4

FIGURE 4-71 *Welding current for carbon electrode (twin torch).*

and for this reason does not have too much use as an industrial welding process.

4-6 THE STUD ARC WELDING PROCESS (SW)

Stud arc welding is a process in which coalescence is produced by heating with an arc drawn between a metal stud or similar part and the other work part until the surfaces to be joined are properly heated when they are brought together under pressure. Partial shielding may be obtained by the use of a ceramic ferrule surrounding the stud. Shielding gas or flux may or may not be used.

Principles of Operation

There are four variations of stud welding. Stud arc welding, also called drawn arc stud welding, is the most popular and will be described in this section; however, the other methods, capacitor discharge stud welding, the drawn arc capacitor discharge stud welding, and the consumable ferrule type stud welding, will be discussed under "Variations of the Process," later in this chapter.

It is questionable if stud welding is a true arc welding process. It has a very specialized field of application and is not a metal joining process in the same manner as the others previously discussed. It end welds prepared studs to the base metal. The process is a combination of arc welding and forge welding. It is based on two steps. First, electrical contact between the stud and the base metal occurs and an arc is established. The heat of the arc melts the surface of the end of the stud and the work surface. As soon as the entire cross section of the stud and an area of equal size on the base metal are melted, the stud is forced against the base metal. The molten end of the stud joins with the molten pool on the work surface and as the metal solidifies the weld is produced. Partial shielding is accomplished by means of a ceramic ferrule that surrounds the arc area and by fluxing ingredients sometimes placed on the arcing end of the stud.

The making of a stud weld is shown by Figure 4-72 and is explained as follows. The stud gun (step A) holds the stud in contact with the work piece until the welder depresses the gun trigger switch. This causes welding current to flow from the power source through the stud (which acts as an electrode) to the work surface. The welding current flow actuates a solenoid within the stud gun which draws the stud away from the work surface (step B) and establishes the arc. The arc time duration is controlled by a timer in the control unit. At the appropriate time the welding current is shut off and the gun solenoid releases its pull on the stud and the spring loaded action plunges the stud into the molten pool of

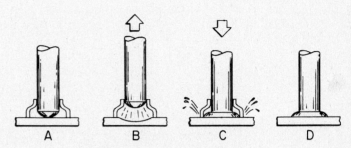

FIGURE 4-72 *Stud welding process.*

the work piece (step C). The molten metal solidifies and produces the weld, plus a small reinforcing fillet. After solidification the gun is released from the stud and the ceramic ferrule is broken off revealing the weld (step D).

Advantages and Major Uses

The stud arc welding process is unique and is a special application process. It offers tremendous cost savings when compared to drilling and tapping for studs, or to manually welding studs to base metal. Stud welding does not destroy the water tightness or weaken the base metal in the way that drilled and tapped holes do. One of the major and earliest uses of the process was to attach wood deck to steel decks of ships. Another use is the installation of insulating material to steel surfaces, specifically the interior of ships. In structural work, the major usage is stud welding of shear connectors to steel work, over which concrete is poured. This insures that the concrete and steel will work together as a composite to withstand the loads imposed. It is also widely used for attaching brackets, hangers, cover plates, and decorative trim to many types of products.

Semiautomatic method of applying is the most common for construction work and for ship work. Automatic feed and automatic location are becoming increasingly popular in the mass production and metalworking manufacturing companies. This information is shown by Figure 4-73.

Method of Applying	Rating
Manual (MA)	No
Semiautomatic (SA)	A
Machine (ME)	B
Automatic (AU)	B

FIGURE 4-73 *Method of applying.*

THE ARC WELDING PROCESSES—USING CC POWER

This process is practically an all-position process; however, the overhead position is difficult. This is shown by Figure 4-74.

Welding Position		Rating
1. Flat		A
Horizontal fillet		—
2. Horizontal		A
3. Vertical		A
4. Overhead		B
5. Pipe—fixed		—

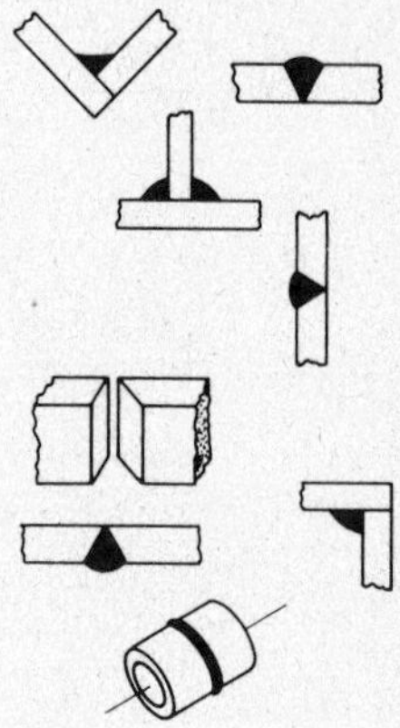

FIGURE 4-74 *Welding position capabilities.*

This process is used most widely for welding studs to mild steels, low alloy steels and some of the austenitic stainless steels. Some of the process variations can be used on nonferrous metals. Figure 4-75 is a chart showing the base metals that can be welded. The stud must have the same analysis as the base metal.

The minimum recommended plate thickness to permit efficient welding without burn-through or excessive distortion is shown by Figure 4-76. As a general rule, the minimum thickness of the plate or base metal is 20% of the stud base diameter. To develop full strength

STUD BASE DIAMETER			BASE METAL THICKNESS		
fraction	inch	mm	inch	gage	mm
3/16	0.187	4.7	0.059	16	1.5
1/4	0.250	6.3	0.075	14	1.9
5/16	0.312	7.9	0.104	12	2.6
3/8	0.375	9.5	0.117	11	2.9
7/16	0.437	11.1	0.135	10	3.4
1/2	0.500	12.7	0.164	—	4.1
5/8	0.625	15.8	0.209	—	5.3
3/4	0.750	19.0	0.250	1/4"	6.3
7/8	0.875	22.2	0.312	5/16"	7.9
1	1.000	25.4	0.375	3/8"	9.5

FIGURE 4-76 *Minimum recommended base metal thickness (steel).*

of the stud the plate thickness should be not less than 50% of the stud base diameter.

Joint Design

Stud welding is a special application process. The joint would be considered a tee type. The weld would be a square groove type with a small reinforcing fillet all around.

Welding Circuit

Figure 4-77 is the circuit diagram for stud welding. It shows the welding power source, which can be a generator or transformer-rectifier, the stud gun, and the special control unit.

Direct current is preferred for stud arc welding and the stud gun (electrode) is connected to the negative terminal (DCEN) or straight polarity. The work or base metal to which the stud is to be welded is attached to the positive pole. AC is not recommended for stud arc welding and direct current constant voltage is usable but not recommended.

	STUD WELDING VARIATIONS			
Base Metal	Conventional Arc Stud	Contact Capacitor Discharge	Drawn Arc Capacitor Discharge	Fusible Cartridge
Aluminum	No	Weldable	Weldable	No
Brass—bronze	No	Weldable	Weldable	No
Copper	No	Weldable	Weldable	No
Low carbon steel	Weldable	Weldable	Weldable	Weldable
Low alloy steel	Weldable	Weldable	Weldable	Weldable
Medium carbon	Limited	Weldable	Weldable	Weldable
Stainless steel	Weldable	Weldable	Weldable	Weldable
Zinc	No	Weldable	Weldable	No
Dissimilar metals	Limited	Weldable	Weldable	No

FIGURE 4-75 *Metals weldable—stud arc welding and variations.*

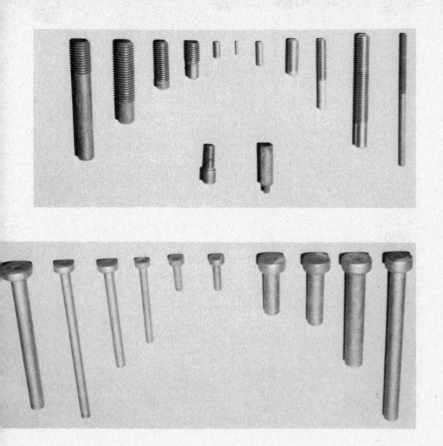

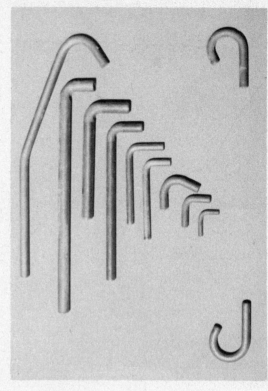

FIGURE 4-77 *Variety of studs available.*

Equipment Required to Operate

Refer to Figure 4-77 regarding equipment required to make stud welds. Figure 4-78 shows stud welding on a bridge frame.

The *power source* for stud welding is normally a direct current motor generator or rectifier constant current welding machine. Batteries can also be used as a source of DC current for stud welding. The size

of the welding machine required is based on the size of the stud to be welded. Welding machines can be paralleled to provide sufficient current for stud welding. Stud welding has an extremely short arc time period, rarely lasting over one second. Therefore, over-capacity currents are drawn from the machine for a short period of time much lower than the normal duty cycle requirements of a machine. For this reason, the welding machine must have sufficient overload capacity. It is

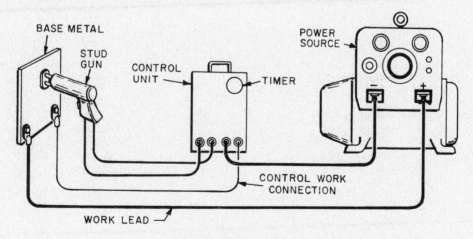

STUD WELDING

FIGURE 4-78 *Circuit diagram—SW.*

THE ARC WELDING PROCESSES—USING CC POWER

FIGURE 4-79 *Making stud welds on bridge girder.*

important to check the specifications of the welding power source that is intended to be used for stud welding to see if it complies. See Figure 4-79 for the recommended welding power sources and cable for welding different sizes of studs. Generator welding machines, either electric or engine driven, have higher overload capacities because of the flywheel effect.

The *stud welding gun* is designed specifically and only for stud welding. The gun is designed to resemble a pistol with a trigger switch for starting the weld cycle. Several different stud guns are shown by Figure 4-80.

The gun contains the solenoid lifting mechanism as well as the spring mechanism for plunging the stud into the molten weld puddle. Because of this the welding current must flow through the gun itself. The gun must also contain adjustments to establish the arc length and to provide for accurate plunge dimensions. The gun is equipped with a stand-off device, to hold it in proper relationship to the work. In addition, the gun allows for the interchange of different types of collets or chucks for each size and type of stud. The stand-off device or arc shield holder must be adjustable to accommodate different types of chucks. Guns are available in different sizes based on the size of the studs to be used. These are also related to the welding current involved.

The *Weld Timer Controller* is a separate unit to establish the arc time required for proper heating. The control unit also contains a contactor capable of breaking the welding current at the end of the arcing period. The timer is calibrated in electrical cycles which are defined as 1/60 of a second, for 60 hertz power. Control units for stud welding are designed specifically to interconnect with the welding stud gun and the power source. Stud guns of one manufacturer should not be used with control panels or studs of another manufacturer. The control cable is between the stud gun and the timer only.

Materials Used

Electrode and filler metal are combined in the stud. Studs are made in many sizes and shapes and their design details are given by catalogs of the companies that manufacture studs. Figure 4-81 shows the wide variety of studs available for welding. The stud gun manufacturers and stud manufacturers offer catalogs with engineering data pertaining to the exact design of the studs they manufacture. Studs up to 1 in. (25 mm) diameter can be welded. The round stud is the most

Stud inch	Diameter mm	Amperes Required	WELD CABLE		POWER SOURCE		ALTERNATE POWER SOURCE	
			Size	No. Reqd.	No.	Rated Size	No.	Rated Size
3/16	4.76	300	2/0	1	1	300 Amp		
1/4	6.35	400	2/0	1	1	400 Amp		
5/16	7.94	500	2/0	1	1	400 Amp		
3/8	9.53	600	2/0	1	1	600 Amp	2	300 Amp
7/16	11.11	700	2/0	1	1	600 Amp	2	300 Amp
1/2	12.70	900	4/0	1	1	1000 Amp	2	400 Amp
5/8	15.87	1150	4/0	1	1	1000 Amp	2	600 Amp
3/4	19.05	1600	4/0	2	2	1000 Amp	2	600 Amp
7/8	22.23	1800	4/0	2	2	1000 Amp		
1	25.40	2000	4/0	2	2	1000 Amp		

Note generator welders have greater overload capacity than rectifiers.

FIGURE 4-80 *Power source requirements for different sizes of studs.*

common and easy to use. Rectangular shaped studs are available. It is recommended that the width to length ratio of rectangular studs is not to exceed a 5-to-1 ratio. Most studs include a method of fluxing. This is accomplished by different means by different manufacturers. In some cases, granular flux is enclosed in the end of the stud by means of a thin metal retaining shield. Other makes utilize a solid flux inset in the end of the stud while others have a coating of flux covering the arcing end of the stud. The design of the arcing end of the stud is slighly different from the different manufacturers of studs. Flux acts as an arc stabilizer and helps protect the molten metal from the atmosphere during welding. Normally studs are made of low-carbon steel having a minimum yield strength of 45,000 psi (20,400 kg) and a 20% minimum elongation in 2 in. (50 mm). Stud types can be threaded fasteners, internally threaded fasteners, flat fasteners with rectangular cross sections, header pins, eye bolts, slotted pins, keys, etc. Studs are also made of different metals.

FIGURE 4-81 *Stud guns.*

A short portion of the stud is melted off during the arcing period so that the finished length is less than the original length of the stud. The amount of burnoff material depends on the diameter of the stud and somewhat on the application. The burn-off value for small studs is 1/8 in. (3.2 mm), 5/32 in. (4 mm) for medium-size studs, and 3/16 in. (4.8 mm) for the larger-size studs.

A ceramic *ferrule* must be used for each stud. This is sometimes called an arc shield. This ferrule is placed over the stud and held in position by a holder or grip on the gun. The ceramic ferrule performs the following functions.

1. It concentrates the heat of the arc in the weld area.

2. Reduces oxidation of the molten metal during welding by restricting contact with the atmosphere.

3. Confines the molten metal.

4. Eliminates the need for a welding headshield or helmet by shielding the arc from the welder.

The inner surface shape of the ferrule is the same shape as the stud being used, usually cylindrical. It has a serrated shape at the base to form vents for escaping shielding gas. Internally it is shaped to help mold the molten metal around the base of the stud to form a small fillet. Specially designed ferrules may be obtained for specific applications. The ferrule is broken off the weld at the completion of the weld and discarded. It can be used only once.

Quality of Welds

The weld quality of properly made stud welds is excellent and usually exceeds the strength of the stud. Properly made welds depend on the weld schedule, the size of the power source, the gun chuck and hold down, the use of correct size of cables, and a good work connection. In addition, the work surface to be welded must be clean of paint, rust, heavy metal scale, etc. It is necessary to hold the gun in the proper position and steady during the weld cycle. Figure 4-82 shows potential quality problems that can occur. Inspection of welds may be made both visually and mechanically.

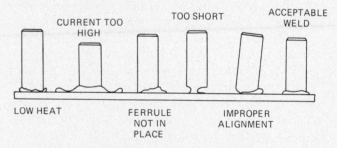

FIGURE 4-82 *Potential quality problems.*

The normal visual inspection is to determine if a uniform fillet is produced completely around the periphery of the stud at its junction to the base metal.

Mechanical tests may be made by shearing the stud from the work or by pulling it from the work. AWS Recommended Practice for Stud Welding[8], the AWS Structural Code and Navy welding codes specify how these tests are to be made.

CHAPTER 4

THE ARC WELDING PROCESSES—USING CC POWER

Weld Tables—Schedules

The welding conditions for stud welding are given in the tables shown by Figure 4-83. This includes all factors necessary to set up procedures for stud welding.

Stud Dia. inches	Amps DCEP	Welding Volts	Time* Cycle	Lift (in.)	Plunge (in.)
3/16	300	30	7	1/16	1/8
1/4	400	30	10	1/16	1/8
5/16	500	30	15	1/16	1/8
3/8	600	28	20	1/16	1/8
7/16	700	28	25	1/16	1/8
1/2	900	28	30	3/32	5/32
5/8	1150	28	40	3/32	5/32
3/4	1600	26	50	1/8	3/16
7/8	1800	24	60	1/8	3/16
1	2000	24	70	1/8	3/16

*Based on 60 Hertz.

FIGURE 4-83 *Stud arc welding conditions.*

Welding Variables and Tips for Using

Stud welding is a relatively simple and foolproof welding process. The apparatus must be properly set up and adjusted for the size and type of studs to be welded. The settings and control schedules must be in agreement with the schedules based on stud sizes. Further, the stud gun must be equipped with the proper size holding devices, etc.

Stud welding does not require manipulative skill necessary for the other arc welding processes.

The most popular method of applying is semiautomatic where the welder holds the stud gun during the welding operation. The gun should be held by both hands, one hand in the same manner as gripping a pistol, and the other hand against the back of the gun to provide stability. It is necessary to place the stud tip against the base metal and push the gun towards the base metal to seat the ferrule firmly against the base metal. This actually compresses the mechanism inside the gun for proper operation. When the trigger is pressed the entire weld cycle is automatic, based on the settings of the control unit. The gun should not be moved during the welding operation. The arc will be noticed and plunging action will be heard at the completion of the welding cycle. After the plunging operation, the gun should be held steady for at least a half-second before withdrawing it from the welding stud.

There are a variety of problems that might occur in making a stud weld. If the current is too low a full-strength weld will not result. If it is too high there will be too much metal discharged from the weld. The time cycle must be set correctly. If the plunge is too short a full weld will not result. If the ferrule is not in place properly the weld will be offside. If the gun is not perpendicular to the work, the stud will be welded at an angle not perpendicular. Other problems can result if there is too much dirt, paint, or foreign material on the base material being welded. Another problem may result if welding leads from the power source to the work and to the stud gun or control are not tight. There is also a problem if these cables are longer than recommended.

Every time a new job is started or a new setup is made the welding procedure should be verified by bending a stud with hammer blows to see that it is welded securely.

Safety Considerations

The ceramic ferrule, or arc shield, shields the arc from the welder and eliminates the need for the normal welding hood. The welder should wear safety glasses, or flash goggles, with a tinted shade. For overhead and vertical position welding above the welder's head protective clothing is required. Gloves are recommended for all arc stud welding.

Limitations of the Process

The arc stud process is limited primarily to the mild and low-alloy steels. High-carbon and alloy steels should not be stud welded unless a heat treating operation is performed. Stud welding can be performed on the austenitic types of stainless steels only. It cannot be used to weld nonferrous metals.

Variations of the Process

There are three variations of stud welding. These are:

The capacitor discharge stud welding method.

The drawn arc capacitor discharge stud welding method.

The fusible cartridge stud welding method.

In the contact capacitor discharge stud welding variation the energy for making the weld is stored at a low voltage in high capacitance capacitors. This method is called the stored energy system or percussive stud welding. The stud is slightly different since it has a small tip on the end of the stud. In operation this small tip is brought into contact with the base material and then pressure is applied by means of a spring or air pressure in the gun. The contactor then closes and the weld circuit is completed. The stored energy is discharged through the small tip or projection at the base of the stud and it presents a high resistance to the electrical energy and rapidly disintegrates. This creates an arc which heats the surface of the stud and base metal. During this arcing period the stud is plunged to the base metal by means of a spring or air pressure. The weld is completed immediately following the high intensity arc, in a short period of about 0.006 seconds. It is done so quickly that the heat effect upon the parts is minimal. Figure 4-84 shows the sequence of operations. For welding mild steel neither flux nor shielding of any kind is used. However, only smaller studs are welded with this method, usually 1/4 in. (6.4 mm) maximum diameter. The power source, stud gun, and controller are designed especially for this stud welding variation.

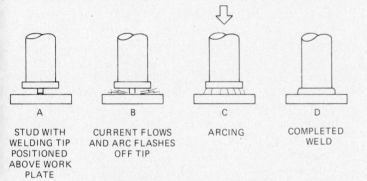

A
STUD WITH WELDING TIP POSITIONED ABOVE WORK PLATE

B
CURRENT FLOWS AND ARC FLASHES OFF TIP

C
ARCING

D
COMPLETED WELD

FIGURE 4-84 *Sequence of operations— contact capacitor discharge.*

In the drawn-arc capacitor discharge stud welding variation arc initiation is obtained in the same manner as arc stud welding. The sequence shown by Figure 4-85 is as follows. The stud is placed against the work, then it lifts from the work to draw an arc. Studs up to 1/4 in. (6.4 mm) diameter are used. The arc time varies from 6 to 15 milliseconds and then the stud is plunged into the molten pool and the weld is completed. Flux is not

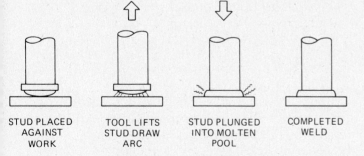

STUD PLACED AGAINST WORK

TOOL LIFTS STUD DRAW ARC

STUD PLUNGED INTO MOLTEN POOL

COMPLETED WELD

FIGURE 4-85 *Sequence of operations— drawn arc capacitor discharge.*

required but shielding gas may be used for welding such metals as aluminum. A special gun, control, and power source are required for this variation of stud welding. This method including power source and associated equipment is very similar to that used for contact capacitor discharge stud welding.

The fusible cartridge stud welding variation of stud welding is used primarily in Europe and does require a longer time period than any of the variations.[9] This stud is square on the welding end without any special preparation, ranging in size from 3/16 in. (5 mm) to 5/8 in. (16 mm) diameter. A fusible cartridge which looks similar to an arc shield is placed on the end of the stud. The fusible cartridge, or ferrule, is made of material similar to coating on an iron powdered electrode. The sequence of operations is shown by Figure 4-86. The arc is initiated by the flow of current through the fusible cartridge. Soon after initiation, full arcing across the face of the stud occurs. The weld cycle is controlled by the cartridge and as the cartridge disintegrates the stud is forced against the work piece by the spring in the gun. The end of the stud and the work piece are molten and when the molten metal solidifies the weld is complete. The weld cycle time is from 1/2 to 2-1/2 seconds. The process is used for carbon steel and stainless steel. A fusible cartridge is used for each weld. The gun is special for this process variation. It has less movable parts since it is not required to withdraw the stud to initiate an arc. The power source is a conventional CC type. Either AC or DC can be used. A line contactor is required, as well as a special control circuit.

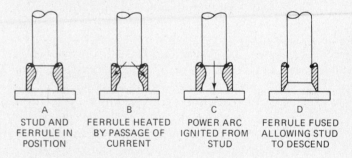

A
STUD AND FERRULE IN POSITION

B
FERRULE HEATED BY PASSAGE OF CURRENT

C
POWER ARC IGNITED FROM STUD

D
FERRULE FUSED ALLOWING STUD TO DESCEND

FIGURE 4-86 *Sequence of operations—fusible cartridge.*

Industries Using and Typical Applications

The construction industry is a major user of stud welding for attaching shear connectors, for attaching conduits, piping, electrical switch boxes, etc., to metal work. The shipbuilding industry uses stud welding for attaching wood decking to metal decking, also for attaching insulation to the interior steel portions of ships. Machinery manufacturers use studs for the attachment of inspection cover plates. The automotive industry uses stud welding for frames and for attaching trim to auto bodies.

4-7 OTHER PROCESSES USING CONSTANT CURRENT POWER

There are a number of welding processes that utilize the constant current welding power. Some of these are included only from an historical point of view since they are of little industrial significance today. Others are mentioned because they may gain industrial importance. Welding processes are developed and if they fulfill a need they will become widely used in industry. As other processes are developed they may replace earlier ones which will gradually fall into disuse. Some of the following processes never became industrially popular but were stepping stones to some of the modern processes now in use.

The following processes will be briefly described:

Atomic hydrogen welding.

Magnetic rotating arc welding.

Automatic welding using:

Continuous covered wire.

Impregnated tape.

Magnetic flux.

Composite electrode.

Atomic hydrogen welding is no longer of industrial significance. It is "an arc welding process where coalescence is produced by heating with an electric arc maintained between two metal electrodes in an atmosphere of hydrogen." Shielding is obtained from the hydrogen. Filler metal may or may not be used. This process was invented by Irving Langmuir of General Electric Company in the mid-1920s. He found that atomic hydrogen was formed when hydrogen gas was passed through an electric arc between two tungsten electrodes at atmospheric pressure. Molecular form is the more stable and when the atoms recombine to form molecular hydrogen intense heat is liberated. Figure 4-87 is a diagram of atomic hydrogen process. The arc is maintained between two tungsten electrodes and the hydrogen gas passes from the electrode holders along the electrodes in through the

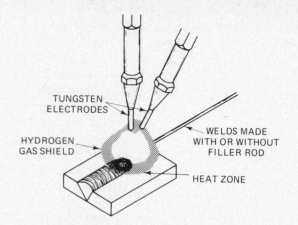

FIGURE 4-87 *Atomic hydrogen welding process.*

arc. The arc stream between the two electrodes assumes a fan shape and is characterized by a sharp singing sound. The arc area is usually 3/8 to 3/4 in. (9 to 20 mm) in diameter. As the hydrogen passes through the arc from the jets around the tungsten electrodes, molecules are separated into atoms and this gives the process its name, "atomic hydrogen." As the gas in its atomic state is forced out of the intense arc heat, it recombines to molecular form giving up its heat of disassociation. This heat produces the extremely high welding temperature flame. Recombining is done in the outer edge of the arc stream, which is the greatest concentration of atomic hydrogen. This extra heat added to the normal heat of the arc produces a considerably higher temperature flame than the ordinary arc. The arc is independent of the work and can be moved closer or further from the work for precise heat control.

There are two other functions of the hydrogen. Hydrogen is a very powerful reducing agent and in the presence of the arc and the molten metal it would tend to reduce any gas forming material in the arc and thus produce a sound porosity free weld. In addition, the hydrogen prevents contamination of the arc and weld puddle by excluding atmospheric oxygen and nitrogen from the weld area. At one time the process was fairly popular for welding certain hard-to-weld metals such as nickel base alloys, molybdenum, high-alloy steels, and steels for making tools and dies.

The arc is visible to the welder and welding is done best in the flat and horizontal positions. Safety precautions must be followed since the open circuit voltage of the power source approaches 300 volts.

The equipment needed for atomic-hydrogen welding is shown by Figure 4-88. The power source has a high open-circuit voltage and for this reason it contains a contactor that is operated by a foot switch. It also contains a solenoid valve for controlling the flow of the hydrogen gas. The atomic-hydrogen welding torch is shown by Figure 4-89.

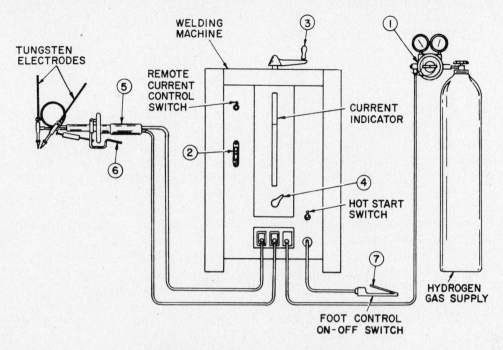

FIGURE 4-88 *Atomic hydrogen circuit diagram.*

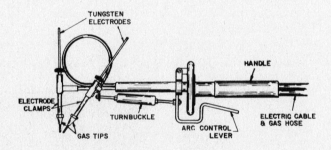

FIGURE 4-89 *Atomic hydrogen torch.*

The process has limited industrial use today. Its main application is tool and die repair welding and for the manufacture of alloy steel chain.

Magnetic Rotating Arc Welding

This is a relatively new welding process that uses the constant current power source. It is being applied for welding tubular or hollow cross-section parts together in Germany. The process is a fully automatic process which utilizes a magnetically rotated arc to heat the surfaces to be welded and then utilizes pressure for consummating the weld.[10] The sequence of operations is shown by Figure 4-90. First two tabular or hollow cross-section parts are clamped in the welding machine. They are moved together until they touch and a hinged magnetic coil is placed around the joint. The welding

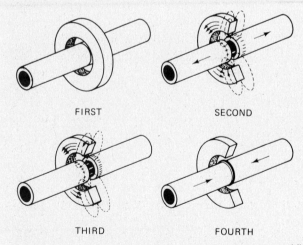

FIGURE 4-90 *Welding schedules for the atomic hydrogen welding process.*

127

current is initiated and the parts moved slightly away from each other. The arc is struck, normally by a high-frequency discharge, and the arc is then caused to rotate by the magnetic field in the coil around the parts. The length of time that the arc rotates is in the order of 1/2 second to 2 seconds depending on the mass of metal to be heated. A shielding gas, normally CO_2, is often used to protect the arc and the molten metal from the atmosphere. Pressure is applied bringing the parts together, extinguishing the arc and making a forge weld. The maximum welding current employed is 1000 amperes and the maximum pressure is on the order of 500 lbs.

The process is capable of welding many different metals; however, mild steel, alloy steels, and stainless steels are the only metals currently being welded. The parts do not need to be circular.

One of the advantages is that it is completely automatic and uses less power than would be required by resistance welding. The rotating arc process does not require as much material to be consumed as does flash welding. It is expected that this process will find increasing use in the mass production or high-volume industries. Currently it is being used to join the cap on the end of a tube to manufacture shock absorbers as shown by Figure 4-91.

FIGURE 4-91 *Magnetic rotating arc weld sample.*

Automatic Welding

Automatic welding began in the early 1920s but the covered electrode which produced higher-quality weld metal soon replaced many bare electrode automatic welding applications. Automatic welding was improved by using lightly coated wires and later by using knurled

wires which held more of the stabilizing and deoxidizing coating materials. The quality was not as good as that produced by covered electrodes.

Various efforts were made to produce an all-position automatic welding process that would produce high-quality weld metal. The coating on the *covered wire* created two problems. One, the coating was fragile and brittle and could not be bent into coils without cracking and falling off. Second, the coatings were insulators and the welding current could not be introduced into the metal core wire in an efficient manner. There were, however, some interesting developments that deserve brief consideration.

One early attempt was made to extrude coatings on large diameter electrode wire and then coil the wire into extremely large diameter coils. At the welding head a cutter removed the coating from one small area of the wire. The welding current was conducted to the core wire through the slot in the coating by multiple pickup shoes. This process had limited success and was abandoned.

Another variation which had greater success was known as the *impregnated-tape* metal arc welding process. Shielding is obtained from decomposition of an impregnated tape wrapped around the bare electrode as it is fed into the arc. The tape which provided the shielding was similar to insect screen wire woven in the same manner but narrow in width sufficient to wrap around the electrode. It was impregnated with stabilizing and deoxidizing materials and then coiled. This

FIGURE 4-92 *Composite flux electrode.*

tape was wrapped around the bare electrode wire below the current pickup jaws and above the arc. It provided arc shielding, deoxidizers, and slag formers. This process was known as Una-Matic and had limited success in the U.S.A.

A similar idea was utilized in Europe; the coating was made by extruding small half-circular sections, which were notched and joined together with two small wires like beads. These were fed in from both sides to the bare electrode wire just above the arc and constituted the covering on the electrode. This process had limited success.

Another system utilized the magnetic field surrounding the electrode wire to carry flux into the arc. The bare electrode wire was fed into the arc, surrounded by carbon dioxide gas which carried powdered magnetic flux. The *magnetic flux* is attracted to the electrode wire by the magnetic field created by the welding current. The magnetic flux covered the electrode wire as it entered the arc zone. The flux performed the normal function. The carbon dioxide gas also aided in shielding the molten weld metal from the atmosphere. This system was utilized for surfacing and fabrication. It never became a major semiautomatic or automatic welding process, largely due to the problems of feeding and handling the magnetic flux.

The one system that became successful and is still in use, utilizes a *composite electrode*. The composite filler metal electrode is of special design. The large-diameter center core wire is wrapped with two strands

of small wires spiraled in one direction and two strands of small wire spiraled in the opposite direction, to create cavities. This composite assembly is run through an extrusion press and flux is forced into the cavities to provide protection in the same manner as the covering on a conventional coated electrode. Figure 9-92 shows the construction of this special electrode. The electrode is placed on rather large coils so that the coating will not crack and separate.

The welding head is fairly conventional except that the current pickup jaws are elongated to provide a large contact area to conduct the welding current to the electrode. Various types of flux coverings are available to provide different characteristics or to match different types of carbon and low-alloy steels.

A recent variation uses CO_2 shielding gas to surround the arc area. This provides for higher welding currents and faster travel speeds. Both the unshielded and CO_2 gas shielded processes are limited to the flat and horizontal fillet weld positions.

This process is not used with semiautomatic equipment since the electrode wire is not sufficiently flexible to be easily manipulated by a welder.

This process is no longer popular in North America, but it is used in Great Britain by the shipbuilding industry.

QUESTIONS

1. Explain the static electrical characteristic curve of a CC power source.

2. A long arc has a _____ voltage. A short arc has a _____ voltage.

3. Explain the difference between a dual-control and single-control machine.

4. Shielded metal arc welding can be used in how many positions? Name them.

5. Name the metals that can be welded with SMAW.

6. What limits the size of electrode and welding current when welding overhead?

7. What is the factor that limits the welders duty cycle or operator factor?

8. What factors are involved in selecting electrodes? Name them and explain each.

9. In GTAW the filler metal is added by a separate rod. How does this minimize spatter?

10. What is the major advantage of gas tungsten arc welding?

11. What are the disadvantages of gas tungsten arc welding?

12. Why is alternating current used for welding aluminum?

13. What determines the size of the tungsten electrode?

14. Why is "stand-off distance" less critical for PAW than GTAW?

15. Explain the difference between "key hole" and melt-in mode.

16. Must the plasma arc torch always be water cooled? Why?

17. Why is carbon arc welding becoming less important industrially?

18. True or false: arc stud welding is a combination of arc and forge welding.

19. Explain the difference between arc stud and the discharge variations.

20. Describe arc shielding used in the older automatic welding methods.

REFERENCES

1. "Specifications for Mild Steel Covered Arc Welding Electrodes," *American Welding Society, (AWS)* A5.1, Miami, Florida.

2. "Specifications for Low-Alloy Steel Covered Arc Welding Electrodes," *American Welding Society, (AWS)* A 5.5, Miami, Florida.

3. U.S. Patent No. 2,269,369, dated January 6, 1942, to George Hafergut, "Firecracker Welding."

4. "Specifications for Tungsten Arc-Welding Electrode," *American Welding Society, (AWS)* A5.12, Miami, Florida.

5. "Recommended Practices for Plasma Arc Welding," *American Welding Society, (AWS)* A5.12, Miami, Florida.

6. "Modern Arc Welding," Hobart Brothers, Troy, Ohio.

7. Military Specification MIL-E-17777B, "Electrode, Cutting and Welding Carbon-Graphite Uncoated and Copper Coated."

8. "Recommended Practices for Stud Welding," *American Welding Society, (AWS)* L5.4-74, Miami, Florida.

9. "Handbook of Stud Welding," Philips Welding, The Netherlands.

10. Osborn, Max R. "Magnetically Controlled Arc Joins Tubular Components," *Metal Progress,* November 1974, p. 79, Cleveland, Ohio.

5

THE ARC WELDING PROCESSES— USING CV POWER

5-1 **WELDING WITH CONSTANT VOLTAGE**

5-2 **SUBMERGED ARC WELDING (SAW)**

5-3 **GAS METAL ARC WELDING (GMAW) OR MIG**

5-4 **FLUX CORED ARC WELDING (FCAW)**

5-5 **ARC WELDING VARIABLES**

5-6 **ELECTRO SLAG WELDING (ESW)**

5-7 **SELECTING THE CORRECT ARC WELDING PROCESS**

5-1 WELDING WITH CONSTANT VOLTAGE

The constant voltage (CV) welding machine and the CV system of automatic arc length control is relatively new. The CV principle of operation was introduced at about the same time as gas metal arc welding. It was the combination of these two inventions that has made gas metal arc welding extremely popular. Prior to the introduction of the constant voltage (CV) welding machine, the drooping characteristic type power source was employed with the voltage-sensing electrode wire feed systems. The reaction time of these systems was not sufficiently fast to avoid burnback and stubbing when using fine wire gas metal arc apparatus. In spite of its widespread use, the CV principle of welding needs a thorough explanation.

The CV electrical system is fairly new to welding but is not new to the electric power industry. It is the basis of operation of the entire commercial electric power system. The electric power delivered to your house and available at every receptacle has a constant voltage. The same voltage is maintained continuously at each outlet whether a small light bulb, with a very low wattage rating, or a heavy-duty electric heater with a high wattage rating is connected. The current that

flows through each of these circuits will be different based on the resistance of the particular item or appliance in accordance with Ohm's law. For example, the small light bulb will draw less than 0.01 amperes of current while the electric heater may draw over 10 amperes. The voltage throughout the system remains constant, but the current flowing through each appliance depends on its resistance or electrical load. The same principle is utilized by the CV welding system.

Welders have long known that when they use a higher current the electrode is melted off more rapidly. With low current the electrode melts off slower. This relationship between melt-off rate and welding current applies to all of the arc welding processes that use a continuously fed electrode. This is a physical relationship that depends upon the size of the electrode, the metal composition, the atmosphere that is surrounding the arc, and welding current. Figure 5-1 shows the melt-off rate curves for different sizes of steel electrode wires in a CO_2 atmosphere. Note that these curves are nearly linear at least in the upper portion of the curve. Similar curves are available for all sizes of electrode wires of different compositions and in different shielding atmospheres. This relationship is definite and fixed, but some variations can occur.

THE ARC WELDING PROCESSES—USING CV POWER

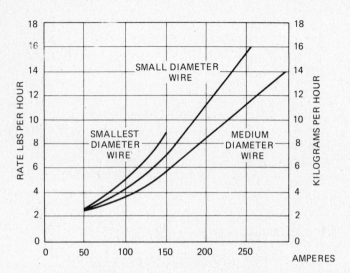

FIGURE 5-1 *Burn-off rates of wire vs current.*

This relationship is the basis of the simplified control for wire feeding using constant voltage. Instead of regulating the electrode wire feed rate to maintain the constant arc length, as is done when using a constant current power source, the electrode wire is fed into the arc at a fixed speed and the power source is designed to provide the necessary current to melt off the electrode wire at this same rate. This concept prompted the development of the constant voltage welding power source.

The volt-ampere characteristics of the constant voltage power source shown by Figure 5-2 was designed to produce substantially the same voltage at no load and at rated or full load. It has characteristics similar to a

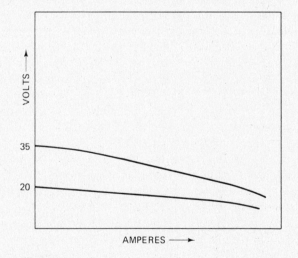

FIGURE 5-2 *Static volt amp characteristic curve of CV machine.*

standard commercial electric power generator. If the load in the circuit changes, the power source automatically adjusts its current output to satisfy this requirement and maintains essentially the same voltage across the output terminals. This system assured a self-regulating arc based on a fixed rate of wire feed and a constant voltage power source. The simplified controls did not require complex circuitry and reversal of the wire feed drive motor.

Another way to understand the constant voltage system is to consider the factors that make up the electrical resistance load in the external portion of the welding circuit. These resistances or voltage drops occur in the welding arc and in the welding cables and connectors, in the welding gun, and in the electrode length beyond the current pickup tip. These voltage drops add up to the output voltage of the welding machine and represent the electrical resistance load on the welding power source. When the resistance of any component in the external circuit changes the voltage balance will be achieved by changing the welding current in the system. The greatest voltage drop occurs across the welding arc. The other voltage drops in the welding cables and connections, etc., are relatively small and constant. The voltage drop across the welding arc is directly dependent upon the arc length. A small change in arc volts results in a relatively large change in welding current. Figure 5-3 shows that if the arc length shortens slightly or 2 volts the welding current increases by approximately 100 amperes. This change in arc length greatly increases the melt-off rate and quickly brings the arc length back to normal.

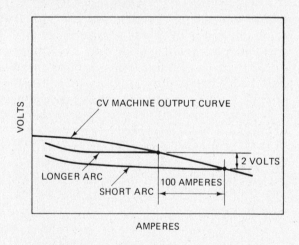

FIGURE 5-3 *Static volt amp curve with arc range.*

The constant voltage power source is continually changing its current output in order to maintain the voltage drop in the external portion of the welding circuit. Changes in wire feed speed which might occur when the welder moves the gun toward or away from the work are compensated for by momentarily changing the current and the melt-off rate until equilibrium is re-established. The same corrective action occurs if the wire feeder has a temporary reduction in speed. The CV power source and fixed wire feed speed system is self-regulating. It is an excellent wire feed system, especially for semiautomatic welding, since movement of the cable assembly often changes the drag or feed rate of the electrode wire. The CV welding power source provides the proper current so that the melf-off rate is equal to the wire feed rate. The arc length is controlled by setting the voltage on the power source. The welding current is controlled by adjusting the wire feed speed.

The characteristics of the welding power source must be designed to provide a stable arc when gas metal arc welding with different electrode sizes and metals and in different atmospheres. Most constant voltage power sources have taps or a means of adjusting the slope of the volt-ampere curve. Experience indicates that a curve having a slope of 1-1/2 to 2 volts per hundred amperes is best for gas metal arc welding with nonferrous electrodes in inert gas, for submerged arc welding, and for flux-cored arc welding with larger-diameter electrode wires. A curve having a medium slope of 2 to 3 volts per hundred amperes is preferred for welding with CO_2 gas shielded metal arc welding and for smaller flux cored electrode wires. A steeper slope of 3 to 4 volts per hundred amperes is recommended for short circuiting arc transfer. These three slopes are shown by Figure 5-4.

unwanted spatter if not controlled. It is controlled by adding reactance or inductance in the circuit. This changes the time factor or response time and provides for a stable arc. In most machines a different amount of inductance is included in the circuit for the different slopes.

The constant voltage welding power system has its greatest advantage when the current density of the electrode wire is high. The current density (amperes/sq in.) relationship for different electrode wire sizes and different currents is shown by Figure 5-5. Note the vast difference between the current density employed for gas metal arc welding with a fine electrode wire compared with conventional shielded metal arc welding with a covered electrode.

Direct current electrode positive DCEP is used for gas metal arc welding. When DC electrode negative DCEN is used the arc is erratic and produces an inferior weld. Direct current electrode negative DCEN can be used for submerged arc welding and flux-cored arc welding.

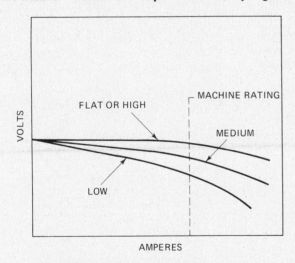

FIGURE 5-4 *Various slopes of characteristic curves.*

The flatter the curve, the more the current changes for an equal change in arc voltage.

The dynamic characteristics of the power source must be carefully engineered. Refer again to Figure 5-3. If the voltage changes abruptly with a short circuit the current will tend to increase quickly to a very high value. This is an advantage in starting the arc but will create

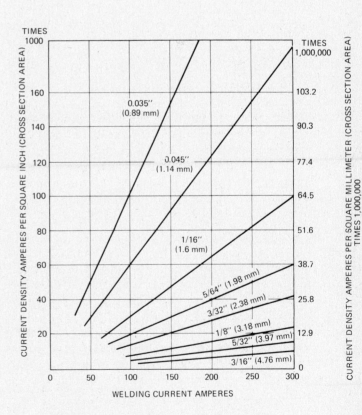

FIGURE 5-5 *Current density—various electrode sizes.*

| Welding Machine Power Output Characteristics | Constant Voltage "Flat" | |
	Direct Current CV-DC	Alternating Current CV-AC
Welding Process		
Shielded metal arc welding (SMAW) stick	No	No
Gas tungsten arc welding (GTAW) or TIG	No	No
Plasma arc welding (PAW)	No	No
Carbon arc welding, cutting and gouging (AAC)	C	No
Stud arc welding (SW)	C	No
Submerged arc welding (SAW)	A	Exp.
Gas metal arc welding (GMAW):	—	—
Inert gas (MIG)—nonferrous	A	No
Spray arc (argon-oxygen)	A	No
CO_2 (big wire)	A	No
Shorting arc (fine wire)	A	No
Flux cored arc welding (FCAW):	—	—
With external shielding gas	A	Exp.
Without shielding gas (self shield)	A	Exp.
Electroslag welding (EW)	A	A

FIGURE 5-6 *Welding process vs welding power type—constant voltage.*

The constant voltage principle of welding with alternating current is normally not used. It can be used for submerged arc welding and for electroslag welding, but is not popular.

The constant voltage power system should not be used for shielded metal arc welding. It may overload and damage the power source by drawing too much current too long. It can be used for carbon arc cutting and gouging with small electrodes and the arc welding processes shown by Figure 5-6.

5-2 SUBMERGED ARC WELDING (SAW)

Submerged arc welding is a process in which coalescence is produced by heating with an arc or arcs between a bare metal electrode or electrodes and the work. The arc is shielded by a blanket of granular fusible material on the work. Pressure is not used and filler metal is obtained from the electrode and sometimes from a supplementary welding rod.

Principle of Operation

The submerged arc welding process is shown by Figure 5-7. It utilizes the heat of an arc between a continuously fed electrode and the work. The heat of the arc melts the surface of the base metal and the end of the electrode. The metal melted off of the electrode is transferred through the arc to the work piece where it becomes the deposited weld metal. Shielding is obtained from a blanket of granular flux, which is laid directly over the weld area. The flux close to the arc melts and inter-

mixes with the molten weld metal and helps purify and fortify it. The flux forms a glasslike slag that is lighter in weight than the deposited weld metal and floats on the surface as a protective cover. The weld is submerged under this layer of flux and slag and hence the name submerged arc welding. The flux and slag normally cover the arc so that it is not visible. The unmelted portion of the flux can be reused. The electrode is fed into the arc automatically from a coil. The arc is maintained automatically and travel can be manual or by machine. The arc is initiated by a fuse type start or by a reversing or retrack system.

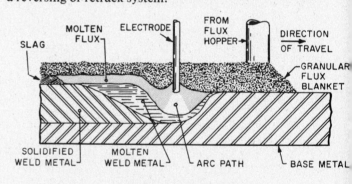

SUBMERGED ARC WELDING

FIGURE 5-7 *Process diagram—submerged arc.*

Advantage and Major Use

The submerged arc welding process, introduced in the early 1930s, is one of the older automatic processes and was originally used to make the longitudinal seam in large pipe. It was developed to provide high-quality deposited weld metal by shielding the arc and the molten metal from the contaminating effects of the air. The major advantages of the process are:

High quality of the weld metal.

Extremely high deposition rate and speed.

Smooth, uniform finished weld with no spatter.

Little or no smoke.

No arc flash, thus minimal need for protective clothing.

High utilization of electrode wire.

Easily automated for high-operator factor.

Manipulative skills normally not involved.

The submerged arc process is widely used in heavy steel plate fabrication work. This includes the welding of structural shapes, the longitudinal seam of larger diameter pipe, the manufacture of machine components for all types of heavy industry, the manufacture of vessels and tanks for pressure and storage use. It is widely used in the shipbuilding industry for splicing and fabricating subassemblies, and by many other industries where steels are used in medium to heavy thicknesses. It is also used for surfacing and buildup work, maintenance, and repair.

Normal Method of Applying and Position Capabilities

The most popular method of applying is the machine method where the operator monitors the welding operation. Second in popularity is the automatic method where welding is a pushbutton operation. The process can be applied semiautomatically; however, this method of application is not too popular. The process cannot be applied manually because it is impossible for a welder to control an arc that is not visible.

The submerged arc welding process is a limited-position welding process. Welding can be done in the flat position and in the horizontal fillet position with

ease. The welding positions are limited because of the large pool of molten metal which is very fluid and the slag is also very fluid and will tend to run out of the joint. Under special controlled procedures it is possible to weld in the horizontal position, sometimes called 3 o'clock welding. This requires special devices to hold the flux up so that the molten slag and weld metal cannot run away. The process cannot be used in the vertical or overhead position.

Metals Weldable and Thickness Range

Submerged arc welding is used to weld low- and medium-carbon steels, low-alloy high-strength steels, quenched and tempered steels, and many stainless steels. Experimentally it has been used to weld certain copper alloys, nickel alloys, and even uranium. This information is summarized by the table shown by Figure 5-8.

Base Metal	Weldability	Chapter
Wrought iron	Weldable	13-3
Low carbon steel	Weldable	11-2
Low alloy steel	Weldable	11-3
High and medium carbon	Possible but not popular	11-4
Alloys steel	Possible but not popular	11-5
Stainless steel	Weldable	11-6

FIGURE 5-8 *Base metals weldable by the submerged arc process.*

Metal thicknesses from 1/16 in. (1.6 mm) to 1/2 in. (12 mm) can be welded with no edge preparation. With edge preparation welds can be made with a single pass on material from 1/4 in. (6.4 mm) through 1 in. (25 mm). When multipass technique is used, the maximum thickness is practically unlimited. This information is summarized by the chart shown in Figure 5-9. Horizontal

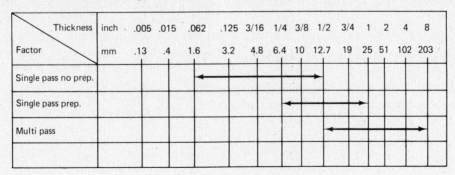

Thickness	inch	.005	.015	.062	.125	3/16	1/4	3/8	1/2	3/4	1	2	4	8
Factor	mm	.13	.4	1.6	3.2	4.8	6.4	10	12.7	19	25	51	102	203
Single pass no prep.														
Single pass prep.														
Multi pass														

FIGURE 5-9 *Base metal thickness range.*

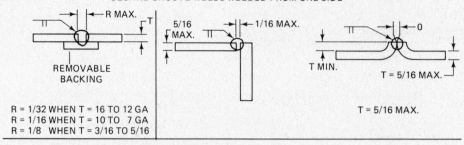

SQUARE-GROOVE WELDS WELDED FROM ONE SIDE

R = 1/32 WHEN T = 16 TO 12 GA
R = 1/16 WHEN T = 10 TO 7 GA
R = 1/8 WHEN T = 3/16 TO 5/16

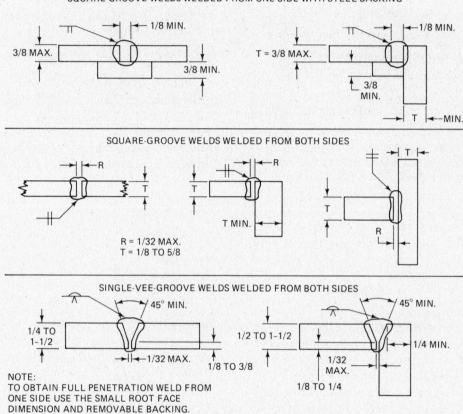

SQUARE-GROOVE WELDS WELDED FROM ONE SIDE WITH STEEL BACKING

SQUARE-GROOVE WELDS WELDED FROM BOTH SIDES

R = 1/32 MAX.
T = 1/8 TO 5/8

SINGLE-VEE-GROOVE WELDS WELDED FROM BOTH SIDES

NOTE:
TO OBTAIN FULL PENETRATION WELD FROM
ONE SIDE USE THE SMALL ROOT FACE
DIMENSION AND REMOVABLE BACKING.

FIGURE 5-10 *Weld joint designs for submerged arc welding.*

fillet welds can be made up to 3/8 in. (9.5 mm) in a single pass and in the flat position fillet welds can be made up to 1 in. (25 mm) size.

Joint Design

The submerged arc welding process can utilize the same joint design details as the shielded metal arc welding process. These joint details are given in Chapter 15. However, for maximum utilization and efficiency of submerged arc welding, different joint details are suggested.

For groove welds, the square groove design can be used up to 5/8 in. (16 mm) thickness. Beyond this thickness bevels are required. Open roots are used but backing bars are necessary since the molten metal will run through the joint. When welding thicker metal, if a sufficiently large root face is used, the backing bar may be eliminated. However, to assure full penetration when welding from one side, backing bars are recommended. Where both sides are accessible, the backing weld can be made which will fuse into the original weld to provide full penetration. Recommended submerged arc joint designs are shown by Figure 5-10.

Welding Circuit and Current

The welding circuit employed for single electrode submerged arc welding is shown by Figure 5-11. This requires a wire feeder system and a power supply.

SINGLE-GROOVE WELDS WELDED FROM ONE SIDE WITH STEEL BACKING

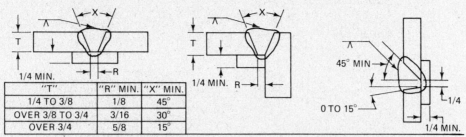

"T"	"R" MIN.	"X" MIN.
1/4 TO 3/8	1/8	45°
OVER 3/8 TO 3/4	3/16	30°
OVER 3/4	5/8	15°

DOUBLE-VEE-GROOVE WELDS WELDED FROM BOTH SIDES

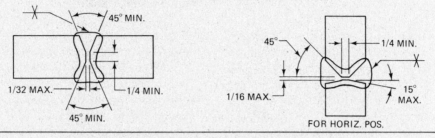

DOUBLE-BEVEL-GROOVE WELDS WELDED FROM BOTH SIDES

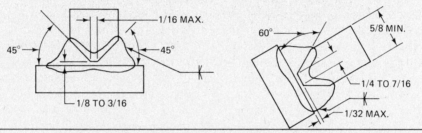

SINGLE-U-GROOVE WELDS WELDED FROM ONE OR BOTH SIDES

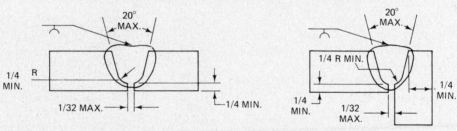

FIGURE 5-10 *(Continued)*

FIGURE 5-11 *Block diagram — SAW.*

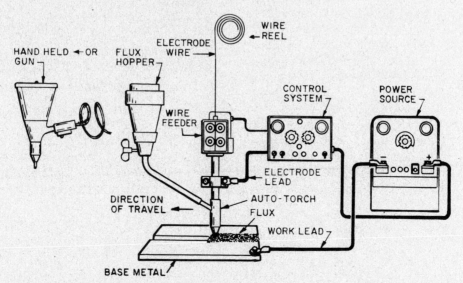

The submerged arc welding process uses either direct or alternating current for welding power. Direct current is used for most applications which employ a single arc. Both direct current electrode positive (DCEP) and electrode negative (DCEN) are used.

The constant voltage type of direct current power is more popular for submerged arc welding with 1/8 inch (3.2 mm) and smaller diameter electrode wires.

The constant current power system is normally used for welding with 5/32 in (4 mm) and large diameter electrode wires. The control circuit for CC power is more complex since it attempts to duplicate the actions of the welder to retain a specific arc length. The wire feed system must sense the voltage across the arc and feed the electrode wire into the arc to maintain this voltage. As conditions change, the wire feed must slow down or speed up to maintain the prefixed voltage across the arc. This adds complexity to the control system and the system cannot react instantaneously. Arc starting is more complicated with the constant current system since it requires the use of a reversing system to strike the arc, retract, and then maintain the preset arc voltage.

For AC welding the constant current power is always used. When multiple electrode wire systems are used with both AC and DC arcs the constant current power system is utilized. The constant voltage system, however, can be applied when two wires are fed into the arc supplied by a single power source. Welding current for submerged arc welding can vary from as low as 50 amperes to as high as 2,000 amperes. Most submerged arc welding is done in the range of 200 to 1,200 amperes.

Equipment Required to Operate

The equipment components required for submerged arc welding are shown by Figure 5-11. These consist of (1) welding machine or power source, (2) the wire feeder and control system, (3) the welding torch for automatic welding, or the welding gun and cable assembly for semiautomatic welding, (4) the flux hopper and feeding mechanism and usually a flux recovery system, and (5) travel mechanism for automatic welding.

The power source for submerged arc welding must be rated for a 100% duty cycle, since the submerged arc welding operations are continuous and the length of time for making a weld may exceed 10 minutes. If a 60% duty cycle power source is used it must be derated according to the duty cycle curve for 100% operation.

When constant current is used, either AC or DC, the voltage sensing electrode wire feeder system must be used. When constant voltage is used the simpler fixed speed wire feeder system is used. The CV system is only used with direct current.

Both generator and transformer-rectifier power sources are used, but the rectifier machines are the more popular. Welding machines for submerged arc welding range in size from 300 amperes to 1,500 amperes. They may be connected in parallel to provide extra power for high-current applications. Direct current power is used for semiautomatic applications but alternating current power is used primarily with the machine or the automatic method. Multiple electrode systems require specialized types of circuits especially when AC is employed.

For semiautomatic application a welding gun and cable assembly are used to carry the electrode and current and to provide the flux at the arc. An application of semiautomatic submerged arc welding is shown by Figure 5-12. A small flux hopper is attached to the end of the cable assembly and the electrode wire is fed through the bottom of this flux hopper through a current pickup tip to the arc. The flux is fed from the hopper to the welding area by means of gravity. The amount of flux fed depends on the height the gun is held above the work. The hopper gun may include a start switch to initiate the weld or it may utilize a "hot" electrode so that when the electrode is touched to the work feeding will begin automatically.

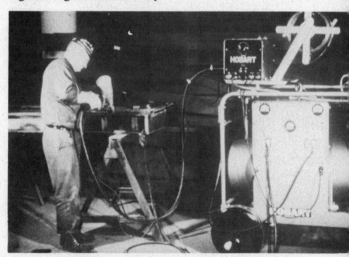

FIGURE 5-12 *Semiautomatic submerged arc welding.*

For automatic welding, the torch is attached to the wire feed motor and includes current pickup tips for transmitting the welding current to the electrode wire. The flux hopper is normally attached to the torch. The flux hopper may have magnetically operated valves so that they can be opened or closed by the control system.

Other pieces of equipment sometimes used may be a travel carriage, which can be a simple tractor or a complex moving specialized fixture. A typical travel carriage is shown by Figure 5-13. A flux recovery unit is normally provided to collect the unused submerged arc flux and return it to the supply hopper.

Materials Used

Two materials are used in submerged arc welding, the welding flux, and the consumable electrode wire.

Submerged arc welding flux shields the arc and the molten weld metal from the harmful effects of atmospheric oxygen and nitrogen. The flux contains deoxidizers and scavengers which help to remove impurities from the molten weld metal. Flux also provides a means of introducing alloys into the weld metal. As this molten flux cools to a glassy slag it forms a covering which protects the surface of the weld. The nonmelted portion of the flux does not change its form, its properties are not affected, and it can be recovered and reused. The flux that does melt and forms the slag covering must be removed from the weld bead. This is easily done after the weld has cooled and in many cases will actually peel without requiring special effort for removal. In groove welds, the solidified slag may have to be removed by a chipping hammer.

Fluxes are formulated for specific applications and for specific types of weld deposits. Submerged arc fluxes come in different particle sizes. Many fluxes are not marked for size of particles because the size is designed and produced for the intended application.

FIGURE 5-13 *Submerged arc head on travel carriage.*

Submerged arc welding systems can become quite complex by incorporating additional devices such as seam followers, weavers, work movers etc.

Wire/Flux Classification Per AWS A5.17-69	TYPICAL DEPOSIT CHEMISTRY					TYPICAL MECHANICAL PROPERTIES					
	C	Mn	Si	P	S	Tensile Strength psi	Yield Strength psi	Elong. % in 2"	Reduction of Area %	Charpy V-Notch Impact Value Ft-lb	°F
EL12	0.09	0.50	0.01	0.020	0.025						
F60-EL12	0.06	0.70	0.75	0.025	0.020	70,300	60,100	27.0	48.0	—	
F63-EL12	0.04	1.18	0.48	0.036	0.011	72,000	56,000	35.0	67.0	30	—40
EH14	0.14	1.85	0.04	0.010	0.018						
F62-EH14	0.08	1.05	0.55	0.020	0.016	72,000	58,000	29.0	58.0	30	—20
F72-EH14	0.08	1.80	0.65	0.016	0.018	88,000	73,000	28.0	56.0	24	—20
F64-EH14	0.12	1.17	0.24	0.022	0.021	71,000	57,000	31.0	59.1	24	—60
EM15K	0.15	1.10	0.25	0.022	0.025						
F70-EM15K	0.09	0.93	0.94	0.027	0.022	81,700	67,000	30.0	61.0	—	
F72-EM15K	0.08	1.54	0.79	0.025	0.021	82,000	58,500	30.0	59.0	26	—20
F64-EM15K	0.11	0.78	0.30	0.022	0.025	70,000	55,000	29.5	56.5	21	—60
(same as above)	0.13	1.95	0.04	0.010	Mo.53						
Weld stress	0.07	1.95	0.70	0.020	Mo.35	99,250	84,000	25.0	57.0	23	—20
relieved	0.08	1.17	0.23	0.017	Mo.38	80,000	65,500	27.0	66.2	22	—60
EM13K	0.11	1.20	0.50	0.020	0.019						
F70-EM13K	0.09	1.74	1.17	0.017	0.026	86,000	66,500	26.0	53.2	—	
F64-EM13K	0.10	0.90	0.54	0.016	0.020	70,500	54,000	31.0	62.8	29	—60

FIGURE 5-14 *Typical analysis and mechanical properties of submerged arc flux-wire combinations.*

Information concerning the manufacture of flux is given in Chapter 10.

There is no specification for submerged arc fluxes in use in North America. A way for classifying fluxes, however, is by means of the deposited weld metal produced by various combinations of electrodes and proprietary submerged arc fluxes. This is covered by the American Welding Society Standard. Bare carbon steel electrodes and fluxes for submerged arc welding.[1] In this way fluxes can be designated to be used with different electrodes to provide the deposited weld metal analysis that is desired. Figure 5-14 shows the flux wire combination and the mechanical properties of the deposited weld metal.

The electrodes used for submerged arc welding are covered more completely in Chapter 10.

Deposition Rates and Weld Quality

The deposition rates of the submerged arc welding process are higher than any other arc welding process. Deposition rates for single electrodes are shown by Figure 5-15. There are at least four related factors that control the deposition rate of submerged arc welding, polarity, long stickout, additives in the flux, and additional electrodes. The deposition rate is the highest for direct current electrode negative DCEN. The deposition rate for alternating current is between DCEP and DCEN. The polarity of maximum heat is the negative pole.

The deposition rate with any welding current can be increased by extending the "stickout." This is the distance from the point where current is introduced into the electrode to the arc. When using "long stickout"

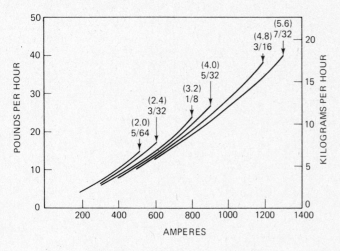

FIGURE 5-15 *Welding deposition rates.*

the amount of penetration is reduced. The deposition rates can be increased by metal additives in the submerged arc flux. Additional electrodes can be used to increase the overall deposition rate.

The quality of the weld metal deposited by the submerged arc welding process is high. The weld metal strength and ductility exceeds that of the mild steel or low-alloy base material when the correct combination of electrode wire and submerged arc flux is used. When submerged arc welds are made by machine or automatically, the human factor inherent to the manual welding processes is eliminated. The weld will be more uniform and free from inconsistencies. In general, the weld bead size per pass is much greater with submerged arc welding than with any of the other arc welding process. The heat input is higher and cooling rates are slower, and, for this reason, gases are allowed more time to escape. Additionally, since the submerged arc slag is lower in density than the weld metal, it will float out to the top of the weld. Uniformity and consistency are advantages of this process when applied automatically.

Several problems may occur when using the semiautomatic application method. The electrode wire may be curved when it leaves the nozzle of the welding gun. This curvature can cause the arc to be struck in a location not expected by the welder. When welding in fairly deep grooves, the curvature may cause the arc to be against one side of the weld joint rather than at the root. This will cause incomplete root fusion and flux will be trapped at the root of the weld. Another problem with semiautomatic welding is the problem of completely filling the weld groove or maintaining exact size, since the weld is hidden and cannot be observed while it is being made. This requires making an extra pass or in some cases too much weld is deposited. Variations in root opening affect the travel speed and if travel speed is uniform the weld may be under- or overfilled in different areas. High operator skill will overcome this problem.

There is another quality problem associated with extremely large single-pass weld deposits. When these large welds solidify the impurities in the melted base metal and in the weld metal all collect at the last point to freeze, which is the centerline of the weld. If there is sufficient restraint and enough impurities are collected at this point, centerline cracking may occur. This can happen when making large single-pass flat fillet welds if the base metal plates are 45° from flat. A simple solution is to avoid placing the parts at a true 45° angle. It should be varied approximately 10° so that the root of the joint is not in line with the centerline of the fillet weld. Another solution is to make multiple passes rather than attempting to make a large weld in a single pass.

Another quality problem has to do with the hardness of the deposited weld metal. Excessively hard weld deposits contribute to cracking of the weld during fabrication or during service. A maximum hardness level of 225 Brinell is recommended. The reason for the hard weld in carbon and low-alloy steels is too rapid cooling, inadequate postweld treatment, or excessive alloy pickup in the weld metal. Excessive alloy pickup is due to selecting an electrode that has too much alloy, or a flux that introduces too much alloy into the weld, or the use of excessively high welding voltages. Caution should be established to avoid these.

In automatic and machine welding defects may occur at the start or at the end of the weld. The best solution is to use runout tabs so that starts and stops will be on the tabs rather than on the product.

Weld Schedules

The submerged arc welding process applied by machine or fully automatically should be done in accordance with welding procedure schedules. Figure 5-16 shows the recommended welding schedules for submerged arc welding using a single electrode on mild and low-alloy steels. These tables can be used for welding other ferrous materials, but were developed for mild steel. All of the welds made by this procedure should pass qualification tests, assuming that the correct electrode

Material Thickness (Gauges, Inches)		Type of Weld See Figure 5-18	Electrode Dia. (2)	Welding Current Amps-DC	Arc Voltage Elec. Pos.	Wire Feed ipm	Travel Speed ipm
16	.063	a Square groove	3/32	300	22	68	100–140
		b Square groove	1/8	425	26	53	95–120
14	.078	a Square groove	3/32	375	23	85	100–140
		b Square groove	1/8	500	27	65	75–85
12	.109	a Square groove	1/8	400	23	51	70–90
		b Square groove	1/8	550	27	65	50–60
		d Fillet	1/8	400	25	51	40–60
10	.140	a Square groove	1/8	425	26	53	50–80
		b Square groove	5/32	650	27	55	40–45
3/16	.188	a Square groove	5/32	600	26	50	40–75
		b Square groove	3/16	875	31	55	35–40
		d Fillet	1/8	525	26	67	35–40
1/4	.250	a Square groove	3/16	800	28	50	30–35
		b Square groove	3/16	875	31	56	22–25
		d Fillet	5/32	650	28	56	30–35
		e Vee groove	3/16	750	30	47	25–40
3/8	.375	b Square groove	3/16	950	32	61	20–25
		f Square groove	3/16	1st pass 500	32	27	30
				2nd pass 750	33	47	30
		e Vee groove	3/16	900	33	57	23–25
		d Fillet	3/16	950	31	61	30–35
1/2	.500	c Vee groove	3/16	975	33	63	12–17
		f Square groove	3/16	1st pass 650	34	40	25
				2nd pass 850	35	54	23–27
		e Vee groove	3/16	950	35	61	18–20
		d Fillet	3/16	950	33	61	14–17
3/4	.750	c Vee groove	7/32	1000	35	49	68
		f Square groove	3/16	1st pass 925	37	59	12
				2nd pass 1000	40	65	11
		e Vee groove	7/32	950	36	46	10–12
		d Fillet	7/32	1000	35	49	6–8
		g Vee groove	7/32	1st pass 950	34	46	15
				2nd pass 750	34	25	22
		h Double vee groove	3/16	1st pass 700	35	42	20–22
				2nd pass 1000	36	65	14–16
1	1.000	g Vee groove	7/32	1st pass 1150	36	58	11
				2nd pass 850	36	40	20
		h Double vee groove	7/32	1st pass 900	36	42	13–15
				2nd pass 1075	36	52	12–14
1-1/4	1.25	h Double vee groove	7/32	1st pass 1000	36	50	13
				2nd pass 1125	37	56	8
1-1/2	1.50	h Double vee groove	7/32	1st pass 1050	36	51	9
				2nd pass 1125	37	56	7

FIGURE 5-16 *Welding procedure schedules.*

THE ARC WELDING PROCESSES—USING CV POWER

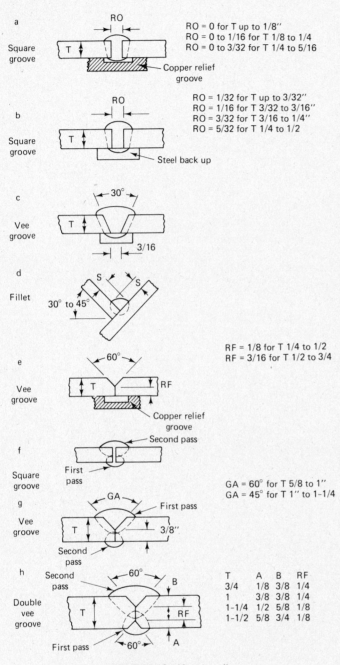

a

Square groove

RO = 0 for T up to 1/8"
RO = 0 to 1/16 for T 1/8 to 1/4
RO = 0 to 3/32 for T 1/4 to 5/16

Copper relief groove

b

Square groove

RO = 1/32 for T up to 3/32"
RO = 1/16 for T 3/32 to 3/16"
RO = 3/32 for T 3/16 to 1/4"
RO = 5/32 for T 1/4 to 1/2

Steel back up

c

Vee groove

30°

3/16

d

Fillet

30° to 45°

e

Vee groove

60°

RF = 1/8 for T 1/4 to 1/2
RF = 3/16 for T 1/2 to 3/4

Copper relief groove

f

Square groove

Second pass

First pass

g

Vee groove

GA

First pass

3/8"

Second pass

GA = 60° for T 5/8 to 1"
GA = 45° for T 1" to 1-1/4

h

Double vee groove

Second pass

60°

First pass

60°

T	A	B	RF
3/4	1/8	3/8	1/4
1	3/8	3/8	1/4
1-1/4	1/2	5/8	1/8
1-1/2	5/8	3/4	1/8

FIGURE 5-16 *(Continued).*

and flux have been selected. If the schedules are varied more than 10%, qualification tests should be performed to determine the weld quality.

Welding Variables

The welding variables for submerged arc welding are much similar to the other arc welding processes, with several exceptions. The variables will be covered in detail in Section 5-5.

In submerged arc welding the electrode type and the flux type are usually based on the mechanical properties required by the weld. The electrode and flux combination selection is based on Figure 5-14 to match the metal being welded. The electrode size is related to the weld joint size and the current recommended for the particular joint. This must also be considered in determining the number of passes or beads for a particular joint. Welds for the same joint dimension can be made in many or few passes; this depends on the weld metal metallurgy desired. Multiple passes are more expensive but usually deposit higher-quality weld metal. The polarity is established initially and is based on whether maximum penetration or maximum deposition rate is required.

The major variables that affect the weld involve heat input and include the welding current, the arc voltage, and the travel speed. Welding current is the most important. For single-pass welds the current should be sufficient for the desired penetration without burn-through. The higher the current, the deeper the penetration. In multipass work the current should be suitable to produce the size of the weld expected in each pass. The welding current should be selected based on the electrode size. The higher the welding current, the greater the melt-off rate (deposition rate).

The arc voltage is varied within narrower limits than welding current. It has an influence on the bead width and shape. Higher voltages will cause the bead to be wider and flatter. Extremely high arc voltage should be avoided since it can cause cracking. This is because an abnormal amount of flux is melted and excess deoxidizers may be transferred to the weld deposit lowering its ductility. Higher arc voltage also increases the amount of flux consumed. The low arc voltage produces a stiffer arc that improves penetration, particularly in the bottom of deep groves. If the voltage is too low, a very narrow bead will result. It will have a high crown and the slag will be difficult to remove.

Travel speed has an influence on both bead width and on penetration. Faster travel speeds produce narrower beads that have less penetration. This can be an advantage for sheet metal welding where small beads and minimum penetration are required. If speeds are too fast, however, there is a tendency for undercut and porosity, since the weld freezes quicker. If the travel speed is too slow, the electrode stays in the weld puddle too long which will create poor bead shape and may cause excessive spatter and flash through the layer of flux.

The secondary variables include the angle of the electrode to the work, the angle of the work itself, the thickness of the flux layer, and, most important, the distance between the current pickup tip and the arc. This latter factor, called electrode "stickout," has a

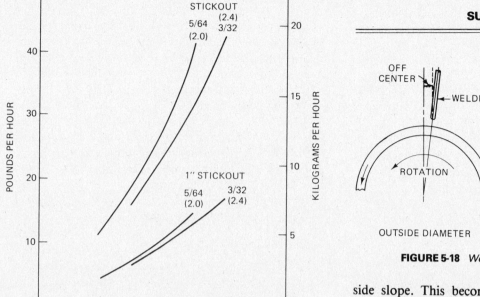

FIGURE 5-17 *Stick out vs deposition rate.*

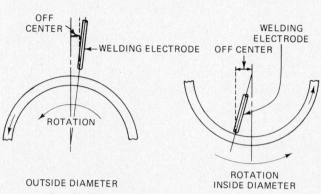

FIGURE 5-18 *Welding on rotating circular parts.*

considerable effect on the weld. Normally, the distance between the contact tip and the work is 1 to 1-1/2 in. (25 to 38 mm). If the stickout is increased beyond this amount it will cause preheating of the electrode wire which will greatly increase the deposition rate. As stickout is increased, the penetration into the base metal decreases. This factor must be given serious consideration because in some situations the penetration is required. The relationship between stickout and deposition rate is shown by Figure 5-17.

The depth of the flux layer must be considered. If it is too thin, there will be too much arcing through the flux or arc flash. This also may cause porosity. If the flux depth is too heavy, the weld may be narrow and humped. On the subject of flux, too many fines (small particles of flux) in the flux can cause surface pitting since the gases generated in the weld may not be allowed to escape. These are sometimes called *pock marks* on the bead surface.

Tips for Using the Process

One of the major applications for submerged arc welding is on circular welds where the parts are rotated under a fixed head. These welds can be made on the inside or outside diameter. Submerged arc welding procudes a large molten weld puddle and molten slag which tends to run. This dictates that on outside diameters the electrode should be positioned ahead of the extreme top, or 12 o'clock position, so that the weld metal will begin to solidify before it starts the down-

side slope. This becomes more of a problem as the diameter of the part being welded is smaller. Improper electrode position will increase the possibility of slag entrapment or a poor weld surface. The angle of the electrode should also be changed and pointed in the direction of travel of the rotating part. When the weld-

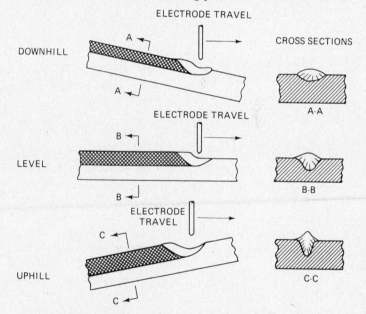

FIGURE 5-19 *Angle of slope of work vs weld.*

ing is done on the inside circumference the electrode should be angled so that it is ahead of bottom center, or the 6 o'clock position. Figure 5-18 illustrates these two conditions.

Sometimes the work being welded is sloped downhill or uphill to provide different types of weld bead contours. If the work is sloped downhill, the bead will have less penetration and will be wider. If the weld is sloped uphill the bead will have deeper penetration and will be narrower. These are based on all other factors remaining the same. This information is shown by Figure 5-19.

The weld will be different depending on the angle of the electrode with respect to the work when the work is level. This is the travel angle which can be a drag or push angle and has a definite effect on the bead contour and weld metal penetration. Figure 5-20 shows the relationship.

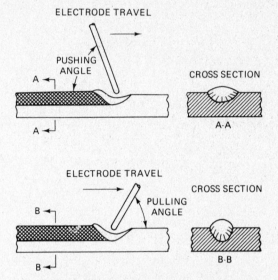

FIGURE 5-20 *Angle of electrode vs weld.*

One side welding with complete root penetration can be obtained with submerged arc welding. When the weld joint is designed with a tight root opening and with a fairly large root face, high current and electrode positive should be used. If the joint is designed with a root opening and a minimum root face, it will be necessary to use a backing bar since there will be nothing to support the molten weld metal. The molten flux is very fluid and will run through narrow openings. If this happens the weld metal will follow and the weld will burn through the joint. Backing bars are needed whenever there is a root opening and a minimum root face.

Copper backing bars are useful when welding thin steel. Without backing bars, the weld would tend to melt through and the weld metal would fall away from the joint. The backing bar holds the weld metal in place until it solidifies. The copper backing bars may be water cooled to avoid the possibility of melting and copper pickup in the weld metal. For thicker materials, the backing may be submerged arc flux or other specialized type flux. More details of weld backing will be given in Chapter 20.

Safety Considerations

Safety precautions for submerged arc welding are somewhat fewer and less exacting than for other arc welding processes, because of the nature of the submerged arc process and because most submerged arc welding is applied automatically. See Chapter 3, "Safety and Health," for details.

The welding arc is normally not visible in the submerged arc welding process. Only small amounts of sparks or flash are produced therefore it is not necessary to wear a welding face helmet. It is good practice, however, to wear tinted safety glasses.

The submerged arc welding process is normally less hazardous than other arc welding processes.

Limitations of the Process

A major limitation of submerged arc welding is its limitation of welding positions. The other limitation is that it is only used to weld primarily mild and low-alloy high-strength steels.

The high-heat input, slow-cooling cycle can be a problem when welding quenched and tempered steels. The heat input limitation of the steel in question must be strictly adhered to when using submerged arc welding. This may require the making of multipass welds where a single pass weld would be acceptable in mild steel. In some cases, the economics may be reduced to the point where flux-cored arc welding or some other process should be considered.

In semiautomatic submerged arc welding, the inability to see the arc and puddle can be a disadvantage in reaching the root of a groove weld and properly filling or sizing.

Variations of the Process

There are a large number of variations to the process that give it additional capabilities. Some of the more popular variations are:

1. Two-wire systems—same power source.
2. Two-wire systems—separate power source.
3. Three-wire system—separate power source.
4. Strip electrode for surfacing.
5. Iron powder additions to the flux.
6. Long stickout welding.
7. Electrically "cold" filler wire.

The multiwire systems offer advantages since deposition rates and travel speeds can be improved by using more electrodes. Figure 5-21 shows the two methods of utilizing two electrodes, one with a single-power source and one with two-power sources. When a single-power source

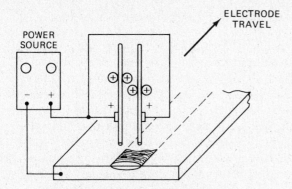

POWER
SOURCE

ELECTRODE
TRAVEL

TRANSVERSE ELECTRODE POSITION

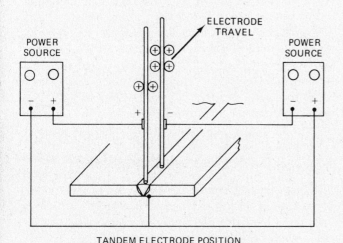

POWER
SOURCE

ELECTRODE
TRAVEL

POWER
SOURCE

TANDEM ELECTRODE POSITION

FIGURE 5-21 *Two electrode wire systems.*

fabricated beams. Extremely high currents can be used with correspondingly high travel speeds and deposition rates.

The strip welding system is used to overlay mild and alloy steels usually with stainless steel. A wide bead is produced that has a uniform and minimum penetration. This process variation is shown by Figure 5-22. Overlay with strip electrodes is popular in Europe and is being used by the nuclear industry in North America. It is used for overlaying the inside of vessels to provide the corrosion resistance of stainless steel while utilizing the strength and economy of the low-alloy steels for the wall thickness. A strip electrode feeder is required and special flux is normally used. When the width of the strip is over 2 in. (50 mm), a magnetic arc oscillationg device is employed to provide for even burnoff of the strip and uniform penetration.

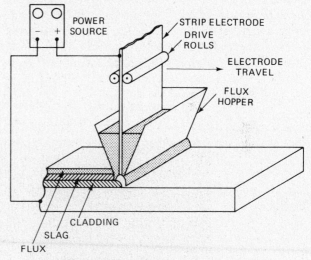

POWER
SOURCE

STRIP ELECTRODE
DRIVE
ROLLS

ELECTRODE
TRAVEL

FLUX
HOPPER

CLADDING

SLAG

FLUX

FIGURE 5-22 *Strip electrode on surfacing.*

is used, the same drive rolls are used for feeding both electrodes into the weld. When two power sources are used, individual wire feeders must be used to provide electrical insulation between the two electrodes. With two electrodes and separate power it is possible to utilize different polarities on the two electrodes or to utilize alternating current on one and direct current on the other. The electrodes can be placed side by side, is what is called *transverse electrode position,* or they can be placed one in front of the other in the *tandem electrode position.*

The two-wire tandem electrode position with individual power sources is used where extreme penetration is required. The leading electrode is positive with the trailing electrode negative. The first electrode creates a digging action and the second electrode will fill the weld joint. When two DC arcs are in close proximity, there is a tendency for arc interference between them. In some cases, the second electrode is connected to alternating current to avoid the interaction of the arc. The three-wire tandem system normally uses AC power on all three electrodes connected to three-phase power systems. The three-wire systems are used for making high-speed longitudinal seams for large-diameter pipe and for

Another way of increasing the deposition rate of submerged arc welding is to add iron base ingredients to the joint under the flux. The iron in this material will melt in the heat of the arc and will become part of the deposited weld metal. This greatly increases deposition rates without decreasing weld metal properties. Metal additives can also be used for special surfacing applications. This variation can be used with single-wire or multiwire installations. Figure 5-23 shows the increased deposition rates attainable.

Another variation is the use of an electrically "cold" filler wire fed into the arc area. The "cold" filler rod can be solid or flux cored to add special alloys to the weld metal. By regulating the addition of the proper

145

THE ARC WELDING PROCESSES—USING CV POWER

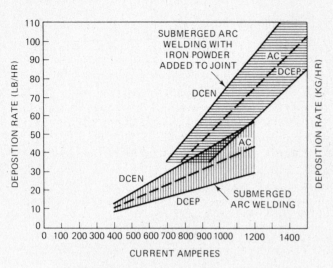

FIGURE 5-23 *Welding with iron powder additives.*

material, the properties of the deposited weld metal can be improved. It is possible to utilize a flux-cored wire for the electrode or for one of the multiple electrodes to introduce special alloys into the weld metal deposit. Each of these variations requires special engineering to ensure that the proper material is added to provide the desired deposit properties.

The submerged arc welding process is widely used in the manufacture of most heavy steel products. These include pressure vessels, boilers, tanks, nuclear reactors, chemical vessels, etc. Another use is in the fabrication of trusses and beams. It is used for welding flanges to the web and this is shown by Figure 5-24. The heavy equipment industry is a major user of submerged arc welding. An unusual application is the simultaneous welding of two joints on a triangular shaped boom of an excavator shown by Figure 5-25. Another application is the welding together of forged halves to make tractor rollers as shown by Figure 5-26. Shipbuilding is another major user of submerged arc welding, using both machine and automatic methods. Fully automatic submerged arc machines splice plates together and weld stiffeners to plates. Much of the machine welding is done using welding heads mounted on small carriages, such as those shown by Figure 5-27. The overlay of surfaces for corrosion resistance and for wear resistance is also a major use for submerged arc welding.

5-3 GAS METAL ARC WELDING (GMAW)

Gas metal arc welding is a process in which coalescence is produced by heating with an arc between a continuous filler metal (consumable) electrode and the work. Shielding is obtained entirely from an externally supplied gas or gas mixture.

FIGURE 5-24 *Welding a bridge girder.*

FIGURE 5-25 *Welding triangular excavator boom—two heads.*

FIGURE 5-26 *Welding track rolls forgings—high volume production.*

FIGURE 5-27 *Using welding head on carriage in shipyard.*

THE ARC WELDING PROCESSES—USING CV POWER

There are a number of variations of the gas metal arc welding process and the process has been given many different trade names which tend to create confusion. For example, variations are called MIG welding, CO_2 welding, fine wire welding, spray arc welding, pulsed arc welding, electrogas welding, and short-circuit arc welding. These variations and methods are of sufficient importance that each will be more clearly defined and explained later in the chapter.

Principles of Operation

The gas metal arc welding process is shown by Figure 5-28. The gas metal arc welding process utilizes the heat of an arc between a continuously fed consumable electrode and the work to be welded. The heat of the arc melts the surface of the base metal and the end of the electrode. The metal melted off the electrode is transferred through the arc to the work where it becomes the deposited weld metal. Shielding is obtained from an envelope of gas, which may be an inert gas, an active gas, or a mixture. The shielding gas surrounds the arc area to protect it from contamination from the atmosphere. The electrode is fed into the arc automatically, usually from a coil. The arc is maintained automatically and travel can be manually or by machine.

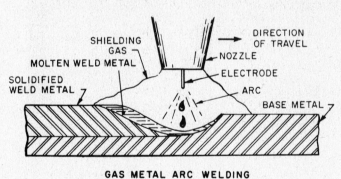

GAS METAL ARC WELDING

FIGURE 5-28 *Process diagram—GMAW.*

The metal being welded dictates the composition of the electrode and the shielding gas. The shielding gas and the size of the electrode wire affects the type of metal transfer. The type of metal transfer is one way of identifying some variation of the process.

The gas metal arc welding process, introduced in the late 1940s, has become one of the most popular arc welding processes. Early development was for welding of aluminum using inert gas for shielding. Hence the name "metal inert gas," or MIG welding. For welding steels inert gases seemed to be too expensive and an active gas, CO_2, was selected. The selection of CO_2 was based on the analysis of gases formed by the disintegration of the coatings of covered electrodes. This variation, named CO_2 welding, was well adapted for welding mild steel in the flat position utilizing relatively large size 1/16 in. (1.6 mm) electrode wires. This variation never became too popular with welders because of the high heat and necessity for high travel speeds. The metal transfer was globular and spatter was greater than desired. Efforts to refine this variation led to an all-position capability still using CO_2 gas shielding but with smaller diameter electrodes in the order of 0.035-in. (0.9 mm) and 0.045-in. (1.1 mm) diameter. This variation is called short circuiting metal transfer or fine wire welding. Efforts to improve weld appearance led to shielding gas mixtures of argon and CO_2, which provided a smoother, nicer appearing weld surface.

Further development with different shielding gases led to the "spray arc" variation, which utilized larger diameter electrodes, with a shielding gas mixture of 95% argon and 5% oxygen (other mixtures were 98-2 and 99-1 argon–oxygen mixtures). This gas mixture produced a spray type metal transfer and welds with an extremely smooth surface. This variation has become extremely popular.

Another variation is electrogas welding which is a "method of gas metal arc welding wherein molding shoes confine the weld metal for vertical position welding. Additional shielding may or may not be obtained. Shielding is obtained from an externally supplied gas or gas mixture." This is a single-pass technique for automatic vertical welding.

Another variation is pulsed arc welding, whereby the welding current is pulsed at regular intervals to create a discrete transfer of metal across the arc rather than a random transfer which occurs in the other variations.

The major advantages of gas metal arc welding are:

High deposition rates when related to shielded metal arc welding.

High operator factor with respect to shielded metal arc welding.

High utilization of filler material.

Elimination of slag and flux removal.

Reduction of smoke and fumes.

May be automated for high-operator factor.

Skill level in the semiautomatic method of application slightly lower than that required for manual shielded metal arc welding.

Extreme versatility and wide and broad application ability.

The gas metal arc welding process is the fastest-growing welding process in use today. Its growth is based on replacing shielded metal arc welding for welding thin metals, and for replacing gas tungsten arc welding for nonferrous metals. It is replacing submerged arc in automatic applications. It has replaced gas welding and torch brazing in many uses. It has also replaced resistance welding in many sheet metal applications and can be utilized for producing arc spot welds, as well as seam welds.

Normal Method of Applying and Position Capabilities

The most popular method of applying is the semi-automatic method where the welder provides manual travel and guidance of a welding gun. Second is the fully automatic method where the welding operation is automated. This process can be applied as machine welding; however, this is of minor popularity. The process cannot be applied manually. The electrogas variation is machine welding since the welding operator monitors the operation continuously.

The gas metal arc welding process is an all-position welding process. However, each of the variations has its own position capabilities, depending on electrode size and metal transfer. The CO_2 welding variation, utilizing large electrode wires, is used primarily in the flat and horizontal fillet position. The spray arc variation is normally used in the flat and horizontal position. It can be used in the vertical and overhead positions if smaller size electrodes are employed. The electrogas method is limited to the vertical position. The short circuiting and pulsed variations can be used in all positions.

Weldable Metals and Thickness Range

The gas metal arc welding process can be used to weld most metals. Carbon dioxide welding is restricted to steels. Electrogas is normally used only for welding steels. The pulsed variation can be used for most metals. Electrodes are matched to the base metals. This information is summarized by Figure 5-29. The process can

also be used for surfacing and for buildup using special metals for bearing surfaces, corrosion-resistant surfaces, and so on.

Metal thickness from 0.005 in. (0.13 mm) upward can be welded. The short circuiting, or fine-wire variation, and the pulsed arc variation are used for welding the thinner materials in all positions. Heavier thicknesses can be welded with large wire CO_2 variation. Weld grooves and multiple-pass technique will allow welding of practically unlimited thickness. This situation is summarized by the chart shown in Figure 5-30. The extreme versatility of the process and its variations allow welding of the thinnest up to the thickest, by choosing the correct electrode wire type and size and gas for shielding.

Joint Design

The gas metal arc welding process can utilize the same joint design details that are used for the shielded metal arc welding process. These joint details will be given in Chapter 15. For maximum economy and efficiency of gas metal arc welding, weld joint details, specifically groove welds, should be modified. The selection of the proper joint detail depends on the variation that is to be used. The diameters of the electrodes employed by gas metal arc welding are smaller than those employed for shielded metal arc welding. Because of this, the groove angles can be reduced as shown by Figure 5-31. Reducing groove angles will still allow the electrode to be directed to the root of the weld joint so that complete penetration will occur.

The different variations require special attention concerning weld design. The CO_2 variation provides extremely deep penetrating qualities and in designing fillet welds, the size of the fillet can be reduced at least one size when converting from shielded metal arc welding to CO_2 welding. The electrogas method requires square groove welds with a wide root opening to accommodate the current pickup tube.

The variation using inert gas on nonferrous metals can use the standard joint details recommended for shielded metal arc welding, except that the groove angle should be reduced. The joint designs used for pipe welding with shielded metal arc welding or gas welding are normally used for gas metal arc welding with the fine wire variation.

Base Metal	Short Circuiting Arc	Spray Arc	CO_2	Pulsing	Chapter
Aluminums	No	Yes	No	Yes	12-1
Bronzes	No	Yes	No	Yes	12-3
Copper	No	Yes	No	Yes	12-3
Copper nickel	No	Yes	No	Yes	12-3
Cast iron	Yes	No	No	—	12-1
Magnesium	No	Yes	No	Yes	12-5
Inconel	No	Yes	No	Yes	12-6
Nickel	No	Yes	No	Yes	12-6
Monel	No	Yes	No	Yes	12-6
Low carbon steel	Yes	Yes	Yes	—	11-2
Low alloy steel	Yes	Yes	Yes	—	11-3
Medium carbon	Yes	Yes	Yes	—	11-4
Stainless steel	Yes	Yes	No	—	11-6
Titanium	No	Yes	No	—	12-9

FIGURE 5-29 *Base metals weldable by the gas metal arc process variations.*

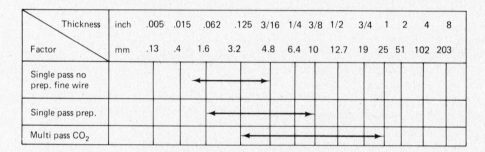

Thickness	inch	.005	.015	.062	.125	3/16	1/4	3/8	1/2	3/4	1	2	4	8
Factor	mm	.13	.4	1.6	3.2	4.8	6.4	10	12.7	19	25	51	102	203
Single pass no prep. fine wire				←———————→										
Single pass prep.					←—————————→									
Multi pass CO_2					←——————————————————→									

FIGURE 5-30 *Base metal thickness range.*

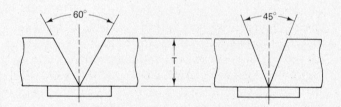

FIGURE 5-31 *Weld joint design differences.*

Welding Circuit and Current

The welding circuit employed for gas metal arc welding is shown by Figure 5-32. It uses a wire feeder system that controls the electrode wire feed and welding arc, as well as the flow of shielding gas and cooling water. The power supply is normally the constant voltage (CV) type. A gun or torch is used for directing the electrode and shielding gas to the arc area. For the electrogas method travel mechanism and retaining shoes are also required.

The gas metal arc welding process uses direct current. Alternating current has not been successfully used. Direct current is normally used with the electrode positive DCEP (reverse polarity). Direct current electrode negative DCEN (straight polarity) can be used with special emissive coated electrode wires which provide for better electron emissions. DCEN is rarely used because the emissive coated electrodes are more expensive.

The fine wire variation became popular when the constant voltage system of welding power was introduced. The constant voltage system reduced the complexity of the wire feed control circuits and eliminated electrode burnback to the contact tip or stubbing to the work. It also provided positive arc starting.

The pulsed current variation requires a special power source which changes from a lower to a higher current at a frequency equal to or double that of the line frequency. This is normally 50 or 60 Hertz and 100 or 120 Hertz.

The welding current varies from as low as 20 amperes at a voltage of 18 volts to as high as 750 amperes at an arc voltage of 50 volts. This broad range of current and voltage encompasses all of the variations.

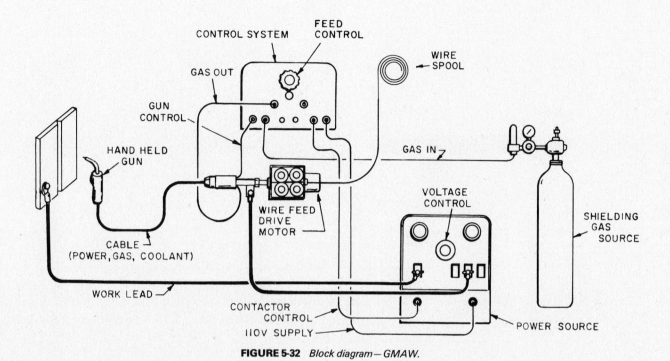

FIGURE 5-32 *Block diagram—GMAW.*

Equipment Required to Operate

The equipment required for gas metal arc welding is shown by Figure 5-32. These consist of (1) the power source, (2) the electrode wire feeder and control system, (3) the welding gun and cable assembly for semiautomatic welding or the welding torch for automatic welding, (4) the gas and water control system for the shielding gas and cooling water when used, and (5) travel mechanism and guidance for automatic welding. There is the extra factor of molding shoes when using the electrogas method.

The power source for gas metal arc welding shall be rated for 100% duty cycle.

Either a generator power source or a transformer-rectifier power source can be used; both are equally satisfactory. For the fine wire variation the 200 ampere size machine is normally used. CO_2 welding and spray arc welding require higher-current power sources up to 650 amperes; 300 and 500 amp machines are the most

popular. For field use, an engine-driven generator power source is used.

The CV type power source is normally used. The slope of the volt-ampere characteristic curve is different for the process variations. Many power sources designed for the fine wire process variation may have adjustable taps or other controls to vary the slope of the output curve. The smaller machines are designed for the fine-wire process.

The best selection is the constant-speed wire feeder with the CV power source since it provides a self-regulating arc. When using the pulsed-current type power source the constant-speed wire feeder is normally employed.

The welding torch or gun assembly is used to carry the electrode, the welding current and the shielding gas to the arc. For the fine-wire variation air-cooled welding guns are employed. When larger currents and larger-diameter electrodes are used, with CO_2 shielding gas, air-cooled guns are also used, since CO_2 is a cooling medium for the welding gun. When inert gas or the argon-oxygen mixture is used at higher currents the gun must be water cooled to avoid overheating. Both pistol type guns and goose neck type guns are employed for gas metal arc welding. Figure 5-33 shows both types of guns which may be air cooled or water cooled.

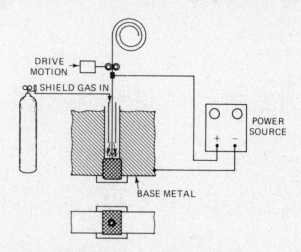

FIGURE 5-34 *Electrogas welding diagram.*

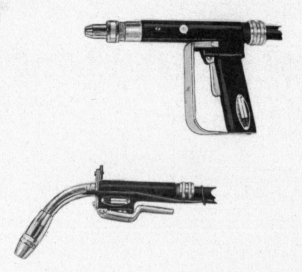

FIGURE 5-33 *Types of welding guns for GMAW.*

For automatic welding the torch can be either air cooled or water cooled. For CO_2 welding the torch is normally air cooled and the CO_2 gas may be introduced by a side delivery type nozzle. When welding nonferrous metals, a concentric type nozzle is utilized. When this type is used, water cooling is normally employed. Water cooling is usually required when utilizing inert gas or argon-oxygen gas mixtures.

Other pieces of equipment sometimes used consist of travel mechanisms and fixtures. More details of fixtures are given in Chapter 9, on automatic welding.

The electrogas method utilizes the molding shoes and traveling mechanism for automatically producing welds in a single pass in the vertical position. This is shown by Figure 5-34.

Materials Used

Two materials are used for the gas metal arc welding process—the filler metal and the shielding gas. The electrode and the shielding gas must be carefully selected with regard to the base metal that is to be welded and also with regard to the process variation that is to be employed.

The filler metal or electrode wire must be selected based on the metal that is to be welded. It is normally related to the strength requirements of the deposited weld metal, as well as to the composition. The electrodes, for most applications, are covered by the American Welding Society filler metal specifications. The following factors govern the selection of the electrode wire.

1. Base Metal To Be Welded: The type or composition and mechanical properties of the base metal must be given major consideration.

2. Thickness and Joint Design: Thicker sections of base metal and complex joint designs usually require filler metals which provide high weld metal ductility.

3. Surface Conditions: The surface of the base metal to be welded whether it is scaly, rusty, etc., has an important effect on the type of electrode wire to be used.

4. Specifications or Service Conditions: Specifications may dictate the type of electrode wire to be used. If specifications are not involved it is important to consider the service requirements that the weldment will encounter.

For mild and low-alloy steels the AWS Specification, "Carbon Steel Electrodes for Gas Metal Arc Welding"[2] should be consulted. This specification provides the composition of the welding electrode and the mechanical properties of the deposited weld metal. It is impossible to list all of the types and analyses of electrode wires available. The proper electrode wire for matching base metals is covered in detail in the chapter on the specific base metal. The size of the electrode wire to be used depends upon the welding position and the variation of the gas metal arc welding process.

The basis for selecting the shielding gas involves the type of electrode, the type of base metal, the welding position, the variation of the process, and the desired weld quality. The recommended shielding gases for different metals and process variations will be covered in Chapter 13 or 14 for the particular metal to be welded.

In order to establish a basis for selection of process variation, it is necessary to know the capabilities and normal applications for each of the process variations. Figure 5-35 is a chart which shows the variations, the type of metal transfer, the base metals that can be welded, the welding position capabilities, and the recommended welding shielding gas. This will simplify the selection of materials necessary to utilize each variation of gas metal arc welding.

Deposition Rates and Weld Quality

Each of the variations has a considerable range of

deposition rates based on the weld procedure employed. Figure 5-36 shows the relationship of deposition rates for the variations and the different electrode sizes that would be used. This chart is based on the utilization of carbon steel base metals and electrodes.

For welding nonferrous metals, deposition rates vary considerably due to the density of the metals being welded.

The deposition rates of gas metal arc welding are higher for the same welding currents than are obtained with shielded metal arc welding. These higher rates occur because there is no electrode coating that must be melted. The current density on the small diameter electrode wires is much higher than with covered electrodes, which contributes to the higher deposition rates for the same welding current. The tip-to-work distance

Process Variation	MIG	CO_2	Fine Wire	Spray Arc	Pulsing
Shielding Gas	Inert Gas	CO_2	CO_2 or CO_2 + Argon (C-25)	Argon + Oxygen (1 to 5%)	Inert Gas
Metal transfer	Various	Globular	Short circuiting	Spray	Discrete
Metals to be welded	Al. & Al. alloys Stainless steels nickel & nickel alloys—copper alloys, Ti, etc.	Low carbon & med. carbon steel—Low alloy high strength steels	Low carbon & medium carbon steels—low alloy high strength steels—some stainless steels.	Low carbon & med. carbon steels—Low alloy high strength steels	Alum. Nickel Steels Nickel Alloys
Metal thickness	12 gage (.109″) to 3/8″ without bevel preparation max. thickness—practically unlimited.	10 gage (.140″) Up to 1/2″ without bevel preparation	20 gage (.038″). Up to 1/4″. Economical in heavier metals for vertical & overhead welding.	1/4″ to 1/2″ with no preparation. Max thickness practically unlimited.	Thin to unlimited thickness
Welding positions	All positions.	Flat & horizontal	All positions (also for pipe welding)	Flat & horizontal with small electrode wire all positions	All position
Major advantages	Welds most nonferrous metals—min. clean-up	Low cost gas-high travel speed deep penetration	Thin material—will bridge gaps—min. clean-up	Smooth surface—deep penetration high travel speed	Uses larger electrode
Limitations	Cost of gases	Spatter removal sometimes required	Uneconomical in heavy thickness-except out of position	Position—min. thickness	Expension power source
Appearance of weld	Fairly smooth convex surface	Relatively smooth some spatter	Smooth surface—relatively minor spatter	Smooth surface—minimum spatter	Smooth surface
Travel speeds	Up to 100 IPM	Up to 250 IPM	Max. 50 IPM-semi-auto.	Up to 150 IPM	Up to 100 IPM
Range of electrode Wire sizes—inches	Dia.-.035, .045, 1/16, 3/32	Dia.-.045, 1/16 5/64, 3/32	Dia.-.030, .035, .045	Dia.-.035, .045, 1/16, 3/32	1/16, 5/64, 3/32, 1/8

FIGURE 5-35 *Variations of the gas metal arc welding process.*

THE ARC WELDING PROCESSES—USING CV POWER

affects deposition rates, and, as the distance is increased, the preheating of the electrode wire contributes to higher deposition rates. Excessive stickout cannot be used without losing the ability to accurately point the electrode wire in the joint.

The quality of welds made with the gas metal arc welding process can be extremely high. Quality of the deposited weld depends on the selection of the electrode wire, the cleanliness of electrode wire, the cleanliness of the welding joint, the welding procedure, the welding position, and the skill of the welder.

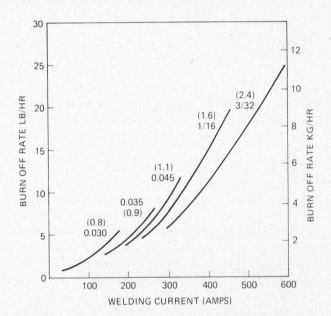

FIGURE 5-36 *Deposition rates—variations and steel electrode size.*

Gas metal arc welding is a "no-hydrogen" or a "low-hydrogen" welding process. There is no hydrogen in the shielding gas atmosphere or in any component involved in making the weld. This is based on the use of welding grade shielding gas and clean dry electrodes on clean, dry surfaces. Another reason for the high quality is the absence of slags and fluxes which might become entrapped in the weld.

Certain factors detract from the quality of the deposited weld metal. One is the possibility of reduced efficiency of the gas shielding envelope. Breezes in the weld area may blow the shielding gas envelope away and allow the atmosphere to come in contact with the molten puddle. The loss of shielding gas is normally noticed by the welder. This will cause a dirty-appearing weld or will create an unstable welding condition or

gross porosity. Another factor is impure gas that contains water vapor, oil, or other impurities. Welding with electrode wire that is dirty, oily, or greasy will also contribute to inferior weld deposits. Anything that reduces the efficiency of the gas shield can contribute to the reduced quality of the weld. Welding on dirty surfaces—wet, oily, or otherwise—will reduce the quality of the weld metal.

Weld Schedules

The type of metal transfer across the arc is related to the particular gas metal arc welding variation. There are four major types of metal transfer. These are (1) spray transfer, (2) globular transfer, (3) short circuiting arc transfer and (4) pulsed metal transfer, and they are shown by Figure 5-37. The type of metal transfer depends upon the welding current density which relates to the electrode wire size and the welding current. The shielding gas type and the arc voltage are important variables in the four modes. Transition areas exist between different types of metal transfer in which both types may occur simultaneously.

Spray transfer occurs when very fine droplets or particles of electrode metal are transferred rapidly through the welding arc from the electrode to the puddle. This type of transfer is experienced when using argon plus small amounts of oxygen gas mixtures. Higher-current density also contributes to spray transfer. The droplets in spray transfer are smaller than the diameter of the electrode. The spray transfer occurs because the argon shielding gas exhibits a pinch effect on the molten tip of the electrode wire. This pinch effect physically limits the size of the molten ball that can form on the end of the electrode wire and therefore only small droplets of metal form.

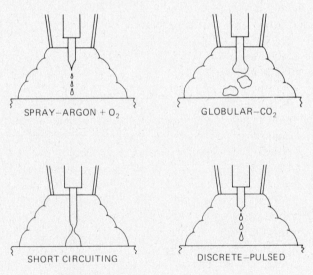

FIGURE 5-37 *The four types of metal transfer.*

The globular type of metal transfer is encountered primarily when CO_2 or CO_2 with argon additions. In the shielding atmosphere containing carbon dioxide an entirely different type of force occurs than with argon shielding gas. The CO_2 arc is longer, since an argon atmosphere tends to reduce arc length, and a force called "cathode jet" occurs. In DCEP the cathode jet originates from the work piece and actually supports a ball on the tip of the electrode. This molten ball can grow in size until its diameter reaches 1-1/2 to 2 times the diameter of the electrode. When it grows to a sufficient size it separates from the electrode and is transferred across the arc. As the globule is transferred across the arc, it is subjected to forces in the arc and takes on an irregular shape and a rotating motion. This irregular shape and motion sometimes causes the globule to reconnect with the electrode, as well as the work. This may cause a short circuit which momentarily extinguishes the arc. With globular metal transfer, the arc is occasionally extinguished. Globular transfer produces spatter which comes from both the molten puddle and the transfer metal.

The third type of metal transfer is known as short circuiting transfer. In this type the molten tip of the electrode wire is supported by the cathode jet mentioned previously. However, the electrode wire is feeding at such a high rate of speed that the molten tip of the electrode will occasionally come in contact with the puddle before the globule can separate from the electrode. At this point, a short circuit occurs and the surface tension of the molten puddle tends to draw the molten metal of the electrode to the puddle. This causes the metal to bridge the gap between the electrode and the puddle. The electrode will then separate from the globule and re-establish an arc. On occasion, the electrode wire will contact the puddle and the electrode wire will act like a fuse and literally explode because of the high-current density. When this explosion occurs, the arc gap is re-established. These conditions continue randomly; metal is transferred by the electrode wire actually touching the puddle, the globule will touch and flow into the puddle, and the fuse action will cause the arc to be re-established. This type of metal transfer allows for all-position welding and is obtained only with specific welding procedures, normally limited to a maximum of 200 amperes on 0.035 in. (0.9 mm) electrode wire. CO_2 or CO_2 gas mixtures are used.

The fourth type of metal transfer is known as the pulsed-transfer. In pulsed-transfer the welding current is rapidly switched from a high or pulsed current to a low or background current level. The rate of switching is equal to the line frequency or to double the line frequency. The high and low currents must be ad-

justed for stable operation. During the high-current pulse the metal is transferred across the arc. During the low level the arc is maintained but there is no metal transfer; thus, the transfer is regulated and matches the pulsing rate. This regular frequency of metal transfer causes discrete droplets to cross the arc. One advantage is that larger-than-normal electrode wires can be utilized with the metal transferred across the arc in small particles at a regular rate. This has advantages for aluminum welding and for welding high-nickel alloy steels. Argon is usually the shielding gas.

The welding schedule for the short circuiting transfer or with fine-wire welding is shown by Figure 5-38. This schedule can be utilized on all types of joint details and is relatively narrow in range based on the electrode size, the welding current, and the voltage. It enables all-position welding of carbon steels on the metal thicknesses shown. The same conditions can be used on thicker metals; however, in the flat position, this variation becomes less economical. For vertical and overhead welding, it is more productive than shielded metal arc welding.

The welding schedules for CO_2 welding are shown by Figure 5-39. Here, welding currents are much higher and deposition rates and productivity are greatly increased. It is normally used in the flat and horizontal position. The basic difference between the two welding procedure schedules is the position capabilities. Note the extremely high currents that can be used on carbon steels.

For spray transfer, high currents can be employed and these data are shown by the welding procedure schedules in Figure 5-40. This is a highly productive variation for welding mild and low-alloy steels. Welding procedure schedules for nonferrous metals will be found in Chapter 13.

The pulsed-current mode of operation is a lower productive variation of the process. The welding procedure schedule depends on the type of equipment used. Pulsed arc transfer is used only if the other types cannot be used.

Welding Variables

Welding procedure variables are essentially the same for all of the continuous electrode feed arc welding processes. They are items within the welding procedures that can be adjusted or changed in order to control the results.

Material Thickness (1)			Electrode Dia.		Welding Current Amps-DC	Arc Voltage Elec. Pos	Wire Feed ipm	Travel Speed ipm	Shielding Gas Flow CFH (3)
Fraction	Decimal	mm	in.	mm					
24 ga.	0.025	0.6	0.030	0.8	30-50	15-17	85 / 100	12-20	15-20
22 ga.	0.031	0.8	0.030	0.8	40-60	15-17	90 / 130	18-22	15-20
20 ga.	0.037	0.9	0.025	0.9	55-85	15-17	70 / 120	35-40	15-20
18 ga.	0.050	1.3	0.035	0.9	70-100	16-19	100 / 160	35-40	15-20
1/16	0.063	1.6	0.035	0.9	80-110	17-20	120 / 180	30-35	20-25
5/64	0.078	2.0	0.035	0.9	100-130	18-20	160 / 220	25-30	20-25
1/8	0.125	3.2	0.035	0.9	120-160	19-22	210 / 290	20-25	20-25
1/8	0.125	3.2	0.045	1.1	180-200	20-24	210 / 240	27-32	20-25
3/16	0.187	4.7	0.035	0.9	140-160	19-22	210 / 290	14-19	20-25
3/16	0.187	4.7	0.045	1.1	180-205	20-24	210 / 245	18-22	20-25
1/4	0.250	6.4	0.035	0.9	140-160	19-22	240 / 290	11-15	20-25
1/4	0.250	6.4	0.045	1.1	180-225	20-24	210 / 290	12-18	20-25

Note: Singlepass flat and horizontal fillet position. Reduce current 10 to 15% for vertical and overhead welding.

(1) For fillet and groove welds. For fillet welds size equals metal thickness. For square groove welds the root opening should equal 1/2 the metal thickness.

(3) Shielding gas is CO_2 or mixture of 75% Argon + 25% CO_2.

FIGURE 5-38 *Short circuiting transfer (fine wire) variation of gas metal arc welding.*

	Material Thickness		Type of Weld (a)	Electrode Dia.		Welding Current Amps-DC	Arc Voltage Elec. Pos.	Wire Feed ipm	Travel Speed ipm	CO_2 Gas Flow CFH
Ga.	in.	mm		in.	mm					
18	0.050	1.3	Fillet	0.045	1.1	280	26	350	190	25
			square groove	0.045	1.1	270	25	340	180	25
16	0.063	1.6	Fillet	0.045	1.1	325	26	360	150	35
			square groove	0.045	1.1	300	28	350	140	35
14	0.078	2.0	Fillet	0.045	1.1	325	27	360	130	35
			square groove	0.045	1.1	325	29	360	110	35
			square groove	0.045	1.1	330	29	350	105	35
11	0.125	3.2	Fillet	1/16	1.6	380	28	210	85	35
			square groove	0.045	1.1	350	29	380	100	35
3/16	0.188	4.8	Fillet	1/16	1.6	425	31	260	75	35
			square groove	1/16	1.6	425	30	320	75	35
			square groove	1/16	1.6	375	31	260	70	35
1/4	0.250	6.4	Fillet	5/64	2.0	500	32	185	40	35
			square groove	1/16	1.6	475	32	340	55	35
3/8	0.375	9.5	Fillet	3/32	2.4	550	34	200	25	35
			square groove	3/32	2.4	575	34	160	40	35
1/2	0.500	12.7	Fillet	3/32	2.4	625	36	160	23	35
			square groove	3/32	2.4	625	35	200	33	35

(1) For mild carbon and low alloy steels on square groove welds backing is required.

FIGURE 5-39 *Globular transfer variation (CO_2) gas metal arc welding.*

Material Thickness in. (1)	mm	Type of Weld	Number of Passes	Electrode Dia. in. (2)	mm	Welding Current Amps-DC	Arc Voltage Elec. Pos.	Wire Feed ipm	Travel Speed ipm	Shielding Gas (3) Flow CFH
1/8	3.2	Fillet or square groove	1	1/16	1.6	300	24	165	35	40-50
3/16	4.8	Fillet or square groove	1	1/16	1.6	350	25	230	32	40-50
1/4	6.4	Vee groove	2	1/16	1.6	325 375	24 25	210 260	30	40-50
1/4	6.4	Vee groove	2	3/32	2.4	400 450	26 29	100 120	35	40-50
1/4	6.4	Fillet	1	1/16	1.6	350	25	230	32	40-50
1/4	6.4	Fillet	1	3/32	2.4	400	26	100	32	40-50
3/8	9.5	Vee groove	2	1/16	1.6	325 375	24 25	210 260	24	40-50
3/8	9.5	Vee groove	2	3/32	2.4	400 450	26 29	100 120	28	40-50
3/8	9.5	Fillet	2	1/16	1.6	350	25	230	20	40-50
3/8	9.5	Fillet	1	3/32	2.4	425	27	110	20	40-50
1/2	12.7	Vee groove	3	1/16	1.6	325 375 375	24 26 26	210 260 250	24	40-50
1/2	12.7	Vee groove	3	3/32	2.4	400 450 425	26 29 27	100 120 110	30	40-50
1/2	12.7	Fillet	3	1/16	1.6	350	25	230	24	40-50
1/2	12.7	Fillet	3	3/32	2.4	425	27	105 110	26	40-50
3/4	19.1	Double Vee Groove	4	1/16	1.6	325 375 350 400 450	24 26 25 26 29	210 260 230 100 120	24	40-50
3/4	19.1	Double Vee	4	3/32	2.4	425	27	110	24	40-50
3/4	19.1	Fillet	5	1/16	1.6	350	25	230	24	40-50
3/4	19.1	Fillet	4	3/32	2.4	425	27	110	26	40-50
1	24.1	Fillet	7	1/16	1.6	350	25	230	24	40-50
1	24.1	Fillet	6	3/32	2.4	425	27	110	26	40-50

Use only in flat and horizontal fillet position.

(1) For fillet welds, material thickness indicates fillet weld size.

(3) Shielding gas is argon plus 1 to 5% oxygen.

FIGURE 5-40 *Spray arc transfer variation—gas metal arc welding.*

The interrelationship of the welding variables will be covered completely in Section 5-5.

Tips for Using the Gas Metal Arc Welding Process

Semiautomatic welding using the short circuiting metal transfer is easy to use. Experienced shielded metal arc welders or people with no welding experience can learn this process variation in a relatively short time. Production welding can be learned in a few days, whereas pipe welding may require 80 to 120 hours of training.

It is important to use the correct welding technique when welding semiautomatically. The electrode wire should be directed to the loading edge of the puddle for optimum results. For out-of-position welding the puddle should remain small for best control.

The gun tip-to-work distance known as **stickout** must be closely controlled. If the stickout becomes too long the electrode will become overheated and will minimize penetration. Also, when the gun nozzle is too far from the arc, the shielding gas efficiency is reduced. Normal nozzle-to-work distance should be approximately one to one-and-one-half times the inside diameter of the gas nozzle being used.

Another important factor is the angle the gun nozzle makes with the work. Two angles are involved. One is known as the *travel angle,* the other is the *work angle.* The work angle is normally half the included angle between the plates forming the joint. When making fillet welds the gun should be at a 45° angle but directed slightly towards the horizontal plate by one electrode wire diameter from the bisecting angle.

The travel angle can be a *drag angle* or a *push angle*. The push angle is used when pure inert gases are employed. The drag angle is used when CO_2 is used—with short circuiting or the globular transfer.

The welding equipment must be in good operating condition. The drive rolls and contact tip must be proper for the electrode size being used. The conduit tube in the gun cable assembly must be kept clean and any centering guides, lineup rolls, etc., must be properly aligned. The nozzle of the gun must be kept clean and all portions of the gas supply system must be tight and operating properly. Finally, the work cable must be tightly connected to the work for trouble-free operation.

The welding parameters must be set in accordance with welding procedure schedules. The correct gas flow rates must be employed for optimum results. The welding polarity must be correct. For almost all gas metal arc welding, direct current electrode positive is employed. The proper inductance or tap should be used and all other adjustments should be in order.

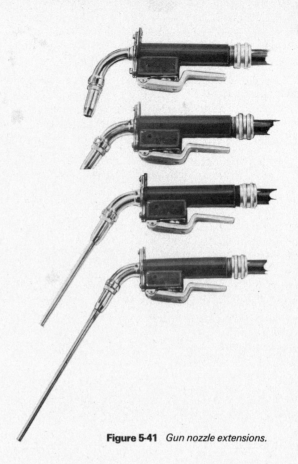

Figure 5-41 *Gun nozzle extensions.*

Safety Considerations

Safety factors and potential hazards involved with the gas metal arc welding process are covered in detail in Chapter 3, "Safety and Health of Welders." In general, gas metal arc welding is a less hazardous process than manual shielded metal arc welding.

Limitations of the Process

There are very few limitations to the overall process if each of the variations is considered. Some of the early limitations are mentioned but most of these have been overcome.

One problem has been the inability to reach inaccessible welding areas with the available guns. Gas metal arc welding guns are not as flexible as the covered electrode used for shielded metal arc welding. However, extensions can be placed on welding guns to reach relatively inaccessible areas. A variety of guns is shown by Figure 5-41.

Another problem was the distance that the gun could be separated from the welding power source and electrode supply. This problem has been overcome with development of the linear type wire feeder in which the gun can be separated from the power source by up to 50 ft. (16 m). An illustration of this is shown by Figure 5-42.

FIGURE 5-42 *Linear wire feed system—field use.*

There has been some concern about the inability to push small-diameter, soft electrode wires through long cable assemblies. There are several solutions to this. One is the spool gun feeder with the wire supply mounted on the welding gun and pushed only a very short distance. This type of wire feeder has been popular for gas metal arc welding of aluminum. Another solution is the use of a linear wire feed motor in the gun handle. The linear booster motor is also used at the wire supply. It is extremely popular for aluminum, but is also used for feeding other electrode wire types.

Often there is the objection of higher-priced equipment that requires additional maintenance. When comparing gas metal arc welding to manual shielded metal arc welding, it is obvious that more equipment is involved. The productivity of GMAW is sufficiently greater to make the extra cost and extra maintenance a minor factor from an overall cost viewpoint.

Another objection to the process has been the problem of wind and drafts affecting the efficiency of the gas shielding envelope around the arc area. This can be a problem when working outdoors or in drafty locations. It can be overcome by establishing wind breaks or shielding the welding area from direct exposure to fans or open doors or the wind. With a little experience welders are able to use their body to shield the arc area from drafts and breezes.

Variations of the Process

The variations of the process are mentioned throughout the chapter. They have a great effect on the selection of the variation for particular applications. There are some applications which can be considered process variations, such as narrow gap welding and arc spot welding. Both are explained in Chapter 20.

FIGURE 5-43 *Welding aluminum bus bar.*

Industrial Use and Typical Locations

The original use for gas metal arc welding was for welding nonferrous metals primarily of the heavier thicknesses. A typical application is the welding of aluminum bus bars in the electrical industry. This is shown by Figure 5-43.

The most aggressive user of the process has been the automotive industry. Many submerged arc welding applications have been changed to gas metal arc welding since it is better for automatic fixturing and avoids the problem of abrasive flux in fixtures. A typical application is the semiautomatic welding of auto bodies shown by Figure 5-44.

FIGURE 5-44 *Welding automobile body.*

Pipe welding also utilizes the gas metal arc welding process. A typical application is shown by Figure 5-45. Here all-position welding is performed at a speed higher than previously obtained with shielded metal arc welding. There are many, many applications of the process.

FIGURE 5-45 *Welding pipe.*

5-4 FLUX-CORED ARC WELDING (FCAW)

The **flux-cored arc welding process** is a process in which coalescence is produced by heating with an arc between a continuous filler metal (consumable) electrode and the work. Shielding is obtained from a flux contained within the electrode. Additional shielding may or may not be obtained from an externally supplied gas or gas mixture.

There are two variations, one using externally supplied shielding gas and the other which relies entirely upon shielding gas generated from the disintegration of flux within the electrode. There is one major method known as flux-cored arc welding electrogas. This is defined as "a method of flux cored arc welding wherein molding shoes confine the molten weld metal for vertical position welding. Additional shielding may or may not be obtained from an externally supplied gas or gas mixture."

Principle of Operation

The flux-cored arc welding process is shown by Figure 5-46. It utilizes the heat of an arc between a continuously fed consumable flux-cored electrode and the work. The heat of the arc melts the surface of the base metal and the end of the electrode. The metal melted off the electrode is transferred through the arc to the work piece where it becomes the deposited weld metal. Shielding is obtained from the disintegration of ingredients contained within the flux-cored electrode. Additional shielding may be obtained from an envelope of gas supplied through a nozzle to the arc area. Ingredients within the electrode produce gas for shielding and also provide deoxidizers, ionizers, purifying agents, and in some cases alloying elements. These ingredients form a glasslike slag, which is lighter in weight than the deposited weld metal, and which floats on the surface of the weld as a protective cover. The electrode is fed into the arc automatically, from a coil. The arc is maintained automatically and travel can be manual or by machine.

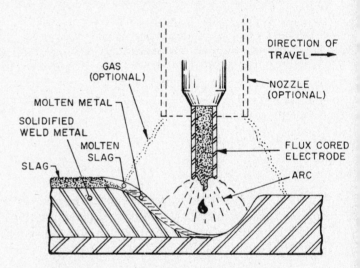

FIGURE 5-46 *Process diagram—flux-cored arc welding.*

Advantages and Major Use

The flux-cored arc welding process introduced in early 1950 is an outgrowth of the gas metal arc welding process. Flux-cored arc welding has many advantages over the manual shielded metal arc welding process. It also provides certain advantages over submerged arc welding and the gas metal arc welding processes. Simply stated, the flux-cored arc welding process provides high-quality weld metal at lower cost with less effort on the part of the welder than shielded metal arc welding. It is more forgiving than gas metal arc welding and is more flexible and adaptable than submerged arc welding. These advantages can be listed as follows:

High-quality weld metal deposit.

Excellent weld appearance—smooth, uniform welds.

Excellent contour of horizontal fillet welds.

Welds a variety of steels over a wide thickness range.

High operating factor—easily mechanized.

High deposition rate—high-current density.

Relatively high electrode metal utilization.

Relatively high travel speeds.

Economical engineering joint designs.

Visible arc—easy to use.

Less precleaning required than gas metal arc welding.

Reduced distortion over shielded metal arc welding.

This process is becoming increasingly popular. It is widely used on medium thickness steel fabricating work where the fine-wire gas metal arc welding process would not apply and where the fitup is such that submerged arc welding would be unsuitable.

Normal Method of Applying and Position Capabilities

The most popular method of applying flux-cored arc welding is by the semiautomatic method. Second is the fully automatic method. The process can also be applied by machine methods, but it cannot be applied manually.

The flux-cored arc welding process is an all-position welding process depending on electrode size.

Weldable Metals and Thickness Range

The flux-cored arc welding process is used to weld low- to medium-carbon steels, low-alloy high-strength steels, quenched and tempered steels, certain stainless steels, and cast iron. The process is also used for surfacing and for buildup. The metals welded by the process are shown by Figure 5-47.

The metal thickness range for the two variations— self-shielding and using CO_2 external gas shielding—is different. With a CO_2 atmosphere weld penetration is considerably deeper and metal thicknesses from 1/16 in. (1.6 mm) to 1/2 in. (13 mm) can be welded with no edge preparation when CO_2 gas is used. When CO_2 gas is not

Base Metal	Weldability	Chapter
Cast iron	Using special electrode	14-1
Low carbon steel	Weldable	12-2
Low alloy steel	Weldable	12-3
High and medium carbon	Weldable	12-4
Alloys steel	Weldable	12-5
Stainless steel—selected	Limited types	12-6

FIGURE 5-47 *Base metals weldable by the flux cored arc process.*

used the maximum is only half or 1/4 in. (6 mm). With edge preparation, welds can be made with a single pass on material from 1/4 in. (6 mm) through 3/4 in. (19 mm), with either variation. With multipass technique and with joint preparation the maximum thickness is practically unlimited. This information is summarized on the chart shown in Figure 5-48. Horizontal fillets can be made up to 3/8 in. (9.5 mm) in a single pass, and in the flat position fillet welds can be made to 3/4 in. (19 mm).

Joint Design

The flux-cored arc welding process can utilize the same joint design details used by the shielded metal arc welding process. These joint details are given in Chapter 15. For maximum utilization and efficiency different joint details are suggested.

For groove welds, the square groove design can be used up to 5/8 in. (16 mm) thickness. Beyond this thickness, bevels are required; however, the included angle of bevel groove welds can be reduced 35-50% over that normally used for shielded metal arc welding. This is because the smaller size electrode wire can get deeper into the joint. Open roots can be used; however, a root face is normally required to avoid burning through. In many structural applications the weld is made with a tight root opening, and the back side is gouged and

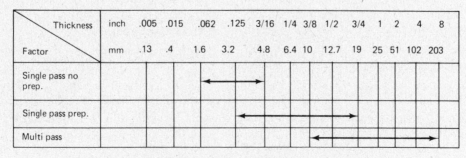

Thickness	inch	.005	.015	.062	.125	3/16	1/4	3/8	1/2	3/4	1	2	4	8
Factor	mm	.13	.4	1.6	3.2	4.8	6.4	10	12.7	19	25	51	102	203
Single pass no prep.				←→										
Single pass prep.					←—				—→					
Multi pass							←—					—→		

FIGURE 5-48 *Base metal thickness range.*

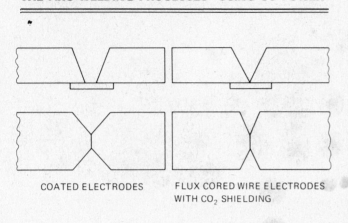

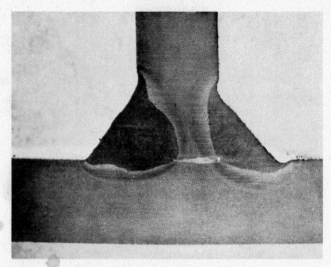

COATED ELECTRODES   FLUX CORED WIRE ELECTRODES
WITH CO_2 SHIELDING

FILLET WELDS OF EQUAL STRENGTH

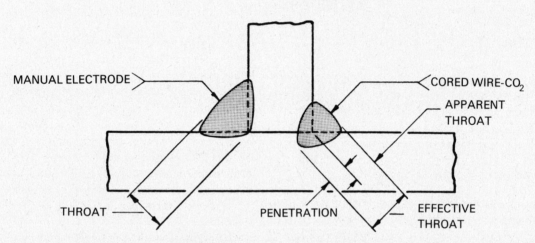

MANUAL ELECTRODE   CORED WIRE-CO_2

APPARENT THROAT

THROAT   PENETRATION   EFFECTIVE THROAT

FIGURE 5-49 *Welding joint design details.*

rewelded. When welding fillet welds using the CO_2 shielded version, the fillet size can be smaller yet will have the same strength as shielded metal arc welds. This is shown by Figure 5-49. The self-shielding type electrode wire does not have the deep penetrating qualities of the CO_2 deposit; therefore, the fillet size cannot be reduced when using this variation.

Welding Circuit and Current

The welding circuit employed for flux-cored arc welding is identical to that used by the gas metal arc welding process. This is shown by Figure 5-50. In the case of self-shielding type electrode wires the gas system can be eliminated.

The flux-cored arc welding process normally uses direct current with the electrode positive DCEP. Some electrodes for the self-shielding variation operate with the electrode negative DCEN. Direct current with con-

stant voltage power is normally employed. This is the same as used for gas metal arc welding.

AC flux-cored arc welding is used in some countries with specially formulated flux-cored electrodes. When AC type electrodes are used the drooping characteristic (CC) type power source and voltage sensing feeders are employed. The welding current for flux-cored arc welding can vary from as low as 50 amperes to as high as 750 amperes. Most flux-cored arc welding is done in the range of 350-500 amperes when the 3/32 in. (2.4 mm) electrode wire is used.

Equipment Required to Operate

The equipment required for flux-cored arc welding is shown by Figure 5-50. These components, when using the externally gas shielded version, are identical to the gas metal arc welding process. The only difference is that higher current power sources and larger welding guns or torches are used.

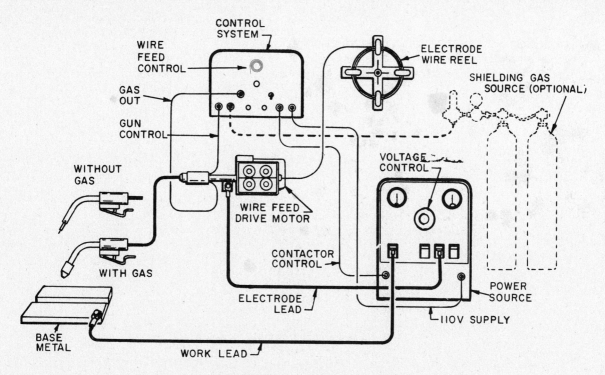

FIGURE 5-50 *Block diagram—FCAW.*

When the gasless version is used, the entire gas supply system is eliminated. This eliminates the gas cylinders, the regulator and flow meter, the hoses, the solenoid valve, and the nozzle on the welding gun. Since the nozzle is removed from the welding guns, the guns can be designed with different tip configurations. The end of the welding gun is smaller and visibility is improved. However, in view of the amount of smoke produced by flux-cored welding, it is becoming increasingly necessary to include smoke suction nozzles surrounding the gun nozzle to reduce smoke and fumes in the shop atmosphere. Guns for self-shielding electrodes normally use special wire guides that include *electrical stickout.* This means that the current is introduced to the electrode before the end of the tip. This preheats the electrode wire and makes it operate more efficiently.

Materials Used

There are two materials required, the electrode and the shielding gas.

The difference between gas metal arc welding and flux-cored arc welding is the electrode. There are several reasons why the cored electrode wires were developed to supplement solid wires of the same or similar analysis. Solid electrode wires are drawn from billets of the proper specific analysis, which may not be readily available and are large and expensive. In the case of cored wires, the special alloying elements are introduced in the core material to provide the proper deposit analysis. Second,

the cored wire production method provides a latitude of composition which is not limited to the analysis of steel billets available. Third, the cored electrode wires are easier for the welder to use than solid wires of the same deposit analysis. This is especially true for alloy type welding on pipe in the fixed position.

A series of flux-cored electrode wires have been developed and specifications established. The information concerning these electrodes based on the "AWS Specification for Carbon Steel Electrodes for Flux-Cored Arc Welding" is shown by Figure 5-51.

Additional electrode wires are available and are specified by different AWS specifications described in Chapter 10.

The self-shielding type flux-cored electrode wires include additional gas forming elements in the core. These are necessary to prohibit the oxygen and nitrogen of the air from contacting the metal transferring across the arc and the molten weld puddle. In addition, self-shielding electrodes include extra deoxidizing and denitrating elements to compensate for oxygen and nitrogen which may contact the molten metal. Self-shielding electrodes are usually more voltage-sensitive and require electrical stickout for smooth operation. The properties of the weld metal deposited by the self-shielding wires are sometimes inferior to those produced by the externally shielded electrode wires, because of the extra amount of deoxidizers included. It is possible for these elements to build up in multipass welds, lower the

Identification AWS Classification	Welding Conditions Shielding Gas (2)	Current & Polarity	X-Ray	Tensile Strength	Yield Strength	Elong. in 2"	Transverse Tension	Guided Bend	Notch Impact Charpy V
				TEST REQUIREMENTS (AS WELDED) All Weld Metal TEnsile					
E60T-7	None	DCEN	Required	67,000 PSI 47.10 KGF/mm^2	55,000 PSI 38.67 KGF/mm^2	22%	—	—	—
E60T-8	None	DCEP	Required	62,000 PSI 43.59 KGF/mm^2	50,000 PSI 33.15 KGF/mm^2	22%	—	—	20 FT/LB at 0° F 2.77 KGF-M at −18° C
E70T-1	CO$_2$	DCEP	Required	72,000 PSI 50.62 KGF/mm^2	60,000 PSI 42.18 KGF/mm^2	22%	—	—	20 FT/LB at 0° F 2.77 KGF-M at −18° C
E70T-2	CO$_2$	DCEP	—	72,000 PSI 50.62 KGF/mm^2	Not required		Required	Required	—
E70T-3	None	DCEP	—	72,000 PSI 50.62 KGF/mm^2	Not required		Required	Required	—
E70T-4	None	DCEP	Required	72,000 PSI 50.62 KGF/mm^2	60,000 PSI 42.18 KGF/mm^2	22%	—	—	—
E70T-5(1)	CO$_2$ None	DCEP	Required	72,000 PSI 50.62 KGF/mm^2	60,000 PSI 42.18 KGF/mm^2	22%	—	—	20 FT/LB at −20° F 2.77 KGF-M at −29° C
E70T-6	None	DCEP	Required	72,000 PSI 50.62 KGF/mm^2	60,000 PSI 42.18 KGF/mm^2	22%	—	—	20 FT/LB at 0° F 2.77 KGF-M at −18° C
E70T-G	Not spec.	Not spec.	—	72,000 PSI (3) 50.62 KGF/mm 72,000 PSI (4) 50.62 KGF/mm^2	Not required 60,000 PSI (4) 42.18 KGF/mm^2	22% (4)	(5)	—	—

(1) Where CO$_2$ and none are indicated as the shielding gases for a given classification, chemical analysis pads and test assemblies shall be prepared using both CO$_2$ and no separate shielding gas. (2) shielding gases are designated as follows: CO$_2$-carbon dioxide, none-no separate shielding gas. (3) requirement for single pass only electrodes. (4) requirement for multiple pass electrodes. (5) if electrode is suitable only for single pass applications, a transverse tension test is required; if electrode is suitable for multiple pass applications, an all-weld-metal tension is required.

FIGURE 5-51 *Summary of specification for flux cored electrodes.*

ductility, and reduce the impact values of the deposit. Many codes prohibit the use of self-shielding wires on steels with the yield strength exceeding 42,000 psi (25.5 kg/mm^2). Other codes prohibit the self-shielding wires from being used on dynamically loaded structures.

The flux-cored welding electrodes are complex in design, but are simple and easy to use. Information concerning the manufacture and types of wires available will be given in Chapter 10.

The other material used with gas shielded flux-cored arc welding is the shielding gas. Carbon dioxide is used most often. However, the CO$_2$–argon mixture (75% argon plus 25% CO$_2$) and the argon-oxygen mixtures are sometimes used. These gas mixtures are used for out-of-position welding, for welding piping or when an extra smooth weld is required. It is important to know if an electrode can be used with a gas mixture since the majority of the electrodes are designed for use with CO$_2$ shielding. Caution should be exercised to determine if this will be detrimental to the weldment. Certain of the deoxidizers included in the electrode wire, such as silicon, manganese, and possible aluminum, will carry across the arc rather than be oxidized in the arc. These elements may build up in multiple-pass welds to a level that could be unacceptable.

Deposition Rates and Weld Quality

The deposition rates for flux-cored electrodes are shown by Figure 5-52. These curves show deposition rates when welding with mild and low-alloy steels using direct current electrode positive (DCEP). Two types of

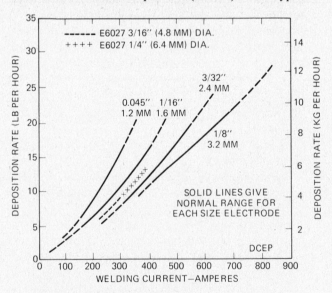

FIGURE 5-52 *Deposition rates of steel flux cored electrodes.*

164

covered electrodes are shown for comparison. Deposition rates of the smaller size flux-cored wires exceed that of the covered electrodes. The metal utilization of the flux-cored electrode is higher. Flux-cored electrodes have a much broader current range than covered electrodes, which increases the flexibility of the process.

The quality of the deposited weld metal produced by the flux-cored arc welding process depends primarily on the flux-cored electrode wire that is used. It can be expected that the deposited weld metal will match or exceed the properties shown for the electrode used. This assures the proper matching of base metal, flux-cored electrode type and shielding gas. Quality depends on the efficiency of the gas shielding envelope, on the joint detail, on the cleanliness of the joint, and finally on the skill of the welder.

The quality level of weld metal deposited by the self-shielding type electrode wires is usually lower than that produced by electrodes that utilize external gas shielding. This is recognized by reviewing the properties of the deposited metal of both types of electrodes. As time goes by, the self-shielding type electrodes will be improved to provide higher levels of quality.

Weld Schedules

The welding procedure schedules for flux-cored arc welding is given in two ways. The first is shown by the table in Figure 5-53 which provides the welding conditions for utilizing the different size electrode wires in the flat, horizontal, and vertical positions. This gives the amperage and voltage range that should be used with the specific electrode diameter. These data are given for direct-current welding with the electrode positive DCEP. In this, or any welding procedure schedule, the ranges can be expanded. Specifically,

higher currents are used when automatic travel is employed. The voltage range can also be expanded and it will increase when longer tip-to-work or stickout distance is increased. Normally the stickout or tip-to-work distance for these tables will be 3/4 in. (19 mm) to 1-1/2 in. (38 mm). When CO_2 gas is used the flow rate would be 35 cu ft/hr (16.5 liters per minute). Wire feed speed is proportional to the welding current.

The second set of tables for welding procedure schedules is shown by Figure 5-54. In this series specific weld details are shown and weld schedules are given for each detail. This is based on generalized conditions using manual travel on carbon steel, and can be altered for specific situations. These tables are starting points and should be verified by qualification tests or production runs. The weld cross-sectional area for flux-cored welding can be reduced over that utilized for coated electrodes and for this reason joint details shown on these charts can be modified to reduce the included angle.

When utilizing the self-shielding type wires that operate with direct current electrode negative DCEN current levels are reduced approximately 20%. Electrical stickout is required for most self-shielding electrodes. The amount varies by electrode type.

Welding Variables

The welding variables involved with flux-cored arc welding are essentially the same as those associated with gas metal arc and submerged arc welding. This information will be more completely covered in Section 5-5, which follows.

ELECTRODE SIZE		FLAT POSITION (1)		HORIZONTAL POSITION (1)		VERTICAL POSITION (1)	
in.	mm	Ampere (2) DC	Voltage (3) EP	Ampere (2) DC	Voltage (3) EP	Ampere (2) CD	Voltage (3) EP
0.045	1.2	150-225	22-27	150-225	22-26	125-200	22-25
1/16	1.6	175-300	24-29	175-275	25-28	150-200	24-27
5/64	2.0	200-400	25-30	200-375	26-30	175-225	25-29
3/32	2.4	300-500	25-32	300-450	25-30	—	—
7/64	2.8	400-525	26-33	—	—	—	—
1/8	3.2	450-650	28-34	—	—	—	—

(1) Applies to groove, bead or fillet welds in position shown.

(2) Ampere range can be expanded. Higher currents can be used, especially with automatic travel.

(3) Voltage range can be expanded. It will increase when larger ''stickout'' tip to work distance is employed.

Shielding gas when employed should be used at 35 cu. ft/hr (16.5 L/M).

FIGURE 5-53 *Welding procedure schedule—electrode size.*

Manual travel—mild steel

Travel speed (top left diagram, 2", 60°, passes 9 7 5 3 2 4 6 8)

Weld pass	IPM
1	11 down
2	3 up
3	3.5 up
4	2.1 up
5	2.7 up
6	2 up
7	1.8 up
8	1.4 up
9	1.3 up

Travel speed (1", 3/32, 1/16, passes 4 3 2 1)

Weld pass	IPM
1	13 down
2	1.4 up
3	2.3 up
4	1.6 up
5	11 down

Travel speed (3/8", passes 3 2 1)

Weld pass	IPM
1	13 down
2	7.7 up
3	5 up

Material Thickness T in.	mm	Type of Joint	Number of Passes	Root Opening in.	mm	Electrode Diameter in.	mm	Volts EP	Amps DC
3/8	9.5	60° single vee	3	0	0	.045	1.1	22	180
1	25.4	60° single vee	5	3/32	2.4	.045	1.1	22	180
2	50.8	60° single vee	9	1/16	1.6	.045	1.1	22	180

(b)

FIGURE 5-54 *Welding procedure schedule— joint details.*

Mild steel with backup

Material Thickness T in.	mm	Type of Joint	Number of Passes	Root Opening R in.	mm	Electrode Diameter in.	mm	Welding Power Volts EP	Amps DC	Travel Speed per pass, ipm
1/8	3.2	square	1	1/32	0.8	3/32	2.4	24-26	325	56
1/4	6.4	60° vee	1	0	0	3/32	2.4	25-27	375	41
1/2	1.27	60° vee	1	0	0	1/8	3.2	27-30	550	14
3/4	19.0	60° vee	3	0	0	1/8	3.2	27-30	550	18
1	25.4	60° vee	6	0	0	1/8	3.2	27-30	550	11

(a)

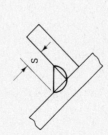

Flat

Horizontal

Weld Size (S) in.	mm	Material Thickness T in.	mm	Number of Passes	Electrode Diameter in.	mm	Welding Power Volts EP	Amps DC	Travel Speed IMP (per pass)
1/8	3.2	1/8	3.2	1	3/32	2.4	24-26	300-350	44-60
1/4	6.4	1/4	6.4	1	3/32	2.4	24-26	350-400	22-24
1/4	6.4	1/4	6.4	1	1/8	3.2	25-27	450-500	26-30
3/8	9.5	3/8	9.5	1	3/32	2.4	26-30	375-500	13-17
3/8	9.5	3/8	9.5	1	1/8	3.2	28-31	500-575	16-20
5/8	15.9	5/8	15.9	3	3/32	2.4	26-31	450-475	12-14
5/8	15.9	5/8	15.9	3	1/8	3.2	27-30	450-500	12-14

(c)

FIGURE 5-55 *Smoke nozzle in use on gun.*

Tips for Using the Process

These are essentially the same as those mentioned for gas metal arc welding (Section 5-3). There is only one major difference; the slag covering on the weld deposit allows the possibility of slag entrapment with flux-cored arc welding. This requires manipulation of the welding arc in the same manner as used with shielded metal arc welding to avoid flux entrapment. The electrode should be directed toward the leading edge of the weld puddle and the tip-to-work or stickout length should be kept uniform. Special guns are available which incorporate electrical stickout, and these increase deposition rates. However, penetration is reduced and therefore the proper current and voltage should be employed to ensure root penetration.

Safety Considerations

Safety considerations for flux-cored arc welding are the same as those for the other arc welding processes, and this was completely covered in Chapter 3.

One safety factor in flux-cored arc welding is the amount of smoke and fumes produced. This process produces more smoke than shielded metal arc welding with covered electrodes; however, much more weld metal is being deposited per hour with this process. Proper positioning of the welder's head and the use of curved front welding hoods will greatly reduce the smoke that will reach the breathing zone. For more efficient collection of smoke, the smoke exhaust weld-

ing guns are recommended. Figure 5-55 shows how the smoke is pulled into the annular suction nozzle which surrounds the conventional nozzle.

Limitations of the Process

The following are some of the limitations to this process:

Flux-cored arc welding is used only to weld ferrous metals, primarily steels.

The process produces a slag covering which must be removed.

Flux-cored electrode wire is more expensive on a weight basis than solid electrode wires.

The equipment is more expensive and complex than required for shielded metal arc welding; however, the increased productivity compensates for this.

The wire feeder and power source must be fairly close to the point of welding. This is being overcome with the new linear wire feeders and repeaters in the cable.

For the gas shielded version, the external shield may be adversely affected by breezes and drafts. This is also true in the self-shielding version, but to a lesser degree.

Variation of the Process

A variation is the electrogas welding method. This was defined and shown in the previous section. The flux-cored electrode wire is fed into the cavity and the weld is made in a single pass between two plates, normally using a square groove weld joint design. The vertical travel of the welding head including the molding shoes is automatic. In some cases two passes may be employed with a special weld joint design. The use of the flux-cored wire is the only difference between this process method and electrogas welding utilizing a solid electrode wire. External shielding gas is still used to protect the weld area. A further variation of the electrogas method is the use of a flux-cored electrode wire that provides sufficient gas for atmospheric protection.

Industrial Use and Typical Applications

The flux-cored arc welding process is replacing shielded metal arc for many applications, replacing gas metal arc welding, primarily the CO_2 version, and replacing submerged arc welding for thinner metal.

THE ARC WELDING PROCESSES—USING CV POWER

The construction equipment industry has used the process to the greatest degree. Figure 5-56 shows the application of the process on a scraper blade which is quite typical; however, frames and other weldments utilize the process.

FIGURE 5-56 *Welding scraper blade.*

The industrial equipment industry that produces machine tool bases, press frames, etc., also is a large user of FCAW. A typical application on a press frame is shown by Figure 5-57. This type of work involves relatively heavy plate fabrication, usually carbon or low-alloy steel.

The tank and vessel industry also utilizes flux-cored arc welding. The process has qualified to the requirements of ASME for pressure vessel work. It is used on many applications that conform to the code. Figure 5-58 is a typical vessel being welded.

The structural steel industry has switched almost entirely to flux-cored arc welding, both for in-plant fabrication and for erection work. For the erection of structural steel it is particularly useful for column splices, beam splices, and beam-to-column connections. Figure 5-59 is a picture of a typical heavy box column being welded with the flux-cored welding process.

The piping industry uses flux-cored arc welding with the smaller-diameter electrode wires that provide all-position capabilities.

The flux-cored arc welding process is gaining in popularity because of its proven economies. Flux-cored arc welding provides high deposition rates. It provides a higher operator factor or duty cycle. The electrode

FIGURE 5-57 *Welding press frame.*

FIGURE 5-58 *Welding pressure vessel.*

FIGURE 5-59 *Welding box column splice.*

wires have a higher rate of utilization, and more economical weld joint details can be employed. This results in lower-cost overall weldment production, which is the continual goal for weldment fabricators.

5-5 ARC WELDING VARIABLES

During the welding operation the welder has control over certain factors that affect the weld. This is particularly true when using manual or the semiautomatic method of application. For example, the welder can increase or decrease the speed of travel along the weld joint. The welder, when welding manually using the shielded metal-arc welding process, can increase the length of the arc and the voltage or decrease the length of the arc and the voltage. In this way the welder is also changing the welding current. With the continuous wire welding processes, the welding voltage can be changed by changing the dial setting on the power source and the welding current can be changed by changing the dial setting on the wire feeder. Additionally, the welder can change the angle of the electrode or the torch to either push or drag, and these changes can be made while welding. These are some of the variables that are involved. When all the variables are in proper balance, the welder will have a smooth-running arc and will deposit high-quality weld metal. This section will explain the effect of each of the welding procedure variables on the characteristics of the weld. It will also explain how these variables interrelate and how some of them are more easily changed and are useful for control.

The effect of changing these variables and the resulting change in the weld is essentially the same for all of the arc welding processes in which the weld metal crosses the arc. The relative amount of change may be slightly different, however. For those arc welding processes in which the arc is used as a heat source and the metal does not cross the arc, the relationship is slightly different. For the electroslag welding process the relationships are completely different. This is brought out in detail in Section 5-6 on electroslag. All welds shown are on steel, however; the basic factors apply to other metals as well.

Welding variables can be divided into three classifications. These are: *primary* adjustable variables, *secondary* adjustable variables, and *pre-selected* or *distinct level* variables.

The *primary* adjustable variables are those most commonly used to change the characteristics of the weld. These are: the travel speed, the arc voltage, and the welding current. They can be easily measured and continually adjusted over a wide range. These primary variables control the formation of the weld by influencing the depth of penetration, the bead width, and the bead height (or reinforcement). They also affect deposition rate, arc stability, spatter level, etc. Specific values are assigned to these primary adjustable variables. They are normally included in welding procedure schedules and can be duplicated time after time.

The *secondary* adjustable variables can also be changed continuously over a wide range. The secondary adjustable variables do not directly affect bead formation; instead, they cause a change in a primary variable which in turn causes the change in weld bead formation. Secondary adjustable variables are more difficult to measure and accurately control. They are assigned values and are usually included in welding procedure schedules. They include tip-to-work distance (stickout) and electrode or nozzle angle.

The third class of variables is known as *distinct level* variables, since they cannot be changed in a continuous fashion, but normally in increments or in specific steps or intervals. Distinct level variables must be preselected and are therefore fixed during a particular weld. They have considerable influence on the weld formation. Distinct level variables that must be preselected are normally included in welding procedure schedules. They included in welding procedure schedules. They include the electrode size, the electrode type, welding current type and polarity, shielding gas composition, and

THE ARC WELDING PROCESSES—USING CV POWER

flux type, among others. These variables are selected with regard to the type of metal to be welded, the thickness of the material, joint design, welding position, deposition rate, and appearance.

Primary Adjustable Variables

The primary adjustable variables are; welding current, arc voltage, and travel speed. To best explain the effect of these variables, bead on plate type welds are shown with the three characteristics involved. These are weld penetration, the weld width, and the weld reinforcement. Each of the variables has a distinct effect on the three weld characteristics.

When making welds to establish a welding procedure or in reviewing welds that did not meet requirements, judgment is based on these three weld characteristics. Welding penetration is usually the most important and it is affected by all three of the variables. Penetration is also affected by the secondary adjustable variable and by the preselected variables.

If in analyzing the weld it is decided that penetration

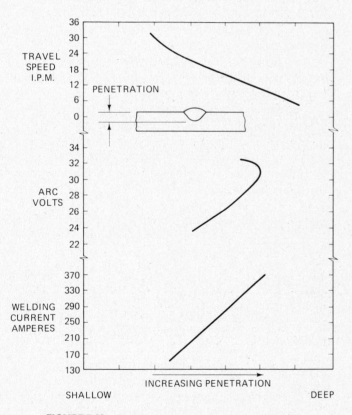

FIGURE 5-60 *Weld penetration related to primary variables.*

must be increased, one of the preselected variables may have to be changed. For example, if the maximum welding current for a particular electrode size is being used, it would be necessary to change to a larger electrode size to further increase the welding current. This same rationale may have to be followed to change the bead width or weld reinforcement if the limit of a primary variable is reached without obtaining the desired results.

Three sets of curves show the effect of the three primary variables on the three weld characteristics. Figure 5-60 shows the effects of the three variables on weld penetration. Penetration is the distance that the fusion zone extends below the original surface of the parts being welded. Joint design is also a factor that must be considered. This curve is based on flux-cored arc welding, but would apply to gas metal arc welding, submerged arc welding, and, to a fairly large degree, to shielded metal arc welding. The values may change, but the relationships are similar. To explain this, Figure 5-61 shows bead appearance and weld cross section of welds made with the flux-cored arc welding process. Welding conditions were 3/32 in. (2.4 mm) electrode, 29 volts, electrode positive, and travel speed 20 ipm (510 mm/min). The depth of penetration increases as the current level increases. The welding current and weld penetration relationship is almost a straight line and is the most effective in controlling this weld characteristic. It should be considered first when a change of penetration is required.

The relationship between travel speed and weld penetration also is a relatively straight line relationship. Penetration is increased as travel speed is decreased. Travel speed should not be used as the major control, since, for economical reasons, it is usually desired to weld at the maximum speed possible.

The relationship of penetration and arc voltage is not a straight line relationship. The curve shows that there is an optimum arc voltage where penetration is maximum. Raising or lowering arc voltage from this point reduces penetration. Thus, a long arc or a short arc will decrease penetration. For a given welding current there is a certain voltage that will provide the smoothest welding arc. It is for this reason that arc voltage is not recommended as a control for penetration.

The weld bead width relationship to the primary variables is shown by Figure 5-62. Bead width is an important characteristic of a weld, particularly when using automatic equipment to fill a weld groove or to produce a specific geometry of a weld. The arc voltage variable, or arc length, is a straight line relationship with weld bead width. As the arc voltage is increased, bead width increases. This can be explained by considering the welding arc. The welding arc has a point-to-plane

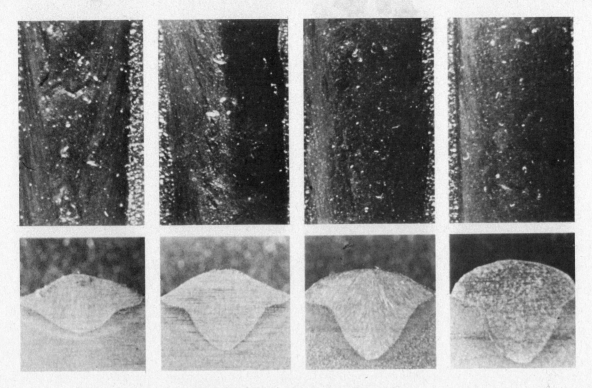

FIGURE 5-61 *Weld penetration vs welding current.*

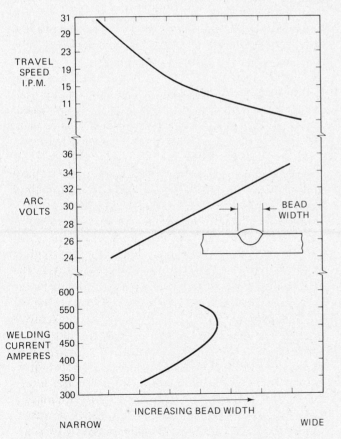

FIGURE 5-62 *Weld bead width related to primary variables.*

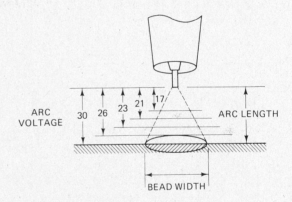

FIGURE 5-63 *Arc length—weld bead with relationship.*

relationship and is thus conical in shape with the point of the cone at the end of the electrode and the wide portion at the surface of the weld. This is shown by Figure 5-63 and explains the relationship between the longer arc with higher voltage and the bead width. This shows the arc voltage at different arc lengths and how the arc spreads out and makes a wider bead. This relationship is also shown by Figure 5-64, which shows the weld surface appearance and cross section of flux-cored arc welds made at different arc voltages. Welding conditions; electrode size and travel speed are the same as previous, current 450 amperes. Since increasing the arc voltage makes the bead wider, the reinforcement is

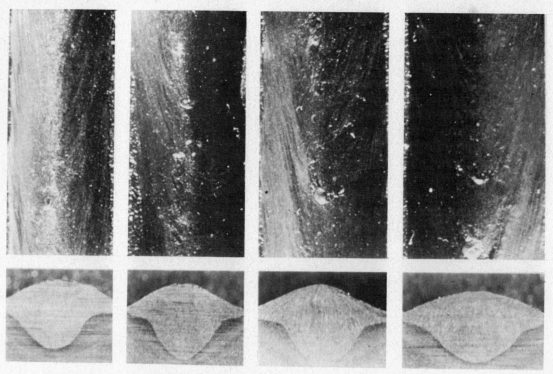

FIGURE 5-64 *Weld bead width related to arc voltage.*

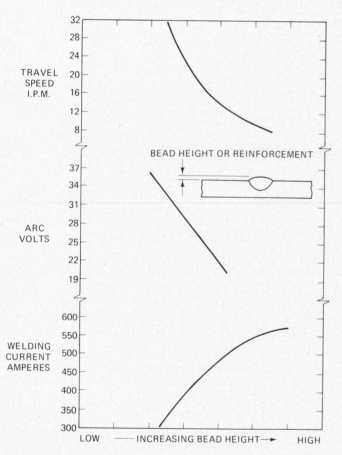

FIGURE 5-65 *Weld bead reinforcement related to primary variables.*

reduced because the same volume of weld metal is involved. Conversely, reducing the arc voltage makes the bead narrower and increases the height of the reinforcement.

Travel speed is the second choice for changing bead width, since it has a relatively straight line relationship. The welding current is not a straight line relationship, as is shown by the curve. Therefore, welding current is not used as a control for weld bead width.

The weld bead reinforcement or height related to the three primary variables is shown by the curve of Figure 5-65. Weld reinforcement is important in automatic welding when considering the requirements for filling a groove weld with the proper amount of metal using the desired number of passes. The weld height is most effectively controlled by travel speed because of the relatively straight line relationship. It should be the first choice for changing weld reinforcement.

The welding current to bead height is a relatively straight line relationship. This is based on the mass or amount of weld metal deposited. The travel speed relationship to the weld characteristics is shown by Figure 5-66, which shows the weld surface appearance and cross section of flux-cored arc welds made at different speeds. Welding conditions: electrode size is the same as previous welding, current 450 amps, voltage 29. At the lower travel speeds the weld is large in mass,

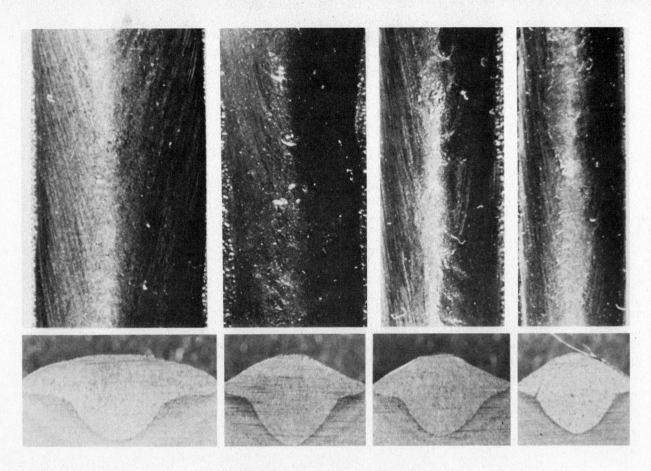

FIGURE 5-66   *Weld bead related to travel speed.*

whereas at the high travel speed it is smaller in mass. This relationship is very easily determined by relating the cross-sectional area of the welding electrode times the wire feed speed to the cross-sectional area of the weld times the travel speed. As more electrode is fed into the arc based on higher welding current, a greater mass of metal is deposited. However, as the speed of travel is increased this mass of metal will be spread out over a longer length. It is possible to establish an exact weld size mathematically based on this relationship. This can be used when establishing a welding procedure, especially for single-pass welds. The relationships shown relate penetration, bead width, and reinforcement to welding current, arc voltage, and travel speed. Notice the interaction that occurs. These relationships can only be varied within limits, since there is a relatively fixed relationship between arc voltage and welding current within the stable operating range. This relationship changes for different processes, shielding gas atmospheres, and electrode sizes. The relationship is shown by Figure 5-67 for flux-cored arc welding.

All of these relationships are relative. Different values would be used for different processes. The

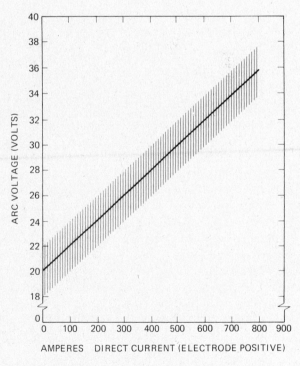

FIGURE 5-67   *Welding voltage—current relationship.*

173

shape of the curves and the changes in weld bead characteristics would be the same.

Secondary Adjustable Variables

The secondary adjustable variables include at least the stickout and torch or electrode angle. These variables do change the weld characteristics because they influence one of the primary variables. When using the CV welding system, welding current is controlled by the electrode wire feed speed. Therefore, penetration is directly influenced by wire feed speed when all other conditions are the same. Since welding current can be easily measured, penetration is normally related to it rather than to wire feed speed. The wire feed speed—current relationship can be changed by changing polarity, shielding media, electrode wire size, and the stickout. Stickout is shown by Figure 5-68. It is in this area where preheating of the electrode occurs and it is sometimes called I^2R heating. This is because the electrode wire extending from the current pickup tip to the arc is heated by the tremendous amount of current being carried by the wire.

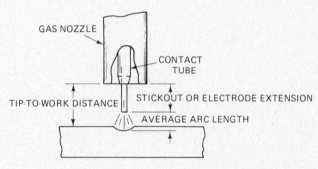

FIGURE 5-68   *Stick out or electrode extension.*

The heat generated in this portion of the electrode wire is equal to I^2R, which is the current squared times the resistance of the electrode wire. This preheats the electrode wire so that when it enters the arc it is at an elevated temperature, which increases the meltoff rate. Increasing stickout increases deposition rate *only if the wire feed speed is increased* sufficiently to maintain the current at a constant value. This is shown by Figure 5-69. Extended type nozzles with insulated guide tubes are used to create *electrical stickout,* which preheats the wire. This factor is used in gas metal arc, submerged arc, and flux-cored arc welding to increase the deposition rate.

The relationship between stickout and welding current is shown by Figure 5-70. Increasing the stickout

will reduce the welding current in the arc by almost 100 amperes when the wire feed speed rate is not changed. This reduces penetration a proportional amount. In semiautomatic welding, the stickout is adjusted by the

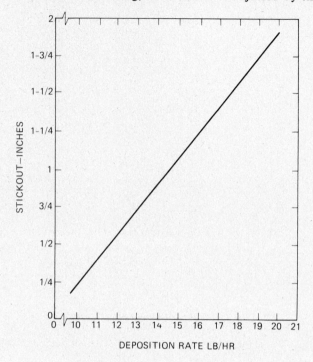

FIGURE 5-69   *Stick out vs deposition rate.*

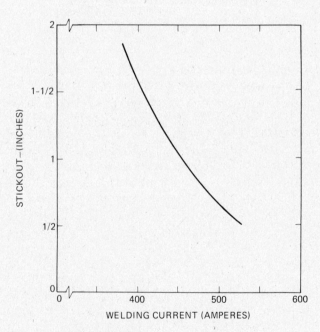

FIGURE 5-70   *Stick out vs welding current.*

welder and is an excellent means of compensating for joint variations without stopping the weld. Stickout exerts an influence on penetration through its effect on welding current as shown by Figure 5-71. Stickout is thus a control during the welding operation.

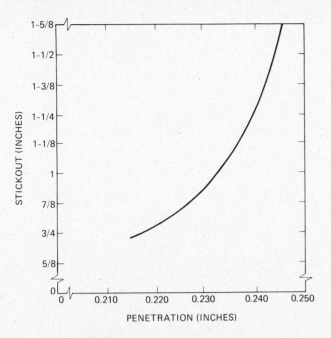

FIGURE 5-71 *Stick out vs penetration.*

Stickout influences the welding current. Increasing the electrode extension increases the resistance in the circuit and with the voltage constant the current will be reduced in accordance with Ohm's law. The voltage between the current pickup tip and the work is the sum of the voltage across the arc and the voltage drop in the electrode extension. As the electrode extension or stickout increases, the circuit resistance increases and the welding current decreases. The output voltage of the power source remains constant; therefore, more voltage occurs across the extension and thus less voltage occurs across the arc. This decrease of both voltage and current will reduce the penetration of the arc.

Conversely, as the stickout (electrode extension) decreases, the preheating effect is reduced and the welding power source furnishes more current. This increase in welding current provides a proportionate increase in penetration.

Another secondary adjustable variable is the electrode or nozzle travel angle, which has an appreciable effect on penetration. Two angles are required to define the position of an electrode or welding gun nozzle: (1) the travel angle and (2) the work angle.

The *work angle* is the angle that the electrode or the centerline of the welding gun makes with the referenced plane or surface of the base metal in a plane perpendicular to the axis of the weld. Figure 5-72 shows the work angle for a fillet weld and a groove weld. For pipe welding, the work angle is the angle that the electrode, or the centerline of the welding gun, makes with the referenced plane or surface of the pipe in a plane extending from the center of the pipe through the puddle.

The *travel angle* is the angle that the electrode, or the centerline of the welding gun, makes with a reference line perpendicular to the axis of the weld in the plane of the weld axis; see Figure 5-73. For pipe welding the travel angle is the angle that the electrode, or the centerline of the welding gun, makes with a reference line extending from the center of the pipe through the arc in the plane of the weld axis. The travel angle is further described as either a *drag angle* or a *push angle*. Figure 5-73 also shows the drag and push travel angles. The drag angle, which points backward from the direction of travel, is also known as *backhand welding*. The *push angle*, which points forward in the direction of travel, is also known as *forehand welding*.

It is found that maximum penetration is obtained when a drag angle of 15–20° is used. If the gun travel angle is changed from this optimum condition, penetration decreases. From a drag angle of 15° to a push angle of 30° the relationship between penetration and travel angle is almost a straight line. Therefore, good

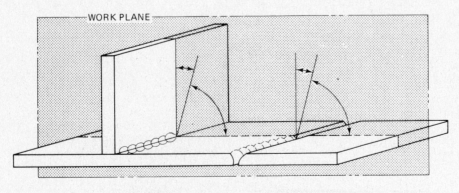

FIGURE 5-72 *Work angle—fillet and groove weld.*

175

THE ARC WELDING PROCESSES—USING CV POWER

control of penetration can be obtained in this range. It is not recommended that a drag angle of greater than 25° be used. The gun travel angle variable can also be used to change bead height and width, since the gun travel angle does affect bead contour. A drag travel angle tends to produce a high, narrow bead. As the drag angle is reduced, the bead height decreases and the width increases. This relationship is shown by Figure 5-74. The push travel angle is used for high travel speeds. These angles vary slightly with different processes and procedures.

Distinct Level Variables

The most important distinct level variable is the selection of the welding process. Once this is done the next important variable can be the polarity of welding. In general direct current electrode positive DCEP produces greater penetration than electrode negative DCEN. Alternating current, which is only used with the

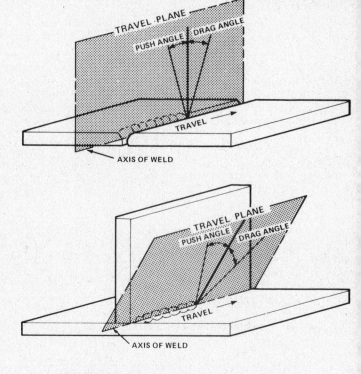

FIGURE 5-73 *Travel angle—fillet and groove weld.*

FIGURE 5-74 *Travel angle vs penetration.*

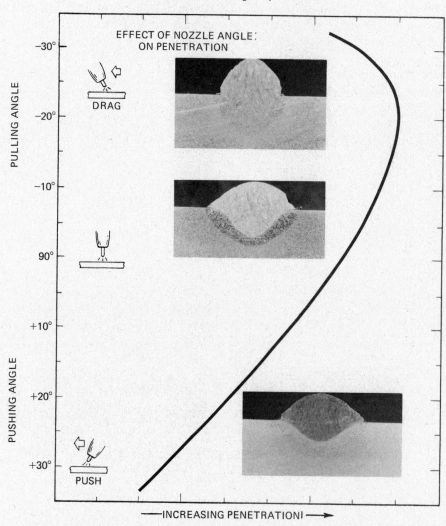

THE ARC WELDING PROCESSES—USING CV POWER

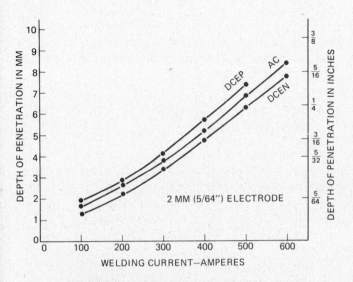

FIGURE 5-75 *Welding polarity vs penetration—curve.*

submerged arc welding, produces penetration between that produced by electrode positive and electrode negative. The polarity of the electrode and its influence on penetration is shown by the three curves in Figure 5-75.

FIGURE 5-76 *Welding polarity vs penetration—cross section.*

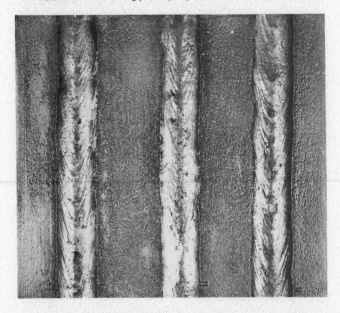

This is also shown by Figure 5-76 which was made using the submerged arc welding process.

The other variable has to do with electrode size. Smaller size electrodes tend to produce deeper penetration. This is related to the geometry of the arc and the point-to-plane relationship. The higher the current density on the electrode wire, the deeper the penetration. Larger-wire electrodes produce wider beads and less penetration, as shown by Figure 5-77.

FIGURE 5-77 *Electrode size vs penetration.*

In gas metal arc and flux-cored arc welding, the use of CO_2 gas shielding provides deeper penetration. It is the characteristic of carbon dioxide to provide deep penetration. The shielding gas relationship to penetration is shown by Figure 5-78, made using a self-shielding flux-cored electrode wire.

Also, in gas metal arc welding the type of shielding gas affects the weld bead shape and penetration pattern. Argon has a characteristic deep center or pointed penetration, while CO_2 provides a wider pattern. The CO_2–argon mixture is between these. The cross-sectional

FIGURE 5-78 *With and without external shielding gas—FCAW.*

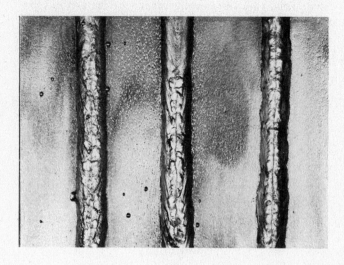

FIGURE 5-79 *Shielding gas vs weld bead shape.*

view and weld surface appearances are shown by Figure 5-79.

By fully understanding the relationship of the variables and their effect on weld characteristics, it is possible to establish a welding procedure to provide the exact type, shape, size of weld, and welding production rate that are required. A summary of the welding variables that can be changed to change the characteristics of the weld is given by the table shown in Figure 5-80.

5-6 ELECTROSLAG WELDING PROCESS (ESW)

The **electroslag welding process** is a process in which coalescence is produced by molten slag, which melts the filler metal and the surfaces of the work to be welded. The weld pool is shielded by this slag which moves along the full cross section of the joint as welding progresses. The conductive slag is maintained molten by its resistance to electric current passing between the electrode and the work. Consumable guide electroslag welding variation is a method of electroslag welding in which filler metal is supplied by an electrode and its guiding member.

Principle of Operation

Electroslag welding is not an arc welding process. It is included here since it utilizes the same basic equipment as the other consumable electrode welding processes described earlier in this chapter.

FIGURE 5-80 *Recommended variable adjustments for continuous wire arc welding.*

Change Required	Welding Variable	Arc Voltage	Welding Current (Feed Speed)	Travel Speed	Travel Angle	Stick-out or Tip-to-Work Distance	Electrode Size
Deeper penetration			[1]Increase		[3]Drag max. 25°	[2]Decrease	[5]Smaller*
Shaller penetration			[1]Decrease		[3]Push	[2]Increase	[5]Larger*
Bead height	Larger bead		[1]Increase	[2]Decrease		[3]Increase*	
	Smaller bead		[1]Decrease	[2]Increase		[3]Decrease*	
and	Higher narrower bead	[1]Decrease			[2]Drag trailing	[3]Increase	
Bead width	Flatter wider bead	[1]Increase			[2]90°	[3]Decrease	
Faster deposition rate			[1]Increase			[2]Increase*	[3]Smaller
Slower deposition rate			[1]Decrease			[2]Decrease*	[3]Larger

Key: 1 First choice, 2 Second choice, 3 Third choice, 4 Fourth choice, 5 Fifth choice.

*It is assumed that the wire feed speed is readjusted to hold welding current constant.

The electroslag welding process is shown by Figure 5-81. The consumable guide variation will be described in this chapter because it is the most popular use of the process.[4] Electroslag welding is done in the vertical position using molding shoes, usually water cooled, in contact with the joint to contain the molten flux and weld metal. The electrode is fed through a guide tube to the bottom of the joint. The guide tube carries the welding current and transmits it to the electrode. The guide tube is normally a heavy wall tube. At the start of the weld, granulated flux is placed in the bottom of the cavity. The electrode is fed to the bottom of the joint and for a brief period will create an arc. In a very short time the granulated flux will melt from the heat of the arc and produce a pool of molten flux. The flux is electrically conductive and the welding current will pass from the electrode through the molten flux to the base metal. The passage of current through the conductive flux causes it to become very hot and it reaches a temperature in excess of the melting temperature of the base metal. The high-temperature flux causes melting of the edges of the joint as well as melting of the electrode and the end of the guide tube. The melted base metal, electrode, and guide tube are heavier than the flux and collect at the bottom of the cavity to form the molten weld metal. As the molten weld metal slowly solidifies from the bottom it joins the parts to be welded. Shielding of the molten metal from atmospheric contamination is provided by the pool of molten flux. Surface contour of the weld is determined by the contour of the molding or retaining shoes. The consumable guide variation of electroslag welding normally uses fixed or nonsliding

molding shoes. The welding head does not move vertically and is normally mounted on the work at the top of the weld joint. Multiple electrodes and guides may be employed for welds of larger cross section. It is also possible to oscillate the electrode and guide tube across the width of the joint.

The surface of the solidified weld metal is covered with an easily removed thin layer of slag. The slag loss must be compensated for by adding flux during the welding operation. A starting tab is necessary to build up the proper depth of flux so that the molten pool is formed at the bottom of the joint. Runoff tabs are required at the top of the joint so that the molten flux will rise above the top of the joint. Both starting and runoff tabs are removed from the ends of the joint after the weld is completed.

The normal electroslag welding variation utilizes a welding head which moves up the joint as the weld is made. Retaining shoes usually slide along the joint and rise with the head. Single- or multiple-electrode systems can be employed and they may be oscillated across the width of the joint. All of the other factors involved in operating the process are the same except that the guide tube is not used and therefore the deposit weld metal is supplied entirely by the electrode.

Advantage and Major Use

The **consumable guide electroslag welding process** is one of the most productive welding processes when it can be applied. Some of its advantages are:

1. Extremely high metal deposition rates. Electroslag has a deposition rate of 35 to 45 lbs/hr per electrode.

2. Ability to weld thick materials in one pass. There is only one setup and no interpass cleaning since there is only one pass.

3. High-quality weld deposit. Weld metal stays molten longer allowing gases to escape.

4. Minimized joint preparation and fitup requirements. Mill edges and square flame-cut edges are normally employed.

5. A mechanized process: once started, continues to completion. There is little operator fatigue since manipulative skill is not involved.

6. Minimized materials handling. The equipment may be moved to the work rather than the work moved to the equipment.

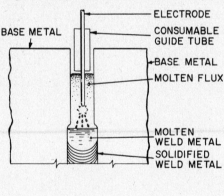

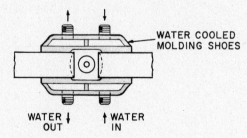

FIGURE 5-81 *Process diagram—electroslag welding.*

THE ARC WELDING PROCESSES—USING CV POWER

7. High filler metal utilization. All the welding electrode is melted into the joint. In addition, the amount of flux consumed is small.

8. Minimum distortion. There is no angular distortion in the horizontal plane. There is minimum distortion (shrinkage) in the vertical plane.

9. Minimal time. It is the fastest welding process for large, thick material.

10. There is no weld spatter and metal finishing of the weld is minimal.

11. There is no arc flash and so the normal welding helmet is not required.

Normal Method of Applying and Position Capabilities

The consumable guide version of electroslag welding is applied as a machine operation. Once the process is started it should be continued until the weld joint is completed. The apparatus should be monitored by the welding operator although little is done in guiding or directing it. Flux is added periodically and the welding operator must monitor the depth of the molten flux pool. When oscillation is used or when sliding shoes are used, closer attention is required.

The electroslag welding process is a limited position process. It can be used only when the axis of the weld joint is vertical or varies from the vertical by not over 15°.

Base Metal	Weldability	Chapter
Aluminum	Limited	13-1
Low carbon steel	Weldable	12-2
Low alloy steel	Weldable	12-3
High and medium carbon steel	Weldable	12-4
Stainless steel—selected	Limited	12-6

FIGURE 5-82 *Base metals weldable by the electroslag process.*

Weldable Metals and Thickness Range

The metals welded by the consumable guide electroslag process are low-carbon steels, low-alloy high-strength steels, medium-carbon steels, and certain stainless steels. Quenched and tempered steels can be electroslag welded; however, a post heat treatment is necessary to compensate for the softened heat-affected zone. The electroslag welding process can also be used to weld aluminum; however, it has been practical only on pure aluminum and not usable for the alloy types. The table shown by Figure 5-82 summarizes the metals that can be welded with electroslag.

Under normal conditions the minimum thickness metal welded with the consumable guide method is 3/4 in.

Plate Thickness		Root Opening		Joint Height		Number of Electrodes	Oscillation	Welding Voltage Elec Pos	Total Current Amp DC	Vert. Speed	
in.	mm	in.	mm	ft.	meters					ipm	mm/min.
3/4	19.0	1	25.4	20	6	1	No	35	500	1.40	3.6
1	25.4	1	25.4	20	6	1	No	38	600	1.20	3.0
2	50.8	1	25.4	20	6	1	No	39	700	1.00	2.5
3	76.2	1	25.4	20	6	1	No	52	700	0.80	20.3
2	50.8	1-1/4	31.8	5	1.5	1	Yes	39	700	0.76	19.3
3	76.2	1-1/4	31.8	5	1.5	1	Yes	40	750	0.64	16.3
4	101.6	1-1/4	31.8	5	1.5	1	Yes	41	750	0.52	13.2
5	127.0	1-1/4	31.8	5	1.5	1	Yes	46	750	0.40	10.2
3	76.2	1	25.4	20	6	2	No	40	850	0.50	12.7
4	101.6	1	25.4	20	6	2	No	41	850	0.44	11.2
5	127.0	1	25.4	20	6	2	No	46	850	0.38	9.7
5	127.0	1-1/4	31.8	10	3	2	Yes	41	1500	0.80	20.3
6	127.0	1-1/4	31.8	10	3	2	Yes	42	1500	0.72	18.2
8	203.2	1-1/4	31.8	10	3	2	Yes	45	1500	0.54	13.7
10	254.0	1-1/4	31.8	10	3	2	Yes	48	1500	0.47	11.9
12	304.8	1-1/4	31.8	10	3	2	Yes	51	1500	0.36	9.1
12-18	304.8-457.2	1-1/2	38.1	6	1.8	3	Yes	55	1800	0.18	4.6
18-24	457.2-609.6	1-1/2	38.1	5	1.5	4	Yes	55	2400	0.18	4.6
24-30	609.6-762.0	1-1/2	38.1	4	1.2	5	Yes	55	3000	0.18	4.6
30-36	762.0-914.4	1-1/2	38.1	3	1	6	Yes	55	3600	0.18	4.6

FIGURE 5-83 *Base metal thickness and height that can be welded.*

(20 mm). Maximum thickness that has been successfully welded with electroslag is 36 in. (950 mm). To weld this thickness, six individual guide tubes and electrodes were used.

A single electrode is used on materials ranging from 1 to 3 in. (25-75 mm) thick. From 2 to 5 in. (50-125 mm) thick, the electrode and guide tube are oscillated in the joint. From 5 to 12 in. (125-320 mm) thick, two electrodes and guide tubes are used and are oscillated in the joint. If oscillation is not employed, additional guide tubes and electrodes are required. This necessitates additional power sources and wire feed systems, and, therefore, oscillation is preferred where it can be used.

The height of the joint has a definite relationship and must be considered. The process can be used for joints as short as 4 in. (100 mm) and as long or high as 20 ft (6.5 m). It is difficult to oscillate extremely long guide tubes since they become heated and flexible. When two guide tubes are used and properly secured together it is possible for oscillation; however, as the number of tubes increases, the height of the joint must be decreased. This is not a disadvantage, since as wide plates become thicker, they become narrower. The relationship of joint thickness and joint length or height is shown by the table of Figure 5-83.

Joint Design

In electroslag welding, there is just one basic type of weld and this is the square groove weld shown by Figure 5-84. The square groove weld can be used to produce many types of joints. The butt joint is by far the most popular. However, tee joints and corner joints and even lap and edge joints can be made. The square groove butt configuration is used for the transition joint where two thicknesses of plate are joined with a smooth

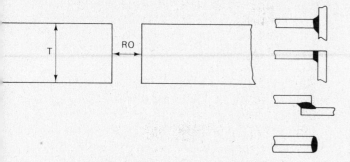

FIGURE 5-84 *Electroslag joint designs.*

contour from one thickness to the other. The transition can be in the plates or in the weld metal. Bead or overlay welds can also be made with electroslag. For tee joints and corner joints a fillet can be incorporated to improve stress flow. The different weld joints made

with the consumable guide system are shown by Figure 5-85.

FIGURE 5-85 *Different weld joints made with electroslag.*

On material over 2 in. (50 mm) thick, the square groove weld is the most economical joint design for full penetration type welds.

In a square groove weld, there are only two dimensions, the thickness of the parts being joined, T and the root opening between the parts, RO. The root opening can be varied but it is relative based on the plate thickness. It is desirable to have the root opening as small as possible to use a minimum amount of weld metal. A limiting factor to the minimum root opening is the size of the consumable guide tube and the insulators that are required to keep it from touching the sides of the joint. The guide tube must be of sufficient size to carry the welding current and structurally strong until it is consumed in the flux pool. If the weld is large and oscillation is necessary, the root opening must be increased to provide for transverse motion. The root opening must be large enough to provide sufficient volume of the molten flux to ensure stable welding conditions. This becomes one of the limiting factors for making small welds, and is particularly true with the consumable guide variation. For this reason, the consumable guide variation is not recommended for welding plates thinner than 3/4 in. (20 mm), nor is it recommended for making welds where the guide tube is unduly restricted.

Welds other than square groove types should not be attempted without extensive prior research.

The water-cooled retaining shoes are designed to accommodate the different types of joints. Shoes are available for the square groove with minor reinforcing. These are used for butt joints and for other joints where the surfaces of the plates to be joined are flush. Other types of welds, particularly fillet and vee groove welds, can be made with the electroslag process. For square groove welds involving corner or tee joints, fillet type shoes are normally used. The fillet is an extension to the groove weld and is normally a one-to-two ratio.

Welding Circuit and Current

The welding circuit used for the consumable guide method of electroslag welding is essentially the same as for the other continuous electrode processes. The block diagram of the welding circuit is shown by Figure 5-86.

For the consumable guide system direct-current welding power is normally employed. The electrode is positive DCEP. The constant-voltage system with the constant adjustable speed wire feeder is used.

In the normal electroslag welding process alternating current is often used, especially for three-wire systems. In these cases the constant-feed electrode wire drive motor is used and the characteristics of the power source are close to flat.

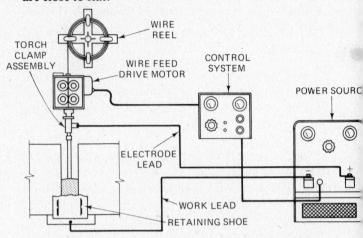

FIGURE 5-86 *Block diagram—electroslag—consumable guide.*

The electrical characteristics of the conductive molten flux are similar to those of a high-current welding arc.

Welding current per electrode wire may range from as low as 400 amperes to as high as 800. The weld voltage will range from 25 to 55 volts. The high voltage is extremely important especially when using long guide tubes.

Equipment Required to Operate

The equipment required for the consumable guide electroslag welding process is shown by Figure 5-86. A system can range from one electrode and guide tube assembly up to systems utilizing 6 electrodes and guide tubes. The systems become more complex as additional electrodes are added. The use of lateral motion or oscillation provides greater latitude of the consumable guide method. All the electrode wires are mounted on one oscillating assembly, so only one oscillating device and control are required.

The power source used for the consumable guide electroslag welding process should be a direct current welding machine of the constant-voltage (CV) type. It must be rated at a 100% duty cycle and should be heavy-duty construction since some electroslag welds take hours to complete. The current capacity of the machine must exceed the current required for a single electrode according to the welding procedure schedule. The power source should have high voltage ratings since starting voltages as high as 55 volts are sometimes required. Transformer-rectifier machines are best suited for electroslag welding, although motor generators can be used. Primary contactors and provisions for remote control including voltage adjustments should be included.

The wire drive feed motor and control system is the same as used for the other consumable electrode wire processes. Normally the wire feed motor is mounted over the welding joint, however, it can be mounted elsewhere and the electrode wire can be conducted to the joint by flexible conduits.

The guide tube torch and clamp assembly is mounted at the top of the joint. This assembly has provisions for clamping the guide tubes directly over the weld joint. The clamp assembly allows variations in the center to center distance of the guide tube torch holders. The distance between the guide tubes must be adjustable for different thicknesses of metal. This assembly is made in such a manner that the entire group of torches can be oscillated across the weld joint. The electrode power cables attached to the torch assemblies and the welding current is transmitted directly to the guide tubes.

The oscillation system is used for certain thicknesses and lengths of joints. This requires a motor control for oscillating the clamp assembly, limit switches for adjusting the width of oscillation, and a control circuit for adjusting the oscillation speed and dwell at each end of oscillation.

The retaining shoes are normally water cooled and are designed specifically for electroslag welding. The shoes must contain a slight relief approximately 3/16 in. (5 mm) deep to provide for reinforcement of the sides of the weld. They must also provide for a full water flow and should not allow entrapment of air which would soon allow steam to collect and block the water circulation. The water circulation must be sufficiently rapid to remove the heat generated in the electroslag weld. Shoes are available in standard lengths.

When the water-cooled shoes are used a system for water circulation and heat removal is required. When running water is available and when it can be easily disposed of, this is the simplest solution. However, water circulating systems, which include heat exchangers, can be used. The size of the heat exchanger must be sufficiently great to remove the heat generated in the weld.

Solid copper or air-cooled retaining shoes can be used; however, the mass of copper required to withstand the heat becomes quite heavy and hard to lift in position. In addition, they build up heat when used continuously to the point that they are difficult to handle.

Materials Used

Three materials are routinely used in making consumable guide electroslag welds: the flux, electrode wire, and the guide tube. There are several other

materials, used under certain circumstances, including runoff tabs and the starting sump. These are reusable and must be the same thickness and composition as the base metal. Insulating material is used for certain applications. Insulators are sometimes required around the bare guide tube to avoid short circuiting the system if the guide tube comes in contact with the retaining shoes or the face of the weld joint. Special round insulators that slip over the bare guide tubes are available. These are designed to be consumed in the molten pool without adversely affecting weld quality. Other reusable items are the strong backs used to hold the retaining shoes against the weld joint. Wedges are used to hold the retaining shoes in place. The strong backs and wedges are reused many times. When more than one electrode is used, a steel wool ball is placed at the bottom of the joint under the electrode wire to aid arc initiation. Steel wool also can be used for single wire applications, although it is not normally required.

When the work surface is irregular it is necessary to install a puttylike material to seal the cracks between the shoes and the work. A compound for this can be made of powdered asbestos and water mixed to a thick consistency. Commercial materials such as furnace sealing compound can also be used. Figure 5-87 is a series of pictures showing the preparation of the joint, the attachment of the strong backs, the guide tube with insulators, and the installation of the retaining shoes.

The *electroslag flux* is a major part of the electroslag welding system. The design of the flux is of utmost importance, since its characteristics determine how well the electroslag process operates. Normal functions for an electroslag flux are:

1. Providing heat to melt the electrode and base metal.
2. Conducting the welding current.
3. Protecting the molten weld metal from the atmosphere.
4. Purifying or scavenging the deposited weld metal.
5. Providing stable operation.

The flux for electroslag welding must possess the following properties:

Electrical Resistance: The molten flux must be electrically conductive but must have resistance high enough to generate a sufficient amount of heat for melting.

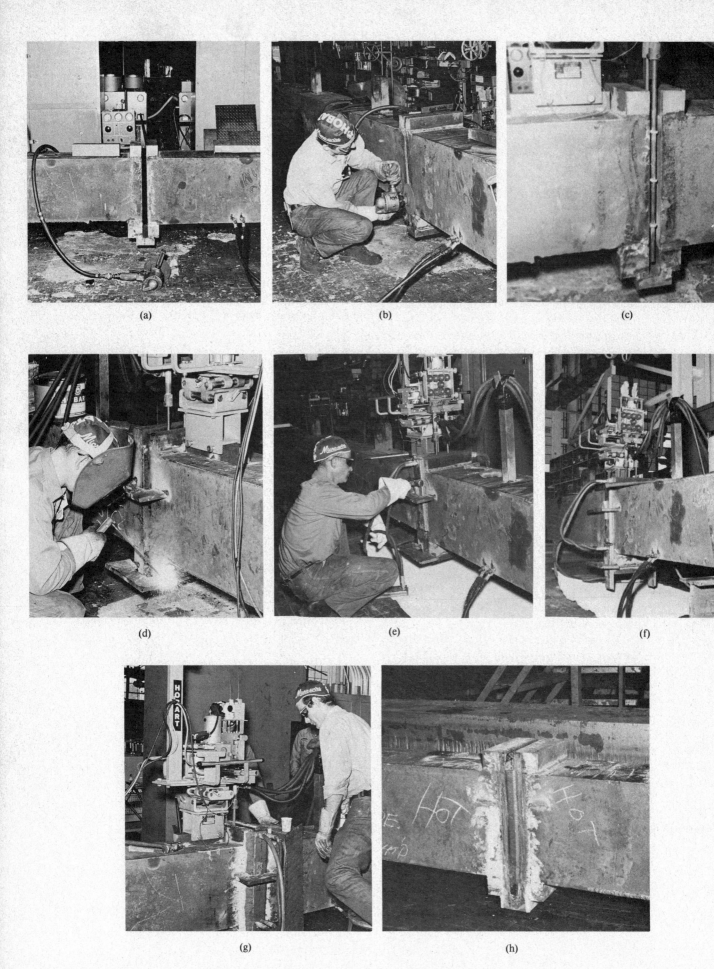

(a)

(b)

(c)

(d)

(e)

(f)

(g)

(h)

FIGURE 5-87 *Sequence of preparing for and making electroslag weld (composite).*

Viscosity: The molten flux at its operating temperature approximately 3,000°F must have the proper viscosity for good circulation to ensure even distribution of heat.

Specific Gravity: The density of the molten flux must be lighter than the material being welded so that the flux rises easily above the molten metal.

Melting Temperature: The melting point of the flux must be well above the melting point of the base metal.

Inertness: The molten flux must be reasonably inert in order not to change the analysis of the deposited weld metal.

Stability: The flux must be stable over a wide range of welding parameters.

There are two types of granular fluxes normally used for electroslag with the consumable guide tube. One is known as a starting flux and the other as a running flux. The starting flux is designed to bring the electroslag process into quick stabilization. It has a low melting point and high viscosity. It melts quickly and wets the bottom of the sump to facilitate starting. It is highly conductive and generates more heat quickly. It is used in small amounts for starting. It is especially useful for starting without the use of a sump. The running flux or operating flux is designed to provide the proper balance for correct electrical conductivity, correct bath temperature and viscosity, and the proper chemical analysis. Running flux will operate over a wide range of conditions.

Only a relatively small amount of electroslag flux is used. Flux solidifies on the cold surface of the water-cooled retaining shoes and results in a thin coating of slag on the exposed edges of the finished weld. Flux consumption is based on the surface area of the shoes exposed to the flux pool. Neglecting losses by leakage, consumption approximates 1/4 lb (100 g) of flux per vertical foot of the joint or height. Another approximation of flux consumed is 5 to 10 lb of flux for each 100 lb (45 kg) of deposited weld metal.

The *electrode* for consumable guide electroslag welding supplies over 80% of the deposited weld metal. The guide tube supplies the remainder. The electrode wire must match the base metal of the parts being welded. This means that the mechanical properties of the deposited weld metal must match the properties of the base metal. Since an electroslag weld deposit is similar to a casting, it is essential that the properties of this *as-cast* metal should overmatch the mechanical properties of the parts being joined. It is important to consider the dilution factor provided by the base metal.

In a consumable guide weld the dilution runs from a low of 25% to a high of 50% base metal. The amount of dilution of base metal depends upon the welding conditions.

The American Welding Society specification for electroslag welding electrodes is based on the use of proprietary fluxes.

The flux adds no alloys and has little effect on the weld deposit in relationship to the analysis of the wire. In general, electrode wires designed for gas metal arc welding and submerged arc welding are employed for electroslag welding. The 3/32 in. (2.4 mm) electrode size is the most common. It is the most easily used to feed through a guide tube and produces the highest deposition rate.

The surface of the electrode wire is extremely important. The copper coating on the wire may form a low temperature brazing material. Capillary action and the heat of the process causes the material to collect between the inside diameter of the guide tube and the electrode wire. If the guide tube momentarily cools, this compound will freeze and seize the electrode wire to the guide tube. This will cause the welding operation to stop. To avoid this, the electrode wire should be specified for use in the consumable guide electroslag welding process. Electrode wire should be procured in large reels or payoff packs since it is essential that sufficient electrode wire be available to complete the entire weld joint.

The *consumable guide tube* melts just above the surface of the molten slag bath. A guide tube must be used whenever the length of the weld is 6 in. (160 mm) or over, assuming that the head is stationary. Guide tubes can be either bare or coated. The bare guide tube is usually a heavy wall seamless tube with an analysis comparable to the base metal being welded. The normal composition is AISI 1018.

When a bare guide tube is used and if the weld is over 12 in. long (304 mm), insulators should be placed on the tube to avoid the guide tube coming in contact with the side wall or face of the joint or the retaining shoes. The insulators are slipped over the guide tubes and spaced approximately 6 in. (160 mm) apart. A tack weld on the guide tube beneath the insulator will prevent it from falling during the welding operation. The insulators, composed of flux and steel, are consumed in the molten weld slag pool and do not adversely affect the quality of the weld metal.

Coated guide tubes are also available and the coating

is an effective insulator particularly when working in tight joints. The composition of the coating on the guide tube is similar to the composition of the running flux. It helps replenish the flux that is lost during the welding operation. Additional flux is normally required since the amount of flux on the guide tube must be kept to the minimum, so that the molten pool depth does not increase when welding.

There are several other variations of the consumable guide tube system and in some cases bars are tack welded to the guide tube or tubes are tacked on edges of bars. These bars contribute metal to the weld deposit. In some cases this entire assembly is coated with the flux material.

Deposition Rates and Weld Quality

Deposition rates of the electroslag welding process are among the highest of any. Figure 5-88 shows the deposition rate versus welding current of the 3/32 in. (2.4 mm) electrode wire and of the 1/8 in. (3.2 mm) electrode wire.

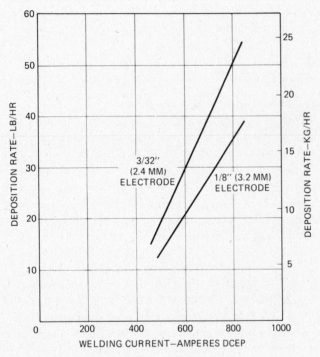

FIGURE 5-88 *Deposition rate.*

The electroslag welding process produces a high quality weld metal deposit. Electroslag welding is similar to the electroslag remelting process which produces

steel of a quality comparable to that produced by the vacuum arc remelting process. The high quality of electroslag weld metal is the result of progressive solidification which begins at the bottom of the joint or cavity. There is always molten metal above the solidifying weld metal and the impurities which are lighter rise above the deposited metal and collect only at the very top of the weld in the area that is normally discarded. Thus the weld itself is free of impurities. The metal transferring from the electrode wire and the deposited molten metal is never exposed to the atmosphere since it is covered by the pool of molten flux.

Electroslag welding is a low-hydrogen welding process, since hydrogen is not present in any of the materials involved in making the weld. Because of the slow-cooling rate any impurities that are in the base metal and are melted during the welding process have time to escape. In spite of the acknowledged high purity and high quality of the weld metal, there is concern about the large grain sizes inherent with the process. Electroslag welds possess properties and characteristics surpassing welds made by many of the arc welding processes. The cooling rate of the electroslag weld is much slower than the cooling rate of welds made by other arc welding processes. Heat input into an electroslag weld is slower due to the temperature of the molten slag bath above the weld metal which performs a preheat for the weld. The slow cooling rate allows large grain growth in the weld metal and also in the heat affected zone of the base metal. The slow cooling rate minimizes the risk of cracking and reduces the high hardness in the heat affected zone sometimes found in conventional arc welds. If excessively high welding current is used, weld cracking may occur. The hardness of the weld is very uniform across the weld and is very similar to the unaffected base metal. Figure 5-89 shows a hardness traverse of a consumable guide weld made in 2-1/2 in. carbon steel. This illustration shows the uniform hardness across the weld and also the lack of a high hardness zone in the adjacent base metal.

Weld metal produced by electroslag welding will qualify under the most strict codes and specifications. When the correct electrode is selected, the yield and tensile strength of the deposit weld metal will equal that of the base metal. Ductility of the weld metal is relatively high in the range of 25–30%. Impact requirements for electroslag welds will meet those required by the AWS structural welding code. Vee-notch Charpy impact specimens producing 5–30 ft-lb at 0°F are normal and expected.

Fatigue strength of the electroslag welds have also been questioned due to the relatively large grain size of the deposit. Research work concluded[5] that unmachined electroslag welds have fatigue strength as high

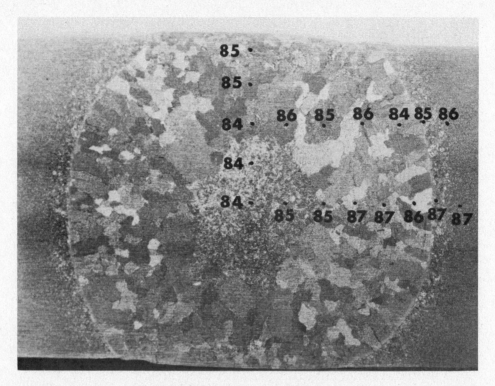

FIGURE 5-89 *Macro graph with hardness survey.*

or higher than those obtainable from butt welds made by other welding processes. Research also indicated that electroslag welds machined flush to the surface of the base metal have fatigue strengths as high or higher than those of the base metals. Therefore, it is apparent that the weld surface contour is a greater factor with respect to fatigue strength than the welding process used to produce the weld.

Weld Schedules

Welding procedure schedules for electroslag welding with the consumable guide method are shown by Figure 5-90. The consumable guide method does not have the broad latitude of operating conditions enjoyed by other arc welding processes. In view of this, the procedure conditions shown are recommended. They may not necessarily be the only conditions that can be used. It is possible that the settings can be adjusted to obtain optimum results, however, a qualification test should be made before utilizing published welding procedure schedules when welding critical work.

Welding procedure schedules are based on welding low-carbon steel under normal conditions using water-cooled copper shoes, and the 5/8 in. (16 mm) outside diameter guide tube with a 1/8 in. (3.2 mm) inside diameter unless otherwise specified. The electrode diameter is 3/32 in. (2.4 mm) and proprietary starting and running fluxes are used. Oscillation speed is based on

the number of seconds per cycle, which is shown as a rate of speed. There is a dwell time at each end of oscillation, which is normally 4 seconds. Any exception to the above is mentioned in the tables.

Welding Variables

Electroslag welding differs from arc welding processes in that the base metal melting results from localized heat generated in the molten slag pool instead of from an arc. The heating involved in an electroslag welding is concentrated in a volume of molten flux which is the product of the metal thickness by the root opening by the slag pool depth. The tremendous amount of heat generated is transferred throughout the weld area by convection (flux motion), conduction, and radiation. This heat generated and transferred is enough to permit 3 in. (75 mm) thick to be welded with a single stationary electrode. If a thicker plate is welded, a method of expanding the heat distribution is used. One way is to add additional electrodes, another is by moving the electrode in the weld cavity. Lateral motion, or oscillation, is used in order to weld thicker material with one electrode. The heat of one oscillated electrode is sufficient to weld material up to 5 in. (125 mm) thick. By adding more electrodes and utilizing lateral motion, it is possible to weld heavier thicknesses. Figure 5-91 shows the principle of electrode and guide tube oscillation to distribute the heat throughout the flux pool and

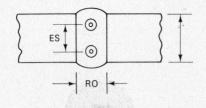 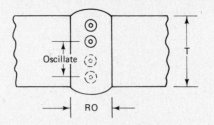

Two Electrodes –Non oscillating

Thickness T		Maximum Height H		Root RO Opening		Electrode Spacing ES		Welding Volts EP	Current Amperes DC
in.	mm	ft.	Meters	in.	mm	in.	mm		
3	75	20	6	1	25	2-1/2	62	41	850
4	100	20	6	1	25	2-1/2	62	44	850
5	125	20	6	1	25	2-1/2	62	47	850

Guide tube size–5/8″ (15.9 mm) O.D. x 1/8″ (3.2 mm) I.D.

Two Electrodes—with Oscillation

Thickness		Maximum Height H		Root RO Opening		Oscill. Length		Traverse Speed (2)		Welding Volts EP	Current Amperes DC
in.	mm	ft.	Meters	in.	mm	in.	mm	ipm	mm		
5	125	10	3	1-1/4	31.4	1	25	20	500	42	1500
6	150	10	3	1-1/4	31.4	2	50	40	1000	43	1500
8	200	10	3	1-1/4	31.4	4	100	80	2000	46	1500
10	250	10	3	1-1/4	31.4	6	150	120	3000	49	1500
12	300	10	3	1-1/4	31.4	8	200	120	3000	52	1500

(a)

Guide tube size–5/8″ (15.9 mm) O.D. x 1/8″ (3.2 mm) I.D.
(1) Electrode diameter 3/32″.
(2) Based on 10 second oscillation cycle.
(3) Water cooled shoes and Porta-slag flux used.

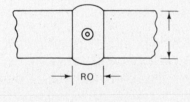 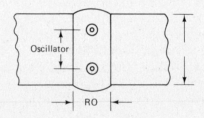

Single electrode—Non oscillating

Thickness T		Maximum Height H		Root RO Opening		Welding Volts EP	Current Amperes DC
in.	mm	ft.	Meters	in.	mm		
3/4	19	20	6	1	25	36	500
1	25	20	6	1	25	39	600
2	50	20	6	1	25	40	700
3	75	20	6	1	25	53	700

(b)

Guide tube size–1/2″ (12.7 mm) O.D. x 1/8″ (3.2 mm) I.D.

Single electrode—with Oscillation

Thickness T		Maximum Height H		Root RO Opening		Oscill. Length		Traverse Speed (2)		Welding Volts EP	Current Amperes DC
in.	mm	ft.	meters	in.	mm	in.	mm	ipm	mm		
2	50	5	1.5	1-1/4	31.4	1-1/4	31.4	25	625	40	700
3	75	5	1.5	1-1/4	31.4	1-1/4	31.4	45	1125	41	750
4	100	5	1.5	1-1/4	31.4	1-1/4	31.4	65	1625	44	750
5	125	5	1.5	1-1/4	31.4	1-1/4	31.4	85	1925	47	750

Guide tube size–5/8″ (15.9 mm) O.D. x 1/8″ (3.2 mm) I.D.
(1) Electrode dia. 3/32″
(2) Based on 10 second oscillation cycle
(3) Water cooled shoes and Porta-slag flux used.

FIGURE 5-90 *Welding procedure schedule.*

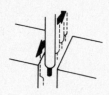

1/2–23/8 inch	23/8–5 inch	5–12 inch
ONE ELECTRODE	ONE ELECTRODE	TWO ELECTRODES
NO OSCILLATION	WITH OSCILLATION	WITH OSCILLATION

FIGURE 5-91 *Electrode oscillation.*

weld joint. In the arc welding processes, the localized heating is confined to the much smaller area of an arc and puddle but the arc is at a much higher temperature. The operation of electroslag welding is thus different from the familiar arc processes. The operating conditions well known in arc welding and their effect on the weld are different in electroslag welding. Weld penetration in the conventional arc welding processes is perpendicular with the axis of the electrode, and is increased by increasing the welding current. However, in electroslag welding the metal surface to be melted (joint side walls) is parallel to the axis of the electrode. Thus, increasing welding current does not increase the depth of penetration of the sidewalls of the base metal. In electroslag welding increasing the electrode wire feed rate and deposition rate causes the CV power source to supply more welding current. More electrode wire is fed into the weld pool but this does not provide for more side wall penetration. Increased welding current does not increase the depth of penetration that occurs with the continuous wire arc welding processes. In fact, increasing welding current beyond reasonable levels can cause problems since the depth of the molten weld metal pool increases, and this adversely affects the form factor and reduces cracking resistance. The higher the welding current, the higher the deposition rate. When welding high-carbon steels or restrained joints, high amperage can cause weld cracking. When welding thin plate, too much current will cause undercutting.

With all arc welding processes, an increase in arc voltage causes the weld bead to widen. In electroslag welding, this same thing is true only now this widening causes an increase in the depth of penetration into the sidewall. The increased voltage raises the slag bath temperature and causes more of the base metal or sidewall to melt. Increase voltage to increase depth of fusion. Excessively high voltage will cause undercutting. Too low a voltage may result in arcing between electrode wire and the molten weld metal at the bottom of the flux pool.

Tips for Using the Process

The electroslag process utilizing the consumable guide tube is completely different in many respects from the arc welding processes. Even though this process is a machine or automatic process, the operator must be continually alert to make various adjustments as required during the welding operation. The operator must have a good operating knowledge of electroslag welding because of the different effects of changing the various parameters. In many respects the effects are exactly opposite those expected with the arc welding processes.

Increasing the wire feed speed of the electrode feed motor does increase the welding current. It does not, however, increase the sidewall penetration of an electroslag weld. Increasing the welding voltage does seemingly increase the gap between the end of the electrode and the molten metal and this increases the sidewall penetration.

The depth of the molten slag pool should be checked if possible. When the pool is accessible to the operator a dipstick can be used to determine its depth. Experience will quickly show that when the pool is quiet and the process is running without sparking or sputtering that the pool depth is correct. If the pool depth becomes too shallow sparks will emit from the surface and can be seen by the operator. Additional flux should be added to the pool. This is simply done with a container, which is shown in Figure 5-92. The pool depth should not become too deep because it will cool and restrict sidewall penetration.

If the retaining shoes do not fit tightly along the joint the molten flux may run out of the opening. When this happens, steps must be taken immediately to stop the leak of flux. This is done by using the putty-like sealing preparation mentioned previously. As soon as flux starts to escape it should be sealed off and additional flux added to the pool to maintain the proper pool depth.

The operator should make a rough calculation to determine the amount of electrode wire required for a specific joint. Sufficient wire should be available on the machine prior to starting a weld. Once the weld is started it should be continuously in operation until it is finished. If the operation stops for any reason, the machine should be turned off. Corrections should then be made and the weld restarted. At the point of stopping and restarting, there is normally an unfused area which must be gouged out and welded by other processes.

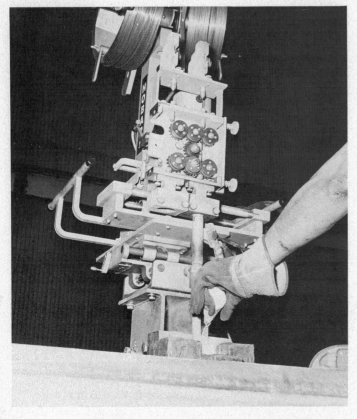

FIGURE 5-92 *Operator adding flux during welding.*

If the retaining shoes leak and water gets into the weld cavity the operation must be stopped. This can create a safety hazard and will create gross porosity in the weld metal. With respect to water-cooled shoes, the operator must ensure that the water flow is uninterrupted during the entire welding operation.

The experienced operator is able to anticipate all of the factors mentioned and will soon be able to make high-quality electroslag welds time after time.

Safety Considerations

The safety factors involved with electroslag welding are much the same as for the other continuous wire processes. There is no arc except at the very beginning; however, safety glasses should be worn because of possible sparking from the molten slag pool.

The major safety factor is the presence of a large mass of molten slag and molten weld metal. The high welding current creates a large mass of metal which must be contained within the weld cavity. If the retaining shoes should fail and allow the molten metal to escape, it is best to evacuate the area, turn off the equip-

ment, and wait for the metal to solidify. Obviously, the surface under the welding operation should be noncombustible.

The work being welded must be securely braced to eliminate the possibility of it falling.

Limitations of the Process

The major limitation is the welding position limitation. It can be used only when the axis of the weld is vertical. A tilt of up to 15° is permitted, but beyond this the process may not function correctly. This is not a serious restriction since many weldments are positioned for flat position welding using the other arc welding processes. It is just as conceivable that they can be positioned for vertical welding to take advantage of the high production rates obtainable with electroslag welding. With the consumable guide system, the welding equipment can be taken to the work.

The second limitation is that the process can be used only on steels. Electroslag welding of aluminum so far is limited to pure aluminum and cannot be used on the aluminum alloys.

There are several other factors including the fact that the process is different from others and does require familiarization and experience before it can be properly applied for maximum productivity. There is also the problem of suspected poor weld metal quality. As research continues, it becomes evident that the quality of the weldment produced by electroslag weld is equal to, or in some cases superior to, that produced by other welding processes.

It is claimed that the high proportion of preparation time is a limitation to the process. Once experience is gained in using the process and the necessary preparation equipment is acquired, this becomes a minor factor, especially in the face of the productivity that can be obtained.

Variations of the Process

The basic process, without the use of the consumable guide, has not become as popular, since more complex equipment is required to provide vertical travel. There are many types of equipment available for the conventional electroslag process. Some that have wider use are the types that are designed for welding thinner materials and will move up the parts being welded.

Industrial Use and Typical Applications

The major user of electroslag welding has been the heavy plate fabrication industry which includes manufacturers of frames, bases, metal-working machinery, etc. A frequent use of the process is the splicing of rolled steel plates to obtain a larger piece for a specific application. A typical application is the splicing of an 8-in.

FIGURE 5-93 *Splicing 8 in. thick 12 ft wide plates.*

(203 mm) thick plate, 12 ft (4 m) wide to make a press frame, shown by Figure 5-93. The weld joint was set up in the vertical position, strong backs were attached, and the two-wire feeder was placed at the top of the joint. Four water-cooled retaining shoes were used, two on either side, which were moved progressively from bottom to top until the weld was completed. Previously with submerged arc welding and turning the plate over after every few passes, the splicing operation required eighty hours. With the consumable guide electroslag process the weld was completed in slightly over four hours.

Another major user of electroslag is the structural steel industry, for making subassemblies for steel buildings. It has also been used for field erection on the building site. A common application is the welding of continuity plates inside box columns. The continuity plate carries the load from one side of the column to the other side at the point of beam-to-column connections. Continuity plates must be welded with complete penetration welds to the two sides of the box column. Figure 5-94 shows an automatic fixture which provides for two welding operations simultaneously. The welding shoes are positioned by clamping devices that also locate the continuity plate and positions two guide tubes. The clamping is automatic and as soon as the fixture is in place the welding operation can be started.

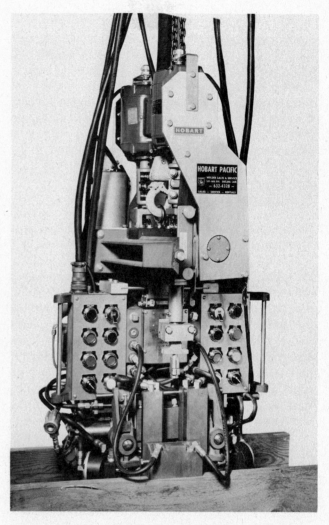

FIGURE 5-94 *Automatic fixture for box column continuity plate.*

Electroslag is also used in steel mills for maintenance operations, for building equipment, and for splicing billets. The billet shown by Figure 5-95 is 30 in. square (760 mm) and a five-wire system utilizing osicllation was required to make the weld. The guide tube torch holders were mounted on an assembly that oscillated over the joint and the wire feed motors were remote but attached to the assembly through flexible conduits. This is the only way that a weld of this type could be made.

The electrical machinery industry also utilizes electroslag welding. Electric motor housings are rolled from a single plate, and a single weld is made to join the two abutting edges. In other cases the material is heavier or the housing might be square. The application shown by Figure 5-96 shows two sides of a motor housing being welded simultaneously. Electroslag reduces total cycle time because the housing was previously welded with submerged arc and only one joint was made at a time.

There are many, many other applications of the consumable guide version, both in the shop and in the field. It is expected that electroslag welding will find more applications as its productivity is better known by more users.

5-7 SELECTING THE CORRECT ARC WELDING PROCESS

The criteria for selecting the correct arc welding process is extremely complex. Before going into detail, it is wise to review the AWS master chart of welding processes shown in Chapter 2 by Figure 2-8. This master chart shows the grouping of all the welding processes. Before developing the method of selecting the correct welding process, it is worth mentioning that welding as a metal joining method competes with the mechanical fasteners and with adhesives that are being increasingly used to join metals. In considering materials joining, we are considering only the joining of metals and are assuming that the joint is to be permanent, not one that may involve occasional disassembly and reassembly.

In view of the complexity of this selection it is well to establish a basis for choosing the correct welding process. There are a number of different factors that must be considered in choosing the best welding method. These can be summarized as follows:

1. The ability to join the metals involved.
2. The quality or reliability of the resulting joint.
3. Capability of the process to join the metals in the thickness and position required.
4. The most economical way of joining metals.
5. The availability of the necessary equipment.
6. The familiarity of the personnel involved in making the joint.
7. Other factors such as the engineering capabilities to design, the user reaction to the method, etc.

Each of these factors will be reviewed with respect to the arc welding processes and also with respect to other welding processes that might satisfactorily fulfill the end product requirement.

FIGURE 5-95 *Five wire unit welding 30-inch joint.*

FIGURE 5-96 *Motor housing—two welds made simultaneously.*

The ability to join the necessary material must be the very first consideration. If the metal in question is not weldable by any of the arc welding processes it is obvious that the non-arc processes must be considered. Most metals can be welded by one or another of the welding processes. In many cases the material can be welded with a number of different welding processes; thus, the selection depends on other factors. Table 1 in Chapter 1 is the starting point. This provides a summary of metals weldable by the various welding processes. Tables in the process chapters (4 and 5) show the metals weldable by that process. These tables also show a reference chapter in which the particular metal is described in detail as well as how to weld it. Chapters 12, 13, and 14 describe the welding processes that can be employed to join that particular metal.

In making a selection analysis, it becomes evident that the basis for initial screening and decisions is to determine which processes can be used to weld the metals to be joined.

The quality or reliability of the joint produced by the different processes is the second basis for process determination. In any design situation the designer must be completely aware of the quality requirement of the product which involves the service requirements, specifications, codes, and environmental exposure that can be expected. The materials to be joined must be selected on the same basis. They must sustain the service requirements with the quality level expected. It is then necessary to determine the metals joining process that will provide a joint of the same quality. The higher the quality requirement or the more severe the service exposure, the more important to select the correct process. Each of the sections on arc welding process discusses the quality aspects of the weld metal deposit produced by that process. This becomes the basis for selection decision. Information in the design chapter aids in making this selection.

The third factor is the necessity to consider the thickness of the metals to be joined to produce the end product and the positions of the work or joints to be welded. These are important factors since welding processes have capabilities and limitations with respect to metal thicknesses. In addition, some processes have all-position capabilities, while others are limited to one or a few welding positions. This information is summarized by Figure 5-97. The position capability may not be as important since many products or assemblies can be turned to place them in the proper position for most advantageous welding. There are some situations, however, in which the position cannot be altered. This is primarily involved in the field erection of large products and in the repair and maintenance of other products that cannot be moved. There are also problems with respect to moving the equipment to the work or moving the work to the equipment.

In all of the arc welding sections a description has been presented showing the capabilities of the process for welding various thicknesses of metals. This information is given in chart form showing the thickness capabilities of the process. In addition, the welding procedure schedules show welding parameters that are established for welding different thicknesses.

The above three factors will narrow down the choice of welding processes available. However, after analyzing these three reasons, there still is the requirement to eliminate all processes but one in order to establish the optimum and most economical welding process. The welding cost factor should then be used. The two major components in welding cost are the cost of labor to apply the welds and the cost of the materials utilized

Welding Position		Welding Process Rating			
		SAW	GMAW	FCAW	ESW
1. Flat		A	A	A	No
Horizontal fillet		A	A	A	No
2. Horizontal		C	A	A	No
3. Vertical		No	A	A	A
4. Overhead		No	A	A	No
5. Pipe-fixed		No	A	A	No

FIGURE 5-97 *Summary of welding positions for each process.*

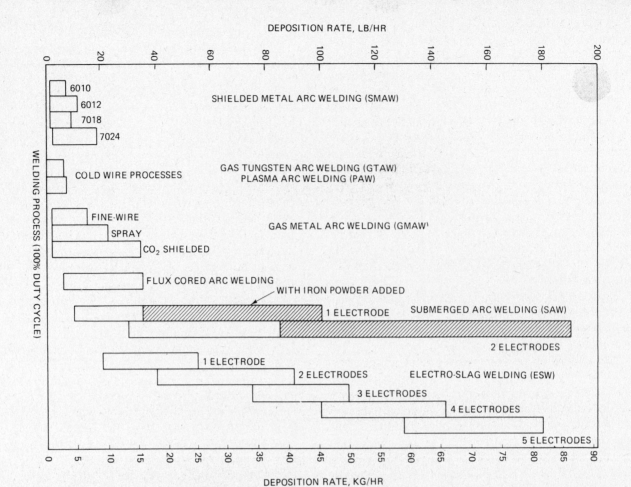

DEPOSITION RATE, LB/HR

WELDING PROCESS (100% DUTY CYCLE)

6010
6012
7018
7024

SHIELDED METAL ARC WELDING (SMAW)

COLD WIRE PROCESSES

GAS TUNGSTEN ARC WELDING (GTAW)
PLASMA ARC WELDING (PAW)

FINE-WIRE
SPRAY
CO_2 SHIELDED

GAS METAL ARC WELDING (GMAW)

FLUX CORED ARC WELDING
WITH IRON POWDER ADDED

1 ELECTRODE   SUBMERGED ARC WELDING (SAW)

2 ELECTRODES

1 ELECTRODE
2 ELECTRODES   ELECTRO-SLAG WELDING (ESW)
3 ELECTRODES
4 ELECTRODES
5 ELECTRODES

DEPOSITION RATE, KG/HR

FIGURE 5-98 *Deposition rate summary by welding process.*

in completing the weldment. The cost of labor has continually increased and will probably increase in the future. It is therefore important to utilize processes that are the most productive and can be used most efficiently. The productivity of a welding process is usually related to its deposition rates. Deposition rate data for each of the processes are presented in the process section; however, a summary of deposition rates based on 100% duty cycle is shown by Figure 5-98. This can only be considered as an indication and is not the entire story. The reason for this is that certain processes provide high deposition rates but may require more weld metal to complete the weld joint. Joint design and the amount of metal required to make the weld joint enter into this situation. The other factor that relates to process productivity also relates to labor cost, since each process may be applied in more than one way. The method of applying each continuous electrode wire process is summarized by Figure 5-99. Each of the methods of applying has a specific operator factor or duty cycle. This is based upon the amount of time that

the process is in actual operation depositing metal, versus the time available. Figure 16-5 in Chapter 16 will show the approximate operator factor related to the method of application.

The other factor has to do with cost of materials. It is obvious that the necessary materials consumed in making the weld must be costed and efforts must be made to determine the amount of each required. To assist in this, it is important to recognize that all filler metals may not be utilized to the same degree. The amount of filler metal purchased may not all be deposited in the weld joint. Figure 16-3 will show the utilization of filler metal related to the welding processes. The

| Method of Applying | WELDING PROCESS USE | | | |
	SAW	GMAW	FCAW	ESW
Manual (MA)	No	No	No	No
Semiautomatic (SA)	B	A	A	No
Machine (ME)	A	A	A	A
Automatic (AU)	A	A	A	A

FIGURE 5-99 *Summary of method of application for each process.*

194

filler metal with the lowest utilization is the covered electrode used for shielded metal arc welding. It is expected that about 65% of the weight of the electrodes purchased actually becomes deposited weld metal in the product. On the other hand, for the cold wire feed processes the amount of filler metal purchased versus that deposited will approach 100%. In gas metal arc welding and electroslag welding the amount of purchased weld metal deposited in the joint is 95% while the amount of flux-cored electrode wire deposited in the joint approaches 80%.

Another factor that affects the cost of welding has to do with the number of pieces or parts or the amount of identical welding that is to be done to meet production requirements. The volume of production has a great bearing on the cost of welding and must be involved in the selection of the process. High volume of like parts will warrant the expense of providing automated equipment. Automated equipment will utilize certain processes and not others; therefore, it becomes an important consideration in process selection. Low-volume work or the production of only a few parts normally involves manual, or possibly semiautomatic, welding. This is less expensive to procure but is also less productive.

The availability of the necessary equipment to produce the product also has a bearing on the process selection. Products similar to those normally produced will utilize the same manufacturing processes. Thus there is no need to consider different equipment since the equipment required is already available. On the other hand, if the new item is sufficiently different from existing products its production will require the addition of more equipment or will possibly utilize existing idle equipment. In any event the cost and the availability of the equipment are important factors. Tooling for high-volume production or specialized precision welding must also be considered.

It is important to be objective at this point of the analysis and question whether or not the availability of equipment overweighs new equipment that might be more productive. It might well be that changing to a more productive welding process will make available equipment obsolete. Perhaps this can be justified by a quick repayment of the cost of the new equipment. The point here is to consider both positions using available equipment or retiring the available equipment to use a more productive process.

The familiarity of the personnel can have an important

bearing on process selection. Many of the factors considered in equipment availability also apply to personnel available and their familiarity with the existing or new equipment. If new processes are adopted, it may be necessary to provide a training program to teach existing personnel new skills. This can be so costly that it might prohibit the change. On the other hand, if the price of the training is such that the overall total expense will be less it is best to provide the training and adopt the new process. A transition of this type can be very difficult because it will affect labor relations in the welding department. An objective point of view should be taken to compare the availability of skills of the existing work force versus what would be required if the process were changed. It is necessary to consider inspection, material preparation, and supervisory and administrative personnel, as well.

There are several other factors that must be considered in the choice of the metals joining process. A very important one has to do with the ability of engineering and design personnel to adopt to a new process. Designers familiar with resistance welding, for example, may have difficulty switching to gas metal arc welding. There is also the reluctance to switch from one well-known process to a new unknown one. Another factor has to do with the user of the end product. Users are familiar and apparently satisfied with the present method of production. Change can upset relationships and cause users to discontinue the product which is made by a new process. For example, the users may be accustomed to the bolted or riveted assembly and are suddenly confronted with a welded assembly and may have questions concerning this new part and its ability to withstand the service that they will subject it to. There is also the question of the reluctance to change versus the necessity to improve products and reduce production costs.

When all of the above factors are considered in the order presented a manageable plan evolves. It will enable the metalworking engineer or executive to chart a course with the known factors and to arrive at the preferred welding process. Alternate selections may result and practical production tests should be made to give the final answer.

QUESTIONS

1. What dictates the relationship between electrode wire meltoff rate and the welding current?

2. How is the CV power system like the electrical system in a house?

3. What welding process most always use the CV power system?

4. What is the main purpose of submerged arc flux? Other purposes?

5. Why isn't submerged arc welding an all-position welding process?

6. If AC is used for submerged arc, is it CC or CV?

7. Is a welding helmet required for submerged arc welding?

8. For gas metal arc welding, what changes are required in weld joint design?

9. Describe the electrogas method of GMAW. What position is it used for?

10. When is cooling water recommended for the GMAW torch or gun?

11. What is the major difference between FCAW and GMAW?

12. What equipment changes are necessary to change processes?

13. What are the three classifications of welding variables?

14. What adjustments change weld penetration? Explain.

15. What adjustments change weld bead width? Explain.

16. What adjustments change weld bead reinforcement? Explain.

17. What is *stickout* and what effect does it have?

18. Define *work angle* and *travel angle*.

19. What makes electroslag welding different from submerged arc welding?

20. What weld groove design is normally used for electroslag welding?

REFERENCES

1. "Bare Mild Steel Electrodes and Fluxes for Submerged Arc Welding," *American Welding Society* A5.17, Miami, Florida.

2. "Mild Steel Electrodes for Gas Metal Arc Welding," *American Welding Society* A5.18, Miami, Florida.

3. "Mild Steel Electrodes for Flux-Cored Arc Welding," *American Welding Society* A5.20, Miami, Florida.

4. "Portaslag Welding," edited by Howard Cary, Hobart Brothers Company, Troy, Ohio.

5. J. D. HARRISON, "Fatigue Tests of Electroslag Welded Joints," *Metal Construction and British Welding Journal,* London, August 1969.

6

THE OTHER
WELDING PROCESSES

6-1 BRAZING

6-2 OXY FUEL GAS WELDING

6-3 RESISTANCE WELDING

6-4 SOLID STATE WELDING

6-5 SOLDERING

6-6 ELECTRON BEAM WELDING

6-7 LASER BEAM WELDING

6-8 THERMIT WELDING

6-9 MISCELLANEOUS WELDING PROCESSES

6-1 BRAZING

Brazing is a group of welding processes which produces coalescence of materials by heating them to a suitable temperature and by using a filler metal having a liquidus above (840°F) 450°C and below the solidus of the base metals. The filler metal is distributed between closely fitted surfaces of the joint by capillary attraction. To avoid confusion it is necessary to explain braze welding which is different since the filler metal is *not* distributed by capillary attraction.

The **solidus** is the highest temperature at which the metal is completely solid, that is, the temperature at which melting starts.

The **liquidus** is the lowest temperature at which the metal is completely liquid, the temperature at which freezing starts. The solidus and liquidus for a particular alloy are definite.

There are five different ways of brazing and they all pertain to the method of applying the heat.

Dip Brazing (DBR)

There are two methods of dip brazing, (1) molten chemical bath dip brazing and (2) molten metal bath

dip brazing. In both cases brazing is accomplished by immersing clean and assembled parts into a molten bath contained in a suitable pot. Dip brazing is used for brazing small parts. The assembly should be self-jigging so the parts will maintain the same assembly relationship during the dipping and removal operation and until the brazing filler metal has completely solidified. By using a high-quality furnace and controller, a close temperature control is obtained with the dip brazing method.

The molten material is contained in a pot type furnace, which is heated by oil, gas, or electricity, or by means of electrical resistance units placed in the bath. Normally the parts to be brazed are first preheated in an air circulating furnace. When the parts have reached the preheat temperature they are then immersed in the molten bath for a suitable length of time.

The difference between the two methods of dip brazing is the molten material in the pot. In the molten chemical bath, the bath is called a flux bath and is made of chemicals which act as a flux. When a chemical bath is used the filler metal must be preplaced in the joints that are to be brazed. Fluxing of the assembly is not required.

The other method of dip brazing utilizes molten metal in the pot. The molten brazing material will flow into the joints to be brazed by capillary attraction. The parts must be clean and fluxed prior to molten metal dip brazing. A flux cover should be maintained over the surface of the molten metal bath.

Dip brazed parts normally distort less than torch brazed parts because of the uniform heating. It is suited for moderate- to high-production runs because the tooling is relatively complex. The method of applying the process may be manual or automatic. The process is suited for brazing small to medium parts with multiple or hidden joints. It can be used for all of the metals that can be brazed and is particularly suited for aluminum and other alloys that have melting points very close to the brazing temperature. The brazing operation can also perform certain of the heat treating operations on aluminum.

Furnace Brazing (FBR)

Furnace brazing is accomplished by placing cleaned parts in a furnace. The parts should be self-jigging and assembled, with filler materials preplaced near or in the joint. The preplaced brazing filler material may be in the form of wire, foil, filings, slugs, powder, paste, tape, etc. The furnaces are usually heated by electrical resistance. Other types of fuel can be used but only for muffle type furnaces. Automatic temperature controllers are required so that they can be programmed for the brazing temperatures and for cooling. The batch type furnace is used for medium-production work. A continuous conveyor type furnace is used for high-volume work. When continuous type furnaces are used several temperature zones may be employed which provide the proper preheat, brazing and cooling temperature. In either type, specialized holding fixtures are required.

Flux is employed except when an atmosphere is specifically introduced in the furnace to perform this function. Flux should not be used where post-braze cleaning is made difficult by the complexity of the design of the brazed parts. Furnace brazing is often done without the use of flux but by the use of special atmospheres in the brazing furnace. Flux is not necessary if the brazing is done in a reducing gas atmosphere such as hydrogen or other special gases. Inert gases—argon or helium—are sometimes employed to obtain special properties. Furnace brazing can also be performed in a vacuum which prevents oxidation and may eliminate the need for flux. Vacuum brazing is widely used in the aerospace and nuclear fields where reactive metals are being joined and where entrapped fluxes would not be acceptable. In vacuum brazing, the vacuum is maintained by continuous pumping which will remove volatile constituents liberated during the brazing operation. There are some base metals and filler metals which cannot be brazed in a vacuum since low boiling point or high vapor pressure constituents would be volatized and lost. The vacuum is a relatively economical method and is an accurately controlled atmosphere. It provides for surface cleanliness and allows the flow of filler metals without the use of fluxes.

It is important to select the correct atmosphere based on the type of base metals and filler metals being employed. For example, copper brazing of steels is normally done in a reducing atmosphere of high-purity hydrogen.

The distortion of furnace brazed assemblies is less than torch brazed parts. Furnace brazing is suitable for joining thin sections to thick sections. It can be used for brazing parts of all sizes having multiple joints and hidden joints. Most metals are completely annealed as a result of furnace brazing. In some cases the heat treating operation can be done in conjunction with the brazing cycle, provided that the two programs are compatible.

Induction Brazing (IBR)

The heat for induction brazing is obtained from the electrical resistance of the work to an electrical current induced in the parts to be brazed. The parts or joint are placed in an alternating current field but do not become a part of the electrical circuit. High-cycle alternating current ranging from 5,000 hertz to 5,000,000 hertz can be used. In general, motor generator alternating current sources operate in the 5,000–10,000 hertz range, spark gap oscillator units operate in the 20,000–300,000 hertz range, and vacuum tube oscillator units operate in the 200,000–5,000,000 hertz range. The output of these power sources is fed into a copper tubing work coil, usually water cooled, designed specifically to fit the shape of the parts to be brazed. They do not touch the parts but are coupled to them by the electrical field. The design of work coils and how they are coupled to the work pieces is quite complex. For more information, refer to the AWS Brazing Manual.[1] The frequency of the power source determines the type of heat that will be induced in the part. High-frequency power sources produce skinheating in the parts. Lower-frequency current results in deeper heating and is used for brazing heavier sections. Heating of the part usually occurs within 10–60 seconds. Sufficient time must be provided for the filler metal to flow through the entire joint and to form good fillets at the interface.

Induction brazing is ideally suited for high-volume manufactured parts. Mechanized systems for moving the parts to and from the coil are quite common. The filler metal is normally preplaced in the joint and the brazing operation can be done in air, in an inert gas atmosphere or in a vacuum. Brazing fluxes may or may not be used, depending on whether the work is done in the vacuum, in air, or in an inert gas. The major advantage of induction brazing is the rapid heating rates that make it suitable for brazing with filler metal alloys that tend to vaporize or segregate. A disadvantage of induction brazing is that the heat may not be uniform. Thin sections tend to heat up quicker than heavy sections and thin sections may tend to overheat. Field or shading coils are sometimes used to reduce the problem of overheating. Induction brazing is applied as an automatic process.

Infra-red Brazing (IRBR)

In infra-red brazing the heat is obtained from infrared heat or a *black* heat source below the red rays in the spectrum. There is some visible light involved but the principle heating is done by the invisible *black* radiation. Heat sources or lamps capable of delivering up to 5,000 watts of radiant energy are used. The lamps do not necessarily need to follow the contour of the parts being brazed even though the heat input varies by the square of the distance from the source. Radiation concentrating reflectors are often used. Sources other than electric lamps, but still supplying infra-red radiation, can be used. Parts to be brazed are positioned so that the radiant energy will impinge on the joint.

With infra-red brazing, as in furnace brazing and induction brazing, the parts can be contained in air, in an inert atmosphere, or in a vacuum. The same comments concerning fluxes and atmospheres apply. Infrared brazing is not as fast as induction brazing; however, the equipment required to provide the heat is much less expensive. Infra-red brazing is designed for automatic application and is not applied manually. Normally the parts to be brazed are self-jigging and the filler material is preplaced in or near the joint.

Resistance Brazing (RBR)

In resistance brazing the heat is obtained from the resistance to the flow of an electrical current through the parts being brazed. The parts become a portion of the electrical circuit. Electrodes may be copper alloys or carbon-graphite material. Resistance-welding machines can be used for supplying the electric current to the parts.

When resistance-welding equipment is used it is used at a lower power input than when it is used for resistance welding. Specially designed machines for resistance brazing are also used. Alternating current is normally employed. Direct current may be used but is not as common. The parts to be brazed are held between the two electrodes while the correct pressure and electrical current are applied. The pressure is maintained until the filler metal has solidified. High amperage current at low voltage is used. The heat is generated at the brazed joint interface and at the electrode to the part interface and depends upon the resistance to the current flow at these locations.

Resistance brazing is normally limited to applications where the brazing filler metal is preplaced; however, face feeding of the filler metal into the joint may be used in some applications. Resistance brazing is normally used for low-volume production where heating is localized at the area to be brazed.

The flux used for resistance brazing must be given specific attention since the conductivity of the flux is important. Normally brazing fluxes are insulators when cool and dry. When they become molten from the heat of the brazing operation they may become conductive. Fluxes are normally employed except when an atmosphere is utilized to perform the same function.

Torch Brazing (TBR)

Torch brazing is done by heating the parts to be brazed with the flame of a gas torch or torches. The temperature and the amount of heat required determine the gases used. The flame can be supplied by acetylene, or other fuel gas which may be burned with air, compressed air, or oxygen. For manual torch brazing normally a single torch and tip are used but for automatic torch brazing multiple tips may be required. Manual torch brazing is probably the most widely used brazing method and is shown by Figure 6-1. Torch brazing is very useful on assemblies that involve heating sections of different mass. The flame can be directed to the heavier part to produce uniform heating. Manual brazing is particularly useful for repair work. Automatic torch brazing is used in manufacturing operations where the rate and volume of production warrants the expense. An automatic brazing operation for an electrical part is shown by Figure 6-2. Normally, the work will move under or between multiple torches, as shown in this application.

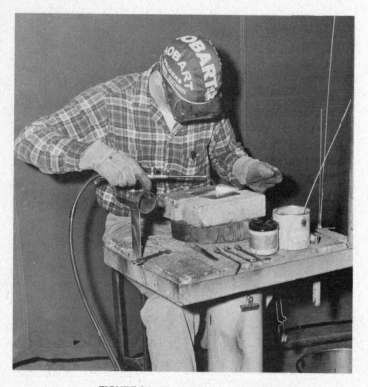

FIGURE 6-1   Torch brazing—manually applied.

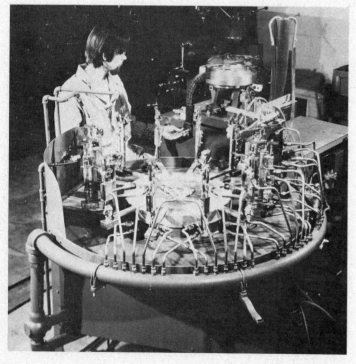

FIGURE 6-2   Automatic torch brazing electrical parts.

The brazing filler metal may be preplaced in or at the joint, or it may be face fed from hand-held filler metal, usually in the form of wire.

The parts must be assembled and self-jigging or held in place by mechanical means during the heating and cooling cycle. The parts must be cleaned, and flux is normally added to assist the capillary flow of the filler material. Position is not too important; however, gravity can be helpful in assisting capillary flow in complex joints.

Torch brazing can be used for a variety of materials. It can be used when parts of unequal mass are being brazed; however, because of the poor temperature control, metals that have a melting point close to the brazing temperature are not normally torch brazed. Torch brazed parts tend to have warpage or distortion when joining parts of different thicknesses. Torch brazing is used when the part to be brazed is too large, is an unusual shape, or cannot be heated by the other methods.

Other Brazing Methods

Other methods of brazing involve different methods of supplying the heat. Exothermic brazing, that utilizes the heat of specific chemical reactions can be used. An exothermic chemical reaction is any reaction between two or more chemicals in which heat is given off due to the free energy of the reaction. The solid state or nearly solid state metal oxide reactions are used. This method is not widely employed.

Other sources of heat can be those used for different welding processes. Any of the arc processes, where the arc is used as a source of heat, can be used for brazing. This includes the carbon arc process, the twin-carbon arc method, the gas tungsten arc process, and the plasma arc process. The laser beam and the electron beam can also be used. The heat input-heat loss relationship must be carefully controlled to avoid melting the parts to be brazed.

Almost all of the common metals can be brazed. Figure 6-3 is a table of these different metals and their **brazability**—that is, the capacity of a metal to be brazed under the fabrication conditions.

The thicknesses of the parts to be joined are dependent upon the ability of providing sufficient heat to the heavier sections without excessive heat to the thinner sections. This can be a problem with some of the brazing methods and is a major consideration in selecting the heating method.

Filler Materials and Fluxing

Filler materials for brazing are covered by an AWS specification.[2] They are classified according to analysis:

Base Metal	Filler Metal Type	Torch Brazing Flame Type	AWS Flux Type
Aluminums	Alum-silicon	Slightly reducing	Flux No. 1
Brasses	Silver alloy	Slightly reducing	No. 3A or 3B
Bronzes	Copper zinc	Slightly reducing	Flux No. 3B
Copper	Copper zinc	Slightly reducing	Flux No. 3B
Inconel	Silver alloy	Slightly reducing	No. 3A or 3B
Iron, cast, nodular	Silver alloy	Neu/slight oxid.	No. 3A or 3B
Iron, wrought	Copper zinc	Slightly reducing	Flux No. 3B
Monel	Silver alloy	Slightly reducing	No. 3A & 3B
Nickel	Silver alloy	Slightly reducing	No. 3A & 3B
Nickel copper	Copper	Slightly reducing	No. 3B
Nickel silver	Silver alloy	Slightly reducing	No. 3A or 3B
Precious metals	Variable	Variable	Variable
Steel low alloys	Copper zinc	Slightly reducing	No. 3B
Steel high carbon	Copper zinc	Slightly reducing	No. 3B
Steel low carbon	Copper zinc	Slightly reducing	No. 3B
Steel medium carbon	Copper zinc	Slightly reducing	No. 3B
Stainless steel	Silver alloy	Slgtly reduc/neu.	No. 4

Note: In many cases, different filler metals may be used. The easiest to use filler metal is shown above. Color matching has been ignored.

FIGURE 6-3 *Brazeability of base metals using the torch brazing.*

aluminum-silicon, copper, copper-zinc, copper-phosphorus, copper-gold, heat-resisting materials, magnesium, and silver are the basic groupings. The composition of the different filler metals, as well as the operating range and recommended uses, are given by the table shown in Figure 6-4. Filler metal selection is based on the metal being brazed.

Certain brazing filler metals contain cadmium in significant amounts. When these are used adequate ventilation is required.

Filler metals are available in many forms; the most common is the wire or rod. Filler metal is also available as thin sheet, powder, paste, or as a clad surface of the part to be brazed.

The placement of the filler metal affects the quality of the joint. For normal lap joints the filler metal should be supplied at only one end and allowed to flow completely through the joint by capillary action. If the filler metal is supplied at both ends gas will be trapped in the joint and will create voids which will drastically reduce the effective area of the braze. Another advantage of supplying filler metal at only one end is for quality control. It will be apparent that the brazed joint is complete if the filler material creates a fillet at the end opposite to where it was introduced. Filler metal cannot be made to flow by means of capillary action into a blind joint. Gas will be trapped and will not allow complete flow of filler metal throughout the faying surfaces. In such cases venting must be provided. This applies also to small tanks or vessels. The gas in these containers will expand as a result of the heat

and will prevent filler metals from penetrating the abutting surfaces.

The correct fluxing material must be used. The American Welding Society has established six different types of fluxes which satisfy most brazing requirements. Figure 6-5 shows the fluxes, as well as the recommended use and types of filler metals and temperature range that they are designed to meet.

Some brazing fluxes contain fluorine compounds. The package will show a precautionary label stating that adequate ventilation is required. The ventilation requirement must be followed.

The placement of the flux also affects the quality of the brazed joint. Paste flux is the most common form and is usually spread over the surfaces to be joined. It is also painted on the preplaced brazing filler materials. Brazing fluxes can be sprayed for high-volume production. In addition liquid flux can be introduced into the fuel gas and supplied to the flame for torch brazing at the point where it is needed. Flux in the flame may not be satisfactory for large, deep, or complex joints. In such cases preplaced paste flux may also be required.

For some of the brazing methods a special atmosphere is used instead of flux. The atmosphere must be selected based on the metals being joined. When atmospheres are used flux may not be required. Atmosphere selection depends upon the base materials being brazed. For torch brazing the atmosphere is the product of the combustion of the flame. The neutral or reducing flame is normally used. A slightly oxidizing flame may be used for certain materials.

AWS Classification	Brazing Temperature Range F°	C°	Nominal Composition (%)
Aluminum-silicon alloys			
BAlSi-2	1110-1150	599-621	92.5 Al, 7.5 Si
BAlSi-3	1060-1120	571-601	86 Al, 10 Si, 4 Cu
BAlSi-4	1080-1120	582-604	88 Al, 12 Si
BAlSi-5	1090-1120	588-604	90 Al, 10 Si
Magnesium alloys			
BMg-1	1120-1160	604-627	89 Mg, 2 Zn, 9 Al
BMg-2a	1080-1130	582-610	83 Mg, 5 Zn, 12 Al
Copper-phosphorus alloys			
BCuP-1	1450-1700	788-927	95 Cu, 5P
BCuP-2	1350-1550	732-843	93 Cu, 7 P
BCuP-3	1300-1500	704-816	89 Cu, 5 Ag, 6 P
BCuP-4	1300-1450	704-788	87 Cu, 6 Ag, 7 P
BCuP-5	1300-1500	704-816	80 Cu, 15 Ag, 5 P
Copper-copper zinc alloys			
BCu-1	2000-2100	1093-1149	99.9 Cu (min)
BCu-1a	2000-2100	1093-1149	99.0 Cu (min)
BCu-2	2000-2100	1093-1149	86.5 Cu (min)
RBCuZn-A	1670-1750	910-954	57 Cu, 42 Zn, 1 Sn
RBCuZn-D	1720-1800	938-982	47 Cu, 11 Ni, 42 Zn
Silver alloys			
BAg-1	1145-1400	618-760	45 Ag, 15 Cu, 16 Zn, 24 Cd
BAg-1a	1175-1400	635-760	45 Ag, 15 Cu, 16 Zn 24 Cd
BAg-2	1295-1550	700-843	45 Ag, 26 Cu, 21 Zn, 18 Cd
BAg-2A	1310-1550	710-843	30 Ag, 27 Cu. 23 Zn, 20 Cd
BAg-3	1270-1500	688-816	52 Ag, 15 Cu, 15 Zn, 15 Cd, 3 Ni
BAg-4	1435-1650	780-899	40 Ag, 30 Cu, 23 Zn, 2 Ni
BAg-5	1370-1550	743-843	45 Ag, 30 Cu, 25 Zn
BAg-6	1425-1600	774-871	53 Ag, 31 Cu, 16 Zn
BAg-7	1205-1400	651-760	56 Ag, 22 Cu, 17 Zn, 5 Sn
BAg-8	1435-1650	780-899	77 Ag, 23 Cu
BAg-8a	1410-1600	766-871	77 Ag, 23 Cu
BAg-13	1575-1775	857-635	54 Ag, 40 Cu, 5 Zn, 1 Ni
BAg-13a	1600-1800	871-982	56 Ag, 42 Cu, 2 Ni
BAg-18	1325-1550	718-843	60 Ag, 40 Cu
BAg-19	1610-1800	877-982	92 Ag, 8 Cu
Precious metals			
BAu-1	1860-2000	1016-1093	37 Au, 63 Cn
BAu-2	1635-1850	890-1010	79.5 Au, 20.5 Cn
BAu-3	1885-1995	1030-1090	34 Au, 62 Cn, 4 Ni
BAu-4	1740-1840	949-1004	82 Au, 18 Ni
Nickel alloys			
BNi-1	1950-2200	1066-1204	14 Cr, 3 Br, 4 Si, 4 Fe, 75 Ni
BNi-2	1850-2150	1010-1177	7 Cr, 3 Br, 4 Si, 3 Fe, 83 Ni
BNi-3	1850-2150	1010-1177	3 Br, 4 Si, 2 Fe, 91 Ni
BNi-4	1850-2150	1010-1177	1 Br, 3 Si, 2 Fe, 94 Ni
BNi-5	2100-2200	1149-1204	19 Cr, 10 Si, 71 Ni
BNi-6	1700-1875	927-1025	11 Br, 89 Ni
BNi-7	1700-1900	927-1038	13 Cr, 10 Br, 77 Ni

FIGURE 6-4 *AWS filler metals for brazing (per AWS A5.8).*

AWS Brazing Flux Type No.	Base Metals Being Brazed	Recommended Filler Metals	Recommended Useful Temp. Range F°	C°	Major Flux Ingredients	Forms Available
1	All brazeable aluminum alloys	BAlSi	700-1190	371-643	Chlorides Fluorides	Powder
2	All brazeable magnesium alloys	BMg	900-1200	482-649	Chlorides Fluorides	Powder
3A	Copper & copper-base alloys (except those with aluminum) iron base alloys; cast iron; carbon & alloy steel; nickel & nickel base alloys; stainless steels; precious metals	BCuP Bag	1050-1600	566-871	Boric Acid Borates Fluorides Fluoborates Wetting Agent	Powder paste liquid
3B	Copper & copper-base alloys (except those with aluminum); iron base alloys; cast iron; carbon & alloy steel; nickel & nickel base alloys; stainless steel; precious metals	BCu BCuP BAg BAu RBCu Zn BNi	1350-2100	732-1149	Boric Acid Borates Fluorides Fluoborates Wetting Agent	Powder paste liquid
4	Aluminum bronze, aluminum brass & iron or nickel base alloys containing minor amounts of Al &/or Ti	BAg (all) BCuP (copper base alloys only)	1050-1600	566-871	Chlorides Fluorides Borates Wetting Agent	Powder paste
5	Same as 3A & B above	Same as 3B (excluding BAg through —7)	1400-2200	760-1204	Borax Boric acid Borates Wetting Agent	Powder paste liquid

FIGURE 6-5 *Fluxes for brazing—AWS brazing manual.*[1]

Joint Designs

When designing a joint for brazing, the following six factors must be considered

The type of joint required.

The clearance between the parts.

The surface finish of the faying surfaces.

Placement of the filler metal.

The placement of the flux when used.

The possibility of gas entrapment.

Brazed joints fall into two general types, *butt joints* and *lap joints*. Butt joints are subjected to tensile or compressive loads and lap joints are normally subjected to shear loading. There are many variations of these two types. Butt joints provide a limited area for brazing. The strength of the filler material is usually less than the strength of the base metal. A butt joint will not provide 100% joint efficiency. If the joint is scarfed to form a bevel, additional area will be provided which will increase the strength of the brazed joint. The bevel area should provide at least three times the area that is obtained with a simple square butt joint. This joint detail is shown by Figure 6-6. Unfortunately, scarf joints are more difficult to hold in alignment than the square butt or lap joints.

Lap joints are more widely used since they can be designed to provide sufficient brazed area so that the joint is as strong as the base metal. Unfortunately, lap joints tend to be unbalanced joints and this produces stress concentrations which adversely affect the joint strength. Every effort should be made to provide a balanced lap joint to properly carry the load. Figure 6-6 shows the different brazed joints and the recommended types.

It is important to compensate for unequal expansion and contraction of a joint design. This can occur when brazing dissimilar metals and when the difference of thermal expansion would create tensile loads on the filler metal during cooling.

The surface finish of the faying surfaces should be between 30 and 80 microinches for best joint strength. The filler metal may not wet the surfaces completely if they are too smooth. Furthermore, the filler metal will not distribute itself throughout the complete joint by capillary attraction if they are too smooth. If the surfaces are too rough, only the high points may be properly brazed. With very rough surfaces the clearance will be too great to provide optimum strength of the brazed joint.

Joint Cleanliness

In addition to selecting the proper joint clearance, the proper filler material, and flux, it is important to have extremely clean surfaces for the brazed joint. Mechanical surface preparations such as grinding, sand blasting, wire brushing, filing, and machining can be used. However, in every case care must be taken to make sure that the surface is clean. For example, grit should not become embedded in the surface. Wire brushing can result in the folding in of oxides and burnishing of the surface. Chemical cleaning can be used to remove dirt and oils. Solvents, alkaline baths, acid baths, salt bath pickling, and ultrasonic cleaning have all been used successfully. When the surfaces have been cleaned, flux is used to protect the surface from oxidation or from other undesirable chemical action during the heating and brazing operation. Fluxes are not designed to clean joints. They are designed to keep cleaned joints clean during the brazing operation. They will combine with, dissolve, or inhibit the formation of chemical compounds which might interfere with the quality of the brazed joint.

Braze Quality

Close adherence to the design factors, filler metal selection, flux selection, and cleanliness will insure quality brazed joints. When the joint does not exhibit the quality required, investigate using the following troubleshooting hints:

1. The brazing filler metal doesn't wet the surface and balls up instead of flowing into the joint:

    Increase the amount of flux used.

    Roughen the surface slightly, especially the surface of cold-drawn or cold-rolled stock.

    Acid pickle parts to remove surface oxides.

    Change work position so gravity will help the filler metal fill the joint.

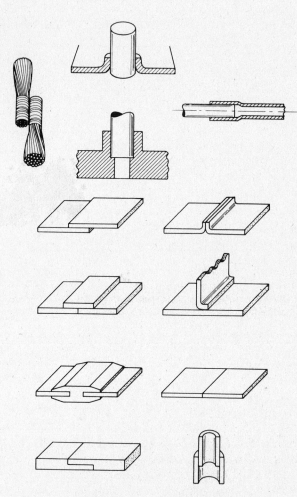

FIGURE 6-6 *Joint details for brazing.*

The clearance between the parts being joined is important. If the joint clearance is too small it will not allow capillary attraction to cause the filler metal to flow uniformly throughout the entire joint. If the clearance is too great, filler metal may not flow throughout the joint, and a low strength joint will result. The brazing filler metal also has an influence on the clearance. Another factor is the length or area of the joint. For smaller areas, a smaller joint clearance can be used. In general, when using an atmosphere system smaller joint clearances can be used. Where fluxes are required the clearances are normally larger. Clearances range from 0.001 in. to 0.025 in. (0.025 mm to 0.635 mm) for clearance when fluxes are involved. The recommended clearances of different groups of brazing filler metals are shown by Figure 6-7.

Filler Metal AWS Classification	Joint Clearance in.	mm	Brazing Conditions
BAlSi group	0.002-0.008	0.051-0.203	For length of lap less than 1/4" (6.4mm)
	0.008-0.010	0.203-0.254	For length of lap greater than 1/4" (6.4mm)
BCuP group	0.001-0.005	0.025-0.127	No flux or mineral brazing fluxes
BAg group	0.002-0.005	0.051-0.127	Mineral brazing fluxes
	0.000-0.002	0.000-0.051	Gas atmosphere brazing fluxes
BAu group	0.002-0.005	0.051-0.127	Mineral brazing fluxes
	0.000-0.002	0.000-0.051	Gas atmosphere brazing fluxes
BCu group	0.000-0.002	0.000-0.051	Gas atmosphere brazing fluxes
BCuZn group	0.002-0.005	0.051-0.127	Mineral brazing fluxes
BMg	0.004-0.010	0.102-0.254	Mineral brazing fluxes
BNi group	0.002-0.005	0.051-0.127	General applications flux or atmosphere
	0.000-0.002	0.000-0.051	Free flowing types, atmosphere brazing

FIGURE 6-7 *Recommended joint clearance at brazing temperatures.*

2. The brazing alloy does not flow through the joint even though it melts and forms a fillet:

Allow more time for heating.

Heat to a higher temperature.

Determine the clearance in the joint and if required rework it to be more loose or tighter.

Apply flux to both the base metals and brazing filler metal.

Do a more thorough cleaning job before assembly.

3. The assembled joint was tight and it opens up during brazing:

The clearance was too small and a load was introduced into one part which causes the opening.

The parts have unequal coefficients of expansion due to dissimilar metals.

Unsupported section might cause improper clearances due to sag from heating.

4. The brazing filler metal melts but does not flow:

Coat the filler metal with flux before using and apply flux generously to the base metal.

Mechanically or chemically clean the filler metal if there are surface oxides present.

5. The brazing filler metal flows away from the joint instead of into the joint:

Provide a reservoir in the joint into which the brazed filler metal can flow.

Reposition the assembly so that gravity will help the filler metal flow into the joint.

Remove burrs, edges, or other obstacles over which the brazing alloy might not flow.

Above all, make sure that the filler metal alloy is compatible with the base metal and that the proper temperatures and fluxes are employed.

To determine the strength of a brazed joint the standard method should be used.[3] The AWS Standard outlines the procedure to be used for making tests that are comparable to others.

For certain work the brazer, or one who performs a manual or semiautomatic brazing operation, must be qualified. Qualification is in accordance with Section IX, of the "ASME Boiler and Pressure Vessel Code."[4] Part C pertains to brazing ferrous and nonferrous materials. This specification must be read carefully. It introduces new uses for positions in flat flow, vertical down flow, vertical up flow, horizontal flow, and special positions. Five types of tests are used and the requirements for "Procedure Qualification and Performance Qualification" are presented.

Disadvantages and Uses

There are several disadvantages to brazing. There is the possibility of lack of *color match* of the parts being brazed and the brazing filler material. The strength of brazed joints may be less than the strength of the parts being joined.

Brazing is widely used throughout industry, and applications are so numerous that it is impossible to list them. Three major industries using brazing are the electrical industry, the utensil-manufacturing industry, and the maintenance industry. In the maintenance field, brazing is widely used for repairing cast iron parts.

6-2 OXY FUEL GAS WELDING

Oxy fuel gas welding is a group of welding processes which produce coalescence by heating materials with a fuel gas flame or flames with or without the application of pressure and with or without the use of filler metal. There are three major processes within this group: oxyacetylene welding, oxyhydrogen welding, and pressure gas welding. There is one process of minor industrial significance, known as air acetylene welding, in which heat is obtained from the combustion of acetylene with air.

The most popular process in this group, **oxyacetylene welding,** is a gas welding process which produces coalescence of metals by heating them with a gas flame or flames obtained from the combustion of acetylene with oxygen. The process may be used with or without the application of pressure and with or without the use of filler metal. Oxyhydrogen welding uses hydrogen as the fuel gas. Pressure gas welding normally uses acetylene. Coalescence is produced simultaneously over the entire area of abutting surfaces by the application of pressure and without the use of filler metal. In this section the oxyacetylene welding process is described in detail with minor attention to the other gas welding processes.

Oxyacetylene Welding (OAW)

The oxyacetylene welding process shown by Figure 6-8 and Figure 6-9 consists of a high-temperature flame produced by the combustion of acetylene with oxygen and directed by a torch. The intense heat of the flame 6,300°F (3,482°C) melts the surface of the base metal to form a molten puddle. Filler metal is added to fill gaps or grooves. As the flame moves along the joint the melted base metal and filler metal solidify to produce the weld.

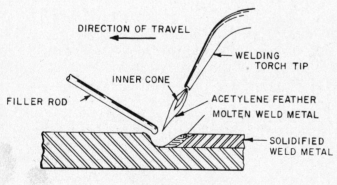

FIGURE 6-8 *Process diagram.*

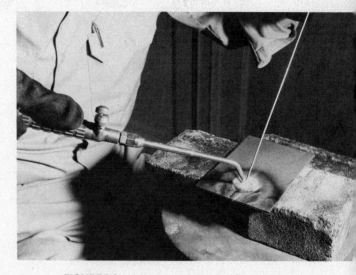

FIGURE 6-9 *Using oxyacetylene welding process.*

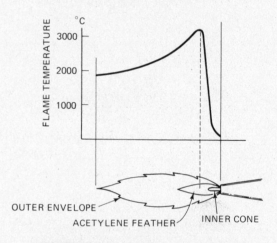

FIGURE 6-10 *The temperature of the flame.*

The temperature of the oxyacetylene flame is not uniform throughout its length and the combustion is also different in different parts of the flame. Figure 6-10 shows the relationship between temperature and the flame and the composition of the gases in different portions of the flame. The temperature is the highest just beyond the end of the inner cone and decreases gradually toward the end of the flame. The chemical reaction for a one-to-one ratio of acetylene and oxygen plus air is as follows:

$$C_2H_2 + O_2 = 2CO + H_2 + \text{Heat}$$

This is the primary reaction; however, both carbon monoxide and hydrogen are combustible and will react with oxygen from the air:

$$2CO + H_2 + 1.5O_2 = 2CO_2 + H_2O + \text{Heat}$$

This is the secondary reaction which produces carbon dioxide, heat, and water.

There are three basic flame types: neutral (or balanced), excess acetylene (carborizing), and excess oxygen (oxidizing). These three flames are shown by Figure 6-11. The neutral flame has a one-to-one ratio of acetylene and oxygen. It obtains additional oxygen from the air and provides complete combustion. It is generally preferred for welding. The neutral flame has a clear, well-defined, or luminous cone indicating that combustion is complete. The carborizing flame has excess acetylene. This is indicated in the flame when the inner cone has a feathery edge extending beyond it. This white feather is called the acetylene feather. If the acetylene feather is twice as long as the inner cone it is known as a 2X flame which is a way of expressing the amount of excess acetylene. The carborizing flame may add carbon to the weld metal. The oxidizing flame

which has an excess of oxygen has a shorter envelope and a small pointed white cone. The reduction in length of the inner core is a measure of excess oxygen. This flame tends to oxidize the weld metal and is used only for welding specific metals. Most welding procedures, utilizing oxyacetylene welding, use the neutral flame. The welder soon learns proper flame adjustment.

Advantages and Major Uses

The oxyacetylene welding process has the following advantages. The equipment is very portable. It is relatively inexpensive, it can be used in all welding positions and the puddle is visible to the welder. The equipment is versatile. It can be used for welding, brazing, soldering, and with proper equipment, for flame cutting. It can also be used as a source of heat for bending, forming, straightening, hardening, etc.

The oxyacetylene welding process is normally used as a manual process. It can be mechanized, but this is not too common. It is rarely used for semiautomatic applications.

Oxyacetylene welding is used for welding most of the common metals as shown by Figure 6-12.

When welding any metal, the appropriate filler material must be selected and used. The filler metal must match the composition of the base metal to be welded and normally contains deoxidizers to aid in producing sound welds. Flux is also required for welding certain materials.

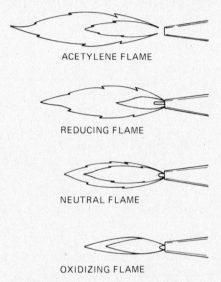

ACETYLENE FLAME

REDUCING FLAME

NEUTRAL FLAME

OXIDIZING FLAME

FIGURE 6-11 *The three types of flame.*

Base Metal	Filler Metal Type	Flame Type	Flux Type	Chapter
Aluminums	Match base metal	Slightly reducing	Al. flux	13-1
Brasses	Navy brass	Slightly oxidizing	Borax flux	13-3
Bronzes	Copper tin	Slightly oxidizing	Borax flux	13-3
Copper	Copper	Neutral	None	13-3
Copper nickel	Copper nickel	Reducing	None	13-3
Inconel	Match base metal	Slightly reducing	Fluoride flux	13-1
Iron, cast	Cast iron	Neutral	Borax flux	13-1
Iron, wrought	Steel	Neutral	None	13-3
Lead	Lead	Slightly reducing	None	13-4
Monel	Match base metal	Slightly reducing	Monel flux	13-6
Nickel	Nickel	Slightly reducing	None	13-6
Nickel silver	Nickel silver	Reducing	None	13-6
Steel, low alloy	Steel	Slightly reducing	None	12-3
Steel, high carbon	Steel	Reducing	None	12-4
Steel, low carbon	Steel	Neutral	None	12-2
Steel, medium carbon	Steel	Slightly reducing	None	12-4
Steel, stainless	Match base metal	Slightly reducing	SS flux	12-6

FIGURE 6-12 *Base metals weldable by the oxyacetylene process.*

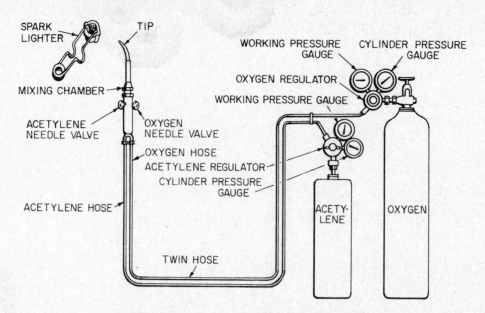

FIGURE 6-13 *The apparatus required for welding with OAW.*

The oxyacetylene welding process is normally used for welding thinner materials up to 1/4 in. (6.4 mm) thick. It can be used for welding heavier material but it is rarely used for thick metals. Its major industrial applications are in the field of maintenance and repair, the welding of small diameter pipe, and for light manufacturing.

Welding Apparatus

The apparatus and equipment employed for oxyacetylene welding are shown by Figure 6-13. This diagram shows the (1) welding torch and tips, (2) oxygen and acetylene hose, (3) oxygen and acetylene regulators, (4) oxygen cylinder, and (5) acetylene cylinder. A spark lighter is normally used. The welding torch, sometimes called a *blow pipe,* is the major piece of equipment for this process. It performs the function of mixing the fuel gas with oxygen and provides the required type of flame, which is directed as desired. The torch consists of a handle or body, which contains the hose connections for the oxygen and the fuel gas. It also contains an oxygen and acetylene valve for regulating gas flow and a mixing chamber. Various-sized tips can be attached. The torch must be well constructed and rugged, since it is in contact with the high temperature flame and is expected to have a long life.

There are two basic types of torches, the medium-pressure torch, which is most popular, and the low-pressure or injector type. When using the medium-pressure torch both oxygen and acetylene are supplied at approximately the same pressure, which may vary from 1–10 psi (0.07–0.7 kg/cm^2) depending on the size of the tip being used. The two gases are mixed together in the mixing chamber in the torch handle. Some torches have the mixing chamber in the tip as shown by Figure 6-14.

The injector type of torch uses acetylene at pressures less than 1 psi (0.07 kg/cm^2) and are designed so that the oxygen at a higher pressure draws the acetylene into the mixing chamber. Any change in oxygen flow will produce a relative change in acetylene flow so that the proportion of the two gases remains constant.

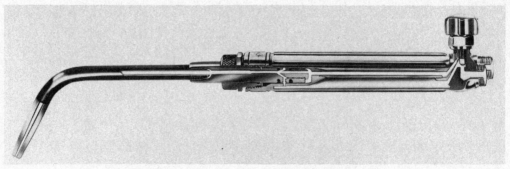

FIGURE 6-14 *Medium pressure oxyacetylene torch.*

The valves on the body of the torch control the amount of oxygen and acetylene or fuel gas which flows to the mixing chamber where they are combined. Different welding tips are available so that the same torch handle can be used for a wide variety of operations.

Welding tips are available in a variety of sizes, which are determined by the drill size of the orifice or hole in the flame end of the tip. The larger tip will have a larger orifice which will produce a larger flame and use more gas. The larger flame supplies greater amounts of heat. The proper tip size must be selected for welding different metals and metal thicknesses. Welding procedure schedules indicate the tip size and gas pressures to be employed. This will determine the volume of gases used. Tips must fit properly and tight to the torch. They must be kept clean for proper operation.

Some welding torches are designed so that they can be converted to an oxyacetylene flame cutting torch by replacing the welding tip with a cutting attachment.

The connections for oxygen and acetylene or fuel gas hoses to the torch handle have special threads so that they cannot be incorrectly attached. The oxygen fittings have right-hand threads and the acetylene or fuel gas fittings have left-hand threads. There is a special sequence for opening and closing valves and lighting the torch, which must be followed for safe operation.

Gas pressure regulators are required, for both the oxygen and acetylene. Regulators reduce the pressure of the gas in the cylinder or supply system to the pressure used in the torch. The pressure in an oxygen cylinder can be as high as 2,200 psi (154 kg/cm^2) and this must be reduced to a working pressure of from 1 to 25 psi (0.07 to 1.75 kg/cm^2). The pressure of acetylene in an acetylene cylinder can be as high as 250 psi (17.5 kg/cm^2) and this must be reduced to a working pressure of from 1 to 12 psi (0.07 to 0.84 kg/cm^2). When gases are piped to the work stations from a central supply the pressure is lower than above, but regulators are still required. A gas pressure regulator will automatically deliver a constant volume of gas to the torch at the adjusted working pressure. The regulators for oxygen and for acetylene and for liquid petroleum fuel gases are of different construction. They must be used only for the gas they are designed for.

There are two types of regulators, the single-stage regulator and the two-stage regulator. The *single-stage regulator* reduces the cylinder pressure of the gas to a working pressure in one step. Single-stage regulators must be readjusted from time to time to maintain the required working pressure. The gas pressure in the cylinder decreases gradually as gas is withdrawn. Single-stage regulators are less expensive than two-stage regulators and are more popular.

The *two-stage regulator* makes the reduction of pressure in two steps. The first step reduces the cylinder pressure to an intermediate pressure. The second step reduces this intermediate pressure to the desired working pressure. A two-stage regulator is simply two single-stage regulators in the same case. The two-stage regulator provides more accurate regulation and eliminates the need to readjust the regulator as the pressure in the supply tank is reduced. Regulators have two pressure gauges; one shows the pressure of the gas inside the cylinder; the other shows the working pressure that is being supplied to the torch. Figure 6-15 shows a cutaway view of a gas regulator.

FIGURE 6-15 *Cut away view of gas regulator.*

The operation of a regulator for controlling gas pressure and providing for uniform gas flow is rather straightforward. The regulator consists of a flexible diaphragm, which controls a needle valve between the high pressure zone and the working pressure zone, a compression spring and an adjusting screw, which compensates for the pressure of the gas against the diaphragm. The needle valve is on the side of the diaphragm exposed to high gas pressure while the compression spring and adjusting screw are on the opposite side in a zone vented to the atmosphere. The spring is compressed by the adjusting knob on the outside of the regulator. In the closed position the diaphragm is flat and the needle valve closes the orifice to the high pressure zone. In this condition the compression spring is

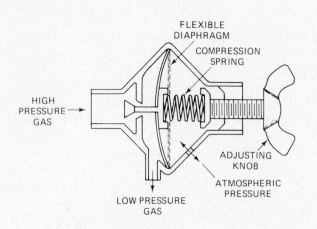

FIGURE 6-16 *Gas regulator operation.*

not loaded and the adjusting knob is backed off. As the knob is turned clockwise it compresses the spring, which in turn presses against the diaphragm to open the needle valve into the high-pressure zone. As the valve opens high-pressure gas enters the chamber and tends to push the diaphragm against the compression spring. This will tend to close the needle valve. When the spring is further compressed it will open the needle valve further allowing more gas through the orifice and into the pressure zone. By balancing the compression spring against the pressure of the gas, the needle valve is kept at the right opening to allow the correct flow of gas through the orifice. The compression spring balanced against the gas pressure keeps the valve at the proper opening to allow the required flow and pressure on the low-pressure side of the regulator. Figure 6-16 shows a diagram of a gas regulator. Single-stage regulators are used for plant piping systems since the pressure in the pipe system is much lower than in the cylinder. Torches, regulators, and other gas apparatuses must be approved by one of the approving agencies. They must also be properly handled and maintained.

Gas hose, used between pieces of apparatus, is specified by its inside diameter. The 1/4 in. (6.4 mm) size is the most common. The hoses may be separate or may be molded together for ease in handling. The hose fittings have the same threaded connections as the connections on the torch and the regulator, i.e., oxygen has right-hand threads and acetylene or fuel gas has left-hand threads. In addition, the acetylene hose connection has a groove around the outside to distinguish it from the oxygen hose connection. There is no common color code for gas hoses; however, in North America green is used for oxygen hose. In Europe, blue is used for oxygen

hose. In North America, red is used for acetylene or fuel gas hose; however, in Europe orange is used. Black is sometimes used for oxygen hose. Hose should never be used for one gas if it was previously used for the other.

The hose should be kept in good repair and if leaks occur they must be repaired or the hose replaced.

A torch must be lighted by a spark lighter consisting of a flint on a lever so it can be moved across a piece of roughened steel. Matches or cigarette lighters or similar items should not be used to light an oxyacetylene torch since it brings the hand too close to the flame. A convenient accessory is a gas saver or economizer, which includes a bracket for hanging the torch. When the torch is hung on the bracket it closes the valves to stop the flow of oxygen and acetylene. The device has a pilot light so that when the torch is lifted from the bracket near the pilot light the torch can be lighted. This apparatus is shown in Figure 6-17.

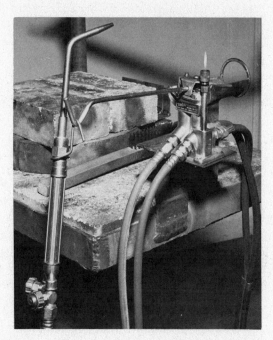

FIGURE 6-17 *Gas valve and pilot light.*

Gas Supply

The oxygen required for oxyacetylene welding can be supplied in several ways. The more common way is the use of high-pressure oxygen cylinders at the welding station. Another way is to manifold oxygen cylinders and run the oxygen through a piping system to the welding stations. This is common where there are a large number of welding or cutting stations that use oxygen. When a piping distribution system is used it can be supplied by liquid oxygen if a large amount of oxygen is used.

Acetylene or fuel gases are often supplied in cylinders taken to the welding station. However, they can be piped throughout the plant in the same manner as oxygen. The acetylene may be supplied to the piping system by manifolding cylinders or by an acetylene generator. The acetylene generator produces acetylene at the plant site by the reaction of carbide and water. In all cases, the installation and operation of piping systems must be in accordance with strict specifications and safety requirements.

Filler Materials

The American Welding Society provides a specification[5] covering the composition of filler metal used with the oxyacetylene or oxy fuel gas welding process. There are three grades, RG 45, RG 60, and RG 65, having a minimum tensile strength of 45,000 psi (3,150 kg/cm^2), 60,000 psi (4,200 kg/cm^2) and 67,000 psi 4,690 kg/cm^2) respectively. There are no chemical composition requirements. Figure 6-12 shows the base metals weldable by the oxyacetylene welding process. This table also shows the type of filler metal required, the flame type, and the type of flux.

Welding flux is required to maintain cleanliness of the base metal, at the welding area, and to help remove the oxide film on the surface of the metal. The welding area should be cleaned by any method. Flux melts at about the melting point of the base metal and helps protect the molten metal from the atmosphere. The molten flux combines with base metal oxides and removes them. There is no national standard for gas welding fluxes. They are categorized according to the basic ingredient in the flux or the base metal for which they are to be used. Fluxes are usually in powder form. These fluxes are often applied by sticking the hot filler metal rod in the flux. Sufficient flux will adhere to the rod to provide proper fluxing action as the filler rod is melted in the flame. Other types of fluxes are of a paste consistency which are usually painted on the filler rod or on the work to be welded. Welding rods with a covering of flux are also available. Fluxes are available from welding supply companies and should be used in accordance with the directions accompanying them.

Quality of Welds

The quality of a weld made with the oxyacetylene process can equal the quality of the base metal being welded. This is based on the use of the proper filler metal, the proper flux and the skill of the welder. The procedure will show the proper tip size, torch adjustment for the proper type of flame, and the travel speed. Figure 6-18 shows a good weld and common welding mistakes.

Welding Schedules

The oxyacetylene welding process is rarely used for joining heavy thicknesses. The table shown by Figure 6-19 is a schedule that can be used for welding material ranging from the thinnest up to the heaviest. The tip size is given by showing the orifice size and the equivalent drill size, since manufacturers utilize different numbering systems for their tips. Each manufacturer relates tip size number to either the drill size or the orifice size. The length of the inner cone is shown, as well as the recommended oxygen and acetylene pressure. The diameter of the filler rod is also shown. This schedule can be used for all-position welding. The major requirement for out-of-position welding is the control of the weld puddle, which relates to the skill of the welder.

This schedule is based on welding of clean mild steel using a neutral flame and not using a flux. Information concerning the welding of the different metals is provided in Chapters 12, 13, and 14.

Safety Considerations

The oxyacetylene process is a safe welding process, provided that proper precautions are taken. Normal precautions involve the respect for open flames, for compressed gases, for combustible gases, and for hot metal. Eye and skin protection is different since the flame is not nearly as bright as an arc. Goggles with the appropriate colored lenses are used and headshields are normally not required. Installation of the equipment and apparatus is extremely important and if piping and manifold systems are used, they must be installed with strict compliance to codes. Torches should always be properly stored when not in use. Hoses and torches should be bled so that gas pressure does not remain in the torch or hoses. Oil or grease should never be used on gas apparatus. Review Chapter 3 for complete information.

Limitations of the Process

The equipment for oxyacetylene welding is the least expensive of all. It is one of the slowest processes due to the heat transfer and temperature involved. For this reason oxyacetylene welding has been largely supplanted for most manufacturing operations. The most popular uses for oxyacetylene are torch brazing and oxygen flame cutting.

A GOOD WELD

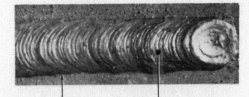

GOOD, EVEN PENETRATION   EVEN EDGES   SMOOTH, EVEN RIPPLE

COMMON WELDING MISTAKES

TOO MUCH OXYGEN

GAS PRESSURE TOO HIGH

IRREGULAR TRAVEL SPEED

TRAVEL SPEED TOO FAST

TOO MUCH HEAT

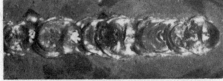

NOT ENOUGH HEAT

INNER CONE TOO CLOSE TO BASE METAL

FIGURE 6-18 *Quality of oxyacetylene welds.*

FIGURE 6-19 *Schedule of oxyacetylene welding of mild steel.*

Material Thickness (Range in in.)	Filler Rod Diameter (in.)	Tip Orifice Size Drill Size	(in.)	Cone Length of Flame (in.)	APPROX. PRESS. OF GAS Acetylene psi	Oxygen psi	APPROX. GAS CONSUMPTION Acetylene cu ft/hr	Oxygen cu ft/hr
22-16 ga	1/16	69	0.029	3/16	1	1	2	2
1/16-1/8	3/32	64	0.036	1/4	2	2	4	4
1/8-3/16	1/8	57	0.043	5/16	3	3	10	10
3/16-5/16	1/8	55	0.052	3/8	4	4	20	20
5/16-7/16	5/32	52	0.064	7/16	5	5	45	45
7/16-1/2	3/16	49	0.073	1/2	6	6	60	60
1/2-3/4	3/16	45	0.082	1/2	7	7	70	70
3/4-1	1/4	42	0.094	9/16	8	8	80	80
over 1 inch	1/4	36	0.107	5/8	9	9	90	90
Heavy Duty	1/4	28	0.140	3/4	10	10	100	100

Note: Based on use of neutral flame. For welding clean steel flux is not normally used. There is no standardized tip size for gas torches —thus table gives data based on tip orifice size in drill size and inch diameter. There are no metric equivalents.

The main variation is gas pressure welding. In this process the entire area of abutting surfaces is heated with gas flames. When the heating is completed the flames are removed and pressure is applied to achieve the weld. This process has been used for joining tubular members such as pipe. It has also been used for joining railroad rails and other parts. It is not of major industrial significance today.

The other variation is the use of hydrogen instead of acetylene. If hydrogen is used the apparatus must be proper for hydrogen and equipment designed for acetylene cannot be used. Oxyhydrogen welding is not too popular and for this reason additional details are not presented.

6-3 RESISTANCE WELDING

Resistance welding is a group of welding processes in which coalescence is produced by the heat obtained from resistance of the work to electric current in a circuit of which the work is a part and by the application of pressure. There are at least seven important resistance-welding processes. These are flash welding, high-frequency resistance welding, percussion welding, projection welding, resistance seam welding, resistance spot welding, and upset welding. They are alike in many respects but are sufficiently different so that each will be explained in this section.

Principles of the Process

The resistance welding processes differ from all those previously mentioned in that filler metal is rarely used and fluxes are not employed. Three factors are involved in making a resistance weld. They are (1) the amount of current that passes through the work, (2) the pressure that the electrodes transfer to the work, and (3) the time the current flows through the work. Heat is generated by the passage of electrical current through a resistance circuit. The maximum amount of heat is generated at the point of maximum resistance, which is at the surfaces between the parts being joined. The high-current generates sufficient heat at this resistance point so that the metal reaches a plastic state. The force applied before, during, and after the current flow, forges the heated parts together so that coalescence will occur. Pressure is required throughout the entire welding cycle to assure a continuous electrical circuit through the work. The amount of current employed and the time period are related to the heat input required to overcome heat losses and raise the temperature of the metal to the welding temperature.

This concept of resistance welding is most easily understood by relating it to resistance spot welding. Resistance spot welding, the most popular, is shown by Figure 6-20. High current at a low voltage flows through the circuit and is in accordance with Ohm's law,

$$I = \frac{E}{R} \text{ or } R = \frac{E}{I}$$

I is the current in amperes, E is the voltage in volts, and R is the resistance of the material in ohms. The total energy is expressed by the formula: Energy equals $I \times E \times T$ in which T is the time in seconds during which current flows in the circuit.

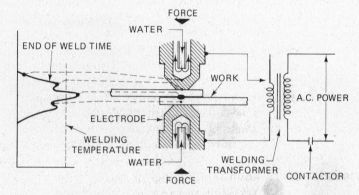

FIGURE 6-20 *Resistance spot welding process.*

Combining these two equations gives H (heat energy) $= I^2 \times R \times T$. For practical reasons a factor which relates to heat losses should be included; therefore, the actual resistance welding formula is

$$H \text{ (heat energy)} = I^2 \times R \times T \times K$$

In this formula I = current squared in amperes, R is the resistance of the work in ohms, T is the time of current flow in seconds, and K represents the heat losses through radiation and conduction.

Welding heat is proportional to the square of the welding current. If the current is doubled the heat generated is quadrupled. Welding heat is proportional to the total time of current flow, thus, if current is doubled the time can be reduced considerably. The welding heat generated is directly proportional to the resistance and is related to the material being welded and the pressure applied. The heat losses should be held to a minimum. It is therefore an advantage to shorten welding time. Mechanical pressure which forces the parts together helps refine the grain structure of the weld.

THE OTHER WELDING PROCESSES

Heat is also generated at the contact between the welding electrodes and the work. This amount of heat generated is lower since the resistance between high conductivity electrode material and the normally employed mild steel is less than that between two pieces of mild steel. In most applications the electrodes are water cooled to minimize the heat generated between the electrode and the work.

Resistance welds are made very quickly; however, each process has its own time cycle.

Resistance-welding operations are automatic. The pressure is applied by mechanical, hydraulic, or pneumatic systems. Motion, when it is involved, is applied mechanically and the current control is completely automatic once the welding operator initiates the weld. Resistance-welding equipment utilizes programmers for controlling current, time cycles, pressure, and movement. The early simple mechanical systems have been improved and now electronic systems are employed. Welding programs for resistance welding can become quite complex. In view of this, quality welds do not depend on welding operator skill but more on the proper set up and adjustment of the equipment and adherence to weld schedules.

Resistance welding is used primarily in the mass production industries where long production runs and consistent conditions can be maintained. Welding is performed with operators who normally load and unload the welding machine and operate the switch for initiating the weld operation. The automotive industry is the major user of the resistance-welding processes, followed by the appliance industry. Resistance welding is used by many industries manufacturing a variety of products made of thinner gauge metals. Resistance welding is also used in the steel industry for manufacturing pipe, tubing and smaller structural sections. Resistance welding has the advantage of producing a high volume of work at high speeds and does not require filler materials. Resistance welds are reproducible and high-quality welds are normal.

The position of making resistance welds is not a factor particularly in the welding of thinner material.

Weldable Metals

Metals that are weldable, the thicknesses that can be welded and joint design are related to specific resistance welding processes. Most of the common metals can be welded by many of the resistance-welding pro-

Base Metal	Weldability	Chapter
Aluminums	Weldable	13-1
Magnesium	Weldable	13-5
Inconel	Weldable	13-6
Nickel	Weldable	13-6
Nickel silver	Weldable	13-6
Monel	Weldable	13-6
Precious metals	Weldable	13-7
Low carbon steel	Weldable	12-2
Low alloy steel	Weldable	12-3
High and medium carbon	Possible but not popular	12-4
Alloys steel	Possible but not popular	12-5
Stainless steel	Weldable	12-6

FIGURE 6-21 *Base metals weldable by the resistance welding process.*

cesses, as shown by Figure 6-21. However, difficulties may be encountered when welding certain metals in thicker sections. Some metals require heat treatment after welding for satisfactory mechanical properties. Weldability is controlled by three factors: (1) resistivity, (2) thermal conductivity, and (3) melting temperature. The metal with a high resistance to current flow and with a low thermal conductivity and a relatively low melting temperature would thus be easily weldable. Ferrous metals all fall into this category. Metals that have a lower resistivity but a higher thermal conductivity will be slightly more difficult to weld and this would include the light metals–aluminum and magnesium. The precious metals comprise the third group, and these are difficult to weld because of the very high thermal conductivity which would be troublesome. The fourth group are the refractory metals which have extremely high melting points and are therefore more difficult to weld.

These three properties can be combined into a formula which will provide an indication of the ease of welding a metal. This formula is

$$W = \frac{R}{FKt} \times 100$$

In this formula W equals weldability, R is resistivity, and F is the melting temperature of the metal in degrees C, and Kt is the relative thermal conductivity with copper equal to 1.00. The metal properties were given in Table 11-1. If weldability (W) is below 0.25 it is a poor rating. If W is between 0.25 and 0.75 weldability becomes fair. Between 0.75 and 2.0 weldability is good and above 2.0 weldability is excellent. In this formula mild steel would have a weldability rating of over 10. Aluminum has a weldability factor of from 1 to 2 depending on the alloy and these are considered having a good weldability rating. Copper and certain brasses have a low weldability factor and are known to be very difficult to weld.[6] This information applies primarily to spot welding but would be an indication for the other resistance-welding processes where arcing does not take place.

Resistance Spot Welding (RSW)

Resistance spot welding is a resistance welding process which produces coalescence at the faying surfaces in one spot by the heat obtained from resistance to electric current through the work parts held together under pressure by electrodes. The size and shape of the individually formed welds are limited primarily by the size and contour of the electrodes. This is shown by Figure 6-20. The equipment for resistance spot welding can be relatively simple and inexpensive up through extremely large multiple spot welding machines. The stationary single spot welding machines are of two general types; the *horn* or *rocker arm* type and the *press* type. The horn type machines have a pivoted or rocking upper electrode arm which is actuated by pneumatic power or by the operator's physical power. They can be used for a wide range of work but are restricted to 50 kva and are used for thinner gauges. Figure 6-22 shows a machine of this type in operation. For larger machines normally over 50 kva, the press type machine is used. In these machines, the upper electrode moves in a slide. The pressure and motion are provided on the upper electrode by hydraulic or pneumatic pressure, or are motor operated. Figure 6-23 shows a press type

FIGURE 6-23 *Press type spot welding machine.*

machine in operation. It can be used for welding medium to the heaviest gauge metals. The Resistance Welder Manufacturers Association has standardized and classified the standard spot welders.[7] This information is shown by Figure 6-24, which gives the size, the kva ratings, and the throat depths.

When the work is too bulky to take to the welding machine a portable machine can be employed. These are used to weld parts assembled in a fixture. The spot welder is moved by the operator from one welding location to the other and a trigger on the gun actuates the welding cycle. Portability is limited by heavy cables connected between the gun and the transformer. In some smaller portable units the transformer is included in the gun. Portable units are normally operated by air or hydraulic pressure; however, extremely small portable spot welding guns are available that utilize manual power to create the pressure. Figure 6-25 shows a portable welding gun in operation.

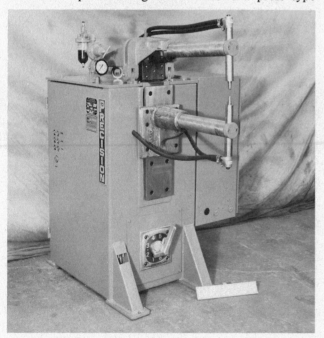

FIGURE 6-22 *Rocker arm spot welding machine.*

Type of Welding Machine	Size RWMA	Rating KVA	Electrode Cooling	Nominal Throat Depth (in.)
Rocker-arm spot welding machines	000	5	Air	8, 12, 16
	00	7.5	Air	8, 12, 16
	0	10	Air	8, 12, 16
	1	15	Air or Water	12, 18, 24
			Water	12, 18, 24, 30, 36
	2	30	Water	12, 18, 24, 30, 36
	3	50		
Press-type spot & project welding machine	000	5	Water	6, 8
	00	20	Water	6, 8
	0	30		
		50	Water	6, 8
	1	30		
		50		
		75	Water	12, 18, 30, 24, 36
	2	100		
		150	Water	12, 18, 24, 30, 36
	3	150		
		200	Water	12, 18, 24, 30, 36
		300		
	4	400		
		500	Water	12, 18, 30

FIGURE 6-24 *RWMA standard spot and projection welding machines.*

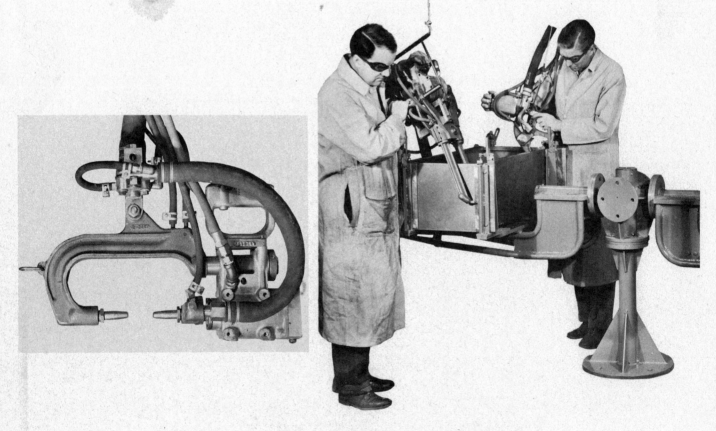

FIGURE 6-25 *Portable resistance welding gun in operation.*

For high-volume production work, such as in the automotive industry, multiple spot welding machines are used. These are in the form of a press on which individual guns carrying electrode tips are mounted. Welds are made in a sequential order so that all electrodes are not carrying current at the same time. Figure 6-26 shows a multiple spot press welding machine for making automotive subassemblies.

(a)

(b)

FIGURE 6-26 *Multiple spot welding machine.*

Several different types of spot welds can be made. Most common is the direct spot weld which is made between the opposing electrode tips. Another method is known as a series type weld where two electrodes are pressed against one side of the material and a common conductor is placed against the opposite side. The current flows from one electrode through the metal being welded, through the common backing conductor, again through the metal being welded, and to the second electrode tip. In this technique two welds are made simultaneously and there is no marking or indentation appearing on the back side of the welded assembly.

Resistance welding electrodes and electrode holders are standardized by the Resistance Welding Manufacturers Association (RWMA). This standard separates electrodes into two basic groups; Group A, which are made of copper-base alloys, and Group B, which are made of copper-tungsten alloys. Group A is broken down into five classes, No. 1 through 5, and Group B is broken down into four classes, No. 10 through 13.

(c)

These different classes identify the analysis of the electrode alloy and relate to the electrode hardness, strength, and conductivity. This information enables the user to select the proper electrode for welding various metals. In addition to alloy numbers the RWMA specification provides design, size information, shank taper, and nose configuration. The electrodes are identified by a code consisting of a letter followed by four digits. The letter indicates the electrode nose configuration and design. A indicates a pointed nose, B a dome shaped nose, C a flat nose, D an eccentric nose, E a truncated taper nose, and F a spherical radius nose. The first digit gives the alloy class as mentioned previously. The second digit shows the taper on the electrode shank. The electrodes fit into the electrode holder, which is also tapered. The taper provides maximum area to conduct the welding current from the holder to the electrode. The connection must be water tight since cooling water circulates through the holder into the electrode. These tapers are given in numbers ranging from 3 through 7. The last two digits of the code are used together and when multiplied by 1/4 provides the overall length of the straight electrode in inches. In this way electrodes can be designated for specific jobs. There are standard electrodes with single and double bends which are used in portable guns. Electrode holders are also identified by codes.

The design of weldments for spot welding is important so that the process can be efficiently utilized. For spot welding a specific joint overlap is required. There is also a specific nugget size which is related to the electrode size. The distance from the center line of the nugget to the edge of the sheet is known as the edge distance and this should be at least 1.5 times the nugget diameter. The separation between sheets being welded should not exceed 10% of the thinnest sheet. Design parameters for spot welding are given in "The Resistance Welding Handbook"[8] and "Resistance Welding Theory and Use."[9] The references also provide the size of welding machines required for welding different thicknesses.

Resistance Spot Weld Quality

The quality of spot welding depends on close adherence to the welding schedule and proper maintenance of equipment. Quality welds require continuous maintenance of the electrode nose contour. The nose must be redressed often to maintain the proper shape. This is more of a problem when welding highly conductive materials and coated materials than when welding mild steel. Special tools are available for dressing electrodes while they are mounted in the resistance spot welding machine.

Another factor involved is the electrode pressure. This can be checked in the machine by portable instruments. Good quality of resistance spot welds depends on the cleanliness of the faying surfaces. These surfaces should be free of scale, oil, etc., for quality resistance spot welds. To maintain the quality of welding, test welds are made on a schedule basis. These welds must be tested according to information in the AWS "Recommended Practices for Resistance Welding."[10] There are standard methods of testing resistance welds. These include the tension shear test, the tension test, the impact test, and the fatigue test.

Projection Welding (RPW)

Projection welding is a resistance welding process which produces coalescence of metals with the heat obtained from resistance to electrical current through the work parts held together under pressure by electrodes.

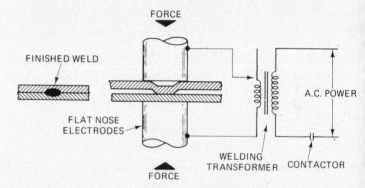

FIGURE 6-27 *Projection welding.*

The resulting welds are localized at predetermined points by projections, embossments, or intersections. Figure 6-27 shows the principles of projection welding. Localization of heating is obtained by a projection or embossment on one or both of the parts being welded. There are several types of projections: (1) the button or dome type, usually round, (2) elongated projections, (3) ring projections, (4) shoulder projections, (5) cross wire welding, and (6) radius projection. The major advantage of projection welding is that electrode life is increased because larger contact surfaces are used. A very common use of projection welding is the use of special nuts that have projections on the portion of the part to be welded to the assembly. These are manufactured with the projections and assist in obtaining quality joints to the parts being welded. Projection dimensions must be properly designed since the height and area have optimum dimensions for welding to specific thicknesses of sheet metal. This data is in the "Resistance Welding Handbook." The press type resistance welding machine is normally used. Flat nose or special electrodes are used.

Resistance Seam Welding (RSEW)

Resistance seam welding is a resistance welding process which produces coalescence at the faying surfaces by the heat obtained from resistance to electric current through the work parts held together under pressure by electrodes. The resulting weld is a series of overlapping resistance spot welds made progressively along a joint by rotating the electrodes. This is shown by Figure 6-28.

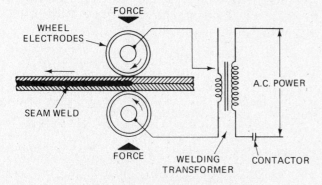

FIGURE 6-28 *Resistance seam welding.*

When the spots are not overlapped enough to produce gastight welds it is a variation known as *roll resistance spot welding*. This process differs from spot welding since the electrodes are wheels. Both the upper and lower electrode wheels are powered. Pressure is applied in the same manner as a press type welder. The wheels can be either in line with the throat of the machine or transverse. If they are in line it is normally called a longitudinal seam welding machine. Welding current is transferred through the bearing of the roller electrode wheels. Water cooling is not provided internally and therefore the weld area is flooded with cooling water to keep the electrode wheels cool. In seam welding a rather complex control system is required. This involves the travel speed as well as the sequence of current flow to provide for overlapping welds. The welding speed, the spots per inch, and the timing schedule are dependent on each other. Welding schedules provide the pressure, the current, the speed, and the size of the electrode wheels. This process is quite common for making flange welds, for making watertight joints for tanks, etc. Another variation is the so-called mash seam welding where the lap is fairly narrow and the electrode wheel is at least twice as wide as used for standard seam welding. The pressure is increased to approximately 300 times normal pressure. The final weld mash seam thickness is only 25% greater than the original single sheet.

Flash Welding (FW)

Flash welding is a resistance welding process which produces coalescence simultaneously over the entire area of abutting surfaces, by the heat obtained from resistance to electric current between the two surfaces, and by the application of pressure after heating is substantially completed. Flashing and upsetting are accompanied by expulsion of metal from the joint. This is shown by Figure 6-29. During the welding operation there is an intense flashing arc and heating of the metal on the surface abutting each other. After a predetermined time the two pieces are forced together and coalescence occurs at the interface, current flow is possible because of the light contact between the two parts being flash welded. The heat is generated by the flashing and is localized in the area between the two parts. The surfaces are brought to the melting point and expelled through the abutting area. As soon as this material is flashed away another small arc is formed which continues until the entire abutting surfaces are at the melting temperature. Pressure is then applied and the arcs are extinguished and upsetting occurs.

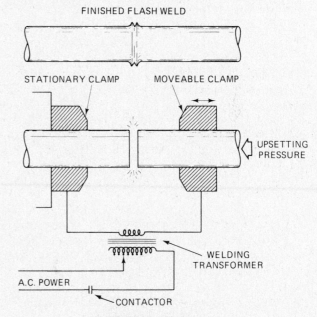

FIGURE 6-29 *Flash welding.*

Flash welding can be used on most metals. No special preparation is required except that heavy scale, rust, and grease must be removed. The joints must be cut square to provide an even flash across the entire surface. The material to be welded is clamped in the jaws of the flash welding machine with a high clamping pressure. The upset pressure for steel exceeds 10,000 psi (700 kg/cm^2). For high-strength materials these pressures may be doubled. For tubing or hollow members the pressures are reduced. As the weld area is more compact, upset pressures are increased. If insufficient upset pressure is used a porous low strength weld will result. Excess upset pressure will result in expelling too much weld metal and upsetting cold metal. The weld may not be uniform across the entire cross section, and fatigue and impact strength will be reduced. The speed of upset—that is, the time between the end of flashing period and the end of the upset period—should be extremely short to minimize oxidation of the molten surfaces. In the flash welding operation a certain amount of material is flashed or burned away. The distance between the jaws after welding compared to the distance before welding is known as the *burnoff*. It can be from 1/8 in. (3.2 mm) for thin material up to several inches for heavy material. Welding currents are high and are related to the following: 50 kva per square inch cross section at 8 seconds. It is desirable to use the lowest flashing voltage at a desired flashing speed. The lowest voltage is normally 2 to 5 volts per square inch of cross section of the weld.

The upsetting force is usually accomplished by means of mechanical cam action. The design of the cams are related to the size of the parts being welded. Flash welding is completely automatic and is an excellent process for mass-produced parts. It requires a machine of large capacity designed specifically for the parts to be welded. Schedules for flash welding are found in references 8 and 9. Flash welds produce a fin around the periphery of the weld which is normally removed.

Upset Welding (UW)

Upset welding is a resistance welding process which produces coalescence simultaneously over the entire area of abutting surfaces or progressively along a joint, by the heat obtained from resistance to electric current through the area where those surfaces are in contact. Pressure is applied before heating is started and is maintained throughout the heating period. The equipment used for upset welding is very similar to that used for flash welding. It can be used only if the parts to be welded are equal in cross-sectional area. The abutting surfaces must be very carefully prepared to provide for proper heating. The difference from flash welding is

that the parts are clamped in the welding machine and force is applied bringing them tightly together. High-amperage current is then passed through the joint which heats the abutting surfaces. When they have been heated to a suitable forging temperature an upsetting force is applied and the current is stopped. The high temperature of the work at the abutting surfaces plus the high pressure causes coalescence to take place. After cooling, the force is released and the weld is completed. There is no arc or flash in upset welding. The area at the joint is usually enlarged over its original dimension. Figure 6-30 shows this process. It is used for welding small wires, tubing, piping, rings, strips, etc., where the cross-sectional areas of both pieces are identical. If intimate contact is not obtained because of improper joint preparation the weld will be defective.

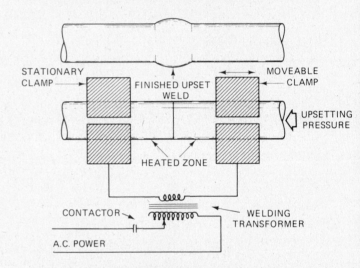

FIGURE 6-30 *Upset welding.*

Percussion Welding (PEW)

Percussion welding is a resistance welding process which produces coalescence of the abutting members using heat from an arc produced by a rapid discharge of electrical energy. Pressure is applied progressively during or immediately following the electrical discharge. This process is quite similar to flash welding and upset welding, but is limited to parts of the same geometry and cross section. It is more complex than the other two processes in that heat is obtained from an arc produced at the abutting surfaces by the very rapid discharge of stored electrical energy across a rapidly decreasing air gap. This is immediately followed by application of pressure to provide an impact bringing the two parts together in a progressive percussive manner. The advantage of the process is that there is an extremely shallow depth of heating and time cycle is very short. It is used only for parts with fairly small cross-sectional areas. It can be used for welding a large number of dissimilar metals. It is used for very specialized applications and the process is entirely automatic.

High Frequency Resistance Welding (HFRW)

High frequency resistance welding is a resistance welding process which produces coalescence of metals with the heat generated from the resistance of the work pieces to a high-frequency alternating current in the 10,000 to 500,000 hertz range and the rapid application of an upsetting force after heating is substantially completed. The path of the current in the work piece is controlled by the proximity effect.

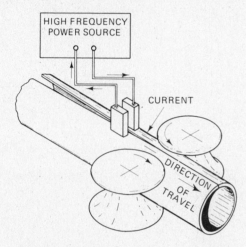

FIGURE 6-31 *High frequency resistance welding.*

This process is ideally suited for making pipe, tubing, and structural shapes. It is used for other manufactured items made from continuous strips of material. Figure 6-31 shows the making of a tubular product. In this process the high frequency welding current is introduced into the metal at the surfaces to be welded but prior to their contact with each other. Current is introduced by means of sliding contacts at the edge of the joint. The high-frequency welding current flows along one edge of the seam to the welding point between the pressure rolls and back along the opposite edge to the other sliding contact. The current is of such high frequency that it flows along the metal surface to a depth of several thousandths of an inch. Each edge of the joint is the conductor of the current and the heating is concentrated on the surface of these edges. At the area between the closing rolls the material is at the plastic temperature, and with the pressure applied, coalescence occurs. The surfaces must be reasonably true with respect to each other and clean. No other special preparation is required. The process can be used to join most common metals and certain

dissimilar metals. The process is entirely automatic and utilizes equipment designed specifically for this process. It is possible to make welds at extremely high speeds approaching 500 ft per minute (150 m per minute) for thin wall tubing.

6-4 SOLID STATE WELDING

Solid state welding is a group of welding processes which produces coalescence at temperatures essentially below the melting point of the base materials being joined, without the addition of a brazing filler metal. Pressure may or may not be used. These processes are sometimes erroneously called solid state bonding processes: this group of welding processes includes cold welding, diffusion welding, explosion welding, forge welding, friction welding, hot pressure welding, roll welding, and ultrasonic welding.

In all of these processes time, temperature, and pressure individually or in combination produce coalescence of the base metal without significant melting of the base metals.

Solid state welding includes some of the very oldest of the welding processes and some of the very newest. Some of the processes offer certain advantages since the base metal does not melt and form a nugget. The metals being joined retain their original properties without the heat-affected zone problems involved when there is base metal melting. When dissimilar metals are joined their thermal expansion and conductivity is of much less importance with solid state welding than with the arc welding processes.

Time, temperature, and pressure are involved; however, in some processes the time element is extremely short, in the microsecond range or up to a few seconds. In other cases, the time is extended to several hours. As temperature increases time is usually reduced. Since each of these processes is different each will be described.

Cold Welding (CW)

Cold welding is a solid state welding process which uses pressure at room temperature to produce coalescence of metals with substantial deformation at the weld. Welding is accomplished by using extremely high pressures on extremely clean interfacing materials. Sufficiently high pressure can be obtained with simple hand tools when extremely thin materials are being joined. When cold welding heavier sections a press is usually required to exert sufficient pressure to make a successful weld. Indentations are usually made in the parts being cold welded. The process is readily adaptable to joining ductile metals. Aluminum and copper are readily cold welded. Aluminum and copper can be joined together by cold welding and an example of this is shown by Figure 6-32. Figure 6-33 shows a cold weld being made.

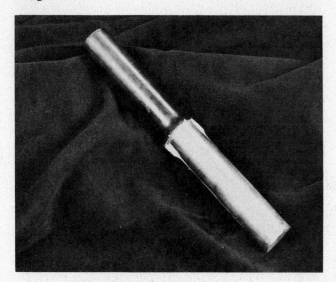

FIGURE 6-32 *Copper welded to aluminum.*

Diffusion Welding (DFW)

Diffusion welding is a solid state welding process which produces coalescence of the faying surfaces by the application of pressure and elevated temperatures. The process does not involve microscopic deformation melting or relative motion of the parts. Filler metal may or may not be used. This may be in the form of electroplated surfaces. The process is used for joining refractory metals at temperatures that do not affect their metallurgical properties. Heating is usually accomplished by induction, resistance, or furnace. Atmosphere and vacuum furnaces are used and for most refractory metals a protective inert atmosphere is desirable. Successful welds have been made on refractory metals at temperatures slightly over half the normal melting temperature of the metal. To accomplish this type of joining extremely close tolerance joint preparation is required and a vacuum or inert atmosphere is used. The process is used quite extensively for joining dissimilar metals. The process is considered diffusion brazing when a layer of filler material is placed between

FIGURE 6-33 *Making a cold weld.*

the faying surfaces of the parts being joined. These processes are used primarily by the aircraft and aerospace industries.

Explosion Welding (EXW)

Explosion welding is a solid state welding process in which coalescence is effected by high-velocity movement together of the parts to be joined produced by a controlled detonation. The explosion welding process is shown by the diagram in Figure 6-34. Even though heat is not applied in making an explosion weld it appears that the metal at the interface is molten during welding.

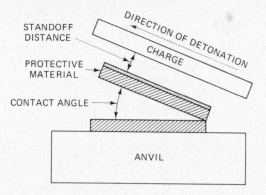

Figure 6-34 *Explosion welding.*

SOLID STATE WELDING

a weld between dissimilar metals. Another application for explosion welding is in the joining of tube-to-tube sheets for the manufacture of heat exchangers. The process is also used as a repair tool for repairing leaking tube-to-tube sheet joints. Another and new application has been the joining of pipes in a socket joint. This application will be of increasing importance in the future.

Forge Welding (FOW)

Forge welding is a solid state welding process which produces coalescence of metals by heating them in a forge and by applying pressure or blows sufficient to cause permanent deformation at the interface. This is one of the older welding processes and at one time was called hammer welding. Forge welds made by blacksmiths were made by heating the parts to be joined to a red heat considerably below the molten temperature. Normal practice was to apply flux to the interface. The blacksmith by skillful use of a hammer and an anvil was able to create pressure at the faying surfaces sufficient to cause coalescence. This process is of minor industrial significance today.

Friction Welding (FRW)

Friction welding is a solid state welding process which produces coalescence of materials by the heat obtained from mechanically-induced sliding motion between rubbing surfaces. The work parts are held together under pressure. This process usually involves the rotating of one part against another to generate frictional heat at the junction. When a suitable high temperature has been reached, rotational motion ceases and additional pressure is applied and coalescence occurs.

This heat comes from several sources, from the shock wave associated with impact and from the energy expended in collision. Heat is also released by plastic deformation associated with jetting and ripple formation at the interface between the parts being welded. Plastic interaction between the metal surfaces is especially pronounced when surface jetting occurs. It is found necessary to allow the metal to flow plastically in order to provide a quality weld. The interface or weld of explosion welded parts is shown by Figure 6-35. Explosion welding creates a strong weld between almost all metals. It has been used to weld dissimilar metals that were not weldable by the arc processes. The weld apparently does not disturb the effects of cold work or other forms of mechanical or thermal treatment. The process is self-contained, it is portable, and welding can be achieved quickly over large areas. The strength of the weld joint is equal to or greater than the strength of the weaker of the two metals joined.

Explosion welding has not become too widely used except in a few limited fields. One of the most widely used applications of explosion welding has been in the cladding of base metals with thinner alloys. The photo micrograph shown by Figure 6-35 is a cross section of

There are two variations of the friction welding process. In the original process one part is held stationary and the other part is rotated by a motor which maintains an essentially constant rotational speed. The two parts are brought in contact under pressure for a specified period of time with a specific pressure. Rotating power is disengaged from the rotating piece and the pressure is increased. When the rotating piece stops the weld is completed. This process can be accurately controlled when speed, pressure, and time are closely regulated.

The other variation is called inertia welding. Here a flywheel is revolved by a motor until a preset speed is reached. It, in turn, rotates one of the pieces to be

FIGURE 6-35 *Interface of explosion weld.*

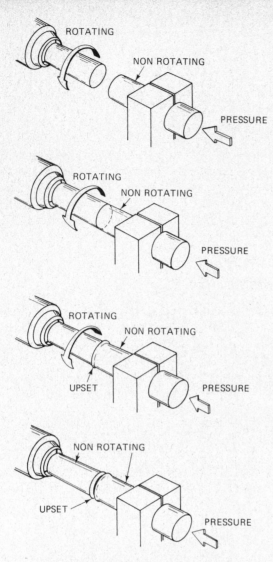

FIGURE 6-36 Friction welding process.

welded. The motor is disengaged from the flywheel and the other part to be welded is brought in contact under pressure with the rotating piece. During the predetermined time during which the rotational speed of the part is reduced the flywheel is brought to an immediate stop and additional pressure is provided to complete the weld.

Both methods utilize frictional heat and produce welds of similar quality. Slightly better control is claimed with the original process. The two methods are similar, offer the same welding advantages, and are shown by Figure 6-36.

Among the advantages of friction welding is the ability to produce high quality welds in a short cycle time. No filler metal is required and flux is not used. The process is capable of welding most of the common metals. It can also be used to join many combinations of dissimilar metals. Friction welding requires relatively expensive apparatus similar to a machine tool. Figure 6-37 is a picture of a friction welding machine.

There are three important factors involved in making a friction weld:

1. The rotational speed which is related to the material to be welded and the diameter of the weld at the interface.

2. The pressure between the two parts to be welded. Pressure changes during the weld sequence. At the start it is very low, but it is increased to create the frictional heat. When the rotation is stopped pressure is rapidly increased so that forging takes place immediately before or after rotation is stopped.

FIGURE 6-37 Friction welding machine.

3. The welding time. Time is related to the shape and the type of metal and the surface area. It is normally a matter of a few seconds. The actual operation of the machine is automatic and is controlled by a sequence controller which can be set according to the weld schedule established for the parts to be joined.

Normally for friction welding one of the parts to be welded is round in cross section; however, this is not an absolute necessity. Visual inspection of weld quality can be based on the flash, which occurs around the outside perimeter of the weld. Normally this flash will extend beyond the outside diameter of the parts and will curl around back toward the part but will have the joint extending beyond the outside diameter of the part. If the flash sticks out relatively straight from the joint it is an indication that the time was too short, the pressure was too low, or the speed was too high. These joints may crack. If the flash curls too far back on the outside diameter it is an indication that the time was too long and the pressure was too high. Between these extremes is the correct flash shape. The flash is normally re-

moved after welding. Figure 6-38 shows a circular friction welded part. Note both the inside flash and the outside flash. Provisions were made in this part so that the inner flash would not extend to inside the chamber where it might interfere with the function of the part.

Hot Pressure Welding (HPW)

Hot pressure welding is a solid state welding process which produces coalescence of materials with heat and the application of pressure sufficient to produce macro-deformation of the base metal.

In this process coalescence occurs at the interface between the parts because of pressure and heat which is accompanied by noticeable deformation. The deformation of the surface cracks the surface oxide film and increases the areas of clean metal. Welding this metal to the clean metal of the abutting part is

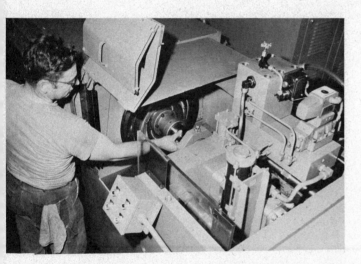

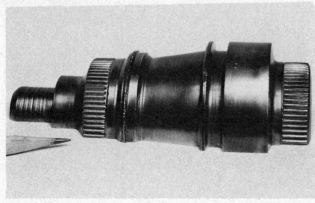

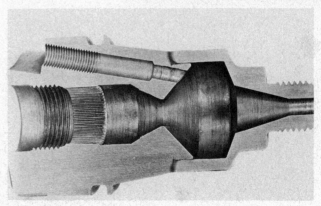

FIGURE 6-38 *Part friction welded.*

accomplished by diffusion across the interface so that coalescence of the faying surface occurs. This type of operation is normally carried on in closed chambers where vacuum or a shielding medium may be used. It is used primarily in the production of weldments for the aerospace industry. A variation is the hot isostatic pressure welding method. In this case, the pressure is applied by means of a hot inert gas in a pressure vessel.

Roll Welding (ROW)

Roll welding is a solid state welding process which produces coalescence of metals by heating and by applying pressure with rolls sufficient to cause deformation at the faying surfaces. This process is similar to forge welding except that pressure is applied by means of rolls rather than by means of hammer blows. Coalescence occurs at the interface between the two parts by means of diffusion at the faying surfaces.

One of the major uses of this process is the cladding of mild or low-alloy steel with a high-alloy material such as stainless steel. It is also used for making bi-metallic materials for the instrument industry. And, it is used to produce the *sandwich* coins used in the U.S.A. Figure 6-39 is a photo micrograph of the weld of a coin made in this manner. This is a weld between copper and a nickel-copper alloy.

FIGURE 6-40 *Ultrasonic welding.*

Ultrasonic Welding (USW)

Ultrasonic welding is a solid state welding process which produces coalescence by the local application of high-frequency vibratory energy as the work parts are held together under pressure. Welding occurs when the ultrasonic tip or electrode, the energy coupling device, is clamped against the work pieces and is made to oscillate in a plane parallel to the weld interface. The combined clamping pressure and oscillating forces introduce dynamic stresses in the base metal. This produces minute deformations which create a moderate temperature rise in the base metal at the weld zone. This coupled with the clamping pressure provides for coalescence across the interface to produce the weld. Ultrasonic energy will aid in cleaning the weld area by breaking up oxide films and causing them to be carried away. The vibratory energy that produces the minute deformation comes from a transducer which converts high-frequency alternating electrical energy into mechanical energy. The transducer is coupled to the work by various types of tooling which can range from tips similar to resistance welding tips to resistance roll welding electrode wheels. The normal weld is the lap joint weld.

The temperature at the weld is not raised to the melting point and therefore there is no nugget similar to resistance welding. Weld strength is equal to the strength of the base metal. Most ductile metals can be welded

FIGURE 6-39 *Interface of roll weld.*

together and there are many combinations of dissimilar metals that can be welded. The process is restricted to relatively thin materials normally in the foil or extremely thin gauge thicknesses.

This process is used extensively in the electronics, aerospace, and instrument industries. It is also used for producing packages and containers and for sealing them.

Figure 6-40 shows an ultrasonic weld being made by continuous seam welder. In this picture a rotating electrode is the active welding tip which delivers ultrasonic energy to the work. The process can also be used for joining plastics and it is finding wider use in this field than in joining metals.

6-5 SOLDERING

Soldering is a group of joining processes which produces coalescence of materials by heating them to a suitable temperature and by using a filler metal having a liquidus not exceeding 840°F (450°C) and below the solidus of the base metals. The filler metal is distributed between the closely fitted surfaces of the joint by capillary attraction. Solder is a filler metal used in soldering which has a liquidus not exceeding 840°F (450°C). The solder or the filler metal is normally a nonferrous alloy. The temperature of 840°F (450°C) is the temperature that differentiates soldering from brazing. This arbitrary number was selected many years ago and is universally accepted. Most of the factors involved with brazing apply to soldering. In fact, slang usage of the terms "soft solder" and "hard solder" or "silver solder" attempts to differentiate between soldering and brazing. Brazing and soldering, the two metal joining processes using filler metals that melt at temperatures below the temperature of the base metal, are much older than the arc welding processes.

There are at least eight soldering methods which are in wide use today. These can be classified into three general groups. One group relates to the means of applying heat and two groups are related to the means of applying the solder with or without flux.

The mechanism for joining by soldering involves three closely related factors: (1) wetting, (2) alloying, and (3) capillary attraction. Wetting is the bonding or spreading of a liquid filler metal or flux on a solid base metal. For soldering it is more specific. When molten solder leaves a continuous permanent film on the metal surface of the base metal it is said to have *wet* that surface. *Wetting* occurs when there is a stronger attraction between certain atoms of the solder and the base metal than between the atoms of the *solder*. Wetting is essentially a chemical reaction. It occurs when one or more elements of the solder react with the base metal being soldered to form a compound.

The ability of a solder to alloy with the base metal is related to its ability of wetting the surface. Heat is applied to facilitate wetting. Alloying is related to the cleanliness of the base metal. The base metal must be oxide free, and this is accomplished by cleaning and using a flux. There must be intimate contact between the solder and the base metal for alloying to occur at the interface. The temperature of wetting may not correspond with the liquidus temperature of the solder alloy. Many metals can be soldered, including the aluminum, copper-base alloys, nickel-base alloys, steels, stainless steel, etc.

The fluidity of the molten solder must be such that it can flow into narrow spaces by capillary attraction. Tests are available to determine this quality. The fluidity of the molten solder is the property that influences the spreading of the solder over the base metal surface. The flowability or spread of solder can also be determined by tests.

The actual application of the solder involves two steps. The wetting of the base metal surface with solder and the filling of the gap between the wetted surfaces with the solder. These two steps are normally carried out together depending on conditions and application; however, for "difficult-to-solder" metals it is desirable to wet the surface of the base metal with solder prior to making the joint. This is called tinning or precoating.

The Strength of Soldered Joints

The strength of a soldered joint depends on the design of the joint and its clearance. Usually joints stressed in tension are not successful. The normal type joint is the lap joint where sufficient overlap occurs to provide enough area for required strength. The clearance between the parts being joined must be held to close limits. The maximum strength of the joint will provide a joint considerably stronger than the strength of the solder alloy. However, if clearance between the joint is excessive, the strength of the joint drops to the strength level of the soldering alloy.

Soldering Procedures

Seven factors are involved in producing a high quality soldered joint. These are: (1) the design and fit of the joint, (2) the precleaning of the base metal, (3) the selection and application of the soldering flux, (4) the selection and application of the solder, (5) the application of heat for soldering, (6) cooling the soldered joint, and (7) postcleaning. These factors will be briefly discussed.

1. Design and Fit of the Joint: The lap joint is the only one that has sufficient strength for most requirements. Clearance between the parts being joined should be sufficient so that the solder will be drawn into the joint by means of capillary attraction. The clearance of from 0.003–0.005 in. (0.07–0.12 mm) is recommended for most applications. The joint design is the same as used for brazing.

2. Precleaning: The joint area must be free of all grease, oil, dirt, oxides, etc. This is best done by mechanical or chemical cleaning. Solder will not wet a dirty surface or a surface covered with oxides. Cleaning can be accomplished by brushing, filing, machining, sanding, and also by the use of chemicals.

3. Selection and Application of Flux: The flux should be selected on the basis of its ability to promote or accelerate the wetting of the base metals. The flux helps remove oxides but the flux must be designed so that it can be removed after the joint is soldered. It should be fluid at the same or at a lower temperature than the liquidus of the solder. It should have a lower specific gravity than the solder so that the solder will displace it in the joint. It should promote wetting of the surface by the solder. Stronger fluxes are the acid fluxes or the inorganic type. Intermediate fluxes are less severe. The organic type fluxes such as resin are the mildest and are often referred to as *pure water, white resin,* or *nonactivated resin.* Resin contains an abietic acid which is very mild, has a long melting point, and remains effective to the highest melting point solder is normally used. Flux should be applied to the base metal to protect it from oxidation.

Solder is available with the flux contained in its core. The amount of flux in the core ranges from about 0.5% to over 3%, 2.2% the most common. Resin cored solder and acid core solders are available. The resin cored type is used for electrical work and the acid core type is used for sheet metal work.

4. Selection and Application of the Solder: Selection of the solder is one of the most important parts of the entire soldering operation. There are two specifications [11,12] that apply to the different types of solder. The chart of Figure 6-41 shows the nominal composition of the solder, its melting range, and typical applications. The 50 A (50% lead, 50% tin) is the most common general purpose solder. Solder selection is based on its ability to wet the surface of the base metals being joined. The grade containing the least amount of tin that provides suitable flowing and wetting action should be used. Special solders such as tin-antimony for food equipment, tin zinc for aluminum, and lead silver for high strength are available.

5. Application of Heat: There are many different methods of applying the heat and each is specifically designed for particular applications. The different methods of applying the heat for soldering will be covered later in this section.

6. Cooling the Solder Joint: The joint is cooled to room temperature after the base metal surfaces have been wetted and the space between them filled with solder. Quite often self-jigging joints are employed, or staking, bending, or other assembly methods are used. Cooling is normally accomplished by removing the heat source and utilizing an air blast.

7. Postcleaning: The final operation is post-cleaning, which is necessary to remove the flux residue which may be corrosive. Certain fluxes are considered noncorrosive and may not be removed unless it affects the appearance or later processing of the part. However, for fluxes identified as corrosive, such as the acid types, it is absolutely necessary that they be removed. They should be neutralized and removed to provide for a successfully soldered joint.

The different methods of soldering are based on the way the heat or the way that the solder is applied. These are briefly, as follows:

Dip Soldering (DS)

This is a soldering process in which the heat required is furnished by a molten metal bath which provides the solder filler metal. The solder may be kept molten by any source of heat.

Furnace Soldering (FS)

This is a soldering process in which the parts to be joined are placed in a furnace and heated to a suitable temperature. In furnace soldering the parts must be assembled and fixed in their proper position. The solder must be preplaced in the joint. The furnace can be fired by any suitable fuel.

Induction Soldering (IS)

This is a soldering process in which the heat required

ASTM Number	Tin	NOMINAL COMPOSITION %			Melting Temp		Freezing Temp		Uses
		Lead	Antimony	Silver	C	F	C	F	
1.55	1	97.5	0.40 max.	1.5	309	588	309	588	For use on copper, brass & similar metals.
2.55	0	97.5	0.40 max.	2.5	304	579	304	579	For use on copper, brass & similar metals. Not in humid environments
5A	5	95	0.12 max.	—	312	594	270	518	For coating & joining metals.
10B	10	90	0.50 max.	—	299	570	268	514	For coating & joining metals
15B	15	85	0.50 max.	—	288	550	227	440	For coating & joining metals
20B	20	80	0.50 max.	—	277	531	183	361	For filling dents or seams in auto bodies
25A	25	75	0.25 max.	—	266	511	183	361	For machine & torch soldering
30A	30	70	0.25 max.	—	255	491	183	361	For machine & torch soldering
35A	35	65	0.25 max.	—	247	477	183	361	General purpose & wiper solder
40A	40	60	0.12 max.	—	238	460	183	361	Wiping solder for auto radiator cores
45A	45	55	0.12 max.	—	227	441	183	361	For auto radiator cores & roofing seams
50A	50	50	0.12 max.	—	216	421	183	361	For general purpose most popular of all
60A	60	40	0.12 max.	—	190	374	183	361	Fine solder where the temp. requirements are critical
63A	63	37	0.12 max.	—	183	361	183	361	As lowest melting (eutectic) solder
70A	70	30	0.12 max.	—	192	378	183	361	For coating metals
20C	20	79	1.0	—	270	517	184	363	For machine soldering & coating of metals, tipping, & like uses
25C	25	73.7	1.3	—	263	504	184	364	For torch & machine soldering
30C	30	68.4	1.6	—	250	482	185	364	For torch soldering or machine soldering
35C	35	63.2	1.8	—	243	470	185	365	For wiping & all uses
40C	40	58	2.0	—	231	448	185	365	Same as (50-50) tin-lead
95TA	95	—	5.0	—	240	464	234	452	For joints on copper, electrical plumbing & heating

Note: The "C" Grades should not be used on galvanized steel.

FIGURE 6-41 *Composition and use of solders per ASTM B32.*

is obtained from the resistance of the work to an induced electric current. This is similar to induction brazing.

Infra-red Soldering (IRS)

This is a soldering process in which the heat required is furnished by infra-red radiation. This is similar to infra-red brazing.

Iron Soldering (INS)

This is a soldering process in which the heat required is obtained from a soldering iron. The part of the soldering iron which is heated and transfers the heat and the solder to the joint is called a *bit,* and is usually made of copper. The bit of a soldering iron may be heated by several different ways. It can be electrically heated by internal resistance coil—hence, the electric soldering iron, it can be heated in a flame, or it can be heated in a furnace. There is also the *soldering gun* which utilizes resistance heating to heat the bit. The bit is the high-resistance part of the electrical circuit. Soldering guns are very popular and widely used for electronic assembly work. The solder is applied manually.

Resistance Soldering (RS)

This is a soldering process in which the heat required is obtained from the resistance to electric current in a circuit of which the work is a part. This is slightly different from resistance brazing and is usually done with a hand-held tool using carbon blocks and introducing a low voltage and relatively high current to the part to be soldered. This is a very common method of manufacturing electrical machinery involving soldered joints. Figure 6-42 shows this soldering method in use,

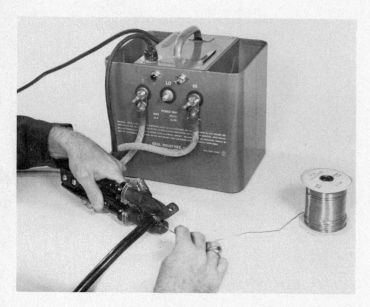

FIGURE 6-42 *Resistance soldering.*

attaching lugs to welding cables. Note that the solder is applied manually. It is also used for soldering copper plumbing fittings.

Torch Soldering (TS)

This is a soldering process in which the heat required is furnished by a fuel gas flame. This is very similar to torch brazing except that lower temperatures are involved and air is used rather than oxygen. Small cylinders of propane are available for this use. The cylinder becomes the handle when a torch head is attached to it. This system is ideal for occasional use. The solder is applied manually. Torch soldering is widely used in the plumbing trade for soldering copper tubing to copper fittings. Torch soldering is shown by Figure 6-43.

Wave Soldering (WS)

This is an automatic soldering process where work parts are passed through a wave of molten solder. This method is used in the production of electronic circuit boards. The circuit boards are assembled with the various electronic components on them with pigtails sticking through the circuit board and crimped over the printed metal circuit on the underside of the board. The boards are then placed over the tank holding the molten solder and the wave of solder touches the metal circuit and joins it to the pigtails of the electronic component with a soldered joint. This is completely automatic and produces high-quality soldered joints. It is widely used in electronics industry. Figure 6-44 shows the wave soldering apparatus.

There are several other methods of soldering. One of these is *ultrasonic soldering.* It is a soldering method in which high-frequency sonic energy is transmitted through molten solder to remove undesirable surface films and promote wetting of the base metal. Flux is not normally used. In this method the ultrasonic vibrations are transmitted to the soldering iron and thereby transmitted to the solder and to the work. A source of ultrasonic energy is required as well as a specialized soldering iron. This method is used for soldering aluminum.

FIGURE 6-43 *Torch soldering.*

FIGURE 6-44 *Wave soldering machine.*

Wipe soldering is a method of producing a joint with the heat supplied by the molten solder poured onto the joint. The solder is manipulated with a hand-held cloth or paddle so as to obtain the required size and contour. The filler metal is also distributed into the joint by capillary attraction. It is similar to an old welding method known as flow welding which produces coalescence of metals by heating them with molten filler metal poured over the surfaces to be welded until the welding temperature is obtained and until the required filler metal has been added. Both wipe soldering and flow welding are of minor industrial importance today.

One other soldering method is known as *sweat soldering*. This is a method in which two or more parts are precoated with solder, assembled into a joint, and reheated without the use of additional solder. This is used in the electrical industry for joining wires to connectors.

Soldering is a very widely used metals-joining method. For more information on this subject consult the AWS soldering manual.[13]

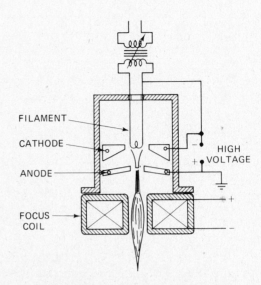

FIGURE 6-45 *Electron beam welding process.*

6-6 ELECTRON BEAM WELDING

Electron beam welding (EBW) is a welding process which produces coalescence of metals with the heat obtained from a concentrated beam composed primarily of high-velocity electrons impinging upon the surfaces to be joined. Heat is generated in the workpiece as it is bombarded by a dense stream of high-velocity electrons. Virtually all of the kinetic energy—the energy of motion—of the electrons is transformed into heat upon impact.

The electron beam welding process had its inception in the 1950s in the nuclear field. There were many requirements to weld refractory and reactive metals. These metals, because of their affinity for oxygen and nitrogen of the air, are very difficult to weld. The development came independently and almost simultaneously from Germany and the U.S.A. The original work was done in a high vacuum. The process utilized an electron gun similar to that used in an X-ray tube. In an X-ray tube the beam of electrons is focused on a target of either tungsten or molybdenum which gives off X-rays. The target becomes extremely hot and must be water cooled. In welding, the target is the base metal which absorbs the heat to bring it to the molten stage. In electron beam welding, X-rays may be produced if the electrical potential is sufficiently high.

As developments continued, two basic designs evolved: (1) the low-voltage electron beam system, which uses accelerating voltages in 30,000 volts or (30 kv) to 60,000-volt (60 kv) range and (2) the high-voltage system with accelerating voltages in the 100,000-volt (100 kv) range. The higher voltage system emits more X-rays than the lower voltage system. In both systems, the electron gun and the workpiece are housed in a vacuum chamber. Figure 6-45 is a diagram showing the principles of the electron beam welding process. There are three basic components in an electron beam welding machine. These are (1) the electron beam gun, (2) the power supply with controls, and (3) a vacuum work chamber with work-handling equipment. The electron beam gun emits electrons, accelerates the beam of electrons, and focuses it on the workpiece. The electron beam gun is similar to that used in a television picture tube. The electrons are emitted by a heated cathode or filament and accelerated by an anode which is a positively-charged plate with a hole through which the electron beam passes. Magnetic focusing coils located beyond the anode, focus and deflect the electron beam.

In the electron beam welding machine the electron beam is focused on the workpiece at the point of welding. The power supply furnishes both the filament current and the accelerating voltage. Both can be changed to provide different power input to the weld. The third component, the vacuum work chamber, must be an absolutely airtight container, which is evacuated by means of mechanical pumps and diffusion pumps to reduce the pressure to a high vacuum. Work-handling equipment is required to move the workpiece under the electron beam and to manipulate it as required to make the weld. The travel mechanisms must be designed for vacuum installations since normal greases, lubricants,

THE OTHER WELDING PROCESSES

and certain insulating varnishes in electric motors may volatilize in a vacuum. Hermetically sealed motors and sealed gearboxes must be used. In some cases the motor and gearboxes are located outside the vacuum chamber with shafts operating through sealed bearings. Electron beam welding equipment is shown by Figure 6-46.

Recent advances in equipment allow the work chamber to operate at a medium vacuum or pressure. In this system, the vacuum in the work chamber is not as high. It is sometimes called a "soft" vacuum. This vacuum range allowed the same contamination that would be obtained in atmosphere of 99.995% argon. Mechanical pumps can produce vacuums to the medium pressure level.

Electron beam welding was initially done in a vacuum because the electron beam is easily deflected by air. The electrons in the beam collide with the molecules of the air and lose velocity and direction so that welding can not be performed. In a high vacuum system, the electron beam can be located as far as 30 in. (662 mm)

away from the workpiece. In the medium vacuum, the working distance is reduced to 12 in. (304.8 mm). The thickness that can be welded in a high vacuum is up to 6 in. (150 mm) thick while in the medium vacuum the thickness that can be welded is reduced to 2 in. (50 mm). This is based on the same electron gun and power in both cases. The advantage of the medium vacuum is that pump down time is reduced, and the vacuum can be obtained by using mechanical pumps only. In the medium vacuum mode the electron gun is in its own separate chamber separated from the work chamber by a small orifice through which the electron beam travels. A diffusion vacuum pump is run continuously, connected to the chamber containing the electron gun, so that it will operate efficiently.

The most recent development is the nonvacuum electron beam welding system. In this system the work area is maintained at atmospheric pressure during welding. The electron beam gun is housed in a high vacuum chamber and there are several intermediate chambers between the gun and the atmospheric work area. Each of these intermediate stages is reduced in pressure by means of vacuum pumps. The electron beam passes from one chamber to another through a small orifice sufficient in size for the electron beam

FIGURE 6-46 *Electron beam welding equipment.*

but too small for a large volume of air. By means of these differential pressure chambers a high vacuum is maintained in the electron beam gun chamber. The nonvacuum system can thus be used for the largest weldments, however the workpiece must be positioned within 1-1/2 in. (37 mm) of the beam exit nozzle. The maximum thickness that can be welded currently is approximately 2 in. (50 mm). The nonvacuum system utilizes the high-voltage power supply.

One of the major advantages of electron beam welding is its tremendous penetration. This occurs when the highly accelerated electron hits the base metal. It will penetrate slightly below the surface and at that point release the bulk of its kinetic energy which turns to heat energy. The addition of the heat brings about a substantial temperature increase at the point of impact. The succession of electrons striking the same place causes melting and then evaporation of the base metal. This creates metal vapors but the electron beam travels through the vapor much easier than solid metal. This causes the beam to penetrate deeper into the base metal. The width of the penetration pattern is extremely narrow. The depth-to-width can exceed a ratio of 20 to 1. As the power density is increased penetration is increased.

The heat input of electron beam welding is controlled by four variables: (1) the number of electrons per second hitting the workpiece or beam current, (2) the electron speed at the moment of impact, the accelerating potential, (3) the diameter of the beam at or within the workpiece, the beam spot size, and (4) the speed of travel or the welding speed. The first two variables, beam current and accelerating potential, are used in establishing welding parameters. The third factor, the beam spot size, is related to the focus of the beam, and the fourth factor is also part of the procedure. Normally the electron beam current ranges from 250 to 1,000 milliamperes, the beam currents can be as low as 25 milliamperes. The accelerating voltage is within the two ranges mentioned previously. Travel speeds can be extremely high and relate to the thickness of the base metal. The other parameter that must be controlled is the gun-to-work distance.

The beam spot size can be varied by the location of the focal point with respect to the surface of the part. Penetration can be increased by placing the focal point below the surface of the base metal. As it is increased in depth below the surface deeper penetration will result. When the beam is focused at the surface there will be more reinforcement on the surface. When the beam is focused above the surface there will be excessive reinforcement and the width of the weld will be greater.

Penetration is also dependent on the beam current.

As beam current is increased penetration is increased. The other variable, travel speed, also affects penetration. As travel speed is increased penetration is reduced.

The heat input produced by electron beam welds is relatively small compared to the arc welding processes. The power in an electron beam weld compared with a gas metal arc weld would be in the same relative amount. The gas metal arc weld would require higher power to produce the same depth of penetration. The energy in joules per inch for the electron beam weld may be only 1/10 as great as the gas metal arc weld. The electron beam weld will be equivalent to the SMAW weld with less power because of the tremendous penetration obtainable by electron beam welding. The power density is in the range of 100 to 10,000 kw/in^2.

Since the electron beam has tremendous penetrating characteristics, with the lower heat input, the heat-affected zone is much smaller than that of any arc welding process. In addition, because of the almost parallel sides of the weld nugget, distortion is very greatly minimized. The cooling rate is much higher and for many metals this is advantageous; however, for high-carbon steel this is a disadvantage and cracking may occur.

The weld joint details for electron beam welding must be selected with care. In high vacuum chamber welding special techniques must be used to properly align the electron beam with the joint. Welds are extremely narrow and therefore preparation for welding must be extremely accurate. The width of a weld in 1/2-in. (12-mm) thick stainless steel, for example, would only be 0.04 in. (0.10 mm) and for this reason small misalignment would cause the electron beam to completely miss the weld joint. Special optical systems are used which allow the operator to align the work with the electron beam. The electron beam is not visible in the vacuum. Welding joint details normally used with gas tungsten arc welding can be used with electron beam welding. The depth to width ratio allows for special lap type joints. Where joint fitup is not precise ordinary lap joints are used and the weld is an arc seam type of weld. Normally, filler metal is not used in electron beam welding; however, when welding mild steel highly deoxidized filler metal is sometimes used. This helps deoxidize the molten metal and produce dense welds.

Almost all metals can be welded with the electron beam welding process. The metals that are most often welded are the super alloys, the refractory metals, the

FIGURE 6-47 *Welding the catalytic converter with EB.*

reactive metals, and the stainless steels. Many combinations of dissimilar metals can also be welded.

One of the disadvantages of the electron beam process is its high capital cost. The price of the equipment is very high and it is expensive to operate due to the need for vacuum pumps. In addition, fitup must be precise and locating the parts with respect to the beam must be perfect.

In the case of the medium vacuum system much larger work chambers can be used. Newer systems are available where the chamber is sealed around the part to be welded. In this case, it has to be designed specifically for the job at hand. The latest uses a sliding seal and a moveable electron beam gun. In other versions of the medium vacuum system, parts can be brought into and taken out of the vacuum work chamber by means of interlocks so that the process can be made more or less continuous. The automotive industry is using this system for welding gear clusters and other small assemblies of completely machined parts. This can be done since the distortion is minimal.

The non-vacuum system is finding acceptance for other applications. One of the most productive applications is the welding of automotive catalytic converters around the entire periphery of the converter. Figure 6-47 shows the electron beam welding of catalytic converters.

The electron beam process is becoming increasingly popular where the cost of equipment can be justified over the production of many parts. It is also used to a very great degree in the atomic energy industry for remote welding and for welding the refractory metals. Electron beam welding is not a cure-all; there are still the possibilities of defects of welds in this process as with any other. The major problem is the welding of plain carbon steel which tends to become porous when welded in a vacuum. The melting of the metal releases gases originally in the metal and results in a porous weld. If deoxidizers cannot be used the process is not suitable. It is expected that the electron beam process will become more popular for welding specialized metals where critical quality standards must be met.

For more information about this process refer to the Welding Research Bulletin entitled "Electron Beam Welding."[14]

6-7 LASER BEAM WELDING

Laser beam welding (LBW) is a welding process which produces coalescence of materials with the heat obtained

234

from the application of a concentrated coherent light beam impinging upon the surfaces to be joined.

The laser beam is a relatively new discovery dating back to the late 1950s. The word laser is obtained from *Light Amplification by Stimulated Emission of Radiation*. The focused laser beam has the highest energy concentration of any known source of energy. The laser beam is a source of electromagnetic energy or light that can be projected without diverging and can be concentrated to a precise spot. The beam is coherent and of a single frequency.

Producing a laser beam is extremely complex. The early laser utilized a solid state transparent single crystal of ruby made into a rod approximately an inch in diameter and several inches long. The end surfaces of the ruby rod were ground flat and parallel and were polished to extreme smoothness. The flat ends were coated with silver except for a small area on one end so that they would reflect light. The ruby rod was closely surrounded by a high-intensity light source, usually a Xenon flash tube. When the Xenon tube was flashed it emitted intense pulses of light which lasted up to two milliseconds. A high-intensity beam of coherent red light was emitted from the opening on the one silvered end of the ruby rod. A burst of light which lasted about two milliseconds occurred each time the Xenon tube was flashed. This device was known as a *pulsed ruby laser*. It was impossible to pulse the ruby crystal too fast or too often because heat built up in the crystal and in the flash lamp. The ruby could not be pulsed or pumped continuously. The pulse durations were very short and there was a relatively long time period between pulses. Continuously operating ruby lasers have been developed but the gas laser offered the possibility of higher power.

As the technology advanced it was found that gases could be made to emit coherent radiation when contained in an optical resonant cavity. It was found also that gas lasers could be operated continuously but originally only at low levels of power. Later developments allowed the gases in the laser to be cooled so that it could be operated continuously at higher power outputs. The gas lasers are pumped by high radio frequency generators which raise the gas atoms to sufficiently high energy level to cause lasing. Currently 2,000-watt carbon dioxide laser systems are in use. Higher powered systems are also being used for experimental and developmental work. A 6-kw laser is being used for automotive welding applications and a 10-kw laser has been built for research purposes. There are other types of lasers; however, the continuous carbon dioxide laser now available with 100 watts to 10 kw of power seems the most promising for metalworking applications.

The coherent light emitted by the laser can be focused and reflected in the same way as a light beam. The focused spot size is controlled by a choice of lenses and the distance from it to the base metal. The spot can be made as small as 0.003 in. (0.076 mm) to large areas 10 times as big. A sharply focused spot is used for welding and for cutting. The large spot is used for heat treating.

The laser offers a source of concentrated energy for welding; however, there are only a few lasers in actual production use today. The high-powered laser is extremely expensive, costing in the order of one-half million dollars. Laser welding technology is still in its infancy so there will be improvements and the cost of equipment will be reduced. Recent use of fiber optic techniques to carry the laser beam to the point of welding may greatly expand the use of lasers in metalworking.

Welding with Lasers

The laser can be compared to solar light beam for welding. The laser can be used in air. The laser beam can be focused and directed by special optical lenses and mirrors. The laser can operate at considerable distance from the workpiece.

When using the laser beam for welding the electromagnetic radiation impinges on the surface of the base metal with such a concentration of energy that the temperature of the surface is melted and volatilized. The beam penetrates through the metal vapor and melts the metal below. One of the original questions concerning the use of the laser was the possibility of reflectivity of the metal so that the beam would be reflected rather than heat the base metal. It was found, however, that once the metal is raised to its melting temperature the surface conditions have little or no effect.

It was found also that the distance from the optical cavity to the base metal had little effect on the laser. The laser beam is coherent and it diverges very, very little and can be focused to the proper spot size at the work with the same amount of energy available, whether it is close or far away.

It was found that with laser welding the molten metal takes on a radial configuration similar to conventional arc welding. However, when the power density rises above a certain threshold level keyholing occurs, the same as with plasma arc welding. Keyholing provides for extremely deep penetration. This provides for a high depth-to-width ratio. Keyholing also mini-

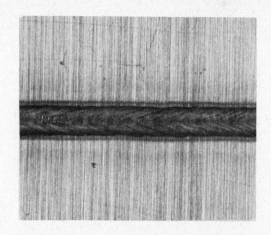

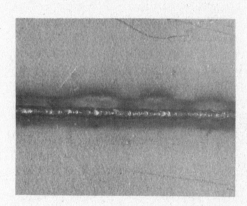

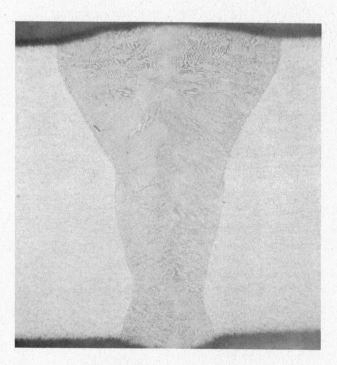

mizes the problem of beam reflection from the shiny molten metal surface since the keyhole behaves like a "black" body and absorbs the majority of the energy. In some applications inert gas is used to shield the molten metal from the atmosphere. It is found that the metal vapor that occurs may cause a breakdown of the shielding gas and creates a plasma in the region of high-beam intensity just above the metal surface. The plasma absorbs energy from the laser beam and can actually block the beam and reduce melting. A solution has been found by use of an inert gas jet directed along the metal surface which eliminates the plasma buildup and also shields the surface from the atmosphere.

The welding characteristics of the laser and of the electron beam are similar. The concentration of energy by both beams is similar with the laser having a power density in the order of 10^6 watts per square centimeter. The power density of the electron beam is only slightly greater. This is compared to a current density of only 10^4 watts per square centimeter for arc welding.

Figure 6-48 shows a laser weld made on stainless steel. It shows the top and underside bead and the cross section. This illustrates the characteristic cross section of the laser weld and also shows the depth-to-width ratio.

Laser beam welding has a tremendous temperature differential between the molten metal and the base metal immediately adjacent to the weld. Heating and cooling rates are much higher in laser beam welding than in arc welding, and the heat-affected zones are much smaller. Rapid cooling rates can create problems such as cracking in high carbon steels.

The laser beam has been used to weld carbon steels, high strength low alloy steels, aluminum, stainless steel and titanium. Laser welds made in these materials are similar in quality to welds made in the same materials by electron beam process. Experimental work using filler metal is being used to weld metals that tend to show porosity when welded with either EB or LB welding.

Experimental work with the laser beam welding process indicates that the normal factors control the weld. Another factor is the location of the focal point of the beam with respect to the surface of the work-piece. Maximum penetration occurs when the beam is focused slightly below the surface. Penetration is less when the beam is focused on the surface or deep within the surface. As power is increased the depth of penetration is increased.

Materials 1/2 in. (12 mm) thick are being welded at a speed of 10 inches (250 mm) per minute.

FIGURE 6-48 *Stainless steel laser weld.*

At this time the efficiency of laser beam welding equipment is in the neighborhood of 5-10%. As time goes on it is expected that this will increase and also that the cost of the equipment will decrease. More applications for laser welding will be found, and this process will challenge the electron beam welding process.

6-8 THERMIT WELDING

Thermit welding (TW) is a welding process which produces coalescence of metals by heating them with superheated liquid metal from a chemical reaction between a metal oxide and aluminum with or without the application of pressure. Filler metal is obtained from the liquid metal. This is one of the older welding processes. It was invented in about 1900, but is still used for specific applications.

The heat for welding is obtained from an exothermic reaction between iron oxide and aluminum. This reaction is shown by the following formula:

$$8Al + 3Fe_3O_4 = 9Fe + 4Al_2O_3 + Heat$$

The temperature resulting from this reaction is approximately 4500°F (2500°C). The superheated steel is contained in a crucible located immediately above the weld joint. The exothermic reaction is relatively slow and requires 20 to 30 seconds no matter how much of the chemicals are involved. The superheated steel runs into a mold which is built around the parts to be welded. Since it is almost twice as hot as the melting temperature of the base metal melting occurs at the edges of the joint and alloys with the molten steel from the crucible. Normal heat losses cause the mass of molten metal to solidify, coalescence occurs, and the weld is completed. The thermit welding process is applied only in the automatic mode. Once the reaction is started it

continues until it goes to completion. Welding utilizes gravity which causes the molten metal to fill the cavity between the parts being welded. It is very similar to the foundry practice of pouring a casting. The difference is the extremely high temperature of the molten metal. The making of a thermit weld is shown by Figure 6-49.

The thermit material is a mechanical mixture of metallic aluminum and processed iron oxide. This mixture may also include various elements for alloying the weld metal. Thermit mixtures can be designed to produce specific weld metal deposits. The normal analysis of thermit employed to weld mild and medium carbon steel is as follows:

Carbon	.20–.30
Manganese	.50–.60
Silicon	.25–.50
Aluminum	.07–.18
Iron	balance

The thermit can be designed for wear resistance and for welding cast iron, as well as for welding carbon steels. The mechanical properties of normal thermit are approximately the same as those of mild steel.

The thermit powder will not ignite until it is brought to an initial temperature of 2400°F (1300°C). It is normally started by using a special ignition powder which creates the temperature. During the exothermic reaction the molten steel will go to the bottom of the crucible. The aluminum oxide will float to the top as a slag, which protects the molten steel from the atmosphere.

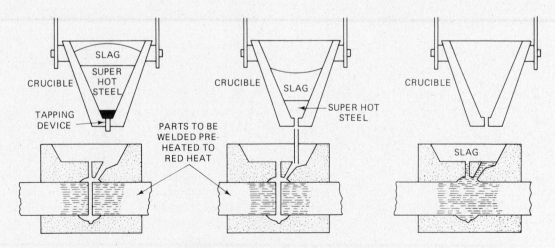

FIGURE 6-49 *Steps in making a thermit weld.*

The molten metal is tapped by a tapping pin at the bottom of the crucible. The superheated molten metal immediately flows into the mold through the pouring gate into the cavity making the weld. Venting must be provided to insure that the cavity is completely filled.

The parts to be welded must be prepared with a square groove joint. The root opening between the parts is related to the cross-sectional area of the weld. The root opening should range from 3/4–1-1/2 in. (19–37 mm) for joining railroad rails. For larger sections the root opening should be greater. Parts to be welded are properly aligned and braced. A mold is then made around the joint. The lost wax technique is sometimes employed for making molds of unusual shapes. The lost wax method involves filling the weld joint or cavity with wax then making the mold around this assembly. A riser must be provided as well as a pouring gate. A heating gate at the lower portion of the joint is provided for preheating which is usually required. This will melt the wax which will run out of the mold and provide the cavity for the molten weld metal. After heating is completed the heating gate is sealed with a plug. This process is shown by Figure 6-50.

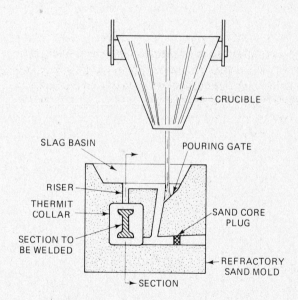

SLAG BASIN

CRUCIBLE

POURING GATE

RISER

THERMIT COLLAR

SECTION TO BE WELDED

SAND CORE PLUG

REFRACTORY SAND MOLD

SECTION

FIGURE 6-50 *Details of the crucible and mold.*

The amount of thermit is calculated to provide sufficient metal to produce the weld. The amount of steel produced by the reaction is approximately one-half the original quantity of thermit material by weight and one-third by volume.

After the weld has cooled, the mold is broken away and discarded, and the gates and risers are removed by oxyacetylene flame cutting. The surface of the completed weld is usually sufficiently smooth and contoured so that it does not require additional metal finishing.

The resulting deposited weld metal is of the composition mentioned above plus any alloying that may have been picked up from the parts being welded. The deposited weld metal is homogenous and quality is relatively high. Distortion is minimized since the weld is accomplished in one pass and since cooling is uniform across the entire weld cross section. There is normally shrinkage across the joint, but little or no angular distortion.

Welds can be made with the parts to be joined in almost any position as long as the cavity has vertical sides. If the cross-sectional area or thicknesses of the parts to be joined are quite large the primary problem is to provide sufficient thermit metal to fill the cavity.

Thermit welding has been used for many special applications. An important use was the welding of stern frames for Liberty ships during World War II. These frames were so large they could not be cast in one piece and were cast in four sections, which were joined together by four thermit welds. Another application has been in the making of large crankshafts. The thermit welds were made through the cross section of the round bearings welded to throws to make the crank. Thermit welds have also been used to weld large thick I beams and railroad and craneway rails. They have also been popular for welding reinforcing bars. Special mixtures of thermit are required for welding alloy steels of this type.

When the thermit process is used for joining rails and reinforcing bars standardized semipermanent molds are used. These molds are made in halves which are clamped around the rail or bar. The molds, which can be reused, are available for various sizes of rails and reinforcing bars.

Thermit welds are relatively inexpensive since little or no equipment is required. The primary cost is the cost of the thermit material which becomes the deposited weld metal. Thermit reinforcing bar welds meet the requirements of the Concrete Institute.

Thermit welds can also be used for welding nonferrous materials. The most popular uses of nonferrous thermit welding is the joining of copper and aluminum conductors for the electrical industry. In these cases the exothermic reaction is a reduction of copper oxide by aluminum which produces molten superheated copper. The high-temperature molten copper flows into the mold, melts the end of the parts to be welded, and, as the metal cools, a solid homogenous weld results. In welding copper and aluminum cables the molds are made of graphite and can be used over and over. When welding nonferrous materials the parts to be joined must be extremely clean and a flux is normally applied to the joint prior to welding. Special kits are available that provide the molds for different sizes of cable, and provide the premixed thermit material. This material also includes enough of the igniting material so that the exothermic reaction is started by means of a special lighter. An example of thermit welds made joining heavy copper cables and for joining cables to lugs is shown by the Figure 6-51.

FIGURE 6-51 *Thermit weld of copper cables.*

6-9 MISCELLANEOUS WELDING PROCESSES

"The other welding processes" include induction welding, flow welding, ionic welding, solar welding, etc. These are processes of limited application, processes of little industrial significance, or processes too new to have been defined by the American Welding Society. Each of these will be briefly described.

Flow Welding (FLOW)

Flow welding is a welding process which produces coalescence of metals by heating them with molten filler metal poured over the surfaces to be welded until the welding temperature is obtained and until the required filler metal has been added. The filler metal is not distributed in the joint by capillary attraction. This process has been called *burning in,* but it is of little industrial significance today.

Plasma MIG Welding

A new welding process developed by Philips in Holland in 1971 is named Plasma-MIG Welding. It is a combination of gas metal-arc welding and plasma arc welding. The standard plasma torch is modified by placing the tungsten electrode slightly to one side of the center line of the torch and nozzle orifice. A contact tip is inserted in the torch on this centerline and a continuous wire is fed to the tip and through the nozzle orifice along with the plasma gas. In other words, the gas metal arc consumable electrode and the arc itself are contained within the plasma column. Separate power supplies are used—one for the plasma and one for the electrode wire. Under certain conditions the gas metal arc rotates within the plasma. Extremely

high temperatures are attained and this results in high deposition rates.

Of necessity the torch is larger than a standard plasma torch and more complex. In addition, an extra power source is employed plus extra circuitry for starting the process. In view of this, it is normal to move the work under a fixed torch. One major application for the process is cladding of steel with dissimilar material such as stainless steel. This process is still very new, has not attained sufficient use and is therefore of little industrial significance today.

Induction Welding (IW)

Induction welding is a welding process which produces coalescence of metals through the heat obtained from resistance of the work to induced electrical current, with or without the application of pressure. Induction brazing and induction soldering are similar in that the heating is done by inducing currents within the workpiece, which, in turn, create heating within the workpiece. High-frequency resistance welding is similar; high-frequency current is transmitted to the workpiece by means of sliding shoes. The high-frequency current creates heating on the surfaces to be welded. Induction welding is a combination of these two. The high-frequency current is introduced into the workpiece by means of induction but the frequency is sufficiently high that it is concentrated on the surface of the work, particularly the surfaces to be welded. Welding is accompanied by pressure and filler metal is not used. The availability of extremely high-frequency generating equipment has made this process possible.

The high-frequency current required is in the range of 200,000–500,000 hertz. The power is generated by means of vacuum tube or solid state oscillators. The generator has a power output of from 1–600 kilowatts depending on the range of material thicknesses and production requirements. The output of the oscillator is fed into coils which have few turns and are water cooled. The work coil induces the high frequency current to flow in the closely coupled workpiece. The coil must be designed for the particular work to be done.

The induced high-frequency current flows on or near the surface of the workpiece and heat is generated only in the section of the base metal that carries the current. A relatively small amount of base metal is brought to the welding temperature. The resistance of the workpiece to the induced current causes the heating. The

THE OTHER WELDING PROCESSES

heated surfaces become sufficiently hot so that with sufficient pressure a weld is made, and, as it cools, coalescence occurs.

One of the most common uses of the process is the making of pipe and tubing. It is also used for joining dissimilar metals and can be used for making butt or lap seams of strip or sheet ranging from the very thinnest to sheet metal gauge thicknesses.

This process is normally automatic and is used for high-volume production of similar joints or parts.

Ion Beam Welding

This process has not been developed to practical use as yet. It is somewhat similar to electron beam welding and heat is created by the bombardment of ions on the surface to be welded. The reason for interest in ion beam welding is that it has different characteristics from electron beam welding. Ion bombardment which, is used in gas tungsten arc welding for the cleaning action, would become part of this welding process. The ion beam is less sensitive to external magnetic fields than the electron beam and there is no X-radiation. To make this process practical it is necessary to have a source of ions, to accelerate the beam of ions to the workpiece, and to be able to focus the beam on the weld area. Efforts to perfect this process are progressing in the laboratories, and perhaps in the not-too-distant future, welding will be performed with ion beams.

Solar Energy Welding

All of the different beam type welding processes are highly concentrated sources of energy. The energy is utilized to heat metals so that welding can be accomplished. The sun is well known as a source of heat and for many years there have been solar furnaces used to melt metals. This is accomplished by focusing the sun's rays with a parabolic mirror to a crucible. The major factor in solar welding is the use of a *heliostat* which is a device utilizing the mirrors moved by clockwork for directing the sun's rays to a constant point. The rays are directed into a parabolic mirror which deflects them to form a small focal spot. The focal spot is then directed to the work and concentrated to the actual welding point. Welding is accomplished by moving the workpiece with respect to the stationary focal spot. Fixturing, travel mechanisms, etc., are similar to those used by the other welding processes. To prevent the

atmosphere from coming in contact with the molten base metal the weld area is shielded by an inert gas. Welds have been successfully made on material from 0.039–0.118 in. (1-3 mm) thick in a single pass. Filler metal was not utilized. Travel speeds up to 2 in. (50 mm) per minute have been accomplished. Welds have been made in stainless steels with qualities equal to the base metal. So far this process is a laboratory curiosity. It may become useful in certain parts of the world depending upon the availability of solar energy.

Spot-Adhesive Welding

A recent development has combined spot welding with adhesives to produce joints that are stronger, more durable, and more resistant to fatigue than joints produced by either method alone. This process has been called *weld bonding* and has been used on aircraft structures.

It is practiced by the application of an adhesive in paste form to the faying surfaces of the weld joint. Resistance spot welds are made through the lap joints. The weldment is then heated in an oven to cure the adhesive. Another variation of the process is accomplished by first making the spot weld joint and then applying the adhesive at the edge of the joint, and, by means of capillary action, the adhesive flows between the faying surfaces and around the resistance spot welded nuggets. After the adhesive dries the assemblies are placed into an oven for curing.

The process has been used on aluminum for aircraft subassemblies.

The strength of the weld bond joints apparently is determined by the strength of the adhesive that is used.

Ultra Pulse Welding

Ultra pulse welding is a resistance type of welding process which uses extremely short weld time. The principle of operation is a capacitor which is charged from the power lines and discharged through a transformer and by means of resistance type electrodes to the work. The charging time is relatively long compared to the discharge or welding time which is in the order of a few milliseconds. The advantage of this system is that it reduces power line draw and makes welds in an extremely short time. Heat does not build up in the workpiece. It is used for welding extremely small and thin parts. The process is used for producing electronic components and instruments.

Undoubtedly there are other welding processes that will be invented and introduced, and some of the currently-used processes will fall into disuse. These changes will ultimately produce weldments at lower cost.

QUESTIONS

1. Does brazing require the use of a filler metal that has a higher or lower melting temperature than the base metal?

2. Is it possible to obtain a brazed joint stronger than the filler metal?

3. What is meant by preplaced filler metal?

4. What are the three types of oxyacetylene flames? How is each produced?

5. Explain how a gas pressure regulator operates.

6. Why shouldn't oil or grease be used on an oxygen apparatus?

7. Is flux required for making oxyacetylene welds on clean mild steel? Why?

8. What creates the heat in a resistance weld?

9. Explain the difference between a resistance spot weld and a projection weld.

10. Resistance welding electrodes are usually made of what materials? Why?

11. Is filler metal required for making a resistance weld?

12. What is the difference between soldering and brazing?

13. Why is flux required for soldering?

14. What is similar between an electron beam welding machine and a television set? Explain.

15. Why is electron beam welding recommended for hard-to-weld metals?

16. Explain the differences between EB welding in a hard vacuum, a soft vacuum, and in the atmosphere.

17. Why is precision joint preparation required for EB welding butt joints?

18. What is the advantage of the laser over the electron beam welding process?

19. Explain how gravity is used in making a thermit weld.

20. Is it possible to thermit weld copper or aluminum?

REFERENCES

1. *AWS Brazing Manual, The American Welding Society,* Miami, Florida.

2. "Specifications for Brazing Filler Metal," *AWS* A5.8, Miami, Florida.

3. "Standard Method for Evaluating the Strength of Brazed Joints," *AWS* C3.2, Miami, Florida.

4. *"ASME Boiler and Pressure Vessel Code,* Section IX Welding Qualifications," New York, New York.

5. "Specifications for Iron and Steel Gas Welding Rods," *AWS* A5.2, Miami, Florida.

6. R. D. ENQUIST, "How Easy Can You Join Metals by Resistance Spot Welding," *The Iron Age,* August 10, 1961.

7. "Resistance Welding Equipment Standards," *ANSI* C-88.2 and Bulletin #16 Sponsored by Resistance Welder Manufacturers Association, Philadelphia, Pennsylvania.

8. "Resistance Welding Manual, Vol. I & II," *Resistance Welder Manufacturers Association,* Philadelphia, Pennsylvania.

9. "Resistance Welding Theory and Use," *The American Welding Society,* Miami, Florida.

10. "Recommended Practices for Resistance Welding," *AWS* C1.1, Miami, Florida.

11. "Specification for Solder Metal," *The American Society for Testing and Materials ASTM* B-32, Philadelphia, Pa.

12. "Rosin Flux Cored Solder," ASTM B-284, Philadelphia, Pennsylvania.

13. *AWS Soldering Manual, The American Welding Society,* Miami, Florida.

14. M. M. SCHWARTZ, "Electron Beam Welding," Bulletin No. 196, July 1974, *Welding Research Council,* New York, New York.

7

PROCESSES RELATED TO WELDING

7-1 **ADHESIVE BONDING**

7-2 **ARC CUTTING**

7-3 **OXYGEN CUTTING**

7-4 **OTHER THERMAL CUTTING PROCESSES**

7-5 **THERMAL SPRAYING**

7-6 **HEAT FORMING AND STRAIGHTENING**

7-7 **PRE- AND POSTHEAT TREATMENT**

7-8 **JOINING OF PLASTICS**

7-1 ADHESIVE BONDING

The AWS Master Chart of Welding and Allied Processes shown by Figure 2-8 includes all of the major groups of metals joining processes. This chapter will present the allied processes.

These process groups are shown by Table 7-1 along with each process name and letter designations. Each process group will be briefly described in this chapter.

Adhesive bonding (AB) is a materials joining process in which an adhesive is placed between the faying surfaces which solidifies to produce an adhesive bond. The adhesive bond is the attractive force, generally physical in character, between an adhesive and the base materials. The two principle interactions that contribute to the adhesion are the van der Waals bond and the diepole bond. The van der Waals bond is defined as a secondary bond arising from the fluctuating-diepole nature of an atom with all occupied electron shells filled. The diepole bond is a pair of equal and opposite forces that hold two atoms together and results from a decrease in energy as two atoms are brought closer to one another.

Adhesive bonding of metal-to-metal applications accounts for less than 2% of the total metal joining

TABLE 7-1 *Processes allied to welding.*

Group	Allied Process	Letter Designation
Adhesive bonding	Dextrin cements	AB-D
	Solvent or rubber cements	AB-RC
	Synthetic resins	AB-SR
	Expoxys	AB-E
Arc cutting (thermal)	Air carbon arc cutting	AAC
	Carbon arc cutting	CAC
	Gas tungsten arc cutting	GTAC
	Metal arc cutting	MAC
	Plasma arc cutting	PAC
Oxygen cutting (thermal)	Chemical flux cutting	FOC
	Metal powder cutting	POC
	Oxygen arc cutting	AOC
	Oxy Fuel gas cutting	OFC
	Oxygen lance cutting	LOC
Other thermal cutting processes	Electron beam cutting	EBC
	Laser beam cutting	LBC
Thermal spraying	Electric arc spraying	EASP
	Flame spraying	FLSP
	Plasma spraying	PSP

requirements. The bonding of metals to nonmetals, especially plastics, is very important and is the major use of adhesive bonding. Most people involved with welding and metalworking are unfamiliar with the branch of chemistry called adhesive bonding. It is unfortunate that there are no uniform industrial standards for adhesives. This causes considerable misunderstanding since the terminology used is confusing and inconsistent.

Adhesive bonding has the problem of process and service life. The parts adhesively bonded using different adhesives require different treatment. Some are toxic, others are flammable, and some require a long cure time with precise temperature and pressure control. Some adhesives are sensitive to the environmental conditions such as humidity and temperature. High temperature service will reduce bond strengths. In spite of this there are many advantages to adhesive bonding and it is important for welding personnel to learn about this joining process.

The following are some of the advantages of adhesive bonding. The stresses on adhesive bonded parts are uniform over the entire bonded area. This may allow the use of thinner materials in an assembly. Adhesive bonding is versatile; materials of all sizes can be bonded. If they have different coefficients of expansion the bond will compensate since it retains some flexibility. Adhesive bonding has a high resistance to fatigue since it provides a damping action and will expand and contract under repeated loads. There is no distortion with adhesive bonding since high temperatures are avoided and holes are not used. Properties of the metals being joined are not affected. Adhesive bonds can have different characteristics as required. Some may provide electrical insulation while others will give electrical conduction depending on the design of the adhesive. Adhesive bonds provide seals to avoid additional sealing operations sometimes required. Adhesive bonds eliminate bulges, gaps, projections, or indentations when compared to mechanical fasteners or welds. Cost savings can be achieved by adhesive bonding when material thicknesses are reduced, or when operations such as drilling or punching are eliminated, and when metal finishing operations can be eliminated.

The description of properties of adhesive bond joints are different from metalworking terms. The joints are judged by such properties as peel strength, cleavage strength, tensile strength, and shear strength. The types of joints used for adhesive bonding are the lap joint and the tongue and groove which can be used for butt,

corner, or T joints. The mortice and tenon are also used for corner joints. These are shown by Figure 7-1. Standardized testing methods have been established by the American Society for Standards and Materials. This includes a test for tensile strength,[1,2] shear strength,[3] cleavage strength for metal-to-metal adhesives,[4] and peel resistance tests.[5].

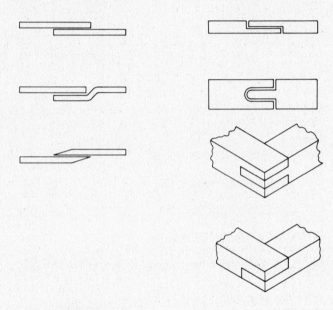

FIGURE 7-1 *Joint types used for adhesive bonding.*

Types of Adhesive for Bonding

The classification of adhesives is difficult because many adhesives are mixtures of basic types. Many are available in different forms. They are also classified with regard to setting characteristics i.e. thermoplastics or thermosetting. Adhesives can be divided into four basic commonly used types: (1) dextrins, (2) rubber base cements, (3) resins, and (4) epoxies.

Dextrin Adhesives

Dextrins belong to the family of starch-derived adhesives ranging in color from white to dark brown and are normally fluid filmy materials. These are glues and pastes used to bond porous materials. They are spread in a thin film. They can be applied by machine at comparatively heavy consistency and generally possess good humidity resistance and stickiness. A prime advantage of dextrins is that they do not set fast. It takes some time for them to establish a bond; thus, they are well suited for many applications. They have a poor resistance to water. They have been used for many years and are recommended for sealing corrugated boxes and attaching labels. They can be applied cold and have good final adhesive.

Rubber cements or **solvent cements** are adhesives that contain organic solvents rather than water. They are based on nitro cellulose or polyvinyl acetate, normally elastomeric products, dispersed in solvent. They are free flowing, thin set materials that dry to hard tack free films. Some others retain a soft tacky, flexible film. They are used in pressure-sensitive labeling operations and in contact bonding for the woodworking industry. They are used for laminating applications, such as veneering. Mastic type adhesives are a solvent cement used in the building and construction industry. They are used to bond wood and dry wall to concrete and other vertical surfaces. Rubber and solvent cements can be sprayed or roller or hand applied. They are usually flammable. The chemistry of the adhesive dictates its characteristics and recommended use.

Resin Adhesives

Synthetic resins are composed of synthetic organic materials and are relatively expensive. They are used when a high-quality bond is required and they are relatively heat and moisture resistant. They can be applied by automatic or semiautomatic equipment, are used for sealing cartons and for wood, and for vinyl film laminations.

One of the major groups is the hot melts which are combinations of waxes and resins that form a bond by applying heat and then cooling. Solid hot melt synthetic resin adhesives contain neither water nor solvents, are available in dry form, and set up quickly. The hot melt type must be heated to between 250 and 400 °F (121 and 204 °C) before they are applied. There are two application methods. (1) The adhesive is applied to one surface in its molten state and the other surface to be bonded is then joined while the adhesive is still liquid. This will cause the hot melt to solidify and form the bond. (2) The hot melt liquid is allowed to cool and form a nontacky film on one surface and then the second surface is added. Heat and pressure is applied to form the bond. Hot melt adhesives resist moisture and can be used on nonporous surfaces. They are used for bonding metal to various surfaces. They are used for packaging and in the furniture and woodworking industry.

Epoxy Adhesives

These are the newest of the adhesives and can be used to bond metal to metal, metal to plastics, and plastics to plastics. They are a family of materials characterized by reactive **epoxy** chemical groups on the ends of resin molecules. They consist of two components, a liquid resin and the hardener to convert the liquid resins to solid. They may contain other modifiers to produce specific properties for special applications. Some epoxies will bond to concrete. One of the newer advances is the *oily metal* epoxy that bonds directly to oily metals "as received" with normal protective films on them. The oily coating need not be removed. Other epoxies can be a one-component type, but these require a heat cure operation. Epoxies are good surface wetters. They achieve intimate molecular contact with the surface to be bonded and will achieve high adhesion on almost any surface. Epoxies are the most expensive of the adhesives; however, they offer more advantages.

There are four factors to consider in using adhesive bonding. These are (1) adhesive selection, (2) surface preparation, (3) application of adhesive, and (4) costs. Each of the four types of adhesives described above has its own advantages and disadvantages. The successful performance of an adhesive depends on surface preparation and, except for the oily type mentioned previously, the surfaces must be clean to allow the adhesive to produce an efficient joint. Application procedures are numerous and in some cases complex. The method of application can be a major expense. The cost of the adhesive should be considered relative to alternate joining methods. In general, the adhesive that assures the most efficient joining, the least amount of preparation and finishing, and provides the best security is the proper selection.

7-2 ARC CUTTING

There are a number of thermal cutting processes that are defined by the American Welding Society. These processes utilize heat and thus differ from mechanical cutting processes such as sawing, shearing, blanking, etc. They are shown on the AWS Master Chart of Welding and Allied Processes. There are four basic processes: arc cutting, oxygen cutting, electron beam cutting, and laser beam cutting.

The arc cutting processes are a group of thermal cutting processes which melt the metals to be cut with the heat of an arc between an electrode and the base metal. Within this group is air carbon arc cutting; carbon arc cutting; gas tungsten arc cutting, shielded metal arc, gas metal arc, and plasma arc cutting. Each will be briefly described.

The thermal cutting processes can be applied by means of manual, semiautomatic, machine, or automatic methods in the same manner as the arc welding processes.

PROCESSES RELATED TO WELDING

Air Carbon Arc Cutting (AAC)

Air carbon arc cutting is "an arc cutting process in which metals to be cut are melted by the heat of a carbon arc and the molten metal is removed by a blast of air."

Principle of Operation: A high velocity air jet traveling parallel to the carbon electrode strikes the molten metal puddle just behind the arc and blows the molten metal out of the immediate area. Figure 7-2 shows the operation of the process. It shows the arc between the carbon electrode and the work and the air stream parallel to the electrode coming from the special electrode holder.

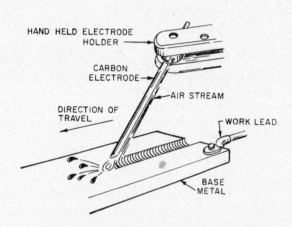

FIGURE 7-2 *Process diagram for air carbon arc cutting.*

Advantage and Major Use: The air carbon arc cutting process is used to cut metal and to gouge out defective metal, to remove old or inferior welds, for root gouging of full penetration welds, and to prepare grooves for welding. Air carbon arc cutting is used when slightly ragged edges are not objectionable. The area of the cut is small, and, since the metal is melted and removed quickly, the surrounding area does not reach high temperatures. This reduces the tendency towards distortion and cracking.

The air carbon arc cutting and gouging process is normally manually operated. The apparatus can be mounted on a travel carriage and thus it is considered machine cutting or gouging. Special applications have been made where cylindrical work has been placed on a lathe-like device and rotated under the air carbon arc torch. This is machine or automatic cutting depending on operator involvement.

The air carbon arc cutting process can be used in all positions. It can also be used for gouging in all positions. Use in the overhead position requires a high degree of skill.

The air carbon arc process can be used for cutting or gouging most of the common metals. This, as well as current and polarity, is given in the table shown by Figure 7-3.

Metal	Current Type	Electrode Polarity
Aluminums	DC	Positive
Copper and alloys	AC	–
Iron, cast, malleable, etc.	DC	Negative
Magnesium	DC	Positive
Nickel and alloys	AC	–
Carbon steels	DC	Positive
Stainless steels	DC	Positive

FIGURE 7-3 *Metals that can be cut with AAC.*

The process is not recommended for weld preparation for stainless steel, titanium, zirconium, and other similar metals without subsequent cleaning. This cleaning, usually by grinding, must remove all of the surface carbonized material adjacent to the cut. The process can be used to cut these materials for scrap for re-melting.

The circuit diagram for air carbon arc cutting or gouging is shown by Figure 7-4. Normally, conventional welding machines with constant current are used. Constant voltage can be used with this process. When using a CV power source precautions must be taken to operate it within its rated output of current and duty cycle. Alternating current power sources having conventional drooping characteristics can also be used for special applications. AC type carbon electrodes must be used.

Equipment required is shown by the block diagram. Special heavy duty high current machines have been made specifically for the air carbon arc process. This

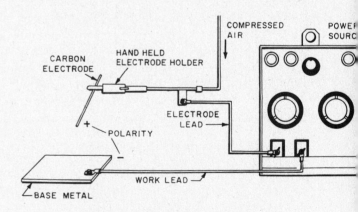

FIGURE 7-4 *Circuit block diagram AAC.*

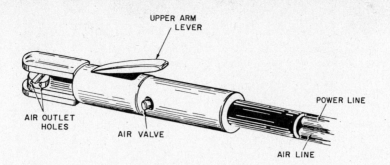

FIGURE 7-4 *(Continued).*

is because of extremely high currents used for the large size carbon electrodes.

The electrode holder is designed for the air carbon arc process. The electrode holder and a copper coated electrode are shown by Figure 7-5. The holder includes a small circular grip head which contains the air jets for directing the compressed air along the electrode. It also has a groove for gripping the electrode. This head can be rotated to allow different angles of electrode with respect to the holder. A heavy electrical lead and an air supply hose are connected to the holder through a terminal block. A valve is included in the holder for turning the compressed air on and off. Holders are available in several sizes depending on the duty cycle of the work performed, the welding current, and size of carbon electrode used. For extra heavy duty work water-cooled holders are used.

The air pressure is not critical but should range from 80 to 100 psi (550 to 690 kPa). The volume of compressed air required ranges from as low as 5 cu ft/min (2.5 liter/min) up to 50 cu ft/min (24 liter/min) for the largest-size carbon electrodes. A one-horsepower compressor will supply sufficient air for smaller-size electrodes. It will require up to a ten-horsepower compressor when using the largest-size electrodes.

The carbon graphite electrodes are made of a mixture of carbon and graphite plus a binder which is baked to produce a homogeneous structure. Electrodes come in several types. The plain uncoated electrode is less expensive, carries less current, and starts easier. The copper-coated electrode provides better electrical conductivity between it and the holder. The copper-coated electrode is better for maintaining the original diameter during operation; it lasts longer and carries higher current. Copper-coated electrodes are of two types, the DC type and the AC type. The composition ratio of the carbon and graphite is slightly different for these two types. The DC type is more common. The AC type contains special elements to stabilize the arc. The AC type electrode is used for direct current electrode negative when cutting cast irons. For normal use, the electrode is operated with the electrode positive. Electrodes range in diameter from 5/32 in. (4.0 mm) to 1 in. (25.4 mm). Electrodes are normally

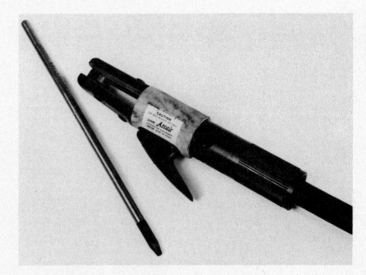

FIGURE 7-5 *Air carbon arc cutting electrode holder.*

12 in. (300 mm) long; however, 6 in. (150-mm) electrodes are available. Copper-coated electrodes with tapered socket joints are available for automatic operation, and allow continuous operation. Figure 7-6 shows the electrode types and the arc current range for different sizes.

Electrode Type	Electrode Size		Current	
	in.	mm	Min.	Max.
	5/32	4.0	90	150
	3/16	4.8	150	200
DC	1/4	6.4	200	400
(Plain)	5/16	7.9	250	450
or	3/8	9.5	350	600
AC	1/2	12.7	600	1000
(Copper	5/8	15.9	800	1200
Covered)	3/4	19.1	1200	1600
	1	25.4	1800	2200

Polarity of electrode is positive (reverse polarity).

Note: For DC copper covered electrodes current can be increased 10%.

FIGURE 7-6 *Electrode type—size and current range.*

Groove Width		Groove Depth		Electrode Dia.		Amperes Direct Current	Volts Electrode Positive	Electrode Feed		Travel Speed	
in.	mm	in.	mm	in.	mm			ipm	mm/min.	ipm	mm/min.
1/4	6.4	1/16	1.6	3/16	4.8	200	43	6.2	155.0	82.0	2050.0
9/32	7.1	1/8	3.2	3/16	4.8	200	40	6.7	167.5	38.2	955.0
5/16	7.9	3/16	4.8	3/16	4.8	190	42	6.7	167.5	27.2	680.0
5/16	7.9	1/4	6.4	3/16	4.8	(To make 1/4" deep groove, make two 1/8" deep passes.)					
5/16	7.9	3/32	2.4	1/4	6.4	270	40	4.0	100.0	54.0	1350.0
5/16	7.9	1/8	3.2	1/4	6.4	300	42	4.0	100.0	51.0	1275.0
5/16	7.9	3/16	4.8	1/4	6.4	300	40	6.7	167.5	38.2	955.0
5/16	7.9	1/4	6.4	1/4	6.4	320	42	6.2	155.0	29.5	737.5
5/16	7.9	3/8	9.5	1/4	6.4	320	46	3.6	90.0	15.0	375.0
3/8	9.5	1/8	3.2	5/16	7.9	320	40	3.0	75.0	65.5	1637.5
3/8	9.5	3/16	4.8	5/16	7.9	400	46	4.3	107.5	46.0	1150.0
3/8	9.5	1/4	6.4	5/16	7.9	420	42	3.8	95.0	31.2	780.0
3/8	9.5	1/2	12.7	5/16	7.9	540	42	5.6	140.0	27.2	680.0
7/16	11.1	1/8	3.2	3/8	9.5	560	42	4.2	105.0	82.0	2050.0
7/16	11.1	1/8	3.2	3/8	9.5	560	42	3.3	82.5	65.0	1625.0
7/16	11.1	3/16	4.8	3/8	9.5	560	42	2.6	65.0	41.0	1025.0
7/16	11.1	1/4	6.4	3/8	9.5	560	42	3.0	75.0	29.5	737.5
7/16	11.1	1/2	12.7	3/8	9.5	560	42	3.2	80.0	15.0	375.0
7/16	11.1	11/16	17.5	3/8	9.5	560	42	3.5	87.5	12.2	305.0
9/16	14.3	1/8	3.2	1/2	12.7	1200	45	3.0	75.0	34.0	850.0
9/16	14.3	1/4	6.4	1/2	12.7	1200	45	3.0	75.0	22.0	550.0
9/16	14.3	3/8	9.5	1/2	12.7	1200	45	3.0	75.0	20.7	517.5
9/16	14.3	1/2	12.7	1/2	12.7	1200	45	3.0	75.0	18.5	462.5
9/16	14.3	5/8	15.9	1/2	12.7	1200	45	3.0	75.0	15.0	375.0
9/16	14.3	3/4	19.1	1/2	12.7	1200	45	3.0	75.0	12.5	312.5
13/16	20.6	1/8	3.2	5/8	15.9	1300	42	2.5	62.5	44.5	1112.5
13/16	20.6	1/4	6.4	5/8	15.9	1300	42	2.5	62.5	29.5	737.5
13/16	20.6	3/8	9.5	5/8	15.9	1300	42	2.5	62.5	20.0	500.0
13/16	20.6	1/2	12.7	5/8	15.9	1300	42	2.5	62.5	14.5	362.5
13/16	20.6	5/8	15.9	5/8	15.9	1300	42	2.5	62.5	13.0	325.0
13/16	20.6	3/4	19.1	5/8	15.9	1300	42	2.5	62.5	11.0	275.0
13/16	20.6	1	25.4	5/8	15.9	1300	42	2.5	62.5	10.0	250.0

Notes: 1 For carbon steel
2 Combination of settings and multiple passes may be used for grooves deeper than 3/4".
3 All values are for DC copper-coated electrodes.
4 Air pressures 80-100 psi is recommended for 1/2" and 5/8" electrodes.
5 A head angle for 45 degrees is used for these settings.

FIGURE 7-7 *Air carbon arc gouging procedure schedule.*

The procedure schedule for making grooves in steel is shown by the table in Figure 7-7. For procedure information beyond these tables, consult the AWS "Recommended Practice for Air Carbon Arc Gouging and Cutting."[6]

To make a cut or a gouging operation the cutter strikes an arc and almost immediately starts the air flow. The electrode is pointed in the direction of travel with a push angle approximately 45° with the axis of the groove. The speed of travel, the electrode angle, and the electrode size and current determine the groove depth. Electrode diameter determines the groove width.

The normal safety precautions similar to carbon arc welding and shielded metal arc welding apply to air carbon arc cutting and gouging. However, two other precautions must be observed. First, the air blast will cause the molten metal to travel a very long distance. Metal deflection plates should be placed in front of the gouging operation. All combustible materials should be moved away from the work area. At high-current levels the mass of molten metal removed is quite large and will become a fire hazard if not properly contained.

The second factor is the high noise level. At high currents with high air pressure a very loud noise occurs. Ear protection, ear muffs or ear plugs should be worn by the arc cutter.

The process is widely used for back gouging, for preparing joints, and for removing defective weld metal.

It is also used in foundries for washing pads, removing risers, and removing defective areas of castings. Another major use is scrap preparation of metals to reduce them to proper size for handling. It is used also for scrapping of metal objects and for maintenance and salvage operations.

Carbon Arc Cutting (CAC)

This is "an arc cutting process in which metals are severed by melting them with the heat of an arc between a carbon electrode and the base metal."

The process is identical to air carbon arc cutting except that the air blast is not employed. The process depends strictly upon the heat input of the carbon arc to cause the metal to melt. The molten metal falls away by gravity to produce the cut. The process is relatively slow, a very ragged cut results and it is used only when other cutting equipment is not available. It has little industrial significance.

Metal Arc Cutting (MAC)

This is "an arc cutting process which severs metals by melting them with the heat of an arc between a metal electrode and the base metal." When covered electrodes are used it is known as shielded metal arc cutting (SMAC).

The equipment required is identical to that required for shielded metal arc welding. When the heat input into the base metal exceeds the heat losses the molten metal pool becomes large and unmanageable. If the base metal is not too thick, the molten metal will fall away and create a hole or cut. The cut produced by the shielded metal arc cutting process is rough and is not normally used for preparing parts for welding. The metal arc cutting process using covered electrodes is used only where a small cutting job is required and other means are not available for the purpose.

For metal arc cutting an electrode with deep penetrating qualities such as an E6010 or an E6011 should be used. A relatively small electrode should be used with DC electrode negative. The current should be set much higher than normally used for welding. This will create a maximum amount of heat in the weld pool which will soon fall away making the cut. This technique can also be used for cutting cast iron. On thick material a sawing action is required to make the cut and to allow the molten metal to fall away. If the electrode coating is made wet by dipping in water the electrode will melt somewhat slower so that more cut can be obtained per electrode. The gas metal arc can also be used for cutting (GMAC).

The metal arc cutting technique can also be used for gouging when utilizing special electrodes. These special electrodes have an insulating type coating which directs the arc. It is sometimes used for back gouging welds prior to making the backing weld. The technique used for gouging is shown by the table of Figure 7-8. The gouging type electrode can be used for cutting.

Gas Tungsten Arc Cutting (GTAC)

This is "an arc cutting process in which metals are severed by melting them with an arc between a single tungsten (nonconsumable) electrode and the work. Shielding is obtained from a gas or gas mixture."

Groove Width		Groove Depth		Electrode Dia.		Amperage Direct Current	Travel Speed	
in.	mm	in.	mm	in.	mm		ipm	mm/min.
3/16	4.8	1/32	0.8	1/8	3.2	210	71	1775
5/32	4.0	1/16	1.6	1/8	3.2	210	58	1450
1/4	6.4	3/32	2.4	1/8	3.2	210	47	1175
1/4	6.4	1/8	3.2	1/8	3.2	210	34	850
3/16	4.8	1/16	1.6	5/32	4.0	300	92	2300
1/4	6.4	3/32	2.4	5/32	4.0	300	71	1775
5/16	7.9	1/8	3.2	5/32	4.0	300	55	1375
5/16	7.9	5/32	4.0	5/32	4.0	300	43	1075
5/16	7.9	1/16	1.6	3/16	4.8	350	66	1650
3/8	9.5	1/8	3.2	3/16	4.8	350	50	1250
3/8	9.5	5/32	4.0	3/16	4.8	350	32	800
3/8	9.5	3/16	4.8	3/16	4.8	350	25	625

Note: (1) Flat and vertical down position

(2) Voltage will vary between 40 and 65 electrode negative (straight Polarity). Set welding machine to maximum open circuit voltage.

FIGURE 7-8 *Shielded metal arc gouging procedure schedule.*

Apparatus used for gas tungsten arc cutting is identical to that used for gas tungsten arc welding. The gas tungsten arc is used to provide sufficient heat input so that the molten puddle will become large and unmanageable and will fall away. Gas tungsten arc cutting is restricted to the thinner metals and can be used for cutting almost any metal in thin sections. The smoothness of the cut in thin materials is largely dependent upon the skill of the arc cutter and the accuracy and uniformity of travel. Under ideal conditions smooth cuts can be made with gas tungsten arc welding on thin materials.

This process has largely been supplanted by plasma arc cutting and is of little industrial significance except for the small jobs when other equipment is not available.

Plasma Arc Cutting (PAC)

This is an arc cutting process which severs metal by melting a localized area with a constricted arc and removing the molten material with a high-velocity jet of hot ionized gas.

There are three major variations: (1) low-current plasma cutting which is a rather recent development, (2) the original relatively high current plasma cutting, and (3) plasma cutting with water added. The low-current plasma variation is gaining in popularity because it can be manually applied.

The principle operation of plasma cutting is almost identical with the keyhole mode of plasma welding. The difference is that the cut is maintained and the keyhole is not allowed to close as in the case of welding. Heat input at the plasma arc is so high and the heat losses cannot carry the heat away quickly enough so that the metal is melted and a hole is formed. The plasma gas at a high velocity helps cut through the metal. The secondary gas can also assist the jet in removing molten metal and limits the formation of drops at the cutting edge. Plasma cutting is ideal for gouging and for piercing. For some operations air is used as the plasma gas. A higher arc voltage is normally used for cutting than for welding.

Plasma arc cutting has less of a detrimental influence on the base metals than conventional oxyacetylene cutting. Welding operations can often be performed directly over plasma cut edges in aluminum and stainless steels. The very thin nitride film on aluminum and the thin oxide film on stainless steel edges are not detrimental and cannot be detected in the weld joint. They do not adversely affect the mechanical properties of the joint and cannot be seen in metallurgical examination.

In the low current arc cutting variation the maximum current is 100 amperes. The torch is relatively small, compared to high-current plasma torches. It is normally used manually and produces high-quality cuts of thin materials.

The plasma arc cutting process can be used to cut metals underwater.

For high current cutting the torch is mounted on a mechanical carriage. Automatic shape cutting can be done with the same equipment used for oxygen cutting, if sufficiently high travel speed is attainable. A water spray is used surrounding the plasma to reduce smoke and noise. Work tables containing water which is in contact with the underside of the metal being cut will also reduce noise and smoke.

The torch can be used in all positions. It can also be used for piercing holes and for gouging.

The metals usually cut with this process are the aluminums and stainless steels. The process can also be used for cutting carbon steels, copper alloys, and nickel alloys.

The arc power source, must have a high open circuit voltage. The high voltage power source requires special protection. The torch and work connections should be provided with special insulation. A power contactor is required so that the open circuit voltage is never present at the machine terminals. Normally a three-phase rectifier power supply with an open circuit voltage of 120 volts is used. Operating range is normally between 75 and 100 volts.

Special controls are required to adjust both plasma and secondary gas flow. Torch-cooling water is required and is monitored by pressure or flow switches for torch protection. The cooling system should be self-contained, which includes a circulating pump and a heat exchanger.

Plasma cutting torches will fit torch holders in automatic flame cutting machines.

The cutting torch is of special design for cutting and is not used for welding.

High purity nitrogen is usually used for the plasma gas, and welding grade carbon dioxide is used for the secondary gas.

The quality of the cut obtained is relatively high and is equal to the cut made with oxyacetylene equipment. The cutting schedules for low-current plasma cutting are shown in the table Figure 7-9.[7]

The amount of gases and fumes generated requires the use of local exhaust for proper ventilation. Cutting should be done over a water reservoir so that the particles removed from the cut will fall in the water to help reduce the amount of fumes released into the air.

Material Type	Material Thickness		Plasma Gas Flow Nitrogen		Shielding Gas Flow CO_2		Current Amperes	Cutting Speed	
	in.	mm	cfh	l/m	cfh	l/m		ipm	mm/min.
Stainless	0.125	3.2	15	7	150	72	80	52	1300
Steel	0.250	6.4	15	7	150	72	80	44	1100
	0.312	7.9	15	7	150	72	85	34	850
	0.500	12.7	15	7	150	72	100	22	550
	0.750	19.1	15	7	150	72	100	20	500
	1.000	25.4	25*	12	150	72	100	8	200
Aluminum	0.093	2.4	15	7	150	72	70	120	3000
	0.250	6.4	15	7	150	72	70	48	1200
	0.375	9.5	25*	12	150	72	80	24	600
	1.000	25.4	25*	12	150	72	100	24	600
Mild steel	0.125	3.2	15	7	150	72	75	52	1300
	0.250	6.4	15	7	150	72	85	36	900
	0.500	12.7	15	7	150	72	100	24	600

*Use 80% argon + 20% hydrogen instead of nitrogen.

FIGURE 7-9 *Cutting procedure schedules low current plasma cutting.*

The noise level generated by the high-powered equipment is uncomfortable. The cutter should be required to wear ear protection. The normal protective clothing to protect the cutter from the arc should be worn. This involves protective clothing, gloves, and helmet. The helmet should be equipped with a shade no. 9 filter glass lens. The other safety considerations as outlined in Chapter 3 should be followed.

There are many applications for low-current plasma arc cutting, including the cutting of stainless and aluminum for production and maintenance. Plasma cutting can also be used for stack cutting and it is more efficient than stack cutting with the oxyacetylene torch. Low-current plasma gouging can also be used for upgrading castings.

The high-current plasma cutting variation is completely described in the booklet, AWS "Recommended Practices for Plasma Arc Cutting."[8] High-current plasma is used with automatic shape cutting equipment and does require relatively high-speed apparatus. The apparatus must also be precise for quality cuts. Hand-held apparatus is not commonly used.

The water-injection variation is an effort to reduce the fumes and smoke produced by the high-current plasma process. The use of water improves the quality of the cut on most materials.

7-3 OXYGEN CUTTING

Oxygen cutting (OC) is a group of thermal cutting processes used to sever or remove metals by means of the chemical reaction of oxygen with the base metal at elevated temperatures. In the case of oxidation-resistant metals the reaction is facilitated by the use of a chemical flux or metal powder. Five basic processes are involved: (1) oxy fuel gas cutting, (2) metal powder cutting, (3) chemical flux cutting, (4) oxygen lance cutting, and (5) oxygen arc cutting. Each of these processes is different and will be described.

Oxyfuel Gas Cutting (OFC)

Oxyfuel gas cutting is used to sever metals by means of the chemical reaction of oxygen with the base metal at elevated temperatures. The necessary temperature is maintained by means of gas flames obtained from the combustion of a fuel gas and oxygen. When an oxyfuel gas cutting operation is described the fuel gas must be specified. There are a number of fuel gases used. The most popular is acetylene. Natural gas is widely used as is propane, methylacetylene-propadiene stabilized and various trade name fuel gases. Hydrogen is rarely used. Each fuel gas has its particular characteristics and may require slightly different apparatus because of these characteristics. The characteristics relate to the flame temperatures, heat content, oxygen fuel gas ratios, and so on. The general concept of oxy fuel gas cutting is similar no matter what fuel gas is used. It is the oxygen jet that makes the cut in steel, and cutting speed depends on how efficiently the oxygen reacts with the steel. For simplicity, we will confine our discussion to the use of acetylene.

The generation of heat by combustion of acetylene and oxygen was described in Chapter 6. This heat is used to bring the base metal steel up to its kindling temperature where it will ignite and burn in an atmosphere of pure oxygen. The chemical formula for three of the oxidation reactions is as follows:

$$Fe + .5O_2 = Fe \quad + Heat$$
$$3Fe + 2O_2 = Fe_3O_4 + Heat$$
$$2Fe + 1.5O_2 = Fe_2O_3 + Heat$$

At elevated temperatures all of the iron oxides are produced in the cutting zone.

The oxyacetylene cutting torch is used to heat the steel by increasing the temperature to its kindling point and then by introducing a stream of pure oxygen to create the burning or rapid oxidation of the steel. The stream of oxygen also assists in removing the material from the cut. This is shown by Figure 7-10.

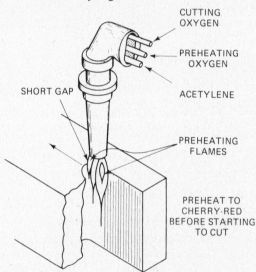

CUTTING OXYGEN

PREHEATING OXYGEN

ACETYLENE

SHORT GAP

PREHEATING FLAMES

PREHEAT TO CHERRY-RED BEFORE STARTING TO CUT

FIGURE 7-10 *Process diagram oxygen cutting.*

Steel and a number of other metals are flame cut with the oxyfuel gas cutting process. The following conditions must apply.

1. The melting point of the material must be above its kindling temperature in oxygen.

2. The oxides of the metal should melt at a lower temperature than the metal itself and below the temperature that is developed by cutting.

3. The heat produced by the combustion of the metal with oxygen must be sufficient to maintain the oxygen cutting operation.

4. The thermal conductivity must be low enough so that the material can be brought to its kindling temperature.

5. The oxides formed in cutting should be fluid when molten so as not to interrupt the cutting operation.

Iron and low-carbon steel fit all of the above requirements and are readily oxygen flame cut. Cast iron is not readily flame cut, because the kindling temperature is above the melting point. It also has a refractory silicate oxide which produces a slag covering. Chrome-nickel stainless steels cannot be flame cut with the normal technique, because of the refractory chromium oxide formed on the surface. Nonferrous metals such as copper and aluminum have refractory oxide coverings which prohibit normal oxygen flame cutting; in addition, they have high thermal conductivity.

FIGURE 7-11 *Manual oxyacetlyene cutting operation.*

When flame cutting, the preheating flame should be neutral or oxidizing. A reducing or carbonizing flame should not be used. Figure 7-11 shows the flame-cutting operation being manually performed. The schedule for flame cutting clean mild steel is shown by the Figure 7-12.

Torches are available for either welding or cutting. By placing the cutting torch attachment on the torch body it is used for manual flame cutting. Figure 7-13 shows a manual oxyacetylene flame-cutting torch. Various sizes of tips can be used for manual flame

Material Thickness		Cutting Orifice Dia. (center hole)			Approx. Press. of Gas		Travel Speed Manual	Travel Speed Automatic
in.	mm	Drill size	in.	mm	Acetylene psi	Oxygen psi	in./min.	in./min.
1/8	3.2	60	0.040	1.0	3	10	20-22	22
1/4	6.4	60	0.040	1.0	3	15	16-18	20
3/8	9.5	55	0.052	1.3	3	20	14-16	19
1/2	12.7	55	0.052	1.3	3	25	12-14	17
3/4	19.0	55	0.052	1.3	4	30	10-12	15
1	25.4	53	0.060	1.5	4	35	8-11	14
1-1/2	38.1	53	0.060	1.5	4	40	6-7-1/2	12
2	50.8	49	0.073	1.9	4	45	5-1/2-7	10
3	76.2	49	0.073	1.9	5	50	5-6-1/2	8
4	101.6	49	0.073	1.9	5	55	4-5	7
5	127.0	45	0.082	2.1	5	60	3-1/2-4-1/2	6
6	152.4	45	0.082	2.1	6	70	3-4	5
8	203.2	45	0.082	2.1	6	75	3	4

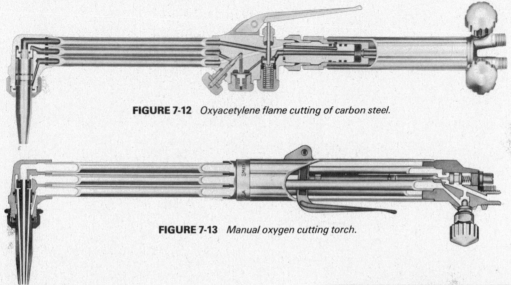

FIGURE 7-12 *Oxyacetylene flame cutting of carbon steel.*

FIGURE 7-13 *Manual oxygen cutting torch.*

cutting. The numbering system for tips is not standardized and most manufacturers use their own tip number system. Each system is, however, based on the size of the oxygen cutting orifice of the tip. These are related to drill sizes. Different tip sizes are required for cutting different thicknesses of carbon steel. The cutting torch and oxygen and acetylene cylinders and regulators are shown by Figure 7-14.

For automatic cutting with mechanized travel, the same types of tips can be used. High-speed type tips, with a specially shaped oxygen orifice, provide for higher-speed cutting and are normally used. The schedule shown provides cutting speeds with normal tips; the speeds can be increased 25 to 50% when using high speed tips.

Automatic shape-cutting machines are widely used by the metalworking industry. These machines can carry several torches and cut a number of pieces simultaneously.

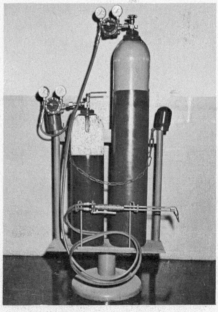

FIGURE 7-14 *Complete manual oxygen cutting equipment.*

FIGURE 7-15 *Automatic shape cutting machine.*

FIGURE 7-16 *Electric eye template system.*

FIGURE 7-17 *Numerical controlled oxygen cutting system.*

Figure 7-15 shows four pieces being cut. Shape-cutting machines have evolved from simple devices that carried one torch that was guided by tracks or cams. Fully automatic machines were guided by photocell type tracing devices. The most modern multitorch cutting machines are directed by numerically-controlled equipment. No matter what type of tracing control system is used, the cutting operation is essentially the same. Figure 7-16 shows an automatic oxygen-cutting machine utilizing the electric eye tracing device. This equipment pioneered chain cutting which led to the newest numerically controlled oxygen cutting equipment shown by Figure 7-17. One of the newer advances in automatic flame cutting is the generation of bevel cuts on contour-shaped parts. This breakthrough has made the use of numerically controlled oxygen cutting equipment even more productive. A bevel cut on medium thickness material is shown by Figure 7-18.

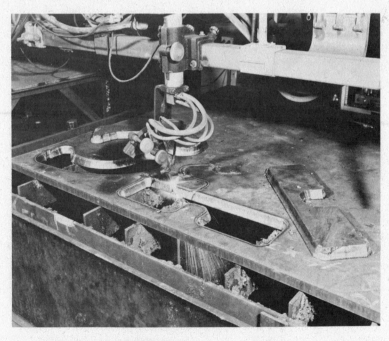

FIGURE 7-18 *Bevel cut procedure.*

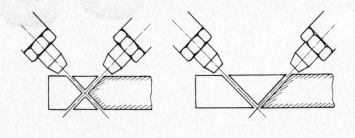

Many specialized automatic oxygen cutting machines are available for specific purposes. Special machines are available for cutting sprockets and other precise items. Oxygen-cutting machines are available for cutting pipe to fit other pipe at different angles and of different diameters. These are quite complex and have built-in contour templates to accommodate different cuts and bevels on the pipe. Other types of machines are designed for cutting holes in drum heads, test specimens, etc. Two or three torches can be used to prepare groove bevels for straight line cuts as shown by Figure 7-19. Extremely smooth oxygen-cut surfaces can be produced when schedules are followed and all equipment is in proper operating condition. The illustration shown by Figure 7-20 shows the oxygen-cut surfaces and gives the probable cause of subquality cuts.

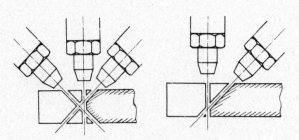

FIGURE 7-19 *Methods of preparing joints.*

FIGURE 7-20 *Surface of flame cut edges.*

CORRECT CUT

Cutting lines are almost vertical and not very pronounced. Edges are square. Little slag evident.

COMMON CUTTING MISTAKES

PREHEAT FLAMES TOO SMALL

Bottom half uneven and wavy.

OXYGEN PRESSURE TOO LOW

Top edge and cutting lines uneven.

PREHEAT FLAMES TOO LARGE

Top edge badly melted. Middle section is smooth. Slag evident at bottom.

OXYGEN PRESSURE TOO HIGH AND/OR NOZZLE SIZE TOO SMALL

Deep gouges into sides of cut due to lack of control.

CUTTING SPEED TOO SLOW

Upper edge melted. Cutting lines rather course.

TORCH TRAVEL UNSTEADY

Cutting lines erratic and uneven.

CUTTING SPEED TOO FAST

Cutting lines curve in opposite direction of travel. Cut edge irregular.

CUTTING WAS LOST

Cut stopped at gouged areas. Usually due to excess travel speed or insufficient preheating.

Metal powder cutting is an oxygen-cutting process which severs metals through the use of powder, such as iron, to facilitate cutting. This process is used for cutting cast iron, chrome nickel stainless steels, and some high-alloy steels. The process uses finely divided material, usually iron powder, added to the cutting-oxygen stream. The powder is heated as it passes through the oxyacetylene preheat flames and almost immediately oxidizes or ignites in the stream of the cutting oxygen. A special apparatus to carry the powder to the cutting tip must be added to the torch. A powder dispenser is also required. Compressed air is used to carry the powder to the torch.

The oxidation or burning of the iron powder provides a much higher temperature in the oxygen stream. This plus the chemical reaction in the flame allows the cutting-oxygen stream to oxidize the metal being cut continuously in the same manner as when cutting carbon steels. By the use of iron powder in the oxygen stream it is possible to start cuts without preheating the base material. Powder cutting has found its broadest use in the cutting of cast iron and stainless steel. It is used for removing gates and risers from iron and stainless steel castings. The same basic process can also be used for scarfing to condition billets, blooms, and slabs, in steel mills.

Cutting speeds and cutting oxygen-pressures are similar to those used when cutting carbon steels. For heavier material over 1 in. thick (25 mm) a nozzle one size larger should be used. Powder flow requirements vary from 1/4–1/2 lb (0.11 to 0.23 kg) of iron powder per minute of cutting. Powder tends to leave a scale on the cut surface which can easily be removed as the surface cools. This is a rather special application process and is used only where required.

Stack cutting is the oxygen cutting of stacked metal sheets or plates arranged so that all the plates are severed by a single cut. In this way, the total thickness of the stack is considered the same as the equivalent thickness of a solid piece of metal. When stack cutting, particularly thicker material, the cut is often lost because the adjacent plates may not be in intimate contact with each other. The preheat may not be sufficient on the lower plate to bring it to the kindling temperature and therefore the oxygen stream will no longer cut through the remaining portion of the stack. One way to overcome this problem is to use the metal powder cutting process. By means of the metal powder and its reaction in the oxygen the cut is completed across separations between adjacent plates.

Chemical Flux Cutting (FOC)

Chemical flux cutting is an oxygen-cutting process in which metals are severed using a chemical flux to facilitate cutting and powdered chemicals are utilized in the same way as iron powder is used in the metal powder cutting process. This process is sometimes called flux injection cutting. Flux is introduced into the cut to combine with the refractory oxides and make them a soluble compound. The chemical fluxes may be salts of sodium such as sodium carbonate. Chemical flux cuttiong process is of minor industrial significance.

Oxygen Lance Cutting (LOC)

Oxygen lance cutting is an oxygen-cutting process used to sever metals with oxygen supplied through a consumable tube. The preheat is obtained by other means. This is sometimes called oxygen lancing. The oxygen lance is a length of pipe or tubing used to carry oxygen to the point of cutting. The oxygen lance is a small (1/8 or 1/4 in. nominal) black iron pipe connected to a suitable handle which contains a shut-off valve. This handle is connected to the oxygen supply hose. The main difference between the oxygen lance and an ordinary flame cutting torch is that there is no preheat flame to maintain the material at the kindling temperature. The lance is consumed as it makes a cut. The principle use of the oxygen lance is the cutting of hot metal in steel mills. The steel is sufficiently heated so that the oxygen will cause rapid oxidation and cutting to occur. For other heavy or deep cuts a standard torch is used to bring the surface of the metal to kindling temperature and then the oxygen lance is brought into play. The end of the oxygen lance becomes hot and supplies iron to the reaction to maintain the high temperature.

There are several proprietary specialized oxygen lance type cutting bars or pipes. In these systems, the pipe is filled with wires which may be aluminum and steel or magnesium and steel. The aluminum and magnesium readily oxidize and increase the temperature of the reaction. The steel of the pipe and the steel wires will tend to slow down the reaction whereas the aluminum or magnesium wires tend to speed up the reaction. This type of apparatus will burn in air, under water, or in noncombustible materials. The tremendous heat produced is sufficient to melt concrete, bricks, and other nonmetals. These devices can be used to sever concrete or masonry walls and will cut almost anything. These bars are known by various trade names.

Oxygen Arc Cutting (AOC)

Oxygen arc cutting is an oxygen-cutting process used to sever metals by means of the chemical reaction of oxygen with the base metal at elevated temperatures. The necessary temperature is maintained by means of an arc between a consumable tubular electrode and the base metal. This process requires a specialized combination electrode holder and oxygen torch. A conventional constant current welding machine and special tubular covered electrodes are used. This process will cut high chrome nickel stainless steels, high-alloy steels, and nonferrous metals. The high temperature heat source is an arc between the special covered tubular electrode and the metal to be cut. As soon as the arc is established a valve on the electrode holder is depressed and oxygen is introduced through the tubular electrode to the arc. The oxygen causes the material to burn and the stream helps remove the material from the cut. Steel from the electrode plus the flux from the covering assist in making the cut. They combine with the oxides and create so much heat that thermal conductivity cannot remove the heat quickly enough to extinguish the oxidation reaction. This process will routinely cut aluminum, copper, brasses, bronzes, Monel, Inconel, nickel, cast iron, stainless steel, and high-alloy steels. The quality of the cut is not as good as the quality of an oxygen cut on mild steel, but sufficient for many applications. Material from 1/4 in.- 3 in. (6.4–75 mm) can be cut with the process. The electric current ranges from 150–250 amperes and oxygen pressure of 3–60 psi (0.21–4.2 kg/cm^2) may be used. Electrodes are normally 3/16 in. (4.8 mm) in diameter and 18 in. (450 mm) long. They are suitable for AC or DC use. This process is used for salvage work, as well as for manufacturing and maintenance operations.

7-4 OTHER THERMAL CUTTING PROCESSES

Electron Beam Cutting (EBC)

This is a thermal cutting process which uses the heat obtained from a concentrated beam composed of high-velocity electrons which impinge upon the workpiece to be cut. The principle of operation of the electron beam was discussed in Chapter 6. The difference between electron beam welding and cutting is the heat input-to-heat output relationship. The electron beam generates heat in the base metal which vaporizes the metal and allows it to penetrate deeper until the depth of the penetration, based on the power input, is achieved. In welding the electron beam actually produces a hole, known as a keyhole. The metal flows around the keyhole and fills in behind. In the case of cutting the heat input is increased so that the keyhole is not closed.

All of the metals that can be welded can also be cut. The quality of the cut surface is equal to the quality of a good oxyacetylene machine cut. The ability to shape cut is limited only by the ability to move the work or the electron gun. The problems of electron beam cutting are greater than those of welding. The work must be in a vacuum. A large amount of volatilized metal will tend to plate out on the inside of the chamber. In addition, the cost of the equipment is very high.

In view of these difficulties the laser beam is replacing the electron beam for cutting.

Laser Beam Cutting (LBC)

Laser beam cutting is a thermal cutting process which severs materials with the heat obtained in the application of a concentrated coherent light beam impinging on the workpiece to be cut. The process can be used without an externally supplied gas. The principle of operation of the laser beam was discussed in Chapter 6. The laser beam can be used for welding or cutting depending upon the heat input-heat output relationship. Laser beam cutting has many advantages over electron beam cutting. The concentrated energy in the laser beam is in the same order as the energy in the electron beam. The ability of both beams to cut materials is essentially the same. The many advantages of the laser beam over the electron beam have practically eliminated the electron beam for cutting. In addition, a gas jet can supplement the laser to remove metal. If oxygen is used the cut can be made faster.

The major advantage of the laser beam is that it can be used in the atmosphere. It can be used with automatic shape cutting equipment normally used for oxygen cutting, as shown by Figure 7-21. Another advantage is that the width of the cut is narrower and the angle of the cut is almost a perfect right angle. The quality of the cut is equal or superior to the best oxygen cut, as shown by Figure 7-22. Furthermore, the laser beam will cut materials other than metals. It has been successfully used for cutting plastics, wood, cloth, etc. One major use of laser beam is cutting plywood and pressed board wood in the woodworking industry. It has also been used to cut cloth for making clothing. The laser beam can also be used for drilling. This is using the same technique but without relative travel. Laser beams can also be used for localized surface heat treating; usually the beam is defocused to provide a larger work area.

FIGURE 7-21 *Laser beam used for automatic shape cutting.*

FIGURE 7-22 *Surface of cut.*

Material	Thickness		Cutting Speed	
	mm	in.	mm/min.	in./min.
ABS Plastic	4	0.157	4.500	0.177
Acrylic	6	0.236	1.700	0.066
Ceramic tile	6.3	0.248	0.300	0.011
Plywood	18	0.708	0.500	0.019
Galvanised steel	1	0.039	4.500	0.177
High carbon steel	3	0.118	1.500	0.059
Mild steel	1	0.039	4.500	0.177
Stainless steel	2.8	0.110	1.200	0.047
Titanium	3	0.188	4.100	0.161

FIGURE 7-23 *Cutting speed on different materials using 400 watt laser.*

The power required for laser cutting is relatively low with respect to the power required for welding. In general, the continuous wave laser with up to 1 kw power and 10% efficiency is sufficient to cut thin gauge metals. The table given by Figure 7-23 shows the travel speed of a 400-watt laser beam for cutting different material. As time goes on, there will be many more applications for laser beam cutting.

7-5 THERMAL SPRAYING

Thermal spraying (THSP) is a group of allied processes in which finely divided metallic or nonmetallic materials are deposited in a molten or semimolten condition to form a coating. The coating material may be in the form of powder, ceramic rod, or wire. There are three separate processes within this group: electric arc spraying, flame spraying, and plasma spraying. These three processes differ considerably, since each uses a different source of heat. The apparatus is different and their capabilities are different. Each of these three processes is described.

The selection of the spraying process depends on the properties desired of the coating. Thermal spraying is utilized to provide surface coatings of different characteristics, such as coatings to reduce abrasive wear, cavitation, or erosion. The coating may be either hard or soft. It may be used to provide high temperature protection. Thermal sprayed coatings improve atmosphere and water corrosion resistance. One of the major uses is to provide coatings resistance to salt water atmospheres. Another use is to restore dimensions to worn parts. The hardness and composition of the deposit are important and dictate whether the part will be machined or ground. Based on this decision it is then necessary to determine the type of material that must be sprayed. If the spray material is available in wire form, the electric arc spray or the flame spray processes can be used. However, if it can be obtained only in a powder form, the flame spray or plasma spraying process must be used. The selection of materials for spraying is beyond the scope of this section. Please see "Recommended Practice for Spraying and Fused Thermal Sprayed Coatings," published by the American Welding Society.[9]

Electric Arc Spraying (EASP)

Electric arc spraying is a thermal spraying process that uses an electric arc between two consumable electrodes of the surfacing materials as the heat source. A compressed gas atomizes and propels the molten material to the workpiece. The principle of this process is shown by Figure 7-24. The two consumable electrode wires are fed by a wire feeder to bring them together at an angle of approximately 30° and to maintain an arc between them. A compressed air jet is located behind and directly in line with the intersecting wires. The wires melt in the arc and the jet of air atomizes the melted metal and propels the fine molten particles to the workpiece. The power source for producing the arc is a direct-current constant-voltage welding machine. The wire feeder is similar to that used for gas metal arc welding except that it feeds two wires. The gun can be hand held or mounted in a holding and movement mechanism. The part or the gun is moved with respect to the other to provide a coating surface on the part.

The welding current ranges from 300 to 500 amperes

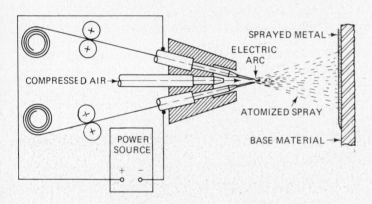

FIGURE 7-24 *Electric arc spraying process.*

direct current with the voltage ranging from 25 to 35 volts. This system will deposit from 15 to 100 lb/hr of metal. The amount of metal deposited depends on the current level and the type of metal being sprayed. Wires for spraying are sized according to the Brown and Sharp wire gage system. Normally either 14 gage (0.064 inch or 1.6 mm) or 11 gage (0.091 inch or 2.3 mm) is used. Larger-diameter wires can be used.

The high temperature of the arc melts the electrode wire faster and deposits particles having higher heat content and greater fluidity than the flame spraying process. The deposition rates are from 3 to 5 times greater and the bond strength is greater. There is coalescence in addition to the mechanical bond. The deposit is more dense and coating strength is greater than when using flame spraying.

Dry compressed air is used for atomizing and propelling the molten metal. A pressure of 80 psi (9 kg/mm^2) and from 30 to 80 cu ft/min (0.85–2.26 cu m/min) is used. Almost any metal that can be drawn into a small wire can be sprayed. Figure 7-25 is a list of metals that are arc sprayed.

Aluminum
Babbitt
Brass
Bronze
Copper
Molybdenum
Monel
Nickel
Stainless steel
Carbon steel
Tin
Zinc

FIGURE 7-25 *Metals in wire form for arc spraying.*

Flame Spraying (FLSP)

Flame spraying is a thermal spraying process that uses an oxy fuel gas flame as a source of heat for melting the coating material. Compressed air is usually used for atomizing and propelling the material to the workpiece. There are two variations: One uses metal in wire form and the other uses materials in powder form. The method of flame spraying which uses powder is sometimes known as powder flame spraying. The method of flame spraying using wire is known as metallizing or wire flame spraying.

In both versions, the material is fed through a gun and nozzle and melted in the oxygen fuel gas flame. Atomizing, if required, is done by an air jet which propels the atomized particles to the workpiece. When wire is used for surfacing material, it is fed into the nozzle by an air-driven wire feeder and is melted in the gas flame. When powdered materials are used they may be fed by gravity from a hopper which is a part of the gun. In another system the powders are picked up by the oxygen fuel gas mixture, carried through the gun

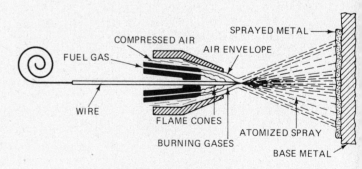

FIGURE 7-26 *Flame spray process.*

where they are melted, and propelled to the surface of the workpiece by the flame.

Figure 7-26 shows the flame spray process using wire. The version that uses wires can spray metals that can be prepared in a wire form. The variation that uses powder has the ability to feed various materials. These include normal metal alloys, oxidation-resistant metals and alloys, and ceramics. It provides sprayed surfaces of many different characteristics.

Plasma Spraying (PSP)

Plasma spraying is a thermal spraying process which uses a nontransferred arc as a source of heat for melting and propelling the surfacing material to the workpiece.

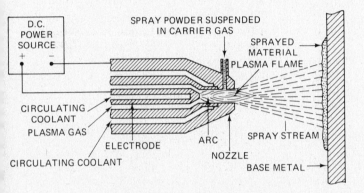

FIGURE 7-27 *Plasma spray process.*

This process is sometimes called plasma flame spraying or plasma metallizing. It uses the plasma arc, which is entirely within the plasma spray gun. The temperature is so much higher than either arc spraying or flame spraying that additional materials can be used as the coating. It appears that most inorganic materials, which melt without decomposition, can be used. The material to be sprayed must be in a powder form and it is carried into the plasma spray gun suspended in a gas. The high-temperature plasma immediately melts the powdered material and propels it to the surface of the workpiece. Since inert gas and extra high temperatures are used, the mechanical and metallurgical properties of the coatings are generally superior to either flame spraying or electric arc spraying. This includes reduced porosity and improved bond and tensile strengths. Coating density can reach 95%. The hardest metals known, and some with extremely high melting temperatures, can be sprayed with the plasma spraying process.

Preparation for Spraying

The most important aspect of thermal spraying is correct preparation of the workpiece. It must be clean.

Machining is normally used to prepare round parts, such as shafting. When thermal spraying is used to correct a dimension, the part is usually machined undersize to allow a sufficient thickness of coating. For large flat areas grit blasting is used. In any case a roughened surface is preferred, but sharp corners should be avoided.

The Spraying Operation

Spraying should be done immediately after the part is cleaned. If the part is not sprayed immediately, it should be protected from the atmosphere by wrapping with paper. If parts are extremely large it may be necessary to preheat the part 200 to 400°F (95 to 205°C). Care must be exercised so that heat does not build up in the workpiece. This increases the possibility of cracking the sprayed surface. The part to be coated should be preheated to the approximate temperature that it would normally attain during the spraying operation. The distance between the spraying gun and the part is dependent on the process and material being sprayed. Recommendations of the equipment manufacturer should be followed and modified by experience. Speed and feed

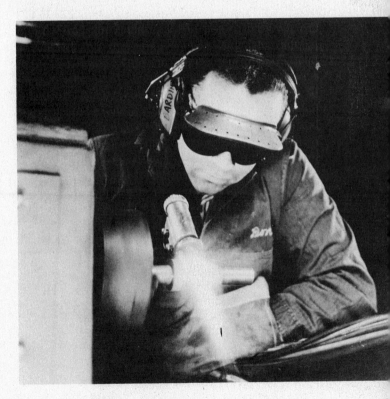

FIGURE 7-28 *The plasma spray process in use.*

PROCESSES RELATED TO WELDING

of spraying should be uniform. The first pass should be applied as quickly as possible. Additional coats may be applied slowly. It is important to maintain uniformity of temperature throughout the part. When there are areas of the part being sprayed where coating is not wanted, the area can be protected by masking it with tape.

Post Treatment and Safety

Certain sprayed surfaces are given an additional treatment to create fusing. This technique involves a gradual and uniform temperature rise to a fusing temperature of 1,850 to 2,370°F (1,010 to 1,300°C), depending on the type of alloy being used. Various methods of applying fusing heat can be used. These include the oxyfuel gas torch, a furnace, or by induction. Temperature control is required to obtain a quality fused coating.

Thermal spraying requires protective clothing, eye and ear protection, and special ventilation. All safety factors with respect to the similar welding process should be observed. Ear protection may be required.

Quality of Coatings

Coatings must be inspected to determine that they are free of cracks, pin holes, blisters, voids, etc. Coatings over sharp corners, such as keyways, require extra attention. The skill of the operator is a major factor in obtaining quality coatings. Written procedures are recommended for each type of application.

7-6 HEAT FORMING AND STRAIGHTENING

Heat, normally applied by the oxyacetylene torch, can be used to bend or straighten metal sections or parts. To best understand heat forming and straightening, the reader should read "Welding Distortion and Warpage" in Chapter 18.

All metals expand when heated and contract when cooled. The amount of expansion depends on the temperature increase, the coefficient of expansion of the metal heated, and size of the heated area. Unrestrained metal expands in all three directions, but metal is normally restrained due to unequal heating. Heat causes plastic deformation or upsetting during heating. This creates dimension changes upon cooling. The strength

of metal decreases rapidly as the temperature of the metal is increased. A thorough understanding of these factors will make it obvious why distortion occurs on weldments. Distortion occurs as a result of forces created by differential heating; therefore, these same forces can be used to bend or to straighten metal pieces.[10]

Flame bending, or heat forming and straightening, is the application of these same principles. It can be used to correct distortion that may occur from welding or from accidents. To best understand the application of heat to bend or straighten members we will first start with a relatively simple example. Consider a flat bar approximately 1/4 in. (6.4 mm) thick by 2 in. (50 mm) wide and approximately 24 in. (650 mm) long. Heat with an oxyacetylene welding torch using a medium-size single-orifice tip. Adjust the torch for a neutral flame. Before beginning the heat pattern use a piece of chalk and mark a triangular area from point A located 1/4 the width of the bar toward edge D and mark a pie-shaped or V-shaped area to point B and C on the E edge of the bar, as shown by Figure 7-29. The angle formed by B, C to A, should be approximately 60°.

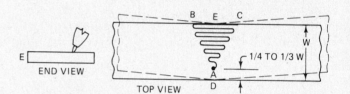

FIGURE 7-29 *The application of heat to vee shaped areas.*

Begin the heating at point A with the flame pointed slightly towards edge E. Hold the flame steady until point A becomes heated to a light red (1,110°F–593°C). Be careful not to melt the surface of the bar at any time. When spot A has reached a light red color, start to move the torch as shown in the figure. Move the torch slowly in a zig-zag fashion bringing it up to the same light red temperature. Continue traveling making sure that the part comes up to temperature before moving further toward the far edge E. Continue the zig-zag line of travel until you have reached points B and C on edge E. This will produce a V-shaped section that has been heated progressively from near edge D to edge E. Cooling will also be progressive in the same direction. After the bar has cooled it will take on the shape as shown by the dotted lines.

The mechanism for creating this action is based on the principles mentioned above. The heat at point A causes the metal to expand in all three directions; however, it is restrained by the adjacent cold metal and it will therefore upset or become slightly thicker. The localized heating continues as the flame is moved from point A toward B and C. The heated area is increased as the flame moves away from point *A*. The metal at point *A* begins to cool and contract as the temperature falls. Normally, points *B* and *C* are reached before the temperature at point *A* cools appreciably. The widened heated area and the contracting metal behind it create the forces that cause the bending. Do not heat the metal at point *D* or beyond point *A* on edge *D*. As the cooling continues from point *A* towards edge *E,* the metal contracts. There is more metal to contract as the heated area becomes wider. This contributes to creating more motion or bending as cooling continues. Point *D* acts as the hinge pin; the material between *B* and *C* contracts the most. The amount of distortion with all factors equal is dependent on the angle between *B* and *C*. For less action use a sharper angle in the order of 45° or for less movement reduce it to 30°.

If more motion is required, additional heated V's can be made adjacent to the first one. It is possible by this method to create a ring out of a straight bar. This requires many applications of heat but it has been done.

The same application of V-shaped heating areas can be used to bend or straighten other types of structural shapes. Figure 7-30 shows the heat application that can be used for other shapes. The large end of the V will be the portion where the maximum contraction occurs. The amount of contraction can actually be calculated; however, since temperature control is not precise, the amount of metal heated is not exactly known due to conduction of heat to the cooler areas. Formulas have been worked out and are available.[11] The best method of understanding the amount required for shrinkage is by experience.

When straightening shapes other than flat bars, it is necessary to consider the relationship between webs and flanges. For example, in Figure 7-30 the rolled angle can be bent with the vertical leg acting as the hinge point. All of the heating is done on the horizontal leg. If, however, the vertical leg of the angle must be shortened it should be done with thorough heating of the vertical leg when the wide portion of the V comes to that leg. Progressive heating should not be done in the vertical leg. Progressive heating is only used when a V-shaped area is to be heated. By closely following Figure 7-30 an angle can be bent in either direction. The same applies to channels, tees, and I or wide flange beams. The basic principles can also be used for box sections and pipe. In all cases, the starting point should be approximately 1/4 from the edge which is to be the hinge point.

The same technique can be used to correct warpage of structures that involve T-type joining utilizing double fillet welds. In this case, the application of heat is linear rather than V-shaped. Warpage is often encountered when fillet welds and double fillets are made on one

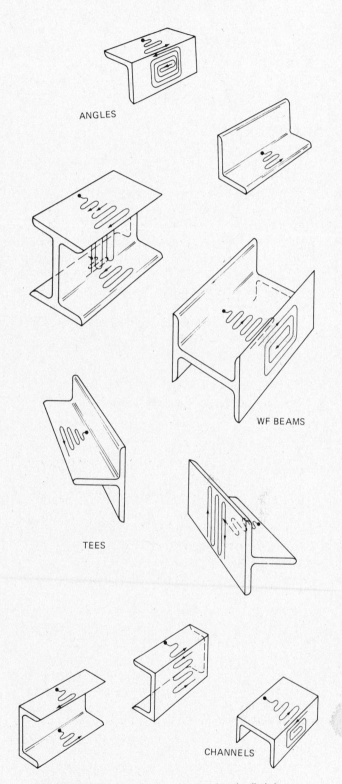

ANGLES

WF BEAMS

TEES

CHANNELS

FIGURE 7-30 *The application of heat to bend volled shapes.*

PROCESSES RELATED TO WELDING

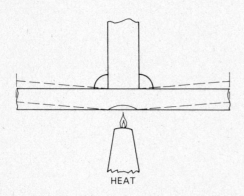

FIGURE 7-31 *The application of heat to correct warpage.*

side of a member forming a T-joint. Applying the torch flame to the center of the top of the Tee will create shrinkage in this point which will tend to bring the top of the tee back into a flat plane. This is shown by Figure 7-31. The same technique of straight line heat application can be used to shorten parts.

Diaphragms, bulkheads, or flat plates in welded assemblies may sometimes buckle as a result of weld distortion. One method of reducing this buckling is to create round heated areas approximately 3 in. (75 mm) in diameter across the surface of the buckled plate. Each individual heated spot will upset and as it cools will create shrinkage in all directions. By adding enough of these round heated spots the buckling can be completely eliminated from the flat plate.

The technique of heat flame straightening and bending can be used to salvage members damaged by accident. It has been successfully applied to structural work of bridges, of large buildings such as aircraft hangars, and of offshore drilling platforms.[12] In these cases, special precautions must be exercised. Force is sometimes applied to assist the heat-forming operation. A careful analysis must be made before attempting such jobs and it must be determined whether the part being straightened is stressed in tension or in compression. When the temperature of the member is increased by flame heating, its strength is greatly decreased. In order to create shrinkage action, it is necessary to have the member loaded in compression. This will assist upsetting and will help create the favorable direction of warpage to straighten the member. If the member is stressed in tension, compression loading should be added, by means of temporary bracing, to accomplish the heat-straightening operation.

The flame bending system is also used for creating camber in beams. Samples of wide flange beams formed to large radiuses for roof structures are given by reference 12. Here 80 ft (24 m) long, 24 in. (288 mm) wide flange beams were formed to curvatures of 135 ft (41.1 m). These were erected and became the arch of a gymnasium. With skill and experience this type of heat forming can be very precise.

There are certain precautions that should be taken in heat forming or bending. It is used on low carbon steels. It should not be performed on medium-, high-carbon or quenched and tempered steels. Precautions should be used in cooling the heated zones. Air quenching or water cooling can be used. This will not appreciably increase the amount of distortion. It will merely decrease the time for the shrinkage action to take place. When steel with high-carbon content is quenched rapidly, it will harden, which can be detrimental to service life.

To avoid damage of the steel it is recommended that 1,200 °F (649 °C) is the maximum temperature utilized. This temperature produces a dark red color on steel in a subdued light.

It has been found that heating the same spot does not adversely affect the steel, provided it is not heated to temperatures above 1,200 °F (649 °C).

Heat forming can be used on stainless steels and on aluminum. For stainless steels the maximum temperature should be 800 °F (427 °C) and for aluminum the maximum temperature should be 400 °F (204 °C).

For thicker materials, larger torch sizes are required. For extremely heavy material multiorifice torches can be used. It is important to maintain the maximum heat differential between the heated area and the adjacent cool area. This provides for more movement during the cooling period.

In straightening materials in two dimensions, it must be remembered that the torch side of the part being heated will be heated to a higher temperature than the underside. For this reason, the torch side will normally have greater reaction than the underside. Thorough heating is required and in many times it is possible to judge the heating by noting the color on the underside.

With sufficient practice and experience some rather amazing feats can be done utilizing the oxyacetylene torch for heat forming and straightening.

7-7 PRE- AND POSTHEAT TREATMENT

Preheating is the application of heat to the base metal immediately before welding, brazing, soldering, or cutting, and postheating is the application of heat to an assembly after welding. This might be better described as a postweld heat treatment, which includes any heat treatment subsequent to welding.

All of the arc welding processes and many of the other welding processes use a high-temperature heat source. These high temperatures exceed the melting temperature of the base metal. This creates the problem of a traveling high temperature, localized heat source, and the effect that it has on the surrounding metal. A steep temperature differential occurs between the localized heat source and the cool base metal. This temperature difference causes differential thermal expansion and contraction, high stresses and hardened areas. By reducing the temperature differential, these problems can be minimized. This will reduce the danger of weld cracking, reduce the maximum hardness, minimize shrinkage stresses, lessen distortion and help gases—particularly hydrogen—escape from the metal. Preheating will reduce the temperature differential. Preheating increases the ability of a weld to withstand service conditions largely because of the reduction of the problems mentioned above. Preheating is also done on highly conductive metals in order to maintain sufficient heat at the weld area.

The preheat temperature depends on many factors, such as the composition and mass of the base metal, the ambient temperature, and the welding procedure. When these factors are more severe the preheat temperature must be increased.

The interpass temperature should also be considered. It is involved in multiple-pass welds and is the temperature, both minimum and maximum, of the deposited weld metal and adjacent base metal before the next pass is started. Usually, the minimum interpass temperature will be the same as the preheat temperature. The weldment temperature should never be allowed to become lower than the preheat or the interpass temperature. If welding is interrupted for any reason, the interpass temperature must be attained before welding is started again. Preheat and interpass temperatures must be attained through the thickness of the area of the weld. The interpass temperature may also be specified as a maximum temperature. When welds are made on a small weldment, its temperature will usually increase due to the heat input from welding. Under certain conditions it is not desirable to allow this heat to build up and exceed a specific temperature; therefore, a maximum interpass temperature will be specified. When heat buildup becomes excessive, the weldment must be allowed to cool, but not below the minimum interpass temperature. The temperature of the welding area must be maintained within the minimum and maximum interpass temperature.

There are many different ways for preheating, including use of gas torches, gas burners, heat-treating furnaces, electrical resistance heaters, low-frequency induction heating, and temporary furnaces. The choice

of the preheating method depends on many factors, such as the preheat temperature, the length of preheating time, the size and shape of the parts, and whether it is a one-of-a-kind or a continuous-production type operation. Any method used for postheating can also be used for preheating.

In certain types of work the preheat temperature must be precisely controlled. In these cases, controllable heating systems are used, and thermocouples are attached directly to the part being heated. The thermocouple will measure the exact temperature of the part and will provide a signal to a controller, which regulates the fuel or electrical power required for heating. By accurate regulation of the fuel or power, the temperature of the part being heated can be held to close limits. Most code work requires precise heat temperature control.

A common method of preheating is by torches or burners utilizing flames. Figure 7-32 shows a natural gas burner used to preheat the lip of a power shovel dipper. This subassembly is a critical part made of quenched and tempered high-alloy steel. The preheat and interpass temperature must be maintained throughout the entire welding cycle. Open flames, such as this, cannot be as precisely controlled as other methods and are therefore used only for preheating and maintaining interpass heat.

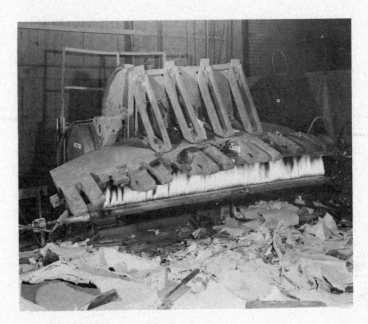

FIGURE 7-32 *Preheating with open natural gas flame.*

Postweld Heating

There are a number of postweld heat treatments for weldments but stress relieving is the most widely used. Some of these other heat treatments are: annealing, normalizing, drawing, and fusing. *Annealing,* or *full annealing* as it is sometimes called, is a heat treatment that increases the temperature of steel above the critical or recrystallization temperature followed by slow cooling. It is normally heated to a point about 100°F (38°C) above the critical temperature line of the steel. Cooling is usually done in a furnace to provide a substantially stress-free condition. Annealing, in addition to removing internal stresses, results in a steel of lower strength and higher ductility. Figure 7-33 shows this in diagrams.

Normalizing is similar to annealing. The heating rate and holding periods are identical but in normalizing the cooling rate is faster and is usually done by allowing the part to cool in still air rather than in the furnace. Due to the higher cooling rate normalizing usually provides a structure with greater strength and less ductility than annealing.

Tempering is a heat treatment done at a much lower temperature than annealing, normalizing, or stress relieving. It is an operation that often follows a quenching operation. It tends to reduce the hardness, the strength of a steel, but it improves ductility and toughness.

Fusing is a specialized process of heating a thermal spray deposit to cause it to coalesce, solidify, and bond metallurgically to the base material. This can be done by almost any heating method.

Stress relieving is of major importance to weldments. It is similar to normalizing except that it is done at a temperature below the critical temperature usually in the 1,050–1,200°F (566–649°C) range.

Both annealing and normalizing relieve residual stresses better than stress relieving. They are carried out above the critical temperature. They involve changes in grain structure and tend to produce heavy scale. They may also produce serious dimensional changes, and require that complex large structures are braced to avoid sagging.

Stress relieving is required by some codes. Refer to the specific portion of the code that is applicable, to decide on the stress-relieving schedule. Many products not built under code are stress relieved for some of the following reasons:

1. To reduce the residual stresses inherent to any weldment, casting, or forging.

2. To improve the resistance to corrosion and caustic embrittlement.

3. To improve the dimensional stability of the weldment during machining operations.

4. To improve the service life of the weldment. Stress relieving should be performed if a weldment is subjected to impact loading or to low temperature service, or if it is exposed to repetitive or fatigue loading.

In all of the heat treatments, with the possible exception of fusion, heating rate and times may be specified. The maximum temperature is related to the composition of the steel, the holding time, at the maximum temperature, is related to the material thickness, and the cooling rate is related to the particular treatment and to the code. The rate of heating is usually in the 300–350°F (149–177°C) per hour rate. The holding temperature is usually one hour for each inch of maximum thickness in order to provide for uniform heating throughout. The cooling rate is also in the range of 300–350°F (149–177°C) per hour down to a specific temperature. In some cases, the cooling rate can be increased when the part has cooled to 500° or 600°F (260–316°C). These rates of heating, holding, and cooling are usually a part of the specification and must be followed explicitly.

Temperature indicators and controlling equipment must be used for postweld heat treatment. Figure 7-34 is a typical car bottom furnace loaded with weldments that have been stress relieved. These furnaces are usually natural gas or fuel oil fired. The complete heat treating cycle may require up to 24 hours. Thermocouples are attached to the weldment and are used to signal the controller. Automatic controllers can be set for a specific rate of heating, a specific holding time, and a specific rate of cooling.

Accurate control is also possible with induction and resistance heating. Figure 7-35 shows the use of resistance heating coils for both pre- and postheating. Figure 7-36 shows the use of low-frequency induction heating coils (400 Hertz) for pre- and postheating welds made on large pipe. Both resistance and induction heating are very popular for preheating and stress-relieving weld joints in power plant piping welds. Temperature record charts are required by some codes for each weld or weldment.

FIGURE 7-33 *Heat treatment cycles.*

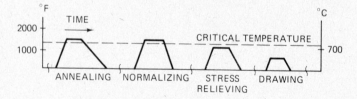

FIGURE 7-34 *Car bottom furnace for stress relieving.*

FIGURE 7-35 *Resistance heating for pre and post heat.*

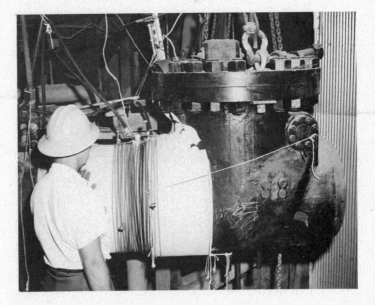

FIGURE 7-36 *Induction heating for pre and post heat.*

267

PROCESSES RELATED TO WELDING

Temperature measurement for preheating can be done by the use of indicating instruments or by temperature-indicating paints and crayons. The temperature-indicating crayons, which melt at specific temperatures, are widely used.

7-8 JOINING OF PLASTICS

The successful use of plastics often requires that parts be securely joined together. Mechanical fasteners can be used; however, for many parts a better joining method is desired. The plastics industry has developed joining methods that fall within two categories: welding and adhesive bonding.

There are at least seven different welding methods in use. Several are similar to processes used for welding metals. These processes are ultrasonic welding, friction welding, hot gas welding, hot wire welding, heated surface welding, and induction heat welding. Each will be briefly described.

Ultrasonic Welding

Ultrasonic welding is a form of heat sealing. Heat is generated by transmitting vibratory energy through the parts being joined from a high-frequency ultrasonic source—frequencies up to 20,000 Hertz. The vibration between the parts causes the two surfaces to move against each other. This causes a temperature rise, as mechanical energy is converted to heat. The heat generated is sufficiently high to melt the surfaces in contact and form a weld. The workpiece transmits the ultrasonic vibration. The area in contact with the sonic probe is not subject to temperature rise and therefore there is no unsightly marking.

Materials which respond best to ultrasonic welding techniques are the rigid thermoplastics with a low modulus of elasticity and a low melting point. Best results are obtained with Acrylonitrile Butadiene Styrene (ABS) and polycarbonate types. It is also possible to weld acetal, nylon, polypropylenes, and high-density polyethylenes.

Friction Welding

Friction welding is a rapid and efficient joining technique which can be applied to most thermoplastics. It offers a particular advantage for those cases where solvent joining is difficult. The major disadvantage is that it can be used only on components having a circular cross section. It can be used to produce welds in similar and dissimilar materials. The typical joint strength on like materials is about 90% of that of the basic material.

The weld is made by holding one section firm and rotating the other part against it under pressure. Frictional heat is developed at the interface. The rotating member is stopped as soon as a melt is formed and the weld is consolidated under pressure. Rotational speed depends on joint diameter. Strength is largely dependent on joint area.

Hot Gas Welding

The hot gas welding technique is normally not employed for joining small parts. Its major use is in the production of very large fabrications made from sheet materials, for example, ducting, pipe work, and ventilator hoods for chemical plant installations.

The apparatus used is relatively inexpensive, consisting of a hot gas torch, a gas pressure regulator, and filler rod. Nitrogen is commonly employed since it helps to eliminate problems associated with oxidation.

Hot Wire Welding

The use of a molded-in resistance wire is a means of heating plastic material to a temperature sufficiently high to produce melting and welding. This method produces high-strength welds, but it has the disadvantage of the added cost of the resistance wire. It can be used anywhere and needs only a source of electric power. The best known application of hot wire welding is the joining of pipe and fittings in which the wire was molded within the fitting.

Heated Surface Welding

The hot plate welding technique is used to join materials which are difficult to join by solvents, adhesives, high-frequency and ultrasonic welding techniques. The technique requires that the two faces to be joined are flat. For optimum strength the area must be commensurate with wall thickness. The two faces of the parts to be joined are held against a heated metal surface. When these faces are melted the pieces are quickly brought together and held under pressure.

When the welding areas are small the surfaces can be joined by the application of direct pressure. When welding thicker sections, the surfaces are slowly moved across each other under pressure to eliminate entrapped air.

In most instances hot plate welding results in a flash around the joint. When large quantities of welds are required the hot plate welding method is easily mechanized.

The welding of plastic parts by induction heating requires a molded-in metal insert. This adds to the cost of the component. The technique has been highly automated. A successful application is the plastic membrane seal in bottle caps. The amount of material to be melted is small and the metal component is part of the product. Induction heating cannot be used to heat thermoplastic materials directly. The heat is obtained by using a piece of metal at the interface of the parts to be joined. Welding is done by putting the components in an alternating current electrical field which will generate sufficient heat in the metal insert to melt the surrounding polymer. During the heating period pressure is applied to the joint and maintained during the cooling period.

Adhesive Bonding

The second basic method of joining plastics is by means of adhesive bonding. In this method chemical action occurs. There are two basic methods, the solvent joining method and the adhesive joining method. They are briefly described below.

Solvent Joining

Solvent joining is applicable to the thermoplastics group of materials which are readily dissolved in a solvent. It cannot be applied to the *inert* materials such as the polyolefines. In the solvent cementing technique the surfaces to be joined are dipped in a solvent and then held together under pressure until the solvent has evaporated to form a seal. Butt, lap, and tongue and groove joints are employed. The joint strength is dependent upon the material and the joint design. Improved bonds are achieved by using specially developed solvent cements which contain a small quantity of the plastics of the type used in the components to be joined. The solvent cements have an advantage over straight solvent welding since they will fill voids in a poorly fitted joint.

Adhesive Joining

Adhesive welding systems, including hot melt types, are used to bond the "inert" thermoplastics and the thermosetting group of materials. Joint strength is equivalent to the strength of the adhesives; however, the bond strength is almost equal to that of the base materials. Most adhesive systems are of the cold curing type but in certain cases cure times can be accelerated by the application of heat.

QUESTIONS

1. Can oily metal surfaces be joined with adhesive bonding?

2. Explain the difference between air carbon arc cutting, and carbon arc cutting.

3. What is the advantage of plasma arc cutting?

4. What is the advantage of using water with plasma arc cutting?

5. What five conditions must apply for successful oxy fuel gas cutting?

6. What is the best way to compare various makes of cutting tips?

7. What methods are used to guide an automatic flame cutting machine?

8. What is meant by stack cutting? What problem is sometimes encountered?

9. Explain the oxygen arc cutting process. What metals can be cut?

10. Will the laser beam cut materials other than metals? If so what ones?

11. What are the three thermal spraying methods?

12. Are powders used for the electric arc spraying process? If so why?

13. Why can plasma spraying apply more materials than flame spraying?

14. Explain how parts are prepared for spraying.

15. Explain how the vee-shaped heated area causes bending.

16. Show how angle iron can be bent with heat—both directions.

17. Is high carbon steel easily straightened with heat? Why?

18. What is meant by interpass temperature?

19. Explain the difference between annealing, stress relieving, and normalizing.

20. How can you measure the temperature of a heated weldment?

REFERENCES

1. "Test for Tensile Properties of Adhesive Bonds," American Society for Testing and Materials (*ASTM*) D897, Philadelphia, Pa.

2. "Test for Tensile Strength of Adhesive by Means of Bar and Rod Specimens," American Society for Testing and Materials (*ASTM*) D2095, Philadelphia, Pa.

3. "Test for Strength Properties of Adhesives in Shear by Tension Loading (Metal to Metal)," American Society for Testing and Materials (*ASTM*) D1002, Philadelphia, Pa.

4. "Test for Cleavage Strength of Metal to Metal Adhesive Bonds," American Society for Testing and Materials (*ASTM*) D1062, Philadelphia, Pa.

5. "Test for Peel or Stripping Strength of Adhesives," American Society for Testing and Materials (*ASTM*) D903, Philadelphia, Pa.

6. "Recommended Practice for Air Carbon Arc Gouging and Cutting," The American Welding Society (*AWS*) C5.3, Miami, Florida.

7. J. J. VISSER, "Low Amp Low Cost Plasma Cutting," *Welding and Metal Fabrication,* October 1974, London, England.

8. "Recommended Practices for Plasma Arc Cutting," The American Welding Society (*AWS*) C5.2, Miami, Florida.

9. "Recommended Practices for Spraying and Fused Thermal Sprayed Coatings," The *American Welding Society,* Miami, Florida.

10. JOSEPH HOLT, *Contraction as a Friend in Need.* Seattle, Washington: Joseph Holt, 1938.

11. RICHARD E. HOLT, "Primary Concepts of Flame Bending," *The Welding Journal,* June 1971, Miami, Florida.

12. J. R. STITT, "Distortion Control During Welding of Large Structures," *SAE-ASME* Paper 844B, April 1964.

8

POWER SOURCES
FOR ARC WELDING

8-1 TYPES OF WELDING POWER SOURCES

8-2 GENERATOR WELDING MACHINES

8-3 TRANSFORMER WELDING MACHINES

8-4 RECTIFIER WELDING MACHINES

8-5 MULTIPLE OPERATOR SYSTEMS

8-6 SELECTING AND SPECIFYING A
     POWER SOURCE

8-7 INSTALLATION AND MAINTENANCE
     OF POWER SOURCES

8-1 TYPES OF WELDING POWER SOURCES

Special electrical power is required to make a weld with the arc welding processes. The power required is from 15 to 35 volts and from 100 to 500 amperes. Voltages and currents higher and lower than these are sometimes used.

There are many ways of describing the electric power used for welding. It can be direct current (DC) or it can be alternating current (AC). Another way is by describing the output characteristics of the power source. It may have constant current (CC), drooping characteristic, or it may have constant voltage (CV), flat characteristic. Figure 8-1 is a diagram showing the principle types of power sources and a classification system. In addition, power sources can be described as: rotating machines, static welding machines, electric motor-driven machines, engine-driven machines, transformer-rectifiers, utility welding machines, single-operator welding machines, multiple-operator power sources, etc.

All of the machines, except in the multioperator power sources section, are the single operator type. These machines are designed to deliver current to only one welding arc.

Electric power for arc welding is obtained in two different ways: (1) generated at the point of use or (2) available power from the utility line is converted. There are two variations of electrical power conversion. First is the transformer which converts the relatively high voltages from the utility line to a lower voltage for AC welding. The second is similar and includes the transformer to lower the voltage but is followed by a rectifier which changes alternating current to direct current for DC welding.

A second way of classifying welding power sources is by the method of adjustment of the welding power. Controlling the power at the arc is done by changing the strength of magnetic fields. In generators, this is done by switches which reconnect the different coils. In static machines the magnetic fields are changed by varying the induction mechanically, electrically or by changing the coupling of coils. It can also be done electronically by feedback signals to control circuits.

A third way of classifying welding power sources is by the welding current provided—whether alternating, or direct, or a combination machine that provides both AC and DC.

271

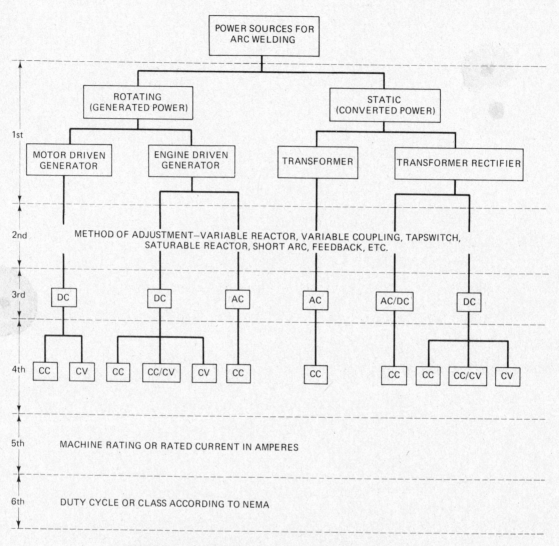

FIGURE 8-1 *A simplified system of classifying welding power sources.*

A fourth way of classifying welding power sources is according to the static volt ampere characteristic output curve. The conventional or *constant current* (CC) welding machine has the *drooping* volt-ampere characteristic curve. This type of output power was described in detail in Section 4-1. The flat or *constant voltage* (CV) sometimes called *constant potential* (CP) power source has a relatively *flat* volt-ampere characteristic curve. This type of output power was described in detail in Section 5-1. For comparison, the normal output curves and the true constant current and true constant voltage curves are shown by Figure 8-2. In both cases the terms are not exact but are accepted and used by the welding industry.

Each power source has particular applications and is ideally suited for specific welding processes; however, alternating current is used for fewer processes than direct current. Constant current (CC) either alternating or direct is used for the shielded metal arc, gas tungsten arc and in some cases for submerged arc welding. Constant voltage (CV) is used for the continuous electrode wire arc welding processes. Normally, constant voltage (CV) machines produce only direct current. Figure 8-3 summarizes the different arc welding processes and shows the type of welding current normally used by each process.

A fifth method of classifying welding power sources is its rating. The rating is the load current obtained from the welding machine without creating excessive temperature rise within the power source. All welding power sources are rated to provide a specific load current at a specific load voltage for a given duty cycle. Ratings in the U.S. are based on the specification of the National Electrical Manufacturers Association (NEMA), "Electric Arc Welding Apparatus."[1]

The sixth way of classifying welding power sources is

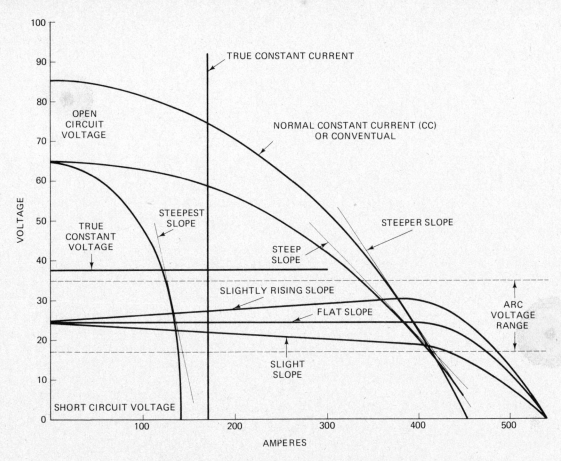

FIGURE 8-2 *Static volt amp characteristic arcs and slopes.*

in accordance with the NEMA class. NEMA has established three classes as follows:

Class I: Rated output at 60%, 80%, or 100% duty cycle.

Class II: Rated output at 30%, 40%, or 50% duty cycle.

Class III: Rated output at 20% duty cycle.

Duty cycle is the ratio of arc time to total time in a 10-minute time period.

No matter what type of classification is used the power source must provide "usability" for the welder, it must give "welder satisfaction." The power source that will enable the welder to produce quality welds using the process specified will be the best machine for the purpose.

FIGURE 8-3 *Welding process and recommended welding power source types.*

Arc Welding Process	Welding Machine Static Volt-Amp Characteristics	Direct Current		Alternating Current
		Constant Current "drooping" CC-DC	Constant Voltage "flat" CV-DC	Constant Current "drooping" CC-AC
Shielded metal arc welding (SMAW) or stick		Yes	Not used	Yes
Gas tungsten arc welding (GTAW) or TIG		Yes	Not used	Yes
Plasma arc welding (PAW)		Yes	Not used	No
Carbon arc welding, cutting and gouging (CAW)		Yes	Possible	Yes
Stud arc welding (SW)		Yes	Possible	No used
Submerged arc welding (SAW)		Yes	Yes	Yes
Gas metal arc welding (GMAW) (inert gas)		Yes	Yes	No used
Gas metal arc welding (GMAW) (spray)		Yes	Yes	Not used
Gas metal arc welding (GMAW) (CO_2)		Usable	Yes	Not used
Gas metal arc welding (GMAW) (shorting arc)		Not used	Yes	Not used
Flux coared arc welding (FCAW)		Yes	Yes	Possible
Electroslag welding (EW)		Possible	Yes	Usable

Welding arcs are subject to severe and rapid fluctuations since the arc length, current and arc voltage constantly change. There is no normal or true steady-state condition, especially for arcs that transfer metal. When metal is not transferred across the arc, for example, in GTAW, much more stable conditions can be obtained. Fluctuations occur in the arc as a result of the welder's inability to hold an exact arc length. Fluctuation also occurs since the arc is moving along the joint. Large drops of molten metal will short circuit the arc for a brief period of time during which the voltage drops below that required to maintain the arc. When the short circuit ends, the arc is re-established. A rapid change of this type is known as a transient. During the short circuit the voltage drops to zero. Immediately following the short circuit the output voltage rises and the arc is re-established. Arc stability refers to the tendency of the arc to burn steady in spite of continued changing electrical conditions. It also implies the differences in arc voltage, from short circuit to open circuit, to arc voltage values. The stability of the welding arc largely depends upon the characteristics of the welding power source.

There are two factors that determine the arc stability. The first is the static volt-ampere characteristic curve. The second is the dynamic characteristics of the machine. This relates the two extreme conditions; the short circuit voltage and the open circuit voltage. The change of the voltage and current provides the dynamic characteristics. At the short circuit the voltage drops to approximately zero. When the short circuit clears, the voltage will rise quickly followed by a stabilization of the voltage to a value below the open circuit voltage, called *recovery voltage.* The current rises rapidly during the short circuit period. When the short is removed, the current quickly stabilizes to the normal value. Arc stability is related to the ability of the power source to smoothly reignite the arc following the short circuit. The quick rising voltage must exceed the arc voltage if arc reignition is to occur. Excessive overshoot of welding current cannot be tolerated. An excessively fast increase in current during the period of initial shorting results in what is called *overshoot* and causes excessive weld metal spatter. Additionally, an excessively fast decrease in the current results in instability which might extinguish the newly formed arc. Current recovery should occur within one or two electrical cycles. The dynamic characteristic of the machine is sometimes called the response time, which is the time required to return to steady or average conditions. The welding machine must contain the proper balance of impedance or inductance to quickly stabilize the output of the machine.

It is normally considered that a soft arc or a harsh arc is related to the magnitude and duration of the current overshoot during the short circuit period. Too much instantaneous current for too long causes a harsh arc. Many old-timer welders would wrap several turns of the welding cable around a keg of nails or a piece of steel to soften a machine with a harsh arc. This action increased the inductance in the welding circuit, which reduced the overshoot.

The rating of the welding machine is determined by tests and is related to the static volt-ampere characteristic curve. Machines are rated according to the duty cycle and the specific load voltage. The load voltage standard changes from 28 to 44 volts depending on the size of the machine. Tests are run at the duty cycle specified to determine that specific temperatures are not exceeded within the machine. The temperature rise allowed is dependent on the class of insulation used.

The NEMA insulation systems for windings are divided into classifications according to the thermal endurance of the system for temperature rating purposes, as follows:

Class A: Cotton, paper, cellulose acetate films, and similar organic materials.

Class B: Mica, glass fiber, asbestos with organic materials for mechanical strength only—slightly higher rise.

Class F: Same as Class B except proper banding materials to permit higher temperature rise.

Class H: Mica, glass fiber, asbestos, silicone elastromer with suitable binders, such as silicone resins.

Sometimes a welding machine is required to produce more than its rated output current. This is possible if the duty cycle is reduced. Sometimes it is necessary to use a machine to weld automatically or for 100% of the 10-minute period even though it has a 60% duty cycle. This is possible if the current is reduced below the rating.

Both of these situations can be resolved by use of the following formula.

$$\text{Desired duty cycle in \%} = \frac{(\text{Rated Current})^2}{(\text{Desired Current})^2}$$

$$\times \text{ Rated duty cycle in \%}$$

For example, a machine rated at 300 amperes and 60% duty cycle must be operated at 350 amperes. What maximum duty cycle can be used?

$$\text{Desired duty cycle in \%} = \frac{(300)^2}{(350)^2} \times .60$$

$$= \frac{90,000}{122,500} \times .6 = 44\%$$

Thus to use this machine at 350 amperes the duty cycle would have to be reduced to 44%. This means welding 4.4 minutes out of every 10 minutes.

In the other situation, the same machine, a 300-ampere 60% duty cycle machine must be used on an automatic-welding application. It must run for a full 10 minutes, or, at a 100% duty cycle. What output current could be safely obtained from the machine?

$$1.00 = \frac{(300)^2}{(\text{Desired current})^2} \times .60$$

$$(\text{Desired current})^2 = \frac{(300)^2}{1.00} \times .60 = 90,000 \times .6$$

$$\text{Desired current} = \sqrt{54,000} = 232 \text{ Amperes}$$

Thus for an automatic operation running 10 minutes continuously the machine output must not exceed 232 amperes without overloading the power source.

This same determination can be made without using the above formula. The curve shown by Figure 8-4, which is the duty cycle versus ampere curve, is a plot of this formula. The sloping lines show typical machine ratings and by drawing a sloping line parallel to those shown different duty cycles or different load current requirements can be determined.

The terms *slope, variable slope* or *slope control,* tend to be confusing and are often misunderstood

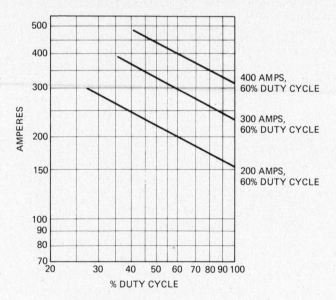

FIGURE 8-4 *Duty cycle vs rated current curve.*

with respect to welding machines. The term slope relates to the static volt-ampere characteristic curve of the machine. It is defined as the output voltage change to the change in the output current expressed in volts per 100 amperes. The term applies to both drooping characteristic machines as well as constant voltage machines. Slope is important within the arc voltage range. Figure 8-2 shows three degrees of slope for both the conventional and the constant voltage type power source. With the dual control constant current (CC) motor generator there are many slopes that can be obtained through the arc voltage range. The slope of the curve depends upon the adjustments of the open circuit voltage rheostat and the current control. In the case of the constant voltage (CV) type machine the three slopes shown are the most widely used. The rising slope is sometimes recommended for GMAW of aluminum but is not too popular. A perfectly flat or true CV curve is normally not used because it has no droop and a small amount of droop improves the stability of the welding arc because it limits the maximum short circuit current.

Most constant voltage (CV) machines have a curve close to flat with several curves having additional slope. Slope on the constant voltage machines is obtained by adding impedance in the output circuit. Most CV machines have three slopes designed for the most widely used welding processes. Adjustments providing for a variable slope or variable inductance are available on some machines. Slope is overrated as a control of machine stability. Stability depends on the transient response of the machine. In general, the correct slope is built into the CV machine for the type of welding to be done. As far as the welder is concerned the slope should be used which provides the best welding arc for the job at hand.

The term power factor is an electrical term which is of high interest to the electrical utility company and to plant engineers. In direct current, power is expressed in watts or kilowatts (kw). In a direct current circuit the product of voltage and amperage entering the circuit is all usable and it all registers on the power meter. With alternating current, however, the kilowatts are used to indicate the usable power while the term kilovolt-ampere (kva) is used to indicate the total product of amperes times volts delivered by the utility company. The term power factor (pf) is the ratio of usable current (kw) to the total current (kva). When the alternating current voltage and current are in phase, the power factor is said to be unity or 1. Thus with a unity power factor the kilowatts (kw) equals the kilovolt-ampere (kva).

The power factor of an industrial company, however, is rarely at unity since most of the electrical loads in the factory consist of motors which are inductive loads that tend to cause the volts and amperes to become out of phase. A single phase AC transformer welding machine is also an inductive electrical load. This causes the current curve of the alternating cycle to lag the voltage curve by a number of degrees and this creates what is known as a lagging power factor. Electrical utility companies monitor their industrial customers and establish a power factor for the company. This is then entered into a formula so that the factory will pay a penalty when the power factor is less than unity. The power factor of an industrial plant can be improved or moved toward unity by correction devices, such as power factor correcting capacitors or synchronous motors. In the case of single phase transformer welding machines, power factor correcting capacitors can be built into the machine to provide a correction factor bringing the power factor close to unity.

8-2 GENERATOR WELDING MACHINES

The generator is the oldest, the most reliable, and perhaps the most versatile of any of the welding power sources. Welding generators can be designed to produce any of the types of static volt-amp characteristic curves desired. Generators that produce constant voltage output have been used for producing power for electrical utilities for many years and are now used for welding. The generator can produce a rising slope curve, a slightly drooping or steeply drooping type output curve. As mentioned previously, it can be used for all the different welding processes.

The rotating generator can be driven by an air motor, an electric motor, normally an induction type, or by an internal combustion engine fueled by gasoline, diesel oil, liquid petroleum gases, or natural gas. The engine can be air cooled or water cooled. The generator can be driven by a power take-off, or by belts and pulleys from a source of rotating power.

The rotating generator is usually used to produce direct current for welding. It can produce alternating current, although this type of power source is not as popular. It has been used to produce high-cycle

AC current in different frequencies, including 180, 240, 360, and 400 Hertz. High cycle AC for welding is of little industrial significance today. The generator can also be used as a source of constant voltage power for multiple operator systems employing a large number of welding arcs.

The most popular use of the engine powered generator is for manual shielded metal arc welding. The most popular type is the dual controlled DC generator welding machine. This type is widely used by the construction industry.

The principle of electrical generation was briefly mentioned in Section 2-5. Electrical generation is based on the principle that "when a conductor moves in a magnetic field so as to cut lines of force an electromotive force (EMF) is generated." An actual generator is much more complex since there are multiple poles and hundreds of loops that cut the magnetic fields.

A generator consists of a stationary frame and a rotating armature. The frame contains an extremely efficient magnetic circuit consisting of multiple magnetic poles. A CV welding generator has four main shunt field poles but a CC generator has two. Each field pole is surrounded by coils of many turns arranged so that two of the poles are north and two of the poles are south. These coils are known as the main field windings, and power to energize them is supplied from an exciter. The output voltage of the welding generator is controlled by a variable resistance or rheostat in the main field circuit. More or less excitation voltage causes an increase or decrease of magnetic lines in the electromagnetic circuit and raises or lowers the open circuit voltage of the welding machine.

Additional magnetic poles known as interpoles are included in the frame of many welding machines. There are usually four interpoles in welding generators. Two of these are called shunt poles and have windings known as shunt field coils. The main welding current flows through the windings of these coils, which are known as series bucking fields or differential series fields. More or less of the main welding current is fed through these coils by means of the diverter or range switches. This serves to reduce the output by setting up magnetic fields in opposition to the main field as welding current is increased. This produces the drooping volt-ampere characteristic necessary for constant current welding. Also by means of range switches more or fewer turns of the differential coils are connected to the output terminals and this determines the amount of current produced by the machine. The other two coils are called the commutating poles which are surrounded by windings known as the commutating field. The commutating poles are the same polarity as the main field poles and provide for sparkless commutation of the

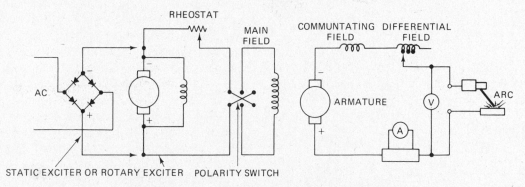

FIGURE 8-5 *Circuit diagram of generator welding machine.*

generator. Figure 8-5 is a circuit drawing of a separately excited differentially compound welding generator circuit.

On constant voltage type welding generators the differential series field is not reversed or bucking but is additive, and this makes the machine produce a constant voltage output.

When generators are driven by engines they will include an exciter generator. This is a small shunt-connected generator mounted on the same shaft as the main generator. This exciter is used to provide current for the field coils of the main generator. It is also used to provide auxiliary power for electric tools, lights, wire feeders, etc. Either alternating or direct current can be obtained from the exciter generator. Some engine-driven machines are called *self-excited* and utilize residual magnetism to provide field coil current. Self-excitation is used only for smaller machines.

For motor driven generators the separate generator exciter may be replaced by a solid state rectifier bridge circuit to provide the DC current necessary for the field coils. In either case the amount of current provided to the field coils determines the open circuit output voltage of the generator.

The rotating part of the welding generator is called an armature or, more accurately, a drum armature. This is mounted on the shaft and consists of many windings which are loops around an iron core that rotates within the magnetic field produced by the poles of the generator. The ends of these winding loops can be connected to slip rings or to a commutator bar. The output of a coil rotating within a magnetic field is an alternating current. When the opposite ends of the winding loop are each connected to a separate slip ring the output of the commutator will be alternating current. This is the normal way that large power generators are constructed. The commutator, which is made of many bars or segments, has the separate ends of the winding loop connected to segments on the opposite side of the circumference of the com-

mutator. By means of carbon brushes on opposite sides of the commutator the current is reversed each quarter revolution. Thus the commutator is a mechanical rectifier since the opposite bars are connected to brushes only when that particular loop winding is generating the maximum EMF as it passes through the magnetic field. An instant later the brushes are connected to a different loop, etc. By means of progressive connections of the winding loops to the segments, direct current is obtained at the brushes in contact with the commutator. The commutator and armature can be seen in Figure 8-6, a cutaway view of a generator welding machine. The construction of the armature can

FIGURE 8-6 *Cut away view of motor generator.*

better be seen by Figure 8-7. Note the large commutator for welding current and the small commutator for the exciter generator. Notice also the sirocco type fan for pulling air through the generator to provide cooling for the coils.

FIGURE 8-7 *Armature of generator welding machine.*

The carbon brushes in contact with the commutator take the generated current through the various coils and to the terminals of the machine. In many of the older machines stabilizing coils or an inductor was placed in series with the output current. In newer machines, sufficient inductance is built into the generator so that external stabilizing coils are not required.

For most welding generators it is possible to reverse the polarity of output. The terminals on a generator are normally marked positive and negative or electrode and work. By utilizing a reversing switch between the output of the exciter, either static or rotating, the field coils change their magnetic polarity. When the polarity of the field coils is changed the polarity of the output of the generator is also changed.

The speed of rotation of a generator welding machine with four poles is 1,800 revolutions per minute. This speed of revolution produces 60 Hertz power from the exciter generator which can be used to run power tools, etc. In order to produce 50 Hertz power the generator would be operated at 1,500 rpm. Engine-driven equipment will run at a slightly higher speed without load, but when under load should approximate 1,800 rpm. When the generator is operated by an electric motor, the motor is normally a squirrel cage type induction motor

designed to run at the above-mentioned speeds. These motors are powered by three-phase current, which provides balance load conditions to the utility line.

The induction motors produce a lagging power factor. The exact amount or degree of lagging power varies with the load. The amount of power factor in percentage is given by the performance curve of a single operator motor generator welding machine shown by Figure 8-8. In some factories, this may be significant and may require the use of power factor corrective devices. These are normally capacitors placed across the input of the electric induction motor.

It is important that the direction of the motor be in accordance with the design of the generator. Most generators show an arrow which indicates the direction of rotation. The three-phase line connected to the power source must have two leads reversed if the rotation is opposite the direction of the arrow. The size of wiring to the machine is based on the current and voltage for which the machine is connected. In North America, welding machines are normally powered by 230 volts or 460 volts, 60 Hertz. Based on this voltage, the different size machines draw different amounts of current. The power input cable should be sized in accordance with current input. This applies also to fuses, switch boxes, and so on.

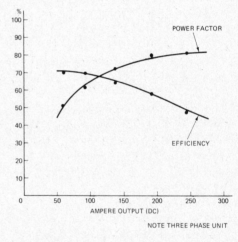

FIGURE 8-8 *Performance curves of a motor generator power source.*

Usually the welding generator motor is equipped with a starting device which connects the motor to the power line without causing excessive current surge when the motor starts. The starter also contains a thermal overload device, which would disconnect the motor in the event of overheating. Special starting switches known as Star-Delta type switches can also be used and are sometimes required by power companies. In this way, the welding machine is started without load at a slower speed for a brief instant and then switched over as speed builds up to normal.

Some engine-driven welding machines will produce only alternating current. Slip rings are used instead of the commutator. Brushes ride against the slip rings taking off the alternating current produced by the revolving loop windings as they pass through the magnetic field. The AC welding generators are usually the utility type. They are normally powered by air-cooled gasoline engines.

The third type of generator is known as the rotating field or alternator type. In this case, the armature is replaced by a rotor, which has four magnetic field coils on an iron core. The frame of the machine, or the *stator*, holds the armature coils in slots. The generation of the EMF is the same since the loop windings cut through the magnetic field. In this case, however, the magnetic field is rotating and the loops are stationary. It is necessary to provide current for the field coils which are rotating. This is done in different ways. In one case, an exciter generator is placed on the same shaft with the rotating fields and its output is taken off as direct current by brushes on the exciter commutator. In some cases, the exciter is an alternator and the output is rectified by diodes. This output is sometimes used for auxiliary power. More importantly, however, this output goes through a control rheostat to adjust the amount of current to the fields and thus adjust the output of the machine. The field current is fed back into the rotating fields by means of slip rings. The output of the alternator is then produced in the armature coils in the stator. This is alternating current, which is then rectified by power diodes for direct current welding.

There is another variation, known as the *brushless generator*. In this machine a small portion of the welding current is controlled and used to excite the field windings of the exciter generator's fixed stator. AC power is then generated in the rotating armature of the exciter, which is fed to a solid state diode rectifier bridge mounted on the rotating shaft. The diodes produce direct current which is used to provide field current for the rotating fields of the main generator, also revolving on the main shaft. The rotating field produces the magnetic lines of force, which generate alternating current in the armature windings of the stator of the main generator. This alternating current is then rectified by power solid state diodes and is the direct current output for welding. This DC welding brushless generator can be powered by engines. It is not very popular, as it does not provide auxiliary power.

Figure 8-8 shows performance curves for a generator welding machine. In addition to the power factor curve, it also shows the efficiency of the generator at different loads. It should be pointed out that a generator is as efficient as a rectifier under load. This can be determined by comparing this curve with those shown in later sections.

8-3 TRANSFORMER WELDING MACHINES

Transformers have been in use ever since alternating current was used for electrical power. The use of the transformer as a welding power source, however, did not occur until about 1920. Holslag is credited with inventing the transformer type welding power source.

Holslag actually discussed the square wave type output which helped stabilize the welding arc by making the current flow through zero quicker. He also covered the phase shifting aspects to help stabilize the arc. It was not, however, until the covered electrode became widely used that the transformer power source became popular.

The principle of operation of a transformer was briefly mentioned previously. It was pointed out that magnetic lines of force build up to the maximum in one direction during one half-cycle of current flow, collapse, then build up in the opposite direction to a maximum during the other half-cycle of current flow, and collapse. If another coil is placed in the same magnetic field the magnetic lines of force will cut across the second coil and induce an EMF in it creating current flow. By changing the magnetic coupling of the two coils the output of the second coil is changed which changes the output of the welding transformer. This coupling can be increased by moving the coils closer together or by increasing the strength of the magnetic field between them. The voltage relation between the two coils is determined by the number of turns in each coil. The transformer also isolates the incoming power in the primary from the welder. This is a safety requirement for transformer power sources.

The transformer welding power source is more complex than an ordinary power transformer used for stepping down voltages. In a power transformer the primary would be, for example, 230 volts. The voltage on the secondary would be 115 volts if it contained half as many turns as the primary. If it had double the amount of turns its voltage would be 460. The secondary voltage is the relationship of the number of turns of the secondary to the number of turns of the primary. The current is inversely proportional to the turns ratio. A power transformer is a constant voltage type of transformer and the open circuit voltage and load voltage would be practically the same. This type of transformer would not be suitable for manual welding.

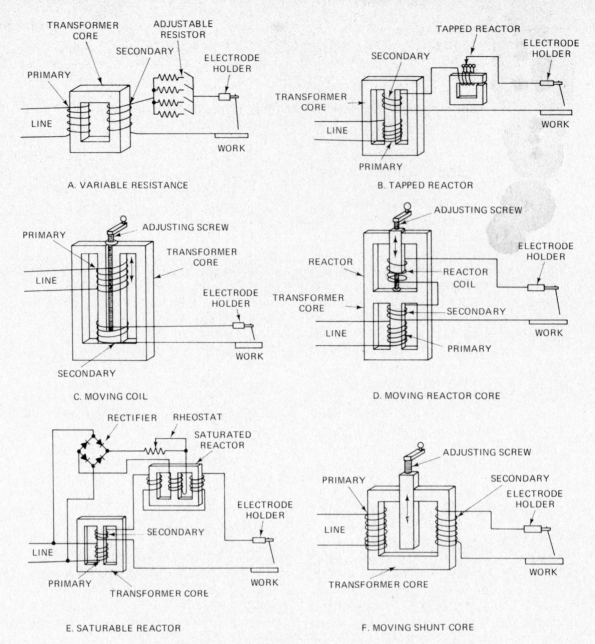

A. VARIABLE RESISTANCE

B. TAPPED REACTOR

C. MOVING COIL

D. MOVING REACTOR CORE

E. SATURABLE REACTOR

F. MOVING SHUNT CORE

FIGURE 8-9 *Methods of controlling the output of transformer power supplies.*

In the early welding transformers the drooping characteristic required for manual welding was obtained by using a resistor in series with the arc. The open circuit voltage would be based on the turns ratio but the output current would be based on the amount of resistance in the output circuit. The voltage across the arc would be suitable for arc welding since part of the voltage would be dropped across the series resistor and the remainder across the arc in accordance with its volt-ampere relationship. This system is not efficient, since the voltage drop in the resistance is wasted. Therefore, better electrical methods for controlling the output were developed.

There are many different ways of varying the output current of a transformer welding machine. The more common ways are as follows and sketches describing each are shown by Figure 8-9.

(a) Placing resistors in series with electrode. This was mentioned above and is inefficient.

(b) Tapping a reactor or placing a variable reactor in the secondary circuit.

(c) Moving the primary or secondary coil with respect to the other coil.

(d) Moving an iron core in a reactor in a secondary circuit.

(e) Using a saturable reactor in the secondary or welding circuit.

(f) Placing a moving iron shunt between the primary and secondary coils.

The methods described in A, B, D, and E adjust the amount of impedance in the transformer secondary circuit. The methods described in C and F adjust the magnetic flux coupling between the primary and secondary of the transformer. There are advantages and disadvantages to each of these systems.

The method described in A has the disadvantage of heat wasted in the resistance in series with arc. It is rarely used today.

Method B uses a tapped reactor in the secondary to change impedance. It does not provide for continuous adjustment of output current. The number of taps restricts the current selection available. This system is relatively efficient and is used for the utility or limited input type welding transformers.

Method C changes the reactance of the transformer by changing the position of the primary and secondary coils with respect to each other. Moving a coil away from the other permits a greater amount of leakage flux to flow between them and increases the reactance of the transformer. This reduces the current output. The coils are moved toward or away from each other by means of a lead screw which provides continuous adjustment. In time the moving coil may loosen and vibrate and cause noise. The connectors attached to the moving coil will be continually flexed in operation, which can create service problems.

Method D uses a moving iron core in the reactor in the secondary circuit. The moving of the core causes the change in the air gap which changes the reactance. The larger the air gap, the smaller the impedance value and the higher the output current. As the air gap is is reduced, impedance increases and the welding current is reduced. Movable parts tend to vibrate as they wear and become loose, which creates undesirable noise. Mechanically moving parts such as lead screws wear and become dirty and become difficult to move. On large machines adjustments that involve movement can be motorized for remote current control.

Method E eliminates the moving parts with their possible service problems, but is more expensive. This

method is made possible by the use of a diode bridge rectifier. In this system the reactor in the secondary circuit is used to regulate the output current. Direct current from the diode bridge is introduced into the reactor which tends to saturate its magnetic field. When there is no DC current flowing in the reactor coil, it has its minimum impedance and thus maximum output of the transformer welding machine. As the DC current is increased, by means of the rheostat in the direct current circuit, the magnetic field contains more direct current or more continuous magnetic lines of force in the magnetic circuit. Impedance of the reactor is increased and the output current of the welding transformer is decreased. This method has a great advantage of eliminating movable parts and flexing conductors.

Method F uses the moving core in a different way to adjust the reactance of the transformer. In this method an iron core called a *shunt core* is moved between the primary and secondary coils. This movement varies the leakage flux between the primary and secondary and adjusts the output current. This method has the same disadvantages as method D.

These brief descriptions are simplified explanations of the transformer welding power source designs. However, by means of these designs a constant current type of AC power source is provided. A certain amount of impedance or droop must be built in the power source to make it usable. Without the secondary impedance the transformer welding machine would have constant voltage characteristics. A constant voltage (CV) or flat characteristic alternating current power source is not practical for any arc welding process.

The alternating current power source, even with proper output characteristics, must rely upon ionizing elements in the arc atmosphere in order to maintain a stable arc. Since the alternating current continuously varies from positive to negative the arc goes out each half-cycle as the voltage passes through the zero point. The ionizing elements added to the flux coating of an electrode help arc reignition each half-cycle which stabilizes the arc. It is impossible to use bare electrodes or bare solid electrode wire with alternating current, without some sort of arc ionizer or stabilizing elements in the arc atmosphere. In the case of gas tungsten arc welding, high-frequency current is sometimes employed to ionize the arc gap.

The reactor core and the transformer core may be one and the same in a transformer welding machine.

POWER SOURCES FOR ARC WELDING

One of the more advanced types of welding machines involves a series of secondary windings which are either closely coupled or loosely coupled to the primary and give different current output ranges. In addition, the same core will include the magnetic shunt adjusting system. However, instead of an iron core moving in the system, this portion of the magnetic circuit will have a direct current coil which will magnetically saturate the core. This will change the impedance which will provide for output current adjustment. A simplified version of this design is shown by Figure 8-10.

Single-operator constant-current welding transformer power sources are available in all three NEMA classes. They are available with current outputs ranging from 100 amperes up through 1,000 amperes. One of the most popular AC transformer welding machines is the NEMA class 3 type known as the *utility* or *farm welder,* and commonly called *limited-input welding machines.* These were originally designed to have a limited input current at their rated output so that they could be used in rural areas on available utility power lines. They are normally rated with a 20% duty cycle. They are used for light-duty repair work and hobbies.

The NEMA class 2 welding transformers are not too popular. The NEMA class 1 welding machines are the industrial type welding power sources that are rated at 60%, or 100% duty cycle. The 60% duty cycle machines are used for manual shielded metal arc welding in indus-

trial applications. The 100% duty cycle machines are used for automatic submerged arc welding.

The transformer welding machine is the most efficient of any welding machine from the electric power point of view. It has the highest efficiency, the lowest no-load losses and, with proper power factor correction, a good power factor. Figure 8-11 shows the performance curves of a typical 300-ampere machine. The efficiency of a transformer welding machine operating on a 40-volt load ranges from 80% to 90%. The no-load losses range from 150–375 watts depending on the machine. The power factor at a 40-volt load is approximately 27% without correction but can be greatly improved by the use of capacitors. The corrective capacity of capacitors should be based on the normal welding range for which the machine is used. Too much correction may be as bad as none. Figure 8-11 also shows the difference in power factor for a transformer welding machine with and without capacitor correction.

There are several disadvantages to transformer power sources. They operate on single-phase input power, which tends to unbalance utility power lines, unless a sufficient number of AC transformer welding machines are used and are properly balanced to the line. In addition, certain types of covered electrodes are not available as alternating current electrodes.

The maintenance expense of transformer welding power sources is the lowest of any type of welding machine; and the use of alternating current reduces the problem of arc blow on complex weldments. Therefore, for many shielded metal arc welding applications the AC transformer power source is the best selection.

FIGURE 8-10 *Modern transformer welding power source.*

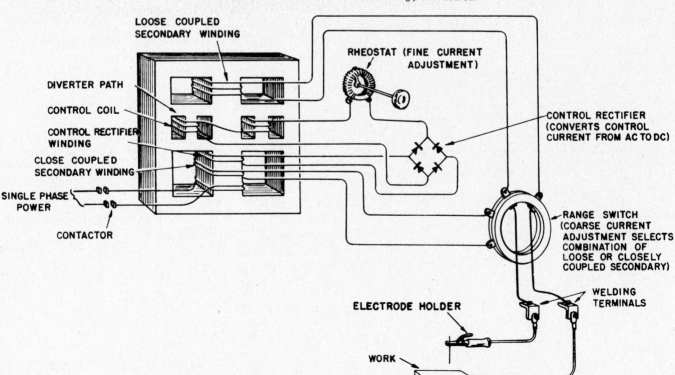

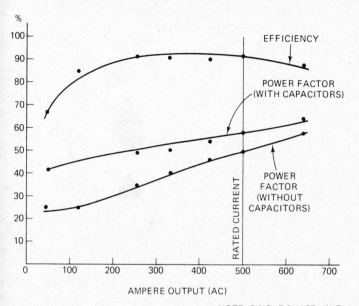

NOTE: SINGLE PHASE UNIT

FIGURE 8-11 *Performance curves of a transformer welding machine.*

8-4 RECTIFIER WELDING MACHINES

The alternating current type power source, while an extremely efficient electrical machine, has limited applications in arc welding. Direct current is used for welding with more processes and for many more applications. This information was summarized by Figure 8-3.

By means of a rectifier the alternating current is changed to direct current. The rectifier is a device that conducts current easier in one direction than the other. The diode vacuum tube was used for many years in radio power supplies for direct current. Six vacuum tube rectifiers were used in the first rectifier welding machine, which was developed in the mid-1930s. This early welding machine called a Weld-O-Tron, which had only a 75-ampere rating, is shown by Figure 8-12. For higher-current output solid state rectifiers were developed. The welding industry adopted the dry disc rectifier, which employs a layer of semiconductor such as selenium between adjacent plates. Selenium rectifiers were used in most early rectifier welding power sources. Today the silicon diode rectifier is used for most welding machines.

Silicon diodes are made of thin wafers of silicon that have small amounts of impurities added to make them semiconductors. The wafers are specially treated and assembled in devices for mounting in welding machines. Figure 8-13 is a picture of a typical power silicon diode. The internal parts, in an exploded view, of a silicon diode are shown by Figure 8-14.

Electrical circuits to utilize rectifiers to change alternating current to direct current are well known. No matter what type rectifier is employed, the results will

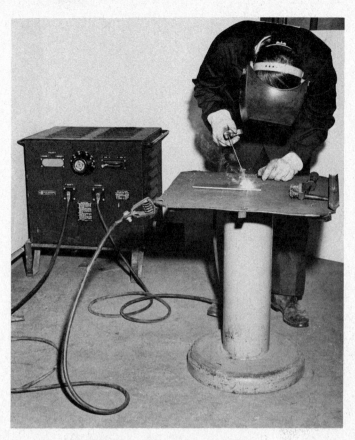

FIGURE 8-12 *First rectifier welding machine.*

FIGURE 8-13 *Silicon power diode rectifier.*

FIGURE 8-14 *Cross section of silicon diode and diagram.*

be essentially the same. The smoothness of the direct current output depends on the rectifier circuit that is used. The rectifier will allow only one polarity of the alternating current sine wave to pass through. The simplest rectifier circuit, known as a half-wave rectifier, will allow only the positive half of each sine wave through and blocks the negative half of the cycle. The AC input and the half-wave rectified DC output are shown by the top diagram of Figure 8-15. This rectified direct current will be very rough rising from 0 to peak voltage, back to 0 then a pause and rising again. This kind of wave shape would be unsuitable for most welding applications. A filter of some sort is necessary but these become extremely large and expensive at power line frequencies. A design which provides a smoother output utilizing four rectifiers instead of one, is known as a full wave rectifier. This is shown by the second line of Figure 8-15. The output is considerably smoother but

still not satisfactory for most welding applications. When the machine must operate on single phase the only way to smooth the output is by the use of inductors in series and capacitors across the DC output of the rectifier circuit. These will tend to smooth the roughness or ripple of the direct current output.

It was previously mentioned that one of the disadvantages of the transformer power source is the fact that it operates only on single-phase AC power. In seeking to provide smoother direct current output it was decided to utilize three-phase AC input. This provided a much smoother DC output, as shown by the third line of Figure 8-15. By using twice as many power diodes it is possible to further smooth out the DC output of a three-phase machine. This is accomplished by using a full-wave rectifier as is shown by the bottom line of Figure 8-15. Most industrial rectifier welding power sources use three-phase input and full-wave rectification to provide smooth DC output for a stable welding arc. These machines provide a balanced load on the three-phase power line which is an advantage over the single-phase transformer welding machines. Some small rectifier welding machines do, however,

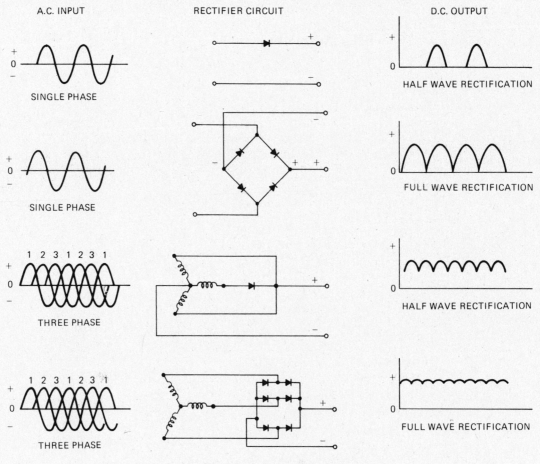

FIGURE 8-15 *Simplified rectifier systems.*

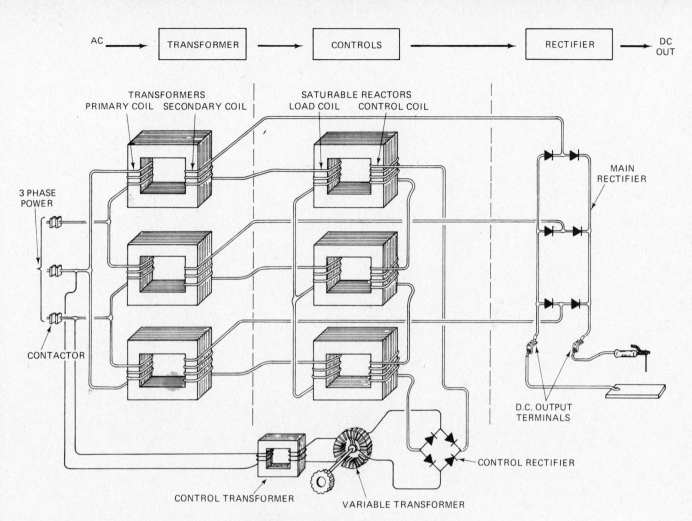

FIGURE 8-16 *Rectifier welding power source.*

operate on a single phase. These machines have a filter circuit but do not provide an arc that is as smooth as the three-phase machines.

All the different transformer welding circuits shown in Figure 8-9 can be converted to direct current power sources by adding rectifiers in the output circuit. The method of control most commonly used is the circuit shown by letter *E,* the saturable reactor type. This is preferred since the output can be controlled electrically through a wide range of current rather than mechanically by moving coils or cores or tap switches. In addition, the same basic types of circuits can be used with three-phase power transformers in place of single-phase transformers. A simplified version of a modern rectifier welding power source is shown by Figure 8-16. Note the block diagram on top of the figure showing alternating current into the transformer through the control circuit then to the rectifier circuit, and the output is direct current suitable for welding. This is the three-phase version of control method *E* saturable reactor shown by Figure 8-9. This type of machine can have drooping static

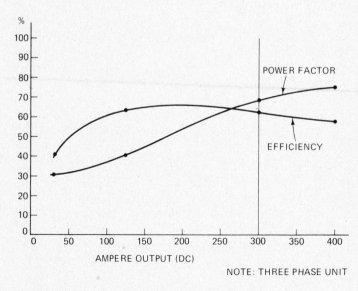

FIGURE 8-17 *Performance curves of a rectifier welding machine.*

285

POWER SOURCES FOR ARC WELDING

volt-amp characteristics or relatively flat or constant voltage output characteristics. The output characteristics depend upon the inductance in the control circuitry. This basic design or similar designs are used for most heavy-duty industrial rectifier welding power sources. The performance curves for this type of machine is shown by Figure 8-17.

Special Power Sources

As the gas tungsten arc welding process became more popular special welding power sources were developed to satisfy the needs of the process. In the early days of gas tungsten arc welding, conventional welding machines for shielded metal arc welding were used. These machines would often burn up because of overloading or saturating the cores. These problems occurred with the transformer power sources. It has been pointed out earlier that rectification can occur in the tungsten arc, particularly when welding on aluminum. When rectification occurred in the arc a portion of the welding current was direct current rather than alternating current. The direct current in the transformer tended to create magnetic saturation of the iron core, which in turn created overheating and eventual burnout of the power source windings. Various efforts were made to eliminate the direct-current component of the tungsten-aluminum welding arc. One system briefly used but rejected, was to connect lead acid automobile type storage batteries in the circuit with the polarity opposite that of the DC component. The battery current was to buck and cancel out the direct current component in the circuit. This solution did not prove to be satisfactory and was quickly abandoned. Another method was to utilize a large number of capacitors in series with the welding circuit. This required a tremendous amount of capacitance which was expensive, and this method was also abandoned. The ultimate solution was found by redesign of the basic transformer power source so that the DC component in the welding circuit would not be harmful to the power source. This was referred to as balanced wave gas tungsten arc welding machines. **Balanced wave** means that both polarities of the AC cycle are in balance. The balanced wave concept created new types of machines which were designed exclusively for the gas tungsten arc welding process. In this same redesign the characteristic output curves of the machines were made steeper so that minor changes in arc length which created small changes in arc voltage would have minor effect on the welding current. It was found that these machines could also be used with shielded metal arc

welding and, therefore, combination machines resulted. Another feature was added to these machines which was brought about by the need of the welder. As the alternating welding current went through the zero point the arc would be extinguished and would have to be reignited. It was found that by superimposing an extremely high frequency current on the welding current that the arc gap would remain ionized so that the arc would immediately be re-established at the arc voltage after going through zero. The superimposed high frequency current also eliminated the rectification in the arc since each half-cycle carried current. The high frequency current was obtained by means of a spark gap oscillator. A high-frequency oscillator consisted of a high-voltage transformer, a spark gap, a high-voltage capacitor, and another high-frequency transformer to couple the high-frequency current to the welding current circuit. The circuit for this is shown in Figure 8-18. The high-frequency current performed another duty in that it allowed arc starting without touching the tungsten electrode to the work. When the tungsten electrode is brought close to the work the spark produced by the oscillator would jump across the gap, ionize the gap, and cause welding current to flow.

This led then to the combination AC-DC welding machines for gas tungsten arc welding. A simplified diagram of such a machine is shown by Figure 8-18. Since this machine provides both AC and DC it necessarily has single-phase power input. For this reason, it does have a filter to help stabilize the DC output when the machine is connected for direct-current welding. The machines are also equipped with an extra terminal so that they can be used for shielded metal arc welding as well as for gas tungsten arc welding. Additional conveniences were provided for the welder by adding solenoid valves for shielding gas and cooling water supplied to the torch. Timers were provided to discontinue the gas and water flow after the arc was stopped. Other refinements included timers to make the high frequency current continuous for AC welding or to make it operate only at the start when welding with direct current. Further refinements provided for remote current adjustment so the welder could change the current during the welding operation by means of a foot pedal rheostat.

The newer machines are available as combination AC-DC machines but also as DC machines utilizing three-phase input and providing a smooth DC output. The DC machines have the most complex control circuitry. A machine of this type is shown by Figure 8-19. This machine offers the welder various options of control including pulsing of the output current, sloping up and down the amount of current, and it provides programs which can be initiated by the welder. In addition, they provide programs which can be used with automatic

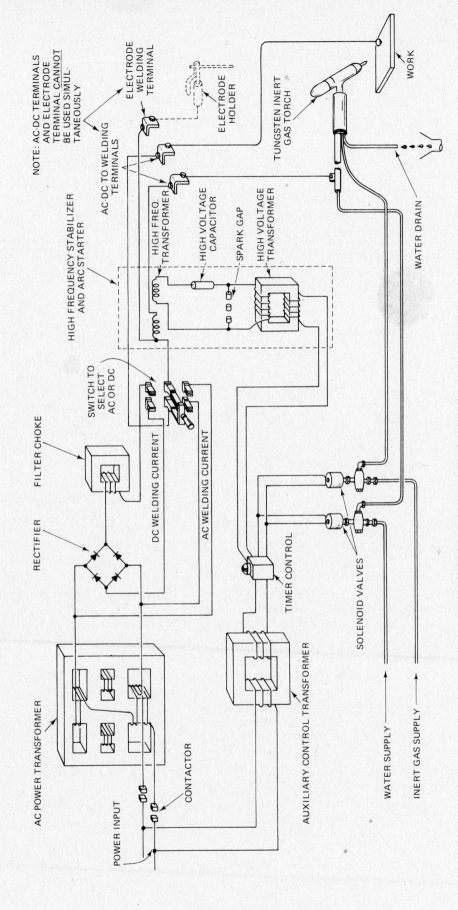

FIGURE 8-18 *Combination AC-DC welding machine for GTAW.*

NOTE: AC-DC TERMINALS AND ELECTRODE TERMINAL CANNOT BE USED SIMULTANEOUSLY

ELECTRODE WELDING TERMINAL

ELECTRODE HOLDER

AC-DC TO WELDING TERMINALS

TUNGSTEN INERT GAS TORCH

WORK

WATER DRAIN

HIGH FREQUENCY STABILIZER AND ARC STARTER

HIGH FREQ. TRANSFORMER

HIGH VOLTAGE CAPACITOR

SPARK GAP

HIGH VOLTAGE TRANSFORMER

SWITCH TO SELECT AC OR DC

FILTER CHOKE

DC WELDING CURRENT

AC WELDING CURRENT

RECTIFIER

TIMER CONTROL

SOLENOID VALVES

AC POWER TRANSFORMER

CONTACTOR

POWER INPUT

AUXILIARY CONTROL TRANSFORMER

WATER SUPPLY

INERT GAS SUPPLY

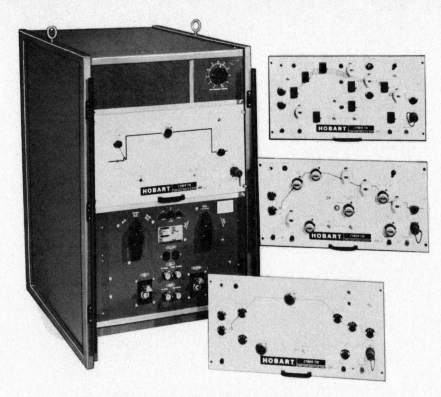

FIGURE 8-19 *Programable DC machine for GTAW.*

welding equipment such as tube-to-tube welding heads. The machines can also be programmed for specific production jobs. A variety of program modules are available that provide simple controls up to extremely complex programs used for automated equipment.

8-5 MULTIPLE OPERATOR SYSTEMS

The multiarc welding system utilizes a high-current constant voltage power source to supply power for many welding arcs. This differs from the conventional single operator power supply system in which an individual welding power source is required for each welding arc. The multiple operator system is used when there is a large concentration of welding arcs in a relatively small area. The system was originally used in shipyards during World War II and has been popular at construction sites for power houses, refineries, chemical plants, and for some manufacturing operations. One of the major advantages of the multiple-operator system is the lower capital investment required per welding arc. This is based on lower cost of installation as well as the cost of the power source and of the individual welding stations. By the use of the multiple operator system it is not necessary to have primary power at each welder's station. See Figure 8-20.

The basis of the multiple-arc system is *load diversity* and *operator factor* or duty cycle. The multiple-operator system uses a single large welding power source with a capacity considerably smaller than the total capacity that would be required when using single operator power sources.[2] Load diversity means that in a large group of welders only a small percentage will be welding at any one time. This is because of the normal welder operator factor or duty cycle. The operator factor is the percentage of time that the welder is actually welding against the total or paid time. The operator factor or duty cycle can vary from 10 to 50% depending on the type of work. Pipe and structural welding tends to be on the low side while heavy plate, flat position welding would tend toward the middle or upper portion of this range. The diversity factor is related to average duty cycle and the laws of probability. It is not expected that all the welders will start to weld at precisely the same time.

The relationship between these factors—the number of arcs to be used, the welding current, and the needed available power—is shown by the following formula.

Total power required = Number of arcs
$$\times \text{ average amperes each arc}$$
$$\times \text{ average duty cycle each arc}$$

To determine the number of arcs that can be employed when the available power is known the formula can be restated as follows:

$$\text{Number of arcs} = \frac{\text{Total power available}}{\text{Ave amp each arc} \times \text{ave duty cycle each arc}}$$

288

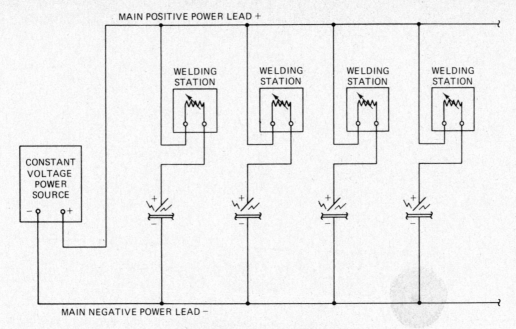

FIGURE 8-20 *Multiple operator system.*

To utilize this formula it is necessary to determine the amperes that will be used at each arc. The current required at an arc can be estimated by referring to Figure 8-21, which shows the average current required based on the welding process and the size of the electrode employed. This chart shows data for shielded metal arc welding, gas tungsten arc welding, air carbon arc cutting and gouging, and stud welding. It is assumed that a normal duty cycle would apply. This excludes stud welding which utilizes much higher currents but the arc duration is normally less than two seconds. These processes utilize direct current electrode positive with the exception of gas tungsten arc welding which utilizes electrode negative.

With these data it is possible to determine the number of welding arcs that can be used with a specific size power source. The power sources available for multiple operator welding range from 500 amperes to 1,500 amperes and larger. Machines can be paralleled to obtain higher currents for larger installations. The data shown in Figure 8-22 are based on the available power of 1,000 amperes at a 100% duty cycle and with the provisions of a 50% overload of not over two minutes.

Power sources used for multiple operator systems have a maximum open circuit voltage of 80 volts and a load voltage of approximately 75 volts at rated output. Motor generators were previously used as power supplies;

Size of Electrode, Carbon or Stud (Dia.)		Average Current Required			
		SMAW DCEP or DCEN	GTAW DCEN	AAC DCEP	SW—2 Sec. Max.—DCEP
in.	mm				
0.035	0.9	—	50	—	—
0.045	1.1	—	75	—	—
1/16	1.6	—	110	—	—
3/32	2.4	75	180	—	—
1/8	3.2	100	235	—	—
5/32	3.9	150	300	150	—
3/16	4.8	200	400	200	300
7/32	5.6	275	450	300	350
1/4	6.4	350	500	400	400
5/16	7.9	—	—	500	500
3/8	9.5	—	—	600	600
1/2	12	—	—	1200	900

FIGURE 8-21 *Average current required—electrode size—by process.*

Ampere Per Arc	25%	20%	25%	30%	35%	40%	45%	50%	55%	60%
					Duty Cycle of Welders					
50	132	100	80	66	56	50	44	40	36	32
75	88	66	56	44	38	32	28	26	24	22
100	66	50	40	32	28	24	22	20	18	16
125	56	40	32	26	22	20	16	16	14	12
150	44	32	26	20	18	16	14	12	12	10
175	38	28	22	18	16	14	12	10	10	8
200	32	24	20	16	14	12	10	10	8	8
225	28	22	16	14	12	10	8	8	8	6
250	26	20	16	12	10	10	8	8	6	6
275	24	18	14	12	10	8	8	6	6	6
300	22	16	12	10	8	8	6	6	6	4
325	20	14	12	10	8	6	6	6	4	4
350	18	14	10	8	8	6	6	4	4	4
375	16	12	10	8	6	6	4	4	4	4
400	16	12	10	8	6	6	4	4	4	4

FIGURE 8-22 *Maximum number of welding arcs with a 1000 amp power source.*

however, the silicon diode three-phase rectifier welding machine is currently used. Machines of this type have an operating efficiency of approximately 90%. These machines do not have controls to change either current or voltage and for this reason are used only with multi-arc systems.

A welding station, or grid, is required at each arc. The welding stations consist of switches and resistance grids, which are in series with the arc. By means of switches the amount of resistance in series with the arc can be adjusted to provide the correct current for that particular arc. Since the normal voltage across a shielded metal arc is 25 volts it would follow that 55 volts would be dropped across the resistance grid of the welding station. If 200-ampere current is used for welding this would amount to 11,000 watts of heat produced in the welding station.

In some welding stations polarity switches are incorporated so that electrode negative can be used. When dual polarity is required two power sources must be used. The positive pole of one and the negative pole of the other power source are connected to the common lead or building frame. Separate conductors are needed for the other leads, both of which are brought to the dual-polarity welding stations. In this way either polarity is available. This is particularly advantageous for gas tungsten arc welding. Normally the systems are set up so that the electrode is positive. The switching arrangement allows for either polarity of welding current and for fine control.

Another advantage of the multiarc system is the volt ampere characteristic curve presented to the arc. This is a straight line drooping relationship based on pure resistance which allows for a very smooth arc. The resistance load is a limiting load and, therefore, transients are minimized. The slope of the curve is determined by the amperage setting of the grids, but in all settings it is a straight line rather than the curved line of a single-operator welding machine.

The multiarc system is often used for air carbon arc cutting and gouging. As with the other applications the duty cycle and current limitations must be observed. The average current based on different size carbon electrodes is shown by the table of Figure 8-21.

The multiarc system is also used for stud arc welding when the current requirement is normally greater than the requirement for shielded metal arc welding. This can be done since all of the components of the system have short-time overload capabilities. The welding power source will provide an overload of up to 50% for two minutes maximum. The time cycle for stud welding is usually less than two seconds. In view of this, overloads for stud welding can be accommodated.

The cables required for MO systems are extremely large in order to minimize power loss. Cable size should be based on a maximum of 3 volts drop per hundred feet. The size of the main conductors from the power source to distribution centers depends on the length and the current carried; normally a 500 MCM cable is used. This size of cable is larger than flexible welding cable and is therefore not of the fine strand flexible cable design. The heavy cable is used with distribution blocks or tap plugs so that smaller cables can be run to the individual welding stations. From the welding stations to the electrode holder conventional welding cable sizes are used. Standard 115-v AC power is usually provided for auxiliary equipment.

For gas tungsten arc welding special high frequency attachments are used in order to avoid touch starting of the tungsten to the work. These are installed at the welding station. The gas tungsten arc process is commonly used for root pass welding of piping in power plants.

It is common practice in power plants to retain the welding system within the building leaving the heavy conductors permanently installed. In this way, future maintenance welding can be done in any part of the power plant.

The multiple-operator or multiarc system has specific advantages and it is necessary to make calculations to determine the installation cost with the multiple-operator system versus single-operator power sources. This would include all equipment and wiring. The operating cost of the multiple-operator system versus individual power source must also be considered or assured as equal. Based on the normal life time of the equipment, the multiple-operator system will often be the least expensive.

8-6 SELECTING AND SPECIFYING A POWER SOURCE

So far in this chapter the different welding power sources have been described and classified according to the output characteristics of the machine, the current type produced, whether it is rotating or static, etc. This brought out the fact that there are two types of characteristic curves, AC and DC current, two types of machines, rotating and nonrotating, and that all of these machines come in different sizes and with different controls. Further, it was learned that there are three basic ways of providing welding current—the generator, the transformer, and the rectifier—and that each has its place. A system of classing welding machines is shown by Figure 8-1.

From the other point of view, if we know the kind of welding work to be done it then becomes a matter of study to determine the type of welding power source that should be used. The following information provides the needed data in order to make the most intelligent selection.

1. Process Selection: This precedes all other factors since it is based on the work to be done. Process selection was covered in Section 5-6. This will generally determine the output characteristics of the power source required.

2. Welding Current: Most arc welding processes utilize direct current; however, some may utilize either direct or alternating current. When the option is available the reason for selecting one or the other may depend on the material being welded. This is particularly true for gas tungsten arc welding. For shielded metal arc welding it might be determined by the type of covered electrodes available for the work.

3. Machine Rating: This is a way of determining the size or capacity of the arc welding power source. Most welding machines are rated by output current in amperes. Larger electrode sizes require more current but larger electrodes can be used only on heavier work. The size is given in amperes but at a rated voltage. The rated voltage for different size machines in amperes has been standardized by the NEMA code.

4. Duty Cycle: The duty cycle is also a measure of the amount of work that the power source can do. A low-duty cycle machine does not have the work capacity of a high-duty cycle machine. Duty cycles range from 20 to 100% and are broken into three classes according to NEMA. The low-duty cycle class has the least capabilities and, therefore, is used for intermittent repair work or for the hobbyist. The high-duty cycle class will be used for heavy-duty industrial applications.

5. Availability of Power: This relates to location and whether electric power is available from the utility company. On construction sites, where electric power is not available engine-driven equipment is required. There are many situations on new construction sites in which power may or may not be available or in which the expense of hooking up the power line is excessive. In these cases, engine-driven equipment is the logical selection. In general, air-cooled engines have lower-duty cycles than water-cooled engines. Diesel-powered machines are required when gasoline might present a fire hazard, such as on an offshore drill rig.

6. Auxiliary Devices: Finally, the selection might depend on auxiliary devices and controls which will make the welder's job easier. This includes high-frequency current for gas tungsten arc welding or remote capabilities for current control. Auxiliary devices include other items such as gas and water controls for processes needing them and timers and programmers for critical work.

Each of the preceding factors should be analyzed. In addition, consider the need of flexibility since a power source is a long term investment and will normally outlast the current job. Will the next job be similar or will it require more power or should it have additional capabilities? Many times it is wise to select a machine with a higher current rating so that it will not operate at maximum output for the present job but will have reserve power for other jobs requiring larger electrodes and higher currents. Also consider auxiliary devices or features which might be used for other jobs. However, as more features and greater output are required the cost of the equipment increases. It then becomes a matter of a tradeoff between the amount of money to be spent versus the extra flexibility that is desired.

In analyzing these six factors, it is wise to investigate newer equipment since they may include features that would be desirable for the current work. It is also desirable to review the welding procedures to determine current levels and duty cycles needed. With this information establish the specification for the equipment to be procured.

Specifying the Equipment

In order to specify a welding power source properly the following data should be provided.

1. Manufacturer's Type Designation: This can be determined by consulting the manufacturer's catalog or data sheets once the other specifications are known.

2. Manufacturer's Identification Number: This also should be determined from the manufacturer's literature and is usually given as a model number.

3. Rated Load Voltage: Most welding machines manufactured in North America are rated in accordance with the NEMA system. In this system minimum load volts are related to the ampere output of the machine. The 200-ampere machine has a minimum load of 28 volts and this increases as the machine ratings are increased.

4. Rated Load Amperes: Rated current is the current that the power source will deliver at the rated volts. This should be related to the welding procedure requirements.

5. Duty Cycle: This is the duty cycle at the rated load that the machine will handle. This is given in percent but may be given in conjunction with the NEMA class designation.

6. Voltage of Incoming Power: This is particularly important so that the machine will match your power line. Various input voltages are available to match the power available in all parts of the world.

7. Frequency of Incoming Power: This is the frequency of the power provided by the power company in Hertz. In North America this is normally 60 Hertz. In other parts of the world it can be 25, 50, or 60 Hertz.

8. Number of Phases of Incoming Power: Normally the incoming power is either single- or three-phase. Single-phase power is used for limited input or low duty cycle machines. Most industrial equipment utilizes three-phase power.

In the case of engine-driven equipment, it is necessary to specify the maximum rated speed in revolutions per minute at no load.

With this information, you can accurately specify a welding power source and be sure of obtaining exactly what is required. Flexibility might again be involved if the machines are to be moved. For example, incoming voltage can be specified with two voltages, such as 230 and 460 volts, which will provide flexibility.

8-7 INSTALLATION AND MAINTENANCE OF POWER SOURCES

The installation, maintenance, and adjustment of welding machines and equipment are usually done by different people. The organizational structure of a company or plant usually determines who performs these different activities. In many plants the original installation may be done by construction people, riggers, and electricians. Once the equipment is operating properly they have completed their work. Maintenance is an on-going operation and is usually done by plant maintenance electricians. Minor adjustments of a routine nature, such as changing tips, nozzles, tungsten electrodes, drive rolls, blowing out cables, etc., are done by welders. In small shops one person may perform all of these operations.

Installation

Installation of welding equipment must be in accordance with the manufacturer's instructions, the company's own standard practices, and all local, state, and national regulations. The manufacturer of welding machines provides information sheets showing the proper method for installing equipment. This usually includes recommended location of the equipment, which should always be in a dry, well-ventilated area, free of excessive dust, moisture, fumes, spray, etc. Information is given about installing primary power to the welding machine, including the wire sizes, the size of the disconnect switch, and the fuse sizes. Most welding machines are manufactured so that different incoming voltages can be used. The voltage is changed by moving links between studs to properly connect the input voltage to match the power incoming voltage. A wiring diagram showing this information is usually posted on the inside of the welding machine. Various safety rules such as the National Electrical Code, the Welding Society's Safety and Health book, and the requirements of the Federal Safety and Health Administration must be met. This means that disconnect and fuses of the correct size are required and that the case of the welding machine must be grounded to earth. For gas tungsten arc welding equipment, special installation requirements must be followed to avoid the radiation of high-frequency current from the spark gap oscillator. This involves extra special grounding to earth of the case and fixtures. This information must be followed explicitly. Other machines may require the installation of gas supply and cooling water supply and drain.

The installation of welding equipment must be inspected with a checklist utilizing the items mentioned above to make sure that it is in accordance with all standards and codes. Motor generator welding machines must be checked for the correct direction of rotation at the time of each installation. Rotation is easily checked since all machines carry an arrow showing the direction of rotation of the armature.

Transformer welding machines must be installed with care. They must be balanced on the three-phase of the power lines and must be phased with respect to adjacent units. This was explained in Chapter 3.

Preventative Maintenance

Preventative maintenance is the routine maintenance performed on equipment while in service so that it does not deteriorate rapidly and fail. This applies to all types of machinery, including welding machines. Preventative maintenance for the different types of welding machines is similar; however, specific types do require specialized attention.

In general, all welding machines and, in fact, all electrical machinery should be kept clean. Dirt and dust from the factory or from the construction site are carried by the ventilating air through the welding machines. This dirt collects in the internal parts of the welding machine and tends to build up on windings and prevent them from cooling efficiently. The dirt may also build up to the point where it blocks the passage of cooling air through the machine. Welding machines should be

inspected every six months and cleaned. In an extra dirty environment with excessive amounts of dust, dirt, lint, or corrosive fumes the inspection period should be shortened.

The welding machine should be cleaned by qualified personnel. For safety reasons, the power should always be disconnected at the wall switch so that live power leads are not inside the welding machine case. The dirt should be removed from the windings by blowing them out with clean, dry compressed air at a pressure of 25 to 30 psi. All foreign material should be removed. High-pressure air should not be used since this will tend to drive the dust and dirt into crevices which will reduce the cooling efficiency of the machine. Vacuum cleaning can be used, particularly if metallic dust is present inside the case. In very dirty environments, filters are recommended to keep dust from entering the machine.

At this inspection be alert for dirt on switch contact points, and check for loose internal and external connections. For machines with cooling fans make sure that the fans are operating and that the fan motors are clean. Look for internal corrosion, internal mechanical damage, and external damage of any type.

Engine-Driven Generators: The normal engine maintenance should be performed on the engines of welding machines. Change oil after every week of continuous operation. This may need to be done more often in especially dirty locations. Inspect and replace, if necessary, oil filters, air filters, and fuel filters. Check coolant and batteries each time fuel is added. The linkage and idling devices should be checked weekly. The radiator should be checked daily to make sure it is full of coolant and free of obstructions. The engine manufacturer's recommendations should be followed.

Generators: Generators should be checked monthly. This includes the windings, contact points, the brushes and brushholders, the commutators, the controls, and the bearings. The brushes should be inspected to make certain that they are bearing properly on the commutator. This applies to the exciter as well as to the main generator. Check also to make sure that the brushes have not worn too short and, if so, new brushes of the same grade should be used for replacement. The pigtail to the brush should be checked and tightened to the connection. The brushes should move freely in the brushholders but should be held against the commutator by the springs. Broken or weak springs should be replaced. The commutator should also be checked for dirt and discoloration. If it is burned or excessively rough, it should be removed and turned on a lathe. For minor discoloration, it can be cleaned with fine sandpaper. Emory paper or cloth should not be used since the dust is conductive and will tend to short out adjacent bars. New brushes should be sanded to make sure they are in full contact with the commutator.

Bearings should be checked monthly: sealed bearings do not normally need grease. However, when grease is required the bearings should be thoroughly cleaned of all old grease and new grease should be installed. Bearings should not be filled full with grease, but should leave sufficient space for the grease to move. All electrical connections should be checked for tightness and all contacts, starters, switches, rheostats, etc., should be inspected and cleaned as required. If the machine is exposed to corrosive or salt atmospheres, special attention should be given to exposed metallic parts. If corrosion is found, the part should be cleaned and repainted and insulation should be replaced.

Static Machines: The nonrotating machines usually include a fan. The fan motor should be checked as well as the bearings to make sure that the fan operates freely. All the foregoing instructions, with respect to contact points, moving switches, or rheostats should be followed. Check for corrosion and repair as required. On machines with mechanical current adjustments and moving cores and coils, the lead screws and guides should be greased for ease of adjustment. Where conductors flex within the machine they should be checked for insulation damage and replaced as required.

On the AC-DC and gas tungsten arc welding power sources, special attention should be given to the spark gap oscillator. The spark gap should be adjusted with a spacing of from 0.006–0.008 in. (0.15–0.2 mm). The contact faces should be cleaned and if they are pitted, they should be redressed. The gas and water valves should also be checked to make sure that they are operating and that there are no leaks in either the gas or cooling water system. Timers and programmers should be checked for calibration and adjusted as required. Solid state circuit boards should be clean and secured in place. Relays and other contact points in control circuits should be checked for proper closing and opening and points dressed if required.

It is advisable to keep a detailed inspection record for each machine by serial number, by inspection and periodic maintenance date, and by extra activities or maintenance that was required. This will help determine if specific machines require extraordinary attention.

Troubleshooting

Troubleshooting is required on a welding machine when it is not operating satisfactorily. It is assumed that the machine had been properly installed and was operating satisfactorily before the problem occurred. Troubleshooting is a matter of solving the problem of the machine and should be done only by qualified personnel. There are two types of troubleshooting. One is a method of diagnosing the problem while the machine is in operation or is energized. This type of work should be done with the case of the machine intact. The other type of troubleshooting is done with the machine completely de-energized and working on the inside of the equipment. Electrical meters can be used in either case.

The following checklist will cover many of the problems that may be encountered with most arc welding machines.

General Troubleshooting Chart for Most Arc Welding Machines

1. Machine will not start (equipped with or without line contactor).

Cause	Remedy
Power lines dead	Check voltage
Broken power lead	Repair lead
Incorrect line voltage	Check power supply
Incorrect connections to welding machine	Check connections against connection diagram
Blown fuse	Replace fuse

2. Machine will not start (equipped with line contactor only).

Overload relay tripped	Reset relay
Open circuit to starter button	Repair
Mechanical obstruction on starter	Remove obstruction
Broken leads at line contactor	Repair

3. Line contactor chatters during welding operation.

Input cable too small	Install larger cable
Low line voltage	Check incoming power

4. Contactor hums.

Dirt on contacting faces of line switch magnet	Clean faces of magnet
Improper alignment of stationary and movable magnet yokes on starter	Correct alignment

5. Contactor operates and blows fuse.

Wrong line voltage	Check line voltage and nameplate of welding machine

Cause	Remedy
Links on line contactor not connected correctly	Check and correct
Fuse too small	Install proper size fuse
Rectifiers burned out (rectifier and AC-DC machines only)	Replace rectifier
Short circuit in primary connections	Remove short circuit
Transformer failure	Repair transformer

6. Welding machine delivers welding current but soon shuts down.

Cause	Remedy
Wrong overload relay elements	Check with renewal part recommendation
Welding machine overloaded	Overload can be carried only for a short time. Reduce welding current
Duty cycle too high	Do not operate continually at overload currents or reduce welding current
Power leads too long or too small	Replace with larger cable
Ambient temperature too high	Operate at reduced loads where temperature exceeds 100 °F (37.8 °C)
Ventilation blocked	Check air inlet and exhaust openings
Fan not operating (fan-cooled machine only)	Disconnect leads and apply motor voltage to check nameplate voltage
Dirty rectifiers (rectifier and AC-DC machines only)	Clean with low pressure air blast—do not use wire brush or abrasives

7. Contactor operates but welding machine will not produce welding current.

Cause	Remedy
Welding terminal shorted	Electrode holder or cable may be shorted
Range switch or polarity switch not centered on arrow or detent.	Set to arrow or detent
Transformer lead open	Have transformer repaired
Transformer secondary failed	Have transformer repaired
Rectifier burned out (rectifier and AC-DC machines only)	Replace rectifier

8. Welding arc is loud and spatters excessively.

Cause	Remedy
Current setting too high	Check setting and output with ammeter, or reduce current
Polarity incorrect	Check polarity, try reversing polarity
Incorrect electrode used on AC	Use AC or AC-DC electrode for AC welding
Filter coil short circuited (AC-DC machines)	Replace filter coil

9. Welding arc sluggish.

Cause	Remedy
Current too low	Check output, and current recommended for electrode being used
Poor connections	Check all electrode holders, leads, and work lead connections
Cables too long or too small	Check cable voltage drop. And change cable

Cause	Remedy
Low line voltage	Check incoming power. Notify power company, if necessary
Power circuit single phased (on 3-phase rectifier machines)	Check for one dead line or fuse

10. Control knob does not control welding current.

Cause	Remedy
Rheostat burned out	Replace
Control rectifier burned out	Replace
Loose connection in control circuit	Check connections at control rectifier and control coil
Control coil failed	Have transformer repaired

11. Range switch or selector switch does not control welding current.

Cause	Remedy
Dial or moving contacts slipping on shaft	Replace dial or moving contact. Tighten on shaft. Clean switch and grease

12. Polarity switch does not control polarity.

Cause	Remedy
Dial or moving contacts slipping on shaft	Replace dial or moving contacts. Tighten on shaft. Clean switch and grease

13. Welding machine operates, but welding current falls off.

Cause	Remedy
Electrode or work lead connections loose	Clean and tighten all connections in welding circuit

14. Receive shock when machine case is touched.

Cause	Remedy
Case of machine not grounded	Ground case to earth

15. Shock when work lead, work, or worktable is touched.

Cause	Remedy
Work table and work not grounded	Ground work and work table to plant ground

The above 15 factors cover the more general problems that can be corrected by routine troubleshooting while the equipment is in service.

It is sometimes difficult to distinguish between a *welding problem* and a *welding equipment problem.* This is particularly apparent when the welder is utilizing unfamiliar equipment or when first using a different welding process. Each of the welding processes have their own specific problem areas and many times the welder may tend to blame the problem on the equipment when it might be a problem with the process. An example of this might be a poor weld due to the lack of shielding gas when using the gas tungsten arc process. The lack of shielding gas can result from a malfunction of the machine such as a solenoid valve, or, on the other hand, it could result from an empty gas cylinder, a stray breeze, a clogged hose, or a loose connection in the torch. In view of this, it is wise to review each of the welding process chapters when troubleshooting equipment used for that welding process. Difficulties can also be encountered when the equipment is not properly adjusted. A very rough arc, for example, when welding with gas metal arc welding using a small electrode can be the result of an out-of-balance adjustment between the welding current and voltage. The equipment might be blamed but it is really a problem of maladjustment. Skillful diagnosis is required to determine the true cause of a welding problem.

The following checklists are provided for welding equipment with specific emphasis on particular welding processes. This information plus a working knowledge of the processes will assist in troubleshooting a problem and making a quick and accurate diagnosis.

1. When using the GTAW process on AC the weld metal *curdles,* and the arc is violent.

Cause	Remedy
Arc is played on weld puddle	Point torch in direction in which you are welding, not directly into the weld puddle
Current too high	Reduce current
Arc length too long	Hold closer arc
Contaminated tungsten electrode	Redress the tungsten electrode
Gas nozzle on torch too small	Increase size of gas nozzle
Too little or too much gas flow	Adjust gas flow according to procedure
High frequency too weak	Check high frequency as outlined below
High frequency rheostat not set properly	Adjust for best welding conditions
Wrong spark gap setting	Adjust spark gap as recommended
Impure inert gas	Change gas cylinders

2. Torch spits tungsten into the weld.

Arc length too long	Hold closer arc
Tungsten too small	Use larger size tungsten electrode
Current too high	Decrease current
Pure tungsten electrodes used at very high currents on AC	Use thoriated tungsten

3. The weld is dirty.

Dirty base metal	Clean the base metal
Dirty filler rod	Keep filler rod clean
High frequency set improperly	Adjust high-frequency rheostat for optimum operating conditions
High frequency too weak	Increase setting of high-frequency rheostat

Cause	Remedy
Too little or too much gas flow	Adjust gas flow according to procedure

4. On DC EN the high frequency jumps the arc gap, but DC power does not follow to initiate the arc.

Use of pure tungsten electrode	Use thoriated tungsten for DC electrode negative
Use of helium gas	Use argon gas for better arc initiation
Tungsten electrode too large	Use smaller tungsten or grind to a point

5. Arc wanders.

Tungsten too large	Use smaller size tungsten
Tungsten contaminated	Redress tungsten electrode
Arc blow	Change position of work lead clamp

6. Tungsten turns purple after weld.

Insufficient gas postflow	Increase setting of postflow timer
Postflow timer sticks	Replace
Gas valve sticks	Replace

7. Water to torch flows too slowly or not at all.

Insufficient water pressure	Increase pressure
Water strainer in circulation system is clogged	Remove and clean or replace
Water valve sticks	Replace

8. Water or gas does not shut off.

Postflow timer set too high	Decrease postflow timer setting
Postflow timer contacts stick	Replace timer
Valves stuck open	Replace valves

If the high-frequency current does not start:

1. Make certain proper line voltage is at the machine and that no fuses are blown or circuit breakers tripped.

2. Make certain power switch is on and fan is running.

3. Check reset button on machine overload trip.

4. Rectifier thermostat may have tripped. Wait 5 minutes with fan running and reset.

5. High-frequency switch may be in the wrong position.

6. Check the remote-local switch to make sure it is in the proper position.

7. Make certain the torch is connected to the "GTAW or TIG torch" terminal and that the work lead is securely connected to the *work* terminal and to the workpiece.

8. Check the spark gaps. They should be set between 0.006 and 0.008 in. (0.15 and 0.2 mm).

9. Check for broken high voltage leads in the spark gap oscillator circuit. Also check components of the circuit.

10. Check for 230 volts to the spark gap oscillator. Use CAUTION—line voltage is present at various terminals.

If the high-frequency circuit is weak:

1. Check tightness of all leads in the external welding circuit.

2. Increase high-frequency rheostat to maximum.

3. Maximum recommended welding cable length is exceeded. Have the machine as close to work as possible.

4. Welding cables should lie in a straight line from the machine to the work for maximum high frequency. Avoid having work and electrode cables touch each other and avoid having them in contact with metallic objects or laying on metal.

5. Check spark gaps—adjust if required.

6. Make certain that the shielding gas is flowing.

7. If above are of no help run checks listed under "If the high-frequency current does not start."

Additional items can be added to your own checklist depending on the complexity of the equipment being used.

The above checklist can be modified slightly and used for gas-metal arc welding, but all reference to high frequency should be eliminated. Refer to the chapters on gas metal arc and flux-cored arc welding for details.

Repairing

If the diagnosis indicates that the machine must be repaired it should be removed from service and taken to the maintenance repair shop. If in-house facilities are not available the equipment should be sent to an authorized repair station. These are usually local electric repair shops that have been approved by the manufacturer of the welding equipment. It is essential that genuine replacement parts are used for all repair work and that the repair mechanics have sufficient knowledge and skill to accomplish this work. After the machine has been repaired it should be tested and checked to make sure that it will fulfill its original function.

QUESTIONS

1. There are six ways of classifying a power source. Name each and explain.

2. What is meant by NEMA Class I, II, and III—how is duty cycle described?

3. What is meant by the static characteristics and dynamic characteristics of a power source?

4. How can you determine the 100% duty cycle rating of a 60% machine? Explain.

5. What is the difference between a static and a rotating exciter?

6. How is the polarity of a generator welding machine changed?

7. What are the principle methods of adjusting the output of a transformer welding machine?

8. How does the saturable reactor system work?

9. What determines the output open circuit voltage of the transformer power supply?

10. What are the disadvantages of a transformer welding machine?

11. What is the function of the rectifier?

12. Why can the rectifier power source be connected to three-phase power?

13. What is a combination AC-DC welding machine? What is its advantage?

14. In a multioperator system what is meant by *load diversity*?

15. Can welding be done simultaneously with DCEP and DCEN using a MO system?

16. How is electrical power wasted in a multioperator system?

17. Why does the MO system provide a smooth arc?

18. What data is necessary to specify a welding power source?

19. Who should do electrical troubleshooting on an electrically hot power source?

20. What is meant by preventative maintenance? How does it apply to welding machines?

REFERENCES

1. "EW-1 Electric Arc-Welding Apparatus," National Electrical Manufacturers Association *ANSI* C87.1, Washington D.C.

2. W. J. LESTER AND F. L. KAESER, "Todd Evaluates Welding Systems for Performance and Cost," *Marine Engineering/Log,* September 1961.

9

WELDING EQUIPMENT

9-1 ELECTRODE HOLDERS AND CABLES

9-2 ELECTRODE WIRE FEEDERS

9-3 ARC MOTION AND STANDARDIZED MACHINES

9-4 FIXTURES AND POSITIONERS

9-5 CUSTOM FIXTURES AND AUTOMATIC WELDING

9-6 AUTOMATED WELDING

9-7 AUXILIARY EQUIPMENT

9-1 ELECTRODE HOLDERS AND CABLES

In all of the arc welding processes some sort of a device is required to transmit the welding current from the welding lead or cable to the electrode. These come in many forms and are given different names. In shielded metal arc welding several types of electrode holders are used to grip the electrode. In carbon arc welding, two different types of electrode holders are used. For gas tungsten arc welding the device is called a welding torch. It holds the tungsten electrode and transmits the current to it. The torch for plasma arc welding is similar. In gas metal arc welding and flux-cored arc welding the device is called a welding gun at least for semiautomatic welding. In semiautomatic submerged arc welding, a welding gun is used which also feeds the welding flux. In stud welding the term welding gun is used and the gun performs the additional function of moving the stud away from and finally into the work to consummate the weld.

All of these devices are different and are designed for a specific welding process. There are no specifications in the United States; however, several countries do provide standards. In view of this, manufacturer's data must be consulted in order to specify the gun, torch, or holder required.

Electrode Holders for SMAW

The electrode holder for shielded metal arc welding is by far the most common of these devices. It transmits electrical current to the electrode and is held by the welder. Electrode holders come in several designs, which include the pincher type and the collet type. A picture in Chapter 4 shows these two types of holders. Each style has its proponents and the selection is largely personal preference. The basis for selecting an electrode holder is shown by Figure 9-1 which shows the classification, the current rating and duty cycle, the maximum electrode size accommodated, and the cable size that may be connected to the holder. A nominal weight range is also provided. In general, the holders range from 8 to 14 in. long based on the rating of the holder. Weights of holders are given without the cable. All electrode holders should be the fully insulated type.

The maintenance of electrode holders is extremely important for efficient welding. Manufacturers supply spare parts so that the holders can be rebuilt and maintained for peak performance.

It is desirable to select the lightest weight holder with the required electrode size and current-carrying capacity.

Electrode Holder Classification	Rating		Electrode Size—Max in.	Cable Size Maximum A.W.G.	Nominal Weight oz
	Max Current amp	Duty Cycle %			
Small	100	50	1/8	1	10 to 12
	200	50	5/32	1/0	10 to 14
Medium	300	60	7/32	2/0	12 to 20
Large	400	60	1/4	3/0	16 to 26
Extra	500	75	5/16	4/0	22 to 30
Large	600	75	3/8	4/0	28 to 36

FIGURE 9-1 *Electrode holders—size and capacity.*

Carbon Arc Welding Electrode Holders

Electrode holders for carbon arc welding are different than those for shielded metal arc welding. There are two types; one is for a single-carbon electrode and the other is for two-carbon electrodes. The single-carbon electrode holder comes in four sizes based on the welding current capacity which relates to the carbon electrode size. These electrode holders are not insulated, and the carbon is usually gripped by collet action or by a set screw. The twin carbon arc holder comes in only one size but will usually accommodate several different sizes of carbon electrodes. Twin carbon arc welding is restricted to lower current levels. Both types of electrode holders were shown in Section 4-5.

Gas Tungsten and Plasma Arc Welding

The torches used for gas tungsten arc welding were described in detail in Section 4-3. Pictures of various sizes of torches, a cross-sectional view, and a table providing capacity information were given.

The torch for plasma arc welding was described and shown in Section 4-4. These torches are similar to the gas tungsten arc welding torches but are more complex and therefore somewhat more delicate. Spare parts are available for rebuilding both plasma and gas tungsten arc welding torches.

Stud Welding

The stud welding guns were described and shown in Section 4-6. The guns for stud arc welding come in several different sizes to accommodate different sizes of studs. These guns are rather complex since movement is involved in making the stud weld. The stud welding gun and the control equipment must be matched and from the same manufacturer. The variations of the stud welding process each require guns for the particular process variation used. Stud guns are available with automatic stud feed equipment. These are used in the automotive industry. The feeding mechanism must be designed for the type of stud used.

Guns for GMAW and FCAW

The guns required for the gas metal arc welding, flux-cored arc welding with self-shielding electrode wire and for semiautomatic submerged arc welding are all different. However, the same basic gun used for gas metal arc can be used for flux-cored arc welding using shielding gas. These guns are broadly classified as pistol grip type and gooseneck type. Figure 9-2 shows a typical gooseneck type gun and a cross-sectional view of a water-cooled gun. Both ambient air and water are used for cooling. The water-cooled guns are used for high-current welding. The CO_2 shielding gas also acts as a coolant for welding guns. Semiautomatic guns include the cable assembly which attaches to the wire feeder. The cable includes the conduit for the electrode wire, the conductor for the welding current, a tube for shielding gas, tubes for cooling water and control cables to the gun switch. The size of the copper conductor is related to the rating of the welding gun. The welding guns have replaceable tips for guiding the electrode wire to the arc and replaceable nozzles for directing the shielding gas. The method of measuring the angle of the gooseneck, the weight or balance point, or size of these guns is not standardized. Manufacturers offer guns in different models according to electrode wire sizes accommodated as well as the welding current level. They provide a duty cycle rating for use with CO_2 and for inert gases. The rating for use with CO_2 is higher than for inert gases due to the cooling effect of the CO_2 shielding gas. The gooseneck type is the most popular; however, some welders prefer the pistol grip or straight line welding gun shown by Figure 9-3. A cross-sectional view shows the water-cooled version. Pistol grip guns are more often used with aluminum since the curve in a gooseneck gun tends to create resistance to the electrode wire which may cause jamming in the cable assembly.

Guns for Gasless FCAW

Special guns are available for flux-cored arc welding using self-shielding electrode wires. This type of gun

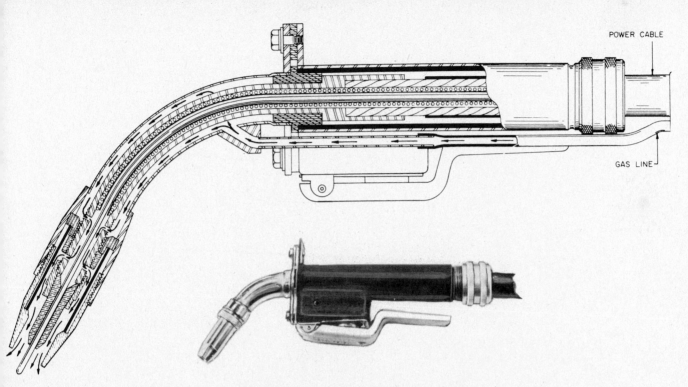

FIGURE 9-2 *Goose neck gun for GMAW and FCAW.*

FIGURE 9-3 *Pistol grip type gun.*

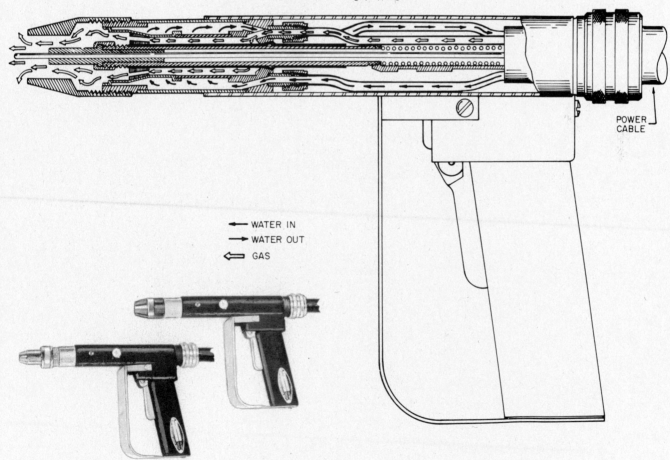

← WATER IN
→ WATER OUT
⇐ GAS

is shown by Figure 9-4. It is simpler since it does not require a gas nozzle or a tube for the shielding gas. The nozzle configuration usually includes an arrangement for *electrical stickout*. This is provided since the self-shielding flux-cored wires normally operate more efficiently with extended stickout. The detail of this nozzle is also shown by Figure 9-4. These guns come in different sizes based on the diameter of the electrode wire used.

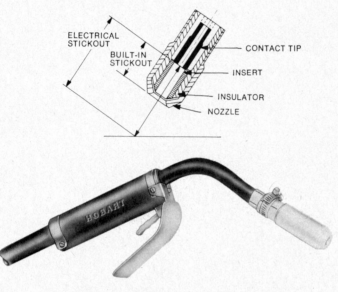

FIGURE 9-5 *Semi automatic gun for submerged arc welding.*

Submerged Arc Welding

Guns for semiautomatic submerged arc welding also feed the submerged arc flux. The more common method of supplying the flux is by means of a cone attached to the gun as shown by Figure 9-5. Guns which supply the flux by means of air pressure are also available.

The cable assemblies provided with the guns are available in different lengths. Consult the manufacturer's literature for details. Adaptors are usually available for attaching gun cable assemblies of one manufacturer to wire feeders of another.

Automatic Weld Torches

Torches used for automatic welding are usually straight line torches; however, there are exceptions. Figure 9-6 shows a variety of automatic torches designed for specific applications. The two left torches are de-

FIGURE 9-4 *Gasless gun and nozzle details for FCAW.*

FIGURE 9-6 *Torches for automatic welding.*

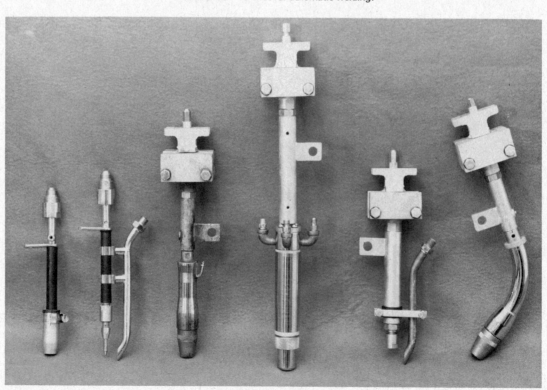

signed for smaller diameter electrode wires and for smaller fixtures where space is at a premium. The torch on the extreme left provides for concentric shielding gas delivery and the next one is for side delivery of CO_2. Side delivery torches should be used only when CO_2 shielding gas is used. The other four torches are for larger electrode wires. The third and fourth torches are for concentric gas delivery but the larger one is water cooled and can be used for inert gas. The fifth torch utilizes larger electrode wires and is designed for the side delivery of CO_2 shielding gas. This same torch can be used without the side delivery nozzle for submerged arc welding and for gasless flux-cored arc welding. The extreme right-hand torch is curved for a specific automatic application.

The torch must be selected to accommodate the welding process. The torch must also be selected with respect to the size of the electrode wire and the duty cycle of the operation. Tips of these torches can be changed to accommodate different wire sizes within limits. For extremely large electrodes, used in submerged arc welding, heavy-duty torches with spring-loaded current contact jaws are usually specified.

Welding Cable

Welding cables are the electrical conductors, normally called the electrode lead and the work lead. These leads carry the welding current from the power source to the arc at the point of welding, and back to the power source. The cables along with the electrode holder and work connection complete an electrical circuit.

The welding circuit can be a source of power waste and an economic loss, as well as a source of weld quality problems due to erratic operation. Power losses might result from the following:

1. Loose connections at the power source.
2. Loose connections at the work connector or electrode holder.
3. Poor quality repair splices in the cable.
4. Use of cable in which strands are broken and the cable is not properly repaired with a splice.
5. Use of cable too small for the amperage or duty cycle being used.
6. Enlarging the hole in a cable lug to fit a larger stud size.
7. Use of excessively long cables which cause abnormal voltage drop.

As the price of electrical energy increases it becomes extremely important to *inspect* and *maintain* welding conductors and the total circuit at peak operating efficiency! A hot point anywhere in the welding circuit is a source of high resistance, a point at which current is being wasted.

It has been found that if 10% of the strands of a cable are broken, the operating temperature of the cable can rise approximately 10°F (5.5°C). Cables will still carry welding current with up to 30% of the strands broken but the operating temperature can rise 30°F (16.6°C). An easy way to check for damaged cables and points where power is wasted is by checking for hot spots by feel.

The work connection, erroneously called a ground clamp, is an important part of the welding circuit. There are several different types of connections available ranging from spring clamps to actual welded-on connections. Figure 9-7 shows different types commercially available. Many styles come in different sizes and are rated according to current-carrying capacity. These connectors should be routinely checked to see that there is not an excessive voltage drop between the cable and the work.

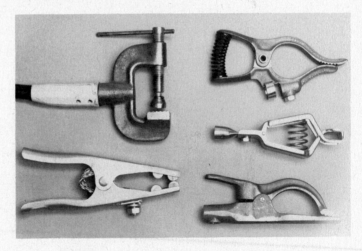

FIGURE 9-7 *A variety of work connection clamps.*

Connectors for splicing additional lengths of cables are commercially available. These allow for quickly increasing or decreasing the lengths of leads. The connectors must be properly attached to the leads and must be well maintained to avoid excessive voltage drops at the connection. Connectors must be fully insulated.

Welding Cable Size Designation

In North America, electrical conductors and cables are specified in size by the American Wire Gauge (AWG). The AWG numbers range from a small size such as

American Wire Gauge (AWG) Size	Nominal Overall Dia. in inches	Approx. Wt. per 1000 ft in Pounds	Resistance DC per 1000 ft @ 68°F in Ohms	Area in Circ. Mils	Metric Nominal Cross-Sectional Area mm^2	Overall Diameter mm	Approx. Wt. per 1000 meters in kg	Resistance Ohms per 1000 m
8	0.340	121	0.688	16510				
				19740	10	10.5	130	1.75
6	0.390	137	0.435	26240				
				31580	16	11.5	235	1.09
4	0.440	194	0.272	41740				
				49350	25	13.0	330	0.70
2	0.550	306	0.173	66360				
				69100	35	14.5	440	0.50
1	0.600	376	0.137	83690				
				98700	50	17.0	610	0.35
1/0	0.660	464	0.109	105600				
2/0	0.715	563	0.087	133100				
				138200	70	19.5	840	0.25
3/0	0.785	708	0.068	167800				
				187500	95	22.0	1120	0.18
4/0	0.875	884	0.054	211600				
				237000	120	24.0	1410	0.146
250 MCM	0.980	1070	0.045	250000				
				296000	150	26.5	1690	0.117
300 MCM	1.060	1260	0.038	300000				
				365000	185	29.0	2100	0.094

FIGURE 9-8 *American and metric cable size comparison.*

number 54 gauge, (ultrafine magnet wire) to cables so large that they are designated by MCM (thousands of circular mils). Wire such as number 12 gauge or number 14 gauge may be encountered in the wiring of a house. Welding cable sizes range from number 6 gauge through number 4/0 (pronounced four ought). The cable sizes for welding are shown by Figure 9-8. The two largest sizes are 250 MCM and 300 MCM. The term mil refers to 0.001 inch. One circular mil equals the area of a circle whose diameter is 0.001 inch.

Outside of North America welding cable is specified in metric size. The relationship between metric and the American Wire Gauge (AWG) is shown by Figure 9-8. Note that these are not soft conversions or exact comparisons since the nominal circular mils areas are not equal. The chart shows the metric size, which is its cross-sectional area in square millimeters. The overall diameter includes the jacket. The dimensions and characteristics given may vary between suppliers because of tolerance and differences in standards.

Welding cables are made of many strands of fine drawn, annealed copper. The copper in this form provides maximum flexibility of the welding cable. A separator, such as paper or mylar-foil is placed between the copper strands and the insulation or jacket. This separator is an aid to jacket removal at terminations. Jacket compounds are designed to be flexible and to

protect the copper conductor from the shop environment. These jackets are made of variations of synthetic rubber which do not melt when they come into momentary contact with sparks or hot metal.

Welding cables are normally made of copper; however, aluminum welding cables are available. Aluminum cables have the advantages of being lighter in weight and are less likely to be pilfered. Disadvantages are that normally two AWG sizes larger must be used than would be used in copper to compensate for the lower conductivity of aluminum. Termination is more difficult and critical, and the flex life of aluminum cables is considerably less than that of copper. Aluminum cables should be used for low-duty cycle welding applications or where the cable is not normally flexed during the welding operation.

The arrangement of the strands within the cable has an influence on flexibility. Rope terminology is used to define these arrangements. A *rope lay* has all the strands, groups of strands, and group layers cabled in the same direction. There are seven groups of fine strands, six of which are cabled around a center group. This will produce a conductor having the correct combination of service life and flexibility at a reasonable cost. For extreme limpness a *hawser lay* is used. In this configuration each layer or group of strands is cabled in the direction opposite to the covering layer. Hawser lay cable provides greater flexibility and can be used for a short portion of the lead at the electrode holder. It is more expensive than the rope lay cable.

What Cable Size to Use

To determine what size of welding cable to use, refer to the chart in Figure 9-9. Three items must be considered:

1. The welding current.
2. The duty cycle or operator factor.
3. The total length of the welding circuit.

This means the total distance from the power source to the work and return. The 100-ft column means that the work is 50 ft from the power source. As the distance increases the cable size should increase. This is to compensate for line loss within the cable due to increased length.

The chart also shows the duty cycle or operator factor which will be involved. This table assumes two categories; (1) up to 60% duty cycle and, (2) from 60% to 100%. Semiautomatic welding is in the top portion of the lower-duty cycle range, whereas automatic welding is in the higher-duty cycle range. The voltage drop in the welding circuit should not exceed 4 volts.

There are three methods to determine the amount of power lost in the welding leads. In the first method, use an accurate voltmeter and measure the voltage at the welding machine terminals and the voltage between the

electrode holder and the work connection while welding. Also, measure the welding current. The difference between the voltage at the power source terminals and at the electrode holder and work connection is the voltage lost in the leads. When multiplied by the welding current, this gives the amount of power lost in the leads. This is in accordance to the following formula:

Power loss = V_1 (at terminals) − V_2 (at holder) × I

or

$$PL = V_1 - V_2 \times I$$

An example would be 35 volts measured at the terminals and 32 volts measured between the electrode holder and the work connector or 3 volts × welding current of 250 amperes = 750 watts lost.

A second way to determine power loss is to find the resistance of the welding cables and multiply this by the welding current squared. The resistance of the different sizes of cables is shown by Figure 9-8. Modify data by total cable length. The formula is

$$PL = I^2R$$

A third way is by the use of the table shown in Figure 9-10. This provides the voltage drop for each cable size, based on a 100-ft long circuit when welding

Weld Type	Welding Current	Length of Welding Cable Circuit in Feet—Cable Size A.W.G.					
		50 ft	100 ft	150 ft	200 ft	300 ft	400 ft
Manual or	75	6	6	4	3	2	1
semiauto-	100	4	4	3	2	1	1/0
matic	150	3	3	2	1	2/0	3/0
welding	200	2	2	1	1/0	3/0	4/0
(up to 60%	250	2	1 2	1/0	2/0	4/0	—
duty cycle)	300	1	1	2/0	3/0	—	—
	350	1/0	1/0	3/0	4/0	—	—
	400	1/0	2/0	3/0	—	—	—
	450	2/0	3/0	4/0	—	—	—
	500	3/0	3/0	4/0	—	—	—
Semi or	400	4/0	4/0	—	—	—	—
automatic	800	2-4/0	2-4/0	—	—	—	—
welding	1200	3-4/0	3-4/0	—	—	—	—
(60% to	1600	4-4/0	4-4/0	—	—	—	—
100% duty							
cycle)							

The length of the cable circuit is the length of the electrode lead plus the length of the work lead.

FIGURE 9-9 *Copper welding cable size guide.*

Welding Current Amps	Voltage Drop per 100 ft of Lead vs. Cable Size (AWG)					
	#2	#1	#1/0	#2/0	#3/0	#4/0
50	1.0	0.7	0.5	0.4	0.3	0.3
75	1.3	1.0	0.8	0.7	0.5	0.4
100	1.8	1.4	1.2	0.9	0.7	0.6
125	2.3	1.7	1.4	1.1	1.0	0.7
150	2.8	2.1	1.7	1.4	1.1	0.9
175	3.3	2.6	2.0	0.7	1.3	1.0
200	3.7	3.0	2.4	2.0	1.5	1.2
250	4.7	3.6	3.0	2.4	1.8	1.5
300	–	4.4	3.4	2.8	2.2	1.7
350	–	–	4.0	3.2	2.5	2.0
400	–	–	4.6	3.7	2.9	2.3
450	–	–	–	4.2	3.2	2.6
500	–	–	–	4.7	3.6	2.8
550	–	–	–	–	3.9	3.1
600	–	–	–	–	4.3	3.4
650	–	–	–	–	–	3.7
700	–	–	–	–	–	4.0

FIGURE 9-10 *Voltage drop for different cable size per 100 ft of welding current.*

at the current shown. In this case the power loss equals the welding current times the voltage drop or $PL = I \times VD$. This data would be factored according to the length of the cable circuit.

Termination Technique

The connection of the cable to the terminal lugs, the electrode holder, and to the work connector are potential sources of high resistance. Therefore, they should be made as efficient as possible. If any of these connections become hot, the joint should be reworked. Soldering the cable to the lugs, etc., is one way to achieve a highly efficient joint. The thermit welding process can also be used for joining cable to special lugs. Mechanical fastenings are also employed; however, these may become loose and must be retightened to maintain a low resistance joint.

Power Cable

The power cable is the conductor used to carry the electrical power from the disconnect or fuse box of the building to the welding power source. Three-conductor cable is usually used for this application; however, four-conductor cable is sometimes used when the welding machine is on a portable mounting. The fourth wire is used to ground the case of the machine to earth.

The basis for determining the size of power conductor cables is the input power required by the welding machine. A factor to consider is whether the machine operates on single-phase or three-phase power. The table shown by Figure 9-11 is the three-conductor power cable size for welding machines. It provides size requirements for motor-driven three-phase welding machines and single-phase transformer-rectifier power sources. The

INPUT AMPERE OF WELDING MACHINE AT RATED OUTPUT			Three Conductor Power Cable Wire Size A.W.G.
Motor Driven Three Phase	Rectifier or Transformer Single Phase	Rectifier or Transformer Three Phase	
up to 24A	up to 30A	up to 24A	10
24 to 32A	30 to 40A	24 to 32A	8
32 to 44A	40 to 55A	32 to 44A	6
44 to 64A	55 to 70A	44 to 64A	4
64 to 76A	70 to 95A	64 to 76A	2
76 to 88A	95 to 110A	76 to 88A	1
88 to 100A	110 to 125A	88 to 100A	1/0
100 to 130A	125 to 165A	100 to 130A	2/0
130 to 155A	165 to 195A	130 to 155A	4/0

FIGURE 9-11 *Copper power cable size guide.*

Size Awg	No. of Conductors	Stranding No. Wires Gauge	Insulation Thickness (in.)	Sheath Thickness	Approx. Outside Dia. in.	I.P.C.E.A.* Amp Rating	Approx. Net. Wt. (1000 ft)
10	3	105 × 30	3/64	Type S	0.700	25	305 lb
8	3	132 × 29	4/64	6/64	0.835	35	440 lb
8	4	132 × 29	4/64	6/64	0.915	35	538 lb
6	3	132 × 27	4/64	6/64	0.900	45	567 lb
6	4	132 × 27	4/64	6/64	1.010	45	705 lb
4	3	259 × 28	4/64	Type W	1.17	65	1050 lb
4	4	259 × 28	4/64	Type W	1.27	55	1295 lb
2	3	413 × 28	4/64	Type W	1.34	90	1275 lb

*Insulated power cable engineers association.

FIGURE 9-12 *Power cable size information.*

power cables are rated at a higher voltage than welding cables since input power to machines can be 480 volts or higher. The name plate of the welding machine will provide the amperage drawn at the rated load and input voltage of the machine. This information is also shown on the data sheets of the machine, available from the manufacturer.

The normal color coding for three-conductor power cables is black, white, and green and for four-conductors black, white, green, and red. The size of cables, their diameter, and weight are presented by Figure 9-12. These cables are flexible tinned copper conductors with paper separators jacketed with insulation suitable for this voltage requirement.

Safety Considerations in the Use of Welding Cable

Welding cables are designed to be used only in conjunction with the relatively low voltages typical of welding equipment.

Welding cable should not be used at power line voltage or for other power applications.

The Occupational Safety and Health Act contains specific requirements which apply to the use of arc welding equipment.

Some OSHA safety requirements:

1. Coiled welding cable must always be spread out before using to avoid overheating during use.

2. Cables must not be spliced within 10 feet of the holder.

3. Welding electrode cable must never be coiled or looped around the body of a welder.

4. Cables with damaged insulation must be repaired or replaced.

5. Welding cables must only be joined together by means of recommended connections.

9-2 ELECTRODE WIRE FEEDERS

Electrode wire feeders are used for all of the semi- and automatic continuous wire arc welding systems. The basic requirement of the wire feeder is to feed the electrode continuously into the arc and to maintain a stable arc at the desired welding current and voltage. The control circuit which regulates the speed of the wire feeder may also control arc initiation, shielding gas, cooling water and other auxiliary equipment.

There are two basic types of automatic electrode wire feeders: (1) the voltage-sensing type and (2) the constant speed type. Figure 9-13 shows the normal uses for the two types of wire feeders.

The *voltage-sensing* type wire feeder utilizes the voltage across the arc and regulates the wire feed speed to maintain a preset arc voltage. This type of unit is used with the Constant Current (CC) "drooping" type power source. The control system replaces the "feeding" action of the welder. It slows down or speeds up the feed rate of the electrode wire to maintain the specified arc voltage. This system uses the arc voltage to control the wire feed motor. As the arc lengthens, the arc voltage increases, which increases the speed of the wire feed motor. This causes the electrode to feed faster and thus shorten the arc. Another system uses a feedback circuit which takes the arc voltage and compares it to a standard, and the difference is used to vary the speed of the wire feed motor. The voltage-sensing system is self-regulating. Welding current is adjusted at the welding power source. A voltage-sensing type wire feeder may be used for either DC or AC welding. It is usually used for feeding large-diameter electrode wires and was originally developed for submerged arc welding. The wire feed system may also include a retract circuit for automatically initiating the arc.

The *constant speed* wire feeder utilizes the fixed burnoff rate versus welding current relationship of the

Power Source Type	ELECTRODE WIRE FEEDER TYPE	
	Voltage Sensing	Constant Speed
"CV" direct current	Difficult to adjust Seldom used Self regulating within limits	Best for gas metal arc welding Best for flux cored arc welding Best for submerged arc when using small diameter electrode wire—self regulating
"CC" direct current	Best for Submerged Arc when using large diameter electrode wire. Used for GMAW on aluminum. Self regulating	Difficult to control Not used Not self regulating
"CC" alternating current	Used for submerged arc (medium and large electrode diameters) Used for flux cored arc welding. Self regulating	Difficult to control Not used Not self regulating

FIGURE 9-13 *Wire feeder types recommended for power source types.*

electrode wire. It must be used with the constant voltage (CV) or "flat" characteristic type power source. The electrode wire feed rate is set by the speed control of the wire feed motor. The CV power source automatically furnishes the correct amount of current to burn it off at the same rate it is fed into the arc. Thus, the wire feed rate controls the welding current. The voltage at the arc is controlled by changing the output voltage of the power source. It is a self-regulating system and is popular for small-diameter electrode wires. It was originally developed for gas metal-arc welding to eliminate stubbing and burnback. The wire feed rates of constant speed wire feeders are adjustable over a wide range of speeds. The range must include the welding conditions involved in the desired welding procedure. A constant speed wire feeder may be used on a CC or "drooping" characteristic power source, but it will be difficult to properly adjust the various controls; furthermore, sudden changes in welding conditions may cause the system to go "out of control."

The wire feeder must have a motor of sufficient power to drive the electrode wire through the cable conduit to the gun. The length of conduit, type of liner or internal surface, type of gun, i.e., *straight thru* or *gooseneck,* and size and surface condition of the electrode must all be considered.

In addition to the different wire feeder control systems there are many different styles and sizes of wire feeders available. These range from the extremely small hand-held type to the large heavy-duty heads used for automatic welding. The type, size, and style depend on the welding application, the type of metal being welded, etc.

The small wire feeders are usually used for semi-automatic welding and the most popular are the *spool guns* shown in Figure 9-14. The spool guns are often used for semiautomatically welding aluminum. The wire feed motor is in the gun handle with a gearbox and drive rolls just behind the gun nozzle. High-speed, low-power permanent magnet motors are used since the wire is fed only a short distance. The control box is remote, usually at the welding power source. Spool guns can be used with either the constant speed or voltage-sensing controls with the appropriate power source. They are awkward to use for certain applications, and the motor life is relatively short. In addition, the filler metal is more expensive when purchased on the small spools.

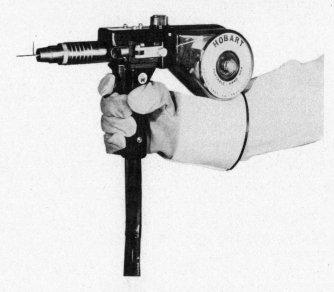

FIGURE 9-14 *Spool guns.*

There are several other wire feeders where the feed mechanism is held by the welder. One type is powered by compressed air and another is a linear type system, which will be described later.

The most popular type of wire feeder used for semiautomatic welding is shown by Figure 9-15. The feeders are designed so that the gun and cable assembly can be quickly disconnected. Most units carry the supply of the electrode wire, sometimes enclosed within the cabinet, sometimes not. The wire feeder shown uses four drive rolls, although many units use two. This type of wire feeder will have a feed motor ranging from 1/8 to 1/4 horsepower. This is sufficient power to feed a small diameter electrode wire through long conduits to curved head guns. In general, these wire feeders are of the constant speed type. The controls can be the auto transformer type or the solid state electronic type. Electronic types may include feedback circuits that maintain constant wire feed speed even though the resistance to feeding the wire may change. This feature is important when using long cable assemblies, which are moved during welding.

These types of wire feeders can be remote from the work or can be separated so that the wire feed motor is close to the work with the wire supply remote. In this manner, the best arrangements can be found to provide the welder with the most satisfactory operating system.

Automatic welding demands heavy-duty wire feed heads that can feed larger electrode wires than are used for semiautomatic welding. Three such wire feed heads are shown by Figure 9-16. Heads of this type are incorporated in automatic fixtures and placed wherever necessary to make the required weld. The largest feeder has a 1/2 horsepower motor for feeding the large electrodes.

FIGURE 9-15 *Popular wire feeders—(two views).*

FIGURE 9-16 *Heavy duty heads for automatic welding.*

All of the wire feeders mentioned above utilize drive rolls which pinch the electrode wire between them. As the drive rolls rotate they feed the wire into the conduit or into the torch and the arc. The surface configuration of the drive rolls must be selected with regard to the size and type of electrode wire. Drive rolls with various surface designs are available. They can be used in different combinations as shown by Figure 9-17. The flat rolls are used for small diameter nonferrous electrode wires. For small steel wires flat rolls with knurled surface to grip the wire are used. For larger diameter wires, knurled V's are used. In some cases a knurled V roller working with a flat roller is recommended. For feeding flux-cored or tubular wires two knurled Vee rolls are used. This reduces the pressure on the electrode and prevents flattening the wire, and allowing the flux to escape. Drive rolls are normally hardened steel although for soft nonferrous wires composition rolls may be used. Drive rolls tend to wear and must be replaced after extended use.

A new concept for feeding wire has recently been introduced. In this concept, called the linear feed system, the drive rolls no longer pinch the wire to feed it but instead revolve around the wire on an angle so that with each revolution it propels the wire the

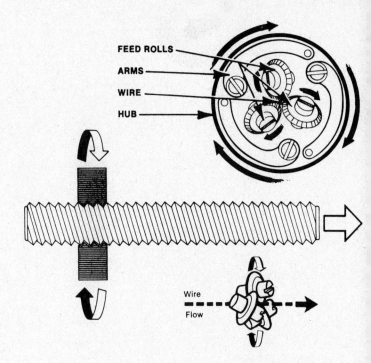

FEED ROLLS
ARMS
WIRE
HUB

Wire
Flow

FIGURE 9-18 *The principle of linear wire feeder.*

amount of the pitch or skew of the drive rolls. This system, shown by Figure 9-18, allows miniaturizing the wire feed motor since the bulky gear train is eliminated. In addition, this system has self-regulating features so that more than one linear feeder can be placed in an extended cable to propel the electrode wire over long distances. The linear feeder can be contained within the welding gun handle or within an enclosure integral with the cable. Figure 9-19 shows the linear feeder in use. Several repeater linear motors can be included in the conduit to give the welder the flexibility formerly achieved only with shielded metal arc welding. The linear feeder can also be used in fully automatic applications. Changing the skew angle of the drive rolls is similar to changing the gearbox ratio of a conventional wire feeder.[1]

Wire feeders are used to feed "cold" filler wire to gas tungsten arcs, plasma arcs, and for submerged arc welding. The *cold* wire does not carry welding current but is fed into and melts in the arc. Cold wire feed rates are much lower than used in the consumable wire systems. The same type of wire feeders can be used; however, the gearbox ratio must be selected for slow feed rates. The constant speed types are used.

Wire feeders have a range of wire feed speeds that can be varied by a control knob. A typical minimum feed rate is 50 in. (127 cm) per minute and the maximum is 400 in. (1,016 cm) per minute. This speed range would be ample for the gas metal arc welding process with the smaller diameter electrode wire. To

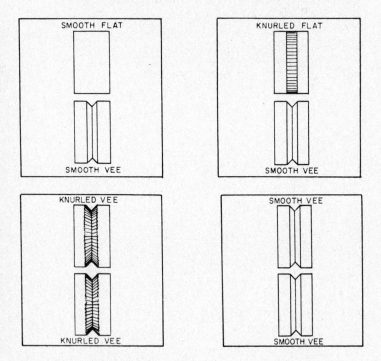

FIGURE 9-17 *Drive roll combinations to feed different size and type electrode.*

FIGURE 9-19 *Welding with the linear system.*

obtain higher or lower feed rates the gear box ratio must be changed. It is important to consider the welding process, the welding current level to be used, the electrode wire size, and type in selecting the wire feed system. The size and type of wire also relate to the type of drive rolls that must be employed and the horsepower of the wire feed motor.

Manufacturers' data sheets provide the maximum and minimum wire feed rates available for their wire feeders with different gearbox ratios. The data sheets also indicate the maximum and minimum size of electrode wire and the different types that can be used with a particular wire feeder and gear ratio. The data sheets may also provide the speed regulation and the length of conduit that can be accommodated by a particular wire feeder. Select the wire feeder that has a speed range that includes the range of wire feed speeds for the welding conditions that are to be used. The speed regulation of the feeder indicates how much the wire feed motor will slow down when extra resistance is placed on the feeding of the electrode wire. Starting and stopping times are of interest as well as the control voltage employed. For semiautomatic welding, low-voltage systems are used so that the trigger switch in the hand-held gun carries a low voltage.

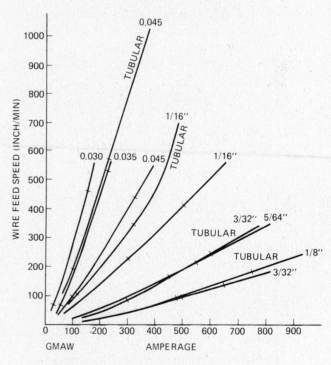

FIGURE 9-20 *Steel electrode wire—GMAW and FCAW.*

311

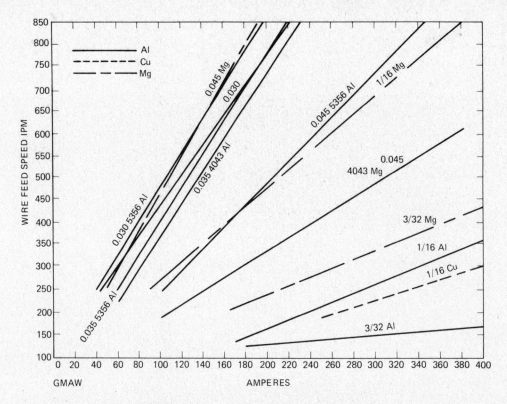

FIGURE 9-21 *Nonferrous electrode wires—GMAW.*

As an aid in selecting a wire feeder and gear ratio consult curves showing the wire feed speeds of different types used with the different welding processes. The following figures show the wire feed speed versus welding current. Figure 9-20 is for the gas metal arc and flux-cored arc welding process using solid and tubular small steel wires. Figure 9-21 is also for GMAW but with nonferrous electrode wires. Figure 9-22 is for steel electrode wires, used with submerged arc welding.

To utilize these curves select the approximate welding conditions, including the electrode type and size, and the welding current range. This will provide a range of wire feed speed requirements. Make sure that this range is within the speed range capabilities of the wire feed system. Hopefully, the midpoint of the welding parameters will be in the midpoint of the wire feed system capability.

There are other features that should be considered when selecting a wire feeder. For semiautomatic welding, quick connection and disconnection of the gun cable conduit to the wire feeder is required. In addition, it is sometimes desirable to have gas purge buttons, jogging buttons, and reversing switches. The wire feeders should include gas valves and for heavy-duty applications, provisions for water cooling. It is necessary to determine if various sizes of spools or coils of electrode wire can be used with the wire feeder. Determine whether the components can be separated for use on boom and mast, or for use in automatic fixtures. Speed regulation should also be checked since it is important for all but the least critical welding requirements.

Wire Feeder Maintenance

Periodic inspection and preventive maintenance will reduce downtime and assure maximum service from the

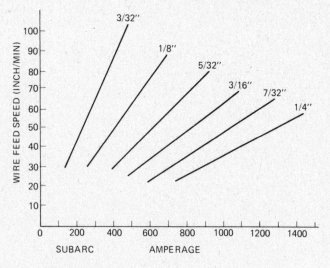

FIGURE 9-22 *Steel electrode wires with sub arc—FCAW.*

wire feed system. The control circuitry should be cleaned by blowing out with dry air at 25–30 psi every three months. Relay contacts and other sliding connections should be checked. The electrical connections, particularly of plugs, that are connected daily should be periodically inspected. The grease in the gear case should be changed at least every 500 hours of operation. The housing should be flushed and new grease of the same type should be installed. The wire feed motors should be inspected and the brushes replaced according to the wear. The commutating surface of the armature should also be checked and if the surface appears rough or worn it should be polished.

The gun cable assemblies should be blown out once a day so there is no accumulation of dirt in the electrode wire conduit. The gun, especially the electrode tip and nozzle, should be checked daily and replaced as required. Drive rolls should be checked weekly and replaced as required. Most manufacturers provide troubleshooting checklists for investigating equipment stoppages. Maintenance work should be done only by qualified people.

9-3 ARC MOTION AND STANDARDIZED MACHINES

High production manufacturers of weldments would like to upgrade semiautomatic applications to fully automatic welding operations. Semiautomatic welding provides the flexibility of following variable and irregular joints not possible with fully automatic welding. However, machine welding which is an intermediate step between semiautomatic welding and fully automatic welding can be the answer. In machine welding the wire feeder maintains the arc and feeds the filler wire and a travel device provides the relative motion. The welding operator will still monitor and adjust the equipment in order to provide a high-quality weld. The individual is still involved, but machines do much of the work. This concept greatly improves productivity since the welder fatigue factor and arc stoppage factors are eliminated. The welding operator, now in a monitoring role, can control high currents and high-duty cycles by means of adjustments rather than doing most of the work. In this way, the operator factor and deposition rate are greatly increased. This increases productivity and reduces welding costs.

This improvement in productivity has been brought about by the use of arc travel or relative motion devices. These fit into four basic categories:

1. Manipulators (boom and mast assemblies).
2. Tractors for flat-position welding.

3. Carriages for all-position welding.
4. Side beam carriages.

Each of these devices is a way of moving an automatic welding head or wire feeder relative to the part being welded. The arc travel devices are widely used for submerged arc welding and they can be used for all of the continuous wire processes and for some of the other arc welding processes.

Manipulators

A welding manipulator consists of a vertical mast and a horizontal boom which will carry an automatic welding head. Figure 9-23 shows a typical welding manipulator being used to weld the circumferential joint of a large tank. Manipulators are usually specified by two dimensions—the maximum height under the arc from the floor and the maximum reach of the arc from the mast. The manipulator can be designated as a 6 × 6 which means that the height weldable is 6 ft high and it can be welded at a distance 6 ft from the face of the mast. Other manipulators range from 4 × 4 to 12 × 12 and larger. There are many variations of manipulators. In some cases, such as the one shown by Figure 9-23, the mast is mounted on a carriage which travels on rails along the shop floor. The length of travel can be unlimited; thus, the same welding manipulator can be used for different weldments by moving the manipulator from one work station to another. In this figure the welding operator actually sits on a seat attached to the boom. The wire supply, flux supply, and wire feeder are all attached. The controls are at the operator's station. This manipulator is a heavy-duty model in order to support this weight. The manipulators usually have power for moving the boom up and down the mast. In some cases, the boom will telescope within the vertical adjusting assembly, whereas in other cases the head may move by power in and out along the boom. In some units the mast will swivel as in the figure. In selecting and specifying welding manipulators it is important to determine the weight to be carried on the end of the boom and how much deflection can be allowed. In addition, the boom or welding head should move smoothly in and out at speed rates that are compatible with the welding process being used. The carriage must also move smoothly at the same speeds. Manipulators are one of the most versatile types of welding equipment available. They can be used for longitudinal

FIGURE 9-23 *Welding manipulator.*

FIGURE 9-24 *Side beam carriage.*

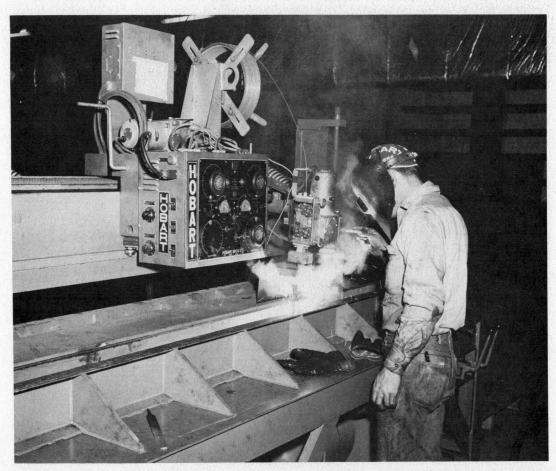

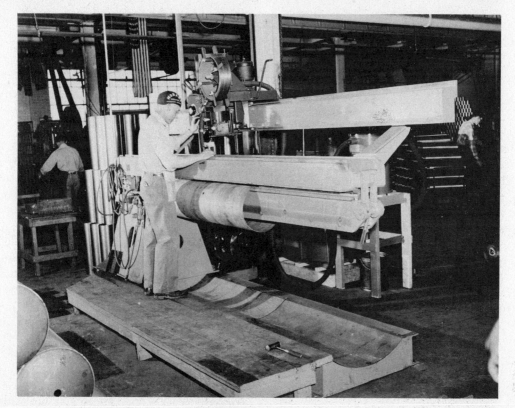

FIGURE 9-27 *Weld seamer.*

FIGURE 9-28 *Spud welder.*

Another type of standardized automatic equipment is used for building up track shoes of crawler type tractors. This is a routine maintenance job and in many mines, quarries and at construction sites track pads are regularly rewelded to build them up to size. This usually includes several heads on side beam carriage with limit switches which stop and start arcs as the carriage progresses from one end to the other.

There are undoubtedly other special but standardized automatic welding machines designed for a particular family of products. These types of machines greatly increase the productivity of welding and should be used whenever possible.

9-4 FIXTURES AND POSITIONERS

Welding Fixtures

A welding fixture is a device that holds the component parts of a weldment in the desired relationship for welding. Fixtures may be used only for tack welding in order to get more work through the fixture and not tie it up while the complete welds are made. Fixtures are used when there is sufficient volume of production to justify their cost or if there is need for extreme dimensional accuracy. Fixtures should be of sufficient mass to hold the parts of the weldment in proper location without distorting under the weight of the parts.

In addition, fixtures must be of sufficient strength to withstand distortion of welding. The fixture must also withstand normal wear and abuse. Fixtures usually include air-powered clamping devices or may be operated with power wrenches. The fixture must locate the parts in their proper relationship but allow for preparation tolerances of each piece. The heat from the arc must not affect the locating points. In addition, these points must be located where spatter from welding will not strike and adhere to them. The fixture-clamping devices must operate quickly and easily and also must be protected from heat and spatter. These location points and clamping devices must be positioned so that they do not introduce distortion or deflection into the parts of the weldment. The holding devices must also withstand loads and distortion of welding. The fixture should be designed so that the parts are easily loaded into it and allow access for welding. Finally, fixtures must be designed so that the finished weldment, with distortion, is easily removed.

The cost of the fixture must be quickly recovered by the time savings of using the fixture versus setting up the parts of the weldment from the blueprint. The design of fixtures is based on the foregoing fundamentals, but it must be tempered with experience gained from making and using fixtures that are successful and economical.

Positioners

A positioner is a device that holds and moves a weldment to the desired position for welding. The welds on a specific weldment are done in different positions assuming that the weldment is not moved. A positioner provides a change of position of the weldment for the most advantageous welding position. In other words, the positioner holds and positions the weldment for flat position welding.

Welding fixtures may be placed on positioners so that parts can be loaded into a fixture and then moved for flat-position welding. For the arc processes, with the exception of electroslag welding, the flat position is the most productive position. The parts can be loaded in the fixture, the weldment can be tack welded and completely welded in the fixture, and when it is completed, it will have been moved many times for flat-position welding. This provides the most economical production rates. There are other situations in which the fixture is used only for tack welding the parts

together. The tack welded assembly is then taken and placed on a positioner for flat-position welding. The difference depends on whether there is sufficient time available or sufficient number of positioners available so that the weldment can be completed in its entirety in the fixture on the positioner. When insufficient fixtures are available it is best to use them only for tack welding and use the positioner only for welding.

The major reason for using positioners is to reduce welding costs. A weld made in the flat position can be made faster because of the use of higher welding currents and easier operation for the welder. Many times the improvement can be as high as 100%. Another reason for positioning is to improve weld quality. If the welder is working in the flat position the weld quality is usually higher. Welders welding in an uncomfortable or cramped position will have a lower-operator factor, the quality may be lower, and the amount of weld deposited in a day may be considerably less. The appearance of the weld deposit is better when it is made in the flat position. Additionally, higher deposition rate processes can be used with higher currents. When using shielded metal arc welding, the high deposition type electrodes are used in the flat position. Studies have been made concerning the advantage of flat position welding. It has been proven that a heavy weldment can be made in approximately one-half the time if all of the welds are made in the flat position.

There are also secondary factors that provide an advantage for welding in the flat position. One is welding with fewer passes, which in turn reduces the amount of distortion encountered. The welding electrodes used in the flat position are usually easier to deslag than those used for vertical or overhead. The weld beads are smoother and this aids in removing the slag.

There are several negative sides to positioning and these should be considered. The foremost is the cost. Positioning equipment is relatively expensive and to be cost effective the savings of flat welding must pay for the cost of the positioning equipment quickly. Weldments must be firmly attached to positioners since overhanging and eccentric loads can be involved as well as welding stresses. The time required for positioner loading and unloading must be considered in the overall utilization of positioners. Another factor is the type of production that will go across a positioner. For example, if like weldments are produced simultaneously in a department there must be sufficient positioners so that all of the weldments welded on positioners can still be completed in the shortest possible cycle time. This means that if there is only one positioner and each weldment must spend a day on the positioner it would take a week to produce five weldments. If there were five positioners, all five weldments could be produced in one day. The total cycle time versus the lot size of production must be carefully considered. Finally, there is the problem of accessibility for welding. Most positioners hold the weldment above the floor so that ladders or scaffolding are involved to bring the welders to the work. This equipment must be considered in the overall capital cost. If automatic welding is utilized the welds must be accessible for the manipulator.

Welding Positioners

There are four types of welding positioners:

Universal or tilt table positioners.

Turning rolls.

Head and tail stock positioners.

Balanced positioners.

The universal tilt table positioners are the most popular. They come in many sizes and types, and an example is shown by Figure 9-29. Powered positioners are available for handling capacities from 100 lb (45.36 kg) to 150 tons (135 mg) or more. The principle of operation is a table which can be tilted from horizontal to the vertical position and beyond the vertical position for certain models. The table rotates about its center by power and the tilt action is also powered. The size of the table which may be round or rectangular is related to the size capacity of the positioner. Tables range from 12 in. in diameter up through 10 ft (3.05 m). The larger positioners are mounted on the floor and they must be securely anchored to withstand the strains imposed by heavy loads. The rotational speed of the table and the tilting speed pertain to the size of the positioner. When selecting a positioner, it is important to consider two factors that relate to capacity. One is the tilt capacity which pertains to the distance from the face of the table to the center of gravity of the weldment placed on the table. The other is the eccentricity capacity.

The positioner tilt capacity is related to the tilt torque required. This is obtained by multiplying the weight of the load in pounds by the distance from the face of the table to the center of gravity of the load in inches. This equals the tilt torque rating in pound inches. Precaution must be used in applying this formula since in some cases the dimension from the center of tilt to the face of the table must be considered in the formula. The tilting mechanism rotates in a bearing and it is the centerline of this shaft that must be considered. To carry this to an extreme, if the center of gravity of the load is 8 ft from the centerline of tilt the weight that can be carried would only be 1/8 the rated capacity.

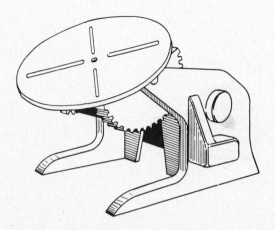

UNIVERSAL GEAR-DRIVEN POSITIONER

FIGURE 9-29 *Table type welding positioners.*

The other factor involved is the eccentricity of the load. This is the distance from the centerline of rotation of the table to the center of gravity of the load. It is usually not possible to precisely align the center of gravity with the center of the table, particularly if different weldments are welded on the same positioner. This involves the rotational torque required to turn the table. The rotational torque required is determined by the weight of the load in pounds by the maximum distance in inches from the centerline of the table to the center of gravity of the load.

It is essential that weldments be very firmly attached to the positioner table. Most positioner tables have slots for bolts and in other cases fixtures on the positioner are actually welded to the positioner and the weldments are loaded and unloaded from the fixture. Whenever fixtures are employed the weight of the fixture must be considered in establishing the capacity of the positioner.

Rotational speeds are extremely important for the positioner, since positioners are often used for rotating parts under fixed welding heads. It is essential for the diameter involved that the travel or circumferential speed be within the parameters of the welding procedure.

Most manufacturers of this type of equipment provide capacity data which includes the rotational torque capacity and tilt torque capacity, as well as tilt and rotational speeds. Positioners must be well manufactured with appropriate bearings so that travel speed is smooth and steady when used for making circumferential welds. Variations of the universal tilt table positioner utilize elevating devices.

Turning Rolls

Turning rolls are used for many positioning applications and are ideally suited for cylindrical parts such as tanks. Turning rolls come in different sizes and ratings based on the weight of the work that is to be rotated. These range in size of rolls that will turn loads of from 1 ton up through 250 tons. For many applications a powered set of rolls and an idler roll are used together. By means of a worm gear and shaft two of the rolls are powered and the other two rolls are idle. Figure 9-30

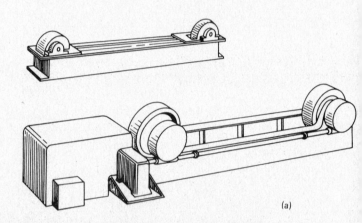

(a)

FIGURE 9-30 *Turning rolls.*

(b)

shows a normal arrangement for turning rolls. For extremely heavy loads or long cylindrical members additional sets of idler rolls are sometimes used. For flexibility of operations sometimes the sets of rolls are placed on tracks so they can be adjusted for long or short cylindrical weldments. The size spacing of the turning rolls are important since they relate to the size of the cylindrical members that can be rotated. In some cases, the spacing of the two rolls on the same frame is adjustable for smaller or larger cylindrical weldments. In other rolls double sets of rollers are used to help distribute the load over more area of the weldment. Rolls that are cylindrical rotating members are also available to provide greater load carrying capacity and for spreading the load on the weldment. The surface of the turning rolls can be metal or rubber. The rubber rollers do not have as high a load capacity as metal rollers. When high-temperature preheating is used the rubber rollers are not recommended. The size of the drive motor is related to the size of the rolls and the weight capacity of the rolls. It also has an effect on the rotational speed of the weldment. These speeds should be selected with care so that the rotational speed of the weldment will be compatible with the parameters of the travel speed of the welding procedure. The speed of rotation must be smooth and steady.

The rolls must be accurately aligned or else the cylindrical weldment will tend to move sideways during rotation. This is not acceptable when making circumferential welds.

Rolls are usually used with cylindrical parts; however, by use of round fixtures, noncylindrical parts can be rotated with conventional positioning rolls. These round fixtures are made in halves which are clamped around the part to be rotated. In this manner, long rectangular, square, or unusual shaped parts can be rotated for ease of welding. This technique has been used for rotating small ocean-going ships.

Head and Tail Stock Positioners

Head and tail stock positioners perform much the same function as positioning rolls. They have the appearance of the tilt table positioners. In fact, a tilt table positioner can be used for the head stock or powered member of a head and tail stock unit. The tilting device can be deactivated so that only the rotation of the table would be employed. Head stock positioners are available without the tilt feature. These are shown by Figure 9-31. The capacity of head and tail stock positioners is similar to that of tilt table positioners and ranges from 5,000 pounds up to 20 tons. Since there are two units the eccentricity or overhang mentioned with respect to tilt table positioners does not

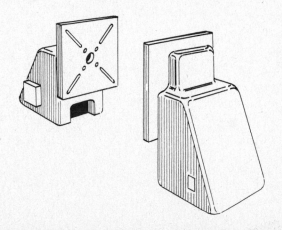

FIGURE 9-31 *Head and tail stock.*

apply. Head and tail stock positioners are used when long rectangular weldments are produced. Head and tail stock positioners are commonly used in railroad freight car building plants to rotate the freight car body during the welding operation.

Universal Balanced Positioners

These positioners are relatively small compared to most powered tilt table positioners. They are balanced and do not contain power; however, power rotation of the table is available. The universal balanced positioner is shown by Figure 9-32. The principle of this type of

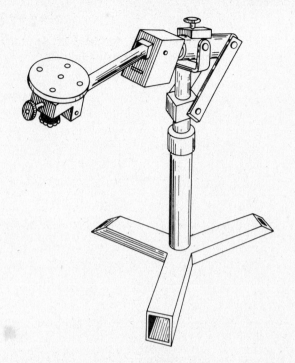

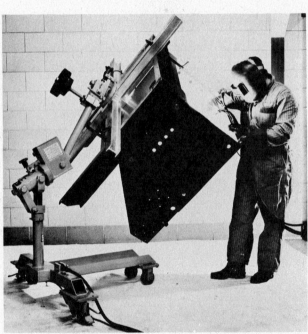

FIGURE 9-32 *Balanced positioner.*

lift but do not require powered tilt table positioners. When sufficient volume is to be produced fixtures can be attached to these positioners so that when the parts are loaded they are accurately located to provide for the balance. In this way the part can be manually moved to put it in the proper position for flat welding. These positioners come in sizes ranging from 100 to 2,000 pounds (45.5 to 890 kg) and have proven to be very efficient for smaller weldments.

9-5 CUSTOM FIXTURES AND AUTOMATIC WELDING

Welding has lagged behind other metalworking operations in the transition from a manual to a mechanized operation. This is because the shielded metal arc welding process does not lend itself to automation since the arc is extinguished each time an electrode is consumed. The first factor that now makes automatic welding possible was the introduction of the continuous electrode wire welding processes. Continuous wire or automatic welding is not new and was practiced in the 1920s using bare electrode wire.[2] Figure 9-33 is a reproduction of the very first picture on record of automatic welding. This picture appeared in the July 1920 issue of the *Welding Engineer* magazine and shows the production of differential housings for automobiles.

FIGURE 9-33 *First automatic welding machine.*

positioner is to accurately determine the center of gravity of the weldment and adjust the angles of the arms of the positioner so that the center of gravity is in line with the main axis of rotation of the arm and also in line with the axis of rotation of the table. Once the part is properly located and balanced on the positioner it can be moved very easily to put it in any position required for welding. This type of unit is extremely popular for weldments that are too heavy to handle or

The second factor that has made automatic welding possible and practical has been the introduction of custom weld fixtures. These are fixtures which are designed to produce one specific part and that part only. Some variations of dedicated fixtures may allow for a family of parts that are similar but vary in size or with some other minor change. Some dedicated fixtures can be adjustable to allow for these differences. The customized welding fixture is applicable and economical only in the high-volume production industries where many hundreds of parts of identical design are produced on a continuous basis.

Automatic welding was given an additional boost with the introduction of gas metal arc welding and flux-cored arc welding. Prior to this time the submerged arc welding process was used for automatic welding, but the problem of flux handling and position restrictions reduced its value for automatic welding.

True automatic welding means welding with equipment which performs the welding operation without constant observation and adjustment of the controls by the welding operator. In manual or semiautomatic welding the weld is under the constant control of the welder, who is continuously and unconsciously, making slight adjustments to insure the quality and continuity of the weld. These adjustments involve speeding up or slowing down the travel, occasionally weaving the arc, lengthening or shortening the arc to assist in controlling penetration, following the weld seam, and maintaining the arc and depositing weld metal. An automatic welding machine must perform all of these functions. One of the requirements for a successful operation is that the parts making up the weldment must come to the automatic machine in consistent and uniform manner with accurate dimensions. Part preparation for automatic welding must be of the highest quality if the welds are to be uniform. In addition, the welding machine and fixture must be maintained so that hold-down clamps, backing structures, locating devices, etc., are the same each and every time the weld is made. With strict quality control, accurate parts, and a well-maintained automatic machine there is still a possibility of nonuniform weldments. This is because of distortion and locked-up stresses in the parts prior to welding. The heat involved with welding may relieve these stresses and when they are different the parts will distort differently and may

create a nonuniform weldment. This type of problem can be held to a minimum or within acceptable tolerance limits; however, parts preparation for automatic welding must be more precise than for manual or semiautomatic welding.

Automatic welding offers the following advantages:

1. Strict cost control—predictable weld time.
2. High uniform quality—predictable and consistant.
3. Low welding cost.
4. High continuous output (no welder fatigue).
5. Minimum operator skill and reduced training time.
6. Can be combined with other automatic processing operations.
7. Increased productivity—higher welding speed and deposition rates.

Figure 9-34 shows the block diagram of an automatic welding system. Each component in the system must be selected with care and with respect to other components in the system. The first item, the power source, is normally the constant voltage type since they are easier to adjust for exact welding current and voltage values. The second item is the wire feeder which must be of the constant speed type matched to the constant voltage power source. The third item required for the system is the welding torch. The torch should be selected to match the welding process, shielding system, the electrode wire type, wire size, the welding current level, and the duty cycle requirement.

FIGURE 9-34 *Block diagram of automatic welding system.*

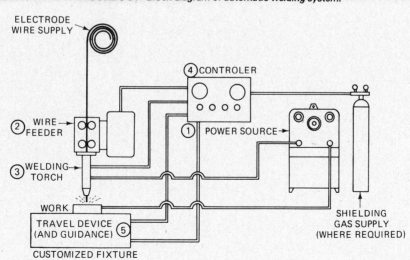

The fourth item is the controller. The controller is required to provide a welding program necessary for all except the most simple automatic welding operations. Controllers may be rather simple or quite complicated depending on the complexity of the welding program. The complexity of the weld program depends on the complexity of the weldment as well as the quantity of parts required, the quality required for the weld, and the cycle time involved.

A typical welding program or sequence of operations is shown by Figure 9-35. In this program the welding procedure operational factors are shown with respect to a time base. The time base can be replaced by motion, which is accomplished by means of limit switches and sensors connected to the control system. At the cycle start point the operation begins and specific activities such as the opening of valves for shielding gas and for cooling water flow are actuated. For many programs there is a preflow shielding gas time period after which the main contactor closes, the arc starts, and the wire feed begins. Travel or motion usually begins at this point or after a brief pause. When the weld is finished there is usually burn-back time and crater fill time

allowed to complete the weld. Next, the contactor will open, the arc stops, and the shielding gas postflow cycle starts. Afterwards, the valve closes and the welding cycle is completed. This cycle can repeat for arc spot welds or for skip welds or can repeat only when new pieces are placed in the machine.

To fully understand a welding program, it is necessary to understand some of the terms and definitions as follows:

Preflow Time: The time interval between start of shielding gas flow and arc starting, (prepurge).

Start Time: The time interval prior to weld time during which arc voltage and current reach a preset value greater or less than welding values.

Start Current: The current value during the start time interval.

Start Voltage: The arc voltage during the start time interval.

Soft Start: Without "hot start."

Hot Start Current: A very brief current pulse at arc initiation to stabilize the arc quickly.

Initial Current: The current after starting, but prior to upslope.

Weld Time: The time interval from the end of start time or end of upslope to beginning of crater fill time or beginning of downslope.

FIGURE 9-35 *Welding program.*

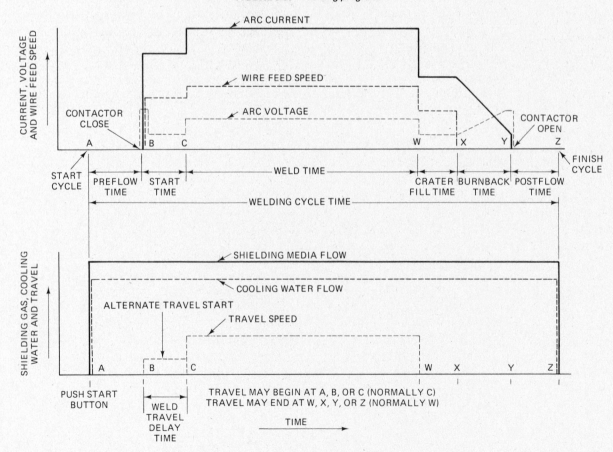

Travel Start Delay Time: The time interval from arc initiation to the start of work or torch travel.

Crater Fill Time: The time interval following weld time, but prior to burn-back time during which arc voltage or current reach a preset value greater or less than welding values. Weld travel may or may not stop at this point.

Crater Fill Current: The arc current value during crater fill time.

Crater Fill Voltage: The arc voltage value during crater fill time.

Burn-back Time: The time interval at the end of crater fill time to arc outage during which electrode feed is stopped. Arc voltage and arc length increase and current decreases to zero to prevent electrode from freezing in the weld deposit.

Downslope Time: The time during which the current is changed continuously from final taper current or welding current to final current.

Upslope Time: The time during which the current changes continuously from initial current value to the welding value.

Postflow Time: Time interval from current shut-off to shielding gas and/or cooling water shut-off, (post purge).

Weld Cycle Time: The total time required to complete the series of events involved in making a weld from beginning of preflow to end of postflow.

Control panels or command circuits are available that can be programmed to provide the weld condition sequences required. A typical system will program welding current or wire feed speed and arc voltage with respect to time as a base or motion with proper sensors. Figure 9-36 shows a control panel with its remote control pendant. This panel can be preprogrammed to control gas preflow, start current, start voltage, welding current, welding voltage, burn-back time, and postflow time. It has provisions for arc spot welding. The normal operator controls are on the outside of the cabinet and controls for job setup are located on the inside. This system utilizes a tachometer feedback of the wire feed motor and allows for precise repeatability of welding conditions. This insures constant weld quality based on consistent parts preparation.

Automatic welding is being used more and more in many industries including automotive, appliance, electrical equipment, just to mention a few. Automatic welding should be used whenever identical parts are manufactured in sufficient number to justify full-time production. The following is a brief description of several automatic welding applications.[3]

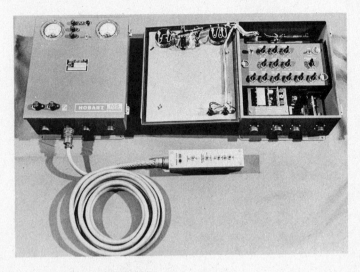

FIGURE 9-36 *Control panel.*

An example of a part made by automatic welding is a compressor housing used for refrigerators, air conditioners, frozen food showcases, etc. These housings are produced in high volume and therefore complex and expensive welding equipment can be utilized. Figure 9-37 shows the extremely complex fixture and welding equipment which makes a fillet lap weld joint around the periphery of the two parts of the housing. These parts are of relatively thin gauge steel and are not round. Since they are not round there is motion in both the head relative to the part as well as rotation of the part. The quality of the weld must be perfect since leaks of a compressor housing cannot be tolerated. This particular machine is adjustable to allow for a limited number of sizes of compressor housings to be made. All of the compressor housings made in this fixture are of the same general shape but the size and dimensions vary. An adjustment of the fixture is required each time the size of the compressor assembly is changed. The gas metal arc welding operation is monitored by the operator who loads and unloads the fixture. The quality of the weld is assured by leak testing later in the sequence of manufacturing operations. Notice the finished welded assemblies.

The next example is less complex but the production requirements are higher. This example is an automobile torque converter part shown by three photographs of Figure 9-38. The fixture is relatively simple because the weld joint is perfectly circular; however, the wire feed head is retractable to allow the load and unload operations. The impeller cover of 1/4-in. (6.4 mm) thick steel

FIGURE 9-37 *Welding compressor housing.*

FIGURE 9-38 *Welding torque converter.*

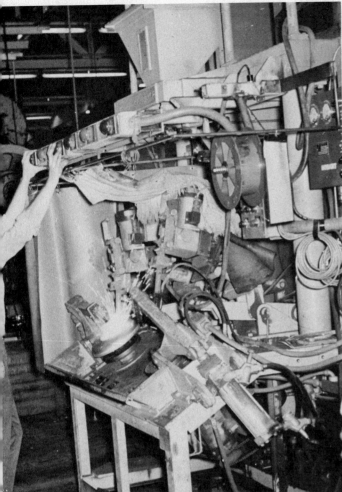

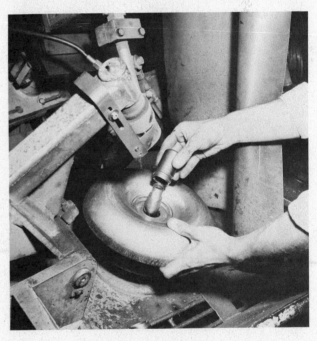

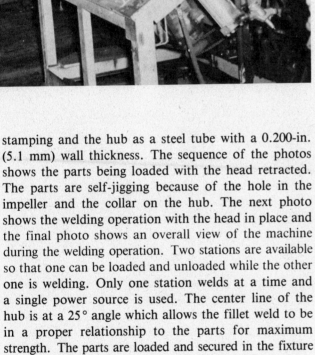

stamping and the hub as a steel tube with a 0.200-in. (5.1 mm) wall thickness. The sequence of the photos shows the parts being loaded with the head retracted. The parts are self-jigging because of the hole in the impeller and the collar on the hub. The next photo shows the welding operation with the head in place and the final photo shows an overall view of the machine during the welding operation. Two stations are available so that one can be loaded and unloaded while the other one is welding. Only one station welds at a time and a single power source is used. The center line of the hub is at a 25° angle which allows the fillet weld to be in a proper relationship to the parts for maximum strength. The parts are loaded and secured in the fixture and the operator then presses two buttons to start the weld sequence. In normal use a curtain comes between the operator and the arc so that a helmet is not required. The gas metal arc welding process with argon-CO_2 shielding gas is used. The resulting part is highly satis-

factory, spatter is minimal, and the quality of the weld is high. The completed torque converter housings are welded together at final assembly and given a hydrostatic pressure test to assure weld quality.

The next example is also from the automotive industry. In this case, a clutch pedal or a brake pedal for a recreational vehicle is being produced. It is a three-piece assembly consisting of a hub, a lever arm, and a pedal,

FIGURE 9-39 *Welding brake and clutch pedal.*

all of mild steel welded together in two operations. The welding of this pedal assembly is shown by the sequence of photos in Figure 9-39. The automatic welding machine consists of two stations which utilize the same welding operator to load and unload one station while the other station is welding. In both stations, however, two arcs are operating simultaneously. On the left side, the hub is being welded to the lever by rotating the assembly approximately 180° and making fillet welds on both sides. The right-hand station uses linear travel, which is accomplished by moving the heads to produce two fillet welds between the lever arm and the foot pedal. All four welds are made using the gas metal arc welding process with CO_2 gas shielding. The cycle time is so fast that the welding operator is continuously loading and unloading the two work stations.

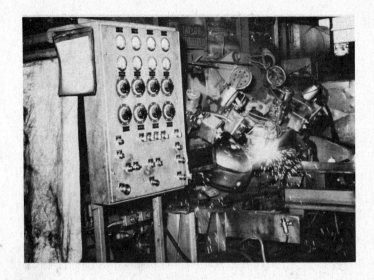

The final example is a considerably larger weldment that requires a much larger fixture. This part which is a heavy duty truck rear axle is shown being welded by the sequence of photos in Figure 9-40. The differential housing weld was made at a previous welding station. The welds being made here are attaching rings to each end of the axle. Four welds are being made simultaneously, two fillet welds joining each ring to the tubular axle. The axle assembly is rotated under the arc and four wire feeders with four power sources are required. The control circuit is integrated into one panel for ease of operator control. The flux-cored arc welding process is being used with CO_2 gas shielding. The quality requirement for these welds is extreme because of normal use and abuse of heavy-duty trucks.

9-6 AUTOMATED WELDING

Automated welding and automatic welding have much in common. There are many similarities but there is a major difference. Automatic welding involves elaborate dedicated fixturing with tooling, work holding devices, accurate part location and orientation. It also involves elaborate welding arc movement devices with predetermined sequences of welding parameter changes and the use of limit switches or timers to accomplish the weld joint. Automatic welding was developed and is being used at a high level of efficiency in the high volume industries where the cost of the equipment is justified by the large number of pieces to be made. Automatic welding reduces manpower requirements, consistently produces high-quality welds, maintains production schedules, and reduces the cost of parts welded. One major disadvantage is the high initial cost of the welding machine involved. Another disadvantage is the need to keep the automatic equipment occupied all of the time. A further disadvantage when used for lower volume of production is the need to provide numerous

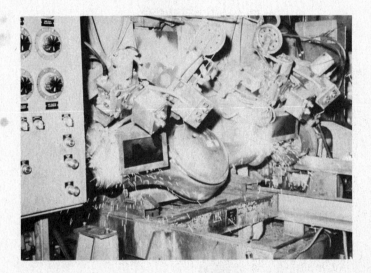

FIGURE 9-40 *Welding track rear axle.*

329

dedicated fixtures for various parts and the fact that none of them are used continuously.

Automated welding eliminates the elaborate, expensive fixture, and the automatic timer or limit switch input necessary to control the arc with respect to the workpiece. In automated welding a program replaces the complex fixturing and sequencing devices. Automated welding provides the same time saving and precision welding of automatic welding, yet can be applied to small-lot production—even to the welding of a single part. In addition, automated welding has quick change capability. It can accommodate changes in the product without the necessity of redesigning and reworking the expensive fixture. These are the basic economic advantages of automated welding.

Automated welding utilizes a program coupled to a welding arc travel device of extreme capabilities instead of elaborate fixtures. The arc motion device can be capable of movement in three directions: longitudinal (X), transverse (Y), and vertical (Z). Other motions, such as rotation and tilt, can also be incorporated. Additional welding positions can be obtained if the work is mounted on a positioner that will rotate and tilt at the command of the program; Figure 9-41 shows the possible motions. The program will thus provide arc travel in all of these directions within the range of the equipment. The program can also provide adjustments to the parameters of the welding procedure. The program is stored in a numerical control tape system or in a minicomputer.

This is automation with flexibility and it will reduce to a minimum the tooling and fixtures required. It will allow the welding of complex parts for small-lot production. It will provide the same welding accuracy and welding speed with repeatable quality and cost. The weldment is located and held at the correct location with respect to the torch. The program is initiated and the weld is made completely without human assistance. The human element is the loading and unloading of the weldment from its locating fixture. By changing the program tape and the locating fixture, the automated welding machine is ready to produce other weldments.

Industries in which limited production of similar parts occurs on an every-month schedule have a need for flexible automation. There is insufficient production to justify many dedicated fixtures since the like weldments required each month can be produced in a day or so. This small-lot production will tie up many pieces of equipment for individual jobs and the utilization of the

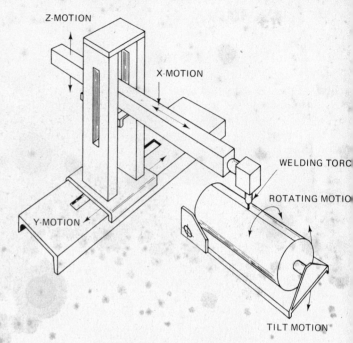

FIGURE 9-41 *Arc travel capabilities.*

equipment is extremely low. Either dedicated fixtures and equipment occupy floor space or the equipment is being continually moved. The solution for this type of production is to have a universal automatic machine which will weld many different weldments. The robot welder coupled to either a numerically controlled or a minicomputer command unit is the answer.

By use of a robot and a moving work-holding fixture a program can be developed for each part for welding that particular part each time it needs to be manufactured. The robot with a work motion device is located at a work station. Various locating fixtures are brought to it in preparation for welding the parts in question. The program is available for each part and is used for welding that part. The setup time is minimal, the machine is kept busy on a full-time basis doing a different type of job every day. This flexible automation is now available in the computer-controlled robot shown by Figure 9-42. It has the capability of changing from one part to another, very quickly. It needs only a positive locating point to align the robot welding torch with the part being welded.

The numerical control utilizes a standard punched tape which contains eight tracks and is one inch wide. A standardized method is used to provide information on the tape. The punched tape system provides for point-to-point movement which can be arranged so as to provide smooth travel for changes of direction. The method of producing and reading tapes is not of primary interest. Tape preparation and tape readers produce the program, which controls all motions and welding parameters.

FIGURE 9-42 *Robot welding machine.*

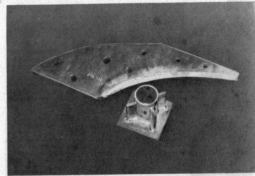

Newer systems utilize minicomputers. The teach mode is used to program the minicomputer by traversing the welding torch along the welding joint by human control of the motions built into the machine. When it is in the teach mode this motion will produce the signals that will reproduce the motion each time. The computer technique has greatly advanced the use of automated welding applications. After the weld is made a tape can be made for use the next time the part is welded. The tape is used only to teach the minicomputer for the next production run.

In addition, the computer-controller is used to change welding parameters during the welding operations. Voltage, current, and travel speed can be changed based on requirements of the weld. Normally, the gas metal arc welding process or the flux-cored arc welding process utilizing small-diameter electrode wires is used so that the automated system has all-position capabilities.

The normal program which provides automated travel and change of parameters is sufficiently capable to maintain high-quality welds on simple parts. However, when parts become larger or when warpage is involved or if the piece parts are not accurately prepared there is the possibility that the weld may not be made in precisely the correct location. To allow for this, sensors of different types are employed to compensate for irregularities and to make sure that the weld is made in the correct location even though the location may not be precisely where it is expected to be. Sensors provide information to the minicomputer by resolving the weld joint centerline and providing control signals to maintain the correct tip-to-work distance or proximity. Other sensors can determine the root opening variations and mismatch of the parts being welded. The control override function provided by the sensors will change torch movement or torch lateral position with respect to the weld centerline. Sensors can also be utilized to control the wire feed rate and current and to change welding travel speed or voltage if required. They can be used to revise or modify the welding parameters originally programmed on the tape.

The robot welder shown by Figure 9-42 has three axis capability. The torch can be tilted in three directions. When this is coupled to work-positioning devices, welds can be made in almost any position. In operation, it is necessary to accurately locate the weldment and to hold it in a precise position. The only limitation is the method of holding and locating the weldment. The machine is placed in the teach mode and the operator causes the welding head to accurately move along the weld joint from the starting point to the finish point. Short intermittent welds can be made in which the program starts and stops the welding operation as required. As the position of the weld changes the parameters of the procedure also change. Thus the welding head can move in a horizontal position and then change to a vertical position with the parameters changing as the direction changes. After the teach mode the machine is tested by running it through the required motions without welding, to determine if the torch is accurately tracing the weld joint. Corrections can be made as required. The next step is to make a weld and determine the joint alignment accuracy and the quality of the resulting weld. If the quality is as expected, the program is then firmed and used on the production run. A tape can be made automatically for future use. The next piece can then be introduced, and the computer is taught a new program in the manner mentioned above. The tape for a particular part is filed and whenever that part is located in its fixture the appropriate tape is used to restore the program in the computer and the weld will always be reproduced.

If there are errors, such as improper part preparation, warpage, etc., the sensing devices will override the program and will cause the arc to follow the weld joint as required.

It is expected that flexible automated welding as described above will be used for more and more short-run applications. It will eliminate the need for complex dedicated fixtures and will find increasing acceptance in the low-volume production plants.

9-7 AUXILIARY EQUIPMENT

There are other pieces of welding equipment that are often required. They are called auxiliary equipment since they do not fall into any of the categories previously described. Many of these devices are extremely useful for certain welding applications. These include electrode oscillators or weavers, welding guidance systems or seam trackers, torch-to-work distance controls, cooling water circulators, heat exchangers, and various devices to provide portability of welding equipment.

Welding oscillators or arc weavers are devices that provide motion to an arc transverse to the direction of travel. These units provide a wider welding bead for surfacing applications and are sometimes used as joints become wider. Oscillators are available as components that can be added to automatic welding equipment. There are several types. One has a transverse or back-and-forth motion and another a pivoting or swinging motion. Motion is controlled either by mechanical devices such as lever arms or cams or by electronic devices containing timing circuits.

The most simple is the mechanical type that provides sinusoidal oscillation using an adjustable crank arm. Similar types utilize cams which can be shaped to provide a specific motion that can include a dwell time at the end of each oscillation stroke. The cam and crank arm types are hard to adjust during operation. Additionally, it is difficult to change the other variables. The newer types utilize electronic controls so that the cycles of oscillation per time period can be very easily changed. The dwell time at either end of the stroke can be lengthened or shortened, and the length of oscillation can be changed during operation. It is possible also to change the centerline of oscillation. The electronic control system can be used for pivoting type motion or for a linear back-and-forth motion. If it is necessary to change any of the oscillation parameters, the electronic type is the best selection. If wide oscillation is required the linear action is preferred to the pivoting action. Oscillators are often used for automatic vertical gas metal arc welding.

Guidance systems are sometimes used with automatic welding equipment. These devices are also available as accessory components. An automatic guidance system normally includes the hardware or adjusters for moving the welding torch transverse to the direction of travel. The mechanical movement is accomplished by an electrical motor and cross slides, which carry the torch. The powered cross slide assembly is attached to the automatic machine. Many of these guidance systems also provide for lateral adjustment of the machine by remote control.

Different methods are used to provide signals to cause the torch to change position. Mechanical probes that ride on the work to be welded, immediately ahead of the welding arc, are often used. In groove welding, the probe is adjusted so that the slightest change in direction will cause the torch to move and follow the joint. The same kind of approach can be used for flat-position fillet welding. Some systems use two cross slides, the second one for vertical adjustment so that as the weld is made the torch will be raised to maintain a fixed torch-to-work distance.

The guidance system can become extremely complex when the mechanical probe is replaced by photoelectric tracking devices. The photoelectric probe can be adjusted to follow the edge of a seam or groove. Other guidance devices, including arc-sensing devices based on changes of arc voltage in the welding circuit, magnetic sensors, and inductive transducers and proximity devices, are used. Each method of sensing errors has its advantages and disadvantages. Mechanical systems that ride the work are difficult to use in high-speed applications. The magnetic systems can be upset by magnetic fields in the parts being welded. Photoelectric devices have the problem of cleanliness in a dirty environment and the proximity, and inductive devices have the problem of reliability. Each type of guidance sensor is best suited to specific applications. The selection should be based on the type of joint to be followed, the metal being welded, the process, and travel speed.

Other torch-to-work control devices are designed specifically for gas tungsten arc welding. These are used for precision welding applications. With this equipment the voltage drop across the welding arc is amplified and used as a signal to the control circuit, which causes the torch to rise or lower by means of a motor. These devices are extremely sensitive and rugged and are widely used when the contour of the weldment may not be flat. Such devices also compensate for irregularities in the work. These devices are sufficiently precise to maintain arc voltage changes within 1% of the selected voltage.

Water cooling equipment is commonly used for heavy-duty gas tungsten arc welding and gas metal arc applications when the welding torch must be cooled. There are two basic types. One is a pump which circulates water through the torch to a reservoir. The volume of water in the reservoir is such that the torch is kept relatively cool. For high-duty cycle welding the water will gradually rise in temperature until it approaches the boiling point. The circulator type system is recommended for light-duty welding. For heavy-duty work, large-volume high-capacity heat exchangers should be used. The same equipment can be used for water-cooled shoes used for electroslag welding. When a large amount of heat is generated for over a long period of time, the amount of heat that must be extracted from the welding torch or retaining shoes should be calculated. This will determine the size of heat exchanger required. Commercially available cooling units are rated in Btu's per hour of heat extracted from the system. For normal gas tungsten arc or gas metal arc welding in medium-current ranges, a 25,000-Btu-per-hour unit is recommended. For heavy-duty work such as high-current gas metal arc or flux-cored arc welding or for electroslag shoes a 50,000-Btu-per-hour unit is recommended. The water cooling circulator should have adjustable flow rate control; it should also have a flow switch and interlock circuit, which can be connected to the welding control circuit. The circulator should allow for adjustable pressures, and for many applications the tank and piping system should be noncorrosive. The minimum flow rate should be at least one-half gallon per minute and adjustable up to four gallons per minute. There should be sufficient water capacity so that if a leak occurs it will not immediately cause a burnout. The heat exchanger should be a radiator with a fan to circulate air through it.

Finally, there are a large number of devices which provide portability to the welding equipment. These are particularly popular for semiautomatic welding equipment. These units usually include a boom or arm which supports the wire feeder so that the gun and its cable are usable over a wide area. This provides flexibility for semiautomatic welding, previously attained with shielded metal arc welding. Many of these units also carry the power supply and the electrode wire supply. Figure 9-43 is an example of a portable mounting for semiautomatic welding. The various types provide different features and the selection is largely a matter of personal preference.

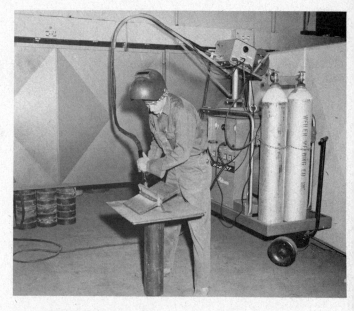

FIGURE 9-43 *Portable boom for semi automatic welding.*

QUESTIONS

1. What are the two types of electrode holders used for SMAW?

2. What are the two types of continuous electrode wire guns?

3. Why are some guns water cooled? Are they used with CO_2 shielding?

4. What factors determine the size of welding cable to be used?

5. What is the maximum voltage drop allowed? How is it determined?

6. Explain the two types of wire feeders. Which type is used with CV machines?

7. What factors must be considered when selecting the speed range of a wire feeder?

8. What are the disadvantages of *spool guns*?

9. What are the four types of travel motion devices?

10. How do you specify the size of a welding manipulator? What other factors should be specified?

11. Explain the difference between a fixture and a positioner.

12. What is the reason for using a welding positioner?

13. What are the four major types of welding positioners?

14. Explain load eccentricity and table-to-load center of gravity with respect to a tilt table positioner.

15. Can turning rolls be used on rectangular parts? Explain.

16. What are the advantages of automatic welding?

17. What factors are included in a welding program?

18. Explain the difference between automatic welding and automated welding.

19. How does a robot welder work?

20. When should a heat exchanger be used in a water cooling circulation system?

REFERENCES

1. C. E. STRAIN AND L. E. WILDENTHALER, "Extended Length Continuous Wire Feed System," *The Welding Journal,* May 1974, Miami, Florida.

2. H. L. UNLAND, "Automatic Electric Arc Welding Machine," *The Welding Engineer,* July 1920, Chicago, Illinois.

3. H. B. CARY, "Automatic Welding in Automotive Production," *The Welding Journal,* March 1973, Miami, florida.

10

FILLER METALS AND
MATERIALS FOR WELDING

10-1 TYPES OF WELDING MATERIALS

10-2 COVERED ELECTRODES

10-3 SOLID ELECTRODE WIRES

10-4 FLUX-CORED OR TUBULAR
      ELECTRODES

10-5 FLUXES FOR WELDING

10-6 OTHER FILLER METALS

10-7 SHIELDING GAS USED IN WELDING

10-8 FUEL GASES FOR WELDING
      AND CUTTING

10-9 GAS CONTAINERS AND
      APPARATUSES

10-1 TYPES OF WELDING MATERIALS

There are many types of materials used to produce welds. These welding materials are generally categorized under the term **filler metals**, defined as "the metal to be added in making a welded, brazed, or soldered joint." The filler metals are used or consumed and become a part of the finished weld. The definition has been expanded and now includes electrodes normally considered nonconsumable such as tungsten and carbon electrodes, fluxes for brazing, submerged arc welding, electroslag welding, etc. The term filler metal does not include electrodes used for resistance welding, nor does it include the studs involved in stud welding.

The American Welding Society has issued 26 specifications covering filler materials, shown by Table 10-1. This table also shows the welding process for which each specification is intended. These specifications are periodically updated and a two-digit suffix indicating the year issued is added to the specification number. Additional specifications are added from time to time.

Most of the industrial countries issue filler metal specifications. The national specifications do not exactly correspond with one another as far as types of filler metals covered. A correlation of national filler metal specifications is shown by the table in Figure 10-1.

In the United States, the American Welding Society filler metal specifications are the primary ones used by industry. The American Society of Mechanical Engineers in their "Pressure Vessel and Boiler Code" issues filler metal specifications that are identical with the AWS specifications. The numbers are changed slightly; ASME adds the prefix letters "SF" to the specification number. These are given in Section 2, "Material Specification" of the ASME code. RDT standards, which have additional requirements over the AWS and ASME specifications are issued by the U.S. Atomic Energy Commission. The U.S. Government Department of Defense also issues standards for welding filler metals. These are known as MIL specifications and are used for procurement by the government. These specifications are usually in agreement with AWS specifications; however, if a military specification is referenced it should be consulted.

TABLE 10-1 *AWS filter metal specification and welding processes.*

AWS Specification	Specifications Title	OAW	SMAW	GTAW	GMAW	SAW	Other
				FOR PROCESS SHOWN			
A5.1	Carbon steel covered arc-welding electrodes		X				
A5.2	Iron & steel gas welding rods	X					
A5.3	Aluminum & aluminum alloy arc welding electrodes		X				
A5.4	Corrosion-resisting chromium & chromium-nickel steel covered welding electrodes		X				
A5.5	Low-alloy steel covered arc welding electrodes		X				
A5.6	Copper & copper alloy covered electrodes		X				
A5.7	Copper & copper alloy welding rods	X		X			PAW
A5.8	Brazing filler metal						BR
A5.9	Corrosion-resisting chromium & chromium-nickel steel bare & composite metal cored & standard arc welding electrodes & rods			X	X	X	PAW
A5.10	Aluminum & aluminum alloy welding rods & bare electrodes	X		X	X		PAW
A5.11	Nickel & nickel alloy covered welding electrodes		X				
A5.12	Tungsten arc welding electrodes			X			PAW
A5.13	Surfacing welding rods & electrodes	X		X			CAW
A5.14	Nickel & nickel alloy bare welding rods and electrodes	X		X	X	X	PAW
A5.15	Welding rods & covered electrodes for welding cast iron	X	X				CAW
A5.16	Titanium & titanium alloy bare welding rods & electrodes			X	X		PAW
A5.17	Bare carbon steel electrodes & fluxes for submerged-arc welding					X	
A5.18	Carbon steel filler metals for gas shielded arc welding			X	X		PAW
A5.19	Magnesium alloy welding rods & bare electrodes	X		X	X		PAW
A5.20	Carbon steel electrodes for flux cored arc welding						FCAW
A5.21	Composite surfacing welding rods & electrodes	X	X	X			
A5.22	Flux cored corrosion-resisting chromium & chromium-nickel steel electrodes						FCAW
A5.23	Bare low-alloy steel electrodes and fluxes for submerged arc welding					X	
A5.24	Zirconium & zirconium alloy bare welding rods and electrodes			X	X		PAW
A5.25	Consumables used for electro-slag welding of carbon & high strength low alloy steels						ES
A5.26	Consumables used for electrogas welding of carbon and high strength low-alloy steels				X (EG)		FCAW (EG)
A5.27	Copper and copper alloy gas welding rods	X					
A5.28	Low-alloy steel filler metals for gas shielded arc welding			X	X		PAW

Note: If GTAW is shown, the specification will also apply to PAW even though not stated.

COUNTRY ISSUING SPECIFICATION	USA	CANADA	INT'L	BRITISH	GERMAN	JAPAN	US MILITARY
Filler metal type	AWS	CSA	ISO	BS	DIN	JIS	US-MIL
Covered electrodes—mild steel	A5.1	W48.1	670 R635 547	639 +1719	1913	Z3210 Z3211	E-15599 E-22200
Covered electrodes—low alloy steel	A5.5	W48.3	1045	2493 1719	1913	Z3212	E-22200/1
Covered electrodes—stainless steel	A5.4	W48.2		2926	8556	Z3221	E-13080 E-22200/2A
Covered electrodes—surfacing	A5.13				8555	Z3251	E-19141
Covered electrodes—for cast iron	A5.15		1163		8573	Z3252	
Covered electrodes—aluminum	A5.3			1616	1732		E-15597
Covered electrodes—copper	A5.6				1733		E-13191 E-21659
Covered electrodes—nickel	A5.11				1736		E-21562
Bare—solid—stainless steel	A5.9		1159	2901	8556	Z3321	E-19933 R-5031
Bare—solid—steel	A5.17 A5.18	W48.4		4165 1453	8559	Z3311	E-18193 E-23765
Flux cored—steel	A5.20 A5.23	W48.5					E-24403

FIGURE 10-1 *National specifications for welding electrodes and wires.*

The Canadian Standards Association issues filler metals specifications which are in general agreement with the AWS specifications. The Canadians have, however, issued additional categories, which are described in detail in the CSA filler metal specifications.

The International Standards Organization (ISO) also issues filler metal specifications. Many of the less industrialized nations utilize specifications of the industrialized countries or ISO standards. The ISO standards are available from the welding or standardization association of each country.

Filler metals can be classified into four basic categories. These are:

Covered electrodes.

Solid (bare) electrode wire or rod.

Fabricated (tubular or cored) electrode wire.

Fluxes for welding.

A category of miscellaneous or others would include the nonconsumable filler metals, such as tungsten and carbon electrodes, consumable backup rings for gas tungsten arc welding of pipe, and possibly strip material for overlay operations.

Included in materials for welding, but a material that cannot be considered a filler metal, would be the gases used in welding. The gases include oxygen and fuel gases for gas welding and cutting and shielding gases for the gas shielded arc welding processes.

The AWS specifications are written in order to provide specific chemical composition of the material and the mechanical properties of the deposited weld metal. The AWS specifications utilize similar methods and testing techniques so that there is consistency between all of the filler metal specifications. The mechanical properties of deposited weld metal are determined based on a standardized welding procedure, in a specified welding joint detail, to produce weld specimens for testing. Specifications also may require other properties such as toughness, quality standards, and, in some cases, standards of porosity. Most specifications include usability factors showing the welding position that the electrode or filler metal is designed for, the type of welding current that should be used, and, in the case of covered electrodes, the type of coating is shown. Size and packaging information is also provided.

The American Welding Society does not test or approve filler metals. The society provides the specifications, which are voluntary conformance standards. It is the manufacturer of the filler material who guarantees that their product conforms to a specific AWS specification and classification.

Weld Metal Certifications

For special applications, filler metals are tested and certified to be in conformance with a specific specification. In some cases filler metals to be used for ships, nuclear reactors, vessels, highway bridges, and certain military products require certification. For military construction, approvals are granted when the specific electrodes are in conformance with the applicable military specification. Tests of this type are usually witnessed by a government inspector. Approved products are then placed on a qualified products list (QPL).

Approval of filler metals for ship construction is similar. A classification society, such as the American Bureau of Shipping, requires that one of their representatives, called a surveyor, witness the welding of test plates based on electrodes selected at random. The surveyor also witnesses the testing of the weld specimens. Classification societies also have special mechanical property requirements and these usually include low-temperature impact data. Approvals are granted for different filler materials based on strength and impact requirements and for welding different classes of steels for ships. The classification society publishes lists of approved electrodes and filler metals manufactured by different companies. Retention of the filler metals on the approved list may be subject to annual tests. Filler metal approvals include covered electrodes, submerged arc electrode wire with flux combinations, and flux-cored arc welding electrodes with gas combinations.

Certification is handled differently from approvals. Certification is required for nuclear work and for certain types of military work. In these cases, a specific batch of electrodes or heat of wire made at one time is tested in accordance with a specification which may be either AWS or MIL specification. The test results must then be certified by the manufacturer and provided to the user. A certification is used only for the specific batch or lot of filler materials made at one time and covered by the test result data. In many cases, the user must previously have approved the manufacturer prior to using filler materials produced by that manufacturer. This will require an audit by the user of the manufacturer to make sure that uniform quality manufacturing procedures are maintained and that quality control procedures provide for strict control and traceability of materials used in the manufacture of the filler material. After the producer has been approved, the certification test may still be required. Traceability of all items is necessary so that the user can provide traceability of all materials used to manufacture the reactor.

Filler metal specifications are of immense value to both producer and user. They allow the user to select the proper filler material to be used to manufacture all types of products. Specifications assure the user that the deposited weld metal, when normally applied, will provide the strength levels indicated by the specification and classification. Specifications are of value to the producer, since they provide standardization of testing methods and procedures and also since they provide categories and classifications to meet the needs of most users.

10-2 COVERED ELECTRODES

The covered electrode is the most popular type of filler metal used in arc welding. The identification of electrode types, the selection of electrodes for specific applications, and the usability of covered electrodes has been discussed in Chapter 4.

The composition of the covering on the electrode determines the usability of the electrode, the composition of the deposited weld metal, and the specification of the electrode. The composition of coatings on covered arc welding electrodes has been surrounded in mystery and little information has been published. The formulation of electrode coatings is very complex and while it is not an exact science it is based on well-established principles of metallurgy, chemistry, and physics, tempered with experience.

The original purpose of the coating was to shield the arc from the oxygen and nitrogen in the atmosphere. It was subsequently found that ionizing agents could be added to the coating which helped stabilize the arc and made electrodes suitable for alternating current welding. It was found that silicates and metal oxides helped form slag, which would improve the weld bead shape because of the reaction at the surface of the weld metal. The deposited weld metal was further refined and its quality improved by the addition of deoxidizers in the coating. In addition, alloying elements were added to improve the strength and provide specific weld metal deposit composition. Finally, iron powder has been added to the coating to improve the deposition rate.

An electrode coating is designed to provide as many as possible of the following desirable characteristics. Some of these characteristics may be incompatible and therefore compromises and balances must be provided and designed into the coating. These desirable characteristics are:

1. Specific composition of the deposited weld metal.
2. Specific mechanical properties of the deposited weld metal.
3. Elimination of weld metal porosity.
4. Elimination of weld metal cracking.

5. Desirable weld deposit contour.

6. Desirable weld metal surface finish, i.e., smooth, with even edges.

7. Elimination of undercut adjacent to the weld.

8. Minimum spatter adjacent to the weld.

9. Ease of manipulation to control slag in all positions.

10. Provide a stable welding arc.

11. Provide penetration control, i.e., deep or shallow.

12. Provide for initial immediate arc striking and restriking capabilities.

13. Provide a high rate of metal deposition.

14. Eliminate noxious odors and fumes.

15. Reduce the tendency of the coating to pick up moisture when in storage.

16. Reduce electrode overheating during use.

17. Provide a strong tough durable coating.

18. Provide for easy slag removal.

19. Provide a coating that will ship well and store indefinitely.

These requirements must be achieved at the minimum possible cost. In addition, the formulation must be manufacturable with conventional extrusion equipment at high production rates. No single electrode type will meet all of the foregoing requirements, and there is no one single "universal electrode." Instead there is a variety of electrode types each having certain desirable characteristics.

The coatings of electrodes for welding mild and low alloy steels may have from 6 to 12 ingredients such as:

Cellulose: to provide a gaseous shield with a reducing agent. The gas shield surrounding the arc is produced by the disintegration of cellulose.

Metal Carbonates: to adjust the basicity of the slag and to provide a reducing atmosphere.

Titanium Dioxide: to help form a highly fluid but quick-freezing slag. It will also provide ionization for the arc.

Ferromanganese and Ferrosilicon: to help deoxidize the molten weld metal and to supplement the manganese content and silicon content of the deposited weld metal.

Clays and Gums: to provide elasticity for extruding the plastic coating material and to help provide strength to the coating.

Calcium Fluoride: to provide shielding gas to protect the arc, adjust the basicity of the slag, and provide fluidity and solubility of the metal oxides.

Mineral Silicates: to provide slag and give strength to the electrode covering.

Alloying Metals: These include nickel, molybdenum, chromium, etc., to provide alloy content to the deposited weld metal.

Iron or Manganese Oxide: to adjust the fluidity and properties of the slag. In small amounts iron oxide helps stabilize the arc.

Iron Powder: to increase the productivity by providing additional metal to be deposited in the weld.

By using combinations and different amounts of these constituents it is possible to provide an infinite variety of electrode coatings. The binder used for most electrode coatings is sodium silicate, which will chemically combine and harden to provide a tough, strong coating. The design of the coating provides the proper balance to give the electrode specific usability characteristics and to provide specific weld deposit chemistry and properties. In general, the different makes of electrodes that meet a particular classification have somewhat similar compositions.

The principle types of electrode coatings for mild steel and low-alloy electrodes are described in the following paragraphs.

Cellulose-Sodium (EXX10): Electrodes of this type have up to 30% cellulosic material in the form of wood flour, or reprocessed paper. The gas shield will contain CO_2 and hydrogen, which are reducing agents. These gases tend to produce a digging arc that produces deep penetration. The weld deposit is somewhat rough and the spatter is at a higher level than other electrodes. It does provide extremely good mechanical properties particularly after aging. This is one of the earliest types of electrodes developed and is widely used for cross country pipe lines using the downhill welding technique. It is normally used with direct current with the electrode positive (reverse polarity).

Cellulose-Potassium (EXX11): This electrode is very similar to the cellulose-sodium electrode with the

FILLER METALS AND MATERIALS FOR WELDING

exception that more potassium is used than sodium. This provides ionization of the arc and makes the electrode suitable for welding with alternating current. The arc action, the penetration, and the weld results are very similar.

In both E6010 and E6011 electrodes, small amounts of iron powder may be added. This will assist in arc stabilization and will also slightly increase the deposition rate.

Rutile-Sodium (EXX12): When the rutile or titanium dioxide content is relatively high with respect to the other constituents, the electrode will be especially appealing to the welder. Electrodes with this coating have a quiet arc, an easily controlled slag, and a low level of spatter. The weld deposit will have a smooth surface and the penetration will be less than with the cellulose electrode. The weld metal properties will be slightly lower than the cellulosic types. This type of electrode provides a fairly high rate of deposition. It has a relatively low arc voltage and can be used on alternating current or on DC with electrode negative (straight polarity).

Rutile-Potassium (EXX13): This electrode coating is very similar to the rutile-sodium type except that potassium is used to provide for arc ionization which makes it more suitable for welding with alternating current. It can also be used with direct current with either polarity. It produces a very quiet smooth running arc.

Rutile-Iron Powder (EXXX4): This coating is very similar to the rutile coatings mentioned above except that iron powder is added. If the addition of iron powder is in the 25–40% category it would result in an EXX14 type electrode. If the iron powder addition is 50% or higher it will result in an EXX24 type electrode. With the smaller amount of iron powder the electrode can be used in all positions, whereas with the high percentage of iron powder it can be used only in the flat position or for making horizontal fillet welds. In both cases the deposition rate is increased based on the amount of iron powder included in the coating.

Low Hydrogen-Sodium (EXXX5): Coatings that contain a high proportion of calcium carbonate or calcium fluoride are termed low hydrogen and in some cases called lime ferritic or basic type electrodes. In this class of coating cellulose, clays, asbestos, and other minerals that contain combined water are not used. This is to insure the lowest possible hydrogen content in the arc atmosphere. In addition, these elctrode coatings are baked at a higher temperature. The low-hydrogen electrode family have superior weld metal properties. They provide the highest ductility of any of the deposits. These electrodes have a medium arc with medium or moderate penetration. They have a medium speed of deposition but do require special welder technique for best results. Low hydrogen electrodes must be stored under controlled conditions. This type is normally used with direct current with electrode positive (reverse polarity).

Low Hydrogen-Potassium (EXXX6): This type of coating is very similar to the low hydrogen-sodium except for the substitution of potassium for sodium to provide arc ionization. This electrode is used with alternating current and can be used with direct current electrode positive (reverse polarity). It has become considerably more popular than the sodium type. The arc action is smoother but penetration of the two electrodes is very similar.

Low Hydrogen-Iron Powder (EXXX8): The coatings in this class of electrodes are similar to the low-hydrogen type mentioned previously; however, iron powder is added to the electrode and if it is added in the amount of 35–40% the electrode will be classed as an EXX18. This electrode produces excellent weld metal.

Low-Hydrogen Iron Powder (EXX28): Similar to the EXX18 but with 50% or more iron powder added to the coating. This electrode is usable only in the flat position or for making horizontal fillet welds. The deposition rate is higher than the EXX18.

Low-hydrogen coatings are used for all of the higher-alloy electrodes. By additions of specific metals in the coatings these electrodes then become the alloy types where suffix letters are used to indicate weld metal compositions. Electrodes for welding stainless steel are also of the low-hydrogen type.

Iron Oxide-Sodium (EXX20): Coatings with high iron oxide content produce a weld deposit with voluminous slag, which is rather difficult to control. This coating type produces high-speed deposition, and it provides medium penetration with a low spatter level. The resulting weld has a very smooth finish. The electrode is restricted to the flat position and for making horizontal fillet welds. The deposit weld metal has good weld metal properties. The electrode can be used with AC or direct current with either polarity. This type of electrode was the original "hot rod" type of electrode. It has become less popular because of the iron powder electrodes.

Iron Oxide-Iron Powder (EXX27): This type of electrode is very similar to the iron oxide-sodium type except that 50% or more iron powder is added to the coating. This greatly increases the deposition rate. This type is quickly supplanting the EXX20 type because although it has similar weld metal deposit characteristics, it is more desirable from the welder's point of view, and it can be used with either alternating or direct current of either polarity.

There are many other types of coatings than those mentioned here, most of which are usually combinations of these types but for special applications such as hard surfacing, cast iron welding, and for nonferrous metals.

Manufacturing

Covered arc welding electrodes are manufactured at machine-gun-like speeds. This is necessary to keep up with the use of covered electrodes throughout the world. There are three basic parts of a covered electrode: the core wire, the chemicals and minerals that comprise the coating, and the liquid binder that hardens and holds it all together. The steps required to manufacture a covered electrode are shown by Figure 10-2 which is a flow chart showing each of the major steps involved.

The core wire for mild-steel and low-alloy steel electrodes is normally a low-carbon steel having a carbon content of about 0.10% carbon, low manganese and silicon content, and the minimum amount of phosphorus and sulphur. Ingots of this composition are produced at the steel mill; they are hot rolled and reduced in size to billets. These are then taken to a bar mill and rolled into small-diameter rods which range in size from 1/4 in. (6.4 mm) to 3/8 in. (9.5 mm) in diameter. This product, which is known as *hot rolled wire rod,* is then taken to the wire drawing mill and drawn into the appropriate diameters for covered electrodes. After the wire has been drawn to the proper diameter it is straightened and cut to the proper length. The lengths vary according to the size and range from 12 to 14 in. in the United States, and from 200 to 500 millimeters in length elsewhere.

The coating is made of different chemicals and minerals obtained throughout the world. They are inspected and ground to the proper mesh size. The specific amount of each chemical is weighed and mixed together in the dry condition. After sufficient dry mixing the proper amounts of liquid, binder, and water are added, and mixing is continued in the wet stage. Mixing is completed when it reaches the proper consistency. This material is then placed in a press where it is formed into large briquets of moist flux coating material.

The briquets of coating material and the cut and straightened lengths of core wire are brought together at the extrusion press. The cut core wires are fed into

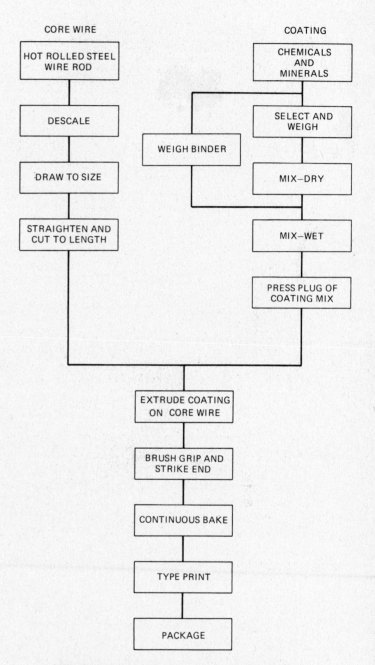

FIGURE 10-2 *Flow chart of operations to manufacture covered electrodes.*

the extrusion die by an automatic feeder. Simultaneously, the press feeds flux into a chamber and extrudes the coating onto the core wire as it passes through the chamber. The extrusion die holder must be adjusted with extreme accuracy so that the flux flows uniformly to make it concentric with the core wire. The coated electrode emerges from the die of the extrusion press at a rate of approximately ten per second.

FIGURE 10-3 *Electrodes being manufactured.*

They drop onto a conveyor where power brushes remove a portion of the coating at the grip end and clean the coating from the strike end. The electrodes move on the conveyor into a drying oven for a sufficient length of time for the coating to solidify and toughen. At the exit end of the oven each electrode is individually printed with the A.W.S. classification number and then inspected and placed in boxes and packed for shipment. The entire operation from start to finish is continuous. Figure 10-3 shows the electrodes during the manufacturing operation.

Care, Storage, and Reconditioning of Covered Electrodes

Covered electrodes can be easily damaged. Each electrode should be treated with care prior to its use. Rough handling in shipment or in storage can cause a portion of the coating to crack loose from the core wire and make the electrode unsuitable. Bending most electrode types will cause the coating to break loose from the core wire. The electrode should not be used where the core wire is exposed.

Electrodes may become unusable if they are exposed to moisture for an extended length of time. The coatings on some types of electrodes absorb moisture when exposed to humid atmospheres. Cellulose, rutile, and acid type electrodes are fairly insensitive to moisture and can tolerate quite high moisture content without the risk of porosity in the weld. The coatings of low-hydrogen type electrodes, particularly those of the EXX16 type, pick up moisture quickly when exposed to a high-humidity atmosphere. Since these electrodes are dried in a high temperature and a low-moisture atmosphere, they are more sensitive to pickup. Stainless steel electrodes are in this same category.

If electrodes, even in unopened cardboard containers, are left outdoors it is possible that they will pick up moisture due to the change of temperature and humidity from day to night. The moisture is absorbed by the packaging and in turn the moisture is gradually absorbed by the coatings of the electrodes inside the package. Efforts to prohibit this are made by wrapping the electrodes in plastic liners or by using vaporproof or metal containers. These provide additional protection to the electrodes.

Once the container is opened, the electrodes should be stored in heated ovens in accordance with the chart shown by Figure 10-4. Low-hydrogen electrodes must not be stored in ovens that hold electrodes of other classifications. Nothing else should be in ovens that hold low-hydrogen electrodes. Food must not be placed in electrode ovens since the moisture given off during cooking will damage the low-hydrogen coatings.

Damp electrodes are difficult to distinguish by the welder. It is easier to recognize the problem based on storage conditions. It is also easy to recognize the problem by reviewing X-rays of weld metal deposited by damp electrodes. The weld metal will be porous if the coatings are damp. It is sometimes possible to shake three or four low-hydrogen electrodes together and listen to the sound as they rattle against each other. If the electrode coatings are dry or contain only small amounts of moisture a clear shrill metallic sound will be heard. Damp electrodes have a hollow sound which is quite different. Experience in testing electrodes in this way will help to distinguish these two different sounds. When welding with an electrode with a damp coating a fierce crackling or explosive sound may be heard. If the electrode is extremely damp condensed vapor may be seen while welding. If the electrode is not completely consumed the coating on the remaining part of the electrode will show longitudinal cracks in the coating.

Electrodes that are only slightly damp can be heated up by shorting them against the work for a few seconds just before beginning to weld. For reconditioning electrodes, special ovens are available. These are set at specific temperatures for specific types of electrode coatings. This is shown by the chart of Figure 10-4.

The baking cycle for reconditioning electrodes should not exceed four hours. The heating rate in the oven is not critical. Electrodes can be taken from room temperature and placed in an oven without affecting the properties of the deposited weld metal. The maximum temperature for any low hydrogen electrode is 800°F (427°C). Some ingredients in the coating tend to oxidize if the temperature is raised above this figure. The holding time at the maximum temperature should be 30 minutes, minimum. This insures that the electrodes are up to the oven temperature. The cooling rate is not critical; however, reconditioned electrodes should not be taken from the oven and allowed to cool until the oven has come down to approximately 300°F (149°C). Electrodes should not be reconditioned by heating more than three times. Going through the extra heating cycle tends to weaken the silicate binder and the coating will eventually become weak and fragile and will chip off easily.

Electrodes should be stored in a special storeroom with controlled atmosphere. The relative humidity should be maintained at 40% or less. This can be accomplished by sealing the room and installing a dehumidifier.

Electrode Classification	Recommended Storage Unopened Boxes	Recommended Storage Open Boxes	Holding Oven	Reconditioning
E-XX10	Dry @ room temp	Dry @ room temp	Not recommended	Not done
E-XX11	Dry @ room temp	Dry @ room temp	Not recommended	Not done
E-XX12	Dry @ room temp	Dry @ room temp	Not recommended	Not done
E-XX13	Dry @ room temp	Dry @ room temp	Not recommended	Not done
E-XX14	Dry @ room temp	150-200°F	150-200°F	250-300°F
E-XX20	Dry @ room temp	150-200°F	150-200°F	1 Hour
E-XX24	Dry @ room temp	150-200°F	150-200°F	
E-XX27	Dry @ room temp	150-200°F	150-200°F	
E-60 or 7015	Dry @ room temp	250-450°F	150-200°F	500-600°F
E-60 or 7016	Dry @ room temp	250-450°F	150-200°F	1 Hour
E-7018	Dry @ room temp	250-450°F	150-200°F	
E-7028	Dry @ room temp	250-450°F	150-200°F	
E-80 & 9015	Dry @ room temp	250-450°F	200-250°F	600-700°F
E-80 & 9016	Dry @ room temp	250-450°F	200-250°F	1 Hour
E-80 & 9018	Dry @ room temp	250-450°F	200-250°F	
E-90-12015	Dry @ room temp	250-450°F	200-250°F	650-750°F
E-90-12016	Dry @ room temp	250-450°F	200-250°F	1 Hour
E-90-12018	Dry @ room temp	250-450°F	200-250°F	
E-XXX-15 or 16	Dry @ room temp	250-450°F	150-200°F	450°F
Stainless	Dry @ room temp	250-450°F	150-200°F	1 Hour

FIGURE 10-4 *Storage and reconditioning of electrodes and filler metals.*

When low-hydrogen electrodes are issued from the controlled atmosphere storeroom, they should be used within two hours. When this cannot be done, individual ovens should be provided for each welder. They can then be left in the heated oven until the electrode is used. All low-hydrogen electrodes not used during a work shift should be returned to the holding oven. For critical work special controls are instituted to maintain dry electrodes.

Electrodes can be damaged by aging. Very old electrodes of most types will have a furry surface on the coating, usually white. This is from the crystallization of the sodium silicate. This surface is normally harmless for mild-steel low-hydrogen electrodes. They should not be used for extremely critical work. If iron powder type electrodes are old, rust may form on the iron powder due to moisture absorbed in the coating. If the core wire is rusty, it is possible that too much moisture may have been absorbed in the coating.

Deposition Rates

The different types of electrodes have different deposition rates, as a result of the composition of the coating. The electrodes containing iron powder in the coating have the highest deposition rates. The percentage of iron powder is confusing when comparing electrodes produced in Europe with those produced in the U.S.A. In the U.S.A. the percentage of iron powder in a coating is in the 10 to 50% range. This is based on the amount of iron powder in the coating versus the coating weight. This is shown in the formula:

$$\% \text{ Iron powder} = \frac{\text{Weight of iron powder}}{\text{Total weight of coating}} \times 100$$

The percentages mentioned above are related to the requirements of the AWS specifications. The European method of specifying iron powder is based on the weight of deposited weld metal versus the weight of the bare core wire consumed or:

$$\% \text{ Iron powder} = \frac{\text{Weight of deposited metal}}{\text{Weight of bare core wire}} \times 100$$

Thus, if the weight of the deposit were double the weight of the core wire it would indicate a 200% deposition efficiency even though the amount of the iron powder in coating represented only half of the total deposit. The 30% iron powder formula used in the U.S. would produce a 100–110% deposition efficiency using the European formula. The 50% iron powder electrode figured on U.S. standards would produce an efficiency of approximately 150% using the European formula.

Quality—Defects

Quality control in manufacturing of covered electrodes starts at the point of receiving chemicals and minerals, the binders and the hot rolled wire rod. The chemicals must meet rigid specifications and are checked when they are received. The wire rod, which is checked on a continuous basis, must also meet stringent specifications. Grind sizes of chemicals, cleanliness of mixing containers, etc., are routinely inspected. The adjustment of the extrusion die to maintain concentricity of the electrodes is checked at each setup. Electrodes are checked after baking for coating concentricity. The surface and structure of the coating is also inspected and each lot is checked by welding to determine that it meets the specifications.

A common complaint of quality of electrodes is *fingernailing* which is the name given to the burning off of an electrode faster on one side than on the other. The welder assumes fingernailing means a nonconcentric electrode, however, other factors might create the fingernailing. Fingernailing is most common when using direct current and is more evident with the smaller electrodes, 1/8 in. (3.2 mm) and 5/32 in. (4.0 mm), when used at low currents. This condition can be aggravated if the coating is not concentric with the core wire. This can be checked by removing the coating from the one side of the core wire and measuring the core wire and covering to the other side, and then removing coating on the opposite side of the electrode and measuring the electrode core wire and covering on the other side. Measuring should be done with a micrometer. Normally, electrodes are concentric within 0.002–0.003 of an inch (0.05–0.07 mm). More often fingernailing results from arc blow, welding current too low, incorrect electrode angle, unbalanced joint preparation, and in some cases, uneven moisture pickup in the coating, which might be greater on one side than the other. A quick check for fingernailing during welding is to stop welding when fingernailing is encountered and rotate the electrode in the holder 180°. Continue to weld and see if fingernailing continues on the same side of the electrode. If it does, the coating is probably off center. If it does not but instead fingernails off the other side, arc blow or one of the other factors mentioned above is the reason. Arc blow is a more frequent cause of fingernailing than off-center electrodes. When welding with lower than normal current fingernailing will appear because there is insufficient arc force to overcome minor arc blow. The other factor is electrode angle which again can be checked by revolving the electrode 180° in the holder. Moisture can be checked as mentioned previously.

Many times the welder or the welding supervisor may wish to compare different brands of electrodes. Many electrode manufacturers provide such charts; however, the most complete chart is published by the American Welding Society and is known as the "Filler Metals Comparison Chart AWS 5.0."

10-3 SOLID ELECTRODE WIRES

Solid metal wires were first used for oxy fuel gas welding to add filler metal to the joint. These wires or rods were provided in straightened lengths approximately 36 in. (1 m) long. The earliest electrodes for arc welding were also solid and bare, usually in the length of 12 to 14 in. long (300–350 mm). Later on, solid wire was provided in coils for "bare wire" automatic arc welding and later for submerged arc and electroslag welding. The latest process to use solid bare wire is gas metal arc welding, which uses relatively small-diameter electrode wires.

The manufacture of wire for welding electrodes or rod is essentially the same except that the straighten and cut operation is added for a welding rod. A simplified flowchart of the manufacturing operations for solid mild steel electrode wire is shown by Figure 10-5. The most complex portion is the drawing operation and this is shown partially by the picture of Figure 10-6. The drawing of steel wires and nonferrous wires is essentially the same; however, different amounts of reduction per drawing die, different drawing lubricants, different heat treatments, etc., are involved.

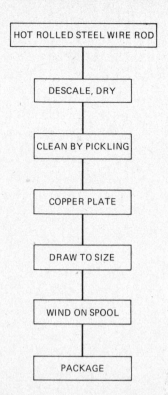

FIGURE 10-5 *Flow chart of operations to manufacture solid electrode wire.*

The solid steel electrode wires may not be "bare." Many suppliers provide a very thin copper coating on the wire. The copper coating is for several purposes. It improves the current pickup between contact tip and the electrode. It aids drawing and helps prevent rusting of the wire when it is exposed to the atmosphere.

Solid electrode wires are also made of various stainless steel analyses, aluminum alloys, nickel alloys,

FIGURE 10-6 *Wire drawing operation.*

CHAPTER 10

FILLER METALS AND MATERIALS FOR WELDING

magnesium alloys, titanium alloys, copper alloys, and other metals. Specifications for these electrodes are shown by Table 10-1.

When the wire is cut and straightened it is called a welding rod, which is a form of filler metal used for welding or brazing which does not conduct the electrical current. If the wire is used in the electrical circuit it is called a welding electrode and is defined as a component of the welding circuit through which current is conducted. A bare electrode is normally thought of as being a wire; however, it can take other forms.

Several different systems are used to identify the classification of a particular electrode or welding rod. In all cases a prefix letter is used.

Prefix R:　　Indicates a welding rod.

Prefix E:　　Indicates a welding electrode.

Prefix RB:　Indicates use as either a welding rod or for brazing filler metal.

Prefix ER:　Indicates either an electrode or welding rod.

The system for identifying bare carbon steel electrodes and rods for gas shielded arc welding is as follows:

ER:　Prefix indicates an electrode or welding rod.

70:　Indicates the required minimum as-welded tensile strength in thousands of pounds per square inch (psi).

S:　Indicates solid electrode or rod.

C:　Indicates composite metal cored or stranded electrode or rod.

1:　Suffix number indicates a particular analysis and usability factor.

Figure 10-7 shows the typical composition and minimum weld metal mechanical properties of the AWS specification, "Carbon Steel Filler Metals for Gas Shielded Arc Welding."

IDENTIFICATION	WELDING CONDITIONS			TEST REQUIREMENTS (AS WELDED)				CHEMICAL COMPOSITION			
Identification	Welding Conditions			Test Requirements (as welded)				Chemical Composition			
AWS Classification	Current Electrode Polarity	External Gas Shield	Radio-graphic Test	All Tensile min psi	Weld Yield min psi	Metal El. % min 2″	Impact Test Charpy V	C	Mn	Si	Other
E-70S-1	DCEP	AO	Rgd.	72,000	60,000	22	Not Rgd.	0.07 to 0.19		0.30 to 0.50	
E-70S-2	DCEP	AO & CO_2	Rgd.	72,000	60,000	22	20@ −20° F	0.6		0.40 to 0.70	Ti-0.05 to 0.15　Zi-0.02 to 0.12　Al-0.05 to 0.15
E-70S-3	DCEP	AO & CO_2	Rgd.	72,000	60,000	22	20@ 0° F	0.06 to 0.15	0.90 to 1.40	0.45 to 0.70	
E-70S-4	DCEP	CO_2	Rgd.	72,000	60,000	22	Not Rgd.	0.07 to 0.15		0.65 to 0.85	
E-70S-5	DCEP	CO_2	Rgd.	72,000	60,000	22	Not Rgd.	0.07 to 0.19		0.30 to 0.60	Al-0.50 to 0.90
E-70S-6	DCEP	CO_2	Rgd.	72,000	60,000	22	20@ −20° F	0.07 to 0.15	1.40 to 1.85	0.80 to 1.15	
E-70S-G	Not specified	Not specified	Rgd.	72,000	60,000	22	Not Rgd.	No chemical requirements			
E-70S-1B	DCEP	CO_2	Rgd.	72,000	60,000	17	20@ −20° F	0.07 to 0.12	1.60 to 2.10	0.50 to 0.80	Ni-0.15　Mo 0.40 to 0.60
E-70S-GB	Not specified	Not specified	Rgd.	72,000	60,000	22	Not Rgd.	No chemical requirements			

Note: P—0.025 Max.　　S—0.035 Max.

FIGURE 10-7　*Summary electrodes for gas metal-arc welding per AWS 5.18-69.*

AWS Classification	Chemical Composition—Percent						Total Other Elements
	C	Mn	Si	S	$	P	
Low manganese classes							
EL8	0.10	0.30 to 0.55	0.05	0.035	0.03	0.30	0.50
EL8K	0.10	0.30 to 0.55	0.10 to 0.20	0.035	0.03	0.30	0.50
EL12	0.07 to 0.15	0.35 to 0.60	0.05	0.035	0.03	0.30	0.50
Medium manganese classes							
EM5K	0.06	0.90 to 1.40	0.40 to 0.70	0.035	0.03	0.30	0.50
EM12	0.07 to 0.15	0.85 to 1.25	0.05	0.035	0.03	0.30	0.50
EM12K	0.07 to 0.15	0.85 to 1.25	0.15 to 0.35	0.035	0.03	0.30	0.50
EM13K	0.07 to 0.19	0.90 to 1.40	0.45 to 0.70	0.035	0.03	0.30	0.50
EM15K	0.12 to 0.20	0.85 to 1.25	0.15 to 0.35	0.035	0.03	0.30	0.50
High manganese class							
EH14	0.10 to 0.18	1.75 to 2.25	0.05	0.035	0.03	0.30	0.50

FIGURE 10-8 *Summary of mild steel electrode wire composition for SAW per AWS A5.17.*

The system for identifying solid bare carbon steel for submerged arc is as follows:

The prefix letter E is used to indicate an electrode. This is followed by a letter which indicates the level of manganese, i.e., L for low, M for medium, and H for high manganese. This is followed by a number which is the average amount of carbon in points or hundredths of a percent. The composition of some of these wires is almost identical with some of the wires in the gas metal arc welding specification.

The electrode wires used for submerged arc welding are given in AWS specification, "Bare Mild Steel Electrodes and Fluxes for Submerged Arc Welding." This specification provides both the wire composition and the weld deposit chemistry based on the flux used. The specification does give composition of the electrode wires. This information is given in the table of Figure 10-8.

When these electrodes are used with specific submerged arc fluxes and welded with prescribed procedures the deposited weld metal will meet mechanical properties required by the specification. This information is presented in Section 10-5 on welding fluxes.

In the case of the filler rods used for oxy fuel gas welding the prefix letter is R, but this is followed by a G indicating that the rod is used expressly for gas welding. These letters are followed by two digits 45, 60, or 65, which designate the approximate tensile strength in 1,000 pounds per square inch.

In the case of nonferrous filler metals, the prefix E, R, or RB is used followed by the chemical symbol of the principal constituent metals in the wire. The initials for one or two elements will follow. If there is more than one alloy containing the same elements a suffix letter or number may be added.

The American Welding Society's specifications are most widely used in the U.S.A. for specifying bare welding rod and electrode wires. There are also military specifications such as the MIL-E or -R types and federal specifications, normally the QQ-R type and AMS specifications. The particular specification involved should be used for specifying filler metals.

The most important aspect of solid electrode wires and rods is their composition which is given by the specification. The chemistry referred to is the composition of the electrode or rod itself. The specifications provide the limits of composition for the different wires and mechanical property requirements. The specification should be referred to for exact information; however, some of the data are summarized in this book.

The bare electrode wires are identified in several different ways. When the wire is coiled on a spool a label is placed on the spool identifying the size and type of the electrode. When it is in coils or in drums, the label is placed on the drum or on a liner on the inside diameter of the coil.

For straight lengths of welding rod, two systems are used. For large-diameter nonferrous rods the classification number may be stamped in the metal. When the diameter is too small, small tags are stuck to each individual rod. These will show the classification number of the piece. An example of this is shown by Figure 10-9. Color coding has been used but is being supplanted by tags showing type numbers. In all cases the container holding the rods carries an identification label. When the rods are not tagged some users have devised various schemes of identifying the welding rod type. An example is shown by Figure 10-10. Here by means of one, two, or three bends, three different types can be identified.

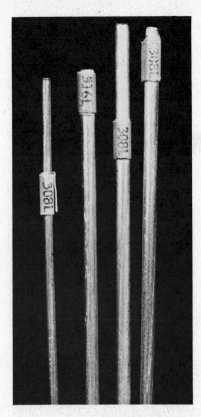

FIGURE 10-9 *Identification of welding rods—tags.*

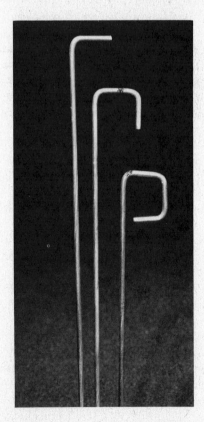

FIGURE 10-10 *Identification of welding rods—bending.*

Other bend configurations can be used; however, this is a limited and relatively expensive identification technique.

In the case of coiled electrode wire a maximum and minimum of cast and helix are specified. These two characteristics of the electrode wire have an important bearing on semiautomatic and automatic welding. The cast is the normal diameter of one loop of the coiled electrode wire when cut from the coil and laid on a flat surface. Normally the cast is greater than the diameter of the package coil. This helps the electrode wire feed through cables and guns. However, there are limits and these are specified. Helix is the distance that one loop will rise from another loop when two loops are laid freely on a flat surface. It should not exceed one inch. The arc will tend to wander as the wire comes out of the welding gun tip if helix is too great.

Occasionally, on copper-plated wires, the copper may flake off in the feed roll mechanism and create problems. It may plug liners, and contact tips. Therefore a light copper coating is desirable. The electrode wire surface should be reasonably free of dirt and drawing compounds. This can be checked by using a white cleaning tissue and pulling a length of wire through it. Too much dirt will clog the liners, will reduce current pickup in the tip, and may create erratic welding operation.

Temper or strength of the wire can be checked in a testing machine. Wire of a higher strength will feed better through guns and cables. The minimum tensile strength recommended by the specification is 140,000 psi (98 kg/mm^2).

Feedability is a measure of the ease with which the wire can be fed through a gun and cable assembly. It depends on all the factors just mentioned.

The continuous electrode wire is available in many different packages. They range from extremely small spools that are used on spool guns through medium-size spools for fine-wire gas metal arc welding. Coils of electrode wire are available which can be placed on reels that are a part of the welding equipment. There are also extremely large reels weighing many hundreds of pounds. The electrode wire is also available in drums or payoffpaks where the wire is laid in the round container and pulled from the container by an automatic wire feeder. Examples of these types of packaging and a summary of size and weights are shown by Figure 10-11.

10-4 FLUX-CORED OR TUBULAR ELECTRODES

The outstanding performance of the flux-cored arc welding process is made possible by the design of the cored electrode. This inside-outside electrode consists of a metal sheath surrounding a core of fluxing and

Type of Package	Diameter	Size of Package (Inch) Height & Width	Arbor Hole or Core	Weight of Solid Steel Wire in lb	Weight of Flux Cored Wire in lb
Spools	4″ O.D.	1-3/4	5/8	—	—
	4 × 2-9/16	1-3/4	5/8	2-1/2	—
	8	2-1/8	2-1/16	10	—
	12 (11-3/16)	4	2-1/16	25	—
	14	4	2-1/16	60	50-60
Coils	12 I.D.	4	—	60	60
	22-1/2 I.D.	4	—	120	—
Big spools or reels	22 O.D.	11	1-5/16	250	250
	30 O.D.	12-1/2 to 13-1/2	1-5/16	750	750
Drums or payoffpaks	20 O.D.	16	13	250	—
	20 O.D.	30	13	500	500-600
	23 O.D.	30	13	750	750
	23 O.D.	34	13	1000	—

FIGURE 10-11 *Summary of packaging for solid and tubuler electrode wire per AWS.*

alloying compounds. The compounds contained in the electrode perform essentially the same functions as the coating on a covered electrode, i.e., deoxidizers, slag formers, arc stabilizers, alloying elements, and may provide shielding gas.

There are three reasons why cored wires are developed to supplement solid electrode wires of the same or similar analysis.

1. There is an economic advantage. Solid wires are drawn from steel billets of the specified analyses. These billets are not readily available and are expensive. Also, a single billet might provide more solid electrode wire than needed.

2. Tubular wire production method provides versatility of composition and is not limited to the analysis of available steel billets.

3. Tubular electrode wires are easier for the welder to use than solid wires of the same deposit analysis, especially for welding pipe in the fixed position.

Figure 10-12 shows cross-sectional views of flux-cored electrodes. The tubular type is the most popular. The sheath or steel portion of the flux-cored wire comprises 75–90% of the weight of the electrode, and the core material represents 10–25% of the weight of the electrode.

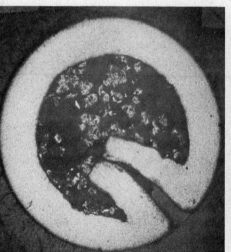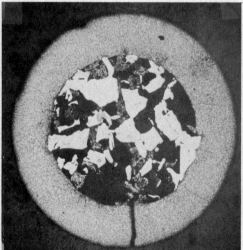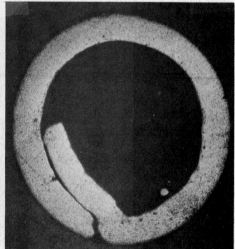

FIGURE 10-12 *Cross section of various types of flux cored electrodes.*

For a covered electrode, the steel represents 75% of the weight and the flux 25%. This is shown in more detail below:

Flux Cored Electrode Wire (E70T-1)			Covered Electrode (E7016)		
By area	Flux	25%	By area	Flux	55%
	steel	75%		steel	45%
By weight	Flux	15%	By weight	Flux	24%
	steel	85%		steel	76%

It is evident that more flux is used on covered electrodes than in a flux-cored wire to do the same job. This is because the covered electrode coating contains binders to keep the coating intact and also contains agents to allow the coating to be extruded.

The manufacture of the flux-cored electrode is an extremely technical and precise operation requiring specially designed machinery. Figure 10-13 shows the simplified flowchart of the manufacturing operation.

FIGURE 10-13 *Flow chart of operations to manufacture tubular electrode wire.*

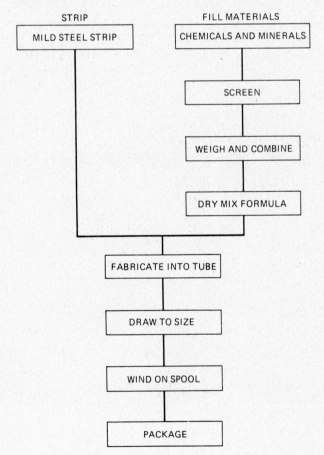

Figure 10-14 shows a simplified version of the apparatus for producing tubular cored electrodes. Thin, narrow, flat, low-carbon steel strip passes through forming rolls which makes it into a U-shaped section. This U-shaped steel passed through a special filling device where a measured amount of the formulated granular core material is added. The flux-filled U-shaped strip then passes through closing rolls which form it into a tube and tightly compress the core materials. This tube is then pulled through drawing dies which reduce its diameter and further compress the core materials. Drawing tightly seals the sheath and secures the core materials inside the tube, thus avoiding discontinuities of the flux. The electrode may or may not be baked during, or between, drawing operations. This depends on the type of electrode and the type of materials enclosed in the sheath.

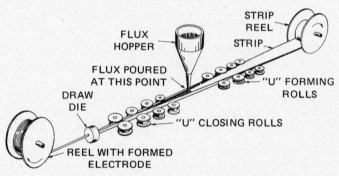

FIGURE 10-14 *Simplified manufacturing operation to make tubular wire.*

The drawing operations produce the various sizes of electrodes. The normal sizes are 1/8-in. (3.2-mm) diameter, 7/64-in. (2.8-mm) diameter, 3/32-in. (2.4-mm) diameter, 5/64-in. (2.0-mm) diameter, 1/16-in. (1.6-mm) diameter, and now 0.045-in. (1.1-mm) diameter. The 3/32-in. (2.4-mm) diameter is the most popular.

The finished cored electrode is packaged as a continuous coil, on spools, or in round drums.

Metal Transfer

Metal transfer from consumable electrodes across an arc has been classified into three general modes. These are spray transfer, globular transfer, and short circuiting transfer. High-speed movies of the arc indicate that the metal transfer of flux-cored electrodes more nearly resembles a fine globular transfer.

On cored electrodes in a CO_2 shielding atmosphere the molten droplets build up around the periphery or outer sheath of the electrode. The core material appears to transfer independently to the surface of the weld puddle.

At low currents, the droplets tend to be larger than when the current density is increased. The transfer from a 3/32-in. (3.2-mm) diameter electrode was observed at 350, 425, and 550 amperes. Transfer was much more frequent with smaller drops when the current was increased. At 550 amperes it appeared that some of the metal may have transferred by the spray mode, although the globular mode prevailed. The larger droplets at the lower currents caused a certain amount of "splashing action" when they entered the weld puddle. This action decreased with the smaller droplet size. This explains why there is less visible spatter, the arc appears smoother to the welder, and the deposition efficiency is higher when the electrode is used at high current rather than at the low end of its current range.

Mild Steel Electrodes

Carbon steel electrodes are classified by the American Welding Society specification, "Carbon steel electrodes for flux-cored-arc welding." This specification includes electrodes having no appreciable alloy content for welding mild and low alloy steels. The table of Figure 5-67 gives a summary of these electrodes, welding polarity, whether external gas shielding is or is not specified, test requirements, and the minimum mechanical properties.

The system for identifying flux-cored electrodes follows the same pattern as electrodes for gas metal arc welding, but is specific for tubular electrodes. As an example: E70T-1

E: Prefix indicates an electrode.

70: Indicates the required minimum as-welded tensile strength in thousands of pounds per square inch (psi).

T: Indicates tubular, fabricated, or flux-cored electrode.

1: Suffix number indicates the chemistry of the deposited weld metal, gas type, and usability factor.

The chemical composition requirements for the cored electrodes are shown by Figure 10-15. These requirements are based on the analysis of weld metal deposited by the electrode when run according to the test procedure.

E60T-7 Electrode Classification: Electrodes of this classification are used without externally applied gas shielding and may be used for single- and multiple-pass applications in the flat and horizontal positions. Due to low penetration, and to other properties, the weld deposits have a low sensitivity to cracking.

E60T-8 Electrode Classification: Electrodes of this classification are used without externally applied gas shielding and may be used for single- and multiple-pass applications in the flat and horizontal positions. Due to low penetration, and to other properties, the weld deposits have a low sensitivity to cracking.

E70T-1 Electrode Classification: Electrodes of this classification are designed to be used with CO_2 shielding gas for single- and multiple-pass welding in the flat position and for horizontal fillets. A quiet arc, high-deposition rate, low spatter loss, flat-to-slightly convex bead configuration, and easily controlled and removed slag are characteristics of this class.

E70T-2 Electrode Classification: Electrodes of this classification are used with CO_2 shielding gas and are designed primarily for single-pass welding in the flat

AWS Classification	Chemical Composition Requirements % (1) (2) (3)						
	Manganese	Silicon	Nickel	Chromium	Molybdenum	Vanadium	Aluminum
E60T-7	1.50	0.90	0.50	0.20 (4)	0.30 (4)	0.08 (4)	1.8
E60T-8	1.50	0.90	0.50	0.20 (4)	0.30 (4)	0.08 (4)	1.0
E70T-1	1.75	0.90	0.30 (4)	0.30 (4)	0.30 (4)	0.08 (4)	—
E70T-2							
E70T-3			No chemical requirements				
E70T-4	1.50	0.90	0.50	0.20 (4)	0.30 (4)	0.08 (4)	1.8
E70T-5	1.50	0.90	0.30 (4)	0.20 (4)	0.30 (4)	0.08 (4)	—
E70T-6	1.60	0.90	0.80	0.20 (4)	0.30 (4)	0.08 (4)	—
E70TG			No chemical requirements				

Note (1) Carbon, phosphorus, and sulphur not specified.

Note (2) Single values shown are maximum.

Note (3) Chemical composition requirements for electrodes are based on the analysis of deposited weld metal.

Note (4) The elements may be present but are not intentionally added.

FIGURE 10-15 *Chemical composition of flux cored electrode deposit per AWS A5.20.*

position and for horizontal fillets. However, multiple-pass welds can be made when the weld beads are heavy and an appreciable amount of admixture of the base and filler metals occurs.

E70T-3 Electrode Classification: Electrodes of this classification are used without externally applied gas shielding and are intended primarily for depositing single-pass, high-speed welds in the flat and horizontal positions on light plate and gage thickness base metals. They should not be used on heavy sections or for multiple-pass applications.

E70T-4 Electrode Classification: Electrodes of this classification are used without externally applied gas shielding and may be used for single- and multiple-pass applications in the flat and horizontal positions. Due to low penetration, and to other properties, the weld deposits have a low sensitivity to cracking.

E70T-5 Electrode Classification: This classification covers electrodes primarily designed for flat fillet or groove welds with or without externally applied shielding gas. Welds made using CO_2 shielding gas have better quality than those made with no shielding gas. These electrodes have a globular transfer, low penetration, slightly convex bead configuration, and a thin, easily removed slag.

E70T-6 Electrode Classification: Electrodes of this classification are similar to those of the E70T-5 classification, but are designed for use without an externally applied shielding gas.

E70T-G Electrode Classification: This classification includes those composite electrodes that are not included in the preceding classes. They may be used with or without gas shielding and may be used for multiple-pass work or may be limited to single-pass applications.

The E70T-G electrodes are not required to meet chemical, radiographic, bend test, or impact requirements; however, they are required to meet tension test requirements. Welding current type is not specified.

The flux-cored electrode wires are considered to be *low hydrogen,* since the materials used in the core do not contain hydrogen. However, certain of these materials are hygroscopic and thus tend to absorb moisture when exposed to a high-humidity atmosphere. Electrode wires are therefore packaged in special containers to prevent this. It is recommended that these electrode wires be stored in a dry room.

Stainless Steel Tubular Wires

Flux-cored tubular electrode wires are available which deposit stainless steel weld metal corresponding to the AISI compositions. These electrodes are covered by the AWS specification, "Flux-Cored Corrosion Resisting Chromium and Chromium-Nickel Steel Electrodes." These electrodes are identified by the prefix E followed by the standard AISI code number. This is followed by the letter T indicating a tubular electrode. Following this and a dash are four possible suffixes as follows:

-1   Indicates the use of CO_2 gas for shielding and DCEP.

-2   Indicates the use of argon plus 2% oxygen for shielding and DCEP.

-3   Indicates no external gas shielding and DCEP.

-G   Indicates that gas shielding and polarity are not specified.

Tubular or flux-cored electrode wires are also used for surfacing and submerged arc welding applications.

10-5 FLUXES FOR WELDING

There are a number of different types of fluxes used in welding, brazing, and soldering. These include fluxes for oxy fuel gas welding, fluxes for brazing, fluxes for soldering, fluxes for oxygen cutting of certain hard-to-cut metals, fluxes for electroslag welding, and fluxes for submerged arc welding. There are no specifications written for any of these fluxes. The American Welding Society provides a specification for weld metal deposited by different combinations of steel electrodes and proprietary fluxes for submerged arc welding. It is included in Table 10-1. For brazing, the American Welding Society provides a brazing flux type number and an indication of the ingredients of the flux. It has also recommended useful temperatures and the forms of the flux. There are various government specifications covering brazing fluxes. The manufacturers of brazing flux provide instructions for the application and use of their fluxes.

Fluxes for gas welding are covered by various government specifications. Manufacturers' recommendations should be followed.

Submerged Arc Flux

The major function of the submerged arc flux is to produce a slag which will protect the molten metal from the atmosphere by providing a mechanical barrier. When it is molten, this slag should provide ionization to permit a stable arc. It should be fluid and of relatively low density so that it will float and cover the top of the deposited weld metal. The melting temperature should be related to that of the molten weld metal and it should have a different coefficient of expansion so that it can easily be removed after cooling. The slag should provide deoxidizers to help cleanse and purify the weld metal. It should also help reduce phosphorous and sulphur that might be present in the base metal. It should not introduce hydrogen into the weld. Finally, the flux should be granular and convenient to handle, should not provide noxious fumes, but should provide for a smooth weld surface.

Submerged arc fluxes consist of mixtures of chemicals and minerals in various combinations to provide the properties just mentioned. Every grain of submerged arc flux should be similar in composition to the others and uniform in size. In the use of submerged arc flux, the granular material is placed over the welding joint and the heat of the arc causes it to melt and produce a molten slag. All of the flux placed over the weld does not melt and the unmelted flux can be removed and reused. Upon cooling the flux that melts transforms to a glasslike slag which must be removed from the weld deposit. The melted slag should not be used for welding since the deoxidizers, and other cleansing elements in the flux, are expended during melting.

There are three types of submerged arc welding fluxes based on the method of manufacturing.[1] The three types are: (1) fused flux, (2) agglomerated flux, and (3) bonded flux.

The ingredients for the flux must be ground, sized, and mixed prior to heating. In the case of *fused fluxes,* the mixture is melted in an electric or gas-fired furnace. The mixture is melted in a temperature range of 2,912°F (1,600°C). After melting, the molten flux is poured into water or onto a chilled plate to produce a glassy material. This material is then dried, crushed, and sized, by means of screens, packaged; and it is then ready for use.

The second method of manufacturing fluxes is the *agglomerated* method. Materials are dry mixed in the same way, except that a binder such as sodium or potassium silicate is added, after which the material is wet mixed. The mixture is then fed into a rotary kiln operating at approximately 1,832°F (1,000°C). Inside the kiln, by means of a tumbling process, the mixture forms into small balls and the ingredients tend to grow together and become larger. When they are properly heated these balls become very tough. After cooling, the balls are ground in the same manner as mentioned above, sized, packaged, and ready for use.

The third method of making fluxes is the *bonded* method, and this is very similar to the agglomerated method, except that the mixture is bonded at a lower temperature. After the pellets are bonded, hardened,

FIGURE 10-16 *Flow chart of operations to manufacture submerged arc flux.*

and cooled they are ground, sized, packaged, and ready for use. Figure 10-16 is a simplified flowchart that applies to either agglomerated or bonded fluxes.

Each manufacturing method produces fluxes that are suitable for submerged arc welding and each has certain advantages and disadvantages. In the case of the fused flux, the high temperatures involved require considerable more energy for production. In addition, many of the elements used for deoxidation are partially expended at the high heat temperature and thus must be enriched to provide sufficient deoxidizing power for welding. The advantage is that all of the grains of the flux are of uniform composition and, in general, the resulting flux is nonhygroscopic; that is, it will not pick up moisture.

In the case of the bonded and agglomerated fluxes the temperatures are lower and therefore less energy is consumed. Additionally, the deoxidizers are not dissipated and are therefore active during the welding operation.

353

The composition of the welding fluxes can be varied to provide ranges in the melting and solidification temperatures, the viscosity, the current-carrying capacity, the arc stability, welding speed capacity, shape and appearance of the weld, and the ease of the slag removal. The slag metal reaction during the welding operation is extremely complex and beyond the scope of this book. For additional information consult reference 1.

Submerged arc flux can be described as neutral, acidic, or basic while it is in the molten stage. If absolute neutrality cannot be obtained it is best then for the flux to be basic, so that it will reduce impurities in the weld metal. The most ideal flux will be metallurgically inactive, which means that the composition of the deposited weld metal will be the same as the composition of the welding electrode. This is not possible over the entire range of welding electrode compositions available and thus a loss or buildup of certain elements may occur. The flux can be used to increase the amount of alloy added to the deposit. Alloying elements can be added via the electrode or the flux. In general, it is more economical to add alloying elements through the flux. This is particularly important when doing surfacing or when welding on alloy base metals.

Sizing is established by means of controlled mesh screens set to both the upper and lower limits of the size allowed for each particle. For example, a 12-mesh screen allows all particle sizes smaller than a certain size to fall through the screen. At the other end would be a 200-mesh screen, which would keep all the particles except the very fine particles from falling through. The resulting flux would have a maximum size allowed through a 12-mesh, a minimum size greater than that allowed to fall through the 200-mesh screen. If there are too many fines in the flux it will tend to reduce the freezing period of the flux and it cannot be used for high-speed or circumferential welding. Larger sizes provide for higher-speed welding but allow arc flash through the layer of flux. Manufacturers provide fluxes between two limits since these sizes are recommended for most welding applications.

Submerged arc fluxes are classified according to the mechanical properties of weld metal made with specific electrode wires whose compositions are given. The information is summarized in the table shown by Figure 10-17.

The flux is identified by a special classification system which uses the prefix letter F to designate a flux. The next digit indicates the minimum tensile strength in 10,000 psi of the weld metal. The next digit is a number or letter code that indicates the lowest temperature at which the impact strength of the weld metal will equal or exceed 20 ft-lb. This code is as follows:

Z: No impact requirement

0   20 ft-lb at 0°F

2   20 ft-lb at −20°F

4   20 ft-lb at −40°F

6   20 ft-lb at −60°F

The code digit is followed by a dash and then the letter E, which designates an electrode. The E is followed by

AWS-Flux* Classification	Strength psi	Yield Strength at 0.2% Offset Min psi	Elongation % in 2 in.	CHARPY V-NOTCH IMPACT STRENGTH	
				Digit Code	Indicating
F6Z-EXXX	62,000			Z	No impact requirement
F60-EXXX	to	50,000	22		
F62-EXXX	80,000			0	20 ft/lb at 0° F
F64-EXXX					
F66-EXXX				2	20 ft/lb at −20° F
F7Z-EXXX	72,000			4	20 ft/lb at −40° F
F70-EXXX	to	60,000	22		
F72-EXXX	95,000			6	20 ft/lb at −69° F
F74-EXXX					
F76-EXXX					

*The letters "EXXX" as used in this table stand for the electrode designations EL8, EL8K, etc. (see Fig. 10-8).

FIGURE 10-17 *Mechanical properties of flux-electrode combination—SAW.*

three digits which indicate the chemical composition of the electrode (see Figure 10-8). An example of this classification system would be: F62-EL12, which would indicate a flux that would produce a weld having a minimum tensile strength of 60,000 psi and an impact strength of 20 ft-lb at −20°F when using a low manganese, approximately 0.12% carbon steel electrode.

In this specification if a specific flux-electrode combination meets the requirements of a given F6X-EXXX classification, the same combination also meets the requirements of all lower numbered classifications in the series. See the specification for more details.

Metallurgically inactive fluxes can be used for high-alloy welding in which the electrode wire matches the chemistry of the base metal. These fluxes are also recommended for submerged arc welding stainless steels. The electrode wire should be selected to match the base metal being welded.

Electroslag Fluxes

Electroslag fluxes are similar to submerged arc fluxes except that they are normally the fused type. Electroslag flux performs differently during the welding operation than submerged arc fluxes. The electrical conductivity of the flux makes the electroslag welding process operate. The flux becomes molten in a pool and the electrode wire melts in the heated bath. It is the resistance of the bath to the welding current flowing between the electrode and the work which maintains the bath at the high temperature. The flux is designed to provide a balance between conductivity and bath temperature for proper electroslag welding. In addition, the flux provides elements to purify and deoxidize the weld metal and also prohibits the oxygen and nitrogen of the air from coming in contact with the molten weld metal. The electroslag flux must have a lower density than steel so that it floats above the molten metal.

The criteria for selecting electroslag fluxes are based on the combination of flux and electrode wire. Tests must be made with standardized electrode wires and proprietary fluxes[2] to qualify procedures.

10-6 OTHER FILLER METALS

There are other filler metals and special items normally consumed in making welds. These include the nonconsumable electrodes—tungsten, carbon, and other materials, including backing tapes, backing devices, and flux additives. Another type of material consumed in making a weld are the consumable rings used for root pass welding of pipe. Additionally, there are ferrules used for stud welding and the guide tubes in the consumable guide electroslag welding method. Other filler materials are solders and brazing alloys.

Nonconsumable Electrodes

There are two types of *nonconsumable electrodes*. The carbon electrode is a nonfiller metal electrode used in arc welding or cutting, consisting of a carbon graphite rod which may or may not be coated with copper or other coatings. The second is the tungsten electrode, defined as a nonfiller metal electrode used in arc welding or cutting made principally of tungsten. The American Welding Society does not provide specification for carbon electrodes but there is a military specification, no. MIL-E-17777C, entitled, "Electrodes Cutting and Welding Carbon-Graphite Uncoated and Copper Coated." This specification provides a classification system based on three grades: plain, uncoated, and copper coated, also copper coated with lock joint ends. It provides diameter information, length information, and requirements for size tolerances, quality assurance, sampling, various tests, etc. Most manufacturers of carbon electrodes provide information indicating the type and size to be used for a specific application. Applications include carbon arc welding, twin carbon arc welding, carbon cutting, and air carbon arc cutting and gouging.

The American Welding Society provides a specification for tungsten electrodes entitled, "Tungsten Arc Welding Electrodes," which is shown by Table 10-1. These electrodes are used for gas tungsten arc welding, plasma arc welding, and atomic hydrogen arc welding. This specification provides for four classes of electrodes in various size diameters and lengths and with two types of finish. The four classifications relate to the composition of the tungsten whether it is pure tungsten or tungsten with small amounts of thorium or zirconium added to improve electron emission. The information concerning tungsten electrodes was covered in Section 4-3.

Backing Materials

Backing materials are being used more frequently for welding. Special tapes exist, some of which include small amounts of flux, which can be used for backing the roots of joints. There are also different composite backing materials, for one-side welding. There are no specifications covering these materials, but more information about them will be provided in Section 20-3.

Consumable rings are used for making butt welds in pipe and tubing. These are rings made of metal that are tack welded in the root of the weld joint and are fused into the joint by the gas tungsten arc. There are three basic types of rings called consumable insert rings and there is no specification covering them. The rings are known as the EB ring, the Grinnel ring, and the Y ring. These names are derived from the cross section of the ring or from the original developer of the ring. The rings are available in different analyses of metal based on normal specifications. More information concerning the consumable rings will be provided in Section 20-1.

Submerged Arc Flux Additives

Specially processed metal powder is sometimes added to the flux used for the submerged arc welding process. Additives are provided to increase productivity or enrich the alloy composition of the deposited weld metal. In both cases, the additives are of a proprietary nature and are described by their manufacturers indicating the benefit derived by using the particular additive. Since there are no specifications covering these types of materials the manufacturer's information must be used.

Electroslag Guide Tubes

There are two types of guide tubes in common use for the consumable guide method of electroslag welding. Guide tubes may be bare or covered. When they are covered they are covered with a coating material that has a composition similar to the composition of the electroslag slux. Both types of tubes are consumed while the weld is being made, and the metal part of the tube becomes a portion of the deposited weld metal. The guide tube is a relatively small percentage of the deposit. The covered guide tube utilizes the flux covering to augment the flux used in the electroslag welding process. There are no AWS specifications for guide tubes; however, they are normally seamless steel tubes of a low-carbon composition. The AISI C1010 composition is often used. Guide tubes are specified by the inside and outside diameter. The flux covering is proprietary and is compatible with the same manufacturer's electroslag flux.

Ceramic Ferrules

Ceramic ferrules are used in the stud welding process. These are small, specially designed short hollow cylinders that fit over the end of the stud and protect the molten metal from the atmosphere during welding. The ferrules also help mold the molten weld metal to an acceptable

weld contour. Ceramic ferrules are available for all different sizes of round studs and for many square or rectangular types. They are available from the stud manufacturer and are made to fit the stud sizes available. A ferrule is used only once and is easily broken away from the weld since it is very brittle. All manufacturers of studs provide the ceramic ferrules. No specifications exist for these items.

Solders

There are many different solder compositions and they are considered filler materials. Specifications for solder are issued by the American Society of Testing and Materials. The information about the different solders was summarized in Section 6-3.

Brazing Alloys

The brazing alloys are covered by a specification issued by the American Welding Society shown in Table 10-1. The information about the different brazing alloys was summarized in Section 6-1.

10-7 SHIELDING GASES USED IN WELDING

In all of the arc welding processes and in many others, the weld metal and adjacent base metal are molten for a short time. When molten metal is exposed to air it becomes contaminated from the gases in the air. By volume, air is made of a mixture of 20.99% oxygen, 78.03% nitrogen, 0.94% argon, and 0.04% other gases, primarily carbon dioxide and hydrogen.

The three gases that cause the most difficulty in welding are oxygen, nitrogen, and hydrogen. When any of the welding processes are used, the molten puddle should be shielded from the air in order to obtain a high quality weld deposit. In the submerged arc and electroslag welding process, the molten metal is shielded from the air by a flux. In the shielded metal arc welding process, shielding from the air is accomplished by gases produced by the disintegration of the coating in the arc. In carbon arc welding, the slow burning away of the carbon electrode produces an atmosphere of carbon monoxide and carbon dioxide which shields the molten metal.

In the gas welding processes and in torch brazing the products of combustion of fuel gas with oxygen shields the molten metal from the atmosphere. As the fuel gas burns, the products of primary combustion are carbon monoxide and hydrogen. The gas flame envelops the welding area, the air is excluded, and the molten metal is exposed only to these two gases. These are reducing gases. The products of second-

Property	Argon	Carbon Dioxide	Helium	Nitrogen	Oxygen	Air	Hydrogen
International symbol and cylinder marking	Ar	CO_2	He	N_2	O_2	Air	H_2
Type of gas	Inert	Active oxidizing	Inert	Not true inert gas	Active oxidizing	Mixture oxidizing	Active reducing
Molecular weight	39.94	44.01	4.003	28.016	32.00	28.98	2.016
Boiling point (at 1 atm) F°	−302.6	−109*	−452.1	−320.5	−297.3	−317.8	−422.9
C°	−184	−178	−269	−196	−182	−194	−252
Specific volume (cu ft/lb) @ 70°F, 1 atm.	9.67	8.76	96.71	13.8	12.1	13.4	192.0
Density (lb/cu ft) @ 70°F and 1 atm.	0.1034	0.1125	0.0103	0.0725	0.0828	0.0749	0.0052
Specific gravity (Air = 1)	1.38	1.53	0.137	0.967	1.105	1.00	0.0695
Thermal conductivity Btu/hr	0.0093	0.0085	0.0823	0.0146	0.0142	0.0140	0.096
Ionization potential (electron volts)	15.7	14.4	24.5	14.48	13.6	−	13.5
Maximum allowable concentration	Non toxic asphyxiant	Non toxic 5,000 ppm	Non toxic asphyxiant	Non toxic 32%	25%	100%	Non toxic asphyxiant

*Sublines directly form a solid to a gas at −109°F (−178°C) and a pressure of 1 atmosphere.

Note: The shipping containers of all of these gases except hydrogen would be marked "Non-Flamable Compressed Gas". The shipping containers of hydrogen would be marked "Flamable Compressed Gas".

TABLE 10-2 *Properties of shielding and active gases.*

ary combustion, which is the reaction of the carbon monoxide and the hydrogen with air, are carbon dioxide and water vapor. These processes could be considered gas shielded processes.

With the gas shielded arc welding processes, shielding from the air is accomplished by surrounding the arc area with a localized gaseous atmosphere provided for this purpose. The air is displaced from the arc area by this gas. It is essential to maintain an efficient shielding atmosphere throughout the welding operation at the molten puddle area. The characteristics of shielding and active gases are given by Table 10-2.

Oxygen

Oxygen is a colorless, odorless, tasteless gas that supports life and makes combustion possible. Oxygen constitutes about one-fifth of the atmosphere and combines with many elements to form oxides. Oxygen is highly reactive and combines with most hot metals at high temperatures. Oxygen combines with iron to form compounds which can remain in the weld metal as inclusions. As the molten weld metal cools free oxygen in the arc area will combine with carbon of the steel and form carbon monoxide. This gas and oxygen may be trapped in the weld metal as it solidifies. The gases collect into pockets which cause pores or hollow spaces. This problem is often overcome by providing deoxidizers such as manganese and silicon

and others which will combine with the oxygen to produce an oxide of manganese or silicon, which will float to the surface of the molten steel.

Nitrogen

Nitrogen comprises the largest single element of the atmosphere. It is colorless, odorless, flavorless, non-toxic and is almost an inert gas. Nitrogen does not burn nor support combustion. In the arc, or at high temperatures, nitrogen will combine with other gases. It is soluble in molten iron but at room temperature the solubility of nitrogen in iron is very low. Therefore, during the cooling and solidification process the nitrogen collects in pockets or precipitates out as iron nitrites. In very small amounts, nitrites can increase the strength and hardness of steel but reduce its ductility. In larger amounts, nitrogen can lead to porosity in the weld deposit. The reduction of ductility due to the presence of iron nitrites may lead to cracking of the weld metal.

Hydrogen

Hydrogen is the lightest gas and is present in the atmosphere in concentrations of about 0.01% at lower altitudes. Hydrogen may also be present in the arc area from water vapor resulting from the products of combustion and also from high temperature reaction with hydrocarbons that might be present.

Hydrogen has a bad effect on the properties of weld metal. Hydrogen is soluble in molten steel but the solubility of hydrogen at room temperature is very low. As molten weld metal cools and solidifies the hydrogen is rejected from the solution and becomes entrapped in the solidifying weld metal. It will collect at grain boundaries or at discontinuities of any type where it will create high pressures, which in turn cause high stresses within the weld. These pressures and stresses lead to minute cracks in the weld metal which can develop into larger cracks. The small concentrations of hydrogen that appear on the fractured surface are known as fish eyes because of their characteristic appearance. Hydrogen also causes underbead cracking in the heat-affected zone. Hydrogen will, however, gradually escape from the solid steel over a period of time. This migration of hydrogen from the weld metal is accelerated if the temperature of the metal is increased.

Gases for Shielding

Various gases are used for arc shielding. These can be inert gases such as helium or argon, or they can be mixtures of these and other gases. All gases have different properties and must be selected for shielding based on the particular metal to be welded, the type of the metal transfer required, and the economics involved. Inert gases will not combine chemically with other elements. There are six truly inert gases: argon, helium, neon, krypton, xenon, and radon. All of these except helium and argon are much too rare and expensive to use for gas shielded welding.

The most commonly used active gas for shielding is carbon dioxide. CO_2 is used when welding steels using the gas metal arc process. It is not an inert gas and compensation must be made for its oxidizing tendencies.

Argon

Argon is colorless, odorless, tasteless, and nontoxic. It is inert and forms no known chemical compound. It is relatively plentiful compared to the other inert gases. A million cubic feet of air contain 93,000 cubic feet of argon. It is separated from air by liquifying the air under pressure and low temperatures and then it is allowed to evaporate by raising the temperature. The argon boils off from the liquid at a temperature of $-302.6\,°F$ ($-184\,°C$). For welding the purity of the argon is approximately 99.995%. Argon is relatively heavy, approximately 23% heavier than air. It is used as a shielding medium for gas tungsten arc welding and for gas metal arc welding of nonferrous metals. Argon has a relatively low ionization potential and the arc voltage of the gas tungsten arc in argon is considerably lower than in helium. The welding arc tends to be stable in argon and for this reason it is used in many shielding gas mixtures. Argon is nontoxic but can cause asphyxiation in confined spaces by replacing the air. Argon is specified by a military specification.[3]

Helium

Helium is the second lightest gas. It is 1/7 as heavy as air. It is inert, has no color, odor, or taste, and is nontoxic. In liquid form it is the only known substance to remain fluid at temperatures near absolute zero. Helium is obtained from natural gas and in the Texas field it represents 2% of the volume. It is found also in natural gas in Canada and the U.S.S.R. Helium has the highest ionization potential of any of the shielding gases and for this reason a gas tungsten arc in helium has an extremely high arc voltage. Because of this, arcs in an atmosphere of helium produce a larger amount of heat. Helium is used in mixtures with argon and with active gases. Because of its light weight helium tends to float away from the arc zone, thus producing an inefficient shield unless higher flow rates are employed. For overhead welding, this can be helpful. It is used for gas tungsten arc welding. It is often mixed with other gases for gas metal arc welding. Helium for welding is sometimes in scarce supply.

Carbon Dioxide

Carbon dioxide is a compound of about 27% carbon to 72% oxygen. It is made of two oxygen atoms joined with a single atom of carbon. At normal atmospheric temperature and pressure it is colorless, nontoxic, and it does not burn. It has a faintly pungent odor and a slightly acid taste. At normal atmospheric temperature and pressure it is considered inert. It is about one and one-half times heavier than air and in confined spaces it will displace the air. At elevated temperatures it will disassociate into oxygen and carbon monoxide. In the welding arc disassociation takes place to the extent that 20–30% of the gas in the arc area is carbon monoxide and oxygen. Thus, CO_2 has oxidizing characteristics in the welding arc. As the carbon monoxide leaves the arc area it quickly recombines with oxygen to form CO_2. Extensive measurements have been made and it has been found that the carbon monoxide level at a distance of 7 in. (175 mm) from the welding arc is 0.01% or 100 ppm, which is regarded as a safe limit for carbon monoxide gas. At a distance of 12 in. from the arc the carbon monoxide concentration is 0.005%. A concentration of 5,000 ppm of carbon dioxide is considered as a safe level and ventilation can be easily provided to keep the CO_2 level below this concentration.

Carbon dioxide can exist simultaneously as a solid, a liquid, and a gas at its triple point. At atmospheric pressure solid CO_2 transforms directly to a gas without passing through the liquid phase; that is, it sublimes. At temperatures and pressures above the triple point and below 87°F in a closed cylinder carbon dioxide liquid and gas exist in an equilibrium. This is normally the way it occurs in cylinders.

Carbon dioxide is manufactured from flue gases, given off by the burning of natural gas, fuel oil, or coke. It is also obtained as a by-product of calcinining operations of lime kilns, from the manufacturing of ammonia, and from the fermentation of alcohol. The CO_2 gas is cleaned, purified, and dried before packaging. The purity of carbon dioxide gas can vary considerably depending upon the process of manufacture. A federal specification governs the degree of purity for welding grade CO_2 gas.[4] The specification covers two classifications of CO_2. Grade B, nonmedical, Type 1, with very little moisture content for special uses covers welding grade CO_2. The purity specified for welding grade CO_2 gas is a minimum dew point temperature of −40°F (−40°C). The table given by Figure 10-18 shows the dew point of CO_2 versus the percent of moisture in the gas. The standard provides a minimum dew point of −40°F (−40°C); however, many manufacturers produce welding grade CO_2 gas with a dew point temperature as low as −70°F (−57°C). This gas has a moisture content of 0.0091% by weight and/or 9 parts per million (ppm). Dew point can be measured using commercial instruments to determine if the gas meets its standard. Too much moisture in the gas will cause weld porosity.

Dew Point			
°F	°C	% Moisture by Weight	ppm Moisture in CO_2
−90	−68	0.00021	2
−80	−62	0.00043	4
−70	−57	0.00091	9
−60	−51	0.00188	19
−50	−46	0.00365	36
−40	−40	0.0066	66
−30	−34	0.0120	120
−20	−29	0.0218	218
−10	−23	0.0354	354
− 0	−17.8	0.0590	590
10	−12.2	0.0980	980

FIGURE 10-18 *Dew point vs percent of moisture in CO_2.*

Welding with the Different Gases

There are several factors that should be considered in selecting gases for shielding arc welds. The selections must relate to the welding process.

When welding with the gas tungsten arc process inert gases only are used. However there are exceptions: one is a mixture of inert gas with a reducing active gas such as hydrogen. Another is the use of an extremely small amount of oxygen. Another is the use of nitrogen alone or mixed with an inert gas.

The following is a review of the gases and their use for shielding in arc welding. This information is summarized by the table shown by Figure 10-19.

Argon Plus Oxygen

The poor bead contour and penetration pattern obtained with argon, when gas metal arc welding on mild steel, is improved with the addition of oxygen. Small amounts of oxygen added to the argon produce significant changes. Oxygen is normally added in amounts of 1, 2, or 5%. The amount of oxygen which can be employed is limited to 5%. Additional oxygen might lead to the formation of porosity in the weld deposit.

Oxygen improves the penetration pattern by broadening the deep penetration finger at the center of the weld. It also improves bead contour and eliminates the undercut at the edge of the weld.

Argon Plus Helium

This gas mixture is used for gas tungsten arc welding of nonferrous materials. The addition of helium in percentages of 50% or 75% raises the arc voltage and increases the heat in the arc. Helium is particularly useful for welding heavier sections of aluminum, magnesium, and copper and for overhead position welding.

Argon Plus Carbon Dioxide

A popular mixture of argon and carbon dioxide is 75% argon and 25% CO_2; however, 80% argon and 20% CO_2 is also used for steel applications. It is used on thinner steels or when deep penetration is not necessary and when bead appearance is important. This gas mixture will provide improved appearance over straight CO_2. It is also helpful for out-of-position welding on extremely thin sheet metal.

Shielding Gas	Composition of the Gas	Welding Process	Gas Reaction	Used For	Remarks
Argon	Ar	GTAW GMAW PAW	Inert	GTAW—all metals GMAW—non ferrous metals	Least expensive; Inert Gas
Argon + helium	50% Ar 50% He	GTAW GMAW	Inert	Al, Mg and copper and their alloys	Higher heat in arc use on heavier thickness-less porosity
Argon + hydrogen	Ar + 2% to 4% H_2	GTAW PAW	Reducing	Nickel and nickel alloys also for austinitic stainless steel (300)	Reduces porosity in nickel and nickel alloys-better appearance
Argon + oxygen	Argon + 1-2% O	GMAW	Oxidizing	Mild, low alloy and stainless steel	Provide spray transfer
Argon + oxygen	Argon + 3-5% O	GMAW	Oxidizing	Stainless steel	Oxygen provides arc stability
Argon + Carbon Dioxide	75% Ar 25% CO_2	GMAW FCAW	Slightly Oxidizing	Mild and low alloy steels (also some stainless with GMAW	Smooth weld surface reduces penetration
Helium	He	GTAW PAW	Inert	Al, Mg and copper and their alloys	High heat and voltage in arc—expensive
Helium + argon + carbon dioxide	90% He + 7.5% Ar + 2.5% CO_2	GMAW	Essentially inert	Stainless steel and some alloy steels	Provides arc stability helpful in out of position
Helium + argon	75% He 25% Ar	GTAW GMAW	Inert	Al and alloys copper and alloys	Higher heat input then Ar. min Porosity
Carbon dioxide	CO_2	GMAW FCAW	Oxidizing	Mild and low alloy steels (also on some stainless with GMAW)	Least cost gas deep penetration
Carbon dioxide + oxygen	CO_2 + up to 20% O_2	GMAW	Oxidizing	Mild steels	Used in Japan not popular in U.S.
Nitrogen	N_2	GTAW	Essentially inert	Copper and copper alloys and purging stainless steel pipe and tubing	Has high heat input not popular in U.S.

FIGURE 10-19 *Summary of shielding gases and mixtures and their use.*

Argon Plus Hydrogen

Small amounts of hydrogen, up to 5%, are added to argon. The hydrogen increases the arc voltage and increases the heat in the arc. This gas mixture should not be used for welding mild or low-alloy steels. Its major usefulness is found for welding nickel and nickel alloys. Some of the heavier stainless steels are also welded with this mixture.

Carbon Dioxide

Carbon dioxide gas eliminated many of the undesirable characteristics that were obtained when using argon. With carbon dioxide a broad, deep penetration pattern is obtained. This makes it easier for the welder to reduce defects such as lack of penetration and lack of fusion. Bead contour is good and there is no tendency toward undercutting. Another advantage of CO_2 shielding is its relatively low cost when compared to inert gases.

The chief drawback of the CO_2 gas is the tendency for the arc to be somewhat violent. This can lead to spatter problems when welding on thin materials. However, for most applications this is not a major problem and the advantages of CO_2 shielding far outweigh its disadvantages.

Nitrogen

Nitrogen is not a true inert gas and should not be used as a shielding gas for welding steels. It is used in some parts of the world for welding copper. It does provide a high heat arc and can be used for copper, although mixtures of argon and nitrogen produce better appearing welds. Nitrogen is often used for purging stainless steel pipe and tubing systems. It is less expensive than argon for purging and it does keep oxygen away from the back side of a weld root bead.

10-8 FUEL GASES FOR WELDING AND CUTTING

Oxygen and sometimes air is used with various hydrocarbon fuel gases for producing heat by means of chemical combustion. These fuel gases, usually with oxygen, are used for soldering, brazing, welding, oxygen cutting, flame spraying, flame hardening, and flame straightening. The major fuel gases are acetylene, natural gas, liquid petroleum gases (propane and propylene), and synthetic gases such as methylacetylene propadiene. The only fuel gases used for welding are compounds of carbon and hydrogen which will react with oxygen to produce a flame having a temperature above the melting point of most metals. Nonhydrocarbon fuel gases should not be used for welding since their products of combustion are toxic.

The fuel gas–oxygen reaction is in two steps. The primary reaction produces carbon monoxide and hydrogen plus heat.

$$\text{Fuel Gas} + O_2 \longrightarrow CO + H_2 + \text{Heat}$$

The secondary reaction which utilizes oxygen from the air, will oxidize the carbon monoxide and hydrogen to carbon dioxide and water vapor plus additional heat.

$$\text{Air} + CO + H_2 \longrightarrow CO_2 + H_2O + \text{Heat}$$

This complete combustion reaction produces a large amount of heat known as the gross heat of combustion (heat of primary reaction plus heat of secondary reaction). This is given in Btu's per pound of fuel gas or Btu's per cubic foot of fuel gas.

The information, shown by Table 10-3, gives the properties of fuel gases. This table also shows the flame temperatures of each of the fuel gases in oxygen and in air. Flame temperature in oxygen is always much higher than in air. The flame temperature and heat of combustion are indications of the amount of work that can be done by the different fuel gases. However, when comparing the cost of using different gases it is also important to consider the ratio of fuel gas to oxygen required for combustion. This is necessary so that the cost of both the fuel gases and the cost of oxygen are combined to obtain the total gas cost. Unfortunately, these data are theoretical since

in actual use a portion of the oxygen required for total combustion comes from the air surrounding the flame. For example, the combination of acetylene and oxygen for the primary reaction is a one-to-one ratio. The additional oxygen required for the secondary reaction requires 1.5 units of extra oxygen; therefore, the total ratio of oxygen to acetylene is 2.5 rather than 1. This is determined by working out the chemistry of both the primary and secondary reactions. The primary reaction is produced in the inner cone and the secondary reaction in the outer envelope of the flame.

Another important consideration in selecting fuel gas is its specific gravity. Hydrogen is the lightest in weight of all gases. Propane, propylene, and methylacetylene are all heavier than air. Acetylene is slightly lighter than air. Methane and natural gas are slightly over half the weight of air. This datum shows that some of the fuel gases would tend to float away into the atmosphere while others would collect in low spots, in enclosed areas of weldments, or in pits and bottoms of tanks. This is a very important safety consideration since fuel gas leakage can occur.

Another safety factor is the flammability limits in air. Acetylene will burn in air with a minimum of 2.5% to a maximum of 81%. This is the widest range of any fuel gas; however, hydrogen is almost as wide. The other gases are much lower. This means that acetylene is the most dangerous, since it will ignite in any percentage with air between these two limits. Table 10-3 also shows the threshold limit values (TLV) of the different gases.[5]

Acetylene (C₂H₂)

Acetylene is a compound of carbon and hydrogen. It is a colorless flammable gas slightly lighter than air. Acetylene of 100% purity is odorless, but gas of commercial purity has a distinctive garlic flavor. Acetylene burns in air with an intensely hot, yellow, luminous, and smoky flame. For safety reasons acetylene is never compressed above 15 pounds per square inch gauge (0.0105 kg/mm²). Acetylene cylinders are made safe by providing a porous mass of material inside the cylinder which is saturated with acetone. Acetylene dissolves in acetone and in this mode can be compressed to 250 psig (0.1750 kg/mm²) without danger.

Acetylene with oxygen produces the highest flame temperature of any of the fuel gases. It also has the most concentrated flame, but it produces less gross heat of combustion than the liquid petroleum gases and the

TABLE 10-3 *Properties of fuel gases.*

Property	Acetylene	Hydrogen	Methane	Methyl Acetylene Propadiene	Propane	Propylene	Natural Gas
International symbol and cylinder marking	C_2H_2	H_2	CH_4	$CH_3C{:}CH$ (MPS)	C_3H_8 (LP Gas)	C_3H_6 (PRY)	MET
Molecular weight	26.036	2.016	16.042	40.07	44.094	42.078	Similar to Methane
Specific gravity of gas (Air = 1)	0.91	0.069	0.55	1.48	1.56	1.48	0.56
Specific volume of gas (at 60°F and 1 atm) cu ft/lb	14.5	192.0	23.6	8.85	8.6	9.5	23.6
Specific gravity of liquid	–	–	–	0.576	0.507	0.527	–
Lb./gal of liquid @ 60°F	–	–	–	4.80	4.25	4.38	–
Density of gas—lb/cu ft	0.0680	0.0052	0.0416	0.113	0.115	0.105	0.0424
Boiling point (at 1 atm) °F	–119.2	–422.9	–258.6	–9.6	–43.8	–53.9	–161
°C	–84	–252	–161	–23.1	–42.1	–47.7	–107
Flame temperature (neutral) in oxygen °F	5600	4800	5000	5300	4600	5250	4600
in oxygen °C	3100	2650	2775	2925	2550	2900	2550
in air °F	4700	4000	3525	3200	3840	3150	3525
in air °C	2600	2200	1950	1760	2100	1730	1950
Ratio of oxygen to fuel gas required for combustion	1 to 1	.5 to 1	1.75 to 1	2.5 to 1	3.5 to 1	4.5 to 1	2 to 1
Ratio of air to fuel gas required for combustion	11.9	2.38	9.52	21.83	24.30	21.83	10.04
Gross heat of combustion Btu per pound	21,600	52,800	23,000	21,000	21,500	22,000	24,000
Btu per cubic feet	1,500	344	1,000	2,500	2,500	2,400	1,000
Flammable limits in air (by volume)	2.5 to 81%	4 to 75%	5.3 to 15%	2.4 to 11.7%	2.2 to 9.5%	2.0 to 10.3%	5.3 to 14%
Max allowable concentration in TLV's	Non toxic asphyxiant	Non toxic asphyxiant	Non toxic up to 9%	Non toxic 1000 PPM	Non toxic asphyxiant	Non toxic	Non toxic up to 25%

Note: The shipping containers of these gases would all be marked "Flammable Compressed Gas."

synthetic gases. Acetylene is manufactured by the reaction of water and calcium carbide. This is sometimes done at plant sites in acetylene generators. Acetylene is nontoxic; however, it is an anesthetic and if present in sufficiently high concentration it is an asphyxiant in that it replaces oxygen and will produce suffocation.

Hydrogen (H_2)

Hydrogen's properties were discussed in the previous section. Hydrogen can be used as a fuel gas and originally was an important commercial fuel gas. Its flammable limits in air range from 4–75%. When hydrogen is burned in either oxygen or air the flame temperature is lower than that of acetylene. It does require less oxygen for complete combustion but does not produce sufficient gross heat of combustion for industrial welding.

Methane (CH_4)

Methane is a colorless, odorless, tasteless, flammable gas. It is generally considered nontoxic, and concentrations of up to 9% can be inhaled without apparent ill effects. Methane is the major component of natural gas. It is normally separated from natural gas and can be obtained from petroleum. It is normally shipped and stored in high-pressure gas cylinders. It can, however, be shipped in liquid form in special insulated tanks at temperatures below its boiling point. It acts in the flame in the manner similar to natural gas.

Natural Gas (essentially CH_4)

Natural gas has much the same characteristics as methane. The composition of natural gas varies in different geographical locations and the gross heat of combustion of natural gas varies from one locality to another. The 1,000 Btu per cubic foot is normally accepted as a minimum. Natural gas is used in oxygen flame cutting. Its flame temperature is relatively low and the gross heat of combustion is also relatively low. It is less expensive than other fuel gases and for this reason has become quite popular. It is not used for gas welding or flame hardening because of its lower flame temperature. It is normally supplied via pipeline to industrial sites and is sold by the cubic foot.

Liquefied Petroleum Gases

The liquefied petroleum (LP) gases are propane and propylene (propene) and butanes. They are by-products of oil refineries and are flammable, colorless, non-corrosive, and nontoxic. They do have an anesthetic effect and when they displace oxygen in the air they act as asphyxiants. This is an important safety factor

since they weigh approximately 1-1/2 to almost 2 times the weight of air.

Pure propane is odorless; however, it is given an artificial odorization while propalene has an unpleasant odor characteristic of refineries. The flame temperature of propane is lower than that of acetylene but its gross heat of combustion is higher, more than 1-1/2 times that of acetylene. Propane is available in pure form and as mixtures. The mixtures contain additives such as ethylene, propylene, or ethyl ether, which increase the flame temperature and the heat of combustion. Additives also increase the price of the gas. Propane base gases are known as acetogen, Chemi-gas, Flamex, Hy-Temp, Chem-O-Lene etc.

Propylene has a higher flame temperature than propane but not as high as acetylene. It also has a gross heat of combustion approximately 1-1/2 times that of acetylene. Propylene is also available as pure gas and with additives and is given trade names as Apachi gas, HPG, B.T.U., Liquifuel, etc.

The liquefied petroleum gases require considerably more oxygen for combustion than acetylene. The liquefied petroleum gases are shipped and stored in the liquefied form in cylinders and tanks. They normally do not have pressures exceeding the 375 psi (0.2625 kg/mm^2). The liquefied petroleum vaporizes in the cylinder and is discharged as a gas. It is usually sold by weight. To determine the cubic feet of gas multiply by the specific volume of the gas shown by Table 10-3.

Synthetic Gases—Methylacetylene Propadiene (MPS)

The most popular synthetic hydrocarbon fuel gas is methylacetylene plus propadiene (allene), sometimes called methylacetylene propadiene stabilized or MPS gas. It is a by-product of the chemical industry. It goes by several trade names including MAPP gas, and Fuel-gas. This gas is colorless, flammable, and slightly toxic. The tentative maximum concentration of 1,000 parts per million has been suggested for its TLV. Methyl-acetylene propadiene stabilized has a flame temperature in oxygen higher than propane but less than acetylene. Its gross heat of combustion is over 1-1/2 times that of acetylene. It is stored and shipped as a liquefied gas in its own vapor pressure of about 60 psi (0.042 kg/mm^2) at 70°F (21.1°C). These gases are usually sold by weight and are available in cylinders and in bulk.

When using these different fuel gases different torches and tips are usually required.

Fuel Gas / Used For	Acetylene	LPG Propane	Natural Gas or Methane	MAAP
Heating	Not preferred	Yes	Yes	Yes
Torch soldering	Yes (in air)	Yes (in air)	Yes (in air)	Yes (in air)
Torch brazing	Yes	Yes	Yes	Yes
Oxygen cutting	Yes	Yes	Yes	Yes
Flame spraying	Yes	Yes	Yes	Yes
Gas welding	Yes	No	No	No
Flame hardening	Yes	No	No	Yes

FIGURE 10-20 *Uses of fuel gases.*

Selecting Fuel Gases

The selection of a fuel gas should be based on the gas that will provide required properties at the least cost. Necessary properties would include the flame temperature, the gross heat of combustion, and the oxygen-to-fuel gas ratio for combustion. This information is shown by Table 10-3. Some of the fuel gases can be used only for heating, for oxygen cutting, or for soldering and brazing. They cannot all be used for gas welding or for flame hardening. The various uses of a particular gas depend on its flame temperature, heat of combustion, heat distribution in the flame, and coupling distance. All fuel gases can be used for flame spraying; however, for spraying high melting temperature metals the higher flame temperature fuel gases must be used. All fuel gases can be used for heating but the type of heating might dictate the fuel gas. For example, acetylene is a more concentrated heat source than the other gases. All fuel gases have a higher temperature when the flame is oxidizing; however, an oxidizing flame cannot be used for welding.

For underwater oxygen cutting, acetylene can be used down to depths of 30 feet, but the methylacetylene propadiene gas can be used to depths of 100 feet. Special torches are required. This information is summarized by Figure 10-20.

The ratios of oxygen to fuel gases have an important bearing on the cost of the total operation since oxygen is expensive. The amount of oxygen required is difficult to determine since it depends on the type of torch employed. A single flame port torch is used for brazing or welding, and a multiflame port torch is used for flame cutting and heating. Multiport torches have a higher oxygen-to-fuel gas ratio since the inner flames are not able to receive oxygen from the air. Another factor to consider is the heat transfer or coupling. This is best done by the practical test.

The gross heat of combustion is an indication of how much work can be done by a given volume of fuel gas—hence, the amounts of oxygen and fuel gas that are required to do work. The measure of comparison here is to establish welding or cutting procedures which will include gas usage and work travel speeds. This information is available from torch and gas manufacturers. Then calculate the time required for a specific cut or weld and compare the results. It is wise to confirm the calculated times by making tests under controlled conditions. In making cutting speed tests make sure that the most efficient cutting tip is used for the conditions being tested. Differences in tips can be more of a determining factor than differences in gases. Remember also it is the oxygen jet that does the cutting.

It is becoming increasingly important to determine the availability of different fuel gases. As energy sources become less available and more expensive, the cost usefulness relationship can change.

10-9 GAS CONTAINERS AND APPARATUSES

The shielding gases and fuel gases must be transported and stored and available at the point of use. The most convenient way for transporting and storing is by portable cylinders which are easily taken to the job site. For installations where a high volume of gas is required the bulk storage system or manufacturing the gas at the site is used. This requires equipment to pipe the gas to the welding or cutting stations. The design of piping systems is complex and should be done only by experts who are familiar with safety regulations and codes. In bulk form, the argon is supplied as a liquid. The capacity of a bulk system is normally between 3,000 and 1 million cubic feet (84,950 to 28,310,000 liters).

Carbon dioxide can also be obtained in bulk containers. With the bulk system, carbon dioxide is usually drawn off as a liquid and heated to the gas state before entering the piping system. The bulk system is normally only used when supplying a large number of welding stations.

There are four basic types of cylinders used for transporting welding gases. They are shown by the table of Figure 10-21. In addition, many of these types of cylinders come in different sizes according to the gas producer. These high-pressure cylinders are commonly used for transporting and storing argon, oxygen, hydrogen, nitrogen, and helium. This same type of cylinder is used for mixtures of these gases and mixtures of argon with CO_2. The cylinder of this type is shown by Figure 10-22. These cylinders are made under very strict manufacturing procedures and are covered by various laws. In the U.S.A., the Department of Transportation provides the regulations. They are made of manganese steel (3A) or chrome molybdenum steel

(3AA) and each must be inspected, numbered, and reinspected at regular intervals, usually every five years.[6]

There are no uniform standards for cylinder sizes even though different gas companies offer standards within their own organization. There is no standard color code in the U.S.A. for the industrial gases. Some gas producers have standardized cylinder color codes within their own organization, however. There is a standardized identification system. Either the total name of the gas or the international symbol of the gas is required on each cylinder. In addition, each cylinder must carry a label showing the hazardous classification of the gas. This information is given in the two charts showing the properties of the gases.

Cylinder Identification	Cylinder Type	Cylinder Contents	Cylinder Capacity (cu ft)	Full Cylinder Pressure at 70° F psi	APPROXIMATE WEIGHT	
					Full lb	Empty lb
Nonflammable compressed gas	DOT Type 3A or 3AA	Argon	244 330	2,200 2,640	158 177	133 143
Nonflammable compressed gas	High Pressure	Argon + Oxygen	330	2,640	177	143
Nonflammable compressed gas		Argon + carbon dioxide	379	2,640	177	143
He non-flammable compressed gas		Helium	213	2,200	135	133
Hydrogen flammable compressed gas	See Fig. 10-22	Hydrogen	191	2,015	134	133
O_2 non-flammable compressed gas		Oxygen	330 244	2,640 2,200	172 153	146 133
CO_2 non-flammable compressed gas	See Fig. 10-23	Carbon dioxide liquid + gas	435	1,000	183	133
LPG or LP gas or PRY or MAAP	DOT B 240	Liquid under vapor pressure	Varies by gas and supplier	94	Varies by gas and supplier	Varies by gas and supplier
Flammable compressed gas						
C_2H_2 flammable compressed gas	DOT 8 AL see Fig. 10-24	Acetylene disolvent in acetone	390	250	207	180

Note: The cylinder capacity and weights will vary by supplier.

FIGURE 10-21 *Gas cylinder types and sizes.*

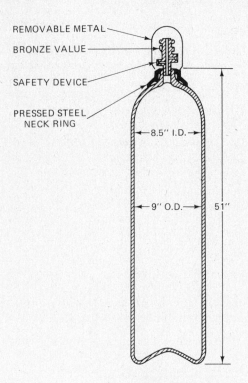

REMOVABLE METAL

BRONZE VALUE

SAFETY DEVICE

PRESSED STEEL
NECK RING

8.5" I.D.

9" O.D.   51"

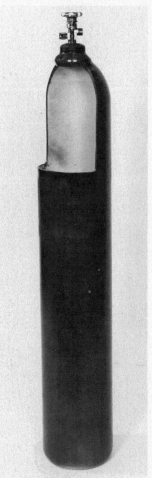

FIGURE 10-22 *High pressure gas cylinders.*

Fortunately, the valve connections of the different gas cylinders have been standardized so that regulators for the same gas can be readily attached to cylinders supplied by different gas producing companies. These standards apply to North America only.

Carbon dioxide (CO_2) welding grade gas is available to the user in high-pressure steel cylinders. Cylinders containing CO_2 are always labeled CO_2 and may be labeled "welding grade." They are usually aluminum colored but no standard code exists.

The standard welding grade carbon dioxide cylinder shown by Figure 10-23 contains approximately 50 lb (22.7 kg) or 435 cubic feet (12,317 liters) of carbon dioxide under a pressure of 1,000 psi (0.7 kg/mm^2). In the CO_2 cylinder, at 70°F (21.1°C) the carbon dioxide is in both a liquid and a vapor form. The liquid carbon dioxide takes up approximately two-thirds of the space in the cylinder. Above the liquid the CO_2 exists as a gas. As the gas is drawn from the cylinder, part of the liquid carbon dioxide vaporizes to replace it. The normal discharge rate of the CO_2 cylinder is from about 4 to 30 cubic feet per hour (2–14L/m). However, a maximum discharge rate of 25 cfh (12 L/m) is normally recommended when welding using a single cylinder. In a cold environment, the maximum discharge rate is reduced. As the CO_2 vapor pressure drops from the cylinder pressure to discharge pressure through the CO_2

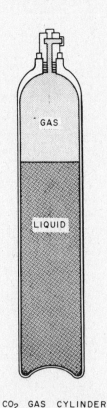

GAS

LIQUID

CO_2 GAS CYLINDER

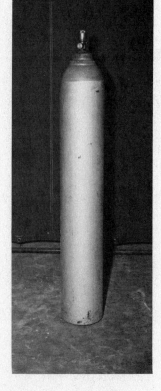

FIGURE 10-23 *Cylinders for carbon dioxide.*

regulator, it absorbs heat. If flow rates are set too high, this absorption of heat can lead to freezing of the CO_2 regulator. When this happens, the gas shield may be interrupted and weld porosity will result. When flow rates higher than 25 cfh are required, normal practice is to manifold two CO_2 cylinders in parallel or to place a heater between the CO_2 cylinder and the pressure regulator. As the carbon dioxide gas is drawn from the cylinder it is replaced with carbon dioxide that vaporizes from the liquid in the cylinder and therefore the overall pressure in the cylinder does not drop. However, as the liquid carbon dioxide is used, a drop in pressure will be indicated by the pressure gauge. When the pressure in the cylinder has dropped to 200 psi (0.1400 kg/mm^2) the cylinder should be replaced with a new one. A positive pressure should always be left in the cylinder in order to prevent moisture and other contaminants from entering.

Pressure is not an accurate measurement of cylinder contents and a partially used CO_2 cylinder should be weighed to determine how much CO_2 it still contains. To do this, first weigh the cylinder then subtract the *tare weight* (weight of cylinder when empty). Tare weight is usually stenciled on the cylinder neck. This gives the weight of the remaining contents. At 70°F, there are 84.7 cubic feet of CO_2 per pound.

The liquefied petroleum gases are transported and stored in a different type of cylinder, one that is made to handle materials at a lower pressure. They are usually larger since the pressure is not so high. They are made to DOT specification B240 and are similar to the high-pressure cylinders except that they are usually larger in diameter.

When a gas is confined to a specific volume, the pressure exerted on the walls of the cylinder will vary in direct proportion with the temperature. Estimating the volume of gas remaining in a cylinder on the basis of gauge pressure is possible only within very broad limits. This is especially true of the liquefied gases.

Acetylene is transported and stored in a very special type of cylinder shown by Figure 10-24. This type of cylinder made to DOT specification 8AL is used only for acetylene. As mentioned previously, it is filled with a porous material soaked with acetone and the acetylene is dissolved in the acetone.

Apparatus

Various pieces of apparatus are required to utilize gas from high-pressure cylinders. These include regulators and flowmeters or combination units. The gas regulator was described in Chapter 5. The function of a regulator is to reduce pressure and provide constant gas flow. Regulators must only be used for the gas for which they are designed.

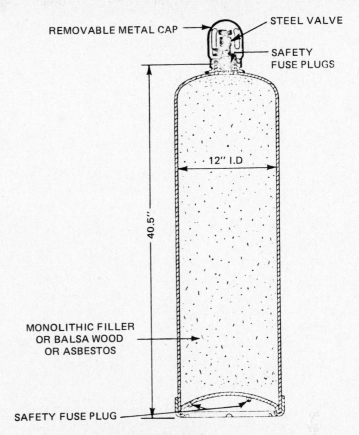

REMOVABLE METAL CAP

STEEL VALVE

SAFETY FUSE PLUGS

12" I.D

40.5"

MONOLITHIC FILLER OR BALSA WOOD OR ASBESTOS

SAFETY FUSE PLUG

FIGURE 10-24 *Cylinders for acetylene.*

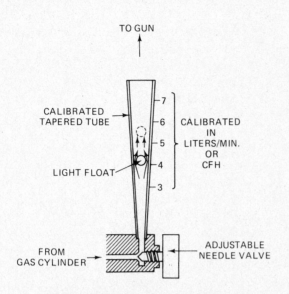

CALIBRATED
TAPERED TUBE

7

6   CALIBRATED
IN
5   LITERS/MIN.
OR
CFH

LIGHT FLOAT

4

3

FROM
GAS CYLINDER

ADJUSTABLE
NEEDLE VALVE

TO GUN

FIGURE 10-25 *Diagram of gas flowmeter.*

Another item of gas apparatus is the *flowmeter,* sometimes called a *rotometer.* The flowmeter contains two components, the adjustable needle valve which allows for very accurate control of gas flow and a slightly tapered transparent tube which contains a float or indicator. The gas enters the flowmeter through the needle valve and then passes upward through the tapered tube. As it passes upward the tube becomes enlarged and the float is suspended in the stream of gas. The higher the flow rate the higher the float will rise in the calibrated tube. This is shown by Figure 10-25. The tapered tube is calibrated in either cubic feet per hour or liters per minute. It is important that the flowmeter used is designed for the gas being measured. Different float weights are used for gases of different specific gravities. For extremely accurate work the discharge head or the resistance of the gas system beyond the flowmeter must be standardized and related to the calibration of the flowmeter. A typical flowmeter attached to a cylinder is shown by Figure 10-26. The flowmeter should be installed with the tube absolutely vertical for accurate measurement.

Permanent type or orifice type flowmeters can also be used; however, they are not adjustable and are installed in the line following the regulator to establish a specific flow rate of a specific gas. These are used when adjustments are not required.

Gas Flow Rates

The flow rates required with the various shielding gases depend upon the density or specific gravity

of the gases. The specific gravity of the gas is a measure of its density relative to the density of air, the density of air is understood to be one. The specific gravity of helium is 0.137. Helium tends to rise quickly from the weld area and higher flow rates are needed to provide adequate shielding. Helium is also greatly affected by cross drafts. The specific gravity of argon is 1.38 and the specific gravity of carbon dioxide is 1.52. Both of these gases tend to blanket the weld area and lower flow rates can be used. Carbon dioxide, being heaviest, gives the best protection. It also has the best resistance to cross drafts or wind currents. Welding procedures will provide the flow rates to be used.

When siphon tube CO_2 cylinders are used external heaters are required. These cylinders are equipped with a plastic siphoning tube extending to the bottom of the cylinder and they discharge liquid CO_2 rather than gaseous CO_2. The liquid must be vaporized in a heater prior to going into the gas shielding system. Siphon tube CO_2 cylinders are not popular in the U.S.A., although they can be obtained on special order. They are used in Europe and in other parts of the world.

The final element in the system is the hose leading from the cylinder to the torch or to the control panel. Different types of hose are available, including rubber and various plastic materials. Each type has different characteristics with respect to pressure and the ability to maintain purity of the gas in the cylinder to the arc. It is best to check manufacturers' literature for this information.

Finally, there are such items as proportioners or mixers which can be used to mix different gases together for a particular application. These are quite involved and should be used based on the manufacturer's recommendations.

FIGURE 10-26 *Gas flowmeter.*

QUESTIONS

1. What are the four basic filler metals? Describe the miscellaneous group.

2. List at least ten functions of a coating on a covered electrode.

3. What is meant by a low-hydrogen coating?

4. How are electrodes reconditioned after they become damp?

5. Explain *fingernailing* and reasons for it.

6. How are bare solid steel electrodes specified? Give an example.

7. How are *cast* and *helix* measured? How do they affect welding?

8. What is the function of the core of a flux-cored electrode?

9. Explain the specification system for flux-cored steel electrodes.

10. What welding processes normally use flux?

11. How is submerged arc flux made? Name two types.

12. What are the two types of consumable guides used for electroslag welding?

13. What is the most common inert gas used for shielding?

14. What happens to CO_2 in an arc? Away from the arc?

15. How is CO_2 gas specified? How is moisture checked?

16. What types of gases are used for GTAW welding?

17. Name the different fuel gases in common use.

18. What is the danger of fuel gases that are heavier than air?

19. Describe the different types of tanks used for transporting welding gases.

20. Explain the principle of operation of a flowmeter.

REFERENCES

1. C. E. Jackson, "Fluxes and Slags in Welding," *Welding Research Council Bulletin* No. 190, December 1973, New York, N.Y.

2. Howard B. Cary, "Porta-slag Welding," Book No. EW-412, Hobart Brothers Company, Troy, Ohio 45373.

3. "Military Specification Argon Technical and DOD," MIL-A-18455B, Washington, D.C.

4. Federal Specification No. BB-C-101A, Washington, D.C.

5. "Threshold Limit Values for Chemical Substances and Physical Agents in the Workroom Environment," The American Conference of Governmental Industrial Hygienists, Lansing, Michigan 48901.

6. "Code of Federal Regulations Support M—Compressed Gas and Compressed Air Equipment," Section 1910.166, "Inspection of Compressed Gas Cylinders," *Federal Register,* Vol. 36, No. 105, Superintendent of Documents, Government Printing Office, Washington, D.C.

11

WELDABILITY AND PROPERTIES OF METALS

11-1 **PROPERTIES OF METALS**

11-2 **METAL SPECIFICATIONS AND GROUPINGS**

11-3 **IDENTIFICATION OF METALS**

11-4 **HEAT AND WELDING**

11-5 **WELDING METALLURGY**

11-6 **WELDABILITY OF METALS**

11-1 PROPERTIES OF METALS

When you look at a piece of bare metal it may appear brownish, it may look bright or dull, it might appear grey; it has color. One of the physical properties of a metal is its color. When you lift a piece of metal it may seem to be heavy or light. When you bend a thin piece of metal it may break or it may bend easily because it possesses ductility. If you attempt to melt the metal with a flame it may become liquid quickly or it may not melt. Metals have different melting temperatures.

These are some of the many different properties of metals which help describe and specify them. Metals have physical properties such as density, melting point, color, conductivity, and others. They also have mechanical properties, which include strength, hardness, ductility, etc., and all of them can be tested. Many mechanical and physical properties of metals determine if and how they can be welded and how they will perform in service.

This section describes the different properties and how they affect the ease of welding. The mechanical properties of the common metals will be described and listed as well as the tests used to determine them. This will provide a better understanding of why some metals are specified for certain uses while others are not, and why some metals are more easily welded than others.

Physical Properties

The physical properties of metals are given in both the metric system and the conventional system. Physicists normally use the metric system while engineers more often use the conventional terms. In time, this will be standardized and all will use the new SI terms which are based on international standards. Table 11-1 shows the more common metals and their physical properties. Each is briefly described below:

Color relates to the quality of light reflected from the metal.

Mass or *density* relates to mass with respect to volume. One of the more common ways of describing this property is by means of specific gravity, which is the ratio of the mass of a given volume of a metal to the mass of the same volume of water. This is at a specified temperature usually 39°F (4°C). The mass of a volume of water is taken as unity and the metals are related to it. In conventional terms this is taken as pounds per cubic foot of the metal or pounds per cubic inch. In the metric system this is taken as grams per cubic millimeter or centimeter.

The melting point of a metal is extremely important with regard to welding. A metal's fusibility is related to its melting point, the temperature when the metal changes from a solid into a molten state. Mercury is

TABLE 11-1 *Physical Properties of Metals.*

Properties Base Metal Or Alloy	Specific Gravity	Density lb/ft³	Density gm/cc	Melting Point °F	Melting Point °C	Boiling Point °F	Boiling Point °C	Relative Thermal Conductivity Copper = 1	Co-efficient of Linear Expansion	Specific Heat Calories per/98 per °C	Electrical Conductivity in % Copper = 100%	Resistivity Microhms per cm
Aluminum and alloys	2.70	166	2.7	1220	660	3270	2480	0.52	13.8	0.22	59.0	2.8
Brass, navy	8.60	532	8.6	1630	885	NA	NA	0.28	11.8	0.09	28.0	6.6
Bronze, alum. (90Cu-9Al)	7.69	480	7.7	1890	1032	NA	NA	0.15	16.6	0.014	12.8	13.5
Bronze, phosphor (90 Cu-10Sn)	8.78	551	8.8	1550	845	NA	NA	0.12	10.2	0.09	11.0	16.0
Bronze, silicon (96Cu-3Si)	8.72	542	8.7	1780	971	NA	NA	0.10	10.0	0.09	7.0	NA
Copper (deoxidized)	8.89	556	8.9	1981	1081	4700	2600	1.00	9.8	0.095	100.0	1.7
Copper nickel (70Cu-30Ni)	8.81	557	8.8	2140	1172	NA	NA	0.07	9.0	0.09	4.6	37.0
Everdur (96Cu-3Si-1Mn)	8.37	523	8.4	1866	1019	NA	NA	0.09	10.0	0.095	NA	NA
Gold	19.3	1205	19.3	1945	1062	5380	2950	0.76	7.8	0.032	71.0	2.2
Inconel (72Ni-16Cr-8Fe)	8.25	530	8.3	2540	1390	NA	NA	0.04	6.4	0.109	1.5	98.1
Iron, cast	7.50	450	7.5	2300	1260	NA	NA	0.12	6.0	0.119	2.9	NA
Iron, wrought	7.80	485	7.8	2750	1510	5500	3000	0.16	6.7	0.115	15.0	NA
Lead	11.34	708	11.3	621	328	3100	1740	0.08	16.4	0.03	8.0	20.6
Magnesium	1.74	108	1.7	1240	671	2010	1100	0.40	14.3	0.246	37.0	5.0
Monel (67 Ni-30Cu)	8.47	551	8.8	2400	1318	NA	NA	0.07	7.8	0.127	3.6	48.2
Nickel	8.8	556	8.8	2650	1452	5250	3000	0.16	7.4	0.105	23.0	7.9
Nickel silver	8.44	546	8.4	1706	930	NA	NA	0.09	9.0	0.09	8.3	1.6
Silver	10.45	656	10.5	1762	963	4010	2210	1.07	10.6	0.056	106.0	–
Steel, low alloy	7.85	490	7.8	2600	1430	NA	NA	0.12	6.7	0.118	14.5	12.0
Steel, high carbon	7.85	490	7.8	2500	1374	NA	NA	0.17	6.7	0.118	9.5	18.0
Steel, low carbon	7.84	490	7.8	2700	1483	NA	NA	0.17	6.7	0.118	14.5	12.0
Steel, manganese (14 Mn)	7.81	490	7.8	2450	1342	NA	NA	0.04	6.7	0.210	NA	72.0
Steel, medium carbon	7.84	490	7.8	2600	1430	NA	NA	0.17	6.7	0.118	15.0	15.0
Steel, stainless (austentic)	7.9	495	7.9	2550	1395	NA	NA	0.12	9.6	0.117	3.0	75.0
Steel, stainless (matensitic)	7.7	485	7.7	2600	1430	NA	NA	0.17	9.5	0.118	3.0	57.0
Steel, stainless (ferritic)	7.7	485	7.7	2745	1507	NA	NA	0.17	9.5	0.334	3.0	60.0
Tantalum	16.6	1035	16.6	5162	2996	7410	5430	0.13	3.6	0.052	13.9	12.5
Tin	7.29	455	7.3	448	231	4100	2270	0.15	12.8	0.125	13.5	11.0
Titanium	4.5	281	4.5	3270	1800	5900	3200	0.04	4.0	0.113	1.1	42.0
Tungsten	18.8	1190	19.3	6170	3420	10600	5600	0.42	2.5	0.034	31.0	5.6
Zinc	7.13	442	7.1	788	419	1660	907	0.27	22.1	0.093	30.0	5.9

the only common metal that is in its molten state at normal room temperature. Metals having low melting temperatures can be welded with lower temperature heat sources. The soldering and brazing processes utilize low-temperature metals to join metals having higher melting temperatures.

The boiling point is also an important factor in welding. The boiling point is the temperature at which the metal changes from the liquid state to vapor state. In welding, some metals, when exposed to the heat of an arc will vaporize.

The thermal conductivity of a metal is its ability to transmit heat throughout its mass. It is of vital importance in welding since one metal may conduct or transmit heat from the welding area much more rapidly than another. It indicates the need for preheating and the size of heat source required. The thermal conductivity of metals is usually related to copper. Copper has the highest thermal conductivity of the common metals exceeded only by silver. Aluminum has approximately half the thermal conductivity of copper, and steels have only about one-tenth the conductivity of copper. Some data use silver as the standard and rate the thermal conductivity with respect to silver. Thermal conductivity is measured in calories per square centimeter per second per degree Celsius. However, since we are using a relative figure these are not used.

Specific heat is a measure of the quantity of heat required to increase the temperature of a metal by a specific amount. Specific heat is important in welding since it is an indication of the amount of heat required to bring the metal to its melting point. A metal having a low melting point but having a relatively high specific heat may require as much heat to bring it to its point of fusion as a metal of a high melting point and low specific heat. Specific heat is expressed in conventional as well as in metric forms. However, metric is more commonly used and it is the number of calories required to raise the temperature of one gram of metal one degree Celsius. It can be stated as a relative specific heat related to a standard. It is usually given at a standard temperature. Table 11-1 provides the specific heat of the different metals based on the calories per gram per degrees C at 20° Celsius.

The coefficient of linear thermal expansion is a measure of the linear increase per unit length based on the change in temperature of the metal. Expansion is the increase in the dimension of a metal caused by heat. The expansion of a metal in a longitudinal direction is known as the linear expansion. The coefficient of linear expansion is expressed as the linear expansion per unit length for one degree of temperature rise.

The expansion of the metal in volume is called the volumetric expansion. Linear expansion is most com-

monly used and the data are available in both the conventional and metric values. The coefficient of linear expansion varies over a wide range for different metals. Aluminum has the greatest, expanding almost twice as much as steel for the same temperature change. This is important for welding with respect to warpage, warpage control and fixturing, and for welding dissimilar metals together.

Electrical conductivity is the capacity of metal to conduct an electric current. A measure of electrical conductivity is provided by the conductance of a metal to the passage of electrical current. The reciprocal of conductivity is resistivity. Electrical resistivity is measured in micro-ohms per cubic centimeter at a standardized temperature, normally 20°C. Electrical conductivity, however, is usually considered as a percentage and is related to copper or silver. Temperature bears an important part in this property. As the temperature of a metal is increased conductivity decreases. This property is particularly important to resistance welding and to electrical circuits.

Mechanical Properties

The mechanical properties of metals determine the range of usefulness of the metal and establish the service that can be expected. Mechanical properties are also used to help specify and identify metals. They are of vital interest to welding since the weld must provide mechanical properties in the same order as the base metals being joined. The adequacy of a weld depends on whether or not it provides properties equal to or exceeding those of the metals being joined.

The most common properties considered are strength, hardness, ductility, and impact resistance. These properties of the common metals are shown by Table 11-2.

Strength

The strength of a metal is its ability to withstand the action of external forces without breaking. *Tensile strength,* also called *ultimate strength,* is the maximum strength developed in a metal in a tension test. The tension test is a method for determining the behavior of a metal under an actual stretch loading. This test provides the elastic limit, elongation, yield point, yield strength, tensile strength, and the reduction in area. Tensile tests are normally taken at standardized room temperatures but may also be made at elevated temperatures. Figure 11-1 shows a tensile testing machine in

TABLE 11-2 *Mechanical Properties of Metals.*

Properties Base Metal Or Alloy	YIELD STRENGTH			TENSILE STRENGTH			Elongation % in 2 in. (50mm)	Hardness BHN
	lb/in.^{2}	MPa	kg/mm^2	lb/in.^{2}	MPa	kg/mm^2		
Aluminum and alloys	5,000	34.5	3.5	13,000	89.6	9.1	35	23
Brass, navy	30,000	206.8	21.0	62,000	427.4	43.6	47	89
Bronze, alum. (90Cu-9Al)	30,000	206.8	21.0	76,000	523.9	53.4	10	125
Bronze, phosphor (90Cu-10Sn)	28,000	193.0	19.7	66,000	455.0	46.4	35	148
Bronze, silicon (96Cu-3Si)	15,000	103.4	10.5	40,000	275.8	28.1	52	119
Copper (deoxidized)	10,000	68.9	7.0	33,000	227.5	23.2	40	30
Copper nickel (70Cu-30Ni)	20,000	137.9	14.0	55,000	379.2	38.6	45	95
Everdur (96Cu-3Si-1Mn)	20,000	137.9	14.0	55,000	379.2	38.6	60	75
Gold	–	–	–	17,000	117.2	11.9	45	25
Inconel (76Ni-16Cr-8Fe)	35,000	241.3	24.6	85,000	586.0	59.7	45	150
Iron, cast	–	–	–	25,000	172.4	17.5	0.5	180
Iron, wrought	27,000	186.1	19.0	40,000	275.8	28.1	25	100
Lead	19,000	131.0	13.4	2,500	17.2	1.7	45	6
Magnesium	13,000	89.6	9.1	25,000	172.4	17.5	4	40
Monel (67 Ni-30Cu)	35,000	241.3	24.6	75,000	517.1	52.7	45	125
Nickel	8,500	58.6	6.0	46,000	317.1	32.3	40	85
Nickel silver	20,000	137.9	14.0	58,000	399.8	40.7	35	90
Silver	8,000	55.2	5.6	23,000	158.6	16.2	35	90
Steel, low alloy	50,000	344.7	35.1	75,000	517.1	52.7	28	170
Steel, high carbon	90,000	620.5	63.2	140,000	965.2	98.4	20	201
Steel, low carbon	36,000	248.2	25.3	60,000	413.6	42.2	35	310
Steel, manganese (14 Mn)	75,000	517.1	52.7	118,000	813.5	82.9	22	200
Steel, medium carbon	52,000	358.5	36.5	87,000	599.8	61.2	24	170
Steel, stainless (austentic)	40,000	275.8	28.1	90,000	620.5	63.2	23	160
Steel, stainless (matensitic)	80,000	551.5	56.2	100,000	68.9	70.3	26	250
Steel, stainless (ferritic)	45,000	310.2	31.6	75,000	517.1	52.7	30	155
Tantalum	–	–	–	50,000	344.7	35.1	40	300
Tin	1,710	11.8	1.2	3,130	21.6	2.2	50	5.3
Titanium	40,000	275.8	28.1	60,000	413.6	42.2	28	–
Tungsten	–	–	–	500,000	3447.0	351.5	15	230
Zinc	18,000	124.1	12.6	25,000	172.35	17.5	20	38

Values depend on heat treatment or mechanical condition or mass of the metal.

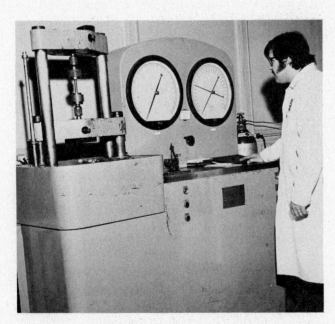

FIGURE 11-1 *Tensile test machine.*

operation. Many tensile testing machines are equipped to plot a curve which shows the load or stress and the strain or movement that occurs during the test operation. A typical curve for mild steel is shown by Figure 11-2. In the testing operation the load is increased gradually and the specimen will stretch or elongate in proportion to the tensile load. The specimen will elongate in direct proportion to the load during the elastic portion of the curve to point *A*. At this point, the specimen will continue to elongate but without an increase in the load. This is known as the yield point of the steel and is the end of the elastic portion. At any point up to point *A* if the load is eliminated, the specimen will come back to its original dimension. Yielding occurs from point *A* to point *B* and this is the area of plastic deformation. If the load were eliminated at point *B* the specimen would not go back to its original dimension but instead take a permanent *set*. Beyond point *B* the load will have to be increased to further stretch the specimen. The load will increase to point *C*, which is the ultimate strength of the material. At point *C* the specimen will break and

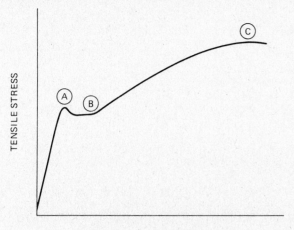

STRAIN OR UNIT DEFORMATION

FIGURE 11-2 *Stress strain curve.*

the load is no longer carried. The ultimate tensile strength of the material is obtained by dividing the ultimate load by the cross-sectional area of the original specimen. This provides the ultimate tensile strength in pounds per square inch or kilograms per square millimeter.

The *yield stress* or *yield point* is obtained by dividing the load at yield or at point *A* by the original area. This provides a figure in pounds per square inch or kilograms per square millimeter. Extremely ductile metals do not have a yield point. They stretch or yield at low loads. For these metals the yield point is determined by the change in elongation. Two tenths of one percent elongation is arbitrarily set as the yield point. The yield point is the limit upon which designs are calculated. Designs of weldments are expected to perform within the elastic limit and the yield point is the measure of that limit.

Ductility

The ductility of a metal is the property that allows it to be stretched or otherwise changed in shape without breaking and to retain the changed shape after the load has been removed. The ductility of a metal can be determined from the tensile test. This is done by determining the percent of elongation. Gauge marks are made two inches apart across the point where fracture will occur. The increase in gauge length related to the original length times 100 is the percentage of elongation. This is done by making center punch marks two inches apart at the reduced section of the test coupon, testing the coupon, tightly holding the two pieces together and remeasuring the distance between the center punch marks. The original two inches is subtracted from the measured length and the difference is divided by two and multiplied by 100 to obtain percentage of elongation.

Ductility of welds or of metals can also be measured by the bend test. In this case, gauge lines are drawn before testing, measured, and measured again after bending. The difference divided by the original length times 100 is the elongation in percentage.

The ductility is of extreme interest to welding since a higher ductility indicates a weld which would be less likely to crack in service. Figure 11-3 shows a typical tensile specimen.

FIGURE 11-3 *Tensile specimen close up.*

Reduction of Area

Reduction of area is another measure of ductility and is obtained from the tensile test by measuring the original cross-sectional area of the specimen and relating it to the cross-sectional area after failure. For a round specimen the diameter is measured and the cross-sectional area is calculated. After the test bar is broken the diameter is measured at the smallest point. The cross-sectional area is again calculated. The difference in area is divided by the original area and multiplied by 100 to give the percentage reduction of area. This figure is of less importance than the elongation but is usually reported when the mechanical properties of a metal are given.

The tensile test specimen also provides another property of metal known as its modulus of elasticity, also called *Young's modulus.* This is the ratio of the stress to the elastic strain. It relates to the slope of the curve to the yield point. For iron or steel the modulus of elasticity is approximately 30,000,000 psi, for aluminum it is approximately 10,300,000 psi, for copper 15,000,000 psi, and for magnesium 6,500,000 psi. The modulus of elasticity is important to designers and is incorporated in many design formulas.

WELDABILITY AND PROPERTIES OF METALS

Hardness

The hardness of a metal is defined as the resistance of a metal to local penetration by harder substance. The hardness of metals is measured by forcing a hardened steel ball or diamond into the surface of the specimen, under a definite weight, in a hardness testing machine. Figure 11-4 shows hardness testing machines. The Brinell is one of the more popular types of machines for measuring hardness. It provides a Brinell hardness number (BHN), which is in kilograms per square millimeter based on the load applied to the hardened ball in kilograms and divided by the area of the impression left by the ball in square millimeters. These metric factors are disregarded and hardness is measured strictly as a Brinell hardness number. The BHN is obtained by an optical magnifier which reads the diameter of the impression and produces a hardness number. Various scales are used in the Brinell hardness testing system based on the load and the diameter of the ball, usually 10 millimeters using a 3,000-kilogram load.

There are several other hardness measuring systems. A popular machine is the Rockwell hardness tester which utilizes a diamond which is forced into the surface of the specimen. Different loads are used to provide different scales. Smaller loads are used for softer materials. The optical unit is not required since the hardness is read from a dial mounted on the machine which relates to the penetration. There is also the superficial Rockwell hardness tester for measuring surface hardnesses of metals. Another method is by means of the Vickers hardness machine, which reads directly, as a diamond is pressed into the surface of the metal. Another way is the Shore sceleroscope which utilizes a small dropped weight which will bounce from the surface of the metal providing a hardness measure. This device is used for field work and is not considered as accurate as those that make impressions. Table 11-3 shows the hardness conversion numbers for these different hardness measuring systems. It is also possible to relate the approximate strength of a metal to its hardness (shown in the table).

Impact Resistance

Resistance of a metal to impacts is evaluated in terms of impact strength. A metal may possess satisfactory ductility under static loads but may fail under dynamic loads or impact. Impact strength is most often determined by the Charpy test. It is sometimes measured

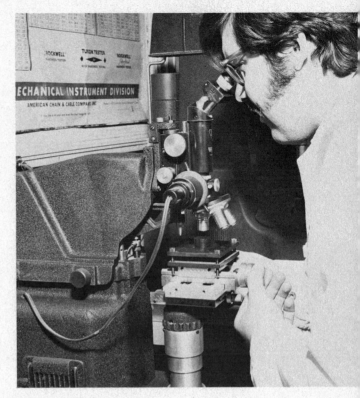

(a)

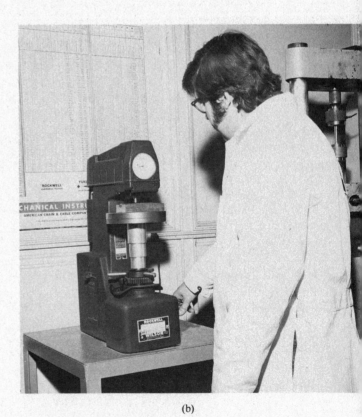

(b)

FIGURE 11-4 *Rockwell and Vickers hardness testing machine.*

TABLE 11-3 *Hardness conversion table.*

BRINELL		Vickers or Firth Hardness No.	ROCKWELL C 150 kg. load 120° Diamond Cone	ROCKWELL B 100 kg. load 1/16 in. dia. ball	Scleroscope No.	Approximate Tensile Strength 1000 psi
Dia. in mm., 3000 kg load 10 mm ball	Hardness No.					
2.05	898					440
2.10	857					420
2.15	817					401
2.20	780	1150	70		106	384
2.25	745	1050	68		100	368
2.30	712	960	66		95	352
2.35	682	885	64		91	337
2.40	653	820	62		87	324
2.45	627	765	60		84	311
2.50	601	717	58		81	298
2.55	578	675	57		78	287
2.60	555	633	55	120	75	276
2.65	534	598	53	119	72	266
2.70	514	567	52	119	70	256
2.75	495	540	50	117	67	247
2.80	477	515	49	117	65	238
2.85	461	494	47	116	63	229
2.90	444	472	46	115	61	220
2.95	429	454	45	115	59	212
3.00	415	437	44	114	57	204
3.05	401	420	42	113	55	196
3.10	388	404	41	112	54	189
3.15	375	389	40	112	52	182
3.20	363	375	38	110	51	176
3.25	352	363	37	110	49	170
3.30	341	350	36	109	48	165
3.35	331	339	35	109	46	160
3.40	321	327	34	108	45	155
3.45	311	316	33	108	44	150
3.50	302	305	32	107	43	146
3.55	293	296	31	106	42	142
3.60	285	287	30	105	40	138
3.65	277	279	29	104	39	134
3.70	269	270	28	104	38	131
3.75	262	263	26	103	37	128
3.80	255	256	25	102	37	125
3.85	248	248	24	102	36	122
3.90	241	241	23	100	35	119
3.95	235	235	22	99	34	116
4.00	229	229	21	98	33	113
4.05	223	223	20	97	32	110
4.10	217	217	18	96	31	107
5.15	212	212	17	96	31	104
4.20	207	207	16	95	30	101
4.25	202	202	15	94	30	99
4.30	197	197	13	93	29	97
4.35	192	192	12	92	28	95
4.40	187	187	10	91	28	93
4.45	183	183	9	90	27	91
4.50	179	179	8	89	27	89
4.55	174	174	7	88	26	87
4.60	170	170	6	87	26	85
4.65	166	166	4	86	25	83
4.70	163	163	3	85	25	82
4.75	159	159	2	84	24	80

TABLE 11-3 *(Continued)*

BRINELL Dis. in mm., 3000 kg load 10 mm ball	Hardness No.	Vickers or Firth Hardness No.	ROCKWELL C 150 kg. load 120° Diamond Cone	B 100 kg. load 1/16 in. dia. ball	Scleroscope No.	Approximate Tensile Strength 1000 psi
4.80	156	156	1	83	24	78
4.85	153	153		82	23	76
4.90	149	149		81	23	75
4.95	146	146		80	22	74
5.00	143	143		79	22	72
5.05	140	140		78	21	71
5.10	137	137		77	21	70
5.15	134	134		76	21	68
5.20	131	131		74	20	66
5.25	128	128		73	20	65
5.30	126	126		72		64
5.35	124	124		71		63
5.40	121	121		70		62
5.45	118	118		69		61
5.50	116	116		68		60
5.55	114	114		67		59
5.60	112	112		66		58
5.65	109	109		65		56
5.70	107	107		64		56
5.75	105	105		62		54
5.80	103	103		61		53
5.85	101	101		60		52
5.90	99	99		59		51
5.95	97	97		57		50
6.00	95	95		56		49

by the Izode test. Both types of tests use the same type of pendulum-testing machine. The Charpy test specimen is a beam supported at both ends and contains a notch in the center. The specimen is placed on supports and struck with a pendulum on the side opposite the notch. The accuracy and location of the notch is of extreme importance. There are several types of Charpy specimens; the V-notch type is the most popular. The specimen is standardized in metric dimensions. Figure 11-5 shows the impact testing machine in action.

The impact strength of a metal is determined by measuring the energy absorbed in the fracture. This is equal to the weight of the pendulum times the height at which the pendulum is released and the height to which the pendulum swings after it has struck the specimen. In conventional terms the impact strength is the foot pounds of energy absorbed. In metric practice, impact resistance is measured two ways. One, the kilogram meter based on energy absorbed and, two, the kilogram meter per square centimeter of the area of the fractured surface or the cross-sectional area under the notch. Both terms are used, but care must be taken to determine which is appropriate. The S.I. system measures energy absorbed in joules. Figure 11-6 shows V-notch Charpy impact bars before and after testing.

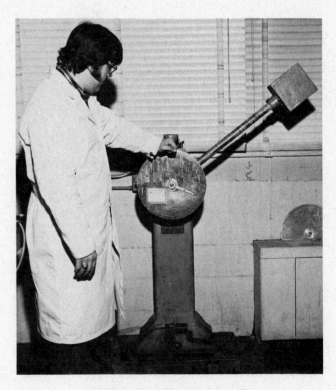

FIGURE 11-5 *Making an impact test.*

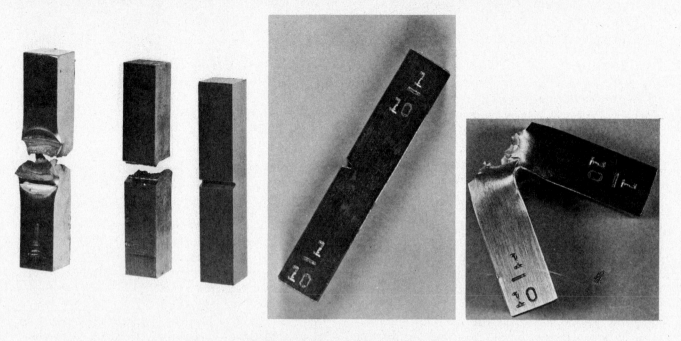

FIGURE 11-6 *Vee notch charpy impact bars.*

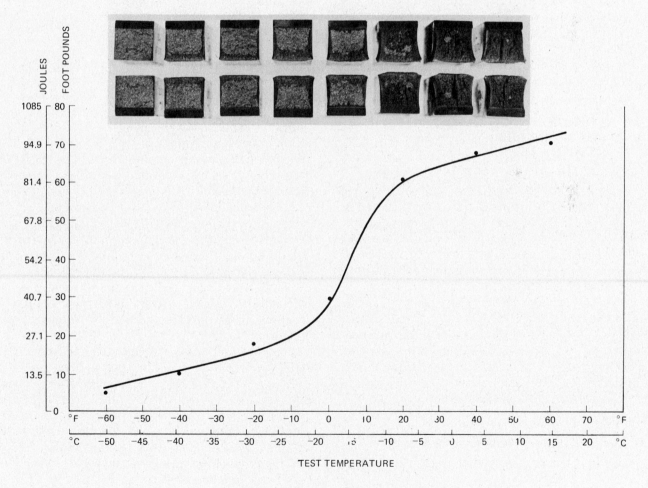

FIGURE 11-7 *Transition chart of impact tests.*

Impact tests are often made at different temperatures, since steels normally become more brittle or will absorb less energy at lower temperatures. Normally seven specimens are broken at each test temperature and the high and low values are discarded. The reported value is the average of the remaining two specimens. Test temperatures are $-60°F$ ($-51°C$), $-50°F$ ($-46°C$), $-40°F$ ($-40°C$), $-20°F$ ($-29°C$), $-14°F$ ($-25°C$), $-4°F$ ($-20°C$), $0°F$ ($-18°C$), $14°F$ ($-10°C$). $32°F$ ($0°C$), $50°F$ ($10°C$), $68°F$ ($20°C$). All temperatures may not be used; however, usually five test temperatures are used so that a transition curve can be drawn. Figure 11-7 shows a temperature transition curve. At the point of *fall off* the transition changes from ductile to brittle. This is known as the transition temperature. The change from ductile to brittle fracture can also be seen by the surface of the broken bars in the figure. Another measure of ductility is also utilized and this is the degree of lateral expansion of the bar at the fracture surface. The greater the degree of change the more ductile the fracture. The fracture surface type is also reported for critical requirements. All these different tests and test specimens are standardized by the American Society for Testing and Materials.

11-2 METAL SPECIFICATIONS AND GROUPINGS

The one rule that must be continually emphasized in welding is that you must know the metal you are welding. There are many different ways that metals are identified and specified. These range from national specifications, professional society specifications, trade association specifications, to trade names of specific metals. To be able to identify metals it is important to know how they are specified.

In North America and in many countries in the world, the most popular method of specifying a metal is by its ASTM number. ASTM stands for the American Society for Testing and Materials which issues an annual book of ASTM standards consisting of at least 33 parts. Seven parts specify metals:

Part 1: Steel piping, tubing, and fittings.

Part 2: Ferrous castings-ferro-alloys.

Part 3: Steel sheet, strip, bar, rod, wire, etc.

Part 4: Structural steel, steel plate, steel rails, wheels, etc.

Part 5: Copper and copper alloys.

Part 6: Die-cast metals, light metals, and alloys.

Part 7: Nonferrous metals and alloys, etc.

Other sections cover basic materials such as concrete, chemicals, insulating materials, papers, petroleum products, fuels, paint, textiles, plastics, rubber, etc.

ASTM issues three parts that relate to testing;

Part 30: General Test Methods.

Part 31: Metals—Physical and Mechanical Nondestructive Tests.

Part 32: Analytical Methods of Analysis.

The ASTM Standards represent a common viewpoint of parties concerned with its provisions: producers, users, and general interest groups. These are voluntary standards written by committees of the membership. The ASTM Standards for metals provide the mechanical properties of the metal and in many cases chemical composition. Specifications for steels usually provide compositions that refer to either the analysis of the steel in the ladle or in its final form. The specifications also provide information concerning the form and size of the products, the size tolerance of products, testing procedures, inspection information, etc.

From the welding point of view the most important information is the mechanical properties and the chemical composition. All specifications, however, do not list chemical compositions.

The ASTM metal specifications are identified by a prefix letter, A indicating ferrous materials and B indicating nonferrous metals. This is followed by a one-, two-, or three-digit number indicating the exact specification number which is then followed by a two-digit number indicating the year that the specification was formally adopted. A suffix letter T, when used, indicates that it is a tentative specification.

There are too many ASTM specifications to be listed here and it is therefore recommended that the first seven parts or volumes be examined for a better understanding of the usefulness of the ASTM specifications.

Each part or volume is updated yearly and is available from the Society. Many libraries have the ASTM Volumes. Individual standards are also available from the Society.

Another widely used specification for steel is the SAE steel specification, which has now been combined with the AISI Standard Steel Compositions. These specifications originally issued by the Society of Automotive Engineers provided a system of indicating the chemical composition of a steel. They are widely used to specify steels for automotive, aircraft, and machinery applications and are used primarily for sheet and bar materials. They provide identification for steels ranging from plain carbon steel to the low-alloy steels and to certain specialty types of steels. Recently the American Iron and Steel Institute has collaborated with the SAE and issued an updated version of the steel specifications which also includes prefix letters indicating the type of manufacture. The detailed compositions and identification numbers of the steels will be shown in Chapter 12.

The AISI and SAE also collaborated on a system of identifying the corrosion-resistant or stainless steels. In this system stainless and heat-resistant steels are classified into four general groups. They are identified by a three-digit number. The first number indicates the group and the last two numbers indicate the type. Modifications of types are indicated by suffix numbers. Information concerning the exact specifications for different AISI steels will be given in Chapter 12.

Different branches of the U.S. federal government also write specifications for metals. These include the MIL specifications issued by the Department of Defense and the QQ specifications issued by the Department of Commerce. Other groups may also issue metal specifications.

Professional societies and trade associations also provide specifications for different metals. For example, the American Petroleum Institute issues a specification covering the mechanical properties and composition of steel for pipe. The American Bureau of Shipping provides specifications for steels used in ship building. The Association for American Railroads provides specifications for wrought and cast steels used by the railroad industry.

Steel castings are identified and specified by ASTM specifications and also by SAE, AAR, and ABS classes. These specifications provide for mechanical properties with various heat treatments and chemical compositions.

Specifications for materials used in the aircraft industry are designated as AMS, which stands for Aerospace Materials Specifications. These are produced by a division of the Society of Automotive Engineers (SAE).

Most industrialized countries have national standards which provide specifications for metals. The British Standards (BS) issue specifications for many different types of metals as do the German Standardization Institute (DIN) and the Japanese Standards Association (JIS). The standardization department of the Soviet Union issues GOST standards. Most of these standards are available in their original language from the standardization society of each of the other industrialized countries. The International Organization of Standards (ISO) also issues specifications for metals.

For nonferrous metals several trade associations are involved. For example, the Aluminum Association provides a system of designating the compositions of aluminum alloys. This system uses four-digit numbers

for wrought aluminum alloys. The AA also provides temper designations as suffix letters and numbers to indicate the temper condition. All of the aluminum producers in North America utilize the Aluminum Association alloy numbers for identifying their different products.

A standard designation for copper and copper alloys has been established by the Copper Development Association, Inc. Their system is used in North America and has been adopted by the U.S. Government, ASTM, SAE, and nearly all producers of copper and copper alloy products. It is not a specification but rather an orderly method using a three-digit number of defining and identifying coppers and copper alloys. The system groups compositions into families including the coppers, the high-copper alloys, the brasses, the bronzes, the copper-nickels, and the copper-zinc alloys. These numbers replace previous tradenames such as copper-nickel, aluminum-bronze, phosphur-bronze, Naval-brass, tough pitch copper, etc.

The various alloys of magnesium are identified by a special designation established by ASTM. Titanium, lead, tin, and other metals are specified by ASTM specifications.

Specifications applicable to each metal will be provided in the chapter concerning each metal.

There are, however, numerous alloys that are identified by trade names or numbers by the producer. Compositions of these alloys can be obtained from the producer. The Society of Automotive Engineers (SAE) also provides specifications for nonferrous metals.

Several books are available[1,2,3,4] that list different alloys and metals by trade name. For those who need to determine the analysis of different proprietory or trade name alloys these books are of immense value.

11-3 IDENTIFICATION OF METALS

Many times, particularly in repair work, it is necessary to weld on a material whose specification or trade name or composition is unknown. In order to produce a successful weld it is necessary to know the composition of the metal being welded. There are a number of ways to determine the composition so that a welding procedure can be developed. The ability to make a rapid identification of the metal will reduce the time required to make a successful weld. If time permits, it is recommended that a piece of the metal be taken to a laboratory for analysis. Since time is rarely

CHAPTER 11

WELDABILITY AND PROPERTIES OF METALS

available, the following basis for identifying the metal should be followed. By using this technique and with experience a fairly accurate identification can be made.

There are seven simple tests that can be performed to help identify metals. They will at least provide sufficient guidance to make a successful weld even though the exact composition may not be learned. Six of the different tests are summarized by the table shown by Figure 11-8. This should be supplemented by Tables 11-1 and 11-2, which present physical and mechanical properties of metals.

Appearance Test

The first test is the appearance of the base metal. This includes features such as color and the appearance of the machined as well as unmachined surfaces. The shape can be descriptive; for example, shape includes such things as cast engine blocks, automobile bumpers, reinforcing rod, I beams or angle irons, pipes, and pipe fittings. Form should be considered and may show how the part was made, such as a casting with its obvious surface appearance and parting mold lines, or hot rolled wrought material, extruded or cold rolled with a smooth surface. Form and shape give definite clues. For example, pipe can be cast, in which case it would be cast iron, or wrought, which would normally be steel. Another example is a hot rolled structural shape in a steel-framed building, which would be mild or low-alloy steel.

Color provides a very strong clue in metal identification. It can distinguish many metals such as copper, brass, aluminum, magnesium, and the precious metals. On oxidized metals, the oxidation can be scraped off to determine the color of the unoxidized metal. This helps to identify lead, magnesium, and even copper. The oxidation on steel, or rust, is usually a clue that can be used to separate plain carbon steels from the corrosion-resisting steels.

Base Metal or Alloy	Color	Magnet	Chisel	Fracture	Flame or Torch	Spark
Aluminum & alloys	bluish-white	non-magnetic	easily cut	white	melts wo/col	non-spark.
Brass, navy	yellow or reddish	non-magnetic	easily cut	not used	not used	non-spark.
Bronze, alum. (90Cu-9Al)	reddish yellow	non-magnetic	easily cut	not used	not used	non-spark.
Bronze, phosphor (90Cu-10Sn)	reddish yellow	non-magnetic	easily cut	not used	not used	non-spark.
Bronze, silicon (96Cu-3Si)	reddish yellow	non-magnetic	easily cut	not used	not used	non-spark.
Copper (deoxidized)	red; 1 cent piece	non-magnetic	easily cut	red	not used	non-spark.
Copper nickel (70Cu-30 Ni)	white 5 cent piece	non-magnetic	easily cut	not used	not used	non-spark.
Everdur (96Cu-3Si-1Mn)	gold	non-magnetic	easily cut	not used	not used	non-spark.
Gold	yellow	non-magnetic	easily cut	not used	not used	non-spark.
Inconel (76Ni-16Cr-8Fe)	white	non-magnetic	easily cut	not used	not used	non-spark.
Iron, cast	dull grey	magnetic	not easily chip.	brittle	melts slowly	see text
Iron, wrought	light grey	magnetic	easily cut	bright grey fib.	melts fast	see text
Lead	dark grey	non-magnetic	very soft	white; crystal	melts quick	non-spark.
Magnesium	silvery white	non-magnetic	soft	not used	burns in air	non-spark.
Monel (67 Ni-30Cu)	light grey	slightly magnet.	tough	light grey	not used	non-spark.
Nickel	white	magnetic	easily cut	almost white	not used	see text
Nickel silver	white	non-magnetic	easily chipped	not used	not used	non-spark.
Silver	white-pre 1965 10¢ pc	non-magnetic	not used	not used	not used	non-spark.
Steel, low alloy	blue-grey	magnetic	depends on comp	medium grey	shows color	see text
Steel, high carbon	dark grey	magnetic	hard to chip	very lgt. grey	shows color	see text
Steel, low carbon	dark grey	magnetic	continuous chip	bright grey	shows color	see text
Steel, manganese (14 Mn)	dull	non-magnetic	work hardens	coarse grained	shows color	see text
Steel, medium carbon	dark grey	magnetic	easily cut	very lgt. grey	shows color	see text
Steel, stainless (austentic)	bright silvery	see text	continuous chip	deps. on type	melts fast	see text
Steel, stainless (matensitic)	grey	slightly magnetic	continuous chip	deps. on type	melts fast	see text
Steel, stainless (ferritic)	bright silvery	slightly-magnet.	-	deps. on type	-	see text
Tantalum	grey	non-magnetic	hard to chip	-	high temp.	-
Tin	silvery white	non-magnetic	usually as plating	usually as plating	melts quick	non-spark.
Titanium	steel grey	non-magnetic	hard	not used	not used	see text
Tungsten	steel grey	non-magnetic	hardest metal	brittle	highest temp.	non-spark
Zinc	dark grey	non-magnetic	usually as plating	at R.T.	melts quick	non-spark

FIGURE 11-8 *Summary of identification tests of metals.*

The use of the metal part is also a clue to identify it. Many machinery parts for agricultural equipment and light and medium-duty industrial equipment are made of cast irons. For heavy-duty work such as brake presses, the castings would probably be steel. A railroad rail obviously can be identified by shape and this gives an immediate clue to its composition.

Hardness Test

A second test that should be used is the hardness test. The *Shore scleroscope* is a portable instrument that can be taken to the work. This gives a hardness indication, which helps determine the type of metal. The other hardness tests can also be used.

Another but less precise hardness test is the file test. A summary of the reaction to filing and the approximate Brinell hardness and the possible type of steel is given in Figure 11-9. A sharp mill file is used. It is assumed that the part is steel and the file test will help identify the type of steel. Experience will help identify steel types with the file test.

File Reaction	Brinell Hardness	Type Steel
File bites easily into metal	100 BHN	Mild steel
File bites into metal with pressure	200 BHN	Medium carbon steel
File does not bite into metal except with extreme pressure	300 BHN	High alloy steel-high carbon steel
Metal can only be filed with difficulty	400 BHN	Unhardened tool steel
File will mark metal but metal is nearly as hard as the file and filing is impractical	500 BHN	Hardened tool steel
Metal is harder than file	600 + BHN	

FIGURE 11-9 *Approximate hardness of steel by the file test.*

Magnetic Test

The magnetic test can be quickly performed using a small pocket magnet. With experience it is possible to judge a strongly magnetic material from a slightly magnetic material. The nonmagnetic materials are easily recognized.

The strongly magnetic materials include the carbon and low-alloy steels, iron alloys, pure nickel, and martensitic stainless steels.

A slightly magnetic reaction is obtained from Monel and high-nickel alloys and the stainless steel of the 18 chrome 8 nickel type when cold worked, such as in a seamless tube.

The nonmagnetic materials are the copper-base alloys, aluminum-base alloys, zinc-base alloys, annealed 18 chrome 8 nickel stainless, the magnesiums, and the precious metals.

With experience it is soon possible to quickly use the magnetic test to help screen the various metals.

Chisel Test

The chip test or chisel test should also be used. The only tools required are a hammer and a cold chisel. The operation is to use the cold chisel and hammer on the edge or corner of the material being examined. The ease of producing a chip is an indication of the hardness of the metal and if the chip is continuous it is indicative of a ductile metal whereas if chips break apart it indicates a brittle material. On such materials as aluminum, mild steel and malleable iron the chips are continuous. They are easily chipped and the chips do not tend to break apart. The chips for grey cast iron are so brittle that they become small broken fragments. On high-carbon steel, the chips are hard to obtain because of the hardness of the material, but they can be continuous.

Fracture Test

This test is simple to use if a small piece of the metal being evaluated is available. The ease of breaking the part is an indication of its ductility or lack of ductility. If the piece bends easily without breaking it is one of the more ductile metals. If it breaks easily with little or no bending it is one of the brittle materials. The surface appearance of the fracture is also an indication. It will have the color of the base metal without oxidation. This will be true of copper, lead, and magnesium. In other cases, the coarseness or roughness of the broken surface is an indication of its structure. A careful study of known metals, how they appear at the fracture will help build experience to judge unknown specimens.

Flame or Torch Test

To perform this test it is necessary to have a high-temperature flame such as the oxyacetylene torch flame. The flame test should be used with discretion since it is possible that it will damage the part being investigated. If at all possible it should be used on a small piece of the metal being checked. The factors learned from this test are the rate of melting, the appearance of the molten metal and slag, and the action of the molten metal under the flame. All these factors provide clues which can aid in making the evaluation. When a sharp corner of a white metal part is heated the rate of melting can be an indication. If the material is aluminum it will not melt until sufficient heat has been used because of the high conductivity of aluminum. If the part is zinc the sharp corner will melt quickly since zinc is not a good conductor. In the case of copper if the sharp corner melts it is normally deoxidized copper. If it does not melt until much heat has been applied it is electrolytic copper. Also, with copper alloys, if lead is the composition, it will boil, thus indicating a lead-bearing alloy. To distinguish aluminum from magnesium apply the torch to filings. Magnesium will burn with a sparkling white flame. Steel will show characteristic colors before melting.

Spark Test

This is a very popular and accurate test for identification of different steels. The test requires the use of high-speed grinding wheel, either fixed or portable. The grinding wheel should have a speed of at least 5,000 surface feet per minute. The surface feet per minute equals the circumference in inches multiplied by the revolutions per minute divided by 12. Spark testing should be done in subdued light since the color of the spark is important. Spark testing is not of much use on nonferrous metals such as coppers, aluminums, and nickel-base alloys, since they do not exhibit spark streams of any significance. This is one way to separate ferrous and nonferrous metals. For example, it can be used to separate stainless steel from high-nickel or copper-nickel materials, such as Monel, since sparks would be produced only by the stainless steel. It is advisable to have specimens or samples of known steels which can be sparked immediately before or after sparking the unknown material to help determine the unknown composition.

Spark testing has been thoroughly studied and data are available with photographs showing the sparks that are produced. The spark resulting from the test should be directed downward and studied. Figure 11-10 shows the sparks resulting from a spark test. The color, shape, length, and activity of the sparks relate to characteristics of the material being tested. The spark stream has specific items which can be identified. The straight lines are called carrier lines. They are usually solid and continuous. At the end of the carrier line they may divide

FIGURE 11-10 *Spark test.*

into three short lines which are called forks. If it divides into more lines at the end it is called a *sprig*. Sprigs also occur at different places along the carrier line. These are sometimes called bursts, either star or fan bursts. In some cases, the carrier line will enlarge slightly for a very short length, continue, and perhaps enlarge again for a short length. When these heavier portions occur at the end of the carrier line they are called spear points or buds. High sulphur creates these thicker spots in carrier lines and the spearheads. Cast irons have extremely short streams, whereas low-carbon steels and most alloy steels have relatively long streams. Steels usually have white to yellow color sparks while cast irons are reddish to straw yellow. By learning to identify the different portions of the spark, and by making tests on known samples it is possible to acquire experience sufficient to make relatively accurate determinations of the metal being investigated. The following is a summary of some metals and the type of spark that is produced.

Cast Iron: Dull red to straw yellow color, short spark stream, many small sprigs short and repeating.

Wrought Iron: Long straw-colored carrier lines, usually whiter away from the grinding wheel. Carrier lines usually end in spearhead arrows or small forks.

Low Carbon or Mild Steel: Long yellow carrier line; occasional forks and lines may end in an arrowhead.

Low-Alloy Steel: Each alloying element has an effect on the spark appearance and very careful observation is required. Type 4130 steel has carrier lines that often end in forks and sharp outer points with few sprigs.

High Carbon Steel: Abundant yellow carrier lines with bright and abundant star bursts.

Manganese Steel: Bright white carrier lines with fan-shaped bursts.

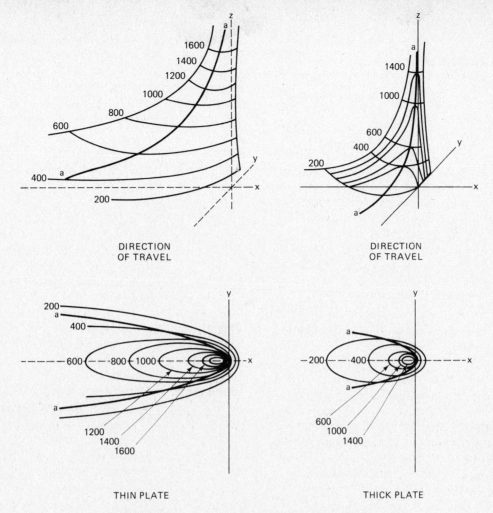

DIRECTION
OF TRAVEL

DIRECTION
OF TRAVEL

THIN PLATE

THICK PLATE

FIGURE 11-12 *Temperature distribution around a moving arc.*

process 70–85% of the heat is utilized in making the weld nugget. For the carbon arc welding process 50–70% of the heat is utilized while in submerged arc welding 80–90% of the heat generated by the arc is utilized in melting the weld metal. A major portion of this lost heat is used to raise the temperature of the base metal adjacent to the weld to near its melting point. Other losses come from weld spatter, heating the electrode and flux, and radiation and convection to the surrounding air.

In analyzing the effects of heat on a weld, a weld joint, or a weldment, it is necessary to determine:

1. The rate of heating.
2. The maximum temperature attained.
3. The length of time at temperature.
4. The rate of cooling.

These factors are difficult to determine; however, a good analysis of the potential damaging effects on the weld can be approximated. This allows precautions or procedure changes to minimize the harmful effects.

The *rate of heating* depends on a number of factors which include: the size and intensity of the heat source, the efficiency of the transfer of the heat to the base metal, the utilization of heat in making the weld, the mass of the base metal, the joint geometry, and thermal conductivity. How much heat is in the heat source and how fast does it flow from the weld area? For example, a low-current gas tungsten arc can be struck on a large thick copper plate and it would be impossible to make a weld since the temperature of the copper plate would not rise sufficiently to cause melting. Not enough heat is produced by the tungsten arc to melt the copper since the conductivity of the copper and its mass causes the heat to flow away so rapidly. The thermal conductivity, the ability to transmit heat throughout its mass, is of major importance when considering the rate of heating. Table 11-1 provides the relative thermal conductivity of the common metals. Numbers cannot be put into formulas to provide exact answers; however, the relationship of thermal conductivity of one metal versus another provides a clue. For example, steel has only about a tenth of the thermal conductivity of copper. With reference to the isotherm diagram, the conductivity of the metal has an effect on the steepness of the curve or change of temperature gradient. Extremely steep temperature gradients occur in welding as a result of the high temperatures of the heat source and the temperature of the base metal. Steepness of the rate of the heating curve is much different for the different processes.

The *maximum temperature* that will be attained in the base metal is also important. The base metal at the weld must be raised to its melting temperature and above. How *much* above is important and depends on the welding process. Efficient welding does not require the base metal to be raised much above its melting temperature. The welding processes that utilize extremely high temperature heat sources such as electron beam or laser beam can raise the base metal temperature so high that it will volatilize the metal. This is why some processes can be used for both welding and cutting, depending on the heat input. When welding thin sheet metal with too much heat input, the material becomes too hot. It melts rapidly and falls away, and holes are produced rather than welds. The maximum temperature reached by the base metal is related to the rate of heat input and to the rate of heat loss. As long as the heat input exceeds the rate of heat loss the base metal will continue to get hotter. This relationship must continue until surface melting of the base metal occurs. More heat is required to make the weld than seems necessary because of the heat losses in the base metal.

Another factor is the *specific heat* of the base metal. This is a measure of the quantity of heat required to increase the temperature of the metal. It relates to the amount of heat required to bring the metal to its melting point. A metal having a low melting temperature but with a relatively high specific heat may require as much heat to cause surface melting as a metal with a high temperature melting point and a low specific heat. This can be seen by comparing aluminum to steel. The specific heat of the common metals is shown by Table 11-1. The temperature required at the weld area should be only slightly greater than the melting temperature of the metal being welded. This is obtained in the base metal by balancing the heat input with the heat losses.

The length of time at the maximum temperature depends upon maintaining a heat balance between heat input and heat losses. There is rarely a true heat balance in any welding situation. During the arcing period, the heat input usually exceeds the heat losses and the base metal becomes hotter. Many times the welder must allow work to cool when the welding puddle becomes too large and unmanageable. The current is reduced or the arc is broken, and the heat input is reduced or ceases. The heat losses continue, and the puddle and base metal begin to cool. Normally the temperature of the work near the arc rises to a maximum. As soon as the arc moves on, the temperature adjacent to the weld begins to fall. The longer the base metal, adjacent to the weld, remains at a high temperature the greater the possibility for grain growth in the weld metal and in the heat affected area. The amount of metal melted, and the heat input and heat loss affect this relationship.

The *rate of cooling* of the weld and adjacent metal is the rate of temperature change from welding temperatures to room temperature. The rate of cooling can be closely controlled and is governed by such conditions as heat transfer, heat losses, and thermal conductivity of the base metal. However, several factors must be considered since they can be used to regulate the cooling rate. The most important one is the initial temperature of the base metal before welding. A higher preheat or the more heat in the weldment the slower it will cool. The second important factor is the heat input that may be given to the weldment after the weld is made. It is usually desirable to reduce the cooling rate if metallurgical problems such as cracking or hard zones occur. Hard zones in or adjacent to the weld usually have lower toughness and ductility and tend to crack when thermal stresses are introduced. By reducing the cooling rate these can be eliminated and quality welds produced. Factors that increase the heat input or the mass of heat, or reduce the heat losses will reduce the cooling rate, which will reduce the possibility of these defects.

11-5 WELDING METALLURGY

The science of joining metals by welding relates closely to the field of metallurgy. Metallurgy involves the science of producing metals from ores, of making and compounding alloys, and the reaction of metals to many different activities and situations. Heat treatment, steel making and processing, forging, and foundry all make use of the science of metallurgy. Welding metallurgy can be considered a special branch, since reaction times are in the order of minutes, seconds, and fractions of seconds, whereas in the other branches reactions are in hours and minutes.

Welding metallurgy deals with the interaction of different metals and the interaction of metals with gases and chemicals of all types. The welding metallurgist is also involved with changes in physical characteristics that happen in short time periods. The solubility of gases in metals and between metals and the effect of impurities are all of major importance to the welding metallurgist.

In a general treatment of welding such as this, many metallurgical factors and practices are found throughout the entire book. This chapter presents a very brief coverage of welding metallurgy.

The structure of metals is complex. When metal is in a liquid state, usually hot, it has no distinct structure or orderly arrangement of atoms. The atoms move freely among themselves within the confines of the liquid. Their mobility allows the liquid metal to yield to the slightest pressure and to conform to the shape of the container. This high degree of mobility of the atoms is due to the heat energy involved during the melting process.

As molten metal cools, the heat energy of the atoms in the liquid state decreases and the atoms move with less mobility. As the temperature is further reduced and the metal cools the atoms are no longer able to move and are attracted together into definite patterns. These patterns consist of three-dimensional lattices known as *space lattices,* which are made of imaginary lines connecting atoms in symmetrical arrangements. These imaginary lines are approximately the same distance from one another and limit the movement. Metals, in a solid state, possess this uniform arrangement,

which is called *crystals.* All metals and alloys are crystalline solids made of atoms arranged in a specific uniform manner.

There are over a dozen types of space lattice possible; however, the majority of common metals fall into only three: (1) the face-centered cubic lattice, (2) the body-centered cubic lattice, and (3) the hexagonal close-packed lattice. The metals and the form of the crystal lattice structure are shown by Figure 11-13. Note that iron has both the face-centered cubic structure and the body-centered cubic structure but at different temperatures. The change from one type of lattice structure to another takes place in the solid state with no change in specific gravity, but a small change in volume. This is known as an *allotropic change.*

The crystal lattices just mentioned are for pure metals that are composed of only one type of atom. Most metals in common use are alloys. In other words, they contain more than one metal. When more than one metal is present the atoms making up the crystals will change. The atoms of the metal making up the minor portion of the alloy will at random replace some of

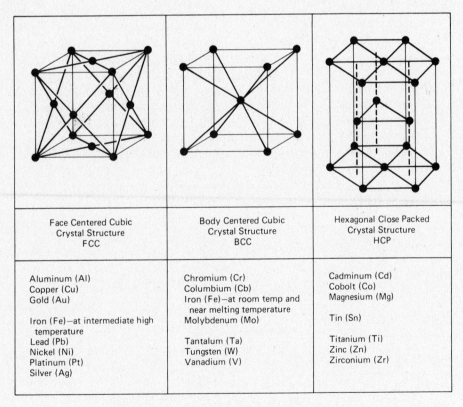

Face Centered Cubic Crystal Structure FCC	Body Centered Cubic Crystal Structure BCC	Hexagonal Close Packed Crystal Structure HCP
Aluminum (Al)	Chromium (Cr)	Cadminum (Cd)
Copper (Cu)	Columbium (Cb)	Cobolt (Co)
Gold (Au)	Iron (Fe)—at room temp and near melting temperature	Magnesium (Mg)
Iron (Fe)—at intermediate high temperature	Molybdenum (Mo)	Tin (Sn)
Lead (Pb)		
Nickel (Ni)	Tantalum (Ta)	Titanium (Ti)
Platinum (Pt)	Tungsten (W)	Zinc (Zn)
Silver (Ag)	Vanadium (V)	Zirconium (Zr)

FIGURE 11-13 *Crystalline structure of common metals.*

the atoms of the metal making up the majority of the alloy. If the crystals are of essentially the same size the minor metal will be considered to be dissolved in the major metal of the alloy. This condition is called a substitutional solid solution. A small amount of nickel added to copper will produce a substitutional solid solution.

If the atoms of the minor metal in the alloy are much smaller than those in the major metal lattice, they do not replace the atoms of the major metal in the lattice but rather locate in points between or in intervening spaces known as *interstices* in the lattice. This type of structure is called an *interstitial solid solution.* Very small amounts of carbon sometimes occur interstitially in iron.

If the minor metal atoms in the alloy cannot completely dissolve either interstitially or substitutionally, they will form the type of chemical compound the composition of which corresponds roughly to the chemical formula. This results in the formation of mixed kinds of atomic groupings consisting of different crystalline structures. These are referred to as *intermetallic compounds* and have a complicated crystal structure.

Each grouping with its own crystalline structure is referred to as a phase in the alloy and the alloy is called a multiphase alloy. The individual phases may be seen and distinguished when examined under a microscope at extremely high magnification.

These different alloys, solid solutions, intermetallic compounds, and phases occur as the molten metal solidifies. Freezing or solidification of a liquid metal does not happen simultaneously throughout the entire melt. Freezing begins at the point of lowest temperature just below the liquidus. At this point a small crystal forms, called a *nucleus.* Different nuclei may be formed almost simultaneously, and each is a point where solidification starts and the solidified metal grows from these points. The growth of solidification advances in all directions that are normal to the main axis of the nuclei crystal. Thus from a cubic crystal, growth progresses in six

directions simultaneously. Growth is simply the adding on of additional crystals as temperature decreases. The growth continues and takes on a treelike pattern with branches and subbranches at right angles to one another. As solidification continues the branches become thicker and larger and fill the spaces between additional branches which are called *dendrites*. This continues until the entire mass has become solid. The dendritic growth of a crystal is shown by Figure 11-14.

The crystals that grow from one nucleus can grow only to the point where they come in contact with another crystal growing from a different nucleus. Since the nuclei occur randomly, growth of dendrites from nuclei crystals are at odd angles with one another and this does not permit the various crystals to merge into a single crystal. The completely solidified metal is made up of individual dendritic crystals which are oriented in different planes but held together by atomic attractive forces at the interface of adjacent dendrites. When this resultant structure is cut in a flat plane, the individual dendritic crystals, which grew until they met adjacent dendritic crystals, form an irregularly shaped area, which is known as a *grain*. The fitting together of the different grains is normally in an irregular outline shape and the interface between grains is known as grain boundaries. This is shown by Figure 11-15. Grains are also very, very small but much larger than the individual crystals.

The size of the crystals and the grains depends on the rate of growth of the crystal. The rate of crystal growth depends on the rate of cooling of the molten solidifying metal. When the rate of cooling is high, the solidification process occurs more rapidly and the crystal size and grain size tend to be smaller. When the rate of cooling is slower, crystal and grain size tend to be larger. With extremely slow cooling or possibly with reheating, grains that have crystal axis almost parallel with one another will tend to grow together and it is possible that two grains will grow into one.

Grains can also grow relatively long and narrow because of the orientation of the nuclei. Grain growths that have proceeded in primarily one direction are those at the edge of a groove weld. The crystal is formed and grows into elongated grains which pro-

FIGURE 11-14 *Dendritic growth from a nuclear crystal.*

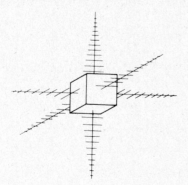

FIGURE 11-15 *Grains are formed by dendrites growing together.*

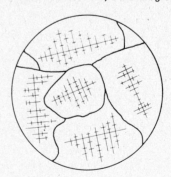

duce a columnar structure. In restricted areas in which nuclei formed close together long grains are not possible and therefore more equiaxed grains result.

The overall arrangement of grains, grain boundaries, and phases present in an alloy is called its *microstructure*. The microstructure is largely responsible for the properties of the metal. It is affected by the composition or alloy content and by other factors such as hot or cold working, straining, heat treating, etc. The microstructure of weld metal and adjacent metal is greatly influenced by the welding process and welding procedure, which influence the properties of the weld.

Anything done to the metal that will disturb or distort the lattice structure causes the metal to harden. Cold working of a metal distorts the structure and thereby hardens it. The presence of foreign atoms in the structure by alloy additions distorts the structure and tends to harden it. When atoms are dissolved in a solid state structure and are then precipitated out, the structure is distorted and is thus hardened.

The grain boundaries contain lower melting point materials since the grain boundaries are the last portion to freeze or solidify. The strength of metals is sometimes determined by the grain boundaries. Grain boundaries increase the strength of some materials at room temperature by inhibiting the deformation of individual grains when the material is stressed. At elevated temperatures the atoms in the boundaries can move more easily and slide past one another thus reducing the material strength. Fine grain materials have better properties at room temperature. Metal structures can be characterized as having large grains (coarse grained) or small grains (fine grained) or a mixture of large and small (mixed grain size). The arrangement of atoms is irregular in the grain boundaries and there are vacancies or missing atoms. The atom spacing may be larger than normal and individual atoms can move more easily in the grain boundaries, and because of this the diffusion of elements, which is the movement of individual atoms through the solid structure, occurs more rapidly at grain boundaries. Odd-sized atoms segregate at the boundaries and this leads to the formation of undesirable phases that reduce the properties of a material by lowering the ductility while making it susceptible to cracking.

Phase Transformation

Some metals change their crystallographic arrangement with changes in temperature. Iron has a crystalline body-centered cubic lattice structure from room temperature up to 1,670°F (910°C), and from this point to 2,535°F (1,388°C) it is face-centered cubic. Above this point, to the melting point of 2,800°F (1,538°C) it is again body-centered cubic. This change in crystalline structure is known as a *phase transformation* or an *allotropic transformation*. Other metals undergoing allotropic transformation at different temperatures are titanium, zirconium, and cobalt.

Another type of transformation occurs when the metal melts or solidifies. When the metal melts the orderly crystalline arrangement of atoms disappears and there is then random movement of atoms. When the metal solidifies the crystalline arrangement re-establishes itself. The change in crystalline structure or the change from liquid to solid is known as phase change. Pure metals melt or solidify at a single temperature while alloys solidify or melt over a range of temperatures with a few exceptions.

The phase changes can be related to alloy composition and temperature when they are in equilibrium, and shown on a diagram. Such diagrams are called phase diagrams, alloy equilibrium diagrams, or constitution diagrams. Metallurgists have developed constitution diagrams for almost every combination of metal alloys.[9] By means of these diagrams it is possible to determine the phases that are present and the percentage of each, based on the alloy composition at any specified temperature. In addition, it is possible to determine what phase changes tend to take place with increasing or decreasing temperature. Most constitutional phase diagrams describe alloy systems containing two elements. Diagrams of more than two elements are complex and difficult to interpret. Phase diagrams are based on equilibrium conditions. This means that the metal is stable at the particular point on the diagram based on relatively slow heating or cooling. In welding this is not true since temperature changes are extremely rapid and equilibrium conditions rarely occur. Even so, the constitution diagram is the best tool available to determine phases.

The Iron-Carbon Diagram

An understanding of the iron-carbon equilibrium diagram shown by Figure 11-16 will provide an insight of the behavior of steels in connection with welding thermal cycles and heat treatment. This diagram represents the alloy of iron with carbon, ranging from 0 to 5% carbon.

Pure iron is a relatively weak but ductile metal. When carbon is added in small amounts the iron acquires a wide range of properties and uses and becomes the most popular metal, steel. It was previously mentioned that iron has either of two crystalline structures, depending on temperature. This can be shown by the iron-carbon diagram by considering pure iron or 0% carbon.

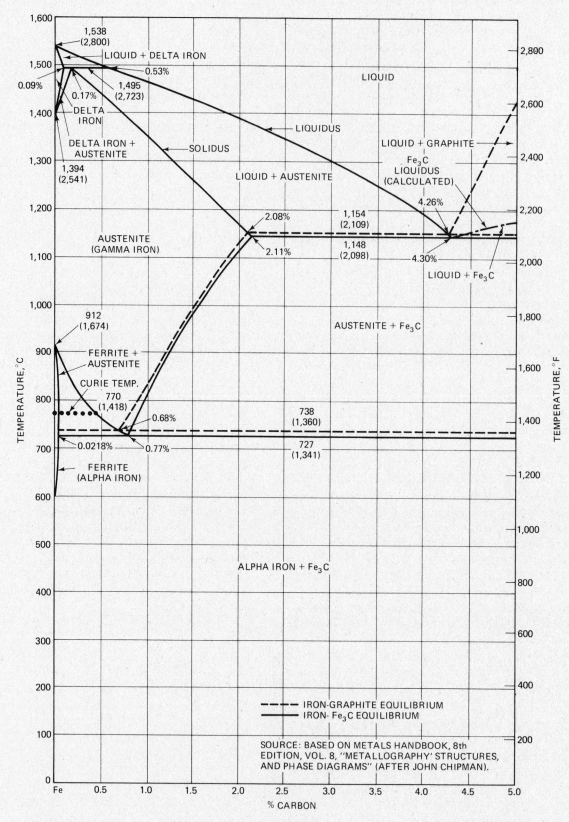

FIGURE 11-16 *Iron-carbon equilibrium diagram (after Metal Progress Data book © American Society of Metals 1977).*

Above 2,800°F (1,540°C) the iron is in liquid state and there is no crystalline structure. Below this temperature, it solidifies and has a body-centered cubic lattice which is known as *delta iron*. As the temperature is further reduced below 2,540°F (1,400°C) a transformation occurs and the crystalline structure changes to a face-centered cubic arrangement known as *gamma iron*. Below 1,670°F (910°C) the iron transforms back to the body-centered structure which is now known as *alpha iron*, and retains this structure down to room temperature. These transformation temperatures establish points on the iron-carbon diagram.

The other points and lines in Figure 11-16 show percentage of carbon involved in solid solution. Iron and carbon form a compound known as iron carbide Fe$_3$C) or cementite. When iron carbide or cementite is heated above 2,100°F (1,115°C) it decomposes into liquid iron saturated with graphite. Graphite is a crystalline form of carbon. Most metals have the ability to dissolve other elements in the solid state and solid solutions are formed. Under suitable temperature and time conditions the dissolved elements will diffuse and homogenity will be obtained. A maximum solubility of carbon in alpha iron occurs at 1,340°F (727°C) and decreases with lower temperature. This establishes a point on the diagram. A solid solution of carbon in alpha iron or delta iron (body-centered cubic) is known as ferrite. A solid solution of carbon in gamma iron (face-centered cubic) is known as austenite. As much as 2.1% carbon can be held in solid solution in gamma iron at a specific temperature and this establishes a point. In fact, the iron-carbon diagram can be divided at this point. Those alloys of iron and carbon less than 2.1% are considered steels, while those containing more than 2.1% are referred to as cast irons. Thus the line on the chart indicates whether or not the carbon is held in solid solution or whether it precipitates out.

To better understand the iron-carbon diagram consider a steel with a composition of 0.25% carbon. This would be indicated by drawing a vertical line between the 0 and 0.5% carbon line. Considering this line it will be seen that above approximately 2,768°F (1,520°C) the steel would be molten. As the temperature decreases delta iron would start to form in the liquid. At just below 2,732°F (1,500°C) it would transform to austenite and molten metal. However, at about 2,696°F (1,480°C) all the liquid metal would be solidified and it would be austenite. At approximately 1,500°F (815°C) the austenite commences to break down and form a new phase at the grain boundaries. This new phase is almost pure iron or ferrite. Ferrite formation would continue until a temperature of 1,340°F (727°C) was reached. At this point, the remaining austenite disappears completely, transforming to a structure known as pearlite

plus ferrite. Pearlite is a mixture of ferrite and cementite and this structure would be retained down to room temperature. Microstructure of pearlite is shown by Figure 11-17. Only the room temperature structures can be seen and analyzed through a microscope.

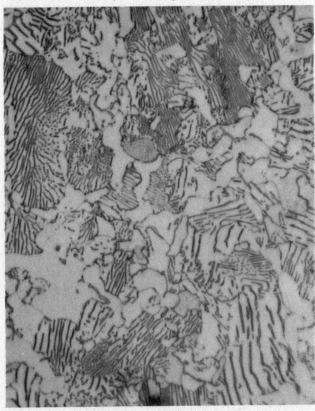

FIGURE 11-17 *Pearlite micro structure.*

By means of alloy additions it is possible to have the other structures at room temperature. The microstructure of pearlite is a lamellar structure which is relatively strong and ductile. The transformation during the cooling cycle will be reversed during the heating cycle. In welding the rise and fall of temperature or the rate of change of temperature is so fast that equilibrium does not occur. Therefore, some of the structures and temperatures mentioned here will be different. For example, if the cooling rate is faster, the austenite-to-ferrite transformation will be appreciably lower in temperature and this will also be true of pearlite. The pearlite will be more finely laminated since the transformation temperature is much lower. With extremely fast cooling rates the austenite might not have sufficient time to transform completely

WELDABILITY AND PROPERTIES OF METALS

to ferrite and pearlite and will provide a different microstructure. In this case, some of the untransformed austenite will be retained and the carbon is held in a supersaturated state. This new structure is known as martensite and is shown by Figure 11-18. Martensite has a needlelike appearance and if the cooling rate is sufficiently fast the austenite might transform completely to martensite. Martensite is harder than pearlite or the pearlite-ferrite microstructure and it has lower ductility. Its hardness depends on the carbon content. Thus, it can be seen that the cooling rate influences the microstructure and causes higher hardness. This is because the crystal lattice is changed or distorted and this in turn hardens the material.

By adding different alloys to the steel the tendency of austenite to transform into martensite upon cooling increases. This is the basis of hardening steels. By proper use of different alloys the amount of martensite produced can be changed. The rate of cooling changes depending on the method of quenching. The more severe quench will create more martensite, and the slower quench will create less martensite and thus a lower hardness. The amount of alloys and their power to create this microstructure transformation is known as *hardenability*. This is an advantage for heat treatment but can be detrimental to welding since high hardness is not desired in welds of softer materials.

The lines on the iron-carbon diagram show the different phases of iron-carbon alloy and indicate

FIGURE 11-18 *Martensite micro structure.*

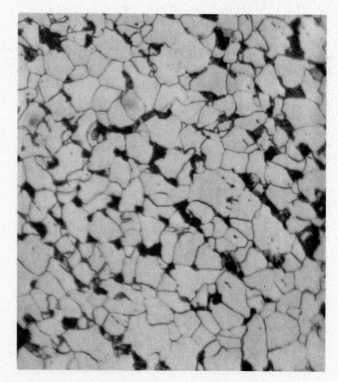

FIGURE 11-19 *Ferrite micro structure.*

the microstructure of each phase. With respect to the iron-carbon alloy only pearlite and ferrite and pearlite and cementite (Fe_3C) are found at room temperature. However, in other iron-alloy systems other microstructures exist at room temperature. Ferrite is an example. Ferrite is a solid solution of carbon in delta or alpha iron. It has a body-centered cubic structure and occurs when less than 0.08% carbon is dissolved in the iron. It is the softest constituent in steel and as the amount of ferrite increases the steel is softer. In alloy steels, certain alloy elements may be dissolved in ferrite as a solid state solution so that it occurs at room temperature. The microstructure of ferrite is shown by Figure 11-19.

The microstructure cementite is a compound of three atoms of iron and one atom of carbon (called iron carbide (Fe_3C) and is shown by Figure 11-20. It is hard and wear resistant and the composition will be varied when other carbide-forming alloys are present. It appears in several different ways, sometimes as a network surrounding the grains in the region of grain boundaries and also within the grains. It may also appear as round, roughly globular-shaped particles in steel that has been specially heat treated.

Austenite is another important constituent. It is the face-centered cubic lattice form and occurs in low-carbon steels in temperatures above 1,333 °F (722 °C). It is not stable at room temperatures in carbon steel; however, in highly alloyed steels and stainless steels it is stable at room temperature. It has good tensile strength and is

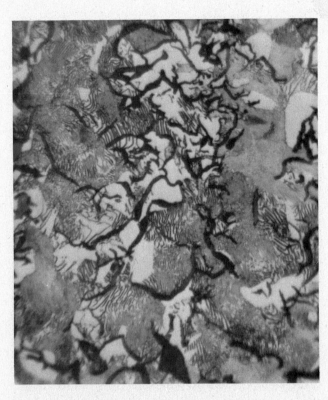

FIGURE 11-20 *Cementite micro structure.*

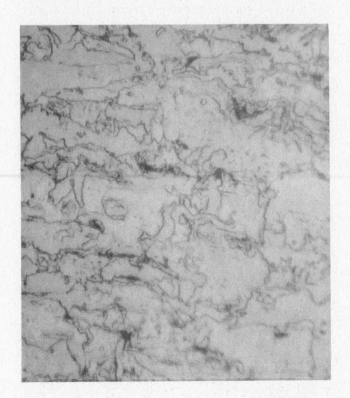

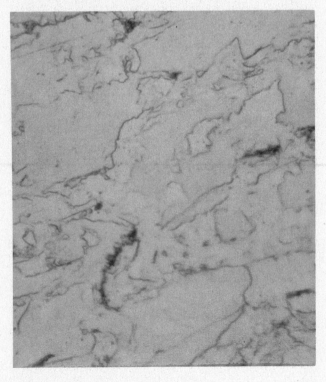

FIGURE 11-21 *Austentite micro structure.*

ductile, but it has a strong tendency to work-harden and is shown by Figure 11-21. There are other micro-structures in steels; however, the foregoing are the most important.

Hardenability

The heat treatment of steels to increase hardness and the metallurgy of welding have much in common. Heat treating to increase hardness is accomplished by heating followed by rapid cooling. The rapid cooling of metal in and adjacent to a weld is in this same order. An understanding of hardening by heat treating will make the metallurgical changes during welding clearer. Most steels possess the property of hardenability, which is defined as the property that determines the depth and distribution of hardness induced by quenching. This property can be measured by the end quench test. In this test, a round bar is heated to a temperature in the austenite range and is then placed vertically in a fixture and a jet of water is directed upwards to the bottom end. The bar rapidly cools to room temperature. Hardness measurements are then made along the bar from the quenched end to the un-quenched end and plotted against distance. This produces a curve with the high hardness at the quenched end and dropping off to normal hardness at the unquenched end. This hardenability curve shows the maximum hardness, the depth of hardness, and so on, under standardized conditions. It is very useful in establishing heat treating procedures. This information also provides data for welding since it indicates the effect of different alloying elements on the hardness of the quenched steel. The microstructure of the quenched steel can also be studied and related to the microstructure of welds.

Grain size and microstructure relate directly to hardness and strength. It was mentioned previously that as crystals deform they become harder and it is the growth of crystals that dictate grain size. In heat treating the steel is heated above the critical temperature, a phase change occurs, and new grains will nucleate and grow within the old grains. Since new grains of austenite were formed within each of the former grains the steel now has more finer grains. Fine grain size promotes both increased strength and hardness.

Alloying elements are added to steel to increase its hardenability. Carbon is the most important and effective, and small amounts will greatly increase hardness up to about 0.65%. Manganese is the next most important. Chromium and molybdenum also increase hardenability. These alloys contribute to other properties as well.

The time required for transformation of steel to begin and end at any constant temperature is a useful measure of the heat-treating characteristics of the steel. This can be shown by diagrams known as isothermal transformation diagrams, which plot temperature against time in seconds. Diagrams are available for steels of different composition and they show the phases that occur at different temperatures and times. The lines are constructed from experimental data and show the process of transformation at each temperature. Isothermal diagrams help to explain the relationship between cooling rates and the microstructure of a specific steel composition. During cooling the steel remains a certain period of time in each phase and the time period is inversely proportional to the cooling rate. With very slow cooling rates the phase changes take place near equilibrium values. As the cooling rate is increased time is reduced so that there is not enough time for completion of the pearlite reaction and some austenite remains below the equilibrium transformation temperature which creates hardening. The cooling rate must be so rapid that much of the austenite transforms to martensite rather than pearlite. These curves are of limited use in welding but do show the types of transformation occurring at subcritical temperatures and their effects.

Welds

When a weld is made all the factors just mentioned occur: the changes of temperature, the changes of dimensions, the growth of crystals and grains, the phase transformations, etc. The type of welding process dictates, in general, how these will occur. In the arc welding processes the heat cycle is of primary importance. The previous section explained the heat input time-temperature relationship or thermal cycle. The rate of cooling or quench is of primary importance and this is controlled by the process, procedure, metal, and mass. The electroslag process has the slowest cooling rate while the gas metal arc has a much faster cooling rate. The rate of change decreases as the distance from the center of the weld increases. This relationship is shown by Figure 11-22. It is obvious that many different cooling rates occur and that different microstructures will result. This is shown by Figure 11-23, which is a multipass groove weld. The different structures that occur in the weld are shown, as well as the different phases shown in the base metal adjacent to the weld.

With any arc process, where metal is transferred across the arc, the metal reaches a superheated temperature much above the melting temperature shown on the iron-carbon diagram. When it is deposited in the weld it is molten or in the liquid phase. Imme-

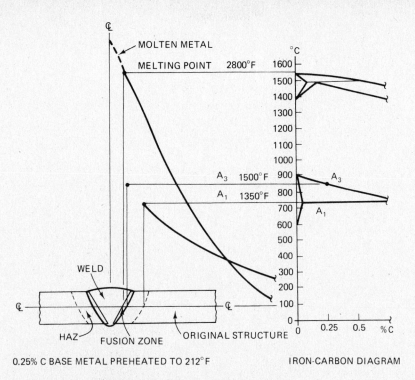

MOLTEN METAL
MELTING POINT   2800°F

A₃   1500°F
A₁   1350°F

WELD

HAZ   FUSION ZONE   ORIGINAL STRUCTURE

0.25% C BASE METAL PREHEATED TO 212°F   IRON-CARBON DIAGRAM

FIGURE 11-22   *Temperature distribution at a weld.*

diately the weld metal starts to freeze or solidify. The heat contained in the molten metal is transmitted to the base metal and its temperature at the weld is raised to the molten stage. Away from the weld the metal is raised to a lower temperature. This creates a multitude of time-temperature curves based on location. As the weld metal freezes, the crystals form into grains which rapidly cool until there is no more liquid metal. The cooling rate is much faster than occurs in a casting or ingot and, therefore, equilibrium, as represented in the iron-carbon diagram, does really not occur.

In addition to the complications created by the rapid cooling there is also the complication of composition variations. As weld metal is deposited on base metal some of the base metal melts and mixes with the weld metal, producing a dilution of weld metal. Unless the composition of the deposited filler metal and the composition of the base metal are identical there will be a variation of composition of the metal at the interface. In multipass welds, the first pass will have a high dilution factor, the second pass less, and the third pass perhaps little or none. When welding on a base metal with a different composition from the deposited metal this variation can be considerable. Variation in composition and the variation in the cooling rates will create variations in microstructure as shown by Figure 11-23. This is the reason that the microstructure of the weld is important and should be studied. The microstructure taken at different locations in the weld is shown by Figure 11-24. Each microstructure has its particular characteristics, as previously discussed. One of the important characteristics is the hardness of the microstructure throughout the weld area. The hardness should not vary over specific limitations. Figure 11-25 shows the macrostructure of a weld-metal—base-metal interface and the hardness at different points across this interface. Note the higher hardness of the weld metal compared to the base metal and that in the heat-affected zone the hardness is between these two values. This macrograph is of a low-carbon base metal joined with slightly alloyed weld metal.

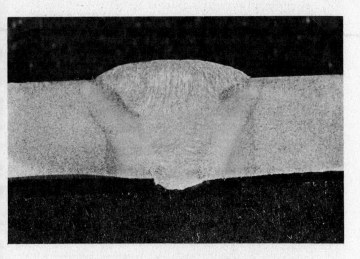

FIGURE 11-23   *Cross section of a multipass arc weld.*

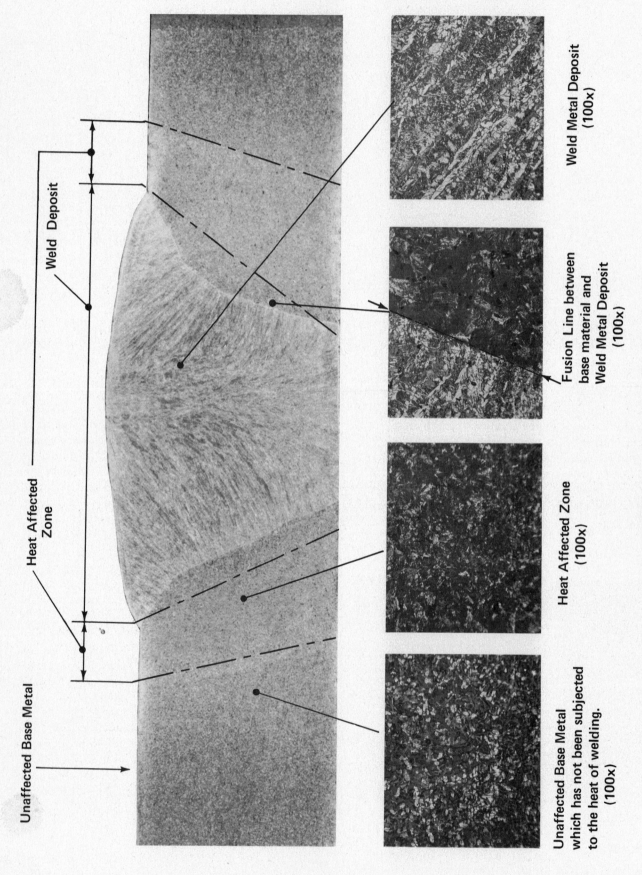

Weld Deposit

Heat Affected Zone

Unaffected Base Metal

Weld Metal Deposit
(100x)

Fusion Line between
base material and
Weld Metal Deposit
(100x)

Heat Affected Zone
(100x)

Unaffected Base Metal
which has not been subjected
to the heat of welding.
(100x)

FIGURE 11-24 Micro structure at different parts of a weld.

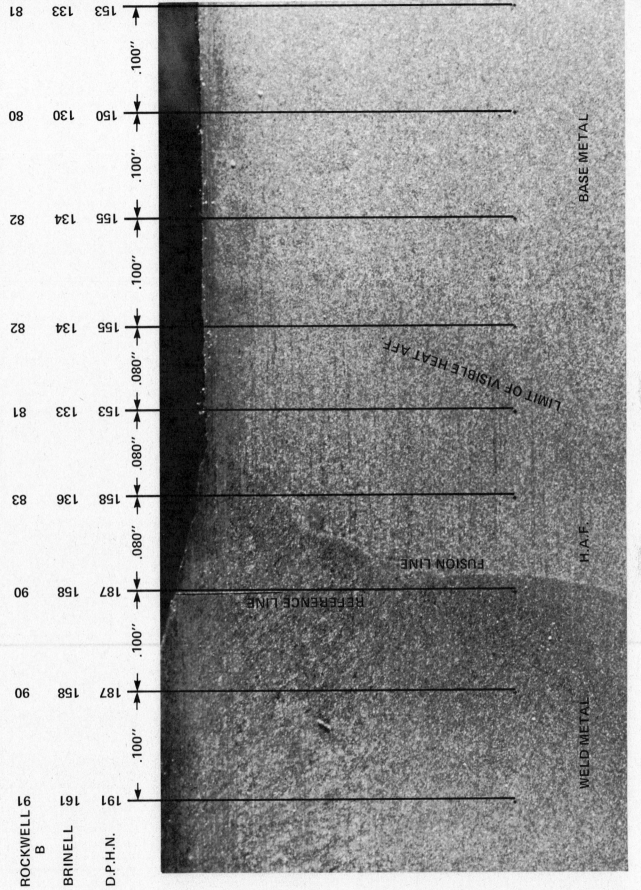

ROCKWELL B	BRINELL	D.P.H.N.		
81	133	153	.100"	
80	130	150	.100"	
82	134	155	.100"	
82	134	155	.080"	
81	133	153	.080"	
83	136	158	.080"	
90	158	187	.100"	
90	158	187	.100"	
91	161	191		

BASE METAL

LIMIT OF VISIBLE HEAT AFF.

FUSION LINE

REFERENCE LINE

H.A.F.

WELD METAL

FIGURE 11-25 Macro structure and hardness across arc weld interface.

The area between the interface of the deposited weld metal, and extending into the base metal far enough that any phase change occurs, is known as the heat affected zone (HAZ). This is shown in several of the pictures. The heat-affected zone, while part of the base metal, is considered to be a portion of the weld joint since it influences the service life of the weld. The heat-affected zone and the admixture zone are the most critical in many welds. For example, when welding a hardenable steel the heat-affected zone can increase in hardness to an undesirable level. On the other hand, when welding a hardened steel the heat-affected zone can become a softened zone since the heat of the weld has annealed the hardened metal. In the case of electroslag welding, the heat-affected zone will contain extremely large grains because of the long time at high temperature and the possibility for grain growth. The hardness of the weld metal, the base metal, and the heat-affected zone of an electroslag weld will be relatively uniform.

When the base metal and weld metal are of completely different analysis the interface zone contains alloys that can be detrimental. Figure 11-26 shows the interface between a low-alloy high-strength steel and a stainless steel weld. Note the heat-affected zone of the base metal and the minute amount of mixing of the lighter colored stainless steel deposit with the base metal. In this case, the alloy mixture produced is of such a small amount that it does not have an appreciable effect on the overall properties of the weld joint.

Another problem of the weld is the segregation during the thermal cycle. Segregation relates to the solubility of elements in metals, particularly alloys. The composition of the first crystals which form as an alloy freezes is different from the composition of the liquid that freezes last. The purer metals have the higher melting point or freezing point and, therefore, freeze first. Metals or elements with lower melting points freeze last. In addition, in weld metal, because of the rapidity of freezing time, very little diffusion occurs and there is a lack of homogeneity in the total weld. Carbon, phosphorus, sulphur, and sometimes manganese are frequently in the segregated state in steel. This can be determined by high-magnification study of the microstructure. Segregation, particularly of carbon, can be partially dispersed by means of heat treatment.

Molten metal has a relatively high capacity of dissolving gases in contact with it. As the metal cools it has less capacity for dissolved gases and when going from liquid to solid state the solubility of gas in metal is much lower. The gas is rejected as the crystals solidify but it may be trapped because of almost instantaneous solidification. Entrapment of the gas causes gas pockets and porosity in the weld. Carbon monoxide which is present in many arc and fuel gas atmospheres is sometimes trapped. Hydrogen that may be present in the arc atmosphere is also trapped. Hydrogen, however, will gradually disperse and escape from the weld metal over a period of time. High temperatures increase the speed for hydrogen migration and removal. The inert gases are not soluble in molten metal and for this reason are used in many gas shielded applications.

The solubility of metals within metals is also of high interest, particularly for dissimilar metal welding. The solubility is determined by the equilibrium diagram of the alloys. The greater the degree of solubility the better the success of welding dissimilar metal combinations.

It is impossible to provide a more thorough coverage of the metallurgy of steel welds in the space available. For more information, please refer to the reference section.[10,11,12]

The metallurgy of nonferrous metals resembles but

FIGURE 11-26 *Interface of dissimilar weld joint.*

is quite different from the metallurgy of iron and steel. The metallurgy of these metals will be briefly discussed in Chapter 13.

11-6 WELDABILITY OF METALS

The American Welding Society defines weldability as "the capacity of a material to be welded under the fabrication conditions imposed into a specific suitably designed structure and to perform satisfactorily in the intended service." This definition includes many qualifying statements, for example, "is the design suitable?" and "is the material suitable?" And what about the welding process and the welding procedures? A more practical definition might be "the ease with which a satisfactory weld can be made that will produce a joint equal to the metal being welded."

It has been stated that all metals are weldable, but some are much more difficult to weld than others. In view of this, it is vitally important that the welding process and welding procedures be considered when determining the weldability of a particular metal.

In any of these definitions it is important to know all about the metals to be welded, the design of the welds and the weldment, and the service requirements including loadings and the environment to which it will be exposed.

Perhaps the best definition is that a weldable material can be welded so that the joint is equal in all respects to the base metal—in other words, a 100% weld joint.

The base metal or metal to be welded must be considered from all points of view. This includes its physical properties, mechanical properties, and chemical composition and structure.

Unfortunately, the physical properties are not always identical in materials of the same composition. This relates to the size of test specimen, method of testing, and the type of microstructure. The mechanical properties can be different for different materials even though they may fall within the same specification or class. For example, hardness is related to structure which is affected by thermal history or heat treatment. The direction of testing has a large effect on strength levels, toughness, and ductility. In addition, composition and microstructure may vary. In heavy material, the composition may have more carbon or alloy to provide the strength called for by the specification, and the structure will change from the outside to the center based on different rates of cooling when the material was produced.

It becomes apparent that materials may not be as uniform or consistent as we have thought. Segregation and changes in metallurgical structure affect the properties. In spite of this, we must still produce the weld or weldment that provides the required service performance.

To better determine weldability, it becomes necessary to make several assumptions:

1. The material to be welded is suitable for the intended use. In other words, it will provide the proper and necessary properties to withstand its service requirements.

2. The design of the weldment is suitable for its intended use. In considering the design for the weldment we should include the design of the welds.

Based on these assumptions, it is necessary then to study the weld joint. The desirable weld joint has uniform strength, ductility, notch toughness, fatigue strength, and corrosion resistance throughout the weld and adjacent material.

Most welds involve the use of heat and the addition of a different metallurgical structure from the unaffected base metal. Welds may also include defects such as voids, cracks, entrapped materials, etc. Two types of problems may occur:

1. Problems of the weld metal deposit or heat-affected zone that occur in connection with or immediately following the welding operation, such as hot cracking, heat-affected zone cracking, hydrogen-induced cracking, etc.

2. Problems in the weld or adjacent to the weld that occur any time during service of the weldment. These can be any kind of defects that will reduce the efficiency of the weld joint under service conditions.

It is our objective to produce a weld that will avoid the problems. Hot cracking may result from any of the following four factors: restraint, weld shape, excessive heat input, or material composition. It can result from any one factor but is much more likely if two or more factors are present.

Restraint is always present in any weld because as the weld solidifies it acquires strength but continues to cool and shrink. It is the degree of restraint that becomes critical. Restraint relates primarily to the weld design, the weldment design, and the thickness of the materials being joined.

Weld shape is also a function of weld design, weldment design, and welding procedure. Weld procedure relates to the placement of welds or beads in the weld, the shape of the beads, and the shape of the finished surface of the weld.

CHAPTER 11

WELDABILITY AND PROPERTIES OF METALS

The third factor is the metal composition, and we must use the nominal composition. Segregation is important, however, since impurities such as sulphur and phosphorus tend to form low melting point films between solidifying grains of the metal. These impurities relate to weld joint detail and the welding process, since they affect the amount of dilution. Lamellar tearing is also associated with base metal impurities and its through-directional strength. When the degree of restraint increases, as it does in thicker metals, this problem becomes more serious.

Hydrogen cracking is considered cold cracking since it occurs soon after the weld is completed, usually within four to eight hours. It usually occurs in the heat-affected zone. It may occur while the weld is cooling down to room temperature. The four main factors that affect heat-affected zone cracking are:

The thickness of the base metal and the type of weld.

The composition of the base metal.

The welding process and filler metal type.

Energy input and preheat temperatures.

The effects of these four factors are interrelated. The thickness and composition of the base metal is established by the design. The weld joint configuration, the welding process, the type of filler metal, and the welding procedure can all contribute to the severity of the factors that cause HAZ cracking. The input energy can be modified by the welding process, welding procedure and the welding preheat temperature. These can also be changed to reduce the cooling rate.

All of the above factors determine the type of microstructure that will occur in the heat-affected zone. The two factors most important to weldability are hardenability and the susceptibility of the hardened structure to cracking. Both are increased by using a higher carbon or higher alloy content in the base metal. Certain alloying elements increase hardenability without a significant increase in the susceptibility to cracking. In this regard the carbon equivalent of the base metal becomes important. The carbon equivalent formula is as follows:

$$CE = \%C + \frac{\%Mn}{6} + \frac{\%Mo}{4} + \frac{\%Cr}{5}$$
$$+ \frac{\%Ni}{15} + \frac{\%Cu}{15} + \frac{\%P}{3}$$

If phosphorus (P) does not exceed 0.06% the value %P can be omitted.

Plain carbon steels which have a carbon equivalent of not over 0.40% are considered readily weldable. This carbon equivalent can be increased up to 0.45% provided that the carbon does not exceed 0.22%, the phosphorus does not exceed 0.06%, and the steel is not over 3/4 in. (19.1 mm) thick.

Usually when the carbon equivalent exceeds 0.40% special controls are required. The low-hydrogen processes or filler metals should be employed. Higher heat input should be employed and preheat may be required.

When the carbon equivalent exceeds 0.60%, low-hydrogen processes are required; preheating is required if the thickness exceeds 3/4 in. (19.1 mm).

Hardenability is related to the cooling rate of metals. The faster cooling rate tends to produce higher hardnesses. The cooling rate depends on the mass of the metal, the welding process, the welding procedure, and preheat temperatures. The welding process and procedure influence the energy input used to make the weld. The greater the energy input, the slower the cooling rate. Heat input is a function of welding current, arc voltage and travel speed. To increase the heat input, increase the welding current or reduce the travel speed. Welding current relates to process and to electrode size. Heat input is calculated by using the formula given in Section 11-4.

Thus, by increasing amperes or voltage, heat input increases but by increasing travel speed heat input decreases. The voltage has minor effect since it varies only slightly compared to the other factors. In general, higher heat input reduces cooling rate. This must be used with caution since on quenched and tempered steels too high a heat input will tend to soften the heat-affected zone and its strength level will be reduced.

In steels of relatively low hardenability it is possible to produce an unhardened heat-affected zone by increasing the heat input. In higher hardenability steels, the tendency toward cracking, and the maximum hardness, will be reduced by a slower cooling rate. There are limits to the amount of heat input that can be used. In this case, preheating is used in order to reduce cooling rates.

There is an interplay of several variables in regard to hydrogen cracking. These are the base metal composition, the heat input, the preheat temperature, the rate of cooling, and restraint. On a nonhardenable thin material the control of hydrogen is not important. As carbon and alloy increase, to provide greater hardenability, and as thickness increases the effect of hydrogen becomes of vital importance.

Heat-affected zone cracking depends on many of the same factors just mentioned. There are, however, general precautions that should be taken with certain types of steels to avoid HAZ cracking. Constructional steels can be grouped into five general classifications depending on whether or not they are hardenable and the nature of the hardened structure. These must be tempered with the effects of thickness which increases restraint and adds to the cracking problem. The five classes are:

1. Nonhardenable low-carbon mild steel.

2. Low hardenability with low susceptibility of cracking when hardened, or low-alloy steels with a carbon equivalent of not over 0.20%. maximum.

3. Low hardenability with high susceptibility of cracking when hardened, normally carbon manganese steels with less than 0.25% carbon and not over 1% manganese.

4. High hardenable steels with low susceptibility of cracking when hardened. This includes most low-carbon low-alloy high-strength steels usually with carbon less than 0.15%, manganese up to 1.5%, nickel up to 1.5%, chromium 0.25%, molybdenum up to 0.25%, and vanadium up to 0.20%.

5. High hardenable steels with high susceptibility of cracking when hardened. This would include alloy steels with carbon not exceeding 0.25% but with alloys.

Precautions that should be taken with the five classifications are:

1. No extraordinary precautions required when welding thin-to-medium thickness materials.

2. Use low-hydrogen processes and filler materials and utilize preheat for thick sections or increase heat input.

3. Low-hydrogen processes are recommended but not essential. High heat input should be used and preheat is not required except in thicker materials and should be in the 480°–660°F (250°–350°C) range.

4. Low-hydrogen processes are required, preheat and interpass are suggested, high heat input processes are recommended and preheat is increased as thickness increases.

5. Low-hydrogen processes are required, preheat and interpass temperature in the range of 300°–480°F (150°–250°C) are necessary and a postweld heat treatment is required.

Recommended preheat and filler metals will be given in Chapter 12 with regard to the welding of steels.

Weldability is extremely complex and all the welding factors interrelate. It is of vital importance for the success of welds to give consideration to all the factors mentioned.

QUESTIONS

1. What are some of the physical properties of a metal? List and define.

2. What are some of the mechanical properties of a metal? List and define.

3. What is ductility and why is it important for welding?

4. Explain the different methods of measuring hardness. Is there a comparison?

5. What types of groups write metal specifications? What is the purpose of a specification?

6. How can you determine the metal composition if you know the specification number?

7. What are seven tests that can be made to identify a metal piece?

8. What can be determined by a metal's appearance?

9. Why is a magnet of value to help identify a piece of metal?

10. What is the spark test? How is it performed?

11. Why is it important to identify a metal before welding it?

12. What different ways is heat generated for welding?

13. What is the weld cooling rate and what importance is it to welding?

14. What is the major difference between welding metallurgy and casting metallurgy?

15. What are the three most common crystal structures of metals? Explain.

16. What is an equilibrium diagram?

17. What is meant by the microstructure of a metal? Name some microstructures.

18. What is the HAZ? What causes it?

19. What is the carbon equivalent formula? How is it used?

20. What is meant by weldability?

REFERENCES

1. CHATE & GRAFF, "Alloys Index," *American Society of Metals,* Metals Park, Ohio.

2. ERIC N. SIMONS, AND FREDERIC MULLER, "A Dictionary of Alloys."

3. WOLDMAN & GIBBONS, *Engineering Alloys,* (5th Edition), Van Nostrand Reinhold Company, New York.

4. SIMMONS & GUNIZ, "Compilation of Trade Name, Specification and Producers of Stainless Alloys and Super Alloys," Data Service DS45, *American Society for Testing and Materials,* Philadelphia, Pa.

5. GERHART TSCHORN, *Spark Atlas of Steels,* A Pergamon Press Book, Macmillan Company, New York.

6. "Non Destructive Rapid Identification of Metals and Alloys by Spot Test," Special Technical Publication 550, American Society for Testing and Materials.

7. IVAN HRIVNAK, *The Theory of Mild Steel and Micro Alloy Steels' Weldability,* Bratislava, Czechoslovakia, 1969.

8. D. ROSENTHAL, "The Theory of Moving Sources of Heat and Its Application to Metal Treatment," Transactions of the ASME, November 1946, p. 849.

9. *Metals Handbook,* Vol. 1, "Properties and Selection of Metals" American Society of Metals, Metals Park, Ohio.

10. G. E. LINNERT, "Welding Metallurgy," *The American Welding Society,* 1967 Miami, Florida.

11. ARTHUR PHILLIPS, Ed., "Introductory Welding Metallurgy, *The American Welding Society,* 1968, Miami, Florida.

12. *Metals Handbook,* Vol. 6, "Welding and Brazing," *American Society of Metals,* Metals Park, Ohio.

12

WELDING THE STEELS

12-1 STEEL CLASSIFICATIONS

12-2 WELDING LOW-CARBON
AND LOW-ALLOY STEELS

12-3 WELDING THE ALLOY STEELS

12-4 WELDING STAINLESS STEELS

12-5 WELDING ULTRA-HIGH-STRENGTH
STEELS

12-1 STEEL CLASSIFICATIONS

Almost 85% of the metal produced and used is steel. The term *steel* encompasses many types of metals made principally of iron. Steel is an alloy of iron and carbon, but steels most often contain other metals such as manganese, chromium, nickel, etc., and nonmetals such as carbon, silicon, phosphorus, sulphur, and others.

The first rule of welding is "know the metal being welded." This theme is repeated throughout this book. It is necessary to consider the different steels, how they are classified and identified.

There are so many different types and kinds of steels that it is sometimes confusing just to be able to identify the steel that is being used. For example, there are structural steels, cast steels, stainless steels, tool steels, hot rolled steel, reinforcing steel, low-alloy high-strength steel, etc. Steels are sometimes given names based on their principal alloy such as carbon steel, chrome-manganese steel, chrome-moly steel, etc. Sometimes steels are identified by numbers, such as C-1020 steel, A36 steel, SAE 1045 steel, Type 304 steel, etc. In other cases, steels may be identified by letters, such as AR steels, T-1 steels, RQC steels, NAX steels, etc. Steels are also called by a trade name given by their manufacturer. Examples of this are Mayari steel, Corten steel, Jalloy steel, Naxtra steel, etc. All these names tend to add to the confusion, but they are clues for finding the true identification of a steel.

In this section information is provided to help identify steels or at least show how their identity, composition, or properties can be determined. The true basis for identifying steel is its composition—mechanical and physical properties.

The method of the manufacture of the steel also enters into the identification system. This would include cast steel, hot or cold rolled steels, forged steels, semi-killed steels, continuous cast steel, etc.

The two best ways of identifying a steel are by its specification number and grade or trade name and number. In the case of the specification, it should be consulted to determine composition and properties, and when a trade name is used the manufacturer's literature should be consulted for composition and properties. In the case of trade names, it is necessary to determine the manufacturer and in the case of specification, it is necessary to determine the sponsoring group of the specification. When this is known, the true identity of the steel can be determined.

The specifications for different metals was covered in the previous chapter, Section 11-2. In this section, we will provide more information concerning steel specifications.

A popular system for classifying steels is the American Iron and Steel Institute Numerical Designation of Standard Carbon and Alloy Steels.[1] This is known as the AISI designation system and sometimes known as the SAE system since it was originated by the Society of Automotive Engineers. The groupings of steels within this numerical system are shown by Figure 12-1. Numbers are used to designate different chemical compositions. A four-digit number series designates carbon and alloy steels according to the types and classes shown in Figure 12-1. This system has been expanded and in some cases five digits are used to designate certain alloy steels. The last two digits are intended to indicate the approximate middle of the carbon range; for example, 0.21 indicates a range of 0.18–0.23% carbon. In a few cases, the system deviates from this rule and some carbon ranges relate to the ranges of manganese, sulphur, phosphorous, chromium, and other elements. Two letters are often used as a prefix to the numerals. The letter C indicates basic open hearth carbon steel, and E indicates electric furnace carbon and alloy steels. The letter H is sometimes used as a suffix to denote steels manufactured to meet hardenability limits. The first two digits indicate the major alloying metals in the steel such as manganese, nickel-chromium, chrome-molybdenum, etc. Thus the system will designate the major elements of a steel and the approximate carbon range of the steel. It will also indicate the manufacturing process used to produce the steel. The complete designation system is shown by Table 12-1.

The other major identifying system is based on specifications written by the American Society for Testing and Materials (ASTM). The ASTM specifications were described in Section 11-2. Other organizations that specify steels include the American Petroleum Institute (API), the American Association of Railroads (AAR), the American Bureau of Shipping (ABS), the Steel Founders Society of America (SFSA), the Society of Automotive Engineers (SAE), which include the AMS specifications as well as many government agencies and private companies.

In view of this it is necessary to learn the different specifying groups and obtain copies of specifications or literature concerning the steels that are to be welded.

Series Designation	Types and Classes
10xx	Non-resulphurized carbon steel grades
11xx	Resulphurized carbon steel grades
13xx	Manganese 1.75%
23xx	Nickel 3.50%
25xx	Nickel 5.00%
31xx	Nickel 1.25%–chromium 0.65 or 0.80%
33xx	Nickel 3.50%–chromium 1.55%
40xx	Molybdenum 0.25%
41xx	Chromium 0.50-0.95%–molybdenum 0.12 or 0.20%
43xx	Nickel 1.80%–chromium 0.50 Or 0.80%–molybdenum 0.25%
46xx	Nickel 1.55 or 1.80%–molybdenum 0.20 or 0.25%
47xx	Nickel 1.05%–chromium 0.45%–molybdenum 0.25%
48xx	Nickel 3.50%–molybdenum 0.25%
50xx	Chromium 0.28 or 0.40%
51xx	Chromium 0.80, 0.90, 0.95, 1.00 or 1.05%
5xxxx	Carbon 1.00%–chromium 0.50, 1.00,or 1.45%
61xx	Chromium 0.80 ro 0.95%–vanadium 0.10 or 0.15% min
86xx	Nickel 0.55%–chromium 0.50 or 0.65%–molybdenum 0.20%
87xx	Nickel 0.55%–chromium 0.50%–molybdenum 0.25%
92xx	Manganese 0.85%–silicon 2.00%
93xx	Nickel 3.25%–chromium 1.20%–molybdenum 0.12%
94xx	Manganese 1.00%–nickel 0.45%–chromium 0.40%–molybdenum 0.12%
97xx	Nickel 0.55%–chromium 0.17%–molybdenum 0.20%
98xx	Nickel 1.00%–chromium 0.80%–molybdenum 0.25%

Prefix letters:
C—Indicates basic open hearth carbon steel.
E—Indicates electric furnace carbon and alloy steels.

FIGURE 12-1 *AISI-SAE numerical designation of carbon and alloy steels.*

TABLE 12-1 *AISI-SAE numerical designation of carbon and alloy steels.*

S.A.E. Steel Specifications

The following numerical system for identifying carbon and alloy steels of various specifications has been adopted by the Society of Automotive Engineers.

COMPARISON
A.I.S.I.–S.A.E. Steel Specifications

The evergrowing variety of chemical compositions and quality requirements of steel specifications have resulted in several thousand different combinations of chemical elements being specified to meet individual demands of purchasers of steel products.

The S.A.E. developed an excellent system of nomenclature for identification of various chemical compositions which have symbolized certain standards as to machining, heat treating and carburizing performance. The American Iron and Steel Institute has now gone further in this regard with a new standardization setup with similar nomenclature but with restricted carbon ranges and combinations of other elements which have been accepted as standard by all manufacturers of bar steel in the steel industry, because it has become apparent that steel producers must concentrate their efforts upon a smaller number of standardized grades. The society of Automotive Engineers have as a result revised most of their specifications to coincide with those set up by the American Iron and Steel Institute.

PREFIX LETTERS–

No prefix for basic open-hearth alloy steel.
(B) Indicates acid Bessemer carbon steel.
(C) Indicates basic open-hearth carbon steel.
(E) Indicates electric furnace steel.

NUMBER DESIGNATIONS–

(10XX series) Basic open-hearth and acid Bessemer Carbon Steel grades, non sulphurized and non-phosphorized.
(11XX series) Basic open-hearth and acid Bessemer Carbon Steel grades, sulphurized but not phosphorized.
(1300 series) Manganese 1.60 to 1.90%
(23XX series) Nickel 3.50%
(25XX series) Nickel 5.0%
(31XX series) Nickel 1.25%–Chromium .60%
(33XX series) (Nickel 3.50%–Chromium 1.60%
(40XX series) Molybdenum
(41XX series) Chromium Molybdenum
(43XX series) Nickel-Chromium-Molybdenum
(46XX series) (Nickel 1.65%–Molybdenum .25%
(48XX series) Nickel 3.25%–Molybdenum .25%
(51XX series) Chromium
(52 XX series) Chromium and High Carbon
(61XX series) Chromium Vanadium
(86XX series) Chrome Nickel Molybdenum
(87XX series) Chrome Nickel Molybdenum
(92XX series) Silicon 2.0%–Chromium
(93XX series) Nickel 3.0%–Chromium–Molybdenum
(94XX series) Nickel–Chromium–Molybdenum
(97XX series) Nickel–Chromium–Molybdenum
(98XX series) Nickel–Chromium–Molybdenum

Carbon Steels

SAE No.	C	Mn	P Max	S Max	AISI Number
–	0.06 max	0.35 max	0.040	0.050	C1005
1006	0.08 max	0.25-0.40	0.040	0.050	C1006
1008	0.10 max	0.25-0.50	0.040	0.050	C1008
1010	0.08-0.13	0.30-0.60	0.040	0.050	C1010
–	0.10-0.15	0.30-0.60	0.040	0.050	C1012
–	0.11-0.16	0.50-0.80	0.040	0.050	C1013
1015	0.13-0.18	0.30-0.60	0.040	0.050	C1015
1016	0.13-0.18	0.60-0.90	0.040	0.050	C1016
1017	0.15-0.20	0.30-0.60	0.040	0.050	C1017
1018	0.15-0.20	0.60-0.90	0.040	0.050	C1018

TABLE 12-1 *(Cont.)*

Carbon Steels (continued)

SAE No.	C	Mn	P Max	S Max	AISI Number
1019	0.15-0.20	0.70-1.00	0.040	0.050	C1019
1020	0.18-0.23	0.30-0.60	0.040	0.050	C1020
–	0.18-0.23	0.60-0.90	0.040	0.050	C1021
1022	0.18-0.23	0.70-1.00	0.040	0.050	C1022
–	0.20-0.25	0.30-0.60	0.040	0.050	C1023
1024	0.19-0.25	1.35-1.65	0.040	0.050	C1024
1025	0.22-0.28	0.30-0.60	0.040	0.050	C1025
–	0.22-0.28	0.60-0.90	0.040	0.050	C1026
1027	0.22-0.29	1.20-1.50	0.040	0.050	C1027
–	0.25-0.31	0.60-0.90	0.040	0.050	C1029
1030	0.28-0.34	0.60-0.90	0.040	0.050	C1030
1033	0.30-0.36	0.70-1.00	0.040	0.050	C1033
1034	0.32-0.38	0.50-0.80	0.040	0.050	C1034
1035	0.32-0.38	0.60-0.90	0.040	0.050	C1035
1036	0.30-0.37	1.20-1.50	0.040	0.050	C1036
1038	0.35-0.42	0.60-0.90	0.040	0.050	C1038
–	0.37-0.44	0.70-1.00	0.040	0.050	C1039
1040	0.37-0.44	0.60-0.90	0.040	0.050	C1040
1041	0.36-0.44	1.35-1.65	0.040	0.050	C1041
1042	0.40-0.47	0.60-0.90	0.040	0.050	C1042
1043	0.40-0.47	0.70-1.00	0.040	0.050	C1043
1045	0.43-0.50	0.60-0.90	0.040	0.050	C1045
1046	0.43-0.50	0.70-1.00	0.040	0.050	C1046
1050	0.48-0.55	0.60-0.90	0.040	0.050	C1050
–	0.45-0.56	0.85-1.15	0.040	0.050	C1051
1052	0.47-0.55	1.20-1.50	0.040	0.050	C1052
–	0.50-0.60	0.50-0.80	0.040	0.050	C1054
1055	0.50-0.60	0.60-0.90	0.040	0.050	C1055
–	0.50-0.61	0.85-1.15	0.040	0.050	C1057
–	0.55-0.65	0.50-0.80	0.040	0.050	C1059
1060	0.55-0.65	0.60-0.90	0.040	0.050	C1060
–	0.54-0.65	0.75-1.05	0.040	0.050	C1061
1062	0.54-0.65	0.85-1.15	0.040	0.050	C1062
1064	0.60-0.70	0.50-0.80	0.040	0.050	C1064
1065	0.60-0.70	0.60-0.90	0.040	0.050	C1065
1066	0.60-0.71	0.85-1.15	0.040	0.050	C1066
–	0.65-0.75	0.40-0.70	0.040	0.050	C1069
1070	0.65-0.75	0.60-0.90	0.040	0.050	C1070
–	0.65-0.76	0.75-1.05	0.040	0.050	C1071
1074	0.70-0.80	0.50-0.80	0.040	0.050	C1074

Alloy Steel

AISI Number	C	Mn	P Max	S Max	Si	Ni	Cr	Other	SAE Number
1320	0.18-0.23	1.60-1.90	0.040	0.040	0.20-0.35	–	–	–	1320
1321	0.17-0.22	1.80-2.10	0.050	0.050	0.20-0.35	–	–	–	–
1330	0.28-0.33	1.60-1.90	0.040	0.040	0.20-0.35	–	–	–	1330
1335	0.33-0.38	1.60-1.90	0.040	0.040	0.20-0.35	–	–	–	1335
1340	0.38-0.43	1.60-1.90	0.040	0.040	0.20-0.35	–	–	–	1340
2317	0.15-0.20	0.40-0.60	0.040	0.040	0.20-0.35	3.25-3.75	–	–	2317
2330	0.28-0.33	0.60-0.80	0.040	0.040	0.20-0.35	3.25-3.75	–	–	2330
2335	0.33-0.38	0.60-0.80	0.040	0.040	0.20-0.35	3.25-3.75	–	–	–
2340	0.33-0.43	0.70-0.90	0.040	0.040	0.20-0.35	3.25-3.75	–	–	2340
2345	0.43-0.48	0.70-0.90	0.040	0.040	0.20-0.35	3.25-3.75	–	–	2345

TABLE 12-1 *(Cont.)*

Alloy Steel (continued)

AISI Number	C	Mn	P Max	S Max	Si	Ni	Cr	Other	SAE Number
E 2512	0.09-0.14	0.45-0.60	0.025	0.025	0.20-0.35	4.75-5.25	–	–	2512
2515	0.12-0.17	0.40-0.60	0.040	0.040	0.20-0.35	4.75-5.25	–	–	2515
E 2517	0.15-0.20	0.45-0.60	0.025	0.025	0.20-0.35	4.75-5.25	–	–	2517
3115	0.13-0.18	0.40-0.60	0.040	0.040	0.20-0.35	1.10-1.40	0.55-0.75	–	3115
3120	0.17-0.22	0.60-0.80	0.040	0.040	0.20-0.35	1.10-1.40	0.55-0.75	–	3120
3130	0.28-0.33	0.60-0.80	0.040	0.040	0.20-0.35	1.10-1.40	0.55-0.75	–	3130
3135	0.33-0.38	0.60-0.80	0.040	0.040	0.20-0.35	1.10-1.40	0.55-0.75	–	3135
3140	0.38-0.43	0.70-0.90	0.040	0.040	0.20-0.35	1.10-1.40	0.55-0.75	–	3140
3141	0.38-0.43	0.70-0.90	0.040	0.040	0.20-0.35	1.10-1.40	0.70-0.90	–	3141
3145	0.43-0.48	0.70-0.90	0.040	0.040	0.20-0.35	1.10-1.40	0.70-0.90	–	3145
3150	0.48-0.53	0.70-0.90	0.040	0.040	0.20-0.35	1.10-1.40	0.70-0.90	–	3150
E 3310	0.08-0.13	0.45-0.60	0.025	0.025	0.20-0.35	3.25-3.75	1.40-1.75	–	3310
E 3316	0.14-0.19	0.45-0.60	0.025	0.025	0.20-0.35	3.25-3.75	1.40-1.75	–	3316
								Mo	
4017	0.15-0.20	0.70-0.90	0.040	0.040	0.20-0.35	–	–	0.20-0.30	4017
4023	0.20-0.25	0.70-0.90	0.040	0.040	0.20-0.35	–	–	0.20-0.30	4023
4024	0.20-0.25	0.70-0.90	0.040	0.035-0.050	0.20-0.35	–	–	0.20-0.30	4024
4027	0.25-0.30	0.70-0.90	0.040	0.040	0.20-0.35	–	–	0.20-0.30	4027
4028	0.25-0.30	0.70-0.90	0.040	0.035-0.050	0.20-0.35	–	–	0.20-0.30	4028
4032	0.30-0.35	0.70-0.90	0.040	0.040	0.20-0.35	–	–	0.20-0.30	4032
4037	0.35-0.40	0.70-0.90	0.040	0.040	0.20-0.35	–	–	0.20-0.30	4037
4042	0.40-0.45	0.70-0.90	0.040	0.040	0.20-0.35	–	–	0.20-0.30	4042
4047	0.45-0.50	0.70-0.90	0.040	0.040	0.20-0.35	–	–	0.20-0.30	4047
4053	0.50-0.56	0.75-1.00	0.040	0.040	0.20-0.35	–	–	0.20-0.30	4053
4063	0.60-0.67	0.75-1.00	0.040	0.040	0.20-0.35	–	–	0.20-0.30	4063
4068	0.63-0.70	0.75-1.00	0.040	0.040	0.20-0.35	–	–	0.20-0.30	4068
–	0.17-0.22	0.70-0.90	0.040	0.040	0.20-0.35	–	0.40-0.60	0.20-0.30	4119
–	0.23-0.28	0.70-0.90	0.040	0.040	0.20-0.35	–	0.40-0.60	0.20-0.30	4125
4130	0.28-0.33	0.40-0.60	0.040	0.040	0.20-0.35	–	0.80-1.10	0.15-0.25	4130
E 4132	0.30-0.35	0.40-0.60	0.025	0.025	0.20-0.35	–	0.80-1.10	0.18-0.25	–
E 4135	0.33-0.38	0.70-0.90	0.025	0.025	0.20-0.35	–	0.80-1.10	0.18-0.25	–
4137	0.35-0.40	0.70-0.90	0.040	0.040	0.20-0.35	–	0.80-1.10	0.15-0.25	4137
E 4137	0.35-0.40	0.70-0.90	0.025	0.025	0.20-0.35	–	0.80-1.10	0.18-0.25	–
4140	0.38-0.43	0.75-1.00	0.040	0.040	0.20-0.35	–	0.80-1.10	0.18-0.25	4140
4142	0.40-0.45	0.75-1.00	0.040	0.040	0.20-0.35	–	0.80-1.10	0.15-0.25	–
4145	0.43-0.48	0.75-1.00	0.040	0.040	0.20-0.35	–	0.80-1.10	0.15-0.25	4145
4147	0.45-0.50	0.75-1.00	0.040	0.040	0.20-0.35	–	0.80-1.10	0.15-0.25	–
4150	0.48-0.53	0.75-1.00	0.040	0.040	0.20-0.35	–	0.80-1.10	0.15-0.25	4150
4317	0.15-0.20	0.45-0.65	0.040	0.040	0.20-0.35	1.65-2.00	0.40-0.60	0.20-0.30	4317
4320	0.17-0.22	0.45-0.65	0.040	0.040	0.20-0.35	1.65-2.00	0.40-0.60	0.20-0.30	4320
4337	0.35-0.40	0.60-0.80	0.040	0.040	0.20-0.35	1.65-2.00	0.70-0.90	0.20-0.30	–
4340	0.38-0.43	0.60-0.80	0.040	0.040	0.20-0.35	1.65-2.00	0.70-0.90	0.20-0.30	4340
4608	0.06-0.11	0.25-0.45	0.040	0.040	0.25 Max	1.40-1.75	–	0.15-0.25	4608
4615	0.13-0.18	0.45-0.65	0.040	0.040	0.20-0.35	1.65-2.00	–	0.20-0.30	4615
–	0.15-0.20	0.45-0.65	0.040	0.040	0.20-0.35	1.65-2.00	–	0.20-0.30	4617
E 4617	0.15-0.20	0.45-0.65	0.025	0.025	0.20-0.35	1.65-2.00	–	0.20-0.27	–
4620	0.17-0.22	0.45-0.65	0.040	0.040	0.20-0.35	1.65-2.00	–	0.20-0.30	4620
X4620	0.18-0.23	0.50-0.70	0.040	0.040	0.20-0.35	1.65-2.00	–	0.20-0.30	X4620
E 4620	0.17-0.22	0.45-0.65	0.025	0.025	0.20-0.35	1.65-2.00	–	0.20-0.27	–
4621	0.18-0.23	0.70-0.90	0.040	0.040	0.20-0.35	1.65-2.00	–	0.20-0.30	4621
4640	0.38-0.43	0.60-0.80	0.040	0.040	0.20-0.35	1.65-2.00	–	0.20-0.30	4640
E 4640	0.38-0.43	0.60-0.80	0.025	0.025	0.20-0.35	1.65-2.00	–	0.20-0.27	–
4812	0.10-0.15	0.40-0.60	0.040	0.040	0.20-0.35	3.25-3.75	–	0.20-0.30	4812
4815	0.13-0.18	0.40-0.60	0.040	0.040	0.20-0.35	3.25-3.75	–	0.20-0.30	4815
4817	0.15-0.20	0.40-0.60	0.040	0.040	0.20-0.35	3.25-3.75	–	0.20-0.30	4817
4820	0.18-0.23	0.50-0.70	0.040	0.040	0.20-0.35	3.25-3.75	–	0.20-0.30	4820
5045	0.43-0.48	0.70-0.90	0.040	0.040	0.20-0.35	–	0.55-0.75	–	5045
5046	0.43-0.50	0.75-1.00	0.040	0.040	0.20-0.35	–	0.20-0.35	–	5046

TABLE 12-1 *(Cont.)*

Alloy Steel (continued)

AISI Number	C	Mn	P Max	S Max	Si	Ni	Cr	Other	SAE Number
—	0.13-0.18	0.70-0.90	0.040	0.040	0.20-0.35	—	0.70-0.90	—	5115
5120	0.17-0.22	0.70-0.90	0.040	0.040	0.20-0.35	—	0.70-0.90	—	5120
5130	0.28-0.33	0.70-0.90	0.040	0.040	0.20-0.35	—	0.80-1.10	—	5130
5132	0.30-0.35	0.60-0.80	0.040	0.040	0.20-0.35	—	0.80-1.05	—	5132
5135	0.33-0.38	0.60-0.80	0.040	0.040	0.20-0.35	—	0.80-1.05	—	5135
5140	0.38-0.43	0.70-0.90	0.040	0.040	0.20-0.35	—	0.70-0.90	—	5140
5145	0.43-0.48	0.70-0.90	0.040	0.040	0.20-0.35	—	0.70-0.90	—	5145
5147	0.45-0.52	0.75-1.00	0.040	0.040	0.20-0.35	—	0.90-1.20	—	5147
5150	0.48-0.53	0.70-0.90	0.040	0.040	0.20-0.35	—	0.70-0.90	—	5150
5152	0.48-0.55	0.70-0.90	0.040	0.040	0.20-0.35	—	0.90-1.20	—	5152
E50100	0.95-1.10	0.25-0.45	0.025	0.025	0.20-0.35	—	0.40-0.60	—	50100
E51100	0.95-1.10	0.25-0.45	0.025	0.025	0.20-0.35	—	0.90-1.15	—	51100
E52100	0.95-1.10	0.25-0.45	0.025	0.025	0.20-0.35	—	1.30-1.60	—	52100
								V	
6120	0.17-0.22	0.70-0.90	0.040	0.040	0.20-0.35	—	0.70-0.90	0.10 Min	—
6145	0.43-0.48	0.70-0.90	0.040	0.040	0.20-0.35	—	0.80-1.10	0.15 Min	—
6150	0.48-0.53	0.70-0.90	0.040	0.040	0.20-0.35	—	0.80-1.10	0.15 Min	6150
6152	0.48-0.55	0.70-0.90	0.040	0.040	0.20-0.35	—	0.80-1.10	0.10 Min	—
								Mo	
8615	0.13-0.18	0.70-0.90	0.040	0.040	0.20-0.35	0.40-0.70	0.50-0.60	0.15-0.25	8615
8617	0.15-0.20	0.70-0.90	0.040	0.040	0.20-0.35	0.40-0.70	0.40-0.60	0.15-0.25	8617
8620	0.18-0.23	0.70-0.90	0.040	0.040	0.20-0.35	0.40-0.70	0.40-0.60	0.15-0.25	8620
8622	0.20-0.25	0.70-0.90	0.040	0.040	0.20-0.35	0.40-0.70	0.40-0.60	0.15-0.25	8622
8625	0.23-0.28	0.70-0.90	0.040	0.040	0.20-0.35	0.40-0.70	0.40-0.60	0.15-0.25	8625
8627	0.25-0.30	0.70-0.90	0.040	0.040	0.20-0.35	0.40-0.70	0.40-0.60	0.15-0.25	8627
8630	0.28-0.33	0.70-0.90	0.040	0.040	0.20-0.35	0.40-0.70	0.40-0.60	0.15-0.25	8630
8632	0.30-0.35	0.70-0.90	0.040	0.040	0.20-0.35	0.40-0.70	0.40-0.60	0.15-0.25	8632
8635	0.33-0.38	0.75-1.00	0.040	0.040	0.20-0.35	0.40-0.70	0.40-0.60	0.15-0.25	8635
8637	0.35-0.40	0.75-1.00	0.040	0.040	0.20-0.35	0.40-0.70	0.40-0.60	0.15-0.25	8637
8640	0.38-0.43	0.75-1.00	0.040	0.040	0.20-0.35	0.40-0.70	0.40-0.60	0.15-0.25	8640
8641	0.38-0.43	0.75-1.00	0.040	0.040-0.060	0.20-0.35	0.40-0.70	0.40-0.60	0.15-0.25	8641
8642	0.40-0.45	0.75-1.00	0.040	0.040	0.20-0.35	0.40-0.70	0.40-0.60	0.15-0.25	8642
8645	0.43-0.48	0.75-1.00	0.040	0.040	0.20-0.35	0.40-0.70	0.40-0.60	0.15-0.25	8645
8647	0.45-0.50	0.75-1.00	0.040	0.040	0.20-0.35	0.40-0.70	0.40-0.60	0.15-0.25	8647
8650	0.48-0.53	0.75-1.00	0.040	0.040	0.20-0.35	0.40-0.70	0.40-0.60	0.15-0.25	8650
8653	0.50-0.56	0.75-1.00	0.040	0.040	0.20-0.35	0.40-0.70	0.40-0.60	0.15-0.25	8653
8655	0.50-0.60	0.75-1.00	0.040	0.040	0.20-0.35	0.40-0.70	0.40-0.60	0.15-0.25	8655
8660	0.50-0.65	0.75-1.00	0.040	0.040	0.20-0.35	0.40-0.70	0.40-0.60	0.15-0.25	8660
8720	0.18-0.23	0.70-0.90	0.040	0.040	0.20-0.35	0.40-0.70	0.40-0.60	0.20-0.30	8720
8735	0.33-0.38	0.75-1.00	0.040	0.040	0.20-0.35	0.40-0.70	0.40-0.60	0.20-0.30	8735
8740	0.38-0.43	0.75-1.00	0.040	0.040	0.20-0.35	0.40-0.70	0.40-0.60	0.20-0.30	8740
8742	0.40-0.45	0.75-1.00	0.040	0.040	0.20-0.35	0.40-0.70	0.40-0.60	0.20-0.30	—
8745	0.43-0.48	0.75-1.00	0.040	0.040	0.20-0.35	0.40-0.70	0.40-0.60	0.20-0.30	8745
8747	0.45-0.50	0.75-1.00	0.040	0.040	0.20-0.35	0.40-0.70	0.40-0.60	0.20-0.30	—
8750	0.48-0.53	0.75-1.00	0.040	0.040	0.20-0.35	0.40-0.70	0.40-0.60	0.20-0.30	8750
—	0.50-0.60	0.50-0.60	0.040	0.040	1.20-1.60	—	0.50-0.80	—	9254
9255	0.50-0.60	0.70-0.95	0.040	0.040	1.80-2.20	—	—	—	9255
9260	0.55-0.65	0.70-1.00	0.040	0.040	1.80-2.20	—	—	—	9260
9261	0.55-0.65	0.75-1.00	0.040	0.040	1.80-2.20	—	0.10-0.25	—	9261
9262	0.55-0.65	0.75-1.00	0.040	0.040	1.80-2.20	—	0.25-0.40	—	9262
E9310	0.08-0.13	0.45-0.65	0.025	0.025	0.20-0.35	3.00-3.50	1.00-1.40	0.08-0.15	9310
E9315	0.13-0.18	0.45-0.65	0.025	0.025	0.20-0.35	3.00-3.50	1.00-1.40	0.08-0.15	9315
E9317	0.15-0.20	0.45-0.65	0.025	0.025	0.20-0.35	3.00-3.50	1.00-1.40	0.08-0.15	9317
9437	0.35-0.40	0.90-1.20	0.040	0.040	0.20-0.35	0.30-0.60	0.30-0.50	0.08-0.15	9437
9440	0.38-0.43	0.90-1.20	0.040	0.040	0.20-0.35	0.30-0.60	0.30-0.50	0.08-0.15	9440
9442	0.40-0.45	1.00-1.30	0.040	0.040	0.20-0.35	0.30-0.60	0.30-0.50	0.08-0.15	9442
9445	0.43-0.48	1.00-1.30	0.040	0.040	0.20-0.35	0.30-0.60	0.30-0.50	0.08-0.15	9445
9747	0.45-0.50	0.50-0.80	0.040	0.040	0.20-0.35	0.40-0.70	0.10-0.25	0.15-0.25	9747
9763	0.60-0.67	0.50-0.80	0.040	0.040	0.20-0.35	0.40-0.70	0.10-0.25	0.15-0.25	9763

TABLE 12-1 *(Cont.)*

Alloy Steel (continued

AISI Number	C	Mn	P Max	S Max	Si	Ni	Cr	Other	SAE Number
9840	0.38-0.43	0.70-0.90	0.040	0.040	0.20-0.35	0.85-1.15	0.70-0.90	0.20-0.30	9840
9845	0.43-0.48	0.70-0.90	0.040	0.040	0.20-0.35	0.85-1.15	0.70-0.90	0.20-0.30	9845
9850	0.48-0.53	0.70-0.90	0.040	0.040	0.20-0.35	0.85-1.15	0.70-0.90	0.20-0.30	9850

12-2 WELDING LOW-CARBON AND LOW-ALLOY STEELS

Low-carbon steels are considered to be steels with a maximum of 0.15% carbon. Mild steels are those that contain 0.15-0.29% carbon. Low-alloy steels are those having a maximum of 0.29% carbon and with their total metal alloy content not exceeding 2%. Different groups have slightly different composition limits. These types of steels will be covered in this section.

Much of the information throughout this book relates to welding processes and provides welding procedures and schedules. The information presented in each section of the arc welding process chapters has been directed toward the welding of mild and low-alloy steels. In addition, the information provided in the filler metal chapter is also related to the mild and low-alloy steels.

For the shielded metal arc welding process, attention was directed toward the selection of the class of covered electrodes based on their usability factors. All the electrodes described in AWS specification A5.1 are applicable to the mild and low-alloy steels. The E-60XX and E-70XX classes of electrodes provide sufficient strength to produce 100% weld joints in the steels. The yield strength of electrodes, in these classes, will overmatch the yield strength of the mild and low-alloy steels. The E-60XX class should be used for steels having yield strengths below 50,000 psi (35 kg/mm²) and the E-70XX class should be used for welding steels having a yield strength below 60,000 psi (42 kg/mm²). Low-hydrogen electrodes should be used and preheat is suggested when welding heavier materials, or restrained joints. The electrode that provides the desired operational features should be selected.

When welding the low-alloy high-strength steels, the operating characteristics of the electrode are not considered since the E-80XX and higher-strength electrodes are all of the low-hydrogen type. There is one exception which is the E-XX10 class. These are shown in the AWS specification for low-alloy steel-covered arc welding electrodes, AWS A5.5.[2] This specification is more complex than the one for mild steel electrodes, even though there are only two basic classes in each strength level. The lower strength level includes the E-8010, E-XX15, E-XX16, and the more popular E-XX18 classes. A suffix code indicates the composition of the weld deposit produced by the specific electrode. A suffix is never applied to an E-60XX class of electrode and may not always be used with the higher-strength levels. However, when the suffix is used, it follows the four- or five-digit classification and designates a specific chemical composition of the deposited weld metal. This information is shown by Figure 12-2.

This new information now allows the selection of the covered electrode to match not only the mechanical properties of the base metal, but also to approximately match the composition of the base metal. For this reason, the base metal composition and the mechanical properties must be known in order to select the correct covered electrode to be used. The only E-80XX or higher-strength electrodes that do not have low-hydrogen coverings are the E-8010 type electrodes which are designed specifically for welding pipe. These high-strength, cellulosic covered, electrodes are matched to specific alloy steel pipe. The deep penetrating characteristics of the cellulosic electrodes make them suitable for cross-country pipe welding. The theory and practice is that alloy steel pipe is relatively thin and it is welded with cellulosic electrodes at relatively high currents. In addition, each welding pass is very thin and the weld metal is aged for a considerable length of time prior to putting the pipeline into service. This allows for hydrogen, which might be absorbed, to escape from the metal and not adversely affect the service life of the pipeline.

Suffix	C	Mn	Si	Ni	Cr	Mo	Va
A1	0.12	0.60 or 1.00*	0.40 or 0.80*	—	—	0.40-0.65	—
B1	0.12	0.90	0.60 or 0.80*	—	0.40-0.65	0.40-0.65	—
B2L	0.05	0.90	1.00	—	1.00-1.50	0.40-0.65	—
B2	0.12	0.90	0.60 or 0.80*	—	1.00-1.50	0.90-1.20	—
B3L	0.05	0.90	1.00	—	2.00-2.50	0.90-1.20	—
B3	0.12	0.90	0.60 or 0.80*	—	2.00-2.50	0.90-1.20	—
B4L	0.05	0.90	1.00	—	1.75-2.25	0.40-0.65	—
C1	0.12	1.20	0.60 or 0.80*	2.0-2.75	—	—	—
C2	0.12	1.20	0.60 or 0.80*	3.0-3.75	—	—	—
C3	0.12	0.40-1.10	0.80	0.80-1.10	0.15	0.35	0.05
D1	0.12	1.25-1.75	0.60 or 0.80*	—	—	0.25-0.45	—
D2	0.15	1.65-2.00	0.60 or 0.80*	—	—	0.25-0.45	—
G	—	1.00 Min.	0.80 Min.	0.50 min.	0.30 min.	0.20 min.	0.10 min.
M	0.10	0.60-2.25*	0.60 or 0.80*	1.4-2.25*	0.15-1.50*	0.25-0.55*	0.05

*Amount depends on electrode classification. Single values indicate max per AWS A5.5.

TABLE 12-2 *AISI stainless steel classification system (Courtesy, The American Iron and Steel Institute).*

The remainder of this section is related to specific types of steels and provides guidance in the selection of filler metal for joining them. For those steels that may not be specifically mentioned, it is possible to relate them to the chemistry of the deposited weld metal in order to establish the proper electrode class. For the other processes, including gas metal arc welding, flux-cored arc welding, and submerged arc welding the basic information provided in the process chapter and in the filler metals chapter must be followed.

Low-Carbon (Mild) Steels

Low-carbon steels include those in the AISI series C-1008 to C-1025. Carbon ranges from 0.10 to 0.25%, manganese ranges from 0.25 to 1.5%, phosphorous is 0.40% maximum, and sulphur is 0.50% maximum. Steels in this range are most widely used for industrial fabrication and construction. These steels can be easily welded with any of the arc, gas, and resistance welding processes.

Medium-Carbon Steels

The medium-carbon steels include those in the AISI series C-1030 to C-1050. The composition is similar to low-carbon steels, except that the carbon ranges from 0.25 to 0.50% and the manganese from 0.60 to 1.65%. With higher carbon and manganese the low-hydrogen type electrodes are recommended, particularly in thicker sections. Preheating may be required and should range from 300–500°F (149–260°C). Postheating is often specified to relieve stress and help reduce hardness that may have been caused by rapid cooling. Medium-carbon steels are readily weldable provided the above precautions are observed. They can be welded with all of the processes mentioned above.

High-Carbon Steels

High-carbon steels include those in the AISI series from C-1050 to C-1095. The composition is similar to medium-carbon steels, except that carbon ranges from 0.50 to 1.03% and manganese ranges from 0.30 to 1.00%. Special precautions must be taken when welding steels in these classes. The low-hydrogen electrodes must be employed and preheating of from 400–600°F (204–316°C) is necessary, especially when heavier sections are welded. A postheat treatment, either stress relieving or annealing, is usually specified. High-carbon steels can be welded with the same processes mentioned previously.

Low-Alloy High-Strength Steels

The low-alloy high-strength steels represent the bulk of the remaining steels in the AISI designation system. These steels are welded with the E-80XX, E-90XX, and E-100XX class of covered welding electrodes. It is also for these types of steels that the suffix to the electrode classification number is used. These steels include the low-manganese steels, the low-to-medium nickel steels, the low nickel-chromium steels, the molybdenum steels, the chromium-molybdenum steels, the nickel-chromium-molybdenum steels, and the other groups shown by Figure 12-1. Not included as an alloy steel but part of the AISI series are the sulphur steels. These are the series designated by 11XX, sometimes known as free-matching steels. Sulphur is 0.08–0.33%. These steels are difficult to weld, because of the high sulphur content, which has a tendency to produce porosity in the weld and cracking. Low-hydrogen electrodes of the minimum-strength level should be used. Welding is tedious on these steels and their use should be avoided when welding is required.

These include those in AISI series 2315, 2515, and 2517. Carbon ranges from 0.12–0.30%, manganese from 0.40–0.60%, silicon from 0.20–0.45% and nickel from 3.25–5.25%. If the carbon does not exceed 0.15% preheat is not necessary, except for extremely heavy sections. If the carbon exceeds 0.15% preheat of up to 500°F (260°C), depending on thickness, is required. On thin material, 1/4 in. (6.4 mm) or less, preheating is unnecessary. Stress relieving after welding is advisable. The electrode suffix C-1 or C-2 would be used depending on the level of nickel in the base material. The strength level would be matched to the base metal. In all cases, the low hydrogen coating is used.

Low-Nickel Chrome Steels

Steels in this group include the AISI 3120, 3135, 3140, 3310, and 3316. In these steels, carbon ranges from 0.14–0.34%, manganese from 0.40–0.90%, silicon from 0.20–0.35%, nickel from 1.10–3.75% and chromium 0.55–0.75%. Thin sections of these steels in the lower carbon ranges can be welded without preheat. A preheat of 200–300°F (93 to 149°C) is necessary for carbon in the 0.20% range, for higher carbon a preheat of up to 600°F (316°C) should be used. The weldment must be stress relieved or annealed after welding. The E-80XX or E-90XX electrodes should be used with the C-1 or C-2 suffix. There is no electrode type that will exactly provide a nickel-chrome deposit the same as the base metal.

Low-Manganese Steels

Included in this group are the AISI type 1320, 1330, 1335, 1340, and 1345 designations. In these steels, the carbon ranges from 0.18–0.48%, manganese from 1.60–1.90%, and silicon from 0.20–0.35%. Preheat is not required at the low range of carbon and manganese. Preheat of 250°–300°F (121–149°C) is desirable as the carbon approaches 0.25%, and mandatory at the higher range of manganese. Thicker sections should be preheated to double the above figure. A stress relief postheat treatment is recommended. The E-80XX or E-90XX electrodes with the A-1, D-1, or D-2 suffix should be used.

Low-Alloy Chromium Steels

Included in this group are the AISI type 5015 to 5160 and the electric furnace steels 50100, 51100, and 52100. In these steels carbon ranges from 0.12–1.10%, manganese from 0.30–1.00%, chromium from 0.20–1.60%, and silicon from 0.20–0.30%. When carbon is at the low end of the range, these steels can be welded without special precautions. As the carbon increases

and as the chromium increases, high hardenability results and a preheat of as high as 750°F (399°C) will be required, particularly for heavy sections. The B suffix should be used. Match the suffix to the chromium content.

The previous examples illustrate how the electrode specifications are used to select the proper covered electrode. They show how the strength level of the deposited weld metal is matched to the strength level of the steel by means of the first two or three digits of the electrode class. The suffix is used to match the composition of the base metal. Compositions are not available to match every base metal, but an effort should be made to come as close as possible. Welding procedure information is then determined by means of the carbon equivalent formula. This formula shown below was explained in the previous chapter (Section 11-6).

$$CE = \%C + \frac{\%Mn}{6} + \frac{\%Mo}{4} + \frac{\%Cr}{5}$$
$$+ \frac{\%Ni}{15} + \frac{\%Cu}{15} + \frac{\%P}{3}$$

The carbon equivalent should be calculated for the exact composition, if known. When only a range of composition is known use the maximum values to be on the safe side. When the carbon equivalent is 0.40% or lower, the material is considered readily weldable. Above 0.40% special controls are required. In all cases, low-hydrogen processes should be used and preheat may also be required. When the carbon equivalent exceeds 0.60%, preheating is required if the thickness exceeds 3/4 in. (19.1 mm). When the carbon equivalent exceeds 0.90%, preheat is absolutely required to a relatively high temperature on all except the thinnest material. This provides the guidance for establishing welding procedures using covered electrodes. For all but the most simple work, the procedure should be qualified according to one of the standard tests to determine whether it produces the quality of weld expected.

When using the submerged arc welding process, it is also necessary to match the composition of the electrode with the composition of the base metal. A flux that neither detracts nor adds elements to the weld metal should be used. In general, preheat can be reduced for submerged arc welding because of the higher heat input and slower cooling rates involved. To make sure that the submerged arc deposit is low hydrogen, the flux must be dry and the electrode and base metal must be clean.

When using the gas metal arc welding process, the electrode should be selected to match the base metal and the shielding gas should be selected to avoid excessive oxidation of the weld metal. Preheating with the gas metal arc welding process should be in the same order as with shielded metal arc welding since the heat input is similar.

When using the flux-cored arc welding process, the deposited weld metal produced by the flux-cored electrode should match the base metal being welded. Preheat requirements would be similar to gas metal arc welding.

When low-alloy high-strength steels are welded to lower-strength grades the electrode should be selected to match the strength of the lower-strength steel. The welding procedure, that is, preheat, heat input, etc., should be suitable for the higher-strength steel.

Weathering Steels

Weathering steels are low-alloy steels that can be exposed to the weather without being painted. The steel protects itself by means of a dense oxide coating (patina) which forms naturally on the steel when it is exposed to the weather. This tight oxide coating reduces continuing corrosion. The corrosion resistance of weathering steels is four to six times that of normal structural carbon steels, and two to three times that of many of the low-alloy structural steels. The weathering steels are covered by the ASTM specification A242. These steels have a minimum yield strength of 50,000 psi (35 kg/mm^2) with an ultimate tensile strength of 70,000 psi (49 kg/mm^2). Two of the better known weathering steels are Corten and Mayari R.

The weathering steels can be welded by all the arc welding processes and by gas welding and resistance welding.

To maintain the weather resistance characteristic of the steel, a special welding procedure should be employed. Use the E-7018 class to within one layer of the top of the joint. The top layer should be made with an E-7018-C1 electrode since the 2% nickel in the weld deposit will cause the weld metal to weather the same as the weathering steel. The C-1 suffix weld deposit should be used for the top layer of any multipass weld.

The same concept can be used for gas metal arc welding, flux-cored arc welding or submerged arc welding. The normal electrode used for a 50,000 psi steel would be employed, but this last layer would include alloy to provide weathering resistance.

12-3 WELDING THE ALLOY STEELS

The following is a definition by the American Iron and Steel Institute. "Steel is considered to be an alloy steel when the maximum of the range given for the content of alloying elements exceeds one or more of the following limits: manganese 1.65%, silicon 0.60%, copper 0.60%, or in which a definite range or a definite minimum quantity of any of the following elements is specified or required within the limits of the recognized field of constructional alloy steels: aluminum, boron, chromium up to 3.99%, cobalt, columbium, molybdenum, nickel, titanium, tungsten, vanadium, zirconium, or any other alloying element added to obtain a desired alloying effect."

It will be noted that some of the steels mentioned in the previous section fall within this range and are considered low-alloy steels. Many of the steels shown in Table 12-1, the AISI steel classification system, are also in the alloy range. If the chromium content exceeds 4% it will be considered a stainless steel, which will be covered in Section 12-4. In this section, we will discuss steels including the quenched and tempered constructional steels, the nickel steels, the chromium-molybdenum steels, and steel castings.

Quenched and Tempered Constructional Steels

The quenched and tempered constructional steels were developed in the early 1950s. The early steels were modifications of quenched and tempered armor plate and were produced by the same steel-making facilities. These steels offer a number of advantages which have made them extremely popular in weldments where a high strength-to-weight ratio is important. The unique properties of these steels are obtained from their special chemical composition plus a quenched and tempered heat treatment. These steels have extremely high strength, in the order of 100,000 psi (70 kg/mm^2) yield strength, combined with good weldability. In addition, they possess good ductility, good notch toughness, good fatigue strength, and corrosion resistance. They can be successfully welded with relatively conventional welding procedures. Little or no preheat is used and for most applications the weldments are used in the as-welded condition.

Initially, these steels were available only as plate since they were produced on the same equipment as armor plate. Later they became available as structural shapes and seamless pipe. The ASTM specification A514 and A517 have been written to cover the quenched and tempered constructional steels produced by different steel manufacturers. The grades and compositions are shown by Figure 12-3. A different grade is given to each trade name steel. Several other ASTM specifications have been written to cover these types of steels, but the welding outlined here will apply to all when the composition is similar. These types of steels are now used extensively in earth moving and mobile construction equipment, bridges, storage tanks, penstocks, ships, buildings, and for other specialized equipment when high strength and minimum weight are essential.

Steels of the same basic composition are also made into castings and forgings.

These steels are water quenched by special techniques from a temperature of 1,500-1,600°F (816-871°C) and tempered at a temperature of from 1,000-1,110°F 538°-593°C). This produces a microstructure of tempered, low-temperature transformation products which have an excellent combination of strength and toughness. The heat treatment greatly increases the yield and tensile strength of the steel. The toughness is improved over the same steel in the hot rolled condition. Additional developments by steel makers have expanded the list to almost a dozen different grades of quenched and tempered steels. Some of the newer ones have the same excellent properties but are less expensive.

To weld the quenched and tempered low-alloy steels of the ASTM A514/A517 class, a low heat input is used in order to obtain the full strength of the welded joint. Three factors must be considered, (1) the use of the correct filler materials, (2) the use of the correct heat input, and (3) following the recommended procedures.

The shielded metal arc welding process, the gas metal arc welding process, and the submerged arc welding process are commonly used for welding these steels. The gas tungsten arc welding process can be used, but is restricted to the thinner sections. Flux-cored arc welding is not normally used because the flux-cored electrodes of the proper composition are not readily available. The electroslag welding process is not recommended for these steels because the long time at high temperature softens the steel and destroys the heat treatment. In effect, it returns the steel to the as-rolled condition and the mechanical properties, in this condition, are appreciably reduced. When using any of the welding processes every effort should be made to keep the process a low-hydrogen process. This means dry electrode coating, dry flux, dry gas, clean joint preparation, etc.

ASTM Grade	Proprietary Steels	COMPOSITION PERCENT							
		C Max.	Mn	Si	Cr	Ni	Mo	Cu Min	Other
A	NAXTRA 100	0.15-0.21	0.80-1.10	0.40-0.80	0.50-0.80	—	0.18-0.28	—	0.035-P, 0.040-S
B	T-1 Type A	0.15-0.21	0.70-1.00	0.20-0.30	0.40-0.65	—	0.15-0.25	—	0.035-P, 0.040-S, 0.01-0.03-Ti, 0.03-0.08-Va
C	Jalloy S-100	0.10-0.20	1.01-1.50	0.15-0.30	—	—	0.20-0.30	—	0.035-P, 0.040-S
D	Armco SSS100A	0.13-0.20	0.40-0.70	0.20-0.35	0.85-1.20	—	0.20-0.25	0.20-0.40	0.035-P, 0.040-S, 0.04-0.10-Ti
E	Armco SSS100	0.12-0.20	0.40-0.70	0.20-0.35	1.40-2.00	—	0.40-0.60	0.20-0.40	0.035-P, 0.040-S 0.04-0.10-Ti
F	USS T-1	0.10-0.20	0.60-1.00	0.15-0.35	0.40-0.65	0.70-1.00	0.40-0.60	0.15-0.50	0.035-P, 0.040-S, 0.03-Va
G	PX-100	0.15-0.21	0.80-1.10	0.50-0.90	0.50-0.90	—	0.40-0.60	—	0.035-P, 0.040-S
H	T-1 Type B	0.12-0.21	0.95-1.30	0.20-0.30	0.40-0.65	0.30-0.70	0.20-0.30	—	0.035-P, 0.040-S, 0.03-0.08-Va
J	RQ 100A	0.12-0.21	0.45-0.70	0.20-0.35	—	—	0.50-0.65	—	0.035-P, 0.040-S
K	CHT-100	0.10-0.20	1.10-1.50	0.15-0.30	—	—	0.45-0.55	—	0.035-P, 0.040-S
L	Armco SSS100B	0.13-0.20	0.40-0.70	0.20-0.35	1.15-1.65	—	0.25-0.40	0.20-0.40	0.035-P, 0.040-S, 0.04-0.10-Ti
M	RQ-100B	0.12-0.21	0.45-0.70	0.20-0.35	—	1.20-1.50	0.45-0.60	—	0.035-P, 0.040-S
N	NAXTRA 100A								
P	RQC-100	0.12-0.20	0.45-0.70	0.20-0.35	0.85-1.20	1.20-1.50	0.45-0.60	—	0.035-P, 0.040-S

B oron of 0.0005 to 0.005 added to each grade.

FIGURE 12-3 Popular quenched and tempered steels, grades and composition per ASTM A514/517.

When using the shielded metal arc welding process, the low-hydrogen electrode types must be used. In order to produce full strength joints, electrodes of the E-11018 or E-12018 classification should be used. If the lower-strength electrodes are used, the strength of the weld joint will not develop the full strength of the base material.

When welding with the gas metal arc welding process, a shielding gas of 98% argon plus 2% oxygen is often used; however, pure carbon dioxide can be used. Proprietary electrode wires are required since there are no AWS specifications for the higher-strength solid electrodes. The electrode wire composition should be approximately the same as the composition of the base metal.

When welding with the submerged arc welding process, a neutral type flux should be used. The electrode wire should be of the same or approximately the same composition as the base metal.

The second factor is to maintain the proper heat input. The heat input depends on the material thickness, the preheat employed, and the interpass temperature. For 1-in. thick plate a minimum preheat of 50°F (10°C) is normally used. When the thickness is increased, a preheat of 200°F (93°C) is recommended. Higher preheats may be required for restrained joints. The allowable heat input is based on the joules per inch of weld joint given by the standard formula. The maximum heat input for different thicknesses of the different grades of the ASTM A514/A517 steels is given in the table shown by Figure 12-4. These are suggested maximum heat input limits and may vary from grade to grade or by different manufacturers. It may be necessary to consult the steel manufacturer's technical data to determine the recommended maximum heat

input limit. When these heat inputs are exceeded there will be a loss of strength of the weld joint. The toughness of the steel in the heat-affected zone is usually excellent and the hardness of the heat-affected zone is normally lower than the base metal or the deposited weld zone. Producers of the T-1 steel have developed a welding heat input calculator. This relates the travel speed, welding current, and arc volts to calculate the heat input in heat units. The heat units are actually kilo joules per inch, the same as calculated by the standard formula. This calculator provides maximum heat inputs for different preheat and interpass temperatures based on different thicknesses of steel. As the plate thickness increases and with lower preheat temperatures the maximum heat units are unlimited.

The third factor is the use of the correct welding procedure. The stringer bead technique is preferred. The use of a full weave reduces the travel speed and in so doing increases the heat input above the maximum limits. If weaving is unavoidable, such as when welding vertical up, the weave should be restricted to two electrode diameters. The base metal should not be allowed to become overheated. Temperatures should be maintained at the interpass temperatures shown by the chart of Figure 12-4. When back gouging of the joint is required, it should be done with the air carbon arc process or by grinding. The surface should be ground after air carbon arc gouging in order to provide a clean surface for welding. The oxyacetylene flame should not be used for back gouging.

Defects in welds made on quenched and tempered constructional steels are more serious than the same defect in mild or low-carbon steels. It is very important that the weld surfaces be smooth with contours that are well blended into the pieces being joined. Each weld should be made so that there is good penetration into the previous weld and no undercut. Complete penetration is essential to utilize the full strength of the quenched and tempered steel.

If the procedures mentioned above are followed, successful welds can be made in these types of steels on a routine production basis.

9% Nickel Steels

The 9% nickel steels are also quenched and tempered but are considered separately because they are intended for different type of service. The 9% nickel steels were developed initially to provide high strength and extreme toughness at very low operating temperatures. The reason for developing the material was to provide a steel that could be used to build tanks and vessels for containing liquified natural gas. The temperature of liquefied natural gas is −262°F (−160°C). The 9% nickel steel will provide good notch toughness at temperatures down to −320°F (−196°C). There is also a

Plate Thickness-in.	Perheat and Interpass Temperature °F			
	70	200	300	400
3/16	17,500	14,000	11,500	9,000
1/4	23,700	19,200	15,800	12,300
1/2	47,400	35,500	31,900	25,900
3/4	88,600	69,900	55,700	41,900
1	Any	110,000	86,000	65,600
1-1/4	Any	154,000	120,000	94,000

Note this applies to Grade F. Check suppliers of other grades for values.

FIGURE 12-4 *Suggested maximum heat input limits.*

low nickel steel in the 5% range, plus 0.25% molybdenum which will provide good properties at temperatures as low as −275°F (−170°C). These steels are welded in the heat-treated condition and do not require a postweld heat treatment to obtain welds that provide properties essentially equal to the base metal.

The 9% nickel steel is supplied in the heat-treated condition. It is specified by ASTM A353 and A553. Two types of heat treatments may be used. One is known as the *double normalized and tempered condition* and the other is accomplished by normalizing at 1,650°F (900°C) then normalizing at 1,450°F (790°C) followed by a tempering at 1,050°F (570°C). In the latter way the steel furnished is water quenched and tempered with water quenching at 1,470°F (800°C) and then tempering at 1,050°F (570°C). The toughness of this steel is obtained by the small amount of austenite which is reformed during the tempering treatment. This phase is stable at the subzero temperatures and contributes to the toughness of the steel.

The 9% nickel steel can be flame cut using normal oxygen fuel gas equipment. The cutting speed is slower than on mild steel. Flame-cut surfaces should be ground to remove any hardened metal and the oxide surface. Welding is done in the fully heat-treated condition and the heat-affected zone has a somewhat different microstructure then the base metal. Welding can be accomplished by the shielded metal arc welding process, the submerged arc welding process and the gas metal arc welding process.

When the shielded metal arc welding process is used the high nickel-chrome-iron type electrodes are used. These are the AWS E Ni Cr Fe-2 type and E Ni Cr Fe-3 type. The higher nickel-chrome electrode will produce slightly higher strength welds which will match the base metal. A preheat or postheat is not required on material 2 in. (50 mm) thick or less. Before welding the base metal should be brought up to normal room temperature of 70°F (21°C). When making Vee or bevel groove welds the minimum included angle should be 70°. The high-nickel electrodes operate differently from mild or stainless steel electrodes. They have low penetration, and do not flow or wash into the side wall of the weld joint. The electrode should be pointed to place the deposited metal where it is desired.

When using the submerged arc welding process, the same basic analysis of electrode wire is used with a neutral type welding flux. For thinner materials, a room temperature preheat of 70°F (21°C) is used. When welding 2 in. (50 mm) and thicker a preheat of 250–300°F (121–149°C) is recommended. The same temperature is used for the interpass temperature. The deposited weld metal will meet the properties of the base metal. In addition, the toughness of the deposited weld metal is in the same order.

When using the gas metal arc welding process, the high nickel chrome electrode, AWS Type ERNiCrFe-6, is used and a shielding gas of 90% helium and 10% argon is recommended. The short circuiting type transfer is used and the properties of the weld are essentially the same as the base metal. The pulsed mode of GMAW is also used for welding 9% nickel steel.

Chromium-Molybdenum Steels

The chromium-molybdenum steels (sometimes called chrome-moly) were originally developed for elevated temperature service. They have been used extensively in power piping where they operate at high pressures and temperatures between 700 and 1,100°F (371–599°C).

The major reason for using chrome-moly steels is that they maintain their strength at high temperatures. In addition, they do not creep which means that they do not stretch or deform under long periods of use at high pressures and temperatures. Also, they do not become brittle after extended periods of high-temperature service. Carbon steels, on the other hand, do tend to stretch at high-temperature service and will become brittle in time.

The chrome-moly steels are used in the normalized and tempered condition and in the quenched and tempered condition. The type of heat treatment dictates the strength level of the steel. The strength levels extend from 85,000 psi (59 kg/mm^2) to 135,000 psi (94 kg/mm^2). There are a number of compositions that have become popular. These are the 1% Cr–1/2% Mo, the 1-1/4% Cr–1/2% Mo, the 2% Cr–1/2% Mo, the 2-1/4% Cr–1% Mo, and the 5% Cr–1/2% Mo. Figure 12-5 is a table showing the more common chrome-molybdenum steels and the nominal compositions. These steels are available as pipe and tubing and also as plate.

The shielded metal arc welding process, the gas tungsten arc welding process, and the gas metal arc welding process are most widely used for joining the chrome-moly steels. The submerged arc welding process and the flux-cored arc welding process can be used. In all cases, it is necessary to closely match the weld metal deposit analysis with the composition of the base metal.

For shielded metal arc welding, the electrode class suffix is the clue to matching the weld metal with the base metal. The suffixes starting with B are the chrome-moly steels ranging from B1 with 1/2% chrome–1/2% moly up through the B4 for the 2-1/2% chrome–1/2%

Popular Name	C	Mn	Si	Cr	Mo	Recommended Electrode Suffix
1/2 Cr-1/2 Mo	0.10 to 0.20	0.30 to 0.60	0.10 to 0.30	0.50 to 0.81	0.44 to 0.65	B1
1 Cr-1/2 Mo	0.15 max.	0.30 to 0.60	0.50 max.	0.80 to 1.25	0.44 to 0.65	B2L
1-1/4 Cr-1/2 Mo	0.15 max.	0.30 to 0.60	0.50 to 1.00	1.0 to 1.50	0.44 to 0.65	B2L
2 Cr-1/2 Mo	0.15 max.	0.30 to 0.60	0.50 max.	1.65 to 2.35	0.44 to 0.65	B4L
2-1/4 Cr-1 Mo	0.15 max.	0.30 to 0.60	0.50 max.	1.90 to 2.60	0.87 to 1.13	B3

From ASTM A199 and A333.

FIGURE 12-5 *Composition of popular chrome-molybdenum steels.*

moly. The higher levels of chromium are not specified by means of a suffix system. Proprietary electrodes are available for the higher chrome-moly steels.

Specifications are not available for the solid electrode wires or rods to provide the chrome-moly compositions required. Proprietary alloys are available to match the different covered electrode suffixes. These solid electrode wires would be used for gas metal arc welding, submerged arc welding, and as filler rods for gas tungsten arc welding. In the case of flux-cored arc welding, proprietary electrodes are available matching many of the chrome-moly compositions.

Much of the welding on these steels is done on pipe. For pipe welding, the gas tungsten arc welding process is often used for making the root pass. The shielded metal arc welding process, gas metal arc, or flux-cored arc welding can be used for the remainder of the weld joint. The submerged arc welding process would be used for roll welding of pipe subassemblies.

The chrome-moly steels are hardenable steels; therefore, it is necessary to provide a welding procedure that includes preheating and postheating. Preheat temperatures range from a minimum of 100°F (37.8°C) to as high as 700°F (371°C). The preheat temperature is dependent on the carbon content, and the thickness of the material being welded. If the carbon content is below 0.20% and the thickness is less than 3/8 in. (9.5 mm), the minimum 100°F preheat can be used. However, if carbon is above this figure and the wall thickness is greater, then the temperature should be increased to 200°F and up to 400°F. For the higher chrome-molys and thicker sections, the preheat will extend up to 700°F (371°C); however, if thickness is less than 3/4 in. (19 mm), the preheat can be reduced to half this value. Details of welding procedures for welding piping is given by the AWS Committee Report, Welding of Chromium-Molybdenum Steel Pipe, B10.8.[3] Specific preheat values are given for different types and wall thicknesses of chrome-moly pipe.

Shielded metal arc welding electrodes, for chrome-moly steels are always of the low-hydrogen type.

Low-hydrogen electrodes are difficult to use with open root joints, therefore the gas tungsten arc welding process is used for making the root pass. Backup rings are not used for welding high-pressure, high-temperature steam pipelines.

A postheat treatment is required when the carbon content exceeds 0.20% or the wall thickness is over 1/2 in. (12 mm). The heat treatment temperature is from 1,150–1,300°F (621–704°C). The lower temperatures are used with the thinner material and the higher temperatures for the heavier wall thickness. Specific recommendations are provided in the above-mentioned AWS booklet.

If for any reason welding is interrupted before the joint is completed, the weld joint must be allowed to slow cool and then be properly preheated before welding continues. The induction heating system and resistance heating are popular for both preheating and postheating chrome-moly piping.

Where different grades of chrome-moly steels are welded together, the preheat and postheat temperatures should be based on the higher-alloy material, but the welding electrode can be based on the lower-alloy material.

Steel Castings

The welding of steel castings is important since they are often incorporated into weldments in which irregular shaped areas are involved. In addition, steel castings may have foundry defects which are repaired by welding. Steel castings are made in many different analyses and it is necessary to know the composition of the casting in order to select the proper process and filler metal. Castings are easily identifiable since they normally carry an imprint of the foundry where they were made. By checking with the foundry it is often possible to determine the exact analysis of the casting. Much of the welding on castings is done at the originating foundry.

In general, steel castings have higher amounts of carbon than rolled steel plates and sections. Many steel castings are heat treated to obtain desired properties. When welding heat treated castings one of the problem areas is that the weld metal deposit usually has a lower carbon content than the casting. In view of this, the heat treatment may not produce the same mechanical properties in the weld metal as in the casting. In these cases, it is best to overmatch the analysis of the weld deposit over the composition of the casting. This will tend to produce a hardness level in the weld metal similar to that in the casting. Another method is to use a welding technique which will melt larger amounts of the base metal to pick up carbon and increase the carbon in the weld deposit. This technique is often used for weld repairing castings prior to heat treatment.

If the casting is to be stress relieved or normalized, the higher carbon requirement in the weld metal may not be necessary.

When using the gas metal arc, the flux cored arc or the submerged arc welding process this problem is reduced because of the higher penetration of these processes. There is more dilution of the weld metal from the base metal and a higher carbon deposit will result. This provides a weld metal deposit more similar to the casting and will provide comparable heat-treated properties.

The flux-cored arc welding process is extremely popular for weld repairing of castings; however, proprietory electrodes must be selected to closely match the analysis of the casting.

Welding procedures for castings should be developed based on the casting analysis. All other factors concerning weldability must be considered in developing the procedure. This would include preheat, heat input, and postheat requirements.

12-4 WELDING STAINLESS STEELS

Stainless steels, or, more precisely, corrosion-resisting steels are a family of iron-base alloys having excellent resistance to corrosion. These steels do not rust and strongly resist attack by a great many liquids, gases, and chemicals. Many of the stainless steels have good low-temperature toughness and ductility. Most stainless steels exhibit good strength properties and resistance to scaling at high temperatures. All stainless steels contain iron as the main element and chromium in amounts ranging from about 11% to 30%. Chromium provides the basic corrosion resistance to stainless steels. A thin film of chromium-oxide forms on the surface of the metal when it is exposed to the oxygen of the air. This film acts as a barrier to further oxidation, rust, and corrosion. Steels which contain only chromium or chromium with small amounts of other alloys are known as straight chrome types. There are about 15 types of straight chrome stainless steels. The straight

chrome steels are the 400 series of stainless steels which are highly magnetic.

Nickel is added to certain of the stainless steels, which are known as *chrome-nickel* stainless steel. The addition of nickel reduces the thermal conductivity and decreases the electrical conductivity. The chrome-nickel steels are in the 300 series of stainless steels. They have austenitic microstructure and are nonmagnetic.

The chrome-nickel stainless steels contain small amounts of carbon. Carbon is undesirable particularly in the 18% chrome, 8% nickel group. Carbon will combine with chromium to form chromium carbides which do not have corrosion resistance. Chromium carbides are formed when the steel is held in a temperature range of 800 to 1,600°F (427–871°C) for prolonged periods, which can happen during welding with slow cooling. The chemical reaction of carbon with chromium to form chromium carbide is called *carbide precipitation*. Carbon can be controlled, however, by the use of stabilizing elements. Carbide precipitation can be reduced or prevented in two ways. The first way is to keep the carbon level at 0.03% or less, which eliminates the formation of chromium carbides. Stainless steels with low carbon in this range are commonly referred to as ELC (Extra Low Carbon) types. The other way of preventing carbide precipitation is to use a stabilizing element. The most popular stabilizers are titanium and columbium (niobium). These elements will combine with carbon to form titanium or columbium carbides, which have corrosion resistance. Both types of stainless steels have equivalent corrosion resistance. These types of stainless steels are identified as ELC type or stabilized type.

Manganese is added to some of the chrome-nickel alloys. Usually these steels contain slightly less nickel since the chrome-nickel-manganese alloys were developed originally to conserve nickel. In these alloys, a small portion of the nickel is replaced by the manganese, generally in a two-to-one relationship. The 200 series of stainless steels are the chrome-nickel-manganese series. These steels have an austenitic microstructure and are nonmagnetic. The 201 and 202 types are used as alternates for 301 and 302.

Molybdenum is also included in some stainless steel alloys. Molybdenum is added to improve the creep resistance of the steel at elevated temperatures. It will also increase resistance to pitting and corrosion in many applications. The different alloy groupings are shown by Figure 12-6.

Series Designation	Metallurgical Group	Principle Elements	Hardenable By Heat Treatment	Magnetic
2xx	Austenitic	Chromium-nickel-manganese	Non-hardenable**	Nonmagnetic
3xx	Austenitic	Chromium-nickel steels	Non-hardenable	Nonmagnetic
4xx	Martensitic	Chromium steels	Hardenable	Magnetic
4xx	Ferritic	Chromium steels	Non-hardenable	Magnetic
5xx*	Martensitic	Chromium-molybdenum steels	Martensitic	Magnetic

*Not stainless.
**Will work harden.

FIGURE 12-6 *Groups of stainless steels.*

TABLE 12-2 *AISI stainless steel classification system (Courtesy, The American Iron and Steel Institute).*

Chemical Analyses of Stainless Steels, per cent

AISI No.	Carbon	Manganese	Silicon	Chromium	Nickel	Other Elements
Chromium-Nickel-Magnesium-Austenitic—Non Hardenable						
201	0.15 Max.	5.5/7.5	1.0	16.0/18.0	3.5/5.5	N_2 0.25 Max.
202	0.15 Max.	7.5/10.	1.0	17.0/19.0	4.0/6.0	N_2 0.25 Max.
Chromium-Nickel-Austenitic-Non-Hardenable						
301	0.15 Max.	2.0	1.0	16.0/18.0	6.0/8.0	—
302	0.15 Max.	2.0	1.0	17.0/19.0	8.0/10.0	—
302B	0.15 Max.	2.0	2.0/3.0	17.0/19.0	8.0/10.0	—
303	0.15 Max.	2.0	1.0	17.0/19.0	8.0/10.0	S 0.15 Min.
303Se	0.15 Max.	2.0	1.0	17.0/19.0	8.0/10.0	Se 0.15 Min.
304	0.08 Max.	2.0	1.0	18.0/20.0	8.0/12.0	—
304L	0.03 Max.	2.0	1.0	18.0/20.0	8.0/12.0	—
305	0.12 Max.	2.0	1.0	17.0/19.0	10.0/13.0	—
308	0.08 Max.	2.0	1.0	19.0/21.0	10.0/12.0	—
309	0.20 Max.	2.0	1.0	22.0/24.0	12.0/15.0	—
309S	0.08 Max.	2.0	1.0	22.0/24.0	12.0/15.0	—
310	0.25 Max.	2.0	1.50	24.0/26.0	19.0/22.0	—
310S	0.08 Max.	2.0	1.50	24.0/26.0	19.0/22.0	—
314	0.25 Max.	2.0	1.5/3.0	23.0/26.0	19.0/22.0	—
316	0.08 Max.	2.0	1.0	16.0/18.0	10.0/14.0	Mo 2.0/3.0
316L	0.03 Max.	2.0	1.0	16.0/18.0	10.0/14.0	Mo 2.0/3.0
317	0.08 Max.	2.0	1.0	18.0/20.0	11.0/15.0	Mo 3.0/4.0
321	0.08 Max.	2.0	1.0	17.0/19.0	9.0/12.0	Ti 5 X C Min.
347	0.08 Max.	2.0	1.0	17.0/19.0	9.0/13.0	Cb+Ta10xC Min.
348	0.08 Max.	2.0	1.0	17.0/19.0	9.0/13.0	Ta 0.10 Max.
Chromium-Martensitic—Hardenable						
403	0.15 Max.	1.0	0.5	11.5/13.0	—	—
410	0.15 Max.	1.0	1.0	11.5/13.5	—	—
414	0.15 Max.	1.0	1.0	11.5/13.5	1.25/2.5	—
416	0.15 Max.	1.25	1.0	12.0/14.0	—	S 0.15 Min.
416Se	0.15 Max.	1.25	1.0	12.0/14.0	—	Se 0.15 Min.
420	Over 0.15	1.0	1.0	12.0/14.0	—	—
431	0.20 Max.	1.0	1.0	15.0/17.0	1.25/2.5	—
440A	0.60/0.75	1.0	1.0	16.0/18.0	—	Mo 0.75 Max.
440B	0.75/0.95	1.0	1.0	16.0/18.0	—	Mo 0.75 Max.
440C	0.95/1.2	1.0	1.0	16.0/18.0	—	Mo 0.75 Max.
Chromium-Ferritic—Non Hardenable						
405	0.08 Max.	1.0	1.0	11.5/14.5	—	Al 1.1/0.3
430	0.12 Max.	1.0	1.0	14.0/18.0	—	—
430F	0.12 Max.	1.25	1.0	14.0/18.0	—	S 0.15 Min.
430FSe	0.12 Max.	1.25	1.0	14.0/18.0	—	Se 0.15 Min.
446	0.20 Max.	1.50	1.0	23.0/27.0	—	N 0.25 Max.
Martensitic						
501	Over 0.10	1.0	1.0	4.0/6.0	—	Mo 0.40/0.65
502	0.10 Max.	1.0	1.0	4.0/6.0	—	Mo 0.40/0.65

Stainless steels are sometimes identified by numbers which refer to the principal alloying elements such as 18/8, 25/20, etc. This identification system has been supplanted by the American Iron and Steel Institute system which utilizes a three-digit number. The first digit indicates the group as shown in Figure 12-6, and the last two digits indicate specific alloys. The AISI numbers refer to the alloys as chrome-nickel stainless steels and chromium stainless steels. They are, however, also identified according to their microstructure which can be austenitic, martensitic, or ferritic. The austenitic chrome-nickel-manganese (200 series) and austenitic chrome-nickel (300 series) steels are shown in the upper portion of Table 12-2. The martensitic types are shown in the center part of the table and represent a portion of the 400 series; the ferritic types are shown in the lower portion of the table and are the remaining alloys in the 400 series.

Welding Stainless Steels

The three most popular processes for welding stainless steels are shielded metal arc welding, gas tungsten arc welding, and gas metal arc welding; however, almost all the welding processes can be used.

Stainless steels are slightly more difficult to weld than mild carbon steels. The physical properties of stainless steel are different from mild steel and this makes it weld differently. These differences are:

1. Lower melting temperature.
2. Lower coefficient of thermal conductivity.
3. Higher coefficient of thermal expansion.
4. Higher electrical resistance.

The properties are not the same for all stainless steels, but they are the same for those having the same microstructure. In view of this, stainless steels of the same metallurgical class have similar welding characteristics and are grouped according to the metallurgical structure with respect to welding.

Austenitic Types

The austenitic stainless steels have about 45% higher manganese steels are not hardenable by heat treatment and are nonmagnetic in the annealed condition. They may become slightly magnetic when cold worked or welded. This helps identify this class of stainless steels. All of the austenitic stainless steels are weldable with most of the welding processes, with the exception of Type 303, which contains high sulphur and Type 303Se, which contains selenium to improve machinability.

The austenitic stainless steels have about 45% higher thermal coefficient of expansion, higher electrical resistance, and lower thermal conductivity than mild-carbon steels. High travel speed welding is recommended, which will reduce heat input, reduce carbide precipitation, and minimize distortion. The melting point of austenitic stainless steel is slightly lower than mild-carbon steel. Because of lower melting temperature and lower thermal conductivity welding current is usually lower. The higher thermal expansion dictates that special precautions should be taken with regard to warpage and distortion. Tack welds should be twice as often as normal. Any of the distortion reducing techniques such as back-step welding, skip welding, and wandering sequence should be used. On thin materials it is very difficult to completely avoid buckling and distortion.

Ferritic Stainless Steels

The ferritic stainless steels are not hardenable by heat treatment and are magnetic. All of the ferritic types are considered weldable with the majority of the welding processes except for the free machining grade 430F, which contains high sulphur. The coefficient of thermal expansion is lower than the austenitic types and is about the same as mild steel. Welding processes that tend to increase carbon pickup are not recommended. This would include the oxyfuel gas process, carbon arc process, and gas metal arc welding with CO_2 shielding gas. The ferritic steels in the 400 series have a tendency for grain growth at elevated temperatures. Grain growth occurs at about 1,600°F (871°C) and increases rapidly at higher temperatures. The lower chromium types show tendencies toward hardening with a resulting martensitic type structure at grain boundaries of the weld area. This lowers the ductility, toughness, and corrosion resistance at the weld. For heavier sections a preheat of 400°F (204°C) is beneficial. To restore full corrosion resistance and improve ductility after welding, annealing at 1,400 to 1,500°F (760–816°C), followed by a water or air quench, is recommended. Large grain size will still prevail, however, and toughness may be impaired. Toughness can be improved only by cold working such as peening the weld. If heat treating after welding is not possible and service demands impact resistance, an austenitic stainless steel filler metal should be used. Otherwise, the filler metal is selected to match the base metal.

Martensitic Stainless Steels

The martensitic stainless steels are hardenable by heat treatment and are magnetic. The low-carbon types

| AWS | | | | Typical Composition % | | | |
Class	C	Cr	Ni	Mo	Mn	Si	Others
E308	0.08	19.5	10.5	—	2.5	0.90	—
E308L	0.04	19.5	10.5	—	2.5	0.90	—
E309	0.15	23.5	13.5	—	2.5	0.90	—
E309Cb	0.12	23.5	13.5	—	2.5	0.90	Cb + Ti —0.85
E309Mo	0.12	23.5	13.5	2.5	2.5	0.90	—
E310	0.20	26.5	21.5	—	2.5	0.75	—
E310Cb	0.12	26.5	21.5	—	2.5	0.75	Cb + Ti —0.85
E310Mo	0.12	26.5	21.5	2.5	2.5	0.75	—
E312	0.15	30.0	9.0	—	2.5	0.90	—
E316	0.08	18.5	12.5	2.5	2.5	0.90	—
E316L	0.04	18.5	12.5	2.5	2.5	0.90	—
E317	0.08	19.5	13.0	3.5	2.5	0.90	—
E318	0.08	18.5	12.5	2.5	2.5	0.90	—
E320	0.07	20.0	34.0	2.5	2.5	0.60	—
E330	0.25	15.5	35.0	—	2.5	0.90	—
E347	0.08	19.5	10.0	—	2.5	0.90	—
E410	0.12	12.5	0.60	—	1.0	0.90	—
E430	0.10	16.5	0.60	—	1.0	0.90	—

Note: Remainder is iron.

FIGURE 12-7 *Stainless steel filler metal alloys per AWS A5.4.*

can be welded without special precautions. The types with over 0.15% carbon tend to be air hardenable and, therefore, preheat and postheat of weldments are required. A preheat temperature range of 450 to 550°F 232–288°C) is recommended. Postheating should immediately follow welding and be in the range of 1,200–1,400°F (649–760°C), followed by slow cooling.

If preheat and postheat are not possible, an austenitic stainless steel filler metal should be used. Type 416Se is the free-machining composition and should not be welded. Welding processes that tend to increase carbon pickup are not recommended. Increased carbon content increases crack sensitivity in the weld area.

The selection of the filler metal alloy for welding the stainless steels is based on the composition of the stainless steel. The various stainless steel filler metal alloys are normally available as covered electrodes and as bare solid wires and are shown by Figure 12-7.[4] Recently flux-cored electrode wires have been developed for welding stainless steels.

Figure 12-8 is a table which gives the recommended filler metal alloy for welding the various stainless steel base metals. The table also shows some alternate alloys. Alternates are provided since there are so many different stainless steel types and there are not electrodes of each type. It is possible to weld several different stainless base metals with the same filler metal alloy.

For shielded metal arc welding, there are two basic types of electrode coatings. These are the lime type indicated by the suffix 15 and the titania type designated by the suffix 16. The lime type electrodes are used only with direct current electrode positive (reverse polarity). The titania-coated electrode with the suffix 16 can be used with alternating current and with direct current electrode positive. Both coatings are of the low-hydrogen type and both are used in all positions; however, the 16 type is smoother, has more welder appeal, and operates better in the flat position. The lime type electrodes are more crack resistant and are slightly better for out-of-position welding. The procedure schedule for using the shielded metal arc welding process is given by Figure 12-9. The width of weaving should be limited to two-and-one-half times the diameter of the electrode core wire.

Covered electrodes for shielded metal arc welding must be stored at normal room temperatures in dry areas. These electrode coatings, of low hydrogen type, are susceptible to moisture pickup. Once the electrode box has been opened, the electrodes should be kept in a dry box until used. If the electrodes do become exposed to moisture they should be reconditioned in accordance with procedures that were presented in Chapter 10.

The gas tungsten arc welding process is widely used for thinner sections of stainless steel. The 2% thoriated tungsten is recommended and the electrode should be ground to a taper. Argon is normally used for gas shielding; however, argon-helium mixtures are sometimes used for automatic applications. Figure 12-10 shows the welding procedure schedule for the gas tungsten arc welding process for stainless steel.

	AISI No.	RECOMMENDED FILLER METAL 1st Choice	2nd Choice	Popular Name	Remarks
Cr-Ni-Mn	201	308	308L		Substitute for 301
	202	308	308L		Substitute for 302
Cr-Ni-Austenitic	301	308	308L		
	302	308	308L		
	302B	308	309		High Silicon
	303	—	—		Free machining—welding not recommended—312
	303Se	—	—		Free machining—welding not recommended—312
	304	308	308L	1818	
	304L	308L	347	1818 Elc	Extra low carbon
	305	308	—		
	308	308	—	19/9	
	309	309	—	25/12	
	309S	309	—		Low carbon
	310	310	—	25/20	
	310S	310	—		Low carbon
	314	310	—		
	316	316	309Cb	18/12Mo	
	316L	316L	309Cb	18/12 Elc	Extra low carbon
	317	317	309Cb	19/14Mo	
	321	347	308L		
	347	347	308L	19/9 Cb	Difficult to weld in heavy sections
	348	347	—	19/9CbLTa	
Cr-Martensitic	403	410	—		
	410	410	430	12Cr	
	414	410	—		
	416	410	—		Use 410-15
	416Se	—	—		Free machining welding not recommended
	420	410	—	12 Cr Hc	High Carbon
	431	430	—		
	440A	—	—		High carbon—welding not recommended
	440B	—	—		High carbon—welding not recommended
	440C	—	—		High carbon—welding not recommended
Cr-Ferritic	405	410	405Cb		
	430	430	309	16Cr	
	430F	—	—		Free machining—welding not recommended
	430FSe	—	—		Free machining—welding not recommended
	446	309	310		
	501	502	—	5Cr-1/2Mo	Chrome-moly steel
	502	502	—	5Cr-1/2Mo	Chrome-moly steel

FIGURE 12-8 *Recommended filler metals for stainless steels (use E or R prefix).*

ga	Material Thickness fraction	in.	Electrode Dia in.	WELDING CURRENT DECP Flat	Vertical	Overhead
26	—	0.018	5/64	20-35	20-25	20-30
22	—	0.030	5/64	30-45	30-40	30-40
18	—	0.048	3/32	50-70	40-55	50-60
14	—	0.075	3/32	60-90	50-65	60-95
11	—	0.120	1/8	90-120	75-90	90-110
—	3/16	0.188	5/32	120-150	90-110	120-140
—	1/4	0.250	3/16	150-200	100-125	—

FIGURE 12-9 *Welding procedure schedule for SMAW of stainless steel.*

| Material Thickness (or Fillet Size) | | | Type of Weld Fillet or Groove | Tungsten Electrode Diameter | | Filler Rod Diameter | | Nozzle Size Inside Dia. | Shielding Gas Flow | Welding Current Amps | No. of Passes | Travel Speed (per pass) |
ga	in.	mm		in.	mm	in.	mm	in.	cfh	DCEN		ipm
24			Square Groove	0.040	1.0	1/16	1.6	1/4	10	20-50	1	26
18			Square Groove	1/16	1.6	1/16	1.6	1/4	10	50-80	1	22
1/16	0.062	1.6	Square Groove	1/16	1.6	1/16	1.6	1/4	12	65-105	1	12
1/16	0.062	1.6	Fillet	1/16	1.6	1/16	1.6	1/4	12	75-125	1	10
3/32	0.093	2.4	Square Groove	1/16	1.6	3/32	2.4	1/4	12	85-125	1	12
3/32	0.093	2.4	Fillet	1/16	1.6	3/32	2.4	1/4	12	95-135	1	10
1/8	0.125	3.2	Square Groove	1/16	1.6	3/32	2.4	5/16	15	100-135	1	12
1/8	0.125	3.2	Fillet	1/16	1.6	3/32	2.4	5/16	15	115-145	1	10
3/16	0.188	4.8	Square Groove	3/32	2.4	1/8	3.2	5/16	15	150-225	1	10
3/16	0.188	4.8	Fillet	1/8	3.2	1/8	3.2	3/8	18	175-250	1	8
1/4	0.25	6.4	Vee Groove	1/8	3.2	3/16	4.8	3/8	18	225-300	2	10
1/4	0.25	6.4	Fillet	1/8	3.2	3/16	4.8	3/8	18	225-300	2	10
3/8	0.375	9.5	Vee Groove	3/16	4.8	3/16	4.8	1/2	25	220-350	2-3	10
3/8	0.379	9.5	Fillet	3/16	4.8	3/16	4.8	1/2	25	250-350	3	10
1/2	0.50	12.7	Vee Groove	3/16	4.8	1/4	6.4	1/2	25	250-350	3	10
1/2	0.50	12.7	Fillet	3/16	4.8	1/4	6.4	1/2	25	250-350	3	10

1 - Increase amperage when backup is used.
2 - Data is for flat position. Reduce amperage 10% to 20% when welding is horizontal, vertical, or overhead position.
3 - For tungsten electrodes—1st choice 2% thoriated EWTh2; 2nd choice 1% thoriated EWTh1.
4 - Argon is used for shielding. The 75% helium 25% argon mixture is used for heavier thickness.

FIGURE 12-10 *Welding procedure schedule for GTAW of stainless steel.*

The gas metal arc welding process is widely used for thicker materials since it is a faster welding process. The spray transfer mode is used for flat position welding and this requires the use of argon for shielding with 2% or 5% oxygen or special mixtures. The oxygen helps produce better wetting action on the edges of the weld. Figure 12-11 shows the welding procedure schedule for gas metal arc welding. The short circuiting transfer can also be used on thinner materials. In this case, CO_2 shielding or the 25% CO_2 plus 75% argon mixture is used. The argon-oxygen mixture can also be used with small-diameter electrode wires. With extra low-carbon electrode wires and CO_2 shielding the amount of carbon pickup will increase slightly. This should be related to the service life of the weldment. If corrosion resistance is a major factor the CO_2 gas or the CO_2-argon mixture should not be used.

Stainless steel can also be welded with the submerged arc welding process. In this case, the electrode wire would be the same as shown in the selection guide table. The submerged arc flux must be selected for stainless steel welding.

For all welding operations, the weld area should be cleaned and free from all foreign material, oil, paint,

| Material Thickness (or Fillet Size) | | | Type of Weld Fillet or Groove | Electrode Diameter | | WELDING POWER | | Wire Feed Speed | Shielding Gas Flow | No. of Passes | Travel Speed (per pass) |
ga	in.	mm		in.	mm	Current Amps DC	Arc Volt EP	ipm	cfh		ipm
16	0.063	1.6	Sq. groove & fillet	0.035	0.9	60-100	15-18	90-190	12-15	1	15-30
13	0.093	2.4	Sq. groove & fillet	0.035	0.9	125-150	18-21	230-280	12-15	1	20-30
				0.045	1.1	125-150	18-21	130-160			20-30
11	0.125	3.2	Sq. groove & fillet	0.035	0.9	130-160	19-24	250-280	12-15	1	20-25
				0.045	1.1	150-225	19-24	160-260			20-30
5/32	0.156	3.9	Vee groove & fillet	0.045	1.1	190-250	22-26	200-290	15-20	1	25-30
1/4	0.250	6.4	Vee groove & fillet	0.045	1.1	225-300	24-30	260-370	25-30	2	25-30

Data is for flat position. Reduce current 10-20% for other positions.

Gas selection—Argon-oxygen (1 to 2% oxygen + argon)—flat position and horizontal fillets.
    Argon-CO_2 (75% Argon-25% CO_2)—all position, some carbon pickup.
    Helium-argon-CO_2 (90%-7.5%-2.5%)—for all position welding.
    Carbon dioxide (CO_2)—where carbon pickup can be tolerated.

FIGURE 12-11 *Welding procedure schedule for GMAW of stainless steel.*

dirt, etc. The welding arc should be as short as possible when using any of the arc processes.

Problem Areas

The most serious problem when welding stainless steels is to avoid carbide precipitation. As previously mentioned, this may occur when the material is held at a relatively high temperature range for a period of time. Electrodes containing extra low carbon, indicated by the suffix L, should be used. Most electrodes are stabilized with columbium or titanium. This also helps to eliminate the carbide precipitation problem. These are definitely required when the weldment will be subjected to high temperature service.

Another factor that affects the quality of austenitic weld joints is the control of ferrite content in the microstructure. Austenitic weld deposits may develop microcracks during welding if ferrite is not controlled. The composition of the filler metal should be selected based on the deposit containing a small percentage of ferrite.

The ferrite content should not become too high or else the weldment will have lower than desired impact strength. For low-temperature service the weld metal should have the ferrite in the range of 4 to 10%. The ferrite content of the weld deposit depends on the composition of the base metal as well as the composition of the deposited filler metal. A special constitution diagram for stainless steel weld metal has been designed by Schaeffler and modified by DeLong.[5] This diagram shown by Figure 12-12 relates the nickel and chromium equivalents to lines which shows the percentage of ferrite. This diagram is useful for estimating the microstructure of the weld deposit and the filler metal composition required to produce the prescribed amount of ferrite in the

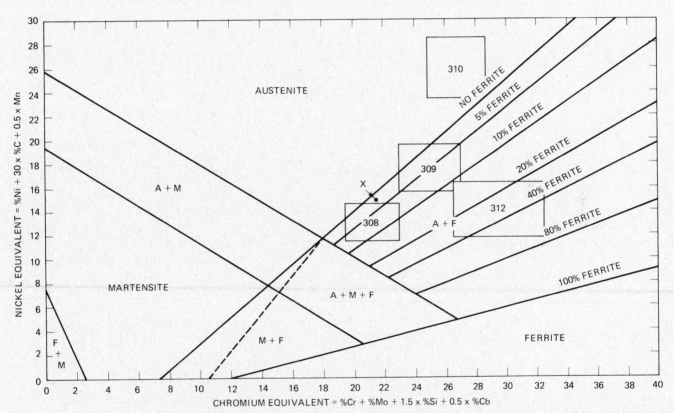

Example: Point X on the diagram indicates the equivalent composition of a type 318 (316 CB) weld deposit containing 0.07 C, 1.55 Mn, 0.57 Si, 18.02 Cr, 11.87 Ni, 2.16 Mo, 0.80 Cb. Each of these percentages was multiplied by the "potency factor" indicated for the element in question along the axes of the diagram, in order to determine the chromium equivalent and the nickel equivalent. When these were plotted, as point X, the constitution of the weld was indicated as austenite plus from 0 to 5% ferrite; magnetic analysis of the actual sample revealed an average ferrite content of 2%.

For austenite-plus-ferrite structures, the diagram predicts the percentage ferrite within 4% for the following stainless steels: 308, 309, 309 Cb, 310, 312, 316, 317, 318 (316 Cb), and 347.

Dashed line is the martensite/M + F boundary modicifation by Eberhard Leinhos, "Mechanische Eigenschaften und Gefügeausbildung von mit Chrom und Nickel legiertem Schweissgut," VEB Deutscher Verlag für Grundstoffindustrie, Leipzig, 1966.

FIGURE 12-12 *Constitution diagram for stainless steel weld metal (from Metal Progress Data Book © American Society for Metals, 1977).*

deposit. The diagram shows how the microstructure of the weld deposit is affected by the alloying elements in the stainless steel, based on those that act like nickel and those that act like chromium. The nickel equivalent group includes nickel and the effect of carbon and manganese. The chromium equivalent group includes chromium and the effect of molybdenum, silicon, and columbium. To estimate the microstructure of a deposit the nickel and the chromium equivalents are calculated using the following formulas.

$$\text{Nickel equiv.} = \%\text{Ni} + 30\%\text{ C} + 0.5\%\text{ Mn}$$
$$\text{Chromium equiv.} = \%\text{Cr} + \%\text{Mo} + 1.5\%\text{ Si} + 0.5\%\text{ Cb}$$

The values obtained are marked on the coordinates of the diagram and a point is located. The microstructure at that point is the one predicted for a deposit of that composition. It is possible to plot the composition of the filler wire and the composition of the base metals and connect them with a line and a resulting weld would be along this line. By the use of the Schaeffler/DeLong diagram, it is possible to select a filler metal that will avoid ferrite or martensite in the stainless steel weld deposit. The diagram can also be used to predict weld deposit composition when welding dissimilar stainless steels.[6]

The stainless steels can be welded by resistance welding and by many of the other specialty welding processes. Stainless steels can also be soldered and brazed.

The PH stainless steels and the higher carbon martensitic stainless steels are covered in the next section.

12-5 WELDING ULTRA-HIGH-STRENGTH STEELS

The term *high-strength steel* is often applied to all steels other than mild low-carbon steels. The steels under discussion in this section are those that have a yield strength of at least 80,000 psi (56 kg/mm²). These are sometimes called the ultra-high-strength steels or super alloys.

The groups of steels that fall into this category are:

1. Medium-carbon low-alloy hardenable steels.
2. Medium-alloy hardenable or tool and die steels.
3. High-alloy hardenable steels.
4. High-nickel maraging steels.
5. Martensitic stainless steels.
6. Semiaustenitic precipitation-hardenable stainless steels.

Each of these groups will be briefly described and welding information will be presented. Nominal compositions of steels in each of these groups is shown by Figure 12-13.

Medium-Carbon Low-Alloy Hardenable Steels

The best known steels in this class are the AISI 4130 and AISI 4140 steels. Also in this class are the higher-strength AISI 4340 steel and the AMS 6434 steel. These steels obtain their high strength by heat treatment to a full martensitic microstructure, which is tempered to improve ductility and toughness. Tempering temperatures greatly affect the strength levels of these steels. The carbon is in the medium range and as low as possible but sufficient to give the required strength. Impurities are kept to an absolute minimum because of high-quality melting and refining methods. These steels are available as sheets, bars, tubing, and light plate. The steels in this group can be mechanically cut or flame cut. However, when they are flame cut they must be preheated to 600°F (316°C). Flame-cut parts should be annealed before additional operations in order to reduce the hardness of the flame-cut edges.

Welding is usually done on these steels when they are in the annealed or normalized condition. They are then heat treated to obtain the desired strength. The gas tungsten arc, the gas metal arc, the shielded metal arc, and the gas welding process are all used for welding these steels. The composition of the filler metal is designed to produce a weld deposit that responds to a heat treatment in approximately the same manner as the base metal. In order to avoid brittleness and the possibility of cracks during welding relatively high preheat and interpass temperatures are used. Preheating is in the order of 600°F (316°C). Complex weldments are heat treated immediately after welding.

Aircraft engine mounts, aircraft tubular frames, and racing car frames are made from AISI 4130 tubular sections. These types of structures are normally not heat treated after welding.

Medium-Alloy Hardenable Steels

These steels are used largely in the aircraft industry for ultra-high-strength structural applications. They have carbon in the low to medium range and possess good fracture toughness at high-strength levels. In addition, they are air hardened which reduces the distortion that is encountered with more drastic quenching methods. Some of the steels in this group are known as hot work die steels and another grade has become known as 5CR-MO-V aircraft quality steel. There are proprietary names for other steels in this class. The

Designation	COMPOSITION, WEIGHT PERCENT						
	C	Mn	Si	Cr	Ni	Mo	Other
Medium carbon low alloy hardenable steels							
AISI 4130	0.28/0.33	0.40/0.60	0.20/0.35	0.80/1.10	—	0.15/0.25	—
AISI 4140	0.38/0.43	0.75/1.0	0.20/0.35	0.80/1.10	—	0.15/0.25	—
AISI 4340	0.38/0.43	0.60/0.80	0.20/0.35	0.70/0.90	1.65/2.00	0.20/0.30	—
AMS 6434	0.31/0.38	0.60/0.80	0.20/0.35	0.65/0.90	1.65/2.00	0.30/0.40	0.17/0.23V
Medium alloy hardenable tool and die steels							
5Cr-Mo-V Aircraft Steel	0.37/0.43	0.20/0.40	0.80/1.20	4.75/5.25	—	1.20/1.40	0.4/0.6 V
H-11 Tool Steel	0.30/0.40	0.20/0.40	0.80/1.20	4.75/5.50	—	1.25/1.75	0.30/0.50 V
H-13 Tool Steel	0.30/0.40	0.20/0.40	0.80/1.20	4.75/5.50	—	1.25/1.75	0.80/1.20 V
High alloy hardenable steels							
HP 9-4-20	0.20	0.30	0.10 max	0.75	9.0	0.75	0.01-S, 0.01-P, 0.10-V, 4.50-Co
HP 9-4-30 (Cr,Mo)	0.30	0.20	0.10 max	1.00	7.5	1.00	0.01-S, 0.01-P, 0.10-V, 4.50-Co
High nickel maraging steels							
18Ni	0.03 max	0.10 max	0.10 max	—	18.0	3.25	0.01-S, 0.01-P, 8.5-Co, 0.20-Ti, 0.10-Al
18Ni	0.03 max	0.10 max	0.10 max	—	18.0	4.90	0.01-S, 0.01-P, 8.0-Co, 0.40-Ti, 0.10-Al
18Ni	0.03 max	0.10 max	0.10 max	—	17.5	4.90	0.01-S, 0.01-P, 9.0-Co, 0.65-Ti, 0.10-Al
18 Ni	0.01 max	0.10 max	0.10 max	—	17.5	3.75	0.01-S, 0.01-P, 12.5-Co, 1.80-Ti, 0.15-Al
Martensitic stainless steels							
AISI 420	0.15 max.	1.0 max	1.0 max	13	—	—	—
AISI 431	0.20 max	1.0 max	1.0 max	16	2.0	—	
12MoV	0.25	0.5	0.5	12	0.5	1.0	0.3-V
17-4PH	0.07 max	1.0 max	1.0 max	16.5	4.0	—	4.0-Cu, 0.3-Cb
PH13-8Mo	0.05 max	0.1 max	0.1 max	12.5	8.0	2.5	1.1-Al
Pyromet X-15	0.03 max	0.1	0.1 max	15	—	2.9	20.0-Co
Custom 455	0.05 max	0.5 max	0.5 max	12	8.5	0.5	2.0-Cu, 0.3-Cb, 1.1-Ti
AFC 77	0.15	—	—	14.5	—	5.0	4.0-Cu, 13.5-Co, 0.5-V, 0.5-N
Semi-austenitic precipitation—hardenable stainless steels							
17-7PH	0.09 max	1.0 max	1.0 max	17.0	7.0	—	1.0-Al
PH15-7Mo	0.09 max	1.0 max	1.0 max	15.0	7.0	2.5	1.1-Al
PH14-8Mo	0.05 max	0.1 max	0.1 max	15.0	8.5	2.5	1.0-Al
AM 350	0.12 max	0.90	0.5 max	16.5	4.5	3.0	0.10-N
AM 355	0.15 max	0.95	0.5	15.5	4.5	3.0	0.09-N

FIGURE 12-13 *Nominal composition of ultra high strength steels.*

steels are available as forging billets, bars, sheet, strip, and plate.

There is another type of steel in this general class which is a medium-alloy quenched and tempered steel known as high-yield or HY 130/150. This type of steel is used for submarines, aerospace applications, and pressure vessels, and is normally available as plate. This steel has good notch toughness properties at 0°F and below. These types of steels have much lower carbon than the grades mentioned previously.

When flame cutting or welding the aircraft quality steels, preheating is absolutely necessary since the steels are air hardening. A preheat of 600°F (316°C) is used before flame cutting and then annealed immediately after the flame-cutting operation. This will avoid a brittle layer at the flame-cut edge which is susceptible to cracking. This type of steel should only be welded in the annealed condition. The steel should be preheated to 600°F (316°C) and this temperature must be maintained throughout the welding operation. After welding, the work must be cooled slowly. This can be done by postheating, or by furnace cooling. The weldment is then stress relieved at 1,300°F (704°C) and air cooled to obtain a fully tempered microstructure suitable for additional operations. It is usually annealed, after all welding is done, prior to final heat treatment. The filler metal should be of the same composition as the base metal. The gas tungsten arc and gas metal arc processes are most widely used. However, shielded metal arc welding, plasma arc, and electron beam welding processes can be used.

The medium-alloy quenched and tempered high-yield strength steels are usually welded with the shielded metal arc, gas metal arc, or the submerged arc welding process. The filler metal must provide deposited metal of a strength level equal to the base material. In all cases, a low-hydrogen or no-hydrogen process is required. For shielded metal arc welding the low-hydrogen electrodes of the E-13018 type are recommended. Electrodes must be properly stored. In the case of the other processes, precautions should be taken to make sure that the gas is dry and that the submerged arc flux is dry. By employing the proper heat input–heat output procedure yield strength and toughness are maintained. Preheating should be at least 100°F (38°C) for thinner materials and double that for heavier materials. The heat input should be such that the adjacent base metal does not become overheated. The heat output is sufficient to maintain the proper microstructure in the heat-affected zone. There may be some softening in the intermixing zone. The properties of welded joints that are properly made will be in the same order as the base metal. Subsequent heat treating is usually not required or desired.

High-Alloy Hardenable Steels

The steels in this group develop high strength by standard hardening and tempering heat treatments. The steels possess extremely high strength in the range of 180,000 psi yield and have a high degree of toughness. This is obtained with a minimum carbon content usually in the range of 0.20%; however, these steels contain relatively high amounts of nickel and cobalt, and they are sometimes called the 9 Ni-4 Co steels. These steels also contain small amounts of other alloys. They are normally welded in the quenched and tempered condition by the gas tungsten arc welding process. No postheat treatment is required. The filler metal must match the analysis of the base metal.

High-Nickel Maraging Steels

This type of steel has a relatively high nickel content but is a low-carbon steel. It obtains its high strength from a special heat treatment called *maraging*. These steels possess an extraordinary combination of ultra-high-strength and fracture toughness and at the same

time are formable, weldable, and easy to heat treat. There are three basic types: the 18% nickel, the 20% nickel, and the 25% nickel types. These steels are available in sheet, forging billets, bars, strip, and plate. Some are available as tubing.

The extra special properties of these steels are obtained by heating the steel to 900°F (482°C) and allowing it to cool to room temperature. During this heat treatment all of the austenite transforms to martensite, which is of the very tough massive type. The time at the 900°F temperature is extremely important and usually is in the range of three hours. The steels derive their strength while aging at this temperature in the martensitic condition and for this reason are known as maraging steels.

These steels are supplied in the soft or annealed condition. They can be cold worked in this condition and can be flame cut or plasma arc cut. Plasma arc cutting is preferred. These steels are usually welded by the gas tungsten arc or the gas metal arc welding process. The shielded metal arc and submerged arc process can also be used with special electrode-flux combinations. The filler metal should have the same composition as the base metal. In addition, the filler metal must be of high purity with low carbon. Preheat or postheat is not required; however, the welding must be followed by the maraging heat treatment which produces weld joints of an extremely high strength.

Martensitic Stainless Steels

These steels are of the straight chromium type essentially the AISI 420 classification. These steels contain 12–14% chromium and up to 0.35% carbon. This composition combines stainlessness with high strength. Numerous variations of this basic composition are available, all of which are in the martensitic classification. This type of steel has been used for compressor and turbine blades of jet engines and for other applications in which moderate corrosion resistance and high strength are required. The strength level of these steels is obtained by a quenching and tempering heat treatment. They can be obtained as sheet, strip, tubing, and plate. The compositions are also used for castings. These steels can be heat treated to strengths as high as 250,000 psi (175 kg/mm^2) yield strength.

These stainless steels can be flame cut by the powder cutting system normally used for flame cutting stainless steels. They can also be cut with the oxy-arc process. Flame cutting should be done with the steel in the annealed condition. Most grades should be preheated to 600°F (316°C) because they are air hardenable. They should be annealed after cutting to restore softness and ductility. These materials can also be cold worked in the annealed condition.

The martensitic stainless steels can be welded in the annealed or fully hardened condition, usually without preheat or postheat. The gas tungsten arc welding process is normally used. The filler metal must be of the same analysis as the base metal. Following welding the weldment should be annealed and then heat treated to the desired strength level.

Semiaustenitic Precipitation-Hardenable Stainless Steels

The steels in this group are chrome-nickel steels that are ductile in the annealed condition but can be hardened to high strength by proper heat treatment. In the annealed condition the steels are austenitic and can be readily cold worked. By special heat treatment the austenite is transformed to martensite and later a precipitant is formed in the martensite. The outstanding extra high strength is obtained by a combination of these two hardening processes. The term *semiaustenitic type* was given these steels to distinguish them from normal stainless steels. They are also called precipitation hardening steels or PH steels. The heat treatment for these steels is based on heating the annealed material to a temperature of 1,700–1,750°F (927–954°C) followed by a tempering or aging treatment in the range of 850–1,100°F (454–593°C). These steels are available as billets, sheet, tubing, and plate.

These steels are normally not flame cut. Welding is

performed using the gas tungsten arc or the gas metal arc welding process. The shielded metal arc welding process is rarely used. The filler metal should have the same composition as the base metal. No preheat or postheat is required if the parts are welded in the annealed condition. Following welding it should be annealed and then heat treated to develop optimum strength levels.

It is possible to weld the PH steels in the heat-treated condition using the gas tungsten arc or gas metal arc welding process. However, there is a loss of joint strength due to heating of the heat-affected zone above the aging temperature. In view of this, it is not possible to produce a 100% efficient joint. Extra reinforcing must be utilized to develop full-strength joints. These steels are also brazed using nickel alloy filler metal.

When welding on any of these high-strength steels, weld quality must be of the highest degree. Root fusion must be complete, and there should be no undercut or any type of stress risers. The weld metal should be free of porosity and any weld cracking is absolutely unacceptable. All precautions must be taken in order to produce the highest weld quality. Arc strikes should be the basis for rejection.

QUESTIONS

1. Explain the AISI-SAE steel classification system.
2. What is the meaning of the last two digits of AISI classification?
3. How much carbon would be in C-1020 steel? In C-1045 steel?
4. What is the significance of the covered electrode class suffix letter?
5. Are all E-80XX, E-90XX, and higher electrodes low-hydrogen types? Why?
6. How is the electrode class suffix letter used to select electrodes for alloy steels?
7. How are the first two (or three) digits useful in selecting electrodes for alloy steels?
8. What is meant by weathering steels? Name two popular brands.
9. How is the carbon equivalent factor used to establish a welding procedure?
10. Why is carbon-moly steel selected for high-temperature service? How is it welded?
11. What is the advantage of high-nickel steel for low-temperature service? How is it welded?
12. What are the advantages of quenched and tempered steels for construction equipment?
13. Why is heat input important when welding quenched and tempered steels?
14. What are the three types of stainless steel? Are any magnetic? Which?
15. Why is the welding of austenitic stainless steel different from mild steel?
16. Are the stainless steel-covered electrodes of the low-hydrogen type? Why?
17. What use is made of the Schaeffler diagram?
18. Can any of the ultra-high-strength steels be welded after heat treatment?
19. What is meant by maraging steel? What processes are used for welding?
20. What is meant by a PH steel? Can it be welded with SMAW?

REFERENCES

1. *Steel Products Manual,* American Iron and Steel Institute, New York.

2. "Specifications for Low Alloy Steel Covered Arc Welding Electrodes," A5.5, *American Welding Society,* Miami, Florida.

3. "Welding of Chromium-Molybdenum Steel Pipe," B10.8, *American Welding Society,* Miami, Florida.

4. "Specifications for Corrosion-Resisting Chromium and Chromium Nickel Steel Covered Welding Electrodes," A5.4, *American Welding Society,* Miami, Florida.

5. ANTON L. SCHAEFFLER, "Constitution Diagram for Stainless Steel Weld Metal," *Metal Progress Data Book,* mid-June 1973, The American Society for Metals, Cleveland, Ohio.

6. LONG, C. AND DELONG W. "The Ferrite Content of Austenitic Stainless Steel Weld Metal," *Welding Journal,* Vol. 52, July 1973, Research Supplement, 281-S to 297-S.

13

WELDING THE NONFERROUS METALS

13-1 ALUMINUM AND ALUMINUM ALLOYS

13-2 COPPER AND COPPER-BASE ALLOYS

13-3 MAGNESIUM-BASE ALLOYS

13-4 NICKEL-BASE ALLOYS

13-5 REACTIVE AND REFRACTORY METALS

13-6 THE OTHER NONFERROUS METALS

13-1 ALUMINUM AND ALUMINUM ALLOYS

The unique combination of light weight and relatively high strength makes aluminum the second most popular metal that is welded. Aluminum is not difficult to join but aluminum welding is different from welding steels.

Many alloys of aluminum have been developed and it is important to know which alloy is to be welded. A system of four-digit numbers has been developed by the Aluminum Association, Inc., to designate the various wrought aluminum alloy types.[1] This system of alloy groups, shown by the table of Figure 13-1, is as follows:

1XXX Series: These are aluminums of 99% or higher purity. They are used primarily in the electrical and chemical industries.

2XXX Series: Copper is the principal alloy in this group. This group provides extremely high strength when properly heat treated. These alloys do not produce as good corrosion resistance and are often clad with pure aluminum or special-alloy aluminum. These alloys are used in the aircraft industry.

3XXX Series: Manganese is the major alloying element in this group. These alloys are non-heat-treatable. Manganese is limited to about 1.5%. These alloys have moderate strength and are easily worked.

Major Alloying Element	Designation
99.0% minimum aluminum and over	1xxx
Copper	2xxx
Manganese	3xxx
Silicon	4xxx
Magnesium	5xxx
Magnesium and silicon	6xxx
Zinc	7xxx
Other element	8xxx

FIGURE 13-1 *Designation of aluminum alloy groups.*

4XXX Series: Silicon is the major alloying element in this group. It can be added in sufficient quantities to reduce substantially the melting point and is used for brazing alloys and welding electrodes. Most of the alloys in this group are non-heat-treatable.

5XXX Series: Magnesium is the major alloying element of this group. These alloys are of medium strength. They possess good welding characteristics, good resistance to corrosion, but the amount of cold work should be limited.

6XXX Series: Alloys in this group contain silicon and magnesium which make them heat treatable. These alloys possess medium strength and good corrosion resistance.

TABLE 13-1 Nominal chemical composition of aluminum wrought alloys (Courtesy, The American Aluminum Association).

AA Destination	Silicon	Iron	Copper	Manganese	Magnesium	Chromium	Nickel	Zinc	Titanium	OTHERS Each	OTHERS Total	Aluminum Min.
1050	0.25	0.40	0.05	0.05	0.05	—	—	0.05	0.03	0.03	—	99.50
1060	0.25	0.35	0.05	0.05	0.03	—	—	0.05	0.03	0.03	—	99.60
1100	1.0 Si + Fe		0.05-0.20	0.05	—	—	—	0.10	—	0.05	0.15	99.00
1145	0.55 Si + Fe		0.05	0.05	0.05	—	—	0.05	0.03	0.03	—	99.45
1175	0.15 Si + Fe		0.10	0.10	0.02	—	—	0.04	0.02	0.02	—	99.75
1200	1.0 Si + Fe		0.05	0.05	—	—	—	0.10	0.05	0.05	0.15	99.00
1230	0.7 Si + Fe		0.10	0.05	0.05	—	—	0.10	0.03	0.03	—	99.30
1235	0.65 Si + Fe		0.05	0.05	0.05	—	—	0.10	0.03	0.03	—	99.35
1345	0.30	0.40	0.10	0.05	0.05	—	—	0.05	0.03	0.03	—	99.45
1350	0.10	0.40	0.05	0.01	—	0.01	—	0.05	—	0.03	0.10	99.50
2011	0.40	0.7	5.0-6.0	—	—	—	—	0.30	—	0.05	0.15	Remainder
2014	0.50-1.2	0.7	3.9-5.0	0.40-1.2	0.20-0.8	0.10	—	0.25	0.15	0.05	0.15	Remainder
2017	0.20-0.8	0.7	3.5-4.5	0.40-1.0	0.40-0.8	0.10	—	0.25	0.15	0.05	0.15	Remainder
2018	0.9	1.0	3.5-4.5	0.20	0.45-0.9	0.10	1.7-2.3	0.25	—	0.05	0.15	Remainder
2024	0.50	0.50	3.8-4.9	0.30-0.9	1.2-1.8	0.10	—	0.25	0.15	0.05	0.15	Remainder
2025	0.50-1.2	1.0	3.9-5.0	0.40-1.2	0.05	0.10	—	0.25	0.15	0.05	0.15	Remainder
2036	0.50	0.50	2.2-3.0	0.10-0.40	0.30-0.6	0.10	—	0.25	0.15	0.05	0.15	Remainder
2117	0.8	0.7	2.2-3.0	0.20	0.20-0.50	0.10	—	0.25	—	0.05	0.15	Remainder
2124	0.20	0.30	3.8-4.9	0.30-0.9	1.2-1.8	0.10	—	0.25	0.15	0.05	0.15	Remainder
2218	0.9	1.0	3.5-4.5	0.20	1.2-1.8	0.10	1.7-2.3	0.25	—	0.05	0.15	Remainder
2219	0.20	0.30	5.8-6.8	0.20-0.40	0.02	—	—	0.10	0.02-0.10	0.05	0.15	Remainder
2319	0.20	0.30	5.8-6.8	0.20-0.40	0.02	—	—	0.10	0.10-0.20	0.05	0.15	Remainder
2618	0.10-0.25	0.9-1.3	1.9-2.7	—	1.3-1.8	—	0.9-1.2	0.10	0.04-0.10	0.05	0.15	Remainder
3003	0.6	0.7	0.05-0.20	1.0-1.5	—	—	—	0.10	—	0.05	0.15	Remainder
3004	0.30	0.7	0.25	1.0-1.5	0.8-1.3	—	—	0.25	—	0.05	0.15	Remainder
3005	0.6	0.7	0.30	1.0-1.5	0.20-0.6	0.10	—	0.25	0.10	0.05	0.15	Remainder
3105	0.6	0.7	0.30	0.30-0.8	0.20-0.8	0.20	—	0.40	0.10	0.05	0.15	Remainder
4032	11.0-13.5	1.0	0.50-1.3	—	0.8-1.3	0.10	0.50-1.3	0.25	0.10	0.05	0.15	Remainder
4043	4.5-6.0	0.8	0.30	0.05	0.05	—	—	0.10	0.20	0.05	0.15	Remainder
4045	9.0-11.0	0.8	0.30	0.05	0.05	—	—	0.10	0.20	0.05	0.15	Remainder
4047	11.0-13.0	0.8	0.30	0.15	0.10	—	—	0.20	—	0.05	0.15	Remainder
4145	9.3-10.7	0.8	3.3-4.7	0.15	0.15	0.15	—	0.20	—	0.05	0.15	Remainder
4343	6.8-8.2	0.8	0.25	0.10	—	—	—	0.20	—	0.05	0.15	Remainder
4643	3.6-4.6	0.8	0.10	0.05	0.10-0.30	—	—	0.10	0.15	0.05	0.15	Remainder
5005	0.30	0.7	0.20	0.20	0.50-1.1	0.10	—	0.25	—	0.05	0.15	Remainder
5050	0.40	0.7	0.20	0.10	1.1-1.8	0.10	—	0.25	—	0.05	0.15	Remainder
5052	0.25	0.40	0.10	0.10	2.2-2.8	0.15-0.35	—	0.10	—	0.05	0.15	Remainder
5056	0.30	0.40	0.10	0.05-0.20	4.5-5.6	0.05-0.20	—	0.10	—	0.05	0.15	Remainder
5083	0.40	0.40	0.10	0.40-1.0	4.0-4.9	0.05-0.25	—	0.25	0.15	0.05	0.15	Remainder
5086	0.40	0.50	0.10	0.20-0.7	3.5-4.5	0.05-0.25	—	0.25	0.15	0.05	0.15	Remainder
5154	0.45 Si + Fe		0.10	0.10	3.1-3.9	0.15-0.35	—	0.20	0.20	0.05	0.15	Remainder
5183	0.40	0.40	0.10	0.50-1.0	4.3-5.2	0.05-0.25	—	0.25	0.15	0.05	0.15	Remainder
5252	0.08	0.10	0.10	0.10	2.2-2.8	—	—	0.05	—	0.05	0.15	Remainder

TABLE 13-1 Nominal chemical composition of aluminum wrought alloys (Courtesy of The Aluminum Association) continued.

AA Designation	Silicon	Iron	Copper	Manganese	Magnesium	Chromium	Nickel	Zinc	Titanium	OTHERS Each	OTHERS Total	Aluminum Min.
5254	0.45 Si + Fe	Si + Fe	0.05	0.01	3.1-3.9	0.15-0.35	—	0.20	0.05	0.05	0.15	Remainder
5356	0.50 Si + Fe	Si + Fe	0.10	0.05-0.20	4.5-5.5	0.05-0.20	—	0.10	0.06-0.20	0.05	0.15	Remainder
5454	0.25	0.40 Si + Fe	0.10	0.50-1.0	2.4-3.0	0.05-0.20	—	0.25	0.20	0.05	0.15	Remainder
5456	0.40 Si + Fe	Si + Fe	0.10	0.50-1.0	4.7-5.5	0.05-0.20	—	0.25	0.20	0.05	0.15	Remainder
5457	0.08	0.10	0.20	0.15-0.45	0.8-1.2	—	—	0.05	—	0.03	0.10	Remainder
5554	0.40 Si + Fe	Si + Fe	0.10	0.50-1.0	2.4-3.0	0.05-0.20	—	0.25	0.05-0.20	0.05	0.15	Remainder
5556	0.40 Si + Fe	Si + Fe	0.10	0.50-1.0	4.7-5.5	0.05-0.20	—	0.25	0.05-0.20	0.05	0.15	Remainder
5652	0.40 Si + Fe	Si + Fe	0.04	0.01	2.2-2.8	0.15-0.35	—	0.10	—	0.05	0.15	Remainder
5654	0.45 Si + Fe	Si + Fe	0.05	0.01	3.1-3.9	0.15-0.35	—	0.20	0.05-0.15	0.05	0.15	Remainder
5657	0.08	0.10	0.10	0.03	0.6-1.0	—	—	0.05	—	0.02	0.05	Remainder
6003	0.35-1.0	0.6	0.10	0.08	0.8-1.5	0.35	—	0.20	0.10	0.05	0.15	Remainder
6005	0.6-0.9	0.35	0.10	0.10	0.40-0.6	0.10	—	0.10	0.10	0.05	0.15	Remainder
6053	—	0.35	0.10	—	1.1-1.4	0.15-0.35	—	0.10	—	0.05	0.15	Remainder
6061	0.40-0.8	0.7	0.15-0.40	0.15	0.8-1.2	0.04-0.35	—	0.25	0.15	0.05	0.15	Remainder
6063	0.20-0.6	0.35	0.10	0.10	0.45-0.9	0.10	—	0.10	0.10	0.05	0.15	Remainder
6066	0.9-1.8	0.50	0.7-1.2	0.6-1.1	0.8-1.4	0.40	—	0.25	0.20	0.05	0.15	Remainder
6070	1.0-1.7	0.50	0.15-0.40	0.40-1.0	0.50-1.2	0.10	—	0.25	0.15	0.05	0.15	Remainder
6101	0.30-0.7	0.50	0.10	0.03	0.35-0.8	0.03	—	0.10	—	0.03	0.10	Remainder
6151	0.6-1.2	1.0	0.35	0.20	0.45-0.8	0.15-0.35	—	0.25	0.15	0.05	0.15	Remainder
6162	0.40-0.8	0.50	0.20	0.10	0.7-1.1	0.10	—	0.25	0.10	0.05	0.15	Remainder
6201	0.50-0.9	0.50	0.10	0.03	0.6-0.9	0.03	—	0.10	—	0.03	0.10	Remainder
6253	—	0.50	0.10	—	1.0-1.5	0.15-0.35	—	1.6-2.4	—	0.05	0.15	Remainder
6262	0.40-0.8	0.7	0.15-0.40	0.15	0.8-1.2	0.04-0.14	—	0.25	0.15	0.05	0.15	Remainder
6351	0.7-1.3	0.50	0.10	0.40-0.8	0.40-0.8	—	—	0.20	0.20	0.05	0.15	Remainder
6463	0.20-0.6	0.15	0.20	0.05	0.45-0.9	—	—	—	—	0.05	0.15	Remainder
6951	0.20-0.50	0.8	0.15-0.40	0.10	0.40-0.8	—	—	0.20	—	0.05	0.15	Remainder
7001	0.35	0.40	1.6-2.6	0.20	2.6-3.4	0.18-0.35	—	6.8-8.0	0.20	0.05	0.15	Remainder
7005	0.35	0.40	0.10	0.20-0.7	1.0-1.8	0.06-0.20	—	4.0-5.0	0.01-0.06	0.05	0.15	Remainder
7008	0.10	0.10	0.05	0.05	0.7-1.4	0.12-0.25	—	4.5-5.5	0.05	0.05	0.15	Remainder
7011	0.15	0.20	0.05	0.10-0.30	1.0-1.6	0.05-0.20	—	4.0-5.5	0.05	0.05	0.15	Remainder
7072	0.7 Si + Fe	Si + Fe	0.10	0.10	0.10	—	—	0.8-1.3	—	0.05	0.15	Remainder
7075	0.40	0.50	1.2-2.0	0.30	2.1-2.9	0.18-0.35	—	5.1-6.1	0.20	0.05	0.15	Remainder
7079	0.30	0.40	0.40-0.8	0.10-0.30	2.9-3.7	0.10-0.25	—	3.8-4.8	0.10	0.05	0.15	Remainder
7178	0.40	0.50	1.6-2.4	0.30	2.4-3.1	0.18-0.35	—	6.3-7.3	0.20	0.05	0.15	Remainder

Note: 1. Composition is percent maximum unless shown as a range, or a minimum.
2. There are sometimes other minor elements percent. See reference for exact data.

7XXX Series: Zinc is the major alloying element in this group. Magnesium is also included in most of these alloys. Together they result in a heat-treatable alloy of very high strength. This series is used for aircraft frames.

The nominal composition of the wrought aluminum alloys is shown by Table 13-1.

Aluminum alloy casting alloys are also designated by the Aluminum Association. Table 13-2 shows the nominal chemical composition of casting alloys. With respect to welding, it is the composition that is most important rather than how the part was made. Castings as well as wrought forms are heat treated, and this is the other factor that must be considered. Otherwise, the welding procedures can be essentially the same.

Temper Designation System

The Aluminum Association has also produced a temper designation system used for wrought and cast aluminum alloys. It is based on the sequence of basic treatments to produce various tempers. In specifying an alloy the temper designation follows the alloy designation separated by a dash. Basic temper designations consist of letters. Subdivisions of the basic tempers, when required, are indicated by one or more digits following the letter.

The basic temper designations and subdivisions are as follows:

F, As fabricated.

O, Annealed, recrystalized (wrought products only); applies to the softest tempers of the wrought products.

H, Strain hardened (wrought products only). This applies to products which have their strength increased by strain hardening with or without supplementary treatment. The H is always followed by two or more digits. The first digit indicates the specific combination of basic operations as follows:

H-1, Strain hardened only.

H-2, Strain hardened and then partially annealed.

H-3, Strain hardened and then stabilized.

The digit following the designation H-1, H-2, and H-3 indicates the final degree of strain hardening. Tempers between 0 (annealed) and 8 (full hard) are designated by Nos. 1 through 7. Numeral 2 indicates quarter hard, the numeral 4 indicates half hard, the numeral 6 indicates three-quarters hard, etc. The numeral 9 indicates extra hard temper.

The third digit, when used, indicates a variation of the two-digit H temper number.

W, Solution heat treated. This is an unstable temper applied only to alloys which age harden at room temperatures after solution heat treatment.

T, Thermally treated to produce stable tempers other than F, O, or H. The T is always followed by one or more digits as follows:

T-1, cooled from an elevated temperature shaping process and naturally aged to a substantially stable condition.

T-2, annealed (cast products only).

T-3, solution heat treated and then cold worked.

T-4, solution heat treated and naturally aged to a substantially stable condition.

T-5, cooled from an elevated temperature shaping process and then artificially aged.

T-6, solution heat treated and then artificially aged.

T-7, solution heat treated and then stabilized.

T-8, solution heat treated and then heat treated, cold worked, and then artificially aged.

T-9, solution heat treated, artificially aged, and then cold worked.

T-10, cooled from an elevated temperature shaping process, artificially aged, and then cold worked.

An additional digit may be used which indicates the variation and treatment that significantly alter the characteristics of the product. For example, TX indicates stress relieving by some process such as stretching, compressing or thermal treatment.

The temper designations are important from a welding point of view since welding which is normally a thermal process can change the characteristics of the metal in the heat-affected zone. Care must be taken when welding on the H, W, or T designations. Metallurgical advice should be obtained to determine treatment required to obtain original properties.

The different temper designations are used for different products such as sheet, plate, pipe, shapes, rod, bar, etc. In addition, the different alloys are available in certain types of mill products. In other words, all products are not available in all compositions nor in all of the different tempers.

The heat-treatable alloys which contain copper or zinc are less resistant to corrosion than the non-heat-treatable alloys. To increase the corrosion resistance of

TABLE 13-2 Nominal chemical composition of aluminum casting alloys (Courtesy, The Aluminum Association).

AA Number	Former Designation	Product	Silicon	Iron	Copper	Manganese	Magnesium	Chromium	Nickel	Zinc	Tin	Titanium	Others Each	Others Total
208.0	108	S	2.5-3.5	1.2	3.5-4.5	0.50	0.10	–	0.35	1.0	–	0.25	–	0.50
213.0	C113	P	1.0-3.0	1.2	6.0-8.0	0.6	0.10	–	0.35	2.5	–	0.25	–	0.50
222.0	122	S&P	2.0	1.5	9.2-10.7	0.50	0.15-0.35	–	0.50	0.8	–	0.25	0.05	0.35
242.0	142	S&P	0.7	1.0	3.5-4.5	0.35	1.2-1.8	0.25	1.7-2.3	0.35	–	0.25	0.05	0.15
295.0	195	S	0.7-1.5	1.0	4.0-5.0	0.35	0.03	–	–	0.35	–	0.25	0.05	0.15
B295.0	B195	P	2.0-3.0	1.2	4.0-5.0	0.35	0.05	–	0.35	0.50	–	0.25	–	0.35
308.0	A108	P	5.0-6.0	1.0	4.0-5.0	0.50	–	–	–	1.0	–	0.25	–	0.50
319.0	319, Allcast	S&P	5.5-6.5	1.0	3.0-4.0	0.50	0.10	–	0.35	1.0	–	0.25	–	0.50
328.0	Red X-8	S	7.5-8.5	1.0	1.0-2.0	0.20-0.6	0.20-0.6	0.35	0.25	1.5	–	0.25	–	–
A332.0	A132	P	11.0-13.0	1.2	0.50-1.5	0.35	0.7-1.3	–	2.0-3.0	0.35	–	0.25	0.05	0.50
F332.0	F132	P	8.5-10.5	1.2	2.0-4.0	0.50	0.50-1.5	–	0.50	1.0	–	0.25	–	0.50
333.0	333	P	8.0-10.0	1.0	3.0-4.0	0.50	0.05-0.50	–	0.50	1.0	–	0.25	–	0.50
355.0	355	S&P	4.5-5.5	0.6	1.0-1.5	0.50	0.40-0.6	0.25	–	0.35	–	0.25	0.05	0.15
C355.0	C355	S&P	4.5-5.5	0.20	1.0-1.5	0.10	0.40-0.6	–	–	0.10	–	0.20	0.05	0.15
356.0	356	S&P	6.5-7.5	0.6	0.25	0.35	0.20-0.40	–	–	0.35	–	0.25	0.05	0.15
A356.0	A356	S&P	6.5-7.5	0.20	0.20	0.10	0.20-0.40	–	–	0.10	–	0.20	0.05	0.15
357.0	357	S&P	6.5-7.5	0.15	0.05	0.03	0.45-0.6	–	–	0.05	–	0.20	0.05	0.15
360.0	360	D	9.0-10.0	2.0	0.6	0.35	0.40-0.6	–	0.50	0.50	0.15	–	–	0.25
A360.0	A360	D	9.0-10.0	1.3	0.6	0.35	0.40-0.6	–	0.50	0.50	0.15	–	–	0.25
380.0	380	D	7.5-9.5	2.0	3.0-4.0	0.50	0.10	–	0.50	3.0	0.35	–	–	0.50
A380.0	A380	D	7.5-9.5	1.3	3.0-4.0	0.50	0.10	–	0.50	3.0	0.35	–	–	0.50
384.0	384	D	10.5-12.0	1.3	3.0-4.5	0.50	0.10	–	0.50	1.0	0.35	–	–	0.50
413.0	13	D	11.0-13.0	2.0	1.0	0.35	0.10	–	0.50	0.50	0.15	–	–	0.25
A413.0	A13	D	11.0-13.0	1.3	1.0	0.35	0.10	–	0.50	0.50	0.15	–	–	0.25
B443.0	43 (0.15 max cu)	S&P	4.5-6.0	0.8	0.15	0.35	0.05	–	–	0.35	–	0.25	0.05	0.15
C443.0	A43	D	4.5-6.0	2.0	0.6	0.35	0.10	–	0.50	0.50	0.15	–	–	0.25
514.0	214	S	0.35	0.50	0.15	0.35	3.5-4.5	–	–	0.15	–	0.25	0.05	0.15
A514.0	A214	P	0.30	0.40	0.10	0.30	3.5-4.5	–	–	1.4-2.2	–	0.20	0.05	0.15
B514.0	B214	S	1.4-2.2	0.6	0.35	0.8	3.5-4.5	0.25	–	0.35	–	0.25	0.05	0.15
518.0	218	D	0.35	1.8	0.25	0.35	7.5-8.5	–	0.15	0.15	0.15	–	–	0.25
520.0	220	S	0.25	0.30	0.25	0.15	9.5-10.6	–	–	0.15	–	0.25	0.05	0.15
535.0	Almost 35	S	0.15	0.15	0.05	0.10-0.25	6.2-7.5	–	–	–	–	0.10-0.25	0.05	0.15
705.0	603, Ternalloy 5	S&P	0.20	0.8	0.20	0.40-0.6	1.4-1.8	0.20-0.40	–	2.7-3.3	–	0.25	0.05	0.15
707.0	607, Ternalloy 7	S&P	0.20	0.8	0.20	0.40-0.6	1.8-2.4	0.20-0.40	–	4.0-4.5	–	0.25	0.05	0.15
A712.0	A612	S	0.15	0.50	0.35-0.65	0.05	0.6-0.8	–	–	6.0-7.0	–	0.25	0.05	0.20
D712.0	D612, 40E	S	0.30	0.50	0.25	0.10	0.50-0.65	0.40-0.6	–	5.0-6.5	–	0.15-0.25	0.05	0.20
713.0	613, Tenzaloy	S&P	0.25	1.1	0.40-1.0	0.6	0.20-0.50	0.35	0.15	7.0-8.0	–	0.25	0.10	0.25
771.0	Precedent 71A	S	0.15	0.15	0.10	0.10	0.8-1.0	0.06-0.20	–	6.5-7.5	–	0.10-0.20	0.05	0.15
850.0	750	S&P	0.7	0.7	0.7-1.3	0.10	0.10	–	0.7-1.3	–	5.5-7.0	0.20	–	0.30
A850.0	A750	S&P	2.0-3.0	0.7	0.7-1.3	0.10	0.10	–	0.30-0.7	–	5.5-7.0	0.20	–	0.30
B850.0	B750	S&P	0.40	0.7	1.7-2.3	0.10	0.6-0.9	–	0.9-1.5	–	5.5-7.0	0.20	–	0.30

Note: 1. Composition is percent maximum unless shown as a range. Aluminum is the remainder.
2. There may be minor elements present. See reference for exact data.
3. Product S = Sand Cast, P = Permanent Mold Cast, D = Die Cast.

these alloys in sheet and plate they are sometimes clad with high-purity aluminum, usually 2 1/2–4% of the total thickness on each side. These are known as *alclad* products.

Welding Aluminum Alloys

Aluminum possesses a number of properties that make welding it different than the welding of steels. These are:

1. Aluminum oxide surface coating.
2. High thermal conductivity.
3. High thermal expansion coefficient.
4. Low melting temperature.
5. The absence of color change as temperature approaches the melting point.

The normal metallurgical factors that apply to other metals apply to aluminum as well.

Aluminum is an active metal and it reacts with oxygen in the air to produce a thin hard film of aluminum oxide on the surface. The melting point of aluminum oxide is approximately 3,600°F (1,926°C) which is almost three times the melting point of pure aluminum, 1,220°F (660°C). In addition, this aluminum oxide film, particularly as it becomes thicker, will absorb moisture from the air. Moisture is a source of hydrogen which is the cause of porosity in aluminum welds. Hydrogen may also come from oil, paint, and dirt in the weld area. It also comes from the oxide and foreign materials on the electrode or filler wire, as well as from the base metal. Hydrogen will enter the weld pool and is soluble in molten aluminum. As the aluminum solidifies it will retain much less hydrogen and the hydrogen is rejected during solidification. With a rapid cooling rate free hydrogen is retained within the weld and will cause porosity. Porosity will decrease weld strength and ductility depending on the amount.

The aluminum oxide film must be removed prior to welding. If it is not all removed small particles of unmelted oxide will be entrapped in the weld pool and will cause a reduction in ductility, lack of fusion, and may cause weld cracking.

The aluminum oxide can be removed by mechanical or chemical means or electrical means. Mechanical removal involves scraping with a sharp tool, sandpaper, wire brush (stainless steel), filing, or any other mechanical method. Chemical removal can be done in two ways. One is by use of cleaning solutions, either the etching types or the nonetching types. The nonetching types should be used only when starting with relatively clean parts. They are used in conjunction with other solvent cleaners. For better cleaning the etching type solutions are recommended but must be used with care. When dipping is employed hot and cold rinsing is highly recommended. The etching type solutions are alkaline solutions. The time in the solution must be controlled so that too much etching does not occur.

Chemical cleaning includes the use of welding fluxes. Fluxes are used for gas welding, brazing, and soldering. The coating on covered aluminum electrodes also contains fluxes for cleaning the base metal. Whenever etch cleaning or flux cleaning is used the flux and alkaline etching materials must be completely removed from the weld area to avoid future corrosion.

The electrical oxide removal system uses cathodic bombardment. Cathodic bombardment occurs during the half cycle of alternating current gas tungsten arc welding when the electrode is positive (reverse polarity). This is an electrical phenomenon that actually blasts away the oxide coating to produce a clean surface. This is one of the reasons why AC gas tungsten arc welding is so popular for welding aluminum.

Since aluminum is so active chemically, the oxide film will immediately start to reform. The time of buildup is not extremely fast but welds should be made after aluminum is cleaned within at least 8 hours for quality welding. If a longer time period occurs the quality of the weld will decrease.

Other reasons that aluminum welding is different are due to its high thermal conductivity and low melting temperature. Aluminum conducts heat from three to five times as fast as steel depending on the specific alloy. This means that more heat must be put into the aluminum even though the melting temperature of aluminum is less than half that of steel. Because of the high thermal conductivity, preheat is often used for welding thicker sections. If the temperature is too high or the period of time is too long it can be detrimental to weld joint strength in both heat-treated and work-hardened alloys. The preheat for aluminum should not exceed 400°F (204°C), and the parts should not be held at that temperature longer than necessary. Because of the high heat conductivity procedures should utilize higher speed welding processes using high heat input. Both the gas tungsten arc and the gas metal arc processes supply this requirement.

The high heat conductivity of aluminum can also be helpful since if heat is conducted away from the weld extremely fast the weld will solidify very quickly. This with surface tension helps hold the weld metal in position and makes all-position welding with gas tungsten arc and gas metal arc welding practical.

					Nominal Composition %					
Type	Si	Fe	Cu	Mn	Mg	Cr	Zn	Ti	Al
1100	1.0 Si + Fe		0.05-0.20	0.05	—	—	0.10	—	99.0 Min.
4043	4.5-6.0	0.80	0.30	0.05	0.05	—	0.10	0.20	Remainder
5154	0.45 Si + Fe		0.10	0.10	3.1-3.9	0.15-0.35	0.20	0.20	Remainder
5254	0.45 Si + Fe		0.05	0.01	3.1-3.9	0.15-0.35	0.20	0.05	Remainder
5652	0.40 Si + Fe		0.04	0.01	2.2-2.8	0.15-0.35	0.10	—	Remainder
5554	0.40 Si + Fe		0.10	0.50-1.0	2.4-3.0	0.05-0.20	0.25	0.05-0.20	Remainder
5356	0.50 Si + Fe		0.10	0.05-0.20	4.5-5.5	0.05-0.20	0.10	0.06-0.20	Remainder
5183	0.40	0.40	0.10	0.50-1.0	4.3-5.2	0.05-0.25	0.25	0.15	Remainder
5556	0.40 Si + Fe		0.10	0.50-1.0	4.7-5.5	0.04-0.35	0.25	0.05-0.20	Remainder
6061	0.40-0.80	0.70	0.15-0.40	0.15	0.80-1.2	0.04-0.35	0.25	0.15	Remainder

Per AWS specification A5.10 "aluminum and aluminum alloy welding rods and bare electrodes".

FIGURE 13-2 *Composition of aluminum filler metals.*

The thermal expansion of aluminum is twice that of steel. In addition, aluminum welds decrease about 6% in volume when solidifying from the molten state. This change in dimension or attempt to change in dimension may cause distortion and cracking.

The final reason why aluminum is different to weld from steels is that it does not exhibit color as it approaches its melting temperature. Aluminum will not show color until it is raised above the melting point, at which time it will glow a dull red. When soldering or brazing aluminum with a torch, flux is used and the flux will melt as the temperature of the base metal approaches the temperature required. The flux first dries out and then melts as the base metal reaches the correct working temperature. When torch welding with oxyacetylene or oxyhydrogen the surface of the base metal will melt first and assume a characteristic wet and shiny appearance. (This aids in knowing when welding temperatures are reached.) When welding with gas tungsten arc, or gas metal arc, color is not too important because the weld is quickly completed before the adjoining area would melt.

When the above factors are taken into consideration and related to the metallurgical factors that apply to other metals, it will allow the making of welded joints in aluminum with little or no more trouble than when welding steels.

Since the gas metal arc and gas tungsten arc welding processes are the most popular of the fusion welding processes they will be given major emphasis. With either process the selection of filler metal is the same. The selection of the filler metal depends on the same factors as when welding other metals. In other words, the base metal composition or alloy must be known. Figure 13-2 provides the nominal composition of the different aluminum filler metals. Refer to AWS specifications A5.3 and A5.10 for details. These provide for bare, solid, straightened electrode wires, coiled wires and covered electrodes. It may not be necessary to make the comparison or selection of the filler metal to weld the different aluminum alloys since this has been fairly well standardized. Table 13-3 is a guide to the choice of filler metals for aluminum welding established by the American Welding Society and is recommended.

Gas Tungsten Arc Welding

The gas tungsten arc welding process is used for welding the thinner sections of aluminum and aluminum alloys. This process was described in detail in Chapter 4. There are several precautions that should be mentioned with respect to using this process.

Alternating current is recommended for general-purpose work since it provides the half-cycle of cleaning action. Figure 13-3 provides welding procedure schedules for using the process on different thicknesses to produce different welds. AC welding, usually with high frequency, is widely used with manual and automatic applications. Procedures should be followed closely and special attention should be given to the type of tungsten electrode, size of welding nozzle, gas type, and gas flow rates. When manual welding the arc length should be kept short and equal to the diameter of the electrode. The tungsten electrode should not protrude too far beyond the end of the nozzle. The tungsten should be kept clean and if it does accidentally touch the molten metal it must be redressed.

Welding power sources designed for the gas tungsten arc welding process should be employed. The newer equipment provides for programming, pre- and postflow of shielding gas, and pulsing. Development work is continuing in this field and advance machines may include square wave output and variable amounts of each polarity.

For automatic or machine welding direct current electrode negative (straight polarity) can be used. Cleaning must be extremely efficient since there is no cathodic bombardment to assist. When DC electrode negative is used, extremely deep penetration and high speeds can

TABLE 13-3 *Guide to the choice of filler metal for welding aluminum.*

Base Metal	319, 333 354, 355 C355	13, 43, 344 356, A356 A357, 359	214, A214 B214, F214	7039, 7005k A612, C612 D612	6070	6061, 6063 6101, 6151 6201, 6951	5456	5454
1060, EC 1100, 3003	4145[e,i]	4043[i,f]	4043[e,i]	4043[i]	4043[i]	4043[i]	5356[c]	4043[e,i]
Alclad 3003	4145[e,i]	4043[i,f]	4043[e,i]	4043[i]	4043[i]	4043[i]	5356[c]	4043[e,i]
2014, 2024	4145[g]	4145	—	—	4145	4145	—	—
2219	4145[g,e,i]	4145[e,i]	4043[i]	4043[i]	4043[f,i]	4043[f,i]	4043	4043[i]
3004								
Alclad 3004	4043[i]	4043[i]	5654[b]	5356[e]	4043[e]	4043[b]	5356[e]	5654[b]
5005, 5050	4043[i]	4043[i]	5654[b]	5356[e]	4043[e]	4043[b]	5356[e]	5654[b]
5052, 5652a	4043[i]	4043[b,i]	5654[b]	5356[e,h]	5356[b,c]	5356[b,e]	5356[b]	5654[b]
5083	—	5356[c,e,i]	5356[e]	5183[e,h]	5356[e]	5356[e]	5183[e]	5356[e]
5086	—	5356[c,e,i]	5356[e]	5356[e,h]	5356[e]	5356[e]	5356[e]	5356[b]
5154, 5254a	—	4043[b,i]	5654[b]	5356[b,h]	5356[b,c]	5356[b,c]	5654[b]	5654[a,b]
5454	4043[i]	4043[b,i]	5654[b]	5356[b,h]	5356[b,c]	5356[b,c]	5356[b]	5654[c,e]
5456	—	5356[c,e,i]	5356[e]	5556[e,h]	5356[e]	5356[e]	5556[e]	
6961, 6063, 6101, 6201, 6151, 6951	4145[e,i]	4043[b,i]	5356[b,c]	5356[b,c,h,i]	4043[b,i]	4043[b,i]		
6070	4145[e,i]	4043[e,i]	5356[c,e]	5356[c,e,h,i]	4043[e,i]			
7039, 7005k A612, C612 D612	4043[i]	4043[b,h,i]	5356[b,h]	5039[e]				
214, A214 B214, F214	—	4043[b,i]	5654[b,d]					
13, 43, 344 356, A356 A357, 359	4145[e,i]	4043[d,i]						
319, 333 354, 355, C355	4145[d,e,i]							

FIGURE 13-3 *Welding procedure schedules for AC-GTAW welding of aluminum.*

Material Thickness (or Fillet Size)			Type of Weld Fillet or Groove	Tungsten Electrode Diameter		Filler Rod Diameter		Nozzle Size Inside Dia.	Shielding Gas Flow	Welding Current Amps AC	No. of Passes	Travel Speed (per pass)
ga	in.	mm		in.	mm	in.	mm	in.	cfh			ipm
3/64	0.046	1.2	Sq. Groove & Fillet	1/16	1.6	1/16	1.6	1/4-3/8	20	40-60	1	14-18
1/16	0.063	1.6	Sq. Groove & Fillet	3/32	2.4	3/32	2.4	5/16-3/8	20	70-90	1	8-12
3/32	0.094	2.4	Sq. Groove & Fillet	3/32	2.4	3/32	2.4	5/16-3/8	20	95-115	1	10-12
1/8	0.125	3.2	Sq. Groove & Fillet	1/8	3.2	1/8	3.2	3/8	20	120-140	1	9-12
3/16	0.187	4.7	Fillet	5/32	3.9	5/32	3.9	7/16-1/2	25	160-200	1	9-12
3/16	0.187	4.7	Vee Groove	5/32	3.9	5/32	3.9	7/16-1/2	25	160-180	2	10-12
1/4	0.250	6.4	Fillet	3/16	4.8	3/16	4.8	7/16-1/2	30	230-250	1	8-11
1/4	0.250	6.4	Vee Groove	3/16	4.8	3/16	4.8	7/16-1/2	30	200-220	2	8-11
3/8	0.375	9.5	Vee Groove	3/16	4.8	3/16	4.8	1/2	35	250-310	2-3	9-11
1/2	0.500	12.7	Vee or U Groove	1/4	6.4	1/4	6.4	5/8	35	400-470	3-4	6

1 - Increase amperage when backup is used.

2 - Data is for all welding positions. Use low side of range for out of position.

3 - For tungsten electrodes—1st choice—pure tungsten EWP; 2nd choice—zirconated EWZr.

4 - Normally Argon is used for shielding, however, mixtures of 10% or more helium with Argon are sometimes used for increased penetration in aluminum 1/4'' thick and over. The gas flow should be increased when helium is added. A mixture of 75% He + 25% Argon is popular. When 100% helium is used, gas flow rates are about twice those used for Argon.

5154 / 5254[a]	5086	5083	5052 / 5652[a]	5005 / 5050	3004 / Alc. 3004	2219	2014 / 2024	1100 / 3003 / Alc. 3003	1060 / EC	Base Metal
4043[e,i]	5356[c]	5356[c]	4043[i]	1100[c]	4043	4145	4145	1100[c]	1260[c,i]	1060, EC 1100, 3003
4043[e,i]	5356[c]	5356[c]	4043[e,i]	4043[c]	4043[e]	4145	4145	1100[c]		Alclad 3003
–	–	–	–	–	–	4145[g]	4145[g]			2014, 2024
4031[i]	4043	4043	4043[i]	4043	4043	2319[e,f,i]				2219 3004
5654[b]	5356[c]	5356[e]	4043[e,i]	4043[e,i]	4043[e]					Alclad 3004
5654[b]	5356[e]	5356[e]	4043[e,i]	4043[d,e]						5005, 5050
5654[b]	5356[e]	5356[e]	5654[a,b,c]							5052, 5652[a]
5356[e]	5356[e]	5183[e]								5083
5356[b]	5356[e]									5086

Note: 1—Service conditions such as immersion in fresh or salt water, exposure to specific chemicals or a sustained high temperature (over 150°F) may limit the choice of filler metals.

Note: 2—Recommendations in this table apply to gas shielded-arc wedling processes. For gas welding, only R1100, R1260, and R4043 filler metals are ordinarily used.

Note: 3—Filler alloys designated with ER prefix are listed in AWS specification A5.10.

[a] Base metal alloys 5652 and 5254 are used for hydrogen peroxide service. ER5654 filler metal is used for welding both alloys for low-temperature service (150°F and below).

[b] ER5183, ER5356, ER5554, ER5556, and ER5654 may be used. In some cases they provide:
  (1) improve color match after anodizing treatment,
  (2) highest weld ductility, and
  (3) higher weld strength, ER5554 is suitable for elevated temperature service.

[c] ER4043 may be used for some applications.

[d] Filler metal with the same analysis as the base metal is sometimes used.

[e] ER5183, ER5356, or ER5556 may be used.

[f] ER4145 may be used for some applications.

[g] ER2319 may be used for some applications.

[h] ER5039 may be used for some applications.

[i] ER4047 may be used for some applications.

[j] ER1100 may be used for some applications.

[k] 7005 extrusions only.

Note: 4—Where no filler metal is listed, the parent alloy combination is not recommended for welding.

*AWS Designation: A5.10-69.

ga	Material Thickness (or Fillet Size)		Type of Weld Fillet or Groove	Tungsten Electrode Diameter		Filler Rod Diameter		Nozzle Size Inside Dia.	Shielding Gas Flow	Welding Current Amps	No. of Passes	Travel Speed (per pass)
	in.	mm		in.	mm	in.	mm	in.	cfh	DCEN		ipm
20	0.032	0.8	Sq. groove & fillet	3/32	2.4	None		3/8	30	65-70	1	52
18	0.046	1.2	Sq. groove & fillet	3/64	1.2	3/64	1.2	3/8	30	35-95	1	45
16	0.063	1.6	Sq. groove & fillet	3/64	1.2	3/64	1.2	3/8	30	45-120	1	36
13	0.094	2.4	Sq. groove & fillet	1/16	1.6	1/16	1.6	3/8	30	90-185	1	32
11	1/8	3.2	Sq. groove & fillet	1/8	3.2	1/8	3.2	3/8	30	120-220	1	20
11	1/8	3.2	Sq. groove & fillet	1/8	3.2	None		3/8	30	180-200	1	24
–	1/4	6.4	Sq. groove & fillet	1/8	3.2	1/8	3.2	1/2	40	230-340	1	22
–	1/4	6.4	Sq. groove & fillet	1/8	3.2	None		1/2	40	220-240	1	22
–	1/2	12.7	Vee groove	3/16	4.8	1/8	3.2	1/2	40	300-450	1	20
–	1/2	12.7	Sq. groove	5/32	3.9	None		1/2	40	260-300	2	20
–	3/4	19.1	Vee groove	3/16	4.8	1/8	3.2	1/2	40	300-450	2	6
–	3/4	19.1	Sq. groove	3/16	4.8	None		1/2	40	450-470	2	6
–	1	25.4	Vee groove	3/16	4.8	1/8	3.2	5/8	40	300-450	2	5

Normally for automatic travel
Use Helium or 75% helium 25% argon

FIGURE 13-4 *Welding procedure schedules for DC-GTAW welding of aluminum.*

WELDING THE NONFERROUS METALS

be obtained. Cleanliness is an absolute necessity. Figure 13-4 provides welding procedure schedules for DC electrode negative welding.

The gases are either argon or helium or a mixture of the two. Argon is the most popular and is used at a lower flow rate. Helium will increase penetration but a higher flow rate is required.

When filler wire is used, either manually or automatically, it must be clean. If the oxide is not removed from the filler wire it may include moisture that will tend to produce posority in the weld deposit.

Gas Metal Arc Welding

The gas metal arc welding process is applicable to heavier thicknesses of aluminum. It is much faster than the gas tungsten arc process and was described in Chapter 5.

Several factors should be mentioned with respect to GMAW welding that apply specifically to aluminum. The electrode wire must be clean. Precautions should be taken when storing electrodes, so that when they are used they will produce quality welds. If porosity occurs, it is possible that it came from moisture absorbed in the oxide coating of the electrode wire.

Pure argon is normally used for gas metal arc welding of aluminum. Precautions should be taken to ensure that the gas shield is extremely efficient. On occasion, leaks in the gas system, in the gun or cable assembly, will allow air to be drawn into the argon which will cause porosity. Gas purge control and post-gas-flow should be used. The angle of the gun or torch is more critical when welding aluminum with inert shielding gas. A 30° leading travel angle is recommended. The electrode wire tip should be oversize for aluminum. Figure 13-5 provides welding procedure schedules for gas metal arc welding of aluminum.

The wire feeding equipment for aluminum welding must be in good adjustment for efficient wire feeding. It is recommended that nylon type liners be used in cable assemblies. Also, proper drive rolls should be selected for the aluminum wire and for the size of the electrode wire. It is more difficult to push extremely small diameter aluminum wires through long gun cable assemblies than steel wires. For this reason, the spool gun or the newly developed guns which contain a linear feed motor are used for the small diameter electrode wires. Water-cooled guns are required except for low-current welding.

Both the constant current power source with matching voltage sensing wire feeder and the constant voltage power source with constant speed wire feeder are used for welding aluminum. In addition, the constant speed wire feeder is sometimes used with the constant current power source. In general, the CV system is preferred

Material Thickness (or Fillet Size)			Type of Weld Fillet or Groove	Electrode Diameter		WELDING POWER		Wire Feed Speed	Shielding Gas Flow	No. of Passes	Travel Speed (per pass)
ga	in.	mm		in.	mm	Current Amps DC	Arc Volt EP	ipm	cfh		ipm
—	0.050	—	Sq. groove & fillet	0.030	0.8	50	12-14	268-308	30	1	17-25
—	0.062	1.6	Sq. groove & fillet	0.030	0.8	55-60	12-14	295-320	30	1	17-25
—	0.062	1.6	Sq. groove & fillet	3/64	1.2	110-125	19-21	175-185	30	1	20-27
—	0.093	2.4	Sq. groove & fillet	0.030	0.8	90-100	14-18	330-370	30	1	24-36
—	0.125	3.2	Fillet	0.030	0.8	110-125	19-22	410-460	30	1	20-24
11	0.125	3.2	Sq. groove	3/64	1.2	110-125	20-24	175-190	40	1	20-24
3/16	0.187	4.7	Sq. groove & fillet	3/64	1.2	160-195	20-24	215-225	40	1	20-25
1/4	0.250	6.4	Fillet	3/64	1.2	160-195	20-24	215-225	40	1	20-25
1/4	0.250	6.4	Vee groove	1/16	1.6	175-225	22-26	150-195	40	3	20-25
3/8	0.375	9.5	Vee groove & fillet	1/16	1.6	200-300	22-26	170-275	40	2-5	25-30
1/2	0.500	12.7	Vee groove & fillet	1/16	1.6	220-230	22-27	195-205	40	3-8	12-18
1/2	0.500	12.7	Double vee groove	3/32	2.4	320-340	22-29	140-150	45	2-5	15-17
3/4	0.750	19.0	Double vee groove	1/16	1.6	255-275	22-27	230-250	50	4-10	8-18
3/4	0.750	19.0	Double vee groove	3/32	2.4	355-375	22-29	155-160	50	4-10	14-16
1	1.000	25.4	Double vee groove	1/16	1.6	255-290	22-27	230-265	50	4-14	6-18
1	1.000	25.4	Double vee groove	3/32	2.4	405-425	22-27	175-180	50	4-8	8-12

1 - For groove and fillet welds—material thickness also indicated fillet weld size. Use vee groove for 3/16″ and thicker.

2 - Use argon for thin and medium material; sue 50% argon and 50% helium for thick material. Increase gas flow rate 10% for overhead position.

3 - Increase amperage 10-20% when backup is used.

4 - Decrease amperage 10-20% when welding out of position.

FIGURE 13-5 *Welding procedure schedules for GMAW welding of aluminum.*

when welding on thin material and using small diameter electrode wire. It provides better arc starting and regulation. The CC system is preferred when welding thick material using larger electrode wires. The weld quality seems better with this system. The constant current power source with a moderate droop of 15 to 20 volts per 100 amperes, and, with a constant speed wire feeder, provides the most stable power input to the weld and also provides the highest weld quality.

Other Welding Processes

The shielded metal arc welding process can be used for welding aluminum. The electrodes are covered similarly to conventional steel electrodes. The covering on these electrodes are hydroscopic and must be protected from the atmosphere prior to use. Welds made with covered electrodes must be cleaned since the residue that remains on the weld will cause corrosion. Arc stability is rather poor and there are a limited number of electrode types. This process is restricted in use.

Plasma welding is used for joining aluminum. It is expected to become more popular.

The atomic hydrogen welding process has also been used for welding aluminum. This process is not too popular and is seldom used today.

The carbon arc welding process was quite popular in the past but is of little use today.

Gas welding has been done using both oxyacetylene and oxyhydrogen flames. In either case, an absolutely neutral flame is required. Flux is used as well as a filler rod. The process also is not too popular because of low heat input and the need to remove flux.

Electroslag welding is used for joining pure aluminum. So far it has not been successful for welding the aluminum alloys. Submerged arc welding has been used in some countries where inert gas is not available. It is not used in North America.

All of the resistance welding processes are used for welding aluminum. In the case of spot and seam welding, extreme cleanliness of surface is required. Different types of power are used but the process is extremely efficient and is widely used in the aircraft industry.

Most of the solid state welding processes including friction welding, ultrasonic welding, and cold welding are used for aluminums. In addition, the stud welding process is used for aluminum. Aluminum can also be joined by soldering and brazing. Brazing can be accomplished by most brazing methods. A high silicon alloy filler material is used.

Electron beam welding process is also used for aluminum welding, as are the plasma process and laser welding. These processes have not, however, been used to a very great degree and will not be covered here.

All of the major aluminum companies provide welding manuals and data for the welding of aluminum.[2] [3] [4] If more detail is required these books are recommended.

13-2 COPPER AND COPPER-BASE ALLOYS

Copper and copper-base alloys have specific properties which make them widely used. Their high electrical conductivity makes them widely used in the electrical industries and corrosion resistance of certain alloys makes them very useful in the process industries. Copper alloys are also widely used for friction or bearing applications.

There are over 300 different alloys commercially available and identifying and classifying them is a problem. All of these different alloys have been used for many years, long before an attempt was made to classify them. The Copper Development Association, Inc., has established an alloy designation system that is widely accepted in North America. It is not a specification system but rather a method of identifying and grouping different coppers and copper alloys. This system has been updated so that it now fits the unified numbering system (UNS). It provides one unified numbering system which includes all of the commercially available metals and alloys. The UNS designation consists of the prefix letter C followed by a space, three digits, another space, and, finally, two zeros. The composition of each UNS number or copper alloy number and its common name is published by the association.[5]

The information shown by Table 13-4 is a grouping of these copper alloys by common names which normally include the constituent alloys. Welding information for those alloy groupings is provided. There may be those alloys within a grouping that may have a composition sufficiently different to create welding problems. These are the exception, however, and the data presented will provide starting point guidelines. There are two categories, wrought materials and cast materials. The welding information is the same whether the material is cast or rolled.

Copper shares some of the characteristics of aluminum, but it is weldable. Attention should be given to its properties that make the welding of copper and copper alloys different from the welding of carbon steels.

Copper alloys possess properties that require special attention when welding. These are:

1. High thermal conductivity.

2. High thermal expansion coefficient.

3. Relatively low melting point.

4. It is *hot short,* i.e., brittle at elevated temperatures.

5. The molten metal is very fluid.

6. It has high electrical conductivity.

7. It owes much of its strength to cold working.

Copper has the highest thermal conductivity of all commercial metals and the comments made concerning thermal conductivity of aluminum apply to copper, to an even greater degree.

Copper has a relatively high coefficient of thermal expansion, approximately 50% higher than carbon steel, but lower than aluminum. One of the problems associated with copper alloys is the fact that some of them, such as aluminum bronze, have a coefficient of expansion over 50% greater than that of copper. This creates problems when making generalized statements about the different copper-based alloys. The physical properties shown by Table 11-1 will show these variations.

The melting point of the different copper alloys varies over a relatively wide range, but is at least 1,000°F (538°C) lower than carbon steel. Some of the copper alloys are *hot short.* This means that they become brittle at high temperatures. This is because some of the alloying elements form oxides and other compounds at the grain boundaries, embrittling the material.

Copper does not exhibit heat colors like steel and when it melts it is relatively fluid. This is essentially the result of the high preheat normally used for heavier sections. Copper has the highest electrical conductivity of any of the commercial metals and this is a definite problem in the resistance welding processes.

All of the copper alloys derive their strength from cold working. The heat of welding will anneal the copper in the heat-affected area adjacent to the weld and reduce the strength provided by cold working. This must be considered when welding high-strength joints.

There is one other problem associated with the copper alloys that contain zinc. Zinc has a relatively low boiling temperature, and under the heat of an arc will tend to vaporize and escape from the weld. For this reason the arc processes are not recommended for the alloys containing zinc.

The grouping of the copper alloys in Table 13-4 is

TABLE 13-4 *Copper and copper alloy designation system.*

Copper Number	Wrought Alloys—Groups
C11X00	Oxygen Free-High Conductivity Copper (99.95 + %)
C11X00	
C12X00	Tough Pitch Copper (99.88 + %)
C13X00	
C19X00	High Copper Alloys (96 + % Copper)
C2XX00	Copper-Zinc-Alloys (Brasses)
C3XX00	Copper-Zinc-Lead Alloys (Leaded Brasses)
C4XX00	Copper-Zinc-Tin Alloys (Tin Brasses)
C50X00	Copper-Tin Alloys (Phosphor Bronzes)
C51X00	
C52X00	
C53X00	Copper-Tin-Lead Alloys (leaded Phospher Bronzes)
C54X00	
C61X00	
C62X00	Copper-Aluminum Alloys (Aluminum Bronzes)
C63X00	
C64X00	Copper-Silicon Alloys (Silicon Bronzes)
C65X00	
C66X00	Copper-Zinc Alloys (Misc. Brasses & Bronzes)
C67X00	
C68X00	
C69X00	
C70X00	
C71X00	Copper-Nickel Alloys
C72X00	
C73X00	
C74X00	
C75X00	
C76X00	Copper-Nickel-Zinc Alloys (Nickel Silvers)
C77X00	
C78X00	
C79X00	
	Cast Alloys—Groups
C80X00	Copper Alloys (99 + % Copper)
C81X00	High Copper Alloys (Beryllium Copper)
C82X00	
C83X00	Copper-Tin-Zinc + Copper-Tin-Zinc-Lead Alloys (Red Brasses and Leaded RB
C84X00	Semi-Red Brasses and Leaded Semi-Red Brasses
C85X00	Yellow Brasses and Leaded Yellow Brasses
C86X00	Manganese and Leaded Manganese Bronze Alloys
C87X00	Copper-Zinc-Silicon Alloys (Silicon Bronzes and Brasses)
C90X00	Copper-Tin Alloys (Tin Bronzes)
C91X00	
C92X00	Copper-Tin-Lead Alloy (Leaded Tin Bronze)
C93X00	Copper-Tin-Lead Alloy (High Leaded Tin Bronze)

for convenience; however, there may be certain alloys within the grouping that are different from the others. In view of these, it is difficult to make generalized statements that apply to all the alloys in a particular grouping. For best results it is wise to know the exact composition of the alloy being welded. If it fits within a particular grouping, the recommended filler metal can be checked by referring to Figure 13-6 which gives the nominal composition of the copper alloy filler metals. The data shown here is for the filler metal whether it is an electrode, a rod, or wire, or for brazing. Refer to AWS specifications A5.6, A5.7, and A5.8 for details. The composition of the filler material should be chosen to match the base metal as closely as possible.

The gas metal arc welding and the gas tungsten arc welding processes are the most popular for welding copper and copper alloys. The different families or groups of copper alloys will be covered along with the recommended welding procedures for these two processes.

The gas tungsten arc welding process normally uses direct current electrode negative (straight polarity), but in some cases alternating current with high frequency is recommended. Gas tungsten arc welding is best for

welding the thinner gauges of copper and copper alloys. It is also recommended for repairing copper alloy castings.

The gas metal arc welding process is used for welding thicker materials. It is faster, has a higher deposition rate, and usually results in less distortion. It can produce high-quality welds in all positions. It uses direct current electrode positive and the CV type power source is recommended.

AWS Class	NOMINAL COMPOSITION IN %										
	Cu	Al	Fe	Mn	Ni	Si	Sn	Pb	Ti	Zn	Other
Cu	98	0.01	—	0.5	—	0.50	1.0	0.22	—	—	0.50
CuAl-A1	rem	6.0 to 9.0	—	—	—	0.10	—	0.02	—	0.20	0.50
CuAl-A2	rem	9.0 to 11.0	1.5	—	—	0.10	—	0.02	—	0.02	0.50
CuAl-B	rem	11.0 to 12.0	3.0 to 4.25	—	—	0.10	—	0.02	—	0.02	0.50
CuNi	rem	—	0.40 to 0.75	1.00	29	0.50	—	0.02	0.15 to 1.00	—	0.50
CuSi	rem	0.01	0.5	1.5	—	2.8 to 4.0	1.5	0.02	—	—	0.50
CuSi-A	94	0.01	0.5	1.5	—	2.8 to 4.0	1.5	0.02	—	1.5	0.50
CuSn-A	rem	0.01	—	—	—	—	4.0 to 6.0	0.02	—	—	0.50
CuSn-C	rem	0.01	—	—	—	—	7.0 to 9.0	0.02	—	—	0.50
CuZn-A	57 to 61	0.01	—	—	—	—	0.25 to 1.0	0.05	—	rem	0.50
CuZn-B	56 to 60	0.01	0.25 to 1.2	0.01 to 0.50	0.2 to 0.8	0.04 to 0.15	0.8 to 1.1	0.05	—	rem	0.50
CuZn-C	56 to 60	0.01	0.25 to 1.2	0.01 to 0.50	—	0.04 to 0.15	0.8 to 1.1	0.05	—	rem	0.50
CuZn-D	46 to 50	0.01	—	—	9.0 to 11.0	0.04 to 0.25	—	0.05	—	rem	0.50

From AWS Specifications A5.6, A5.7, and A5.8.

AWS class filler metals may have the prefix letter E = Electrode, R = Rod, RB = Rod or Brazing, B = Brazing.

FIGURE 13-6 *Composition of copper alloy filler metals.*

Welding Copper

There are three basic groups in the C100 series of copper designations. The C10X is the oxygen-free type which has a copper analysis of 99.95% or higher. The second subgroup are the tough pitch coppers which have a copper composition of 99.88% or higher and some high copper alloys which have 96% or more copper. The ECu or RCu class filler metal is recommended.

The oxygen-free high-conductivity copper contains no oxygen and is not subjected to grain boundary migration. Adequate gas coverage should be employed to avoid oxygen of the air coming into contact with the molten metal. Welds should be made as quickly as possible since too much heat or slow welding can contribute to oxidation. The deoxidized coppers are preferred because of their freedom from embrittlement by hydrogen. Hydrogen embrittlement occurs when copper oxide is exposed to a reducing gas at high temperature. The hydrogen reduces the copper oxide to copper and water vapor. The entrapped high temperature water vapor or steam can create sufficient pressure to cause cracking. In common with all copper welding, preheat should be used and can run from 250–1,000 °F (121–538 °C), depending on the mass involved.

The tough pitch electrolytic copper is difficult to weld because of the presence of copper oxide within the material. During welding the copper oxide will migrate to the grain boundaries, at high temperatures, which reduces ductility and tensile strength. The gas-shielded processes are recommended since the welding area is more localized and the copper oxide is less able to migrate in appreciable quantities.

The third copper subgroup is the high-copper alloys which may contain deoxidizers such as phosphorus. The ECuSi filler wires are used with this material. The preheat temperatures needed to make the weld quickly, apply to all three grades.

Welding Copper-Zinc Alloys (Brasses)

These are the C2XX family of copper alloys. Within this group there are many different types of brasses. These alloys contain zinc, and zinc vaporization can be reduced by decreasing or eliminating preheat and by using lower welding currents. The various filler metals, copper-silicon, copper-tin and aluminum-bronze, can all be used.

For lighter sections argon shielding is used. For thicker sections preheat should be employed at approximately 400 °F (200 °C). Helium and helium gas mixtures are recommended. The weld joint should be opened up sufficiently to allow root penetration.

The Copper-Zinc-Lead Alloys (Leaded Brasses)

The leaded alloys in this C3XX group are not suitable for welding since the lead will create excessive porosity and promote cracking in the weld area.

Copper-Zinc-Tin Alloys (Tin Brasses)

This is the C4XX subgroup. These are the yellow brasses and are generally welded with the CuA1-A2 aluminum bronze filler metal. The same comments made concerning the copper-zinc alloys apply here.

Copper-Tin Alloys (Phosphur Bronzes)

This is the C5XX series which also includes the leaded phosphur bronzes. Except for the leaded bronzes, both the gas metal arc and gas tungsten arc processes can be used with the normal recommendations concerning thickness. These alloys do have a tendency to be hot short and they have high thermal conductivity. High current density and a high travel speed should be used. Helium is recommended for shielding with gas tungsten arc welding. The level of tin in the filler wire should be selected to match the tin in the base metal. Preheat should be used in the 300–400 °F (150–200 °C) range. When groove angles are used wide angles should be used. To reduce stresses and distortion, hot peening of the weld deposit is recommended. The CuSn filler metal should be used with the higher amount of tin to match the higher amount of tin in the base metals. The leaded phosphur bronzes should not be welded.

Copper-Aluminum Alloys (Aluminum Bronzes)

These are a subgroup in the C6XX class representing the lower numbers. Both the gas metal arc and gas tungsten arc processes are used with the gas metal arc used for the heavier thicknesses. Filler metal should be the CuA1-A2 type. Argon-helium mixtures are recommended. Preheat is required only for the heavier thicknesses. Full-penetration welds are recommended.

Copper-Silicon Alloys (Silicon Bronzes)

This is also a subgroup of the C6XX series. Both gas metal arc and gas tungsten arc welding can be used for this family of copper alloys. This alloy is free of volatile alloying elements and has a lower conductivity than many of the others. Preheat is recommended for heavier thicknesses. The leaded grade in this class is not suitable for welding.

These alloys are in the low C7XX class of alloys. Both gas metal arc and gas tungsten arc welding processes can be used for these alloys. The filler metal should be the CuNi 70/30 type. Argon is normally used but for heavy thicknesses argon and helium mixtures can be employed. Preheating is normally not used and the interpass temperatures should not be allowed to rise above 150°F (65°C).

Copper-Nickel-Zinc Alloys (Nickel-Silver)

These alloys are in the high C7XX class of alloys. These alloys are not normally welded and would be joined by brazing instead. This is because of the relatively high amount of zinc included in each of these compositions.

The analysis of cast alloys similar to wrought alloys would be welded the same way. The filler metal should be selected to most closely approximate the analysis of the base metal.

Figure 13-7 provides recommended welding conditions for both gas metal arc and gas tungsten arc welding of copper alloys. This is a summary of material just covered and is a starting point for establishing a welding procedure. Welding procedure schedules are provided for welding the different copper alloys with both gas tungsten arc welding and gas metal arc welding. These are shown by Figures 13-8 and 13-9.

| Material Type | Filler Metal | GTAW | | | GMAW | | Notes |
		Shielding Gas	Welding Current	Electrode Type	Electrode Class	Shielding Gas	
Copper* (E1xx)	RCu	Helium Argon or Mixture	DCEN AC-HF	EWTh-2	ECu	Argon plus helium	Preheat higher temp for thicker materials
Brasses* (C-2xx) (copper-zinc)	RCuZn-B RCuZn-C RCuZn-D	Argon Helium Mixture	DCEN AC-HF	EWTh-1	–	–	Preheat—open up joint—do not weld leaded types
Tin brasses* (C4xx)	RCuZn-A RCuZn-C	Argon Helium Mixtures	DCEN AC-HF	EWTh-1	ECuAl-A2	Argon plus Helium	Preheat—open up joint—higher temp for thicker materials.
Phosphor bronze (C5xx) (copper-tin)	RCuSn-A RCuSn-C RCuSn-D	Argon Helium Mixture	DCEN	EWTh-2	ECuSn-A ECuSn-C	Argon	Weld quickly hot short do not weld leaded types
Aluminum bronze (C61x, 62x and 63x) (copper-aluminum)	RCuAl-A2 RCuAl-B	Argon Helium Mixture	AC-HF DCEN	EWTh-1	ECuAl-A1 ECuAl-A2 ECuAl-B	Argon	Relatively easy to weld
Silicon bronze (C64x and 65x) (copper-aluminum)	RCuSi-A	Argon	DCEN	EWTh-1	ECuSi	Argon	Relatively easy to weld. Do not weld leaded types
Copper nickel (C7xx)	RCuNi	Argon	DCEN	EWTh-1	ECuNi	Argon	Relatively easy to weld

*Preheat.

FIGURE 13-7 *Recommended welding condition for GMAW and GTAW of copper alloys.*

Material Thickness (or Fillet Size)			Type of Weld	Tungsten Electrode Diameter		Filler Rod Diameter		Nozzle Size Inside Dia.	Shielding Gas Flow	Welding Current Amps	No. of Passes	Travel Speed (per pass)
ga	in.	mm	Fillet or Groove	in.	mm	in.	mm	in.	cfh	DCEN		ipm
16	0.063	1.6	Square Groove	1/16	1.6	1/16	1.6	3/8	20	100-150	1	10-12
16	0.063	1.6	Fillet	1/16	1.6	1/16	1.6	3/8	20	85-125	1	10-12
11	0.125	3.2	Square Groove	3/32	2.4	3/32	2.4	3/8	20	170-235	1	8-11
11	0.125	3.2	Fillet	3/32	2.4	3/32	2.4	3/8	20	115-165	1	10-12
3/16	0.187	4.7	Square Groove	1/8	3.2	1/8	3.2	5/8	35	185-255	1	8-12
3/16	0.187	4.7	Fillet	3/32	2.4	3/32	2.4	3/8	25	170-230	1	8-12
1/4	0.250	6.4	Fillet	1/8	3.2	1/8	3.2	1/2	40	220-275	1	7-10
1/4	0.250	6.4	Single Vee	1/8	3.2	1/8	3.2	1/2	40	220-275	2	7-10
1/4	0.250	6.4	Edge	1/8	3.2	1/8	3.2	1/2	25	160-225	1	7-10
1/4	0.250	6.4	Double Vee	1/8	3.2	1/8	3.2	1/2	20	180-220	3	8-12
3/8	0.375	9.5	Fillet	3/16	4.8	3/16	4.8	1/2	45	275-325	3	8-12
3/8	0.375	9.5	Single Vee	1/8	3.2	1/8	3.2	1/2	25	225-290	3	8-12
3/8	0.375	9.5	Double Vee	5/32	3.9	1/8	3.2	1/2	20	200-250	3	8-12
1/2	0.500	12.7	Fillet	1/4	6.4	1/4	6.4	5/8	45	370-500	4	8-12
1/2	0.500	12.7	Single Vee	1/8	3.2	1/8	3.2	1/2	30	280-330	7	7-10
1/2	0.500	12.7	Double Vee	5/32	3.9	1/8	3.2	1/2	30	180-250	4	7-10

1. Increase amperage 100% when backup is used.
2. Data is for flat position. Reduce amperage 10%-20% when welding in horizontal, vertical, or overhead position.
3. For tungsten electrodes: 1st choice 1% thoriated EWTh1; 2nd choice 2% thoriated EWTh2.
4. For copper use helium for shielding, however, a mixture of 75% He + 25% Argon is very popular on copper and somce copper alloys. Argon is usually for bronzes.
5. Preheat 3/16'' copper 200°F, 1/4'' −300°F, 3/8'' −500°F, Preheat '/4'' and up −900°F.
6. Deoxidized copper and copper alloys use DCEN−aluminum bronze uses ACHF and argon for shielding.

FIGURE 13-8 *Welding procedure schedules for GTAW of copper alloys.*

Material Thickness (or Fillet Size)			Type of Weld	Electrode Diameter		WELDING POWER		Wire Feed Speed	Shielding Gas Flow	No. of Passes	Travel Speed (per pass)
ga	in.	mm	Fillet or Groove	in.	mm	Current Amps DC	Arc Volt EP	ipm	cfh		ipm
16	0.063	1.6	Deoxidized copper	3/64	1.2	150-170	22-24	210-220	35	1	20-23
14	0.078	1.9	Sq. groove & fillet	3/64	1.2	180-200	22-25	240-270	40	1	20-25
12	0.109	2.8	Sq. groove & fillet	3/64	1.2	200-230	23-27	270-290	40	1	20-25
11	0.125	3.2	Sq. groove & fillet	3/64	1.2	210-240	23-27	280-300	40	1	20-25
1/4	0.250	6.4	Sq. groove & fillet	1/16	1.6	380-410	23-29	260-270	40	1	12-15
1/4	0.250	6.4	Vee groove & fillet	1/16	1.6	300-330	23-27	190-210	40	1-3	14-17
3/8	0.375	9.2	Vee groove & fillet	1/16	1.6	340-360	24-28	220-240	40	1	12-15
1/2	0.500	12.7	Double Vee groove	3/32	2.4	400-440	24-30	270-290	50	2	8-10
3/4	0.750	19.0	Double Vee groove	3/32	2.4	420-460	24-30	290-315	50	3	7-9
1	1.000	25.4	Double Vee groove	3/32	2.4	420-460	24-30	270-300	50	4	7-9
1/8	0.125	1.2	Silicon Bronze	3/64	1.2	130-160	25-28	220-230	35	1	25-32
1/4	0.250	6.4	Fillet & Vee groove	1/16	1.6	270-290	27-30	170-190	40	1-3	26-33
1/4	0.250	6.4	Fillet & Vee groove	1/16	1.6	450-465	25-28	220-250	50	1	30-34
1/2	0.500	12.7	Fillet & Vee groove	1/16	1.6	335-350	27-30	180-200	50	3-5	15-20
1/8	0.125	3.2	Aluminum bronze	3/64	1.2	190-225	22-25	280-300	40	1	18-24
1/4	0.250	6.4	Vee groove & fillet	1/16	1.6	275-300	23-29	170-190	50	2	16-22
3/8	0.375	9.2	Vee groove & fillet	1/16	1.6	300-340	23-29	190-210	50	3-6	16-22
1/2	0.500	12.7	Double Vee groove	1/16	1.6	320-350	23-29	200-220	50	6-8	11-15
5/8	0.625	15.9	Double Vee groove	1/16	1.6	320-345	23-29	220-240	50	6-8	9-13
3/4	0.750	19.0	Double Vee groove	1/16	1.6	340-370	23-29	220-240	50	6-8	9-12

1 - If preheating is required, a range of 500 to 900°F may be used for aluminum, bronze, and deoxidized copper and 400 to 600°F on silicon bronze.
2 - If porosity is encountered, it can be eliminated by adding an equal amount of helium to the argon flow.
3 - Speeds and currents for fully automatic welding are approximately 15% higher.

FIGURE 13-9 *Welding procedure schedules for GMAW of copper alloys.*

Many of the other welding processes can be used to join copper and copper alloys. Soldering is widely used for joining most of the copper alloys; however, the high aluminum content and aluminum-manganese bronzes are not readily soldered. Both corrosive and resin type fluxes are used for soldering copper. There is one precaution. Solders containing more than 1.0% antimony or more than 0.02% arsenic should not be used to solder the copper-zinc alloys. They will produce brittle joints or have poor bonding. The soldering process does not utilize sufficiently high heat to cause annealing of the copper base alloys.

Brazing is widely used for joining copper. The copper-phosphorus filler material (BCuP) and some of the silver alloy (BAg) types are used. The copper phosphorus is much less expensive but is not used for copper alloys that contain more than 10% nickel. In addition, the copper phosphorus alloy does not provide as high an electrical conductivity as the silver alloys.

Several of the other arc welding processes can be used. Plasma arc welding is becoming more popular for welding copper alloys. The same comments made concerning gas tungsten arc welding previously will apply. The submerged arc welding process has been used for copper alloy welding and overlaying. Specialized fluxes for copper alloys must be used.

The cold welding process is widely used on coppers; so are high-frequency welding and the electron beam welding process. Additional information for the weld-

ing of copper and copper alloys can be obtained from the various bulletins published by the Copper Development Association, Inc.

13-3 MAGNESIUM-BASE ALLOYS

Magnesium is the lightest structural metal. It is approximately two-thirds as heavy as aluminum and one-fourth as heavy as steel. Magnesium alloys containing small amounts of aluminum, manganese, zinc, zirconium, etc., have strengths equaling that of mild steels. They can be rolled into plate, shapes, and strip. Magnesium can be cast, forged, fabricated, and machined. As a structural metal it is used in aircraft. It is used by the materials-moving industry for parts of machinery and for hand-power tools due to its strength to weight ratio. Magnesium can be welded by many of the arc and resistance welding processes, as well as by the oxy-fuel gas welding process, and it can be brazed.

The more popular magnesium alloys are shown by the table of Figure 13-10. This chart shows the ASTM designations per ASTM B275, "Codification of Light Metals and Alloys." Magnesium like aluminum is produced with different tempers. These are based on heat treatment and work hardening. They are listed following the alloy classification and use the prefix

FIGURE 13-10 *Composition of magnesium alloys.*

Alloy	NOMINAL COMPOSITION—PERCENT						
	Aluminum	Manganese	Zinc	Zirconium	Rare earths	Thorium	Magnesium
sand and permanent mold castings							
AZ92A	9.0	0.15	2.0	—	—	—	Balance
AZ63A	6.9	0.25	3.0	—	—	—	Balance
AZ81A	7.6	0.13 min.	0.7	—	—	—	Balance
AZ91C	8.7	0.20	0.7	—	—	—	Balance
EK30A	—	—	—	0.35	3.0	—	Balance
EK41A	—	—	—	0.6	4.0	—	Balance
EZ33A	—	—	2.7	0.7	3.0	—	Balance
HK31A	—	—	—	0.7	—	3.0	Balance
HZ32A	—	—	2.1	0.7	—	3.0	Balance
die castings							
AZ91A							
AZ91B	9.0	0.20	0.6	—	—	—	Balance
extrusions							
AZ31B							
AZ31C	3.0	0.45	1.0	—	—	—	Balance
AZ61A	6.5	0.30	1.0	—	—	—	Balance
M1A	—	1.50	—	—	—	—	Balance
AZ80A	8.5	0.25	0.5	—	—	—	Balance
ZK60A	—	—	5.7	0.55	—	—	Balance
sheet and plate							
AZ31B	3.0	0.45	1.0	—	—	—	Balance
HK31A	—	—	—	0.7	—	3.0	Balance

Per ASTM B275 magnesium alloys (abridged).

AWS Classification	NOMINAL COMPOSITION %										
	Mg	Al	Be	Mn	Zn	Zi	Rare earth	Cu	Fe	Ni	Si
AZ61A	rem	5.8 to 7.2	0.0002 to 0.0008	0.15	0.40 to 1.5	—	—	0.05	0.005	0.005	0.05
AZ101A	rem	9.5	0.0002 to 0.008	0.13	0.75	—	—	0.05	0.005	0.005	0.05
AZ92A	rem	8.3 to 9.7	0.0002 to 0.0008	0.15	1.75	—	—	0.05	0.005	0.005	0.05
EZ33A	rem	—	—	—	2.0 to 3.1	0.45 to 1.0	2.5 to 4.0	—	—	—	—

Use suffix letter E = Electrode or R = Rod.

FIGURE 13-11 *Composition of magnesium filler metals per AWS A5.19.*

FIGURE 13-12 *Guide to the choice of filler metal for welding magnesium.*[6]

Filler Alloy \ Base Alloy	AM100A	AZ10A	AZ31B&C	AZ61A	AZ63A	AZ80A	AZ81A	AZ91C	AZ92A	EK41A	EZ33A
AM100A	1. AZ92A 2. AZ101										
AZ10A	AZ92A	1. AZ61A 2. AZ32A									
AZ31B&C	AZ92A	1. AZ61A 2. AZ92A	1. AZ61A 2. AZ92A								
AZ61A	AZ92A	1. AZ61A 2. AZ92A	1. AZ61A 2. AZ92A	1. AZ61A 2. AZ92A							
AZ63A	X	X	X	X	AZ92A						
AZ30A	AZ92A	1. AZ61A 2. AZ92A	1. AZ61A 2. AZ92A	1. AZ61A 2. AZ92A	X	1. AZ61A 2. AZ92A					
AZ81A	AZ92A	AZ92A	AZ92A	AZ92A	X	AZ92A	1. AZ92A 2. AZ101				
AZ91C	AZ92A	AZ92A	AZ92A	AZ92A	X	AZ92A	AZ92A	1. AZ92A 2. AZ101			
AZ92A	AZ92A	AZ92A	AZ92A	AZ92A	X	AZ92A	AZ92A	AZ92A	AZ101		
EK41A	AZ92A	AZ92A	AZ92A	AZ92A	X	AZ92A	AZ92A	AZ92A	AZ92A	EZ33A	
EZ33A	AZ92A	AZ92A	AZ92A	AZ92A	X	AZ92A	AZ92A	AZ92A	AZ92A	EZ33A	EZ33A
HK31A	AZ92A	AZ92A	AZ92A	AZ92A	X	AZ92A	AZ92A	AZ92A	AZ92A	EZ33A	EZ33A
HM21A	AZ92A	AZ92A	AZ92A	AZ92A	X	AZ92A	AZ92A	AZ92A	AZ92A	EZ33A	EZ33A
HM31A	AZ92A	AZ92A	AZ92A	AZ92A	X	AZ92A	AZ92A	AZ92A	AZ92A	EZ33A	EZ33A
HZ32A	AZ92A	AZ92A	AZ92A	AZ92A	X	AZ92A	AZ92A	AZ92A	AZ92A	EZ33A	EZ33A
K1A	AZ92A	AZ92A	AZ92A	AZ92A	X	AZ92A	AZ92A	AZ92A	AZ92A	EZ33A	EZ33A
LA141A	0	0	E233A	X	X	X	X	X	X	0	0
M1A MG1	AZ92A	1. AZ61A 2. AZ92A	1. AZ61A 2. AZ92A	1. AZ61A 2. AZ92A	X	1. AZ61A 2. AZ92A	AZ92A	AZ92A	AZ92A	AZ92A	AZ92A
QE22A	0	0	0	0	X	0	0	0	0	EZ33A	EZ33A
ZE10A	AZ92A	1. AZ61A 2. AZ92A	1. AZ61A 2. AZ92A	1. AZ61A 2. AZ92A	X	1. AZ61A 2. AZ92A	AZ92A	AZ92A	AZ92A	1. EZ33A 2. AZ92A	1. EZ33A 2. AZ92A
ZE41A	0	0	0	0	X	0	0	0	0	EZ33A	EZ33A
ZX21A	AZ92A	1. AZ61A 2. AZ92A	1. AZ61A 2. AZ92A	1. AZ61A 2. AZ92A	X	1. AZ61A 2. AZ92A	AZ92A	AZ92A	AZ92A	AZ92A	AZ92A
ZH62A ZK51A ZK60A ZK61A	X	X	X	X	X	X	X	X	X	X	X

letter T followed by a number ranging from 1 to 10, the higher numbers indicating the higher hardness. The letter F is also used indicating as fabricated. The letter H is used to indicate the heat treat condition. The strength of a weld joint is lowered in base metal, in the work-hardened condition, as a result of recrystalization and grain growth in the heat-affected zone. This effect is minimized with gas metal arc welding because of the higher welding speed utilized. This is not a factor in the base metals that are welded in the soft condition.

Welding Magnesium Alloys

Magnesium possesses properties that make welding it different than the welding of steels. Many of these are the same as for aluminum. These are:

1. Magnesium oxide surface coating.
2. High thermal conductivity.
3. Relatively high thermal expansion coefficient.
4. Relatively low melting temperature.

5. The absence of color change as temperature approaches the melting point.

The normal metallurgical factors that apply to other metals apply to magnesium as well.

Magnesium is a very active metal and the rate of oxidation increases as the temperature is increased. The melting point of magnesium is very close to that of aluminum, but the melting point of the oxide is very high. In view of this, the oxide coating must be removed.

Magnesium has high thermal heat conductivity and a high coefficient of thermal expansion. The thermal conductivity is not as high as aluminum but the coefficient of thermal expansion is very nearly the same. The absence of color change is not too important with respect to the arc welding processes.

FIGURE 13-12 *(Cont.)*

HK31A	HM21A	HM31A	HZ32A	K1A	LA141A	M1A MG1	QE22A	ZE10A	ZE41A	ZX21A	ZH62A ZK51A ZK60A ZK61A
EZ33A											
EZ33A	EZ33A										
EZ33A	EZ33A	EZ33A									
EZ33A	EZ33A	EZ33A	EZ33A								
EZ33A	EZ33A	EZ33A	EZ33A	EZ33A							
0	EZ33A	0	0	0	1. LA141A 2. EZ33A						
AZ92A	AZ92A	AZ92A	AZ92A	AZ92A	0	1. AZ61A 2. AZ92A					
EZ33A	EZ33A	EZ33A	EZ33A	EZ33A	EZ33A	0	EZ33A				
1. EZ33A 2. AZ92A	1. EZ33A 2. AZ92A	1. EZ33A 2. AZ92A	1. EZ33A 2. AZ92A	1. EZ33A 2. AZ92A	EZ33A	1. AZ61A 2. AZ92A	1. EZ33A 2. AZ92A	1. AZ61A 2. AZ92A			
EZ33A	EZ33A	EZ33A	EZ33A	EZ33A	0	0	EZ33A	0	EZ33A		
AZ92A	AZ92A	AZ92A	AZ92A	AZ92A	0	1. AZ61A 2. AZ92A	AZ92A	1. AZ61A 2. AZ92A	AZ92A	1. AZ61A 2. AZ92A	
X	X	X	X	X	X	X	X	X	X	X	EZ33A

The welds produced between similar alloys will develop the full strength of the base metals; however, the strength of the heat-affected zone may be reduced slightly. In all magnesium alloys the solidification range increases and the melting point and the thermal expansion decrease as the alloy content increases. Aluminum added as an alloy up to 10% improves weldability since it tends to refine the weld grain structure. Zinc of more than 1% increases hot shortness which can result in weld cracking. The high zinc alloys are not recommended for arc welding because of their cracking tendencies. Magnesium, containing small amounts of thorium, possesses excellent welding qualities and freedom from cracking. Weldments of these alloys do not require stress relieving.

Certain magnesium alloys are subject to stress corrosion. Weldments subjected to corrosive attack over a period of time may crack adjacent to welds if the residual stresses are not removed. For weldments intended for this type of service stress relieving is required.

The gas tungsten arc welding process and the gas metal arc welding process are the two recommended for joining magnesium. Gas tungsten arc is recommended for thinner materials and gas metal arc is recommended for thicker materials; however, there is considerable overlap. The equipment for applying these processes has been previously described.

The filler metal alloys used for joining magnesium are shown by Figure 13-11. It is based on AWS specification A5.19. The composition of filler metals should match the composition of the base materials; however, there are many cases in which this cannot be done. In all cases the recommendations shown by Figure 13-12 should be used.

Gas Tungsten Arc Welding

Welding procedure schedules for gas tungsten arc welding of magnesium are given by Figure 13-13. All the precautions mentioned for welding aluminum should be observed. A short arc should be used and the torch should have a slight leading travel angle. The cold wire filler metal should be brought in as near to horizontal as possible (on flat work). The filler wire is added to the leading edge of the weld puddle. High-frequency current should be used for starting the direct current arc and with alternating current high frequency should be used continuously. Runoff tabs are recommended for welding any except the thinner materials. Uniform travel speed and weld beads are recommended. The shielding gas is normally argon. However, a mixture of 75% helium plus 25% argon is used for thicker materials. For heavy thicknesses 100% helium can be used; more helium is required than argon to do the same job.

Direct current with electrode positive can be used but only for machine or automatic welding. In this case, materials must be perfectly clean prior to welding. Additional details are given by the welding procedure schedule.

FIGURE 13-13 *Welding procedure schedules for GTAW of magnesium.*

Material Thickness (or Fillet Size)			Type of Weld Fillet or Groove	Tungsten Electrode Diameter		Filler Rod Diameter		Nozzle Size Inside Dia.	Shielding Gas Flow	Welding Current Amps DCEN	No. of Passes	Travel Speed (per pass)
ga	in.	mm		in.	mm	in.	mm	in.	cfh			ipm
20	0.038	0.9	Square Groove	1/16	1.6	3/32	2.4	1/4	15	25-40	1	20
20	0.038	0.9	Fillet	1/16	1.6	3/32	2.4	1/4	15	30-45	1	20
16	0.063	1.6	Square Groove	1/16	1.6	3/32	2.4	1/4	15	45-60	1	20
16	0.063	1.6	Fillet	1/16	1.6	3/32	2.4	1/4	15	45-60	1	20
14	0.078	1.9	Square Groove	1/16	1.6	3/32	2.4	1/4	15	60-75	1	17
14	0.078	1.9	Fillet	1/16	1.6	3/32	2.4	1/4	15	60-75	1	17
12	0.109	2.8	Square Groove	3/32	2.4	1/8	3.2	5/16	15	80-100	1	17
12	0.109	2.8	Fillet	3/32	2.4	1/8	3.2	5/16	15	80-100	1	17
11	0.125	3.2	Square Groove	3/32	2.4	1/8	3.2	5/16	25	95-115	1	17
11	0.125	3.2	Fillet	3/32	2.4	1/8	3.2	5/16	25	95-115	1	17
3/16	0.187	4.7	Vee Groove	1/8	3.2	1/8	3.2	3/8	25	95-115	2	26
1/4	0.250	6.4	Vee Groove	1/8	3.2	3/16	4.8	1/2	25	110-130	2	24
3/8	0.375	9.5	Vee Groove	1/8	3.2	3/16	4.8	1/2	30	135-165	2	20

(1) Increase amperage when backup is used.
(2) Data is for flat position. Reduce amperage 10%-20% when welding in horizontal, verticla or overhead positions.
(3) Tungsten electrode: 1st choice zirconated EWZr; 2nd choice pure tungsten EWP.
(4) Select filler metal in accordance to selection chart.
(5) Shielding gas is normally argon. A mixture of 75% helium + 25% argon is used for heavier thickness. For heavy thickness 100% helium is used. Gas flow rates for helium are approximately twice those used for argon.

| Material Thickness (or Fillet Size) | | | Type of Weld | Electrode Diameter | | WELDING POWER | | Wire Feed Speed | Shielding Gas Flow | No. of Passes | Travel Speed (per pass) |
ga	in.	mm	Fillet or Groove	in.	mm	Current Amps DC	Arc Volt EP	ipm	cfh		ipm
0.025	–	–	Sq. groove & fillet	0.040	1.0	26-27	13-16	180	40-60	1	24-36
0.040	–	–	Sq. groove & fillet	0.040	1.0	35-50	13-16	250-340	40-60	1	24-36
0.063	1/16	1.6	Sq. groove & fillet	0.063	1.6	60-75	13-16	140-170	40-60	1	24-36
0.090	3/32	2.4	Sq. groove & fillet	0.063	1.6	95-125	13-16	210-280	40-60	1	24-36
0.125	1/8	3.2	Sq. groove & fillet	0.094	2.4	110-135	13-16	100-130	40-60	1	24-36
0.160	5/32	3.9	Sq. groove & fillet	0.094	2.4	135-140	13-16	130-140	40-60	1	24-36
0.190	3/16	4.8	Vee groove & fillet	0.094	2.4	175-205	13-16	160-190	40-60	2	24-36
0.250	1/4	6.4	Vee groove & fillet	0.063	1.6	240-290	24-30	550-660	50-80	2	24-36
0.375	3/8	9.5	Vee groove & fillet	0.094	2.4	320-350	24-30	350-385	50-80	2	24-36
0.500	1/2	12.7	Vee groove & fillet	0.094	2.4	350-420	24-30	385-415	50-80	2	24-36
1.00	1	25.4	Vee groove & fillet	0.094	2.4	350-420	24-30	385-415	50-80	4	24-36

Note: Values are for flat position

(1) For groove and fillet welds—material thickness also indicates fillet weld size. Use vee groove for 1/4″ and thicker.

(2) Shielding gas is argon or for heavier thicknesses use helium-argon mixtures.

(3) Above 200 amps and 20 volts, metal transfer is spray type—below 200 amps and 20 volts, metal transfer is short circuiting type.

FIGURE 13-14 *Welding procedure schedules for GMAW of magnesium.*

Gas Metal Arc Welding of Magnesium

The gas metal arc welding process is used for the medium to thicker sections. It is considerably faster than gas tungsten arc welding. Special high-speed gear ratios are usually required in the wire feeders since the magnesium electrode wire has an extremely high meltoff rate. The normal wire feeder and power supply used for aluminum welding will be suitable for welding magnesium. The different types of arc transfer can be obtained when welding magnesium. This is primarily a matter of current level or current density and voltage setting. The short-circuiting transfer and the spray transfer are recommended. Argon is usually used for gas metal arc welding of aluminum; however, argon-helium mixtures can be used. In general, the spray transfer should be used on material 3/16 inch and thicker and the short-circuiting arc used for thinner metals. Figure 13-14 provides welding procedure schedules.

Welding Problems

Magnesium is usually delivered with an oil preservative. The oil must be removed with a solvent and the material should be cleaned either by mechanical or chemical methods. Scraping and brushing is often used and a stainless steel brush should be used.

If cracking persists in magnesium welds, check the welding technique. Craters must always be filled and runoff tabs should be used. Preheating is recommended for complex weldments. Preheating in the 200–400°F (93–204°C) should be employed.

Stress relieving is recommended when the weldment is exposed to corrosion.

Other Welding Processes

The resistance welding processes can be used for welding magnesium, including spot welding, seam welding, and flash welding. Magnesium can also be joined by brazing. Most of the different brazing techniques can be used. In all cases, brazing flux is required and the flux residue must be completely removed from the finished part. Soldering is not too popular since the strength of the joint is relatively low.

Magnesium can be stud welded, gas welded, and plasma welded. Finely divided pieces of magnesium such as shavings, filings, etc., should not be in the welding area since they will burn. Magnesium castings, or wrought materials do not create a safety hazard since the possibility of fire caused by welding on these sections is very remote. The producers of magnesium provide additional data for welding magnesium.[6]

13-4 NICKEL-BASE ALLOYS

Nickel and the high-nickel alloys are commonly used when corrosion resistance is required. They are used in the chemical industry and the food industry. Nickel and nickel alloys are also widely used as filler metals for joining dissimilar materials and cast iron.

There is no standard specification or designation system for the nickel alloys. In general, they are identified by trademark names and suffix numbers. The trademark names are:

Monel, which is a nickel-copper alloy.

Inconel, which is a high nickel-chromium alloy with iron.

451

WELDING THE NONFERROUS METALS

Incoloy, which is a nickel-iron-chromium alloy plus others.

There are other nickel alloys such as Hastealloy, which is a nickel-molybdenum-iron alloy, and others. In view of this, the trademark names of the International Nickel Company or the Huntington Alloy Products Division will be used. The more common nickel alloys are shown by Figure 13-15.

When welding, the nickel alloys can be treated much in the same manner as austenitic stainless steels with a few exceptions. These exceptions are:

1. The nickel alloys will acquire a surface oxide coating which melts at a temperature approximately 1,000°F (538°C) above the melting point of the base metal.

2. The nickel alloys are susceptible to embrittlement at welding temperatures by lead, sulphur, phosphorus, and some low-temperature metals and alloys.

3. Weld penetration is less than expected with other metals.

When compensation is made for these three factors the welding procedures used for the nickel alloys can be the same as those used for stainless steel. This is because the melting point, the coefficient of thermal expansion, and the thermal conductivity are similar to austenitic stainless steel.

Alloy Designation	Ni	C	Mn	Fe	S	Si	Cu	Cr	Al	Ti	Cb[c]	Others
Nickel 200	99.5[b]	0.08	0.18	0.2	0.005	0.18	0.13	—	—	—	—	—
Nickel 201	99.5[b]	0.01	0.18	0.2	0.005	0.18	0.13	—	—	—	—	—
Nickel 205	99.5[b]	0.08	0.18	0.10	0.004	0.08	0.08	—	—	0.03	—	Mg 0.05
Nickel 211	95.0[b]	0.10	4.75	0.38	0.008	0.08	0.13	—	—	—	—	—
Nickel 220	99.5[b]	0.04	0.10	0.05	0.004	0.03	0.05	—	—	0.03	—	Mg 0.05
Nickel 230	99.5[b]	0.05	0.08	0.05	0.004	0.02	0.05	—	—	0.003	—	Mg 0.06
Nickel 270	99.98	0.01	<0.001	0.003	<0.001	<0.001	<0.001	<0.001	—	<0.001	—	Mg <0.001, Co <0.001
Duranickel alloy 301	96.5[b]	0.15	0.25	0.30	0.005	0.5	0.13	—	4.38	0.63	—	—
Permanickel alloy 300	98.5[b]	0.20	0.25	0.30	0.005	0.18	0.13	—	—	0.40	—	Mg 0.35
Monel alloy 400	66.5[b]	0.15	1.0	1.25	0.012	0.25	31.5	—	—	—	—	—
Monel alloy 401	42.5[b]	0.05	1.6	0.38	0.008	0.13	Bal.	—	—	—	—	—
Monel alloy 404	54.5[b]	0.08	0.05	0.25	0.012	0.05	44.0	—	0.03	—	—	—
Monel alloy R-405	66.5[b]	0.15	1.0	1.25	0.043	0.25	31.5	—	—	—	—	—
Monel alloy K-500	66.5[b]	0.13	0.75	1.00	0.005	0.25	29.5	—	2.73	0.60	—	—
Monel alloy 502	66.5[b]	0.05	0.75	1.00	0.005	0.25	28.0	—	3.00	0.25	—	—
Inconel alloy 600	76.0[b]	0.08	0.5	8.0	0.008	0.25	0.25	15.5	—	—	—	—
Inconel alloy 601	60.5	0.05	0.5	14.1	0.007	0.25	0.50	23.0	1.35	—	—	—
Inconel alloy 617	54.0	0.07	—	—	—	—	—	22.0	1.0	—	—	Co 12.5, Mo 9.0
Inconel alloy 625	61.0[b]	0.05	0.25	2.5	0.008	0.25	—	21.5	0.2	0.2	3.65	Mo 9.0
Inconel alloy 671	Bal.	0.05	—	—	—	—	—	48.0	—	0.35	—	—
Inconel alloy 702	79.5[b]	0.05	0.50	1.0	0.005	0.35	0.25	15.5	3.25	0.63	—	—
Inconel alloy 706	41.5	0.03	0.18	40.0	0.008	0.18	0.15	16.0	0.20	1.75	2.9	—
Inconel alloy 718	52.5	0.04	0.18	18.5	0.008	0.18	0.15	19.0	0.50	0.90	5.13	Mo 3.05
Inconel alloy 721	71.0[b]	0.04	2.25	6.5	0.005	0.08	0.10	16.0	—	3.05	—	—
Inconel alloy 722	75.0[b]	0.04	0.50	7.0	0.005	0.35	0.25	15.5	0.70	2.38	—	—
Inconel alloy X-750	73.0[b]	0.04	0.50	7.0	0.005	0.25	0.25	15.5	0.70	2.50	0.95	—
Inconel alloy 751	72.5[b]	0.05	0.5	7.0	0.005	0.25	0.25	15.5	1.20	2.30	0.95	—
Incoloy alloy 800	32.5	0.05	0.75	46.0	0.008	0.50	0.38	21.0	0.38	0.38	—	—
Incoloy alloy 801	32.0	0.05	0.75	44.5	0.008	0.50	0.25	20.5	—	1.13	—	—
Incoloy alloy 802	32.5	0.35	0.75	46.0	0.008	0.38	—	21.0	0.58	0.75	—	—
Incoloy alloy 804	41.0	0.05	0.75	25.4	0.008	0.38	0.25	29.5	0.30	0.60	—	—
Incoloy alloy 825	42.0	0.03	0.50	30.0	0.015	0.25	2.25	21.5	0.10	0.90	—	Mo 3.0
Ni-span-C alloy 902	42.25	0.03	0.40	48.5	0.02	0.50	0.05	5.33	0.55	2.58	—	—

[b]Not for specification purposes.

[c]Cobalt included.

Plus tantalum.

FIGURE 13-15 *Composition of nickels and nickel alloys.*

AWS Class	Trade Name	NOMINAL COMPOSITION %											
		Ni	C	Mn	Fe	S	Si	Cu	Cr	Al	Ti	Cb	Mo
ERNi-3	Nickel 61	96.0	0.06	0.30	0.10	0.005	0.40	0.02	–	–	3.0	–	–
ENi-1	Nickel 141	96.0	0.03	0.30	0.05	0.005	0.60	0.03	–	0.25	2.5	–	–
ERNiCu-7	Monel 60	65.0	0.03	3.5	0.20	0.005	1.00	27.0	–	–	2.2	–	–
ERNiCu-8	Monel 64	65.0	0.15	0.60	1.00	0.005	0.15	30.0	–	2.8	0.5	–	–
ERCuNi	Monel 67	31.0	0.02	0.75	0.50	0.005	0.10	67.5	–	–	0.30	–	–
ENiCuAl-1	Monel 134	64.0	0.20	2.50	1.00	0.005	0.30	30.0	–	1.8	0.75	–	–
ECuNi	Monel 187	32.0	0.02	2.00	0.60	0.01	0.15	65.0	–	–	–	–	–
ENiCu-2	Monel 190	65.0	0.01	3.10	0.30	0.007	0.75	30.5	–	0.15	0.55	–	–
ERNiCrFe-5	Inconel 62	74.0	0.02	0.10	7.50	0.005	0.10	0.03	16.0	–	–	2.25	–
ERNiCrFe-7	Inconel 69	73.0	0.04	0.55	6.50	0.007	0.30	0.05	15.2	0.70	2.5	0.85	–
ERNiCr-3	Inconel 82	72.0	0.02	3.00	1.00	0.007	0.20	0.04	20.0	–	0.55	2.55	–
ERNiCrFe-6	Inconel 92	71.0	0.03	2.30	6.60	0.007	0.10	0.04	16.4	–	3.2	–	–
–	Inconel 601	60.5	0.05	0.50	14.1	0.007	0.25	0.50	23.0	1.35	–	–	–
–	Inconel 625	61.0	0.05	0.25	2.5	0.008	0.25	–	21.5	0.2	0.2	3.65	9.0
–	Inconel 718	52.5	0.04	0.20	18.5	0.007	0.30	0.07	18.6	0.40	0.90	5.0	3.1
–	Inconel 112	61.0	0.05	0.3	4.0	0.010	0.40	–	21.5	–	–	3.6	9.0
ENiCrFe-1	Inconel 132	73.0	0.04	0.75	8.5	0.006	0.20	0.04	15.0	–	–	2.1	–
ENiCrFe-3	Inconel 182	67.0	0.05	7.75	7.5	0.008	0.50	0.10	14.0	–	0.40	1.75	–
ENiCrFe-2	Inco-weld A	70.0	0.03	2.0	9.0	0.008	0.30	0.06	15.0	–	–	2.0	1.5
–	Inco-weld B	70.0	0.13	2.0	9.0	0.008	0.30	0.06	15.0	–	–	2.5	2.0
–	Incoloy 65	42.0	0.03	0.70	30.0	0.007	0.30	1.70	21.0	–	1.0	–	3.0
–	Incoloy 135	36.0	0.05	2.00	26.0	0.008	0.40	1.80	29.0	–	–	–	3.75
ENi-C1	Ni-rod	95.0	1.00	0.20	3.0	0.005	0.70	0.10	–	–	–	–	–
ENiFe-C1	Ni-rod 55	53.0	1.50	0.30	45.0	0.005	0.50	0.10	–	–	–	–	–

Prefix E = Electrode, R = Rod

From AWS specification A5.6, A5.7, A5.11, A5.14 and A5.15.

FIGURE 13-16 *Composition of nickel alloy filler metals.*

It is necessary that each of these precautions be considered. The surface oxide should be completely removed from the joint area by grinding, abrasive blasting, machining, or by chemical means. When chemical etches are used they must be completely removed by rinsing prior to welding. The oxide which melts at temperatures above the melting point of the base metal may enter the weld as a foreign material, or impurity, and will greatly reduce the strength and ductility of the weld.

The problem of embrittlement at welding temperatures also means that the weld surface must be absolutely clean. Paints, marking crayons, grease, oil, machining lubricants, and cutting oils may all contain the ingredients which will cause embrittlement. They must be completely removed from the weld area to avoid embrittlement.

Finally, with respect to the minimum penetration, it is necessary to increase the opening of groove angles and to provide adequate root openings when full-penetration welds are used. The bevel or groove angles should be increased to approximately 40% over those used for carbon steel.

Almost all the welding processes can be used for welding the nickel alloys. In addition, they can be joined by brazing and soldering. The filler metals to be used for joining nickel alloys are shown in Figure 13-16. These are based on AWS filler metal specifications A5.11, A5.14, and A5.15 and include covered electrodes as well as bare solid wire for gas metal arc welding or for cold wire applications with other processes. The recommended filler metals for joining different alloys are shown by Figure 13-17.

Welding Nickel Alloys

The most popular processes for welding nickel alloys are the shielded metal arc welding process, the gas tungsten arc welding process, and the gas metal arc welding process. Process selection depends on the normal factors. When shielded metal arc welding is used the procedures are essentially the same as those used for stainless steel welding.

The welding procedure schedule for using gas tungsten arc welding is shown by Figure 13-18. The welding procedure schedule for gas metal arc welding is shown by Figure 13-19. The procedure information set forth

Alloy Designation	ELECTRODE Inco Name	AWS—Spec	ROD Inco Name	AWS—Spec
Nickel 200	141	ENi-1	61	ERNi-3
Nickel 201	141	ENi-1	61	ERNi-3
Monel alloy 400	190	ENiCu-2	60	ERNiCu-7
Monel alloy 404	190	ENiCu-2	60	ERNiCu-7
Monel alloy K-500	190*	ENiCu-2	60*	ERNiCu-7
Monel alloy 502	190*	ENiCu-2	60*	ERNiCu-7
Inconel alloy 600	132 (182)	ENiCrFe-1	62 (82)	ERNiCrFe-5
Inconel alloy 601	132 (182)	ENiCrFe-1	601 (82)	ERNiCrFe-5
Inconel alloy 625	112	—	625	—
Inconel alloy 706	N.A.	—	718	—
Inconel alloy 718	N.A.	—	718	—
Inconel alloy 722	N.A.	—	69	ERNiCrFe-7
Inconel alloy X-750	N.A.	—	69	ERNiCrFe-7
Incoloy alloy 800	182 (A)	ENiCrFe-3	82 (625)	ERNiCr-3
Incoloy alloy 801	182 (A)	ENiCrFe-3	82 (625)	ERNiCr-3
Incoloy alloy 825	135 (112)	—	65 (625)	—

Note: Electrodes are covered electrode for SMAW. Rods are solid bare for GTAW, GMAW, and SAW.

*will not age harden

N.A.—not available—use GTAW only

FIGURE 13-17 *Guide for selecting filler metal for welding nickel alloys.*

Material Thickness (or Fillet Size) ga	in.	mm	Type of Weld Fillet or Groove	Tungsten Electrode Diameter in.	mm	Filler Rod Diameter in.	mm	Nozzle Size Inside Dia. in.	Shielding Gas Flow cfh	Welding Current Amps DCEN	No. of Passes	Travel Speed (per pass) ipm
24	0.024	0.6	Sq. groove & fillet	1/16	1.6	None		3/8	15	8-10	1	8
16	0.063	1.6	Sq. groove & fillet	3/32	2.4	1/16	1.6	1/2	18	25-45	1	8
1/8	0.125	3.2	Sq. groove & fillet	1/8	3.2	3/32	2.4	1/2	25	125-175	1	11
1/4	0.25	6.4	Vee groove & fillet	1/8	3.2	1/8	3.2	1/2	30	125-175	2	8

1 - Tungsten used, 1st choice 2% thoriated EWTh2—2nd choice 1% thoriated EWTh1.

2 - Adequate gas shielding is a must not only for the arc but also heated metal. Backing gas is recommended at all times. A trailing gas shield is also recommended. Argon is preferred but for higher heat input on thicker material use Argon-Helium mixture.

3 - Data is for flat position. Reduce amperage 10% to 20% when welding is horizontal, vertical, or overhead position.

FIGURE 13-18 *Welding procedure schedules for GTAW of nickel alloys.*

Material Thickness (or Fillet Size) ga	in.	mm	Type of Weld Fillet or Groove	Electrode Diameter in.	mm	WELDING POWER Current Amps DC	Arc Volt EP	Wire Feed Speed ipm	Shielding Gas Flow cfh	No. of Passes	Travel Speed (per pass) ipm
1/16	0.062	1.6	Sq. groove & fillet	3/64	1.2	200-250	23-27	200-250	50	1	55-65
1/8	0.125	3.2	Sq. groove & fillet	1/16	1.6	290-340	25-35	150-175	60	1	30-35
1/4	0.250	6.4	Double Vee & fillet	1/16	1.6	300-350	28-38	170-200	80	3	20-25

1 - Use 50% helium and 50% argon for thin metal and 100% helium for thick—higher voltage is for helium.

2 - Increase amperage 10-20% when backup is used.

Data is for flat position. Reduce current 10-20% for other positions.

FIGURE 13-19 *Welding procedure schedules for GMAW of nickel alloys.*

No postweld heat treatment is required to maintain or restore corrosion resistance of the nickel alloys. Heat treatment is required for precipitating hardening alloys and stress relief may be required to meet certain specifications to avoid stress corrosion cracking in applications involving hydrofluoric acid vapors or caustic solutions. Additional welding information is available from the producers of the base materials.[7]

13-5 REACTIVE AND REFRACTORY METALS

The reactive and refractory metals were originally used in the aerospace industry and are now being welded for more and more requirements. These metals share many common welding problems and are, therefore, grouped together in this section. Reactive metals have a strong affinity for oxygen and nitrogen at elevated temperatures and when combined form very stable compounds. At lower temperatures they are highly resistant to corrosion. Refractory metals have extremely high melting points. They may also exhibit some of the same characteristics of reactive metals.

The reactive metals are:

Zirconium

Titanium

Beryllium

The refractory metals are:

Tungsten

Molybdenum

Tantalum

Columbium (Niobium)

A summary of some of the physical properties of these metals is shown by Figure 13-20. Note that the refractory metals all have extremely high melting points, relatively high density, and thermal conductivity. The reactive metals have lower melting points, lower densities, and, except for zirconium, have higher coefficient of thermal expansion.

The reactive metals are becoming increasingly important because of their use in nuclear and space

technology. They are considered in the difficult-to-weld category. These metals have a high affinity for oxygen and other gases at elevated temperatures and for this reason cannot be welded with any process that utilizes fluxes, or where heated metal is exposed to the atmosphere. Minor amounts of impurities cause these metals to become brittle.

Most of these metals have the characteristic known as the *ductile-brittle* transition. This refers to a temperature at which the metal breaks in a brittle manner rather than in a ductile fashion. The recrystalization of the metal during welding can raise the transition temperature. Contamination during the high temperature period and impurities can raise the transition temperature so that the material is brittle at room temperatures. If contamination occurs so that transition temperature is raised sufficiently it will make the weldment worthless. Gas contamination can occur at temperatures below the melting point of the metal. These temperatures range from 700°F (371°C) up to 1,000°F (538°C).

At room temperature the reactive metals have an impervious oxide coating that resists further reaction with air. The oxide coatings melt at temperatures considerably higher than the melting point of the base metal and create problems. The oxidized coating may enter molten weld metal and create discontinuities which greatly reduce the strength and ductility of the weld. Of the three reactive metals, titanium is the most popular and is routinely welded with special precautions.

All of the refractory metals incur internal contamination or surface errosion when exposed to the air at elevated temperatures. Molybdenum has an extremely high rate of oxidation at high temperatures above 1,500°F (816°C). Tungsten is much the same. Tantalum and columbium form pentoxides that are not volatile

Element	Crystal Structure	Melting Point F°	Melting Point C°	Density lb/ft^3	Density gr/cc	Thermal Conductivity cal/cm^2/cm/sec/°C	Thermal Expansion micro-in./in./°F
W-Tungsten	BCC	6170	3410	1190	19.3	0.397	2.55
Ta-Tantalum	BCC	5425	2996	1035	16.6	0.130	3.6
Mo-Molybdenum	BCC	4730	2610	650	9.0	0.34	2.7
Cb-Columbium	BCC	4474	2567	524	8.4	0.125	4.06
Zr-Zirconium	HCP	3366	1796	*402	6.4	—	3.2
Be-Beryllium	HCP	2332	1377	*114	1.8	0.35	6.4
Ti-Titanium	HCP	3035	1668	281	4.5	—	4.67

FIGURE 13-20 *Physical properties of the refractory and reactive metals.*

WELDING THE NONFERROUS METALS

below 25°F (−3.9°C), but these provide little protection because they are nonadherent. Molybdenum and tungsten both become embrittled when a minute amount of oxygen or nitrogen is absorbed. Columbium and tantalum can withstand larger amounts of oxygen and nitrogen.

Titanium can withstand much more oxygen or nitrogen before becoming embrittled; however, small amounts of hydrogen will cause embrittlement. Zirconium can withstand about as much oxygen but much less nitrogen or hydrogen. Beryllium is similar to zirconium in this regard.

Welding Refractory Metals

It is obvious that these metals must be perfectly clean prior to welding and that they must be welded in such a manner that air does not come into contact with the heated material. Cleaning is usually done with chemicals. A water rinse is necessary to remove all traces of chemicals from the surface. After the parts are cleaned they must be protected from reoxidation. This is best done by storing in an inert gas chamber or in a vacuum chamber.

Molybdenum is welded by the gas tungsten arc welding process and the electron beam process. The gas metal arc process can be used but sufficient thickness of molybdenum is rarely available to justify this process. It has been welded by other arc processes but results are not too satisfactory. Welding with the gas shielding processes is accomplished in an inert gas chamber or dry box. This is a chamber that can be evacuated and purged with inert gas until all active gases are removed. A typical dry box welding chamber is shown by Figure 13-21. Welding is done in the pure inert atmosphere with normally good results. The filler metal compositions should be the same as the base metal. The base metal in the heat-affected zone becomes embrittled by grain growth and recrystallization as a result of the welding temperatures. Recrystallization raises the transition temperature so that molybdenum welds tend to be brittle. Molybdenum is highly notch sensitive, craters and notch effects such as undercutting must be avoided. Molybdenum can also be welded with the resistance welding processes and by diffusion welding.

Tungsten is welded in the same manner as molybdenum and has the same problems, only more intensely so. It has greater susceptibility to cracking because the ductile-to-brittle transition temperatures are higher.

FIGURE 13-21 *Welding in a dry box.*

The preparation of tungsten for welding is more difficult. The gas tungsten arc welding process is used with direct current electrode negative. Welding should be done slowly to avoid cracking. Preheating may assist in reducing cracking but must be done in the inert gas atmosphere.

Commercially pure tantalum is soft and ductile and does not seem to have a ductile-brittle transition. There are several alloys of tantalum commercially available. Even though the material is easier to weld, it should be well cleaned and for best results should be welded in the inert gas chamber. The gas tungsten arc welding process is recommended. Some tantalum products are produced by powder metallurgy technology and this may result in porosity in the weld. The arc cast product does not have porosity. Filler wire is normally not used when welding tantalum and for best results direct current electrode negative is used. High frequency should be used for initiating the arc. Helium is recommended for welding tantalum to provide for maximum penetration since joints are designed to avoid using filler metal.

There are several different alloys of columbium (niobium) available. Some are ductile and others brittle since the transition temperature is near room temperature. The gas tungsten arc welding process is used for the pure columbium and for the lower strength commercial alloys. In certain alloys the welding can be done outside of an inert gas chamber but special precautions should be taken to provide extremely good inert gas shielding coverage. In certain of the alloys preheating is recommended to provide for a crack-free weld. Electron beam welding is used and columbium can also be resistance welded.

Reactive Metals

Beryllium has been welded with the gas tungsten arc welding process and with the gas metal arc welding process. It is also joined by brazing. Beryllium should not be welded without expert technical assistance. Beryllium is a toxic metal and extra special precautions should be provided for proper ventilation and handling.

Zirconium and zirconium-tin alloys are ductile metals and can be prepared by conventional processes. Cleaning is extremely important and chemical cleaning is preferred over mechanical cleaning. Both the gas tungsten arc welding and the gas metal arc welding processes are used for joining zirconium. The inert gas chamber should be employed to maintain an efficient gas shield. Argon or argon-helium mixtures are used. The zircalloys are alloys of zirconium which contain small amounts of tin, iron, and chromium. These alloys can be welded in the open in much the same manner as titanium. The electron beam process and the resistance welding processes have been used for joining zirconium.

The secret to the successful welding of titanium is *cleanliness*. Small amounts of contamination can render a titanium weld completely brittle. Contamination from grease, oils, paint, fingerprints, or dirt, etc., can have the same effect. If the material is cleaned thoroughly before welding and well protected during welding there is little difficulty in the welding of titanium.

The gas tungsten-arc and gas metal-arc welding processes can be used for welding titanium. Special procedures must be employed when using the gas-shielded welding processes. These special procedures include the use of large gas nozzles and trailing shields to shield the face of the weld from air. Backing bars that provide inert gas to shield the back of the welds from air are also used. Not only the molten weld metal, but the material heated above 1,000°F by the weld must be adequately shielded in order to prevent embrittlement.

When using the GTAW process a thoriated tungsten electrode should be used. The electrode size should be the smallest diameter that will carry the welding current. The electrode should be ground to a point. The electrode may extend 1-1/2 times its diameter beyond the end of the nozzle. Welding is done with direct current, electrode negative (straight polarity).

Selection of the filler metal will depend upon the titanium alloys being joined. When welding pure titanium, a pure titanium wire should be used. When welding a titanium alloy, the next lowest strength alloy should be employed as a filler wire. Due to the dilution which will take place during welding the weld deposit will pick up the required strength. The same considerations are true when GMAW welding of titanium.

Argon is normally used with the gas-shielded process. For thicker metal use helium or a mixture of argon and helium. The purity of welding grade gases is satisfactory. They should have a dew point of minus 65°F. Welding grade shielding gases are generally free from contamination; however, tests can be made before welding. A simple test is to make a bead on a piece of scrap, clean titanium, and notice its color. The bead should be shiny. Any discoloration of the surface indicates a contamination.

Extra gas shielding provides protection for the heated solid metal next to the weld metal. This shielding is provided by special trailing gas nozzles shown by Figure 13-22 or by chill bars laid immediately next to the weld. Backup gas shielding should be provided to protect the underside of the weld joint. Protection of the back side of the joint can also be provided by placing chill bars in intimate contact with the backing strips. If the contact is close enough, backup shielding gas is not required. For critical applications use an inert gas welding chamber.

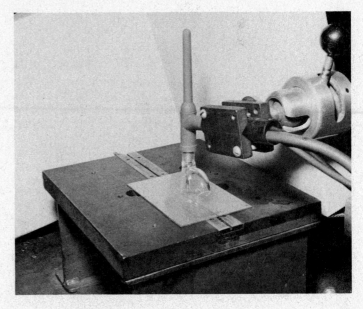

FIGURE 13-22 *Trailing shield on welding torch.*

Material Thickness (or Fillet Size)			Type of Weld Fillet or Groove	Tungsten Electrode Diameter		Filler Rod Diameter		Nozzle Size Inside Dia.	Shielding Gas Flow	Welding Current Amps	No. of Passes	Travel Speed (per pass)
ga	in.	mm		in.	mm	in.	mm	in.	cfh	DCEN		ipm
24	0.024	0.6	Sq. groove & fillet	1/16	1.6	None		3/8	18	20-35	1	6
16	0.063	1.6	Sq. groove & fillet	1/16	1.6	None		5/8	18	85-140	1	6
3/32	0.093	1.6	Sq. groove & fillet	3/32	2.4	1/16	1.6	5/8	25	170-215	1	8
1/8	0.125	3.2	Sq. groove & fillet	3/32	2/4	1/16	1.6	5/8	25	190-235	1	8
3/16	0.188	4.8	Sq. groove & fillet	3/32	2.4	1/8	3.2	5/8	25	220-280	2	8
1/4	0.25	6.4	Vee groove & fillet	1/8	3.2	1/8	3.2	5/8	30	275-320	2	8
3/8	0.375	9.5	Vee groove & fillet	1/8	3.2	1/8	3.2	3/4	35	300-350	2	6
1/2	0.50	12.7	Vee groove & fillet	1/8	3.2	5/32	3.9	3/4	40	325-425	3	6

1 - Tungsten used, 1st choice 2% thoriated EWTh2—2nd choice 1% thoriated EWTh1.

2 - Use filler metal one or two grades lower in strength than the base metal.

3 - Adequate gas shielding is a must not only for the arc but also heated metal. Backing gas is recommended at all times. A trailing gas shield is also recommended. Argon is preferred for higher heat input on thicker material use Argon-Helium mixture.

4 - Without backup or chill bar, decrease current 20%.

FIGURE 13-23 *Welding procedure schedule for GTAW of titanium.*

These can either be flexible, rigid, or vacuum-purge chambers.

To guarantee that embrittlement of the weld will not occur, proper cleaning steps must be taken. Solvents containing chlorine should not be used. Recommended solvents would be tri alcohol or acetone. Titanium can be ground with discs of aluminum oxide or silicon carbide. Wet grinding is preferred, however, if wet grinding cannot be used the grinding should be done slowly to avoid overheating the surface of the titanium.

Figure 13-23 gives procedure schedules for welding titanium. Joint types that are satisfactory for stainless steel should be used.

13-6 THE OTHER NONFERROUS METALS

This section covers the welding of the low melting point metals, lead and zinc, and the precious metals, silver, gold, and platinum. Lead has one of the lowest melting temperatures. It melts at 621°F (328°C), and boils at 3,092°F (1,700°C). It is very soft and very ductile. When freshly cut its surface has a bright silvery luster which almost immediately oxidizes to a dull grey. It is available in sheet form and is used as a liner for tanks because of its corrosion resistance, particularly to sulphuric acid. It is also available in pipe since it is widely used in the chemical industry.

Lead is normally joined by the oxyacetylene or oxy-fuel gas torch and for this reason has erroneously been called lead burning. Lead can also be welded with the gas tungsten arc welding process. Since lead is usually in thin sheets the square butt and lap joints are the most commonly used. The surface of the lead at the welding area should be cleaned. Filler metal can be obtained by shearing strips of the base metal.

Sufficient heat is obtained to melt the surface of the base metal and the filler rod when used. When using gas tungsten arc a rather long arc is recommended to reduce the actual heat in the joint. There are a few difficulties that can be encountered when welding lead. Its popularity and use is declining and for this reason the joining of lead is of minor importance.

Zinc is much similar to lead except that it is not as heavy. Zinc is widely used as a coating on steel which is called galvanized steel. Zinc also has a tough oxide coating which must be removed prior to welding. The major use of zinc where welding is involved is as high-zinc die castings. Die castings of this type are commonly used for automotive grills and for decorative trim on automobiles. In addition, the carburetor bodies of most automobiles are made of the zinc die castings. For decorative trim and grills, the zinc may be chromium plated. It is important to distinguish zinc die castings from aluminum and magnesium alloy castings. Zinc has a weight in the same order as steel. In addition, the zinc has a light grey appearance which is rather easy to identify.

The oxyacetylene or oxyfuel gas torch method is used to weld zinc alloys, particularly the die castings. The technique is essentially the same as for lead. The filler metal for matching zinc die castings is difficult to obtain and for this reason it may be necessary to manufacture it. One easy way is to obtain scrap carburetors and melt down the bodies of them and pour into a groove formed when an angle iron is positioned with the point down. This will make a mold and the zinc alloy can be poured into it to produce a rod of filler metal. These alloys contain approximately 10% aluminum, 1 to 2-1/2% copper, a trace of magnesium, with the remainder zinc.

For welding die castings the joint must be prepared

with an extra wide root angle approaching 90 degrees. The part should be positioned and braced in position. If the part is chrome plated the welding should be done from the nonplated side. If not, the plating should be removed adjacent to the welding area. The surfaces to be welded should be cleaned by wire brushing, sanding, filing, etc. In addition, the filler wire should be cleaned by sanding. A relatively small size torch should be used and only sufficient heat is required to melt the surface and the filler rod. The filler rod is handled in the same manner as with lead welding.

The gas tungsten arc welding process can also be used for welding zinc die casting alloys. When using the gas tungsten torch a small tungsten and low current should be employed. Sufficient heat should be maintained to melt the surface of the work being welded. The filler rod is moved in and out of the molten puddle, normally on the leading edge. If the part is unusually shaped it may be necessary to back it up with asbestos paper to maintain the shape of the part.

Welding the Precious Metals

Silver is welded in much the same manner as copper since it has a high conductivity and a low affinity for oxygen and nitrogen. Silver is usually used as an alloy with copper. Filler metal normally used for brazing can be used. Silver is used for jewelry and tableware, but also for industrial applications where tanks are lined with sheet silver for the chemical industry.

The oxyacetylene or oxyfuel gas torch has been widely used for welding silver. Silver is also clad to other metals for chemical vessels. The gas tungsten arc welding process can also be used for welding silver and in this case direct current electrode negative is employed. The torch or tungsten must be sufficiently small to match the welding job. Procedure data for welding pure copper can be used to establish starting points for a silver welding procedure. Silver is also brazed using silver alloy filler metal.

Gold is one of the most expensive metals and, therefore, parts to be welded are of unusually thin or small intricate shape. Gold can be soldered or brazed with brazing alloys and it may be cold or pressure welded. It is normally welded with the oxyfuel gas process using the small torch.

Platinum is used in the chemical industry and also in the glass industry for making filaments for fiberglass. Welding is often required and is normally accomplished by the oxyacetylene or oxyfuel gas and the gas tungsten arc welding process. The other precious metals of the platinum group can be welded in the same manner. All the precious metals can be resistance welded, in spite of their high conductivity. The plasma arc welding process can also be used.

QUESTIONS

1. Explain the aluminum designation system of alloys and tempers.

2. What properties of aluminum make it different from welding steel? Explain.

3. What are the two most popular processes for welding aluminum? Where is each used?

4. Of all the factors involved in welding aluminum, which one is the most important?

5. What are the properties of copper and its alloys that make welding difficult?

6. What is the effect of zinc in copper alloys on welding?

7. What problem does magnesium oxide present when welding magnesium?

8. Welding nickel and high-nickel alloys is similar to welding what other metals?

9. Why should the bevel or groove angles be increased for nickel alloys?

10. What welding process is most widely used for welding nickel alloys?

11. What are the reactive metals? Why are they difficult to weld?

12. What precautions should be taken when welding beryllium or its alloys?

13. What welding processes are used to join titanium *in the open*?

14. What is the most important factor to consider when welding titanium?

15. What precautions must be taken when welding lead?

16. What filler metal is used for lead? Where is it obtained?

17. Can zinc die castings be welded? How?

18. What precautions must be taken when welding zinc?

19. Silver is welded in much the same way as what other metal?

20. What processes are commonly used to join gold?

REFERENCES

1. "Aluminum Standards and Data," The Aluminum Association, 1976, 750 Third Avenue, New York.

2. "Welding Kaiser Aluminum," Kaiser Aluminum and Chemical Sales, Inc., 1967, Kaiser Center, Oakland, Calif.

3. "Welding Alcoa Aluminum," Aluminum Company of America, Pittsburgh, Pa.

4. "Welding Aluminum," Reynolds Metals Company, Richmond, Va.

5. "Standards Handbook Wrought Copper and Copper Alloy Mill Products, Part 2-Alloy Data," Copper Development Association, Inc., 405 Lexington Avenue, New York.

6. "Joining Magnesium," The Dow Metal Products Company, Midland, Mich.

7. "Joining Huntington Alloys," The International Nickel Company, Huntington, W.Va.

14

WELDING SPECIAL AND
DISSIMILAR METALS

14-1   CAST IRON AND OTHER IRONS

14-2   TOOL STEELS

14-3   REINFORCING BARS

14-4   COATED STEELS

14-5   OTHER METALS

14-6   CLAD METALS

14-7   DISSIMILAR METALS

14-1   CAST IRON AND OTHER IRONS

The term *cast iron* is a rather broad description of many types of irons which are castings but which may have different properties and serve different purposes. In general, a cast iron is an alloy of iron, carbon, and silicon in which more carbon is present than can be retained in solid solution in austenite at the Eutectic temperature. The amount of carbon is usually more than 1.7% and less than 4.5%. There are many types of cast iron; in fact, pig iron, which is the product of the blast furnace, can be considered cast iron since it is iron cast into pigs or ingots for later remelting and casting into the final form. The most widely used type of cast iron is known as gray iron. Its tonnage production exceeds that of any other cast metal.

Gray iron has a variety of compositions but it is usually such that the matrix structure is primarily pearlite with many graphite flakes dispersed throughout. It is these graphite flakes that provide the characteristic "gray" appearance of the fracture. The graphite flakes promote machinability which is one of the primary advantages of gray cast iron. Another advantage is the ability to cast material into complex shapes with relatively thin walls. Gray cast iron is the least expensive metal for making many parts. Gray cast iron also has good resistance to wear and seems to have a damping effect for vibration.

Gray cast iron is used in the automotive industry for: engine blocks and heads, automatic transmission housings, differential housing, water pump housing, brake drums, and engine pistons. There are exceptions to this but the exceptions are usually aluminum, which is readily identifiable from cast iron.

There are also alloy cast irons which contain small amounts of chromium, nickel, molybdenum, copper, or other elements added to provide specific properties. These usually provide higher strength cast irons. One of the major uses for the higher strength irons is casting automotive crankshafts. These are sometimes called semisteel or proprietory names.

Another alloy iron is the austenitic cast iron which is modified by additions of nickel and other elements to reduce the transformation temperature so that the structure is austenitic at room or normal temperatures. Austenitic cast irons have a high degree of corrosion resistance.

Another type of cast iron is known as white cast iron in which almost all the carbon is in the combined form. This provides a cast iron with higher hardness which is used for abrasion resistance.

WELDING SPECIAL AND DISSIMILAR METALS

Another class of cast iron is called malleable iron. This is made by giving white cast iron a special annealing heat treatment to change the structure of the carbon in the iron. By so doing, the structure is changed to pearlitic or ferritic which increases its ductility.

There are two other classes of cast iron which are more ductile than gray cast iron. These are known as nodular iron and ductile cast iron. These are made by the addition of magnesium or aluminum which will either tie up the carbon in a combined state or will give the free carbon a spherical or nodular shape rather than the normal flake shape in gray cast iron. This structure provides a greater degree of ductility or malleability of the casting.

Cast irons are used in many different industries and they are popular because of their ease of casting, the ease of machining, and the relative low cost compared to other metals. They are widely used in agricultural equipment including farm tools, garden tools, etc. They are also used for bases, brackets, covers, and so on, on machine tools. Cast iron is used for pipe fittings and for cast iron pipe which is widely used in the chemical industry. Cast iron is also used for automobile engine blocks, heads, and manifolds, and for water pumps, etc. Cast iron is rarely used in structural work except for compression members. It is widely used in construction machinery for counterweights and in other applications in which weight is required.

Cast irons, particularly gray cast irons, are covered by ASTM specification A48, which establishes seven classes based on the tensile strength of the material. These range from 20,000 psi (14.5 kg/mm^2) tensile and 150 Brinell hardness number to 40,000 psi (28.5 kg/mm^2) and 250 Brinell hardness number. The cast irons above 40,000 psi (28.5 kg/mm^2) tensile strength are considered high-strength irons and are more expensive and more difficult to machine. Cast iron is much stronger in compression than in tension with the compression strength being about double the tension value. Other ASTM specifications are used to describe other classes of cast iron but these are primarily with respect to the end use of the material.

Welding of the Cast Irons

The gray cast iron has a very low ability to bend and low ductility. Possibly a maximum of 2% ductility will be obtained in the extreme low carbon range. The low ductility is due to the presence of the graphite flakes which act as discontinuities. In most welding processes the heating and cooling cycle creates expansion and contraction, which sets up tensile stresses during the contraction period. For this reason, gray cast iron is difficult to weld without special precautions. On the other hand, the ductile cast irons such as malleable iron,

FIGURE 14-1 *Welding processes and filler metals for cast iron.*

Welding Process & Filler Metal Type	Filler Metal Spec[1]	Filler Metal Type[1]	Color Match	Machineable Deposit
SMAW (Stick)				
Cast iron	E-CI	Cast iron	Good	Yes
Copper-tin[2]	ECuSn A & C	Copper—5 or 8% tin	No	Yes
Copper-aluminum[2]	ECuAl-A2	Copper-10% aluminum	No	Yes
Mild steel	E-St	Mild steel	Fair	No
Nickel	ENi-CI	High nickel alloy	No	Yes
Nickel-iron	ENiFe-CI	50% Nickel plus iron	No	Yes
Nickel-copper	ENiCu-A & B	55 or 65% Ni + 40 or 30% W	No	Yes
Oxy Fuel Gas				
Cast iron	RCI & A & B	Cast iron—with minor alloys	Good	Yes
Copper zinc[2]	RCuZn B & C	58% Copper—zinc	No	Yes
Brazing[3]				
Copper Zinc	RBCuZn A & D	Copper-zinc & copper-Zinc-nickel	No	Yes
GMAW (MIG)				
Mild steel	E60S-3	Mild steel	Fair	No
Copper base[2]	ECuZn-C	Silicon bronze	No	Yes
Nickel-copper	ENiCu-B	High nickel	No	Yes
FCAW				
Mild steel	E70T-7	Mild steel	Fair	No
Nickel type	No spec	50% Nickel plus iron	No	Yes

Note: (1) See AWS Specification for Welding Rods and Covered Electrode for Welding Cast Iron.[1]
(2) Would be considered a brass weld.
(3) Heat source any for brazing also carbon arc, twin carbon arc, gas tungsten arc or plasma arc.

ductile iron, and nodular iron can be successfully welded. For best results, these types of cast irons should be welded in the annealed condition.

Welding is used to salvage new iron castings, to repair castings that have failed in service and also to join castings to each other or to steel parts in manufacturing operations. The table in Figure 14-1 shows the welding processes that can be used for welding cast, malleable, and nodular irons. The selection of the welding process and the welding filler metals depends on the type of weld properties desired and the service life that is expected. For example, when using the shielded metal arc welding process different types of filler metal can be used. The filler metal will have an effect on the color match of the weld compared to the base material. The color match can be a determining factor, specifically in the salvage or repair of castings, where a difference of color would not be acceptable.

No matter which of the welding processes is selected certain preparatory steps should be made. It is important to determine the exact type of cast iron to be welded, whether it is gray cast iron or a malleable or ductile type. If exact information is not known it is best to assume that it is gray cast iron with little or no ductility. This would err on the side of safety so that a successful repair weld would be obtained.

In general it is not recommended to weld repair gray iron castings that are subjected to heating and cooling in normal service. At least it should not be used when heating and cooling vary over a range of temperatures exceeding 400°F (204°C). The reason is that unless cast iron is used as the filler material, the weld metal and base metal may have different coefficients of expansion and contraction. This will contribute to internal stresses which cannot be withstood by gray cast iron. Repair of these types of castings can be made, but the reliability and service life on such repairs cannot be predicted with accuracy.

Preparation for Welding

In preparing the casting for welding it is necessary to remove all surface materials to completely clean the casting in the area of the weld. This means removing paint, grease, oil, and other foreign material from the weld zone. It is desirable to heat the weld area for a short time to remove entrapped gas from the weld zone of the base metal. Additionally, the skin or high silicon surface should be removed adjacent to the weld area on both the face and root side.

Where grooves are involved a V groove from a 60–90° included angle should be used. Complete penetration welds should always be used since a crack or defect not completely removed may quickly reappear under service conditions.

Preheating is desirable for welding with any of the welding processes. It can be reduced when using extremely ductile filler metal. Preheating will reduce the thermal gradiant between the weld and the remainder of the case iron. Preheat temperatures should be related to the welding process, the filler metal type, the mass and the complexity of the casting.

Preheating can be done by any of the normal methods. Torch heating is normally used for relatively small castings weighing 30 pounds (13.6 kg) or less. Larger parts may be furnace preheated and in some cases temporary furnaces are built around the part rather than taking the part to a furnace. In this way, the parts can be maintained at a high interpass temperature in the temporary furnace during welding. Preheating should be general since it helps to improve the ductility of the material and will spread shrinkage stresses over a large area to avoid critical stresses at any one point. It tends to help soften the area adjacent to the weld; it assists in degassing the casting and this in turn reduces the possibility of porosity of the deposited weld metal; and it also increases welding speed.

Slow cooling or post heating improves the machinability of the heat-affected zone in the cast iron adjacent to the weld. The post cooling should be as slow as possible. Many times this is done by covering the casting with insulating materials to keep the air or breezes from it.

Arc Welding

The shielded metal arc welding process can be utilized for welding cast iron. There are four types of filler metals that may be used. These are cast iron covered electrodes, covered copper base alloy electrodes, covered nickel base alloy electrodes and mild steel covered electrodes. There are reasons for using each of the different specific types of electrodes as follows: The machinability of the deposit, the color match of the deposit, the strength of the deposit, and the ductility of the final weld.

When arc welding with the cast iron electrodes (ECI) preheat to between 250° and 800°F (121° and 425°C), depending on the size and complexity of the casting and the need to machine the deposit and adjacent areas. The higher degree of heating, the easier it will be to machine the weld deposit. In general, it is best to use small-size electrodes and a relatively low current setting. A medium arc length should be used and if at all possible welding should be done in the flat position. Wandering or skip welding procedure should be used, and peening will help reduce stresses and will minimize distortion. Slow cooling after welding is recommended. These electrodes provide an excellent color match on gray iron. The strength of the weld will equal the strength of the base metal.

There are two types of copper-base electrodes, the copper tin alloy (ECuSn-A and C) and the copper aluminum (ECuA1-A2) types. The copper zinc alloys cannot be used for arc welding electrodes because of the low boiling temperature of zinc. Zinc will volatilize in the arc and will cause weld metal porosity. The copper tin electrodes will produce a braze weld having good ductility. The ECuSn-A has less amount of tin. It is more of a general purpose electrode. The ECuSn-C provides a stronger deposit with higher hardness.

The copper aluminum alloy electrode (ECuA1-A2) provides much stronger welds and is used on the higher strength alloy cast irons.

When the copper base electrodes are used, a preheat of 250–400°F (121–204°C) is recommended and small electrodes and low current should be used. The welding technique should be to direct the arc against the deposited metal or puddle to avoid penetration and mixing the base metal with the weld metal. Slow cooling is recommended after welding. The copper-base electrodes do not provide a good color match.

There are three types of nickel electrodes used for welding case iron. The ENiFe-CI contains approximately 50% nickel with iron, the ENiCI contains about 85% nickel and the ENiCu type contains nickel and copper. The ENiFeCI electrode is less expensive and provides results approximately equal to the high-nickel electrode. These electrodes can be used without preheat; however, heating to 100°F (38°C) is recommended. These electrodes can be used in all positions; however, the flat position is recommended. The welding slag should be removed between passes. The nickel and nickel iron deposits are extremely ductile and will not become brittle with the carbon pickup. The hardness of the heat-affected zone can be minimized by reducing penetration into the cast iron base metal. The technique mentioned above, that is, playing the arc on the puddle rather than on the base metal, will help minimize dilution. Slow cooling and if necessary postheating will improve machinability of the heat-affected zone. The nickel-base electrodes do not provide a close color match.

The copper nickel type comes in two grades; the ENiCu-A with 55% nickel and 40% copper and the ENiCu-B with 65% nickel and 30% copper. Either of these electrodes can be used in the same manner as the nickel or nickel iron electrode with about the same technique and results. The deposits of these electrodes do not provide a color match.

Mild steel electrodes (E St) are not recommended for welding cast iron if the deposit is to be machined. The mild steel deposit will pick up sufficient carbon to make a high-carbon deposit which is impossible to machine. Additionally, the mild steel deposit will have a reduced level of ductility as a result of increased carbon content. This type of electrode should be used only for small repairs and should not be used when machining is required. Minimum preheat is possible for small repair jobs. Here again, small electrodes at low current are recommended to minimize dilution, and to avoid the concentration of shrinkage stresses. Short welds using a wandering sequence should be used and the weld should be peened as quickly as possible after welding. The mild steel electrode deposit provides a fair color match.

Oxy-Fuel Gas Welding

The oxy-fuel gas process is often used for welding cast iron. Most of the fuel gases can be used. The flame should be neutral to slightly reducing. Flux should be used. Two types of filler metals are available: The cast iron rods (RCI and A and B) and the copper zinc rods (RCuZn-B and C).

Welds made with the proper cast iron electrode will be as strong as the base metal. The RCI classification is used for ordinary gray cast iron. The RCI-A has small amounts of alloy and is used for the high strength alloy cast irons and the RCI-B is used for welding malleable and nodular cast iron. Good color match is provided by all of these welding rods. The optimum welding procedure should be used with regard to joint preparation, preheat, and post heat.

The copper zinc rods produce braze welds. There are two classifications: RCuZn-B, which is a manganese bronze and RCuZn-C, which is a low-fuming bronze. The deposited bronze has relatively high ductility but will not provide a color match.

Brazing and Braze Welding

Brazing is used for joining cast iron to cast iron and steels. In these cases, the joint design must be selected for brazing so that capillary attraction causes the filler metal to flow between closely fitting parts. The torch method is normally used; however, any of the other heating methods mentioned in Chapter 6 can be used. In addition, the carbon arc, the twin carbon arc, the gas tungsten arc, and the plasma arc can all be used as sources of heat. Two brazing filler metal alloys are normally used; both are copper zinc alloys; see Figure 14-1 for specification for the brazing alloys.

Braze welding can also be used to join cast iron. In braze welding the filler metal is not drawn into the joint by capillary attraction. This is sometimes called bronze welding. The filler material having a liquidous above 850°F (454°C) should be used. Braze welding will not provide a color match.

Braze welding can also be accomplished by the shielded metal arc and the gas metal arc welding processes. High temperature preheating is not usually required for braze welding unless the part is extremely heavy or complex in geometry. The bronze weld metal deposit has extremely high ductility which compensates for the lack of ductility of the cast iron. The reason for not requiring high temperature preheat is the desire to avoid intermix of base metal with the filler metal. The heat of the arc is sufficient to bring the surface of the cast iron up to a temperature at which the copper base filler metal alloy will make a bond to the cast iron. Since there is little or no intermixing of the materials the zone adjacent to the weld in the base metal is not appreciably hardened. The weld and adjacent area is machinable after the weld is completed. In general, a 200°F (93°C) preheat is sufficient for most applications. The cooling rate is not extremely critical and a stress relief heat treatment is not usually required. This type of welding is commonly used for repair welding of automotive parts, agricultural implement parts, and even automotive engine blocks and heads. It can only be used when the absence of color match is not objectionable.

Gas Metal Arc Welding

The gas metal arc welding process can be used for making welds between malleable iron and carbon steels. Several types of electrode wires can be used,[2] including:

Mild Steel (E70S-3) using 75% Argon + 25% CO_2 for shielding.

Nickel Copper (ENiCu-B) using 100% Argon for shielding.

Silicon Bronze (ECuZn-C) using 50% Argon +50% Helium for shielding.

In all cases small diameter electrode wire should be used at low current. With the mild steel electrode wire the Argon-CO_2 shielding gas mixture is used to minimize penetration. In the case of the nickel base filler metal and the copper base filler metal the deposited filler metal is extremely ductile. The mild steel provides a fair color match. A higher preheat is usually required to reduce residual stresses and cracking tendencies.

Flux-Cored Arc Welding

This process has recently been used for welding cast irons. The more successful application has been using a nickel base flux-cored wire which produces a weld metal deposit very similar to the 50% nickel deposit provided by the ENiFe-CI covered electrode. This electrode wire is normally operated with CO_2 shielding gas but when lower mechanical properties are not objectionable it can be operated without external shielding gas. The minimum preheat temperatures can be used. The technique should minimize penetration into the cast iron base metal. Postheating is normally not required. A color match is not obtained.

Flux-cored self-shielding electrode wires (E60T-7), operating with electrode negative (straight polarity), have also been used for certain cast iron to mild steel applications. In this case, a minimum penetration type weld is obtained and by the proper technique penetration should be kept to a minimum. It is not recommended for deposits that must be machined.

Figure 14-1 provides a summary of welding processes for joining cast iron. For manufacturing operations it is highly recommended that a welding procedure be developed utilizing the process selected and that service type tests be made prior to using the process on a particular product. In general, those cast irons having maximum ductility are those that can be most successfully welded. Thus malleable iron, nodular iron, and ductile iron can be welded for many applications. The most successful welds would be those that provide an extremely ductile weld deposit. These types should be used for repair work.

Other welding processes can also be used for cast iron. Thermit welding has been used for repairing certain types of cast iron machine tool parts. The procedure is identical to that used for welding steel except that a special thermit mixture is required.

Soldering can be used for joining cast iron and is sometimes used for repairing small defects in small castings. Soldering is done with an iron or with a torch. Flash welding can also be used for welding cast iron.

14-2 TOOL STEELS

Steels used for making tools, punches, and dies are perhaps the hardest, the strongest, and toughest steels used in industry. It is obvious that tools used for working steels and other metals must be stronger and harder than the steels or material they cut or form. The metallurgical characteristics of various compositions of tool steels are extremely complex and beyond the scope of this chapter. Tools and dies do wear and are damaged but by means of welding they can be repaired and reworked and returned to service. In addition, certain kinds of tools and dies can be fabricated by welding. The repairing of damaged tools and dies and the fabrication by welding of dies will save money. For this reason, information is provided for welding these types of material.

There are hundreds of different makes and types of tool steels available and each may have a specific composition and end use. The Society of Automotive Engineers, in cooperation with the American Iron and Steel Institute, has established a classification system which relates to the use of the material and its composition or type of heat treatment. This classification system divides the tool and die steels into separate categories that are shown by Figure 14-2. In general, tool steels are basically medium- to high-carbon steels with specific elements included in different amounts to provide special characteristics. The carbon in the tool steel is provided to help harden the steel to greater hardness for cutting and wear resistance. Other elements are added to provide greater toughness or strength. In some cases elements are added to retain the size and shape of the tool during its heat treat hardening operation or to make the hardening operation safer and to provide *red hardness* so that the tool retains its hardness and strength when it becomes extremely hot. Various alloying elements in addition to carbon are chromium (Cr), cobalt (Co), manganese (Mn), molybdenum (Mo), nickel (Ni), tungsten (W), and vanadium (V). Iron is the predominant element in the composition and represents by far the majority of the metal in the alloy but is not normally specified in a composition of a tool steel.

The addition of elements produces different effects on the resultant composition as follows: Chromium produces deeper hardness penetration in heat treatment and contributes wear resistance and toughness. Cobalt is used in high-speed steels and increases the red hardness so that they can be used at higher operating temperatures. Manganese in small amounts is used to aid in making steel sound and further additions help steel to harden deeper and more quickly in heat treatment. It also helps to lower the quenching temperature necessary to harden steels. Larger amounts of manganese

in the 1.20–1.60% range allow steels to be oil quenched rather than water quenched. Molybdenum increases the hardness penetration in heat treatment and reduces quenching temperatures. It also helps increase red hardness and wear resistance. Nickel adds toughness and wear resistance to steel and is used in conjunction with hardening elements. Tungsten added to the steel increases its wear resistance and provides red hardness characteristics. Approximately 1.5% increases wear resistance and about 4% in combination with high carbon will greatly increase wear resistance. Tungsten in large quantities with chromium provides for red hardness. Vanadium in small quantities increases the toughening effect and reduces grain size. Vanadium in amounts over 1% provides extreme wear resistance especially to high-speed steels. Smaller amounts of vanadium in conjunction with chromium, and tungsten, aid in increasing red hardness properties.

The tool or die steels are designed for special purposes that are dependent upon composition. Certain tool steels are made for producing die blocks; some are made for producing molds, others are made for hot working, and still others for high-speed cutting applications. The other way for classifying tool steels is according to the type of quench required to harden the steel. The most severe quench after heating is the water quench, (water-hardening steels). A less severe quench is the oil quench obtained by cooling the tool steel in oil baths (oil-hardening steels). The least drastic quench is cooling in air (air-hardening steels).

Tool steels and dies can also be classified according to the work that is to be done by the tool. This is based on class numbers. Class I steels are used to make tools that work by a shearing or cutting action, such as cutoff dies, shearing dies, blanking dies, trimming dies, etc. Class II steels are used to make tools that produce the desired shape of the part by causing the material being worked, either hot or cold, to *flow* under tension. This includes drawing dies, forming dies, reducing dies, forging dies, etc. This class also includes plastic molds and die cast molding dies. The Class III steels are used to make tools that act upon the material being worked by partially or wholly reforming it without changing the actual dimensions. This includes bending dies, folding dies, twisting dies, etc. Class IV steels are used to make dies that work under heavy pressure and that produce a flow of metal or other material compressing it into the desired form. This includes crimping dies, embossing dies, heading dies, extrusion dies, staking dies, etc. It is important to understand and have sufficient information concerning the composition of the tool or die, the type of heat treatment that it has received, and the type of work that it performs.

As far as welding is concerned there are four basic

AISI-SAE Types	Classification of Tools Steels	COMPOSITION %					
		C	Cr	V	W	Mo	Other
W1	Water hardening	0.60	—	—	—	—	—
W2		0.60	—	0.25	—	—	—
S1		0.50	1.50	—	2.50	⊢	—
S5	Shock resisting	0.55	—	—	—	0.40	0.80 Mn 2.00 Si
S7		0.50	3.25	—	—	1.40	—
O1	Oil hardening	0.90	0.50	—	0.50	—	—
O6		1.45	—	—	—	0.25	1.00 Si
A2	Cold work	1.00	5.00	—	—	1.00	—
A4	Medium alloy air hardening	1.00	1.00	—	—	1.00	2.00 Mn
D2	Cold work High carbon High chromium	1.50	12.00	—	—	1.00	—
M1	Cold work	0.80	4.00	1.00	1.50	8.00	—
M2	Molybdenum	0.85	4.00	2.00	6.00	5.00	—
M10		0.90	4.00	2.00	—	8.00	—
H11	Hot work	0.35	5.00	0.40	—	1.50	—
H12	Chromium	0.35	5.00	0.40	1.50	1.50	—
H13		0.35	5.00	1.00	—	1.50	—
P20	Die casting mold	0.35	1.25	—	—	0.40	—

FIGURE 14-2 *Abridged chart of tool steel types.*

types of die steels that are weld repairable. These are water-hardening dies, oil-hardening dies, air-hardening dies, and hot work tools. High-speed tools can also be repaired. In tool and die welding it is not always necessary for the electrode used to provide a deposited weld metal that exactly matches the analysis of the tool steel being welded. It is necessary, however, that the weld metal deposited match the heat treatment of the tool or die steel as closely as possible. Thus, selection of the proper electrode is based on matching the heat treatment of the tool or die steel.

There are no specifications covering the composition of tool and die welding electrodes. However, all manufacturers of these types of electrodes provide information concerning each of their electrodes showing the type of tool or die steels for which it is designed. They also provide the properties of the weld metal that is deposited. Welding electrodes are not available to match the composition of each and every tool steel composition or to match the specific heat treatment of each tool or die steel. However, tool and die electrodes are available that match the various different categories of tool and die steels given in Figure 14-2. Assistance can be obtained from the catalogs of electrodes for this type of welding or by consulting with representatives of the companies that manufacture these electrodes. If the identification of the electrodes is lost it is possible to use the spark test in matching the electrode to the tool steel. A comparison is made of the sparks from the

tool or die steel to be welded and compared with the spark pattern of the welding electrode. The matching spark patterns will be the guide or basis for selection of the electrode.

Successful tool and die welding depends on the selection or development of a welding procedure and welding sequence. It is beyond the scope of this chapter to outline specific welding procedures to be used for each classification of tool steel using each type of tool and die welding electrodes. Normally the manufacturer of electrodes will provide specific procedure sheets pertaining to the different electrodes they offer. These should be carefully followed.

In general, weld deposits of tool and die electrodes are sufficiently hard in the as-welded condition. If the welded tools or dies lend themselves to grinding, treatment other than tempering is not required. However, if machining is required the weld deposits should be annealed and heat treated after machining. The hardness of the weld deposit will vary in accordance with the following:

Preheat temperature if used.

Welding technique and sequence.

Mixture or dilution of the weld metal with base metal.

Rate of cooling, which depends on the mass of the tool being welded.

Tempering temperature of the welded tool or die after welding.

WELDING SPECIAL AND DISSIMILAR METALS

Uniform hardness of the as-welded deposit is obtained if the temperature of the tool or die is maintained constant during the welding operation. The temperature of the tool or die being welded should never exceed the maximum of the *draw range* temperature for the particular class of tool steel being welded. Manufacturer's recommendation should be followed with respect to these temperatures.

The welding procedure for repair welding tools and dies should consist of at least the following factors:

Identification of the tool steel being welded.

Selection of the electrode to match the same class of material or heat treatment.

Establishing the correct joint detail for the repair and preparing the joint.

Preheating the workpiece.

Making the weld deposit in accordance with manufacturer's recommendations.

Postheating to temper the deposit or the repaired part.

One of the major problems is proper preparation of the part for repair welding. When making large repairs to worn cutting edges or surfaces the damaged area should be ground sufficiently under size to allow a uniform depth of finished deposit of at least 1/8 in. (3.2 mm).

In some cases very small weld deposits are made using the gas tungsten arc welding process to build up a worn or damaged edge or corner. It is important to provide a uniformly thick weld deposit which will be refinished to the original dimensions. This insures a more uniform hardness throughout the deposit. For inlay deposit or other overlay work a thickness of 3/16 in. (4.8 mm) is required.

When preheating the part to be repaired observe the "draw temperature range" of the base metal. The preheat temperature should slightly exceed the minimum of the draw range and the interpass temperature should never exceed the maximum of the draw range of the particular tool steel. Exceeding the maximum draw range will reduce the hardness of the tool by softening it. The table shown by Figure 14-3 provides recommended preheat temperatures to be employed on some of the popular types of tool steels.

Most of the tool and die welding electrodes are used with DC electrode positive or with alternating current. The recommended currents for each different size should be provided with the electrode manufacturer's technical data. For making tool and die welds a slow travel is recommended in order to maintain an even deposit and to assure uniform weld penetration. The work should be positioned for flat position welding except that it is recommended that welding be done with the work positioned for slightly uphill travel, in the order of 5–15°. This causes the deposit to build up evenly and helps keep the slag free of the weld puddle. Uniform motion without weaving is recommended. When welding

AISI-SAE Type	Preheat Temperature °F	Interpass Temperature °F	Draw or Tempering Temperature °F
W1	250/450	250/450	300/650
W2	250/450	250/450	300/650
S1	300/500	300/500	300/500
S5	300/500	300/500	500 Min.
S7	300/500	300/500	425/400
O1	300/400	300/400	300/450
O6	300/400	300/400	300/450
A2	300/500	300/500	350/400
A4	300/500	300/500	350/400
D2	700/900	700/900	925/900
M1	950/1100	950/1100	1000/1050
M2	950/1100	950/1100	1000/1050
M10	950/1100	950/1100	950/1050
H11	900/1200	900/1200	1000/1150
H12	900/1200	900/1200	1000/1150
H13	900/1200	900/1200	1000/1150
P20	400/800	400/800	1000

FIGURE 14-3 *Preheat and interpass temperature when welding on tool steels.*

on tool cutting edges, position the work so that the deposit will flow or roll over the cutting edge.

Peening should be done immediately on all weld deposits. Peening, however, should be controlled. Peening should be used to provide sufficient mechanical work to help improve the properties of the deposit and help refine the metallurgical structure. It will also assist in relieving shrinkage stresses and possibly assist in correcting distortion. Peening can be done manually or small pneumatic power hammers can be used. The welding technique should avoid craters. In all cases, craters should be filled by reversing the direction of travel and pausing slightly. This will insure a more uniform deposit.

When welding deeply damaged cutting edges that require multiple passes it is necessary to start at the bottom and gradually fill up damaged areas. The current for the first or second beads can be higher than used on the final bead. It is important to peen the weld metal while hot to help eliminate shrinkage, warpage, and possibly cracks. The random or wandering welding technique should be used when welding circular parts, such as on the inner edge of a die. Warpage or distortion can be reduced by preheating, which expands the part, and peening during the contraction period will reduce stresses. On parts such as a long shear blade where welding is done all on one side it is recommended that the parts be reverse formed. This will help keep the part straight during welding. It is recommended to weld only short lengths of 2 to 3 inches and then to peen to reduce stresses and warpage.

After the repair welds are completed the part may be allowed to cool to room temperature. It is then tempered by reheating to the recommended temperature, as specified by the type of tool steel being welded or by the welding electrode manufacturer's technical data. The draw temperature would always be used. For small or light-duty work parts the draw temperature should be on the minimum side of the draw range. On larger or heavy-duty parts the draw temperature should be on the maximum.

Composite dies manufactured by welding are becoming more popular. Tool steel is used for cutting or working surfaces and medium-carbon steel is used for the remainder of the part. This type of construction greatly reduces the cost of the composite die. The electrode is selected based on the type of tool steel employed. The weld preparation, preheat, and welding sequence would be the same as mentioned previously.

Experience with tool and die welding is very helpful and will avoid the possibility of failures. The procedure development, including identification of material, selection of electrodes, and welding techniques should follow the tool steel manufacturer's data and the welding electrode manufacturer's information.

14-3 REINFORCING BARS

Concrete reinforcing bars, or as they are more technically known, *deformed steel reinforcing bars,* are used in reinforced concrete construction. This includes buildings, bridges, highways, locks, dams, docks, piers, etc. The principle applications of reinforcing bars include reinforcement of columns, girders, beams, slabs, pavements, as well as precast and prestressed concrete structures. Concrete is strong in compression and shear but is weak in tension. By using deformed steel reinforcing bars imbedded in the concrete, tensile stresses can be accommodated, thus reinforced concrete provides compression strength of concrete and tensile strength of steel. It is necessary, however, that the concrete and steel work together. This is accomplished by means of a bond between the bar and the concrete which is achieved by means of deformations which are rolled into the bars. These deformations keep the bars from slipping through the concrete.

Concrete reinforcing bars come in different sizes which are designated by a series of numbers. There are eleven standard sizes known as *reinforcing bar* No. 3 through No. 11 and No. 14 and No. 18. The numbers assigned to bars are based on the number of 1/8 inches included in the nominal diameter. The nominal diameter of a deformed reinforcement bar is equivalent to the diameter of a plain steel bar having the same weight per foot as the deformed bar. Lengths up to 60 ft long are available.

There are three ASTM specifications for reinforcing bars: A 615, plain billet steel bars; A 616, rail steel reinforcing bars; and A 617, axle steel reinforcement bars. Information concerning these different specifications is shown by Figure 14-4.

All of the reinforcing bars produced in the U.S.A. are identified by markings rolled into the bar. These markings will show, by means of raised letters, the code for the manufacturer of the steel bar. The different code letters have been identified by the Concrete Reinforcing Steel Institute.[3] This is then followed by the letter identifying the specific steel mill where the bar was produced, based on standard designations. The next symbol indicates the bar size by the bar number. The next symbol indicates the type of steel as follows: N indicates new billet steel, A indicates axle steel, and the third symbol which is a cross section of a railroad rail indicates that the bar was rerolled from used railroad rails.

The next identification symbol is a number indicating

ASTM Specification	Specification Identification	Grades Produced	Grade Identification	Size Designation Bar Number	Strength		Composition (1)	
					Tensile min psi	Yield min psi	Carbon	Manganese
A-615 (New billet steel)	N	40	Blank	#3 thru #11 and #14 and #18	70,000	40,000	– –	– –
		60	60		90,000	60,000	– –	– –
		75	75		100,000	75,000	– –	– –
A-616 (Made from A-1)	 I	50	Blank	#3 thru #11	80,000	50,000	0.55–0.82	0.60–1.00
		60	60		90,000	60,000		
A-617 (Made from A-21)	A	40	Blank	#3 thru #11	70,000	40,000	0.40–0.59	0.60–0.90
		60	60		90,000	60,000		

Note: See specific ASTM specification for additional information—
(1) composition based on A-1 and A-21

FIGURE 14-4 *Reinforcing bars—summary of information.*

the grade of the steel. If there is no number it normally means it is the minimum grade within the specification. Grades are also identified by a single or double continuous longitudinal line through at least five spaces offset from the center of the bar. A single line indicates the middle-strength grade and a double line indicates the highest-strength grade. It is important to determine the type of steel and the grade since this will be valuable information in establishing the welding procedure.

The specifications do not include chemical requirements for the different classes; however, when bars are purchased from the mill, the mill will provide a chemical analysis report of the bars, if requested. The grade number is the indication of the strength of the bars and the numbers indicate the yield point in thousand pounds per square inch minimum. All bar sizes are not made in all grades and it is only the specification A 615 that provides the large No. 14 and No. 18 size bars.

It is necessary to splice concrete reinforcing bars in all but the most simple concrete structures. In the past, splicing was done by overlapping the bars from 20 to 40 diameters and wiring them together and relying on the surrounding concrete to transmit the load from one bar to the other. This method is wasteful of the steel and is sometimes impractical. An example of this occurs when splices must be made in heavily reinforced columns. The close spacing of the bars makes lap splicing particularly difficult and often requires a larger column diameter to provide sufficient concrete between bars and covering the bars. Welding is thus finding increasing importance for splicing concrete reinforcing bars. Three welding processes are used for the majority of welding splices; however, several of the other processes can be used. There is a mechanical splice similar to welding which utilizes medium-strength metal cast *metallic grout* around the ends of the bars enclosed within a steel sleeve having internal grooves.

The welding processes most commonly used are the shielded metal arc welding process, the gas metal arc welding process, and the thermit welding process. It was mentioned previously that there are no chemistry requirements for the three ASTM specifications. However, the reinforcing bars of specification A 616 rail steel are produced from used railroad rails which were originally made to specification A 1. Old railroad rails are salvaged and heated and cut into three parts, the flange, the web, and the head. The heads are then rolled into the deformed reinforcing bars. ASTM specification A 1 has chemical requirements for steel rails and they contain relatively high amounts of carbon and manganese.

The reinforcing bars made to specification A 617 are made from salvaged carbon steel axles used for railroad cars. These axles when originally produced were made to specification A 21. In this ASTM specification the carbon and manganese is relatively high. These are both considered in the hard-to-weld category of steels. The bars produced to A 615 have only a maximum for phosphorous content; however, based on the strength level of steels the alloy content should not be too high. For quality welding it is best to assume that they, too, are in the hard-to-weld category. If at all possible, the analysis of the reinforcing bars should be determined from mill reports. If this is not possible the bars could be analyzed for exact composition. In lieu of this, it is recommended that the bars be considered to have a carbon equivalent of 0.75, thus in the hard-to-weld category.

The American Welding Society has provided a specification entitled, "Reinforcing Steel Welding Code."[4] This code provides a table of carbon equivalents which relates to the bar size and then presents recommended preheat and interpass temperature. The standard formula for the determination of carbon equivalent is used.

There are six carbon equivalents which can only be calculated if the analysis of the reinforcing bars are known.

The code also provides joint design information for making direct butt splices, for making indirect butt splices, and for making lap splices. A butt splice is a direct end-to-end splice of bars with their axis approximately in line and of approximately the same size. An indirect butt splice is one in which an intermediary piece such as a steel plate or rolled angle is used with each reinforcing bar welded directly to the same piece. The lap welded splice is made by overlapping the two bars alongside each other and welding together. These different splices are shown by Figure 14-5. Direct splices can be made between bars of different sizes by providing a transition type configuration to aid stress flow.

For butt splices when the bars are in the horizontal position the single groove weld is most often used with a 45° to 60° included angle. Double groove welds can be made in the larger size bars. When the bars are to be welded with the axis vertical a single or double bevel groove weld is used with the flat side or horizontal side on the lower bar. On occasion, the reinforcing bar may need to be welded to other steel members and a variety of weld joints can be used.

This code provides filler metal selection information based on the grade number of the steel. When welding

FIGURE 14-5 *Types of reinforcing bar splices.*

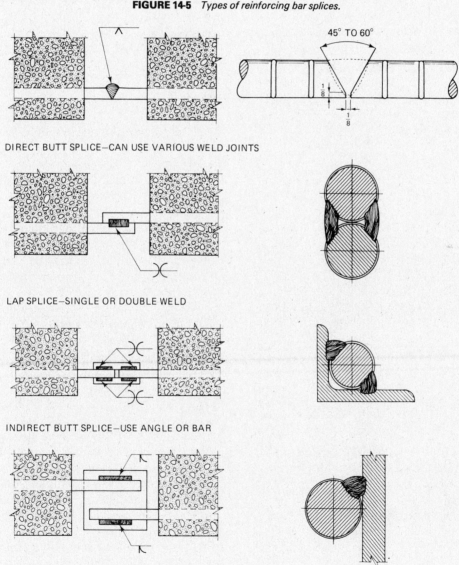

DIRECT BUTT SPLICE—CAN USE VARIOUS WELD JOINTS

LAP SPLICE—SINGLE OR DOUBLE WELD

INDIRECT BUTT SPLICE—USE ANGLE OR BAR

INDIRECT LAP SPLICE—AVOID EXCESSIVE ECCENTRICITY

using the shielded metal arc welding process, Grade 40, the AWS E-7018 would be recommended, for Grade 50 the AWS E-8018 would be recommended, for Grade 60 and the low-alloy A706 the AWS E-9018 electrode is recommended, and for the Grade 75 the AWS E-10018 electrode would be recommended. If the XX18 is not available the XX15 or XX16 can be used. In the case of gas metal arc welding, the E-70S electrode would be used and for flux-cored arc welding the E70T type would be used when welding Grade 40 bars. There are no filler metal specifications for the higher levels of gas metal arc welding electrodes or flux-cored arc welding electrodes at this time. If these processes are to be used the filler metal would need to meet the same mechanical properties as the equivalent shielded metal arc welding electrode mentioned.

The code also provides minimum preheat and innerpass temperatures based on the carbon equivalent of the reinforcing bars. It also relates to the size of the bar. It is therefore important to determine the composition of the bar so that carbon equivalent can be determined. This establishes the heat requirement which ranges from 50°F (10°C) up through 500°F (260°C) based on the size of the bar and the carbon equivalent. In the case of large bars and if the carbon equivalent is not known the 500° preheat would be recommended. Consult the code for further information.

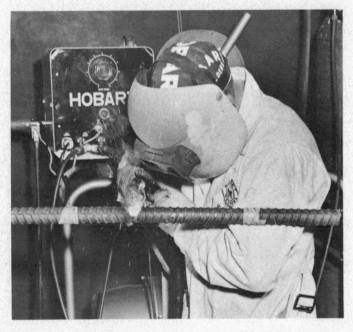

FIGURE 14-6 *Joining bars with gas metal arc welding.*

The code further requires that joint welding procedures should be established based on the welding process, filler metal type and size, and welding technique, which involves position, joint detail, etc. Welders must be qualified. A direct butt splice or indirect butt splice specimen is used. The test bars are tested in tension. Figure 14-6 shows a reinforcing bar being welded using the gas metal arc welding process.

The gas metal arc welding process will make the weld in approximately one-half the time required for shielded metal arc covered electrodes. In either case, however, or with the flux-cored arc welding process, the welds will develop strengths equal or exceeding the specified yield strength of the reinforcing steel bars. Welding is highly recommended as the way to splice reinforcing bars since the concrete will fail at values substantially below the yield strength of the reinforcing bars. This means that the strength of the welds will exceed the requirements for most applications. In any case, the welded splices will exceed the strength of lapped and wired splices. It will also exceed a strength level of the cast metal splices which are sufficiently strong to withstand the strength level of the reinforced concrete composite structure.

14-4 COATED STEELS

In this chapter, the only coated steel that will be discussed in detail is galvanized or zinc-coated sheet steel. There are many other types of coated steels and some of them are welded. For example, tern plate steel, which is tin-lead coated, and aluminized steel. Steels coated with tin, cadmium, copper, and lead may be soldered. The soldering chapter should be referred to for the proper solder alloy and flux selection.

Galvanized steel is widely used and is becoming increasingly important. Manufacturers of many items such as truck bodies, buses, and automobiles are increasingly concerned with the effects of corrosion particularly when chemicals are used on roads for ice control. Galvanized metal is also used in many appliances such as household washing machines and driers and in many industrial products such as air conditioning housings, processing tanks, etc. Other uses for galvanized products are for high tension electrical transmission towers, highway sign standards, and protective items.

There are two basic methods of galvanizing steel. One is by coating sheet metal and the other is by hot dipping the individual item. The coated sheet metal is produced by the continuous hot dip process, but traffic control devices, high-tension transmission tower parts, etc., are made by dipping each item. The continuous hot dip or zinc coated sheet comes in eight classes based on the thickness of the zinc coating. The coating varies from 1 to 1.75 ounces of zinc per square foot of the surface, based on coating both sides of the sheet. Hot-dipped individual parts usually have coatings exceeding the thickness mentioned above. Welding of zinc-coated steel can be done, but specific precautions should be taken. When galvanized steel is arc welded the heat of the welding arc vaporizes the zinc coating in the weld area. This is because the boiling point of zinc (1,600°F, 871°C) is below the melting point of steel (2,800°F, 1,538°C). The zinc volatilizes and leaves the base metal adjacent to the weld. The extent to which the coating is disturbed depends on the heat input of the arc and the heat loss from the base metal. The disturbed area is greater with the slower welding speed processes such as oxyacetylene welding or gas tungsten arc welding.

When galvanized sheet is resistance welded, the welding heat causes less disturbance of the zinc coating than the arc processes. The resistance to corrosion or rather the protection by the zinc is not disturbed since the zinc forced from the spot weld will solidify adjacent to the spot weld and protect the weld nugget. Resistance welding of galvanized steel is, however, more of a problem than arc welding. This is because of the zinc pickup of the welding tips and tools. This is quite complex and beyond the scope of this section. Refer to reference 5 for additional information on resistance welding.

Weld Quality

The low boiling temperature of the zinc of the coating causes it to volatilize in the heat produced by an arc or by an oxyacetylene flame. The zinc in the gaseous state may become entrapped in the molten weld metal as it solidifies. If this occurs there will be porosity in the weld metal and if sufficient zinc is available it will cause large voids in the surface of the deposit. The presence of the zinc in stressed welds can cause cracking and it may also cause delayed cracking due to stress corrosion. To eliminate this, the weld joint must be designed to allow the zinc vapor to completely escape from the joint. Fixturing, backing straps, etc., should be arranged to allow for the zinc to completely escape. Other ways to avoid zinc entrapment in weld metal is to use sufficient heat input when making the weld. It is also important to secure complete and full penetration of the joint. The ultimate precaution would be to remove the zinc from the area to be welded.

When welding on galvanized steel or any coated steel, particularly those with coatings that produce noxious fumes, positive ventilation must be provided. Positive ventilation involves the use of a suction hose at the weld area. When using the gas metal arc process or the flux-cored arc process, the suction type gun nozzles should be used. Welding on zinc or other coated steels should never be done in confined areas.

For corrosion resistance of the weld it is sometimes advisable to use a corrosion-resistant weld metal. This can be done by using a bronze deposit such as a copper-zinc alloy, or a stainless steel electrode. In any case, when arc or oxyacetylene welding is used the area adjacent to the weld will lose the protective zinc coating which must be repaired.

Arc Welding

When using covered electrodes, the electrode selection should be based on the thickness of metal and the position that will be used when welding galvanized steel. The E-XX12 or 13 will be used for welding thinner material, the E-XX10 or 11 will be used for welding galvanized pipe and for welding hot-dipped galvanized parts of heavier thickness. The low-hydrogen electrodes can also be used on heavier thicknesses. The welding technique should utilize slow travel speed to permit degassing of the molten metal. The electrode should point forward to force the zinc vapor ahead of the arc. The quality of welds will be equal to those of bare metal, assuming the weldability of the steel is equal.

The gas metal arc welding process is becoming more widely used for joining galvanized steel. For the thinner gauges the fine-wire short-circuiting method is recommended. In this case, the technique would be similar to that used for bare metal. The shielding gas can be 100% CO_2 or the 75% argon and 25% CO_2 mixture. The selection is dependent primarily on the material thickness and position of welding. For certain applications, the argon-oxygen mixture is used. The amount of spatter produced when welding galvanized steel is slightly greater than when welding bare steel. For this reason, the gun tip and nozzle should be cleaned more often. The electrode wire should be of the highly deoxidized type; however, a stainless steel or bronze type can be employed. This will produce a weld deposit that will be corrosion resistant.

The flux-cored arc welding process can be used as easily as gas metal arc welding for galvanized steel. It is recommended for the heavy gauges and on hot-dipped galvanized parts. The highly deoxidized type of welding electrode should be used.

The gas tungsten arc welding process is not popular for welding galvanized steel. It is not popular since it is one of the slower welding processes and does cause a larger area of zinc adjacent to the weld to be destroyed.

In addition, the volatilized zinc is apt to contaminate the tungsten electrode and require frequent redressing of the electrode. In an effort to overcome this, extra high gas flow rates are sometimes used, which can be expensive. Other techniques may be used. If a filler rod is used it may be of either the highly deoxidized steel type or of the bronze type previously mentioned. In this case the arc is played on the filler rod and zinc contamination of the tungsten electrode is avoided.

The carbon arc welding process has been widely used for welding galvanized steel. Both the single carbon torch and twin carbon torch can be used. The twin carbon torch is used as a source of heat much the same as the oxyacetylene flame; however, when the single carbon is used the carbon can be played on the filler rod and extremely high rates of speed can be accomplished. Normally in this situation the filler rod is the 60 copper–40 zinc alloy, Type RBCuZn-A. By directing the arc on the filler rod it melts and sufficient heat is produced in the base metal for fusion but not sufficient to destroy the zinc coating. This process and technique is widely used in the sheet metal duct work industry.

Torch Brazing

The oxyacetylene torch is also widely used for brazing galvanized steel. The technique is similar to that mentioned with the carbon arc. The torch is directed toward the filler rod which melts and then fills the weld joint. A generous quantity of brazing flux is used and this helps reduce the zinc loss adjacent to the weld.

Repairing the Zinc Coating

The area adjacent to the arc or gas weld will be free of zinc because of the high temperature of the weld. To produce a corrosion-resistant joint, the zinc must be replaced in this area and on the weld itself unless the nonferrous filler metal was used. There are several ways of replacing the zinc. One is by the use of zinc base paste sticks sometimes called zinc sticks or galvanized sticks sold under different proprietary names. These sticks are wiped on the heated bare metal. With practice a very good coating can be placed which will blend with the original zinc coating. This coating will be thicker than the original coating, however. Another way of replacing the depleted zinc coating is by means of flame spraying using a zinc spray filler material. This is a faster method and is used if there is sufficient zinc coating to be replaced. The coating should be two to two-and-one-half times as thick as the original coating for proper corrosion protection.

Other Coated Metals

One other coated metal that is often welded is known as tern plate. This is sheet steel hot dipped with a coating of a lead-tin alloy. The tern alloy is specified in thicknesses based on the weight of tern coating per square foot of sheet metal. This ranges from .35 to 1.45 ounces per square foot of sheet metal based on both sides being coated. Tern plate is often used for making gasoline tanks for automobiles. It is thus welded most often by the resistance welding process. If it is arc welded or oxyacetylene welded the tern plating is destroyed adjacent to the weld and it must be replaced. This can be done similar to soldering.

Aluminized steel is also widely used in the automotive industry particularly for exhaust mufflers. In this case, a high silicon-aluminum alloy is coated to both sides of the sheet steel by the hot dip method. There are two common weights of coating, the regular is 0.40 ounces per square foot and the lightweight coating is 0.25 ounces per square foot based on coating both sides of the sheet steel. Here again, if an arc or gas weld is made on aluminum-coated steel the aluminum coating is destroyed. It is relatively difficult to replace the aluminum coating and, therefore, painting is most often used.

14-5 OTHER METALS

This section includes special steels not covered in Chapter 12. They are metals that are often welded, and information is presented here to help establish welding procedures to successfully join them. These metals are abrasion-resisting steel, free-machining steel, manganese steel, silicon steel, and wrought iron. Each of these metals is briefly described and welding information is provided.

Abrasion-Resisting Steel

Abrasion-resisting (AR) steel is carbon steel usually with a high-carbon analysis used as liners in material-moving systems and for construction equipment where severe abrasion and sharp hard materials are encountered. Abrasion-resisting steels are often used to line dump truck bodies for quarry service, for lining conveyors, chutes, bins, etc. Normally the abrasion-resisting steel is not used for structural strength purposes, but only to provide lining materials for wear resistance. Various steel companies make different proprietary alloys that all have similar properties and, in general, similar compositions. Most AR steels are high-carbon steel in the 0.80–0.90% carbon range; however some are low carbon with multiple alloying elements. These steels are strong and have a hardness up to 40 Rockwell C or 375 BHN. Abrasion-resisting bars or plates are welded to the structures and as they wear they are removed by oxygen cutting or air carbon arc and new plates installed by welding.

Low-hydrogen welding processes are required for welding abrasion-resisting steels. Local preheat of 400°F (204°C) is advisable to avoid underbead cracking of the base metal or cracking of the weld. In some cases this can be avoided by using a preheat weld bead on the carbon steel structure and filling in between the bead and the abrasion-resisting steel with a second bead in the groove provided. The first bead tends to locally preheat the abrasion-resisting steel to avoid cracking and the second bead is made having a full throat. Intermittent welds are usually made since continuous or full-length welds are usually not required. Efforts should be made to avoid deep weld penetration into the abrasion-resisting steel so as not to pick up too much carbon in the weld metal deposit. If too much carbon is picked up the weld bead will have a tendency to crack.

When using the shielded metal arc welding process the E-XX15, E-XX16, or E-XXX8 type electrodes are used. When using gas metal arc welding the low penetrating type shielding gases such as the 75% argon–25% CO_2 mixture should be used. The flux-cored arc welding process is used and the self-shielding version is preferred since it does not have the deep penetrating quality as the CO_2 shielded version.

During cold weather applications, it is recommended that the abrasion-resistant steel be brought up to 100°F (38°C) temperature prior to welding.

Free Machining Steels

The term *free machining* can apply to many metals but it is normally associated with steel and brass. Free machining is the property that makes machining easy because small cutting chips are formed. This characteristic is given to steel by sulphur and in some cases by lead. It is given to brass by lead. Sulphur and lead are not considered alloying elements. In general, they are considered impurities in the steel. The specifications for steel show a maximum amount of sulphur as 0.040% with the actual sulphur content running lower, in the neighborhood of 0.030%. Lead is usually not mentioned in steel specifications since it is not expected and is considered a "tramp" element. Lead is sometimes purposely added to steel to give it free-machining properties.

Free-machining steels are usually specified for parts that require a considerable amount of machine tool work. The addition of the sulphur makes the steel easier to turn, drill, mill, etc., even though the hardness

is the same as a steel of the same composition without the sulphur.

The sulphur content of free-machining steels will range from 0.07–0.12% as high as 0.24–0.33%. The amount of sulphur is specified in the AISI specifications for carbon steels. Sulphur is not added to any of the alloy steels. Leaded grades comparable to 12L14 and 11L18 are available.

Unless the correct welding procedure is used, the weld deposits on free-machining steel will always be porous and will not provide properties normally expected of a steel of the analysis but without the sulphur or lead.

The basis for establishing a welding procedure for free-machining steels is the same as that required for carbon steels of the same analysis. These steels usually run from 0.010% carbon to as high as 1.0% carbon. They may also contain manganese ranging from 0.30% to as high as 1.65%. Therefore, the procedure is based on these elements. If the steels are free-machining and contain a high percentage of sulphur the only change in procedure is to change to a low-hydrogen type weld deposit. In the case of shielded metal arc welding this means the use of low-hydrogen type electrode of the E-XX15, E-XX16, or the E-XXX8 classification. In the case of gas metal arc welding or flux-cored arc welding the same type of filler metal is specified as is normally used since these are no-hydrogen welding processes. Submerged arc welding would not normally be used on free-machining steels. Gas tungsten arc welding is not normally used since free-machining steels are used in thicker sections which are not usually welded with the GTAW process.

The welding procedure should minimize dilution of base metal with the filler metal. Efforts should be made to reduce penetration so as to melt less sulphur of the base metal. This means using lower welding currents.

Free-machining steel can be successfully welded and quality welds made; however, the procedures are slower and, for this reason, free-machining steel should not be specified for weldments, unless absolutely necessary.

Manganese Steel

Manganese steel is sometimes called austenitic manganese steel because of its metallurgical structure. It is also called Hadfield manganese steel after its inventor. It is an extremely tough, nonmagnetic alloy. It has an extremely high tensile strength, a high percentage of ductility, and excellent wear resistance. It also has a high resistance to impact and is practically impossible to machine.

Hadfield manganese steel is probably more widely used as castings but is also available as rolled shapes. Manganese steel is popular for impact wear resistance. It is used for railroad frogs, for steel mill coupling housings, pinions, spindles, and for dipper lips of power shovels operating in quarries. It is also used for power shovel track pads, drive tumblers, and dipper racks and pinions.

The composition of austenitic manganese is from 12–14% manganese and 1–1.4% carbon. The composition of cast manganese steel would be 12% manganese and 1.2% carbon. Nickel is oftentimes added to the composition of the rolled manganese steel.

A special heat treatment is required to provide the superior properties of manganese steel. This involves heating to 1,850°F (1,008°C) followed by quenching in water. In view of this type of heat treatment and the material toughness, special attention must be given to welding and to any reheating of manganese steel.

Manganese steel can be welded to itself and defects can be weld repaired in manganese castings. Manganese steel can also be welded to carbon and alloy steels and weld surfacing deposits can be made on manganese steels.

Manganese steel can be prepared for welding by flame cutting; however, every effort should be made to keep the base metal as cool as possible. If the mass of the part to be cut is sufficiently large it is doubtful if much heat will build up in the part sufficient to cause embrittlement. However, if the part is small it is recommended that it be frequently cooled in water or, if possible, partially submerged in water during the flame cutting operation. For removal of cracks the air carbon arc process can be used. The base metal must be kept cool. Cracks should be completely removed to sound metal prior to rewelding. Grinding can be employed to smooth up these surfaces.

There are two types of manganese steel electrodes available. Both are similar in analysis to the base metal but with the addition of elements which maintain the toughness of the weld deposit without quenching. The EFeMn-A electrode is known as the nickel manganese electrode and contains from 3–5% nickel in addition to the 12–14% manganese. The carbon is lower than normal manganese ranging from 0.50 to 0.90%. The weld deposits of this electrode on large manganese castings

will result in a tough deposit due to the rapid cooling of the weld metal.

The other electrode used is a molybdenum-manganese steel type EFeMn-B. This electrode contains 0.6–1.4% molybdenum instead of the nickel. This electrode is less often used for repair welding of manganese steel or for joining manganese steel itself or to carbon steel. The manganese nickel steel is more often used as a buildup deposit to maintain the characteristics of manganese steel when surfacing is required.

Stainless steel electrodes can also be used for welding manganese steels and for welding them to carbon and low-alloy steels. The 18-8 chrome-nickel types are popular; however, in some cases when welding to alloy steels the 29-9 nickel is sometimes used. These electrodes are considerably more expensive than the manganese steel electrode and for this reason are not popular.

When welding manganese steel with manganese type electrodes the welds should be made with relatively low current and they should be peened with a pneumatic hammer as quickly as possible. This helps spread or deform the deposited weld metal and avoids triaxial shrinkage stresses which can cause cracking. The base metal should be kept as cool as possible and for small parts it is recommended that they be frequently cooled or partially submerged in water. The manganese steel electrodes are available both as covered electrodes and as bare electrodes. Bare electrodes are not popular; however, they are used for large repair welds, etc. Covered electrodes and bare electrodes are operated with the electrode positive on direct current. Preheating is never employed when welding manganese steels and should the part become heated to over 500°F (260°C) it must be reheat-treated to retain its toughness.

Silicon Steel

Silicon steels, or, as they are sometimes called, electrical steels, are steels that contain from 0.5% to almost 5% silicon but with low carbon and low sulphur and phosphorous. Silicon steel is primarily provided as sheet or strip so that it can be punched or stamped to make laminations for electrical machinery. The silicon steels are designed to have lower hysteresis and eddy current losses than plain steel when used in magnetic circuits. This is a particular advantage when used in alternating magnetic fields. Their magnetic properties make silicon steels useful in direct current fields for most applications. Silicon steel stampings are used in the laminations of electric motor armatures, rotors, and generators. They are widely used in transformers for the electrical power industry and for transformers, chokes, and other components in the electronics industry.

Welding is important to silicon steels since many of

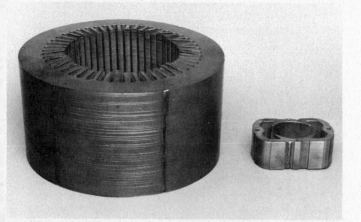

FIGURE 14-7 *Welded lamination stack of silicon steel.*

trode processes are used the stampings are usually indented to allow for deposition of filler metal. For gas tungsten arc and plasma arc the filler metals are not used and the edges are fused. The size of the weld bead should be kept minimum so that eddy currents are not conducted between laminations in the electrical stack.

One precaution that should be taken in welding silicon steel laminations is to make sure that the laminations are tightly pressed together and that all of the oil used for protection and used in manufacturing is at a minimum. Oil can cause porosity in the welds which might be detrimental to the lamination assembly.

Wrought Iron

Wrought iron is a ferrous material made of highly-refined iron with slag minutely and uniformly distributed throughout. The slag is a form of stringers that are in a longitudinal arrangement in the finished product. Wrought iron has been used for structural applications and for pipe. It provides good corrosion resistance and has been used for closed loop piping systems such as hot water coils for radiant heating and brine coils, for cooling ice rinks, etc. It is also used for certain architectural applications.

Many applications of so-called "wrought iron" are actually made of mild low-carbon steel. Very little wrought iron is manufactured in the United States today; however, welding is sometimes required for repair or modifications of existing systems.

Wrought iron should be treated exactly the same as low-carbon mild steel and the same welding processes, procedures, and filler metals should be employed. For welding small wrought iron pipe the oxyacetylene process should be used. For larger pipe use the shielded metal arc process with E-60XX electrodes or the gas metal arc welding process using E70S3 electrode wire. The flux-cored arc welding process using E70T-1 electrodes can also be used.

14-6 CLAD METALS

Most clad metals are composites of a cladding metal such as stainless steel, nickel and nickel alloys, and copper and copper alloys welded to a backing material of either carbon or alloy steel. The two metals are welded together at a mill in a roll under heat and pressure. The clad composite plates are usually specified in a thickness

the laminations are assembled in packs which are welded together. Figure 14-7 is an example of welding a stack of laminations. Welds are made on the edge of each sheet to hold the stack together. Welding is done instead of punching holes and riveting the laminations in order to reduce manufacturing costs. Almost all of the arc welding processes are used, submerged arc, shielded metal arc, gas metal arc, gas tungsten arc and plasma arc welding. The more popular processes are gas metal arc using CO_2 for gas shielding and the gas tungsten arc process. The plasma arc process is used for some of the smaller assemblies. When the consumable elec-

of the cladding which ranges from 5% to 20% of the total composite thickness. The advantage of composite material is to provide at relatively low cost the benefits of an expensive material which can provide corrosion resistance, abrasion resistance, and other benefits with the strength of the backing metal. Clad metals were developed in the early 1930s and one of the first to be used was nickel bonded to carbon steel. This composite was used in the construction of tank cars. Other products made of clad steels are heat exchangers, tanks, processing vessels, materials-handling equipment, storage equipment, etc.

Clad or composites can be made by several different welding manufacturing methods. The most widely used process is roll welding which employs heat and roll pressure to weld the clad to the backing steel. Explosive welding is also used and weld surfacing or overlay is another method of producing a composite material.

Clad steels can have as the cladding material chromium steel in the 12–15% range, stainless steels primarily of the 18/8 and 25/12 analysis, nickel base alloys such as Monel and Inconel, copper-nickel, and copper. The backing material is usually high-quality steel of the ASTM-A285, A212, or similar grade. The tensile strength of clad material depends on the tensile strength of its components and their ratio to its thickness. The clad thickness is uniform throughout the cross section, and the weld between the two metals is continuous throughout.

A slightly different procedure is used for oxygen cutting of clad steel. All of the clad metals mentioned above can be oxygen flame cut with the exception of the copper-clad composite material. The normal limit of clad plate cutting is when the clad material does not exceed 30% of the total thickness. However, higher percentage of cladding may be cut in thicknesses of 1/2 in. (12 mm) and over. The oxygen pressure is lower when cutting clad steel; however, larger cutting tips are used. The quality of the cut is very similar to the quality of the cut of carbon steel. When flame cutting clad material the cladding material must be on the underside so that the flame will first cut the carbon steel. The addition of iron powder to the flame will assist the cutting operation. Schedules of flame cutting are provided by clad steel producers as well as flame-cutting equipment producers. For oxygen flame cutting copper and copper-nickel clad steels the copper clad surface must be removed and the backing steel cut in the same fashion as bare carbon steel. Copper and brass clad plate can

be cut using iron powder cutting. Clad steels can be fabricated by bending and rolling, shearing, punching, and machining in the same manner as the equivalent carbon steels. Clad materials can be preheated and given stress relief heat treatment in the same manner as carbon steels. However, stress relieving temperatures should be verified by consulting with the manufacturer of the clad material.

Welding Clad Steels

Clad materials can be successfully welded by adopting special joint details and following welding procedures presented later in this section. Special joint details and welding procedures are established in order to maintain uniform characteristic of the clad material. Inasmuch as the clad material is utilized to provide special properties it is important that the weld joint retain these same properties. It is also important that the structural strength of the joint be obtained with the quality welds of the backing metal.

The normal procedure for making a butt joint in clad plate is to weld the backing or steel side first with a welding procedure suitable for the carbon steel base material being welded. Then the clad side is welded with the suitable procedure for the material being joined. This sequence is preferable in order to avoid the possibility of producing hard brittle deposits, which might occur if carbon steel weld metal is deposited on the clad material. Different joint preparations can be used to avoid the possible pickup of carbon steel in the clad alloy weld. Any weld joint made on clad material should be a full-penetration joint. When designing the joint details it is wise to make the root of the weld the clad side of the composite plate. This may not always be possible; however, it is more economical since most of the weld metal can be of the less expensive carbon steel rather than the expensive alloy clad metal. This is shown in Figure 14-8 as the preferred type of joint. Dimensions for the groove angles and root face are presented and are recommended for welding clad materials. If the material is of sufficient thickness that a double groove type weld is required it is recommended that the smaller groove side be the clad side. Again, this is in the interest of utilizing the least amount of the alloy clad metal. For thin material 3/16 in. (4.8 mm) or thinner a square groove joint detail should be used.

The selection of the welding process or processes to be used would be based on that normally used for welding the material in question in the thickness and position required. Shielded metal arc welding is probably used more often; however, submerged arc welding is used for fabricating large thick vessels and the gas metal arc welding process is used for medium thicknesses; the flux-cored arc welding process is used for the steel side,

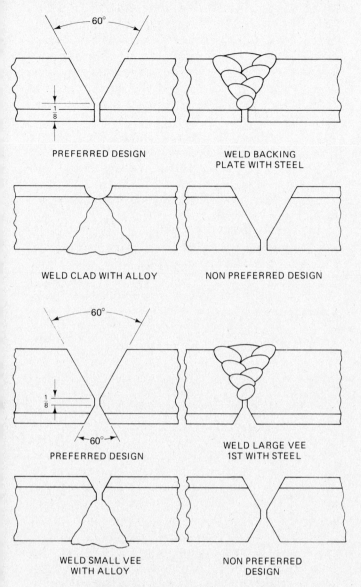

PREFERRED DESIGN

WELD BACKING
PLATE WITH STEEL

WELD CLAD WITH ALLOY

NON PREFERRED DESIGN

PREFERRED DESIGN

WELD LARGE VEE
1ST WITH STEEL

WELD SMALL VEE
WITH ALLOY

NON PREFERRED
DESIGN

FIGURE 14-8 *Weld joint design for clad plate.*

and gas tungsten arc welding is sometimes used for the thinner materials, particularly the clad side. The selection of process should be based on all factors normally considered. It is important to select a process that will avoid penetrating from one material into the other. The welding procedure should be designed so that the clad side is joined using the appropriate process and filler metal to be used with the clad metal and the backing side should be welded with the appropriate process and filler metal recommended for the backing metal. Exceptions to this will be covered later. For code work the welding procedure must be qualified in accordance with the specification requirements.

The normal procedure, assuming that the material is properly prepared and fitted is as follows. The backing side or steel side would be welded first. The depth of

penetration of the root pass must be closely controlled by selecting the proper procedure and filler metal. It is desirable to produce a root pass which will penetrate through the root of the backing metal weld joint into the root face area yet not come in contact with the clad metal. A low-hydrogen deposit is recommended. If penetration is excessive and the root bead melts into the clad material because of poor fitup or any other reason the deposit will be brittle. If this occurs the weld will have to be removed and remade. However, if the penetration of the backing steel root bead is insufficient the amount of back gouging will be excessive and larger amounts of the clad material weld metal will be required. The steel side of the joint should be welded at least half way prior to making any of the weld on the clad side. If warpage is not a factor, the steel side weld can be completed before welding is started on the clad side.

The clad side of the joint is prepared by gouging to sound metal or into the root pass made from the backing steel side. This can be done by air carbon arc gouging or by chipping. The gouging should be sufficient to penetrate into the root pass so that a full penetration of the joint will result. This will determine the depth of the gouging operation. It is also a measure of the depth of penetration of the root pass. Grinding is not recommended since it tends to wander from the root of the joint and may also cover up an unfused root by smearing the metal. If the depth of gouging is excessive, weld passes made with the steel electrode may be required to avoid using an excessive amount of clad metal electrode.

On thin materials the gas tungsten arc welding process may be used, on thicker materials the shielded metal arc welding process or the gas metal arc process may be used. The filler metal must be selected to be compatible with the clad metal analysis. There is always the likelihood of diluting the clad metal deposit by too much penetration into the steel backing metal. Special technique should be used to minimize penetration into the steel backing material. This is done by directing the arc on the molten puddle instead of on the base metal. When welding copper or copper nickel clad steels a high nickel electrode is recommended for the first pass (ECuNi or ENi-1). The remaining passes of the joint in the clad metal should be welded so that the copper or copper nickel electrode matches the composition of the clad metal.

When the clad metal is stainless steel the initial pass which might fuse into the carbon steel backing

WELDING SPECIAL AND DISSIMILAR METALS

should be of a richer analysis of alloying elements than necessary to match the stainless cladding. This same principle is used when the clad material is Inconel or Monel. The remaining portion of the clad side weld should be made with the electrode compatible with or having the same analysis as the clad metal. The procedure should be designed so that the final weld layer will have the same composition as the clad metal.

On heavier thicknesses where the weld of the backing steel is made from both sides it is important to avoid allowing the steel weld metal to come in contact or to fuse with the clad metal. This will cause a contamination of the deposit which may result in a brittle weld.

When welding thinner gauge clad plate and inside clad pipe it may be more economical to make the complete weld using the alloy weld metal compatible with the clad metal instead of using two types of filler metal. The alloy filler metal must be compatible with the steel backing metal. The expense of the welding filler metal may be higher, but the total weld joint may be less expensive because of the more straightforward procedure. Joint preparation may also be less extensive using this procedure. For medium thickness, the joint preparation is a single vee or bevel without a large root face. The root face is obtained by grinding the feather edge to provide a small root face. If possible the face of the weld will be the steel or backing side of the joint. The backing side or steel side is welded first using the

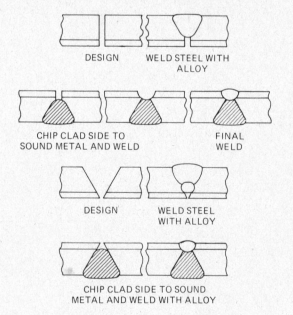

FIGURE 14-9 *Alloy weld from both sides.*

small diameter electrode for the root pass to insure complete penetration. The remainder of the weld is made on the steel side. The weld is completed by chipping the back side or clad side of the joint and making a final pass from that side. This is shown by Figure 14-9.

If the composite is a pipe or if it must be welded from one side the *buttering* technique should be used. In this case the filler metal must provide an analysis equal to the clad metal and be compatible with the backing steel. Weld passes are made on the edge of the composite to butter the clad and backing metal. The buttering pass must be smoothed to the design dimensions prior to fitup. The same electrode can be used to make the joint.

When the joint is welded from the clad side the procedure is the same. The filler metal deposit must match the clad metal but be compatible with the backing metal. This is shown by Figure 14-10. The same basic joint procedure can be used for flat, vertical, horizontal, or overhead weld joints.

ALLOY WELD METAL TO MATCH CLAD

FIGURE 14-10 *Alloy weld from clad side.*

When welding heavy, thick composite plate the U groove weld joint design is recommended instead of the vee groove in order to minimize the amount of weld metal. The same principles mentioned previously are used.

When the submerged arc welding process is used for the steel side of the clad plate caution must be exercised to avoid penetrating into the clad metal. This same caution applies to automatic flux-cored arc welding or gas metal arc welding. A larger root face is required and fitup must be very accurate in order to control root bead penetration.

The submerged arc process can also be used on the clad side when welding stainless alloys. However, caution must be exercised to minimize dilution of a high-alloy material with the carbon steel backing metal. The proper filler metal and flux must be utilized. To minimize admixture of the final pass it is recommended that the clad side be welded with at least two passes so that dilution would be minimized in the final pass.

When making T joints, corner joints, or lap joints special precautions must be taken so as to have the weld metal proper relationship to the clad and the backing material. This is illustrated by Figure 14-11.

Special quality control precautions must be established when welding clad metals so that undercut, in-

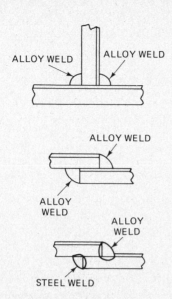

ALLOY WELD   ALLOY WELD

ALLOY WELD

ALLOY
WELD

ALLOY
WELD

STEEL WELD

FIGURE 14-11 *Tee and lap joints.*

complete penetration, lack of fusion, etc., are not allowed. In addition, special inspection techniques must be incorporated to detect cracks or other defects in the weld joints. This is particularly important with respect to the clad side which may be exposed to a corrosive environment.

14-7 DISSIMILAR METALS

The procedures described for obtaining quality welds so far in this book have been between identical metals or at least of metals of similar composition and properties. There are many applications, however, in which weldments are made from metals of different compositions. For example, a small portion of a weldment may be subjected to a corrosive atmosphere. It is, therefore, desirable to make the area subjected to corrosion out of a material that withstands the corrosive action; the remainder could be made of mild steel. The same is true of a mechanical wear problem, a high-temperature situation, or other conditions in which different properties are required from different parts of the same weldment. This brings about the need for joining dissimilar metals. Welding dissimilar metals is being done more often; therefore, this chapter is devoted to providing information for successfully joining some of the more common metal combinations.

A successful weld between dissimilar metals is one that is as strong as the weaker of the two metals being joined, i.e., possessing sufficient tensile strength and ductility so that the joint will not fail in the weld. Such joints can be accomplished in a variety of different metals and by a number of the welding processes. The problem of making welds between dissimilar metals relates to the transition zone between the metals and the

intermetallic compounds formed in this transition zone. For the fusion type welding processes it is important to investigate the phase diagram of the two metals involved. If there is mutual solubility of the two metals the dissimilar joints can be made successfully. If there is little or no solubility between the two metals to be joined the weld joint will not be successful. The intermetallic compounds that are formed, between the dissimilar metals, must be investigated to determine their crack sensitivity, ductility, susceptibility to corrosion, etc. The microstructure of this intermetallic compound is extremely important. In some cases, it is necessary to use a third metal that is soluble with each metal in order to produce a successful joint.

Another factor involved in predicting a successful service life for a dissimilar metals joint relates to the coefficient of thermal expansion of both materials. If these are widely different, there will be internal stresses set up in the intermetallic zone during any temperature change of the weldment. If the intermetallic zone is extremely brittle service failure may soon occur. The difference in melting temperatures of the two metals that are to be joined must also be considered. This is of primary interest when a welding process utilizing heat is involved since one metal will be molten long before the other when subjected to the same heat source. When metals of different melting temperatures and thermal expansion rates are to be joined the welding process with a high heat input that will make the weld quickly has an advantage.

The difference of the metals on the electrochemical scale is an indication of their susceptibility to corrosion at the intermetallic zone. If they are far apart on the scale, corrosion can be a serious problem.

In certain situations, the only way to make a successful joint is to use a transition material between the two dissimilar metals. An example of this is the attempt to weld copper to steel. The two metals are not mutually soluble, but nickel is soluble with both of them. Therefore, by using nickel as an intermediary metal the joint can be made. Two methods are used: (1) use a piece of nickel, or (2) deposit several layers of nickel alloy on the steel, i.e., buttering or surfacing the steel with a nickel weld metal deposit. The nickel or nickel deposit can be welded to the copper alloy using a nickel filler metal. Such a joint will provide satisfactory properties and will be successful.

Another method of joining dissimilar metals is the use of a composite insert between the two metals at the weld joint. The composite insert consists of a transition joint between dissimilar metals made by a welding process that does not involve heating. By selecting the proper materials for the composite insert like metals can be welded to like metals in making the fusion weld joint.

The following is a brief description of some of the welding processes that can be used for composite inserts that include transition joints and that do not employ filler metals.

Explosion welding is used to join many so-called incompatible metals. In explosion welding the joint properties will be equal to those of the weaker of the two base materials. Since minimum heat is introduced there is no melting, no heat-affected zone, and no thermal compounds are formed. The characteristic sine wave pattern of the interface greatly increases the interface area. This process is used for cladding, but is also used to make composite transition inserts used for the fusion welding of dissimilar metals. Composites containing a transition joint are commercially available between aluminum and steel, aluminum and stainless steel, aluminum and copper, and other materials.

Cold welding is used for making dissimilar metal transition joints. This process does not use heat, thus avoids the heat-affected zone and the intermetallic fusion alloy. Little or no mixing of the base metals takes place. It is commonly used to join aluminum to copper.

Ultrasonic welding is also used for welding dissimilar metals since very little heat is developed at the weld joint. Ultrasonic welding can be used only for very thin materials or small parts.

Friction welding is also used for joining dissimilar metals and for making composite transition inserts. Various dissimilar combinations of metal have been welded, including steel to copper base alloys, steel to aluminum, stainless to nickel base alloys, etc. In friction welding only a very small amount of the base metal is heated and that which is melted is thrown from the joint, therefore, the intermetallic material is kept to a very minimum. The heat-affected zone is also minimal.

The high-frequency resistance process is also widely used for dissimilar metal welding. Here the heat is concentrated on the very surface of the parts being joined and pressure applied is sufficient to make welds of many dissimilar materials. It can be used for joining copper to steel at very high speeds.

Diffusion welding is widely used for aerospace applications of dissimilar metals welding. Percussion welding is also used but this process is restricted to wires or small parts. The laser beam welding process has also been used but it is restricted, at this time, to very thin materials.

The electron beam welding process has had wide application for joining dissimilar metals. Electron beam uses high-density energy and fast welding speed. It seems to overcome the difference of thermal conductivity when welding metals together having wide variation of thermal conductivity. In addition, the weld zone is extremely small and filler metal is not introduced. Since there is such a small amount of intermetallic compound formed electron beam does offer an advantage for many dissimilar combinations.

The flash butt welding process will make high-quality welds between copper and aluminum. With proper controls all or most of the molten metal is forced out of the joint and the weld is complete as a solid state process. Flash butt welds are made in rods, wires, bars, and tubes.

Arc Welding Dissimilar Metals

There are many welding requirements to join dissimilar metals in which the above processes cannot be used. In these cases, the three popular arc welding processes are most often utilized. These are the shielded metal arc welding process, the gas tungsten arc welding process, and the gas metal arc welding process. The remainder of the chapter will be devoted to utilizing these processes to join some of the more common combinations of dissimilar metals.

The popular combinations of dissimilar metals that are joined are shown by the table of Figure 14-12. This table summarizes the requirement to join aluminum to different metals, copper and copper alloys to different metals, nickel alloys to different metals, stainless steel to carbon steels, and the welding together of various types of steels. All these combinations can be successfully welded using the following procedures.

Welding Aluminum to Various Metals

There is a wide difference of melting temperatures, for aluminum, approximately 1,200°F (649°C), and for steel, approximately 2,800°F (1,538°C). The aluminum will melt and flow away well before the steel has melted. The aluminum iron phase diagram shows that a number of complex brittle intermetallics are formed. It is found that iron aluminum alloys containing more than 12% iron have little or no ductility. In addition, there is wide difference in the coefficient of linear expansion, in thermal conductivity, and in specific heats of aluminum and steel. This will introduce thermal stresses of considerable magnitude.

The most successful method is to use an aluminum-

Base Metal Combinations	WELDING PROCESS and FILLER METAL		
	SMAW	GTAW	GMAW
Aluminum to mild and La steel	Use a transition insert of these metals or coat the surface of the steel and GTAW		
Aluminum to stainless steel	Use a transition insert of these metals or coat the surface of the SS and GTAW		
Aluminum to copper	Use a transition insert of these metals		
Copper to mild and La steel	ECu	RCu	ECu
Copper to stainless steel	ECuAl-A2	RCuAl-A2	ECuAl-A2
Brass to mild and La steel	ECuAl-A2	RCuAl-A2	ECuAl-A2
Aluminum bronze to La steel	ECuAl-A2	RCuAl-A2	ECuAl-A2
Inconel to mild and La steel	ENiCrFe-3	RNiCrFe-3	ENiCrFe-3
Inconel to austinitic stainless steel	ENiCrFe-3	RNiCrFe-3	ENiCrFe-3
Inconel to ferritic stainless steel	ENiCrFe-3	RNiCrFe-3	ENiCuFe-3
Monel to mild steel and La	EniCu-2	RNiCu-2	ENiCu-2
Monel to Austinitic Stainless steel	ENiCu-2	RNiCu-2	ENiCu-2
Monel to ferritic stainless steel	ENiCu-2	RNiCu-2	ENiCu-2
Ferritic stainless steel to mild and La steel	ENiCrFe-3	RNiCrFe-3	ENiCrFe-3
Austinitic stainless steel to mild and La steel	ENiCrFe-3	RNiCrFe-3	ENiCrFe-3
Alloy steel to mild and La steel	E7018	E70S-X	E70S-X
Q and T steel to mild and La steel	E7018	E70S-X	E70S-X

FIGURE 14-12 *Popular dissimilar metal combinations.*

steel transition insert with each metal welded to its own base metal using any of the three arc welding processes.[6]

The other way is to coat the steel surface with a metal compatible with aluminum. The success of this type of joint depends on metal used to coat the steel, the thickness of the coating, and the bond between the coating and the steel surface. A coating of zinc on steel can be used and the aluminum welded to it by the gas tungsten arc welding process. A high-silicon aluminum filler wire should be used. Direct the arc toward the aluminum; pulsing will assist the welder.

For welding aluminum to stainless steel transition inserts are available. It is also possible to use the coating technique. A coating for the stainless steel is pure aluminum coating, which can be applied by dipping clean stainless steel into molten aluminum. Another way to obtain a compatible coating is by tinning the stainless steel with a high-silicon aluminum alloy. The aluminum surface can then be gas tungsten arc welded to the aluminum. The arc should be directed toward the aluminum; pulsing will assist the welder.

The welding of aluminum to copper is accomplished by using a copper-aluminum transition insert piece.

Welding Copper to Various Metals

Copper and copper-base alloys can be welded to mild and low-alloy steels and to stainless steels. For thinner sections, in the gauge metal thickness, the gas tungsten arc welding process can be used with a high-

copper-alloy filler rod. The pulsed mode makes it easier to obtain a quality weld. The arc should be directed to the copper section to minimize pickup of iron. In the heavier thicknesses first overlay or butter the steel with the same filler metal and then weld the overlayed surface to the copper. It is important to avoid excessive penetration into the steel portion of the joint since iron pickup in copper alloys creates a brittle material. The copper must be preheated.

Another method is to overlay the copper with a nickel-base electrode. A second overlay or layer is recommended on thicker materials. When making the overlay welds on thick copper, the copper should be preheated to 1,000°F (538°C). The overlay or buttered surface of the copper part should be smoothed to provide a uniform joint preparation. The copper part should be preheated to 1,000°F (538°C) when making the weld. Effort should be made to minimize dilution of the copper with the nickel electrode. The shielded metal arc process, the gas tungsten arc or gas metal arc welding processes can all be used. The selection will depend on equipment available and the thickness of the material being joined. Copper can also be joined to stainless steel, and brass can be joined to mild and low-alloy steels. The proper filler metal in each case is shown by Figure 14-12.

Welding Nickel-Base Alloys to Steels

Nickel-base alloys such as Monel and Inconel can be successfully welded to low-alloy steel by using the

Monel analysis of filler material when using any of the arc welding processes. In the case of Inconel to mild or low-alloy steel the Inconel base electrode would be used. The same situation applies also to the welding of Inconel or Monel to stainless steels. Here the Inconel or Monel type electrode is used.

Welding Stainless Steels to
Various Metals

Most stainless steels can be successfully welded to mild and low-alloy steels. Consideration must be given to the effects of dilution of the weld metal with the two base metals and the different coefficients of thermal expansion of stainless steel and mild or low-alloy steels. The deposited weld metal will tend to pick up alloys from both parts of the joint being welded. The effect of this dilution can be controlled by buttering or overlaying the surface of one of the metals being joined and by selecting the correct electrode or filler material. The weld joint between a slightly ferritic stainless steel and a mild or low-alloy steel would be hard and brittle if made with a slightly ferritic stainless steel electrode. However, if a fully austenitic stainless steel electrode or filler rod is used the amount of ferrite in the weld metal would be reduced to a tolerable level. The electrode or filler wires normally used would be an E310 electrode or filler rod corresponding to the 310 composition. The stabilized stainless steel electrodes and filler wires should be used. The best selection would be a E NiCr-1 which is a 15% chromium high nickel composition. This analysis has an austenitic composition and the weld deposit can tolerate considerable dilution before becoming crack sensitive.

Stainless steel has a coefficient of thermal expansion about twice that of mild or low-alloy steel. During the cooling cycle of the weld the stainless steel side will tend to contract more than the mild steel or low alloy steel side. This difference in contraction will set up stresses in the weld joint. If the weld joint is subjected to repetitive thermal cycles the resulting stress cycling could cause premature failure of the joint in a manner similar to fatigue fracture. The use of stainless steel for buttering will not solve this situation. The buttering technique is not recommended for those situations in which repetitive thermal cycles are involved. The best solution is to use a high-nickel electrode such as the E NiCr-1 or the E NiCrFe-2 electrodes or the ERNiCrFe-5 or ERNiCr-3 filler wires. The relatively low iron content

of these alloys allows minimum iron in the weld which reduces cracking. High-nickel deposits have a thermal expansion coefficient similar to that of the low-alloy steel. Thus the thermal stress will be set up in the weld metal and stainless steel rather than the mild or low-alloy steel. Both the weld metal and the stainless steel have good ductility and can absorb the stress cycles without premature failure.

When using the stainless steel or the high-nickel overlay, the joint design must be altered to provide space for the buttering layer.

Welding Steel to Different Steels

Steels which have similar metallurgical structures are normally welded with electrodes matching the composition or behavior of the lower-strength material. This applies not only to various grades or strength levels of carbon and low-alloy steels but also to various grades of stainless steels. For example, a 316 stainless steel should be welded to a 304 stainless steel with a 308 composition filler metal. The 308 would slightly overmatch the 304 composition. More data on the subject of welding different stainless steels is covered in Chapter 12.

Another example would be the welding of a quenched and tempered steel to a low-alloy high-strength steel. The electrode normally used for joining the low-alloy high-strength steel to itself should be used for welding it to the quenched and tempered steel. The heat input requirements of the quenched and tempered steel should be followed and in all cases a low-hydrogen deposit is required. Another example is the welding of a low-chrome-moly steel to a plain carbon mild steel. In this case, the standard E7018 type electrode or filler metal designed for the carbon or low-alloy steel would be used.

Conclusion

When welding dissimilar metals together it is important to consider the problem areas. These relate to the solubility of the metals with one another and the formation of brittle alloys. Second, it relates to the difference in thermal expansion and contraction and the recommended use of ductile weld metals to help absorb these stresses. The buttering technique is most often used. Other techniques involve the plating or coating of one of the base materials with a material compatible to both metals and then making the weld. Bimetallic, composite transition inserts, which can be welded to each type of base metal, are used for certain combinations. Information about joining less common metals can be found in *Welding Research Bulletin* No. 210.[7]

QUESTIONS

1. What property of cast iron makes it difficult to weld? Why?

2. Why is a full-penetration weld necessary for repairing cast iron?

3. Identify the four types of covered electrodes for cast iron welding. What is the advantage of each?

4. Why is preheat usually specified for welding cast iron?

5. Tool and die welding is very complex. What are the important factors?

6. How are deformed reinforcing bars identified? Why is this important?

7. Describe the different types of splices of reinforcing bars.

8. Low-hydrogen electrodes are required. What else is required for successfully welding rebars?

9. What safety precaution should be taken when welding galvanized steel?

10. Why is the galvanized coating damaged adjacent to the weld?

11. Why is GTAW not recommended for welding galvanized steel? How can the difficulty be reduced?

12. How can the galvanized coating be repaired alongside the weld?

13. What welding technique should be utilized when welding free-machining steel?

14. Why shouldn't preheat be used when welding Hadfield manganese steel?

15. When welding clad metals which side should have the vee groove? Why?

16. Which side should be welded first, the clad side or the steel side?

17. When making a tee joint of clad metal should alloy or steel electrode be used?

18. What makes welding different types of metal together more difficult? Why?

19. What is a transition piece? Where is it placed in the joint?

20. What welding processes are used to produce transition pieces?

REFERENCES

1. "Specification for Welding Rods and Covered Electrodes for Welding Cast Iron," A5.15-69, *American Welding Society,* Miami, Florida.

2. JOHN ENTENMANN AND HOWARD CARY, "Welding Malleable Iron," *Modern Castings,* November 1963.

3. "Manual of Standard Practice," Concrete Reinforcing Steel Institute, January 1, 1973, Chicago, Illinois 60601.

4. "Reinforcing Steel Welding Code," D12.1-75, *American Welding Society,* Miami, Florida.

5. ELDON BRANDON, "Techniques for Welding and Brazing Galvanized Steel," *Metals Progress,* June 1966.

6. F. R. BAYSINGER, "Welding Aluminum to Steel Using Transition Insert Pieces," *The Welding Journal,* February 1969, Miami, Florida.

7. M. M. SCHWARTZ, "The Fabrication of Dissimilar Metal Joints Containing Reactive and Refractory Metals," *Welding Research Council,* Bulletin No. 210, October 1975, New York.

15

DESIGN FOR WELDING

15-1 ADVANTAGES OF WELDED CONSTRUCTION

15-2 WELDMENT DESIGN FACTORS

15-3 WELDING POSITIONS, JOINTS, WELDS, AND WELD JOINTS

15-4 WELD JOINT DESIGN

15-5 WELDMENT DESIGN AND INFLUENCE OF SPECIFICATIONS

15-6 WELDMENT CONVERSION FROM OTHER DESIGNS

15-7 WELDMENT DESIGN PROBLEMS

15-8 WELDING SYMBOLS

15-1 ADVANTAGES OF WELDED CONSTRUCTION

A weldment is an assembly whose component parts are joined by welding, thus a weldment can be an all-welded ship, a skyscraper, an automobile body, a bicycle frame, or a coffee pot, but in any case it must be designed as a weldment. The weldment must meet its service life performance requirements at a minimum cost and have a pleasing appearance. To do this, it should be designed initially as a weldment and not designed as a riveted structure or as a casting or forging and then switched. In order to properly utilize the tremendous cost savings potential of welding, the structure or part must be designed as a weldment from the beginning. Too often designers remembering the principles learned for riveted or bolted design may use lap type joints in weldments, or the designer remembering the principles of casting design may incorporate thicker sections than are required for the strength of the part. The weldment must be designed strictly to satisfy the functional requirements of the part.

Designers of welded parts and structures must break from past design limitations and design weldments to perform the required function with the least amount of material and the minimum amount of production labor.

This requires some ingenuity, but the results obtained from intelligently utilizing the advantages of weld design are worth this effort.

The most economical solution to a design requirement will be accomplished if the designer intelligently takes into consideration these ten points.

1. The total service requirements of the product.
2. The types of loadings and methods of accurately calculating stresses.
3. The allowable working stresses.
4. The mechanical and physical properties of base materials to be used.
5. The capabilities of the welding processes to be used and the weld deposit properties.
6. Joint types and weld types—their design and limitations.
7. Fabrication methods available—the advantages and potential problems and costs.
8. The cost of welding when using different processes and procedures.
9. Clear-cut communications of weld designs including the use of welding symbols.
10. Quality specifications and inspection techniques.

Weldments offer many advantages over other design concepts.

1. The weldment is normally lighter in weight than cast or mechanically fastened structures; thus, it requires less material.

2. The weldment design can be readily and economically modified to meet changing product requirements.

3. The production time for a weldment is usually less than that of other manufacturing methods. (This is a hidden cost savings benefit.)

4. The weldment will be more accurate with respect to dimensional tolerances than a casting (another hidden cost savings advantage).

5. Weldments are more easily machined than castings.

6. Weldments are tight and leakproof and will not shift or give like riveted structures.

7. The capital investment for producing weldments is much lower than for producing castings. Additionally, environmental controls are more easily adopted to the welding shop than to the foundry.

8. Weldments can be more pleasing to the eye than castings. They are cleaner in their lines and usually smoother and more easily prepared for final use.

Some of the examples mentioned in the beginning of this book should be re-examined. Compare the welded and riveted splice mentioned in Section 1-1 and prove again to yourself the reduction in material and the reduction in production labor. There is also the advantage of splicing pipe by means of welds versus mechanical methods. This may allow the entire structure to be made of thinner wall pipe with a great savings in the cost of the pipe. It also provides a smoother surface for the material traveling through the pipe. Also consider the advantage of welding when different types and strengths of metals can be placed strategically within the weldment to meet design requirements.

Observe the savings obtained with the welded truck brake shoe when compared to the cast design shown by Figure 15-1. The weight of the rough casting is 23 lb (10.5 kg) and the weldment is 13 lb (5.8 kg). The extra material and machine work penalize most casting designs. It is often claimed by proponents of cast design

FIGURE 15-1 *Welded truck brake shoe.*

that if quantity production is involved the casting will be the least expensive. This is definitely not true in the case of this truck brake shoe. Quantity production is required to justify the blanking dies for the web plates and also for the two welding operations. This is a good example of tooling for high-volume production of a weldment.

Normally, under ideal conditions, the weldment is less expensive than the casting or mechanically assembled structure. Whenever this is not true it is an indication that the weldment design is less than optimum or there is some other factor or circumstance involved.

Ultimately the success or lack of success of a specific weldment lies with the designer of the weldment. The designer is responsible for the design of the weldment and it is essential that the designer be fully aware of these responsibilities. All of the following comments concerning weldment design are for products made of carbon steel or low-alloy steels, which are most often used for making weldments. The principles involved normally apply for other metals.

The designer must have a clear understanding of the service requirements and expected service life of the product being designed. It is the designer's responsibility for developing a product that will properly function under the service conditions that it will encounter. In addition, the designer must be completely aware of the properties of the materials involved and how the materials must be treated and handled in fabricating and welding. The design, the materials to be used, and the production procedures are interrelated and must be considered when making the design. Design factors are sometimes covered by specifications or codes. In a sense, the specifications relieve the designer of some of these responsibilities, specifically with respect to working stresses, minimum weld sizes, etc. However, even with codes and specifications the interrelationships of materials and manufacturing methods must still be considered.

Design of weldments can be extremely difficult because weldments cover such a wide variety of parts ranging from miniature pieces in precision equipment through heavy massive machinery. Weldments are used to build transportation equipment such as railroad cars, locomotives, truck frames, airplane frames, automobiles, and ships. Weldments become storage tanks, pressure vessels, and missile cases. Weldments also become simple single-span bridges and complicated structures such as offshore drill rigs, ore unloaders, giant stripping power shovels, and the tallest building in the world. It is difficult to provide detailed information for each of these types of weldments, and, therefore, only general comments will be made concerning design systems and design concepts. It is the designer's added responsibility to select and choose the correct weld configurations to properly carry the loads to sustain the service life of each weldment designed. It is fundamental that the designer give attention to the following principles:

The economic factors must be considered. The first costs of the weldment may not always be the determining factor. Maintenance and repair cost over the life of the weldment may be more important than the initial cost. The initial cost must be considered concurrently with economic benefits to be derived from welded construction.

Weight reduction is desirable for moving structures such as railroad rolling stock, construction equipment, highway trucks, etc. Increased payloads and reduced energy required to move such structures plus cost of materials makes a strong case for designing to minimum weight. Welding allows for weight reduction by utilizing the full section of members.

Utilization of high-strength metals becomes increasingly important in reducing weight and cost. The designer must select the strength level and the strength-density relationship of the metal to be used in the weldment.

Dynamic loading is most common in moving equipment. This factor has an effect on the fatigue life of the structure. It is based on reversal of loading but is also involved with varying loads either tensile or compression. Impact loadings can increase the normal static stress by factors as high as ten times. This has a great effect on weldment design.

In conjunction with dynamic and impact loading, the service temperature is important. Structures exposed to extreme low temperatures, like those encountered with construction equipment operating in the Arctic, and severe dynamic loads, must be very carefully

designed to sustain these loads. Cold temperatures magnify the problems of stress concentrations, design notches, and make notch toughness more important.

In some cases, the loading of the weldment is strictly of a static nature and no movement is involved. Under these conditions fatigue loading or impact loading can be ignored.

In some weldments, corrosion resistance might be a major factor with respect to service life. Corrosion exposure largely determines the metal that will be used for the weldment and also is a major factor in the selection of the welding process and procedure.

There are other problems such as exposure to high temperatures that may be continuous or intermittent. Continuous high temperature poses a design problem but temperature changes can be more difficult for the designer.

Coupling high temperatures or intermittent high and low temperatures with corrosive atmospheres compounds the designer's problems.

Another important factor is resistance to the abrasion that can occur from use in construction equipment or the interaction of moving parts. Special metals can be placed at wear points. There are many other service related factors that should be considered, such as exposure to nuclear radiation, exposure to potential vandalism, plus exposure to a combination of many of the factors mentioned previously.

In addition, the designer must have a clear-cut understanding of the fabrication procedures and techniques that may be involved in the production of the weldment. This would include such factors as the ability to oxygen flame cut, shear, or machine the material involved. The designer must also know whether it is possible to form the material cold or whether hot forming is required. The designer must know if the metals can be welded and if special precautions are required. The weldability of the material to be used for the product is of utmost importance in the design of any weldment.

The designer must provide the inspector with guidance to select the class of inspection required and indicate special requirements for specific weld joints. The type of nondestructive testing should be specified and tolerances for dimensions should be given.

Finally, the designer must transmit to the welding shop the exact specification for each weld by means of welding symbols. At the last, or perhaps it should be first, the designer must make sure that every weld joint is accessible to the welder and can be made.

15-2 WELDMENT DESIGN FACTORS

The design of a weldment involves the efficient and economical use of the metals selected. It also involves the design of all weld joints used to make the weldment. We have previously mentioned the service conditions to which the weldment will be exposed and have covered the factors that the designer must consider in designing the weldment to meet these service requirements. In order to satisfy the service requirements the designer must have reliable information of the performance characteristics of the metals that are used. Most designs are based on an analysis of the stresses imposed on the part or structure. The designer must completely and accurately determine the loads that will be encountered. The **load** is the force that stresses a part.

It is beyond the scope of this book to develop how loads imposed on structures are calculated and how these loads develop specific stresses within the part. Since we are concerned with the welding phase or the design of the welds rather than the total design of the part, no extended treatment of stress analysis will be given. There are many excellent handbooks on machine design and structural design that may be consulted. It is important to point out, however, that the more accurately these loads and stresses can be determined the more efficient the design of the weldment will be. It is important that designers be fully cognizant of loads and stresses and utilize techniques such as electric strain gauges, electronic measuring devices, elastic examination of transparent models, and the use of brittle lacquers which crack under strain on actual parts under load. Stress analysis must include not only static loadings, but also the dynamic loads, including impact and fatigue that are involved. The structure must support its own weight, the dead loads, and the superimposed loads and forces produced by all service conditions, including all live loads and forces, centrifugal or accelerating and decelerating action, transmission of loads, erection loads, loads due to thermal stresses, etc. The term *load* is used here in all inclusive sense to cover all the external forces acting on the weldment that cause internal stresses.

In the design of a weldment it is first necessary to establish the outlines of the proposed part or structure and then to determine all the different loads and forces and environmental conditions that will be imposed upon it.

The weldment must be designed to resist these imposed forces and this partially depends upon the mechanical and physical properties of the metal pieces that become the weldment. The important properties of the base metals that the designer must know are the values of the yield strength of the metal, the ultimate strength of the metal, and its modulus of elasticity. The strength properties of the material to be used must be selected. The allowable working stress must be determined, the factor of safety of the weldment must be established, and the cross-sectional area of the parts can then be calculated.

The *allowable working stress,* sometimes called allowable unit stress, is the maximum stress level that is allowed anywhere within the weldment. The stresses within the weldment, calculated by the various formulas based on the type of loading, must not exceed this value. The allowable working stress, for many types of weldments, is dictated by a code or specification. Different values are given for different materials or different grades of steels and for different kinds of loadings. Handbooks provide this information. When a code or specification does not apply, the allowable working or unit stress is calculated by dividing the yield strength of the material by a factor so that the part would not be subjected to stresses above its yield point. If this were to happen the part would deform and possibly fail. This factor is known as the *factor of safety.*

The factor of safety has an extremely important bearing on the economics of a design. In order to produce the lightest-weight item or weldment utilizing the least amount of material the factor of safety should be the lowest. For example, in the design of missile cases, the factor of safety may be as low as 1.25. The selection of the factor of safety must be based on several factors, and the accuracy with which loads are selected to represent service conditions has a great bearing. When these loads can be determined with great accuracy, the factor of safety can be lower. If there is doubt or there are numerous assumptions about the loads, the allowable working stress should be more conservative or lower which makes the safety factor higher. Approximations are often used in determining loads on members and their accuracy is based on the designer's experience. However, with accurate experimental work to determine loads the expected service life can be attained with a lower factor of safety. Finally, it is important for the designer to thoroughly consider the importance of the structure or weldment and the possibility of damage that might occur if it were to fail. The factor of safety used in designing many weldments is based on experience from previous application of weldments in similar service situations. In general, it is possible to reduce the factor of safety when a strict quality control program is utilized to insure that the materials and welds are precisely designed. In most structural work and machine design, a factor of safety of 1.8 is commonly used. However, it is the responsibility of the designer to select the proper factor of safety and allowable stresses for materials utilized, based on the service requirements that will be encountered by the end product.

The internal forces normally called stresses that occur within the structure can then be calculated. In simple tension this would be done by dividing the load or force by the cross-sectional area of the component parts that resist these forces. Dividing the force by the cross-sectional area of the part will determine the unit stress. The cross-sectional area is proportioned to provide sufficient area so that the internal stress does not exceed the allowable working stress for the type of loading and the metal used in the weldment. There are special cases such as long, slender columns, in which the length to slenderness ratio is the controlling factor since such members may fail by buckling. Bending loads also need special analysis.

For the designer to better analyze the stresses in a weldment it is common practice to classify the types of loads that are imposed and from these to determine the types of stresses in the structure. The five basic types of loads on a weldment are as follows:

1. Tensile load.
2a. Compression-load (short section).
2b. Compression-load (long section).
3. Bending loads.
4. Torsion load.
5. Shear load.

Most all loadings can be categorized into one of these. In many cases it is necessary to consider combinations of these loads. It must be remembered in considering loads and stresses in a weldment that the weldment is a monolithic structure and that all pieces work together. In other words, a load imposed on one particular piece of a weldment is transmitted through the weld joints to the other pieces of the weldment. For simplicity's sake, it is usually assumed that the parts are connected by welds and that the weld transmits the load from one part to another.

Most design work is classified as either structural design or machine design. In general, structural design refers to large structures usually made of hot rolled steel sections such as angles, bars, plates, beams, etc., connected at intersecting points by welds. Machine design on the other hand normally involves smaller, more compact weldments made of parts cut from steel plates and sometimes castings welded together at all interfacing points. As an example, compare a roof truss which is a structural assembly to a machine tool base,

which is made of many pieces of steel plates. The design analysis techniques are somewhat different for these two types of weldments.

A *tensile* load is considered the simplest type of load. The axial load is indicated by the letter *F*, indicating force or tensile force. The load, parallel to the axis, is divided by the cross-sectional area, *A*, which is the product of the thickness *t*, and width *l*, to give the unit stress indicated by *S*. Thus the equation

$$S = \frac{F}{A} \text{ or } \frac{F}{tl}$$

Such stresses and loading are considered to be uniform over the entire cross-sectional area. The cross-sectional area can be taken anywhere along the member or through the throat of the butt weld joint. In normal weldment calculations the weld reinforcement is ignored and a unit stress is based on the theoretical thickness by the length of the weld as shown by Figure 15-2.

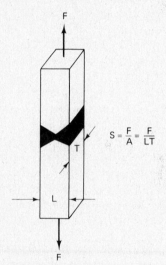

$$S = \frac{F}{A} = \frac{F}{LT}$$

FIGURE 15-2 *Tension loading.*

A *compression* load is much the same as a tensile load if it is imposed on a short member. Figure 15-3 illustrates this situation with the loading and equation identical to that of the tensile load, which produces tension; however, in this case the load results in compression of the weld. In most metals the mechanical properties in compression are not as clearly defined as those in tension. There is no definite ultimate strength, although there is a yield point which is generally considered to be equal to the tension yield point. Beyond this yield point there is plastic flow and lateral expansion

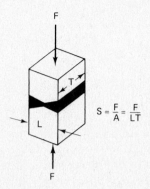

$$S = \frac{F}{A} = \frac{F}{LT}$$

FIGURE 15-3 *Compression loading.*

of ductile metals. For short compression members the basic working stress in compression is assumed to be equal to the working unit stress in tension. Hence the same formula applies except that the load is compressive rather than tension.

The *compression load* on long members requires a different analysis. As the length of a member increases, failure takes place by buckling at an average compressive unit stress less than the yield point. By definition a column is a straight member subjected to compressive forces acting in the direction of its axis as shown by Figure 15-4. If the length of the column is sufficiently great compared to its cross section, failure will take place by lateral bending or buckling. Unlike tension and short-compression members, the fiber stress in columns is not directly proportional to the load. Loading of columns is a technical subject too complex to be treated here. If the loading on the column becomes eccentric or not exactly on the axis of the column the tendency to buckle increases. Long slender members subjected to axially compressive loads should be designed with utmost caution. Flat plates can also be exposed to compressive loads on the edge. In such cases, local buckling will occur in short lengths of a section due to its elastic instability under the action of a compressive

FIGURE 15-4 *Compression loading—long member.*

load. The buckling of a flat plate of a ductile material does not necessarily mean failure of the entire member since it may occur while the member is still capable of carrying additional load.

Bending is an extremely common type of loading, both in structural and machinery applications. Many machine and structural parts act as beams and their primary function is to support various loads. These loads tend to produce bending in the part, as shown by Figure 15-5. Internal stresses due to bending are of two types: one called *fiber stresses* (also known as normal or bending stresses), which act perpendicular to a transverse cross section of the beam, and the other called *shearing stresses,* which act in both the transverse and longitudinal directions.

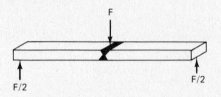

FIGURE 15-5 *Bending loading.*

Maximum fiber stress in a beam occurs at the fibers furthest from the neutral axis. For beams with a symmetrical cross section about the neutral axis, the section modulus to either the top or bottom extreme fibers is the same. For beams with an unsymmetrical cross section the section modulus to each extreme fiber is different. The fiber stress resulting from bending is given by the formula

$$s = \frac{M}{S}$$

where

s = the fiber stress,

M = the external bending moment, and

S = the section modulus of the beam under consideration.

Most handbooks on design provide values of the section modulus (S) for rolled structural steel shapes and commonly used built up sections. The section modulus can also be expressed as I divided by c where I is the moment of inertia of the section of the beam about its neutral axis and c is the distance from the neutral axis to the extreme fiber. The basic allowable unit stress for extreme fibers tension due to bending is the same as that allowed in simple tension.

The shearing stress at any point in a beam is given by the formula

$$v = \frac{V}{dt}$$

where

v is the shearing stress and V is the vertical shear load, d is the depth of the beam and t is the thickness of the web or vertical portion of the beam.

Normally, the shearing stress in bending is a maximum at the neutral axis and zero at the extreme fibers.

There are many different types of loading of beams and there are many different support systems for beams. In view of this, it is recommended that the designer consult engineering design handbooks for making these calculations.

Torsion, the next type of loading, is based on forces attempting to twist and forces resisting the tendency to twist. The internal stress that results from these two twisting forces is actually a shear stress. This is shown by Figure 15-6, in which a twisting motion is attempting to shear the weld joining the rod to the plate. The equation for torsion is

$$S_s = T \times \frac{r}{J}$$

where

Ss is the shear stress and T is the torque expressed in force multiplied by distance from the centerline of the axis of the part being twisted, r is the radius of the tube being twisted, and J is the polar moment of inertia of the circular part being twisted.

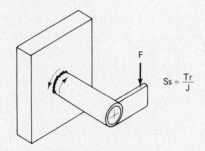

FIGURE 15-6 *Torsion loading.*

The polar moment of inertia is available in design handbooks for round type members. In the case of the weld shown in Figure 15-6, the polar moment of inertia of the failure area or throat of the fillet weld would have to be calculated. This formula should only be used for circular cross-sectional areas subjected to torsion. Formulas for other cross sections subject to twisting loads are very involved and must be individually calculated.

Shearing stress is the force acting in the plane of the cross section. This is in contrast to tensile and compressive stresses which act perpendicular to the plane of the cross section of a member. The failure area in shear is parallel to the load and the stress is caused by two equal and opposite parallel loads. It can be thought of

as the load attempting to slide two pieces apart. The equation in this case is

$$s = \frac{F}{A}$$

where

s is the shear stress, F is the force, and A is the area, but as mentioned above it is an area parallel to the loads.

It is common practice to assume that the weld is stressed equally over its entire area, although this is not precisely true. The shear is shown by Figure 15-7. Shear is extremely important with respect to welding. Fillet welds fail through the throat, which is considered a shear failure. The controlling stress in a fillet weld is shear instead of tension or compression. Failure usually occurs by shear at 45° along or through the throat section.

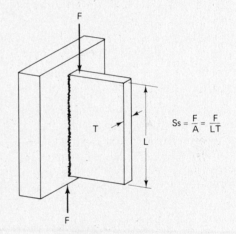

FIGURE 15-7 *Shear loading.*

Fillet welds are the most widely used welds. Since it is somewhat controversial to compare longitudinally versus transversely stressed fillets it is interesting to review Figure 15-8. This figure shows that the shear stress is identical in either situation providing that bending does not occur. Even so, it is felt that stress distribution is more uniform in the transverse type of weld than in the longitudinal fillet which is parallel to the line of force. This type of connection is widely used for the welding of structural angles to plates. The different types of loading and the stresses that result can act singly but in many weldments they may act in combination. Combined stresses should be given special attention.

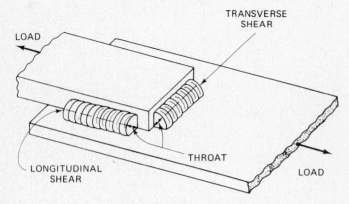

FIGURE 15-8 *Stresses in a fillet weld.*

These are defined as stresses acting simultaneously at a point in more than one principal direction. Under combined stresses, in steel and other ductile metals, yielding may occur at a stress that is less than the *uniaxial* (stress in one direction only) yield stress. Combined stress will more often occur in extremely thick sections.

The foregoing information should give you an idea of the calculations necessary to determine stress in sections based on applied loads. It is not intended to cover the more complex loadings of machine parts or structural members. The designer should consult a design handbook for additional information concerning calculation of stresses based on different types of loading.

Stress Concentrations: When considering the design criteria of a weldment special emphasis must be placed on the factor of stress concentration or the distribution of stress at notches or discontinuities. It was mentioned previously that when the parts of a weldment are welded together they all act as a monolithic structure. This means that stresses are spread throughout the entire structure when transmitting a force from one point to another within the structure. Designers may claim that a particular part is not expected to carry any of the load. However, once that piece is welded to the total weldment it becomes a part of the weldment and will usually carry a portion of the load. Tensile stresses are normally thought to be uniformly distributed throughout the entire cross section of a member. This is usually a safe assumption for simple statically loaded structures. However, complex structures exposed to static dynamic loading, repetitive loading, and impact loading will not have a uniform stress pattern across the structure. The effect of the nonuniformity is more important in fatigue and cold weather impact.

Under some service requirements external loads may fluctuate or may be applied *repetitively* thousands of times. Most metals exhibit a lower ultimate strength under the application of repetitive loads and it is normal practice to reduce the working unit stress when cyclic or repeated loading is applied. This involves the fatigue life of the metal and depends upon the stress cycle imposed. The types of stress cycle refer to whether there is a complete reversal of stress from tension to compression or whether it is a loading from a minimum to a maximum amount of either tension or compression. Most codes specify an allowable fatigue stress.

Suddenly applied loads are called *impact loads* and the suddenness of the application of the load is a matter of degree of the impact. The effect of impact is to immediately increase the internal stress in the weldment. These internal stresses may be localized and cause problems. In a weldment the stresses remote from the point of application of the load will be considerably less than at the point of loading. The design calculations are made on the basis of equivalent static load conditions. It is normal practice to allow for the effect of impact-producing stresses by the use of factors based on static loading. An alternate is the complete calculation of the stresses. The factor is an across-the-board cut in the allowable unit stress of the total structure or weldment.

A stress concentration is a point within the structure at which the stresses will be more concentrated than throughout the remaining cross-sectional area of the weldment. Stresses at a notch, for example, can be two to four times as great as the stresses throughout the remaining portion of the structure. This may not be harmful to the statically-loaded weldment but to weldments loaded dynamically, repetitively, or by impact, it can create a point of premature failure. A simple example of a stress concentration based on the presence of a notch is shown by Figure 15-9. On the left the bar stressed in tension will have uniform stress throughout its cross section. The bar on the right has a sharp

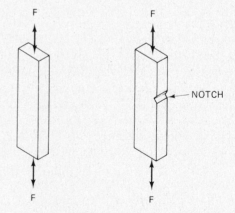

FIGURE 15-9 *Bar with and without notch.*

notch in the edge and the stresses will be concentrated at the root of the notch. Even though the stresses were uniformly distributed at the ends of the bar, they cannot be transmitted through the notch; therefore, they are concentrated at the root of the notch. The bar with the notch will have service life considerably less than half that of the bar not containing the notch.

There are three basic types of notches that result in stress concentration. They are:

1. Design notches both in weldments and in welds.
2. Workmanship notches of weldments but more importantly of welds.
3. Metallurgical notches.

A design notch is designed into a weldment such as in an abrupt change of section. A practical example is a square hatch opening in the deck of a ship. The section at the point where the hatch opening occurs changes abruptly and there is a concentration of stresses. Another similar design notch would be the attachment of a deckhouse welded to the deck of a ship. The section changes drastically at the point the deckhouse is welded to the deck. It might be argued that the deckhouse is not expected to carry any of the load or stress. It will, however, since it is welded and becomes integral to the other parts of the hull. The intersection of members, re-entrant angles, abrupt reinforcing structures, and square holes are all examples of notches in the design of weldments.

Notches also occur in welds.[1] A design notch in a weld joint would be the incomplete fusion area at the center or root of a weld. Any weld joint design which

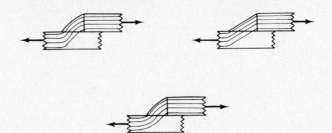

FIGURE 15-10 *Single fillet lap weld in tension.*

provides partial penetration or incomplete fusion includes a notch. Fillet welds used for lap joints are notch prone. See Figure 15-10 for examples of stress concentration when the joint is loaded in tension. Tee joints made with fillets are also notch prone. The worst notch is the single fillet welded tee joint. Figure 15-11 shows these tee weld joint details and how different designs contain notches. This figure shows three joint designs, the stress paths, and a comparison of their relative static tensile strength, resistance to fatigue, and impact strength. The static strength is determined by the area of the weld, the other factors by tests. Butt joints are less notch prone because of their geometry. Full-penetration welds produce highly efficient butt joints. However, partial penetration welds contain notches at the center of the weld or at the outer surface. Four examples are shown by Figure 15-12. It is easy to determine whether a notch is designed into the joint by drawing a cross section of the weld joint indicating the paths that the lines of stress must follow. Notches can also occur in weld joints when the section changes abruptly. Figure 15-13 shows three butt joints between a thick and a thin

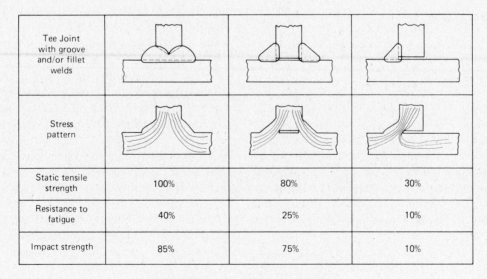

Tee Joint with groove and/or fillet welds			
Stress pattern			
Static tensile strength	100%	80%	30%
Resistance to fatigue	40%	25%	10%
Impact strength	85%	75%	10%

FIGURE 15-11 *Tee joints—stress pattern.*

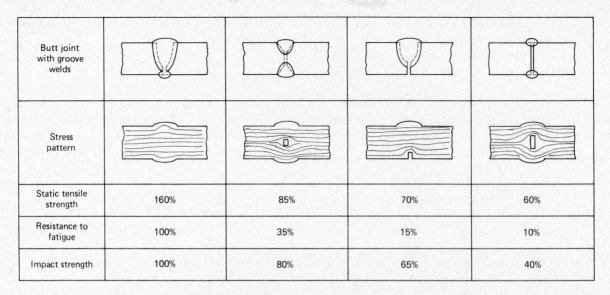

Butt joint with groove welds				
Stress pattern				
Static tensile strength	160%	85%	70%	60%
Resistance to fatigue	100%	35%	15%	10%
Impact strength	100%	80%	65%	40%

FIGURE 15-12 *Butt joints—stress pattern.*

member. There are several ways to provide a smooth stress flow through the weld. There is also a way to make two stress concentrations. The left joint in Figure 15-13 would create two points of stress concentration, one at the toe of the fillet and one at the unfused root. The right joint would be a better solution and would provide a relatively smooth stress flow. The joint on the bottom would also provide a relatively smooth stress flow and would be less expensive since less weld metal is required.

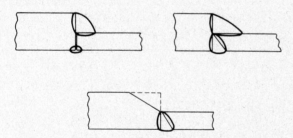

FIGURE 15-13 *Thick to thin butt.*

Workmanship notches can be troublesome and difficult to control. These occur when the welds are not full-penetration welds, even though they are designed to be. The root opening may have been eliminated by an accumulation of tolerances, or back gouging and rewelding may have been omitted. A fillet weld at the point of joining a thin section to a thick section may not be full size and will create a notch. Examples of these defects are given in Chapter 17.

The third notch factor is in the minority and creates the least trouble. These are metallurgical notches which may be caused by joining metals of different yield strengths, or by welding on hardenable steels and creating extremely hard spots in welds. Abrupt change in strength along a cross section can have the effect of a notch even though there is no abrupt change in section.

In the design and construction of weldments subjected to dynamic loads and low-temperature service, every effort must be taken to provide a smooth flow of stress lines throughout the weldment. This becomes more important when high-strength materials are used.

Rigid Frame Construction

This type of construction, also called continuous construction, is a design system incorporating *plastic analysis.* This type of design and construction is specifically suited to welded construction. In this type of work the members are welded directly together rather than through connection plates, gussets, or filler plates. The material in a rigid or continuous welded frame is utilized more efficiently because the bending moments are distributed better. This type of construction provides the maximum degree of end restraint. Rigid frame design allows large savings in steel of the members involved. It also eliminates the steel of the connecting plates, gussets, etc. It requires a more thorough analysis and should be done only by designers well qualified to perform this type of work.

In machine members the rigid connection concepts or plastic analysis provides for greater dimensional accuracy. The weldment will maintain correct alignment and dimensional accuracy throughout its service life. This type of design and construction should be used for all welded machine parts, unless they must be disassembled.

Welding Positions

Welding cannot always be done in the most desirable position. Often the welding must be done on the ceiling, in the corner, or on the floor. Welding must be done in the position in which the part will be used. We must be able to describe and define these different welding positions. This is necessary since welding procedures must indicate the position in which the welding is to be performed, and the selection of welding processes is necessary since some have all-position capabilities whereas others may be used in only one or two positions. The American Welding Society has defined the four basic welding positions as follows (see Figure 15-14).

Flat: "When welding is performed from the upper side of the joint and the face of the weld is approximately horizontal."

Horizontal: "The axis of the weld is approximately horizontal, but the type of the weld dictates the complete definition. For a fillet weld, welding is performed on the upper side of an approximately horizontal surface and against an approximately vertical surface. For a groove weld the face of the weld lies in an approximately vertical plane."

Vertical: "The axis of the weld is approximately vertical."

Overhead: "When welding is performed from the underside of the joint."

There are at least two foreign definitions that are different from those of the American Welding Society.

FIGURE 15-14 *Welding position—fillet and groove welds.*

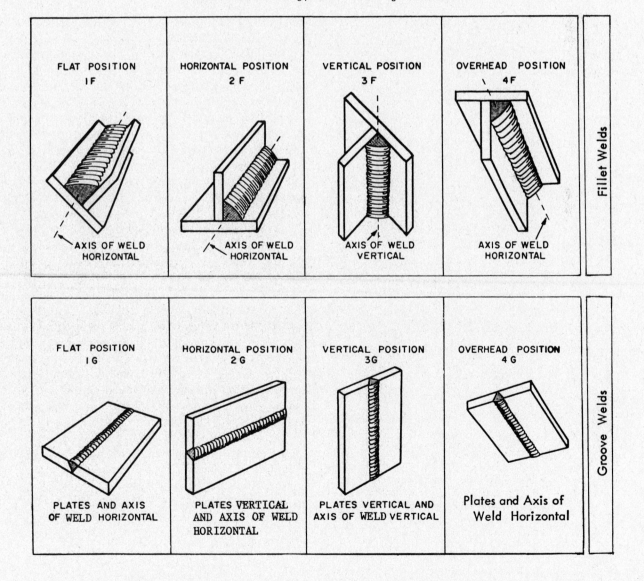

The British and others use the term *downhand* to describe the flat position. This term is quite descriptive and is also used in the U.S.A. They also use the term *horizontal-vertical* to describe welds between plates, one in approximately the horizontal plane and one in the vertical plane when the axis of the weld is approximately horizontal. The positions are identified as follows: F indicates fillets, 1 indicates flat, 2 indicates horizontal,

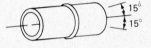

A—TEST POSITION 1G
PIPE HORIZONTAL AND ROTATED. WELD FLAT (± 15°). DEPOSIT FILLER METAL AT OR NEAR THE TOP.

B—TEST POSITION 2G
PIPE OR TUBE VERTICAL AND NOT ROTATED DURING WELDING. WELD HORIZONTAL (± 15°).

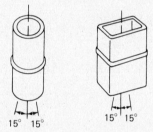

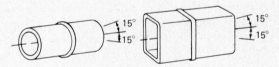

C—TEST POSITION 5G
PIPE OR TUBE HORIZONTAL FIXED (± 15°). WELD FLAT, VERTICAL, OVERHEAD

RESTRICTION RING

TEST WELD

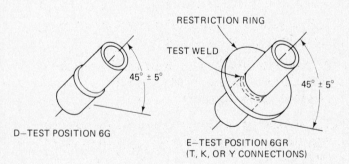

45° ± 5°

45° ± 5°

D—TEST POSITION 6G

E—TEST POSITION 6GR
(T, K, OR Y CONNECTIONS)

PIPE INCLINED FIXED (45° ± 5°) AND NOT ROTATED DURING WELDING.

FIGURE 15-15 *Welding position—pipe welds.*

3 indicates vertical, and 4 indicates overhead. The same numbers apply to groove welds, which are designated with the letter G.

Pipe weld joints are a special case and are identified as test positions as shown by Figure 15-15. They are normally groove welds and they are indicated by the letter G. Test position 1G is roll welding with the axis of pipe horizontal, but with the welding done in flat position with the pipe rotating under the arc. Test position 2G is known as horizontal welding but the axis of the pipe is in the vertical position with the axis of the weld in the horizontal position. There is no 3G or 4G test position on pipe welding. Test position 5G is known as horizontal fixed position. Here the axis of the pipe is horizontal, but the pipe is not to be turned or rolled during the welding operation. Test position 6G for pipe has the axis of the pipe at 45° and the pipe is not to be turned while welding. For qualification work a 6G restricted position is often used. Restricted accessibility is provided by a restriction ring placed near the weld. It is called 6GR. The axis of the pipe may vary ± 15° for 1G, 2G, and 5G test positions, but only ± 5° for the 6G position. Square and rectangular tubing is accommodated in the 2G and 5G test positions as shown by Figure 15-15.

The official AWS diagrams for welding positions are precise. They utilize the angle of the axis of the weld which is "a line through the length of the weld perpendicular to the cross section at its center of gravity." Figure 15-16 shows the fillet weld and the limits of the various positions. It is necessary to consider the inclination of the axis of the weld as well as the rotation of the face of the fillet weld.

Figure 15-17 shows the groove weld positions in the same manner. The inclination of the axis of the groove and fillet weld is the same as far as limits are concerned. For the flat position the rotation of the face of the weld is the same for fillet and groove welds. However, the rotation of the face of the horizontal, vertical, and overhead groove welds are different.

The design of a joint is often changed whenever the welding position or type of backing is changed. In general, narrower included angles are used for other than flat-position groove welds. Welds made in the horizontal position usually have a flat face on the bottom member and a beveled face on the upper member. When backing strips are used the root opening is usually wider. Specific joint details for different thicknesses and positions are shown in the next section.

The welding position must always be accurately described. It is an important variable in any welding procedure. It is especially important with respect to training and qualifying welders and must always be given consideration when selecting a welding process.

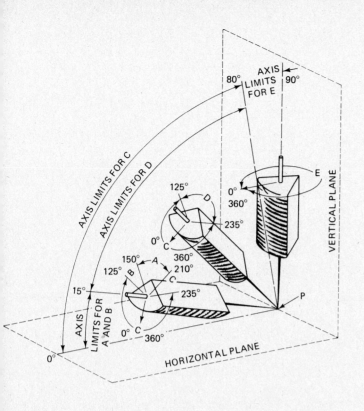

FIGURE 15-16 *Welding position—fillet welds.*

TABULATION OF POSITIONS OF FILLET WELDS			
POSITION	DIAGRAM REFERENCE	INCLINATION OF AXIS	ROTATION OF FACE
FLAT	A	0° TO 15°	150° TO 210°
HORIZONTAL	B	0° TO 15°	125° TO 150°
			210° TO 235°
OVERHEAD	C	0° TO 80°	0° TO 125°
			235° TO 360°
VERTICAL	D	15° TO 80°	125° TO 235°
	E	80° TO 90°	0° TO 360°

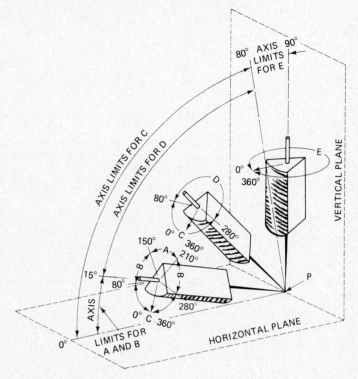

FIGURE 15-17 *Welding position—groove welds.*

TABULATION OF POSITIONS OF GROOVE WELDS			
POSITION	DIAGRAM REFERENCE	INCLINATION OF AXIS	ROTATION OF FACE
FLAT	A	0° TO 15°	150° TO 210°
HORIZONTAL	B	0° TO 15°	80° TO 150°
			210° TO 280°
OVERHEAD	C	0° TO 80°	0° TO 80°
			280° TO 360°
VERTICAL	D	15° TO 80°	80° TO 280°
	E	80° TO 90°	0° TO 360°

Joint Types

Welds are made at the junction of the various pieces that make up the weldment. These junctions of parts are called joints defined as "the location where two or more members are to be joined." Parts being joined to produce the weldment may be in the form of rolled plate, sheet, shapes, pipe, or they may be castings, forgings, or billets. It is the placement of these members that helps us define the joints. There are five basic types of joints for bringing two members together for welding (see Figure 15-18):

B, Butt Joint: "a joint between two members lying approximately in the same plane."

C, Corner Joint: "a joint between two members located approximately at right angles to each other in the form of an angle."

T, Tee Joint: "a joint between two members located approximately at right angles to each other in the form of a T."

L, Lap Joint: "a joint between two overlapping members."

499

E, Edge Joint: "a joint between the edges of two or more parallel or mainly parallel members."

When more than two members are brought together other designs may be used, or they may be a combination of the five basic joints. The most popular is the *cross* or *cruciform joint*—"a joint between three members at right angles to each other in the form of a cross." It is actually a double T joint. The joint describes the geometry in cross section of the members to be welded.

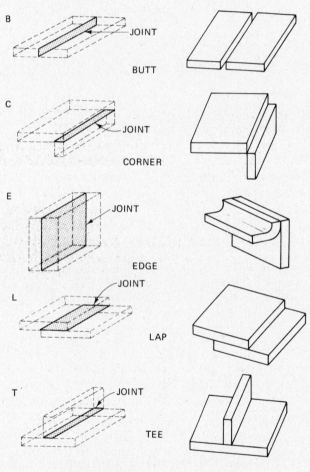

FIGURE 15-18 *The five basic joint types.*

A weld is "a localized coalescence of metals suitably produced either by heating the materials to temperatures with or without the application of pressure or by the application of pressure alone and with or without the use of filler metal."

To describe or specify a weld joint it is necessary to describe both the joint and the weld. As mentioned

before there are five basic types of joints: butt, corner, tee, lap, and edge. A fairly large number of different types of welds and variations are used to join the members to make the *weld joint*.

Weld Types

There are eight types of welds shown by Figure 15-19 which are separate and distinct. However, some of these weld types have many variations. In addition, weld types can be combined. The eight types of welds are:

1. The fillet weld.
2. The plug or slot weld.
3. The spot or projection weld.
4. The seam weld.
5. The groove weld (seven types).
6. The back or backing weld.
7. The surfacing weld.
8. The flange weld (edge and corner).

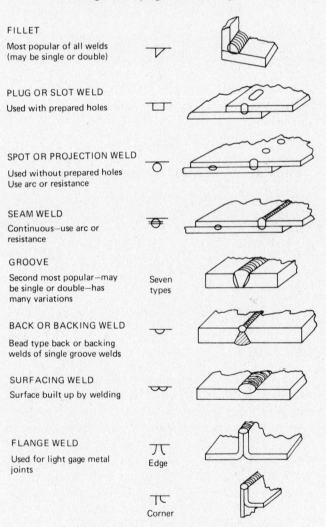

FIGURE 15-19 *The eight types of welds.*

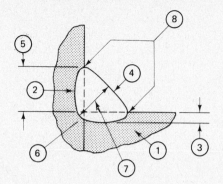

(1) BASE METAL: Metal to be welded.

(2) BOND LINE: The junction of the weld metal and the base metal.

(3) DEPTH OF FUSION: The distance that fusion extends into the base metal.

(4) FACE OF WELD: The exposed surface of a weld on the side from which the welding was done.

(5) LEG OF A FILLET WELD: The distance from the root of the joint to the toe of the fillet weld.

(6) ROOT OF WELD: The point or points, as shown in cross-section, at which the bottom of the weld intersects the base metal surface or surfaces.

(7) THROAT OF FILLET WELD: The shortest distance from the root of the fillet weld to its face.

(8) TOE OF A WELD: The junction between the face of a weld and the base metal.

FIGURE 15-20   *The fillet weld.*

The most commonly used weld type is the "fillet" weld. The fillet weld is so named because of its cross-sectional shape. The fillet is regarded as being *on the joint* and is defined as "a weld of approximately triangular cross section joining two surfaces approximately at right angles to each other." Details of the fillet weld are shown by Figure 15-20. The design of the fillet weld will be covered in Section 15-4.

Plug or Slot Welds: These are used with prepared holes. They are considered together since the welding symbol to specify them is the same. The important difference is the type of hole in the prepared member being joined. If the hole is round, it is considered a plug weld; if it is elongated, it is considered a slot weld. The design of plug and slot welds will be given in Section 15-4.

The Spot and Projection Weld: This is shown by the same weld symbol. These types of welds can be

applied by different welding processes which change the actual weld. For example, when the resistance welding process is used, the weld is at the interface of the members being joined. If the electron beam, laser, or an arc welding process is used the weld melts through one member into the second member. More information about arc spot welds will be given in Chapter 20.

Seam Weld: This weld in cross section looks similar to a spot weld. The weld geometry is influenced by the welding process employed. With resistance welding, the weld is at the interface between members being joined, but with an electron beam, laser, or arc welding process, the weld melts through the one member to join it to the second member. There are no prepared holes in either the spot or the seam weld.

Groove Weld: This is the second most popular weld type employed. It is defined as "a weld made

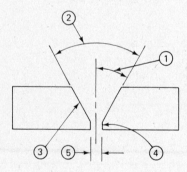

(1) BEVEL ANGLE: The angle formed between the prepared edge of a member and a plane perpendicular to the surface of the member.

(2) GROOVE ANGLE: The total included angle of the groove between parts to be joined by a groove weld.

(3) GROOVE FACE: The surface of a member included in the groove.

(4) ROOT FACE: That portion of the groove face adjacent to the root of the joint.

(5) ROOT OPENING: The separation between the members to be joined at the root of the joint.

FIGURE 15-21   *The groove weld.*

Weld Type	Symbol	The Five Basic Joint Types				
		B Butt	C Corner	E Edge	L Lap	T Tee
Fillet		Special	Yes	Special	Yes	Yes
Plug or slot		–	–	–	Yes	Yes
Spot or projection		–	–	–	Yes	Special
Seam		–	Special	–	Yes	Special
Square groove		Yes	Yes	Yes	–	Yes
Vee groove		Yes	Yes	Yes	–	Yes
Bevel groove		Yes	Yes	Yes	Yes	Yes
U groove		Yes	Yes	Yes	–	–
J groove		Yes	Yes	Yes	Yes	Yes
Flare V groove		Yes	Yes	–	–	–
Flare bevel groove		Yes	Yes	–	Yes	Yes
Backing weld		Combin.	Combin.	–	–	Combin.
Surfacing		–	–	–	–	–
Flange edge		–	–	Yes	–	–
Flange corner		–	Yes	–	–	–

FIGURE 15-22 *The welds applicable to the basic joints—combining table.*

in the groove between two members to be joined." The groove weld is regarded as being "in the joint." There are seven basic groove weld designs, and they can be used as single or double welds. The details of the groove weld are shown by Figure 15-21. The design of the groove weld will be given in Section 15-4.

Back or Backing Weld: This is a special type of weld made on the back side or root side of a previously made weld. The root of the original weld is gouged, chipped, or ground to sound metal before the back or backing weld is made. This will improve the quality of the weld joint by assuring complete penetration.

Surfacing Weld: This is a type of weld composed of one or more stringer or weave beads deposited on base metal as an unbroken surface. It is used to build up surface dimensions, to provide metals of different properties, or to provide protection of the base metal from hostile environment. It is not used for joining and will be covered in more detail in Chapter 19, which is devoted to surfacing.

Flange Weld Edge: This is used primarily for light gauge or sheet metal joints.

Flange Weld Corner: This is also used for light gauge or sheet metal parts. In both cases, parts must be prepared for these specific joint details.

Weld Joints

In order to produce weldments it is necessary to combine the joint types with weld types to produce weld joints for joining the separate members. Each weld type cannot always be combined with each joint type to make a weld joint. Figure 15-22 shows the welds applicable to the basic joints. Since fillet welds and groove welds have more possibilities and their use is more complex, more information regarding their use on different joints will be given in Section 15-4.

15-4 WELD JOINT DESIGN

The purpose of the weld joint is to transfer the stresses between the members and throughout the weldment. Forces and loads are introduced at different points and are transmitted to different areas throughout the weldment. The amount of stress to be transferred across the joint is estimated using calculations, experience, and so on, as previously discussed. The type of loading and service of the weldment have a great bearing on the joint design that should be selected. All weld joints can be classified in two basic categories. These are the *full-penetration joints* and *partial-penetration joints*. The names are sufficiently descriptive; however, a full-penetration weld joint has weld metal throughout the entire cross section of the weld joint. The partial-penetration joint is designed to have an unfused area. The weld does not completely penetrate the joint. The rating of the joint is based on the percentage of weld metal depth to the total joint. If the weld metal penetrated a quarter of the way from both sides it would still leave half of the joint unfused. A 50% partial-penetration joint would have weld metal halfway through the joint. It was previously mentioned that weldments subjected to static loading need only sufficient weld metal to transfer the static loads. When joints are subjected to dynamic loading, reversing loads, and impact loads the weld joint must be more efficient. This is more important if the weldment is subjected to cold-temperature service. With this type of service the full-penetration weld is required. Designs that increase stresses by the use of partial-penetration joints would not be acceptable for this type of service.

The strength of the weld joint depends not only upon the size of the weld but also upon the strength of the weld metal. When using mild and low-alloy steels little thought is given to the strength of the weld metal because the strength of weld metal is normally stronger than the materials being joined. The yield strength of normal structural steel is about 36,000 psi (25.3 kg/mm^2). The yield strength for a normal E-60XX type of electrode deposit is a minimum of 50,000 psi (35.2 kg/mm^2), thus the weld metal is considerably stronger than the base metal. A properly made weld between structural steel members using an E-6010 electrode is stronger than the base metal. When this joint is pulled in tension it will always break outside of the weld. The base metal will yield first. The weld metal strength over matches the base metal in strength and in other mechanical properties. For this reason weld reinforcement should be kept to a minimum since it is not required and is wasteful. In welding high-alloy steel, heat-treated steels, or other heat-treated metals this situation may not apply. Many materials obtain their strength by heat treatment. The weld metal does not have this same heat treatment; therefore, it might have lower-strength properties than the base metal. The welding operation might nullify the heat treatment of the base metal causing it to revert to its lower strength adjacent to the weld. When welding high-alloy or heat-treated materials special investigation and precaution must be taken.

Design of the Weld Joint

There are many factors that must be considered in designing a weld joint. Many have an influence on the economics of the weldment design as well as on the strength of the weld joint and the ability of the welder to make it. The designer must take into consideration the strength requirements mentioned here and the penetration requirements dictated by loading and service.

The joint design must accommodate these requirements in the most economical way. This is done by analyzing the following factors.

The weld joint should be designed so that its cross-sectional area is the minimum possible. This is a matter of cost since deposited weld metal is much more expensive than the base metal. The cross-sectional area is a measure of the amount or weight of weld metal that needs to be used to make the joint. The minimum cross-sectional area, however, may not provide a practical weld joint as discussed later.

Another factor of economics has to do with the amount of edge preparation required to produce the particular weld joint design. Weld joints are normally prepared by three methods: shearing, thermal cutting—primarily oxygen cutting, and machining. They should be considered in this same cost relationship. Shearing is by far the most economical way to cut metals; however, there are limitations to thickness that can be sheared and the sheared edge is a square cut edge without bevels. Flame cutting is the most popular method of preparation and is used for most work above the gauge metal thicknesses. It can be used for cutting square edges but also for adding bevels. The root face or square edge, the bevel—both front and back, can all be accomplished with one passage of torch assemblies, especially on straight line cuts. Machining is the third method of preparation but this involves more expensive equipment. Machining is used normally for preparing the J type and U type of preparation. It is popular for preparing weld joints on circular parts. Joint design and metal thickness dictate the type of preparation tools required. Sometimes compromises are made based on the type of tooling available versus the amount of weld metal required to complete the joint.

The individual joint details must also relate to the welding process to be employed. The joint details given later in this section are all related to the shielded metal arc welding process applied manually. This is a good starting point, since designs can be altered to accommodate other processes. Chapters 4 and 5 gave information concerning those differences from the basic weld joint that should be considered in utilizing the specific process in question.

The welding position must also be considered when designing the weld joint. A good example of this is the design of weld joints to splice columns in structural steel buildings. It is good practice to have the bottom side of the joint flat with the bevel on the top piece.

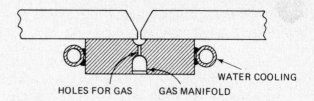

HOLES FOR GAS   GAS MANIFOLD

WATER COOLING

WATER COOLED COPPER BACKING WITH GAS SHIELDING

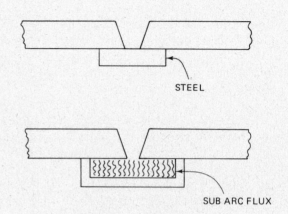

STEEL

SUB ARC FLUX

FIGURE 15-24 *Weld backing system.*

This has become more or less standard practice for horizontal welds. Several examples of this will be shown in the joint details later.

One factor that is sometimes overlooked by designers is the matter of accessibility to the weld joint. The weld joint must be accessible to the welder using the process that is employed. Too often, weld joints are designed for welds that cannot be made. Figure 15-23 is an illustration of several types of inaccessible welds. Welds cannot be made on the inside of small-diameter pipe or inside of box columns. The welds in the channel would be very difficult to make in the position shown. Sequence of assembly has an effect on accessibility, but it is advantageous if the weldment can be designed so that all weld joints are accessible for welding after the weldment has been completely assembled and tack welded.

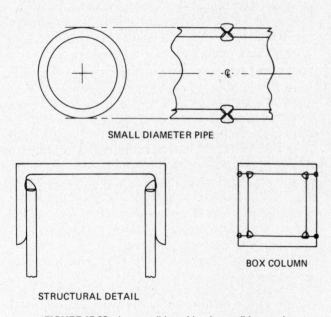

SMALL DIAMETER PIPE

STRUCTURAL DETAIL

BOX COLUMN

FIGURE 15-23 *Inaccessible welds—impossible to make.*

Another factor related to the accessibility problem is the backing or back welding that can be performed on a particular joint. The term *one-side welding* has become very popular since it indicates that the weld is to be completely made on one side of the joint. Section 20-3 will provide more information on this subject. An example of a one-side weld is the weld on small-diameter pipe. It is impossible to do any work to the back side of such a joint; therefore, the joint must be made completely from the outside of the pipe or a *one-side weld.*

There are many different types of backing. The consumable inserts mentioned previously become part of the joint. Backing straps are beyond the joint but when fused to the root become part of the weldment. Backing straps must be continuous. Fluxes, flux-coated tape, and flux-filled bars are all used for weld backing. Various nonmetal items such as ceramic materials and ceramic coated bars can be used as backing materials. Water-cooled copper bars are also used; sometimes they are a part of the fixturing. When used with gas shielded processes they may incorporate root shielding gas ports. A variety of backing systems are shown by Figure 15-24. If the back side of the joint is accessible it is possible to specify a gouging operation and a backing weld. This is normally a more economical way to weld, but cannot always be utilized.

Consideration should also be given to the problem of weld distortion. Study the possibilities of distortion control by means of the use of double welds or welds made on both sides of the center line of the joint to minimize angular distortion. Weld sizes also have an effect on distortion. Use the smallest sizes possible. Weld distortion will be covered more completely in Section 18-2.

On the same general subject is the problem of weld stresses and the possible damage that might occur. This is important when welding high-strength materials in which brittle fracture may be involved. Consideration must be given to lamellar tearing of the base metal directly under the weld, which might be harmful. These subjects will be covered in Section 18-3.

Fillet Weld Design

SECTION 15-4

WELD JOINT DESIGN

The fillet weld is the most popular of all types of welds because there is normally no preparation required for it. In some cases, the fillet weld is the least expensive, even though it might require more filler metal than a groove weld since the preparation cost would be less. It can be used for the lap joint, the tee joint, and the corner joint without preparation. Since these are extremely popular the fillet has wide usage. On corner joints the double fillet can actually produce a full-penetration weld joint. The use of the fillet for making all five of the basic joints is shown by Figure 15-25. Fillet welds are also used in conjunction with groove welds, particularly for corner and tee joints.

The fillet weld is expected to have equal-length legs and thus the face of the fillet is on a 45° angle. This is not always so since a fillet may be designed to have a longer base than height in which case it is specified by the two leg lengths. On the 45° or normal type of fillet the strength of the fillet is based on the shortest or throat dimension which is 0.707 × the leg length. In

North America, fillet welds are specified by the leg length, whereas in many European countries they are specified by the throat dimension. For fillets having unequal legs the throat length must be calculated and is the shortest distance between the root of the fillet and the theoretical face of the fillet. In calculating the strength of fillet welds the reinforcement is ignored. Also the root penetration is ignored unless a deep penetrating process is used. If semi- or fully-automatic application is used the extra penetration can be considered. See Figure 15-26 for details about the weld.

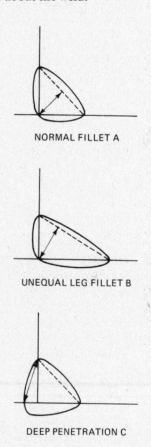

NORMAL FILLET A

UNEQUAL LEG FILLET B

DEEP PENETRATION C

FIGURE 15-26 *Fillet weld throat dimension.*

Under these circumstances the size of the fillet can be reduced yet equal strength will result. Such reductions can be utilized only when strict welding procedures are enforced. The strength of the fillet weld is determined by its failure area, which relates to the throat dimension. Doubling the size (leg length) of a fillet will double its strength, since it doubles the throat dimension (and area). However, doubling the fillet size will increase its cross-sectional area and weight four

JOINTS	SINGLE FILLET	DOUBLE FILLET
BUTT (B)	F OR F	F OR F
CORNER (C)	F OR F	F
TEE (T)	F	F
LAP (L)	F	F
EDGE (E)	F	EDGE WELD OF LAPPED STRIPS

FIGURE 15-25 *The fillet used to make the five basic joints.*

505

times. This is illustrated by Figure 15-27, which shows the relationship to throat-versus-cross-sectional area or weight of a fillet weld. For example, a 3/8-in. fillet is twice as strong as a 3/16-in. fillet; however, the 3/8-in. fillet requires four times as much weld metal.

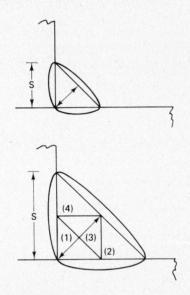

FIGURE 15-27 *Fillet weld size vs strength.*

In design work the fillet size is sometimes governed by the thickness of the metals joined. And in some situations the minimum size of the fillet must be based on practical reasons rather than the theoretical need of the design. Intermittent fillets are sometimes used when the size is minimum, based on code, or for practical reasons, rather than because of strength requirements. Many intermittent welds are based on a pitch and length so that the weld metal is reduced in half. Large intermittent fillets are not recommended because of the volume-throat dimension relationship mentioned previously. For example, a 3/8-in. fillet 6 in. long on a 12-in. pitch (center to center of intermittent welds) could be reduced to a continuous 3/16-in. fillet, and the strength would be the same, but the amount of weld metal would be only one half as much.

Single fillet welds are extremely vulnerable to cracking if the root of the weld is subjected to tension loading. This applies to tee joints, corner joints, and lap joints. The simple remedy for such joints is to make double fillets, which prohibit the tensile load from being applied to the root of the fillet. This is shown by Figure 15-25. Notice the F (force) arrowhead.

Groove Weld Design

There are seven types of variations of the groove weld. The square, V, bevel, U, J, flare V, and flare bevel are shown by Figure 15-28. They can all be used singly or as double welds. Three of them, the square groove, the flare V, and the flare bevel groove weld, can be made without extra preparation of the joint detail. The square groove is the simplest, since it requires only a square cutoff and for thinner metals this is accomplished by shearing. The flare V and flare bevel welds are normally used for thinner materials in which a bent section joins another section or in which a round section is involved. Two of the other groove welds require preparation on only one of the members of the joint and these are the bevel groove weld and the J groove weld. The remaining two, the V groove and the U groove welds, require preparation of both members of the joint.

There are several names given to groove weld designs that are not standard with AWS. These names, used mostly in Europe, are descriptive and deserve mention. Some of the more common names are X, K, and Y grooves. These letters describe the weld in cross section. The X weld is a double vee weld without a root face. The K weld is a double bevel weld and the Y weld is a single vee with a relatively large root face. Draw these in cross section and you will see the resemblance to these letters.

Preparation can become a major factor in deciding which of the groove types to use. For example, the square groove can be prepared by shearing if the metal is relatively thin. The square groove preparation is also used on thick materials for the electroslag process and other *narrow gap* processes. When the thickness is too great, preparation would be by flame cutting. The flare V joints can be made with thin material flanged by a brake press or with square or rectangular mechanical tubing. They are also used when two intersecting round sections are welded together such as the welding of reinforcing bars or rods together. The flare bevel weld is similar, except that only one member has the radius. The bevel and V groove welds are normally used for medium to thicker materials and flame cutting would be required. For the heavier plate the double V or double bevel groove would be used and flame cutting would be employed. The choice between single and double groove welds is shown by Figure 15-28. In both the V groove and bevel groove a root face may or may not be involved. The normal practice is to have a small root face which helps provide dimensional control of the parts during parts preparation operations. The J groove design requires the J type preparation on one of the parts whether a single or a double J is used. In the case

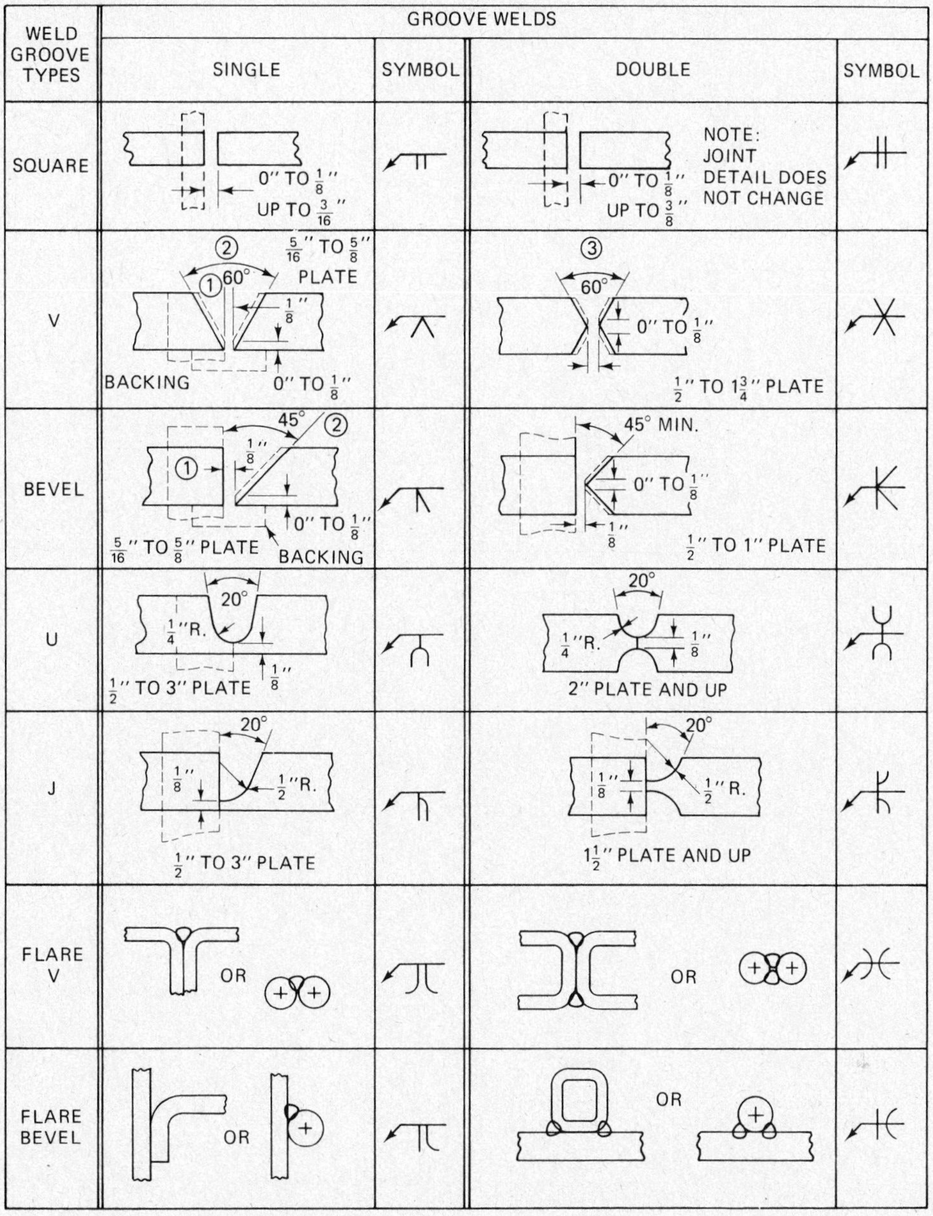

WELD GROOVE TYPES	GROOVE WELDS			
	SINGLE	SYMBOL	DOUBLE	SYMBOL
SQUARE	0″ TO 1/8″ UP TO 3/16″		0″ TO 1/8″ UP TO 3/8″ NOTE: JOINT DETAIL DOES NOT CHANGE	
V	5/16″ TO 5/8″ PLATE 60° ① ② 1/8″ BACKING 0″ TO 1/8″		③ 60° 0″ TO 1/8″ 1/2″ TO 1 3/4″ PLATE	
BEVEL	45° ② 1/8″ ① 0″ TO 1/8″ 5/16″ TO 5/8″ PLATE BACKING		45° MIN. 0″ TO 1/8″ 1/8″ 1/2″ TO 1″ PLATE	
U	20° 1/4″ R. 1/8″ 1/2″ TO 3″ PLATE		20° 1/4″ R. 1/8″ 2″ PLATE AND UP	
J	20° 1/8″ 1/2″ R. 1/2″ TO 3″ PLATE		20° 1/8″ 1/2″ R. 1 1/2″ PLATE AND UP	
FLARE V	OR		OR	
FLARE BEVEL	OR		OR	

FIGURE 15-28 *The seven basic types of groove welds.*

of the U groove both members must have the special curved shape which involves either machining or special gouging and cutting. The other groove designs are also easier to make on circular parts.

The root face mentioned above is used primarily to assure dimensional control of the parts during the parts preparation operation. When large plates are flame cut to feather edges, either V or bevel joint preparations, it is more difficult to hold dimensions than if a root face is involved. However, to obtain two surfaces on the edge requires two passages of the cutting torch unless an existing square edge is used. The root face should be kept to a minimum where full joint penetration is required. When partial penetration is required the root face can be quite large but rarely over 50% of the thickness of the part being beveled. In weldments for which stiffness and weight are the primary criteria, and if dynamic loading is not involved, large root faces will save a considerable amount of weld metal and make the joint less expensive.

There are two other factors that must be considered with respect to the V and bevel groove welds. They must be considered together since they do affect the welder's ability to make or place a weld bead at the root of the joint. These are the included angle and the root opening. In full-penetration welds it is absolutely neces-

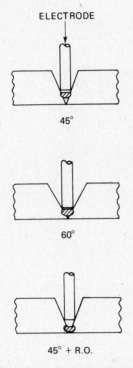

FIGURE 15-29 *Groove weld root opening—included angle relationship.*

sary that the welder have sufficient room and accessibility to place the weld at the root of the joint. If the root opening is too tight or if the included angle is too narrow it will be impossible for the welding electrode to deposit the weld metal at the root of the joint. This is shown by Figure 15-29. It is obvious that one or the other must be widened to allow the root weld to be made. The illustration shows what is accomplished by increasing the included angle, but the best solution is accomplished by increasing the root opening. There are optimum included angles and root openings for shielded metal arc welding and these are based on producing a completely fused root pass. The sample designs shown later utilize these optimum or standardized angles. For different processes these vary.

The J and U groove welds have been fairly well standardized in design. This means that the radius at the root and the included angle have been optimized and these are shown by the examples later in this chapter. Finally, for certain metals and for certain applications special joint details are used. For example, for aluminum pipe a broad single-U preparation is used so that the root is similar to a thinner member which can readily be fused together when making the root pass. In heavy

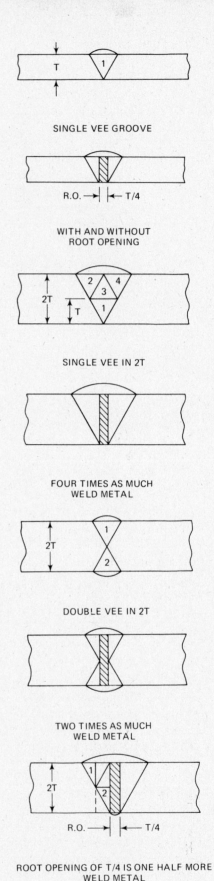

FIGURE 15-30 *Summary of groove weld design dimensions.*

wall pipe compound angles are often used when the joints are prepared by machining. Special joints will be covered in the appropriate sections. Figure 15-30 shows the relationship of cross-sectional area for single and double welds and when the root opening is increased.

Finally, when using groove welds on corner and tee joints, fillets may be used for reinforcement to avoid sharp changes of direction and stress concentrations. Examples of these are shown in weld joint details later.

The problem of joint design can sometimes be compounded in the weldment, if fit is not as designed. Unfortunately, in many shop situations root openings are sometimes used to accommodate tolerances of flame cut parts. If the part is larger than it should be the root openings will disappear and the ability for the welder to make a full penetration weld is reduced. This is why inspection of tacked up weldments can detect potential problems and eliminate defective welds. If the parts come too small the root openings will be excessively large and extra weld metal is required to make the weld

joint. This is expensive and sometimes it is more economical to remake a piece than to fill in a larger than designed groove. If the root opening and the overall joint design is adversely affected by these manufacturing tolerances it will be necessary to retrim the parts according to their original design or remake the part to its original design so that the weld joint preparation is in accordance with the design.

Other Weld Designs

Plug and slot welds and spot and seam welds can be used for lap joints, tee joints, and corner joints. Seam welds and spot welds can be made with resistance welding processes and arc welding processes. In this chapter we are primarily interested in the weld joints

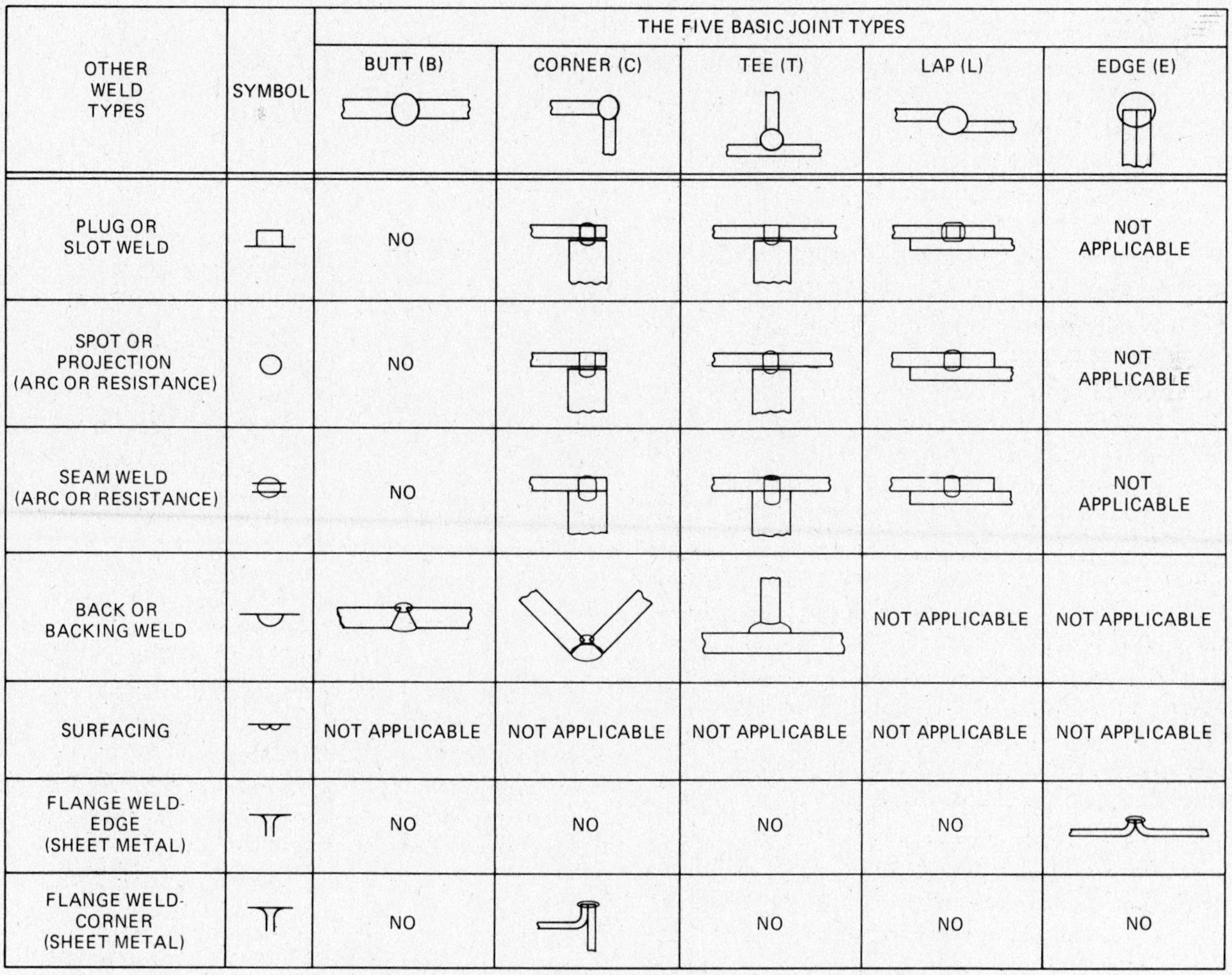

OTHER WELD TYPES	SYMBOL	BUTT (B)	CORNER (C)	TEE (T)	LAP (L)	EDGE (E)
PLUG OR SLOT WELD		NO				NOT APPLICABLE
SPOT OR PROJECTION (ARC OR RESISTANCE)		NO				NOT APPLICABLE
SEAM WELD (ARC OR RESISTANCE)		NO				NOT APPLICABLE
BACK OR BACKING WELD					NOT APPLICABLE	NOT APPLICABLE
SURFACING		NOT APPLICABLE	NOT APPLICABLE	NOT APPLICABLE	NOT APPLICABLE	NOT APPLICABLE
FLANGE WELD-EDGE (SHEET METAL)		NO	NO	NO	NO	
FLANGE WELD-CORNER (SHEET METAL)		NO		NO	NO	NO

THE FIVE BASIC JOINT TYPES

FIGURE 15-31 *The other type of welds related to the five basic joints.*

509

that are designed for the shielded metal arc or other arc welding processes. The seam welds and the spot welds, when using the arc processes, are made by melting through the top member of the joint into the other members. These are sometimes called *burn-through* welds. They are restricted to thin materials and are largely dependent on the depth of penetration of the process involved. They are quite popular for the CO_2 welding process, which has deep penetrating qualities. The joint strength is based on the area of the weld at the interface between the two members. The way these and the other weld types are applied to the five basic joint types is shown by Figure 15-31.

Plug and slot welds have holes prepared in one member so that the welding can be done through the hole into the other member. Plug welds are round and slot welds are elongated. The holes are completely filled in making the plug or slot welds. Plug and slot welds are becoming less popular. Plug welds were made in early design transitions from riveted structures to welded structures. As designs have improved, the plug welds have been discontinued. They are, however, used for certain blind applications in which the other side of the weld joint may be inaccessible.

Standard dimensions have been established for plug or slot welds. This information is shown by Figure 15-32. The strength of these welds is obtained by calculating the area of the plug hole or slot hole where it interfaces the other member.

Surfacing welds are normally bead welds made on a surface to provide special properties or dimensions of that surface. They are commonly used to build up areas for remachining, to provide corrosion-resistant surfaces, or to provide abrasion resistant or hard surfaces for wear. Surfacing welds can be made using most of the arc processes and with strip electrode shielded by submerged arc flux to provide wider beads. They do not involve the making of joints.

The flange weld edge and corner joints are used primarily for lighter gauge metals, usually called sheet metals. Flanging of parts at welds is common practice to improve stiffness, reduce distortion, and to provide an area for welding. The only preparation involved is shearing and bending.

The last of the weld types is the back or backing weld which is used primarily for improving the properties of the root of the single groove welds. To perform the back or backing weld the back side of the weld must be accessible. It is used to ensure that root fusion is

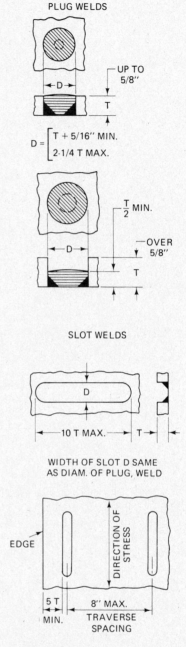

FIGURE 15-32 *Plug and slot weld designs.*

complete and that any potential stress risers are eliminated. Gouging, chipping, or grinding is performed before the backing weld is made.

A summary of formulas used to determine stresses in welds for different welds and different joints and loadings is given by Figure 15-33.[3]

Weld Joint Details

The following illustrations of joint designs will provide a ready reference for the designer. These joint

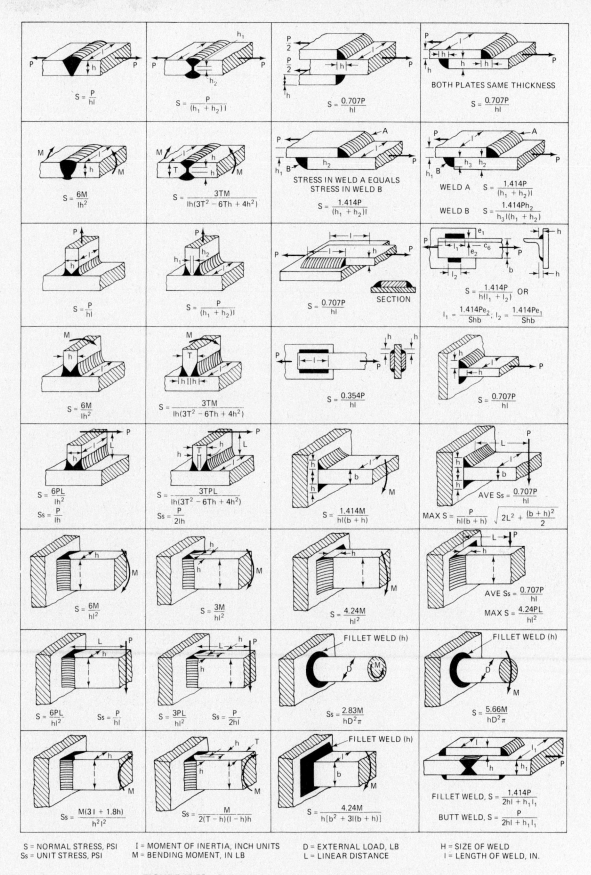

$$S = \frac{P}{hl}$$

$$S = \frac{P}{(h_1 + h_2)l}$$

$$S = \frac{0.707P}{hl}$$

BOTH PLATES SAME THICKNESS

$$S = \frac{0.707P}{hl}$$

$$S = \frac{6M}{lh^2}$$

$$S = \frac{3TM}{lh(3T^2 - 6Th + 4h^2)}$$

STRESS IN WELD A EQUALS
STRESS IN WELD B

$$S = \frac{1.414P}{(h_1 + h_2)l}$$

WELD A $S = \dfrac{1.414P}{(h_1 + h_2)l}$

WELD B $S = \dfrac{1.414Ph_2}{h_3 l(h_1 + h_2)}$

$$S = \frac{P}{hl}$$

$$S = \frac{P}{(h_1 + h_2)l}$$

$$S = \frac{0.707P}{hl}$$

SECTION

$$S = \frac{1.414P}{h(l_1 + l_2)} \text{ OR}$$

$$l_1 = \frac{1.414Pe_2}{Shb}; \ l_2 = \frac{1.414Pe_1}{Shb}$$

$$S = \frac{6M}{lh^2}$$

$$S = \frac{3TM}{lh(3T^2 - 6Th + 4h^2)}$$

$$S = \frac{0.354P}{hl}$$

$$S = \frac{0.707P}{hl}$$

$$S = \frac{6PL}{lh^2}$$

$$Ss = \frac{P}{lh}$$

$$S = \frac{3TPL}{lh(3T^2 - 6Th + 4h^2)}$$

$$Ss = \frac{P}{2lh}$$

$$S = \frac{1.414M}{hl(b + h)}$$

AVE $Ss = \dfrac{0.707P}{hl}$

MAX $S = \dfrac{P}{hl(b+h)} \sqrt{2L^2 + \dfrac{(b+h)^2}{2}}$

$$S = \frac{6M}{hl^2}$$

$$S = \frac{3M}{hl^2}$$

$$S = \frac{4.24M}{hl^2}$$

AVE $Ss = \dfrac{0.707P}{hl}$

MAX $S = \dfrac{4.24PL}{hl^2}$

$$S = \frac{6PL}{hl^2} \qquad Ss = \frac{P}{l}$$

$$S = \frac{3PL}{hl^2} \qquad Ss = \frac{P}{2hl}$$

FILLET WELD (h)

$$Ss = \frac{2.83M}{hD^2\pi}$$

FILLET WELD (h)

$$S = \frac{5.66M}{hD^2\pi}$$

$$Ss = \frac{M(3l + 1.8h)}{h^2 l^2}$$

$$Ss = \frac{M}{2(T - h)(l - h)h}$$

FILLET WELD (h)

$$S = \frac{4.24M}{h[b^2 + 3l(b + h)]}$$

FILLET WELD, $S = \dfrac{1.414P}{2hl + h_1 l_1}$

BUTT WELD, $S = \dfrac{P}{2hl + h_1 l_1}$

S = NORMAL STRESS, PSI   I = MOMENT OF INERTIA, INCH UNITS   D = EXTERNAL LOAD, LB   H = SIZE OF WELD
Ss = UNIT STRESS, PSI   M = BENDING MOMENT, IN LB   L = LINEAR DISTANCE   l = LENGTH OF WELD, IN.

FIGURE 15-33 *Summary of formulas used to determine allowable stresses.*

designs are based on the requirements of the American Welding Society's structural code and include the details of the joint, the joint limitations, its use, and the welding symbol.[4] These joint designs are for use with the shielded metal arc welding process and have been prequalified for this process. The joint designs are suitable for oxyacetylene welding, gas tungsten arc welding, gas metal arc welding, and flux-cored arc welding. However, modifications can be made on these joint designs if they are to be used exclusively for one specific welding process. These changes would make the joints more economical when they are used with a specific process. To obtain efficient, economical weld joint designs the designer must follow the recommended limitations on material thickness. One joint design may be optimum for thin material but poor on thick material.

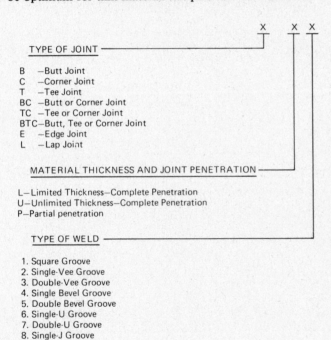

TYPE OF JOINT

B —Butt Joint
C —Corner Joint
T —Tee Joint
BC —Butt or Corner Joint
TC —Tee or Corner Joint
BTC—Butt, Tee or Corner Joint
E —Edge Joint
L —Lap Joint

MATERIAL THICKNESS AND JOINT PENETRATION

L—Limited Thickness—Complete Penetration
U—Unlimited Thickness—Complete Penetration
P—Partial penetration

TYPE OF WELD

1. Square Groove
2. Single-Vee Groove
3. Double-Vee Groove
4. Single Bevel Groove
5. Double Bevel Groove
6. Single-U Groove
7. Double-U Groove
8. Single-J Groove
9. Double-J Groove
10. Single Fillet
11. Double Fillet
12. Edge

FIGURE 15-34 *Weld joint identification number.*

The weld joint details are indexed by a code system which will identify the type of joint, the material thickness limitation, joint penetration requirement, and the type of weld. Similar weld designs are then expanded alphabetically to provide variations. This is shown by

Figure 15-34. This system is similar to that of "Welded Joint Design" MIL-STD-0022B (SHIPS) but sufficiently different so that the numbering system is not interchangeable.[5] A similar system is presented by the American Institute of Steel Construction.[6]

Figure 15-35 presents the series of weld joint details indexed by this code system. In this identification system, the joint type is first. The types of joints are listed alphabetically; thus, the butt joints (B) come first followed by corner joints (C), and so on. Combination joints are also included since the weld detail can be the same for more than one type of joint.

The second factor is the material thickness and penetration requirements. There are three categories. L indicates limited thickness. A maximum nominal thickness is shown for each particular joint. In these cases the thickness maximum must be adhered to. U indicates unlimited thickness. This would be used for materials thicker than the L category. P indicates partial penetration. In these joint designs a sufficiently large face is used to avoid complete penetration. Caution must be exercised when using partial penetration joints in which dynamic loading and cold temperature service are involved.

The third factor is the type of weld. These are based on the types of weld groove and welds. When using these joint designs, the following allowable variations of dimensions should be observed. The root face (RF) of joints is zero unless dimensioned otherwise. If the dimension is zero or otherwise the root face should not exceed the dimension shown by more than 1/16 in.

The root opening (RO) of joints shown is minimum. Root opening should not exceed the dimension shown by more than 1/16 in. The groove angle (A) shown is minimum. The angle dimension should not be exceeded by more than +10° or −0°. The radius (R) dimension of J and U grooves is minimum. This minimum dimension should not be exceeded by more than 1/8 in. Single-bevel and single-J groove weld joints should not be used for butt joints except in the horizontal position and in this case the beveled or grooved part must be the upper member of the joint.

Double-groove welds may have grooves of unequal depths. The depth of a shallower groove should not be less than 1/4 the thickness of the thinner member joined.

In composite welds where fillet welds are used to reinforce groove welds, particularly in tee and corner joints, the size of the fillet shall equal T/4 of the thinner member but should not exceed 3/8 in. maximum. For more detailed information and additional joints refer to the American Welding Society's structural welding code.

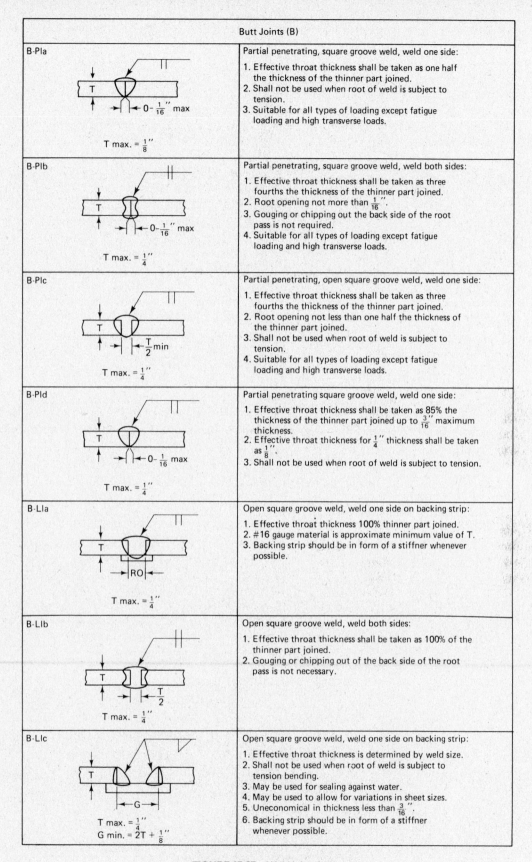

Butt Joints (B)	
B-P1a T max. = $\frac{1}{8}$"	Partial penetrating, square groove weld, weld one side: 1. Effective throat thickness shall be taken as one half the thickness of the thinner part joined. 2. Shall not be used when root of weld is subject to tension. 3. Suitable for all types of loading except fatigue loading and high transverse loads.
B-P1b T max. = $\frac{1}{4}$"	Partial penetrating, square groove weld, weld both sides: 1. Effective throat thickness shall be taken as three fourths the thickness of the thinner part joined. 2. Root opening not more than $\frac{1}{16}$". 3. Gouging or chipping out the back side of the root pass is not required. 4. Suitable for all types of loading except fatigue loading and high transverse loads.
B-P1c T max. = $\frac{1}{4}$"	Partial penetrating, open square groove weld, weld one side: 1. Effective throat thickness shall be taken as three fourths the thickness of the thinner part joined. 2. Root opening not less than one half the thickness of the thinner part joined. 3. Shall not be used when root of weld is subject to tension. 4. Suitable for all types of loading except fatigue loading and high transverse loads.
B-P1d T max. = $\frac{1}{4}$"	Partial penetrating square groove weld, weld one side: 1. Effective throat thickness shall be taken as 85% the thickness of the thinner part joined up to $\frac{3}{16}$" maximum thickness. 2. Effective throat thickness for $\frac{1}{4}$" thickness shall be taken as $\frac{1}{8}$". 3. Shall not be used when root of weld is subject to tension.
B-L1a T max. = $\frac{1}{4}$"	Open square groove weld, weld one side on backing strip: 1. Effective throat thickness 100% thinner part joined. 2. #16 gauge material is approximate minimum value of T. 3. Backing strip should be in form of a stiffner whenever possible.
B-L1b T max. = $\frac{1}{4}$"	Open square groove weld, weld both sides: 1. Effective throat thickness shall be taken as 100% of the thinner part joined. 2. Gouging or chipping out of the back side of the root pass is not necessary.
B-L1c T max. = $\frac{1}{4}$" G min. = 2T + $\frac{1}{8}$"	Open square groove weld, weld one side on backing strip: 1. Effective throat thickness is determined by weld size. 2. Shall not be used when root of weld is subject to tension bending. 3. May be used for sealing against water. 4. May be used to allow for variations in sheet sizes. 5. Uneconomical in thickness less than $\frac{3}{16}$". 6. Backing strip should be in form of a stiffner whenever possible.

FIGURE 15-35 *Weld joint designs.*

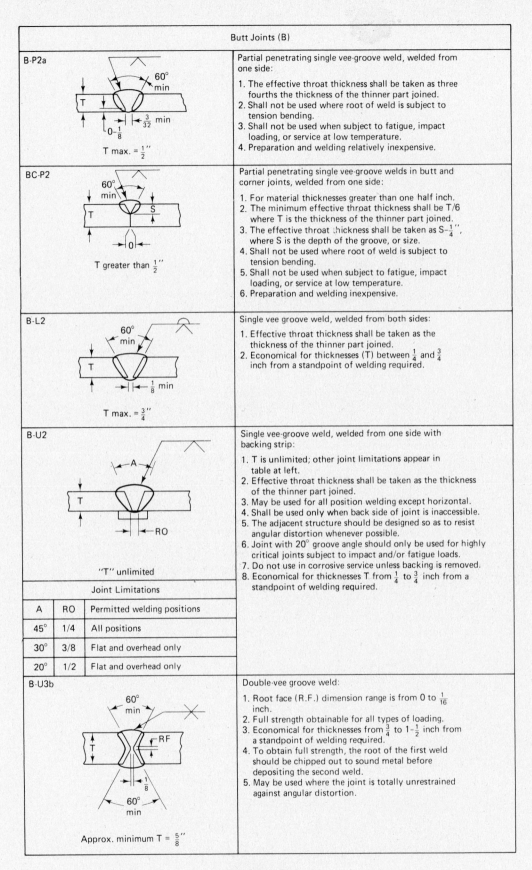

Butt Joints (B)	
B-P2a T max. = $\frac{1}{2}$''	Partial penetrating single vee-groove weld, welded from one side: 1. The effective throat thickness shall be taken as three fourths the thickness of the thinner part joined. 2. Shall not be used where root of weld is subject to tension bending. 3. Shall not be used when subject to fatigue, impact loading, or service at low temperature. 4. Preparation and welding relatively inexpensive.
BC-P2 T greater than $\frac{1}{2}$''	Partial penetrating single vee-groove welds in butt and corner joints, welded from one side: 1. For material thicknesses greater than one half inch. 2. The minimum effective throat thickness shall be T/6 where T is the thickness of the thinner part joined. 3. The effective throat thickness shall be taken as $S-\frac{1}{4}$'', where S is the depth of the groove, or size. 4. Shall not be used where root of weld is subject to tension bending. 5. Shall not be used when subject to fatigue, impact loading, or service at low temperature. 6. Preparation and welding inexpensive.
B-L2 T max. = $\frac{3}{4}$''	Single vee groove weld, welded from both sides: 1. Effective throat thickness shall be taken as the thickness of the thinner part joined. 2. Economical for thicknesses (T) between $\frac{1}{4}$ and $\frac{3}{4}$ inch from a standpoint of welding required.

Single vee-groove weld, welded from one side with backing strip:

1. T is unlimited; other joint limitations appear in table at left.
2. Effective throat thickness shall be taken as the thickness of the thinner part joined.
3. May be used for all position welding except horizontal.
4. Shall be used only when back side of joint is inaccessible.
5. The adjacent structure should be designed so as to resist angular distortion whenever possible.
6. Joint with 20° groove angle should only be used for highly critical joints subject to impact and/or fatigue loads.
7. Do not use in corrosive service unless backing is removed.
8. Economical for thicknesses T from $\frac{1}{4}$ to $\frac{3}{4}$ inch from a standpoint of welding required.

B-U2

"T" unlimited

Joint Limitations

A	RO	Permitted welding positions
45°	1/4	All positions
30°	3/8	Flat and overhead only
20°	1/2	Flat and overhead only

B-U3b

Double-vee groove weld:

1. Root face (R.F.) dimension range is from 0 to $\frac{1}{16}$ inch.
2. Full strength obtainable for all types of loading.
3. Economical for thicknesses from $\frac{3}{4}$ to $1-\frac{1}{2}$ inch from a standpoint of welding required.
4. To obtain full strength, the root of the first weld should be chipped out to sound metal before depositing the second weld.
5. May be used where the joint is totally unrestrained against angular distortion.

Approx. minimum T = $\frac{5}{8}$''

FIGURE 15-35 *(Cont.)*

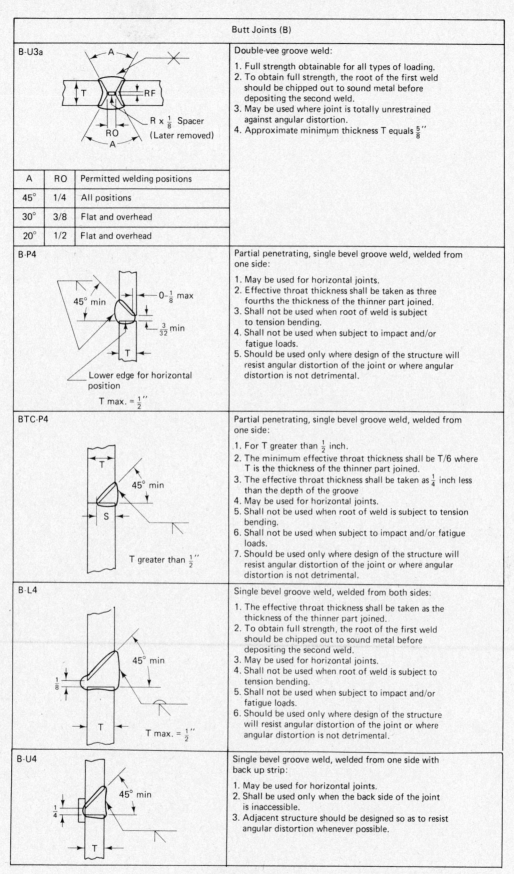

Butt Joints (B)	
B-U3a	Double-vee groove weld: 1. Full strength obtainable for all types of loading. 2. To obtain full strength, the root of the first weld should be chipped out to sound metal before depositing the second weld. 3. May be used where joint is totally unrestrained against angular distortion. 4. Approximate minimum thickness T equals $\frac{5}{8}$ ''

A	RO	Permitted welding positions
45°	1/4	All positions
30°	3/8	Flat and overhead
20°	1/2	Flat and overhead

B-P4 — 45° min, $0-\frac{1}{8}$ max, $\frac{3}{32}$ min, T, Lower edge for horizontal position, T max. = $\frac{1}{2}$ ''

Partial penetrating, single bevel groove weld, welded from one side:

1. May be used for horizontal joints.
2. Effective throat thickness shall be taken as three fourths the thickness of the thinner part joined.
3. Shall not be used when root of weld is subject to tension bending.
4. Shall not be used when subject to impact and/or fatigue loads.
5. Should be used only where design of the structure will resist angular distortion of the joint or where angular distortion is not detrimental.

BTC-P4 — T, 45° min, S, T greater than $\frac{1}{2}$ ''

Partial penetrating, single bevel groove weld, welded from one side:

1. For T greater than $\frac{1}{2}$ inch.
2. The minimum effective throat thickness shall be T/6 where T is the thickness of the thinner part joined.
3. The effective throat thickness shall be taken as $\frac{1}{4}$ inch less than the depth of the groove
4. May be used for horizontal joints.
5. Shall not be used when root of weld is subject to tension bending.
6. Shall not be used when subject to impact and/or fatigue loads.
7. Should be used only where design of the structure will resist angular distortion of the joint or where angular distortion is not detrimental.

B-L4 — 45° min, $\frac{1}{8}$, T, T max. = $\frac{1}{2}$ ''

Single bevel groove weld, welded from both sides:

1. The effective throat thickness shall be taken as the thickness of the thinner part joined.
2. To obtain full strength, the root of the first weld should be chipped out to sound metal before depositing the second weld.
3. May be used for horizontal joints.
4. Shall not be used when root of weld is subject to tension bending.
5. Shall not be used when subject to impact and/or fatigue loads.
6. Should be used only where design of the structure will resist angular distortion of the joint or where angular distortion is not detrimental.

B-U4 — 45° min, $\frac{1}{4}$, T

Single bevel groove weld, welded from one side with back up strip:

1. May be used for horizontal joints.
2. Shall be used only when the back side of the joint is inaccessible.
3. Adjacent structure should be designed so as to resist angular distortion whenever possible.

FIGURE 15-35 *(Cont.)*

Butt Joints (B)	
B-U5a 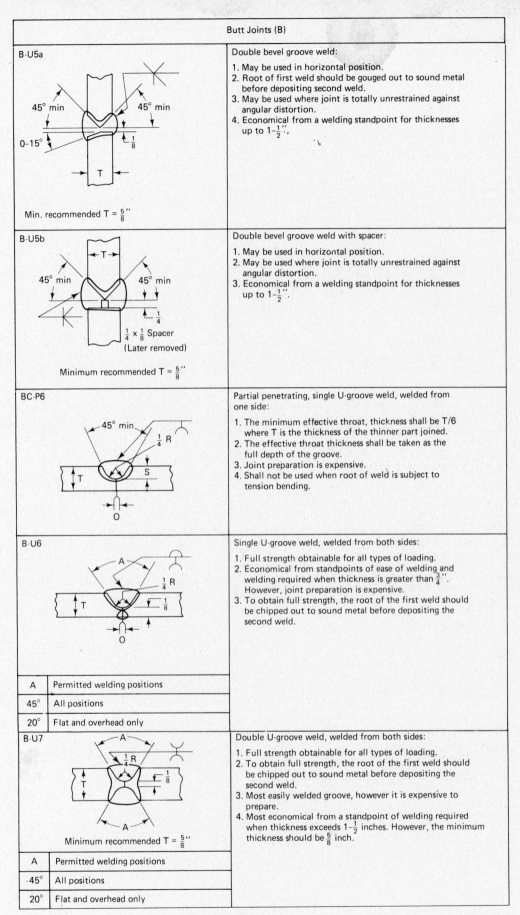 45° min 45° min 0–15° $\frac{1}{8}$ T Min. recommended T = $\frac{5}{8}$″	**Double bevel groove weld:** 1. May be used in horizontal position. 2. Root of first weld should be gouged out to sound metal before depositing second weld. 3. May be used where joint is totally unrestrained against angular distortion. 4. Economical from a welding standpoint for thicknesses up to 1-$\frac{1}{2}$″.
B-U5b T 45° min 45° min $\frac{1}{4}$ $\frac{1}{4} \times \frac{1}{8}$ Spacer (Later removed) Minimum recommended T = $\frac{5}{8}$″	**Double bevel groove weld with spacer:** 1. May be used in horizontal position. 2. May be used where joint is totally unrestrained against angular distortion. 3. Economical from a welding standpoint for thicknesses up to 1-$\frac{1}{2}$″.
BC-P6 45° min $\frac{1}{4}$ R T S O	**Partial penetrating, single U-groove weld, welded from one side:** 1. The minimum effective throat, thickness shall be T/6 where T is the thickness of the thinner part joined. 2. The effective throat thickness shall be taken as the full depth of the groove. 3. Joint preparation is expensive. 4. Shall not be used when root of weld is subject to tension bending.
B-U6 A $\frac{1}{4}$ R T $\frac{1}{8}$ O	**Single U-groove weld, welded from both sides:** 1. Full strength obtainable for all types of loading. 2. Economical from standpoints of ease of welding and welding required when thickness is greater than $\frac{3}{4}$″. However, joint preparation is expensive. 3. To obtain full strength, the root of the first weld should be chipped out to sound metal before depositing the second weld.
A Permitted welding positions	
45° All positions	
20° Flat and overhead only	
B-U7 A $\frac{1}{4}$ R T $\frac{1}{8}$ A Minimum recommended T = $\frac{5}{8}$″	**Double U-groove weld, welded from both sides:** 1. Full strength obtainable for all types of loading. 2. To obtain full strength, the root of the first weld should be chipped out to sound metal before depositing the second weld. 3. Most easily welded groove, however it is expensive to prepare. 4. Most economical from a standpoint of welding required when thickness exceeds 1-$\frac{1}{2}$ inches. However, the minimum thickness should be $\frac{5}{8}$ inch.
A Permitted welding positions	
·45° All positions	
20° Flat and overhead only	

Figure 15-35 *(Cont.)*

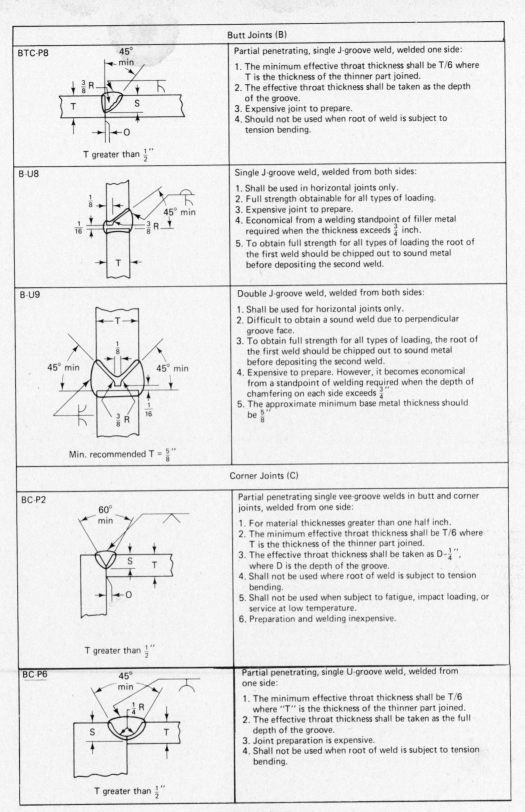

Butt Joints (B)	
BTC-P8 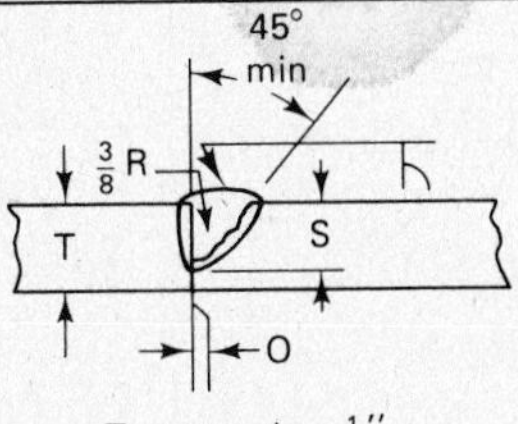 45° min $\frac{3}{8}$ R T S O T greater than $\frac{1}{2}''$	Partial penetrating, single J-groove weld, welded one side: 1. The minimum effective throat thickness shall be T/6 where T is the thickness of the thinner part joined. 2. The effective throat thickness shall be taken as the depth of the groove. 3. Expensive joint to prepare. 4. Should not be used when root of weld is subject to tension bending.
B-U8 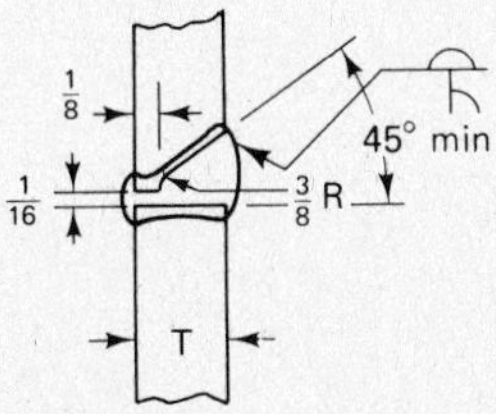 $\frac{1}{8}$ 45° min $\frac{1}{16}$ $\frac{3}{8}$ R T	Single J-groove weld, welded from both sides: 1. Shall be used in horizontal joints only. 2. Full strength obtainable for all types of loading. 3. Expensive joint to prepare. 4. Economical from a welding standpoint of filler metal required when the thickness exceeds $\frac{3}{4}$ inch. 5. To obtain full strength for all types of loading the root of the first weld should be chipped out to sound metal before depositing the second weld.
B-U9 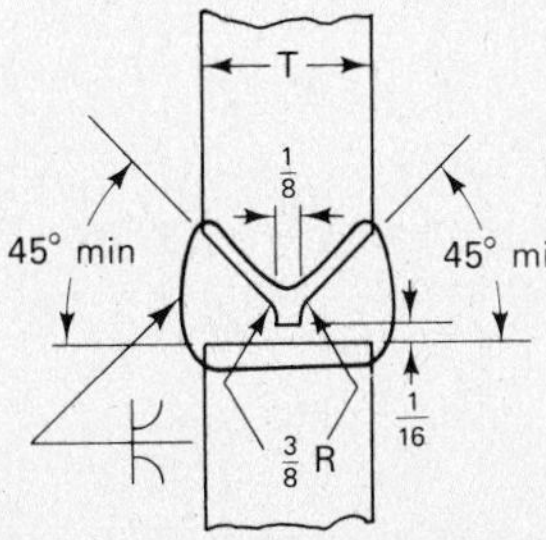 T $\frac{1}{8}$ 45° min 45° min $\frac{1}{16}$ $\frac{3}{8}$ R Min. recommended T = $\frac{5}{8}''$	Double J-groove weld, welded from both sides: 1. Shall be used for horizontal joints only. 2. Difficult to obtain a sound weld due to perpendicular groove face. 3. To obtain full strength for all types of loading, the root of the first weld should be chipped out to sound metal before depositing the second weld. 4. Expensive to prepare. However, it becomes economical from a standpoint of welding required when the depth of chamfering on each side exceeds $\frac{3}{4}''$ 5. The approximate minimum base metal thickness should be $\frac{5}{8}''$
Corner Joints (C)	
BC-P2 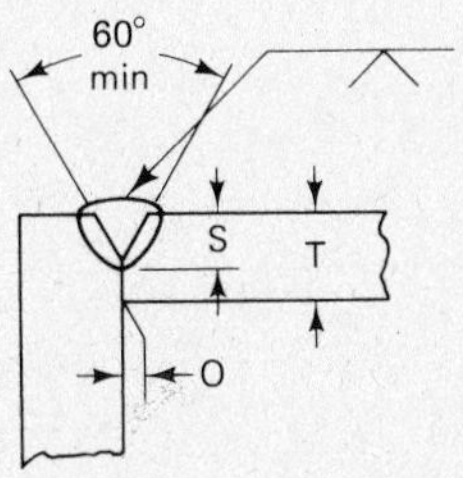 60° min S T O T greater than $\frac{1}{2}''$	Partial penetrating single vee-groove welds in butt and corner joints, welded from one side: 1. For material thicknesses greater than one half inch. 2. The minimum effective throat thickness shall be T/6 where T is the thickness of the thinner part joined. 3. The effective throat thickness shall be taken as D-$\frac{1}{4}''$, where D is the depth of the groove. 4. Shall not be used where root of weld is subject to tension bending. 5. Shall not be used when subject to fatigue, impact loading, or service at low temperature. 6. Preparation and welding inexpensive.
BC-P6 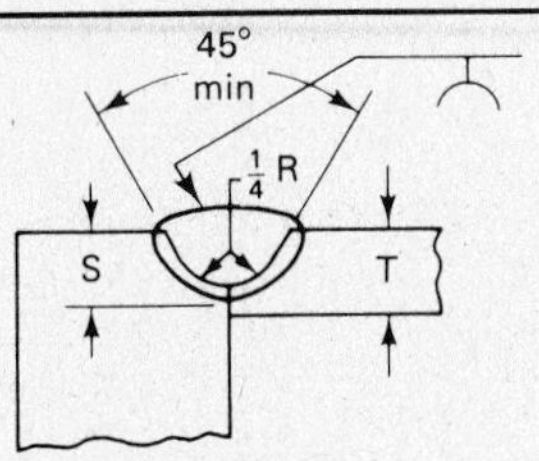 45° min $\frac{1}{4}$ R S T T greater than $\frac{1}{2}''$	Partial penetrating, single U-groove weld, welded from one side: 1. The minimum effective throat thickness shall be T/6 where "T" is the thickness of the thinner part joined. 2. The effective throat thickness shall be taken as the full depth of the groove. 3. Joint preparation is expensive. 4. Shall not be used when root of weld is subject to tension bending.

FIGURE 15-35 *(Cont.)*

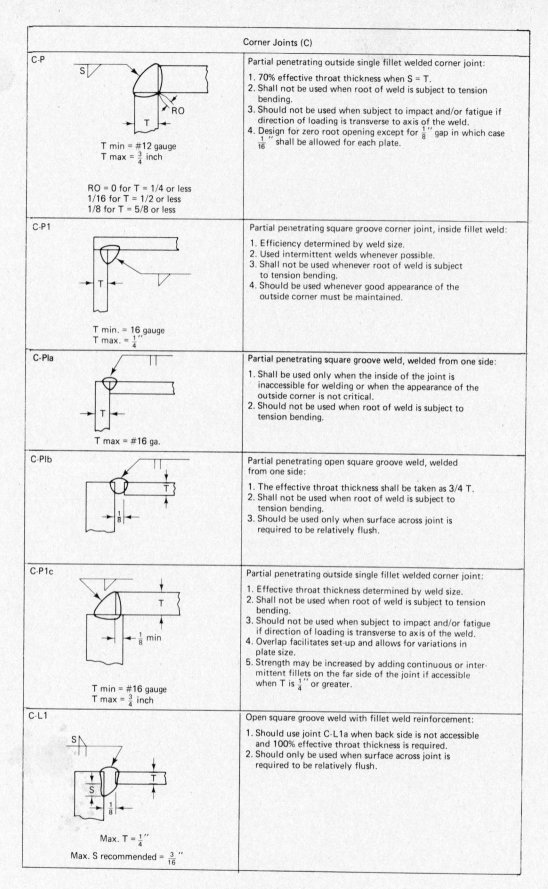

Corner Joints (C)	
C-P	Partial penetrating outside single fillet welded corner joint: 1. 70% effective throat thickness when S = T. 2. Shall not be used when root of weld is subject to tension bending. 3. Should not be used when subject to impact and/or fatigue if direction of loading is transverse to axis of the weld. 4. Design for zero root opening except for $\frac{1}{8}$" gap in which case $\frac{1}{16}$" shall be allowed for each plate.
T min = #12 gauge T max = $\frac{3}{4}$ inch RO = 0 for T = 1/4 or less 1/16 for T = 1/2 or less 1/8 for T = 5/8 or less	
C-P1	Partial penetrating square groove corner joint, inside fillet weld: 1. Efficiency determined by weld size. 2. Used intermittent welds whenever possible. 3. Shall not be used whenever root of weld is subject to tension bending. 4. Should be used whenever good appearance of the outside corner must be maintained.
T min. = 16 gauge T max. = $\frac{1}{4}$"	
C-PIa	Partial penetrating square groove weld, welded from one side: 1. Shall be used only when the inside of the joint is inaccessible for welding or when the appearance of the outside corner is not critical. 2. Should not be used when root of weld is subject to tension bending.
T max = #16 ga.	
C-PIb	Partial penetrating open square groove weld, welded from one side: 1. The effective throat thickness shall be taken as 3/4 T. 2. Shall not be used when root of weld is subject to tension bending. 3. Should be used only when surface across joint is required to be relatively flush.
$\frac{1}{8}$	
C-P1c	Partial penetrating outside single fillet welded corner joint: 1. Effective throat thickness determined by weld size. 2. Shall not be used when root of weld is subject to tension bending. 3. Should not be used when subject to impact and/or fatigue if direction of loading is transverse to axis of the weld. 4. Overlap facilitates set-up and allows for variations in plate size. 5. Strength may be increased by adding continuous or intermittent fillets on the far side of the joint if accessible when T is $\frac{1}{4}$" or greater.
$\frac{1}{8}$ min T min = #16 gauge T max = $\frac{3}{4}$ inch	
C-L1	Open square groove weld with fillet weld reinforcement: 1. Should use joint C-L1a when back side is not accessible and 100% effective throat thickness is required. 2. Should only be used when surface across joint is required to be relatively flush.
Max. T = $\frac{1}{4}$" Max. S recommended = $\frac{3}{16}$"	

FIGURE 15-35 *(Cont.)*

518

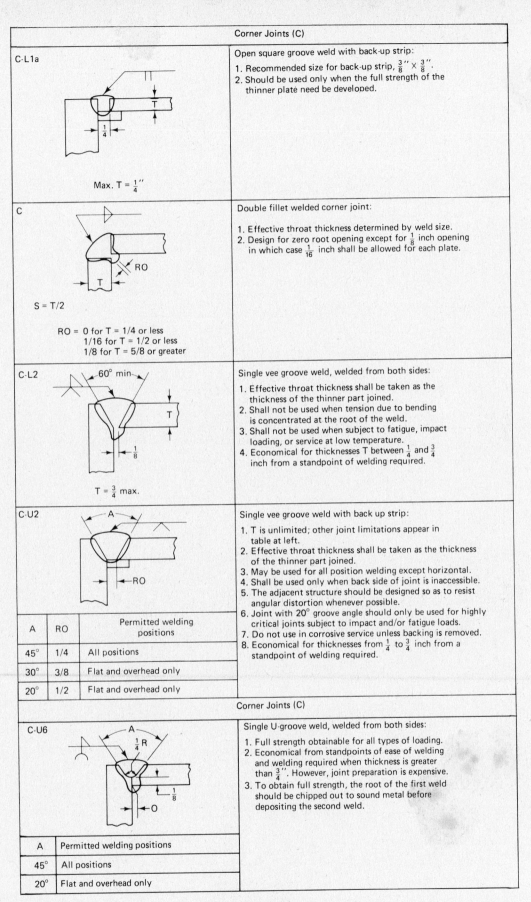

Corner Joints (C)	
C-L1a Max. T = $\frac{1}{4}''$	**Open square groove weld with back-up strip:** 1. Recommended size for back-up strip, $\frac{3}{8}'' \times \frac{3}{8}''$. 2. Should be used only when the full strength of the thinner plate need be developed.
C S = T/2 RO = 0 for T = 1/4 or less 1/16 for T = 1/2 or less 1/8 for T = 5/8 or greater	**Double fillet welded corner joint:** 1. Effective throat thickness determined by weld size. 2. Design for zero root opening except for $\frac{1}{8}$ inch opening in which case $\frac{1}{16}$ inch shall be allowed for each plate.
C-L2 60° min T = $\frac{3}{4}$ max.	**Single vee groove weld, welded from both sides:** 1. Effective throat thickness shall be taken as the thickness of the thinner part joined. 2. Shall not be used when tension due to bending is concentrated at the root of the weld. 3. Shall not be used when subject to fatigue, impact loading, or service at low temperature. 4. Economical for thicknesses T between $\frac{1}{4}$ and $\frac{3}{4}$ inch from a standpoint of welding required.
C-U2 A RO	**Single vee groove weld with back up strip:** 1. T is unlimited; other joint limitations appear in table at left. 2. Effective throat thickness shall be taken as the thickness of the thinner part joined. 3. May be used for all position welding except horizontal. 4. Shall be used only when back side of joint is inaccessible. 5. The adjacent structure should be designed so as to resist angular distortion whenever possible. 6. Joint with 20° groove angle should only be used for highly critical joints subject to impact and/or fatigue loads. 7. Do not use in corrosive service unless backing is removed. 8. Economical for thicknesses from $\frac{1}{4}$ to $\frac{3}{4}$ inch from a standpoint of welding required.

A	RO	Permitted welding positions
45°	1/4	All positions
30°	3/8	Flat and overhead only
20°	1/2	Flat and overhead only

Corner Joints (C)	
C-U6 A $\frac{1}{4}$ R $\frac{1}{8}$ O	**Single U-groove weld, welded from both sides:** 1. Full strength obtainable for all types of loading. 2. Economical from standpoints of ease of welding and welding required when thickness is greater than $\frac{3}{4}''$. However, joint preparation is expensive. 3. To obtain full strength, the root of the first weld should be chipped out to sound metal before depositing the second weld.

A	Permitted welding positions
45°	All positions
20°	Flat and overhead only

FIGURE 15-35 *(Cont.)*

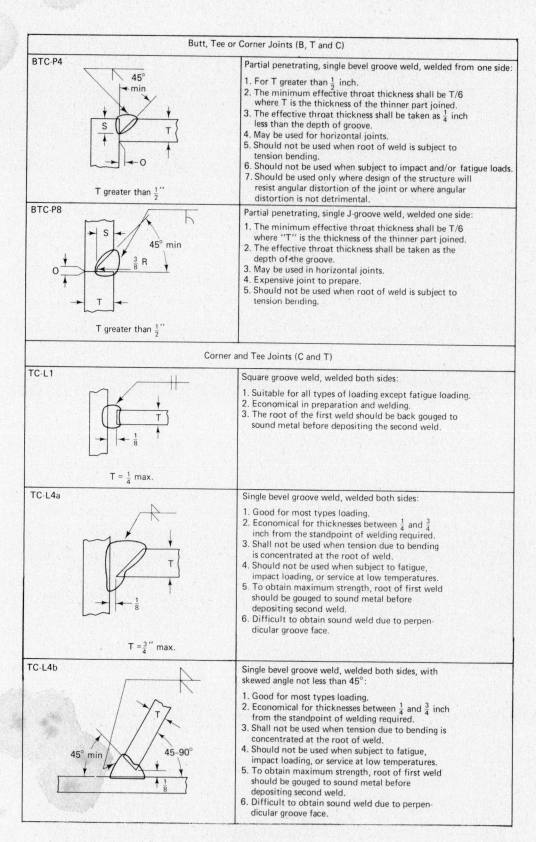

Butt, Tee or Corner Joints (B, T and C)	
BTC-P4 T greater than $\frac{1}{2}''$	Partial penetrating, single bevel groove weld, welded from one side: 1. For T greater than $\frac{1}{2}$ inch. 2. The minimum effective throat thickness shall be T/6 where T is the thickness of the thinner part joined. 3. The effective throat thickness shall be taken as $\frac{1}{4}$ inch less than the depth of groove. 4. May be used for horizontal joints. 5. Should not be used when root of weld is subject to tension bending. 6. Should not be used when subject to impact and/or fatigue loads. 7. Should be used only where design of the structure will resist angular distortion of the joint or where angular distortion is not detrimental.
BTC-P8 T greater than $\frac{1}{2}''$	Partial penetrating, single J-groove weld, welded one side: 1. The minimum effective throat thickness shall be T/6 where "T" is the thickness of the thinner part joined. 2. The effective throat thickness shall be taken as the depth of the groove. 3. May be used in horizontal joints. 4. Expensive joint to prepare. 5. Should not be used when root of weld is subject to tension bending.
Corner and Tee Joints (C and T)	
TC-L1 T = $\frac{1}{4}$ max.	Square groove weld, welded both sides: 1. Suitable for all types of loading except fatigue loading. 2. Economical in preparation and welding. 3. The root of the first weld should be back gouged to sound metal before depositing the second weld.
TC-L4a T = $\frac{3}{4}''$ max.	Single bevel groove weld, welded both sides: 1. Good for most types loading. 2. Economical for thicknesses between $\frac{1}{4}$ and $\frac{3}{4}$ inch from the standpoint of welding required. 3. Shall not be used when tension due to bending is concentrated at the root of weld. 4. Should not be used when subject to fatigue, impact loading, or service at low temperatures. 5. To obtain maximum strength, root of first weld should be gouged to sound metal before depositing second weld. 6. Difficult to obtain sound weld due to perpendicular groove face.
TC-L4b	Single bevel groove weld, welded both sides, with skewed angle not less than 45°: 1. Good for most types loading. 2. Economical for thicknesses between $\frac{1}{4}$ and $\frac{3}{4}$ inch from the standpoint of welding required. 3. Shall not be used when tension due to bending is concentrated at the root of weld. 4. Should not be used when subject to fatigue, impact loading, or service at low temperatures. 5. To obtain maximum strength, root of first weld should be gouged to sound metal before depositing second weld. 6. Difficult to obtain sound weld due to perpendicular groove face.

FIGURE 15-35 *(Cont.)*

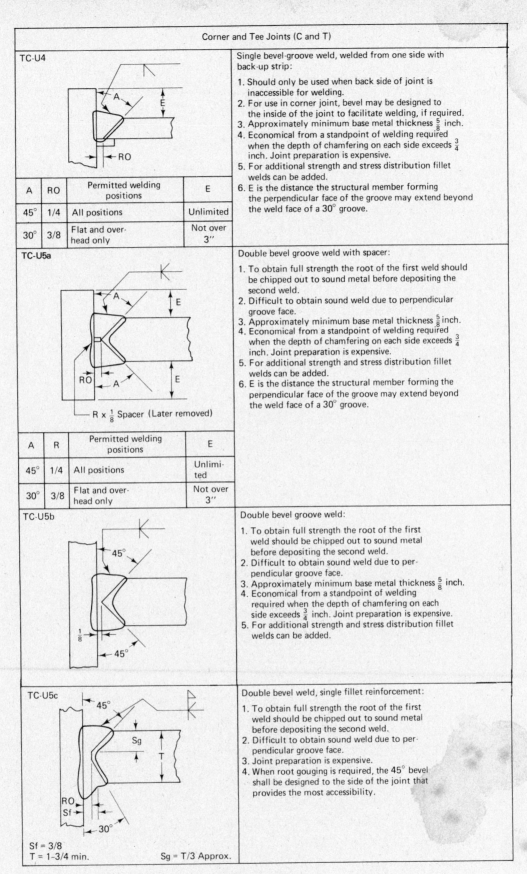

	Corner and Tee Joints (C and T)

TC-U4

Single bevel-groove weld, welded from one side with back-up strip:

1. Should only be used when back side of joint is inaccessible for welding.
2. For use in corner joint, bevel may be designed to the inside of the joint to facilitate welding, if required.
3. Approximately minimum base metal thickness $\frac{5}{8}$ inch.
4. Economical from a standpoint of welding required when the depth of chamfering on each side exceeds $\frac{3}{4}$ inch. Joint preparation is expensive.
5. For additional strength and stress distribution fillet welds can be added.
6. E is the distance the structural member forming the perpendicular face of the groove may extend beyond the weld face of a 30° groove.

A	RO	Permitted welding positions	E
45°	1/4	All positions	Unlimited
30°	3/8	Flat and over-head only	Not over 3''

TC-U5a

Double bevel groove weld with spacer:

1. To obtain full strength the root of the first weld should be chipped out to sound metal before depositing the second weld.
2. Difficult to obtain sound weld due to perpendicular groove face.
3. Approximately minimum base metal thickness $\frac{5}{8}$ inch.
4. Economical from a standpoint of welding required when the depth of chamfering on each side exceeds $\frac{3}{4}$ inch. Joint preparation is expensive.
5. For additional strength and stress distribution fillet welds can be added.
6. E is the distance the structural member forming the perpendicular face of the groove may extend beyond the weld face of a 30° groove.

R x $\frac{1}{8}$ Spacer (Later removed)

A	R	Permitted welding positions	E
45°	1/4	All positions	Unlimited
30°	3/8	Flat and over-head only	Not over 3''

TC-U5b

Double bevel groove weld:

1. To obtain full strength the root of the first weld should be chipped out to sound metal before depositing the second weld.
2. Difficult to obtain sound weld due to perpendicular groove face.
3. Approximately minimum base metal thickness $\frac{5}{8}$ inch.
4. Economical from a standpoint of welding required when the depth of chamfering on each side exceeds $\frac{3}{4}$ inch. Joint preparation is expensive.
5. For additional strength and stress distribution fillet welds can be added.

TC-U5c

Double bevel weld, single fillet reinforcement:

1. To obtain full strength the root of the first weld should be chipped out to sound metal before depositing the second weld.
2. Difficult to obtain sound weld due to perpendicular groove face.
3. Joint preparation is expensive.
4. When root gouging is required, the 45° bevel shall be designed to the side of the joint that provides the most accessibility.

Sf = 3/8
T = 1-3/4 min.   Sg = T/3 Approx.

FIGURE 15-35 *(Cont.)*

521

Corner and Tee Joints (C and T)

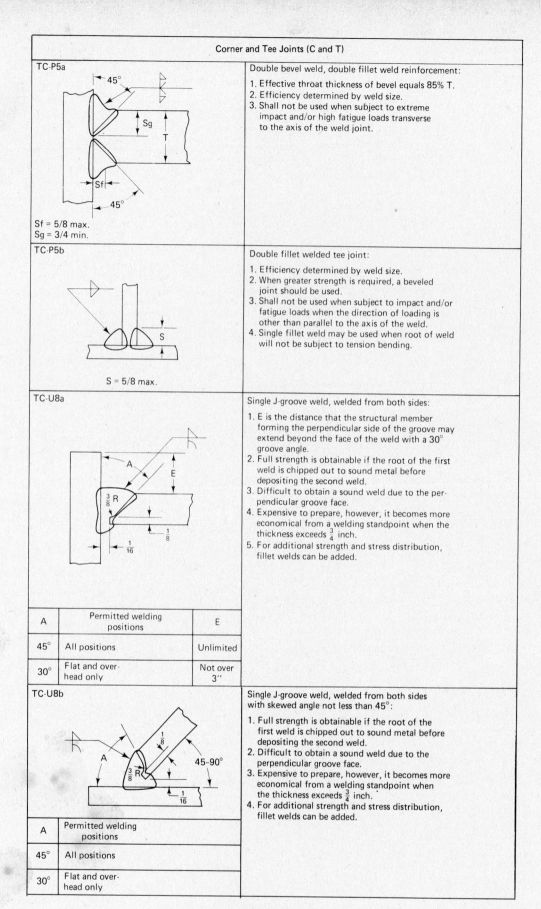

TC-P5a

$Sf = 5/8$ max.
$Sg = 3/4$ min.

Double bevel weld, double fillet weld reinforcement:

1. Effective throat thickness of bevel equals 85% T.
2. Efficiency determined by weld size.
3. Shall not be used when subject to extreme impact and/or high fatigue loads transverse to the axis of the weld joint.

TC-P5b

$S = 5/8$ max.

Double fillet welded tee joint:

1. Efficiency determined by weld size.
2. When greater strength is required, a beveled joint should be used.
3. Shall not be used when subject to impact and/or fatigue loads when the direction of loading is other than parallel to the axis of the weld.
4. Single fillet weld may be used when root of weld will not be subject to tension bending.

TC-U8a

Single J-groove weld, welded from both sides:

1. E is the distance that the structural member forming the perpendicular side of the groove may extend beyond the face of the weld with a 30° groove angle.
2. Full strength is obtainable if the root of the first weld is chipped out to sound metal before depositing the second weld.
3. Difficult to obtain a sound weld due to the perpendicular groove face.
4. Expensive to prepare, however, it becomes more economical from a welding standpoint when the thickness exceeds $\frac{3}{4}$ inch.
5. For additional strength and stress distribution, fillet welds can be added.

A	Permitted welding positions	E
45°	All positions	Unlimited
30°	Flat and over-head only	Not over 3''

TC-U8b

Single J-groove weld, welded from both sides with skewed angle not less than 45°:

1. Full strength is obtainable if the root of the first weld is chipped out to sound metal before depositing the second weld.
2. Difficult to obtain a sound weld due to the perpendicular groove face.
3. Expensive to prepare, however, it becomes more economical from a welding standpoint when the thickness exceeds $\frac{3}{4}$ inch.
4. For additional strength and stress distribution, fillet welds can be added.

A	Permitted welding positions
45°	All positions
30°	Flat and over-head only

FIGURE 15-35 *(Cont.)*

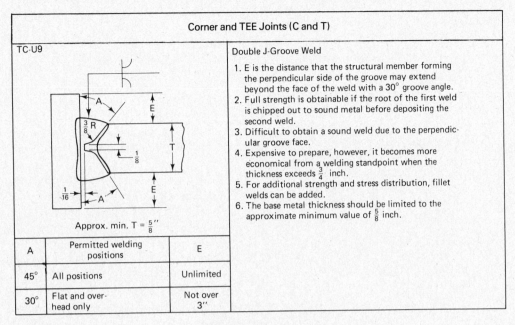

	Corner and TEE Joints (C and T)	
TC-U9	**Double J-Groove Weld**	
	1. E is the distance that the structural member forming the perpendicular side of the groove may extend beyond the face of the weld with a 30° groove angle.	
	2. Full strength is obtainable if the root of the first weld is chipped out to sound metal before depositing the second weld.	
	3. Difficult to obtain a sound weld due to the perpendicular groove face.	
	4. Expensive to prepare, however, it becomes more economical from a welding standpoint when the thickness exceeds $\frac{3}{4}$ inch.	
	5. For additional strength and stress distribution, fillet welds can be added.	
	6. The base metal thickness should be limited to the approximate minimum value of $\frac{5}{8}$ inch.	

Approx. min. $T = \frac{5}{8}''$

A	Permitted welding positions	E
45°	All positions	Unlimited
30°	Flat and over-head only	Not over 3''

FIGURE 15-35 (Continued)

15-5 WELDMENT DESIGN AND INFLUENCE OF SPECIFICATIONS

We have previously discussed the factors involved in designing a weldment. This included the study of the service requirements of the part and the various loadings that would be imposed on the part as well as the severity and repetition of such loadings. It was shown how these loadings and service conditions affect the calculation of stresses that must be withstood by the weldment or structure. It was brought out how these stresses are related to the allowable stress and the strength of the metals used in the weldment.

We have also discussed the design of welds, the design of joints, and how these are combined as weld joints. It was also discussed how these are designed to sustain specific loadings of a static and dynamic nature. Finally, examples of specific weld joint details were presented with their specifications.

In the course of these discussions, it has been assumed that the weld metal is equal in strength to the base metal. This is normal for mild and low-alloy steels. Further, it was mentioned that the different codes and specifications provide the allowable stresses, safety factors, etc., that must be employed on certain weldments for particular types of applications. The allowable stress levels of welds and weld metals are covered in much detail by different specifications and standards. We will attempt to show this information and relate it to the specific industries for which the weldment is intended. Prior to this, there are specific weldment design considerations that should be discussed.

Design of Weldments

It has been mentioned previously that the weld metal portion of a weldment is the most expensive metal in the weldment. The amount of weld metal used to produce the weldment should be kept at an absolute minimum. In the design of a weld joint it was brought out that weld joint cross section is a measure of the amount of weld metal required. Data was presented showing how adequate joints can be obtained with a minimum amount of weld metal. There are other situations in the weldment in which joints, or rather, changes in the placement of parts, can be achieved without making welds. To accomplish this, the designer should utilize as much as possible hot rolled shapes or sections such as standard I beams, wide flange beams, standard channels, ship building channels, angles, pipe, and tubing—circular, square, and rectangular. To use these types of members, the designer must be familiar with tolerances on hot rolled parts, including size tolerance, warpage tolerance, curvature tolerance, etc. These data are provided in standard handbooks.

Cold rolled steel products should be used, when required, since they have closer tolerances than hot rolled sections. Cold rolled parts that can be used include bars and mechanical tubing—round, square, and rectangular. The properties of the metals used to produce these sections are much the same as the hot rolled members.

Another way of eliminating weld joints is the use of formed shapes. These are shapes formed on machines such as brake presses, bending rolls, roll forming equipment, etc. Many standardized types and shapes can be obtained, but many shapes can be produced on a brake press with the proper tooling. Most metal of this type is cold formed; however, for larger or thicker sections, hot forming may be used. In these cases hydraulic presses, bulldozers, forging hammers, bending dies, etc., are used. Special shapes can be formed with special tools and dies for certain requirements. The designer must be familiar with the processes involved since knowledge of tolerances, spring back, sizes, and available shapes is invaluable.

There are many points in a weldment when extra thick parts may be required; for example, at load-bearing points, housings for antifriction bearings, counterweights, etc. In these cases extra heavy plate or billets can be flame cut to the approximate finished size and shapes. Final sizing can be done by machining using regular machine tools. In this case the parts are usually called premachined parts that go into the weldment. Forgings can also be included in this category.

Another way of obtaining thicker sections for specific areas in a weldment can be done by the use of relatively small and simple steel castings. The smaller castings are easily obtained, they are less prone to foundry defects, and pattern costs are reduced. For simple cylindrical parts, patterns can be eliminated. Many castings can include transition sections and may even have weld joint preparation incorporated in the pattern. Small castings are available from foundries and can be economically utilized in complex weldments.

The weldment can be a composite of different types of base materials. It can include clad metal parts, high-strength, medium-strength, and mild steel. Areas of the weldment can be overlayed with special types of alloys to resist wear or corrosive service which might occur in only specific parts of the finished structure. One of the strong features of weldment design is that the type of metal required for the service intended can be placed at the point where it is required and it need not be placed throughout the entire weldment.

Finally, larger weldments and weldments which are too big to ship assembled can be made of subassemblies, which will be welded together at the point of field erection or final assembly. All welded ships are made of many subassemblies which are moved to the final assembly basin where the entire ship is finish welded. Large construction machinery is also field erected because of shipping restrictions. All welded steel buildings are made of large shop fabricated beams, trusses, columns,

etc., which are taken to the erection site for field welding. This type of approach is abslutely necessary when parts are too large to transport.

Specifications Applicable to Weldment Design

Codes and specifications apply to weldments that are designed for certain types of service. Among the more popular specifications are those covering structural work, including buildings and bridges, ships—both navy and commercial, pressure vessels and pressure piping, storage tanks, aircraft, and machinery of many types. We will attempt to indicate the specifications that apply to these different types of structures or weldments and the important requirements each might represent or require.

The designer is responsible for obtaining and studying the latest edition of all specifications that apply to the product being designed. The points brought out in this chapter are some of the more important, but are not all inclusive. Therefore, this chapter cannot be used as a substitute for the specification.

Structural Welding

The welding of structures, both buildings and bridges—either highway or railroad, are covered by the "Structural Welding Code AWS D1.1," published by the American Welding Society, by the "Specification for the Design, Fabrication and Erection of Structural Steel for Buildings," published by the American Institute of Steel Construction, the "Standard Specifications for Highway Bridges," adopted by the American Association of State Highway officials, and the "Specifications for Steel Railroad Bridges," published by the American Railway Engineering Association.

These specifications all refer to welding and usually they refer welding details to the AWS structural welding code. Specific information regarding welding is provided in the other three specifications. This includes such things as types of filler metals to be used, the allowable unit stresses with various types of loading, and the relationship of working stresses to various grades of steel. The specifications covering bridges also refer to fatigue loading and the stress factors that should be employed. Information concerning specific design data is provided. The AWS structural welding code provides for the use of prequalified weld joints when made by specific welding processes. This is particularly advantageous since it avoids qualification of many joints under normal conditions.

Most of the sources regarding welding in these four specifications are in general agreement. One point of agreement is the minimum size of fillet welds. Weld sizes should not exceed the thickness of the thinner part joined unless a larger size is required by calculated stress.

There is also a maximum effective size of fillet welds. The maximum size that may be used along the edges of connected parts should be:

1. Along edges of material less than 1/4 in. thick the maximum size may be equal to the thickness of the material.

2. Along edges of material 1/4 in. or more in thickness the maximum size shall be 1/16 in. less than the thickness of the material unless the weld is especially designed on the drawings to be built up to obtain full throat thickness.

There are other regulations concerning the length of fillet welds, the use of intermittent fillet welds, the use of lap joints, fillet welds in holes, and the end return of fillet welds. Finally, there is specific information concerning plug and slot welds. The data presented in the Section 15-4 are in substantial agreement with the requirements of these four specifications.

Ships

The welding on ships is covered by different specifications and codes. For naval vessels the specifications for welding are provided by the government. In the United States, publications of Navships of the Department of Defense apply. The U.S. Coast Guard has similar requirements, which are published in their Marine Engineering Regulations. The Coast Guard and the Navy requirements are substantially in agreement.

For nonnaval ships the classifications and specifications are established by the U.S. Maritime Commission or by the classification society that is surveying the particular ship. This can be the American Bureau of Shipping or other similar organizations. The classification societies based in Europe have uniform specifications accepted by all classification societies. These specifications and those of the American Bureau of Shipping are in substantial agreement. These organizations all have design requirements for the ships, and they approve ship designs. They also have special requirements for steels used and for the welding filler metals that are employed. The classification societies make tests on filler metals produced by most companies and issue certificates of approval. Designers involved with marine type structures and ships must follow these specifications.

Pressure Vessels

The general term *pressure vessel* includes unfired pressure vessels, fired vessels, nuclear reactors, and other items that enclose pressurized vapors. This would include processing tanks, vessels, heat exchangers, etc., for the chemical industry, for refineries, and for certain manufacturing processes. In North America, the prominent specification is the "ASME Boiler and Pressure Vessel Code," which is sponsored by the American Society of Mechanical Engineers. In most of the states and most of the provinces of Canada this code is referenced or reprinted and thus has become the law of that particular state or province. The ASME code includes eleven sections, covering design, inspection rules, operating and maintenance rules, nondestructive examination procedures, material specifications, and welding procedure qualification standards. The design requirements for power boilers, heating boilers, and nuclear power plant components must be followed by the designer in accordance with the applicable section of the specification.

Closely allied to the pressure vessel code is the code for piping. This code originated by the ASME is sponsored by the American National Standards Institute and is known as the "American National Standard Code for Pressure Piping, Number ANSI B 31.X." The X is for various numbers following the decimal point indicating the particular type of service to which the pipe is intended for use. The ANSI code covers pressure piping, refrigeration piping, transmission piping, service piping, etc. The design parameters and welding requirements must be followed.

Storage Tanks

Large storage tanks are covered by codes and specifications. The two more prominent ones are the code for elevated storage tanks prepared by the American Waterworks Association and the code for petroleum storage tanks prepared by the American Petroleum Institute. The elevated storage tank code is predominantly the same as the American Welding Society requirements as far as welding is concerned, but the design is specific to the product. The same is true of the API storage tank specification.

Machinery

Various types of machinery are covered by specifications. The more popular standards issued by the American Welding Society cover the following products:

Welding industrial and mill cranes (D14.1)

Metal cutting machine tool weldments (D14.2)

Welding on earth-moving and construction equipment (D14.3)

Welding press and press components (D14.4)

In many regards these codes are similar to the AWS structural welding code. They do recommend specific types of joints and they provide the strength levels of different types of welds, for both static and fatigue stresses. These specifications establish common acceptance standards for weld performance and process application.

Aircraft and Spacecraft

Weldments intended for use in aircraft and spacecraft are covered primarily by U.S. Government specifications, either by the National Aeronautics and Space Administration or the Department of Defense. Welds are given attention in various different codes and a listing is not possible here. All of the above codes and specifications are available to designers working on the products for these particular industries.

Many products are not covered by codes or specifications. For guidance, designers should consult specifications or codes covering products similar to those being designed. When designers are not governed by a specific code or specification the following twenty welding joint design guidelines may be used.

Twenty Welding Joint Design Guidelines

For mild and low-alloy steel fabrication the following must apply (when codes or specifications do not apply).

1. Designed Strength: Each weld joint shall be designed or selected to meet the strength requirements for the intended application. Consideration shall be given to stress concentrations due to abrupt changes in section, especially when impact, fatigue loads, or low-temperature service is involved.

2. Standardized Joints: Use the standard welding joints shown by Figure 15-34. These have been designed to require the least amount of weld metal. The design of additional joints not covered shall be designed on the same basis.

3. Complete Penetration Joints: Highest efficiencies for all types of loadings result from joints designed for full penetration. Joints requiring the highest efficiency must specify CP in the tail of the weld symbol. For welds

made from both sides the designation CP in the tail of the weld symbol means that the back side of the joint shall be gouged out to sound metal before depositing the backing weld. Bevel joints welded from one side to a backing strip with the specified root opening shall be full-penetration welds, even though CP is not specified.

4. Joint Preparation Time: Weld joints shall be selected or designed to require the least amount of edge preparation with respect to the welding time required to fill the joint. It is generally less expensive to bevel and weld thinner plates from one side; however, for thicker plate it is less time consuming to bevel and weld from both sides. When comparing fillet welds to groove welds, the effective throat of a fillet weld is less than three-fourths the weld size while that of the groove weld is normally equal to the weld size.

5. J and U-Groove Preparation: J and U-grooves shall be used only on parts that are readily prepared by machining. In general, machining is more expensive than flame cutting.

6. Reduce Overwelding: Overwelding increases welding costs and causes extra distortion. Joints designed for 100% efficiency may be subjected to all types of loadings; however, when stiffness is the principal requirement, joints with efficiencies as low as 50% may be satisfactory.

7. Fillet Weld Size: The designed size might be smaller than shown but for general appearance and for fabricating reasons, the minimum fillet weld size to be used on a given thickness of plate for tee joints is as follows: The design size might be larger in some cases.

Thickness of Thinner Plate (in inches)	Minimum Fillet Weld Size (in inches)
Up to 1/4 incl.	1/8
3/8 to 3/4 incl.	1/4
7/8 to 1-1/4 incl.	1/2
1-3/8 to 2 incl.	1/2
2-1/4 to 4 incl.	3/4

8. Intermittent Fillet Welds: Intermittent fillet welds shall only be used for strength when the smallest fillet sizes, as given in No. 7 above, are too large for continuous welds. Exceptions are metallurgical or warpage reasons. Intermittent welds shall be used whenever possible on sheet metal and structural parts when stiffness is their prime purpose.

9. Length of Intermittent Fillet Welds: For material 1/4 inch thick and greater the minimum length

of intermittent fillet welds shall be 8 times their nominal size, but not less than 2 inches; the maximum length shall be 16 times their nominal size, but not more than 6 inches.

10. **Pitch of Intermittent Fillet Welds:** For material 1/4 inch thick and greater, the maximum center to center dimension shall be 32 times the thickness of the thinner plate, but in no case shall the clear spacing between intermittent fillets be greater than 12 inches.

11. **Reduce Welds:** Eliminate a weld joint by making simple bends wherever possible.

12. **Butt Joints:** For butt joints of unequal thickness, smooth the transition by removing metal rather than by adding weld metal.

13. **Double Tee Joints:** Avoid double tee or cruciform joints whenever possible. Such joints have maximum locked up stresses.

14. **Corner Joints:** For corner joints when bevels are used, prepare the thinner member whenever possible.

15. **Plug and Slot Welds:** Plug or slot welds, or fillet welds in holes or slots, shall not be used in highly stressed members unless absolutely necessary. They shall be used where subjected principally to shearing stresses or where needed to prevent buckling of lapped parts.

16. **Groove Weld Preparation:** Whenever possible require only one member of a joint to have bevel joint preparation.

17. **Welding Position:** Weldments shall be designed so that the position in which the welds are made shall have the following order of preference:

Fillet welds, flat, horizontal fillet, horizontal, vertical, overhead.

Groove welds, flat, vertical, horizontal, overhead.

18. **Enclosed Welding:** Whenever welding is required in enclosed areas or pockets, the enclosed area must contain sufficient openings for access and ventilation for the welders.

19. **Accessibility:** All joints shall be located so that the welder will have sufficient room to weld, gouge, peen, and clean slag. There shall be no obstructions which prevent the welder from seeing to the root of the joint.

20. **Weld Symbols:** All weld joints shall be specified by weld symbols. Symbols shall be conspicuously placed on all drawings and shall refer to views

that show the most joint detail, which is normally the profile view.

15-6 WELDMENT CONVERSION FROM OTHER DESIGNS

The trend to welded design is accelerating. Many of these welded products are redesigns of parts or structures manufactured by other methods. Enthusiasm for redesigned welded structures continues since most redesign projects are extremely successful.

Most companies have value engineering groups that review all practices within a company to find better ways to do things to provide the same value but for less cost. Value engineering programs are responsible for cost savings resulting from weld redesigns.

Many changes are coming about in the structural field such that parts that were previously designed for riveting are now being redesigned for welded connections. The types of assemblies involved are roof trusses, girders, and storage tanks, as well as the structural parts of heavy engineering products. In process chemical plants and refineries old installations are being modernized by replacing threaded pipe work with all welded piping systems. Old riveted tanks and process vessels are being discarded for new welded ones. In all types of machinery many of the components are being changed from castings to weldments. Composite, cast-riveted structures used in this equipment are also being changed as they loosen and wear out. Damaged or worn parts are being redesigned as weldments since replacements for old castings are no longer available or would take too long to obtain. This trend is taking place right now. In addition many new manufactured items are being changed from cast components to welded components.

The redesign from older manufacturing methods to weldments is being done for many different reasons. These reasons can be classified into the following categories:

Economics or reduced cost.

Quality improvement.

Appearance improvement.

Design or product improvement.

Easier to machine.

Reduced production cycle time.

Environmental and other reasons.

DESIGN FOR WELDING

An investigation into these reasons for redesign soon shows the basic facts that are involved. By taking each of the above reasons and investigating them we are able to establish the following factors which should assist in making a decision to redesign the present part or structure to a weldment. The economic or *cost reduction* reasons consist of at least the following.

Hundreds of cost studies show that weldments are less expensive than the previously used steel casting. When the weldment is not able to be produced at a lower cost, the design of the weldment should be reviewed. One of the primary cost reduction reasons is the ability to make the product lighter. This occurs since the metal thickness of castings is often thicker than the stresses require but is needed to assist metal flow in the mold. Those parts with excessive thickness due to the metal casting flow requirements are obviously wasteful and would be reduced in the weldment. Often extremely generous radiuses are provided at intersecting planes to assist in metal flow during the casting process. Many surfaces are made extra thick to allow for machining and the machining allowance may be oversize to allow for potential warpage, pattern shifts, etc. The lap joint used for riveting is also excess metal which can be eliminated with the welding operation. This applies not only to structural work but to tanks and vessels of all types. Weight can be reduced. This reduces the initial cost of the weldment because less metal is involved. This also reduces other charges, for example, shipping charges of the finished product. Reducing the weight allows for more payload in many types of structures.

A machining cost problem that plagues many users of cast parts is the detection of internal defects in the casting after it has been partially or completely machined. Sometimes these castings can be salvaged by welding, many times they must be scrapped with the consequent loss of the time spent machining them prior to finding the defect. Another important cost factor is that weldments frequently simplify and streamline production flow of manufactured parts. Many companies acquire their castings from outside sources. An abnormally long time is usually required to place the order and get the casting produced and received. This cycle time can be greatly reduced by having an in-plant welding department. In comparing the cost of a foundry and cost of a welding shop and metal-preparation shop, studies show that the capital investment to produce the same quantity of product is much lower for the weldment than the casting. The direct labor required to produce

the same part in the foundry is usually higher. Many structures made of castings are of such size that the castings must be connected together. These require machined surfaces and bolt holes for fitted bolts. Providing for this extra machining can run up the cost of the final product. Many foundries do not want short-run jobs. Many foundries also do not have their own pattern shops and this adds more cycle time whenever changes are involved or patterns need to be repaired. Pattern damage and damage to core boxes are relatively common for larger castings and these items are expensive to maintain and repair. Many foundries are discontinuing business due to the air pollution problems associated with older iron foundries.

The *quality* of castings is another reason why many companies are making a change from castings to weldments. It was mentioned above that considerable machining may be performed on a casting before it is found to be defective. There are many casting designs that are extremely difficult to manufacture from the foundry's point of view, and these usually contribute to high casting costs. The difficulty in casting certain designs creates a high scrappage rate in the foundry and consequently higher selling prices to the user.

Riveted joints in tanks and vessels, especially those subjected to heating and cooling, will in time become loose and will leak. Riveted vessels often have corrosion problems at the edges and at riveted joints. These problems have been overcome by replacing machined surfaces and gaskets using bolted joints with welded joints. An example of this is the hermetically sealed compresser case for refrigeration equipment. Mechanical fasteners often contribute to other problems such as leaks of threaded screw joints in piping systems. In large structures in which castings and structural members are riveted together the joints become loose and start to flex. In time the rivets shear, the holes elongate, and the working stresses cause the parts to fail. The service problems of composite structures are the reason why they are being changed to weldments. The frames of large coal stripping power shovels were changed from composite structures to weldments. Figure 15-36 is the composite design, and Figure 15-37 is the welded design of a lower frame. The welded frame has proved to be far superior in service.

Design improvement is another reason for redesign. This can be not only for more reliability, reduced cost, etc., but also for several other reasons. One of the most important is the design freedom that welding provides the designer. Changes in the ultimate product are relatively easy to make if it is designed of weldments. Design changes are more often required in low-volume production equipment than in mass-produced items. On some types of machinery the production volume is low

FIGURE 15-36 *Riveted lower frame—large stripping shovel.*

and pattern costs are absorbed by only a limited number of pieces and thus become an excessive cost. Changes on patterns are very expensive. Another reason for changing is to design the product to reduce space requirements. Weldments can be designed to be more compact than castings. This is particularly important in the transportation industry, especially in aircraft. Another design improvement is the ability for composite weldment construction. Composite construction is a weldment produced by joining steel castings, hot rolled plate, rolled shapes, stamped items, forged, and formed pieces. Composite construction can also involve the welding together of parts of different types of steel. The type of metal may be placed at a specific spot to provide the necessary strength, corrosion resistance, abrasion resistance, etc., whereas a casting or forging must be completely of the same material. Design improvement can also be made with regard to deflection resistance or stiffness and vibration control. In general, since the modulus of elasticity of steel is at least double that of

FIGURE 15-37 *Welded lower frame—large stripping shovel.*

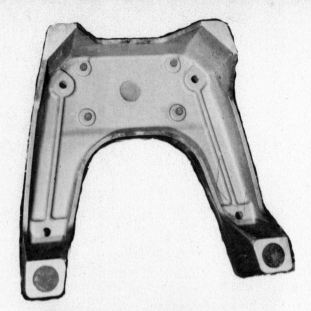

cast iron, the weight of items of similar stiffness can be reduced.

Improved appearance is a reason for redesign. This is very similar to design improvement, but can be considered as styling, which is becoming increasingly important for industrial machinery as well as for consumer products. Weldments have cleaner, crisper lines than castings. Figure 15-38 illustrates this point. It shows the top and bottom view of a cast food machinery base and the part redesigned as a weldment. The weldment utilized square mechanical tubing and thin plate. The cost was reduced and cycle time improved. A casting gives an appearance of an old-fashioned design and thus detracts from its appeal to the buyer. Weldments eliminate the cracks and crevices in riveted and bolted construction, and also eliminate the problem of rivets, rivet heads, etc. The surface of hot rolled steel is vastly superior to that of cast steel or cast iron and allows reduced finishing costs. Cold rolled steel is easy to plate with a minimum amount of finishing and polishing. It is sometimes difficult to get a cast surface free enough of porosity to meet sanitation standards. The welds when made with semiautomatic or automatic equipment have a smooth appearance and a surface that can be painted with little or no finishing.

Another reason for the change to a weldment has to do with manufacturing or machining requirements. Machining a large casting requires large machine tools that are extremely expensive. In many cases small castings can be incorporated into composite design weldments. The small castings can be premachined on smaller machine tools. By proper fixturing and using balanced procedures the parts can be incorporated into weldments with minimum warpage, which will eliminate much of the machining requiring large machine tools. The use of small castings eliminates much of the foundry quality problems since they have simpler design and are easier to produce. Scrappage rates and machining time are reduced.

An important factor that enters into this picture has to do with maintenance or rebuilding of equipment. In repair work parts must often be replaced. If they are castings which are no longer available or require excessive procurement time it is wise to redesign the part as a weldment. The weldment can be made immediately and machined so that the equipment can be returned to service more quickly. In preparing welded steel versus cast iron parts the ease of weld repair of the steel is much greater than the repair by welding of cast iron parts.

FIGURE 15-38 *Cast and welded food machinery base—top and bottom views.*

Another reason to use welded steel is the fact that cast iron parts may without warning suddenly fail in a brittle manner. This rarely occurs with welded steel structures.

A cost-savings reason for redesign is the reduced cycle time required by the weldment versus other methods of manufacture. This has been previously mentioned, but it is an important cost factor in manufacturing. It is extremely important in the maintenance of equipment and machinery. An effective way for a company to reduce costs is to reduce the cycle time from start to finish of the manufactured product. With an in-house welding shop and an in-house parts preparation department, this is obtainable. Reduced cycle time is also important for companies supplying spare parts for existing machinery. Most often when a spare part is ordered the user is experiencing an emergency and needs assistance to return to production as quickly as possible. If the extra time required by the foundry is included the time to supply the spare part becomes excessive and the user may decide to fabricate the part.

There are a few other factors that may be the reason for redesign. One of these has to do with the environmental pollution problems experienced by foundries. Older foundries when faced with the expense of installing pollution control devices may decide to discontinue business. When this happens to a foundry supplying you with castings you may suddenly find the patterns delivered to your door rather than the castings. Redesigns made under these conditions may not be the best.

A new problem sometimes experienced in structural shops is the fact that equipment for riveting may have been discarded. Rivet heaters and rivet squeezers or guns may have been scrapped as a result of lack of use. If a riveted design structure is required a quick redesign may be in order. It appears that as time goes on there are fewer people who have the ability to hot rivet structures, and riveting may become a lost art due to the high noise level and the efforts to reduce high noise levels in plants.

These are valid reasons for changing to weldments from other types of construction. If you make an analysis of some of the castings or riveted or other mechanically-fastened products in your plant, you might find that some of these reasons apply and that you can reap some of the economic advantages by making the change.

How to Redesign to a Weldment

There are three basic methods for converting from the existing design to a weldment. These are known as (1) direct copy redesign, (2) redesign from the existing part, and (3) new design based on loads and stresses.

The optimum economic advantage is obtained when the part is designed from the beginning as a weldment based on the loads and stresses applied to it. This type of design is covered elsewhere in this chapter. The second best method, from an economic point of view, would be the redesign of the part based on the existing part. The quickest method, but the least economical, is to copy as closely as possible the existing part as the new weldment. The latter two methods will be discussed briefly.

Direct Copy

A direct copy of an existing part to a weldment is not the recommended design technique to use if you expect to make the maximum cost reduction. Direct copy redesign is done quickly, usually when time is not available for a complete study. It is also done by less experienced designers. The following is a review of how direct copy redesigns can be done.

Storage tanks, vessels, and similar items can be very easily redesigned as a direct copy of an existing structure. Usually the lap joints required for the riveted design are eliminated and butt joints for welding are substituted. In this case, the full strength of the metal should be utilized. This means a weld joint having full penetration should be used. The direct copy redesign will produce a stronger tank as well as a leakproof tank. It will be stronger unless the thickness of the metal is reduced. An analysis should be made before reducing metal thickness. If the thickness can be safely reduced it should be done by compensating for the areas used for the rivet holes. This can be done by determining the vertical percentage of the area through the metal and rivet holes compared to the area through the metal alone. The rivet holes can reduce this area by as much as 25%. One half of this percentage can be used to reduce the thickness of the plate. If the plates become too thin corrosion or pitting may be the controlling factor. Before reducing the plate thickness a thorough analysis should be made by a qualified designer.

Piping work can also be redesigned by the direct copy method. This is simply a matter of eliminating screw-threaded joints and substituting welded joints or welding type fittings. If the same size and wall thickness of pipe are used the welded design will have a higher safety factor than the screw-threaded design. This is because of the material cut away when making the threads on the pipe. The pipe wall thickness or schedule should not be reduced to the thickness at the thinnest point, which is at the root of the threads. This could shorten

Rivet Diameter inch	Rivet Shear Value at 15,000 psi Shear Stress	Equivalent Length of Fillet Weld Based on 19,000 psi Deposit—in inches		
		1/4 Fillet	3/8 Fillet	1/2 Fillet
1/2	2950 lb	1	5/8	1/2
5/8	4600 lb	1-3/8	1	3/4
3/4	6630 lb	2	1-3/8	1
7/8	9020 lb	2-3/4	1-7/8	1-3/8
1	11780 lb	3-1/2	2-3/8	1-3/4

FIGURE 15-39 *Length of fillet weld (AWS E70XX electrode) to replace rivets (ASTM A502-1 rivets).*

the life of the piping system as a result of corrosion, erosion, etc.

If wall thickness is reduced it should be done by a qualified designer after reviewing the entire design. A conservative way would be to reduce wall thickness by one schedule rather than attempting to provide a wall thickness equal to the thickness at the root of the thread. In any event, a leakproof system will result which will be superior to the threaded system.

In structural work the direct copy conversion can be used for certain types of structures. Riveted roof trusses, for example, can be changed to a welded roof truss merely by eliminating the rivets in certain of the members, and replacing them with plug welds. The number of holes required would be cut in half. However, the section sizes would remain the same and gusset plates, spacers, etc., would still be required. Another way for the direct copy method would be to eliminate all holes and use fillet welds at the toes and heels of angles to join the members to gusset plates rather than using plug welds. The equivalent length of weld to provide the same strength as rivets is shown in Figure 15-39.[7] Direct copy method can be used for many other similar type structures as well.

Castings can be redesigned as weldments also utilizing the direct copy method. However, to intelligently make these conversions it is necessary to consider the metal from which the casting is made. In the case of redesigning a cast iron part the section thicknesses can be reduced by one-half and still produce a weldment stronger than the replaced cast iron part. This is because the strength of the steel is at least twice as great as the strength of the cast iron. The stiffness of the steel part will be as stiff as the cast iron part even though the section thicknesses are reduced by one half. Stiffness is related to deflection and the deflection formulas involve the modulus of elasticity of the material. If all other factors are equal, the deflection of the steel part would be only one-half the deflection of the cast iron part. This is because the modulus of elasticity in steel is approximately twice that of cast iron. Therefore, by reducing the section thickness by one-half, the deflection of the redesigned part and the original part will be

the same. There is sometimes the question of vibration of a steel weldment versus a cast iron part. The damping capacity of steel and cast iron are about the same. In steel plate construction, the design should be such that the natural frequencies that might occur are outside of the operating range. This is a complex problem and should be placed in the hands of an experienced designer.

In the case of redesigning steel castings to steel weldments, the dimensions of the sections should be left roughly the same. It is difficult to determine when sections of the steel casting are heavier than they need to be unless the design history for the casting is known. Since this is normally not known it is wise to make the sections of the same thickness. The welds that are taking the full load should be designed for complete penetration. Many times various gussets and large radiuses can be eliminated or reduced without affecting the service capabilities of the weldment.

Aluminum sand castings are also sometimes redesigned as weldments. The ratio of strength of steel versus strength of aluminum should be considered. However, there are many different strength levels of aluminum; therefore, it is impossible to provide ratios unless tests are performed. Many times aluminum castings can be replaced by steel stampings of fairly thin parts. An example of this type of redesign is shown by Figure 15-40. This is almost a direct copy of the casting although several dimensions are changed. The weldment made of the stampings is about the same weight as the aluminum casting.

Analysis of Part

The redesign by analysis of the existing part to determine its strength is more economically advantageous than the direct copy technique. In this case efforts are made to analyze the part that is being redesigned without actually analyzing the loads and stresses that are imposed on it. In this type of redesign it is usually possible to reduce the size of members or thickness of sections appreciably to reduce the total weight of the weldment versus the part it replaces. An example of the redesign by part analysis would be the redesign of a structural

FIGURE 15-40 *Redesign aluminum casting to steel stampings.*

member such as a roof truss. In a roof truss there are certain members subjected to tensile loading and other members subjected to compressive or possibly column loading. In the members loaded in tension the part can be reduced in cross-sectional area by the amount or by the size of the rivet holes that were punched in the member. In other words, the replacement member would have the same strength since its entire cross section would be utilized. In such a substitution it is necessary to utilize standard rolled sections and the section that provides the cross-sectional area equal to that of the original less the rivet hole area should be used. For safety sake it might be wise to go up one size for the member. This technique must not be used for long members subjected to compressive loads. It cannot be used because such members are probably designed based on the slenderness ratio and the possibility of buckling rather than the unit stress produced in compression. For those members subjected to bending, particularly outer fibers where the bending load produces tensile stresses, the size of the members should not be reduced; the reason is that the stress is not consistent through the entire length of the member, and the stress level could be extremely low at the joint or where the rivet holes are located. The allowable stresses may be proper at some other point; therefore, this member should not be reduced in size. A designer with experience in this field should be involved.

In the redesign of existing pipe or vessel systems, the remarks made concerning the direct copy method should be considered, but more critically in light of maximum section thickness reduction to achieve maximum savings.

In redesigning castings the previous comments concerning castings should be considered. If it is possible

to change the shape to incorporate rolled sections, or to avoid unnecessary bends or curves, those changes should be made while retaining the general shape of the original part. Much experience is required to redesign, using the analysis of the existing structure. The more experience one has had in this work, the more impressive will be the savings that result.

For maximum cost savings, the complete redesign based on the fundamental factors involved should be done. This would mean following the general concepts outlined in the earlier part of this chapter.

15-7 WELDMENT DESIGN PROBLEMS

Welding is the principal joining method for the manufacture of large heavy engineering structures. It is fortunate that welding is a tolerant and forgiving process and relatively few serious weldment failures occur. However, there are service failures of large structures and of many smaller structures and parts whether they are welded or produced by other methods. There is an opportunity to learn from failures so that future designs will be improved. For this reason, every effort should be made to objectively analyze the problems or situations that in any way contributed to the failure of any product. This section will attempt to relate those factors involved with failures that seem to reflect poor design of the weldment. Later in section 18-5 we shall delve more deeply into this subject and shall cover all factors that influenced the service failure. It is recommended that a clinical approach be taken in the analysis of welded structure failures in order to develop a positive design philosophy that will overcome discrepancies of design that caused weldment failures.

There are certain basic fundamentals that should be repeated, particularly while analyzing the failure of a weldment. One is the fact that a welded structure or weldment is one piece of metal or a monolithic structure. It may be made of many pieces, and, for economical reasons, the joints between the pieces may contain unfused areas. The weldment may contain internal and external notches that are potential crack starters. Static loading of such a weldment would not create a service failure problem. Dynamic loading, impact loading, and low-temperature service might cause these notches to initiate a service failure.

Some problems occur because designers may not pay sufficient attention to the weld joint details that are used in the structure. This may be a carryover from earlier days when detailers were given the task of drawing rivets, rivet holes, bolt holes, etc., and this became a routine matter of detailing the mechanical connection. This may be why joint designs, with internal unfused areas, are sometimes specified for highly loaded dynamic stresses. Weld joints, particularly of plastic analysis or rigid frame design, must be designed with every bit as much care as any other portion of the design.

The weldability factor is sometimes overlooked in the specifications for the base metal that is to be employed. The designer may be thinking of the large amount of machine work that has to be done on a weldment and may therefore specify a free-machining steel, which is a poor choice for a weldment. Free-machining steel has a poor weldability rating even though it does have properties that are suitable for many uses. Materials can be selected and specified for a particular need such as hardness or machinability without taking into consideration the need for weldability.

It is also important to analyze weldments that create manufacturing problems during their construction. Such weldments may also be prone to service failures. The factors that create fabrication welding problems during manufacture should be analyzed and corrected so that they do not create service problems of the final product.

Inaccessibility is involved with field failures and has to do with the ability of the welders to make quality welds in the locations required by the design. It is easy for the designer to make a weld on paper in almost any location. It may be difficult or even impossible for the welder to make the weld on the actual part in the shop. Extra thought should be given to the problem of accessibility to make it easy for the welder. Too often the designer's attitude is, "Let's see the welder make that one." If quality weld metal cannot be deposited in the joint as designed, it is the designer's responsibility to redesign so that a quality weld can be made.

There is also the problem of welds being covered up by additional parts of the weldment. These are hidden welds that cannot be inspected from the outside of the product or from the inside where disassembly is possible. Such internal welds can easily be overlooked and omitted. The remaining parts of the weldment are assembled and the missing welds are not found until the service failure is analyzed. It is easy to blame the welder, but it is really the designer's responsibility to provide for an inspection method of determining that such hidden welds were made.

This is not an attempt to blame the designer for all field service failures of weldments. The welder has responsibilities that are sometimes not fulfilled that can create problems that cause service failures. This chapter on design will concentrate on those factors involving design.

In the design of a weldment one of the first problems or questions is, "What metal is it to be made of?" The base metal selection is of paramount importance, to insure satisfactory service life of the structure. The major question relates to the weldability. Is the material weldable with the process that is to be utilized to manufacture the product? These factors must be considered and analyzed in the failure investigation. Consider the hardenability of the base metal, the hydrogen cracking that might occur, the cooling rate of the weld metal, the base metal and the joint detail, the number of cooling paths, the mass of the metal, etc. It is also important to consider the joint with respect to the weld stresses involved and the stresses that are expected to be transmitted by the joint. Remember that any piece welded to the main weldment or structure will carry loads and stresses whether designed to do this or not. If a hard-to-weld material must be used, special precautions should be taken to completely develop the welding procedure that will produce weld joints acceptable to the service requirements of the weldment. Under-bead cracking, extremely high hardness in the heat-affected zone, or too much porosity in the welds, all of which can be determined at the failure investigation, relate to the weldability of the metal selected.

Another reason for field failure is caused by brittle fracture. Brittle fracture involves cleavage of individual crystals with little or no yielding, except at the tip of the crack. Ductile fracture, on the other hand, occurs by deformation of crystals which slip relative to one another. There is yielding at the point of failure when ductile fracture occurs. Brittle fracture can occur with little or no external application of load to the structure. It can occur when corrective measures are being taken to overcome warpage while still in the factory. It can also occur when the temperature of the structure is changed quickly. It has to do with the design and also with the toughness of the base metal used. Residual stresses in the weldment can cause fracture if the material has insignificant toughness and a defect of a suitable size is present. Low temperatures aggravate the problem, as does the size of the defect.

Service failures often occur because load fluctuations may be more severe or more rapid than anticipated. These create stress changes which may occur from a pulsating live load, from vibration, or from continuous temperature changes. Normally the designer considered fatigue loading by avoiding gross stress concentrations but may have miscalculated the severity or the frequency of the load changes. Stress concentrations resulting from design or workmanship accelerate the time of failure due to stress concentration factors being higher than anticipated.

In some types of structures, particularly those produced in volume, it is possible to analyze many identical parts. An analysis of the weldments will help develop a history of failures. It is often found that the failures will occur in the same areas on different pieces of the same design. If such an analysis can be made, it is soon discovered that the failures or cracks are starting at points of stress concentration which may have been a design factor, an overloaded area, or a point at which quality weld metal could not be deposited. This type of analysis can be very rewarding so that design changes can quickly and easily be made to overcome the reason for the premature service failures.

Another failure mode is known as stress corrosion cracking. This type of failure requires both a tensile stress and a corrosive atmosphere. Sometimes sufficient stress is supplied by residual stresses in the weldment. Stress relief *can* overcome this problem, but was it specified? The corrosive atmosphere, the presence of crevices where moisture can collect, the presence of sharp corners, and abrupt changes of section, all contribute to this problem.

Abrupt changes of section create stress risers that cause weldment failure. Backing bars might not be considered a load-carrying part of the member by the designer. However, once it is welded into the member it does carry part of the load. Backing bars should therefore be continuous throughout the stressed area. If bars are short and merely placed end to end without being welded together, they do create a discontinuity which can contribute to premature failure.

Finally, we should learn from each failure. We should remember that the designer is responsible for the design of the product. The designer must have complete information about the service environment, as well as the required performance. With this knowledge and the acceptance of the basic fundamentals brought out earlier in the chapter, designers will learn from the analysis of field failures and improve their designs for the future.

15-8 WELDING SYMBOLS

During the 1930s the American Welding Society developed and established a system of welding symbols. Their purpose was to help identify the lcoation of welds and transmit the information by means of engineering drawings from the designer to the welding shop. The symbols developed in the 1930s used hatch marks, cross hatch marks, etc., and were quite different from

those we use today. In 1942, the society issued its welding symbols in a form not too different from what we have and use today. Since that time, various revisions have occurred until we have a set of welding symbols that are being used increasingly by progressive fabricators and users of welding. At the same time, welding symbols were established by other welding groups in many countries around the world. In the last few years, through the efforts of the International Institute of Welding and the Welding Committee of the International Standards Organization, these symbols are all being unified so that we will have a common system of weld symbols throughout the world. The most recent edition of the American Welding Society's Symbols Book is in substantial agreement with the ISO Standards and those of other countries throughout the world.[8] With increased international trade, the standardization of welding symbols throughout the world is an important accomplishment. Several more steps must be taken to make them truly identical throughout the world. The first is the conversion of the American measuring system to the metric system. This is in process at this time but has not yet been finalized. The second is to resolve the difference between drafting room practices in North America and in Europe. In North America the third-angle projection is normally used on orthographic projection of engineering drawings. In Europe the first-angle projection is normally used. These are defined as the angle of projection of the dihedral angle which lies in the first quadrant or the third quadrant of the Cartesian coordinate system.[9] This system provides the planes of the coordinate system and provides the three-dimensional characteristic. These three planes become the planes of projection used in an orthographic projection drawing. To draw an object in a third angle of the projection the object is placed in the space defined by the three planes and the third dihedral angle. The first-angle projection system places the object in the space defined by three planes and the first dihedral angle. Both systems give the same point of observation for the front and top view. In the first-angle projection the side viewpoint of observation is the position from which you observe the object that is a left side view. In the third angle projection the side view is the right side view on the third plane. This is perhaps not too important with respect to weld symbols but can cause confusion in interpreting them.

The purpose of welding symbols is to provide the means of placing complete welding information on engineering drawings. The symbols are used to accurately and completely describe the desired weld joint for every weld required. It is the means of transmitting information from the designer to the factory. In their more practical application, the welding symbols are used more by the welding department than by the material preparation department. Detailed drawings of individual parts are normally used and cross-sectional diagrams of edge preparation are often included on these drawings. The symbols come into their widest use at the setup operation. They are also widely used for field welding shop fabricated structures. The welding symbol can be used to transmit additional information such as specifications, and procedures. This is done by means of special information, in the tail of the arrow, which is understood by the designer and the welder. In addition to welding symbols the American Welding Society also provides nondestructive testing symbols and symbols for metalizing.

Welding Symbols

The welding symbol consists of the following eight elements which may or may not all be used in each symbol.

1. Reference line. (Always show horizontally)*
2. Arrow.*
3. Basic weld symbol.**
4. Dimension and other data.
5. Supplementary symbol.
6. Finish symbol.
7. Tail.**
8. Specification, process, or reference.

The first and second elements and either the third or seventh must be used to make an intelligible welding symbol. The others may or may not be used in accordance with the necessity of passing along the information or the standard practice of the organization that is using them.

The foundation for constructing a welding symbol is the reference line. The reference line is always shown in the horizontal position, and it should be drawn near the weld joint that it is to identify. The other parts of the welding symbol are constructed on the reference line. Each of the other elements of the welding symbol must

* Mandatory.
**One or the other mandatory.

be placed in proper location with respect to the reference line and in accordance with the details shown by Figure 15-41. The elements that describe the basic weld, the dimensions, and other data, the supplementary, and the finish symbol are always located with the same relationship to the reference line no matter which end of the reference line carries the arrow.

The next important element of the welding symbol is the arrow. This is a line connected to one end of the reference line and connected at the other end with an arrowhead to the arrow side or arrow side member of the weld joint. When the symbol is used for joints which require the preparation of one member only the arrowhead should point with a definite break in the arrow line toward the member of the joint that is to be prepared.

The other end of the reference line carries the tail of the arrow. The area in the tail is used to provide references to specifications, processes, or other specific information. When no specification, process, or other information is used with the welding symbol, the tail is omitted.

Possibly the most important part of the welding symbol is the weld symbol or the *ideograph* which is used to indicate the desired type of weld. These basic weld symbols are shown by Figure 15-42. The distinctive symbol for each type of weld is similar to the cross section of the weld. We have noted previously that there are eight specific types of welds and they have been described earlier in the chapter by Figure 15-18. Since the groove welds all have different symbols they have been shown by Figure 15-41, which shows the seven types of groove welds.

If the basic weld symbol is placed under the reference line that symbol is to define the weld on the *arrow* side of the joint or the arrow side member of the joint. If the basic weld symbol is placed above the reference line it is to define the weld made on the *other* side or other side member of the joint. When the symbol is placed on both sides, it would indicate that the weld is made on *both* sides. Figure 15-43 shows the identification of the arrow side and the other side of the joint and the arrow side and other side member of the joint.

Different basic weld symbols can be placed on the arrow side and the other side of the reference line. Basic weld symbols, as well as the standard location of elements of a welding symbol, should be learned by all who use weld symbols: the designer, the draftsman, the detailer, the layout man, the setup man, the welder, and the welding inspector. The various dimensions that help describe and define the weld have a specific location relationship to the weld symbol. The size of the weld is to be placed at the left of the weld symbol. In groove

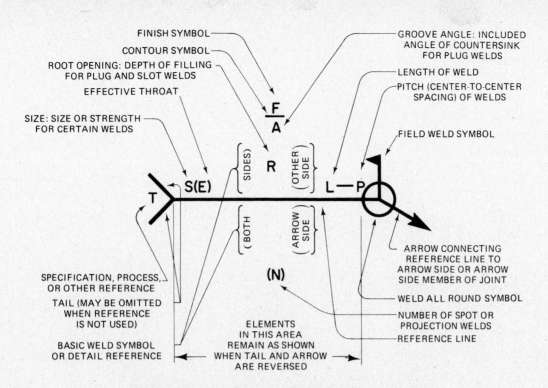

FINISH SYMBOL

CONTOUR SYMBOL

ROOT OPENING: DEPTH OF FILLING
FOR PLUG AND SLOT WELDS

EFFECTIVE THROAT

SIZE: SIZE OR STRENGTH
FOR CERTAIN WELDS

GROOVE ANGLE: INCLUDED
ANGLE OF COUNTERSINK
FOR PLUG WELDS

LENGTH OF WELD

PITCH (CENTER-TO-CENTER
SPACING) OF WELDS

FIELD WELD SYMBOL

ARROW CONNECTING
REFERENCE LINE TO
ARROW SIDE OR ARROW
SIDE MEMBER OF JOINT

WELD ALL ROUND SYMBOL

NUMBER OF SPOT OR
PROJECTION WELDS

REFERENCE LINE

SPECIFICATION, PROCESS,
OR OTHER REFERENCE

TAIL (MAY BE OMITTED
WHEN REFERENCE
IS NOT USED)

BASIC WELD SYMBOL
OR DETAIL REFERENCE

ELEMENTS
IN THIS AREA
REMAIN AS SHOWN
WHEN TAIL AND ARROW
ARE REVERSED

FIGURE 15-41 *Standard location of elements of a welding symbol.*

	GROOVE						FILLET	PLUG OR SLOT	SPOT PROJECTION	SEAM	BACK OR BACKING	SUR-FACING	FLANGE	
SQUARE	V	BEVEL	U	J	FLARE-V	FLARE-BEVEL							EDGE	CORNER
‖	∨	⋁	Y	⊬	⟨	⟨	◺	▭	◯	⊖	⌣	⌣	⌐⌐	‖

FIGURE 15-42 *Basic weld symbols.*

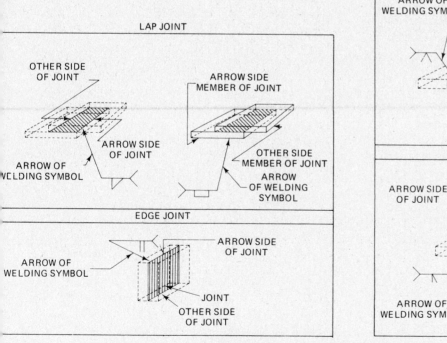

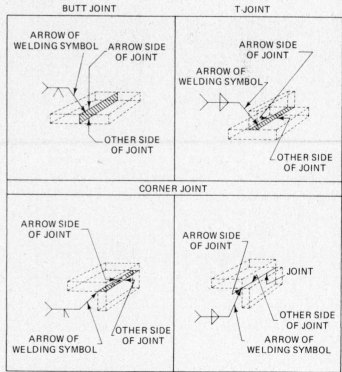

FIGURE 15-43 *Identification of arrow side and other side.*

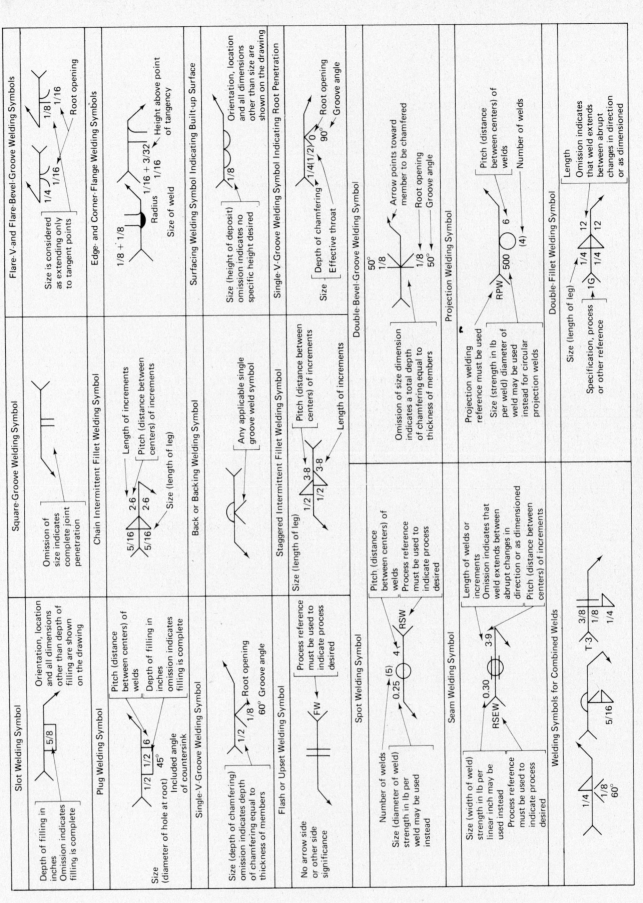

FIGURE 15-44 *Typical welding symbols.*

welds, if the size is not shown, it indicates complete joint penetration. The root opening or depth of filling for plug and slot welding is to be placed directly in the weld symbol. The groove angle, that is the included angle, for groove welds and the included angle of countersink for plug welds are placed above or below the weld symbol. This dimension is often omitted if there is a company standard or all-inclusive note on the drawing. To the immediate right of the weld symbol will be the dimension indicating the length of the weld and if required a dash and the next number will indicate the pitch which is the center-to-center spacing of intermittent type welds. These are all given by Figure 15-44, showing the standard location of elements of a welding symbol.

More than one basic weld symbol can be used to specify a weld joint. For example, in a tee joint a fillet weld may be included in addition to a groove weld. In this case, the basic groove weld symbol would be made to touch the reference line and the basic fillet weld symbol would be added on top. This is also shown by Figure 15-44 and is called combined welds.

The next element of the welding symbol is known as supplementary symbols. These are to be used in conjunction with welding symbols and have a specific location. The supplementary symbols are shown by Figure 15-45. The supplementary symbols may not all be used and are available for situations that require them. Supplementary symbols include: the weld all-around symbol, the field weld symbol, and the melt-through symbol. Additionally, they include symbols indicating the contour of the finished weld which may be flush, convex, or concave. Surface finish symbols are used for very exacting requirements.

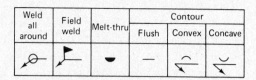

FIGURE 15-45 *Supplementary symbols.*

The following letters are used to indicate the method of finishing but not the degree of finish.

C, chipping.

G, grinding.

M, machining.

R, rolling.

H, hammering.

The use of supplementary symbols is shown by Figure 15-46. They are: weld all-around, field weld, melt-through, flush contour, convex contour, and concave contour.

This chapter has utilized welding symbols to help describe the welds under discussion. The weld joint details all show the specific weld symbol for that particular weld joint design. Numerous other applications and clarifications of the welding symbols are shown by Figure 15-44. The AWS "Welding Symbols Standard" (A2.4) should be available in every drafting room, engineering department, and weld shop.

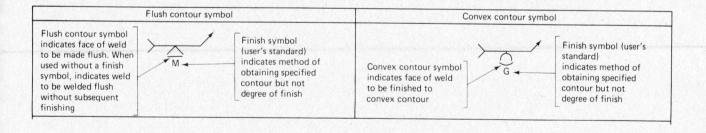

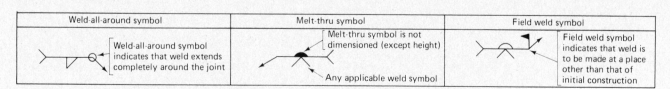

FIGURE 15-46 *The use of supplementary welding symbols.*

QUESTIONS

1. List all the advantages of welded construction.
2. The designer is responsible for weldment design. List the factors that the designer must consider.
3. Define and explain factor of safety.
4. Explain tensile loading, compression loading, bending loading, torsion loading, and shear loading.
5. What is a stress concentration? What causes it? How can it be eliminated?
6. How does a stress concentration affect the life of a weld?
7. What are the five basic types of joints? Describe each.
8. What are the eight types of welds? Describe or sketch each.
9. Define the eight parts of a fillet weld.
10. Define the five parts of a groove weld.
11. If the fillet weld size is doubled, how much more weld metal is required?
12. What are the seven types of groove welds? Are they used as single or double?
13. Explain the weld joint identification number system. What is a BTC-P8 joint?
14. Name the fields of welding in which weldment design is governed by specifications.
15. What is meant by weldment conversion? How is it done?
16. Is the welder responsible for not making an inaccessible weld? Why?
17. Which is the arrow side of a weld symbol reference line?
18. What does the triangle symbol indicate? How do you specify a double one?
19. What is the symbol for *weld all around*? What does it mean?
20. What symbol indicates welding at field erection? Describe it.

REFERENCES

1. "Steel Plates and Their Fabrication" Lukens Steel Company, Coatsville, Pa.
2. ROGER W. BOLZ, "Production Processes—Welding Design Factors," *Welding Arcs,* February 1950, General Electric Company, Schenectady, New York.
3. CHARLES M. JENNINGS, "Welding Design," Transaction of the American Society of Mechanical Engineers, October 1936, New York.
4. "Structural Welding Code," D1.1, *American Welding Society,* Miami, Florida.
5. "Welded Joint Design," Military Standard MIL-STD-00228 (SHIPS), U.S. Naval Ship Engineering Center, May 1969, Hyattsville, Md.
6. *Manual of Steel Construction,* American Institute of Steel Construction, 101 Park Avenue, New York.
7. "Welding Symbols, USA Standard," *AWS* A2.0, *American Welding Society,* Miami, Florida.
8. GEORGE A. RANDALL, "First Angle of Projection and Other Procedures for Converting to SI Metrics," School Shop, New York.

16

THE COST OF WELDING

16-1 THE ELEMENTS OF WELDMENT COST

16-2 WELD METAL REQUIRED FOR WELD JOINTS

16-3 FILLER METAL AND MATERIALS REQUIRED

16-4 TIME AND LABOR REQUIRED

16-5 POWER AND OVERHEAD COSTS

16-6 TOTAL WELD COST FORMULAS AND EXAMPLES

16-7 REDUCING WELDING COSTS

16-1 THE ELEMENTS OF WELDMENT COST

The cost of welding in common with the cost of any other industrial process includes the cost of labor, the cost of materials, and the overhead expense. Welding costs must be accurately determined since they are used as a part of the total cost of products being manufactured. Welding costs are also used to determine cost estimates for bidding on welding work, for setting rates for incentive programs, and for comparing welded construction and competing processes.

The cost of a weld is not necessarily the only cost that must be determined in establishing the cost of a weldment. It is necessary to differentiate between the cost of weldments versus the cost of a specific weld. For example, in pipe welding, we are interested only in the cost of making each butt weld rather than the cost of the pipeline. This is done in order to determine the cost advantage of one process versus another or between one joint design versus another. The cost comparison will reveal the least expensive way to make the joint.

There are other cases in which the cost of the weldment is of major importance. This includes the cost of the weld, but would also include the cost of the material required for the weldment, the preparation of the parts prior to welding, and the postweld treatment that is required. The cost of the weldment is required in order to determine the competitive advantage of a weldment versus a casting or some other type of construction.

The costing of welds and weldments must fit into the general accounting practices of a particular company or activity. This can be easily done since all cost systems must include the same basic factors: labor, material, and overhead. In this chapter methods are presented which will allow welds to be costed when using any of the arc welding processes or joint designs. These methods can be used to determine the cost of competing welding processes in order to select the most economical one. The basis for this system is the amount of weld metal that must be deposited to produce the weld joint. This system can be used even if weld metal is not deposited. It requires the input of data concerning labor costs related to the time required to accomplish the weld, plus the materials required, plus the overhead involved. Normally, additional overhead such as factory administrative costs and company overall sales and administrative costs are prorated according to direct labor involved in manufacturing the part. Thus the total costs can be charged against the particular weld or weldment since the basis for apportioning the overhead expenses is included.

The basis for labor costs is time: time per weld, or time per increment of the length of a weld, or the time required to weld a part. Time is universal and in many situations time represents the cost of the labor. When labor costs are prorated in another fashion this can quickly be related to time per part or parts per unit of time. Welding materials are also related to the total cost. For those processes in which weld metal is deposited in the joint, the amount of metal deposited becomes the basis of the calculation of material costs. The amount of metal required is then related to process to determine the amount of filler metal required to accomplish the weld joint. The rate of depositing weld metal relates to time, which is the basis for labor costs.

This same basis can be used for establishing time standards, and for constructing charts of data that can be used for estimating purposes, rate setting purposes, etc.

The welding procedure is the starting point in establishing welding costs. Welding procedure schedules presented throughout this book can all be used with confidence in establishing welding costs. To be of value, the welding procedure schedule must include the welding joint details, which in turn establish the amount of weld metal required to accomplish a weld joint. The procedure must also include the welding process and the type of filler metal involved since this relates to the cost of the filler metal. The welding procedure must specify the welding current since welding current, filler metal, and process relate to the rate of depositing weld metal and the utilization of filler metal purchased and deposited in the weld. The welding procedure must include the method of application since this influences the operator factor or duty cycle which in turn determines the amount of labor used in depositing weld metal. Finally, the travel speed should be included in the procedure since it determines the amount of time required to produce weld joints. Travel speed is vitally important when weld metal is not deposited in the joint.

The procedure provides the data needed to calculate the cost of a weld. The labor rate and the cost of filler metals must be included to get actual costs. In this regard, filler metals include fluxes, shielding gases, and any other material consumed in making the weld.

The cost of the weldment includes the cost of welding; however, closely related to the cost of welding is the cost of joint preparation. This is included in total weldment cost since joint preparation is required in order to weld and it adds to the total weldment cost. Joint preparation varies according to material thickness and to joint design. To minimize welding cost it is wise to consider alternate joint designs and alternate welding processes. Less expensive joint preparation, based on joint design, can be used with certain processes; for example, with the electroslag welding process a square cut or a mill edge can be used for the joint preparation. For deep groove welds one or two cuts are required for the sub-merged arc welding process. The tradeoff involves the amount of weld metal required to produce the joint. In some cases more weld metal might be required but it can be deposited at a higher rate. This might offset a different joint design that uses less weld metal but is applied with a slower process. These factors must all be studied in order that the optimum weld joint can be made for the least cost.

The cost of postweld treatment will be considered but without the same detail with respect to welding costs. Postweld treatment includes final machining, grinding or polishing, heat treating, shot blasting, and possibly straightening. They are important, since some welding processes require more or less postweld treatment which influences the total cost of the weld and the weldment.

Welding, particularly manual welding, is a high labor intensive manufacturing process. As plants change from manual to semiautomatic, the labor input is reduced, and in changing to automatic it is further reduced. In view of this, calculations for weld costs must be extremely accurate and based on practical procedure data. Comparisons of different welding processes or methods of application can be the basis for investments in automatic equipment. Data in the book and in the remainder of this chapter will provide the basis for making these cost determinations which can greatly affect capital expenditures.

Field welding costs more than shop welding. Welding in the horizontal, vertical, or overhead positions costs more than welding in the flat position. Local working conditions in the shop or on the job definitely affect costs. Available equipment for fitting up or handling the work, experience and skill of the welders, local power rates, special code requirements, weather and temperature conditions, industrial regulations, and many other variables exert so much influence on final costs that tables, graphs, charts, etc., can be offered only as general guide lines applying to work handled under average conditions.

The tables and graphs that provide final costs when labor rates are inserted should be used with caution. These data must be based on many assumptions which might not apply in your situation. For example, the weld designs can vary in root opening, groove angle, etc. The reinforcement may be different. The filler metal yield varies from electrode to electrode and the operator factor can also vary widely. In addition, the data might be slanted to favor one process over another. The data presented in the remainder of this chapter will enable you to construct your own tables and charts based on your standard weld designs, the processes you employ, the filler metals used, and the operator factors you expect. When these data are modified by your pay rates, overhead rates and materials prices, they will apply to your operation. It is wise to include only the standard data in charts, since rates and prices continuously change.

The material cost, for most weld joints, is based on the amount of weld metal deposited in the joint. The exception to this is the autogenous welds in which weld metal is not deposited. The same method of calculating weld metal in joints can also be used for surfacing and overlay applications. The procedure involved applies to all the arc welding and other welding processes in which metal is deposited.

This system utilizes standardized joint designs. Most welding organizations and many welding codes and standards provide standardized designs. Standardized designs were also presented in Chapter 15, entitled "Design for Welding." This same information is presented here with area and weight data calculated for these joint designs in various thicknesses of material. This information is based on the use of steel base metal and weld metal. However, the information is presented so that data for other metals can be easily calculated. The cross-sectional area is related to the standard joints and can be modified for different metals based on their density. In this chapter only the conventional measurements are provided since to include metric measurements would unduly complicate the tables.

Every weld design has a cross-sectional area which can be determined by straightforward geometric calculations. If weld details are standardized it is a simple matter to calculate the cross-sectional area. The formulas for the different welds are shown by Figure 16-1. In these figures the letter designation for the different parts of the weld are the same as Chapter 15. These are as follows:

A = Groove angle

CSA = Cross-sectional area

D = Diameter plug weld, arc spot weld

L = Length of slot weld

R = Radius (used in J and U grooves)

RF = Root face

RO = Root opening

S = Fillet size, bead size, or size of groove weld if not complete penetration

T = Thickness

W = Width of surfacing

In addition, the weld designs are the same as those shown in Chapter 15.

Information shown in Figure 16-1 can be applied in different ways. For example, the bead weld can be used for a backing bead or a sealing bead or for thinner metals for the flange, edge, and corner weld. The amount of metal required for the bead weld may sometimes be added to the full-penetration groove welds for root reinforcing.

When welds of different design configuration are used, the cross-sectional area may be calculated using geometric area formulas. The formulas for each weld presented in Figure 16-1 provide theoretical cross-sectional area values with a flush surface. Weld reinforcement is not included in these values. For practical purposes, however, reinforcement must be added to a weld. The standardized weld designs are shown again by Figure 16-2. In this figure, the different weld designs are related to material thicknesses where the specific weld would normally be applied. The theoretical cross-sectional area in square inches is shown for each weld in its normal range thickness. In addition, the theoretical weight of the weld deposit is shown related to the design and thickness. The weight is based on 12 inches or one foot of joint of steel weld metal in pounds per linear foot of weld. This is calculated using Equation (16-1).

The constant 0.283 is the weight of a cubic inch of steel in pounds per cubic inch. The data can be used for any metal by using its density or weight per cubic inch.

To make this datum more practical a reinforcement factor is added. A value of reinforcement of 10% is added to single groove welds and 20% reinforcement is added to the double groove welds. Ten percent is also added as reinforcement for fillet welds. These are arbitrary figures but they are sufficiently accurate for most calculations. For greater accuracy it may be desirable to make representative welds, cross section them, and measure the reinforcement. This would be accurate for a particular weld, but since reinforcement varies it may not be worth the effort. The amount of reinforcement on a weld must be held to a minimum since it increases the amount of weld metal required. Figure 16-2 therefore includes two additional columns. One provides the cross-sectional area of the welds in square inches in the different material thicknesses with the reinforcement added. The final column provides the weight of the weld deposit with reinforcement in pounds per foot of weld. These data are the basis for many of the calculations that follow.

$$\text{Weight of deposit (lb/ft)} = \text{cross-sectional area (in.}^2) \times 0.283 \text{ (lb/in.}^3) \times 12 = 3.396 \qquad (16\text{-}1)$$

WELD	DESIGN	FORMULA FOR CROSS SECTION AREA
FILLET (EQUAL LEGS)	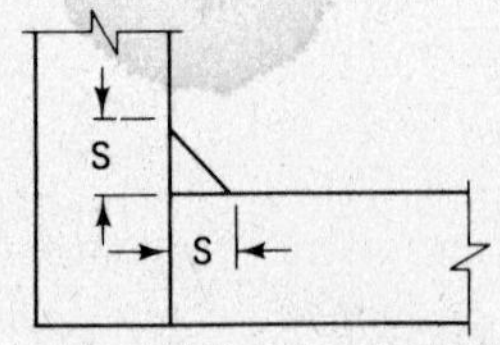	$CSA = 1/2(S)^2$
FILLET (UNEQUAL LEGS)	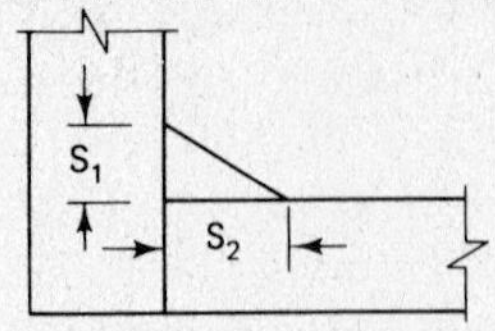	$CSA = 1/2 (S_1 \times S_2)$
SURFACE	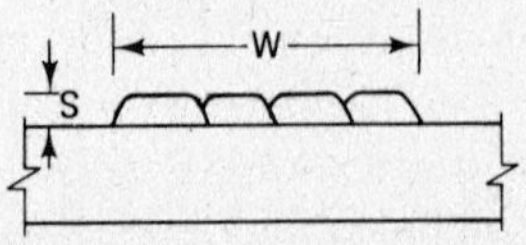	$CSA = S \times W$
PLUG	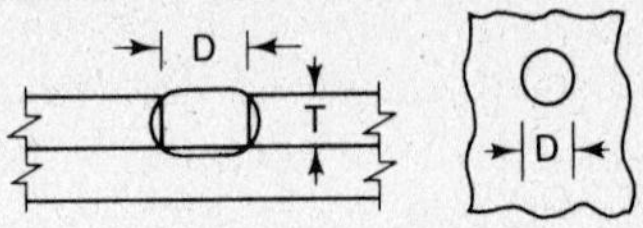	$VOL = \pi \left(\dfrac{D}{2}\right)^2 \times T$ FORMULA PROVIDES VOLUME OF WELD METAL PER WELD
SLOT	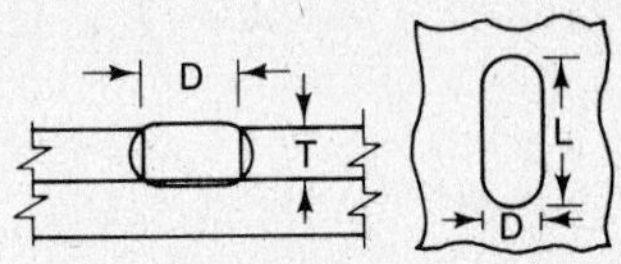	$VOL = \left[\pi \left(\dfrac{D}{2}\right)^2 + (L-D)D\right] T$ FORMULA PROVIDES VOLUME OF WELD METAL PER WELD
ARC SPOT	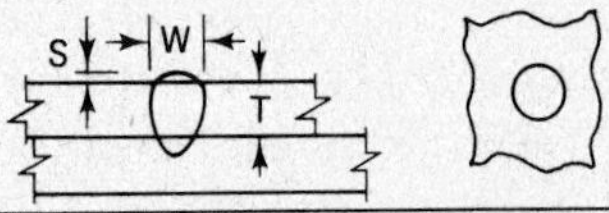	$VOL = 1/2S \left(\pi \dfrac{W}{2}\right)^2$ (VOLUME PER WELD) FORMULA PROVIDES VOLUME OF WELD METAL PER WELD
ARC SEAM	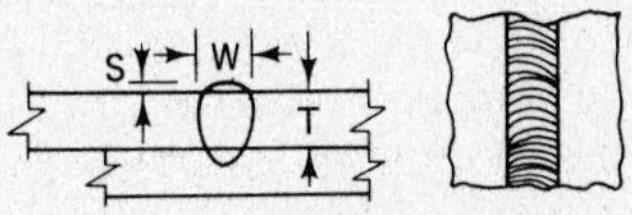	$CSA = 1/2WS$
BEAD	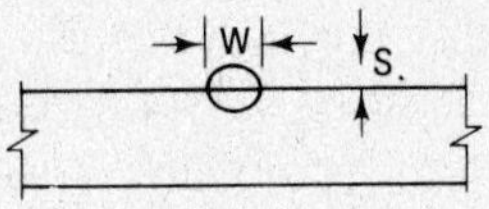	$CSA = 1/2WS$
SQUARE	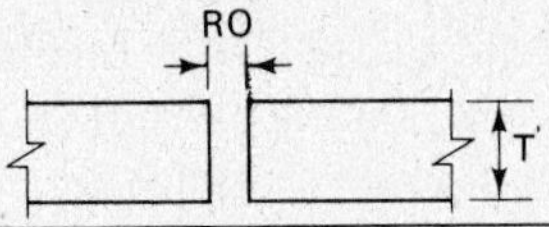	$CSA = RO \times T$
SINGLE V	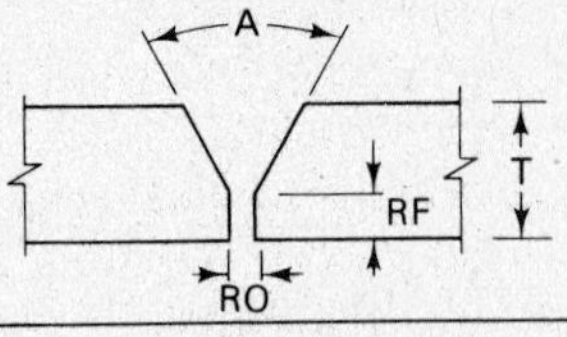	$CSA = (T-RF)^2 \tan\left(\dfrac{A}{2}\right) + RO \times T$
DOUBLE V	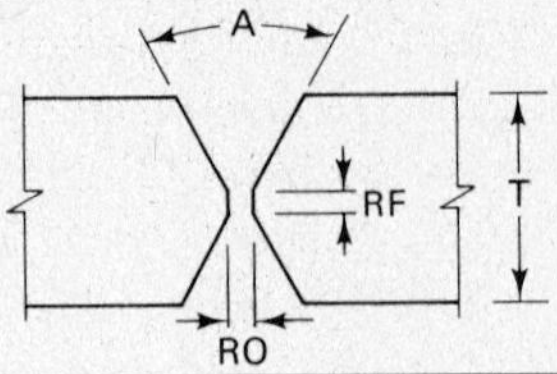	$CSA = 1/2(T-RF)^2 \tan\left(\dfrac{A}{2}\right) + RO \times T$

FIGURE 16-1 *Cross sectional area of welds.*

WELD	DESIGN	FORMULA FOR CROSS SECTION AREA
SINGLE BEVEL	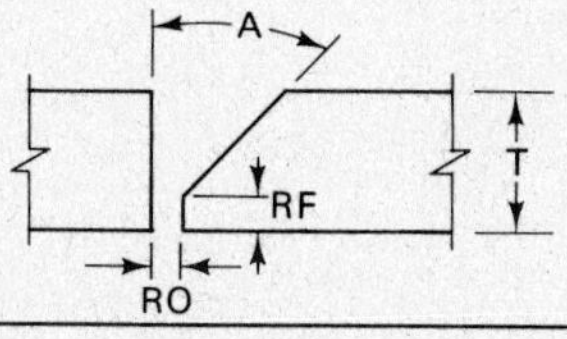	$CSA = 1/2(T - RF)^2 \tan A + RO \times T$
DOUBLE BEVEL	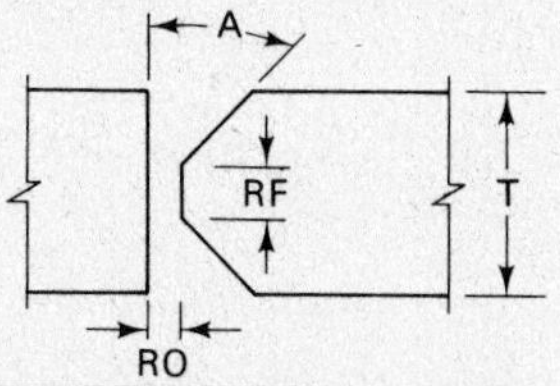	$CSA = 1/4(T - RF)^2 \tan A + RO \times T$
SINGLE U	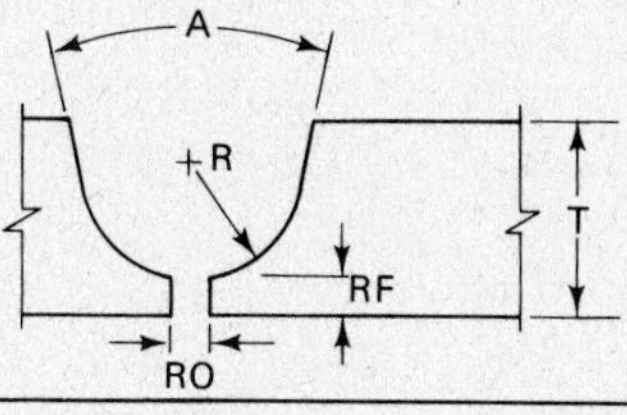	$CSA = (T - R - RF)^2 \tan\left(\dfrac{A}{2}\right) + 2R(T - R - RF)$ $+ 1/2\pi R^2 + RO \times T$
DOUBLE U	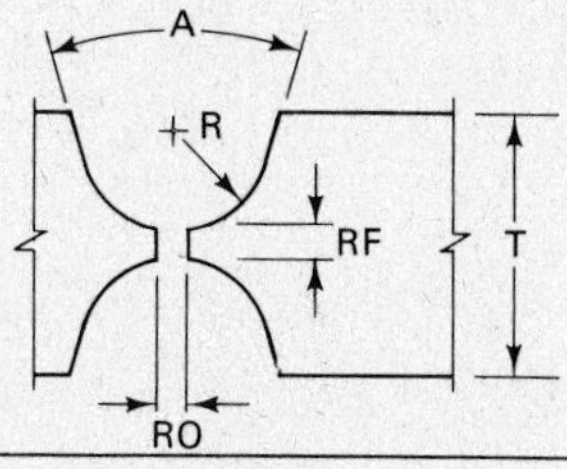	$CSA = 1/2(T - 2R - RF)^2 \tan\left(\dfrac{A}{2}\right) + 2R(T - 2R - RF)$ $+ \pi R^2 + RO \times T$
SINGLE J	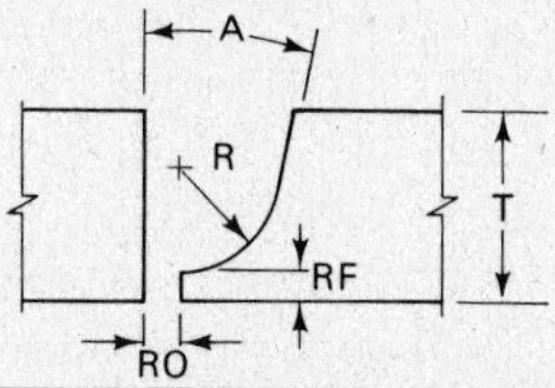	$CSA = 1/2(T - R - RF)^2 \tan(A + R)(T - R - RF)$ $+ 1/4\pi R^2 + RO \times T$
DOUBLE J	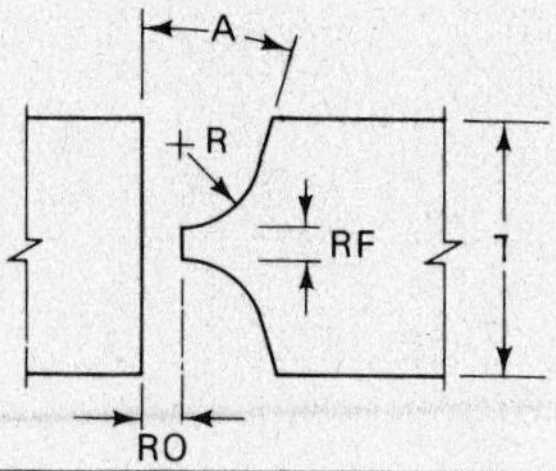	$CSA = 1/4(T - 2R - RF)^2 \tan(A + R)(T - 2R - RF)$ $+ 1/2\pi R^2 + RO \times T$
FLARE V	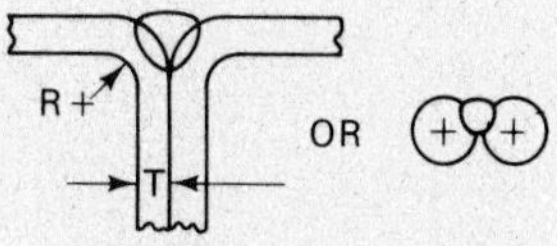 OR	$CSA = \dfrac{(2 \times R + T)^2 - \pi(R + T)^2}{2}$
FLARE BEVEL	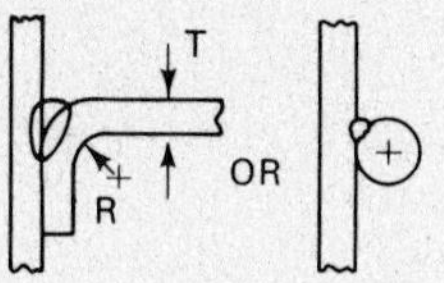 OR	$CSA = \dfrac{(2 \times R + T)^2 - \pi(R + T)^2}{4}$

FIGURE 16-1 *(Cont.) Cross sectional area of welds.*

Weld	Design	T in inch	CSA Theor. in.²	Weld Deposit Theoretical lb/ft	CSA w/ref. in.²	Weld Deposit w/ Reinforcement lb/ft
Fillet (equal legs)	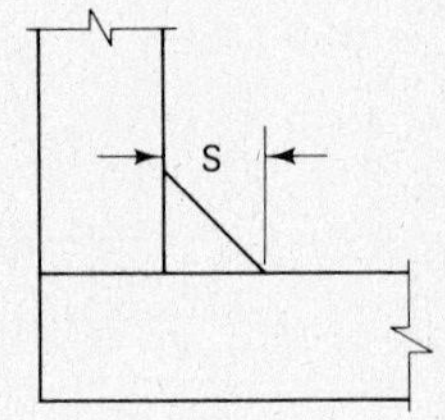	1/8	0.008	0.027	0.009	0.030
		3/16	0.018	0.061	0.020	0.067
		1/4	0.031	0.106	0.034	0.117
		5/16	0.049	0.167	0.054	0.184
		3/8	0.070	0.238	0.077	0.262
		7/16	0.096	0.326	0.106	0.360
		1/2	0.125	0.425	0.138	0.468
		9/16	0.158	0.537	0.174	0.591
		5/8	0.195	0.663	0.215	0.729
		3/4	0.281	0.956	0.309	1.052
		7/8	0.383	1.503	0.421	1.653
		1	0.500	1.700	0.550	1.876
Fillet (unequal legs)	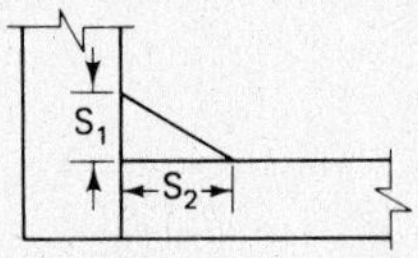	1/4 × 3/8	0.047	0.160	0.052	0.176
		3/8 × 1/2	0.094	0.319	0.103	0.351
		1/2 × 5/8	0.156	0.530	0.172	0.583
		5/8 × 3/4	0.234	0.795	0.258	0.875
		3/4 × 1	0.375	1.274	0.413	1.401
Square	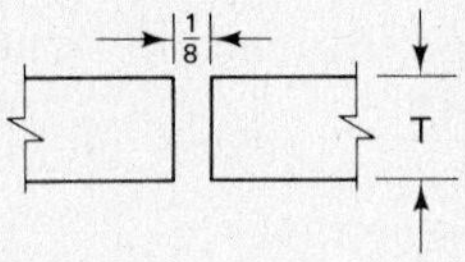	1/8	0.016	0.054	0.019	0.065
		5/32	0.019	0.065	0.023	0.078
		3/16	0.023	0.078	0.027	0.094
		7/32	0.027	0.092	0.032	0.110
		1/4	0.031	0.105	0.037	0.126
		9/32	0.035	0.119	0.042	0.143
		5/16	0.039	0.132	0.047	0.158
Single V	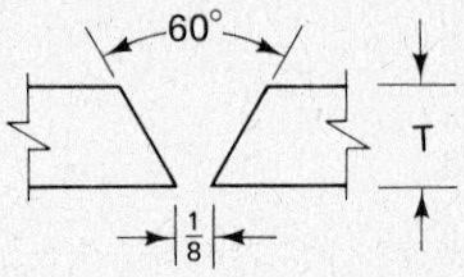	1/4	0.067	0.228	0.074	0.251
		3/8	0.128	0.384	0.141	0.422
		1/2	0.206	0.702	0.227	0.772
		5/8	0.305	1.040	0.336	1.144
		3/4	0.418	1.430	0.460	1.573
		1	0.702	2.395	0.772	2.635
Single bevel	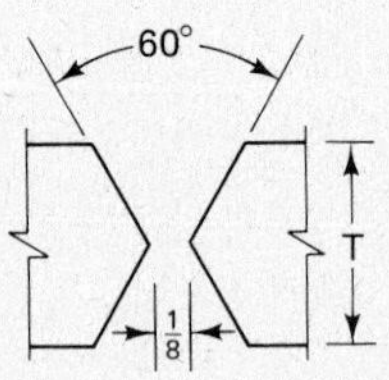	3/4	0.256	0.874	0.307	1.049
		1	0.414	1.420	0.497	1.704
		1-1/4	0.608	2.075	0.730	2.490
		1-1/2	0.838	2.860	1.006	3.432
		1-3/4	1.105	3.765	1.326	4.518
		2	1.405	4.780	1.686	5.736
		2-1/4	1.742	5.945	2.090	7.134
		2-1/2	2.210	7.530	2.652	9.036
		2-3/4	2.530	8.620	3.036	10.344
		3	2.978	10.150	3.574	12.180
		3-1/2	3.970	13.530	4.764	16.236
		4	5.620	19.130	6.744	22.956
Single bevel	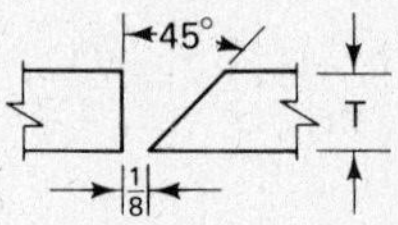	1/4	0.063	0.215	0.069	0.237
		3/8	0.117	0.364	0.129	0.400
		1/2	0.188	0.641	0.207	0.705
		5/8	0.301	1.025	0.331	1.128
		3/4	0.375	1.280	0.413	1.408
		1	0.625	2.135	0.687	2.349
Double bevel	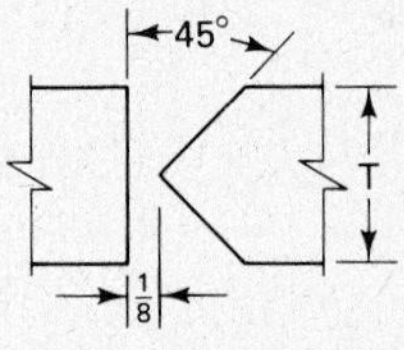	5/8	0.176	0.600	0.211	0.720
		3/4	0.234	0.798	0.281	0.958
		7/8	0.301	1.025	0.361	1.230
		1	0.375	1.279	0.450	1.535
		1-1/4	0.547	1.862	0.656	2.234
		1-1/2	0.750	2.560	0.900	3.072
		1-3/4	0.984	3.360	1.181	4.032
		2	1.250	4.260	1.500	5.112
		2-1/2	1.875	6.398	2.250	7.677
		3	2.625	8.950	3.150	10.740

FIGURE 16-2 *Area and weight of weld metal deposit.*

Weld	Design	T in inch	CSA Theor. in.²	Weld Deposit Theoretical lb/ft	CSA w/ref. in.²	Weld Deposit w/ Reinforcement lb/ft
Single U		1/2	0.163	0.555	0.179	0.611
		3/4	0.310	1.058	0.341	1.164
		7/8	0.392	1.338	0.431	1.472
		1	0.479	1.635	0.527	1.799
		1-1/4	0.671	2.288	0.738	2.517
		1-1/2	0.885	3.020	0.974	3.322
		1-3/4	1.120	3.820	1.232	4.202
		2	1.376	4.680	1.514	5.148
		2-1/2	1.961	6.680	2.157	7.348
		3	2.631	8.960	2.894	9.856
Double U		1	0.396	1.350	0.475	1.620
		1-1/4	0.543	1.852	0.652	2.222
		1-1/2	0.701	2.390	0.841	2.868
		1-3/4	0.870	2.968	1.044	3.562
		2	1.151	3.922	1.381	4.706
		2-1/4	1.242	4.235	1.490	5.082
		2-1/2	1.444	4.925	1.732	5.910
		2-3/4	1.658	5.650	2.000	6.780
		3	1.879	6.410	2.255	7.692
		3-1/2	2.389	8.150	2.867	9.780
		4	2.951	10.070	3.541	12.084
		4-1/2	3.459	11.790	4.151	14.148
Single J		1/2	0.180	0.614	0.198	0.675
		3/4	0.261	0.890	0.287	0.979
		1	0.409	1.395	0.450	1.535
		1-1/4	0.580	1.978	0.638	2.176
		1-1/2	0.774	2.640	0.851	2.904
		1-3/4	0.989	3.375	1.088	3.713
		2	1.229	4.190	1.352	4.609
		2-1/4	1.491	5.080	1.640	5.589
		2-1/2	1.774	6.050	1.951	6.655
Double V		1	0.360	1.228	0.432	1.474
		1-1/4	0.437	1.490	0.524	1.788
		1-1/2	0.589	2.010	0.707	2.412
		1-3/4	0.728	2.482	0.874	2.978
		2	0.875	2.983	1.050	3.580
		2-1/4	1.029	3.510	1.235	4.212
		2-1/2	1.191	4.065	1.429	4.878
		2-3/4	1.360	4.640	1.632	5.568
		3	1.535	5.235	1.842	6.282
		3-1/2	1.909	6.510	2.291	7.812
		4	2.313	7.880	2.776	9.456

FIGURE 16-2   *(Cont.) Area and weight of weld metal deposit.*

Another value of the data presented in Figure 16-2 is their usefulness in visualizing how welding costs are related to weld designs. They will illustrate the amount of weld metal required for different weld designs. For example, they will show the increase in cross-sectional area or weight of weld metal required when the size of a fillet weld is increased. Other comparisons can be made such as the difference in the cross-sectional area or weld metal required between a bevel to a V groove weld or between a single- and a double-groove weld. The data presented by Figure 16-2 apply to all of the arc welding processes in which metal is deposited. They can also be used for torch brazing and for oxy-fuel gas welding. These

data can be the basis for a standard cost system when standard weld design as shown is used. If the weld designs are different from those used in the charts the data must be recalculated to reflect these changes.

16-3 FILLER METAL AND MATERIALS REQUIRED

Electrodes.

Section 16-2 provided information necessary to calculate the "weight of weld metal deposited" in a weld joint or on a weldment. It also provides tables that give the weight of weld metal deposit for standard weld designs made in different thicknesses of material. The total weight of weld metal deposited in the joint or required to produce the weldment can easily be calculated using these tables.

The weight of filler metal purchased to make the weld or the weldment is greater than the weight of the weld metal deposit. This is true for most of the arc welding processes, but not for all. Stated another way, it means that more filler metal must be purchased than is deposited because of stub end losses, coating or slag losses, spatter losses, and so on. This can be shown by Equation (16-2).

These losses are sometimes represented as a ratio and called deposition efficiency, filler metal yield, or recovery rate.

Deposition efficiency is the ratio of the weight of deposited weld metal in the weld divided by the net weight of filler metal consumed, exclusive of stubs. As shown by Equation (16-3) filler metal yield is the ratio of the weight of deposited weld metal divided by the gross weight of the filler metal used. Thus yield relates to the amount of filler metal purchased. The filler metal yield for the different types of electrodes and filler metals will range from as low as 50% to 100%. Yield is the better term to use since stubs occur only when using covered electrodes.

The covered electrode has the lowest yield, that is, it has the highest losses. These losses are made up of the stub end loss, the coating or slag loss, and the spatter loss. Considering a 2-in. stub, the 14-in. long electrode has a 14% stub loss, the 18-in. electrode has an 11% stub loss, and the 28-in. electrode has a 7% stub loss. Unfortunately, electrodes are not always melted to a 2-in. stub. The coating or slag loss of a covered electrode can range from 10-50%. The thinner coatings on an E6010 electrode are at the lower end of the scale approximating

10% while the heavy coating on an E7024 type electrode will approach 50%. This can apply even when iron powder is incorporated in the coating. For accurate results measure this factor for the electrodes to be used. The spatter loss depends on the welding technique, but normally ranges from 5-15% loss. Thus it is easy to see why the coated electrode has such a low yield.

The solid bare electrode or filler rod has the highest yield since the losses are minimized. Normally, in the continuous electrode wire processes the entire spool or coil of electrode wire is consumed in making the weld. Scrap ends of coils are discarded, but this is usually negligible compared to the total weight. The spatter loss relates to welding process and welding technique. In the submerged arc and electroslag welding processes virtually all of the electrode is deposited in the weld. There is no spatter and, therefore, deposition efficiency or yield approaches 100%. In gas metal arc welding there is a loss from spatter and this amounts to approximately 5% of the electrode melted which provides a deposition efficiency or yield of 95%. In the case of the cold wire processes, gas tungsten arc, plasma arc, and carbon arc, the *cold wire* filler metal is completely used and has a 100% yield.

The flux cored electrode has slightly higher losses because of the fluxing ingredients within the tubular wire which are consumed and lost as slag. The fluxing materials in the core represent from 10-20% of the weight of the electrode. Different types and sizes have different core to steel weight ratios. For accurate results measure yield of the electrode to be used. There can be up to 5% loss as spatter; therefore, the deposition efficiency or yield for flux cored electrodes ranges from 75-85%.

The deposition efficiency or yield of filler material and the process have an important bearing on the cost of the deposited weld metal.

The filler metal cost can be calculated several different ways. The most common is based on cost per foot of weld, as shown by Equation (16-4).

$$\text{Weight of filler metal required (lb)} = \frac{\text{weight of weld metal deposited (lb)}}{1 - \text{total electrode loss}} \qquad (16\text{-}2)$$

$$\text{Weight of filler metal required (lb)} = \frac{\text{weight of weld metal deposited (lb)}}{\text{filler metal yield (\%)}} \qquad (16\text{-}3)$$

$$\text{Electrode cost (\$/ft)} = \frac{\text{electrode price (\$/lb)} \times \text{weld metal deposited (lb/ft)}}{\text{filler metal yield (\%)}} \qquad (16\text{-}4)$$

The electrode price is the delivered cost to your plant. Electrode price can be reduced by purchasing in large lot sizes. In this formula the yield can be taken from the table shown by Figure 16-3. These are average figures and should be sufficient for most calculations; however, for more accuracy actual measurements should be made using the filler metals that are to be employed.

Electrode Type and Process	Yield %
Covered Electrode for:	
SMAW 14″ manual	55 to 65%
SMAW 18″ manual	60 to 70%
SMAW 28″ automatic	65 to 75%
Solid Bare electrode for:	
Submerged arc	95 to 100%
Electroslag	95 to 100%
Gas metal arc welding	90 to 95%
Cold wire use	100%
Tubular-flux cored electrode for:	
Flux cored arc welding	80 to 85%
Cold wire use	100%

Note - Does not include shielding except for covered electrodes.

FIGURE 16-3 *Filler metal yield—various types of electrodes.*

A different method can be used for calculating the amount of weld metal required when the continuous wire processes are used. It is particularly advantageous for single-pass welding. Three simple calculations are required but the end result is the electrode cost per foot of weld. The first step is to determine the amount of electrode used expressed as pounds per hour, using Equation (16-5).

The pounds per hour of electrode used disregards the yield or deposition rate factor since we are measuring the actual filler material consumed. The factor 60 is the minutes in an hour which converts minutes to hours. The weight per length of the electrode wire is a physical property of wire based on the size of the wire and the density of the metal of the wire. This is shown in the chart of Table 16-1 for various sizes and types of bare metal wire. The wire feed speed in inches per minute is the same as the melt off rate of an electrode wire. It is not a true deposition rate since the spatter losses and slag losses are not considered. The wire feed speed can be determined from charts that relate the welding current to the wire feed speed, according to the size of the electrode wire, the composition of the electrode wire, and the welding process. Charts showing this data are given by Figures 9-20, 9-21, and 9-23 in Chapter 9 on wire feeders. To use this formula it is essential to know the welding current or to measure the wire feed speed. In some instances wire feed speed is called for in the welding procedure. For very accurate work it is best to make a measurement of wire feed speed. This can be done simply by setting the wire feeder, making a test weld to determine that the weld procedure is satisfactory, and then without welding, actually allow the electrode wire to feed through the gun and measure the amount of wire fed per minute. Since this can be an extremely large amount of wire it can be simplified by feeding for five

$$\text{Weight of filler metal required (lb/hr)} = \frac{\text{wire feed speed (in./min)} \times 60}{\text{length of wire per weight (in./lb)}} \qquad (16\text{--}5)$$

TABLE 16-1 *Length vs weight (inches per pound) of bare electrode wire of type and size shown.*

WIRE DIAMETER			INCHES OF METAL OR ALLOY PER POUND							
Decimal Inches	Fraction Inches	Aluminum	Alum. 10% Bronze	Silicon Bronze	Copper (deox.)	Copper Nickel	Magnesium	Nickel	Steel, Mild	Steel, Stainless
0.020		32400	11600	10300	9800	9950	50500	9900	11100	10950
0.025		22300	7960	7100	6750	6820	34700	6820	7680	7550
0.030		14420	5150	4600	4360	4430	22400	4400	4960	4880
0.035		10600	3780	3380	3200	3260	16500	3240	3650	3590
0.040		8120	2900	2580	2450	2490	12600	2480	2790	2750
0.045	3/64	6410	2290	2040	1940	1970	9990	1960	2210	2170
0.062	1/16	3382	1120	1070	1020	1040	5270	1030	1160	1140
0.078	5/64	2120	756	675	640	650	3300	647	730	718
0.093	3/32	1510	538	510	455	462	2350	460	519	510
0.125	1/8	825	295	263	249	253	1280	252	284	279
0.156	5/32	530	189	169	160	163	825	162	182	179
0.187	3/16	377	134	120	114	116	587	115	130	127
0.250	1/4	206	74	66	62	64	320	63	71	70

Electrode Size-Dia. (Inch)	Length by Weight (Inches/Pound)
0.045	2400
1/16	1250
5/64	1000
3/32	650
7/64	470
0.120	380
1/8	345
5/32	225

FIGURE 16-4 *Length per pound of steel flux-cored electrode wire.*

seconds and multiplying the amount of wire fed by 12 to relate it to inches per minute. Instruments are available for making this measurement.

The second part of this calculation is to determine or measure the weld travel speed and arrive at this rate in feet per hour. Normally, welding procedures provide weld travel speed in inches per minute. If these data are not in the welding procedure, tests should be made to determine travel speed while making the required weld. This is then converted to feet per hour by using Equation (16-6).

The 60 represents minutes in an hour and the 12 represents inches in a foot which is the factor 5 previously mentioned.

The third part of this calculation is to determine the weight of weld metal required per foot of weld, as shown by using Equation (16-7).

The above information would then be multiplied by the electrode price ($/lb) to obtain electrode cost in $/ft.

This system can also be used for flux-cored arc welding, but in this case the length of electrode wire per pound is a little more difficult to determine since different types of flux-cored wire have different amounts of core material of different densities. See the table shown by Figure 16-4. The length per pound of steel flux-cored electrode wire can be used for normal calculations. For better accuracy, actual tests should be made to determine the number of inches of the wire that weighs a pound.

Flux.

When flux is used the cost of flux must be included in the cost of materials used. The cost of flux in submerged arc welding and in electroslag welding and even in oxy-fuel gas welding can be related to the weight of weld metal deposited and may be calculated using Equation (16-8).

In the submerged arc welding process normally one pound of submerged arc flux is used with each pound of electrode wire deposited. This is a flux-to-steel ratio of one. This ratio may change for different welding procedures and for different types of flux.

The flux ratio of one can be used for costing; however, for more accuracy, tests should be run with the particular flux used. The flux ratio can rise as high as 1.5.

For electroslag welding the ratio of flux to electrode wire deposited is 5 to 10 pounds of flux per 100 pounds of electrode consumed. This is a flux-to-steel ratio of 0.05 to 0.10. The exact amount of flux used is based on the surface area of retaining shoes exposed to the flux pool. The rule of thumb is 1/4 pound of flux per vertical foot of joint height. The 10% ratio mentioned above is ample to cover the cost of electroslag flux.

In oxy-fuel gas welding and torch brazing the amount of flux used can also be related to the amount of filler wire consumed. This ratio is usually in the order of 5–10%. Accurate checks can be made for a more precise flux ratio if desired.

Shielding Gas.

When shielding gas is used it must be included in material cost. The cost of the gas is related to the time required to make the weld. Shielding gas is normally used at a specified flow rate and measured in cubic feet per hour. The amount of shielding gas used would be the time required to make the weld, times the rate of gas usage. The cost of shielding gas can be figured two ways.

$$\text{Travel speed (ft/hr)} = \frac{\text{travel speed (in./min)} \times 60}{12} = \text{travel speed (in./min)} \times 5 \quad (16\text{-}6)$$

$$\text{Weight of filler metal required (lb/ft)} = \frac{\text{deposition rate (lb/hr)}}{\text{weld travel speed (in./min)} \times 5} \quad (16\text{-}7)$$

$$\text{Flux cost (\$/ft)} = \text{flux price (\$/lb)} \times \text{weld metal deposit (lb/ft)} \times \text{flux ratio} \quad (16\text{-}8)$$

Normally the gas cost is based on the cost per foot of the weld and it is calculated using Equation (16-9).

The gas flow rate is provided in the welding procedure or it can be measured with a flow meter. The price of gas is the delivered price at the welding station. The time length factor can be replaced by the numeral 5.

Shielding gas cost per minute of operation is used when calculating the cost of making an arc spot weld, a small joint, or a small part. This is based on the time required to make the weld and it may be calculated using Equation (16-10).

Miscellaneous Material Costs.

To obtain a total cost of a particular weld other items should be included. These include the guide tube used in the consumable guide electroslag welding process, the cost of ferrules and studs in arc stud welding, and so on. In stud welding the price of each stud must be considered, even though studs are not strictly filler metals. They relate to the number of welds made and can be calculated in this manner. Also, since a ceramic ferrule is required for each weld they must also be included in the cost of making each stud weld.

16-4 TIME AND LABOR REQUIRED

The cost of the labor required to make a weld is perhaps the single greatest factor in the total cost of a weld. Section 16-3 provided the weight of metal deposit for different welds of different sizes. With this information, the cost of the filler metals required is determined. The amount of weld deposit or the amount of filler metal required is one basis for determining the amount of time required to make a weld or weldment. Time is normally the basis for pay for welders since many are paid by the per hour rate. The data that follows will be used to determine the cost of welding when welders are paid an hourly rate.

Welders are sometimes paid on the basis of welds made. This may be on a per footage basis for different size welds or on the number of pieces welded per hour. This type of pay is usually involved with incentive systems. In order to establish costs on this basis it may be neces-

sary to determine time required for making welds of different types or, conversely, the speed of making certain welds. In some cases, time studies are used to determine the normal welding time for making a particular weld or weldment. In other cases, standard cost data are used and quite often these are based on weight of weld metal deposited. In developing the cost of welds, these data can be used in many different ways according to the pay systems and accounting systems employed. The basis for accurate weld costing is a welding procedure. The welding procedure may not be available at the time of setting cost standards or estimating costs, therefore, welding procedure schedules like those presented throughout this book should be employed.

The basis for calculating the cost of labor on a dollars-per-foot basis is given in Equation (16-11).

The operator factor shown here is the same as duty cycle, which is the percentage of arc time against the total paid time.

Each of the elements of this formula requires considerable analysis. The welder's hourly pay rate can be entered into the formula; however, in most cases companies prefer to factor the pay rate to cover fringe benefits such as cost of insurance, cost of holidays, cost of vacations, and so on. This is a factor that must be determined and must be in line with the company accounting policies. It should be the same as the method used for determining machining costs and other direct labor costs used in the plant.

The above assumption applies primarily to single-pass welds, since weld travel speed is available in the welding procedure schedules. It should be used when metal is not deposited, specifically GTAW and PAW. Travel speeds relate to the welding job, the weld type, and the welding process employed. This is relatively easy to determine when single-pass welds are made but it is more difficult with multipass welds and for this reason a different system is used for large multipass welds.

$$\text{Gas cost (\$/ft)} = \frac{\text{price of gas (\$/ft}^3) \times \text{flow rate (ft}^3/\text{hr})}{\text{weld travel speed (ipm)} \times 5} \qquad (16\text{-}9)$$

$$\text{Gas cost (\$/weld)} = \frac{\text{price of gas (\$/ft}^3) \times \text{flow rate (ft}^3/\text{hr}) \times \text{weld time (min)}}{60} \qquad (16\text{-}10)$$

$$\text{Labor cost (\$/ft)} = \frac{\text{welder pay rate (\$/hr)}}{\text{travel speed (ipm)} \times \text{operator factor (\%)} \times 5} \qquad (16\text{-}11)$$

Duty cycle or operator factor also requires analysis since it varies considerably from job to job and from process to process. The shielded metal arc welding process has the lowest operator factor while semiautomatic welding is much higher, often double. The type of work dictates the operator factor. Construction work, in which small welds are made in scattered locations, has a low operator factor. Heavy production work in which large welds are made on heavy weldments can have much higher operator factors. Time study is sometimes used to determine operator factor based on work similar to the job being costed. Automatic time recorders are sometimes used to accurately determine operator factor on repetitive jobs. The results of this calculation provide labor costs in dollars per foot of weld joint. This can be added to the cost per foot of materials, filler metals, etc.

As an estimate for operator factor, when other data are not available, refer to Figure 16-5. This shows the operator factor related to the method of applying the weld. The arrangement of the work, the use of positioners and fixtures, whether the work is indoors or outdoors, all have an effect on the operator factor. Construction would tend to be the lower end of the range with heavy production toward the high end of the range. The operator factor will vary from plant to plant, and it will also vary from part to part if the amount of weld is much different. The operator factor is higher for the continuous electrode wire processes since the welder does not have to stop every time a covered electrode is consumed. Slag chipping, electrode changes, and moving from one joint to another all reduce the operator factor.

When the welding procedure schedule is not available or when travel speed involves more than one pass, Equation (16-12) is used.

The weight of weld metal deposit in pounds per foot is the datum presented in the table Figure 16-2. As previously mentioned, this can be changed for different metals if the density of the metal is used in the formula. This datum is universal when the proper density factor is used.

The new factor introduced by this formula is deposition rate in pounds per hour. Deposition rate is the weight of the filler metal deposited in a unit of time. It is expressed as pounds per hour. Charts and graphs presented in Chapters 4 and 5 in the sections for particular processes provide deposition rate information. Shielded metal arc welding deposition rate data are shown by Figure 4-20, Figure 4-42 is for gas tungsten arc welding, Figure 4-59 is for plasma arc welding, Figure 5-17 is for submerged arc welding, Figure 4-42 is for gas metal arc welding, Figure 5-84 is for the electroslag welding process, and Figure 16-6 is a composite chart covering most of these processes using steel electrodes. The deposition rate has a tremendous effect on welding costs. The greater the deposition rate, usually the less time required to make a weld. This should be tempered, however, since certain high-deposition rate processes cannot be applied to smaller jobs. For accurate results it is necessary to calculate deposition rates for specific weld procedures. This can be done by weighing the filler metal used, weighing the weld metal deposited, and measuring the arc time. This will provide the deposition rate and also the filler metal yield or utilization.

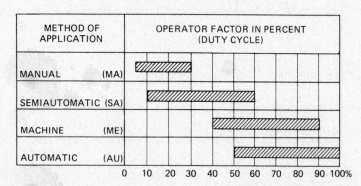

FIGURE 16-5 *Welder operator factor related to method of application.*

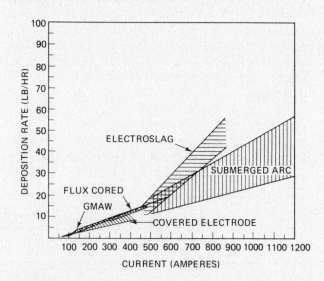

FIGURE 16-6 *Deposition rates—various processes—steel electrodes.*

$$\text{Labor cost (\$/ft)} = \frac{\text{welder pay rate (\$/hr)} \times \text{weight of weld metal deposit (lb/ft)}}{\text{deposition rate (lb/hr)} \times \text{operator factor (\%)}} \qquad (16\text{--}12)$$

Deposition rate can also be calculated. This requires the use of Figures 9-20, 9-21, and 9-22. These curves show the electrode wire feed speed per minute versus welding current. The meltoff rate and the wire feed speed are the same. The curves can be used for these calculations. The meltoff rates are given in inches per minute based on the type and size of electrode, the welding process and current. Table 16-1 shows the length of bare electrode wire per pound of different metals. The relationship between meltoff rate and weight of filler metal melted can be determined by using Equation (16-13).

Use the wire feed speed graphs for meltoff rate and use Table 16-1 for the length and weight of electrode wire.

The deposition rate, which is pounds of metal deposited per hour, is related to the meltoff rate by dividing meltoff rate by the filler metal yield or deposition efficiency. For solid wire systems the yield is high, as shown in Figure 16-3.

16-5 POWER AND OVERHEAD COSTS

Overhead cost consists of many, many things, both in the factory and in the office. It consists of the salaries of plant executives, supervisors, inspectors, maintenance personnel, janitors, and others for whom their time cannot be directly charged to the individual job or weldment. These costs are apportioned pro rata among all work going through the plant. Another important overhead cost is rent or depreciation of the plant and the general maintenance of the building, grounds, etc. Depreciation of plant equipment includes welding machines, materials-handling equipment, overhead cranes, and all other equipment that is not charged directly to individual or specific weldments. In addition, all the taxes on the plant, the real estate, equipment, payroll, and any other taxes applying to the operation of the plant are considered as overhead. Another item

of overhead is the small tools such as chipping hammers, electrode holders, safety equipment, etc., that are not charged to specific jobs. In addition, in most plants the cost of heat, lights, maintenance and repair of the building and equipment are also charged to overhead accounts.

Almost all plants or factories have similar systems for handling overhead expenses but they vary in detail and for this reason cannot be described here. In all cases, however, the overhead charges must be distributed to the welding jobs in one manner or another. In some cases this is based on a per ton of steel fabricated or on the weight of steel consumed. Usually, however, the overhead costs are prorated in accordance to the direct labor charges against the different welding jobs. When this system is used accurate labor costs for weldments are essential.

Overhead rates are sometimes separate and not included in the welder pay rate. When this is the case the same formulas as described before, i.e., nos. 10 and 11, are used but the overhead rate is substituted for the welder's pay rate. Both are in dollars per hour.

For single-pass welding use Equation (16-14).

Duty cycle and operator factor are the same.

For multipass welding Equation (16-15) should be used.

The cost of electrical power is sometimes considered part of the overhead expense. On the other hand, when it is necessary to compare competing manufacturing processes or competing welding processes it is wise to include the cost of electric power in the calculations. In some plants electric power is considered a direct cost and is charged against the particular job. This is more often the case for field welding than for large production weldment shops. In this case, Equation (16-16) is used.

$$\text{Deposition rate (lb/hr)} = \frac{\text{meltoff rate (in./min)} \times 60}{\text{length of electrode wire per weight (in./lb)} \times \text{FM yield (\%)}} \qquad (16\text{-}13)$$

$$\text{Overhead cost (\$/ft)} = \frac{\text{overhead rate (\$/hr)}}{\text{travel speed (ipm)} \times \text{operator factor (\%)} \times 5} \qquad (16\text{-}14)$$

$$\text{Overhead cost (\$/ft)} = \frac{\text{overhead rate (\$/hr)} \times \text{weight of weld metal deposit (lb/ft)}}{\text{deposition rate (lb/hr)} \times \text{operator factor (\%)}} \qquad (16\text{-}15)$$

$$\text{Power cost (\$/ft)} = \frac{\text{local power rate (\$/kwh)} \times \text{volts} \times \text{amperes} \times \text{weld metal deposit (lb/ft)}}{1000 \times \text{deposition rate (lb/hr)} \times \text{operator factor (\%)} \times \text{power source efficiency (\%)}} \qquad (16\text{-}16)$$

The local power rate is based on the rate charged to the factory by the local utility company. If time penalties, power factor penalties, etc., are involved, they should be included. The volts and amperes are the values used when making the weld. The weight of the weld metal is the weight of weld metal deposited. The deposition rate is that used for a particular weld, as is the operator factor. The final factor is the efficiency of the power source, and this can be found from the machine performance curve. Performance curves were presented in Chapter 8.

Special fixtures for tack welding and holding may be considered an indirect labor cost and are classed as overhead. These items are considered as capital expenditures and are depreciated over a period of five years or so. These costs are also added to the welding department or plant overhead.

When this information is calculated it is added to overhead costs, the cost of labor, and the cost of filler metal to arrive at the total weld cost.

16-6 TOTAL WELD COST FORMULAS AND EXAMPLES

Welding costs are obtained by adding the major cost elements:

Materials cost—filler metals, flux, gas, etc.

Labor cost—direct labor

Overhead cost—normally prorated to direct labor

One method of calculating cost is using the basis of cost per weld. This is used for arc spot welds (sometimes cost is based on per 100 arc spot welds), for plug welds, and for small welds used to make a small part.

The more common method of calculating welding cost is on the basis of cost per foot of weld. Each of the cost elements mentioned above can be determined on this basis. There are several ways of determining the cost of the principal elements. One method is best for single-pass welding with continuous wire processes and the other is best for multipass welding.

The different cost formulas are again presented in this section but are given the same numbers as originally. The reference illustration number is also given.

Material Cost

For all welding processes that utilize deposited metal, the weight of deposited metal is the basis for material costs. The weight of each type of weld in the weldment is calculated and the results are added together to determine the total weight. Use Equation (16-3). The weight of weld metal deposit for each weld type is obtained from Figure 16-2. Filler metal yield is based on the type of filler metal and the data of Figure 16-3. These results can be factored by the electrode price to obtain the cost of filler metal required.

To determine the electrode cost per foot use Equation (16-4). Obtain the weld metal deposit from Figure 16-2. The electrode price is the delivered cost to the plant, and the filler metal yield is from Figure 16-3.

Another method of determining the weight of filler metal required, when using a continuous electrode wire process, is to use Equation (16-5). The wire feed speed is taken from the welding procedure, or can be obtained from the process schedule charts in Chapter 9, based on welding current, or it can be measured. The length of wire per weight is taken from Table 16-1.

$$\text{Weight of filler metal required (lb)} = \frac{\text{weight of weld metal deposited (lb)}}{\text{filler metal yield (\%)}} \qquad (16\text{-}3)$$

$$\text{Electrode cost (\$/ft)} = \frac{\text{electrode price (\$/lb)} \times \text{weld metal deposit (lb/ft)}}{\text{filler metal yield (\%)}} \qquad (16\text{-}4)$$

$$\text{Weight of filler metal required (lb/hr)} = \frac{\text{wire feed speed (in./min)} \times 60}{\text{length of wire per weight (in./lb)}} \qquad (16\text{-}5)$$

$$\text{Weight of filler metal required (lb/ft)} = \frac{\text{deposition rate (lb/hr)}}{\text{weld travel speed (in./min)} \times 5} \qquad (16\text{--}7)$$

$$\text{Deposition rate (lb/hr)} = \frac{\text{melt-off rate (in./min)} \times 60}{\text{length of electrode wire per weight (in./lb)} \times \text{FM yield (\%)}} \qquad (16\text{--}13)$$

$$\text{Flux cost (\$/ft)} = \text{flux price (\$/lb)} \times \text{weld metal deposit (lb/ft)} \times \text{flux ratio} \qquad (16\text{--}8)$$

$$\text{Gas cost (\$/ft)} = \frac{\text{price of gas (\$/cu ft}^3） \times \text{flow rate (cu ft/hr)}}{\text{weld travel speed (ipm)} \times 5} \qquad (16\text{--}9)$$

$$\text{Gas cost (\$/weld)} = \frac{\text{price of gas (\$/cu ft}^3) \times \text{flow rate (cu ft}^3\text{/hr)} \times \text{weld time (min)}}{60} \qquad (16\text{--}10)$$

$$\text{Labor cost (\$/ft)} = \frac{\text{welder pay rate (\$/hr)}}{\text{travel speed (ipm)} \times \text{operator factor (\%)} \times 5} \qquad (16\text{--}11)$$

$$\text{Labor cost (\$/ft)} = \frac{\text{welder pay rate (\$/hr)} \times \text{weight of weld metal deposit (lb/ft)}}{\text{deposition rate (lb/hr)} \times \text{operator factor (\%)}} \qquad (16\text{--}12)$$

A third method of determining filler metal required is to use equation (16-7). The weld travel speed can be obtained from the welding procedure, or from process schedule charts, or by measurement. The deposition rate is arrived at from process deposition rate charts based on the current or by Equation (16-13).

The wire feed rate is obtained from the welding procedure, from charts based on welding current, or by measurement. The other two factors were mentioned above.

The cost of welding flux used for SAW is determined by using Equation (16-8). The flux price is the delivered cost to the plant. The weld metal deposit is computed using Equation (16-1), or from Figure 16-2). The flux ratio is 1, 1.1, or 1.2. For accuracy make a test and measure the weld deposit and flux consumed and establish the ratio.

The cost of shielding gas per foot for GMAW, FCAW or GTAW is determined by using Equation (16-9). The price of gas is the cost of the shielding gas delivered to the plant. The flow rate is the usage and is from the welding procedure, from process schedule charts, or by measurement. The welding travel speed is obtained the same way.

The cost of shielding gas per weld is used when costing arc spot welds, plug welds, or small weldments and is determined by using Equation (16-10). The price of gas and the flow rate are obtained as mentioned above. The weld time is established by the welding procedure, or from process schedule charts or by measurement.

Labor and Overhead Cost

The labor and overhead cost for single pass welding is determined by using Equation (16-11). The welder pay rate is normally factored to cover fringe benefits. The travel speed is taken from the welding procedure, process schedule charts, or is measured. The operator factor, or duty cycle, is obtained from Figure 16-4; however, for better accuracy it should be measured, or related to similar jobs.

The labor and overhead cost for multipass welding is determined by using Equation (16-12). The data for each factor involved in this formula has been explained above.

$$\text{Overhead cost (\$/ft)} = \frac{\text{overhead rate (\$/hr)}}{\text{travel speed (imp)} \times \text{operator factor (\%)} \times 5} \qquad \text{(16-14)}$$

$$\text{Overhead cost (\$/ft)} = \frac{\text{overhead rate (\$/hr)} \times \text{weight of weld metal deposit (lb/ft)}}{\text{deposition rate (lb/hr)} \times \text{operator factor (\%)}} \qquad \text{(16-15)}$$

The overhead cost must be calculated separately if all of the overhead is not included in the welder pay rate. The equations to use are similar to the two equations above. They are Equations (16-14) and (16-15). The only difference is that the overhead rate is substituted for the welder pay rate and it should be obtained from the accounting calculations, which is beyond the scope of this book. For simplicity the overhead rate can be assumed to be equal to the welder pay rate. In some plants it is one and one-half times the pay rate or even higher.

Sample Cost Calculations

The following examples will show how the different equations are used. The following rates will be used in these examples.

Welder pay rate = $5.00 per hour

Overhead rate = $5.00 per hour

Power cost = $0.02 per kwh

Argon gas cost = $0.11 per cubic feet

CO_2 gas cost = $0.02 per cubic feet

Covered electrode price = $0.28 per pound

Steel electrode wire price = $0.32 per pound

Flux-cored electrode price = $0.44 per pound

All other data is taken from referenced information elsewhere in this book.

Example 1: In this example, we wish to establish the cost of a weldment that contains 120 ft of 1/4-in. fillet, 300 ft of 3/16-in. fillet, 40 ft of 1/2-in. vee groove, and 60 ft of 3/8-in. vee groove weld. To determine the amount of weld metal deposited consult Figure 16-2. Obtain the weight of weld metal of each type of weld, multiply it by the length of that type of weld and add the total:

120 ft of 1/4-in. fillet	= 12.7 lb
300 ft of 3/16-in. fillet	= 18.3 lb
40 ft of 1/2-in. vee groove	= 28.1 lb
60 ft of 3/8-in. vee groove	= 23.0 lb
Total of weld metal deposited	= 82.1 lb

What is the cost of the filler metal required when covered electrodes are used? Use Equation (16-4) (modified) and a filler metal yield of 60% from Figure 16-3.

$$\text{Electrode cost (\$)} = \frac{0.28 \times 82.1}{0.60} = \$38.31$$

What is the cost of filler metal required if solid steel electrodes are used? Use Equation (16-4) (modified) and a filler metal yield of 92.5% from Figure 16-3.

$$\text{Electrode cost (\$)} = \frac{0.32 \times 82.1}{0.925} = \$28.40$$

Note that the total cost is less when using a higher priced electrode.

Example 2: What is the cost of a foot of 1/4-in. fillet weld made manually with SMAW using E-6024 electrode 3/16-in. size? The operator factor is 30% and is from Figure 16-5. The filler metal yield is 55% and is from Figure 16-3. The weight of weld metal deposited is 0.117 lb/ft and is from Figure 16-2. Use Equation (16-4).

$$\text{Electrode cost (\$/ft)} = \frac{0.28 \times 0.117}{0.55} = 0.060 \ \$/ft$$

The labor cost is calculated using Equation (16-11). The travel speed of 15 IPM is from the process schedule chart Figure 4-22, in Chapter 4.

$$\text{Labor cost (\$/ft)} = \frac{5.00}{15 \times 0.30 \times 5} = 0.222 \ \$/ft$$

The overhead cost is the same as the labor cost. The total cost is the total of: $0.060 + 0.222 + 0.222 = 0.504 $/ft of weld.

Example 3: What is the cost of a foot of 1/4-in. fillet weld made semi-automatically using GMAW with CO_2 gas shielding using an E70S-1 electrode 0.035 in. size? The operator factor is 50% and is from Figure 16-5. The filler metal yield is 95% and is from Figure 16-3. Use Equation (16-4).

$$\text{Electrode cost (\$/ft)} = \frac{0.32 \times 0.117}{0.95} = 0.039 \ \$/ft.$$

The gas cost is calculated using Equation (16-9). The shielding gas flow rate is 25 cubic feet per minute and the travel speed is 15 in. per minute obtained from the process schedule chart Figure 5–38, in Chapter 5.

$$\text{Gas cost (\$/ft)} = \frac{0.02 \times 25}{15 \times 5} = 0.008 \text{ \$/ft.}$$

The labor cost is determined using Equation (16-11). All of the factors involved were mentioned above.

$$\text{Labor cost (\$/ft)} = \frac{5.00}{15 \times .5 \times 5} = 0.133 \text{ \$/ft.}$$

The overhead cost is determined using Equation (16-14). All of the factors involved were mentioned above.

$$\text{Overhead cost (\$/ft)} = \frac{5.00}{15 \times .5 \times 5} = 0.133 \text{ \$/ft.}$$

The total cost is the total of: $0.039 + 0.008 + 0.133 + 0.133 = 0.313$ \$/ft of weld. Note that the gas metal arc welding is less expensive than shielded metal arc welding for the same weld size.

Example 4: How many pounds of E70S-1 electrode wire 0.035-in. diameter should be purchased to make a tank requiring 25,500 ft of 1/8-in. square groove butt weld? The weld deposit, with reinforcement, is 0.065 lb/ft from Figure 16-2 times 25,500 ft, which is 1657.5 lb. To determine the pounds of filler metal needed, use Equation (16-3). The filler metal yield of 90% is from Figure 16-3.

$$\text{Weight of filler metal required (lb)} = \frac{1,657.5}{0.90} = 1,842 \text{ lb}$$

By using an electronic calculator these computations are quickly made.

16-7 REDUCING WELDING COSTS

Weld metal is the most expensive metal involved in normal steel fabrication work. Every effort should be made to reduce the amount of weld metal required for each weld joint and each weldment. The value engineering concept should be used. A cost analysis should be made and it should touch on four specific areas: weld design, welding procedure, manufacturing operations, and the welding department.

The following checklist can be used in making an analysis.

Weld Design.

1. Eliminate weld joints whenever possible. Use rolled sections, use small steel castings for complex

areas, and use formed or bent plates whenever possible.

2. Reduce the cross-sectional area of welds. Do this by utilizing small root openings, small groove angles, double- instead of single-groove welds, etc.

3. Utilize fillet welds with caution. If the size is doubled the strength is doubled, but the cross-sectional area and the weight increase four times.

4. Intermittent fillet welds should be studied. It is sometimes possible to reduce the fillet size and make the weld continuous and save weld metal.

5. Provide easy accessibility for all welds. Failure to do so may mean that extra special attention and time are required.

6. Select weldable materials. Difficult-to-weld metals involve complex and expensive procedures.

7. Use weld symbols with size notations for all welds.

Welding Procedure.

1. Written welding procedures or job sheets should be provided to the welding department for all jobs.

2. Select the welding process with the maximum deposition rates that can be used.

3. Select the method of application to provide the maximum operator factor.

4. Filler metal must be related to the welding process. When using covered electrodes, select the type for maximum deposition rate for the job at hand.

5. When using covered electrodes, the stub losses are less for 18-in. electrodes than for 14-in. electrodes. Select the longer electrodes when possible.

6. Polarity increases deposition rate with the submerged arc welding process. Electrode negative has higher deposition rate but has less penetration.

7. Consider the filler wire packaging. Solid electrode wire is more expensive when it is purchased on small spools than on large reels or in payoffpaks.

*Manufacturing Operations—Other Than
the Welding Department.*

1. Parts preparation must be accurate. This applies particularly to bending and forming.

2. Parts preparation of edges. Shearing or blanking is more economical than thermal cutting.

3. Parts preparation when flame cutting. Select the optimum fuel gas for oxy-fuel gas cutting. Utilize high-speed tips and an oxidizing flame.

4. Avoid designs that require machining for weld preparation.

5. Utilize automatic shape-cutting equipment with templates or numerical control. Avoid hand layout and manual flame cutting.

6. Keep all material moving to the welding department to reduce cycle time. Provide all parts to the welding department at the same time.

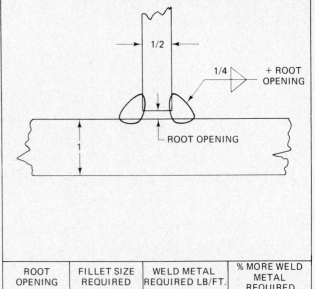

ROOT OPENING	FILLET SIZE REQUIRED	WELD METAL REQUIRED LB/FT.	% MORE WELD METAL REQUIRED
0	1/4″	0.212	—
1/16″	5/16″	0.334	157%
1/8″	3/8″	0.476	224%
3/16″	7/16″	0.652	308%

FIGURE 16-7 *Extra weld metal required for poor fit up.*

Fillet Size Overwelded Size	Theoretical wt/ft	Overwelding Requires This Much More Weldmetal
1/4 × 1/4 design	0.106	—
1/4 × 5/16	0.133	25%
1/4 × 3/8	0.159	50%
5/16 × 5/16 design	0.166	—
5/16 × 3/8	0.199	20%
5/16 × 7/16	0.232	40%
3/8 × 3/8 design	0.239	—
3/8 × 1/2	0.318	33%
3/8 × 9/16	0.358	50%
1/2 × 1/2 design	0.425	—
1/2 × 9/16	0.477	12%
1/2 × 5/8	0.531	25%

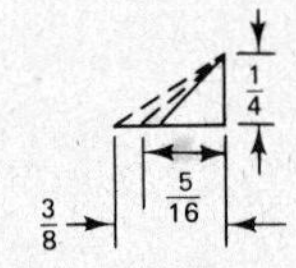

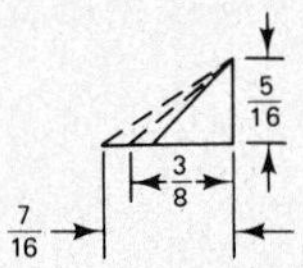

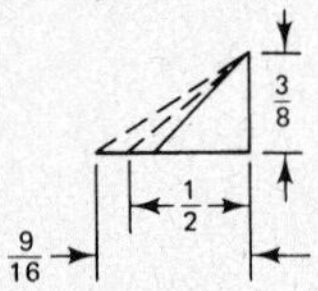

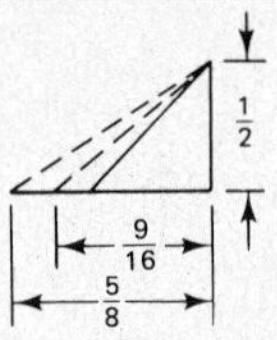

FIGURE 16-8 *Extra weld metal required for overwelded fillets.*

1. Provide accurate fitup of all weldments. Utilize fixtures for setup whenever possible. Tack weld only in weld setup fixtures. Inspect fitup prior to welding. Figure 16-7 shows the effect of fitup on extra weld metal and time required.

2. Avoid overwelding. Observe and obey sizes shown on weld symbols. Figure 16-8 shows the additional weld metal and time required when overwelding.

3. Eliminate excessive reinforcing on all welds. Reinforcing is not required to obtain weld strength.

4. Utilize subassemblies whenever possible. This will minimize distortion and reduce cycle time.

5. Follow welding procedures to minimize distortion. Distortion will create poor fitup which may require additional welding.

6. Utilize positioning equipment. When rotating positioning equipment is not available large weldments can be positioned on the shop floor with cranes. Forty-five degree fixed positioners can be used.

7. Maintain proper arc length and welding current. Excessively long arcs increase spatter losses, as does excessive high current. Low current may result in lower than standard deposition rates.

8. Completely utilize the filler metal purchased. This involves using covered electrodes to minimum stub lengths. Avoid discarding excessive lengths of continuous electrode wire.

9. Provide power tools for slag removal, and for weld finishing.

10. Provide for welder comfort and safety by utilizing scaffolding, worker positioners, guard rails, etc.

11. Retain high welding power source efficiency through routine maintenance procedures.

12. Check for cable and connector efficiency in the welding circuit. Hot spots waste power.

Continued use of this list will help reduce welding costs.

QUESTIONS

1. Why is the cross-sectional area of a weld joint important to cost?

2. What is the formula for the cross-sectional area of a regular fillet weld?

3. If the cross-sectional area is known, how do you determine the weight of a foot of weld?

4. What is the meaning of filler metal yield? Why is the yield lower for covered electrodes?

5. Which has the better yield, a 14-in. electrode or an 18-in. electrode? Why?

6. Why is deposition rate so important for welding costs?

7. What different materials are included in gas metal arc welding? In submerged arc welding? In shielded metal arc welding?

8. What is meant by operator factor when calculating welding cost?

9. Why is operator factor so important?

10. What major cost elements must be totaled to obtain total welding cost?

11. How does poor fitup affect welding costs?

12. How does one's welding affect welding costs? How does excessive reinforcement affect costs?

13. Why is a positioner used to reduce welding costs?

14. What is the most expensive metal of a weldment? Why?

15. What is the formula for the cross section area of a corner joint utilizing a single bevel weld reinforced by a fillet weld?

16. What is the theoretical weld deposit for a 3/8 in. fillet weld 72 in. long?

17. What is the weight of weld deposit for a single vee butt joint in a 30 in. diameter pipe having a 3/8 in. wall thickness?

18. How many pounds of E6010 electrodes should be purchased for making a 5 mile long pipe line using the pipe size in question 17? Assume each length of pipe is 20 ft.

19. A weldment requires 40 ft of 1/2 in. vee, 60 ft of 3/8 in. vee, 120 ft of 1/4 in. fillet and 300 ft of 3/16 in. fillet welds. What is the weight of weld metal deposited including reinforcement?

20. In question 19 how many pounds of electrode should be purchased to make 10 weldments using the shielded metal arc process? How many pounds of electrode wire should be purchased for the 10 weldments using the flux-cored arc welding process?

17

WELDMENT EVALUATION AND QUALITY CONTROL

17-1 WELD RELIABILITY

17-2 WELDING CODES AND SPECIFICATIONS

17-3 WELDING PROCEDURES AND QUALIFYING THEM

17-4 WELD TESTING—DESTRUCTIVE TESTING

17-5 VISUAL INSPECTION

17-6 NONDESTRUCTIVE EXAMINATION

17-7 WELD DEFECTS—CAUSES AND CORRECTIVE ACTIONS

17-8 WORKMANSHIP SPECIMENS, STANDARDS, AND INSPECTION SYMBOLS

17-1 WELD RELIABILITY

The demand for more reliable products, our increasingly complex technology, and the need to conserve resources make weld quality increasingly important. Quality in the aerospace or nuclear field means one hundred percent reliability of each and every weld. In other fields, it means producing welds that are satisfactory to successfully sustain the service encountered. In many fields product liability is a major factor. Quality welding on any product can only be judged with respect to a specific quality standard. This standard must be based on the anticipated service of the product. It must be balanced between the service requirements and the consequence of failure versus economic cost factors. For many products, and in many industries, weld quality requirements are controlled by rigid codes and specifications. However, for some products there are no applicable codes or specifications; for these products, the producer must maintain high product quality in order to survive.

The success of maintaining the delicate balance between high-quality requirements and cost factors is decided in the field and in the marketplace, where quality and price determine the producer's continuing success.

In recent years weld joints have been produced that provide the utmost in quality to satisfy design and safety requirements of many high-performance applications. The welding encountered in space vehicles and nuclear vessels is exposed to environments unheard of in the past. Weld perfection demanded by this super control class of work has been possible thanks to extensive training, stringent quality control and inspection procedures, and the development of better filler metals. This quality level is attained because of extensive preparation and time-consuming procedures which contribute to its high cost. Perfect welds, however, are not required on every type of weldment produced. We must guard against establishing super quality requirements when they are not required.

The responsibility for producing quality products rests on many people. It is the responsibility of management to create the proper cooperative spirit among designers, materials managers, welders and other production workers, supervisors, and quality control and inspection personnel to make sure that the quality requirement is reasonable and in agreement with the service expected for the weldment in question. The responsibility for producing quality welds rests upon the welder. Each welder must accept this responsibility. The welding supervisor has the responsibility of the welders and for their performance. The welding inspectors must guarantee that quality standards are maintained. The welding standards or specifications and procedures are the basis for weld quality and these factors coupled with the weldment design are the responsibility of designers, material managers, and others. It is a total responsibility with all involved and this interrelationship is so complex that it is impossible to clearly state that one person or another has specific single responsibilities for quality.

The guarantee for quality welding is eternal vigilance. The designers, the specification writers, material specifiers, and others must keep close contact with field requirements and problems. They must be sensitive to needs for change and they must be able to relax standards as well as tighten standards when the need occurs. Welding supervisors and other production managers must be continually alert for evidence of substandard workmanship. Weldments properly designed, utilizing correct specifications and controls, will perform satisfactorily in the service environment for which they were designed. The fact that this is true is a tribute to the people involved in their selection of design specifications and weld quality control programs.

The need to differentiate between the adequate and the perfect weld has led to research concerning the acceptability of weld imperfection and how these imperfections affect service life. This, in turn, has led to investigations of the degree of imperfection and what may or may not be allowable. Through the years much of these data have been translated into codes and specifications for special types of equipment or weldments. The knowledge gained from field experience and experience of producing weldments is reflected in the different codes.

Codes are written for specific classes or types of products. This occurs since products tend to follow different types of service, and it is the service life that affects the quality requirements of the specification.

To produce quality weldments it is necessary to look at the total picture, which involves the service requirement of the product, the basic design to produce the product, the design of the welds, the materials selected, and the code or specification to be utilized. This leads to procedures which are normally qualified in accordance with the code. It is then necessary to qualify the welders to see that they adequately comply with the qualified procedure. Finally, it is necessary to determine that the quality level of the welding on the product is in accordance with the specification. This final factor is the human requirement for continually maintaining quality through inspection techniques and testing. One of the major problems encountered in weldment production is the suspicion of the designer that the weldment will not be manufactured as expected. Many engineers and designers feel uneasy with weldment design for this reason. They feel sure that the weldment they have designed will withstand the service requirements.

They also feel that the joint details and procedures are such that the quality required will be attained. The suspicion occurs when they consider workmanship factors that are seemingly beyond their control. They feel, yes, the welder can produce joints equal to the design requirements under ideal conditions and, yes, the welder did produce a quality weld when the performance qualification test was passed; however, what assurance do they have that each and every weld in the weldment will be of this same quality. The only assurance that can be given to designers is the implementation of a complete quality control program. It is acknowledged that it costs money to implement such a program. Such programs save money in the long run as they eliminate the problem of premature field failures, catastrophic disasters, or continual repair work to maintain the product in service.

The implementation of a total quality control program involves a review of the design, of the material involved, very close attention to the weld joints, a review of the qualified procedure, and the adequacy of the procedure to fulfill the needs of service. This becomes an economic factor since the type of service and service life have a very definite effect on the quality demands of the welds. Such factors as low-temperature service, 100% reliability, impact, and fatigue loading are certainly different from low-level stresses and static loading that might be encountered. In the overall review of the program these factors must be considered. The selection of base metals, welding processes, welding filler materials, and procedures all have an influence on the total quality of the end product. Finally, it is important that the welders themselves be adequately trained and qualified to apply the selected process on the base metals with the proper filler materials in the joints and in the positions encountered. This must be determined and established and continual vigilance must be maintained on the inspection of the welding on the product.

For certain classes of work, particularly in the nuclear field, special quality control requirements have been established. These requirements make it necessary to consider all factors and write a quality control program. These more strict requirements are found in the nuclear code of ASME, [1] the Atomic Energy Commission Specifications, [2] and the Department of Defense Nuclear Code. [3] These standards require the producer of nuclear products to institute a quality assurance program. The design of the program is based upon the technical and manufacturing aspects of the product. The program must insure adequate quality throughout from the design, acquisition, and manufacture, to final shipment. Such programs must have authority and responsibility clearly defined for each portion of the work. The quality assurance plan must include many factors, among which are the following:

1. **Organization:** The organization for quality must be clearly prescribed. It should define and show charts of responsibility and authority and the organizational freedom to identify and evaluate quality problems. Quality control personnel should not report to production personnel.

2. **Quality Assurance Program:** The producer must conduct a review of the requirements of the quality required of the product. The various factors such as specialized controls, processes, testing equipment, skills, etc., for assuring product quality must be identified. This program must be documented by written policies, procedures, and instructions.

3. **Design Control:** The design control must provide for verifying or checking the adequacy of the design, via performance testing, and independent review. It should include qualification and testing of prototypes and must conform to specifications.

4. **Procurement Document Control:** The program requires that specifications be written for each item required to be purchased and that the specification insure the quality required by the end product. These types of specifications may also include various approvals and quality control programs that may be required from vendors.

5. **Instructions, Procedures, and Drawings:** The quality program must insure that all work affecting quality must be prescribed in clear and complete documented instructions of a type appropriate to the work. Compliance with instructions must be monitored.

6. **Document Control:** The quality program must include a procedure for maintaining the completeness and correctness of drawings and instructions, and the like, showing dates, control,

effective point, etc. These drawings, procedures, and instructions must be maintained and continuity explained by change notices.

7. **Control of Purchased Material, Equipment, and Services:** The program must include a control system for purchasing those requirements from qualified vendors. This means that vendors must have similar quality programs for producing their items. The need for a qualified products list is indicated and only those vendors having adequate quality programs and providing quality parts will be included. The program also requires the use of inspection systems at receiving so that purchased parts can be checked against the specifications. Raw materials, purchased parts, and the like, will be inspected by means of instruments, laboratory procedures, etc., to insure that the products meet the specification.

8. **Identification and Control of Materials:** The program must provide for identification of all parts, materials, components, etc., from receipt throughout all processing to the final item. Records shall provide traceability of all materials, components, etc.

9. **Control of Special Processes:** The quality program must insure that all manufacturing operations including welding are accomplished under controlled conditions. These controlled conditions involve the use of documented work instructions, drawings, special equipment, etc. It further requires that such instructions be properly recorded and maintained. Detailed work documentation is required for all complex processing operations.

10. **Inspection:** The quality assurance program shall insure a system of inspection and testing for all products. Such testing shall simulate the product service and records must be maintained of the adequacy of the product to meet these specifications.

11. **Test Control:** The program must assure that all tests are performed according to written instructions. Instructions must provide requirements and acceptance limits. Test results must be documented and evaluated to assure that test requirements are met.

12. Control of Measuring and Testing Equipment: The program shall provide for methods of maintaining the accuracy of gauges, testing devices, meters, and other precision devices showing that they are calibrated against certified measurement standards on a periodic basis.

13. Handling, Storage, and Delivery: The program shall provide for adequate instructions for handling, storage, preservation, packaging, shipping, etc., so that the product is protected from its time of manufacture until its time of use.

14. Inspection Test and Operating Status: The program must include methods of identifying parts to determine its status as far as inspection and approval are concerned.

15. Nonconforming Materials, Parts, or Components: There shall be a procedure established to maintain an effective and positive system for controlling nonconforming material. It may include and allow for rework; however, records must be maintained of such work.

16. Corrective Actions: The quality program must establish methods of dealing promptly with any conditions that are adverse to quality, including design, procurement, manufacturing, testing, etc. The program should also include methods of overcoming defects taking corrective action to produce a part to meet the required quality.

17. Quality Assurance Records: Program requires that records be maintained, including all data essential to the economical and effective operation of the quality program. Records must be complete and reliable and include measurements, inspections, observations, etc., and these records must be available for review.

18. Cost Related to Quality: The program should allow for maintenance and use of cost data for identifying the cost of the program and for the prevention and correction of defects encountered.

19. Production Tooling and Inspection Equipment: Various items of tooling, including fixtures, templates, patterns, etc., may be used for inspection purposes provided that their accuracy be checked at periodic intervals.

20. Audits: The program must include a system of planned and periodic audits to verify compliance with all aspects of the quality assurance program. The audit must be done by personnel not normally involved in the areas being audited. Audits must be documented and reviewed, and action must be taken to correct any deficiencies found.

The preceding list is an abbreviated outline of the requirement of a quality assurance program necessary for critical products.

As time goes on and as requirements for higher quality continue similar types of programs may be required for other products.

It is the responsibility of a company's management to assure that designs and designers are adequate for the type of product produced, that they have knowledge of the service life, materials, and manufacturing processes involved. They must also assure that correct specifications are used and that the necessary inspection techniques are followed. They must insure that the welders are qualified for the work in question and that the quality program is used; inspections are made to insure that the quality expected will be obtained. Finally, it is necessary to maintain the balance between the quality requirements and the economic requirements of the products to insure a successful product that can be sold in the marketplace.

17-2 WELDING CODES AND SPECIFICATIONS

There are many different welding codes and specifications and additional ones are being written every year. Figure 17-1 is a picture of the front of some of the representative codes and specifications that involve welding. In order to properly understand these different codes and where they are used it is best to consider the various industries that employ welding specifications. Certain products are more thoroughly regulated by codes and specifications than others, especially regarding welding. For the purposes of classifying the various codes, they are listed according to the product produced.

The products that utilize welding specifications can be listed as follows.

Pressure vessels.

Nuclear reactors.

Pressure piping.

Pipelines and piping in general.

Bridges and buildings.

Ships, commercial and naval.

Storage tanks and vessels.

FIGURE 17-1 *Picture of codes and specifications.*

Railroad rolling stock.

Aircraft and spacecraft.

Construction equipment.

Miscellaneous machinery.

Miscellaneous products such as radio and television towers, tank trucks, compressed gas cylinders, etc.

It is fortunate that specifications applying to similar products are very similar with regard to welding. In some cases the qualifications of one specification may be acceptable by another one for the same products. Also, it is fortunate that most manufacturers and contractors usually produce weldments that are similar or come under the same general types of specifications. Efforts are continuing to make specifications more uniform and, it is hoped, to make the interchange of qualifications easier between specifications for similar products.

To provide maximum assistance we are listing the welded product followed by the various specifications that apply to those products. Included also are important points of the specifications with respect to welder qualification and procedure qualifications.

Pressure Vessels.

In North America the manufacturer of pressure vessels and all other items defined as pressure vessels by various laws comes under the general specification of the ASME boiler and pressure vessel code.[1] This code consists of eleven sections as follows:

Section I: Power boilers.

Section II: Material specifications—ferrous.

                Material specifications—nonferrous.

                Material specifications—welding rods, electrodes, and filler metals.

Section III: Nuclear power plant components.

Section IV: Heating boilers.

Section V: Nondestructive examination.

Section VI: Recommended rules for care and operation of heating boilers.

Section VII: Recommended rules for care of power boilers.

Section VIII: Pressure vessels division I and II (two sections).

Section IX: Welding qualifications.

Section X: Fiberglas reinforced plastic pressure vessels.

Section XI: Rules for in-service inspection of nuclear reactor coolant system.

WELDMENT EVALUATION AND QUALITY CONTROL

All products manufactured under the requirements of these codes may also be manufactured under the rules and regulations of the different states and provinces which either reference or reprint various sections of the ASME boiler and pressure vessel code. From the welding point of view, the following sections of the ASME code are the most important: Section I, power boilers; Section II, material specifications for welding materials; Section III, nuclear power plant components; Section V, non-destructive testing; and Section IX, welding qualifications. In general Section IX is universally used throughout North America and in other parts of the world as the method of qualifying welders for work on pressure vessels and similar items.

Nuclear Reactors.

The nuclear reactors and associated components and materials used in nuclear power plants are covered by the provisions of Section III of the ASME pressure vessel code. Any part that is utilized in a nuclear plant should be manufactured under the jurisdiction of this code. The exceptions are those components for navy ship use which are covered by a similar but different code issued by the Department of Defense Naval Ship Division. This is known as Navships 250-1500-1 Standard for Welding of Reactor Coolant and Associated Systems and Components for Naval Nuclear Power Plants.[3] This is a specialized code that includes additional restrictions. It requires the certification of materials and traceability of all materials including welding filler metals to the point of origin. It also includes strict control systems of inspection during the manufacture of nuclear power plant components.

Pressure Piping.

Work on pressure piping is covered by the American National Pressure Piping Code issued by the American National Standard Institute and known as ANSI B31.1 Code for Pressure Piping. [4] This code is similar, as far as welding requirements are concerned, to Section IX of the ASME pressure vessel code mentioned above. Pressure piping is also regulated by many states and provinces and in most cases the basis for the local code is B31.1. In some states, welders are tested and certified by the individual states.

Pipelines.

For many years the specification controlling the welding on cross-country pipelines was American Petroleum Institute's Standard 1104. [5] This has become a worldwide specification used for transmission pipelines, particularly by the petroleum industry. Through the years it has gradually become accepted for other types of pipe welding, but recently additional codes have been issued for other types of pipe welding. Many of these codes are in substantial agreement with API 1104. This code does provide the welding procedure information and welder qualification information for work on cross-country pipelines. It has been adopted for distribution piping, even though this was not intended. The Office of Pipeline Safety of the U.S. Dept. of Transportation in their "Regulations for the Transportation of Natural Gas and other Gas by Pipeline" [6] and "Transportation of Liquids by Pipeline" [7] refers to API Standard 1104 for welding.

The newer codes involved are American Standard Pipe Codes issued by American National Standards Institute and are as follows:[4]

B31.1 "Power Piping."

B31.2 "Fuel Gas Piping."

B31.3 "Petroleum Refinery Piping."

B31.4 "Liquid Petroleum Transportation Piping Systems."

B31.5 "Refrigeration Piping."

B31.6 "Chemical Plant Piping."

B31.7 "Nuclear Power Piping."

B31.8 "Gas Transmission and Distribution Piping Systems."

Bridges and Buildings.

Structural welding is done under the requirements of certain large cities and for bridges under the jurisdiction of state or provincial highway departments. The basis for these codes either by reference or by direct copy is the "Structural Welding Code" published by the American Welding Society.[8] This code incorporates the requirements of the Department of Transportation Bureau of Public Roads of the U.S. Government. The Bureau of Public Roads has interest in all state structural specifications; therefore, they are all similar. However, most states publish their own welding code. Welding on highway bridges is under the jurisdiction of the state highway departments and in many states welders are examined yearly and certified by the state to work on bridges. Many state highway departments also require certification of welding electrodes and filler metals. The "Standard Specifications for Highway Bridges," [9] adopted by the American Association of State Highway Officials, and the "Specifications for Steel Railroad Bridges," [10] published by the American Railway Engineering Association, are in substantial agreement with the AWS structural welding code from the welding procedure qualification and welder qualification point of view.

Large steel buildings welded in the major cities in North America are covered by city codes and specifications. These codes and specifications are also in substantial agreement with the AWS structural welding code, at least from the welder's point of view. Only the larger cities publish welding codes. Some cities require certification of filler metals for structures welded under their jurisdiction.

Ships.

Welding on ships is well covered by different specifications and codes. In the United States, all government vessels are covered by codes issued by the U.S. Coast Guard [11] or the Navships Division of the Department of Defense. [12] These requirements, as far as welding qualification and welder procedure qualification, are identical. They are also very similar to the requirements of the Maritime Administration on commercial ships.[13] Qualification of welders is usually transferable among these three organizations. The American Bureau of Shipping has similar requirements for welding on ships that they survey. [14] Lloyds and other classification societies also publish specifications that cover welding. Certification of weld metal is usually required by all of these codes.

Storage Tanks and Vessels.

There are two major codes for the welding of storage tanks. One is for the welding of elevated storage tanks and is published by the American Water Works Association, "Standard for Welded Steel Elevated Tanks, Standpipes, and Reservoirs for Water Storage."[15] The other one is for oil or petroleum products storage tanks published by American Petroleum Institute, "Standard for Welded Steel Tanks for Oil Storage."[16] Both of these codes refer to Section IX of the ASME boiler code as far as welding qualification is concerned.

Railroad Rolling Stock.

Specifications for the manufacture of rolling stock for North American railroads is under the jurisdiction of the Department of Transportation (formerly Interstate Commerce Commission), at least in the United States. However, as far as welding qualification and welding design requirements are concerned, the controlling specifications are issued by the Association of American Railroads. Various specifications are involved including "Specifications for Tank Cars" standard[17] and "Specifications for Design, Fabrication, and Construction of Freight Cars" standard.[18] These specifications provide information concerning the design of welds and the

qualification of welders' manufacturing these products. They are in substantial agreement with requirements of the AWS structural welding code.

The Department of Transportation also has codes covering the manufacture of tanks for transporting gas under high pressure[19] and for tanks carrying liquid petroleum and similar products.[20]

Aerospace and Aircraft.

Weldments intended for use in aircraft and spacecraft are welded to the requirements of U.S. government specifications. There are other groups that write specifications for materials that might be utilized, including the Society of Automotive Engineers [21] and the Aircraft Manufacturing Industry Association. [22] Welding codes or requirements are covered by specifications of the National Aeronautics and Space Administration (NASA) and of the Department of Defense Military (Mil) Standards and Specifications. The one pertaining primarily to welding on aircraft is "Test, Aircraft Welding Operators Certification" MIL-T-5021D. [23] This standard covers many welding processes, metals, and levels of proficiency for testing welders and must be strictly adhered to when welding on aircraft. Qualification under this standard is done under the jurisdiction of government inspectors.

Construction Equipment.

Much construction equipment is made to company standards which have been found acceptable based on the product acceptance in the field. Most major manufacturers of construction equipment have their own specifications. Recently the American Welding Society has issued specifications that establish common acceptance standards for weld performance and process application for this industry. This is known as "Welding on Earth Moving and Construction Equipment." [24] Qualification of welders is not a major issue in this standard.

Machinery.

Most industrial machinery utilizing weldments is not covered by code or specification. Here again, however, the American Welding Society has issued some specifications which establish common accepted standards for weld performance and process application. Two of these are:

"Welding Industrial and Mill Cranes." [25]

"Metal Cutting Machine Tool Weldments." [26]

The welder qualification requirements are similar to the requirements of AWS structural code or ASME Section IX of the pressure vessel code.

17-3 WELDING PROCEDURES AND QUALIFYING THEM

A welding procedure is "the detailed methods and practices including all joint welding procedures involved in the production of a weldment." The joint welding procedure is the "materials, detailed methods, and practices employed in the welding of a particular joint." The written welding procedure is a "manner of doing" or "the detailed elements (with prescribed values or ranges of values) of a process or method used to produce a specific result."

In other words, a welding procedure is the step-by-step directions for making a specific weld. It usually consists of three parts, (1) a written description, (2) drawings of the weld joint and tables showing the welding schedule or conditions, followed by (3) a test data sheet showing the results of testing the weld made using the prescribed conditions. The successful test results show that the weld has passed the requirements and thus the procedure is qualified.

Procedures can also be step-by-step directions for making a specific weldment. These procedures are usually written in order to reduce weld distortion or to show how a weldment should be made to avoid the possibility of missing welds.

Sometimes there is confusion concerning the qualification of welding procedures and the qualification of welders. The subject of qualifying welders was covered in Chapter 3, "Qualifying and Certifying Welders." In general, a welder who makes the welds to qualify a welding procedure is automatically qualified. Certain codes require written and qualified procedures before welding can be done on items made under the code. Other codes may not have this requirement and in some cases have prequalified procedures based on specific welding processes and prequalified joint design details.

The majority of the welders working on structural work, piping, and pressure vessels must follow qualified welding procedures. This is because the three most popular codes used in the United States require written and qualified welding procedures. These are the ASME "Boiler and Pressure Vessel Code," the AWS "Structural Welding Code," and the API standard 1104, "Standard for Welding Pipelines and Related Facilities." The API code makes the following statement.

"Prior to the start of production welding, a detailed procedure specification shall be established and qualified to demonstrate that welds having suitable mechanical properties and soundness can be made by this procedure. The quality of the weld shall be determined by destructive testing."

The ASME Code makes the following statement.

"Each manufacturer or contractor shall record in detail and shall qualify the procedure specification for any welding procedure followed in the construction of weldments built in accordance with this code. All welding used in qualifying welding procedures shall be performed in accordance with the procedure specification."

The AWS structural welding code states:

"Each manufacturer or contractor shall conduct the tests required by this code to qualify the welding procedure and the welders, welding operators, tackers who will apply these procedures."

There are other codes that relate directly to some of the above. In these cases the requirements of the referenced code must be followed. The writing and qualifying of procedures have not always been strictly enforced. However, as more codes are accepted and adopted by states, provinces, and cities, enforcement is becoming more stringent. As mentioned above, the manufacturers or contractors qualify the procedures. There are two other groups that can be included. These are owners of plants who may wish to qualify their own employees for installation and maintenance work and associations of manufacturers or contractors who qualify procedures in the name of the association. The procedures are qualified in the name of the association and the welders are also qualified in the name of the association. This practice is fairly common in metropolitan areas where welders, particularly in the plumbing and pipefitting trade, change employers often. By qualifying the procedures and welders in the name of the association, qualification tests are not required and repeated every time the welder works for a different contractor.

The welding procedure specification, abbreviated WPS, must be signed by the person making the tests on the welds and by the person responsible for the qualified procedure. The procedure includes the sentence, "We certify that the statements in this record are correct and that the test welds were prepared, welded, and tested in accordance with the requirements of this code." Signing this, certifies that the qualification is in order. This is known as self-certification and can be done by manufacturers, contractors, owners, users, or associations. Some states issue regulations which have specific requirements in this regard. These requirements supplement the specification.

In all three of the specifications mentioned, the qualified welding procedures are usually monitored by others. In the case of pressure vessels, qualification tests are usually monitored by insurance company inspectors, who provide boiler insurance, and possibly also by inspectors representing the industrial commission of the state. In the case of structural welding, the qualification tests may be monitored by state highway department inspectors or city building inspectors. On pipelines, qualification tests are monitored by inspectors employed by the owner of the pipeline and perhaps by inspectors representing the state or the Department of Transportation.

In the case of piping, there is an organization known as the National Certified Pipe Welding Bureau, which is a division of Mechanical Contractors Association. This bureau has chapters in the various metropolitan areas throughout North America. The piping contractors in the area are members of these chapters. The National Certified Pipe Welding Bureau has a large number of procedures which have been qualified by different members and adopted by the association. It is therefore possible for member contractors to adopt these standardized qualified welding procedures and use them in their own operations. In some cases, the member companies also qualify the welders on a periodic basis using the different welding procedures. The welders are qualified in the name of the local chapter of the association.

Section IX of the ASME pressure vessel code covers the details of the procedure qualification. It also provides sample specimens of the necessary writeup for a qualified welding procedure and the forms necessary for complying with the specification. The following is a brief discussion of the preparation of an ASME qualified welding procedure. This is abridged and for those required to produce a qualified welding procedure according to ASME it is absolutely necessary to obtain and refer to Section IX "Welding Qualifications." The following will make it easier to understand the code, which is somewhat complex.

The ASME boiler and pressure vessel code holds the manufacturer, contractor, owner, or user responsible for the welding that is done. Since it is impossible for the manufacturer to personally produce or inspect each weld that is made, it is necessary to establish some standard that each welder can follow so that consistent quality can be maintained. The standard which is employed is the "Qualified Welding Procedure."

The theory of a written welding procedure is that there are certain known variables that can affect the quality of a weld and that if those variables are covered in a written procedure specification, anyone who has demonstrated the necessary ability to produce sound welds and who has followed that written procedure would produce a weld of predictable quality. Based on this theory, Section IX of the ASME code, "Welding Qualifications, states that each manufacturer shall record in detail the procedure that is followed in construction of weldments built in accordance with the code." Suggested forms showing the information required are given in the appendix to Section IX. This procedure is then tested to determine if it is "qualified" to produce the quality welds required.

To qualify a procedure, the code requires that the welder prepare test specimens following the written welding procedure in all details. These test specimens are subjected to two types of tests for groove welds and one for fillet welds. Qualification used in groove welds also qualifies for fillet welds, but not vice versa. Tension tests are used to measure the tensile strength of groove weld joints, and guided bend tests are used to check the soundness and ductility. The geometry and number of test specimens required and the criteria by which they are judged acceptable or unacceptable are clearly indicated in the text and figures in Section IX.

The written procedure should be titled and should be given an identifying number and should be dated. The procedure should identify the materials to be welded, preferably by ASME or ASTM designation. The chemistry and physical properties can be used but this should be done only when special material permitted by a code case is used. The filler materials are identified as conforming to the AWS classification number. The position of the work during welding must be designated. A single procedure can cover more than one position but if that is done, it is necessary to spell out the technique for each position. The electrical characteristics of the welding circuit must be designated by making a statement as to whether AC or DC will be used and, in the case of DC, whether the electrode will be positive or negative.

Joint welding procedure or welding technique should be clearly shown by a drawing or may be shown on the same drawing with the joint preparation. These drawings should show the number of passes or layers and whether they are weaving or stringer beads. They should show electrode sizes for each pass and the mean voltage and current for each size of electrodes. In the case of vertical welds, progression upward or downward should be indicated. "Mean" is intended to refer to a single specified average of the specified upper and lower limits, neither of which should be more than 15% above or below the average.

ABC PIPE FABRICATORS CO., ANYTOWN, U.S.A.

PROCEDURE SPECIFICATION FOR SEMIAUTOMATIC GAS METAL-ARC WELDING OF 381-Y-52 PIPE PER SECTION IX ASME "WELDING QUALIFICATIONS"	SPECIFICATION NO. 1 DATE 8-11-75 SHEET 1 of 5

WELDING PROCESS. The welding shall be done by the gas metal-arc welding process using semiautomatic equipment. The welding shall be done using a consumable electrode of steel.

BASE METAL. The base metal shall conform to the following chemical composition: Carbon 0.32 max., Manganese 1.32 max., Phosphorus 0.05 max., Sulphur 0.05 max.

BASE METAL THICKNESS. This procedure will allow welding of 381-Y-52 steel pipe from 3/16" through 1-1/8" thickness.

FILLER METAL. No code number has been assigned. The chemical composition of the filler metal shall be within the following limits: 0.11/0.19% Carbon, 0.95/2.10% Manganese, 0.45/0.85% Silicon, 0.025% max. Phosphorus, 0.035% max. Sulphur. 0.15% max., Nickel, 0.60% max. Molybdenum.

GAS FOR SHIELDING ARC WELDING. Welding grade carbon dioxide gas (dew point −40° F or better) shall be used for gas shielding.

POSITION. The welding shall be done in all positions.

BACKING STRIP. The welded joints shall not utilize a backing strip.

PREHEATING AND TEMPERATURE CONTROL. The joint shall be preheated to 100° F before welding.

POSTHEATING. No stress relief is required by this specification.

PREPARATION OF BASE METAL. The edges or surfaces of the parts to be joined by welding may be prepared by flame cutting or machining as shown on the attached sketches and must be cleaned of oil or grease and excessive amounts of moisture, scale, rust, or other foreign material.

NATURE OF THE ELECTRIC CURRENT. Direct current reverse polarity (electrode positive).

WELDING TECHNIQUE. The welding techniques, such as filler wire sizes, and mean voltages and currents for each electrode, shall be substantially as shown on the attached sketches (Fig. No. 1, 2, & 3).

APPEARANCE OF WELDING LAYERS. The welding current and manner of depositing the weld metal shall be such that there shall be practically no undercutting of the side walls of the welding groove or the adjoining base metal.

CLEANING. None required.

DEFECTS. Any cracks or holes that appear on the surface of weld beads shall be removed by chipping, grinding, or gouging before depositing the next successive bead.

PEENING. None required.

TREATMENT OF THE UNDERSIDE OF WELDING GROOVE. None required.

All tacks shall be ground to a feather edge. All starts and stops shall be ground to sound metal.

ABC PIPE FABRICATOR COMPANY

APPROVED BY _____   PREPARED BY _____

DATE _____   DATE _____

FIGURE 17-2 *ASME qualified procedure.*

Next, a statement regarding cleaning should be given. The cleaning of the base material is usually covered in the paragraph on preparation of base material, so this paragraph need only mention the extent and manner of cleaning each deposited weld bead or layer. A statement of the treatment of defects should also be included.

Peening should not be mentioned at all in a welding procedure specification except in those rare cases in which the desired result cannot be achieved by any other means, and then only in such a manner as to clearly describe the details of the peening operation. Treatment of the underside of the welding groove is intended to indicate whether a backing strip will be used and, if not, whether the underside will be back-gouged or not.

Preheating and temperature control should be mentioned only if they are to be used. Be specific regarding base material and clearly indicate the extent to which preheating and temperature control are necessary. The same comments apply to heat treatment, which includes stress relief.

Other variables will have to be covered when other processes are used. For example, in submerged arc welding, flux and rate of travel must be specified. For gas tungsten arc welding, the size of the tungsten electrode, the kind and rate of flow of shielding gas, and in the case of all automatic welding, the rate of travel, should be indicated. The basic principle that should be kept in mind is that any variable which would affect the quality or characteristics of the finished weld should be spelled out sufficiently to make the intent clear and still allow sufficient leeway for practical application.

An example of an ASME qualified welding procedure is shown by Figures 17-2, 17-3, 17-4, 17-5, 17-6. These show the writeup, the weld joint drawings and weld schedule, and the test results. The ASME forms change with different editions of the code. The information required remains the same as shown on these illustrations.

The welder who prepares the procedure qualification test specimens that pass the code requirements is personally qualified under the code. This means that the specific welder is not required to take the performance qualification test.

In revising or altering welding procedure specifications, it is necessary to refer to the code to learn the essential variables and what changes must be made or what changes are allowed to be made when requalifying the procedure. In general, the code must be requalified if there is a change from one base metal type grade to another, if there is a change in the filler metal used, if there is a change of 100 °F or more in preheat temperatures, if there is a change in type of welding current, and, of course, if there is a change in welding process. These are completely spelled out in the code and must be strictly followed.

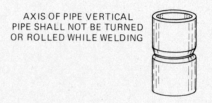

ABC PIPE FABRICATORS CO., ANYTOWN, U.S.A.

SPECIFICATION NO. 1
DATE   8-11-75
SHEET   2 of 5

AXIS OF PIPE VERTICAL
PIPE SHALL NOT BE TURNED
OR ROLLED WHILE WELDING

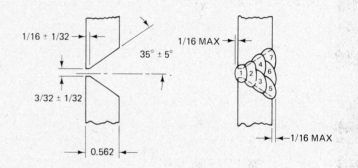

BEAD	WIRE SIZE IN.	AMPS DCRP	VOLTS	GAS FLOW CFH	PRE-HEAT TEMPERATURE
1	0.035	160	22	20	100°F
2	0.035	160	22	20	
3	0.035	160	22	20	
4	0.035	160	22	20	
5	0.035	160	22	20	
6	0.035	160	22	20	
7	0.035	160	22	20	

NOTE: GAS FLOW RATE MAY BE INCREASED 50% TO 100% WHEN WELDING OUTSIDE.

FIGURE 17-3 *ASME qualified procedure (cont'd).*

ABC PIPE FABRICATORS CO., ANYTOWN, U.S.A.

RECORD OF WELDING PROCEDURE QUALIFICATION TESTS

Welding Process   Gas Metal-Arc   Method   Semi-automatic
Material Specification: 381-Y-52   To   381-Y-52
of Material Group "P" Number   No code number   to "P" Number   No code number
Thickness (if pipe, diameter and wall thickness)   24" dia. Schedule 30 (0.562 wall) pipe
Thickness Range this test qualifies   3/16" through 1-1/8"
Filler Metal Group No. F-   No code number assigned
Weld Metal Analysis No. A-   See filler metal composition   Flux or Atmosphere
Describe Filler Metal if not included in Table Q-11.2 or QN-11.2   Flux Trade Name or Composition   None
    (See page 2).   Inert Gas Composition   Welding grade CO_2
                                                                Trade Name CO_2 Flow Rate   20 c.f.h.
                    Welding Procedure   Is Backing Strip Used?   None required
Single or Multiple Pass   Multiple   Preheat Temperature Range   Preheat 100°F
Single or Multiple Arc   Single   Postheat Treatment   None required
Position of Groove   Horizontal 2G   Welding Direction   Not applicable

                                      For Information Only

Filler Wire—Diameter   0.035 diameter   Welding Techniques
Trade Name   Hobart MIG 18   Joint Dimensions Accord with   Figure 1
Type of Backing   None required   amps   160   volts   22   inches per min.   Avg. 11 IPM
Forehand or Backhand   Not applicable

Reduced Section Tensile Test (Fig. Q-6 and QN-6)

| Specimen No. | Dimensions | | Area | Ultimate Total Load, lbs | Ultimate Unit Stress, psi | Character of Failure and Location |
	Width	Thickness				
2G1	0.752	0.377	0.283	26,100	92,500	Fractured in base metal
2G2	0.754	0.377	0.282	25,400	89,500	Fractured in base metal

Guided Bend Tests (Figs. Q-7.1, Q-7.2, QN-7.1, QN-7.2, QN-7.3)

Type and Figure No.	Result	Type and Figure No.	Result
Q-7.1 Side Bend	No defects	Q-7.1 Side Bend	No defects
Q-7.1 Side Bend	No defects	Q-7.1 Side Bend	1 minor defect

Welder's Name   Clock No.   3506   Stamp No.
Who by virtue of these tests meets welder performance requirements.

Test Conducted by   Hobart Procedure Laboratory   Laboratory—Test No.   T-376

                per

We certify that the statements in this record are correct and that the test welds were prepared, welded and tests in accorance with the

requirements of   Section IX of the ASME Code

                                              ABC PIPE FABRICATORS

                                      Signed   _____

                                          By   _____

FIGURE 17-4   *ASME qualified procedure.*

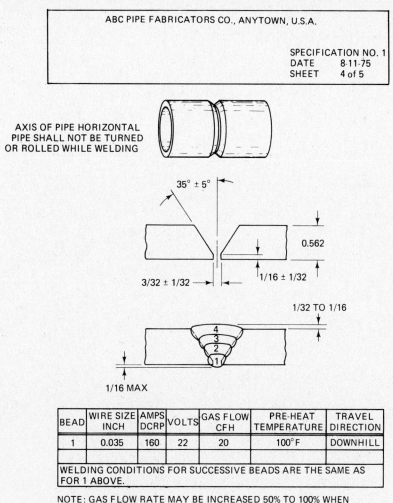

ABC PIPE FABRICATORS CO., ANYTOWN, U.S.A.

SPECIFICATION NO. 1
DATE   8-11-75
SHEET   4 of 5

AXIS OF PIPE HORIZONTAL
PIPE SHALL NOT BE TURNED
OR ROLLED WHILE WELDING

35° ± 5°

0.562

3/32 ± 1/32

1/16 ± 1/32

1/32 TO 1/16

1/16 MAX

BEAD	WIRE SIZE INCH	AMPS DCRP	VOLTS	GAS FLOW CFH	PRE-HEAT TEMPERATURE	TRAVEL DIRECTION
1	0.035	160	22	20	100°F	DOWNHILL
WELDING CONDITIONS FOR SUCCESSIVE BEADS ARE THE SAME AS FOR 1 ABOVE.						

NOTE: GAS FLOW RATE MAY BE INCREASED 50% TO 100% WHEN
WELDING OUTSIDE.

FIGURE 17-5 *ASME qualified procedure (cont'd).*

The requirements for manual and semiautomatic procedures are somewhat different from those which are applied by machine or fully automatic means. This is identified as done by a welder for manual or semiautomatic welding or by a welding operator if performed by machine or automatic equipment.

Manufacturers or contractors who regularly build or install pressure vessels or pressure piping will usually have an ASME *symbol stamp.* This means that the particular contractor or manufacturer has been approved by the American Society of Mechanical Engineers as an authorized manufacturer or installer of the type of equipment specified. Various stamps are used to mark the installation or the product manufactured. Some of the symbol stamps are:

N, Nuclear vessel.

PP, Pressure piping.

U, Pressure vessel, Division I.

U2, Pressure vessel, Division II, plus others.

To obtain an ASME symbol stamp a manufacturer or contractor must contact the American Society of Mechanical Engineers, Boiler and Pressure Code Committee and make application for the code symbol required. The actual mechanics are quite involved but include obtaining a contract with an authorized inspection agency, normally one of the states or provinces or a casualty insurance company. The American Society of Mechanical Engineers will advise of the various requirements. The requirements include at least the need to write up welding procedures and qualify them and submit them to the inspection agency and possibly also to the National Board of Boiler and Pressure Vessel Inspectors. It is necessary to qualify the welders who will work on the pressure vessel. The ASME will send a survey team to inspect your facilities and review qualified procedures and records. If everything is in order ASME will issue a certificate of authorization and code symbol stamp.

ABC PIPE FABRICATORS, CO. ANYTOWN, U.S.A.	SPECIFICATION: 1 REVISION: DATE: 8-11-75 SHEET 5 OF 5
RECORD OF WELDING PROCEDURE QUALIFICATION TESTS	

Welding Process Gas Metal-Arc — Method Semi-automatic

Material Specification: 381-Y-52 — To 381-Y-52

of Material Group "P" Number No code number — to "P" Number No code number

Thickness (if pipe, diameter and wall thickness) 24" dia. Schedule 30 (0.562" wall) pipe

Thickness Range this test qualifies 3/16" through 1-1/8"

Filler Metal Group No. F- No code number assigned

Weld Metal Analysis No. A- See filler metal composition

Describe Filler Metal if not included in Table Q-11.2 or QN-11.2 (See page 2).

Flux or Atmosphere

Flux Trade Name or Composition None

Inert Gas Composition Welding grade CO_2

Trade Name CO_2 — Flow Rate 20 c.f.h.

Is Backing Strip Used? None required

Preheat Temperature Range Preheat 100°F

Postheat Treatment None required

Welding Direction Root pass vertical down
Remaining pass vertical up

Welding Procedure

Single or Multiple Pass Multiple

Single or Multiple Arc Single

Position of Groove Horizontal Fixed 5G

For Information Only

Filler Wire—Diameter 0.035 diameter

Trade Name Hobart MIG 18

Type of Backing None required

Forehand or Backhand Not applicable

Welding Techniques

Joint Dimensions Accord with Figure 3

amps 160 volts 22 inches per min. 9
120 19 Average 5 IPM

Reduced Section Tensile Test (Fig. Q-6 and QN-6)

Specimen No.	Dimensions: Width	Dimensions: Thickness	Area	Ultimate Total Load, lbs	Ultimate Unit Stress, psi	Character of Failure and Location
5GU2	0.753	0.378	0.284	21,800	77,000	Fractured in weld
5GU4	0.754	0.378	0.284	25,000	88,000	Fractured in base metal

Guided Bend Tests (Figs. Q-7.1, Q-7.2, QN-7.1, QN-7.2, QN-7.3)

Type and Figure No.	Result	Type and Figure No.	Result
Q-7.1 Side Bend	No defects	Q-7.1 Side Bend	No defects
Q-7.1 Side Bend	No defects	Q-7.1 Side Bend	No defects

Welder's Name — Clock No. 3506 — Stamp No.

Who by virtue of these tests meets welder performance requirements.

Test Conducted by Hobart Procedure — Laboratory—Test No. T-376

per

We certify that the statements in this record are correct and that the test welds were prepared, welded and testes in accordance with the requirements of Section IX of the ASME Code

ABC Pipe Fabricators

Signed ______________________

By ______________________

FIGURE 17-6 *ASME qualified procedure (cont'd).*

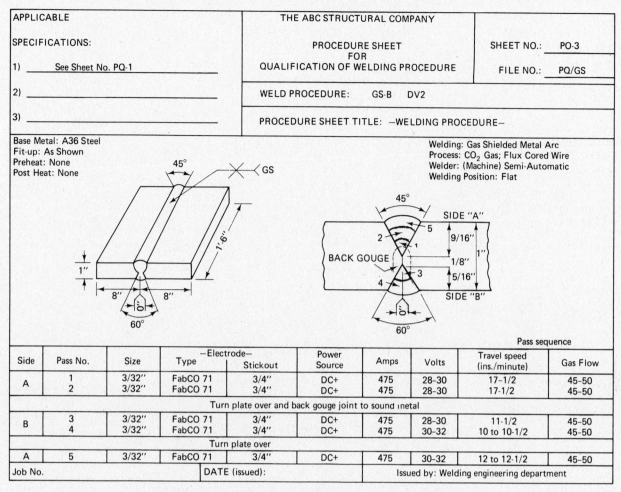

APPLICABLE SPECIFICATIONS:	THE ABC STRUCTURAL COMPANY	
	PROCEDURE SHEET FOR QUALIFICATION OF WELDING PROCEDURE	SHEET NO.: PO-3 FILE NO.: PQ/GS

1) _____See Sheet No. PQ-1_____

2) _____

3) _____

WELD PROCEDURE: GS-B   DV2

PROCEDURE SHEET TITLE: —WELDING PROCEDURE—

Base Metal: A36 Steel
Fit-up: As Shown
Preheat: None
Post Heat: None

Welding: Gas Shielded Metal Arc
Process: CO_2 Gas; Flux Cored Wire
Welder: (Machine) Semi-Automatic
Welding Position: Flat

Pass sequence

Side	Pass No.	Size	Electrode Type	Electrode Stickout	Power Source	Amps	Volts	Travel speed (ins./minute)	Gas Flow
A	1	3/32"	FabCO 71	3/4"	DC+	475	28–30	17-1/2	45–50
	2	3/32"	FabCO 71	3/4"	DC+	475	28–30	17-1/2	45–50
Turn plate over and back gouge joint to sound metal									
B	3	3/32"	FabCO 71	3/4"	DC+	475	28–30	11-1/2	45–50
	4	3/32"	FabCO 71	3/4"	DC+	475	30–32	10 to 10-1/2	45–50
Turn plate over									
A	5	3/32"	FabCO 71	3/4"	DC+	475	30–32	12 to 12-1/2	45–50
Job No.		DATE (issued):			Issued by: Welding engineering department				

FIGURE 17-7 *Welding procedure sheet for AWS prequalified joint.*

Structural Welding: Requirements for the AWS Structural Welding Code are not quite as involved as the ASME Pressure Vessel Code. In fact, manual shielded metal arc, submerged arc, gas metal arc, and flux-cored arc welding procedures which conform with provisions outlined are considered prequalified. These processes are approved for use without performing procedure qualification tests. Electroslag and electrogas welding may also be used providing the procedures are performed to specific provisions and stud welding may be used providing it conforms to applicable provisions. These provisions are outlined in detail in the code. Provisions are with regard to details of welded joints. If the joint designs conform with those presented in the code this is one step toward prequalification. The procedure must comply with the workmanship requirements. These requirements of technique are primarily quality control requirements used on new buildings, new bridges, and new tubular structures. Even though testing is not

required the welding procedure specification must be prepared as a written procedure specification and available for inspection by authorized personnel. A sample of this written welding procedure is shown by Figure 17-7.

When the procedure does not fall within the prequalified welding procedures an entire written procedure similar to that required for the pressure vessel and boiler code is necessary. A sample of this type of procedure is shown by Figures 17-8, 17-9, 17-10, 17-11, and 17-12. These procedures are based primarily on the minimum specified yield strength of the base material. It sets forth a schedule showing the variables and the necessity of requalifying when any of these variables are changed beyond that allowed by the specification. This includes such items as electrode type, the size of the electrode, a major change in voltage or current, a major change in the weld groove design, a change in position, etc. The specification itself must be referred to when writing qualified welding procedures.

WELDMENT EVALUATION AND QUALITY CONTROL

ABC STRUCTURAL STEEL ERECTING CO., ANYTOWN, U.S.A.	SPECIFICATION 2 REVISION NO. 3 DATE 11-12-75 SHEET 1 OF 5

PROCEDURE SPECIFICATION FOR SEMI-AUTOMATIC CARBON DIOXIDE (CO_2)
GAS METAL-ARC WELDING OF CARBON STEEL PER AWS STANDARDS*

WELDING PROCESS This welding shall be done by the gas metal-arc welding process using semi-automatic equipment and a consumable electrode of steel.

BASE METAL This procedure will allow welding of carbon steel having a yield strength of not over 60,000 p.s.i.

BASE METAL THICKNESS This procedure will allow welding of carbon steel through 3/4'' thickness (Minimum of 20-gauge recommended).

FILLER METAL No code number has been assigned. The chemical composition of the filler metal shall be within the following limits: 0.11/0.19% Carbon, 0.95/2.10% Manganese, 0.45/0.85% Silicon, 0.025% max. Phosphorus, 0.35% max. Sulphur, 0.15% max. Nickel, 0.60% max. Molybdenum (Hobart MIG-25).

GAS FOR SHIELDING ARC WELDING Welding grade carbon dioxide (CO_2 with a dew point of $-40°$ F or better) shall be used for gas shielding.

POSITION The welding shall be done in all positions.

BACKING STRIP The welded joints shall not utilize a backing strip.

PREHEATING AND TEMPERATURE CONTROL No preheating is required by this specification.

POSTHEATING No stress relief is required by this specification.

PREPARATION OF BASE METAL The edges or surfaces of the parts to be joined by welding shall be prepared by flame cutting or machining as shown on the attached sketches, (Figs. 1 & 2) and must be cleaned of oil or grease and excessive amounts of moisture, scale, rust, or other foreign material.

NATURE OF THE ELECTRIC CURRENT Direct current, reverse polarity (electrode positive).

WELDING TECHNIQUE The welding techniques, such as filler wire sizes, and mean voltages and currents for each electrode, shall be substantially as shown on the attached sketches (Figs. 1 & 2).

APPEARANCE OF WELDING LAYERS The welding current and manner of depositing the weld metal shall be such that there shall be practically no undercutting of the side walls of the welding groove or the adjoining base metal.

DEFECTS Any cracks or holes that appear on the surface of weld beads shall be removed by chipping, grinding, or gouging before depositing the next successive bead.

APPROVED BY: _____   PREPARED BY: _____

DATE: _____   DATE: _____

*Applicable Standards

FIGURE 17-8 *AWS qualified procedure.*

ABC STRUCTURAL STEEL ERECTING CO., ANYTOWN, U.S.A.

SPECIFICATION   2
REVISION   NO. 3
DATE   11-12-75
SHEET   2 OF 5

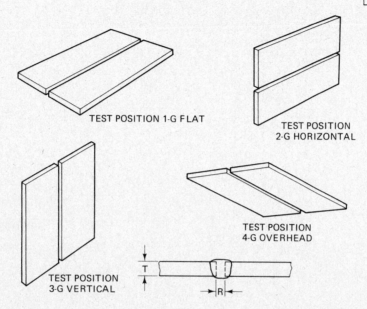

TEST POSITION 1-G FLAT

TEST POSITION
2-G HORIZONTAL

TEST POSITION
4-G OVERHEAD

TEST POSITION
3-G VERTICAL

MATERIAL THICKNESS (T) INCH	ROOT OPENING (R) INCH	SIZE OF WIRE INCH	AMPS D.C.R.P.	VOLTS	TRAVEL SPEED I.P.M.	WIRE FEED SPEED I.P.M.	GAS FLOW C.F.H.
20 GA.	0	0.035	50	18	20	90	12–15
18 GA.	0	0.035	70	20	20	105	12–15
16 GA.	0	0.035	80	20	30	125	12–15
1/8	1/16	0.035	120	21	20	200	12–15

FIG. 1A—JOINT DESIGN AND WELDING CONDITIONS FOR 20GA. TO 1/8IN. MILD STEEL FOR ALL POSITIONS. WIRE STICKOUT (TIP-TO-WORK DISTANCE) SHOULD BE ABOUT 1/4 IN.

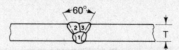

MATERIAL THICKNESS INCH	SIZE OF WIRE INCH	AMPS* D.C.R.P.	VOLTS	TRAVEL SPEED I.P.M.	WIRE FEED SPEED I.P.M.	GAS FLOW C.F.H.	NO. OF PASSES
3/16	.0.035	140	23	15	240	12–15	1
1/4	0.035	120	21	20	200	12–15	3
1/2	0.035	140	23	15	240	12–15	6
3/4	0.035	140	23	15	200	12–15	10

FIG. 1B—JOINT DESIGN AND WELDING CONDITIONS FOR 3/16IN. TO 3/4 IN. MILD STEEL, FOR ALL POSITIONS. WIRE STICKOUT (TIP-TO-WORK DISTANCE) SHOULD BE ABOUT 1/4 IN.

*NOTE: THESE CONDITIONS ARE FOR ALL POSITIONS, BUT HIGHER CURRENT CAN BE USED IN THE FLAT POSITION.

FIGURE 17-9   *AWS qualified procedure (cont'd).*

RECORD OF WELDING PROCEDURE QUALIFICATION TESTS
(GROOVE WELDS)

Specification No. 83.0-41T   Date 5-11-72   Sheet 3   Of 6
Welding Process Gas Metal-Arc   Manual ( ); Semi-Automatic (X); Automatic ( ).
Material Specification A-7   To A-7   Of P-No. 1   To P-No. 1
Thickness (If Pipe, Diameter and Wall Thickness) 3/8" Plate
Thickness Range This Test Qualifies All Thickness Up To 3/4 in. (20 Gage Minimum Recommended)

Reduced Section Tensile Test Table 1

| Specimen No.* | Dimensions | | Area | Ultimate Total Load Lbs | Ultimate Unit Stress psi | Location of Failure |
	Width in.	Thickness in.				
F-1	1.499	0.378	0.578	37,800	66,800	Base Metal
H-8	1.504	0.378	0.569	37,600	66,100	Base Metal
H-1	1.503	0.380	0.572	38,400	67,200	Base Metal
H-8	1.498	0.377	0.565	38,900	68,900	Base Metal
V-1	1.498	0.377	0.565	37,600	66,600	Base Metal
V-8	1.502	0.381	0.571	37,700	66,000	Base Metal
O-1	1.445	0.327	0.472	34,600	73,400	Base Metal
O-8	1.430	0.327	0.468	34,500	73,700	Base Metal

Free Bend Test Table 2

Specimen No.*	Initial Gauge Length in.	Final Gauge Length in.	Elongation in.	Elongation %	Description of Defects
F-3	5/8	55/64	15/64	37.5%	No Defects
F-6	11/16	1-3/64	23/64	52.3%	Tore At Edge
H-3	1/2	1-1/32	17/32	106.3%	Tore Full Width
H-6	17/32	1-1/16	17/32	100.0%	No Defects
V-3	5/8	61/64	21/64	52.5%	No Defects
V-6	11/16	1-1/64	21/64	47.8%	No Defects
O-3	21/32	31/32	5/16	47.5%	Tear, Center Punch Mark
O-6	5/8	59/64	19/64	47.5%	No Defects

Guided Bend Tests Table 3

Specimen No.*	Result	Specimen No.	Result
F-2	No Defect	V-2	No Defect
F-4	No Defect	V-4	No Defect
F-5	2 Minor Defects	V-5	No Defect
F-7	No Defect	V-7	No Defect
H-2	No Defect	O-2	2 Minor Defects
H-4	No Defect	O-4	No Defect
H-5	No Defect	O-5	1 Minor Defect
H-7	No Defect	O-7	No Defect

*Note: F means flat (1-G); H means horizontal (2-G); V means vertical (3-G); O means overhead (4-G).

Weldor's Name _____   Clock No. _____   Stamp No. R.D.F.

Who By Virtue Of These Tests Meets Weldor Performance Requirements.   Lab-Test No. 870

Test Conducted By _____   Per _____

We certify that the statements in this record are correct and that the test welds were prepared, welded, and tested in accordance with the requirements of American Welding Society standard qualification procedure B3.0-41T.

Date _____   Revised to Correct   Signed _____
                               11-12-75
                                                           By _____

FIGURE 17-10 *AWS qualified procedure (cont'd).*

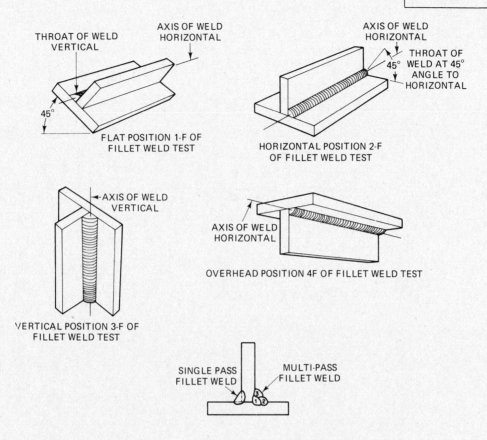

FILLET SIZE	SIZE OF WIRE IN.	AMPS.* D.C.R.P.	VOLTS	TRAVEL SPEED I.P.M.	WIRE FEED SPEED I.P.M.	GAS FLOW C.F.H,	NUMBER OF PASSES
1/8	0.035	140	20	30	235	12–15	1
1/4	0.035	180	21	20	330	12–15	1
3/8	0.035	180	21	20	330	12–15	3
1/2	0.035	180	21	15	330	12–15	3

FIG. 2—JOINT DESIGN AND WELDING CONDITIONS FOR ALL POSITION FILLET WELDS. WIRE STICKOUT (TIP-TO-WORK DISTANCE) SHOULD BE ABOUT 1/4 IN.

*NOTE: THESE CONDITIONS ARE FOR ALL POSITION WELDING, BUT HIGHER CURRENTS SHOULD BE USED FOR FILLETS IN THE FLAT AND HORIZONTAL POSITIONS.

FIGURE 17-11 *AWS qualified procedure (cont'd).*

579

ABC STRUCTURAL STEEL ERECTING CO., ANYTOWN, U.S.A.

RECORD OF WELDING PROCEDURE QUALIFICATION TESTS
(FILLET WELDS)

SPECIFICATION   2
REVISION   NO. 3
DATE   11-12-75
SHEET   5 OF 5

Specification No.   B3.0-41T   Date   5-11-72   Sheet   5   Of   6
Welding Process   Gas Metal-Arc   Manual ( ); Semi-Automatic (X); Automatic ( ).
Material Specification   A-7   To A-7   Of P-No.   1   To P-No.   1
Fillet Sizes This Test Qualifies   All Sizes Of Fillet Welds

Longitudinal Shear Test Table 4

Specimen and Welding Position*	Fillet		Total Length in.	Total Area Sq. in.	Ultimate Total Load lbs	Ultimate Shear Strength lb/lin in.	Ultimate Shear Stress psi	Character of Failure and Location
	Length in.	Size in.						
F-13	1.330	0.155	5.32	0.584	62,900	11,823	107,972	Weld
F-14	1.490	0.245	5.86	1.015	90,300	15,409	88,966	Weld
H-13	1.390	0.156	5.59	0.872	72,050	12,889	112,273	Weld
H-14	1.405	0.249	5.62	0.992	92,730	16,500	93,325	Weld
V-13	1.433	0.175	5.74	0.710	70,250	12,239	98,940	Weld
V-14	1.457	0.227	5.83	0.935	86,050	14,759	91,784	Weld
O-13	1.380	0.180	5.53	0.705	72,850	13,173	103,479	Weld
O-14	1.450	0.248	5.80	1.016	95,300	16,431	93,891	Weld

Free Bend Test Table 5

Specimen No.*	Initial Gauge Length in.	Final Gauge Length in.	Elongation in.	Elongation %	Description and Size of Defects
F-9	1-3/32	1-33/64	27/64	38.6%	No Defect
F-11	1-1/8	1-7/16	5/16	27.8%	No Defect
H-9	1	1-7/16	7/16	43.8%	Tore One Edge
H-11	1-1/4	1-23/32	15/32	37.5%	No Defect
V-9	1	1-29/64	29/64	45.3%	No Defect
V-11	1-3/32	1-33/64	27/64	38.6%	No Defect
O-9	1-3/32	1-19/32	1/2	45.7%	No Defect
O-11	1-1/8	1-39/64	31/64	43.0%	No Defect

Soundness Test Table 6

Specimen No.	Result	Specification No.	Result
F-10	No Defect	V-10	No Defect
F-12	No Defect	V-12	No Defect
H-10	5 Minor Defects	O-10	No Defect
H-12	1 Minor Defect	O-12	1 Minor Defect

*Note: F means flat (1-F); H means horizontal (2-F); V means vertical (3-F); O means overhead (4-F).

Weldor's Name   Clock No.   Stamp No.   R.D.F.

Who By Virtue Of These Tests Meets Weldor Performance Requirements.

Test Conducted By   _____   Laboratory—Test No.   870

              Per   _____

We certify that the statements in this record are correct and that the test welds were prepared, welded, and tested in accordance with the requirements of American Welding Society standard qualification procedure B3.0-41T.

                    Revised to Correct   Signed   _____
Date   _____
                    11-12-75   By   _____

FIGURE 17-12 *AWS qualified procedure (cont'd).*

Cross-country Pipeline: The use of the API standard 1104 for welding pipelines and related facilities includes the procedure qualification. The procedure specification includes the process, the base metal material, the size of pipe, diameter and wall thickness, the joint detail, the filler metal type and number of passes, and the electrical characteristics utilized. For gas welding and other processes, the flame characteristics, direction of welding, welding position, type of flux, type of shielding gas, etc., must be made known. In the event that the essential variables are changed, the welding procedure must be re-established and completely requalified when there is a change in welding process, a change in the pipe material, a change in the joint design, a change in the position, a change in pipe size and wall thickness, a change in filler metal, a change in filler metal size, etc. Some of these changes are described in detail so that the code must be referred to in writing a qualified welding procedure. The sample of an API type qualified welding procedure is given by Figures 17-13, 17-14, and 17-15.

Other codes may reference the three mentioned above and in these cases the same provisions would apply. These codes have been mentioned. It must be re-emphasized that the code in question must be obtained and studied in order to intelligently write a procedure qualification and to qualify it.

ABC CROSS COUNTY PIPELINE CO., ANYTOWN, U.S.A.

SHIELDED METAL-ARC WELDING (E-6010 and E-7010) OF STANDARD LINE PIPE
GRADE X-52 PER API STANDARD 1104 SEVENTH EDITION

SPECIFICATION NO. 3
DATE: 11-30-72
SHEET NO. 1 of 3

A. *PROCESS:* The Shielded Metal-Arc Welding process shall be employed.

B. *PIPE MATERIAL:* API Standard Line pipe grade X-52 shall be used.

C. *DIAMETER AND WALL THICKNESS:* The outside diameter of the pipe shall be over 4-1/2" to 12-3/4" inclusive. The wall thickness of the pipe shall be 3/16" to 3/4" inclusive.

D. *JOINT DESIGN:* A single Vee groove joint design shall be used. The included angle shall be from 60° to 75°. The root face shall be 1/16", plus or minus 1/32". The root opening shall be 1/16", plus or minus 1/32".

E. *FILLER METAL, ELECTRODE AND NUMBER OF BEADS:* The following three electrodes shall be used with this procedure: 1/8" E-6010 (Hobart 10), 1/8" E-7010 (Hobart 885), and 5/32" E-7010 (Hobart 885). The minimum number of beads employed shall be three. The root bead shall be made with 1/8" E-6010 electrodes. The second bead shall be made with 1/8" E-7010 electrodes. The third bead shall be made with 5/32" E-7010 electrodes.

F. *ELECTRICAL CHARACTERISTICS:* Direct current reverse polarity (electrode positive) shall be employed. The voltage shall be between 25 and 28 volts. The current shall be between 110 and 130 amperes.

G. *POSITION:* "Position Welding" shall be employed. This is also known as horizontal fixed position welding.

H. *DIRECTION OF WELDING:* The direction of welding shall be vertical down (downhill welding).

I. *NUMBER OF WELDERS:* The minimum number of root bead welders shall be one. The minimum number of second bead welders shall be one.

J. *TIME LAPSE BETWEEN PASSES:* The time lapse between passes is unlimited.

K. *TYPE OF LINE-UP CLAMP:* No line-up clamp is required by this specification.

L. *REMOVAL OF LINE-UP CLAMP:* Not required.

M. *CLEANING:* All slag or flux remaining on any weld bead shall be removed before laying down the next successive bead.

N. *PREHEAT, STRESS RELIEF:* No preheat of stress relief is required.

PREPARED BY: _____ DATE: _____

APPROVED BY: _____ DATE: _____

FIGURE 17-13 *API qualified procedure.*

ABC CROSS COUNTY PIPELINE CO., ANYTOWN, U.S.A.

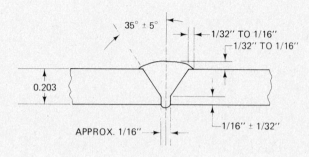

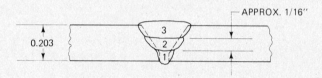

ELECTRODE SIZE AND NUMBER OF BEADS

NUMBER OF BEADS

PIPE WALL THICKNESS	1/8" E-6010 ELECTRODE	1/8" E-7010 ELECTRODE	5/32" E-7010 ELECTRODE	TOTAL NUMBER OF BEADS
0.203	1	1	1	3

NOTE: USE 1/8" E-6010 ELECTRODE ON FIRST PASS ONLY. THE HOT PASS SHALL
BE MADE WITH 1/8" E-7010 ELECTRODE. REMAINING PASS SHALL BE MADE
WITH 5/32" E-7010 ELECTRODE.

VOLTAGE AND AMPERAGE RANGE

ELECTRODE DIAMETER	AMPERAGE	ARC VOLTS
1/8" (E-6010)	110	27
1/8" (E-7010)	110	27
5/32" (E-7010)	130	26

FIGURE NO. 1

TOTAL TIME FOR WELDING JOINT 23 MIN.

NOTE: FOR HEAVIER THICKNESS MORE PASSES SHOULD BE
USED WITH SAME WELDING CONDITIONS AS FOR ABOVE.

FIGURE 17-14   *API qualified procedure (cont'd).*

ABC CROSS COUNTY PIPELINE CO., ANYTOWN, U.S.A.

SPECIFICATION NO. 3
DATE 11-30-72
SHEET 3 of 3

COUPON TEST REPORT

Test No. T242

Location Troy, Ohio — Date 11-30-72
Contractor ABC Cross Country Pipe Line Co. — Sub-contractor
Schedule — Gang — Inspector
Date 11-23-72 — State Ohio — Roll weld No — Fixed position weld Yes
Welder John C. Hickman — Mark 3509
Welding time 23 min. — Time of day 10:00 A.M. — Temperature 70°F.
Weather condition Welding was done indoors.
Wind break used None — voltage 25-28 — amperage 110-130
Make of welding machine Hobart DC Gen. — Size 300 Amp
Brand of Electrode Hobart #10, Hobart #885
Size of reinforcement 1/32 to 1/16
Pipe mfg'r — Kind 5LX Grade X-52
Wall thickness 203 — Dia. O.D. 8" — W./ft. 18.27 — Joint length Nipple

Bead No.	1	2	3	4	5	6	7
Size of Electrode	1/8	1/8	5/32				
No. of Electrode	4	3	4				

Coupon stenciled	1	2	3	4	5	6	7
Original	0.935	1.015					
Dimension: of plate	0.203	0.203					
Orig. area of plate in ^{2}	0.190	0.206					
Maximum load	15,200	16,300					
Tensile S/in. plate area	80,000	79,200					
Fracture location	Base Metal	Base Metal					

☒ Procedure — ☒ Qualifying Test — ☒ Qualified
☐ Welder — ☐ Line Test — ☐ Disqualified

Max. tensile 80,000 — Min. tensile 79,200 — Avg. tensile 79,500

Remarks on tensile

1. Failed in base metal 2" from weld
2. Failed in base metal 1-1/2" from weld
3.
4.

Remarks on Bend Tests

1. Root bend, no defects, passed
2. Root bend, one minor defect, passed
3. Face bend, no defects, passed
4. Face bend, minor defects, passed

Remarks on Nick Tests

1. No defects
2. No defects
3.
4.

Test made at Hobart Technical Center — Date 11-23-72
Tested by — Supervised by

FIGURE 17-15 *API qualified procedure (cont'd).*

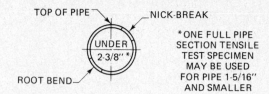

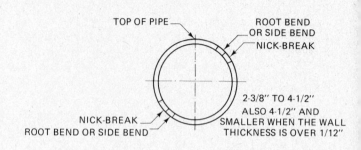

17-4 WELD TESTING—
DESTRUCTIVE TESTING

Welds and weld metal are probably subjected to more different types of tests than any other metal produced. Weld metal can be tested in the same manner as any other form of metal. Mechanical tests are used to qualify welding procedures and welders. Mechanical tests are also used to qualify welding processes and to determine if electrodes and filler metals meet the requirements of the specification covering them. Welds in weldments are often tested for soundness, strength, and toughness by means of mechanical tests. Mechanical tests are used for proving prototype weldments prior to establishing welding procedures.

Mechanical tests are destructive tests since the weldment or weld joint is destroyed in making the test specimen. They are also expensive since they involve the preparation of material, the making of welds, the cutting and often machining of weld test specimens, and finally the mechanical or destructive testing of these specimens.

Procedure Qualification.

To qualify a welding procedure you must make specific welds, cut them into standardized sizes and shapes, and test them to destruction. These tests are spelled out in detail by the specification being followed.

The purpose of the procedure qualification test is to prove that a weld made under prescribed conditions will provide the necessary mechanical properties. This involves making test plates according to the welding procedure and then testing the weld by mechanical means to determine that it has the mechanical properties to meet the acceptance standards of the specification. The welding process, filler metals, and the welding schedule are selected to make the weld in the position required on the base metal that is to be used. Welding joint details and material thickness may be specified and may not be exactly as will be utilized in making the production weldment. Requirements vary from code to code and it is therefore essential that the code be studied when making the test welds.

The procedure qualification welds may be tested by a number of different mechanical tests. The tests to be used and the design of the test specimens are specified by the code. Unfortunately, the different codes have somewhat different requirements and the test specimens are not always exactly the same dimensions, nor are they taken from the same positions of a test plate. In general,

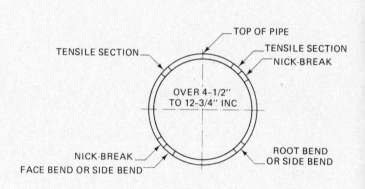

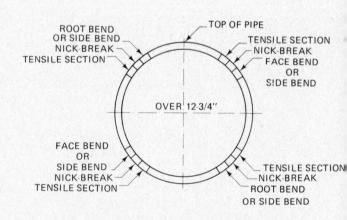

NOTE: AT THE COMPANY'S OPTION, THE LOCATIONS MAY BE ROTATED 45 DEGREES COUNTER-CLOCKWISE OR THEY MAY BE EQUALLY SPACED AROUND THE PIPE EXCEPT SPECIMENS SHALL NOT INCLUDE THE LONGITUDINAL WELD. ALSO, AT THE COMPANY'S OPTION, ADDITIONAL SPECIMENS MAY BE TAKEN.

FIGURE 17-16 *API test specimen locations.*

the codes utilize the same type of weld specimens, including the fillet break test, the nick break test, the tensile test, and the guided bend test. These are by far the most commonly used; however, other different types of specimens may sometimes be required. It is therefore absolutely essential that the code or specification in question be continually checked when making test welds, test specimens, and making the tests of the weld. This is doubly important since both procedure qualification and welder qualification tests are involved.

The American Petroleum Institute standard 1104, "Standard For Welding Pipeline and Related Facilities," is extremely popular and used for much of the field pipe welding. The important thing about this code is that the test specimens be taken from the test weld pipe in the proper location. Figure 17-16 shows the location of test specimens based on the size of pipe. It also indicates the number of test specimens that should be taken. The detail of the nick break test specimen is shown by Figure 17-17. The bend test specimens are shown by Figure 17-18 and 17-19. Note that root and face bend specimens are used for thinner material and the side bend specimen is used when the wall thickness of the pipe is over 3/8 in. (9.5 mm).

The API tests are most often made in field locations and for this reason it is permissible to use flame-cut edges and grinding to prepare specimens. This eliminates all machining operations and makes it possible to quickly prepare test specimens and to make tests in the field with a portable bend test machine. The basis for acceptance of the test welds is spelled out in detail in the code.

The American Society of Mechanical Engineers Boiler and Pressure Vessel Code Section IX, "Welding and Brazing Qualifications," is also very widely used for pressure vessel work and is the reference specification for much of the pressure piping work. This code is more complex than the API standard 1104. This is because it covers many more types of metals and welding processes. This code does make use of the guided bend test and the fillet weld test. The fillet weld performance test is similar to the AWS test which will be detailed later. The ASME also requires reduced section tensile test for certain requirements. The location of test specimens and the details of them are shown in the code. For the guided bend test, special jig dimensions are provided for different thicknesses of test specimens. It is recommended that the code be referred to whenever making any test for this code.

The American Welding Society's "Structural Welding Code," D1.1 is very popular and widely used for qualifying procedures and welders. This code was primarily intended for plate and structural work, but in view of tubular sections now being used in structures, it includes specimens for tubular or pipe welded joints.

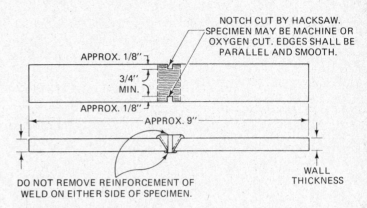

FIGURE 17-17 *API code nick break test specimen (Courtesy, The American Petroleum Institute).*

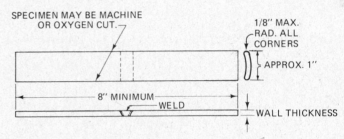

FIGURE 17-18 *API code root and face bend specimen (Courtesy, The American Petroleum Institute).*

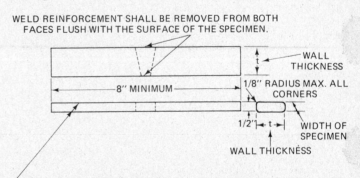

FIGURE 17-19 *API code root side bend specimen (Courtesy, The American Petroleum Institute).*

WELDMENT EVALUATION AND QUALITY CONTROL

One of the tests from the structural code, that is often used for preliminary evaluation, is the fillet weld break specimen used for qualifying weld tackers. This specimen, and the way it is fractured is shown by Figure 17-20. According to AWS code the thickness of the plates shall be 1/2 in. (12.6 mm) and the fillet weld shall be 1/4 in. (6.3 mm) out. Variations of this specimen use thinner plates and smaller fillet sizes. This specimen can be used for each welding position and can be used for all of the arc welding processes and any electrode type.

TEST SPECIMEN DETAIL:

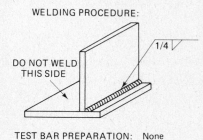

TACK WELD AT EACH END

BILL OF MATERIAL
Item 1 — 1 piece 3/8 x 5 x 6 Mild Steel — ASTM A-37
Item 2 — 1 piece 3/8 x 4 x 6 Mild Steel — ASTM A-37

WELDING PROCEDURE:

DO NOT WELD THIS SIDE

Position	Electrode size (in)	Fillet size (in)
Flat	3/16	1/4
Horizontal	3/16	1/4
Vertical	5/32	1/4 and 3/8
Overhead	5/32	1/4 and 3/8

TEST BAR PREPARATION: None

TEST SPECIFIED

Force

PLACE IN PRESS OR HIT WITH SLEDGE HAMMER UNTIL BROKEN OR FLATTENED.

STANDARD OF ACCEPTABILITY:
(a) Contour—The exposed face of the weld shall be reasonably smooth and regular. There shall be no overlapping or undercutting. The weld shall conform to the required cross section for the size of weld specified per gage.
(b) Extent of Fusion—There shall be complete fusion between the weld and base metal and full penetration to the root of the weld.
(c) Soundness—The weld shall contain no gas pocket, oxide particle or slag inclusion exceeding 3/32 in. in greatest dimension. In addition, no square inch of weld metal area shall contain more than 6 gas pockets exceeding 1/16 inch in greatest dimension.

FIGURE 17-20 *AWS—fillet break test.*

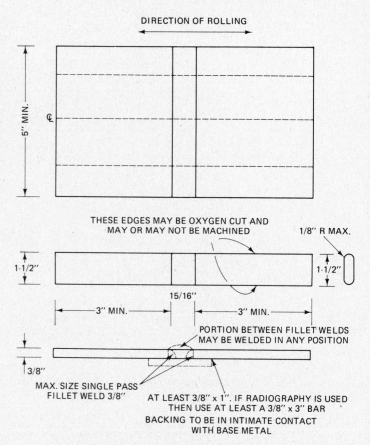

DIRECTION OF ROLLING

5″ MIN.

THESE EDGES MAY BE OXYGEN CUT AND MAY OR MAY NOT BE MACHINED

1/8″ R MAX.

1-1/2″   1-1/2″

15/16″

3″ MIN.   3″ MIN.

PORTION BETWEEN FILLET WELDS MAY BE WELDED IN ANY POSITION

3/8″

MAX. SIZE SINGLE PASS FILLET WELD 3/8″

AT LEAST 3/8″ x 1″. IF RADIOGRAPHY IS USED THEN USE AT LEAST A 3/8″ x 3″ BAR BACKING TO BE IN INTIMATE CONTACT WITH BASE METAL

Weld reinforcement and backing shall be removed flush with base metal, flame cutting may be used for the removal of the major part of the backing, provided at least 1/8″ of its thickness is left to be removed by machining or grinding

FIGURE 17-21 *AWS—fillet weld root bend test.*

The structural code includes another fillet type test that has been used for many years for welder qualification. In this test, fillet welds are made between two plates and a backing bar. The plates are separated 15/16 in. (24 mm), and the fillet welds are made between each plate and the backing bar. The remaining area between the fillet welds is filled in like a groove weld. The difficulty with this test is that the backing bar must be removed for testing and this requires machining. This specimen can be tested with the root in tension, or the face in tension. This test can be very critical when it is tested with the root in tension, since this shows root penetration of the fillets. The details of this test are shown by Figure 17-21. The examination of this test and the requirements are listed in the code. For welder qualification, the code also provides for groove weld test specimens. The weld joint detail is a single vee groove weld with a 45° included angle and a 1/4-in. (6.3-mm) root opening. A backup bar is used. The specimen can be welded in different positions. However, when it is used in the horizontal position it is usually a single bevel weld

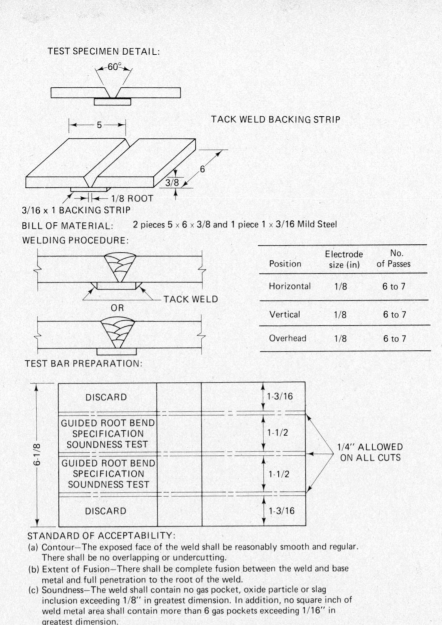

TEST SPECIMEN DETAIL:

60°

TACK WELD BACKING STRIP

5

6

3/8

1/8 ROOT

3/16 x 1 BACKING STRIP

BILL OF MATERIAL: 2 pieces 5 x 6 x 3/8 and 1 piece 1 x 3/16 Mild Steel

WELDING PROCEDURE:

TACK WELD

OR

Position	Electrode size (in)	No. of Passes
Horizontal	1/8	6 to 7
Vertical	1/8	6 to 7
Overhead	1/8	6 to 7

TEST BAR PREPARATION:

DISCARD			1-3/16
GUIDED ROOT BEND SPECIFICATION SOUNDNESS TEST			1-1/2
GUIDED ROOT BEND SPECIFICATION SOUNDNESS TEST			1-1/2
DISCARD			1-3/16

6-1/8

1/4" ALLOWED ON ALL CUTS

STANDARD OF ACCEPTABILITY:
(a) Contour—The exposed face of the weld shall be reasonably smooth and regular. There shall be no overlapping or undercutting.
(b) Extent of Fusion—There shall be complete fusion between the weld and base metal and full penetration to the root of the weld.
(c) Soundness—The weld shall contain no gas pocket, oxide particle or slag inclusion exceeding 1/8" in greatest dimension. In addition, no square inch of weld metal area shall contain more than 6 gas pockets exceeding 1/16" in greatest dimension.

FIGURE 17-22 *AWS—groove weld.*

with the bottom horizontal and the 45° bevel on the top piece. These specimens are shown by Figure 17-22. These specimens are then cut and given a guided bend test. For plate heavier than 3/8 in. (9.5 mm) the side bend specimen is required. Similar test specimens are required for pipe and tubing welded for structural applications. Since different strength level materials have different bending characteristics, the design detail of the guided bend fixture is altered. The details of the guided bend fixture are shown by Figure 17-23. Testing a specimen in a guided bend fixture is shown by Figure 17-24.

Other destructive type tests are required for other reasons. For example, to check the compliance of deposited weld metal a special joint design and *all-weld*

metal test specimen is required. This joint design and test specimen are specified in the various different filler metal specifications of the American Welding Society. Figure 17-25 shows the joint detail for making *all-weld* metal test specimens. In many cases impact properties are also specified on a particular filler metal. When this is so, impact test specimens must also be made and they are made with the same joint detail as shown in Figure 17-25. The detail of the *all-weld* metal test specimen Type 505 is shown by Figure 17-26 and the detail of the Charpy vee notch impact test specimen is shown by Figure 17-27. These test specimens are universally used and the detail dimensions of them are identical in most specifications.

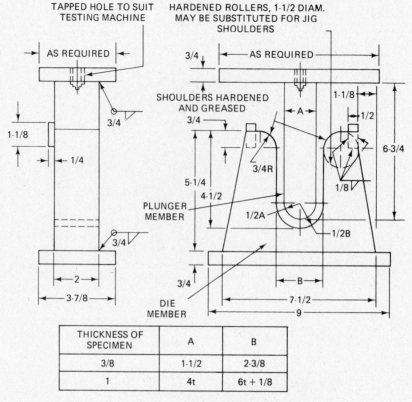

THICKNESS OF SPECIMEN	A	B
3/8	1-1/2	2-3/8
1	4t	6t + 1/8

ALL DIMENSIONS ARE IN INCHES.

NOTES:

1. The ram shall be fitted with an appropriate base and provision for attachment to the testing machine. The ram shall also be designed to minimize deflection and misalignment.

2. The specimen shall be forced into the die by applying the load on the plunger until the curvature of the specimen is such that a 1/8 in. (3.2 mm) diam. wire cannot be placed between the specimen and any point in the curvature of the plunger member of the jig.

FIGURE 17-23   *AWS—guided bend test jig.*

FIGURE 17-24   *Making a guided bend test.*

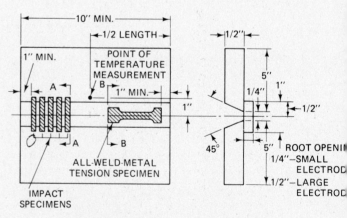

FIGURE 17-25   *AWS all weld make test.*

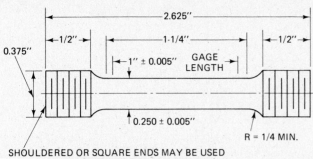

FIGURE 17-26 *Tensile test specimen—505.*

SHOULDERED OR SQUARE ENDS MAY BE USED
IF DESIRED; DIMENSIONS SHOWN ARE FOR
THREADED ENDS.

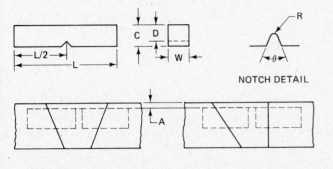

NOTCH DETAIL

Dimension	in.	mm
A—Distance of sample below specimen	0.06 ± 0.002	1.6 + 0.05 − 0
L—Length of specimen	2.165 ± 0.002	55.0 + 0, −2.5
/2—Location of notch	1.082 ± 0.002	27.5 ± 1.0
C—Cross section (depth)	0.394 ± 0.001	10.000 ± 0.025
W—Cross section (width)	0.394 ± 0.001	10.000 ± 0.025
D—Bottom of notch to base	0.315 ± 0.001	8.000 ± 0.025
R—Radius of notch	0.010 ± 0.001	0.250 ± 0.025
θ—Angle of notch 45 deg ± 1 deg	Adjacent sides shall be 90 deg ± 10 minutes	

FIGURE 17-27 *Impact test specimen—charpy vee notch.*

FIGURE 17-28 *Transverse fillet weld test.*

t = SPECIFIED SIZE OF FILLET WELD + 1/8

ALL DIMENSIONS ARE IN INCHES.

NOTE: SEE SPECIFICATION FOR ADDITIONAL DETAILS.

There are several other different weld test specimens that may be used. This includes such specimens as the transverse fillet weld specimen, which is shown by Figure 17-28. Another similar specimen is the longitudinal fillet weld shear specimen which is shown in Figure 17-29. Many other weld specimens are used for development and research work. However, these types of specimens are beyond the scope of this book. For further information on these, please refer to the AWS "Standard Method of Mechanical Testing of Welds." [28] This document shows the detail dimensions of many weld specimens in addition to those previously mentioned.

FIGURE 17-29 *Longitudinal fillet weld test.*

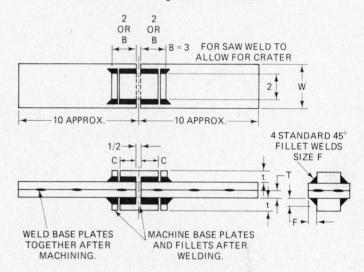

WELD BASE PLATES TOGETHER AFTER MACHINING.

MACHINE BASE PLATES AND FILLETS AFTER WELDING.

ALL DIMENSIONS ARE IN INCHES.

DIMENSION	IN.	MM
SIZE OF WELD, F	1/8	3.2
	1/4	6.4
	3/8	9.5
	1/2	12.7
THICKNESS, t, MIN.	3/8	9.5
	1/2	12.7
	3/4	19.1
	1	25.4
THICKNESS, T, MIN.	1/4	6.4
	3/8	9.5
	1/2	12.7
	5/8	15.9
WIDTH, W	3	76.2
	3	76.2
	3	76.2
	3-1/2*	88.9
FILLET LENGTH, C	1-1/2	38.1
	1-1/2	38.1
	1-1/2	38.1
	1	25.4

NOTE: SEE SPECIFICATION FOR ADDITIONAL DETAILS.

17-5 VISUAL INSPECTION

Visual inspection is a nondestructive testing technique or evaluation method. It is by far the most popular and most widely used of the nondestructive testing techniques. It is used particularly for noncritical weldment inspection and is extremely effective. It is the least expensive of the various inspection methods, requires the least investment in tools and equipment and usually requires less time. The welding inspector can utilize visual inspection throughout the entire production cycle of a weldment. It is an effective quality control method which will ensure procedure conformity and will also catch errors at early stages. In the long run this will allow for the production of higher quality weldments at less cost. The work of the welding inspector utilizing visual inspection methods can be subdivided into three main divisions: (1) Visual inspection prior to welding, (2) visual inspection during welding operations, and (3) visual inspection of the finished weldment.

Visual Inspection Prior to Welding.

There are a great many items that must be reviewed and checked prior to welding of the weldment. These include the following:

1. Review all applicable drawings, specifications, procedures, welder qualifications, etc. This helps the inspector to become familiar with the job and all specifications, etc., that apply to it.

2. Review the material specifications of the parts comprising the weldment and determine that the materials are according to specifications.

3. Compare the edge preparation of each joint with the drawings. At the same time check edge preparation for surface condition.

4. Check the dimensions of each item since they will effect weldment fitup.

5. At the fitup operation check assembly dimensions and fitup with special emphasis on root openings of the weld joints.

6. At the fitup operation check the backing system, bars, rings, copper, flux, etc., to be sure that they are in accordance with requirements.

7. Last, at the fitup operation, check the cleanliness of the welding joints and the conditions of tack welds, if used.

At the fitup and tack weld operations, many weldments are completely fitted and ready for production welding. In other cases, however, certain welds may later be hidden and these welds must be completed before fitup is finished. It is recommended good practice for the fitter to mark in chalk the weld symbols showing weld sizes for all welds to be made by the production welders. On high-volume production work this may not be necessary, especially if samples of production parts are available for reference by production welders.

At the fitup station, the welding inspector should check the tack welds to determine that the correct electrode types are being used for the base metal that is being welded. Also, to see if any special precautions such as preheat, etc., are required. If preheating is specified local preheating may be the answer at this point of production.

Visual Inspection During the Welding Operation

At the point of production welding and before welding begins, there are several factors that should be checked by the welding inspector. These relate to the welding procedures employed. Make sure that they are in order, applicable to the weldment in question, and available to the people doing the welding. Various items that must be checked are the following:

1. Determine that the designated welding process and method of application to be employed are in accordance with procedures.

2. Determine that the designated electrodes or filler metal, proper for base metals to be welded, will be employed. In some cases, it is necessary to determine the storage factors, the condition of electrodes, and, for critical work, to record the heat numbers of the electrodes utilized in specific joints or weldments.

3. The welding inspector should make a survey of the welding equipment to make sure that it is in good operating condition. This should also include clamping devices, fixtures, locating devices, etc.

4. The welding inspector should determine that the correct type welding current and the proper polarity are being used.

5. It is necessary to determine that preheat requirements are adhered to at the time of welding. This involves the checking of temperatures of base metal and determining that base metal temperatures are heat-soaking temperatures instead of merely surface heat. The time of preheating can help establish whether through-heating is accomplished. Preheat temperatures can be

checked by the use of temperature-indicating devices. The same equipment can be used for checking interpass temperatures.

6. The inspector should identify all welders assigned to the particular weldment or job or joint in question. Their qualification level must be in accordance with the requirements of the job. Qualification papers should be reviewed to determine that they are in order and have not expired.

7. The welding inspector should actually observe welders making welds. This has a rather startling effect on welders, especially when they know that their welds are being watched as they are being made. If a welder does not appear to have the necessary skill for the job in question, the inspector can, in consultation with the welder's supervisor, request that the welder make requalification tests. This requirement is not always in all codes but is common practice for high-quality work.

8. The inspector should determine that interpass temperatures are being maintained during the welding operations and, if welding operations are discontinued for a period of time, that the interpass temperatures be obtained before welding is resumed.

9. The inspector should determine also that interpass cleaning by chipping, grinding, gouging, etc., is being done in accordance with the requirements of the procedure or specification or in accordance with good practice.

With the welding inspector on the site during welding operations, it is possible for any unusual activities or disassembly or repair to be noticed. This type of work is often required but should have special attention and supervision to determine that the quality requirements are maintained. In many situations repair work, at least on certain types of existing structures such as pressure vessels, must be described and approved prior to doing the work.

1. The inspector must document all repair welding, why it was required, the extent of the work, and how it was done. This should be recorded in an inspection notebook or on applicable report forms, whichever is required.

2. The inspector should determine that any type of postheat treatment performed is done in accordance with the procedure or other requirements.

3. Finally, the inspector should check any warpage corrective activities such as press work, thermal bending, etc., that might be employed. It is also necessary that these types of activities be recorded in the inspection notebook.

Visual Inspection After Welding on the Completed Weldment.

The inspector is expected to determine that the weldment conforms to the drawings and specifications for which it is designed and constructed. This includes many factors with respect to the weldment but more importantly to the welds. The welds must all be made to the size specified.

1. It is important to check the weld size of all welds. This is not as difficult as it might sound. By means of weld gauges, the size of fillet welds can easily be determined. Figure 17-30 shows the use of a standard fillet weld gauge used in North America. Figure 17-31 shows the size of fillet welds and the method of checking fillet welds to determine that they do meet the size specifications. There are many other types of gauges. Figure 17-32 shows the use of a U.S. Navy type gauge for checking fillets. There are other gauges primarily from Europe that are used throughout the world including North America. One of the most popular is the British gauge shown by Figure 17-33. It has many capabilities for checking many sizes of different types of welds.

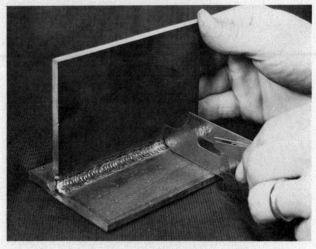

FIGURE 17-30 *Use of standard fillet gage.*

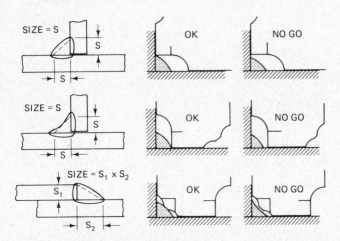

FIGURE 17-31 *Fillet weld size and method of checking.*

FIGURE 17-32 *Navy type weld gage.*

FIGURE 17-33 *British welding gage.*

2. All welds should be inspected to see that they do not have any of the defects listed below.

Surface cracks.

Crater cracks (or unfilled crater).

Surface porosity.

Incomplete root penetration.

Undercut

Underfill on face, groove or fillet (concave).

Underfill of root (suck back).

Excessive face reinforcement, groove or fillet (convex).

Excessive root reinforcement (or drop-through).

Overlap.

Misalignment (high-low).

Arc strikes.

Excessive spatter.

Each of these defects is described in detail in Section 17-7.

3. The weldment must also be checked by the inspector. The following are considered weldment defects.

Warpage: Warpage of weldments can be a reason for rejection or repair work. If warpage is beyond the allowable or acceptable limits, corrective action should be initiated. This can include mechanical methods such as the use of clamps, strong backs, presses, etc., or thermal methods such as the use of torches. Judgment of the inspector is important. If it is felt that damage can be done to the weldment by means of these remedial actions, records of action taken must be made in the inspection notebook.

Base Metal Defects: The inspector must also be on the lookout for these, which can appear as lamination in edges of steel plates. They can also be scabs or seams in the base metal. Caution is particularly important when steel plate is stressed in the through or Z direction.

Backing welds: These must be utilized whenever there is a question about the quality of root fusion of groove welds and corner type fillet weld joints.

Nondestructive Testing Recommendations.

The welding inspector should determine if nondestructive testing symbols are on weldment drawings. If so, it is important that the appropriate joints be marked for nondestructive testing and the inspector determine that such tests are made. These tests can involve the use of ultrasonic inspection, magnetic particle inspection, radiographic inspection, or dye-penetrant inspection. Dye-penetrant inspection is sometimes used for root pass inspection, especially on high-quality pipe weldments.

It is the privilege of the welding inspector, utilizing visual inspection, to call for any of the nondestructive techniques if there is reason to be suspicious of a specific joint, welder, or weldment. Good practice allows such weldments to be taken and subjected to at least one of the NDT methods. It is expected that the inspector will have reason for making such requests. If such weldments, joints, etc., do not show defects it is then wise to determine the competence of the inspector. Visual inspection of the surface of welds or of any metal part will not reveal internal flaws or problems. Surface inspection cannot show the lack of fusion at the root of the weld if the root of the weld is inaccessible for visual inspection. Internal porosity cannot be seen from the surface or internal cracks and other internal defects. It is, therefore, necessary to allow the inspector the privilege of requiring occasional internal examinations in order for the inspector to maintain credibility with welders.

The welding inspector must have certain qualifications. Under certain circumstances on certain types of work, it is necessary that the welding inspector be qualified and possibly certified. With or without this requirement it is necessary that the inspector have good eyesight corrected to 20-20 vision. The inspector must also be given the necessary tools and instruments for making inspections. This would include welding gauges, measuring instruments, temperature indicators, welding shields, flashlights, notebooks for recording data, and last, but not least, a marking crayon normally yellow or red, which can be used to mark those welds or weldments that must be repaired or rejected. Finally, it is necessary that the welders cooperate with the inspector, that weld slag be removed for adequate inspection, and that critical welds are never covered over before they can be inspected. This type of cooperation will insure that quality weldments are produced by the team of welders, supervisor, and inspector involved.

17-6 NONDESTRUCTIVE EXAMINATION

Nondestructive testing is the application of physical principles for the detection of flaws or discontinuities in materials without impairing their usefulness. The growth of nondestructive testing has been greatly accelerated with the need for higher quality and better reliability of manufactured products. This in part is due to the need of subjecting materials and welds to environmental conditions never encountered before. Nondestructive testing properly used will give assurance that each part or weld produced is of acceptable quality.

In the field of welding there are four nondestructive tests which are most widely used. These are dye-penetrant testing and flourescent-penetrant testing, magnetic particle testing, ultrasonic testing, and radiographic testing. Each of these techniques has specific advantages and limitations. A comparison of the different techniques compared with visual testing is shown to provide guidance in selection of the different tests.

Dye-Penetrant Testing. (DPT)

Liquid-penetrant inspection is a highly sensitive, nondestructive method for detecting minute discontinuities (flaws) such as cracks, pores, and porosity, which are *open to the surface* of the material being inspected. This method may be applied to many materials, such as ferrous and nonferrous metals, glass, and plastics. Although there are several types of penetrants and developers, they all employ common fundamental principles, as shown in Figure 17-34.

One of the most important aspects of liquid-penetrant inspection is the preparation of the part before the penetrant is applied. The surface must be cleaned with

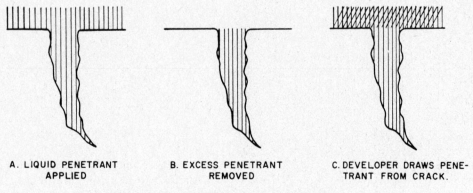

A. LIQUID PENETRANT APPLIED   B. EXCESS PENETRANT REMOVED   C. DEVELOPER DRAWS PENETRANT FROM CRACK.

FIGURE 17-34 *Principle of dye penetrant inspection.*

a solvent cleaner to remove any dirt or film. Discontinuities must be free from dirt, rust, grease, or paint, to enable the penetrant to enter the surface opening. The solvent cleaner, used to remove the excess penetrant, is excellent for precleaning of the part surfaces.

A liquid penetrant is applied to the surface of the part to be inspected. The penetrant remains on the surface and seeps into any surface opening. The penetrant is drawn into the surface opening by capillary action. The parts may be in any position when tested. After sufficient penetration time has elapsed, the surface is cleaned and excess penetrant is removed. When the surface is dry a powdered, absorbent material or a powder suspended in a liquid is applied. The result is a blotting action which draws the penetrant from any surface opening. The penetrant is usually a red color; therefore, the indication shows up brilliantly against the white background of the developer. The indication is larger than the actual defect. Thus, even small defects may be located.

Equipment: Portable inspection kits are available for visible dye penetrant. Some of the kits have liquid materials packaged in pressurized cans so that liquids may be sprayed upon the parts to be inspected. The use of pressurized liquids is shown by Figure 17-35.

Stationary equipment usually consists of a tank in which the penetrant is applied either by dipping, pouring, or by brushing. Other tanks provide drain, wash, and developer stations. The size of the tanks and individual stations is governed by the sizes and quantities of the parts to be inspected. Specialized, high-volume units

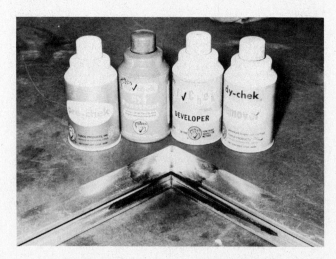

FIGURE 17-35 *Using penetrant on finished weld.*

are available which provide for parts inspection at production rate speeds. This equipment utilizes moving conveyors. The operator places the parts on the conveyor and they proceed unattended throughout the entire processing operation. At the end, an inspector examines the parts and interprets the indications.

Applications: In the field of welding, liquid-penetrant inspection is used to detect surface defects in aluminum, magnesium, and stainless steel weldments when the magnetic particle inspection method cannot be used. It is very useful for locating leaks in all types of welds. Welds in pressure and storage vessels and in piping for the petroleum industry are inspected for surface cracks and for porosity, using this method.

Fluorescent-Penetrant Inspection. (FPT)

The fluorescent-penetrant inspection technique is almost identical with the dye-penetrant technique. There are two basic differences, however. The penetrant is fluorescent and when it is exposed to ultraviolet or black light it shows as a glowing fluorescent type of readout. It provides a greater contrast than the visible dye penetrants. It is considered to have greater sensitivity.

For the use of the fluorescent-penetrant system, the other extra piece of equipment required is the ultraviolet or black light source. It is recommended that the inspection be done in a darkened area with the black light as the predominant source of light available.

Kits similar to the dye-penetrant equipment are also available for fluorescent-penetrant inspection; however, it is probably more popularly used with liquid penetrants in tanks similar to that mentioned for dye penetrant but with the fluorescent-penetrant material and the black light for inspection.

Inspection is made using the ultraviolet or the black light. Sound areas appear deep violet in color while the defects will glow a brilliant yellowish green color. The width and brightness of the fluorescent indication depends upon the size of the crack or defect.

Applications: Probably one of the most useful applications of fluorescent-penetrant testing is for leak detection in magnetic and nonmagnetic weldments. A fluorescent penetrant is applied to one side of the joint and a portable ultraviolet light (black light) is then used on the reverse side of the joint to examine the weld for leaks. Fluorescent-penetrant inspection is also widely used to inspect the root pass of highly critical pipe welds.

Interpretation: When visible dye penetrants are used, defects are indicated by the presence of a red color against the white background of the developer. A crack appears as a continuous line indication. The width and brightness of the dye indication depend upon the

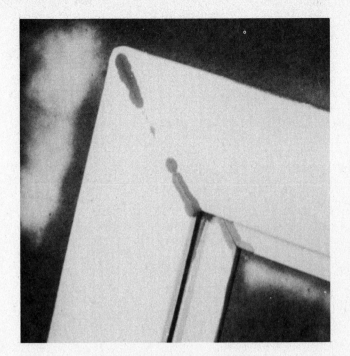

FIGURE 17-36 *Penetrant indication.*

volume of the crack or defect. Figure 17-36 shows a typical indication.

A *cold shut* (lap) which is caused by imperfect fusion is smooth in outline and continuous. Penetrant indications of gas holes appear round with definite color contrast.

The most effective aid for identifying and recognizing defects is a collection of parts containing defects. These can be compared to parts containing unknown indications. Extreme care and judgment must be exercised in interpreting indications. Consult the specification involved for standards of acceptability and qualification of operators. The most widely used standard for the qualification of operators is SNT ''Nondestructive Testing Personnel Qualification and Certification Supplement D.'' [29] See also ASTM ''Liquid Penetrant Inspection.'' [30]

Magnetic Particle Testing. (MT)

Magnetic particle inspection is a nondestructive method of detecting cracks, porosity, seams, inclusions, lack of fusion, and other discontinuities in ferromagnetic materials. Surface discontinuities and shallow subsurface discontinuities can be detected by using this method. There is no restriction as to the size and shape of the parts to be inspected; *only ferromagnetic materials can be inspected by this method.*

This inspection method consists of establishing a magnetic field in the test object, applying magnetic particles to the surface of the test object, and examining the surface for accumulations of the particles which are the indications of defects.

Ferromagnetism is the property of some metals, mainly iron and steel, to attract other pieces of iron and steel. A magnet will attract magnetic particles to its ends or poles, as they are called. Magnetic lines of force or flux flow between the poles of a magnet. Magnets will attract magnetic materials only where the lines of force enter or leave the magnet at the poles.

If a magnet is bent and the two poles are joined so as to form a closed ring, no external poles exist and hence it will have no attraction for magnetic materials. This is the basic principle of magnetic particle inspection. As long as the part to be inspected is free of cracks or other discontinuities, magnetic particles will not be attracted. When a crack is present, north and south magnetic poles are set up at the edge of the crack. The magnetic particles will be attracted to the poles which are the edges of the crack or discontinuity. This is shown by Figure 17-37.

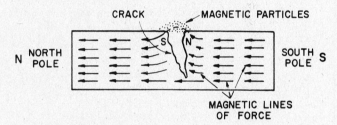

FIGURE 17-37 *Principle of magnetic particle inspection.*

Electric currents are used to induce magnetic fields in ferromagnetic materials. An electric current passing through a straight conductor creates a circular magnetic field. For reliable inspection, the magnetic lines of force should be at right angles to the defect to be detected. Hence, for a straight conductor with a circular field, any defect parallel to the conductor will be detected.

If the part is too large to run current through it, the part can be circularly magnetized by using probe contacts. This is shown by Figure 17-38. Direct current is the most desirable type of current for subsurface discontinuities. It is most commonly used for wet magnetic particle inspection. For dry particles, pulsating DC is used for both surface and subsurface defects. This current causes the particles to pulse, which gives them mobility and aids in the formation of indications. Alternating current tends to magnetize the cracks of the metal only, and hence it is used only for surface discontinuities such as fatigue or cracks caused by grinding.

Ferromagnetic parts that have been magnetized retain a certain amount of residual magnetism. Certain parts

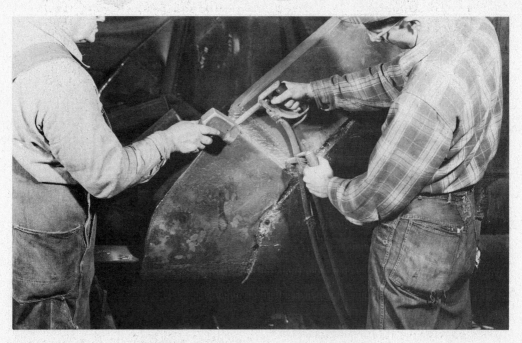

FIGURE 17-38 *Using magnetic particle inspection.*

may require demagnetization if they are to function properly. The attraction of small chips or particles caused by the residual magnetism may cause excessive wear and failures with rotating parts such as bearings and bearing surfaces.

Equipment: The most necessary piece of equipment for magnetic particle inspection is the *specialized power source.* Small portable units are available which supply AC while operating from 115 volts, AC power lines. These units generally use dry powder, but portable magnetic particle units which employ a pressurized spray may also be used.

Stationary units are widely used for the inspection of small manufactured parts. These units usually contain a built-in tank with a pump which agitates the wet particle bath and pumps the fluid through a hose for application to the test parts. These stationary units are usually provided with an inspection hood; ultraviolet or black light can be used so that fluorescent particles can be used and viewed.

Applications: The iron particles can be applied as dry powder or suspended in a liquid. Magnetic particle inspection may be applied to all types of weldments. On multipass welds, it is sometimes used to inspect each pass immediately after it has been deposited. An indication using the dry powder method is shown by Figure 17-39.

The majority of steel weldments in the aircraft industry are inspected by the magnetic particle method. If the weldments are thin enough, this method may provide sufficient sensitivity to detect any subsurface defects. Consult the specification involved for standards of acceptability and qualifications of equipment and operators. The most widely used standard for the qualification of the operator is SNT Supplement B. [31] See also ASTM wet method [32] and dry method. [33] Parts may have to be demagnetized after testing.

FIGURE 17-39 *Magnetic powder indication.*

596

Radiography is a nondestructive test method which uses invisible, X-ray, or gamma radiation to examine the interior of materials. Radiographic examination gives a permanent film record of defects which is relatively easy to interpret. Although this is a slow and expensive method of nondestructive testing, it is a positive method for detecting porosity, inclusions, cracks, and voids in the interior of castings, welds, and other structures.

X-rays, generated by electron bombardment of tungsten, and gamma rays emitted by radioactive elements, are penetrating radiation whose intensity is modified by passage through a material. The amount of energy absorbed by a material depends upon its thickness and density. Thus a thinner part will absorb less energy than a thick part and a heavy dense metal, like steel, will absorb more energy than a light metal such as aluminum. Energy not absorbed by the material will cause exposure of the radiographic film. These areas will be dark when the film is developed. Areas of the film exposed to less energy remain lighter. Therefore, areas of the material where the thickness has been changed by discontinuities, such as porosity or cracks, will appear as dark outlines on the film. Inclusions of low density, such as slag, will appear as dark areas on the film while inclusions of high density, such as tungsten, will appear as light areas. All discontinuities are detected by viewing shape and variations in the density of the processed film.

The X-ray or gamma ray source and penetrameter are placed above the piece to be radiographed and the film is placed on the opposite side of the part. A diagram of this is shown by Figure 17-40. Figure 17-41 shows an

FIGURE 17-41 *Setting up to take radiograph.*

oil-cooled and shielded head, which encloses the X-ray tube, being positioned to make a radiograph of a weld test plate.

Equipment: X-rays are produced by electrons hitting a tungsten target inside an X-ray tube. In addition to the X-ray tube, the apparatus consists of a high-voltage generator with necessary controls. X-rays are produced when a high-speed stream of electrons collides with a piece of tungsten. The electrons are produced in the X-ray tube by a hot cathode. They are accelerated towards the anode by means of an electron gun in the vacuum of the X-ray tube. The anode is a piece of tungsten and when the electrons hit it X-rays are produced, which are directed through a window to the part being inspected.

Gamma rays are produced by radioactive decay of certain radioisotopes. The radioisotopes normally used are Cobalt-60, Iridium-192, Thulium-170, and Cesium-137. These isotopes are contained in a lead or spent uranium vault or capsule to provide safe handling. They have a relatively short half-life and the strength of the radiation decreases with time.

Isotopes must be handled in such a way that radiographic sources can be positioned and yet produce mini-

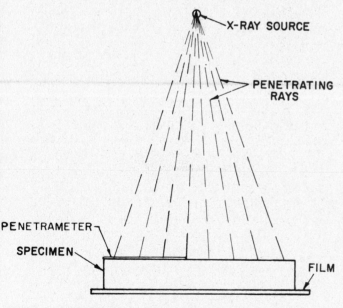

FIGURE 17-40 *Principle of radiographic inspection.*

mum radiation hazards to operating personnel. Remote handling equipment is employed when the radioactive source is drawn from the shielded container to the material to be radiographed. In the United States a license from the U.S. Atomic Energy Commission is required for use of radioisotopes.

The radiation intensity or output from an X-ray machine or from radioisotope sources will vary. Common materials such as concrete and steel are used to house the X-ray machine and protect the operator from exposure. The thickness of the shielding enclosure walls should be sufficient to reduce exposure in all occupied areas to a minimum value. If the work is too large or too heavy to be brought into the shielded room, special precautions such as lead-lined booths and portable screens are used to protect personnel. In the field, radiography protection is usually obtained from distance alone, since radiation intensity decreases as distance increases.

Penetrameters are used to determine the sensitivity of the radiograph. They are made of the same material that is being inspected and are usually 2% of the thickness of the part being tested. Therefore, if the penetrameter can be seen clearly on the radiograph, any change in thickness of the part (2% or more) will be seen clearly.

Radiographic film consists of a transparent plastic sheet coated with a photographic emulsion. When X-rays strike the emulsion, an image is produced. The image is made visible and permanent by a film-processing operation. Most processing equipment consists of tanks which contain a developer, a fixer, and rinse solutions. Film-processing operations are just as critical as the film exposure. Unsatisfactory radiographs can sometimes be attributed to errors in the processing technique or from mishandling of materials.

Application: Radiography is one of the most popular nondestructive testing methods for locating subsurface defects. It is used for inspection of weldments in all types of materials: steel, aluminum, magnesium, etc. Radiography is used in the pipeline industry to insure proper weld quality.

Interpretations: Most indications will show up as dark regions against the light background of the sound weld. Radiographs should be examined with a film illuminator providing a strong light source. Figure 17-42 shows a radiograph being examined.

It is essential that qualified personnel conduct X-ray interpretations since false interpretation of radiographs

FIGURE 17-42 *Examination of a radiograph.*

causes a loss of time and money. Radiographs for reference are extremely helpful in securing correct interpretations. Consult the specification involved for standards of acceptability and qualifications of equipment and operators. The most widely used standard for the qualification of the operator is SNT Supplement A. [34] See also ASTM "Radiographic Testing, [35] "Terminology," [36] and "Reference Radiographs," [37] also IIW "Reference Radiographs," [38] and "Hobart Pipeline Radiographs and Interpretation." [39]

Ultrasonic Testing. (UT)

Ultrasonic testing is a nondestructive inspection method which employs mechanical vibrations similar to sound waves but of a higher frequency. A beam of ultrasonic energy is directed into the specimen to be tested. This beam travels through a material with only a small loss, except when it is intercepted and reflected by a discontinuity or by a change in material.

Ultrasonic testing is capable of finding *surface and subsurface* discontinuities. The ultrasonic contact pulse reflection technique is used. This system uses a transducer, which changes electrical energy into mechanical energy. The transducer is excited by a high-frequency voltage which causes a crystal to vibrate mechanically. The crystal probe becomes the source of ultrasonic mechanical vibrations. These vibrations are transmitted into the test piece through a coupling fluid, usually a film of oil, called a *couplant*. When the pulse of ultrasonic waves strikes a discontinuity in the test piece, it is reflected back to its point of origin. Thus the energy returns to the transducer. The transducer now serves as a receiver for the reflected energy. The initial signal or main bang, the returned echoes from the discontinuities, and the echo of the rear surface of the test material are all displayed by a trace on the screen of a cathode-ray oscilloscope. Video type tapes may be used for permanent records.

The basic principles of ultrasonic testing are shown by Figure 17-43. The transducer is sending out a beam of ultrasonic energy. Some of the energy is reflected by the internal flaw and the remainder is reflected by the back surface of the specimen. Figure 17-44 shows the equipment in use.

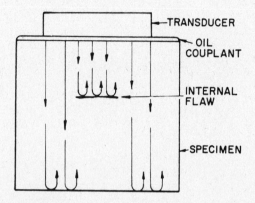

FIGURE 17-43 *Principle of ultrasonic inspection.*

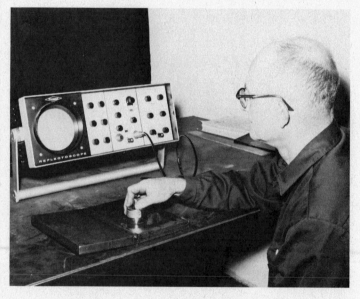

FIGURE 17-44 *Making an ultrasonic inspection of a weld.*

Figure 17-45 shows a typical display as presented on the oscilloscope screen. Signal strength is indicated by a vertical deflection of the pip on the screen and transmitted time is indicated by horizontal deflection. By measuring the height of the pip, the size of the flaw can be determined. The depth of the flaw from the surface is found by use of horizontal base. The front reflection and the rear reflection are at the extreme ends of the screen. The echo from the flaw is between the two.

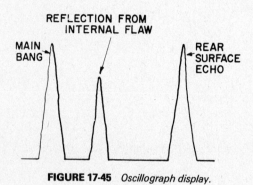

FIGURE 17-45 *Oscillograph display.*

In order to determine the size and depth of flaws, calibration techniques and reference standards must be used. A set of test blanks which have holes of known diameter and depth is used to calibrate the test instrument. The reference standard is described in detail by the AWS structural code.

Testing can be done with one transducer which serves both as a transmitter and receiver, or with two transducers, one transmitting only, and the other receiving only. The single transducer is used for portable equipment.

For testing of irregularly shaped parts, the immersion testing method is often used. With this method, the part and the transducer are submerged in water. The water transmits and couples the ultrasonic beam and the part. Actual contact is not required, thus irregular surfaces can be scanned.

The application of ultrasonic testing methods to weldments is shown by Figure 17-46. A 45° angled beam transducer is used to inspect the weld area. This search unit directs the beam toward the weld from a position on one side of the weld.

Butt joint welds in plate are usually tested with the angled search unit. The reflection obtained by the use

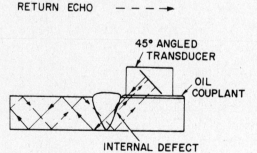

FIGURE 17-46 *45° Use of ultrasonic inspection.*

TABLE 17-1 *Guide to weld quality control techniques.*

Technique	Equipment	Defects Detected	Advantages	Disadvantages	Other Considerations
Visual Inspection VT	Pocket magnifier, welding viewer, flashlight, weld gage, scale, etc.	Weld preparation, fit-up cleanliness, roughness, spatter, undercuts, over-laps, weld contour and size. Welding procedures.	Easy to use; fast, inexpensive, usable at all stages of production.	For surface conditions only. Dependent on subjective opinion of inspector.	Most universally used inspection method.
Dye penetrant or Fluorescent Inspection DPT, FPT	Fluorescent or visible penetrating liquids and developers. Ultra violet light for the fluorescent type.	Defects open to the surface only. Good for leak detection.	Detects very small, tight, surface imper-fections. Easy to apply and to interpret. Inexpensive. Use on magnetic or non-magnetic materials.	Time consuming in the various steps of the processes. Normally no permanent record.	Often used on root pass of highly critical pipe welds. If material is improperly cleaned some indications may be misleading.
Magnetic Particle Inspection MT	Iron particles, wet or dry, or fluorescent. Special power source. Ultra violet light for the fluorescent type.	Surface and near-surface discontinuities, cracks, etc. Porosity, slag, etc.	Indicates discontinuities not visible to the naked eye. Useful in checking edges prior to welding, also, repairs. No size restriction.	Used on magnetic materials only. Surface roughness may distort magnetic field. Normally no permanent record.	Testing should be from two perpendicular directions to catch discontinuities which may be parallel to one set of magnetic lines of force.
Radiographic Inspection RT	X-ray or gamma ray. Source. Film processing equipment. Film viewing equipment. Penetrameters.	Most internal discontinui-ties and flaws. Limited by direction of dis-continuity.	Provides permanent record. Indicates both surface and internal flaws. Applicable on all materials.	Usually not suitable for fillet weld inspection. Film exposure and processing are critical Slow and expensive.	Most popular technique for sub-surface inspec-tion. Required by some codes and specifications.
Ultrasonic Inspection UT	Ultrasonic units and probes. Reference and comparison patterns.	Can locate all internal flaws located by other methods with the addition of exceptionally small flaws.	Extremely sensitive. Use restricted only by very complex weldments. Can be used on all materials.	Demands highly developed interpretation skill. No permanent record is obtained.	Required by some code and specifications.

of the 45° head is similar to Figure 17-45. The joint welds of heavy plate are tested with a straight beam transducer through the top of the joint, or with an angled beam transducer from one side of the bottom. Fillet welds are more difficult to inspect.

Equipment: Equipment required for this process consists of a transducer, pulse rate generator, amplifier, timer and cathode-ray oscilloscope. These devices are electronic, quite small in size for portability, yet rugged. Instant cameras are available to photograph oscilloscope displays for permanent records.

Application: Ultrasonic testing can be used to test practically any metal or material. Its use is restricted only by very complex weldments. The process is increasing in popularity and widely used. Consult the specification involved for standards of acceptability and qualification of equipment and operators. The most widely used standard for the qualification of the operator is SNT Supplement C. [40] See also ASTM "Ultrasonic Contact Inspection of Weldments." [41]

Leak Testing (LT)

Leak testing can be accomplished in many different ways. It can be used only when the weldment can be made to contain a gas or liquid. The most common leak test is the soap bubble test, which can be applied to external joints if internal gas pressure is present. Another is the use of internal liquids and the maintaining of high pressure over an extended period. The same test can utilize a vacuum. Halogen gases with sensitive detection meters are used to inspect production parts.

Do not use toxic or flammable gases or air for internal pressure testing. The internal pressure will store energy and if the part should fail it could cause an explosion that might cause bodily harm. Use internal liquids instead, or test inside a safety chamber.

Proof Testing (PRT)

Proof testing is controversial. Years ago proof testing was used on vessels and mechanical components. It consisted of loading the part to a load 50% or 100% greater than the designed load. It was felt that if the part passed this test and if it was never loaded above its designed load that it would never fail in service. Recently, however, it has been thought that proof testing might cause internal damage to the parts being tested that might reduce the service life. Also proof testing cannot provide assurance if the part is subjected to corrosion, fatigue, low-temperature impacts, etc. Consider proof testing but only in light of the above.

Guide to NDT Techniques: Table 17-1 is a guide to welding quality control comparing the different non-

destructive testing techniques. This table shows the equipment required, the defects that can be detected, the advantages and disadvantages of each technique, and other factors.

17-7 WELD DEFECTS—CAUSES AND CORRECTIVE ACTIONS

The problem of weld defects has become quite complex in the last few years, partially because of the great number of different words and definitions that are being used. For example, a welding flaw is a synonym for discontinuity; *discontinuity* is the preferred term. A discontinuity is "an interruption of the typical structure of a weldment such as a lack of homogeneity in the mechanical or metallurgical or physical characteristics of the material or weldment." A discontinuity is not necessarily a defect. A defect is "a discontinuity or discontinuities which by nature or accumulated effect render a part or product unable to meet minimum applicable accepted standards or specifications." This term designates rejectability. A defective weld then becomes "a weld containing one or more defects." For our purposes, we will consider defects as anything undesirable in a weld. It may or may not be cause for rejection or cause for repair. This is normally a matter left up to the specifications or codes involved. It is important that we learn to recognize the various different types of weld defects and learn enough about them so that we can recognize them, repair them, and avoid them.

Shown below will be the following: a collection of photographs and drawings of the possible different welding defects that can occur. It is an effort to present, in an organized manner, the various defects that can occur, the description of the particular defect or problem, and an indication of how or what caused the problem and the action that should be taken to correct the specific problem. The presentation may not describe all possible weld defects. There are others and some of the defects described may occur on different types of welds, but they would generally resemble those presented.

In this collection, an effort is made to indicate the responsibility for each of the defects. This breakdown is broad, but will indicate whether it is the fault of the welder, that is, a problem of welding technique, the fault of the designer, that is, a drawing or design error, or a fault of some manufacturing or shop function, such as materials preparation. In the tables that follow, these are designated as a *welder* responsibility, *designer* responsibility, or a *shop* responsibility.

An indication of how the particular defect was detected will also be given. This is based on the inspection techniques used to find the defect. The five most popular nondestructive testing techniques are considered as follows:

VT, Visual inspection.

MT, Magnetic particle inspection.

PT, Dye-penetrant inspection (including flourescent).

RT, Radiographic inspection.

UT, Ultrasonic inspection

The tables show information concerning each defect and probable causes for the defect. If there is more than one probable cause they will be represented in the order of importance or frequency. The corrective action for the particular defect is briefly mentioned. Corrective action means the way to correct the specific defect in order to make the weldment suitable for service. Efforts will also be made to provide an explanation to prevent a recurrence of the same type of defect. This may also be covered in the general information concerning the different classifications of defects.

Each particular type of defect is not shown with each different type of weld. Efforts are made to show the more common types of welds with the defects that may occur most often in them. When a defect is not detectable with visual inspection, that is, the defect is not visible from the surface of the weld, additional information is shown to help describe the defect. For subsurface type defects, which must be found by internal type inspection methods, cross sections are shown.

Certain welding processes have specific defects peculiar to those processes. For example, tungsten inclusions in a weld are defects associated with the gas tungsten arc welding process. These are mentioned in the appropriate section. The defects are arranged according to the classification of weld defect system established by Commission V of the International Institute of Welding. Their document no. IIS/IIW-340-69 and Commission V Document classifies the defects into six groups as follows.[42]

Series 100, Cracks.

Including longitudinal, transverse, radiation, crater, etc.

Series 200, Cavities.

Including gas pockets, internal porosity, surface porosity, shrinkage, etc.

Series 300, Solid inclusions.

Including slag, flux, metal oxides, foreign material, etc.

Series 400, Incomplete fusion or penetration.

Including incomplete fusion, incomplete penetration, etc.

Series 500, Imperfect shape or unacceptable contour.

Including undercut, excessive reinforcement, underfill, fillet shape, overlap, etc.

Series 600, Miscellaneous defects not included above.

Including arc strikes, excessive spatter, rough surface, etc.

This International Institute of Welding document catalogs all welding defects and provides index numbers of three digits for groups and four digits for specific types of defects within the groups. It also cross indexes with the International Institute of Welding Radiographic Reference Radiographics [38].

The various examples of defects presented here are shown to illustrate the problems, and no reference is made to what is acceptable or allowable. To determine the acceptability limits of these different defects you must refer to the specification or code that is involved. Certain defects of a minor nature may be acceptable under specific conditions in some codes, whereas they may be unacceptable in other codes. In general, defects that can propagate under stress are not acceptable in any code; for example, cracks are not allowed in any of the major codes but porosity of a specific magnitude and spacing may be acceptable. Detailed information is found in the book, *Welding Inspection.* [43]

There are other problem areas encountered in welding that are not specifically included. These would include such factors as distortion and warping, brittle heat-affected zones, brittle weld metal, arc blow, etc. These topics are covered elsewhere in the book.

Series 100: *Cracks* are the first category of weld defects. A crack is "a fracture-type discontinuity characterized by a sharp tip and high ratio of length and width to opening displacement." Cracks are perhaps the most serious of the defects that occur in the welds or weld joints in weldments. However, cracks are defects that can be found in other metal products such as forgings, castings, and even hot rolled steel products. Cracks are considered dangerous because they create a serious reduction in strength. They can propagate and cause sudden failure. They are most serious when impact loading and cold-temperature service are involved. Cracks must be repaired.

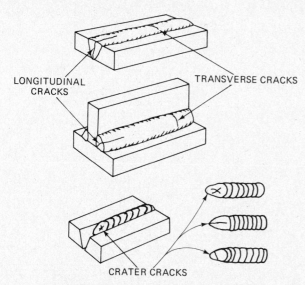

FIGURE 17-47 *Types of surface cracks of welds.*

There are many different types of cracks. One way of categorizing them is as surface or subsurface cracks. *Surface cracks* can be seen on the surface of the weld using the visual testing technique. There are several types of surface cracks—transverse, longitudinal, and crater cracks. There are also toe cracks in adjacent parent metal which normally come to the surface. These cracks are shown by Figure 17-47. The subsurface or internal cracks are also of many types. Some may be in the weld, some in the heat-affected zone—sometimes called *underbead cracks,* shown by Figure 17-48, some at the interface between weld metal and base metal and some completely in the base metal, sometimes called *lamanar tearing.* There can also be microsize cracks as well as macrosize cracks. Sometimes the smaller cracks are called *fissures,* or if the cracks are extremely small they are called *microfissures,* which require special techniques to find.

There is another way of classifying cracks and this is with respect to temperature. There are hot cracks which occur during or immediately after the weld is made or during the cooling cycle. Cold cracks are cracks which occur sometime after the weld is finished, after it has cooled to room temperature. Cold cracks may be delayed hours or even days after the weld is finished. There are also fatigue type cracks which may occur months or years after fabrication as a result of an initiating point and fatigue loading. There are also stress corrosion cracks which are caused by a corrosive atmosphere and a high stress condition.

In general, cracks in welds or cracks adjacent to welds indicate that the weld metal or the base metal has low ductility and that there is high restraint. Any factor that contributes to low ductility of the weld and adjacent metal and high restraint will contribute to cracking. Some of these factors are rapid cooling, high-alloy composition, insufficient heat input, poor joint preparation, incorrect electrode type, etc. The examples that are shown in Figure 17-49 will help further explain the different types of weld cracks, the probable cause, and corrective action.

Series 200: *Cavities* are the second category of defects listed in the IIW document. The most common type of cavity is called porosity, defined as "cavity type discontinuities formed by gas entrapment during solidification." Specific defects can be called *gas pockets* which are cavities caused by entrapped gas. These are sometimes called *blow holes.* Porosity can also be divided into two types: surface porosity, which can be seen by the naked eye and detected by visual inspection technique; and subsurface porosity, which can be found only by the internal detecting techniques. The gas pockets can occur as extremely large holes in the weld metal or extremely small holes scattered throughout the cross section of a weld. Some types of porosity are called *worm holes* when they are long and continuous. Others are called *piping,* usually long in length and parallel to the root of the weld. Some types may occur exclusively at the root and others almost at the surface. Porosity is not as serious a defect as cracks primarily because porosity cavities usually have rounded ends, and will not propagate like cracks. Many codes and specifications provide comparison charts showing the amount of porosity that may be acceptable. Figure 17-50 shows an example of comparison charts used by the API 1104 code. The AWS structural welding code has taken a slightly different point of view and has a sliding scale shown by Figure 17-51. This takes into account size and spacing of the porosity and relates to the size of the weld. For more information refer to these codes.

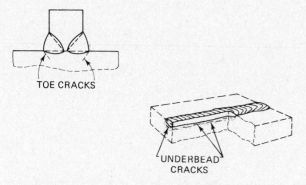

FIGURE 17-48 *Weld toe cracks and underbead cracks.*

Weld Defect Class — Cracks Appearance or Cross Section or Radiograph Class No. 100		How Defected		Respon-sibility	Probable Cause	Corrective Action
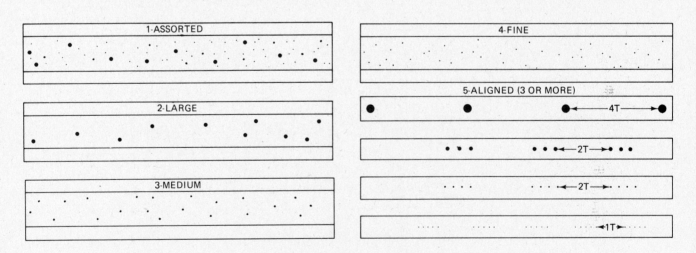A	Longitudinal Crack	VT	X	design	General 1. Incorrect electrode. 2. High restraint of joint. 3. Rapid cooling of weld. 4. Improper joint pre-paration. 5. Fillet weld longitud-inal crack.	1. Use proper or matched electrode. 2. Reduce rigidity of weld-ment or change welding sequence. Use higher ductility welding filler metal. 3. Use preheat and/or inner pass heat to reduce cooling rate. 4. Use proper joint for welding process. 5. Change center line of weld to avoid interface between parts.
		MT	X	X		
		PT	X	welder		
		RT	X	X		
		UT	X	shop		
B	Longitudinal Crack	VT		design		
		MT	X	X		
		PT		welder		
		RT	X	X		
		UT	X	shop		
C	Crater Crack	VT	X	design	Crater Crack 1. Unfilled crater. 2. Crater crack in sub-merged arc welding.	1. Filler crater with proper technique. 2. Utilize run-out-tab.
		MT	X			
		PT	X	welder		
		RT	X	X		
		UT	X	shop		
D	Transverse Crack	VT	X	design	Transverse Crack 1. Incorrect electrode. 2. Rapid cooling. 3. Welds too small for size of parts joined.	1. Use proper electrode. 2. Use larger electrode, higher welding current or preheat. 3. Use larger weld possibly larger welding electrode.
		MT	X			
		PT	X	welder		
		RT	X	X		
		UT	X	shop X		

FIGURE 17-49 *Collection of weld defects—cracks.*

1-ASSORTED

2-LARGE

3-MEDIUM

4-FINE

5-ALIGNED (3 OR MORE)

4T

2T

2T

1T

WALL THICKNESS 1/2" OR LESS
MAXIMUM DISTRIBUTION OF GAS POCKETS

FIGURE 17-50 *Porosity chart for pipe welds—API (Courtesy, The American Petroleum Institute).*

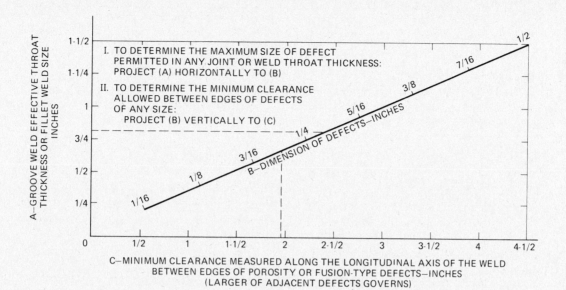

I. TO DETERMINE THE MAXIMUM SIZE OF DEFECT
PERMITTED IN ANY JOINT OR WELD THROAT THICKNESS:
PROJECT (A) HORIZONTALLY TO (B)

II. TO DETERMINE THE MINIMUM CLEARANCE
ALLOWED BETWEEN EDGES OF DEFECTS
OF ANY SIZE:
PROJECT (B) VERTICALLY TO (C)

A—GROOVE WELD EFFECTIVE THROAT THICKNESS OR FILLET WELD SIZE INCHES

B—DIMENSION OF DEFECTS—INCHES

C—MINIMUM CLEARANCE MEASURED ALONG THE LONGITUDINAL AXIS OF THE WELD
BETWEEN EDGES OF POROSITY OR FUSION-TYPE DEFECTS—INCHES
(LARGER OF ADJACENT DEFECTS GOVERNS)

FIGURE 17-51 *AWS structural code weld quality requirements.*

FIGURE 17-52 *Collection of weld defects—cavities.*

Weld Defect Class – Cavities Appearance or Cross Section or Radiograph Class No. 200	How Defected		Responsibility	Probable Cause	Corrective Action
A	VT	X	design	**General** Surface Porosity 1. Welding over foreign material on surface such as rust, oil, moisture, paint, etc. 2. Damp electrodes. 3. Improper base metals such as free machining or high sulphur. 4. Welding current too low.	1. Clean weld bevels and area adjacent to weld and keep clean. 2. Use fresh dry electrodes or rebake electrodes that have been exposed to dampness. 3. Utilize correct base metal possibility to use low hydrogen type electrodes. 4. Increase welding current.
	MT				
	PT	X	welder		
	RT		shop		
	UT	X			
B	VT	X	design	**Gas Shielded Welding Processes** 1. Incorrect shielding gas type. 2. Incomplete gas coverage due to breeze, defective gas system, clogged nozzle, etc. 3. Moisture in the shielding system. 4. Poor gas coverage. 5. Welding over tack weld made with shielded metal arc process.	1. Use specified shielding gas. 2. Provide windshields, check efficiency of gas system such as broken hoses, gas valves, empty tanks, clean nozzle. 3. Check to make sure gas is dry or welding grade. Check for water leaks in water cooled systems. 4. Use proper nozzle to work distance. Check gas flow rate, may be too high or too low. 5. Utilize gas metal arc for tack welding.
	MT				
	PT	X	welder		
	RT		shop		
	UT	X			
C	VT	X	design		
	MT	X			
	PT	X	welder		
	RT	X	shop		
	UT	X	X		
D	VT	X	design	**Submerged Arc Welding** 1. Damp submerged arc flux. 2. Contaminated surface of electrode wire, dirt and/or moisture. 3. Too many fines in flux.	1. Utilize fresh dry flux or dry damp flux. 2. Clean surface of electrode wire. 3. Use fresh flux and discard fines.
	MT				
	PT		welder		
	RT	X	shop		
	UT	X	X		

There are other types of cavities and some are called shrinkage voids which are defined as cavity discontinuities normally formed by shrinkage during solidification.

Cavities, voids, and porosity are caused by gases that are present in the arc area, or may be present in base metal, that are trapped in the molten weld during the solidification process. Common causes of porosity are inclusion of gases from the atmosphere, or generated in the weld. Common causes for porosity are high sulphur in the base metal, hydrocarbons on the surface of the metal such as paint, water, oil, moisture from damp electrodes, wet submerged arc flux, or wet shielding gas. When the porosity exceeds that acceptable by the code it must be removed and repair welds made. In general, surface porosity is an indication that subsurface porosity may have been in the weld before it became noticeable as surface porosity. In these cases extra inspection should be done to determine the extent of the subsurface porosity. The following examples of cavities will help further explain this type of weld defect. These are shown by Figure 17-52.

Series 300: *Solid inclusions* are the next type of defect considered by the IIW Document. Solid inclusions are normally expected to be a subsurface type of defect and would include any foreign material entrapped in the deposited weld metal. The most common type of solid inclusion is a slag inclusion defined as "nonmetallic solid material entrapped in weld metal or between weld metal and base metal." Another and very similar type of inclusion is a flux inclusion which is an entrapment of flux from an electrode, from submerged arc flux, or from another source of flux that for one reason or another did not float out of the weld metal as it solidified. Slag inclusions and flux inclusions can be continuous, intermittent, or very randomly spaced. In general, flux or slag inclusions are rounded and do not possess sharp corners like cracks and for this reason are not quite as serious as cracks. The applicable code or specification indicates how much entrapped slag or flux is acceptable.

On certain metals, particularly those that have high-temperature oxide coatings, there is the possibility of oxide inclusions in the weld metal. This is a troublesome problem when welding aluminum. Aluminum oxide will form rapidly in the atmosphere and can be entrapped in the weld metal very easily if cleaning and other precautions are not taken. Oxide inclusions are detected by the internal inspection techniques. There are also other metallic inclusions such as tungsten inclusions which may occur in the gas tungsten arc welding process. Small pieces of tungsten are deposited in the weld metal deposit and can only be found by internal inspection techniques, particularly radiographic testing. Oxide inclusions or tungsten inclusions are not acceptable for high-quality work. When copper backing bars are used local melting may occur and pieces of copper can be entrapped in the weld metal. This can be detected from the underside surface of a weld or by internal detection techniques. All such inclusions are defects that must be evaluated in accordance with the code or specification in question or with respect to good practice. The examples shown by Figure 17-53 will help further explain these types of defects.

Series 400: *Incomplete fusion* or *penetration* is the next defect category of the IIW document. This is sometimes called lack of fusion or lack of penetration; however, the preferred term is *incomplete fusion,* defined as "fusion which is less than complete." It is shown by Figure 17-54. This can be inadequate joint penetration, defined as "joint penetration which is less than specified." The word penetration is not preferred and the term should be joint penetration or root penetration. *Root penetration* is defined as "the depth a groove weld extends into the root of a joint measured on the center line of the root cross section" and is shown by Figure 17-55. These illustrations help show the difference between complete and incomplete fusion and complete joint versus partial joint penetration and root penetration. Incomplete fusion as a defect means that the weld deposited did not completely fill the joint preparation or there is space in between the beads or passes or a space at the root of the joint. Penetration is a slightly different term. The term joint penetration is the minimum depth of the joint; the groove or flange weld extends from its face into the root, exclusive of reinforcement. The term penetration means the depth that the groove weld extends into the root of a joint measured on the centerline of the cross section. These terms are often used interchangeably but do have a different meaning. The defect is the absence of complete fusion of a joint and this provides a stress riser, which is undesirable for welds loaded in fatigue or subject to impacts or low-temperature service. Figure 17-56 helps illustrate different types of this defect. The cause of such defects can be dirty surfaces such as heavy mill scale, heavy rust, grease, the failure to remove slag from previous beads, the fact that the root opening may not be sufficiently large or that the welder technique is not satisfactory. The danger of the defect is the serious reduction in static strength and the production of a stress riser as mentioned above.

Weld Defect Class — Solid Inclusions Appearance or Cross Section or Radiograph Class No. 300	How Defected	Responsibility	Probable Cause	Corrective Action
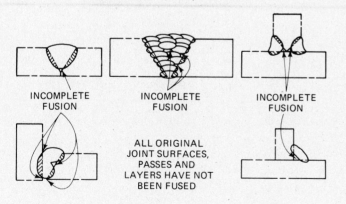 A	VT MT PT RT X UT X	design welder X shop	General 1. Slag inclusion-between passes. 2. Intermittent slag inclusion at edge of bead (wagon tracks). 3. Irregular surface of bevels. 4. Incorrect welding technique or wrong current or voltage. 5. Submerged arc welding —flux inclusion.	General 1. Remove solidified slag after each pass. 2. Remove slag at bead edge. Utilize proper technique to avoid high crowned bead contour. 3. Provide for smooth bevel surface, grind if necessary. 4. Utilize correct welding technique for electrode type and joint design. 5. Improper direction of electrode wire. Welding current too low. Electrode wire misdirected possible correction use wire straightener. Improper joint detail.
B	VT MT PT RT X UT X	design welder X shop		
C	VT MT PT RT X UT X	design welder X shop X		

FIGURE 17-53 *Collection of weld defects—solid inclusions.*

FIGURE 17-54 *Incomplete fusion.*

FIGURE 17-55 *Root penetration and joint penetration.*

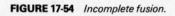
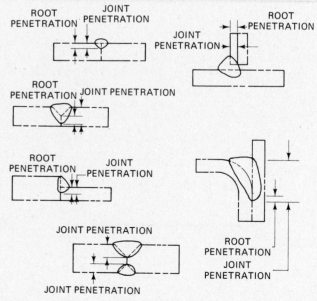

Weld Defect Class — Incomplete Fusion Appearance or Cross Section or Radiograph Class No. 400	How Defected		Responsibility	Probable Cause	Corrective Action
A	VT	X	design	General	1. Reduce welding speed.
	MT		X	1. Welding speed too fast.	2. Utilize correct size electrode.
	PT	X	welder	2. Electrode too large for joint detail.	3. Increase welding current for more penetration.
	RT	X	X	3. Welding current too low.	4. Utilize correct joint detail.
	UT	X	shop	4. Improper joint design such as excessive root face or minimum root opening..	
B	VT	X	design	5. Improper joint fit-up such as root opening too small.	5. Make setup correct to agree with joint design detail.
	MT	X	X		
	PT		welder	Shielded Metal Arc Welding	1. High speed will reduce complete fusion, lower speed will cause complete fusion.
	RT	X	X	1. Irregular travel speed.	
	UT	X	shop	2. Irregular arc length	2. Maintain proper arc length.
C	VT	X	design	Gas Metal Arc Welding	1. Direct arc at leading edge of puddle. Current too low, voltage too low, adjust for proper procedure. Pause too short at dwell when weaving. Increase pause to allow melting of base metal.
	MT	X	X	1. Incomplete fusion— (cold shut)	
	PT		welder		
	RT	X	X		
	UT		shop		
D	VT	X	design		
	MT			Submerged Arc Welding—Semiautomatic	1. Failure to direct welding electrode to root of weld joint.
	PT		welder	1. Incomplete root fusion.	
	RT	X	X		
	UT	X	shop		

FIGURE 17-56 *Collection of weld defects—incomplete fusion.*

Series 500: *Imperfect shape,* or *unacceptable contour,* is the next category of defect from the IIW document. One of the most serious of these defects is undercut, which is shown by Figure 17-57. Undercut occurs not only on fillet welds but on groove welds as well. Undercut also produces stress risers that create problems under impact, fatigue, or low-temperature service. It is normally caused by excessive currents, incorrect manipulation of electrode, incorrect electrode angle or type of electrode. Undercut actually refers more to the base metal adjacent to the weld, whereas imperfect shape is a defect of the weld itself. This can include such things as excessive reinforcement on the face of a weld, which can occur on groove welds as well as fillet welds. There is also the problem of excessive reinforcement on the root of the weld, primarily open root groove welds. Excessive reinforcement is an economic waste. It can also be a stress riser and is objectionable from an appearance point of view. It is normally a factor involved with fitup, welder technique, welding current,

type of electrode, etc. Another similar flaw is the concave type contour or lack of fill on the face of the weld or a suckback on the root of a groove weld. The proper term in both cases is *underfill,* defined "as a

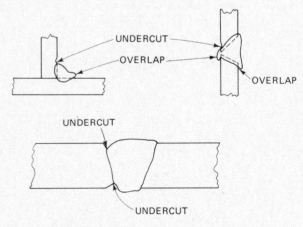

FIGURE 17-57 *Undercut fillet and groove and overlap.*

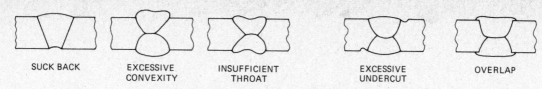

SUCK BACK EXCESSIVE CONVEXITY INSUFFICIENT THROAT EXCESSIVE UNDERCUT OVERLAP

FIGURE 17-58 *Groove welds and various defects.*

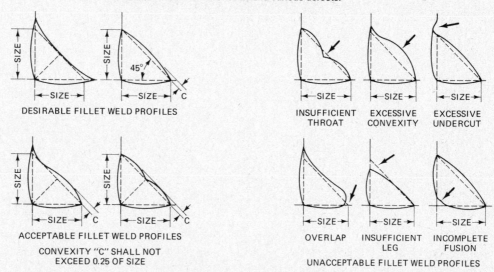

DESIRABLE FILLET WELD PROFILES

INSUFFICIENT THROAT EXCESSIVE CONVEXITY EXCESSIVE UNDERCUT

ACCEPTABLE FILLET WELD PROFILES

CONVEXITY "C" SHALL NOT EXCEED 0.25 OF SIZE

OVERLAP INSUFFICIENT LEG INCOMPLETE FUSION

UNACCEPTABLE FILLET WELD PROFILES

FIGURE 17-59 *Contour of fillet welds.*

depression on the face of a weld or root surface extending below the surface of the adjacent base metal," as shown by Figure 17-58. Underfill does reduce the cross-sectional area of the weld below the designed amount and therefore is a point of weakness and potentially a stress riser where failure may initiate. The fillet weld is particularly vulnerable to the problem of imperfect shape. Figure 17-59 shows some different fillet weld

FIGURE 17-60 *Collection of weld defects—incorrect shape.*

Weld Defect Class — Imperfect Shape Appearance or Cross Section or Radiograph Class No. 500		How Defected	Respon- sibility	Probable Cause	Corrective Action
A	VT	X	design	Undercutting	1. Use uniform weave in groove welding pause at edges.
	MT	X		1. Faulty electrode manipulation.	2. Use prescribed welding current for electrode size.
	PT	X	welder X	2. Welding current too high.	3. Use correct electrode size for size weld being made.
	RT	X	shop	3. Incorrect electrode size (usually too large).	4. Use correct electrode with position capabilities.
	UT	X		4. Incorrect electrode for welding position.	5. Adjust electrode angle to fill undercut area.
B	VT		design	5. Incorrect electrode angle.	
	MT	X			
	PT		welder X		
	RT	X	shop		
	UT	X			
C	VT	X	design	Incorrect Profile	1. Root opening too wide.
	MT	X		1. Excessive root penetration.	
	PT	X	welder X	2. Travel speed too slow.	2. Welding current too high.
				3. Excessive crown or reinforcement.	3. a. Welding speed too slow. b. Welding voltage too high arc length too long, correct arc length.
	RT	X	shop		
	UT	X		4. Incomplete root or negative root reinforcement or "suck back" (internal concavity).	4. a. Voltage too high. b. Travel speed too fast. c. Root opening too wide.
D	VT	X	design	5. Improper fillet contour usually wide on horizontal and not sufficient on vertical leg.	5. Improper welding technique. Welding current too high. Reduce welding current. Use smaller electrode.
	MT	X			
	PT	X	welder X		
	RT	X	shop	6. Incorrect electrode type.	6. Use proper electrode type.
	UT	X			

Weld Defect Class — Miscellaneous Defects Appearance or Cross Section or Radiograph Class No. 600	How Defected		Responsibility	Probable Cause	Corrective Action
	VT	X	design	**Poor Appearance** 1. Welding current too high or too low. 2. Improper technique. 3. Faulty electrode. 4. Irregular travel speed.	1. Use prescribed procedure. 2. Provide additional welding training. 3. Use fresh electrode or determine correct type to be used. 4. Allow for additional practice and experience.
	MT				
	PT		welder X		
	RT	X			
	UT		shop		
	VT	X	design	**Excessive Weld Spatter** 1. Arc blow—see welding current. 2. Excessive welding cur- for type and size electrode. 3. Excessive long arc— high voltage. 4. Improper electrode type.	1. Reduce arc blow—use alternating current. 2. Adjust for proper welding current for size elect-rode used. 3. Hold proper arc length and use correct arc voltage. 4. Utilize proper electrode type for location.
	MT				
	PT		welder X		
	RT				
	UT		shop		
	VT	X	design	**Poor Tie-In** 1. Incorrect electrode angle. 2. Improper technique for restriking arc.	1. Use correct electrode angle. Imrpove training. 2. Provide additional train-ing and experience.
	MT				
	PT		welder X		
	RT	X			
	UT		shop		
	VT	X	design	**Whiskers** 1. Root opening too wide.	1. Use correct root opening. Use weaving motion and direct arc on weld puddle.
	MT				
	PT		welder X		
	RT	X			
	UT	X	shop X		

FIGURE 17-61 *Collection of weld defects—miscellaneous.*

contours, both acceptable and unacceptable. Those that reduce the throat of the fillet weld actually reduce the strength of the weld so that premature failure may occur. Examples shown by Figure 17-60 help show the different types of possible defects in this series.

Series 600: *Miscellaneous defects* are the final category considered and actually cover all defects which may not be categorized in any of the other classifications given above. One type of defect in this classification is *arc strikes,* which are unacceptable in certain types of work. These are defects where the welder accidentally struck the electrode on the base metal adjacent to the weld. This creates problems particularly on hardenable steel and on critical types of applications and is not acceptable. For certain types of work protective wrappings are made around the part, especially pipe adjacent to the weld, to avoid stray arc strikes. *Excessive spatter* adjacent to the weld is also a defect and is unacceptable. This may be caused by arc blow, by the selection of the incorrect electrode or welding current, or the technique of the welder. There are several other types of miscellaneous defects and those defects that apply specifically to particular processes are covered in the examples shown by Figure 17-61.

The above explanation and illustrations of different weld defects are based on opinions of various experts in the welding industry. Unfortunately, it is not universally possible to indicate which are acceptable and which are unacceptable. Some defects can be critical under certain service conditions or on specific metals. The same defect on less demanding service on mild steel would be acceptable. Unfortunately, there is no common agreement as to what is acceptable and unacceptable in different types of service. It is hoped that, by establishing categories and providing illustrations of the different defects within these categories, more common agreement will eventually be reached and will be defined in codes and specifications in a uniform manner.

17-8 WORKMANSHIP SPECIMENS, STANDARDS, AND INSPECTION SYMBOLS

A *workmanship specimen* is an actual weld with each weld bead showing for a short length. Such a specimen provides the welder with an example of what is expected and also gives the welding supervisor and the inspector an example. Figure 17-62 shows a vertical up V-groove weld made of medium thickness plate. The workmanship specimen is a quality control tool and also an instruction device for conveying the type of welding expected in the production of weldments. The workmanship specimen concept originated with the U.S. Army Ordnance Department to insure the quality of weldments produced during World War II. Workmanship specimens were made for each of the various weld joints of an ordnance weldment. Cross-sectional samples were cut from the weld joint, polished, etched, and tack welded to the specimen to provide additional information. A weld schedule or procedure chart which shows the welding joint design, welding conditions, current, voltage, etc., for each layer of the particular joint made. The schedules were followed in making the specific workmanship specimens. This technique provides a good tool for quality control. The information should be posted and made available to welders, welding supervisors, inspectors, engineers, and others at the point where the welding is done. Workmanship specimens can be made for any weld joint welded in any position using almost any of the welding processes. The principle is to show the joint fitup detail and each bead or weld as it is made in producing the total weld joint.

Complete parts are sometimes used instead of welding drawings or blueprints in the manufacturing department. These welded parts sometimes are posted in the welding booth where that part is normally manufactured. For companies making relatively small parts on a production basis this is an excellent technique of informing the welder and others of exactly the type of welds expected

on the finished weldment. Many times these parts are painted with the welds highlighted in a different color of paint to make them stand out. Welding schedule and weld size information can be posted adjacent to the weldment specimen.

The concept of workmanship specimens is used extensively by major structural steel contracting companies. Many of these companies operate erection crews in widely separated areas, yet expect to have welds made the same way by the different crews. These companies produce workmanship specimens and provide the welding schedules for producing these workmanship specimens. They go a step further, and run qualification tests on each of the different types of joints that are normally employed. By providing code numbers for these different weld joints they are able to provide information for the designer, the welding supervisor,

FIGURE 17-62 *Workmanship specimen—vertical up SMAW.*

FIGURE 17-63 *Workmanship specimen and test bars.*

FIGURE 17-64 *Welding schedule for workmanship specimen.*

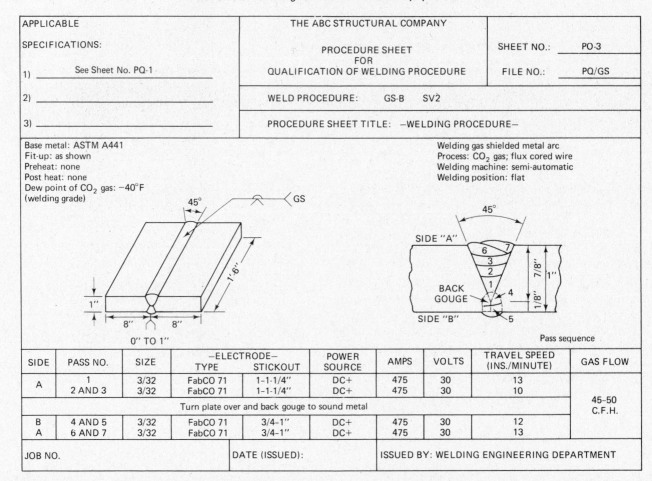

| APPLICABLE SPECIFICATIONS: | THE ABC STRUCTURAL COMPANY
PROCEDURE SHEET
FOR
QUALIFICATION OF WELDING PROCEDURE | SHEET NO.: PO-3 |
| | | FILE NO.: PQ/GS |

1) _____ See Sheet No. PQ-1 _____

2) _____

3) _____

WELD PROCEDURE: GS-B SV2

PROCEDURE SHEET TITLE: —WELDING PROCEDURE—

Base metal: ASTM A441
Fit-up: as shown
Preheat: none
Post heat: none
Dew point of CO_2 gas: $-40°F$
(welding grade)

Welding gas shielded metal arc
Process: CO_2 gas; flux cored wire
Welding machine: semi-automatic
Welding position: flat

SIDE	PASS NO.	SIZE	—ELECTRODE— TYPE	STICKOUT	POWER SOURCE	AMPS	VOLTS	TRAVEL SPEED (INS./MINUTE)	GAS FLOW
A	1	3/32	FabCO 71	1-1-1/4''	DC+	475	30	13	
	2 AND 3	3/32	FabCO 71	1-1-1/4''	DC+	475	30	10	45-50 C.F.H.
	Turn plate over and back gouge to sound metal								
B	4 AND 5	3/32	FabCO 71	3/4-1''	DC+	475	30	12	
A	6 AND 7	3/32	FabCO 71	3/4-1''	DC+	475	30	13	

JOB NO.

DATE (ISSUED):

ISSUED BY: WELDING ENGINEERING DEPARTMENT

welders, and inspectors. The tail of the arrow of the welding symbol will show a specific joint detail specification. This detail specification will refer to a particular weld schedule and workmanship specimen that has been qualified in accordance with the code. By this means, the welding crew will always make the weld joint in the same manner and will utilize the procedure that is known to produce quality welds. This assists in consistency of welding but also in consistency of weld quality and further provides cost control since the weld joint is always made the same way. Figure 17-63 shows a typical welding specimen and the test bars produced from the specimen. The schedule for making this weld is shown by Figure 17-64. Structural companies operating with many crews provide welding supervisors and others with a book of weld schedules indexed by the detail specification of the weld joint.

Companies producing fairly heavy weldments can also use workmanship specimens but in these cases they may only show the various different joints in a small section. These joints serve to show a proper way for making a particular weld, and also provide the appearance expected for making these welds. Companies that manufacture construction equipment often use these types of welding workmanship specimens.

Other companies have used the same concept to produce examples of welds that are acceptable and unacceptable. Originally the construction equipment company produced good and questionable welds of different types and posted the actual welds in the manufacturing department. However, they found that there was sufficient variation in the different specimens to create confusion. To overcome this, they produced one set of acceptable welds and one set of unacceptable welds and then made plastic replicas of each weld from these molds. The plastic weld replicas were posted in the department and in this case each and every impression was identical. Figure 17-65 shows the plastic replicas of acceptable and unacceptable welds. These types of exhibits are also very useful in the training of welders. The Department of Defense has used a similar technique with their military specification, which covers the smoothness of flame-cut surfaces. To actually portray the surface as expected and surfaces not acceptable they have made a plastic replica of the actual flame-cut surfaces.

FIGURE 17-65 *Plastic replicas of acceptable and unacceptable welds.*

WELDMENT EVALUATION AND QUALITY CONTROL

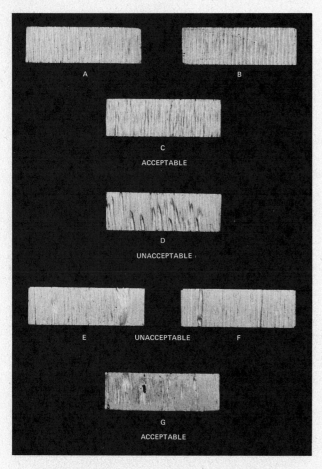

FIGURE 17-66 *Plastic replicas of flame cut surface.*

This is reproduced and is a part of a military specification. This is shown by Figure 17-66.[44] For the purpose of welder training, workmanship specimens have been used and specimens have been developed that include typical problem areas for training welders. Some of the problem areas are welding into and out of corners, changing of electrodes and making fillet welds on outside corners. These welding problems can be put together into one particular workmanship specimen and displayed in the welding department. The automotive industry uses a similar approach for controlling the weld quality on automobile frames. The frame design of different automobiles is usually different; however, the welding involved is similar. In order to provide control, a workmanship specimen display has been made and posted in the department. Figure 17-67 is a typical type of display to assist welders, supervisors, and inspectors in producing the types of welds required for this application.

The use of *workmanship standards* is another technique for maintaining weld quality and can be used as a quality control tool. Workmanship standards are a list of requirements that should be followed in producing quality weldments. A list of 20 requirements has been compiled and is shown by Figure 17-68. This list was made by reviewing the major welding codes and specifications and adopting those rules that repeat in the different codes. In many cases, welding codes and specifications are written in legal terms that may not be easily understood by welders. In compiling these 20 requirements an effort has been made to utilize terminology more understandable by welders. The result of this is a collection of 20 rules that can and should be used when a code is not involved. It is suggested that companies that do not ulitize established codes can adopt these 20 requirements as their own company standard. The following is a discussion concerning these 20 rules to indicate why they are included and involved and why they are important in producing quality welds. The basis for this standard is stated in the first sentence. "Welders are responsible for their own work."

1. "Surfaces to be welded must be reasonably free from scale, grease, paint, water, etc." This is qualified by the word *reasonably* to allow welders and inspectors to exercise judgment. For example, mill scale on new steel just received is usually tight and it would be needless to remove mill scale from new steel; however, if the steel has excessive mill scale or rust it should be removed. The basis here is to provide a good surface for welding. This is good practice and is mentioned in most codes. The surface must be sufficiently clean so that there is nothing that might contain hydrocarbons, which break down in the heat of the arc providing hydrogen which can be absorbed in the weld and cause cracks. Welding over paint is a special case and is discussed elsewhere.

2. "If it is necessary to trim adjacent parts for proper fit-up, the designed bevel and root opening must be maintained." Too often tolerances accumulate in the setting up of a weldment and these are absorbed in weld joints. This can eliminate root openings so that incomplete penetration will result. On the other hand, it can result in excessively large gaps which will require too much weld metal to make the joint and may also introduce undesired warpage. Welders should be cautioned to review the requirements of the weld joint and take corrective action when required if the joint is not as designed.

SPECIFICATION OF FILLET ARC WELDS FOR PASSENGER CAR FRAME FABRICATION

THIS SPECIFICATION IS ISSUED TO DEFINE THE QUALITY LEVEL OF FILLET WELDS IN ORDER THAT THE REQUIRED WELD QUALITY IN MILD STEEL MAY BE DESIGNATED TO INSURE PROPER PERFORMANCE OF EACH AND EVERY INDIVIDUAL PASSENGER CAR FRAME WELDMENT. ALL FILLET WELDS IN THIS SPECIFICATION WILL BE MADE WITH 60,000 P.S.I. MINIMUM FILLER METAL

DEFINITIONS

GAP (G)
THE DISTANCE BETWEEN TWO BASE METAL COMPONENTS AT THE ROOT OF THE JOINT TO BE WELDED.

MATERIAL THICKNESS—T_1-T_2
T_1—THINNER MATERIAL IN WELD JOINT
T_2—THICKEST MATERIAL IN WELD JOINT

FILLET WELD
A WELD OF APPROXIMATELY TRIANGULAR CROSS SECTION JOINING TWO SURFACES APPROXIMATELY AT RIGHT ANGLES TO EACH OTHER IN A LAP JOINT, TEE JOINT, OR CORNER JOINT.

WELD LENGTH
WELD LENGTH SPECIFIED ON THE ENGINEERING DRAWING REFERS TO THE MINIMUM EFFECTIVE DIMENSION. EFFECTIVE LENGTH IS MEASURED WHERE THE FILLET MEETS THE MINIMUM THROAT DIMENSIONS LISTED UNDER JOINT CHARACTERISTICS. WELD ARC STARTING OR FINISHING LOSSES ARE NOT INCLUDED IN THE EFFECTIVE LENGTH.

SIZE OF FILLET WELD
THE LEG LENGTH OF THE LARGEST ISOSCELES RIGHT TRIANGLE WHICH CAN BE INSCRIBED WITHIN THE FILLET WELD CROSS SECTION.

THROAT OF A FILLET WELD (TH)
THE SHORTEST DISTANCE FROM THE ROOT OF A FILLET WELD TO ITS FACE.

UNDERCUT
A GROOVE MELTED INTO THE BASE METAL ADJACENT TO A WELD AND LEFT UNFILLED BY WELD METAL.

SKIP
AN UNWELDED PORTION OF THE DESIGNATED WELD.

BURN THROUGH
AN UNFILLED HOLE MELTED COMPLETELY THROUGH THE BASE METAL BY THE HEAT OF THE WELDING ARC.

SLAG INCLUSION
NON-METALLIC SOLID MATERIAL ENTRAPPED IN THE WELD METAL OR BETWEEN WELD METAL AND BASE METAL.

VISIBLE POROSITY
GAS POCKETS OR VOIDS IN THE WELD DEPOSIT EXPOSED TO THE WELD SURFACE NOT GREATER THAN 1/16" DIAMETER.

CRATER
A DEPRESSION AT THE TERMINATION OF A WELD BEAD.

CRACK
A BREAK OR RUPTURE APPEARING IN THE WELD DEPOSIT OR ADJACENT TO THE WELD JUST AFTER WELD HAS COOLED TO ROOM TEMPERATURE.

WORKMANSHIP SAMPLES
WELD QUALITY WILL BE JUDGED ON DIMENSIONAL AND VISUAL APPEARANCE CONSIDERATIONS AS DEFINED BY THIS SPECIFICATION. WORKMANSHIP SAMPLES WHICH SHOW EXAMPLES OF ACCEPTABLE QUALITY WELDS SHALL BE RETAINED AT THE MANUFACTURING SOURCE.

JOINT CHARACTERISTICS

FILLET SIZE
THE WELD LEG LENGTH OF EACH SIDE OF THE JOINT SHALL BE EQUAL TO THE THICKNESS OF THE THINNER MATERIAL BEING WELDED. WHERE GAPS ARE PRESENT IN THE WELD JOINT, THE LEG LENGTH ON BOTH LEGS OF THE WELD MUST BE INCREASED BY THE WIDTH OF THE GAP, AND THROAT REQUIREMENTS WILL BE PROPORTIONATELY INCREASED.

T_1 = THICKNESS OF THINNER STOCK
G = WIDTH OF GAP
LEG LENGTH = T_1 + G
(WHEN GAP IS PRESENT)

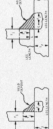

SATISFACTORY

SATISFACTORY WITH GAP

UNSATISFACTORY WITH GAP

MINIMUM THROAT OF THE FILLET WELD TO BE 0.60 OF THINNER MATERIAL BEING JOINED.

SATISFACTORY
THROAT - 0.61T

UNSATISFACTORY
THROAT - 0.50T

MAXIMUM ALLOWABLE UNDERCUT OF BASE METAL
UNDERCUTTING WILL BE PERMITTED PROVIDED THAT IT DOES NOT EXCEED A DEPTH 10% OF THE THINNER STOCK THICKNESS WITHIN 1/2" OF THE EXTREMITIES OF THE WELD, AND THAT THE DEPTH OF THE UNDERCUT DOES NOT EXCEED AN AVERAGE OF 15% OF THE THINNER STOCK THICKNESS ON EITHER TOE OVER THE BALANCE OF THE WELD, NOR REACH A DEPTH EXCEEDING 30% OF THE THINNER STOCK THICKNESS AT ANY POINT.

SATISFACTORY
AVERAGE UNDERCUT - 15%
MAXIMUM UNDERCUT - 30%

UNSATISFACTORY
AVERAGE UNDERCUT - 22%
MAXIMUM UNDERCUT - 26%

BURN THROUGHS
NONE PERMITTED

UNSATISFACTORY

VISIBLE POROSITY
SCATTERED SURFACE POROSITY WILL BE TOLERATED PROVIDED IT DOES NOT EXCEED AN AVERAGE OF ONE PIN HOLE PER INCH OF WELD NOR MORE THAN THREE PIN HOLES IN ANY ONE INCH OF WELD. THE LARGEST PIN HOLES SHALL BE NO GREATER THAN 1/16" IN DIAMETER.

SATISFACTORY

UNSATISFACTORY

ALLOWABLE SKIPS OR SLAG INCLUSIONS
ONE 3/8" DEFECT PERMISSIBLE IN ANY 4" OF WELD.

SATISFACTORY

UNSATISFACTORY

PRESENCE OF CRACKS
NONE ALLOWED.

FIGURE 17-67 *Automobile frame welding requirements.*

Every welder is responsible for welding done. The following workmanship standards must be followed in order to produce high quality welds. They may be employed as a company standard. (Specific codes or specifications take priority over these standards.)

1. Surfaces to be welded must be reasonably free from scale, paint, grease, water, etc.

2. When it is necessary to trim adjacent parts for proper fit-up, the designed bevel and root opening must be maintained. Trim to provide designed weld grooves.

3. Where the clearance between members to be joined in a T type joint is greater than 1/16" the size of the fillet weld shall be the size specified plus the amount of the gap.

4. Preheat or interpass temperature requirements must be adhered to.

5. The specified electrode or electrode matching the base metal must be used.

6. Welding procedures must be followed explicitly. This means using the correct welding electrode type, and size, the proper preheat and interpass temperature and compliance with any specific insturctions with regard to the joint or weldment.

7. Cracked or defective tack welds must be removed before weld is made.

8. Cracked or welds having surface irregularities must be repaired before welding continues.

9. Welds showing excessive, surface porosity must be removed and rewelded.

10. All weld craters shall be completely filled before depositing the next weld bead or pass.

11. Each bead or pass of a multi-pass weld must be cleaned before the next bead is made.

12. The specified size and length of weld as shown on blueprint and drawings, is the minum acceptable. Weld size tolerances should be + 1/6" − 0.

13. Weld reinforcement or crown shall not exceed 1/16" for manual welds.

14. Under-cut is not permissible on highly stressed or dynamically loaded members.

15. Root fusion must be complete on all joints designed with a root opening. This applies whether a backing strip is, or is not, employed.

16. Welds showing subsurface slag or voids, by nondestructive inspection, must be gouged out to sound metal and rewelded. Small amounts of slag or small voids may be permitted.

17. Welds showing subsurface cracks by NDT must be gouged out to sound metal and rewelded.

18. All work should be positioned for flat position welding whenever possible.

19. Weldments or specific welds may be taken at random and submitted to a 100% visual inspection or to any of the nondestructive testing methods to determine weld quality.

20. Welders may be required to requalify if in the opinion of the inspector and the supervisor the work is of questionable quality.

FIGURE 17-68 *Table of workmanship standards.*

3. "Where the spacing between members to be joined by a T type fillet welded joint is greater than 1/16 inch (1.6 mm) the size of the fillets should be the size specified plus the amount of the opening." This is best shown by Figure 17-69. The problem involved here is the fact that a weld fillet size as specified can be made and the inspection after the weld is made would indicate that it is acceptable. However, in view of the opening between the parts, the effective, or failure, area of the fillets would be much less than designed. This type of problem is relatively common, particularly when small fixtures are used to locate premachined parts such as lugs, gussets, etc. Premature failure can occur, particularly if these lugs are load-carrying parts.

4. "The preheat or interpass temperature specified must be adhered to." This is obvious, but it is occasionally ignored. Or, it may be that the preheat is only a surface heat instead of a through heat. If at all possible the temperature of the part should be checked on the side opposite that to which heat is being applied. Specifications recommend a soak type of heating, not merely a surface heating, which would rapidly disappear as soon as the heat source is removed.

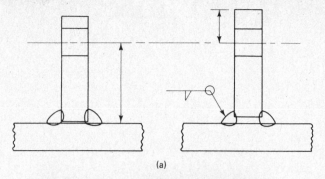

(a)

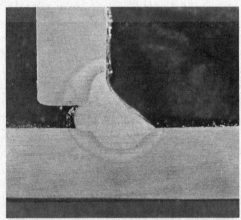

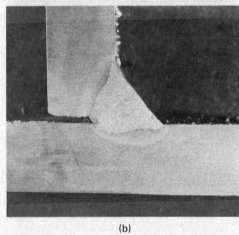

(b)

FIGURE 17-69 *Equal size fillets with different strength.*

6. "The written welding procedure provided must be followed explicitly." This is an obvious requirement; however, in construction jobs and in large manufacturing plants often the written procedures are not available to the welders making the weld. The cost of producing written procedures and especially qualified procedures can be quite expensive. It is prudent to provide specialized information and it must be followed.

7. "Cracked or defective tack welds must be removed before the weld is made." This is an obvious instruction; however, one that is ignored quite often. Tack welds frequently crack. They crack because they are small with respect to the loads that can be imposed on them. Some welders feel that they have sufficient skill to burn through or melt out a cracked tack weld. This may be so, but the average welder normally does not have this skill and the crack is still in the final weld where it can cause problems or will be detected by NDT techniques.

8. "Cracked welds or welds having surface irregularities must be repaired before welding is continued." An obvious requirement, but one sometimes ignored with the feeling that the crack or the irregular weld metal can be covered up with the next pass or bead. This might be so, but it is poor practice since the crack is still in the weld joint and may propagate and create problems later or will be detected by NDT techniques. The best policy is to repair a weld at its earliest possible time of repair.

9. "Welds showing excessive surface porosity must be removed and rewelded." Again, an obvious instruction. Surface porosity is usually an indication that there is subsurface porosity. For this reason, porous welds should be removed and a determination made if sub-surface porosity occurs. The cause of the porosity should be corrected before welding continues.

10. "All weld craters should be completely filled before depositing the next weld bead or pass." A crater is the depression at the end of the weld. However, if welding is to be immediately continued, after changing an electrode, the crater may be allowed to stay, provided that it is filled quickly with the next electrode when the bead is continued. If the crater is at the end of a weld

5. "The specified electrode and electrodes matching the base metal must be used." This is an obvious requirement but one that may be ignored. On some weldments two, three, or more types of electrodes may be employed. It is extremely important that the proper electrode or filler metal be used where specified. The quality of deposited weld metal from different shielded metal arc welding electrodes varies and the properties may not be satisfactory for the service life. For example, the electrode that would be used for welding sheet metal guards or guard rails is not proper for welding major joints of heavy thickness.

or at a change of direction or at a corner it must be filled. This is a matter of technique and can be done by the welder by hesitating shortly before breaking the arc. Craters are prone to produce cracks which might propagate and reduce the service life of the weldment.

11. "Each bead or pass of multipass welds must be cleaned before the next bead is made." This requirement is primarily for the shielded metal arc welding process. It applies also to submerged arc welding and flux-cored arc welding. It is important to remove the slag since it can be trapped in an undercut area and will reduce the strength of the joint. Entrapped slag will be detected by X-ray or other NDT tests. In the case of horizontal multipass fillets judgment can be exercised and cleaning may not be required if in the opinion of the inspector the welder has sufficient skill to avoid entrapping the slag.

12. "The specified size and length of welds as shown on blueprints and drawings is the minimum acceptable. Weld size tolerances should be -0 and $+ 1/16$ inch (1.6 mm)." This is important since many drawings have standardized tolerances printed on the drawing form. Such standardized statements might state that all fractional dimensions shall be $+$ or $-1/16$ inch or similar. Assuming that there are many 3/16-inch fillets, this would mean that fillet welds could all be 1/8 inch, which would in effect greatly weaken them. It is necessary to review such standardized statements and modify them if they could be interpreted to involve dimensions on welds.

13. "Weld reinforcement or crown should not exceed 1/16 inch (1.6 mm) for manual welds or 1/8 inch (3.2 mm) for automatic welds." There are several reasons for this requirement. The most important is that the extra reinforcement is an economic waste since it is not required. Secondly, some welders may put extra reinforcement on a weld to camouflage internal flaws they may know about in a groove weld. Additionally, extra reinforcement causes stress concentration since the stresses will spread to occupy the entire cross section. If the toe of the weld or the edge of the reinforcement is abrupt this can create a concentration. Finally, it is absolutely unnecessary for mild steel and low-alloy steels;

the weld metal is stronger because it has a higher yield strength than the base metal.

14. "Undercut is not permissible on highly stressed or dynamically loaded members." Undercut is a defect that is not allowed by most codes. Some codes allow a small amount of undercut, provided that it is within specified limits. Undercut areas tend to concentrate stresses and will in time create field problems. This can occur on both tensile and compressive loaded weld joints.

15. "Root fusion must be complete on all joints designed with a root opening. This applies whether a backing strip is or is not employed." Root fusion is absolutely necessary for any weld with a root opening. The root face of both sides of the joint must be fused into the weld. It is also the reason why NDT technique is used for root pass inspection.

16. "Welds showing subsurface slag or voids by nondestructive testing must be removed to solid metal and rewelded. Small amounts of slag or voids may be permitted." Some codes do allow small amounts of slag or porosity provided they are small and not continuous. Usually slag and porosity voids have rounded edges and will not propagate under designed loads.

17. "Welds showing subsurface cracks by nondestructive testing must be removed to sound metal and rewelded." In this case the requirement is absolutely necessary since cracks have sharp corners or sharp ends and do propagate under load. They are stress risers and will cause premature failure or shortened service life. Internal cracks are not acceptable by any of the major codes.

18. "All work should be positioned for flat position welding whenever possible." This is good economical practice. A welder who is working in a comfortable position will produce higher quality welds.

19. "Weldments or specific welds can be taken at random and submitted to a 100% visual inspection or to any of the nondestructive testing methods to determine weld quality." This rule and the next are provided to give the inspector the necessary authority that is needed to maintain quality.

20. "Welders may be required to requalify if in the opinion of the inspector and the welder's supervisor the work produced is of questionable quality." Quite often a welder's work may deteriorate for one reason or another. This rule

provides the mechanism of having the welder retested to determine the cause of the problem. This can be a temporary problem or a problem correctible by training or by improved eyesight, etc. This is a tool that the inspector can use whenever a questionable weld quality situation results from the work of a specific welder.

Companies that are not required to produce weldments by a national code should adopt a set of workmanship standards similar to the above. This will provide a better understanding between the designer, the welder, the welding supervisor, and the inspector to maintain weld quality.

Nondestructive testing symbols have been established by the American Welding Society.[45] They are used by the designer to convey information to the inspector concerning joints, welds, or weldments that need special attention. These symbols are very similar to welding symbols and can be used in conjunction with welding symbols. Figure 17-70 shows the elements of the testing symbol and the standard location with respect to each other and only those elements required to provide the needed information are used. The testing symbol or designated letters are shown by Figure 17-71. One special symbol is used to show the direction of radiation which is used in conjunction with the radiographic testing symbol. This is shown by Figure 17-72. Typical testing symbols are shown by Figure 17-73.

For complete information on nondestructive testing symbols it is recommended that the reader consult the AWS standard mentioned above.

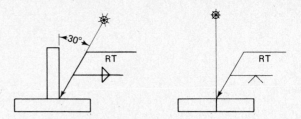

FIGURE 17-72 *Radiation location symbol.*

FIGURE 17-73 *Typical testing symbol.*

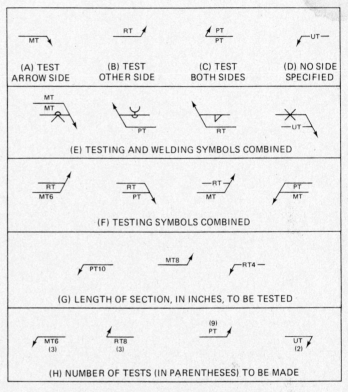

When nondestructive testing symbols have no arrow or otherside significance, the testing symbols shall be centered on the reference line as follows:

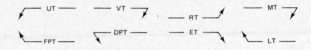

Nondestructive testing symbols and welding symbols may be combined as follows:

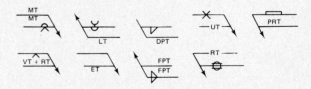

Nondestructive testing symbols may be combined as follows:

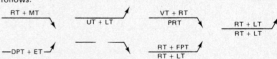

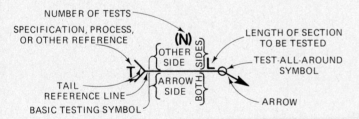

FIGURE 17-70 *Elements of testing symbols.*

Type of Test	Symbol
Visual	VT
Penetrant	PT
Dye penetrant	DPT
Fluorescent Penetrant	FPT
Magnetic Particle	MT
Eddy Current	ET
Ultrasonic	UT
Acoustic Emission	AET
Leak	LT
Proof	PRT
Radiographic	RT
Neutron Radiographic	NRT

FIGURE 17-71 *Nondestructive testing symbol.*

QUESTIONS

1. Why are perfect welds required for some classes of work and not others?

2. There are twenty factors included in a quality assurance plan. Name as many as you can.

3. Who is responsible for a quality product? A quality weld?

4. Codes and specifications are related to industries. What industries use welding specifications?

5. What is a welding procedure? Name the three parts of a welding procedure.

6. What is meant by self-certification? Explain.

7. How is a welding procedure qualified?

8. What is meant by prequalified welding processes? Explain.

9. What is a destructive test? What is a *nick break* test?

10. What is the difference between a root bend and a face bend test?

11. What is the most widely used nondestructive testing technique? What are its three divisions?

12. Explain the use of a fillet weld gage. Show typical fillet size problems.

13. Explain the principle of dye-penetrant testing. What are its limitations?

14. Explain the principle of magnetic particle testing. Why is it not used on aluminum?

15. Explain the principle of radiographic testing. How are radiographs developed?

16. Explain the principle of ultrasonic testing. Can it be used for field inspection?

17. Why is leak testing with compressed air dangerous?

18. Name the six classes of defects.

19. Why are cracks more dangerous than porosity to weld service life?

20. Can testing symbols be combined with weld symbols?

REFERENCES

1. *Boiler and Pressure Vessel Code,* Section III, "Nuclear Power Plant Components," American Society of Mechanical Engineers, 1971, New York.

2. *Code of Federal Regulations,* Section 10, Energy Part 50, Appendix B (10CFR50-B), Quality Assurance Criteria for Nuclear Power Plants and Fuel Reprocessing Plants.

3. Navships 250-1500-1, Standard for Welding of Reactor Coolant and Associated Systems and Components for Naval Nuclear Power Plants.

4. "Code for Pressure Piping," ANSI B 31, American National Standards Institute, New York.

5. API Standard 1104, "Standard for Welding Pipe Lines and Related Facilities," 13th edition, American Petroleum Institute, July 1973, Washington, D.C.

6. "Regulations for the Transportation of Natural and Other Gas by Pipeline," Part 192, Title 49 *Code of Federal Regulations,* Department of Transportation, Office of Pipeline Safety, October, 1, 1973, Washington, D.C.

7. "Transportation of Liquids by Pipeline," Part 195 Department of Transportation, Hazardous Materials Regulations Board, Washington, D.C.

8. *Structural Welding Code,* AWS D1.1-75, 1975, *American Welding Society,* Miami, Florida.

9. "Standard Specifications for Highway Bridges," *American Association of State Highway Officials,* Washington, D.C.

10. "Specification for Steel Railway Bridges," *American Railway Engineering Association,* 1969, Chicago, Illinois.

11. *Marine Engineering Regulations,* Sub Chapter F, Part 57, "Welding and Brazing," U.S. Coast Guard, Department of Transportation, Code of Federal Regulations, Washington, D.C.

12. Navships 0900-000-1000, Fabrication, Welding and Inspection of Ship Hulls, Dept. of the Navy, *Naval Ship Systems Command,* October 1, 1968, Washington, D.C.

13. "Standard Specification for Merchant Ship Construction," Reference ABS, U.S. Maritime Administration.

14. "Rules for Building and Classing Steel Vessels," American Bureau of Shipping, 1974, New York.

15. AWWA "Standard for Welded Steel Elevated Tanks, Standpipes and Reservoirs for Water Storage," D100-73, *American Waterworks Association,* New York.

16. "Welded Steel Tanks for Oil Storage," API Standard 650, American Petroleum Institute, Washington, D.C.

17. "Specifications for Tank Cars," Standard 1972, *Association of American Railroads,* Washington, D.C.

18. "Specifications for Design, Fabrication and Construction of Freight Cars," Standard 1964, *Association of American Railroads,* Chicago, Illinois.

19. *Code of Federal Regulations,* Title 49, Transportation Section 178.340, Part D, "General Design and Con-

struction Requirements," Superintendent of Documents, Washington, D.C.

20. *Code of Federal Regulations,* Title 49, Transportation Section 178.337, "Specification for Cargo Tanks" (ML 331), Superintendent of Documents, Washington, D.C.

21. "Aerospace Material Specifications," Society of Automotive Engineers, Inc., New York.

22. "National Aerospace Standards," *Aerospace Industries Association of America,* Inc., available from National Standards Association, 1315 14th Street, N.W., Washington, D.C.

23. MIL-T-5021D, "Test, Aircraft and Missile Welding Operators Qualification," Department of Defense, Washington, D.C.

24. "Welding on Earthmoving and Construction Equipment," *AWS* D14.3, *American Welding Society,* Miami, Florida.

25. "Welding Industrial and Mill Cranes," *AWS* D14.1, *American Welding Society,* Miami, Florida.

26. "Metal Cutting Machine Tool Weldments," *AWS* D14.2, *American Welding Society,* Miami, Florida.

27. "Welding Procedure and Performance Qualification," AWS B3.0-77, *American Welding Society,* Miami, Florida.

28. "Standard Method of Mechanical Testing of Welds," *AWS* B4.0-74, *American Welding Society,* Miami, Florida.

29. "Nondestructive Testing Personnel Qualification and Certification Supplement D Liquid Penetrant Testing Method." SNT Recommended Practice No. SNT-TC-1A.

30. "Standard Method for Liquid Penetrant Inspection," *American Society for Testing and Materials* (ASTM) E-165.

31. "Nondestructive Testing Personnel Qualification and Certification Supplement B Magnetic Particle Method," SNT Recommended Practice No. SNT-TC-1A.

32. "Standard Method for Wet Magnetic Particle Inspection," ASTM E 138.

33. "Standard Method for Dry Powder Magnetic Particle Inspection," ASTM E 109.

34. "Nondestructive Testing Personnel Qualification and Certification Supplement A Radiographic Testing Method," SNT Recommended Practice No. SNT-TC-1A.

35. "Recommended Practice for Radiographic Testing," ASTM E 94.

36. "Industrial Radiographic Terminology," ASTM E 52.

37. "Standard Reference Radiographs for Steel Welds," ASTM E 390.

38. IIW Collection of Reference Radiographs of Welds, available from the American Welding Society, Miami, Florida.

39. A. G. Barkow, *Interpretation of Radiographs of Pipeline Welding Defects,* Hobart Brothers Company, Troy, Ohio.

40. "Nondestructive Testing Personnel Qualification and Certification Supplement C Ultrasonic Testing Method," SNT Recommended Practice No. SNT-TC-1A.

41. "Standard Method for Ultrasonic Contact Inspection of Weldments," ASTM E 164.

42. IIS/IIW-340-69 (ex doc V-360-67), "Classification of Defects in Metallic Fusion Welds with Explanation," *Metal Construction and British Welding Journal,* February 1970, London, England.

43. "Welding Inspection," *American Welding Society,* Miami, Florida.

44. Navships 0900-999-9000, Acceptance Standards for Surface Finish on Flame or Arc-Cut Material, Department of the Navy, Naval Engineering Center, Washington, D.C., 20360.

45. "Symbols for Welding and Nondestructive Testing," AWS 2.4-76, *American Welding Society,* Miami, Florida.

18

WELDING PROBLEMS
AND SOLUTIONS

18-1 **ARC BLOW**

18-2 **WELDING DISTORTION AND WARPAGE**

18-3 **WELD STRESSES AND CRACKING**

18-4 **IN-SERVICE CRACKING**

18-5 **WELD FAILURE ANALYSIS**

18-6 **OTHER WELDING PROBLEMS**

18-1 ARC BLOW

Arc blow is the deflection of an electric arc from its normal path because of magnetic forces. Deflection of the arc or blow as it is called can be extremely frustrating to a welder. It will usually adversely affect appearance of the weld, will cause excessive spatter, and can also impair the quality of the weld. It is well known to the welder using the shielded metal arc welding process with covered electrodes. It is also a factor in semiautomatic and fully automatic arc welding processes. Arc blow occurs primarily when welding steels or ferromagnetic materials but it can also be encountered when welding nonmagnetic materials. The welding arc is usually deflected forward or backward of the direction of travel; however, it may be deflected from one side to the other. It can become so severe that it is impossible to make a satisfactory weld. Arc blow is one of the most troublesome problems encountered by the welder and one that is the least understood. [1]

The laws of physics, specifically electricity and magnetism, can help explain the problem of arc blow. When an electric current passes through an electrical conductor it produces a magnetic flux in circles around the conductor in planes perpendicular to the conductor and with their centers in the conductor. The *right-hand rule* is used to determine the direction of the magnetic flux. It states that when the thumb of the right hand points in the direction in which the current flows (conventional flow) in the conductor, the fingers point in the direction of the flux. The direction of the magnetic flux produces polarity in the magnetic field, the same as the north and south poles of a permanent magnet. This magnetic field is the same as produced by an electromagnet. The rules of magnetism which state that like poles repel and opposite poles attract apply in this situation. Welding current is much higher than the electrical current normally encountered. Likewise, the magnetic fields are also much stronger.

The welding arc is an electrical conductor and, therefore, the magnetic flux is set up surrounding it in accordance with the *right-hand rule*. The important magnetic field in the vicinity of the welding arc is the field produced by the welding current which passes through it from the electrode and to the base metal or work. This is a self-induced circular magnetic field which surrounds the arc and exerts a force on it from all sides according to the electrical-magnetic rule. As long as the magnetic field is symmetrical there is no unbalanced magnetic force and, therefore, there is no arc deflection. Under these conditions the arc is parallel or in line with the centerline of the electrode and it takes the shortest path to the base plate. If the symmetry of this magnetic field is disturbed the forces on the arc are no longer equal and the arc is deflected by the strongest force.

This electrical-magnetic relationship is used in certain welding applications for magnetically moving or oscillating the welding arc. The gas tungsten arc is rather easily deflected by means of magnetic flux. It can be oscillated by transverse magnetic fields or it can be made to deflect in the direction of travel. To move the arc magnetically is to rely upon the flux field surrounding the arc and to move this field by introducing an external-like polarity field. Oscillation is obtained by reversing the external transverse field to cause it to attract the field surrounding the arc. As the self-induced field around the arc is attracted and repelled, it tends to move the arc column, which tries to maintain symmetry within its own self-induced magnetic field. Magnetic oscillation of the gas tungsten welding arc is used to widen the deposition. Arcs can also be made to rotate around the periphery of abutting pipes by means of rotating magnetic fields. Longer arcs are more easily moved than short arcs. The amount of magnetic flux to create the movement must be of the same order as the flux field surrounding the arc column. This is purposely done and mentioned here only to describe the use of magnetic fields and their relationships to arcs.

The preceding data are helpful in explaining the phenomena of magnetic arc blow. Whenever the symmetry of the field is disturbed by some other magnetic force, it will tend to move the self-induced field surrounding the arc and thus deflect the arc itself.

Except under the most simple conditions the self-induced magnetic field is not symmetrical throughout the entire electric circuit and changes direction at the arc. Furthermore, there is always an unbalance of the magnetic field around the arc because the arc is moving and thus the current flow pattern through the base material is not constant.

There is another factor that helps produce the non-symmetrical or unbalanced relationship of the magnetic forces. The magnetic flux will pass through a magnetic material such as steel much easier than it will pass through air. In fact, the magnetic flux path will tend to stay within the steel and will be more concentrated and stronger than in air. Welding current passes through the electrode lead, the electrode holder to the welding electrode, then through the arc into the base metal (assume steel). At this point the current changes direction to pass to the work lead connection, then through the work lead back to the welding machine. This is shown by Figure 18-1. At the point the arc is in contact with the work, the change of direction is relatively abrupt, and the fact that

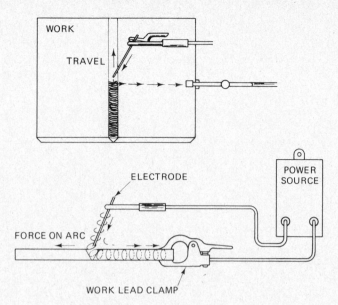

FIGURE 18-1 *Unbalanced magnetic force due to current direction change.*

the lines of force are perpendicular to the path of the welding current creates a magnetic unbalance at this point. The lines of force are concentrated together on the inside of the angle of the current path through the electrode and the work and are spread out on the outside angle of this path. Consequently, the magnetic field is much stronger on the side of the arc toward the work lead connection than on the other side. This unbalance of magnetic forces produces a force on the stronger side which deflects the arc to the left. This is toward the weaker force and is oppostie the direction of the current path. The direction of this force is the same whether the current is flowing in one direction or the other. In other words, if the welding current is reversed the magnetic field is also reversed but the direction of the magentic force acting on the arc is always in the same direction away from the path of the current through the work.

The second factor that keeps a magnetic field from being symmetrical is the fact that the arc is moving and depositing weld metal. As a weld is made joining two plates, the arc moves from one end of the joint to the other. The magnetic field in the plates will constantly change. If we assume the work lead is immediately under the arc and moving with the arc the magnetic path in the work will not be concentric about the point of the arc. This is because the lines of force do not take the shortest path but the easiest path. Near the start end of the joint the lines of force are crowded together since the lines of force will tend to stay within the steel. Toward the finish end of the joint the lines of force will be separated since there is more area. This is shown by Figure 18-2. In addition, where the weld has been made

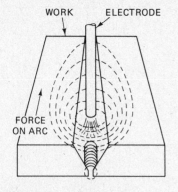

WORK ELECTRODE

FORCE
ON ARC

FIGURE 18-2 *Unbalanced magnetic force due to unbalanced magnetic path.*

the lines of force go through steel. Where the weld is not made the lines of force must cross the air gap or root opening. The magnetic field is more intense on the short end and the unbalance produces a force which deflects the arc to the right or toward the long end.

When welding with direct current the total force tending to cause the arc to deflect is a combination of these two forces. Sometimes these forces tend to add and sometimes they subtract from each other, and at times they may meet at right angles. The polarity or direction of flow of the current does not affect the direction of these forces nor the resultant force. By analyzing the path of the welding current through the electrode and into the base metal to the work lead, and analyzing the magnetic field within the base metal it is possible to determine the resultant forces and predict the resulting arc deflection or arc blow.

The use of alternating current for welding greatly reduces the magnitude of deflection or arc blow. Alternating current for welding does not completely eliminate arc blow. The reason for the reduction of arc blow is that the alternating current sets up other currents that tend to neutralize the magnetic field or greatly reduce its strength. Alternating current varies between maximum value of one polarity and the maximum value of the opposite polarity, and the magnetic field surrounding the alternating current conductor does the same thing. The alternating magnetic field is a moving field which induces current in any conductor through which it passes, according to one of the basic laws of electricity, the *induction principle.* This means that currents are induced in nearby conductors in a direction opposite to that of the inducing current. These induced currents are called eddy currents. They, in turn, produce a magnetic field of their own which tends to neutralize the magnetic field of the arc current. These currents are alternating currents of the same frequency as the arc current and are in the part of the work nearest the arc. They always flow from the opposite direction as shown by Figure

18-3. When alternating current is used for welding, eddy currents are induced in the workpiece, which produce magnetic fields which reduce the intensity of the field acting on the arc. Unfortunately, alternating current cannot be used for all welding applications and for this reason changing from direct current to alternating current may not always be possible to eliminate or reduce arc blow.

With an understanding of the factors that affect arc blow, it is now possible to explain the practical factors involved with arc blow and provide solutions for overcoming them. Arc blow is caused by magnetic forces.

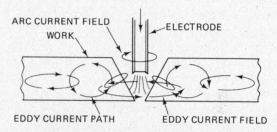

ARC CURRENT FIELD ELECTRODE
WORK

EDDY CURRENT PATH EDDY CURRENT FIELD

FIGURE 18-3 *Reduction of magnetic force due to induced fields.*

The induced magnetic forces are not symmetrical about the magnetic field surrounding the path of the welding current. One factor is the nonsymmetrical location of magnetic material with respect to the arc. This creates a magnetic force on the arc which acts toward the easiest magnetic path and is independent of electrode polarity. The location of the easiest magnetic path changes constantly as welding progresses; therefore, the intensity and the direction of the force changes.

The second factor is the change in direction of the welding current as it leaves the arc and enters the workpiece. Welding current will take the easiest path but not always the most direct path through the work to the work lead connection. The resultant magnetic force is opposite in direction to the current from the arc to the work lead connection. It is independent of welding current polarity.

The third factor explains why arc blow is much less with alternating current. This is because the induction principle creates current flow within the base metal, which creates magnetic fields, that tend to neutralize the magnetic field affecting the arc.

The greatest magnetic force on the arc is caused by the difference in resistance of the magnetic path in the base metal around the arc. The location of the work lead connection is of secondary importance but may have an appreciable effect on reducing the total magnetic force on the arc. It is best to have the work lead connection at the starting point of the weld. This is particularly true in electroslag welding where the work lead should be connected to the starting sump. On occasion, the work lead can be changed to the opposite end of the joint. In some cases, leads can be connected to both ends.

The conditions that affect the magnetic force acting on the arc vary so widely that it is impossible to do more than make generalized statements regarding them. The following suggestions may help reduce arc blow under some conditions.

The magnetic forces acting on the arc can be modified by changing the magnetic path across the joint. This can be accomplished by runoff tabs, starting plates, large tack welds, and backing strips, as well as the welding sequence. An external magnetic field produced by an electromagnet may be effective. This can be accomplished by wrapping several turns of welding lead around the workpiece. Arc blow is usually more pronounced at the start of the weld seam. In this case a magnetic shunt or runoff tab will reduce the blow. In addition, it is always wise to use as short an arc as possible so that there is less of an arc for the magnetic forces to control.

The welding fixture can be a source of arc blow; therefore, an analysis with respect to fixturing is important. The hold-down clamps and backing bars must fit closely and tightly to the work. In general, copper or nonferrous metals should be used. Magnetic structure of the fixture can affect the magnetic forces controlling the arc.

Another major problem can result from magnetic fields already in the base metal particularly when the base metal has been handled by magnet lifting cranes. Residual magnetism in heavy thick plates handled by magnets can be of such magnitude that it is almost impossible to make a weld. The solution here is to attempt to demagnetize the parts, wrap the part with welding leads to help overcome their effect, or, if all else fails, stress relieve or anneal the parts.

18-2 WELDING DISTORTION AND WARPAGE

Most of the welding processes involve heat. High-temperature heat is largely responsible for welding distortion, warpage, and stresses. When metal is heated it expands and it expands in all directions. When metal cools it contracts and again contracts in all directions.

To better understand this consider an extremely small piece of metal the shape of a cube as shown in Figure 18-4. When it is exposed to a temperature increase it will expand in all three directions, the *x*, the *y*, and the *z* direction. There is a direct relationship between the amount of temperature change and the change in dimension. This is based on the coefficient of thermal expansion. This is a measure of the linear increase per unit length based on the change in temperature of the material. The coefficient of expansion is different for the various metals. Aluminum has one of the greatest coefficient of expansion ratios, and changes in dimension almost twice as much as steel for the same temperature change. Table 11-1 showed the coefficient of expansion of the common metals. A metal expands or contracts by the same amount when heated or cooled the same temperature, if it is not restrained. This can be proved; take a straight wire and accurately measure its length cold, then heat it, and measure its length hot. Allow

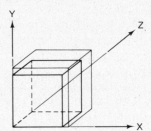

FIGURE 18-4 *Cube of metal showing expansion.*

it to cool, measure it again, and notice that it has returned to the original length. Unfortunately, in the case of welding the metals that are heated and cooled are not unrestrained. They are restrained because they are a part of a larger piece of metal which is not heated to the same temperature. This pinpoints the problem. If we have uniform heating and unrestrained parts, the heating and cooling are relatively distortion free. In actual practice, particularly with respect to welding, the heating is rarely uniform across the cross section of a part. There is always restraint, because the parts not heated or heated to a lesser amount tend to restrain that portion of the same piece of metal that is heated to a higher temperature. This differential or nonuniform heating, which always occurs in welding, and the partial restraint resulting from the part's being nonuniformly heated is the principal cause for thermal distortion and warpage that occur in welding.

The coefficient of expansion is an important factor when considering warpage. It is the factor responsible for the different degrees of warpage between different metals. Of the common structural metals, aluminum has the highest coefficient of expansion; it is approximately twice as great as plain carbon steel. This is significant when trying to relate the warpage that might occur when welding steel compared to when welding aluminum. The coefficient of expansion for steel is 6.7 × 10 to the minus 6th; written out, this would be 0.000,0067, which is the amount in inches that steel expands for every degree Fahrenheit the temperature rises. As an example, if a piece of steel is taken from 100 °F (38 °C) up to a dull red heat of say 1,100° (593 °C) we would have a temperature change of 1000 °F (538 °C). By multiplying these two figures it would show that there is a change in dimension per inch of 0.0067 in. (0.17 mm) or 6.7 thousandths of an inch. This may appear to be a relatively small change in dimension but it is significant. If we would select aluminum for this same temperature change the results would be larger. Aluminum has a coefficient of expansion of 13.8×10^{-6} inches per degree change of temperature. Again, if we expose it to 1,000 °F (538 °C) change in temperature (which, incidentally would almost cause aluminum to melt), we would have a change of $0.000,0138 \times 1,000$ or a dimensional change of 0.0138 in. (0.35 mm) which is a change of almost 1/64 of an inch (0.4 mm) per inch. To make this a little more meaningful, assume a 10-in. (250-mm)-long round rod. Each inch of the aluminum will expand the 0.0138 in. (0.035 mm), but the entire bar, which is 10 in. (250 mm) long, would expand 0.138 in. (3.5 mm) or slightly over 1/8 of an inch (3.2mm) and this is certainly significant with respect to warpage.

In practical application, we do not have the free expansion and contraction and uniform heating. Consider a round rod placed in a vise or in some device that is absolutely unmovable. This is shown by Figure 18-5A. With this rod in between two unmovable surfaces we will uniformly heat the bar a 1,000 °F (538 °C) or the difference above room temperature to say 1,070 °F (577 °C). In the case of the steel bar it would try to expand 0.067 in. (1.7 mm) per inch or in the case of aluminum would try to expand 0.138 in. (3.5 mm) per inch. However, the restraining surfaces will not move. In other words, the bar is restrained in its ability to expand in the length or x direction. However, each of the small cubes within this bar still will expand according to the laws of the temperature and expansion. Therefore, all of the expansion will be in the y and z directions because it will be unable to expand in the x direction. This means that the bar will become slightly larger but not longer, as shown by Figure 18-5 B. This is the principle of upsetting or deformation. Molecules have rearranged themselves and have expanded in two directions, but not in the third direction. Now we will allow the bar to cool down to room temperature. The small cubes within the bar will tend to contract in the x, y, and z directions and will contract the same amount that they expanded. This means that after cooling back to room temperature the bar will be slightly shorter

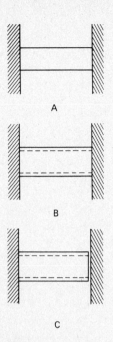

FIGURE 18-5 *Round rod in vice.*

than it was originally, as shown by Figure 18-5C. It will be slightly larger from the upsetting force that occurred during the heating cycle. This illustrates the effect of restraint and shows that the heated portion will not return to its original shape. This illustration, however, may be too elementary, at least when we are considering differential heating with respect to welding. To be a little more practical, we'll assume the same round rod between two immovable surfaces but in this case will include a compression spring. This is shown by Figure 18-6A. We will now go through the same heating cycle. We will heat the rod the same 1,000° (538 °C) but in this situation

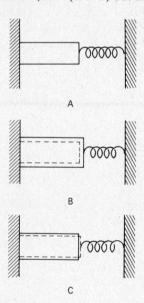

FIGURE 18-6 *Round rod in vice with spring.*

it will expand in length (x direction) as well as in the y and z directions; however, there is a degree of restraint in the length or x direction. The rod does actually become longer and will cause the spring to compress slightly. Therefore, because of the restraint, by the spring, the bar length increase will be less than if it were completely unrestrained. The spring exerts force against the rod, which restrains the rod from expanding as much as when free. There will be expansion in the x and y and z directions so that the bar will actually become somewhat larger as well as longer, as shown by Figure 18-6B. When the heat is removed and the part comes back to its original room temperature it will be somewhat shorter than originally and somewhat larger in diameter, as shown by Figure 18-6C. In other words, the deformation will not be quite so great because the restraint was not so great. This is more similar to what actually happens to weld metal in a differentially heated and cooled cycle.

Another example should be given to illustrate this. Again use a round rod, but in this case considerably longer, and again place it between the immovable surfaces as shown by Figure 18-7A. In this situation, however, we will heat only the center portion of the rod. In the heated area, the temperature rise will cause expansion in all three directions. Restraint will be exercised by jaws and the compressibility of unheated metal; but, this is relatively small. In the heated area there will be expansion in the y and z directions so that the diameter of the bar in the heated area will become larger, as shown by Figure 18-7B. This is a perfect example of plastic deformation or of upsetting. When the bar is allowed to cool contraction will occur uniformly in the x, y, and z

directions and, in effect, the length of the bar will be slightly reduced when it returns to room temperature, as shown by Figure 18-7C.

One final example should be discussed to bring the problem of differential heating and upsetting into focus with regard to practical applications. This is illustrated by the use of a piece of flat rectangular bar stock fairly thin and fairly wide; use 1/4 in. (6.4 mm) thick, 2 in. (50 mm) wide, and 12 in. (300 mm) long. Pass a high temperature heat source such as a gas tungsten arc torch along one edge, as shown by Fig. 18-8A. This creates differential heating across the width of the bar. The top edge of the bar will be heated to the molten stage. Approximately 1/2 in. (50 mm) below the edge in the bar it will remain at room temperature. At the bottom edge of the bar it will also be at room temperature. Each small increment of metal in the upper edge of the bar is being heated to approximately 2,000 °F (1,093 ° C) and it will expand in all three directions. Slightly below the top edge, the bar will be heated to a smaller degree or to a lower temperature; however, even at this point there will be a degree of expansion in all three directions. Further down with little or no change of temperature there will be little or no change in dimension. This will cause the top edge of the bar to tend to expand as shown by Fig. 18-8B, but it is intimately a part of the lower portion of the bar, which has no tendency to expand because it is not heated. The restraint is therefore from the lower portion of the bar, which will restrain the upper portion from expanding to the amount to which it would expand if it were free of the remaining portion of the bar. Plastic deformation will occur and the bar

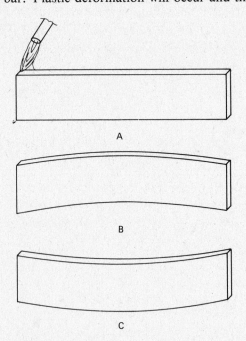

FIGURE 18-8   Long rectangular bar heat on one edge.

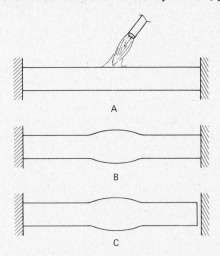

FIGURE 18-7   Long rod in vice.

will become slightly thicker at the heated edge. When the bar cools contraction will occur in all three directions and this will cause the upper edge of the bar to shorten. Shortening the upper edge of the bar without shortening the center or lower edge of the bar will cause warpage, and this is shown by Figure 18-8C. Shortening one edge of a bar and not shortening the other edge will in effect create a curved bar.

There is another factor that must be considered. This is the fact that metals have lower strengths at high temperatures. In other words, as the temperature of a metal increases, its strength decreases. This can be shown by a plot curve on which we plot the yield strength of the metal against temperature. In the case of steel, which might have a yield strength of approximately 36,000 psi, (25.3 kg/mm²), we have a curve shown by Figure 18-9. As the temperature rises the strength decreases at approximately somewhere between 1,000°F (538°C) and 1,500°F (816°C) depending on its composition. For low-carbon mild steel when the temperature is above 1,500°F (816°C) the strength is reduced drastically. This factor is involved because in welding operations a portion of the base metal goes above this temperature since surface melting is involved.

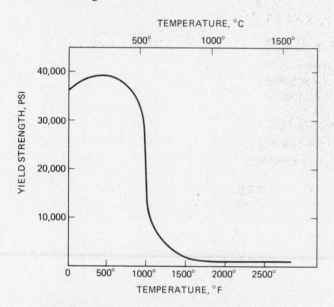

FIGURE 18-9 *Temperature—yield strength relationship.*

One other factor must be considered with regard to the temperature differential in different parts of a metal structure. In the case of the rectangular bar in the previous example, the bar was made of steel. When the temperature at the top edge was practically at the molten stage, a quarter of the way down it would be at a relatively low temperature, and halfway down the width of the bar and at the bottom edge, it would be at room temperature. If the bar is copper or aluminum another factor

becomes important: that of *thermal conductivity,* which has been previously defined. The property of conductivity is shown by Table 11-1, for the more common metals. The thermal conductivity of copper is the highest, that of aluminum is approximately half this figure, but that of steels is only about one-fifth as much. If the bar were made of copper instead of steel the temperature at the top would be practically at the molten stage but the heat would quickly move within the bar to the lower portions so that the temperature differential would not be nearly so great. This would also be true of aluminum and other high thermal conductivity metals. This is important in discussions of warpage since the higher the thermal conductivity the less effect differential heating will have. This physical property should be considered, along with the fact that arc temperatures are very similar but the metal melting points are somewhat different.

When welds are made we must take into consideration all the factors mentioned above and try to determine how each one reacts alone but also how they react with one another. To better understand these relationships consider a weld bead made longitudinally on a relatively thin rectangular plate, as shown by Figure 18-10. When making a weld bead on the plate the deposited weld metal is momentarily at a temperature of about 3,000°F (1649°C), slightly above its melting point. The base metal immediately under the weld bead is also brought to the molten stage. As the weld metal cools and fuses to the base metal immediately under it, it takes shape and forms a bead. The cooling solidification pattern is such that the molten base metal will cool and freeze first. The lower portion of the weld bead then freezes and finally

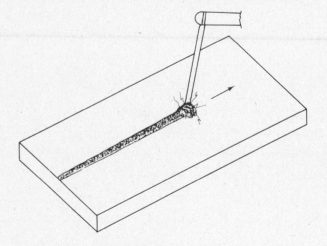

FIGURE 18-10 *Bead on plate.*

629

the total, including the upper portion of the weld bead, freezes. At the point of solidification the molten metal has little or no strength. As it cools it acquires strength, as shown by the graph of Figure 18-9. At this time it is also in its expanded form because of its high temperature. Additionally, the weld metal is now intimately fused to the base metal and they work together, not independently. As the metal continues to cool it acquires higher strength and is now contracting in all three of the x, y, and the z directions. These factors are further complicated because the arc depositing molten metal is a moving source of heat. In addition, the cooling differential is also a moving factor, but does tend to follow the travel of the arc. With the temperature further reducing and each small increment of heated metal tending to contract, contracting stresses will occur and there will be movement in the metal adjacent to the weld. The unheated metal tends to resist the cooling dimension changes of the previously molten metal. Temperature differential has an effect on this. If it is a low conductivity metal the changes will occur over a relatively small distance. If the metal has high thermal conductivity the heating differential will be less and the change in dimensions will be spread over a larger area. In this bead on plate example the cooling shrinkage changes are above the centerline of the thickness of the plate. The tendency of the weld bead to shorten in length, in thickness, and in width tends to warp the plate by shortening the weld area and shortening the top surface in both directions. An exaggeration of how a plate of this type will warp is shown by Figure 18-11.

Running a weld bead on the edge of a bar would be similar to the example of a heat source on the edge of the

bar. The effect might be slightly more since additional molten or high-temperature metal is deposited on the edge of the bar. The deposited metal becomes integral with the bar and provides a greater mass of metal solidifying, gaining strength and cooling. This shortening of the heated edge without a similar change in the nonheated edge would create the warpage which would be the same or similar to that shown by Figure 18-8.

When making a weld joint, specifically a butt joint, between two narrow and relatively thin plates, another factor becomes important. This factor has to do with heat input or the speed of welding or travel. Refer to Figure 18-12, which shows the weld joint partially made.

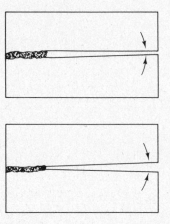

FIGURE 18-12 *Butt joint showing warpage.*

If the same joint is welded with covered electrodes the unwelded end of the joint tends to close or the parts become closer together. If the same joint is welded with submerged arc, the unwelded end of the joint tends to open or the parts become farther apart. The explanation is complicated since it involves the geometry of the pieces being joined, the thermal coefficient of expansion, the thermal conductivity, the mass of molten metal, but most importantly, the travel speed of the heat source or arc. If the travel speed is relatively fast the effect of the heat of the arc will cause expansion of the edges of the plates and they will bow outward and open up the joint. If the travel speed is relatively slow the effect of the arc temperature and the cooling down will cause contraction of the edges of the plates and they will bow inward and close up the joint. This is the same as running a bead on the edge of the plate. In either case it is a momentary situation which continues to change as the weld progresses. By experimenting with current and travel speed, the exact speed can be found for a specific joint design so that the root will neither open up or close together. This is one of the advantages of fine wire gas metal arc welding of sheet metal. The heat input balance, using normal procedures, approaches this travel speed relationship and

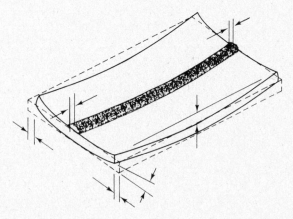

FIGURE 18-11 *Warpage produced by bead on plate.*

for this reason warpage is minimized. Conversely when using the gas tungsten arc process on sheet metal the travel speed is inherently slower and greater warpage usually results.

All these factors must be considered, yet they are not the entire story for more complicated type welds. For example, in a fairly large fillet weld we have all of the factors mentioned previously plus more.

Consider a single-pass fillet weld used to make a corner joint with the fillet weld on the inside, as shown by Figure 18-13. A fillet weld, by definition, has a triangular cross section. In making the weld, the base metal immediately under the fillet in the horizontal plate and the vertical plate are molten. The fillet itself is completely molten and soon after it is deposited it begins to cool and to freeze. Initially it has little strength, but the strength rapidly increases as the temperature decreases. From the geometry of the joint we have absolute restraint where the edge of the vertical plate is in contact with the surface of the horizontal plate. The deposited metal is now integral with the material of the two parts being joined and as it cools its strength is increasing but it is also decreasing in volume. Each and every small increment of the heated metal is decreasing in volume in its *x, y,* and *z* directions. As each increment shrinks and as the surfaces between the two plates are restrained and cannot shrink, we see the evidence of warpage. Normally the fillet weld freezes from the root to the face and the strength would increase in the cooling metal in this same relationship. However, one reason why fillet welds are a special case is the fact that at the root of the fillet there are less small increments of metal that are contracting. But as we progress towards the face of the fillet there are more and more increments that are contracting. Therefore there is more shrinkage or actual movement at the face of the fillet than there is at the root of the fillet, not only because of the larger volume of metal at the face but also because of the restraint between the plates at their interface. If the plates had not been tightly placed together, there would be less total angular warpage at the joint. The fillet weld is particularly vulnerable to warpage because of these two factors. Figure 18-13 shows how this warpage occurs. Note that all the

warpage occurs on one side of the centerline of the vertical plate and on the top side of the centerline of the horizontal plate.

One practical method of eliminating the angular distortion produced by a single-fillet weld is to utilize the T-type joint or the corner joint with the one plate inset so that double-fillet welds would be made. If the fillet welds could be made on both sides simultaneously it would greatly reduce the angular warpage since the two would be working against each other and their shrinkage stresses and their shrinkage dimensions would balance out. However, the longitudinal shrinkage will still occur and in this case we will have highly stressed welds that might be stressed to the yield point of the weld metal at least at the initial freezing period.

If one fillet is made and then the second fillet is made, there will be a certain amount of angular distortion because the initial fillet will freeze first and create some distortion. However, if the second fillet is made larger than the first fillet this can be somewhat overcome. In common practice it is normal to place one pass of the fillet on one side then to place two passes of the second fillet on the other side and finish up on the first side with its second pass. This type of procedure tends to reduce angular warpage of fillet welded T joints.

The cross-section geometry of other welds and the technique of making such welds also contribute to the warpage problem. Consider a weld joint that utilizes a single-V weld. A single-V weld in thinner material can be made with one pass. Figure 18-14 shows a single-pass V-groove butt joint and the warpage that would occur.

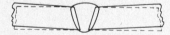

FIGURE 18-14 *Single pass single vee groove butt joint.*

As the weld is deposited in the groove the adjacent base metal is raised to its melting temperature and becomes molten. The entire weld is molten but immediately begins to cool and freeze. As it freezes it increases its strength, but it still has considerable temperature in the order of 1,000°F (538°C). At the bottom of the root of the V groove, each increment of weld metal shrinks in all three directions. At the center of the weld each increment of metal shrinks in all three dimensions, and at the face of the weld the same is true. A weld in thinner material may freeze throughout its cross section almost simultaneously. Since there is more volume of metal freezing

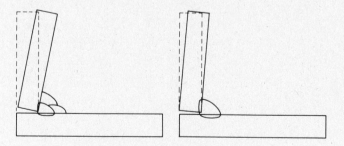

FIGURE 18-13 *Inside fillet corner joint.*

and contracting at the top or wide portion of the joint there will be more total dimensional contraction at this point. The bottom of the joint will contract a small amount. Angular distortion will occur because of the greater amount of weld metal at the top of the joint.

A change in weld joint design has an effect on warpage. If the root opening is increased and the bevel angle is reduced the difference in the amount of weld metal at the root of the weld and at the face of the weld would be more similar and therefore angular distortion would be less. This is one reason for reducing the groove angles to change the ratio of the metal at the root and at the face. Another approach at reducing angular distortion is to use the double V preparation, in which one portion of the weld would be made on one side of the plate and the other portion on the other side. This is applicable only to thicker plates, where it can be justified to have the double preparation. On a thinner material the square groove weld has an advantage since the shrinkage dimension at the root of the weld would be the same as that at the face of the weld. The differential in the cooling rate might create some change which would lead to some angular distortion. In all of these cases, however, the longitudinal distortion or shrinkage will still occur and create the shortening of the total weld and its effect on distortion.

One of the interesting results of electroslag welds is the fact that they produce little or no angular distortion. Normally in electroslag welding process square groove preparation is used. This means that the root and face of the weld has the same dimensions. The weld is made in one pass, thus the root and the face are made simultaneously. As the electroslag weld cools there is some distortion from the start of the weld to the finish of the weld, but this is relatively small. The angular distortion which would tend to warp plates out of line is nonexistent. This is attributed to the fact that the incremental portions of weld metal solidifying and contracting are the same at the root and the face of the weld, and thus we have equal contraction, which eliminates the angular distortion.

Another type of weld that has almost square sides is the electron beam weld with its high depth-to-width ratio. The cross section of an EB weld is almost rectangular, since the width at the face and the width close to the root are almost the same. Angular distortion for electron beam welds is also very small since the sides of the weld are nearly parallel.

The problem of distortion in fillet welds and in V-groove welds increases as the size of the welds increase. The groove weld shown by Figure 18-15 will require many passes to complete the joint or fill the groove. In most processes the number of passes depends on the technique, the size of the electrode, the welding current, etc. With shielded metal arc welding the first or root pass would be placed in the joint and in effect would create a homogeneous structure between the two parts being joined. Very little angular distortion would result from the root pass weld. The next pass would cause some angular distortion. In the second pass the first pass acts as a restraining force, assuming that it is not completely melted by the second pass as it is deposited. The freezing of the second pass would create shrinkage of the deposited metal but the root pass would offer restraint and there would be shrinkage closer to the upper surface of the plates and less at the bottom surface. When the next pass is made there would be more restraint at the root of the joint and more shrinkage at the surface of the weld. Successive passes are larger and wider and there is a greater mass of weld metal cooling and shrinking. This condition continues until the joint is completed. Each new pass creates its heating and cooling and shrinking cycle with previous passes acting as restraint. In

FIGURE 18-15 *Multipass single vee groove butt joint.*

effect, it is like a hinge with the root, or the first pass, acting as the hinge pin, and each additional pass tending to bring the edges of the joint closer together. For this reason multipass single-V-groove welds are particularly susceptible to angular distortion.

It has been proven by many tests and experiments that the larger number of passes used increases the angular distortion in single-groove welds. This disproves the theory that by making many small passes with low currents the distortion will be reduced. This is not true because each pass produces molten metal and in each pass the weld metal cools and shrinks. The mass of molten metal and the mass of metal restraining the shrinkage does affect the relationship, however. The additional applications of heat make for more angular distortion of multipass single fillets and multipass single-V-groove welds. A solution to this is the use of larger passes, larger electrodes, or processes that provide larger passes.

The obvious solution to such situations is to go to the double- weld, the double-V, or double-bevel preparation for thicker materials. The double fillet should be used if at all possible, and if not, a combination of fillet and groove welds. The principle of using double welds is to equalize the shrinkage on both sides of the centerline of the weld joint. If one side of the centerline contracts more than the other it will create angular distortion.

Angular distortion is greatly reduced by balancing the welding on either side of the centerline of the joints being made. The distortion that occurs in the joint is based on the shrinkage which follows thermal cooling versus restraint conditions previously described. Even though there may be no visible evidence of warpage there are probably high stresses approaching yield points of the metals in and near the joint. If a square groove butt weld were made between two edges of two flat plates and if the welds were made from both sides properly, there would be little evidence of warpage. However, if we would saw through the throat of the weld its entire length the warpage would become immediately evident. The result would be the same as two bars with welds on one edge, as previously described. This is shown by Figure 18-16.

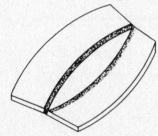

FIGURE 18-16 *Square groove butt weld cut apart.*

The principle of balanced welding must be considered in designing and controlling warpage of weldments. In the case of the bar stock just mentioned the centerline of the bar is actually its neutral axis. The neutral axis is the geographical center of gravity of the cross section of the part. Weldments also have a center of gravity or neutral axis which can be calculated based on the thickness and size of the component parts. An example of a symmetrical cross section part is a fabricated wide flange or H-type beam. To build such a beam will require the welding of two flange plates to a web plate. If the parts are all equal size and thickness the center of gravity would be the exact center of this assembly. This is shown by Figure 18-17. The welds to join the flanges to the web would be equally spaced about the center of gravity. If all four of these fillet welds could be made simultaneously it would be possible to produce the beam without longitudinal distortion. The edges of the flanges will pull in slightly, because of the angular distortion produced by each fillet weld. If the welding can be balanced around the neutral axis of the weldment the distortion can be greatly reduced. This is the principle of balanced welding.

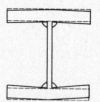

FIGURE 18-17 *Fabricated beam.*

If each application of heat is done in a logical manner about the center of gravity of the weldment, it tends to keep the weldment to true shape. A box section, such as a box column, presents the same type of opportunity. The weldment is symmetrical around its neutral axis and the applications of the welds are also symmetrical. If all four welds could be made simultaneously a straight part would result. In practice it is rarely possible to make all four welds simultaneously. Some of these types of structures have been welded in the vertical position to make the welds simultaneously. A more common practice, however, is to make two of the welds at one time and then make the remaining two welds. Sometimes it is possible to vary the sizes of the welds to produce balanced stresses to create a straight member. Unfortunately, most weldments are not symmetrical around their neutral axis, and more often than not most welding has to be done on one side or another of the neutral axis, and this creates the warpage.

The following factors should be taken into consideration in order to reduce welding warpage.

(a) The location of the neutral axis and its relationship in both directions.

(b) The locations of welds, size of welds, and distance from the neutral axis in both directions.

(c) The time factor of welding and cooling rates when making the various welds.

(d) The opportunity for balancing welding around the neutral axis.

(e) Repetitive identical structures and varying the welding technique based on measurable warpage.

(f) The use of procedures and sequences to minimize weldment distortion.

It is impossible to provide more than general guidelines in warpage control. It is not a true science and even when identical parts are made with identical procedures the warpage of each assembly will be different. This can result from stresses within parts themselves prior to

WELDING PROBLEMS AND SOLUTIONS

welding or seemingly insignificant changes in technique by different welders. However, based on certain rules it is possible to minimize distortion on most weldments. For example, when making the welds on a large box structure such as a ship hull, the sequence of welding can be varied from side to side and from top to bottom to minimize distortion. At shipyards exact measuring devices are used to determine the amount of distortion produced by welding. Corrective action is taken to maintain straightness of the structure. This same technique can be used for any large engineering structures. If the structure tends to warp to one side welding can be increased on the other side to compensate and by measuring continuously and altering procedures the structure can be made to come true. On large weldments it is important to establish a procedure to minimize warpage. The order of joining plates in a deck or on a tank will affect stresses and distortion. As a general rule transverse welds should be made before longitudinal welds. Figure 18-18 shows the order in which the joints should be welded.

Platens and bases that are large and thin and usually made of "egg crate" design can also be regulated by making certain joints first, particularly the joints joining the web sections together prior to joining them to the top and bottom or flange sections. The joining of the web sections to the flanges should be regulated and con-

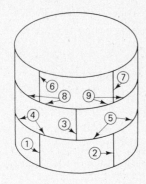

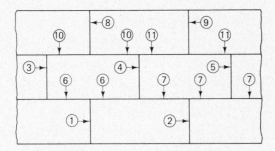

FIGURE 18-18 *The order of making weld joints.*

stant measurements made to determine that the weldment remains true.

The design and type of weldment have much to do with the warpage that is encountered. Large weldments made of relatively thin materials have a tendency to warp. Control is possible with proper procedures. Large weldments made of extremely heavy or thick members have a greater degree of restraint and the amount of warpage may be less. Balancing the welding and monitoring the effect of changes in technique and procedure are the best ways of controlling warpage in large structures.

Warpage can be minimized in smaller structures by different techniques. These include:

(a) The use of restraining fixtures, strong backs, or many tack welds.

(b) The use of heat sinks or the fast cooling of welds.

(c) The predistortion or prebending of parts prior to welding.

(d) Balancing welds about the weldment neutral axis or using wandering sequences or back step welding.

(e) Intermittent welding to reduce the volume of weld metal.

(f) Proper joint design selection and minimum size.

(g) As a last resort use preheat or peening.

Each of these techniques has advantages and can be used in certain applications. No one of them is a cure-all for the problem of weld warpage. Fixtures can be very effective, but fixtures must be extremely strong if they are to resist warpage. Many fixtures are not sufficiently strong for this purpose and even with fixtures balanced welding is necessary. Strong backs or face plates can be used to physically restrain the parts. Massive tack welds and heavy bracing can also be used. If the weldment is stress relieved with bracing in place warpage will be minimized. Figure 18-19 shows a dipper for a large power shovel with massive bracing. Each half was stress relieved separately, because of furnace size limitations, with the bracing in place. Distortion was minimized.

Prebending or predistortion of parts or providing special dimensions for warpage or prewarping with the hope that the welds will bring the parts into proper alignment, can be helpful. This type of technique is most useful when repetitive products are being welded and a history of the amount of warpage involved can be determined. This technique is particularly useful in field erection of large structures. It is also used for welding premachined parts together to avoid finish machining of a total weldment.

Rapid cooling by means of heat extractors or heat sinks has been used successfully in the aircraft industry. By means of hydraulic or pneumatic clamps the parts of the weldment are put in intimate contact with large

FIGURE 18-19 *Heavy braces in weldment.*

masses of highly conductive metal. These are known as heat sinks which pull the heat away from the weld quicker than normal. This creates a more uniform heat distribution and reduces the heat differential and distortion. Some weld fixtures have water-cooled heat sinks to help reduce distortion.

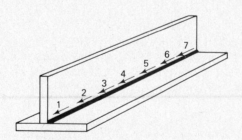

FIGURE 18-20 *Back step technique.*

The use of different types of techniques, such as the back-step technique shown by Figure 18-20, has an advantage in that each small increment will have its own shrinkage pattern which then becomes insignificant to the total pattern of the total weldment. This technique may be rather time consuming and questionable for certain types of applications.

The use of intermittent welding can be helpful; however, intermittent welding is merely a method of reducing the amount of weld metal in certain areas to avoid war-

page at that specific area. Using the smallest possible weld size helps reduce distortion.

As a last resort use peening. Peening actually works the weld metal and expands it which counteracts the shrinkage that occurred upon cooling. Peening is hard to regulate, it is noisy, and usually not the best solution. If the weldment is warped so that it cannot be used it may be salvageable by mechanical or thermal methods. Mechanical methods involve the use of force such as a straightening press, jacks, clamps, etc. This can be expensive but justifiable. Thermal methods involve local heating to relieve stresses in some cases or to cause corrective warpage. Heat is usually applied by torches. Caution should be exercised to not overheat the metal, especially heat-treated materials. Local torch heating should not be used on highly thermal conductive metals such as aluminum or copper. The heat will be conducted away so quickly that local upsetting will not occur. This was covered in Section 11-4.

Finally, one of the best approaches to weldment distortion is the intelligent design of the weldment itself. With intelligent review it is possible to design or redesign a weldment to better place the welds in a balanced geometry around the neutral axis of the weldment. Many times this can change the location of a weld. It might involve the use of a bend, it might involve the use of two welds, but if it can be accomplished it will greatly enhance the distortion control of the weldment.

18-3 WELD STRESSES AND CRACKING

The subjects of weld stresses, cracking, weld distortion, lamellar tearing, brittle fracture, fatigue cracking, weld design, and weld defects are so interrelated that it is almost impossible to treat them separately. All of these factors relate to weldment failure and it is weldment failure that should be eliminated. For clarification and ease of understanding, these factors are segregated and discussed in an orderly fashion. Weld distortion and warpage, which are greatly related to weld stresses, were treated in the previous section. In this section, the problem of welding stress and its effect on weld cracking is explained. The next section will cover brittle fracture, lamellar tearing, and fatigue cracking, which are more or less service oriented, and the final section, weld failure analysis, will attempt to relate all these factors together to the overall problem of producing quality weldments to withstand their designed service requirements.

Residual Stresses.

It was just pointed out in Section 18-2 that metals expand and contract the same amount when heated and cooled the same amount, if the metal is in no way restrained. It was also explained that the heating and cooling that occur in welding are not uniform and there is a temperature difference between the area at the weld and areas adjacent to the weld which in effect creates a restraint on the higher temperature heated metal. The amount of nonuniform heating and the partial, and in some cases almost total, restraint creates stresses within the metal in the weld area and including the weld metal. It was also brought out that as metal is heated to a higher temperature its strength is greatly reduced; however, as the metal cools it regains its strength up to the yield point. If further temperature change occurs the stresses will be greater than the yield point of the metal; however, yielding will occur so that the retained or residual stress will be at the yield point of the metal involved. It was further explained that heated metal expands in all three directions: the *x*, *y*, and *z* directions. Also that as metal cools it contracts in the same three directions but that restraint is always involved. This means that yield point stresses within the weldment may occur in all three directions simultaneously. These internal or remaining stresses are known as residual stress, the "stress remaining in a structure or member as a result of thermal or mechanical treatment or both."

When stresses applied to a member exceed the yield strength the member will yield in a plastic fashion so that the stresses will be reduced to yield point. This is normal in simple structures with stresses occurring in one direction on parts made of ductile materials. Shrinkage stresses due to normal heating and cooling do occur in all three dimensions, however. For example, in a thin flat plate there will be tension stresses at right angles, in other words, in the *x* and *y* directions. As the plate becomes thicker or in extremely thick materials the stresses occur in the *x*, *y*, and *z*—or through—directions as well.

When simple stresses are imposed on a thin brittle material, the material will fail in tension in a brittle manner, the fracture will exhibit little or no ductility. In such cases there is no yield point for the material since the yield strength and the ultimate strength are practically the same. The failures that occur without plastic deformation are known as brittle failures. When two or more stresses occur in a ductile material and particularly when three stresses occur in the *x*, *y*, and *z* directions in a thick material, brittle fracture may occur which is similar to the fracture of a brittle material.

Residual stresses are not peculiar to weldments. They also occur in other types of metal structures, such as castings and forgings, and even hot rolled shapes. Several examples of spontaneous fracturing of rolled structural shapes under conditions of zero external load have been reported. [2] In one case an I beam fractured spontaneously under a condition of zero external load when the beam was lying flat on the ground and under normal temperature conditions. The failure occurred through the center of the web splitting the beam its entire length. High residual stresses also occur in castings and forgings as a result of the differential cooling that occurs. The outer portion of the part cools first and the thicker and the inner portion considerably later. As the parts cool, they contract and pick up strength in accordance with the strength temperature relationship, so that the earlier portions that cool go into a compressive load and the latter portions that cool go into a tensile stress mode. In complicated parts the stresses may cause warpage.

Residual stresses must not always be considered as detrimental. They may have no effect or may have a beneficial effect on the service life of parts. Normally the outer fibers of a part are subjected to tensile loading and thus with residual compression loading there is a tendency to neutralize the stress in the outer fibers of the part. An example of the use of residual stress is in the shrink fit assembly of parts. A typical example is the cooling of sleeve bearings to insert them into machined holes, then allow them to expand to their normal dimension to retain them in the proper location. Sleeve bearings are used for heavy slow machinery and are subjected to compressive residual loading, keeping them within the hole. Also, on the subject of bearings, large roller bearings are usually assembled to shafts by heating them to expand them slightly so that they will fit on the shaft, then allowing them to cool, to produce a tight assembly. One of the most dramatic uses of shrink fit assembly for heavy-duty service is the shrink fitting of steel tires on wheels for railroad locomotives. In this case, the tire is made of relatively high-carbon steel with the required flange. These tires are heated to an elevated temperature and then placed on the locomotive wheel and allowed to cool. This differential heating and cooling allow the tire to shrink around the wheel and make a very strong contact or mechanical connection. Even with the tremendous loads encountered in use, the residual stresses continue to hold the tire on the wheel and thus there is no relaxing of the stresses resulting from mechanical working. It seems rather certain that normal operating loads cannot be expected to reduce the magnitude of internal residual stresses.

Many investigations have been made and techniques established to measure residual stresses. Residual stresses occur in all arc welds but it is only in the more simple joints that accurate measurements have been made. Probably the most common method of measuring has been to produce weld specimens and then to machine away specific amounts of metal which in effect are resisting the tensile stress in and adjacent to the weld and then to measure the movement that occurs. This can be done to produce data showing the magnitude of the residual stresses. Another method is the use of grid marks or data points on the surface of weldments that can be measured in multiple directions. Cuts are made to reduce or release residual stresses from certain parts of the weld joint and the measurements are taken again. The amount of movement relates to the magnitude of the stresses. Another technique is to utilize extremely small strain gauges and gradually mechanically cut the weldment from adjoining portions to determine the change in internal stresses. In this way, experts have been able to establish patterns and actually determine amounts of stress within parts that were caused by the thermal effect of welds.

Based on this data, it is possible to establish a pattern of residual stresses that occurs in a simple weld.

An example of this the residual stresses in an edge weld is shown by Figure 18-21. The metal close to the weld tends to expand in all directions when heated by the welding arc. This metal is restrained by adjacent cold metal and is slightly upset or its thickness slightly increased during this heating period. When the weld metal starts to cool the upset area attempts to contract but is again restrained by cooler metal. This results in the heated zone (upset zone) becoming stressed in tension. When the weld has cooled to room temperature the weld

metal and the adjacent base metal are under tensile stresses close to the yield strength. Therefore there is a portion that is compressive and beyond this another tensile stress area. Thus the two edges are in tensile residual stress with the center in compressive residual stress as shown by the figure.

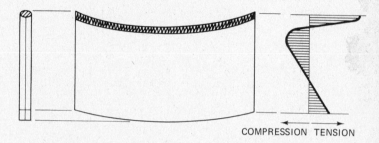

COMPRESSION   TENSION

FIGURE 18-21   *Edge welded joint—residual stress pattern.*

The residual stresses in a butt weld joint made of relatively thin plate are more difficult to analyze. This is because stresses occur in the longitudinal direction of the weld and perpendicular to the axis of the weld. The residual stresses within the weld are tensile in the longitudinal direction of the weld and the magnitude is at the yield strength of the metal. The base metal adjacent to the weld is also at yield stress parallel to the weld and along most of the length of the weld. When moving away from the weld into the base metal the residual stresses rapidly fall to zero and in order to maintain equilibrium change to compression. This is shown by Figure 18-22. The residual stresses in the weld at right angles to the axis of the weld are tensile at the center of the plate and are compressive at the ends. For thicker materials when the welds are made with multipasses, the relationship is different because of the many passages of the heat source. Except for single-pass simple joint designs the compressive and tensile residual stresses can only be estimated. In order to determine at least generally the mode and type of stresses it is important to remember that as each weld is made it will contract as it solidifies and gains strength as the metal cools. As it contracts it tends to pull, and this creates tensile stresses at and adjacent to the weld. Further from the weld or bead the metal must remain in equilibrium and therefore compressive stresses occur. In heavier weldments when restraint is involved, movement is not possible and therefore residual stresses are of a higher magnitude. For example, in a multipass single-groove weld the first weld or root pass originally created a tensile stress. The second, third,

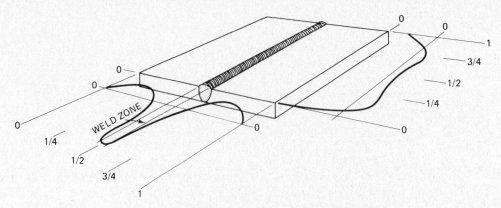

0
0
0
1
3/4
1/2
1/4
0
WELD ZONE
0
0
0
1/4
1/2
3/4
1

FIGURE 18-22 *Butt welded joint—residual stress pattern.*

and fourth passes contract and cause a compressive load in the root pass. As more passes are made until the weld is finished the top passes will be in tensile load, the center of the plate in compression, and the root pass will have tensile residual stress again.

Residual stresses can be decreased in several ways. If the weld is stressed by a load to beyond its yield strength plastic deformation will occur and the stresses will be made more uniform but still at the yield point of the metal. This will not eliminate residual stresses but at least will create a more uniform stress pattern. Another way to reduce high or peak residual stresses is by means of loading or stretching the weld by heating adjacent areas causing them to expand. The heat reduces the yield strength of the weld metal and the expansion will tend to reduce peak residual stresses within the weld. This technique will also make the stress pattern at the weld area more uniform. The more positive way of reducing high residual stresses is by means of the stress relief heat treatment. Here the weldment is uniformly heated to an elevated temperature at which the yield strength of the metal is greatly reduced. The weldment is then allowed to cool slowly and uniformly so that the temperature differential between parts is minor and thus the cooling will be uniform and a uniform low stress pattern will develop within the weldment. High-temperature preheating will also reduce residual stresses since the entire weldment is at a relatively high temperature and will cool more or less uniformly from that temperature and thus reduce peak residual stresses.

Weld Cracking.

Residual stresses do contribute to weld cracking. Often weld cracking will occur during the manufacturing operation of the weldment or shortly after the weldment is completed. Cracking may occur for many reasons and may occur years after the weldment is completed. It is our purpose to describe weld cracking that is a result of the residual stresses and weld cracking that occurs during the manufacturing operation or shortly thereafter.

Cracks are the most serious of the defects that occur in welds or weld joints in the weldments. Cracks are not permitted in most weldments, particularly those subjected to low-temperature service, impact loading, reversing stresses, or when the failure of the weldment will endanger life. It is therefore important to understand the mechanism of weld cracking so as to avoid weld cracks in all welds and weldments. Weld cracking that occurs during or shortly after the fabrication of the weldment can be classified as hot cracking or cold cracking. In addition, welds may crack in the weld metal or in the base metal adjacent to the weld metal, usually in the heat-affected zone. Welds crack for a variety of reasons; such as:

Insufficient weld metal cross section to sustain the loads involved.

Insufficient ductility of weld metal to yield under stresses involved.

Under-bead cracking due to hydrogen pick up in a hardenable type of base material.

Restraint and residual stresses are among the main reasons for weld cracking during the fabrication of a weldment. Weld restraint can come from several factors. One of the most important is the stiffness or rigidity of the weldment itself. For example, if the weldment is made of thick material and it is of a highly restrained nature there will be little chance for yielding or movement in the weld joint. If the weld metal does not have sufficient ductility cracking will occur. Weld metal shrinks as it cools and if the parts being welded cannot move with respect to one another and if the weld metal has insufficient ductility a crack will result. Additionally, movement of welds may impose high loads on other welds and cause them to crack during the fabrication. In such cases, the best solution is to use a more ductile filler material or to make the weld with sufficient cross-sectional area so that as it cools it will have sufficient strength to withstand cracking tendencies. The typical weld crack can occur in the root pass when the parts are unable to move.

Another factor involved is the rapid cooling of the weld deposit. If the base metal being joined is cold and the weld is relatively small it will cool extremely rapidly. Shrinkage will occur quickly and cracking can occur. If the parts being joined are preheated even slightly the cooling rate will be lower and cracking can be eliminated. One of the advantages of preheat is to reduce the cooling rate of the weld and of the adjacent material. In addition, if the base metal is at an elevated temperature it will have lower yield strength and will not be as restrictive as far as its restraint on the weld is concerned.

Another reason for this type of cracking can be the alloy or carbon content of the base material. When a weld is made with higher-carbon or high-alloy base material a certain amount of the base material is melted and mixed with the electrode to produce the weld metal. The resulting weld metal has higher carbon and alloy content; it may have a higher strength but it has less ductility and as it shrinks it may not have sufficient ductility to cause plastic deformation and therefore cracking may occur. This can be eliminated by using a more ductile weld metal or by reducing the cooling rate of the weld and also by reducing the amount of base metal picked up and mixed in with the weld metal.

Another factor involved can be hydrogen pickup in the weld metal and in the heat-affected zone of the base metal. When using cellulose covered electrodes or when hydrogen is present because of damp gas, damp flux, hydrocarbon surface materials, etc., the hydrogen in the arc atmosphere will be absorbed in the molten weld metal and in the adjacent high-temperature base metal. As the metal cools it will reject the hydrogen and if there is sufficient restraint cracking will occur. This type of cracking can be reduced by increasing preheat, reducing restraint, and, of course, eliminating the hydrogen from the arc atmosphere.

As a general rule to eliminate weld cracking during fabrication it is wise to follow these principles.

Use ductile weld metal.

Avoid extremely high restraint or residual stresses.

Revise welding procedures to reduce restraint.

Utilize low-alloy and low-carbon materials.

Reduce cooling rate by use of preheat.

Utilize low-hydrogen welding processes and filler metals.

When cracking is in the heat-affected zone or if cracking

is delayed the culprit is possibly hydrogen pickup in the weld metal and heat-affected zone of the base metal. Another important factor is the presence of higher-carbon materials or high alloy in the base metal. It is important to utilize low-hydrogen filler metals, to reduce cooling rates by means of preheat, etc., and to use ductile filler materials.

One possible solution when welding high-alloy or high-carbon steels is to use the buttering technique. This involves surfacing the weld face of the joint with a weld metal that is much lower in carbon and alloy content than the base metal. The weld is then made between the deposited surfacing material and avoids the carbon and alloy pickup in the weld metal and thus allows a more ductile weld metal deposit. Care must be used, however, so that the total joint strength is sufficient to meet design requirements. Underbead cracking is greatly reduced by use of low-hydrogen type processes and filler metals. The use of preheat reduces the rate of cooling which tends to decrease the possibility of cracking. Reducing the cooling rate will materially reduce the chance of cracks. When welds are too small for the service intended they will probably crack. This is common in tack welds where a small weld is expected to carry extreme loads which are impossible for the strength level of the tack welds. Many specifications list minimum size of fillet welds that can be used to join different thicknesses of steel sections. If these minimum sizes are used cracking will be eliminated.

18-4 IN-SERVICE CRACKING

Our objective is always to design and build weldments that perform adequately in service. The risk of failure of a weldment is relatively small, but it can occur in structures such as bridges, pressure vessels, storage tanks, ships, penstocks, etc. Welding has sometimes been blamed for the failure of large engineering structures, but it should be noted that failures have occurred in riveted and bolted structures and in castings, forgings, hot rolled plate and shapes, as well as other types of construction. Failures of these types of structures occurred before welding was widely used and still occur in un-welded structures today. In spite of this, it is important from the welding point of view to make weldments and welded structures as safe against premature failure of any type as we possibly can. There are at least four specific types of failures that we should be aware of so that proper steps can be taken to avoid them. These types of failures are:

Brittle fracture.

Fatigue fracture.

Lamellar tearing.

Stress corrosion cracking.

Most of these problems have been previously mentioned in other sections of this book. In the design section, designers were alerted to the problems involved and how they can be avoided. In the quality control section, attention was focused on workmanship errors that might contribute to failures of these types. It is important to understand the background of these types of failure so that design, workmanship, and every other factor that might contribute can be considered and avoided. Each of these failure modes will be covered in detail.

Brittle Fracture.

The fracture of metals is a very complex subject which is beyond the scope of this book; however, fracture can be classified into two general categories, *ductile* and *brittle.*

Ductile fracture occurs by deformation of the crystals and slip relative to each other. There is a definite stretching or yielding. There is a reduction of cross-sectional area at the fracture (see Figure 18-23). Brittle fracture occurs by cleavage across individual crystals and the fracture exposes the granular structure; there is little or no stretching or yielding. There is no reduction of area at the fracture (see Figure 18-24).

It is possible that a broken surface will display both ductile and brittle fracture over different areas of the surface. This means that the fracture which propagated

FIGURE 18-24 *Brittle fracture surface.*

across the section changed its mode of fracture. There are four factors that should be reviewed when analyzing a fractured surface. They are: (1) growth marking, (2) fracture mode, (3) fracture surface texture and appearance, and (4) amount of yielding or plastic deformation at the fracture surface. Growth markings are one way to identify the type of failure. Fatigue failures are characterized by a fine texture surface with distinct markings produced by erratic growth of the crack as it progresses. The *chevron* or *herringbone* pattern occurs with brittle or impact failures. The apex of the chevron appearing on the fractured surface always points toward the origin of the fracture and is an indicator of the direction of crack propagation. The second factor is the fracture mode. Ductile fractures have a shear mode of crystalline failure. The surface texture is silky or fibrous in appearance. Ductile fractures often appear to have failed in shear as evidenced by all parts of the fracture surface assuming an angle of approximately 45° with respect to the axis of the load stress. Brittle or cleavage fractures have either a granular or a crystalline appearance. There is usually a point of origin of brittle fractures. The chevron pattern will help locate this point. The necking down of the surface of the fractured part is an indication of the amount of plastic deformation. There is little or no deformation for a brittle fracture and usually a considerably necked down area in the case of a ductile fracture.

One characteristic of brittle fracture is that the steel breaks quickly and without warning. The fractures propagate at very high speeds and the steels fracture at stresses below the yield strength normal for the steel. Mild steels, which show a normal degree of ductility, when tested in tension as a normal test bar, may fail in a brittle manner. In fact, mild steel may exhibit good toughness characteristics at room temperature. Brittle fracture is therefore more similar to the fracture of glass than fracture of normal ductile materials. A combination of conditions must be present simultaneously for brittle fracture to occur. This is reassuring since some of these factors can be eliminated and thus reduce

FIGURE 18-23 *Ductile fracture surface.*

the possibility of brittle fracture. The following conditions must be present: (1) low temperature, (2) a notch or defect, (3) a relatively high rate of loading, (4) triaxial stresses normally due to thickness or residual stresses, and finally (5) the microstructure of the metal.

Temperature is an important factor. However, temperature must be considered in conjunction with microstructure of the material and the presence of a notch. Impact testing of steels using a standard notched bar specimen at different temperatures shows a transition from a ductile type failure to a brittle type failure based on a lowered temperature. This was shown and explained in Section 11-1. The change from ductile to brittle fracture is known as the transition temperature. Unfortunately, notched specimens are different from large engineered weldments. However, notched specimen results do provide a correlation that is useful in selecting the better material.

The notch that can result from faulty workmanship or from improper design produces an extremely high stress concentration which prohibits yielding in the normal sense. A crack, for example, will not carry stress across it and the load is transmitted to the end of the crack. It is concentrated at this point and little or no yielding will occur. Metal adjacent to the end of the crack which does not carry load will not undergo a reduction of area since it is not stressed. It is in effect a restraint which helps set up triaxial stresses at the base of the notch or the end of the crack. Stress levels much higher than normal occur at this point and contribute to starting the fracture.

The rate of loading is the time versus strain rate. The high rate of strain, which is a result of impact or shock loading, does not allow sufficient time for the normal slip process to occur. The material under load behaves elastically allowing a stress level beyond the normal yield point. When the rate of loading, from impact or shock stresses, occurs near a notch in heavy thick material, the material at the base of the notch is subjected very suddenly to very high stresses. The effect of this is often complete and rapid failure of a structure and is what makes brittle fracture so dangerous.

Triaxial stresses are more likely to occur in thicker material than in thin material. The z direction acts as a restraint at the base of the notch and for thicker material the degree of restraint in the through direction is higher. This is why brittle fracture is more likely to occur in thick plates or complex sections than in thinner materials. In addition, thicker plates usually have less mechanical working in their manufacture than thinner plates and are more susceptible to lower ductility in the z axis. The microstructure and chemistry of the material in the center of thicker plates have poorer properties than the thinner material which receives more mechanical working.

The microstructure of the material is of major importance with respect to the fracture behavior and transition temperature range. Microstructure of a steel depends on the chemical composition and production processes used in manufacturing it. A steel in the as-rolled condition will have a higher transition temperature or lower toughness than the same steel in a normalized condition. Normalizing, that is, heating to the proper temperature and slow cooling, produces a grain refinement which provides for higher toughness. Unfortunately, fabrication operations on steel such as hot and cold forming, punching, and flame cutting affect the original microstructure. This raises the transition temperature of the steel.

Unfortunately, welding tends to accentuate some of the undesirable characteristics that we wish to avoid in order to avoid brittle fracture. The thermal treatment resulting from welding will tend to reduce the toughness of the steel or possibly to raise its transition temperature in the heat-affected zone. The monolithic structure of a weldment means that more energy is locked up and there is the possibility of residual stresses which may be at yield point levels. Additionally, the monolithic structure causes stresses and strains to be transmitted throughout the entire weldment, and defects in weld joints can be the nucleus for the notch or crack that will cause fracture initiation. More information on brittle fracture is found in reference 3.

The problem of brittle fracture can be greatly reduced in weldments by selecting steels that have sufficient toughness at the service temperatures. The transition temperature should be below the service temperature to which the weldment will be subjected. Heat treatment or normalizing or any method of reducing locked-up stresses will reduce the triaxial yield strength stresses within the weldment. Design notches must be eliminated and notches resulting from poor workmanship must not occur. This requires the elimination of internal cracks within the welds and of unfused root areas either by design or by accident. By closely following these conditions the possibility of brittle fracture will be eliminated or greatly reduced.

Fatigue Failure.

Structures sometimes fail at calculated nominal stresses considerably below the tensile strength of the materials involved. The materials involved were ductile in the normal tensile tests but the failures generally exhibited little or no ductility. Most of these failures developed after the structure had been subjected to a large number of cycles of loading. This type of failure is called a fatigue failure. Fatigue failure is the formation of and development of a crack by repeated or fluctuating loading. When sudden failure occurs it is because the crack has propagated sufficiently to reduce the load-carrying capacity of the part. Fatigue cracks may exist in some weldments but they will not fail until the load-carrying area is sufficiently reduced. It is the repeated loading which causes progressive enlargement of the fatigue cracks through the material. The rate at which the fatigue crack propagates depends upon the type and intensity of stress and a number of other factors involving the design, the rate of loading, type of material, etc.

The fracture surface of a fatigue failure has a typical characteristic appearance. It is generally a smooth surface and frequently shows concentric rings or areas spreading from the point where the crack initiated. These rings show the propagation of the crack which might be related to periods of high stress followed by periods of inactivity. The fracture surface also tends to become rougher as the rate of propagation of the crack increases. Figure 18-25 shows the characteristic fatigue failure surface.

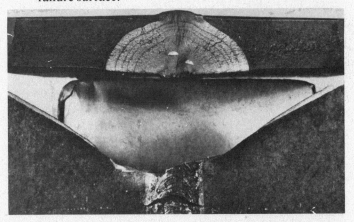

FIGURE 18-25 *Fatigue fracture surface.*

In Chapter 15, it was pointed out that many structures are designed to a permissible static stress based on the yield point of the material in use and the safety factor that has been selected. This is based on statically loaded structures, the stress of which remains relatively constant with respect to time. Many structures, however, are subjected to other than static loads in service. They are loaded by various live loads applied in different ways, for example, that caused by cyclic loading as in the case

of a rotating device or of a bridge carrying varying traffic, or dynamic loads from machinery, or loads based on temperature changes or vibrations, etc. These changes may range from simple cyclic fluctuations to completely random variations. In this type of loading the structure must be designed for dynamic loading and considered with respect to fatigue stresses.

The varying loads involved with fatigue stresses can be categorized in different manners. These can be alternating cycles from tension to compression. Or they can be pulsating loads with pulses from zero load to a maximum tensile load, or from a zero load to a compressive load, or loads can be high and rise higher, either tensile or compressive. In addition to the loadings, it is important to consider the number of times the weldment is subjected to the cyclic loading. For practical purposes, loading is considered in millions of cycles. Fatigue is a cumulative process and its effect is in no way healed during periods of inactivity. Testing machines are available for loading metal specimens to millions of cycles and the results are plotted on stress vs cycle curves, which show the relation between the stress range and the number of cycles, for the particular stress used. Fatigue test specimens are machined and polished and the results obtained on such a specimen may not correlate with actual service life of a weldment. It is therefore important to determine those factors which adversely affect the fatigue life of a weldment.

The possibility of a fatigue failure depends on four factors.

The material used.

The number of loading cycles.

The stress level and nature of stress variations.

Total design and design details.

It is this last factor that is controllable in the design and manufacture of the weldment. The problem of uniform distribution of stress throughout the cross-sectional area of a section was discussed in Chapter 15. It was brought out that weld joints can be designed for uniform stress distribution utilizing a full-penetration weld, but in other cases joints may not have full penetration because of an unfused root and this prohibits uniform stress distribution. Even with a full-penetration weld if the reinforcement is excessive a portion of the stress will flow through the reinforced area and will not be uniformly distributed. It was brought out previously that welds designed for full penetration might not have complete penetration because of workmanship factors such as cracks, slag inclusions, incomplete penetration, etc. and therefore contain a stress concentration. One reason why fatigue failures in welded structures occur is because the welded design can introduce more severe stress concentrations than other types of design. The weld defect

as mentioned previously and including excessive reinforcement, undercut, or negative reinforcement will all
contribute to the stress concentration factor. In addition,
a weld forms an integral part of the structure and when
parts are attached by welding they may produce sud-
den·changes of section which contribute to stress concentrations under normal types of loading.

Anything that can be done to smooth out the stress
flow in the weldment will reduce stress concentrations
and make the weldment less subject to fatigue failure.
Total design with this in mind and careful workmanship
will greatly eliminate this type of a problem. See reference
4 for more information on fatigue failure.

Lamellar Tearing.

Lamellar tearing is a relatively new term which has come
into prominence because of some failures in structural
steel work in buildings and in offshore drill rigs and
platforms. Lamellar tearing is a cracking which occurs
beneath welds and is found in rolled steel plate weldments. The tearing always lies within the base metal,
usually outside the heat-affected zone and generally
parallel to the weld fusion boundary. Lamellar tearing
is not new, but the term is. This type of cracking has
been found in corner joints where the shrinkage across
the weld tended to open up in a manner similar to lamination of plate steel. In these cases, the lamination type
crack is removed and replaced with weld metal. Unfortunately, before the advent of ultrasonic testing this
type of failure was probably occurring and was not found.
It is only when welds subjected the base metal to tensile loads in the z or through direction of the rolled
steel that the problem is encountered. For many years
the lower strength of rolled steel in the through direction
was recognized and the structural code prohibited z-
directional tensile loads on steel spacer plates. Figure
18-26 shows how lamellar tearing will come to the surface
of the metal. Figure 18-27 showing a tee joint is a more
common type of lamellar tearing, which is much more
difficult to find. In this case, the crack does not come
to the surface and is under the weld. This type of crack
can only be found with ultrasonic testing or if failure
occurs the section can actually come out and separate
from the main piece of metal.

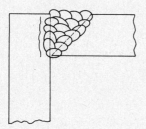

FIGURE 18-26   *Corner joint.*

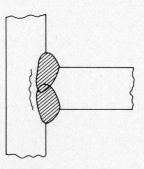

FIGURE 18-27   *Tee joint.*

Three conditions must occur to cause lamellar tearing.
These are:

1. Strains must develop in the through direction
   of the plate. These strains are caused by weld
   metal shrinkage in the joint and can be increased
   by residual stresses and by loading.

2. The weld orientation must be such that the stress
   acts through the joint across the plate thickness
   or in the z direction. The fusion line beneath
   the weld is roughly parallel to the lamellar
   separation.

3. The material will have poor ductility in the
   through or z direction.

Lamellar tearing can occur during flame-cutting operations and also in cold-shearing operations. It is primarily
the low strength of the material in the z or through direction that contributes to the problem. It is a stress placed
in the z direction that triggers the tearing. The thermal
heating and the stresses resulting from weld shrinking
create the fracture. Lamellar tearing is not associated
with the under-bead hydrogen cracking problem. It can
occur soon after the weld has been made but on occasion
will occur at a period months later. Also, the tears are
under the heat-affected zone and it is seemingly more
apt to happen in thicker materials and in higher-strength
materials.

Only a very small percentage of steel plates are susceptible to lamellar tearing. There are only certain plates
where the concentration of inclusions are coupled with
the unfavorable shape and type that presents the risk of
tearing. These conditions rarely occur with the other
two factors mentioned previously. In general, the three
situations must occur in combination: structural restraint,
joint design, and the condition of the steel. The experience gained to date indicates that joint details can be
changed to avoid the possibility of lamellar tearing. In

the case of tee joints it appears that double-fillet weld joints are less susceptible than full-penetration welds. In addition, balanced welds on both sides of the joint appear to present less risk than large single-sided welds.

Corner joints are common in box columns. Lamellar tearing at the corner joints is readily detected on the exposed edge of the plate. The easy way of overcoming the problem of corner joints is to place the bevel for the joint on the edge of the plate that would exhibit the tearing rather than on the other plate. This is shown by Figure 18-28.

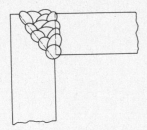

FIGURE 18-28 *Redesigned corner joint to avoid lamellar tearing.*

Butt joints rarely are a problem with respect to lamellar tearing since the shrinkage of the weld does not set up a tensile stress in the thickness direction of the plates.

Experience has indicated that the arc welding processes having the higher heat input are less likely to create lamellar tearing. This may be because of the fewer number of applications of heat and the lesser number of shrinkage cycles involved in making a weld. It is also found that the deposited filler metal with lower yield strength and high ductility seemingly reduces the possibility of lamellar tearing. It does not appear that preheat is specifically advantageous with respect to lamellar tearing. Also, stress relief heat treatment does not appear to have any beneficial effect. As a last resort, the buttering technique of laying one or more layers of low-strength high-ductility weld metal deposit on the surface of the plate stressed in the z direction will materially reduce the possibility of lamellar tearing. This is perhaps an extreme solution and should only be used as a last resort. The steel companies are attempting to initiate improvements in steel processing to avoid lamellar tearing. However, by observing the design factors just mentioned, the lamellar tearing problem is reduced. More information regarding lamellar tearing can be found in reference 5.

Stress Corrosion Cracking.

Stress corrosion cracking and delayed cracking due to hydrogen embrittlement can both be troublesome when the weldment is subjected to the type of environment that accentuates this problem. Delayed cracking is caused by hydrogen absorbed in the base metal or weld metal at high temperatures. Liquid or molten steel will absorb large quantities of hydrogen. As the metal solidifies it cannot retain all of the hydrogen and it is forced out of solution. The hydrogen coming out of the solution sets up high stresses and if sufficient amount of hydrogen is present it will cause cracking in the weld or the heat-affected zone. These cracks develop over a period of time after the weld is completed. The concentration of hydrogen and the stresses resulting from it when coupled with residual stresses promote cracking, cracking will be accelerated if the weldment is subjected to thermal stresses due to repeated heating and cooling.

Stress corrosion cracking in steels is sometimes called *caustic embrittlement.* This type of cracking takes place when hot concentrated caustic solutions are in contact with steel that is stressed in tension to a relatively high level. The high level of tension stresses can be created by loading or by high residual stresses. Stress corrosion cracking will occur if the concentration of the caustic solution in contact with the steel is sufficiently high and if the stress level in the weldment is sufficiently high. This situation can be reduced by reducing stress level and reducing the concentration of the caustic solution. Various inhibitors can be added to the solution to reduce the concentration. Another solution is to maintain close inspection on highly stressed areas.

Another type of cracking is called *graphitization.* This is caused by long service life exposed to thermal cycling, that is, repeated heating and cooling. This may cause a breakdown of carbides in the steel into small areas of graphite and iron. This formation of graphite, in the edge of the heat-affected area, exposed to the thermal cycling causes cracking. It will more often occur in carbon steels deoxidized with aluminum. The addition of molybdenum to the steel tends to restrict graphitization and for this reason carbon molybdenum steels are normally used in high-temperature power plant service. These steels must be welded with filler metals of the same composition.

18-5 WELD FAILURE ANALYSIS

Fortunately it is only rarely that there are failures of welded structures. However, failures of large engineered structures of all types of construction do occur occasionally. Catastrophic failures of major structures are usually reported whenever they might occur. The details and results of investigations of these failures are usually reported. These reports are useful since they often provide information that is helpful in avoiding future similar problems. In the same manner, there are occasional failures of noncritical welds and weldments that should also be investigated. The determination of the cause of these failures can be fruitful since once the reason is determined it can then be avoided.

It is important to make an objective study of any failure of parts or structures to determine the cause of the failure. This is done by investigating the service life, the conditions that led up to the failure, and the actual mode of the failure. An objective study of failure should utilize every bit of information available, should investigate all factors that could remotely be considered, and should attempt to evaluate all this information to arrive at the reason for the failure.

Failure investigation quite often will uncover facts that will lead to changes in design, manufacturing, or operating practice that will eliminate similar failures in the future. In this way the advantage and usefulness of the investigation will be realized. Failures of insignificant parts can also lead to advances in knowledge and should be done objectively, just as with a large expensive structure. Each failure and subsequent investigation will lead to changes that will assure a more reliable product in the future.

More and more often failure analysis is required in order to establish cause of the failure and determine the responsibility for the failure. In view of this, it is absolutely necessary that an objective position be maintained throughout the investigation and that the information and data speak for themselves in establishing the cause of the failure. The investigator should use extreme care and should present the facts in a logical order. The following four areas of interest should be investigated to determine the cause of the failure and the interplay of factors involved.

1. **The Initial Observation:** The detailed study by visual inspection of the actual component that failed should be made at the failure site as quickly as possible. Photographs should be taken, preferably in color, of all parts, structures, failure surfaces, fracture texture appearance, final location of component debris, and all other factors. Witnesses to the failure should all be interviewed and all information determined from them should be recorded.

2. **Background Data:** Investigators should gather all information concerning specifications, drawings, component design, fabrication methods, welding procedures, weld schedules, repairs in and during manufacturing and in service, maintenance, and service use. Efforts should be made to obtain facts pertinent to all possible failure modes. Particular attention should be given to environmental details, including operating temperatures, normal service loads, overloads, cyclic loading, abuse, etc.

3. **Laboratory Studies:** Investigators should make tests to verify that the material in the failed parts actually possesses the specified composition, mechanical properties, dimensions, etc. Studies should also be made microscopically in those situations where it would lead to additional information. Each failed part should be thoroughly investigated to determine what bits of information it can add to the total picture. Fracture surfaces can be extremely important. Original drawings should be obtained and marked showing failure locations. This should be coupled to design stress data originally used in designing the product. Any other defects in the structure that are apparent, even though they might not have contributed to the failure, should also be noted and investigated.

4. **Failure Assumptions:** The investigator should list not only all positive facts and evidence that may have contributed to the failure, but also all negative responses that may be learned about the failure. It is sometimes as important to know what specific things did not happen or what evidence did not appear to help determine what happened. These data should be tabulated. The actual failure should be synthesized to include all available evidence. This might lead to the need for collecting additional data or asking more questions.

The true cause of failure will emerge by means of this study. Assumptions must be challenged by every bit of information available until it stands up as the one and only plausible cause for the failure. Failure cause can usually be classified in one of the following three classifications:

(a) Failure due to faulty design or misapplication of material.

(b) Failure due to improper processing or improper workmanship.

(c) Failure due to deterioration during service.

The following is a summary of these three situations.

Failure due to faulty design or misapplication of the material involves failure due to inadequate stress analysis, or a mistake in design such as incorrect calculations on the basis of static loading instead of dynamic or fatigue loading. Ductile failure can be caused by a load too great for the section area or the strength of the material. Brittle fracture may occur from stress risers inherent in the design, or the wrong material may have been specified for producing the part.

Failures can be due to faulty processing or poor workmanship that may be related to the design of the weld joint, or the weld joint design can be proper but the quality of the weld is substandard. The poor quality weld might include such defects as undercut, lack of fusion, or cracks. Failures can be attributed to poor fabrication practice such as the elimination of a root opening, which will contribute to incomplete penetration. There is also the possibility that the incorrect filler metal was used for welding the part that failed.

One of the major problems with failure due to deterioration during service is the problem of overload, which may be difficult to determine. Normal wear and abuse to the equipment may have resulted in reducing sections to the degree that they no longer can support the load. Corrosion due to environmental conditions and accentuated by stress concentrations will contribute to failure. In addition, there may be other types of situations such as poor maintenance, poor repair techniques involved with maintenance, and accidental conditions beyond the user's control. Or, the product might be exposed to an environment for which it was not designed.

Analysis of the data, if done thoroughly and objectively, will undoubtedly provide the exact cause of the failure. The next step is to take corrective action with regard to other similar structures or in the design and manufacture of like products.

There have been investigation reports made of many, many failures. Thorough investigations are always made if there is a loss of life. Various governmental agencies are required to investigate failures of specific products; for example, the Federal Aviation Agency always investigates aircraft accidents, the Office of Pipeline Safety determines the cause of pipeline failures, and the Department of Transportation investigates serious railroad and highway catastrophes. Some of the most complex and informative investigations have helped advance the level of knowledge in a particular field. An example of this is the documentation of the Comet airplane failures. [6] This represented an outstanding example of the tremendous amount of study and experimentation sometimes necessary to track down the initial cause of failure. This investigation provided knowledge regarding the fatigue problems involved with aircraft structures. Another important failure analysis report, particularly

with respect to welding, is the "Brittle Fracture in Carbon Steel Plate Structures Other Than Ships."[7] This Welding Research Council bulletin provided insight into the failure mode of large welded structures. It brought home the fact that weldments are monolithic structures. This report provided an appreciation that a welded structure is one piece of metal and this one piece of metal, for economical reasons, may have designed into it internal and external notches, stress risers, and crack starters. The report brought forth the fact that stresses are distributed throughout weldments because they are monolithic structures whether the designer had this in mind or not.

The investigation that brought this to a point of indisputable fact was "The Design and Method of Construction of Welded Steel Merchant Vessels." [8] This is a report of a board of investigation of the merchant ships built during World War II that failed in service. This is one of the most exhaustive reports produced with respect to the failure of weldments.

Prior to describing this investigation in detail, it should be brought out that failures of the most insignificant weldments deserve a thorough investigation so that the products made in the future will be more reliable. Figure 18-29 is an excellent example of a small welded part that failed and how it was modified and improved after an investigation pinpointed the problem. This part is a guard for a vee belt sheave of a gasoline-engine-driven garden tractor. Vibration and loading created premature failure of the weld joining the sheet metal part to the curved collar that clamped around the main shaft bearing. The curved collar was in intimate contact with the bearing and clamped solidly to it. The sheet metal section, which was flanged at the edge and subjected to a vibration loading both up and down, would tend to concentrate the stress at the toe of the fillet weld joining the two parts. The cross-section drawing also shown by Figure 18-29 shows the detail of the weld joint. It was soon realized that the vibration created the concentration at the toe of the weld based on the relative motion between the curved collar and the sheet metal guard. The solution of this problem was to form the sheet metal part so that the junction between it and the curved collar was extended over a greater distance so that the stresses would not be concentrated at one point of the weld. This was done by making the center portion of the sheet metal guard extend straight to the curved collar while allowing the remaining portion to be formed down to where it meets the curved collar in the same manner as the initial design. This meant that the weld was extended about 3/4 in. (19.0 mm); the stress concentration was not at the toe of the weld in a single plane. By making this change, also shown by Figure 18-29, the problem was solved and the reliability of the product was very

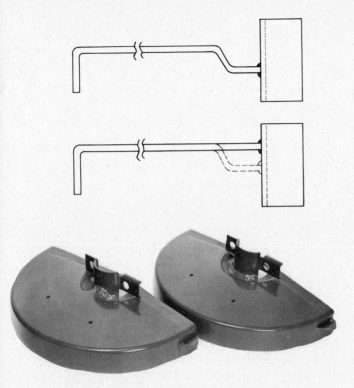

FIGURE 18-29 *Redesigned part to overcome failure.*

FIGURE 18-30 *Power shovel boom failure.*

(a)

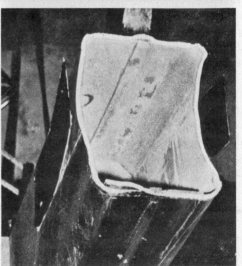

(b)

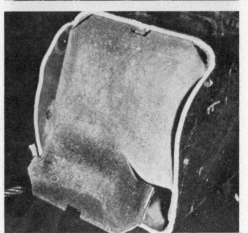

(c)

greatly improved. This is a simple problem and solution, but it illustrates the type of changes that can be made to overcome design problems through an objective investigation and analysis.

This next example, which was investigated by the author, is of a failure in a larger and much more expensive weldment. A diesel-driven power shovel with a three-cubic-yard dipper was digging taconite from an open pit mine in the Mesabi range in northern Minnesota. On a mid-January day when the temperature was $-20°F$ ($-29°C$) the boom failed without warning while the shovel was digging. The failure was complete since the point of the boom broke off completely about midway between the shipper shaft bearing and the outer end of the boom. This failure is shown by Figure 18-30A. The lower portion of the boom and its fracture surface is shown by Figure 18-30B, and the point end and its fracture surface are shown by Figure 18-30C. This was a brittle failure since there was no deformation or necking down of the material thickness. The normal working stresses on this portion of the boom are compressive since the normal working load involved is the load transmitted by the hoist cable over the point sheave on the end of the boom support cables attached at about the same place. When the boom *swings,* that is, the machine rotates, there is a bending moment on the boom point section. Possibly the severest load is caused by the pull of the dipper as it goes through the heavy blasted ore. During this period the boom shakes and vibrates violently. Even so, the calculated stresses when combined were within the allowable limit.

The boom is a rectangular cross section varying in dimensions by about 16 in. (424 mm) by 20 in. (526 mm) at the location of the fracture. It was made of low-carbon mild steel. Investigation revealed that the notch bar impact properties of this steel were poor. The boom is made of two formed half-sections joined longitudinally by submerged arc welds. A 3/8 in. (9.5-mm) by 1-in. (25.4-mm) backup bar was used under the submerged arc welds. The point section was made as a subassembly and butt welded to the main section using the manual shielded metal arc process using E6012 electrodes. A diaphragm was located very near this butt welded joint. The boom was not stress relieved.

The fracture was initiated at the fillet weld attaching the diaphragm to the box section and at a point of poor root fusion in the transverse butt weld. The termination of the backup bars contributed to the stress concentration in the failure area. It was concluded that the stresses were concentrated at this area because of the abrupt end of the backing bars, the fillet weld joining the diaphragm to the box section, and the unfused root of the butt transverse weld. The base metal and the E6012 weld metal both have relatively poor low-temperature impact resistance. The ambient temperature of $-20\,°F$ ($-29\,°C$) and the shock loading plus the stress concentration and the poor low-temperature toughness of the steel contributed to the failure.

The steps taken to avoid future similar failures were (1) to change to a low-alloy steel with better low-temperature impact toughness, (2) to change to E6015 electrodes, which possess better properties than the E6012, (3) to taper the ends of the submerged arc backup bars, and (4) to initiate root pass inspection on the transverse butt weld. These measures apparently solved the problem which did not reoccur on later booms.

The remaining portion of this section is a review of the findings of the board of investigation that inquired into the failures as well as the design and methods of construction of welded steel merchant vessels.

Early in World War II, welded merchant vessels built in the United States experienced difficulties in the form of fractures which could not be explained. Many of these fractures occurred with explosive suddenness and exhibited a quality of brittleness which was not associated with the behavior of normally ductile material used in the building of these ships. It was evident that these failures on welded ships had far-reaching consequences and, therefore, immediate steps were taken to investigate and solve the problem. A board of investigation was appointed to solve the problem by making a complete investigation and reporting the facts as established. The board was expected to investigate the design and manufacture of the ships and if they found defects in design or methods, they were to submit recommendations for correcting these defects.

The board of inquiry was composed of the engineering chief of the United States Coast Guard, the chief of the bureau of ships of the United States Navy, the vice-chairman of the United States Maritime Commission, and the chief surveyor of American Bureau of Shipping. The investigation took over three years, starting in April 1943 and concluding in July 1946. The investigation involved a total of 4,694 ships of which 970 sustained some type of structural casualty. Eight ships were lost at sea, four others broke in two but were not lost. Twenty six lives were lost. This was a tremendous effort to analyze the failures. It involved design studies of each type of vessel involved, loading and ballasting conditions, convoy routes with accompanying sea and weather conditions, and extensive laboratory research aimed at studying fabrication and materials used in construction of the welded ships. The results of this effort to control and eliminate the occurrence of hull fractures were successful. The number of fractures decreased sharply after remedial measures were taken based on the findings of the board of inquiry.

The following is a summary of the findings of the board:

The highest incidence of fracture occurred under the combination of low temperatures and heavy seas.

The age of the vessel had no appreciable influence on the tendency to fracture.

The loading and ballasting system did not create abnormal bending moments.

There was no marked correlation between the incidence of fracture on the ships and the construction practices of shipyards. It was found, however, that ships constructed in yards utilizing subaverage construction practices showed a higher-than-average incidence of fractures.

The bulk of failures were reported on liberty ships with only relatively fewer serious fractures on the victory ships. (Victory ships were designed with fewer structural notches.)

The steel supplied for ship construction complied with the applicable specification for ship steel.

Locked-in stresses in the decks of completed vessels were not appreciably reduced in service.

Welding sequence in general had no effect upon the magnitude of residual welding stresses.

Every fracture examined started at a geometrical discontinuity or notch resulting from unsuitable design or poor workmanship.

There is a large variation in the notch sensitivity of steels used in ship construction. Steel removed from fractured vessels showed high notch sensitivity.

The investigators researched failures of riveted ships that had previously been reported. They found that when a crack starts in a riveted ship it generally progresses only to the first break in the continuity of the metal; that is, a riveted seam. There it awaits reloading to a stress which will give it a fresh start. In a welded structure, however, the crack will continue to propagate as long as sufficient energy is available.

A particularly bewildering phenomenon in the welded ship casualties was the appearance and nature of the fracture. It had been generally believed that medium ship steel incorporated in ship structures would deform elastically when loaded within the elastic limit and that if it were loaded beyond that point plastic flow would take place and a permanent deformation would result as evidenced by a reduction in thickness or area. It was previously believed that if the load were increased sufficiently material would fail only after considerable elongation. It was found on examining the ship fractures that the fractured surface appeared crystalline, rather than silky as it would in a ductile failure. The break was square and the line of separation normal to the surface of the place, rather than at a 45-degree angle, as would be the case of a failure on the plane of maximum shear. Very little ductility was evidenced as indicated by practically zero reduction in the thickness of the plate at the fracture. This type of fracture is termed cleavage fracture denoting a separation of the surface of the crystal lattice rather than sliding action along slip planes.

Early in the investigation, the designs were recalculated. The calculation showed that the hull girder strength was ample and that the margin of strength in the structure was over that required by the design standards. The monolithic character of the welded ship produced specific areas that have high-stress concentrations and severe restraint that inhibit plastic flow. This condition did not exist generally in riveted ships. The danger of high concentration at points of structural discontinuities in the welded ship is further aggravated by the welds usually present at such points. The welding produces a complex metallurgical condition, which is sometimes aggravated by discontinuities in the form of defects in the welds. The design of a ship hull requires numerous openings, machinery foundations, deck houses, etc. At each of these points of structural discontinuity the section modules change abruptly, and when under a bending load, a stress concentration occurs. Stress concentrations of dangerous magnitude exist at structural discontinuities such as the hatch corners, shear strake cutouts, and at the point where foundation and deck houses are welded to the deck. Investigation found that most of the serious fractures started at hatch corners and many started in the shear strake cutout for the accommodation ladder. This strongly emphasized that insufficient attention was paid to the elimination of discontinuities or notches whether they be large or small and that the effect of these discontinuities is aggravated by the monolithic character of welded construction. At the time when the investigation started the mechanism of metal fracture was not well understood. The incidence of serious failures of large welded steel structures, both during construction and during service, indicated the need for a better understanding of the fundamental factors affecting steel performance. Lack of reliable information had led designers to overdesign in the interest of safety, which in some cases enhanced the possibility of failure. Impact tests of steel samples taken from vessels which suffered fractures indicated that in many cases the steel was notch sensitive. That is, its ability to absorb energy in the notched condition especially at low temperatures was low. The research investigators explored the behavior of ship steel in the welded and unwelded condition and under the influence of multiaxial stress in the presence of discontinuities such as notches, especially at low temperatures. These studies found that *notch sensitivity* was an important factor in the occurrence of *brittle* failures, that is, failures that exhibit a low degree of ductility.

Construction methods and workmanship were believed to be responsible to a large degree for the fractures experienced. The welding subcommittee made a survey of shipyards and found varying degrees of quality in workmanship and in methods of construction. The analysis of structural failures did not indicate a marked correlation between the incidence of fractures in welded ships in shipyard construction practice; however, ships produced in those yards utilizing below-average practices showed a higher-than-average incidence of failure. It was concluded that high-quality workmanship is an important need in the building of welded ships. They felt that welds should be identified as to who made the weld and that there should be an improvement of welder training and upgrading. They also felt that welding sequences and procedures must be prepared and the work must follow them. They did conclude that evidence was found to indicate that locked-in stresses or residual welding stresses were important in causing the fractures of welded ships.

The board concluded that the fractures in welded ships were caused by notches and by steel that was notch sensitive at operating temperatures. When an adverse combination of these occur the ship may be unable to resist the bending stresses of normal service. The notch means a structural discontinuity such as occurs at hatch openings, shear strake cutouts, foundations, vent openings, bilge keels, and the abrupt termination of structural members. Notches also occur at imperfections in the structure resulting from fabrication such as peened over cracks, undercut welds, porosity and inclusion in the welds, and incomplete penetration which leaves voids inside the joint.

Figure 18-31 shows the liberty ship and details with abnormal frequency of fractures. Figure 18-32 shows the tragic results of the S.S. Schenectady breaking in two at the outfitting dock prior to being placed in service.

The epidemic of fractures was greatly reduced through the combined effect of corrective measures taken on the structure of the ship during construction and after completion, improvements in new design, and improved construction practices in the shipyards. The first remedial step taken was to eliminate stress concentration of cargo hatch openings. This was done by modifying the corners to provide rounded corners rather than sharp or square corners. Various types of *crack arrestors* were installed. These consisted of riveted seam straps placed over slots made in the decks. Other changes were made and these are discussed in the report.

This exhaustive examination of catastrophic and major failures led the welding industry to properly appreciate the fact that weldments are monolithic in character, that anything welded onto a structure will carry part of the load whether intended or not; also, that abrupt changes in section, either because of adding a deck house or removing a portion of the deck for a hatch opening, create stress concentration. Under normal loading, if the steel at the point of stress concentration is notch sensitive at the service temperature, failure can result.

It was reported by the board that the results of the investigation have vindicated the all-welded ship. The

LIBERTY SHIP
DETAILS WITH ABNORMAL FREQUENCY OF FRACTURES
THESE DATA INCLUDE 2504 FRACTURES OF KNOWN
ORIGIN, OCCURING BEFORE 1 AUG. 45.

BUTT WELDS OF
BULWARK RAIL
154 FRACTURES 6.2%

FIGURE 18-31 *Liberty ship—location of fractures.*

FIGURE 18-32 *S.S. Schenectady after splitting in two at dock.*

statistics show that the percentage of vessels sustaining serious fracture is small. With proper design, high-quality workmanship, and steels that have good notch sensitivity at operating temperatures, a satisfactory all-welded ship can be obtained. This was further reinforced by the fact that the victory ships which were designed to reduce stress concentration sustained fewer and less serious fractures.

18-6 OTHER WELDING PROBLEMS

There are two other welding problems that require some explanation and solutions. These are welding over painted surfaces and painting of welds.

In general, the practice of welding over paint should be discouraged. In every code or specification it is specifically stated that welding should be done on clean metal. In some industries, however, welds are made over paint and in others flame cutting is done on painted base metal. Both of these require discussion.

In the shipbuilding industry and in several other industries, steel, when it is received from the steel mill, is shot blasted, given a coating of prime paint, and then stored out of doors. Painting is done to preserve the steel during storage, and also to identify it. In some shipyards a different color paint is used for different classes of steel. When this practice is used every effort

should be made to obtain a prime paint that is compatible with welding. There are at least three factors involved with the success of the weld when welding over painted surfaces.

The compatibility of the paint with welding.

The dryness of the paint.

The paint film thickness.

Paint compatibility varies according to the composition of the paint. Certain paints contain large amounts of aluminum or titanium dioxide and these paints are usually compatible with welding. Other paints may contain zinc, lead, vinyls, and other hydrocarbons, and are not compatible with welding. The paint manufacturer or supplier should be consulted. Anything that contributes to deoxidizing the weld such as aluminum, silicon, or titanium will, in general, be compatible. Anything that is a harmful ingredient such as lead, zinc, and hydrocarbons will be detrimental. The fillet break test can be used to determine compatibility. The surfaces should be painted with the paint under consideration. The normal paint film thickness should be used, and the paint must be dry. The fillet break test should be run using the proposed welding procedure over the painted surface. It should be broken and the weld examined. If the weld breaks at the interface of the plate with the paint it is obvious that the paint is not compatible with the weld. The paint supplier should be consulted since there are some paints that are compatible with welding.

The dryness of the paint should be considered. Many paints employ an oil base which is a hydrocarbon. These paints dry slowly since it takes a considerable length of time for the hydrocarbons to evaporate. If welding is done before the paint is dry hydrogen will be in the arc atmosphere and can contribute to underbead cracking. The paint will also cause porosity if there is sufficient oil present. Water based paints should also be dry prior to welding.

The thickness of the paint film is another important factor. Some paints may be compatible if the thickness of the film is in the neighborhood of 3 to 4 mils maximum. If the paint film thicknesses are double that amount, such as occurs at an overlap area, there is the possibility of weld porosity. Paint films that are to be welded over should be of the minimum thickness possible.

Tests should be run with the maximum film thickness to be used, but dry, with the various types of paints to determine which paint has the least harmful effects on the weld deposit.

Cutting painted surfaces with arc or flame processes should be done with caution. Demolition of old structural steel work that had been painted many, many times with flame-cutting or arc-cutting techniques can create health problems. Cutting through many layers of lead paint will cause an abnormally high lead concentration in the immediate area and will require special precautions such as extra ventilation or personnel protection. Please refer to Chapter 3 for more information on safety precautions.

Painting over welds is also a problem. The success of any paint film depends on its adherence to the base metal and the weld. This is influenced by surface deposits left on the weld and adjacent to it. The metallurgical factors of the weld bead and the smoothness of the weld are of minor importance with regard to the success of the paint. Paint failure occurs when the weld and the immediate area are not properly cleaned prior to painting. Deterioration of the paint over the weld also seems to be dependent upon the amount of spatter present. Spatter on or adjacent to the weld leads to rusting of the base material under the paint. It seems that the paint does not completely adhere to spatter and some spatter does

fall off in time leaving bare metal spots in the paint coating.

The success of the paint job can be insured by observing both preweld and postweld treatment. Preweld treatment found most effective is to use antispatter compounds, as well as cleaning of the weld area, before welding. The antispatter compound extends the paint life because of the reduction of spatter. The antispatter compound must be compatible with the paint to be used. This treatment thus reduces spatter, which insures a better success for the paint film.

Postweld treatment for insuring paint film success consists of mechanical and chemical cleaning. Mechanical cleaning methods can consist of hand chipping and wire brushing or power wire brushing, or sand or grit blasting. Sand or grit blasting is the most effective mechanical cleaning method. If the weldment is furnace stress relieved and then grit blasted it is prepared for painting. When sand or grit blasting cannot be used power wire brushing is the next most effective method. In addition to the mechanical cleaning, a chemical bath washing is also recommended. Slag coverings on weld deposits must be thoroughly removed from the surface of the weld and from the adjacent base metal. Different types of coatings create more or less problems in their removal and also with respect to paint adherence. Weld slag of many electrodes is alkaline in nature and for this reason must be neutralized to avoid chemical reactions with the paint which will cause the paint to loosen and deteriorate. For this reason, the weld should be scrubbed with water, which will usually remove the residual coating slag and smoke film from the weld. If a small amount of phosphoric acid up to a 5% solution is used it will be more effective in neutralizing and removing the slag. However, if this is used it should be followed by a water rinse. If water only is used, it is advisable to add small amounts of phosphate or chromate inhibitors to the water to avoid rusting which might otherwise occur.

It has been found that the method of applying paint is not an important factor in determining the life of the paint over welds. The type of paint employed must be suitable for coating metals and proper for the service intended.

Successful paint jobs over welds can be obtained by observing the following:

1. Minimize weld spatter using a compatible antispatter compound.

2. Mechanically clean the weld and adjacent area.

3. Wash the weld area with a neutralizing bath and rinse.

QUESTIONS

1. What causes arc blow?

2. Why is alternating current welding less likely to encounter arc blow?

3. What is the final solution to arc blow when parts are magnetized?

4. If a metal piece is not restrained will it come back to its original dimension after heating?

5. Is heating uniform in metal during welding? Does this cause distortion?

6. Is a weldment restrained? Does this cause warpage?

7. What is plastic deformation? How is it affected by heat?

8. How can angular distortion be reduced?

9. Explain the reason for the special order of making weld joints on a tank.

10. What is the stress level of residual stresses?

11. How do residual stresses build up and change in a multipass groove weld?

12. How can residual stresses be reduced?

13. What is the danger of brittle fracture?

14. How is the type of failure determined?

15. What is the characteristic of a fatigue fracture? What four factors are involved?

16. What are the four areas of interest when making a failure analysis?

17. Are all failures caused by poor welding? If not, what other factors are involved?

18. What was a major cause of welded ship failures?

19. What is a monolithic structure? Is a weldment a monolithic structure?

20. What is required to obtain a good paint job over welds?

REFERENCES

1. Charles H. Jennings and Alfred B. White, "Magnetic Arc Blow," *The Welding Journal,* October 1941, Miami, Florida.

2. F. Campus, "Effects of Residual Stresses on the Behavior of Structures" in *Residual Stresses in Metals and Metal Construction,* W. R. Osgood, editor, Rineholt Publishing Corporation, 1954, New York, New York.

3. M. E. Shank, "Control of Steel Construction to Avoid Brittle Failure" Welding Research Council, New York.

4. W. H. Munse, "Fatigue of Welded Steel Structures" Welding Research Council, New York.

5. J. C. M. Farrar and R. E. Dolby, "Lamellar Tearing in Welded Steel Fabrication," The Welding Institute, Cambridge, England.

6. Civil Aircraft Accident Report of the Court of Inquiry into the Accidents to the Comet G, HMSO, London, England, 1955.

7. M.E. Shank, "Brittle Failure in Carbon Plate Steel Structures Other than Ships," Bulletin No. 17, January 1954, Welding Research Council, New York.

8. "The Design and Methods of Construction of Welded Steel Merchant Vessels," First Report of a Board of Investigation, *The Welding Journal,* July 1974.

19

WELDING FOR REPAIR
AND SURFACING

19-1 THE NEED FOR WELD REPAIR AND SURFACING

19-2 ANALYZE AND DEVELOP REWORK PROCEDURE

19-3 REPAIR WELDING

19-4 REBUILDING AND OVERLAY

19-5 SURFACING FOR WEAR RESISTANCE

19-6 SURFACING FOR CORROSION RESISTANCE

19-7 OTHER SURFACING REQUIREMENTS

19-1 THE NEED FOR WELD REPAIR AND SURFACING

There are probably more welders employed doing maintenance and repair welding than there are in any other industry grouping. It was pointed out in Chapter 1 that welders doing repair work are in the ninth ranking group, entitled ''Repair Services—Auto and Miscellaneous,'' and fifth ranking group entitled ''Primary Metal Industries,'' and in the eighth ranking group entitled ''Transportation, Communication, and Utility Services.'' The work done in the primary metal industry is primarily maintenance and repair. This is true also of the utility services category and by combining these with repair services you find that approximately 18% of the welders are engaged in this type of work. It is felt that some of the welders in almost every other group are also employed doing maintenance and repair work. It is for this reason this subject deserves an entire chapter. In addition, it has prime importance to welding since the earliest use of welding was for repair work. The most famous incident happened at the outbreak of World War I when German ships were interned in New York harbor. Their crews, hoping to make the ships inoperable, sabotaged the

engines and machinery. However, by means of welding, repairs were quickly made and the ships were placed in transatlantic service to deliver material from the U.S. to Europe.

Repair welding and surfacing are both considered in the field of maintenance welding and are covered together since they are both done by the same welders. Often it is extremely difficult to separate what is considered repair welding from maintenance welding, and surfacing can be included in both situations. The same basic factors apply to both weld repair and surfacing. For example:

1. Why did the part break or wear out?
2. What results do we expect of the repaired or resurfaced part?
3. What metal are we planning to work on?

Parts break and wear out continually. It may be impossible to obtain another part exactly like the one that broke or wore out. This is particularly true of older industrial machinery, construction machinery, agricultural machinery, machine tool parts, and even automobiles. Repaired parts may be more serviceable than the original part, since they can be reinforced and the weaknesses of the original part corrected. It is often more economical to weld repair since the delay in obtaining the replacement part could be excessive and the cost of the new part would normally exceed the cost of repairing the damaged part.

Weld repair is commonly used to improve, update, and rework parts so that they equal or exceed the usefulness of the original part. This is normally attained, with the possible exception of weld-repaired cast iron parts that are subjected to heating and cooling. Weld repairs on cast iron parts subjected to repetitive heating and cooling may or may not provide adequate service life. The problem is that cast iron parts subjected to high-temperature heating and cooling, such as machinery brakes, furnace sections, etc., fail originally from this type of service and due to metallurgical changes the weld may fail again without providing adequate service life. Except for emergency situations, it is not wise to repair cast iron parts of this type.

The metal that the part to be repaired is made of has a great influence on the service life of the repaired parts. Parts made of low-carbon and low-alloy steels can be repaired without adversely affecting the service life of the part. On the other hand, high-carbon steels may be weld repaired but must be properly heat treated if they are to provide adequate service life.

It is absolutely essential that we know the type, specification, or composition of the metal that we are planning to weld. Chapter 11 on the weldability and property of metals provides information so that the metal can be identified prior to welding. Whenever a repair job is to be done it is wise to review this information so that we can identify the metal. As mentioned above, it may be unwise to weld repair certain metals. But we should not weld on any metal unless we know its composition.

The economics of weld repairing are usually very favorable and this applies to the smallest or the largest weld repair job. Some weld repair jobs may take only a few minutes and others may require weeks for proper preparation and welding. Even so, the money involved in a repair job may be less than the cost of a new part.

A part made of any metal that can be welded can be repair welded or surfaced. In fact, some of the metals that are not normally welded can be given special surfacing coatings by one process or another. All the arc welding processes are used for repair and maintenance

work. In addition the brazing processes, the oxy-fuel gas welding processes, soldering, thermit welding, electroslag welding, electron beam welding, and laser beam welding are also used. The thermal spraying processes are all widely used for surfacing applications. In addition, the various thermal cutting processes are used for preparing parts for repair welding.

The selection of the appropriate preparation process and welding process depends on the same factors that are considered in selecting a welding or cutting process for the original manufacturing operation. In the case of repair welding, there are usually limitations, such as the availability of equipment for a one-time job and the necessity of obtaining equipment quickly for emergency repair work. This limits the selection and it is for this reason that the shielded metal arc welding process, the gas metal arc welding process, the gas tungsten arc welding process, and oxyacetylene welding and torch brazing are most commonly used. These processes will be given the maximum attention in this chapter. However, for many routine and continuous types of repair work some of the other welding processes may be the most economical. For example, submerged arc welding is widely used for building up the surface of worn parts. The electroslag process has been used to repair and resurface parts for hammer mills, for construction equipment, and for rebuilding rolls for steel mills. Thus there is a difference in the selection of the welding process for the routine, continuing types of repair and surfacing work versus the one-of-a-type or breakdown emergency repair job.

19-2 ANALYZE AND DEVELOP REWORK PROCEDURE

The success of a repair or surfacing job depends on the thought and preparation prior to doing any actual work on the project. Many factors must be considered in making a thorough analysis. A thorough analysis as outlined may not be required in many situations. This is due to experience gained by welders and others in analyzing jobs, making repairs, and then checking on the service life of the repaired part. As experience is gained many short cuts can be taken, but it is the intent to provide a detailed method of analyzing jobs so that the repair will be as successful as possible.

Prior to studying this chapter it would be well to review Section 18-6, ''Weld Failure Analysis.'' Chapter 18 presented in detail how a failure investigation should be made. One of the reasons for such an investigation is to establish the cause of the failure in the case of a broken part or the cause of wear or erosion in the case of a part to be surfaced. The four points outlined are:

1. Make a detailed study of the actual parts that failed.

2. Learn the background information concerning the specifications and design.

3. Make an investigation of the materials used.

4. Make a listing of all of the facts so that at the conclusion the reason of failure will be as accurate as possible.

This type investigation is necessary for any except the most simple and minor repair jobs.

There is another factor that should also be considered that relates to the type or classification of repair work required. Is it a fairly well standardized and repetitive type of routine? An example of this would be the routine surfacing of dipper teeth on a power shovel or the rebuilding of track shoes on a crawler tractor. These are parts that routinely wear and must be repaired by weld surfacing. The same repair procedure is used and almost any part subjected to wear can be considered on this basis. On the other hand, there are those weld repairs which are strictly one-of-a-kind and often of an emergency nature. An example would be the broken power shovel boom as shown in Figure 18-29. This is the type of repair work that must be quickly analyzed and a procedure must be established as fast as possible so that the repair weld can be made to return the equipment back to service. Construction equipment such as power shovels, particularly the extremely large ones, create delays in an entire operation, such as a mine, and cost extremely large amounts of money while they are out of operation. This same thought applies to oil-drilling rigs, offshore platforms, cement kilns, steel rolling mills, electric power generators and other production equipment. These are the types of repair work where *return to service* is most important and there is no time to obtain a replacement part.

There are certain situations and certain types of equipment for which repair welding may not be done or may be done only with prior approvals.

If a failure occurs when equipment is relatively new and within the manufacturer's warranty it is necessary to contact the manufacturer of the equipment. The manufacturer must be made aware of the problem or failure and the repair that is planned. Failure to do this may cancel the warranty from the manufacturer.

Civil aircraft may be repaired by welding but only under most stringent controls. Every welder doing repair welding on aircraft should be qualified in accordance with MIL-T-5021D "Tests; Aircraft and Missile Welding Operators Certification," on the type of metal being welded, using the process for which the welder is qualified and on the category of parts involved. Furthermore,

the welder should be certified in accordance with requirements of the Department of Transportation Federal Aviation Administration. The FAA issues two documents, "Advisory Circular Acceptable Methods, Techniques and Practices—Aircraft Inspection and Repair," [1] and the "Air Frame and Power Plant Mechanics Air Frame Handbook." [2] Both of these booklets provide precautionary information and provide techniques, practices, and methods that may be used for repair welding. They further state that techniques, practices, and methods other than those prescribed may be used provided that they are acceptable to the administrator of the Federal Aviation Administration. It is mentioned that extensive damage must not be weld repaired on items such as engine mounts, landing gear, fuselage components unless the method of repair is specifically approved by an authorized representative of the FAA or the repair is accomplished in accordance with the FAA-approved instructions furnished by the aircraft manufacturer. The reason for these regulations is that many such parts are of high-strength material and the strength is obtained by heat treatment or post weld working of one type or another. Furthermore, certain parts are not to be welded if the damage is beyond a specific amount. In view of this, consult with Federal Aviation Administration authorities or the manufacturer of the particular aircraft to be repaired. Welding must not be done on aircraft, inside hangars, unless all fuel is completely removed from the aircraft and it is made inert.

Certain types of containers and transportation equipment must not be weld repaired or may be welded only with special permission and approval. These include railroad locomotive and car wheels, high-alloy high-strength truck frames, and compressed gas cylinders.

Most pieces of power-generating machinery, including turbines, generators, and large engines, are covered by casualty insurance. Weld repair on such machinery can be done only with the prior approval of the welding procedure by the insurance underwriters. In some cases, approval may not be granted. An example of this can be cast iron crankshafts in large stationary diesel engines. Certain weld repairs may be made but it is necessary to develop a written procedure which must be approved in writing by the underwriting company's representative.

Repairs by welding to boilers and pressure vessels require special attention. Pressure vessels that carry an ASME stamp or are under the jurisdiction of any state or province or government agency must be repaired in accordance with the National Board Inspection Code. [3]

CHAPTER 19

WELDING FOR REPAIR AND SURFACING

Chapter VI covers repairs and alteration to boilers and pressure vessels. This code includes special rulings which must be followed. These are briefly as follows: Repairs by welding are limited to steels having known weldable quality. It provides a maximum carbon content of 0.35% for carbon steels and a carbon content of 0.25% for low-alloy steels. For welding high-alloy materials and nonferrous materials the work must be done in accordance with the ASME code. Welders making such repairs must be qualified based on the thickness of the material and the type of material being welded. Full-penetration welds are required with welding recommended from both sides. Permissible welded repairs are defined as cracks, corroded surfaces, and seal welding, patches, and the replacement of stays. A repair is the work necessary to return a boiler or pressure vessel to a safe and satisfactory operating condition. Alterations are also permitted and this is a change in a boiler or pressure vessel that substantially alters the original design and in this case work can be done only by a manufacturer possessing a valid certificate authorization from ASME. All alterations must comply with the section of code to which the original boiler or pressure vessel was constructed. A written repair procedure is required for doing either repair work or alterations. In the case of an alteration a record must be made and all alteration work must be approved. These records must be filed with the inspection agency or the jurisdictional agency, the National Board of Boiler and Pressure Vessel Inspectors, and all work must be inspected. It is recommended that in any repair work or alteration work that the National Board Inspection Code be referred to prior to doing any work. Alterations on bridges, large steel frame buildings, and ships may be done only with special authorization. The alteration work must be designed and approved. The welders must be qualified according to the code used and the work must be inspected. Written welding procedures are required.

Once the decision has been made to make a weld repair it is then necessary to establish why the part failed or wore out. This relates to the type of repair job since it also determines whether reinforcing may be required. Reasons for the part to fail or wear out can be among the following:

Accident.

Misapplication.

Abuse.

Overload.

Poor design.

Incorrect material.

Poor workmanship.

If the part failed because of an accident or an overload, it may be returned to service with the weld repair made to bring it back to its original strength. The same consideration applies if the part has been abused or misapplied. It may be necessary to reinforce the part so that it will stand temporary overloads, misapplication, or abuse. This decision should be made prior to the weld repair.

In the case of poor workmanship, poor design, or incorrect material the weld repair should eliminate the poor workmanship that was responsible for the failure. In this case, the part would be returned to its original design. If failure is due to poor design, design changes may be required and reinforcement may be added. In a case of wrong material it will be assumed that the material was of a lower strength level which contributed to the failure. In this case reinforcing would be required. If the repair or alteration job is to modify the part, it is necessary that the modification be designed by competent designers who have the knowledge of the design conditions of the original part. This may require reinforcing to make sure the modification or alteration is satisfactory. Drawings and additional parts will be required. Another important factor that must be considered is what results are expected of the repaired or reworked part. Should it be reinforced or should it be redesigned and altered to provide necessary service life? Finally, in the case of surfacing, what better surface could be provided to withstand the service that caused the premature wear or failure?

Rework Procedure.

A written repair procedure is required for all but the most simple jobs. It is absolutely necessary that the type of material being welded is known. This can be found in several ways. If possible, refer to the drawing of the part and the specifications that are shown for the part or parts to be welded. If this is not possible, particularly in the field or at the maintenance shop, look for clues as to the type of metal involved. Analyze the application of the metal for clues. For example, an automobile engine block is normally cast iron except for some which might be cast aluminum. Aluminum and iron are easily distinguishable. The spring of an automobile or truck would normally be high-carbon steel. The body structure of a car or truck would be mild steel. The appearance often helps provide clues. See Chapter 11 for more identification systems. As a final resort it may be necessary to obtain a laboratory analysis of the metal. Filings or a piece of the metal must be sent to a laboratory capable of making such determinations.

The normal method of selecting the welding process will be followed once the material to be welded has been identified. This involves the type of metal, the thickness of the metal, the position of welding, etc. This also leads into the question of filler metal to be used. After this, the normal method of filler metal selection is followed. This involves matching base metal composition, matching the base metal properties, particularly strength, and providing weld metal that will withstand the service involved.

In surfacing, the surface characteristics desired for the finished job depend entirely on the service to which the surface will be exposed. This is based on knowledge and experience and on the fact that the surface has deteriorated to the point that it needs to be reworked or resurfaced. When wear is involved, surfaces can be rebuilt many times without reducing the strength of the part and the service life will be greatly extended.

The repair procedure should be very similar to a procedure developed for welding a critical part. It should include the process and filler metal and the technique to be used in making welds. The format utilized by Section IX of the pressure vessel code can be utilized for repair procedures.

The written weld repair procedure for complex jobs should be qualified to determine that it will provide a repair weld that is equal in strength to the original part. This is done in the same manner as qualifying a procedure by one of the more strict codes, such as Section IX of the ASME pressure vessel code. When necessary, the repair procedure should be approved by the proper authority: the inspector of a casualty insurance company, the inspector of the National Board, the representative of the

manufacturer of the original equipment, particularly if the equipment is under warranty, and possibly the review and approval by the governmental representative, such as the state piping or boiler inspector. In other situations approvals might be required from a structural engineer or architect. In work on ships and vessels, the shipping rating agency should be consulted. In every case consider the specification or code under which the product was built. Consider also if an insurance or government activity is involved. In any case, on extensive repairs on critical items make sure that the procedure is practical and will provide the necessary strength for the service intended. There may be other situations in which repair procedures must be approved and filed with the original manufacturer or with the government agency or the insurance underwriter. It is well to investigate this possibility prior to making any repair welds. After the procedure has been approved it is time to make the repair weld.

19-3 REPAIR WELDING

A repair weld is to be made. All the factors have been reviewed and analyzed, and decisions have been made. The analysis indicated that the failure was caused by an overload. The material that contains the failure is low-carbon steel. A repair welding procedure has been prepared and all involved have agreed. The weld repair may be as simple as the removal and replacing of a body panel in an automobile like the one shown in Figure 19-1 or as complex as the repair of a rolling mill frame,

FIGURE 19-1 *Automobile body repair welding.*

FIGURE 19-2 *Complex weld repair—rolling mill frame.*

as shown in Figure 19-2. In either case, there are three separate phases to the job. They are:

Preparation for welding.

Welding.

Postweld operation.

The amount of detail that must be considered depends entirely on the complexity of the job.

Preparation for Welding.

A large number of factors should be considered and decisions made before starting to weld.

1. Safety: The repair welding location or area must be surveyed and all safety considerations satisfied. This can include the posting of the area required by certain regulations, removal of all combustible materials from the area, the draining of fuel tanks of construction equipment, aircraft, boats, trucks, etc. The removal or inerting of any fuel pipeline, tanks, blind compartments, etc. If electrical cables are involved they should be removed or made inactive. Other precautions include the elimination of toxic materials such as thick coats of lead paint, plastic coverings of metals, etc. If heights are involved, proper scaffolding with safety devices should be used. If welding is enclosed, preparations for proper ventilation and personnel removal should be made. If these hazards cannot all be removed, special safeguards should be established such as fire watch, wetting down, or protecting combustible wooden floors, etc. Traditionally, repair welding creates more safety problems than production welding, and for this reason, extra special precautions must be taken.

2. Cleaning: The immediate work area must be clean from all contaminants and this includes removal of dirt, grease, oil, rust, paint, plastic coverings, etc., from the surface of the parts being welded. The method of cleaning depends on the material to be removed and the location of the workpiece. For most construction and production equipment, steam cleaning is recommended. When this is not possible solvent cleaning can be used. Blast cleaning with abrasives is also used. For small parts pickling or solvent dip cleaning can be used and, finally, power tool cleaning with brushes, grinding wheels, disc grinding, etc., can be employed. The time spent cleaning a weld repair area will pay off in the the long run.

3. Disassembly: Except for the most simple repair jobs disassembly may be required. This can be related to items mentioned above but also applies to lubrication lines, instrument tubing, wiring, etc. Sometimes it is necessary to disassemble major components such as machinery from machinery frames, etc. Experience with similar jobs is important, since it is expensive to disassemble and remove machinery when not required.

4. Protection of Adjacent Machinery and Machined Surfaces: When repair welding is done on machinery many parts that are not removed should be protected from weld spatter, flame cutting sparks, and other foreign material generated by the repair process. Sheet metal guards or baffles are used to protect adjacent machinery. For machined surfaces, asbestos cloth can be employed. It is wise to secure protective material with wire, clamps, or other temporary bracing. Machined surfaces within five feet of the welding operation should be protected.

5. Bracing and Clamping: On complex repair jobs bracing or clamping may be required. This is because of the heavy weight of parts or the fact that loads may be exerted on the part being weld repaired. If main structural members are to be cut the load must be carried by temporary braces. The braces can be temporarily welded to the structure being repaired. The braces can be strong backs or pieces welded on both sides of the repair area to maintain alignment of the part while the repair weld is being made. If strong backs or bracing are used they should be prepared and located so that they do not interfere with the repair welding.

6. Lay Out Repair Work: In most repair jobs it is necessary to remove metal so that a full-penetration weld can be made. A layout should be made to show the metal that is to be removed by cutting or gouging to prepare the part for welding. The minimum amount of metal should be removed to obtain a full-penetration weld. The layout should be selected so that welding can be balanced, if possible, and that the bulk of the welding can be made from the more comfortable welding position. The root opening should be specified, and if the welding can be done on the back side it should be gouged for full-penetration welding. If the back side cannot be reached for welding, backing straps should be employed. The groove angle should be the minimum possible for use but should be of sufficient size so that

the welder has room to manipulate the arc at the root.

7. Preheating: The preheating and flame cutting or gouging is part of the preparation for welding but can be considered part of the welding operation. When flame cutting or gouging is required, preheating should be the same as when welding. It might not be quite as important since stresses are much smaller; however, the thermal shock on the metal can occur in gouging as well as in welding. It is wise to preheat prior to cutting or gouging to at least one-half the temperature that will be used for the repair welding operation. Preheating should be based on the mass of the metal involved. If the mass is great heating should be slow so that thorough heating throughout the section occurs. Surface heating is not acceptable. Preheating can be done by any of the normal methods; however, the slower processes would be advantageous. The equipment for preheating and sufficient fuel should be available prior to starting.

8. Cutting and Gouging: The oxygen fuel gas cutting torch is most often used for this application. Special gouging tips are available and they should be selected based on the particular geometry of the joint preparation. It is possible, by closely watching the cut surface, to find and follow cracks during the flamegouging operation. The edges of the cracks will show since they become slightly hotter. The air carbon arc cutting and gouging process is also widely used for weld repair preparation. Proper power sources and carbons should be selected for the volume of metal to be removed. The technique should be selected to avoid carbon deposit on the prepared metal surface. For some metals the torch or carbon arc might not be appropriate and in these cases mechanical chipping and grinding may be employed. Chipping is preferable to grinding and air power tools should be employed. The resulting groove should be as close as possible to the layout and should be smooth without re-entrant gouges or notches.

9. Grinding and Cleaning: The resulting surfaces may not be as smooth as desired and may include burned areas, oxide, etc. Grind the surfaces to clean bright metal prior to starting to weld. For

critical work or where there is a suspicion of additional cracks it is wise to check the surface by magnetic particle inspection to make sure that all cracks and defects have been removed.

The above nine steps constitute a listing of steps for weld preparation. Some of these may be eliminated but they should all be considered to properly prepare the joint for welding.

Repair Welding.

Successful repair welding also involves following a logical sequence to make sure that all factors are considered and adequately provided for.

1. Welding Procedure: The welding procedure must be available for the use of the welders. It must include the process to be used, the specific filler metals, the preheat required, and any other specific information concerning the welding joint technique. This procedure must be understood by all concerned.

2. Welding Equipment: Sufficient welding equipment should be available so that there will be no delays. Standby equipment might also be required. For large jobs sufficient equipment must be provided for each welder. This not only includes welding equipment but includes sufficient electrode holders, grinders, wire feeders if required, cables, etc. Sufficient power must be available at the site to run all of the equipment required. In addition, if the job runs around the clock, provisions for lighting, personnel comforts such as wind breaks or covers, etc., should be provided.

3. Materials: Sufficient materials must also be available for the entire job. This includes the filler metals stored properly for use on the repair. It also includes materials such as insert pieces, reinforcing pieces, etc. Materials also include fuel for maintaining preheat and interpass temperature, shielding gases—if used, and fuel for engine powered welding machines.

   If inspection equipment is required for intermediate checking this equipment must also be available.

4. Alignment Markers: Prior to making the weld alignment markers are sometimes used. These can be nothing more than center punch marks made across the joint in various locations. With precise measuring equipment such marks are useful in maintaining proper dimensional control and alignment during the welding operation. This is more important when repairing mechanical equipment than for structural applications.

5. Welding Sequences: The welding sequence should be well described in the welding procedure and can include block welding, back-step sequence welding, wandering sequence welding, and peening. All these techniques are useful to reduce distortion and to help maintain alignment and dimensional control. By means of these procedures and by means of making precision measurements from the check points the technique can be varied to maintain the alignment required.

6. Manpower: Finally, there should be a sufficient number of welders assigned to the job so that the job can be completed quickly. Individual welders should be rotated so that they will be able to produce quality welds. It is impossible for welders to continue to work excessive hours on precise jobs and still maintain the quality necessary. Many jobs require three shifts of welders when the need to return to service is paramount.

7. Safety: Finally, safety cannot be overlooked throughout the welding operation. For example, ventilation must be provided when fuel gases are used for preheating, etc.

8. Weld Quality: The quality of the weld should be continually checked. The final weld should be smooth, there should be no notches, and reinforcing, if used, should fair smoothly into the existing structure. If necessary, grinding should be done to maintain smooth flowing contours.

Postweld Operation.

After the weld has been completed, it should be allowed to slow cool. It should not be exposed to winds or drafts, nor should the machinery loads be placed on the repaired part until the temperature has returned to the normal ambient temperature.

1. Inspection: The finished weld should be inspected for smoothness and quality. This can include nondestructive testing such as magnetic particle, ultrasonic, or X-ray. The repair weld should be of high quality since it is replacing original metal of high quality.

2. Clean Up Operation: This includes the removal of strongbacks and the smooth grinding of the points where they were attached. It also involves

the removal of other bracing and protective covers, etc. In addition, all weld stubs, weld spatter, weld slag, and other residue should be removed from the repair area to make it cleaner than it was originally. Grinding dust is particularly troublesome and every effort should be made to remove it entirely since it is abrasive and can get into working joints, bearings, etc., and create future problems.

3. **Repainting:** After the weld and adjacent repair area has been cleaned it should be repainted and other areas should be regreased in preparation for the reoperation of the machinery.

4. **Reassembly:** After cleaning and painting, etc., the pieces of machinery that were taken away are returned. This involves the reassembly of machinery. Particular attention should be paid to the fit of machinery. If necessary, remachining or redressing should be done to assure proper fit. All other items such as grease lines, cables, conduits, etc., should be reassembled and once this has been done, the machinery should be ready for operation.

There is a gratifying sense of accomplishment of a successful repair job. Most complex repair jobs are done under much pressure due to time out of service, the inaccessibility of welds, the need for quality, etc. This sense of satisfaction can be extremely gratifying for all who are involved.

19-4 REBUILDING AND OVERLAY

Rebuilding and overlaying with weld metal or spray metal are both considered surfacing operations. Surfacing is the deposition of filler metal on a base metal to obtain desired dimensions or properties. Overlay is considered to be a weld or spray metal deposit that has specific properties sometimes unlike the original surface properties. Rebuilding is used to bring parts back to their original dimensions and properties, such as the rebuilding of worn shafts, repair of parts that were machined undersize, etc. Overlay surfacing is used to return the part to original dimensions but with weld metal having particular properties to reduce wear, erosion, corrosion, etc.

Rebuilding and overlay, or the all-embracing term, surfacing, can be done by many of the welding processes and by the thermal spraying processes. The selection of the process is based on the same factors that are used to select a welding process for fabricating or repairing. There are some situations in which the thermal spray processes should be selected. The thermal spray processes do not introduce as much heat into the work as the welding processes. Where this is an important require-

FIGURE 19-3 *Surfacing a dredge cutterhead with SMAW.*

ment, the thermal spray method should be used. It is possible to thermal spray certain materials that cannot be deposited with the welding processes. This applies particularly to the ceramic sprayed coatings or other nonmetallic materials. The capabilities of the thermal spray processes are covered in Chapter 7.

The selection of the welding process and the welding procedure and technique is as important as the selection of the deposit alloy. Almost all the arc welding processes and several others can be used for the application of hardfacing weld metal. The various factors that have been previously discussed should be considered in making this selection; however, some additional factors must be considered. Whether the job is to be done in the field or in the shop has a definite bearing on process selection. In addition, if it cannot be moved and must be welded in place this prohibits the use of some processes. The properties and analysis of the base metal also have an important bearing as does the cost factor, which cannot be ignored.

The shielded metal arc welding process is probably the most commonly used of any of the welding processes for hardfacing. It can be used in the field and in the shop and can be applied to small and large parts in any position. Figure 19-3 shows the shielded metal arc process being used to surface a dredge cutter head.

Submerged arc welding is also used for many applications but it is restricted to welding in the flat position. Most often it is used for plant operations and not used in the field. It is often used for repeating applications when the same part is surfaced on a routine basis.

FIGURE 19-4 *Surfacing a ring gear with submerged arc.*

Rollers, track shoes, drums are commonly hardfaced with submerged arc welding. Figure 19-4 shows a power shovel ring gear being surfaced to bring the worn area up to its original dimension. Over 400 pounds of weld metal were deposited in this operation. Submerged arc using strip electrodes is sometimes used for overlaying surfaces in nuclear vessels with stainless steel to improve service life.

FIGURE 19-5 *Surfacing a dipper lip with flux cored arc.*

Flux-cored arc welding with and without shielding gas is a popular semiautomatic welding process. It can be used in the field or in the shop and is not restricted to the flat position. Conventional CV power sources and wire feeding equipment are used. Figure 19-5 shows the process being used in the field to build up a dipper lip.

The gas metal arc welding process is also used. There is not as wide a selection of solid electrode wires available for hardsurfacing applications. It is used more for buildup applications. It can be used either semiautomatically or fully automatically. Figure 19-6 shows the process being used on a small shaft.

The gas tungsten arc welding process is used for many smaller applications, usually for shop work in which the part can be brought to the shop and manipulated and moved for ease of welding. Gas tungsten arc can be used manually or in an automatic mode with automatic wire feeders, oscillators, etc. It is more expensive than the other processes and for this reason is restricted to the more technical type jobs.

Plasma arc welding is also used much in the same manner as gas tungsten arc. It does have a higher temperature and for this reason can be used in certain cases where gas tungsten arc welding is not applicable. It is again restricted to the smaller types of jobs.

The electroslag welding process is also used for certain special applications. It has been widely used for rebuilding crusher hammers. These can be rebuilt with special fixturing and done quite rapidly with the electroslag process.

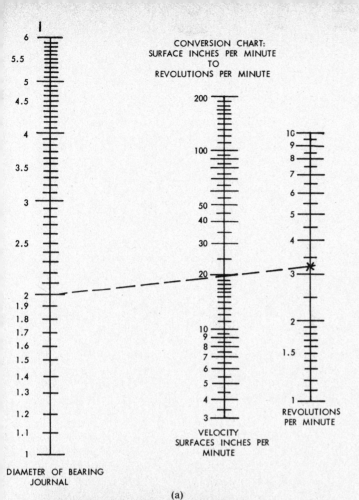

CONVERSION CHART:
SURFACE INCHES PER MINUTE
TO
REVOLUTIONS PER MINUTE

DIAMETER OF BEARING
JOURNAL

VELOCITY
SURFACES INCHES PER
MINUTE

REVOLUTIONS
PER MINUTE

(a)

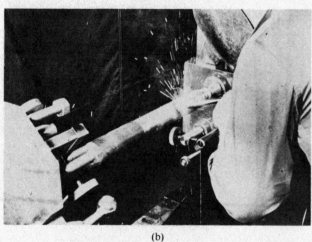

(b)

FIGURE 19-6 *Surfacing a shaft with gas metal arc.*

Oxyacetylene welding is also used for certain applications. It is widely used for application of specialized cobalt alloys on relatively thin edges.

In general, the process is selected based on normal process selection factors and modified by some of the above comments. Once the process is selected, the next requirement is the selection of the deposited metal to provide the necessary properties. This will be covered in the next few sections.

Salvaging of Shafts.

The rebuilding of round shafting is an important application of surfacing. Worn shafts can be salvaged and restored to original condition by surfacing with the gas metal arc welding process. The same basic procedure can be used for rebuilding the shaft to its original surface or providing an overlay with specific properties to improve its service life. As an example, a pump shaft that is exposed to a corrosive atmosphere can be surfaced with stainless steel to improve its service performance. Another important use of this procedure is to salvage incorrectly machined shafts. Shoulders and undersized areas can be quickly built up for remachining.

The gas metal arc welding process utilizing small-diameter electrodes is preferred to thermal spraying since it produces a weld to the shaft. This is important since splines and keyways can be cut in weld deposits without harming the overlay.

The welding procedure must be closely followed when making circumferential weld deposits. Close adherence to the procedure is more important when the diameter of the part being welded is small. Figure 19-6 shows the GMAW process being applied for surfacing a relatively small diameter generator shaft. An old lathe or similar device to provide rotation can be used. Precautions should be taken to avoid welding current from passing through roller bearings. A rotaty connection should be used. The welding gun can be mounted in the lathe tool holder if desired. The gun should be properly located for quality welding. It should be offset approximately 1/4 of the diameter of the part being welded or at the one-thirty or two o'clock position. The offset is always toward the direction of rotation and the electrode should point to the centerline of the rotating part. The travel of the gun with respect to the longitudinal axis of the part should be fast enough so that each weld bead blends smoothly into the preceding one, yet is deposited on the original surface. If the offset distance is insufficient the molten metal will not solidify before it reaches the top or twelve o'clock position and may form a high crown bead. If the offset distance is too large the molten weld metal may run down the shaft ahead of the arc. The angle of the gun can be adjusted to point slightly ahead up to 5° to improve shielding gas coverage. Experience will assist in setting these exact distances based on different diameter parts.

The welding procedure must include the welding travel speed. On rotating parts such as a shaft this is known as surface inches per minute. The correct speed will insure a smooth weld deposit that will require a minimum amount of machining. The graph shown in Figure 19-6 provides the method of determining the surface speed based on the diameter of the part being welded and related to the rotational speed of the turning equipment. The 20 inch per minute surface speed is proper when using the 0.035-in. diameter electrode wire in the 120–150 ampere range. For bigger jobs, larger electrodes can be used and the procedure variables will be different.

Welding procedures presented in the different welding process chapters are used to develop the proper procedures for surfacing, overlaying, and rebuilding. Experience soon provides the reasonable parameters that can be used for each process.

Please refer to the graph of Figure 19-6 to determine the proper rotational speed. To determine the revolutions needed for the desired travel speed draw a straight line between points that indicate the diameter of the shaft being welded and the desired travel speed. The desired travel speed is the 20 ipm. Where the line intersects revolutions per minute read the rpm required. For example, with a 2-in. diameter shaft to be welded draw the line through 20 of the surface travel speed to the intersection with revolutions per minute which would be 3.3. The lathe or other device should be set for this rotational speed.

19-5 SURFACING FOR WEAR RESISTANCE

Wear.

The deterioration of surfaces is a very real problem in many industries. Wear is the result of impact, erosion, metal-to-metal contact, abrasion, oxidation, and corrosion, or a combination of these. The effects of wear, which are extremely expensive, can be repaired by means of welding. Surfacing with specialized welding filler metals using the normal welding processes is used to replace worn metal with metal that can provide more satisfactory wear than the original. Hardfacing applies a coating for the purpose of reducing wear or loss of material by abrasion, impact, erosion, oxidation, cavitation, etc. It can be used to extend the usable life of wear parts and can save money since the replacement of worn parts is costly, particularly when the downtime and repair labor is considered.

The selection of the hardfacing material is extremely complex and will be the subject of this section. This information will aid in selecting the optimum weld deposit for hardfacing application.

In order to properly select a hardfacing alloy for a specific requirement it is necessary to understand the wear that has occurred and what caused the metal deterioration. The various types of wear can be categorized and defined as follows:

Impact wear is the striking of one object against another. It is a battering, pounding type of wear that breaks, splits, or deforms metal surfaces. It is a slamming contact of metal surfaces with other hard surfaces or objects. A good example is the impact encountered by a shovel dipper lip or tamper.

Abrasion is the wearing away of surfaces by rubbing, grinding, or other types of friction. It usually occurs when a hard material is used on a softer material. It is a scraping or grinding wear that rubs away metal surfaces. It is usually caused by the scouring action of sand, gravel, slag, earth, and other gritty material.

Erosion is the wearing away or destruction of metals and other materials by the abrasive action of water. This type of action gouges or grooves out metal surfaces. This is also caused by steam and slurries that carry abrasive materials. Pump parts are subject to this type of wear.

Compression is a deformation type of wear caused by heavy static loads or by slowly increasing pressure on metal surfaces. Compression wear causes metal to move and lose its dimensional accuracy. This can be damaging when parts must maintain close dimensional tolerances.

Cavitation wear results from turbulent flow of liquids, which may carry small suspended abrasive particles.

Metal-to-metal wear is a seizing and galling type of wear that rips and tears out portions of metal surfaces. It is often caused by metal parts seizing together because of lack of lubrication. It usually occurs when the metals moving together are of the same hardness. Frictional heat helps create this type of wear.

Corrosion wear is the gradual eating away or deterioration of unprotected metal surfaces by the effects of the atmosphere, acids, gases, alkalies, etc. This type of wear creates pits and perforations and may eventually dissolve metal parts.

Oxidation is a special type of wear indicated by the flaking off or crumbling of metal surfaces, which takes place when unprotected metal is exposed to a combination of heat, air, moisture. Rust is an example of oxidation.

Corrosion (erosion) wear takes place at the same time. This can happen when corrosive liquids flow over unprotected surfaces.

Thermal shock is a problem indicated by cracking or splintering, which is caused by rapid heating and cooling cycles. While not exactly a wear problem it is a deterioration problem and is thus considered here.

Many of the above types of wear occur in combination with one another. It is wise to consider not only one factor, but to look for a combination of factors that create the wear problem in order to best determine the type of hardfacing material to apply. This is done by studying the worn part, the job it does, how it works with other parts of the equipment and the environment in which it works. With these factors in mind it is then possible to make a hardfacing alloy selection.

Hardfacing Alloy Selection.

Unfortunately, there is no standardized method of classifying and specifying the different surfacing weld rods and electrodes. The American Welding Society has issued two specifications, A5.13, "Specifications for Surfacing Weld Rods and Electrodes," and A5.21, "Specifications for Composite Surfacing Weld Rods and Electrodes." There is some overlap between these two specifications and also A5.6 and A5.7, "Copper and Copper Alloy Welding Electrodes and Rods." Many of the hardfacing electrodes commercially available are not covered by any of these specifications. Various filler metal suppliers provide data setting forth classes of service and have categorized their own products within these classes. Many suppliers also provide complete information for using their specific products for various applications and for different industries such as quarrying, steel mills, foundries, etc. This information is extremely valuable and should be consulted.

The best system of classification has been established by the American Society for Metals Committee on Hardfacing. This data is found in the Metals Handbook, Section 3, "Hardfacing by Arc Welding." [4] In this system, there are five major groups classed according to total alloy content other than iron, with subdivisions based on the major alloying elements. These data have been abridged and simplified by Spencer who added the AWS Specifications where they apply.[5] This information is shown by Figure 19-7, "Typical Compositions of Hardfacing Materials." Most of these alloys are available as solid bare filler rod in straightened lengths or in coils or covered electrodes. Some of the materials are available as powder for special applications.

The following is a brief description of the five major groups, what they contain as alloys, and where they are recommended.

Group 1 is the low-alloy steels that, with few exceptions, contain chromium as the principal alloying element. The subgroup 1A has from 2–6% alloy including carbon. These alloys are often used as buildup materials under higher-alloy hardfacing materials. The Group 1B is similar except that they have a higher-alloy content ranging from 6–12%. Several alloys in the group have higher carbon content exceeding 2%, and include several

alloy cast irons. The alloys of Group 1 have the greatest impact resistance of all hardfacing alloys except the austenitic manganese steels (Group 2D) and have better wear resistance than low or medium carbon steels. They are the least expensive of the alloy surfacing materials and are extremely popular. They are machinable and have a moderate improvement over the wear properties of the base metal to which they are welded. They have a high compressive strength and fair resistance to erosion and scratch abrasion.

Group 2 contains higher alloyed steels. Group 2A has chromium (Cr) as the chief alloying element with total alloy content of 12–25%.

Many of these alloys also contain molybdenum. Those with over 1.75% carbon are medium-alloy cast irons. Group 2B has molybdenum (Mo) as the principal alloying element but many of these also contain appreciable amounts of chromium. The hardfacing alloys of Groups 2A and 2B are more wear resistant, less shock resistant, and more expensive than those in Group 1.

Groups 2A and 2B are quite strong and have relatively high compressive strengths. They are effective for rebuilding severely worn parts and are used for buildup prior to using higher alloy facing materials. They provide high impact resistance and good abrasion resistance at normal temperatures.

Group 2C contains tungsten and modified high-speed tool steels. They are excellent choices at service temperatures up to 1,100°F (593°C) and when good resistance coupled with toughness is required. They are not considered as good high abrasion-resistant types but are resistant to hot abrasion up to 1,100°F and exhibit good metal-to-metal wear at elevated temperatures.

Group 2D are the austenitic manganese steels, which contain either nickel or molybdenum as stabilizers. The alloys in Group 2D are highly shock resistant but have limited wear resistance unless subjected to work hardening. The total alloy content ranges from 12–25%. This group is excellent for metal-to-metal wear and impact when the deposit is work hardened in use. The as-welded deposit hardness is low, from 170 to 230 BHN, but will work harden to 450–550 BHN. The deposit may deform under battering but it will not crack. The deposit should not be heated to above 500°F (260°C), which would cause embrittlement.

Group 3 contains higher-alloyed compositions ranging from 25–50% total alloy. They are all high-chromium alloys and some contain nickel, molybdenum, or both. The carbon can range from slightly under 2% to over

Group	AWS Class	C %	Mn %	Si %	Cr %	Ni %	Mo %	W %	V %	Co %
1A	—	0.10	1.3	0.75	2.0		1.0			
		0.25	0.8	0.50	0.4	0.70	0.6			
		0.20	0.25	0.40	3.25		1.0			
		0.35	1.2	0.10	4.0		0.5			
		0.55	1.0		1.8			2.25		
1B	—	0.70	0.9	0.3	6.5		0.8			
		0.70	1.0	0.7	3.0		4.0			
		0.70	1.2	1.0	5.0		0.5			
		2.2	0.4	0.5	5.0				5.0	
		3.0	0.7	1.0	3.0					
		3.4			4.8				2.4	
2A	—	0.6	0.4	0.7	7.0		0.9	3.5	1.0	
		0.5	2.0	1.0	9.0		1.7			
		3.0	2.5	1.0	12.0		1.5			
		1.0	4.0		12.0					
		3.8			15.0	2.0	8.0			
		3.0			16.0	6.0	8.0			
2B		0.80			4.0		9.0		1.5	
		1.0			0.9		15.3			
		1.4			4.2		9.7			
		3.5			5.0		4.0			
	RFeMoC	3.6					10.0			
2C	EFe5A	0.85	0.5	0.7	4.0		5.0	6.0	2.0	
	EFe5B	0.70	0.5	0.7	4.0		8.0	2.0	1.0	
	EFe5C	0.40	0.5	0.7	4.0		8.0	2.0	1.0	
2D	EFeMnA	0.80	16.0	0.3	0.4	4.0				
	EFeMnB	0.80	14.0	0.8	0.5		1.0			
		1.2	12.0	0.6		4.8				
3A		2.7	1.0	1.0	26.0					
		3.0			18.0		16.0		1.5	6.0
		3.7	1.0	0.8	28.0					3.0
	EFeCrA1	4.0	6.0	1.7	29.0					
	EFeCrA2	4.0	1.0	1.3	29.0	3.5				
3B	—	2.5			25.0	12.0	8.0			
		4.0		4.5	16.0				0.5	
		4.0		1.0	17.0	6.0			0.5	
		3.4	4.5	0.8	30.0					
3C	—	2.3			16.0	6.0				20.0
		3.6	0.6	1.6	15.5		3.0			23.5
4A	ECoCrA	1.0	2.0	0.5	29.0	3.0	1.0	4.0		Rem.
	ECoCrB	1.3	2.0	1.0	29.0	3.0	1.0	8.0		Rem.
		2.5			32.0		17.0			Rem.
	ECoCrC	2.5	2.0	1.0	30.0	3.0	1.0	12.0		Rem.
		0.3			27.0	2.7	5.0			Rem.
4B	ENiCrA	0.35		3.5	12.0	Rem.				
	ENiCrB	0.40		4.0	15.0	Rem.				
		0.10			16.0	Rem.	17.0	4.5		
	ENiCrC	0.75		4.5	15.0	Rem.				1.0
4C	—	2.5			29.0	39.0		14.0		8.0
		2.5			25.0	15.0	8.0			25.0
		3.7			16.0	4.0	6.5			20.0
5	EWC RWC	Tungsten carbide particles (38 to 60+ % encased in matrix most wear resistant of materials.								

FIGURE 19-7 *Typical composition of hardfacing materials.*

4%. The alloys in this group exhibit better impact, erosion resistance, metal-to-metal wear, and shock resistance than the previous groups. The 3B grouping will withstand elevated temperatures of up to 1,000°F (538°C). The 3C group is high in cobalt which improves high-temperature properties. The Group 3 alloys are more expensive than Groups 1 and 2.

The compositions within Group 4 are nonferrous alloys—either cobalt base or nickel base with total content of nonferrous metals from 50 to 99%.

The Group 4A alloys are the high-cobalt-based alloys with high percentage of chromium. These alloys are used exclusively for applications subjected to a combination of heat, corrosion, erosion, and oxidation. They are considered the most versatile of the hardfacing materials. The alloys with higher carbon are used for applications requiring high hardness and abrasion resistance but when impact is not as important. These alloys are excellent when service temperatures are above 1,200°F (649°C). They resist oxidation temperatures of up to 1,800°F (982°C).

The Group 4B alloys are the nickel-based alloys which contain relatively high percentages of chromium. This group of alloys is excellent for metal-to-metal resistance, exhibits good scratch abrasion resistance, and corrosion resistance. They will retain hardness to 1,000°F (538°C). The alloys with higher carbon content provide higher hardnesses but are more difficult to machine and provide for less toughness. These alloys show good oxidation resistance up to 1,750°F (954°C).

The Group 4C alloys are the chrome-nickel cobalt alloys and all are recommended for elevated temperatures. The high-nickel alloy has excellent resistance to hot impact, abrasion, and corrosion and moderate resistance to wear and deformation at elevated temperatures. The medium-nickel alloy has high-temperature wear resistance and impact resistance. It also provides resistance to erosion, corrosion, and oxidation. The low-nickel alloy is used for moderate high temperatures and provides good edge strength, corrosion resistance, and moderate strength.

The Group 5 alloys provide a tungsten carbide weld deposit. This deposit consists of tungsten carbide particles distributed in a metal matrix. The matrix metals include iron, carbon steel, nickel-based alloys, cobalt-based alloys, and copper-based alloys. The tungsten carbide particles are crushed to mesh sizes varying from 8 to 10 down to 100 and have excellent resistance to abrasion and corrosion, and moderate resistance to impact. The matrix material determines the resistance to corrosion and high-temperature resistance. The finish of the deposit depends on the tungsten carbide particle size. The finer the particles the smoother the finish. The deposits are not machinable and are very difficult to grind.

There is another class of surfacing materials which is used primarily as overlays to provide corrosion or oxidation resistance surfaces. These will be covered in Section 19-6.

In an effort to aid the selection of the proper hardsurfacing alloy to provide the overlay with desired properties refer to Figure 19-8. This shows the same hardsurfacing alloy classes as previously shown and provides the properties of each with the welding processes that can be used, the method of finishing, and the suggested application for the different alloy classes. Admittedly, this is generalized information and is presented here as a starting point for making the final selection. This is done since there are many proprietary alloys that can be considered in the different classes but which might have specific properties or advantages to make them more suitable for the application.

For the successful hardsurfacing or overlaying operation a welding procedure should be established. The procedure should be related to the particular part being surfaced and the composition or analysis of the part. It should specify the welding process to be used, the method of application, the prewelding operations such as cleaning, undercutting, etc. The welding procedure should also give the preheat and interpass temperature and any special techniques that should be employed, such as the pattern of hardsurfacing, the method of welding—whether beading or weaving, the interface between adjacent beads, and finally, any postwelding operations such as peening and the method of cooling. When a properly developed procedure is followed the service life of the job will be predictable.

The selection of the surfacing alloy was discussed. In many cases two separate materials may be required—the buildup alloy, which is used when the part is to be reclaimed or is excessively worn, and the hardfacing alloy. In general, over three layers of hardfacing alloys are not deposited. The hardsurfacing alloys are considerably more expensive than buildup alloys. The hardsurfacing should be replaced when the hardfacing alloy is worn away. When deposit exceeds three layers other problems may be encountered such as cracking, etc., which will influence the service life of the deposit. The other factor to be considered is dilution. This is the diluting of the hardfacing alloy with base metal. Excessive dilution will reduce the effectiveness of the hardfacing material. Excessive penetration and poor tie-in of adjacent beads should be avoided.

In common with all welding operations, fabrication, repair, or surfacing, the base metal to be welded must

Class & No.	Cold Abrasion	Impact	Erosion	Metal-to-Metal Wear	Corrosion	Hot Abrasion	Rockwell Hardness, Layers & Process	Finishing	Applications
1A	F	Ex	No	No	No	No	RC 30 to 40 FCAW, SAW SMAW, 2-3 layers.	Machinable, with carbide tools.	Used as built-up for hardfacing or by itself. Fair-to-good strength, toughness and moderate abrasion resistance;
1B	F	G	F	F	No	No	RC 50-57 2 layers SAW, SMAW, FCAW	Use carbide tools or grind	Hardfacing or heavy duty built-up application involving heavy impact
2A	F to G	F to G	F	F	No	No	RC 50-55 2 layers SMAW & SAW	Use grinding practices	High strength, low crack sensitivity deposits for severe abrasion and compression, moderate to heavy impact, good resistance to erosion and mild corrosion.
2B	Ex	No	Ex	F	No	No	RC 65 Gas	Use grinding practices	Excellent for cold abrasive wear— also for metal to metal wear and mild impact.
2C	F	F	G	F	F	Ex	RC 55 to 60 anneal to RC 30	Machines if softened.	Hardness up to 1100° F (593° C). Good wear resistance and toughness use for tools.
2D	F to G	Ex	F to G	G to Ex	No	F to G	SMAW & Gas	Use grinding practices	Work hardness—build up and hard facing—austenitic manganese, steel
3A	Ex	F to G	G to Ex	F to G	No	No	RC 47 to 62 SMAW, gas	Machine if softened or grind	Holds hardness up to 800° F (427° C) or 1150° F (621° C) depending on alloy, good wear and oxidation resistance. Moderate impact and severe abrasion.
3B	G to Ex	F to G	F to G	G to Ex	F to G	F	RC 35 to 65 SMAW, gas	Use carbide tool or grind	Excellent abrasion resistance and metal to metal wear at moderate temperatures.
3C	Ex	Ex	F	Ex	G	F	RC 45-55 SMAW, gas	Machinable with carbide tools	Good edge strength—use for tools.
4A	V	V	G	G	G	G	RC 35-50 SMAW, gas	Use carbide tools or grind	Very good for metal to metal wear. Good for hot and cold abrasion— depends on specific alloy. Impact varies.
4B	V	V	G	G	G to E	G	RC-30-40 SMAW, gas	Machinable with carbide tools	Best for corrosion for erosion. Also good for metal to metal wear other properties depend on specific alloy.
4C	Ex	G	F	F	G	Ex	SMAW, gas	Use carbide	Excellent wear resistance. Good high temperature properties.
5	Ex	F	G	No	V	V	RC 90-95 SMAW, gas	Use grinding practices	Severe abrasion. Limited to 1200° F (649° C). Moderate resistance to impact.

Ex-Excellent, G-Good, F-Fair, No-do not use, V-Varies

FIGURE 19-8 *Hardfacing alloys vs service conditions.*

be clean. Shot blasting, grinding, machining, and brushing are all methods for cleaning the base metal prior to welding. A major consideration is the location of finished surface with respect to the worn surface. In many cases, the first layer of surfacing may have sufficient dilution of base metal so that it is unsuitable for the desired service. In this case, the worn surface should be further removed so that there is sufficient room for two layers of surfacing metal. This will provide a better service life. There are other situations in which the part is to be remachined after surfacing and it is not recommended that the machining surface be at the interface between weld surfacing metal and the base metal. Here again, premachining may be required. This is particularly important when the base metal is of hardenable material.

Preheating, interpass temperature, and cooling of the part being surfaced are as important in surfacing as they are in repair welding or in original fabrication. The factors that apply to the welding of the base metal in normal fabrication should be followed when overlaying with surfacing weld metal. Preheating is used to minimize distortion, to avoid thermal shock, and to prevent surfacing cracking. The temperature of preheat depends on at least two factors, the carbon and the alloy content of the base metal and the mass of the part being surfaced. A soak type preheat should be used and sufficient time must be allowed for the preheat temperature to stabilize throughout the part. If it is extremely complex in shape, preheat should be increased. If the ambient temperature is low, preheat should also be increased. Any part that is preheated should be maintained at that temperature throughout the entire welding operation and should then be allowed to slow cool.

The base metal composition must be known in order to provide proper preheat temperatures. Certain materials such as austenitic manganese steel should be treated in accordance with the requirements of the steel. In this case preheat should not exceed 500°F (260°C). Cast iron should be given sufficient preheat. Cast iron is crack sensitive and normally it is not hardsurfaced because it is relatively inexpensive compared to the cost of surfacing metal and it may be more economical to replace the part than to surface it. Cast iron should be preheated to at least 800°F but less than 1100°F, should be slow cooled, and often the weld deposit should be peened to avoid cracking.

Welding should be done in the flat position if at all possible. If not possible, the correct electrode and procedure must be specified.

The thickness of the surfacing deposit is extremely important. If the deposit is too heavy, problems can be encountered. Hardfacing alloys should be restricted to two layers. The first will include dilution from the base metal, but the second layer should provide the properties expected. Some types of alloys can be used in three layers

and others are recommended for use of one layer only. When edges are being built up with surfacing material make sure that sufficient material is removed so that the edge has at least two layers of surfacing metal prior to remachining or grinding. Consult the manufacturer's data for the particular product involved.

The technique for buildup should be to within 1/4 in. (6 mm) of the final surface. This will then allow two layers of surfacing material to bring the part to final dimension. A weaving technique is recommended instead of stringer bead welding. In addition, the pass thickness or layer thickness should not exceed 3/16 in. (5 mm). The adjacent beads must fair into the previous bead to provide as smooth a surface as possible. There is considerable controversy concerning the exact pattern of welds that should be made when applying the surfacing deposits. In general, the direction of welding should not be transverse to the load on the part. This can create stress concentrations and may affect service life of the part. Diagonally-shaped welds have an advantage in this regard. In certain types of metal, peening is recommended but this is based on the metal. The manufacturer's instructions should be followed.

Hardfacing by welding is an excellent method of reclaiming parts and will save considerable time and money. It will often reduce downtime of equipment and may keep equipment going without as much downtime. It is considerably cheaper than replacing original parts and should be used whenever possible. It is now becoming popular for original equipment manufacturers to actually hardface wear parts on new equipment to provide better service life of the equipment.

19-6 SURFACING FOR CORROSION RESISTANCE

The corrosion of metals is one of the less known but more expensive of the factors that cause premature failures of many things. These range from automobile bodies to chemical plant equipment to ship hulls. The cost of corrosion is difficult to measure but should include the loss of efficiency of operating equipment such as pumps, mixers, valves, etc., as well as total failures. Fortunately, corrosion can be prevented or at least substantially reduced so that metal parts will have a longer life cycle. One of the best ways to reduce corrosion is to protect the metal with an overlay or surface of a material less susceptible to corrosion in a specific environment. Coated metals such as galvanized steel and clad metals with nonferrous facings have long been used to reduce the effect of corrosion. In more and more applications, surfaces that are sprayed or welded are contributing to longer service life of parts exposed to corrosive atmospheres.

It was mentioned previously that the deterioration of metal surfaces is caused by the combination of factors, such as, corrosion and oxidation, corrosion and erosion, or cavitation. In repairing corroded or deteriorated surfaces it is necessary to analyze the reason for the deterioration. These factors should be considered in designing new surfaces for specific types of service. That is, consideration should be made in selecting a material for overlays to prevent corrosion. There is considerable confusion in this field concerning the proper terms to use for the weld surface that is applied. The general term *surfacing* is sometimes used but more often the term *cladding* is used. The terms *weld buildup* and *buttering* have no official status; however, the term *corrosion-resistant weld-overlay cladding* does provide an understanding of what is being covered in this section.

Cladding of this type is applied for many reasons: (1) to produce a corrosion-resistant surface as on the inside of a nuclear pressurizer vessel, (2) to produce a corrosion-resistant material to replace a higher-priced or unavailable high-alloy material, (3) to produce a metallurgical structural composition that is more weldable, (4) to deposit weld metal which would later be used as a filler metal such as in a tube-to-tube sheet weld, or (5) to produce a wear- or erosion-resistant surface.

Various methods or techniques can be used to provide these surfaces, such as explosive clad metal, roll bond clad, and loose cladding liners, plug and seam welded to the inside of a vessel or tank. Many of the welding processes can be used for applying liners. When attaching liner plates or sheets to carbon mild steel the problems of welding dissimilar metals must be considered. This involves the metallurgical requirements of the clad material and the compatibility of the two materials. When the solid solubility, that is, ability of one element to be dissolved in another, is exceeded, cracking may occur. In addition, the effects of elements such as sulfur and phosphorous from dilution can be a source of trouble. The welding technique and procedure involving the selection of filler metals, coatings, fluxes, etc., must be considered as with any dissimilar welding operations. These same factors apply whether the material is being applied as a weld surfacing or as separate sheets or plates welded to the carbon steel structure.

There are a number of alloys that are used for overlays or clads for corrosion and oxidation resistance. These are usually standardized compositions commonly used by themselves for the same requirements. These are summarized as follows:

The copper-based alloys are used for certain corrosion requirements. The copper silicon alloys and the copper tin alloys are used for certain corrosion-resistance requirements.

The austenitic stainless steels, which include the standard alloy types 308, 309, 310, 316, and 347 are all used for corrosion-resistant surfaces. These alloys exhibit moderate resistance to high-stress abrasion and have excellent oxidation-resistance and impact properties.

The nickel-base alloys are also used for this purpose. This includes 100% nickel, the Monel (67 Ni-30 Cu) and Inconel (72 Ni-7 Fe-16 Cr). These alloys are frequently used as overlays on carbon and low-alloy steels for cladding of tanks and vessels.

The high-cobalt chromium alloys are used for specific overlays when corrosion is a major problem. These are used quite often in refineries where high pressures, high temperatures, and corrosive materials are pumped and stored. These alloys can be applied in several ways; as a powder applied by the plasma process, as a *cold* wire, or by covered electrodes with the shielded metal arc welding process. The selection of the overlay is based entirely on the requirements of the materials to which the product is exposed. The selection must be based on normal metallurgical factors.

A unique but rather typical application of weld overlay is used in the repairing of digesters used in pulp and paper mills. Digesters are tanks or pressure vessels ranging in height from 25–50 ft and in diameter from 8–12 ft. They are used for the first chemical processing step of converting wood chips into pulp for paper manufacturing, primarily in the sulphate or kraft paper process. The wood chips are placed in the digester and are cooked in a highly corrosive alkaline solution. The mixture of wood chips and alkaline liquor is under pressure and operates at a relatively high temperature.

The digester is made of carbon steel of from 1 to 2 inches thick. The internal surfaces of digesters corrode at a high rate at the surface of the liquor due to the corrosive action of the alkaline solution. The steel walls of the vessel will gradually deteriorate until they become so thin that pressures and temperatures must be reduced for safety. Unless the metal is replaced by welding the digester will eventually become unsafe and will have to be abandoned.

Welding has been employed to repair the pitted or corroded areas and to rebuild wall thickness to original dimension. Originally, carbon steel weld metal was used. It was found, however, that stainless steel electrodes provide a surface that is less subject to the corrosive action. Tests revealed that stainless overlay outlasts the original carbon steel many times.

Recently the gas metal arc welding process has been used for this overlaying operation. Automatic methods have been used to make the overlay welds more rapidly than with manual application. The automatic application will deposit weld metal in horizontal beads on the vertical inside circumference of the tank. The automatic welding heads are mounted on a boom that rotates about the centerline of the tank and deposits metal as it revolves inside the tank. It is possible to utilize two or even three automatic heads that automatically travel around the inside circumference of the tank. This work is done starting at the lower portion to be welded and moves upwards as it revolves. The most popular procedure uses either 316 or 310 stainless alloy in the 0.035-in. diameter electrode wire with argon for shielding.

In normal applications the inside diameter of the digester is prepared for welding by grit blasting the entire surface to be welded. This may be followed by an acid wash and water rinse. The welding opeation, once it is begun, is usually continuous to eliminate any voids in the surface. Each pass must fair smoothly into the previous one and the depth of the surface should be from 1/8 to 3/16 of an inch thick.

This technique is used occasionally for new digesters to reduce the rate of corrosion and the length of time between maintenance repair work. Penetration must be closely controlled so that dilution will not appreciably lower the alloy content of the deposit.

Other procedures for accomplishing an overlay on the inside diameter of smaller tanks are done by rotating the tank and doing the welding in the flat position. In this case, the welding is done by the submerged arc process using one or more electrode wires. There are some situations in which the strip overlay method is used. The submerged arc welding process increases the speed of making the overlay. In some cases the GTAW or plasma hot wire process is used. The process should be selected which is most appropriate for the position and the job to be done.

Normally single layers are used; however, for certain applications such as pump linings and wear areas a second layer of surfacing is applied. The second layer can be made with an electrode of lower alloy content since the dilution factor is drastically reduced.

Figure 19-9 shows the strip overlay method being used to surface an extremely large area. This is an automatic method in which the work is moving under the welding head.

Welding procedures must be developed and qualified for these applications.

FIGURE 19-9 *Strip overlay.*

19-7 OTHER SURFACING REQUIREMENTS

There are other surfacing requirements which cannot be considered as hardfacing for wear resistance or cladding for corrosion resistance. These may involve wear, however.

One popular use of surfacing is the overlaying of metal parts with bronze to provide wearing surfaces for metal-to-metal contact. This includes guides and ways for reciprocating and sliding motion. In order to avoid making a part completely of brass or bronze it is possible that it can be made of steel and that the steel is then overlaid with a bronze. The copper-base alloys provide a relatively soft deposit for metal-to-metal wear. The AWS Class ECuAl types, which are the aluminum bronzes, are well suited for overlay for bearing surfaces. The different classes such as ECuAl A-2, B, C, D, and E can all be used. The hardness is greater with the higher suffix letters.

Aluminum bronze overlays can also be used for plungers in pumps, rams in extrusion presses, and for rings on hydraulic rams. They will wear faster than the hardened steel with which they make contact. It is advantageous to concentrate the wear on the bronze which can be easily replaced rather than cause wear on the inside diameter of a hardened steel cylinder.

The aluminum bronzes can also be used for repair welding of worn bronze bearings used in heavy slow moving machinery, and for overlaying worn cast iron gears and sheaves. By overlaying with bronze and remachining, the part can be made better than new.

Tool and die welding, as described in Chapter 14, is often a form of surfacing. In many cases, the tool steel electrodes are used to build up wearing surfaces or cutting edges. These applications are covered in Section 14-2.

Overlay of bronze is sometimes used for decorative purposes, particularly for architectural metal applications. By judicious use of the bronze and stainless steel the color contrast can be made very attractive.

Another use for bronze surfacing is for projectiles where a brass overlay is welded around the projectile. This causes it to fit the rifling of the gun barrel tightly to avoid loss of pressure and also to give the desired spin to the projectile.

The use of weld surfacing can be extended to provide safety surfaces. When metal flooring becomes smooth from wear it is possible to run stringer beads on the smooth surface to produce a rougher surface. The weld surface will be safer and eliminate slipping. This is used on treads of metal steps, walkways, and other places where smooth metal surfaces can be hazardous.

Undoubtedly there are other applications for surfacing and overlays since it is an ideal method to utilize less expensive materials and provide specialized materials only at specific points.

QUESTIONS

1. More welders are employed doing maintenance and repair welding than any other class of welders. True or False?

2. Which arc welding processes can be used for maintenance and repair welding?

3. What are the four points of a failure analysis?

4. When should repair welding *not* be performed?

5. Approval of a repair procedure is required on what type of products?

6. Why is it wise to prepare a written procedure for repair welds?

7. List the factors involved in preparation for repair welding.

8. List the factors involved in repair welding.

9. List the factors involved in repair postweld treatment.

10. Explain rebuilding, overlaying, and surfacing. How do they differ?

11. What is the best process to use to rebuild small shafts?

12. Define and give an example of impact wear.

13. Define and give an example of abrasion wear. How is it different from erosion?

14. Define and give an example of corrosion wear. How is it different from oxidation?

15. What is the basis for selecting hardfacing alloys?

16. What is a buildup surfacing material? Where is it used?

17. What is corrosion-resistant weld-overlay cladding? Where is it used?

18. What is the advantage of stainless steel cladding over carbon steel?

19. Why are wear parts surfaced with bronze or brass?

20. How is weld surfacing used to provide a safety surface?

REFERENCES

1. "Acceptable Methods, Techniques and Practices—Aircraft Inspection and Repair," AC43.13-1A-1972, Federal Aviation Administration, Department of Transportation, Washington, D.C.

2. *Air Frame and Power Plant Mechanics Air Frame Handbook,* AC65-15, Federal Aviation Administration, Department of Transportation, Washington, D.C.

3. "National Board Inspection Code," 12th Edition, The National Board of Boiler and Pressure Vessel Inspectors, 1972, Columbus, Ohio.

4. "Welding and Brazing," *Metals Handbook,* Vol. 6, 8th Edition, Chapter on Hardfacing by Welding, The American Society for Metals, Metals Park, Ohio.

5. LESTER F. SPENCER, "Hardfacing, Picking the Proper Alloy," *The Welding Engineer,* November 1970, Chicago, Illinois.

20

SPECIAL APPLICATIONS
OF WELDING

20-1 PIPE WELDING

20-2 ARC SPOT WELDING

20-3 ONE-SIDE WELDING

20-4 UNDERWATER WELDING

20-5 NARROW GAP WELDING

20-6 GRAVITY WELDING

20-7 REMOTE WELDING

20-8 SHEET METAL WELDING

20-1 PIPE WELDING

In the United States over 10% of the steel produced is made into tubular products, essentially pipe. Except for the small sizes the majority of the pipe is installed by welding.

The piping industry is roughly divided into three major categories:

Pressure or power piping

Transmission and distribution piping

Subcritical piping

Today, all high-pressure piping used in the thermal and nuclear power generating stations, refineries, chemical plants, on ships, etc., is welded. The welding of pressure piping is usually based on specifications issued by various organizations, but normally, as far as welding is concerned, it is based on Section IX of the ASME pressure vessel code. The pipe employed normally has a medium to thick wall in medium to large sizes. Welds are made using many of the welding processes.

The transmission and distribution pipeline industry uses welding to construct pipelines which transmit gas and petroleum products from the producing field to the consumers. Welding on this type of pipe has become highly developed and utilizes special techniques and procedures. Welding performed on such pipelines is governed by the API specification 1104. The pipe employed is usually of medium to high strength and has a relatively thin wall section and is usually in medium to large diameters. Gas distribution piping and other similar types of pipe installations are made to the same basic standard.

The subcritical piping field embraces many different types of piping ranging all the way from domestic hot water supply systems through low-pressure heating systems, sprinkler systems, sanitary systems, compressed air lines, fuel distribution systems, and many, many more. Welding has not been universally adopted in this field. Screw thread pipe, soldered copper pipe, and plastic pipe are used for certain of these applications. Steel pipe is usually of mild steel in small to medium sizes and in standard wall thickness. Qualification tests may not be required, although they will utilize qualified welders and qualified procedures for much of this type of work.

The piping industry has continually sought ways of joining pipe to produce more efficient joints and to reduce installation costs. Welding has responded and has provided many, different welding procedures based

SPECIAL APPLICATIONS OF WELDING

on the requirements of the pipe material, the size, the wall thickness, the service requirements, etc. This has led to the use of most of the arc welding processes with continued emphasis on adopting the newer processes.

Welding of pipe and tubular products is done in accordance with established procedures,[1] which are usually written but may be based on the welder's experience. Pipe welds are made under many different requirements and in different welding situations. For this reason, a list of welding procedures used by the different industries would be extremely long. It would also be difficult to list procedures based on a particular code or specification or on welding different base metals. Pipe welds that will meet the requirements of one specification or code will usually meet requirements of the others. The only exception would be the nuclear piping which must have extremely high quality welds.

Welding procedures are selected and designed based on the pipe diameter, and wall thickness, and also the pipe material. The welding position that must be

used and the code that applies must be considered. These factors are dictated by the job. Once they are known, the welding process and the progression of travel are selected. The method of application of the process depends on the process selected and the equipment availability. The filler metal is selected based on the specifications or composition of the base metal being welded. Process selection is based on the factors mentioned throughout this book.

In an effort to help establish pipe welding procedures a listing of pipe welding schedules is given by Figure 20-1. This list is based on the pipe or tubing size (see Table 20-1) which is categorized as small, 4 in. (100 mm) and smaller; medium, 4 in. (100 mm) to 12 in. (300 mm); and large, 12 in. (300 mm) and larger. The wall thickness is categorized as thin (less than standard), standard (schedule 40) and heavy (greater than standard). This information is followed by the position of welding that must be used as dictated by the job. In general, the position is fixed but in some cases it can be rolled for flat-position work. Knowing this information allows selection of the process and several options are shown. The type of service involved and the code or specification may have an influence on process selection. Several methods of applica-

Tube & Pipe Diameter & Wall Thickness	Welding Position	Welding Process	Method of Applying	Schedule No.
Small tubing-thin wall	All position	GTAW	MA	1
Small tubing-thin wall	All position	PAW	MA	2
Small tubing-thin wall	All position	GTAW-No FM	AU	3
Small pipe-std. wall	All position	OFG	MA	4
Small pipe-std. wall	All position	GTAW-with FM	AU	5
Small to medium-std. wall	All position	GMAW	SA	6
Small to medium-std. wall	All position	SMAW	MA	7
Medium-std. & heavy wall	All position	Comb. GTAW & SMAW	MA	8
Medium & large-thin & std.	All downhill	SMAW	MA	9
Medium & large-thin & std.	All up or down	GMAW	SA	10
Medium & large-thin & std.	All up or down	FCAW	SA	11
Medium & large-all wall	All uphill	SMAW	MA	12
Medium & large-all wall	Flat roll (1G)	SAW	AU	13
Medium & large-all wall	Flat roll (1G)	Comb. GMAW & SAW	SA or AU & AU	14
Medium & large-all wall	Flat roll (1G)	Comb. GMAW & FCAW	SA or AU & AU	15
Small, medium & large-std. to thick	All position	FCAW	SA	16
Small, medium & large-std. to thick	All position	Comb. GTAW & FCAW	MA & SA	17
Small, medium & large-std. to thick	All position	Comb. GTAW & GMAW	MA & SA	18

Note: Small is 4 in. and down, Medium is 4 to 12 in. Large is 12 in. and up—See Table 20-1 Standard wall thickness is Schedule 40, Thin wall is less than standard, Heavy wall is thicker than standard.

FIGURE 20-1 *Listing of pipe welding schedules.*

TABLE 20-1 *Pipe size and wall thickness.*

Nominal Pipe Size	Outside Dia.	NOMINAL WALL THICKNESS FOR													
		Sched. 5*	Sched. 10*	Sched. 20	Sched. 30	Standard +	Sched. 40	Sched. 60	Extra Strong	Sched. 80	Sched. 100	Sched. 120	Sched. 140	Sched. 160	XX Strong
1/8	0.405	—	0.049	—	—	0.068	0.068	—	0.095	0.095	—	—	—	—	—
1/4	0.540	—	0.065	—	—	0.088	0.088	—	0.119	0.119	—	—	—	—	—
3/8	0.675	—	0.065	—	—	0.091	0.091	—	0.126	0.126	—	—	—	—	—
1/2	0.840	—	0.083	—	—	0.109	0.109	—	0.147	0.147	—	—	—	0.187	0.294
3/4	1.050	0.065	0.083	—	—	0.113	0.113	—	0.154	0.154	—	—	—	0.218	0.308
1	1.315	0.065	0.109	—	—	0.133	0.133	—	0.179	0.179	—	—	—	0.250	0.358
1-1/4	1.660	0.065	0.109	—	—	0.140	0.140	—	0.191	0.191	—	—	—	0.250	0.382
1-1/2	1.900	0.065	0.109	—	—	0.145	0.145	—	0.200	0.200	—	—	—	0.281	0.400
2	2.375	0.065	0.109	—	—	0.154	0.154	—	0.218	0.218	—	—	—	0.343	0.436
2-1/2	2.875	0.083	0.120	—	—	0.203	0.203	—	0.276	0.276	—	—	—	0.375	0.552
3	3.5	0.083	0.120	—	—	0.216	0.216	—	0.300	0.300	—	—	—	0.438	0.600
3-1/2	4.0	0.083	0.120	—	—	0.226	0.226	—	0.318	0.318	—	—	—	—	—
4	4.5	0.083	0.120	—	—	0.237	0.237	—	0.337	0.337	—	0.438	—	0.531	0.674
5	5.563	0.109	0.134	—	—	0.258	0.258	—	0.375	0.375	—	0.500	—	0.625	0.750
6	6.625	0.109	0.134	—	—	0.280	0.280	—	0.432	0.432	—	0.562	—	0.718	0.864
8	8.625	0.109	0.148	0.250	0.277	0.322	0.322	0.406	0.500	0.500	0.593	0.718	0.812	0.906	0.875
10	10.75	0.134	0.165	0.250	0.307	0.365	0.365	0.500	0.500	0.593	0.718	0.843	1.000	1.125	—
12	12.75	0.156	0.180	0.250	0.330	0.375	0.406	0.562	0.500	0.687	0.843	1.000	1.125	1.312	—
14	14.0	—	0.250	0.312	0.375	0.375	0.438	0.593	0.500	0.750	0.937	1.093	1.250	1.406	—
16	16.0	—	0.250	0.312	0.375	0.375	0.500	0.656	0.500	0.843	1.031	1.218	1.438	1.593	—
18	18.0	—	0.250	0.312	0.438	0.375	0.562	0.750	0.500	0.937	1.156	1.375	1.562	1.781	—
20	20.0	—	0.250	0.375	0.500	0.375	0.593	0.812	0.500	1.031	1.281	1.500	1.750	1.968	—
22	22.0	—	0.250	0.375	—	0.375	—	—	0.500	—	—	—	—	—	—
24	24.0	—	0.250	0.375	0.562	0.375	0.687	0.968	0.500	1.218	1.531	1.812	2.062	2.343	—
26	26.0	—	—	0.500	0.562	0.375	—	—	0.500	—	—	—	—	—	—
30	30.0	—	0.312	0.500	0.625	0.375	—	—	0.500	—	—	—	—	—	—
34	34.0	—	—	—	—	0.375	—	—	0.500	—	—	—	—	—	—
36	36.0	—	—	—	—	0.375	—	—	0.500	—	—	—	—	—	—
42	42.0	—	—	—	—	0.375	—	—	0.500	—	—	—	—	—	—

* Thicknesses shown in italics are for Schedules 5S and 10S which are available in stainless steel only.
+ Thicknesses shown in italics are available also in stainless steel, under the designation Schedule 40S.
: Thicknesses shown in italics are available also in stainless steel, under the designation Schedule 80S.

FIGURE 20-2 Pipe welding schedules.

No.	Process	Joint Design (See Fig. 20-3)	Method of Application	Pass	Electrode or Rod Dia. in.	mm	Amperes	Voltage	Shielding Details	Travel	Other Information
1	GTAW	A	MA	1	1/16	1.6	35-45 DC	10-12 EN	Argon @ 12-15 CFH	Downhill	Use purge gas inside for high quality
2	PAW	A or D	MA	1	None		60-70 DC	— EP	Argon @ 12-15 CFH	10 IPM	Plasma gas is 95 argon + 5 H_2 @ 1 CFH
3	GTAW	A or D	AU	1	None		40 DC	10 EN	Argon @ 20 CFH	4-1/4 IPM	1/16″ Tungsten use purge gas
4	OFG	B or E	MA	1 / 2	1/8	3.2	—	—	Product of Combustion	either / 3 IPM	Forehand—Oxygen and Acetylene
5	GTAW	E	MA	1	3/32	2.4	90-100 DC	EN	Argon @ 15-20 CFH	Uphill	Use purge gas
				2+	1/8	3.2	100-140 DC	EN	Argon @ 20-25 CFH	Uphill	No purge gas
6	GMAW	C or E	SA	1	0.035	0.8	150-170 DC	22 EP	CO_2 @ 12-15 CFH	Downhill	Travel 11 in/min
				2+	0.035	0.8	120-130 DC	20 EP	CO_2 @ 12-15 CFH	Up or Down	Travel 4 in/min
7	SMAW	C	MA	1	1/8	3.2	90-100 DC	EP	Coating	Uphill	E6010
				2+	1/8	3.2	110-140 DC	EP	Coating	Uphill	E6010
7 Alt	SMAW	E	MA	1	1/8	3.2	80-120 DC	EP	Coating	Downhill	E6010
				2+	1/8	3.2	150-200 DC	EP	Coating	Downhill	E6010
8	GTAW	E or F	MA	1+2	3/32	2.4	90-100 DC	EN	Argon @ 15-20 CFH	Uphill	Alt use insert.
	SMAW			3+	5/32	4.0	120-200 DC	20-22 EP	Coating	Uphill	Low Hydrogen
9	SMAW	C	MA	1	5/32	4.0	140-160 DC	24-26 EP	Coating	Downhill	E6010 or E7016
				2+	3/16	4.8	160-190 DC	24-28 EP	Coating	Downhill	E6010 or E7016
10	GMAW	C or E	SA	1+	0.035	0.8	180-200 DC	21-23 EP	CO_2 @ 20-30 CFH	Downhill	Less passes for Downhill
11	FCAW	C or E	SA	1+	0.035	0.8	120-150 DC	19-20 EP	CO_2 @ 20-30 CFH	Uphill	Use purge for high quality
12	SMAW	E or F	MA	1	1/8	3.2	135-145 DC	20-21 EP	Coating	Uphill	E-XX10 electrode
				2+	5/32	4.0	325-350 DC	25-28 EP	Coating	Uphill	Low hydrogen elec.
13	SAW	E or F	AU	1	5/53	4.0	375-425 DC	23-25 EP	Sub Arc Flux	20 IPM	Back up ring rqd.
				2+	5/32	4.0	480-505 DC	26-27 EP	Sub Arc Flux	26 IPM	
14	GMAW	C, E, F	SA or AU	1	0.035	0.8	150-170 DC	20-22 EP	CO_2 @ 20-35 CFH	12 IPM	Flat-Roll
	SAW		AU	2+	5/32	4.0	480-505 DC	26-27 EP	Sub Arc Flux	26 IPM	Flat-Roll
15	GMAW	C, E, F	SA or AU	1	0.035	0.8	150-170 DC	20-22 EP	CO_2 @ 20-35 CFH	12 IPM	Double Ending
	FCAW		AU	2+	3/32	2.4	350-400 DC	25-28 EP	CO_2 @ 30-35 CFH	10 IPM	Flat-Roll
16	FCAW	C or E	SA	1	0.045	0.9	135-145 DC	20-21 EP	CO_2 @ 20-30 CFH	Uphill	Double Ending
				2+	3/32	2.4	325-350 DC	25-28 EP	CO_2 @ 25-30 CFH	Uphill	
17	GTAW	C or E	MA	1+2	3/32	2.4	90-100 DC	EN	Argon @ 15-20 CFH	Uphill	Use purge for high quality
	FCAW		SA	3+	3/32	2.4	325-350 DC	25-28 EP	CO_2 @ 25-30 CFH	Uphill	Alt use insert
18	GTAW	C or E	MA	1+2	3/32	2.4	90-100 DC	EN	Argon @ 15-20 CFH	Uphill	Alt use insert
	GMAW		SA	3+	0.035	0.8	120-150 DC	19-20 EP	CO_2 @ 20-30 CFH	Either	

tion options are also given. In some cases, combinations of welding processes and methods of applying are given. These can be varied to meet conditions. Usually the schedules can be used for more than one method of application, with minor variations. The same is true with regard to position. The economic and job location may require additional variations. Filler metal information is not given since this relates specifically to the base metals. However, the basic schedule information will still apply, but variations may be required. These schedules shown by Figure 20-2 should be used as a starting point.

With these data welding procedures can be designed to meet the job requirements. The writing of welding procedures was covered in Chapter 17. The information given in the different schedules will allow the making of test welds, which, if properly welded, will allow the procedure to be qualified to meet the most strict code or specification requirements.

Joint Design.

The joint designs for pipe welding have been fairly well standardized. For the thinner wall pipe the joint design is the square groove weld. As thickness increases a single vee type joint is used. The included angle of the vee groove has been standardized at 60° and 75°. The 75° included angle is more common in pressure piping and the 60° included angle is common in cross-country transmission line piping. The root face and root opening are approximately the same. As the wall thickness increases the joint design will change so that a lesser amount of weld metal will be required. This means that the included angle changes to a narrower angle partially up the joint. These types of joint designs are more commonly used in power plant piping where extremely heavy wall thickness pipe is used. Other variations in joint design depend on the composition of the pipe. Backing

FIGURE 20-3 *Pipe welding joint designs.*

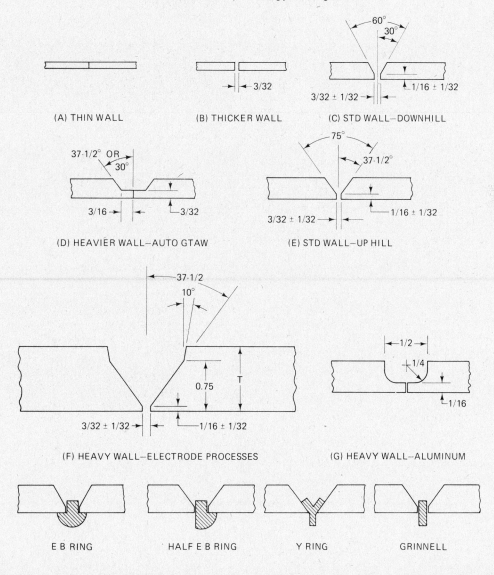

(A) THIN WALL (B) THICKER WALL (C) STD WALL—DOWNHILL

(D) HEAVIER WALL—AUTO GTAW (E) STD WALL—UP HILL

(F) HEAVY WALL—ELECTRODE PROCESSES (G) HEAVY WALL—ALUMINUM

E B RING HALF E B RING Y RING GRINNELL

diameter pipe external type clamps are normally employed. Figure 20-6 shows an assortment of these. Some of these clamps have sufficient power to reform the pipe into a perfect circle to facilitate fitup. This is possible on the

rings are not normally used, but consumable insert rings may be used under certain conditions.

For aluminum pipe special joint details have been developed and these are normally associated with combination-type procedures. This allows a root weld to be made in much the same manner as the weld on thin wall tubing. In all cases, it is assumed that a backing ring is not employed. The rectangular backing ring is rarely used when fluids are transmitted through the piping system. It may be used for structural applications in which pipe and tubular members are used to transmit loads rather than materials. Socket joints and bell and spigot joints which utilize fillets are sometimes used but they are becoming less popular. The different pipe welding joint designs are shown by Figure 20-3.

Pipe Joint Alignment and Fitup.

The most important factor in obtaining a high quality pipe joint is to make sure that the fitup of the joint prior to welding is perfect. It must be in accordance with the joint detail and uniform throughout the circumference of the pipe joint. This is relatively difficult to obtain because pipe is not always exactly round and the diameter may change within limits of the pipe size. Tubing is made to closer size tolerances and is easier to fit up. The nonroundness or ovality of pipe presents a major welding problem. This is particularly true in the larger diameter pipe. The most common method of preparing joints for welding is by oxygen flame cutting. Specialized torches that revolve around the pipe circumference are used. Preparation of this type is shown by Figure 20-4. Mechanical cutters are also used to make a uniform joint preparation on pipe. Weld bevels are prepared at the pipe mill for welding and in the case of forged fittings the joint details are prepared at the manufacturer's plant. Bevels are also prepared by mechanical means such as machining and also by flame cutting when the pipe is rotated under a torch.

Fitting of the pipe becomes the most difficult part of piping installations. For small assemblies this is relatively easy and much of this work is done in pipe-fabricating shops where assemblies of different fittings and pipe sections are welded together under ideal conditions. A typical shop fabrication welded with semiautomatic equipment is shown by Figure 20-5. These assemblies are then transported to the field erection site for field welding. Assembly in the field is usually more difficult since dimensional variations are less easy to control.

A variety of alignment devices are available and used for many pipe application installations. For small-

FIGURE 20-4 *Preparation of bevel on pipe.*

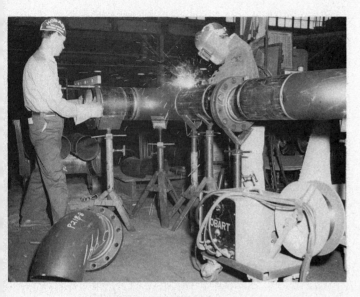

FIGURE 20-5 *Shop fabrication of pipe assembly.*

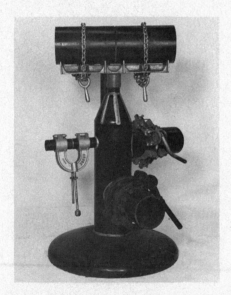

FIGURE 20-6 *A variety of external line up clamps.*

FIGURE 20-7 *A pneumatic internal line up clamp.*

thin wall pipe but becomes increasingly difficult as the pipe wall thickness increases.

The cross-country transmission pipeline welding techniques have become extremely sophisticated. Normally the *stove pipe* method of installing pipe is used. This means simply that each section or length of pipe is added on to the existing pipe installation. The crew for doing this moves consistently along the right of way from the beginning to the end of the pipeline. Welding procedures and techniques vary somewhat based on the diameter of the pipe.

For the large-diameter pipe welds an internal lineup clamp is utilized. Figure 20-7 shows a typical and complex internal lineup clamp. They are inserted in the end of the last section of the pipeline and are remotely operated by air pressure. The air pressure clamps the internal lineup clamp to the section already welded to the pipeline and then as the new section is being placed in position it clamps, locates, and spaces the new section. It helps round out the pipe due to the pressures and strength of the clamp. Many of these clamps also include a copper backing ring. Normally the clamps are left in the pipe joint until the first or stringer pass of the weld is made. In some cases, the second pass is also made before the clamp is released and removed from the joint. In welding large-diameter pipes there is usually one welding crew that makes the root pass and second pass commonly known as the *stringer* and the *hot* pass. They move on with the lineup clamp crew and work on the next pipe joint and other welding crews come in to finish the weld. These crews make the so-called *filler* passes, which are those that fill the weld joint, the *stripper* passes, which are usually made in the vertical portion of the pipe joint, and the last pass, known as the *cap* pass. The stringer crew and the other crews may represent three or four pipe welding groups that are progressing along the pipeline during its construction. The equipment is carried on tractors or trucks so it can move along the pipeline right of way and progressively add sections and build the pipeline. Different sized crews and equipment are required for small-diameter or large-diameter pipelines. This is shown by the two photographs of Figure 20-8.

Pipe Welding Materials.

The steel for pipe is specified by several different specification bodies, primarily the American Society for Testing and Materials (ASTM) and the American Petrolem Institute (API). The ASTM standards are used for small-diameter pipe, the medium-diameter pipe, and the heavy-wall pipe. The API specifications are used for line pipe normally of the larger diameters and thinner wall. These specifications are described in detail in Chapter 12.

The steel used for small and medium grade pipe is usually carbon steel. In the thicker sections, also carbon steel, but for high-temperature high-pressure power plant applications low-alloy steels are employed. The API specs are for carbon steels but as the strength levels increase they become low-alloy high-strength steels. Quenched and tempered steels are also used for piping. For thin wall tubing, low-alloy steels and stainless steels are used. Aluminum, titanium alloys, copper, brass, and high-nickel alloys are also used to make pipe and tubing. The welding process must be selected for the base metal being welded and the filler metal must be selected to meet the specific properties of the material.

Automatic Welding.

Automatic welding systems have been perfected for joining tubular members. The most successful type of equipment utilizes the gas tungsten arc welding process and is used for automatically joining smaller-diameter thin wall pipe and tubing. This type of equipment clamps the pipe or tube in alignment and rotates a tungsten electrode around the weld joint to make an autogenous type weld without the addition of the filler metal. This type of equipment can be used for welding pipe and tubing up to 8-in. (200-mm) diameter with a wall thickness of up to 0.109 in. (2.8 mm). For heavier wall thickness up to 0.500 in. (12 mm) filler metal is used to

FIGURE 20-8 *Pipe line welding of small and large pipe.*

FIGURE 20-9 *Tube to tube automatic GTAW welding machine.*

completely fill the weld joint. The equipment is similar but in this case filler metal is added as the automatic head revolves around the pipe. Figure 20-9 shows this type of equipment in operation.

The gas metal arc welding process has also been employed for fully automatic pipe welding. Wire feeders using small diameter electrode wires are made to rotate around the weld joint and deposit weld metal. The major problem with this type of equipment is the making of a quality root pass. This is difficult since the root opening, the root faces, and pipe alignment are not uniform around the joint because of pipe ovality and tolerances of joint preparation. Once the root pass has been made, the remaining passes are easier to make. This type of equipment is being used on barges for welding under water pipelines on the surface prior to lowering them to the ocean floor. These automatic welding machines are also used for cross-country pipelines under certain conditions. Figure 20-10 shows a fully automatic gas metal arc welding machine for automatic welding 30-in. (750 mm) diameter pipe in the field.

FIGURE 20-11 *Automatic roll welding using submerged arc.*

FIGURE 20-10 *Automatic GMAW welding machine for large pipe.*

The flux-cored arc welding process, as well as gas metal arc and submerged arc welding, is used to a great degree for automatically welding pipe when the pipe can be rotated. This is commonly used in fabricating shops and for double joining or welding together two pipe lengths prior to installing the pipe in the pipeline. This allows for higher efficiency since welds can be made more rapidly in the flat or roll position than in the fixed position. Figure 20-11 shows a typical fab shop application using the sumberged arc welding process. In some cases, the first or root pass is made by gas metal arc

welding or even gas tungsten arc welding. Schedules show the different types of possible welding procedures that can be used. If the pipe is sufficiently large backing welds can be made. These can be made automatically with the submerged arc welding process or they can be made with the gas metal or flux-cored arc welding process automatically or semiautomatically on the inside of the pipe.

There is one other type of automatic pipe welding commonly employed and this is for welding pipe automatically when the axis of the pipe is vertical. This type of automatic welding is used for well drilling and also for piling installations when tubular sections are used. This automatic welding normally will use a backing ring or bar and can be more easily automated since the welding conditions are fixed throughout the entire circumference of the weld joint. This is different than in the fixed position where the welding conditions change continuously around the circumference of the pipe. The flux-cored arc welding process and gas metal arc welding are most commonly used.

Another type of automatic welding that should be mentioned is the tube-to-header or tube-to-tube sheet welding required to manufacture heat exchangers. A

FIGURE 20-12 *Tube to tube sheet automatic welding machine.*

heat exchanger consists of many pipes or tubes between two headers. One fluid is transmitted through the tubes and another occupies the space outside of these tubes. Heat exchangers are used extensively in refineries, chemical plants, power plants, and for other process manufacturing. Each heat exchanger may involve hundreds of tube-to-header welds. Figure 20-12 shows a typical heat exchanger head with the tubes being welded. The majority of this welding is done with small automatic machines that rotate the tungsten around the joint between the tube and tube sheet or header.

Welding procedures for these joints are shown by Figure 20-13. The three more common joint designs are shown. These are the flush tube design, the extended tube design, and the recess tube design, and each has a

FIGURE 20-13 *Tube to tube sheet schedules.*

A. Extended Tube   B. Flush Tube   C. Recessed Tube

Tube O.D.		Wall Thickness		Joint Type	Weld Current (Amperes)	Filler Rod Type	Weld Time (sec)
in.	mm	in.	mm				
Stainless steel tube-to-stainless steel tube sheet							
0.75	19.1	0.062	1.6	A	140	E304	18
0.75	19.1	0.080	2.0	A	130	None	18
0.75	19.1	0.090	2.3	A	140	E304	31
Mild steel to mild steel							
0.75	19.1	0.062	1.6	B	140	E705-3	28
1.00	25	0.125	3.2	B	140	E705-3	184
1.00	25*	0.125	3.2	C	180	E705-3	49
2.50	62	0.187	4.7	B	175	E705-3	245
Stainless steel to mild steel							
0.62	15.7	0.062	1.6	A	120	E309	14
1.00	25	0.083	2.1	A	155	E309	40
CuNi–CuNi							
0.75	19.1	0.062	1.6	A	160	ERCuNi	28
CuNi–mild steel							
0.75	19.1	0.062	1.6	A	160	ERNiCrFe-6	28
Cu–mild steel							
0.62	15.7	0.025	0.6	A	140	ERCuSi–A	16

* 2 passes

variation. For some applications it is necessary only to seal or fuse the tube to the tube sheet. In other situations it is required to add filler metal. For the extended tube or the recessed tube a fillet weld can be produced; however, a fillet weld will not have as much strength as the tube because of the throat dimension of the fillet throat. Nuclear specifications require that the filler metal thickness be equal to the thickness of the tube; thus a J groove is required. In this design sufficient weld metal can be added to make the weld stronger through its shortest dimension than the wall of the tube. The joint design must be based on the specifications involved.

The automatic tube-to-header welding machine uses a 3/32-in.-diameter tungsten electrode of the EWTh-2 classification (2% thoriated). When filler metal is being added the arc length should be held to 3/32–1/8 in. (2.4–3.2 mm). If filler metal is not used the smaller arc length should be used. For the extended tube design the tungsten should be angled 10° and a sharper taper should be used on the point. This should approach 20°, whereas for standard work the normal 60° point can be used.

Pulsing is used for making automatic tube-to-tube sheet welds. In addition, preflow, postflow, upslope, and downslope, and the various delays are employed to produce a high-quality weld. These data are presented in qualified welding procedures manuals available from manufacturers of the tube-to-header automatic welding machines.

20-2 ARC SPOT WELDING

An arc spot weld is a spot weld made by an arc welding process. A spot weld "is a weld made between or upon overlapping members wherein coalescence may start and occur on the same surfaces or may have proceeded from the surface of one member." The weld cross section in plan view is approximately circular. Figure 20-14 shows the different spot welds. An arc spot weld differs from the resistance spot weld since in resistance welding coalescence is located at the faying surfaces of the parts being joined.[2] The arc spot weld does not require a hole in either member. Therefore, it differs from the plug weld, which requires a hole to be prepared for the weld.

The arc spot welding is performed by melting through one of the members, usually the top member. Therefore, the thickness of this member is the limiting factor. This limiting factor also depends on the welding process used.

The gas tungsten arc welding process and the gas metal arc welding process are most commonly employed

FIGURE 20-14 *Spot welds—top, bottom and cross section.*

Material	Mild Steel	Mild Steel	Mild Steel	Stainless	Aluminum
Thickness	20 ga (0.0359 in.)	11 ga (0.1196 in.)	3/16 in.	22 ga (0.0299 in.)	11 ga (0.0907 in.)
Process	GMAW (CO_2)	GMAW (CO_2)	FCAW (CO_2)	GTAW (Argon)	GMAW (Argon)
Electrode size	0.035 in.–.9 mm	1/16 in.–1.6 mm	7/64 in.–2.8 mm	– – –	3/64 in.–1.2 mm
Electrode type	E70S-3	E70S-3	E70T-1	– – –	5356
Top side					
Bottom side					
Cross section					

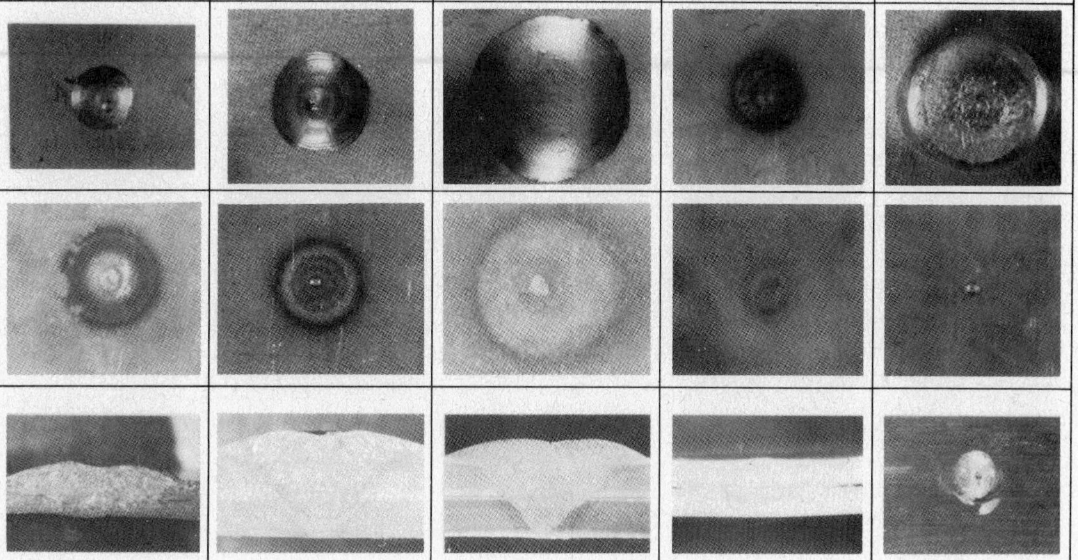

for making arc spot welds. Flux-cored arc welding, and shielded metal arc welding using covered electrodes can be used for making arc spot welds.

The principal operation of arc spot welding is to strike and hold an arc without travel at a point where the two parts to be joined are held tightly together. The heat of the arc melts the surface of the top member and the depth of the melting is dependent on the welding process, the electrode size and type, the welding current and the time, and, in the case of gas metal arc welding, the type of shielding gas employed. Timing is usually done automatically by means of a timer or some device which can be varied to provide different times for different size welds. With the exception of the timer and special electrode holders or guns, the process utilizes the same equipment that is normally used for that process.

Arc spot welding is an extremely adaptable process that offers many metal joining advantages. One advantage of the process is that it can be used without exercising manipulative skills. Thus, the training required to make arc spot welds is minor. In addition, the welder making arc spot welds does not have to use a welding helmet since the arc is contained within a gun or gun nozzle. Arc spot welding is an extremely fast process, can be used for a variety of joining requirements, and can be fully automated.

The arc spot welds can be made in the flat position and in the horizontal position, but it is almost impossible to make arc spot welds in the overhead position. The reason for this is that the sheet or member closest to the welding gun must melt completely through and this can result in a fairly large mass of molten metal which tends to fall away from the weld in the overhead position.

The metals normally welded with arc spot welding are the mild, low-alloy and stainless steels. However, with special precautions other metals can be welded. Aluminum can be arc spot welded if a clean interface between the parts to be joined is maintained.

Thickness range of metals that can be welded by the different arc processes is shown in the schedules for GTAW and GMAW welding. This information involves the thickness of the top of the member which must be melted through. The thickness of the other member is not important. Lap joints are the most common types of joints used for arc spot welding; however, tee joints can also be made.

Gas Tungsten Arc Spot Welding.

Arc spot welding with the gas tungsten arc process is an extremely efficient and simple tool to use for making weld joints. All of the advantages listed above apply to gas tungsten arc welding. The process is limited, however, to a maximum thickness of 1/16 in. (1.6 mm) of the top sheet or the sheet closest to the arc. Gas tungsten arc welding can be used for mild steels, low-alloy steels, stainless steels, and aluminum alloys.

The basic difference in equipment is the welding gun or torch, which should be designed for arc spot welding. The nozzle of the gun is used to apply pressure to hold the parts being welded in close contact. A gun for gas tungsten arc spot welding is shown by Figure 20-15. The gun nozzles are made of copper or stainless steel and are normally water cooled since the arc is contained entirely within the nozzle. The nozzle design controls the critical distance between the tungsten electrode and the surface of the work; it should have ports for the shielding gas to escape. The inside diameter should be related to the size of the tungsten electrode being used. The 0.500-in. (12-mm) inside diameter is most common. The nozzles can also be designed to help locate the arc spot weld, especially with respect to corners or edges of the top sheet. Arc spot welding equipment is also used to make tack welds at inside or outside corner joints, etc. The gas tungsten arc spot gun will also include a trigger switch which will actuate the arc spot operation.

FIGURE 20-15 *Making a GTAW arc spot weld.*

The second extra part in the system is the timer. Some gas tungsten arc power sources include timers, which are preferred. However, if these are not included, a separate timer can be used provided that the power source includes a contactor. The timer should have the capability of being adjusted from 0.5 second up to 5 seconds.

The tungsten electrode type would be the same as would be normally selected for the material to be welded. The 1/8-in. (3.2-mm) electrode diameter is recommended for all work. The point of the tungsten should be the standard ground point but then squared off or blunted on the very end. This tends to improve the depth of penetration obtained with the process. The tungsten must be kept clean and should have sufficient distance between it and the work surface to avoid contamination during the arc spot welding period. An arc length of 1/16 in. (1.6 mm) is recommended. If the arc length is too short the weld area or spot weld size will be small. As the arc length increases the size of this arc spot weld will be larger. If the arc length is too long, an unstable arc will result and there will be a lack of uniformity between the different arc spot welds.

The normal sequence of events is as follows: The nozzle of the torch or gun is placed on the joint and sufficient pressure is applied to bring the parts in intimate contact. The trigger on the gun is depressed, which starts the welding cycle. Gas flow is initiated to purge the system and the area within the gun nozzle. If water cooling is employed the cooling water will also start to flow at this time. The arc will be initiated by the high-frequency discharge supplied by the gas tungsten arc power source. The arc will continue for the period of time established and will be extinguished as the power

contactor opens. The shielding gas will continue to flow for a predetermined postflow time. This completes the normal cycle.

Normally the thinnest metals joined are 24 gauge, which is 0.022 in. (0.56 mm). If both top and bottom sheets are of the same thickness, it is best to utilize a copper backup to prevent the weld from falling through the joint and having a depression on the top and excessive penetration on the bottom. A schedule for gas tungsten arc welding is shown by Figure 20-16. It is best to use high current and short time cycle rather than lower currents with longer time period. The amount of current increases the size of the nuggets and the thickness of the materials that can be welded. Nugget diameter and strength is also influenced by the welding current. The time factor tends to increase penetration but to a much lesser degree than the current. If too much current is used it can cause splashing. This blowing away of the molten metal from the top member may contaminate the tungsten and will result in an unsatisfactory weld. If the current is too low the nuggets may not penetrate completely through the top sheet to the bottom sheet.

FIGURE 20-16 *Schedule of arc spot welds using GTAW.*

Material Type	Metal Thickness (top sheet)			Welding Conditions Amperes		Arc Spot Time	Shielding Gas Argon	
	Gage	in.	mm	DCEP + HF	AC-HF	Seconds	CFH	L/M
Stainless Steel	24	0.025	0.64	125	175	1.0	10	4.5
	24	0.025	0.64	110	175	1.25	10	4.5
	24	0.025	0.64	100	150	1.5	10	4.5
	22	0.031	0.79	125	175	1.5	10	4.5
	22	0.031	0.79	100	175	1.75	10	4.5
	18	0.050	1.27	140	200	1.5	12	5.6
	18	0.050	1.27	110	150	2.5	12	5.6
	16	0.062	1.57	170	250	3.0	12	5.6
	16	0.062	1.57	140	—	3.25	12	5.6
	16	0.062	1.57	115	—	5.25	12	5.6
		0.064	1.62	160	250	2.25	12	5.6
Mild Steel	22	0.031	0.79	170	250	1.5	8	3.6
	22	0.031	0.79	140	200	2.0	8	3.6
	22	0.031	0.79	120	175	2.25	8	3.6
	18	0.050	1.27	170	250	1.75	10	4.5
	18	0.050	1.27	140	200	2.0	10	4.5
	18	0.050	1.27	135	200	2.5	10	4.5
	16	0.062	1.57	170	250	3.0	12	5.6
	16	0.062	1.57	155	225	3.5	12	5.6
Aluminum		0.022	0.56	—	170	1.1	8	3.6
		0.032	0.81	—	200	1.5	8	3.6
		0.048	1.21	—	220	1.7	8	3.6
		0.064	1.63	—	250	2.2	8	3.6

Note: 1. The electrode is 2% thoriated tungsten, except for aluminum pure tungsten electrodes are used, 1/8 in. diameter.
   2. High-frequency is used to start the arc when using direct current and alternating current.
   3. Arc length electrode to work 1/16.

The shielding gas will be either argon or helium with a flow rate of from 6–10 cubic feet per hour (2.5–4.5 liters per minute). The nozzle should make good contact with the work to contain the gas inside the nozzle. Vents are contained in the nozzle to allow shielding gas to flow and escape. If the nozzle-to-work contact is not tight, higher flow rates will be required. Helium will provide a smaller weld nugget with a greater depth of penetration. Argon produces a larger weld nugget with penetration not quite so deep.

Direct current should be used for all materials, except aluminum, with the electrode negative (straight polarity). Alternating current with continuous high frequency should be employed on aluminum. If aluminum is well cleaned the electrode negative (straight polarity) can be used. Parts to be welded should be clean of oil, dirt, grease, scale, etc. This is especially true at the interface and absolutely necessary when welding aluminum. The weld diameter is the basis for the shear strength of arc spot welds. The shear strength of arc spot welds will be similar to resistance spot welds made in the same material.

Gas tungsten arc spot welding has become widely used in the manufacture of automotive parts, parts for appliances, precision metal parts, and parts for electronic components. It is normally applied as a semiautomatic process; however, it can be mechanized and used for high-volume production work. It can also be controlled by numerical tape-controlling devices.

Gas Metal and Flux-Cored Arc Spot Welding.

The gas metal arc and the flux-cored arc welding processes can both be used for making arc spot welds. The equipment used for the two processes is identical. The only difference is in the type of electrode wire employed.

The arc welding equipment used for normal semiautomatic welding can be used for gas metal and flux-cored arc spot welding by the addition of a timer and a special torch or gun nozzle. Time range for the timer should be the same as used for gas tungsten arc spot welding. The nozzle or torch should be sufficiently strong so that it can transmit the necessary force to hold the parts to be welded together in intimate contact.

The sequence of events to produce an arc spot weld are almost the same as used for gas tungsten arc welding. The only difference is that when the arc starts the wire feeder will feed electrode wire into the arc.

The gas metal and flux-cored arc processes can weld through greater thicknesses of metal than gas tungsten arc spot welding. With proper conditions, welds can be

routinely made through the top plate of 1/4 in. (6.4 mm) thickness. At the other end of the scale, welds can be made in 24-gauge material 0.022 in. (0.56 mm) thick. Welding is done primarily on the mild steels and low-

FIGURE 20-17 *Making a GMAW or FCAW arc spot weld.*

alloy steels. Figure 20-17 shows an arc spot weld being made.

The shielding gas normally used for gas metal arc spot welding is CO_2. Carbon dioxide is selected since it has the highest penetrating qualities of any of the shielding gases. If high penetration is not required, the 75% argon–25% CO_2 mixture can be used.

The size and type of electrode wire have a large effect on the depth of penetration and on the diameter of the weld nugget at the interface. Normally for maximum strength of the arc spot weld it is desirable to have a large nugget at the interface. The composition of the electrode wire should be of a de-oxidized type; normally the E70S1 electrode wire is used with solid wires or the E70T1 electrode wire is used for the flux-cored arc welding process.

The weld schedule for making GMAW and FCAW arc spot welds is shown by the table of Figure 20-18. The electrode wire changes from a small diameter of 0.030-in. (0.8 mm) solid wire up through 1/16-in. (1.6 mm) solid wire and to 3/32-in. (2.4 mm) diameter flux-cored wire to accommodate all the weld thicknesses shown. The larger diameter wires provide higher strength per arc spot weld, since they produce larger nuggets at the interface.

When welding thinner materials, a backup bar should be used behind the second sheet. It is also possible to

Gage	Material Thickness in.	mm	Electrode Wire Dia. in.	mm	Electrode Type	Current DC Amperes	Arc Voltage EP Volts	Arc Spot Time Seconds	Wire Consumed Per Spot in.	mm	Typical Shear Strength Per Spot lbs	kg
24	0.022	0.56	0.030	0.76	E60S-1	90	24	1.0	4-5/8	115	625	283.60
22	0.032	0.81	0.030	0.76	E60S-1	120	27	1.2	5	125	730	331.13
20	0.037	0.94	0.030	0.76	E60S-1	120	27	1.2	10-1/8	253	1337	606.46
22	0.032	0.81	0.035	0.89	E60S-1	140	26	1.0	6	150	800	362.88
20	0.037	0.94	0.035	0.89	E60S-1	140	26	1.0	6	150	1147	520.33
18	0.033	0.84	0.035	0.89	E60S-1	190	27	1.0	8-1/2	212	1507	683.58
16	0.059	1.50	0.035	0.89	E60S-1	190	28	2.0	17-1/4	431	1434	641.46
14	0.072	1.82	0.035	0.89	E60S-1	190	28	5.0	40-1/4	1006	2600	1179.96
18	0.039	0.99	0.045	1.14	E60S-1	200	27	0.7	4	100	1414	641.39
16	0.059	1.50	0.045	1.14	E60S-1	260	29	1.0	6	150	2070	938.95
14	0.072	1.82	0.045	1.14	E60S-1	300	30	1.5	12-3/4	319	3224	1462.41
12	0.110	2.79	0.045	1.14	E60S-1	300	30	3.5	28-1/2	712	4300	1950.48
11	0.124	3.15	0.045	1.14	E60S-1	300	30	4.2	34	850	4114	1866.11
16	0.059	1.50	1/16	1.6	E60S-1	250	29	1.0	2-3/4	69	1654	750.25
14	0.072	1.82	1/16	1.6	E60S-1	360	31	1.0	5-1/2	137	3340	1515.02
12	0.110	2.79	1/16	1.6	E60S-1	440	32	1.0	7-1/4	181	5000	2268.00
11	1/8	3.18	1/16	1.6	E60S-1	490	32	1.0	8-1/2	212	5634	2556.58
	5/32	4.0	1/16	1.6	E60S-1	490	32	1.5	9	225	5447	2460.76
	3/16	4.76	1/16	1.6	E60S-1	490	32	2	16-3/4	419	6834	3179.90
	1/4	6.4	1/16	1.6	E60S-1	490	34	3.5	28-1/8	703	8667	4721.35
16	0.062	1.57	3/32	2.4	E70T-2	400	30	0.6	1-3/8	34	2550	1156.68
11	1/8	3.18	3/32	2.4	E70T-2	500	34	0.8	3	75	3400	1442.24
	3/16	4.76	3/32	2.4	E70T-2	650	38	1.6	8-1/4	206	7050	3197.88
	1/4	6.4	3/32	2.4	E70T-2	750	40	2.2	15-3/4	394	10300	4672.08

Note: Contact tip-to-work distance: 1/4 to 3/8 for fine wire, 7/8 for large wire and flux-cored wire, CO_2-7/8 Gasless—1-1/2.

Note: CO_2 when used: 35 cubic feet per hour (6.5 L/M).

FIGURE 20-18 *Schedule of arc spot welds using FCAW and GMAW.*

make arc spot welds through more than two sheets, provided that the combined thickness of the top two sheets is within the range of the welding schedule. Intimate contact is required. The same equipment with timers can be used for tack welds and for plug welds when the top sheet is too thick.

The advantages of arc spot welds over resistance spot welds are as follows:

1. Access is required only to the top or front side of the joint.

2. The amount of pressure is not excessive, assuming that the parts are fitted properly.

3. Operator skill is minimum and a welder's helmet is not required.

4. The consistency and reproducibility of arc spot welds are excellent.

5. The amount of weld spatter, smoke, and flash is minimum and metal finishing can be eliminated for many products.

6. Distortion is minimum.

7. Close cost control is obtained using arc spot welding.

8. It is readily adaptable to designs originally used for bolting or riveting.

Arc spot welding is being used in automobile body assemblies, the attachment of brackets to body assemblies, frame assemblies, etc., the assembly of industrial products, the assembly of lattice structural beams, and for innumerable other applications. The strengths are in the same relationship as resistance spot welds on the same materials in the same thicknesses.

Variations of the Process.

There is one variation that would be more accurately described as a plug weld. It is used to join dissimilar metals. It has been used for joining aluminum to copper and galvanized steel to aluminum. Aluminum filler wire is used with inert gas through a hole in the copper or galvanized part. Copper terminals can be joined to aluminum cables and galvanized steel brackets can be joined to aluminum pans. [3]

Another variation of arc spot welding is done with the shielded metal arc welding process using covered electrodes. Special spot welding guns or holders are used. Small diameter electrodes are used. The special holder causes the arc to strike and holds a short arc for several seconds without manual assistance. An arc shield surrounds the arc area and a welding helmet is not required. This process variation will weld through 16 gauge steel in the flat, horizontal, and vertical positions. It is used primarily in autobody repair shops.

Applications.

One of the more interesting applications of arc spot welding is the welding of door hinges of a compact automobile to the door post of the body and to the door. Three spot welds are used and they are made automatically. A series of pictures in Figure 20-19 shows the weld and the apparatus for making these welds.

20-3 ONE-SIDE WELDING

The term *one-side welding* has become very popular in the welding industry within the last few years. One-side welding is certainly not new, but it has been popularized and given a high degree of importance in the shipyards of Japan. The term one-side welding means the production of a butt weld from one side of the joint that achieves a 100% efficiency and produces a back side of the joint that is acceptable from an appearance and quality viewpoint.

It is obvious that all welds made on medium-and-small-diameter pipe and tubes are one-side welds, since the back side of the weld is inaccessible. It is also true that at the early stages of the use of the submerged arc welding process techniques for backing the joint with flux were developed and produced welds made entirely from the top side of the joint. Also welds made on small tanks, particularly the closure welds, welds made on other containers, and tubular products are all one-side welds. Why then is there so much interest in one-side welding today?

In the shipbuilding industry, particularly in Japan where large tankers are being built, it is customary to handle welded assemblies that measure up to 60 ft (20 m) by 60 ft (20 m). Previously with smaller assemblies, automatic welding was done by a single pass from each side of the joint. This meant that the assemblies had to be turned over to complete the weld. Turning these large assemblies over involved extremely heavy-duty overhead cranes with sufficient height to handle the large weldments. The heavy-duty cranes and high bays in the buildings are extremely expensive. Elimination of this turning over and the crane-handling time represented a substantial cost savings in the building of large tankers. As

FIGURE 20-19 *Arc spot welding door hinge on autobody.*

ship assemblies became larger, research work was intensified to avoid the turning-over operation. It was felt also that there would be another advantage of welding from one side. It would fit in with the use of automatic conveyors and material-moving systems for better assembly–line welding. (4)

One-side welding can be divided into two distinctly different methods of operation. The first method uses a weld-backing apparatus located in a production line area. Flat plates are set on this apparatus for one-side welding. This method is suitable for long straight joints normally in flat plates welded in the flat position.

The second method utilizes portable backing materials that are taken to and applied to the back side of the weld joint, wherever it is located. This method of operation is an attempt to eliminate overhead welding of short joints in assemblies that cannot be positioned for flat or downhand welding.

One-side welding is difficult when it is necessary to obtain full penetration and a 100% joint efficiency over the entire length of the joint. With conventional welding equipment, automatic or machine, it is difficult to produce a 100% joint from one side of the joint. Welds made from both sides usually have a varying degree of overlap of penetration from the welds made from each side. The extent of this overlap or penetration has not been considered important as long as the overlap occurred. The problem of joining a large assembly from one side or both sides is complicated by the material preparation tolerances. The proper alignment and flatness and the warpage that is inherent to any welding operation create fitup problems. Weld penetration is related to heat input which in turn relates to welding current and travel speed, as well as the heat losses of the weldment. Fitup, which depends on the root opening, the root face, and the angular dimension of grooves is never perfect on large weldments. It is possible that the root opening can vary from zero to an amount equal to or greater than the diameter of the electrodes being used. The root face can vary in the same degree. With welding conditions relatively constant the variations of fitup rapidly change the penetration of the weld. Changes in the welding schedule while making the joint affect the penetration of the weld. The research done in the Japanese shipyards was directed toward accepting the normal fabricating tolerances, yet producing a weld with a fixed amount of reinforcement on the back side of the weld joint.

Another problem inherent to welding of large structures is the straightness or fairness of the parts at the joint. Misalignment is very common and the so-called *high-low* where the surfaces of the plates are not in the same planes complicate the problem of welding. Methods of applying pressure and straightening weldments and plates were adopted as part of the overall program.

One of the earliest methods of making one-side welds was developed during the early use of the submerged arc welding process. This method utilized submerged arc flux on the underside of the weld joint. The flux was brought into intimate contact with the back of the joint by means of air pressure in a hose. This system has been called the FB method. FB indicates flux backing method and is shown by the diagram of Figure 20-20.

Another method utilized in many fixtures and used in seam welding machines is known as the copper backing method indicated by the letters CB. This has been used for many welding processes including gas metal arc, gas tungsten arc, submerged arc, as well as shielded metal arc welding. The copper backing method utilized fairly heavy copper bars brought into intimate contact to the back side of the weld joint. Often a recessed

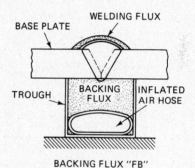

BACKING FLUX "FB"

FLUX-COPPER BACKING "FCB"

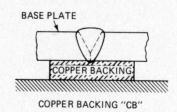

COPPER BACKING "CB"

FIGURE 20-20 *Fixed position weld backing methods.*

groove was placed in the copper bar immediately below the weld joint to allow for root penetration. For heavy-duty welding or for high-duty cycle work, the copper bar would be water cooled. For some of the more high-quality work on nonferrous metals using gas metal arc welding, backing gas was introduced into the recessed groove on the underside of the joint. This method is also shown by Figure 20-20.

Both of these methods of backing were highly successful for thinner materials or for relatively short welds. For heavier plates, when warpage was involved, or for long joints neither of these methods was entirely satisfactory. A development combining the two and retaining the advantages of both while eliminating their disadvantages is known as the flux-copper backing method abbreviated FCB. The FCB backing method utilizes a layer of granulated flux of consistent thickness in positive contact with the underside of both plates at the joint and in contact with the upper side of a copper backing bar. The copper backing bar helps control the uniform size of the reinforcement bead. The Japanese have named the backing bead *Uranami,* which describes the root bead viewed from the back side of the joint. In the FCB method the width of the flux layer is approximately 4 in. (100 mm). This width will accommodate normal variation in the line of the joint so that special accuracy is not required. The copper bar which backs up the flux is approximately 1/2 in. (50 mm) thick and 5 in. (125 mm) wide and as long as the joint. The thickness of the backing flux is about 1/4 in. (6 mm). The backing flux and the welding flux are normally the same flux. In some cases special backing flux is used. The back pressure keeps the flux in intimate contact with the parts being welded. This system is also shown by Figure 20-20.

There are variations to all three of these methods. For example, with the flux backing (FB) method, sometimes the flux is encased in a paper container, which is burned during the welding operation. This makes the fitup quicker. In the case of the copper backing (CB) method different kinds of coatings are placed on the copper bar. Some users place fiberglas tape or other inorganic fibers on the copper bar to maintain pressure against the back side of the weld. In the flux copper backing (FCB) method, different systems are used to keep the backing against the work and different types of flux are used.

There is also a variation when the weld groove is filled with iron powder to improve welding productivity. This was previously described in the submerged arc section of the book. The Japanese have called this the IP method, indicating iron powder. The IP method can be combined with the different backing methods.

One-side welding tries to utilize single-pass welding, particularly with the submerged arc welding process. The FCB method offers the advantage of a smaller groove area which reduces the amount of filler metal required and allows for a higher travel speed. It also provides for lower consumption of submerged arc flux and backing flux. An improvement of the FB method utilizes special backing flux, called RF, indicating refractory flux. This flux helps to form a uniform backing or uranami bead. The flux contains phenol resin binders which undergo thermal hardening from the heat of the arc. The flux will become solidified while maintaining intimate contact with the underside of the joint.This reaction will proceed with the arc and provides a solid and continual support for the back side of the joint as the welding progresses. The iron powder and deoxidizers help the formation of the uranami bead and a uniform backing bead is produced even though there are wide variations in fitup.

The second basic type of one-side welding utilizes portable backing material applied to the back side of the joint. The portable system utilizes short lengths of assemblies containing the backing materials. One of these, known as the KL method, is shown by Figure 20-21. This backing assembly is approximately 2 ft long. The top layer of flux is the bead-forming type, underneath which is a solid refractory material. Both are enclosed in a thin sheet metal trough. This trough is sufficiently flexible so that it can be formed to fit the changing contour of a ship structure.

Another portable method utilizes the FAB backing assembly also shown by Figure 20-21. This assembly is

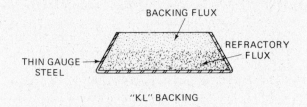

"KL" BACKING

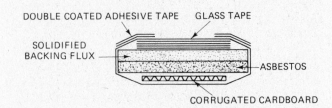

FLEXIBLE "FAB" BACKING

FIGURE 20-21 *Portable type weld backing methods.*

composed of several layers of fiberglas tape, under this is a layer of asbestos paper and below this is a corrugated cardboard pad. This entire assembly is enclosed in a cardboard container, which holds double-coated adhesive tape used to keep the backing assembly in intimate contact with the underside of the weld joint. This assembly is relatively flexible and can be formed to the contour of the part being welded. There are several other configurations of portable backing devices. One of particular interest is known as the CRB method, for coated rod backing. It is similar to a large covered electrode except that the flux covering on the rod does not melt. When the root opening is excessive two rods may be used. Magnetic brackets are used to hold the backing assembly in place.

A type of portable backing has become popular in North America. This method utilizes an adhesive tape, which contains a layer of backing flux along with several layers of aluminum foil. This material can be placed against the underside of the joint and will adhere to the parts being joined because of the adhesive on the tape. The flux will help form the backing bead. Tape is being used with gas metal arc welding, flux-cored arc welding, submerged arc welding, and shielded metal arc welding. When using submerged arc welding, sufficient root face must be allowed so that the welding arc does not come in contact with the backing flux in the tape. Figure

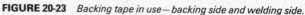

FIGURE 20-23   *Backing tape in use—backing side and welding side.*

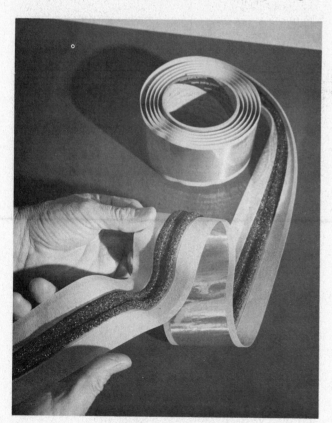

FIGURE 20-22   *Backing tape with flux.*

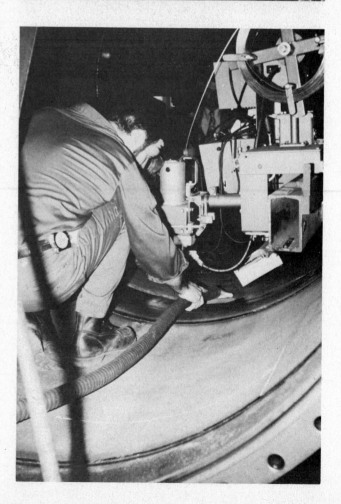

20-22 shows this tape with flux which will be in contact with the root side of the weld joint. This tape can be used in a variety of ways. Figure 20-23 shows the tape in use joining cylindrical members. The tape is on the outside of the structure which is the root side of the joint and the weld is made from the inside with an automatic submerged arc welding head.

Tape has been found to be very useful for welding vertical joints using the gas metal arc process.In this case the tape is used to eliminate a draft of air blowing through the root opening, which tends to reduce the efficiency of the shielding gas.

20-4 UNDERWATER WELDING

Underwater welding began during World War I when the British Navy used it to make temporary repairs on ships. These repairs consisted of welding around leaking rivets on ship hulls. With the introduction of covered electrodes it was possible to successfully weld under water and to produce welds having approximately 80% of the strength and 40% of the ductility of similar welds made in air. For this reason underwater welding has been restricted to salvage operations and emergency repair work. In addition, it was limited to depths below the surface of not over 30 ft (10 m).

In recent years there has been an increased interest in underwater welding because of the offshore exploration, drilling, and recovery of oil and gas. This activity has led to the laying of underwater pipelines and the necessity to sometimes repair them and the underwater portion of drill rigs and production platforms. Specialized techniques have been developed so that weld metal quality meeting the American Petroleum Institute standard 1104 can be made under water. Efforts are now going forward to increase the depth at which quality welding can be performed.

Underwater welding can be subdivided into two major categories—welding in a *wet environment* and welding in a *dry environment*. Welding in the wet or the wet environment is used primarily for emergency repairs or salvage operations in relatively shallow water.

Wet Welding.

The poor quality of welds made *in the wet* is due primarily to the problem of heat transfer, welder visibility, and the presence of hydrogen in the arc atmosphere during the welding operation.

When the base metal and the arc area are surrounded entirely by water there is no temperature or heat buildup of the base metal at the weld. This creates a high-temperature gradient or quench effect, which reduces the quality of the weld metal. The arc area is composed of a high concentration of water vapor. The arc atmosphere of hydrogen and oxygen of the water vapor is absorbed in the molten weld metal and contributes to porosity and hydrogen cracking. In addition, welders working under water are restricted in their efforts to manipulate the welding arc in the same manner as possible on the surface. They are further restricted by low visibility, because of their equipment and the contaminants in the water, plus those generated in the arc. Under the most ideal conditions, the welds that are produced in the wet with covered electrodes are at best marginal. They may be used for short periods of time but should be replaced with quality welds as quickly as possible.

Various efforts have been made to produce a bubble of gas in which the weld can be made. This technique has not yet been sufficiently successful to insure quality welds when made with covered electrodes in-the-wet. Efforts have also been made to utilize the gas metal arc welding process in-the-wet, again welding inside a gas bubble. These efforts also have been less than successful.

In spite of the fact that deposit weld metal may not have the same quality as weld metal made in air it is often necessary to do in-the-wet underwater welding. The general arrangements for underwater in-the-wet welding are shown by Figure 20-24. The power source for underwater welding should always be a direct current machine rated at 300 or 400 amperes. Motor generator welding machines are most often employed for underwater welding in-the-wet. The frame of the welding machine must be grounded to the ship. In addition, the welding circuit must include a positive type of switch, usually a knife switch that is operated on the surface upon the command of the welder-diver. The knife switch in the electrode circuit must be capable of breaking the full welding current. This switch is required for safety reasons. The welding power should be connected to the electrode holder only during the period that the welder is welding. Direct current with electrode negative (straight polarity) is used. Special welding electrode holders with extra insulation against the water are employed. Figure 20-25 is an example of this type of holder. The underwater welding electrode holder utilizes a twist type head for gripping the electrode. It will accommodate two sizes of electrodes. The electrode size normally used is 3/16 in. (4.8 mm); however, 5/32-in. (4-mm) electrodes can also be employed. The electrode types used conform to AWS E6013 classification. The electrodes must be waterproofed prior to using for underwater welding. This can be done by wrapping them with waterproof tape or by dipping them in special sodium silicate mixes and

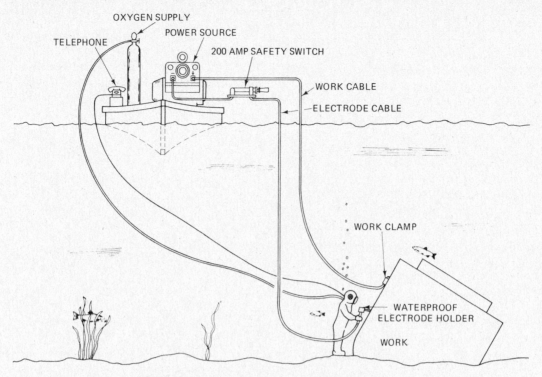

FIGURE 20-24 *Arrangements for underwater welding.*

allowing them to dry. Commercial electrodes for underwater welding are available.

The welding lead and work lead should be at least 2/0 size, and the insulation must be perfect. If the total length of the leads exceeds 300 ft. (100 m) they should be paralleled. With paralleled leads to the electrode holder the last 3 ft. (1 m) should be a single cable. All connections must be thoroughly insulated so that the water cannot come in contact with the metal parts. If the insulation does leak the sea water will come in contact with the metal conductor and part of the current will leak away and will not be available at the arc. In addition there will be rapid deterioration of the copper cable at the point of the leak. The work lead should be connected to the piece being welded within 3 ft. (1 m) of the point of welding.

A special underwater cutting torch which utilizes the electric arc-oxygen cutting process using tubular steel covered electrodes is also available. This torch is fully insulated and utilizes the twist type collet for gripping the electrode. It includes an oxygen valve lever and connections for attaching the welding lead and an oxygen hose. It is equipped to handle up to a 5/16-in. (7.9 mm) tubular electrode. The arc-oxygen cutting process is explained in Chapter 7. Briefly, it utilizes a special tubular covered insulated and waterproof electrode. In this process the arc is struck in the normal fashion and oxygen is fed through the center hole of the electrode to provide the cutting action. The electric arc-oxygen cutting torch is shown by Figure 20-25. The same electrical connections mentioned above are employed.

The welding techniques for underwater welding are unique and involve signaling the surface helper to close the knife switch when the welder begins welding. The drag travel system is used and a bead technique is employed. When the electrode is consumed the welder signals *current off* to the helper who opens the knife switch. "Current on" is signaled when a new electrode is in position against the work and ready for welding.

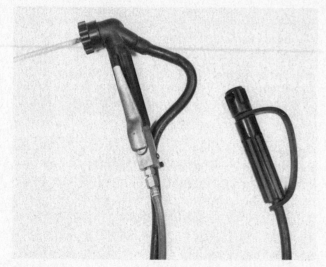

FIGURE 20-25 *Underwater electrode holder for welding and cutting.*

The current must be connected only when the electrode is against the work ready for welding, or when welding.

Complete information concerning underwater welding in-the-wet with covered electrodes is given by the "U.S. Navy Underwater Welding and Cutting Manual." [5] This same manual provides information on oxy-arc cutting, as well as other cutting processes. This manual should be utilized by those interested in underwater welding.

The need to produce high-quality welds under water has increased as the oil and gas industry moved off shore into deeper water. Most of the offshore exploration, drilling, and production, until recently, was in shallow water ranging from 30-50 ft. (10-16 m) deep. When a pipeline needed to be repaired, it was raised to the surface, repaired on a barge and lowered back to the ocean bottom. These repairs were made by welding or by mechanical means. Exploration, drilling, and now production are moving into deeper water up to the 1,000-ft. (305 mm) depth. In deeper water it is impossible to raise the pipeline to the surface for repair work. There is too much unsupported pipe and the pipeline would fail in such an operation. For this reason the repair work must be done on the ocean floor. In addition it appears that more and more pipelines are damaged and there is more activity off shore. Also, there is the necessity of making tie-ins as new wells are brought to production. Tie-ins must also be made on the ocean floor. The repairs and tie-ins must be high-quality weld joints to prohibit possibility of leaks or oil spills. The majority of this type of work is now being done in depths of 200–600 ft. (61–182 m).

Dry Welding.

The new welding developments for welding *in-the-dry* or with a dry environment make it possible to produce high-quality weld joints that meet X-ray and code requirements. In addition, welding can be accomplished much quicker, which provides for major savings.

Three welding processes are currently in use for welding in-the-dry. These are the shielded metal arc welding process, the gas tungsten arc welding process, and gas metal arc welding. Research is going forward to utilize the plasma process for in-the-dry underwater welding. The shielded metal arc stick welding process is not too popular for welding in-the-dry environment, because of the large amount of smoke and fumes produced by the electrode as it is consumed. When covered electrodes are used, an extensive air moving, filtering,

and refrigeration system must be employed. This is necessary since the dry environment area will quickly fill with the smoke fumes, making it impossible for the welder to function properly. The gas tungsten arc welding process is being used to produce weld joints in pipe that meet the quality requirements of API standard 1104. It is being used at depths of up to 200 ft. (61 m) for joining pipe and the resulting welds meet the X-ray requirements and weld metal requirements of the API code. The gas tungsten arc welding process is a relatively slow welding process, but it is acceptable since the welding operation is only a part of the total repair operation.

In view of the success with the gas tungsten arc welding process, efforts are going forward to more fully develop plasma arc welding for similar operations. It has not been used to a very great degree, however, at this time. The gas metal arc welding process seems to offer the greatest potential for underwater welding in-the-dry. Gas metal arc welding is relatively fast, it is an all-position process and can be adopted for welding the metals involved in underwater work. Successful application of gas metal arc welding has been made to depths as great as 180 ft. (51 m). Development work is continuing to overcome problem areas and to increase the depth of operating.

There are two basic types of in-the-dry underwater welding. One involves a welding chamber or habitat and is known as hyperbaric welding. The habitat or large welding chamber provides the welder-diver with all the necessary equipment for welding and associated operation in a dry environment. The weld chamber or habitat is made so that it can be sealed around the part to be welded. Since the majority of this type of work is on pipe, arrangements are made to seal the habitat to the pipe. The bottom of the chamber is exposed to open water and is covered by a grating. The pressure of the atmosphere inside the chamber is equal to the water pressure at the operating depth. Figure 20-26 shows a typical welding habitat. The deep-diving chambers are welded to conform to a special ASME PVHO code.

Life-support equipment involves two-way telephone communications apparatus, TV video camera for continuous observation, life support atmosphere for the welder-diver, which may be different from the atmosphere in the habitat, electric power for tools and for welding, along with the gas supply for the welding atmosphere. Habitats usually include an atmosphere-conditioning system used to both cool and filter the atmosphere in the chamber. Filtering is especially important since vaporized metal is released during welding. Air conditioning is also employed since considerable heat is generated by the welding operation and the temperature of the atmosphere in the chamber would gradually increase to an intolerable level. For gas metal arc welding the welding power source is normally on the

FIGURE 20-26 *Underwater pressurized habitat for pipe welding.*

surface with the other support equipment. Welding cables are lowered to the habitat to provide power for the arc. The electrode wire and wire feeder and control unit are located in the habitat. The wire feeder must be protected from the high pressure and high humidity conditions in the chamber. Two or more wire feeders may be employed, depending on the size of the pipe being welded. The electrode wire must also be protected from the humidity.

The shielding gas system for welding is designed for use at the high-atmosphere pressures which are involved. To appreciate the pressures involved consider that the underwater pressure increases one atmosphere or 14.7 lbs/in^2 (1 kg/cm^2) for each 33 ft (10 m) of depths. The water pressure must be equalized by the pressure of the atmosphere within the habitat. This high pressure creates problems for welding as the depth increases. The problems associated with shielded metal arc welding involve the removal and filtering of the atmosphere in the habitat. With gas tungsten arc welding very little smoke and fumes are generated but the inert gas used for welding disrupts the atmosphere. The habitat atmosphere must be regulated to maintain the proper gas mixture balance. Helium is normally used for arc shielding, since helium is also used in the atmosphere of the habitat. For most saturated diving work the breathing atmosphere of the welder-diver is approximately 90–95%

helium and 5–10% oxygen. This balance must be maintained.

Welding with the gas metal arc in a habitat presents specific problems. In shallow depths the welding in the habitat is essentially the same as welding on the surface. CO$_2$ gas is used for shielding. When welding at depths of 125 ft (35 m), the pressure is 4 to 5 atmospheres (approximately 50 psi gauge). The weld metal quality is essentially the same as surface made welds; however, the welding voltage increases and weld bead penetration increases. As the depth is increased and the atmospheric pressure increases the arc becomes more constricted. At depths beyond 125 ft (35 m) the weld puddle becomes increasingly difficult to control and an increased amount of smoke is generated which affects visibility. Smaller electrode wires are recommended. At greater depths argon is added to the carbon dioxide for shielding. Best results seem to be obtained with 95% argon and 5% carbon dioxide gas mixture. Increasing pressure causes the gas metal arc weld to become more constricted and this leads to increased arc voltage and increased penetration and higher burnoff rates, which makes the weld puddle increasingly difficult to handle.

Wet-Dry Welding.

Recent improvements have been made to provide more flexibility to the gas metal arc welding process when welding underwater. The dry hyperbaric chambers or habitats are extremely expensive and must be built for specific applications. For example, for repairing or making tie-ins on horizontally laid pipe, or for repairing an almost verticle leg of a drilling platform. In view of this, efforts have been made to take the gas metal arc welding process outside of the habitat and give it the flexibility of in-the-wet welding. This has resulted in the development of special nozzles or domes or shrouds which surround the torch used for GMAW welding. This is similar to the bubble technique but with specialized nozzles or domes or miniature chambers. In using this type of apparatus the welder-diver is in the wet or in the water but the nozzle of the welding gun and material to be welded is in the dry atmosphere. The gun and nozzle are in this small chamber, but the wire feed apparatus and the wire supply must be in a watertight pressurized enclosure. The pressure of the shielding gas coming through the system is greater than the pressure of the water at the operating level. The gas flows through the wire feeder enclosure through the electrode wire conduit to the torch, where it provides the shielding for welding.

It also provides the pressure to evacuate the small chamber or shroud to provide the dry atmosphere. These types of shrouds or localized dry gas environment chambers are relatively inexpensive, small and light-weight. They can be provided with flexible seals to be used against the part being welded. They can be hand held or made with clamps for quick attachment to the part to be welded. The wire feeder is connected to the small welding chamber by means of the normal flexible wire conduit and cable assembly. The gun can be hand manipulated inside the small chamber, the same way as on the surface. The larger boxes are usually shaped to fit the joint structure being welded and the bottom is normally open. The welder can then insert the gun into the chamber and evacuate the water from it by the use of the shielding gas. The pressure of the atmosphere in the chamber will equal that of the water surrounding it. The chambers are made of transparent material or have sufficient number of windows so that the welder can see inside to properly manipulate and direct the welding gun. This technique can be utilized for welding with gas metal arc up to 125 ft. (35 m) below the surface of the water. Quality welds can be produced that meet code requirements. It is expected that this type of technique will be utilized for underwater repair work whenever possible.

Special safety precautions must be followed when doing underwater welding. These include all precautions normally employed by divers plus those required for welding. Welder-divers must be aware of the possibilities of entrapped gases in parts being welded or cut. These gases are usually rich in hydrogen and oxygen and may explode when ignited. Only experienced, well-trained personnel should do underwater welding.

20-5 NARROW GAP WELDING

Narrow gap welding is a term applied to the method of making welds in a square groove weld joint in thick materials and utilizing a very small or narrow root opening. The root opening or gap between the parts being welded is only 1/4 in. (6.4 mm) to 3/8 in. (9.5 mm) wide, and welds can be made with this opening in material up to 8 in. (200 mm) thick. The gas metal arc welding process is normally used for making the welds. Welding can be done in all positions. The power source is a constant voltage type with a constant speed wire feeder. The head, including nozzles, is designed specifically for the narrow gap method.

Figure 20-27 is a cross-sectional view of a narrow

gap weld made on thick material. One of the major reasons for interest in narrow gap welding is the reduction in the amount of weld metal required to make the joint. This can be visualized by drawing the cross section of

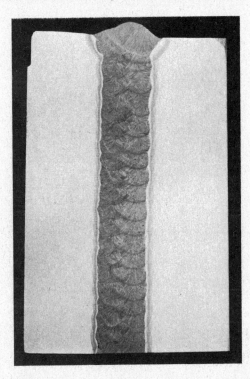

FIGURE 20-27 *Cross sectional view of narrow gap weld.*

joints normally used for welding thick sections and comparing the cross-sectional areas of the different welds. The competing processes would be the submerged arc welding process, which would utilize the double vee joint preparation, also the gas metal arc or flux-cored arc welding process that would use a similar joint preparation. The electroslag welding process which utilizes a square groove design but with a larger root opening, also competes. The economic advantage is based on using less weld metal to produce the joint. The welding time would be greatly reduced, thus the labor cost would be reduced. Potentially the narrow gap welding method can join material of heavy thickness more economically than other welding processes. The other major advantage relates to the low heat input used with this method. The heat-affected zone is much lower and the total heat input is also lower. The weld metal properties are extremely good, largely for this reason. It is possible to weld high-alloy steels with this method and with the lower heat input. [6]

The method also reduces the distortion that would occur with vee-groove welds. Finally, it is a fully automatic method and can be used in all positions. It can be used for welding carbon steels, high-strength quenched and tempered steels, aluminum, and titanium, and possibly other metals. The economic advantage of the method is obtained when welding a thickness of approximately 2 inches, and increases above that thickness.

Narrow gap welding cannot be considered a welding process since it is a variation of gas metal arc welding. It utilizes the spray transfer range of operating parameters. The weld is a multipass weld made by using contact tubes which are inserted into the joint. Normally, two electrode wires and contact tubes are used simultaneously. One is directed toward one side wall and the other toward the opposite side wall. The contact tubes are mounted on a carriage with a fixed distance between them and the respective side walls. This assures proper side wall fusion, regardless of variations in the joint opening. The narrow gap method can be used with one electrode wire. The shielding gas is introduced into the joint by means of a special nozzle, which extends into the joint. The welding is done from one side of the joint. The electrode diameter is either 0.035 in. (0.9 mm) or 0.045 in. (1.1 mm). With an 0.035-in. diameter electrode the welding current would be 220–240 amperes direct current electrode positive at 25–26 volts. The travel speed would be from 40–45 in. per minute and the shielding gas would be 75% argon plus 25% CO_2. This results in a heat input of from 7,500-10,500 joules per inch per pass for each electrode. The tip-to-work distance is maintained at 1/2 in. A backing strip is used for making the first pass, subsequent passes are deposited on previous layers. Approximately ten passes are required for each inch of thickness of the parts being joined.

Each electrode requires its own constant voltage DC power supply and wire feed system. The piece of equipment that is different is the welding head which carries the contact tubes into the joint. This head must be accurately arranged to travel in the joint and maintain the proper distance from both side walls. The contact tubes are arranged so that they can be retracted as the weld is made. This is necessary to maintain the 1/2 inch between the end of the contact tip and the arc. The major factor in this method is the necessity to direct the electrode wire to the proper joint side wall. This is normally done by introducing a cast into the electrode wire immediately before it goes to the contact tube. This insures that as the wire leaves the contact tube, it will divert to the side to which it is directed. The electrode wires are directed to opposite sides. In this way, side wall penetration is maintained and lack of penetration is avoided. The contact tubes are normally water cooled. They are insulated to minimize the possibility of shorting out against the base metal, and are coated to avoid the adherence of spatter to the tube. The head-carrying contact tube tips must be accurately located and directed to keep the contact tubes in the proper relationship to the edges of the joint. The control system is special-

ized in order to provide automatic sequence circuits to start and stop the electrode wires at the same location in the joint.

The major problem involved with the narrow gap welding method is the preparation of the weld joint so that the root opening or gap is uniform. This is relatively easy on test plates but becomes progressively more difficult as the structure being welded becomes larger. Some of the economic advantage may be lost because of the necessity of providing extremely close tolerances of fitup for the weld. For certain types of work this is not a problem, but for the average fabrication shop it may be insurmountable.

There is another system very similar to the narrow gap welding that uses a root opening or spacing approximately double in size. [7] This method utilizes a root opening of from 1/2-5/8 in. and can be used for welding material from 2-1/2 in. (62 mm) to 6-1/2 in. (162 mm) thick. The economics is similar, even though the amount of metal required is greater since the deposition rate can be increased. There are several major differences between the two methods, even though both employ the gas metal arc welding process. In the second method a larger single electrode wire is employed and the welding current is considerably higher. The electrode used is 1/8 in. (3.2 mm) diameter. The welding current ranges from 400–450 amperes, and the voltage ranges from 30–37 volts. The travel speed is in the order of 15 in. per min. The other difference is that direct current electrode negative (straight polarity) is used while the first method uses electrode positive (reverse polarity). Shielding gas is varied; however, the 33% helium, 33% argon, 33% CO_2 mixture has proven successful. The other difference is the fact that the contact tube does not extend inside the joint. This allows for considerable stickout which creates heating of the electrode wire. The phenomenon that makes this process practical is the fact that the walls of the joint seem to help direct the electrode wire down the center of the joint. Metal transfer is the globular type, but the spread of the wire is sufficient to make a pass as wide as the groove. In this method an 18° drag travel angle of the electrode wire is used.

This method utilizes a constant voltage power source and matching wire feeder but with a special carriage to obtain the electrode wire relationship with the bottom of the joint and the side walls. Heat input is greater with this method than with narrow gap welding and it shares the problem of requiring precise joint dimensions for successful application.

20-6 GRAVITY WELDING

Gravity welding, which utilizes heavy-coated electrodes, was first described in 1938 by K. K. Madsen of Denmark. [8] Gravity welding is considered an automatic method of applying the shielded metal arc welding process. It utilizes a low-cost mechanism that includes an electrode holder attached to a bracket which slides down an inclined bar arranged along the line of the weld. Heavy-coated electrodes are maintained in contact with the workpiece by the weight of the electrode holder bracket and the electrode. Once the process is started, it continues automatically until the electrode is consumed. When the electrode has burned to a short stub the bracket and electrode holder are automatically kicked up to break the arc.

This welding method received considerable publicity in the early 1960s, based on work done in Japanese shipyards. Equipment for gravity welding was made during the 1940s in the Scandinavian countries, in England, in Japan, and in the U.S.A. An inclined arm automatic stick feeder was available in the late 1940s in the United States. A *contact-er* electrode holder was developed in the U.S. to utilize contact type electrodes in approximately 1950. Credit should be given, however, to the Japanese shipbuilders for perfecting and utilizing the gravity welding system on a large scale. In recent years the gravity system is being used in shipyards throughout the world. It is also used in railroad car shops, barge yards, and other applications in which a large amount of horizontal fillet welds are to be made within a relatively small area.

The gravity welding system provides welding economies over manual welding since the operator can use a number of the gravity feeders simultaneously. This increases the productivity of the welders, reduces welder fatigue, operator training is minimized, and there is a substantial savings in welding labor cost. Initially, electrodes of standard length 18 in. (550 mm) were used; however, the Japanese designed extra-long electrodes to make the process more productive. The most common length is now 28 in. (800 mm). This reduces the stub end loss ratio over that of normal length electrodes.

Special electrode feeders are required to utilize gravity welding. There are two types of feeders in common use: one is the tripod or gravity feeder, which has an electrode holder mounted on a bracket or carriage that slides down an incline. It is used to produce horizontal fillet welds. The second type is a spring-loaded holder, used for making flat-position groove welds, as well as horizontal fillet welds. The tripod type feeder is shown by Figure 20-28 and the spring-loaded holder by Figure 20-29.

FIGURE 20-28 *Gravity feeder.*

FIGURE 20-29 *Spring loaded feeder.*

The tripod type feeder seems to be the most popular since it provides a more uniform weld throughout its entire length. On the other hand, the spring-loaded feeder has the advantage of being usable in less accessible spaces.

The power source for gravity welding is a conventional constant current welding machine normally used for manual shielded metal arc welding. The current rating and duty cycle of the machine must be considered, since gravity welding can attain a 90% duty cycle. Currents of up to 400 amperes can be used based on the electrode type and size. The conventional 60% duty cycle welding power source must be derated to allow for the 90% duty cycle. Both alternating current and direct current power sources are used.

Heavy-coated welding electrodes must be used for gravity welding. The E6027 type and the E7024 type are most often used; however, the E7028 type can also be used. The most common size electrodes used are the 7/32 in. (5.6 mm) and the 1/4 in. (6.4 mm) diameters. The most commonly used length is 28 in. (800 mm); however, other lengths can be used. Shorter electrodes do not provide the economies and longer electrodes are somewhat flexible and not as efficient as the 28 in. length. The size and length of fillet welds produced by these electrodes can be varied with the gravity feeders based on changing the angle of the inclined track. It is possible to obtain welds from 20 to 40 in. (500–1000 mm) long with the 28 in. electrode, depending on the adjustment of the feeder. Horizontal fillet welds with a leg length of from 7/32 in. (5.6 mm) to 3/8 in. (9.5 mm) can be obtained. The properties of the weld metal deposited when using the gravity feed system are the same as when electrodes are used manually. Welding procedures are slightly different for the different types and makes of gravity feeders.

There are two variations of the tripod gravity feeder. One is preset and allows for no adjustment of the incline of the track, whereas the other type allows the tripod legs to be adjusted. The preset feeder will produce welds in lengths that vary from approximately 27–31 in. (675–775 mm) long, using the 28 in. (800 mm) electrode. The table shown by Figure 20-30 shows the welding procedure when using the preset type gravity feeder. With two different sizes of electrodes available two different sizes of fillet welds can be obtained. Normally to fulfill the requirements of a particular weldment design, a specific fillet size is specified. The electrode type and size must be selected to fulfill this requirement. The welding current will determine the specific weld fillet size and weld length. Two different levels of current can be

FIGURE 20-30 *Weld procedures for gravity welding.*

Feeder	Fillet Size inch	Electrode	Electrode Size inch	Weld Length inch	Weld Current
	9/32	E6026	7/32	27	260 AC
	11/32	E6027	1/4	27	290 AC
Preset	1/4	E6027	7/32	31	260 AC
Gravity	5/16	E6027	1/4	31	290 AC
Type					
(Fixed	9/32	E7024	7/32	27	290 AC
Legs)	11/32	E7024	1/4	27	350 AC
	1/4	E7024	7/32	31	290 AC
	5/16	E7024	1/4	31	350 AC
	1/4	E6027	7/32	27	240 AC
Spring	5/16	E6027	1/4	27	270 AC
Loaded					
Type	1/4	E7024	7/32	27	270 AC
	5/16	E7024	1/4	27	300 AC

employed with the two different types of electrodes. The higher current will produce the larger fillet size. Some of the nonadjustable type feeders have a base plate variation, which slightly changes the track angle and will increase the length of the resulting weld. It is important to relate the exact type of feeder to this table for consistent results.

The second variation of the tripod gravity feeder provides for leg length adjustments. This in turn provides for a different angle of the inclined track. With this type of feeder a variety of fillet sizes can be produced with the same electrode size operating at the same welding current. The manufacturer's instruction book provides information for leg length adjustments. The fillet weld size is determined by the electrode diameter and by the weld length setting used with the adjustable feeder. The weld length may be varied depending on angle adjustments which are established between the axis of the weld, the centerline of the electrode, and the inclined track of the gravity feeder. The actual range of weld sizes obtainable will depend on the specific feeder being used. It is recommended that the instruction manual for the feeder be consulted in setting up these angles. It may be necessary to do some experimental work to establish the proper relationships to produce the weld size and length with the electrodes to be used at the current level recommended.

For production work of the same size of fillet the preset or nonadjustable tripod feeder is preferred.

The spring-loaded feeder produces welds 24–27 in. (600–675 mm) long using a 28 in. (800 mm) long electrode. A normal stub length of 2 in. (50 mm) results. The weld size is established by the electrode diameter. Figure 20-30 is a table that shows the fillet weld size and is related to the electrode size and welding current.

When using the spring-loaded feeders you will find a small size variation of the fillet from the beginning of the weld to the end of the weld. There is also a change in penetration from start to end. This is due to the change in the approach angle that occurs during the melting of the electrode. There is also a variation in weld appearance and spatter levels, depending on whether the weld is progressing toward or away from the work lead connection. It may be necessary to refer to the feeder instruction manual when making a particular setup. It may also require some experimental work to establish exact size and length of weld for the electrodes being used.

When direct current is used the electrode is negative (straight polarity). When alternating current is used the

change in bead smoothness and spatter level is less dependent on welding direction related to the work lead connection.

The economic advantage of using gravity feeders depends on one welding operator utilizing two or more feeders simultaneously. It is the labor cost that is largely involved in using more than one feeder. When using long electrodes in gravity feeders, the welding current is less than normally used for manual welding with standard length electrodes. Because of this, it is necessary to use at least two gravity feeders to obtain an economic advantage. Additional cost reduction is possible by using three or four automatic feeders. The most economical operation is to have a sufficient number of feeders in a small area so that the operator can move from one to another and reload the holder and re-establish the arc of one feeder while all of the other feeders are welding. In this way, maximum utilization is obtainable and all but one of the feeders are actually welding simultaneously. Figure 20-31 shows a comparison of the deposition rate based on pounds per hour when using one electrode manually versus two, then three or four or five gravity feeders.

Method of Applying To Make 5/16" Fillet	Deposition Rate (E6027)
Manual-one arc	10 lb/hr
Gravity-two arcs	17 lb/hr
Gravity-three arcs	26 lb/hr
Gravity-four arcs	34 lb/hr
Gravity-five arcs	43 lb/hr

FIGURE 20-31 *Comparison of deposition rates.*

Gravity welding is an excellent method of applying shielded metal arc welding when a sufficient number of horizontal fillets are to be made in a concentrated area. The feeders must be properly set up and adjusted to obtain the fillet weld size desired. The work must be laid out and a logical progression established so that the welding operator can move from feeder to feeder to minimize the nonarcing time of any one feeder. When this is possible the economic advantage is considerable and for this reason gravity welding has a very important place in production welding.

20-7 REMOTE WELDING

Remote welding and fully automatic welding have much in common. In both cases, the welding is done without the immediate presence of a human welding operator. In the case of automatic welding, the operator may be only a few feet away from the welding operation, but the welder could just as well be many feet away. This is because monitoring and adjustments are not required during the operation. In the previous chapter it was brought out how automated welding is done in mass production industries without monitoring by the operator. In many cases, the welding operation is performed behind curtains so that the operator cannot even see the operation or is not affected by the arc.

Remote welding is much similar in that the welding operator is not at the welding location and may be a great distance from it. The difference, however, is that automated welding is designed normally for making the same identical weld time after time. Remote welding usually involves maintenance operations where each weld may be different from the previous one. Where the same weld is performed time after time remote welding becomes similar to automated welding.

Remote welding is becoming more widely used. Remote welding is performed where humans cannot be present because a hostile atmosphere, such as a high level of radioactivity, exists. Maintenance on radioactive units must therefore involve remote work including welding. Special protective equipment, necessary for the welder, seriously hampers visibility and flexibility. The radiation level greatly reduces the time that the welder can work in such locations. Therefore, remote welding operations must be performed.

The nuclear industry uses remote welding for the sealing of radioactive materials into metal containers. Remote sealing of some fuel elements and target rods is also performed. This is done by specialized techniques and newly developed equipment. These involve ingenuity, mechanical gadgetry, and sound knowledge of joining processes and problems that are encountered in the growing nuclear industry. Figure 20-32 shows an operator monitoring a welding operation inside a heavily shielded hot cell. Figure 20-33 shows the welding operation. Note the remote movement and adjustment device.

Remote welding is performed in some radio chemical processing plants where highly corrosive solutions are handled. Remote welding is also done around nuclear reactors where service conditions require the highest weld quality obtainable.

One of the maintenance problems of a nuclear power plant is the repair of heat exchangers. Occasionally, one tube of a heat exchanger becomes defective and allows leakage. The solution is to weld plugs into the faulty tube at the tube sheet at both ends of the heat exchanger. This is accomplished as a remote welding operation using a version of the automatic gas tungsten arc welding machine normally used for making the original tube-to-tube sheet welds. A plug with a flange is inserted into each end of the tube and the periphery of the flange is welded to the tube sheet. The flange on

FIGURE 20-32 *Remote welding.*

FIGURE 20-33 *Remote welding.*

the plug provides metal for the weld and also covers the leak path that may occur between the tube and tube sheet. An automatic gas tungsten arc welding head was modified to reduce its size for use inside the heat exchanger. The spindle was modified and a hole was made in the plug, to mount the spindle of the head. The spindle guides the welding head around the periphery of the flange of the plug. When the plug is seated into the end of the tube the machine is started and performs its normal cycle of operations. This procedure has been successfully used in many applications.

Pipe weld joints are also made remotely. The automatic gas tungsten arc pipe welding heads can be remotely operated. In these cases the pipe joints must be properly aligned and prepared. The pipe welding head is then assembled to the joint. The welders can only work in the radioactive atmosphere for a short time period. Therefore the setup must be minimized to reduce exposure of the welders. Schedules that are normally employed for pipe or tube welding of the wall thickness and size being joined are employed. The power source and operator pendant are remote from the welding operation and shielded from the area of radioactivity. Remote welds are made in pipe and tubing, as they would be made with the equipment under normal conditions. The weld joints are then X-rayed or inspected by ultrasonic inspection and the required quality is achieved.

Welding is used to seal radioactive materials in metal containers. Welding is preferred as the joining method, particularly when the radioactive source is to be used at high temperatures or in corrosive liquid mediums. Automated equipment, which is remotely operated, is used for this operation. Manipulators place the container in a chuck for positive location. The chuck rotates the container under the welding arc. Programmed welds are used to insure the high quality required for these applications. In some cases welding is monitored remotely by means of television cameras.

The more difficult jobs are the one-of-a-kind type, in which specialized equipment may not be available. The ingenuity of the personnel and the use of gadgetry and special welding equipment is a must. More and more of this type of work will be required as the necessity for repair welding in radioactive hot areas increases.

Remote cutting using the thermal processes is also performed. The plasma method seems most appropriate for automation and preparation in the radioactive atmospheres. Another requirement is the necessity to cut under water. This is required, for certain maintenance operations, when the area is flooded to improve the shielding from radioactivity. In this case, the plasma process would be utilized.

20-8 SHEET METAL WELDING

Sheet metal is metal having a thickness of 1/8 in. (3.2 mm) or less. This usually means that it has a gauge number of 11 and higher, the higher numbers indicating thinner thicknesses of metal. The thickness and gauge number relationship are shown by Table 20-2. Welding can be performed on the thinnest metal produced but extra special fixturing and automatic travel are required. Under normal conditions using a manual or semiautomatic process sheet metal of approximately 0.035 inch (0.9 mm) in thickness or roughly 20 gauge can be welded. Of the gauges shown by Table 20-2, the Steel Sheets Manufacture Standard is the most commonly used in the U.S.A.

Thin sheets of stainless steels, aluminum alloys and nickel alloys are also welded. The processes most commonly used are shielded metal arc welding for the thicker sections, gas metal arc welding and oxy-fuel gas welding for the thin sections, and gas tungsten arc welding and plasma arc welding for the thinnest metals. The medium and thin sections are soldered and brazed. When brazing sheet metal, the torch method can be used. Brazing can also be done with the twin carbon arc and the single carbon arc method using the bronze type filler rod.

There are two major problems involved with the welding of sheet metal. These are (1) minimizing the distortion, and (2) avoiding burn-through when using the arc processes.

The problems of burn-through and distortion can both be minimized by the use of tight fitup, clamping, and aligning fixtures and backup bars. These are all recommended for production welding; however, for maintenance and repair welding the accurate fitup, clamping, and the use of backup bars may not be possible.

Of all the welding processes mentioned above, the gas metal arc welding process, utilizing the short-circuiting arc transfer and using a small-diameter electrode wire, is the most suitable. This process has replaced shielded metal arc welding for almost all sheet metal applications. It is used by the furniture industry, metal office equipment industry, appliance industry, and the automotive body industry for manufacturing. The main reason for the preference of this process is its ability to operate at a wide range of current levels. Its relative high-speed travel, which balances the heat loss-buildup problem, greatly reduces weld distortion. Brazing is used when color match is not important. It is also used for galvanized steel applications and for joining nonferrous metals such as the copper alloys. The single carbon arc method and the gas torch method are both popular.

The problem of burn-through occurs with many of

TABLE 20-2 *Sheet metal gages.*

Gauge Number	Aluminum and Brass Brown and Sharp (inch)	Steel Sheets Mfrs. Std.* (inch)	Strip and Tubing and Copper Birmingham or Stubs (inch)	Nearest Metric Thickness***	Stress Wire Gauge**
6/0's	0.5800				0.4615
5/0's	0.5165		0.500		0.4305
4/0's	0.4600		0.454		0.3938
4/0's	0.4096		0.425		0.3625
3/0's	0.3648		0.380	10.0	0.3310
2/0's	0.3249		0.340	9.0	0.3065
1	0.2893		0.300	8.0	0.2830
2	0.2576		0.284	7.0	0.2625
3	0.2294	0.2391	0.259	6.0	0.2437
4	0.2043	0.2242	0.238	5.5	0.2253
5	0.1819	0.2092	0.220	5.0	0.2070
6	0.1620	0.1943	0.203	4.8	0.1920
7	0.1443	0.1793	0.180	4.5	0.1770
8	0.1285	0.1644	0.165	4.2	0.1620
9	0.1144	0.1495	0.148	3.8	0.1483
10	0.1019	0.1345	0.134	3.5	0.1350
11	0.0907	0.1196	0.120	3.0	0.1205
12	0.0808	0.1046	0.109	2.8	0.1055
13	0.0720	0.0897	0.095	2.2	0.0915
14	0.0641	0.0747	0.083	2.0	0.0800
15	0.0571	0.0673	0.072	1.8	0.0720
16	0.0508	0.0598	0.065	1.6	0.0625
17	0.0453	0.0538	0.058	1.4	0.0540
18	0.0403	0.0478	0.049	1.2	0.0475
19	0.0359	0.0418	0.042	1.1	0.0410
20	0.0320	0.0359	0.035	1.0	0.0348
21	0.0285	0.0329	0.032	0.090	0.0317
22	0.0253	0.0299	0.028	0.080	0.0286
23	0.0226	0.0269	0.025	0.070	0.0258
24	0.0201	0.0239	0.022	0.060	0.0230
25	0.0179	0.0209	0.020	0.055	0.0204
26	0.0159	0.0179	0.018	0.045	0.0181
27	0.0142	0.0164	0.016	0.040	0.0173
28	0.0126	0.0149	0.014	0.035	0.0162
29	0.0113	0.0135	0.013		0.0150
30	0.0100	0.0120	0.012		0.0140
31	0.0089	0.0105	0.010		0.0132
32	0.0080	0.0097	0.009		0.0128
33	0.0071	0.0090	0.008		0.0118
34	0.0063	0.0082	0.007		0.0104
35	0.0056	0.0075	0.005		0.0095
36	0.0050	0.0067	0.004		0.0090
37	0.0045	0.0064			0.0085
38	0.0040	0.0060			0.0080

*Replaces U.S. standard (revised) gauge.
**Replaces Washburn and Moen gauge.
***ANSI B32.3.

SPECIAL APPLICATIONS OF WELDING

the processes, and steps to avoid burn-through include the use of close tolerance cutting to provide tight, even fitup between parts. The preparation of sheet metal for welding is normally by shearing, which produces straight edges that can easily be aligned properly. A back up bar of copper and the ample use of clamps to hold the sheet metal in alignment against the backing bar will aid in making quality weld joints. The travel speed should be as high as possible; this is a matter of welder skill and practice. The welder should try to travel at a uniform high speed but must be able to follow the joint accurately.

Fitup and distortion are closely related since distortion ahead of the arc will cause the fitup to vary and this in turn creates the likelihood for burn-through. A large number of small tack welds should be used. They should be relatively short but closely spaced. This will help maintain close or tight fitup and will reduce weld distortion. It is also helpful to use the push-travel angle when using any of the arc processes. Another assist is to make the weld in a downhill position. If the work can be tilted so that welding can be done downhill, approximately 45°, a flatter bead will result, travel speed will be higher, and this in turn will reduce distortion.

In utilizing the gas metal arc welding process the argon CO_2 mixture (75% argon and 25% CO_2) helps improve the welding operation since it tends to reduce penetration into the joint, reduces the spatter level, and produces a smoother weld bead. The fine wire gas metal arc

Welding Process	Sheet Metal Gage (Note 1)			Filler Rod or Electrode Dia.		Volt	Amps DC	Shielding Gas & Flow	Travel Speed
	Gage	in.	mm	in.	mm				
PAW	25	0.020	0.5	—	—	18	12	20	21
	20	0.030	0.8	—	—	18	34	20	17
	16	0.062	1.6	—	—	20	65	20	14
	3/32	0.093	2.4	—	—	17	85	20	16
	1/8	0.125	3.2	—	—	18	100	20	16
GTAW	20	0.032	0.8	1/16	1.6	11	75-100	10	13
	18	0.040	1.0	1/16	1.6	12	90-120	10	15
	16	0.063	1.6	1/16	1.6	11	95-135	10	15
	3/32	0.094	2.4	3/32	2.4	12	135-175	10	14
	1/8	0.125	3.2	1/8	3.2	12	145-205	12	11
GMAW	24	0.025	0.6	0.030	0.8	16	30-50	20	12-20
	22	0.031	0.8	0.030	0.8	16	40-60	20	15-22
	20	0.037	0.9	0.035	0.9	17	55-85	20	35-40
	18	0.050	1.3	0.035	0.9	18	70-100	20	35-40
	16	0.063	1.6	0.035	0.9	18	80-110	20	30-35
	5/64	0.078	1.9	0.035	0.9	19	100-130	20	25-30
	1/8	0.125	3.2	0.035	0.9	20	120-160	20	15-25
SMAW	24	0.025	0.1	3/32	2.4	25	40	—	20
	22	0.031	0.8	3/32	2.4	25	40	—	21
	20	0.038	0.9	3/32	2.4	25	50	—	28
	18	0.050	1.3	3/32	2.4	27	65	—	30
	16	0.063	1.6	3/32	2.4	27	75	—	30
	5/64	0.078	1.9	3/32	2.4	28	100	—	35
	1/8	0.125	3.2	1/8	3.2	26	120	—	40
CAW See Note	24	0.025	0.1	3/16	4.8	18	25	—	8
	22	0.031	0.8	3/16	4.8	18	35	—	8
	20	0.038	0.9	3/16	4.8	18	45	—	10
	18	0.050	1.3	3/16	4.8	19	50	—	10
	16	0.062	1.6	3/16	4.8	20	50	—	12
	5/64	0.078	1.9	1/4	6.4	20	80	—	13
	1/8	0.125	3.2	1/4	6.4	19	95	—	15

Note 1—Parameters are suitable for square groove butt or for filler welds.
Note 2—Bronze cold wire (1/8 in. 3.2mm) used with carbon arc welding.

FIGURE 20-34 *Sheet metal welding schedules.*

process can be used at extremely low currents and the current level should match the thickness of the metal being welded. This information is summarized on the procedure schedule, Figure 20-34.

When utilizing the shielded metal arc welding process use the smallest size electrodes possible. This is the 3/32-in. diameter size or for heavier sheet metal the 1/8 in. diameter size. Either AC or DC can be used; however, this will dictate the electrode type that should be selected. If AC equipment is available the E-6011 or E-6013 electrodes should be used. AC is preferred over direct current for welding thinner sheet metal. If DC equipment is to be used the selection would be AWS E-6010 or AWS E-6012 type electrodes.

The E-6012 or E-6013 type is used for butt, edge, and corner joints. When using the electrode negative (straight polarity), penetration is reduced and the metal transfer is more of a spray type. A short arc length should be used and the arc length should equal one core wire diameter.

For welding T fillets, inside corner fillets or fillets for lap joints the AWS E-6012 or E-6013 electrodes are recommended. The pull travel angle is preferred and, if possible, the weld joint should be positioned at approximately 45° so that the welding can be done in the downhill direction. This will tend to flatten the bead, allow for higher travel speed, and minimize distortion.

When plasma arc welding or gas tungsten arc welding

is employed, use high current and maximum travel speed. The gas tungsten arc process employs the slowest travel speed of the arc processes and this tends to increase distortion. The plasma process can be used at a higher speed and this will reduce distortion.

When utilizing the oxy-fuel gas welding process the forehand technique or forward pushing travel angle should be used. When using the single carbon arc process and a brazing rod the arc is played on the rod and allowed to melt onto the sheet metal joint. This reduces heat into the joint, reduces distortion, and allows for rapid travel speed.

When using the gas tungsten arc, the plasma arc or the oxy-fuel gas processes the filler rod must be selected to match the requirements of the base metal. When using the gas metal arc welding process the electrode wire also must be matched to the base metal. The shielded metal arc welding process is restricted primarily to mild steel or possibly stainless steel; however, stainless steel electrodes only have the low-hydrogen type electrode coating and this is a limiting factor.

Procedure information using all of these processes is given in a schedule shown by Figure 20-34.

QUESTIONS

1. What code applies to most high-pressure pipe welding? Cross-country pipe?

2. Can more than one welding process be used to make a pipe weld? Explain.

3. Why are different pipe weld joint designs used? Where is each used?

4. Explain the difference between internal and external lineup clamps. What determines the type to be used?

5. Explain the difference between uphill and downhill pipe welding. Will both techniques qualify for cross-country pipe welding? For pressure pipe welding?

6. What are the three joint types for tube-to-tube sheet welds?

7. Why is a timer used for arc spot welding?

8. Is it necessary to wear a welding helmet when arc spot welding?

9. Can arc spot welding be used to join aluminum?

10. What is meant by one-side welding?

11. Describe two different backing methods.

12. Why is one-side welding important in shipbuilding?

13. Explain the difference between underwater *wet welding* and *dry welding*.

14. Can pipe welds be made under water to meet code requirements?

15. What design of weld joint is used for narrow gap welding?

16. What welding processes are used for narrow gap welding?

17. What industry is a major user of gravity welding? Why?

18. How many gravity feeders must be used to make the method economical?

19. What is the major reason for making welds remotely?

20. What process is recommended for steel sheet metal welding?

REFERENCES

1. HOOBASAR RAMPAUL, "Pipe Welding Procedures," Industrial Press, Inc., 1973, New York.

2. T.W. SHEARER, "Arc Spot Welding," Bulletin No. 105, Welding Research Council, May 1965, New York.

3. R.A. STOEHR AND F.R. COLLINS, "Gas Metal Arc Spot Welding Joins Aluminum to Other Metals," *The Welding Journal,* April 1963, Miami, Florida.

4. J.T. BISKUP, "One Side Welding Is Strong In Japan," *Canadian Machinery and Metal Working,* May 1970, Toronto, Ontario, Canada.

5. "Underwater Cutting and Welding," Navy Ships 0929-000-08010, U.S. Navy Supervisor of Welding, Naval Ship Systems Command, Washington, D.C. 20360.

6. BUTLER, MEISTER, AND RANDALL, "Narrow Gap Welding, A Process for All Positions," *The Welding Journal,* February 1969.

7. JACKSON AND SARGENT, "Straight Polarity Groove Welding of Fixed Plate," *The Welding Journal,* November 1967.

8. P.T. HOULDCROFT, "Letter to Editor," *Metal Construction and British Welding Journal,* p. 101, March 1974, London, England.

APPENDIX

APPENDIX

A-1 GLOSSARY OF WELDING TERMS

A-2 ORGANIZATIONS INVOLVED WITH WELDING

A-3 CONVERSION INFORMATION

A-4 WEIGHTS AND MEASURES

A-1 GLOSSARY OF WELDING TERMS

Air*: Slang for oxygen, should not use.

All-Weld-Metal-Test-Specimen: A test specimen with the reduced section composed wholly of weld metal.

Alternating current or AC*: Electricity which reverses its direction periodically. For cycle current, the current goes in one direction and then in the other direction 60 times in the same second, so that the current changes its direction 120 times in one second.

Ammeter*: An instrument for measuring either direct or alternating electric current (depending on its construction). Its scale is usually graduated in amperes, milliamperes.

Anode*: The positive terminal of an electrical source.

Arc Blow: The deflection of an electric arc from its normal path because of magnetic forces.

Arc Length*: The distance from the end of the electrode to the point where the arc makes contact with work surface.

Arc Spot Weld: A spot weld made by an arc welding process.

Arc Voltage: The voltage across the welding arc.

Arc Wander*: Wander or drifting of arc in various directions. (*See* Arc Blow).

As-Welded: The condition of weld metal, welded joints, and weldments after welding but prior to any subsequent thermal, mechanical, or chemical treatments.

* Asterisks indicate term is not effective AWS one.

Autogenous Weld: A fusion weld made without the addition of filler metal.

Backfire: The noise associated with a momentary recession of the flame into the welding tip or cutting tip followed by immediate reappearance or complete extinction of the flame.

Backing: A material (base metal, weld metal, carbon, or granulated flux) placed at the root of weld joint for the purpose of supporting molten weld metal.

Back-Step Welding: A welding technique wherein the increments of welding are deposited opposite the direction of progression.

Base Metal: The metal (material) to be welded, brazed, soldered, or cut. *Also see* Substrate.

Bell Hole Welding*: A pipeline term whereby the pipe sections are welded together to the end of the transmission line in position.

Bevel: An angular type of edge preparation.

Blacksmith Welding: *See* preferred term, Forge Welding.

Bottle*: *See* preferred term, Cylinder.

Boxing: The continuation of a fillet weld around a corner of a member as an extension of the principal weld.

Braze: A weld produced by heating an assembly to suitable temperatures and by using a filler metal, having liquidus above 450°C (842°F) and below the solidus of the base materials.

The filler metal is distributed between the closely fitted surfaces of the joint by capillary attraction.

Buckling*: Distortion of sheet metal due to the forces of expansion and contraction caused by the application of heat.

Burner: *See* preferred term, Oxygen Cutter.

Butt Joint: A joint between two members aligned approximately in the same place.

Butt Weld: An erroneous term for a weld in a butt joint. (*See* Butt Joint.)

Button Weld*: *See* Arc Spot Weld.

Cap Pass*: A pipe line term that refers to the final or reinforcing pass of the weld joint.

Carbon Steel*: Carbon steel is a term applied to a broad range of material containing: Carbon 1.7% max., Manganese 1.65% max., Silicon 0.60% max. Carbon Steel is subdivided as follows:

Low-Carbon Steels	0.15% C max.
Mild-Carbon Steels	0.15—0.29% C.
Medium-Carbon Steels	0.30—0.59% C.
High-Carbon Steels	0.60—1.70% C.

Cast Iron*: A wide variety of iron base materials containing Carbon 1.7 to 4.5%; Silicon 0.5 to 3%; Phos. 0.8% max.; Sulphur 0.2% max.; molybdenum, nickel, chromium, and copper can be added to produce alloyed cast irons.

Chamfer: *See* preferred term, Bevel.

Covered Electrode*: A filler-metal electrode, used in arc welding, consisting of a metal core wire with a relatively thick covering which provides protection for the molten metal from the atmosphere, improves the properties of the weld metal, and stabilizes the arc.

Crater: In arc welding, a depression at the termination of a weld bead or in the molten weld pool.

Cup: *See* preferred term, Nozzle.

Cylinder: A portable container used for transportation and storage of a compressed gas.

Depth of Fusion: The distance that fusion extends into the base metal or previous pass from the surface melted during welding.

Direct Current or DC*: Electric current which flows only in one direction. It is measured by an ammeter.

Double Ending*: A pipeline term meaning welding two lengths of pipe together, usually roll welding in the flat position.

Downhand: *See* preferred term Flat Position.

Downhill Welding*: A pipe welding term indicating that the weld progresses from top of the pipe to the bottom of the pipe. The pipe is not rotated.

Electrode: Tungsten Electrode: A non-filler metal electrode used in arc welding or cutting, made principally of tungsten.

Elongation*: Extension produced between two gauge marks during a tensile test. Expressed as a percentage of the original gauge length.

FabCO Welding*: Tradename; *See* Flux-Cored Arc Welding.

Face of Weld: The exposed surface of a weld on the side from which welding was done.

Filler Bead*: A pipeline term referring to the passes laid over the hot pass but not the next to last or final pass.

Fillet Weld: A weld of approximately triangular cross section joining two surfaces approximately at right angles to each other in a lap joint, T-joint, or corner joint.

Firing Line Welder*: A pipeline term—a welder in the hot pass crew.

Flat Position: The welding position used to weld from the upper side of the joint; the face of the weld is approximately horizontal.

Flux: Material used to prevent, dissolve, or facilitate removal of oxides and other undesirable surface substances.

Flux-Cored Arc Welding (FCAW): An arc welding process which produces coalescence of metals by heating them with an arc between a continuous filler metal (consumable) electrode and the work. Shielding is provided by a flux contained within the tubular electrode. Additional shielding may or may not be obtained from an externally supplied gas or gas mixture. (*See* flux-cored electrode.)

Flux Cored Electrode: A composite filler metal electrode consisting of a metal tube or other hollow configuration containing ingredients to provide such functions as shielding atmosphere, deoxidation, arc stabilization and slag formation. Alloying materials may be included in the core. External shielding may or may not be used.

Forge Welding: A solid state welding process which produces coalescence of metals by heating them in air in a forge and by applying pressure or blows sufficient to cause permanent deformation at the interface.

Friction Welding: A solid state welding process which produces coalescence of materials by the heat obtained from a mechanically induced sliding motion between rubbing surfaces. The work parts are held together under pressure.

Furnace Brazing: A brazing process in which the parts to be joined are placed in a furnace heated to a suitable temperature.

Gas Metal Arc Welding (GMAW): An arc welding process which produces coalescence of metals by heating them with an arc between a continuous filler metal (consumable) electrode and the work. Shielding is obtained entirely from an externally supplied gas or gas mixture. Some methods of this process are called MIG or CO_2 welding.

Gas Pocket*:   Pipeline term for porosity.

Gas Shielded Metal Arc Welding: A general term used to describe gas metal arc welding, gas tungsten arc welding, and flux-cored arc welding when gas shielding is employed.

Gas Tungsten Arc Welding (GTAW):   An arc welding process which produces coalescence of metals by heating them with an arc between a tungsten (nonconsumable) electrode and the work. Shielding is obtained from a gas or gas mixture. Pressure may or may not be used and filler metal may or may not be used. (This process has sometimes been called TIG welding.)

Groove Weld: A weld made in the groove between two members to be joined. The standard types of groove welds are as follows: double-bevel-groove weld, double-flare-bevel-groove weld, double-flare-V-groove weld, double-J-groove weld, double-U-groove weld, double-V-groove weld, single-bevel-groove weld, single-flare-bevel-groove weld, single-flare-V-groove weld, single-J-groove weld, single-U-groove weld, single-V-groove weld, and square-groove weld.

Ground Connection: An electrical connection of the welding machine frame to the earth for safety. (*See also* Work Connection and Work Lead.)

Ground Lead: *See* preferred term, Work Lead.

Heat-Affected Zone: That portion of the base metal which has not been melted, but whose mechanical properties or microstructure have been altered by the heat of welding, brazing, soldering, or cutting.

Heliarc*:   Tradename; *See* Gas Tungsten Arc Welding.

Hollow Bead*:   Pipeline term for porosity in the root reinforcement.

Horizontal Position: Fillet weld; the position in which welding is performed on the upper side of an approximately horizontal surface and against an approximately vertical surface. Groove weld; the position of welding in which the axis of the weld lies in an approximately horizontal plane and the face of the weld lies in an approximately vertical plane.

Hot Pass*:   A pipeline term that refers to the second pass or the pass over the stringer bead. The hot pass is usually made at high currents to practically remelt the entire stringer bead.

Impact Resistance*:   Energy absorbed during breakage by impact of specially prepared notched specimen, the result being commonly expressed in foot-pounds.

Induction Brazing: A brazing process in which the heat required is obtained from the resistance of the work to induced electric current.

Inertia Welding*:   *See* Friction Welding.

Joint Penetration: The minimum depth a groove or flange weld extends from its face into a joint, exclusive of reinforcement. Joint penetration may include root penetration.

Joint Welding Procedure: The materials, detailed methods and practices employed in the welding of a particular joint.

Kerf: The width of the cut produced during a cutting process.

Lap Joint: A joint between two overlapping members.

Lead Burning: An erroneous term used to denote the welding of lead.

Leg of a Fillet Weld: The distance from the root of the joint to the toe of the fillet weld.

Low-Alloy Steel*:   Low-alloy steels are those containing low percentages of alloying elements.

Melting Rate: The weight or length of electrode melted in a unit of time.

Micro-Wire Welding*:   Tradename; *See* Gas Metal Arc Welding.

MIG Welding*:   *See* preferred terms, Gas Metal Arc Welding, Flux-Cored Arc Welding.

Molten Weld Pool: The liquid state of a weld prior to solidification as weld metal.

Nose*:   *See* Root Face.

Nozzle: A device which directs shielding media.

Off-Center Coating*:   When flux coating on a covered electrode is thicker on one side than the opposite side.

Open-Circuit Voltage: The voltage between the output terminals of the welding machine when no current is flowing in the welding circuit.

Overfill*:   Excessive reinforcement.

Overhead Position: The position in which welding is performed from the underside of the joint.

Overlap: The protrusion of weld metal beyond the toe, face, or root of the weld.

Oxygen Cutter: One who performs a manual oxygen cutting operation.

Parent Metal: *See* preferred term, Base Metal.

Pass: A single progression of a welding or surfacing operation along a joint, weld deposit, or substrate. The result of a pass is a weld bead, layer, or spray deposit.

Peening: The mechanical working of metals using impact blows.

Penetration: *See* preferred terms, Joint Penetration and Root Penetration.

Performance Qualification*: Methods, tests, and acceptable standards used to qualify a welding procedure.

Porosity: Cavity discontinuities formed by gas entrapment during solidification.

Postheating: *See* Postweld Heat Treatment.

Postweld Heat Treatment: Any heat treatment subsequent to welding.

PSI*: Pounds per square inch.

Preheating: The application of heat to the base metal immediately before welding, brazing, soldering, or cutting.

Procedure Qualification: The demonstration that welds made by a specific procedure can meet prescribed standards.

Procedure Specification*: A very complete and formal welding procedure written in accordance with code.

Puddle: *See* preferred term, Molten Weld Pool.

Quarter Weld*: Pipe welding making the joint in four segments rotating the pipe 90° between each segment.

QWP*: Qualified welding procedure.

Radiography*: The use of radiant energy in the form of X-rays or gamma rays for the nondestructive examination of metals.

Reduction of Area*: The difference between the original cross-sectional area and that of the smallest area at the point of rupture; usually stated as a percentage of the original area.

Returning (Boxing)*: The practice of continuing a weld around a corner as an extension of the principal weld.

Reverse Polarity: The arrangement of direct current arc welding leads with the work as the negative pole and the electrode as the positive pole of the welding arc. A synonym for direct current electrode positive.

Rheostat*: A variable resistor which has one fixed terminal and a movable contact (often erroneously referred to as a "two-terminal potentiometer"). Potentiometers may be used as rheostats, but a rheostat cannot be used as a potentiometer, because connections cannot be made to both ends of the resistance element.

Root Face: That portion of the groove face adjacent to the root of the joint.

Root Gap: *See* preferred term, Root Opening.

Root of Weld: The points, as shown in cross section, at which the back of the weld intersects the base metal surfaces.

Root Opening: The separation between the members to be joined at the root of the joint.

Root Penetration: The depth that a weld extends into the root of a joint measured on the centerline of the root cross section.

Seal weld: Any weld designed primarily to provide a specific degree of tightness against leakage.

Shielded Metal Arc Welding (SMAW): An arc welding process which produces coalescence of metals by heating them with an arc between a covered metal electrode and the work. Shielding is obtained from decomposition of the electrode covering. Pressure is not used and filler metal is obtained from the electrode.

Shoulder: *See* preferred term, Root Face.

Silver Soldering: Nonpreferred term used to denote brazing with a silver-base filler metal. *See* preferred terms, Furnace Brazing, Induction Brazing, Torch Brazing.

Size of Weld: Groove weld—the joint penetration (depth of bevel plus the root penetration when specified). The size of a groove weld and its effective throat are one and the same. Fillet weld—for equal leg fillet welds, the leg lengths of the largest isosceles right triangle which can be inscribed within the fillet weld cross section. For unequal leg fillet welds, the leg lengths of the largest right triangle which can be inscribed within the fillet weld cross section. When one member makes an angle with the other member greater than 105°, the leg length (size) is of less significance than the effective throat which is the controlling factor for the strength of a weld.

Slag Inclusion: Nonmetallic solid material entrapped in weld metal or between weld metal and base metal.

Slugging: The act of adding a separate piece or pieces of material in a joint before or during welding that results in a welded joint, not complying with design, drawing, or specification requirements.

Spatter: In arc and gas welding, the metal particles expelled during welding and which do not form a part of the weld.

Squirt Welding*: Semiautomatic and submerged arc welding.

Stick Welding*: Welding using shielded metal arc welding.

Stinger*: Electrode holder.

Stove Pipe Welding*: A pipeline term whereby each length of pipe is joined to the transmission line in a progressive fashion with each joint made in position.

Straight polarity: The arrangement of direct current arc welding leads in which the work is the positive pole and the electrode is the negative pole of the welding arc. An alternate term is direct current electrode negative.

Stress Relief Heat Treatment: Uniform heating of a structure or a portion thereof to a sufficient temperature to relieve the major portion of the residual stresses, followed by uniform cooling.

Stringer Bead: A type of weld bead made without appreciable weaving motion. *See also* Weave Bead.

Stripper*: A pipeline term referring to the pass that brings the weld groove flush with the surface of the pipe.

Substrate: Any base material to which a thermal sprayed coating or surfacing weld is applied.

Tack Weld: A weld made to hold parts of a weldment in proper alignment until the final welds are made.

Tensile Strength*: The maximum load per unit of original cross-sectional area obtained before rupture of a tensile specimen. Measured in pounds per square inch.

Throat of a Fillet Weld: Theoretical throat; the distance from the beginning of the root of the joint perpendicular to the hypotenuse of the largest right triangle that can be inscribed within the fillet weld cross section. Actual throat; the shortest distance from the root of a fillet weld to its face. Effective throat; the minimum distance minus any reinforcement from the root of a weld to its face.

TIG Welding*: *See* Gas Tungsten Arc Welding.

Toe of Weld: The junction between the face of a weld and the base metal.

Torch Brazing: A brazing process in which the heat required is furnished by a fuel gas flame.

Tungsten Electrode: *See* Electrode: Tungsten Electrode.

Ultimate Tensile Strength*: The maximum tensile stress which will cause a material to break (usually expressed in pounds per square inch).

Underbead Crack: A crack in the heat-affected zone generally not extending to the surface of the base metal.

Undercut: A groove melted into the base metal adjacent to the toe or root of a weld and left unfilled by weld metal.

Underfill: A depression on the face of the weld or root surface extending below the surface of the adjacent base metal.

Uphill Welding*: A pipe welding term indicating that the welds are made from the bottom of the pipe to the top of the pipe. The pipe is not rotated.

VAE*: Visually Acceptable External—inspection of a weld.

Vertical Position: The position of welding in which the axis of the weld is approximately vertical.

Weave Bead: A type of weld bead made with transverse oscillation.

Weaving*: A technique of depositing weld metal in which the electrode is oscillated.

Weld: A localized coalescence of metals or nonmetals produced either by heating the materials to suitable temperatures, with or without the application of pressure, or by the application of pressure alone, and with or without the use of filler material.

Welder: One who performs a manual or semiautomatic welding operation. (Sometimes erroneously used to denote a welding machine.)

Weld Metal: That portion of a weld which has been melted during welding.

Welding Ground: *See* preferred term, Work Connection.

Welding Procedure: The detailed methods and practices including all joint welding procedures involved in the production of a weldment. (*See* Joint Welding Procedure.)

Welding Rod: A form of filler metal used for welding or brazing which does not conduct the electrical current.

Weldment: An assembly whose component parts are joined by welding.

Weldor: *See* preferred term, Welder.

Whipping*: A term applied to an inward and upward movement of the electrode which is employed in vertical welding to avoid undercut.

Wire Welding*: Gas Metal Arc Welding.

Work Connection: The connection of the work lead to the work.

Work Lead: The electric conductor between the source of arc welding current and the work.

A-2 ORGANIZATIONS INVOLVED WITH WELDING

AIA: Aerospace Industries Association of America, 1725 De Sales Street, N.W., Washington, D.C. 20036. A national industry association of companies in U.S. engaged in the research, development, and manufacture of aerospace systems, missiles and astronautical vehicles, their propulsion or control units or associated equipment.

AA: Aluminum Association, 750 Third Avenue, New York, New York 10017. An industry association of producers of aluminum. The association aim is to increase understanding of the aluminum industry, to provide technical, statistical, and marketing information.

AASHT: American Association of State Highway Transportation, Suite 341, National Press Building, Washington, D.C. 20045. An association of state transportation officials. The AASHT issues various standards and specifications.

ABS: American Bureau of Shipping, 45 Broad Street, New York, New York 10004. A nonprofit classification society.

* Indicates slang terms.

Classification is a service for shipowners to establish that the ship has been built to recognized standards. ABS provides rules for building ships and issues approvals for welding filler metals.

AFS: American Foundrymen's Society, Golf and Wolf Roads, Des Plains, Illinois 60016. A technical society devoted to the advancement of manufacture and utilization of castings through research, education, and dissemination of technology.

AISC: American Institute of Steel Construction, 1221 Avenue of the Americas, New York, New York 10020. An industry association of fabricated structural steel producers. AISC provides design information and standards pertaining to structural steel.

AISI: American Iron and Steel Institute, 1000 Sixteenth Street, N.W., Washington, D.C. 20036. An industry association of the iron and steel producers. It provides statistics on steel production and use and publishes the steel products manuals.

ANSI: American National Standards Institute, 1430 Broadway, New York, New York 10018. ANSI formerly the United States of America Standards Institute (USASI) formerly the American Standard Association (ASA). ANSI is the United States representative to ISO. It is a nonprofit corporation that publishes National Standards in cooperation with technical and engineering societies, trade associations, and government agencies.

API: American Petroleum Institute, 1801 K Street, Washington, D.C. 20006. An association of the petroleum industry. It publishes various standards involved with welding including cross-country pipeline welding, storage tanks, line pipe, etc.

AREA: American Railway Engineering Association, 59 East Van Buren Street, Chicago, Illinois 60605. An engineering society which, through committees, develops standards applicable to railroads. Several of these involve welding.

ASME: American Society of Mechanical Engineers, 345 East 47th Street, New York, New York 10017. An engineering society which, among other things, publishes the boiler and pressure vessel code. Section IX is the welding qualification section of this code.

ASM: American Society for Metals, Metals Park, Ohio 44073. A technical society that seeks to advance the knowledge of metals and materials, their engineering, design, processing, and fabricating, through research education and dissemination of information.

ASNT: American Society for Nondestructive Testing, Inc., 3200 Riverside Drive, Columbus, Ohio 43221. The purpose of this engineering society is scientific and educational, directed toward the advancement of theory and practice of nondestructive test methods for improved product quality and reliability.

ASQC: American Society for Quality Control, Inc., 161 West Wisconsin Avenue, Milwaukee, Wisconsin 53203. An engineering society that seeks to create, promote, and stimulate interest in the advancement and diffusion of knowledge of the science of control and its application to the quality of industrial products.

ASTM: American Society of Testing & Materials, 1916 Race Street, Philadelphia, Pennsylvania 19103. A scientific and technical organization for standards, materials, products, systems. It is the world's largest source of voluntary consensus standards.

AWWA: American Water Works Association, 6666 West Quincy Avenue, Denver, Colorado 80235. An industry association of water companies and companies serving the water supply industry. It publishes numerous standards, several in cooperation with AWS.

AWS: American Welding Society, 2501 N.W. 7th Street, Miami, Florida 33125. A nonprofit technical society organized and founded for the purpose of advancing the art and science of welding. The AWS publishes codes and standards concerning all phases of welding and *The Welding Journal.*

AAR: Association of American Railroads, 1920 L Street, N.W., Washington, D.C. 20036. An industry association of railroads. Among other things, it publishes specifications for rolling stock, welding qualifications, etc.

AWI: Australian Welding Institute, 2nd Floor, Eagle House, 118 Alfred Street, Milsons Point, N.S.W., Australia 2061. The national welding society of Australia.

BSI: British Standards Institution, 2 Park Street, London, England. A nonprofit concern. The principal object is to coordinate the efforts of producers and users for the improvement, standardization, and simplification of engineering and industrial material.

CSA: Canadian Standards Association, 178 Rexdale Boulevard, Rexdale, Ontario, Canada M9W 1R3. A National Association of Technical Committees to provide a national standardizing body for Canada. It publishes many standards involving welding.

CWB: Canadian Welding Bureau, 254 Merton Street, Toronto, Ontario, Canada M4S 1A9. A division of the CSA, its purpose is to provide the necessary codes and standards covering all phases of welding, for the guidance of fabricators, designers, architects, consulting engineers, and governmental departments.

CWS: Canadian Welding Society, Inc., 6 Milvan Drive, Weston, Ontario, Canada M9L 1Z2. The national welding society of Canada.

CWDI: Canadian Welding Development Institute, Institute Canadian de Development de La Soudure, 254 Merton Street, Toronto, Ontario, Canada M4S 1A9. A division of the Canadian Standards Association.

CGA: Compressed Gas Association, 500 Fifth Avenue, New York, New York 10036. A nonprofit membership association and technical organization interested in both adequate data and sound utilization for gases.

CDA: Copper Development Association, 57th Floor, Chrysler Building, 405 Lexington Avenue, New York, New York 10017. A trade association of copper producers. Publishes standards of commercial copper mill products and standard designations for copper and copper alloys.

DVS: Deutscher Verband fur Schweisstechnik, Schadowstrasse 42, Postfach 2725, 4 Dusseldorf, Germany. The national welding society of West Germany.

IIW: International Institute of Welding. Contact your local welding society. An international society of national associations to promote the development of welding and assist in the international standards for welding in collaboration with the ISO.

ISO: International Organization for Standardization, Paris, France, is a worldwide federation of national standards institutes.

JIS: Japanese Industrial Standards, 1-24 Akasaka, 4-chome, Minato-ku, Tokyo 107, Japan. The Japanese Standards Association publishes standards including metals, welding filler metals, etc.

Lloyd's Register of Shipping Trust Corp. Inc., 71 Fenchurch Street, London EC3M 4BS, England. This society was established for the purpose of obtaining for the use of merchants, shipowners, and underwriters a faithful and accurate classification of mercantile shipping. The society approves design, surveys, apparatus, material, etc.

MCAA: Mechanical Contractors Association of America, 5530 Wisconsin Avenue N.W., Washington, D.C. 20015. Formerly Heating, Piping and Air-Conditioning Contractors National Association. A trade association of contractors in the piping, heating and air-conditioning business. It sponsors the National Certified Pipe Welding Bureau.

NBBPVI: National Board of Boiler and Pressure Vessel Inspectors, 1055 Crupper Drive, Columbus, Ohio 43229. An organization of chief boiler inspectors of the states and cities in the U.S. and provinces of Canada. The national board enforces the various sections of the ASME boiler code.

NCPWB: National Certified Pipe Welding Bureau, 5530 Wisconsin Avenue, Suite 750, Washington, D.C. 20015. A division of the Mechanical Contractors Association of America, Inc. Its purpose is to develop and test procedures and, through its local chapters, to establish pools of men qualified to weld under these procedures.

NEMA: National Electrical Manufacturers Association, 2101 L. Street N.W. Washington, D.C. 20037. An industry association of manufacturers of electrical machinery. Publishes standards and industry statistics including welding.

NFPA: National Fire Protection Association, 470 Atlantic Avenue, Boston, Massachusetts 02210. An organization dedicated to promote the science and improve the methods of fire protection. NFPA publishes the National Electrical Code.

The code provides safety installation information for welding machines.

NWSA: National Welding Supply Association, 1900 Arch Street, Philadelphia, Pennsylvania 19103. An industry association of welding supply distributors.

PFI: Pipe Fabrication Institute, 1326 Freeport Road, Pittsburgh, Pennsylvania 15238. An association of the pipe-fabricating industry.

PLCA: Pipe Line Contractors Association, 2800 Republic National Bank Building, Dallas, Texas 75201. An industry association of contractors that build underground pipelines, especially cross-country pipelines.

RWMA: Resistance Welder Manufacturer Association, 1900 Arch Street, Philadelphia, Pennsylvania 19103. An industry of manufacturers of resistance welding equipment. Establishes standards for welding equipment and procedure information.

SAE: Society of Automotive Engineers, Inc., 400 Commonwealth Drive, Warrendale, Pennsylvania 15086. An engineering society with the objective to promote the arts, sciences, standards, and engineering practices connected with the design, construction and utilization of self-propelled mechanisms, prime movers, components thereof, and related equipment.

SFSA: Steel Founders' Society of America, 20611 Center Ridge Road, Rocky River, Ohio 44116. The Steel Founders' Society is an association of companies engaged in the manufacture of steel castings. They publish technical bulletins, the "Steel Castings Handbook" and the "Journal of Steel Castings Research."

SPFA: Steel Plate Fabricators Association, 15 Spinning Wheel Road, Hinsdale, Illinois 60521. A nonprofit industry association of metal plate fabricators.

UL: Underwriters Laboratories, Inc., 207 East Ohio Street, Chicago, Illinois 60611. The Underwriters' Laboratories, Inc., is a non-profit organization that operates laboratories for the examination and testing of devices, systems and materials. They publish standards for safety for oxy-fuel gas torches, regulators, gauges, acetylene generators, transformer type arc welding machines and for many other items.

WI: Welding Institute, Abington Hall, Cambridge, England. A professional institute to further the exchange of technical knowledge by meetings, publications, a library and information service, and courses in its School of Welding Technology. Its research division provides research on welding.

WRC: Welding Research Council, 345 E. 47th Street, New York, New York 10017. A nonprofit association organized to provide a mechanism for conducting cooperative research work in the welding field.

A-3 CONVERSION INFORMATION

Length, or Distance and Speed Conversion

Approximate Conversion

Flow Rate—Liquid Measure

Approximate Conversion

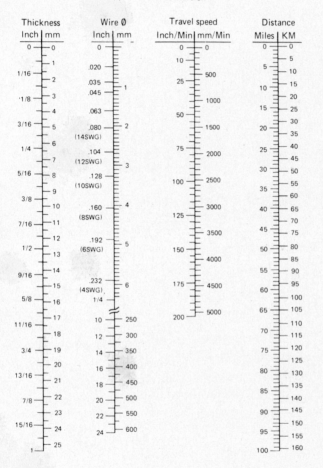

Exact Conversion

For use with electronic calculators—First enter the conversion constant number. Press the x button. Enter the known quantity of the dimension. Press the =. The desired value of the dimension will appear on the display.

25.40	☒ ___ inch	▣ ___ mm
304.8	☒ ___ feet	▣ ___ mm
.0393	☒ ___ mm	▣ ___ inch
.00328	☒ ___ mm	▣ ___ feet
.0621	☒ ___ km	▣ ___ miles
1.609	☒ ___ miles	▣ ___ Kilometers
.4233	☒ ___ in/min	▣ ___ mm/second
2.362	☒ ___ mm/s	▣ ___ in/min.

Exact Conversion

For use with electronic calculators. First enter the conversion constant number. Press the ☒ button. Enter the known quantity of the dimension. Press the ▣. The desired value of the dimension will appear on the display.

0.4719	☒ _____ cu ft/hr	▣ _____ litre/min.
2.119	☒ _____ L/min	▣ _____ cu ft/hr.
3.785	☒ _____ gal/min	▣ _____ litre/min.
0.264	☒ _____ Litre/min	▣ _____ gal/min.
645.2	☒ _____ in²	▣ _____ mm²
0.00155	☒ _____ mm²	▣ _____ in²

Weight–Pressure–Load

Approximate Conversion

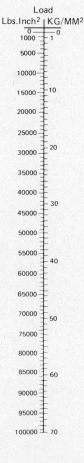

Exact Conversion

For use with electronic calculators—First enter the conversion constant number. Press the ⊠ button. Enter the known quantity of the dimension. Press the ⊟. The desired value of the dimension will appear on the display

.000703	⊡	lb/in²	⊟	kg/mm²
6,894.7	⊡	lb/in²	⊟	Kilo Pascal (KPa)
0.006895	⊠	lb/in²	⊟	Mega Pascal (MPa)
0.07030	⊠	lb/in²	⊟	kg/cm²
14.2234	⊠	kg/cm²	⊟	lbs/in²
1422.34	⊠	kg/mm²	⊟	lb/in² (PSI)
	⊠	kg/mm²	⊟	Pascal (Pa)
	⊠	kg/mm²	⊟	Kilo Pascal (KPa)
0.00145	⊠	Pa	⊟	lb/in² (PSI)
0.145	⊠	KPa	⊟	lb/in² (PSI)
	⊠	Pa	⊟	kg/mm²
	⊠	KPa	⊟	kg/mm²
4.448	⊡	lbs	⊟	Newton (N)
9.807	⊡	kg	⊟	Newton (N)
.2248	⊡	N	⊟	lbs
.1009	⊡	N	⊟	kg
0.4536	⊠	lbs	⊟	kg
2.205	⊠	kg	⊟	lbs

Temperature–Impact Values

Approximate Conversion

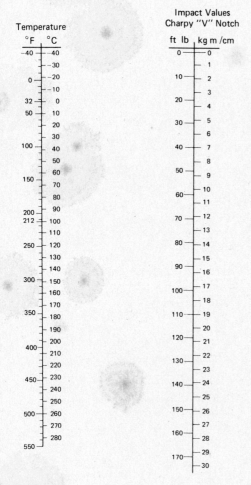

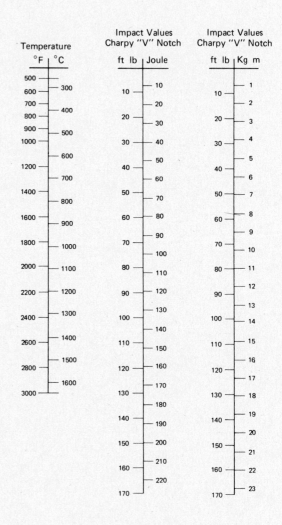

Exact Conversion

For use with electronic calculators—First enter the conversion constant number. Press the ☒ button. Enter the known quantity of the dimension. Press the ☐. The desired value of the dimension will appear on the display.

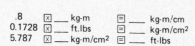

___ °F ☐ 32 ☐ ☒ .555 ☐ ___°C
___ °C ☒ 1.8 ☐ ☐ 32 ☐ ___°F

0.1383 ☒ ___ ft-lbs ☐ ___ kg-m
7.233 ☒ ___ kg-m ☐ ___ ft-lbs

1.356 ☒ ___ ft-lbs ☐ ___ Joule
.7376 ☒ ___ Joule ☐ ___ ft-lbs

.8 ☒ ___ kg-m ☐ ___ kg-m/cm
0.1728 ☒ ___ ft.lbs ☐ ___ kg-m/cm^2
5.787 ☒ ___ kg-m/cm^2 ☐ ___ ft-lbs

A-4 WEIGHTS AND MEASURES

Impact values: sometimes metric values are kg-M/cm^2 if so multiply by 0.8 (area under notch).

Foot-Pounds	Kilogram-Meters	Joules	Foot-Pounds	Kilogram-Meters	Joules	Foot-Pounds	Kilogram-Meters	Joules
1	0.14	1.36	35	4.84	47.46	69	9.54	93.56
2	0.28	2.71	36	4.98	48.82	70	9.68	94.92
3	0.42	4.07	37	5.12	50.17	71	9.82	96.28
4	0.55	5.42	38	5.25	51.53	72	9.95	97.63
5	0.69	6.78	39	5.39	52.88	73	10.09	98.99
6	0.83	8.14	40	5.53	54.24	74	10.23	100.34
7	0.97	9.49	41	5.67	55.60	75	10.37	101.70
8	1.11	10.85	42	5.81	56.95	76	10.51	103.06
9	1.24	12.20	43	5.95	58.31	77	10.65	104.41
10	1.38	13.56	44	6.08	59.66	78	10.78	105.77
11	1.52	14.92	45	6.22	61.02	79	10.92	107.12
12	1.66	16.27	46	6.36	62.38	80	11.06	108.48
13	1.80	17.63	47	6.50	63.73	81	11.20	109.84
14	1.94	18.98	48	6.64	65.09	82	11.34	111.19
15	2.07	20.34	49	6.78	66.44	83	11.48	112.55
16	2.21	21.70	50	6.91	67.80	84	11.61	113.90
17	2.35	23.05	51	7.05	69.16	85	11.75	115.26
18	2.49	24.41	52	7.19	70.51	86	11.89	116.62
19	2.63	25.76	53	7.33	71.87	87	12.03	117.97
20	2.77	27.12	54	7.47	73.22	88	12.17	119.33
21	2.90	28.48	55	7.60	74.58	89	12.31	120.68
22	3.04	29.83	56	7.74	75.94	90	12.44	122.04
23	3.18	31.19	57	7.88	77.29	91	12.58	123.40
24	3.32	32.54	58	8.02	78.65	92	12.72	124.75
25	3.46	33.90	59	8.16	80.00	93	12.86	126.11
26	3.60	35.26	60	8.30	81.36	94	13.00	127.46
27	3.73	36.61	61	8.43	82.72	95	13.13	128.82
28	3.87	37.97	62	8.57	84.07	96	13.27	130.18
29	4.01	39.32	63	8.71	85.43	97	13.41	131.53
30	4.15	40.68	64	8.85	86.78	98	13.55	132.89
31	4.29	42.04	65	8.99	88.14	99	13.69	134.24
32	4.42	43.39	66	9.13	89.50	100	13.83	135.60
33	4.56	44.75	67	9.26	90.85			
34	4.70	46.10	68	9.40	92.21			

WIRE GAGES DIAMETER

U.S. Steel Wire Gage No.	in.	mm	U.S. Steel Wire Gage No.	in.	mm	U.S. Steel Wire Gage No.	in.	mm
7/0's	0.4900	12.447	11	0.1205	3.0607	28	0.0162	0.4115
6/0's	0.4615	11.7221	12	0.1055	2.6797	29	0.0150	0.381
5/0's	0.4305	10.9347	13	0.0915	2.3241	30	0.0140	0.3556
4/0's	0.3938	10.0025	14	0.0800	2.032	31	0.0132	0.3353
3/0's	0.3625	9.2075	15	0.0720	1.8389	32	0.0128	0.3251
2/0's	0.3310	8.4074	16	0.0625	1.5875	33	0.0118	0.2997
0	0.3065	7.7851	17	0.0540	1.3716	34	0.0104	0.2642
1	0.2830	7.1882	18	0.0475	1.2065	35	0.0095	0.2413
2	0.2625	6.6675	19	0.0410	1.0414	36	0.0090	0.2286
3	0.2437	6.1899	20	0.0348	0.8839	37	0.0085	0.2159
4	0.2253	5.7226	21	0.0317	0.8052	38	0.0080	0.2032
5	0.2070	5.2578	22	0.0286	0.7264	39	0.0075	0.1905
6	0.1920	4.8768	23	0.0258	0.6553	40	0.0070	0.1778
7	0.1770	4.4958	24	0.0230	0.5842	41	0.0066	0.1678
8	0.1620	4.1148	25	0.0204	0.5182	42	0.0062	0.1575
9	0.1483	3.7668	26	0.0181	0.4597	43	0.0060	0.1524
10	0.1350	3.429	27	0.0173	0.4394	44	0.0058	0.1473

Impact values: sometimes metric values are kg-M/cm^2; if so, multiply by 0.8 (area under notch).

Metric Conversion.

Inch. Fraction	Inch. Decimal	Millimeter	Inch. Fraction	Inch. Decimal	Millimeter	Inch. Fraction	Inc. Decimal	Millimeter
1/64	0.0158	0.3969	11/32	0.3437	8.7312	43/64	0.6719	17.0656
1/32	0.0312	0.7937	23/64	0.3594	9.1281	11/16	0.6875	17.4625
3/64	0.0469	1.1906	3/8	0.375	9.525	45/64	0.7031	17.8594
1/16	0.0625	1.5875	25/64	0.3906	9.9219	23/32	0.7187	18.2562
5/64	0.0781	1.9844	13/32	0.4062	10.3187	47/64	0.7344	18.6532
3/32	0.0937	2.3812	27/64	0.4219	10.7156	3/4	0.750	19.050
7/64	0.1094	2.7781	7/16	0.4375	11.1125	49/64	0.7656	19.4469
1/8	0.125	3.175	29/64	0.4531	11.5094	25/32	0.7812	19.8433
9/64	0.1406	3.5719	15/32	0.4687	11.9062	51/64	0.7969	20.2402
5/32	0.1562	3.9687	31/64	0.4844	12.3031	13/16	0.8125	20.6375
11/64	0.1719	4.3656	1/2	0.500	12.700	53/64	0.8281	21.0344
3/16	0.1875	4.7625	33/64	0.5156	13.0968	27/32	0.8437	21.4312
13/64	0.2031	5.1594	17/32	0.5312	13.4937	55/64	0.8594	21.8281
7/32	0.2187	5.5562	35/64	0.5469	13.8906	7/8	0.875	22.2250
15/64	0.2344	5.9531	9/16	0.5625	14.2875	57/64	0.8906	22.6219
1/4	0.25	6.35	37/64	0.5781	14.6844	29/32	0.9062	23.0187
17/64	0.2656	6.7469	19/32	0.5937	15.0812	59/64	0.9219	23.4156
9/32	0.2812	7.1437	39/64	0.6094	15.4781	15/16	0.9375	23.8125
19/64	0.2969	7.5406	5/8	0.625	15.875	61/64	0.9531	24.2094
5/16	0.3125	7.9375	41/64	0.6406	16.2719	31/32	0.9687	24.6062
21/4	0.3281	8.3344	21/32	0.6562	16.6687	63/64	0.9844	25.0031

Geometric Formula.

Areas

Paralleologram = base × altitude.

Triangle = half base × altitude.

Trapezoid = half the sum of the two parallel sides × the perpendicular distance between them.

Regular polygon = half of perimeter × the perpendicular distance from the center to any one side.

Circle = square of the diameter × 0.7854.

Sector of circle = number of degrees in arc × square of radius × 0.008727.

Segment of circle = area of sector with same arc minus area of triangle formed by radii of the arc and chord of the segment.

Octagon = square of diameter of inscribed circle × 0.828.

Hexagon = square of diameter of inscribed circle × 0.866.

Sphere = area of its great circle × 4; or square of diameter × 3.1416 (π).

Volumes

Prism = area of base × altitude.

Wedge = length of edge plus twice length of base × one-sixth of the product of the height of the wedge and the breadth of its base.

Cylinder = area of base × altitude.

Cone = area of base × one-third of altitude.

Sphere = cube of diameter × 0.5236.

Miscellaneous

Diameter of circle = circumference × 0.31831.

Circumference of circle = diameter × 3.1416 (π).

INDEX

INDEX

A

Abrasion, 666
Abrasion resistance, 669
AC arc welding machine, 279
AC-DC welding machines, 286
AC or DC, choosing, 273
AC Tig welding, 96
Acetone, 361
Acetylene:
   adjusting working pressures, 209
   cylinder, 361
   fittings, 59
   formula, 361
   hose, 59
   properties, 362
   regulator, 209
   safety rules for, 59
Adhesive bonding, 243
Air acetylene brazing, 199
Air acetylene welding, 206
Air carbon arc cutting, 246
Air contamination, 55
Air filter, welding machine, 293
Air pressure test, 601
Alloy fluxes, 354
Alloy steels:
   definition, 411

Alloy steels *(cont.):*
   identifying, 381
   properties, 371
   weldability, 401
Alloying elements in steel, 406
Alternating current, 36, 271
Alternating current welding:
   arc welding machines, types, 271
   covered electrodes, 340
   high frequency stabilizion, 286
   transformer adjustment, 280
   transformer power factor, 275
   transformers, 38, 279
Aluminum and aluminum alloys, 431
American Iron and Steel Institute, 405
American Welding Society:
   address, 714
   procedure qualifications, 575
   specification for filler metal, 335
   welding symbols, 536
Amperes, 35
Annealing, 266
Aptitude for welders, 48
Arc air cutting, 246
Arc blow, 623
Arc carbon electrode, 116, 355
Arc characteristics, 38
Arc cutting, 245

Arc gouging, 248
Arc heating, 118
Arc length, 40
Arc plasma welding, 250
Arc spot welding, 685
Arc stream, 41
Arc voltage, 40
Arc welding:
  AC, 271
  accessories, 332
  booth, 56, 67
  carbon arc, 115
  DC, 271
  electrodes, 335
  flux cored, 28
  fundamentals, 21
  gas metal, 28
  gas tungsten, 27
  generator, 276
  machine maintenance, 293
  plasma, 28
  pulsed, 105
  safety, 49
  stud, 28
  submerged, 27
  tubular wire, 28
Argon, 358
Argon—CO_2 shielding, 359
Argon—helium shielding, 359
Associations and Societies, 713
Atomic hydrogen arc welding, 13, 126
Austenite, 393
Austenite manganese steels, 475
Austenite stainless steel, 421
Automated welding, 329
Automatic tungsten arc welding, 107
Automatic welding, 31, 129, 322
AWS (see American Welding Society)

B

Backfire, 62
Backhand welding, 175
Backing rings, 678
Backing strip/straps, 504, 691
Backstep welding, 635
Backup methods, 504, 691
Bare electrode welding, 12, 36
Base metal, 23
Beads, 90
Bend test, 585
Bending moment, 492
Beryllium, 455

Bevel groove weld, 501
  double, 507
  single, 507
Blacksmith welding (see Forge welding)
Blowholes, 603
Blowpipe, 209
Body-centered lattice, 389
Brass, 442
Braze welding, 30
Brazing:
  alloys, 202
  aluminum, 201
  carbon arc, 118
  cast iron, 201
  copper, 201
  filler metal, 200
  fluxes, 200
  heating methods, 197
  joint types, 204, 500
  safety, 55, 59, 201
  silver brazing alloys, 202
  silver brazing flux, 201
  stainless steel, 201
  temperatures, 197, 202
Brinell hardness test, 376
Bronze, 442
Brushes, adjusting, 293
Brushes, generator, 277
Butt joint:
  double bevel, 516
  double U, 517
  double vee, 514
  single bevel, 514
  single U, 516
  single vee, 514
  square, 513
Butt welder (see Flash welding)
''Buttering'', 483, 672

C

Cable connectors, 303
Cables, size, 303
Cables, welding, 303
Calcium carbide, 363
Carbide precipitation, 419
Carbon arc
  brazing, 118, 200
  cutting, 117, 249
  electrodes, 116
  equipment, 116
  gouging, 246
  twin arc, 118
  welding, 115
Carbon dioxide, 358, 360
Carbon electrode, 116, 355
Carbon equivalents, 402
Carbon steels:
  compositions, 407

Carbon steels *(cont.)*:
  high-strength, low alloy, 412
  low carbon, 411
    medium and high carbon, 412
    quenched and tempered, 414
Carburizing flame, 207
Cast iron:
  arc welding, 463
  brazing, 464
  cutting, 246, 252, 257
  gas welding, 464
  identifying, 384, 461
  malleable, 462
  preheating, 463
  repair, 463
  types, 461
  weldability, 462
Cast steel welding, 418
Casting conversion, 2, 531
Cementite, 395
Certification, 69
Charpy test, 376
Chemical analysis, 385
Chemistry of welding, 43
Chrome molybdenum steel, 415
Chrome nickel steels, 413
Chromium steel, 413
Circumferential welding, 143, 665
Clad steels, welding, 478
Classification of welds, 500
Cleaning:
  aluminum surfaces, 436
  joints in carbon steel, 614
  joints in cast iron, 463
  joints in stainless steel, 424
  tungsten electrodes, 101
  welded surfaces, 652
  weldment, 652
Clothing protection, 64
Coalescence, 26
Coated electrodes, 82, 338
Coated steels, 472
Codes and specifications, 564
Coefficient of expansion, 373, 626
Cold cracking, 638
Cold welding, 221
Cold working, 391, 434
Color codes, 83, 100
Compounds, 44
Compression loading, 492
Compression strength, 491
Conductors, 35, 303
Constant potential, 75
Constant voltage, 75, 131
Consumables, welding, 335
Containers, 61
Contours, weld, 608
Control of heat input, 385
Conversion factors, 716

Cooling rates, 386, 390
Cooling systems, 333
Copper and alloys:
  copper weldability, 441
  GMAW welding, 445
  GTAW welding, 445
Corner joint, 21, 509, 520
Corrosion, 644, 671
Cost, welding, 541
Cover glass (lens), 53
Covered electrodes, 82, 338
Cracking, 617, 638
Craters, welds, 617
Creep, 417
Crystalline structure, 389
Current, 35, 75, 132
Cutting:
  air carbon arc, 246
  arc-air, (*see* Air carbon arc cutting)
  arc-oxygen, 258
  automatic, 253
  carbon arc, 249
  cast iron, 246, 257
  containers, 61
  electrode sizes, 247
  flame, 252
  flux, 257
  fuel gases, 59, 361
  gas, 251
  gas tungsten arc, 249
  machine, 253
  metal arc, 249
  oxyacetylene, 252
  oxy-fuel gas, 251, 364
  oxygen, 251
  oxygen lance, 257
  oxy-propane, 363
  plasma arc, 250
  powder, 257
  safety, 59
  stainless steel, 257
  tips, 252
  torch, 252
  underwater, 695
Cylinders:
  acetylene, 365
  argon, 365
  carbon dioxide, 366
  compressed gas, 62
  fuel gas, 62
  helium, 365
  oxygen, 62
  safe handling, 62

D

DCEN, 36
DCEP, 36
DCRP (see DCEP)
DCSP (see DCEN)
Defects, 601
Definitions, 21, 709
Density, 371
Deposition efficiency, 548
Deposition rate, 85, 194, 344, 552
Design:
    joints, 499
    procedures, 527
    weldments, 490
Design factors, 490
Destructive testing, 584
Diagram, iron carbon, 391
Dip transfer arc welding, 155
Direct current welding machines, 271
Distortion:
    cause of, 626
    control, 504, 633
Double ending, 683
Double joints, 506
Drafting symbols, 535
Drooping type, welding machine, 272
Dry box welding, 456
Dry welding, 696
Drying low hydrogen electrodes, 342
Drying ovens, 343
Ductile iron, 462
Ductility, 375
Duty cycle, 274
Dye penetrant inspection, 593

E

Edge joint, 21, 500
Edge preparation, 506
Edge welds (see Edge joint)
Elastic limit, 373
Electrical conductivity, 373
Electrical leads, 303
Electrical resistance, 35
Electrical units, 35
Electrode angle, 175
Electrode consumption, 549
Electrode holders, 82, 299
Electrode position, 84, 175
Electrodes:
    alloy steel, 411
    aluminum, 437
    carbon, 116, 355

Electrodes (cont.):
    carbon arc cutting, 247, 335
    cast iron, 463
    classification, 83, 335
    composite, 128
    copper and copper alloy, 443
    covered, 82, 338
    covering analysis, 339
    flux cored, 348
    gas metal arc, 346
    holder, 82, 299
    identifying, 83
    iron powder, 340
    low hydrogen, 340
    markings, 83
    oxygen arc, 258
    resistance welding, 217
    selecting, 83
    spot welding, 217
    stainless steel, 422
    storage, 342
    submerged arc, 347
    tungsten, 100, 355
    weight, 549
Electro-gas welding, 152
Electromotive force, 35
Electron beam welding, 231
Electron gun, 231
Electroslag welding, 29, 178
Elongation, 375
Engine driven generators, 277
Equal pressure type torch, 208
Equilibrium diagram, 392
Expansion of metals, 373
Expansion, thermal, 373
Explosion welding, 222
Eye protection, 52

F

Face centered latticed, 389
Face shields, 52
Factor of safety, 490
Fatigue allowables, 494
Fatigue cracking, 641
Fatigue strength, 494
Ferrite, 393
Ferrite stainless steel, 421
Filler metal, 335
Fillet lap joint:
    double, 505
    single, 505
Fillet weld:
    application, 22
    contour, 609
    size, 505, 592
    unequal, 505
Filter glass, 53
Fire cracker welding, 92

Fire prevention, 58
Fitup, 614
Fixtures, 317
Flame characteristics, 207
Flame cutting, 251
Flame oxyacetylene, 206
Flame shrinking, 264
Flame straightening, 262
Flange joint, 510
Flash welding, 219
Flashback, 62
Flashthrough, 354
Flat position, 23
Flow meters, 370
Flux:
    aluminum, 441
    brazing, 201
    cast iron, 464
    magnetic, 128
    oxygen cutting, 257
    soldering, 228
    welding, 352
Flux cored arc welding, 160
Flux cored electrode, 163, 348
Flux, magnetic, 128
Flux recovery, 139
Flux, submerged arc, 139, 352
Forehand welding, 176
Forge welding, 11, 225
Friction welding, 223
Fuel gases, 62, 361
Fumes, 55

       G

Galvanized steel, 472
Gas arc welding:
    see Gas metal arc welding
    see Gas tungsten arc welding
Gas cost, 550
Gas flow, 369
Gas metal arc welding:
    application, 148
    automated, 329
    cabon and low alloy steels, 152
    current, 150
    cutting, 249
    equipment, 151
    gas used, 153
    metal transfer, 154
    power supply, 151
    shielding gas, 356
    stainless steel, 424
Gas mixtures, 359
Gas regulators, 209
Gas shielded flux cored welding, 160
Gas spot welding, 648
Gas tungsten arc welding:
    application, 95

Gas tungsten arc welding *(cont.):*
    automatic, 95
    current, 96
    equipment, 97
    position capabilities, 96
    power supply, 97
    shielding gas, 101
    speed, 102
    technique, 103
    torch, 98
Gas welding apparatus, 208
Gases:
    cutting, 361
    explosive, 361
    shielding, 63, 356
Gauge:
    American, wire, 719
    Brown and Sharpe, 705
    pressure, 209
    sheet metal, 705
    U.S. Standard, 705, 719
    wire, 719
Generator:
    acetylene, 363
    engine driven, 277
    welding machine, 276
Globular metal transfer, 154
Gloves, 54
Goggles, 52
Gouging:
    air carbon arc, 246
    with cutting torch, 661
Grains (metal), 390
Gravity welding, 92, 700
Gray iron, 461
Groove-fillet welds, 521
Groove weld, 22
Ground clamp (*see* Work clamp)
Grounding, 51
Guided bend test, 585
Gun:
    electron, 231
    mig (GMAW), 152
    stud, 123
    tig (GTAW), 686

       H

Hard surfacing, 666
Hardness of metals, 376
Hardenability, 394, 402
Hard-facing, 666
Hard soldering, 227

Hardness testing:
  Brinell, 376
  Rockwell, 376
  scleroscope, 376
  table, 379
  Vickers, 376
Hazards, 50
Heat treatment, 226
Heliarc (*see* Gas tungsten arc welding)
Helium, 358
Helmets, 52
Hexagonal lattice, 389
High frequency welding, 221
History:
  American Welding Society, 12
  covered electrodes, 12
  welding processes, 11
Holder, electrode, 81
Hopkins process, 15
Horizontal position, 23
Hose:
  acetylene, 59, 210
  connections, 59
  oxygen, 59, 210
  safe practices, 65
  welding, 210
Hot cracking, 638, 603
Hot shortness, 442
Hot Tapping, 61
Hot wire welding, 107
Hot work permit, 60
Hydrogen, 357
Hydrogen oxygen welding, 251
Hydrostatic testing, 601

I

Identification of iron, steel, 331
Impact:
  data conversion table, 718
  strength, 376
  testing, 380
Inclusions, 606
Inertia welding, 223
Initial current, 324
Inspection:
  liquid penetrant, 593
  magnetic particle, 595
  nondestructive, 593
  pneumatic, 601
  ultra sonic, 598
  visual, 590

Inspection *(cont.)*:
  X-ray, 597
Installation, welding equipment, 292
Intermittent welds, 506
Interpass temperatures, 265
Iron:
  cast, 461
  critical temperature, 391
  ductile, 462
  gray cast, 461
  malleable, 462
  nodular, 462
  white cast, 461
  wrought, 477
Iron-carbide alloy, 391
Iron-carbon diagram, 392
Iron-powder electrodes, 340
Izode test, 378

J

Jigs and fixtures, 317
Joint:
  backing, 504
  design, 498
  position, 497
  preparation, 506
  types, 499

K

Kilovolt-ampere (KVA), 275

L

Lamellar cracking, 643
Lancing, 258
Lap joint:
  double fillet, 505
  joint, 21, 500
  single fillet, 505
Laser welding, 234
Lattice space:
  body centered cubic, 389
  face centered cubic, 389
  hexagonal close packed, 389
Leads:
  aluminum, 304
  copper, 304
Leak detection, 601
Lenses, 52
Lighters, friction, 210
Lighting torch, 65, 210
Liquid penetrant, 593
Loads:
  fatigue, 494

Loads *(cont.):*
  impact, 494
  types of, 490
Long stickout, 174
Low alloy steels, 411
Low carbon steels, 411
Low hydrogen electrodes, 340

M

Machine welding, 31
Magnaflux *(see* Magnetic particle inspection)
Magnesium, 447
Magnetic arc blow, 623
Magnetic field, 37
Magnetic particle inspection, 595
Magnetic rotating arc welding, 127
Maintenance welding equipment, 292
Malleable iron, 462
Manipulators, welding, 313
Manual cutting, 33
Manual welding, 31
Mapp gas, 363
Maraging nickel steel, 428
Marking electrodes, 83
Martensitic stainless steel, 421
Master chart of welding processes, 25
Mechanical properties, 373
Mechanical testing, 584
Mechanized travel units, 313
Mechanized welding *(see* Machine welding *and* Automatic welding)
Melt off rates, 131
Melting point, 371
Metal arc cutting, 249
Metal fume hazard, 55
Metal inert gas arc welding *(see* Gas metal arc welding)
Metal transfer, 154
Metallizing, 259
Metallurgy, 388
Metals:
  alloy, 388
  coefficient of expansion, 373
  density, 371
  identifying, 381
  properties, 371
  source of information, 380
  spraying, 259
  structure, 370
  surfacing, 663
Meters, electrical, 35
Meters, flow, 370
Metric conversion, 716
MIG welding *(see* Gas Metal Arc Welding)
Modulus of elasticity, 375
Molecular theory, 43
Molybdenum, 455
Moment, bending, 492

Motor generator, 276
Multiple arc welding, 288
Multiple electrodes, 145, 217
Multiple pass welding, 80

N

Narrow gap welding, 698
Nick-break test, 585
Nickel alloys:
  composition, 452
  weldability, 453
Nickel steels, 413, 426
Nitrogen, 357
Nodular cast iron, 462
Nondestructive testing:
  dye penetrant, 593
  leak, 601
  magnetic particle, 595
  radiographic, 597
  ultrasonic, 598
  visual, 590
Nonferrous metal, 431
Normalizing, 266
Notch impact strength, 376
Nozzles, 99, 152, 163
Nugget, weld, 218, 687
Number, tip sizes, 209, 253
Numerical control, 255, 330

O

One-side welding, 690
Open circuit voltage, 75, 132, 273
Operator factor, 552
Operator, qualification, 69, 568
OSHA, 50
Oven drying, 343
Oven electrode, 343
Overhead position, 23, 497
Overwelding, 558, 609, 618
Oxidizing flame, 207
Oxyacetylene:
  brazing, 199
  cutting attachments, 252
  cutting cast iron, 257
  cutting pressures, 253
  cutting safety, 59
  cutting torch, 253
  cutting torch tip, 253
  equipment, 252
  hose, 210

Oxyacetylene *(cont.):*
   regulators, 209
   safety, 59
   shape cutting machine, 253
   torches, 208
   welding, 206
Oxyacetylene flame:
   adjustment, 207
   preheat adjustment, 252
   temperature, 206
Oxyacetylene welding
   aluminum, 441
   cast iron, 464
   pipe, 677
Oxy-fuel gas welding, 30
Oxygen
   acetylene welding, 206
   arc cutting, 258
   arc underwater cutting, 257
   cutting chemistry, 251
   cutting equipment, 252
   cutting machine, 253
   cylinder, 62, 210, 365
   cylinder pressure, 365
   fuel gas underwater cutting, 364
   lance, 257
   lance cutting, 257
   manifolds, 210
Oxy-hydrogen welding, 206
Oxy-propane cutting, 251, 363

P

Painting over welds, 652
Paralleling welding machines, 51
Pearlite, 393
Peening, 635
Penetrameters, 597
Penetration, 606
Percussion welding, 220
Physical properties of metals, 371
Pipe beveling machine, 680
Pipe clamps, 680
Pipe cutting, 680
Pipe joints, 678
Pipe size, 679
Pipe welding, 675
Pipe welding procedures, 676
Pipeline welding, 683
Placement of arrows, 536
Plasma arc cutting, 250
Plasma arc spraying, 261
Plasma arc welding:
   keyhole welding, 108
   position capabilities, 109

Plasma arc welding *(cont.):*
   torch, 111
   transferred arc, 108
Plastic welding, 268
Plate preparation, 506
Plug welding, 501, 509
Polarity:
   reverse DCRP *(see* DCEP)
   straight DCSP *(see* DCEN)
Porosity, 603
Positioner, 318
Positioning work, 319
Positions of fillet welds, 499
Positions of groove weld, 499
Positions, welding, 497
Post heating, 266
Postflow time, 325
Postweld heat treating, 266
Powder cutting, 257
Power sources:
   classification, 271
   selecting, 291
Prebending, 634
Precautions and safe practices, 64
Precipitation hardening, 429
Preheating, 264
Pressure regulators, 209
Pressure time, 213
Pressure vessels, 564
Procedure qualification, 568
Procedures, welding:
   basic guidelines, 33
   see specific metal
   see specific process
Processes, welding:
   atomic hydrogen, 126
   carbon arc, 11, 26, 115
   electrogas, 152
   electroslag, 15, 29, 178
   gas metal arc, 14, 148
   gas shielded flux cored, 14, 28, 160
   gas tungsten arc, 13, 27, 95
   history of development, 11, 25
   laser, 16
   master chart of, 25
   plasma arc, 27, 107
   selection, 192
   self-shielded flux cored, 160, 1528
   shielded metal arc, 12, 26, 79
   submerged arc, 13, 27, 134
Programmed welding, 324
Projection welding, 218
Prooftesting, 601
Properties of fuel gases, 362
Properties of metals, 372, 374
Properties of shielding gases, 357
Proportions for joints, 507
Protection:
   ear, 63
   eye, 52

Protective:
  clothing, 54
  shield, 53
Pulse current, high, low, 105
Pulse time, high, low, 105
Pulsed arc transfer, 154
Pulsed GMA welding, 154
Pulsed GTA welding, 105

Q

Qualification codes, 564
Qualification test for welders, 69
Qualifying personnel, 69
Quality, weld, 561
Quenched and tempered steels, 414

R

Radiographic inspection, 597
Rebuilding worn surfaces, 663
Rectifier, 283
Redesign, 527
Regulators, 209
Reinforcing bars, welding, 469
Remote welding, 702
Repair shop, 6
Repair welding, 665
Repairing welding machines, 294
Residual stress, 636
Resistance welding, 213
Rings, backing, 678
Robot, welding, 330
Rockwell hardness test, 376
Rods, welding, 23, 211, 335
RWMA, 215, 715

S

SAE Numbers, 406
Safety:
  arc welding, 50, 52
  containers, 61
  cutting, 61
  cylinders, 62
  fire, 58, 60
  hot work permit, 60
  lighting torch, 65
  OSHA, 50
  oxyacetylene, 62
  particulate matter, 55
  precautions, 64
  rules, 64
  testing leaks, 64
Safety equipment:
  clothing, 54
  filter lens, 52

Safety equipment *(cont.):*
  gloves, 54
  goggles, 53
  helmet, 52
  leather jacket, 54
  ventilation, 56
Safety factor, 490
Schaeffler diagram, 425
School for welding, 67
Scleroscope, 376
Seam welding, 219
Self-shielded flux cored, 160, 351
Semiaustenitic steels, 426, 429
Semiautomatic guns, 152, 300
Semiautomatic welding, 31
Shades of lenses, 53
Shape cutting, 253
Shear loading, 493
Shear strength, 493
Sheet metal welding, 704
Shield, head, 52
Shielded metal arc welding, 79
Shielding gas, 356
Ship construction, 567
Shore scleroscope hardness, 376
Short arc welding (*see* Short circuiting transfer)
Short circuiting transfer, 155
Shrinkage, 626
Silver brazing, 201
Silver soldering, 227
Single joints, 506
Sizes of welds, 505, 592
Slag removal, 139
Slope control, 78, 133
Smoke extracting torch, 57
Solder:
  flux, 228
  lead tin alloys, 229
  melting temperatures, 229
  silver, 227
  tinning, 227
Solder composition, 229
Soldering:
  alloys, 229
  aluminum, 230
  dip bath, 228
  flux, 228
  gun, 229
  hard, 227
  heating, 228
  iron, 229
  principals, 227
  procedure, 227
  silver, 227
  resistance, 229

Soldering *(cont.)*:
soft, 227
torch, 230
wave, 230
Solenoid valve, 286
Solid state welding, 31
Solution heat treating, 434
Space lattice, 389
Spark test, 384
Spatter, 610
Specific heat of metals, 373
Specifications:
in design, 524
for qualification, 564
Specimens, test, 584
Speed, travel, 551
Spot welding:
electrode, 217
KVA rating, 215
portable, 215
resistance, 213
Spray arc transfer, 154
Spraying metal (*see* Thermal spraying)
Stabilized stainless steel, 419
Stack cutting, 257
Stainless steel:
austenitic, 421
composition, 420
ferritic, 421
martensitic, 421
Standards and codes, 524, 564
Steel, classifying, 405
Steel production, 4
Steel specification, 405
Steels:
AISI-SAE designation system, 381
alloys, 413
high carbon, 412
identifying, 381
low alloy, 412
low carbon, 412
maraging, 428
numbering system, 406
properties, 372
reinforcing bar, 469
stainless, 419
tool steels, 465
Stick electrode welding, 79
Stickout, electrical, 174
Strength, 373
Stress:
allowable, 490
analysis, 491
beams, 492
concentration, 494

Stress *(cont.)*:
distribution, 491
raisers, 494
relief, 266
residual, 636
Stress strain curve, 375
Strongbacks, 634
Stud welding:
equipment, 121
gun, 123
process, 119
sizes, 122
Submerged arc welding:
angles, 144
equipment, 138
flux, 139, 352
process, 134
surfacing, 146
Surface condition, 614
Surface defects, 601, 617
Surfacing, 666
Symbols:
nondestructive testing, 619
welding, 535

T

Tack welds, 617
Tandem arc welding, 145
Tanks, 525, 567, 634
Tee joint, 21, 23, 500
Temperature:
measurement, 266
scales, 718
Temperature conversion table, 718
Temperatures, preheat and interpass, 264
Tempering, 266
Tensile strength, 491
Tensile testing, 373
Tension loading, 491
Terms and definitions, arc welding, 709
Tests:
air pressure, 601
bend, 588
Brinell, 376
Charpy, 376
destructive, 584
fracture, 586
hardness, 376
impact, 376
Izode, 378
magnetic partical, 595
nick break, 585
nondestructive, 593
pneumatic, 601
qualification, procedure, 568
qualification, welder, 69
Rockwell, 376
scleroscope, 376

Tests *(cont.):*
    tensile, 373
    torch, 383
    welder qualification, 69
    X-ray, 597
Thermal conductivity, 373
Thermal spraying:
    electric arc spraying, 259
    flame spraying, 260
    plasma spraying, 261
Thermit reaction chemistry, 237
Thermit welding, 237
Tig welding (*see* Gas tungsten
    arc welding)
Time, welding, 324
Timers, 324
Tinning, 227
Tips, torch:
    cutting, 253
    welding, 209
Titanium, 455
T-Joint (*see* Tee joint)
Tool and die welding, 465
Tool steel welding, 465
Torch:
    adjustments, 207
    angles, 175
    backfire, 62
    brazing, 199
    cutting, 253
    gas tungsten arc, 98
    injector type, 208
    lighter and economizer, 210
    lighters, 210
    lighting, 65
    medium pressure, 208
    position, 175
    soldering, 230
    tests for metal, 383
    tip drill size, 209, 253
    tips, 252
    types of, 210
Torsion, 493
Total cost per foot, 554
Toughness, 376
Training, 48
Tracers, shape cutting, 255
Transformer, 38, 279
Transition curve, 379
Travel speed, 551
Trouble shooting, 294
Tube welding, 682
Tubular welding (*see* Pipe welding)
Tungsten, 455
Tungsten electrode:
    color code, 100
    current carrying capacity, 101
    holder, 98
    nozzle sizes, 99
    point, 101

Tungsten electrode *(cont.):*
    sizes, 100
    torch, 98
Tungsten inert-gas welding (*see* Gas tungsten arc welding)
Twin-electrode welding, 117, 145
Two-stage regulators, 209

U

Ultrasonic inspection, 598
Ultrasonic welding, 226
Underbead cracking, 638
Undercut, 608
Underwater cutting, 257
Underwater welding, 694
Upset welding, 220
Upslope time, 325

V

Vacuum furnace, 198
Vee joint:
    double, 506
    single, 507
Ventilation, smoke exhaust system, 57
Vertical down welding, 89
Vertical up welding, 89
Vertical welding, 23
Visibility, arc, 52
Voltage, 35
Voltage drop, 305
Volt ampere curve, 75

W

Warpage, 626
Washing, 652
Water cooled torch, 300
Weathering steel, 414
Weight of filler metal, 549
Weights of steel weld metal, 543
Weld contours, 608
Weld reinforcement, 608, 618
Weld types, 22
Weldability, 401
Welded design:
    advantages, 487
    conversion, 531
    formulas, 491, 511
    influence of specifications, 523
    joint types, 500
    problems, 533

weld joint details, 511
welds, 500
Welder qualifications:
procedure, 34
records, 69
testing specimens, 584
tests, 588
Welder, rectifier, 283
Welders, 31
Welding:
booths, 56
cables, 304
definition of, 21, 709
electrodes, 82, 338
flames, 207
fluxes (*see* type)
gases (*see* type)
goggle, shade numbers, 52
goggles, 52
gun (*see* type)
habitat, 696
hose, 59
lead, 304
multiple passes, 80
pipe, 675
plastic, 268
positions, 497
procedures, 33
processes, 25
rod, 23, 211, 335
safety, 49
symbols, 535
torch, 210
torch tip sizes, 209, 252

Welding arc, 38
Welding circuit, 35
Welding current:
choice of, 76
constant current, 75
for different processes, 76
Welding defects, 601
Welding heads, 307
Welding machines, 271
Welding operators, 31
Welding processes, 25, 30
Welding rods, 23, 211, 335
Weldment, 21
Wet welding, 694
White iron, 461
Wire gauges, 719
Wire data, 305, 719
Wire feeder, 307
Wire weight, 549
Work clamp, 303
Workmanship, 611
Wrought iron, 477

X

X-Ray inspection, 597

Y

Yield point, 374
Yield strength, 375

Z

Zirconium, 455

welds, transverse welds, and, when in a fixed position with a rotating device, for circular welds.

Side Beam Carriage

The side beam carriage is less versatile than the manipulator and is less expensive. The side beam carriage performs only one function and that is to provide longitudinal travel of the welding head. A typical side beam carriage is shown by Figure 9-24. In this case, the carriage is mounted on an I beam that has been modified with bars to provide for powered travel. Side beam carriages can be of medium to high precision depending on the accuracy used in the manufacture of the beam and the accuracy and speed regulation of the travel motor system. The carriage will carry the welding head, wire supply, flux, if required, and the controls for the operator. In Figure 9-24 the flux-cored arc welding process is being used. The side beam carriage is teamed up with a work-holding device for making special trailer axles. The welding head on the carriage can be adjusted for different heights and for transverse dimensional requirements. The welding head is monitored by the welding operator and can be adjusted to follow joints that might not be in perfect alignment. Travel speed of the side beam carriage is adjustable to accommodate

different welding procedures and processes.

Welding Tractor

The welding tractor is an inexpensive way of providing arc motion. Tractors of this type were designed for mechanized flame cutting. In fact, the same tractor could be used for either. Some tractors are designed to ride on the material being welded, and others may require special tracks. The tractor should have sufficient stability to carry the wire feeder head as well as the wire supply, flux if used, and the welding controls. A typical welding tractor is shown by Figure 9-25. This tractor using the submerged arc welding process carries the welding head and controls, the electrode supply, and the flux supply. It rides on a track and has sufficient adjustment so that the head can be made to follow the weld joint. This type of equipment is extremely popular in shipyards and in plate-fabricating shops. The travel speed of the tractor must be related to the welding process, closely regulated and smooth.

FIGURE 9-25 *Welding tractor.*

All-Position Welding Carriage

The welding tractor mentioned previously is used for flat-position work only. There are many requirements, especially for gas metal arc welding and flux-cored arc welding, for all-position welding. A similar tractor but one that utilizes a track that holds the tractor in place is used. Figure 9-26 shows an all-position welding carriage that utilizes semiautomatic welding equipment and only carries the welding gun. The gun is attached to the wire feeder by means of the standard cable assembly. In this particular case an oscillator is employed to provide lateral arc motion. This type of welding carriage can be used in the vertical, horizontal, or overhead positions. Adjustments are available for aligning the torch to the joint and for maintaining this alignment. A variation of this type of carriage is used for welding the horizontal girth welds on large storage tanks. In this case, the carriage may straddle the top course of plates and make welds simultaneously on the inside and the outside of the tank shell. Other units are designed for vertical welding that may carry the wire feeder and sometimes

FIGURE 9-26 *All position welding carriage.*

the welding operator. Such devices are used for the vertical joints on storage tanks and large ships. Another variation is known as a skate welder which can be designed to follow irregular joint contours. Usually skate welder travel devices are extremely compact and may carry a miniaturized type of wire feeder or merely the torch. These types of devices are used for cylindrical tanks, aircraft assemblies, etc. The travel speed must be within the parameters of the welding procedures and closely regulated.

Standardized Machines

Standardized welding machines are available and are made with special work-holding devices plus some of the travel devices mentioned above. These can include devices for making the longitudinal welds on tanks, welding the heads in tanks, making wide flange beams, welding small-diameter spuds in plates, and for performing specialized maintenance work, such as building up track pads for track type tractors.

The most popular are tank seamers, shown by Figure 9-27. In this type of equipment a side beam carriage is employed but positioned over a special holddown fixture which can be adjusted for making various diameters and lengths of tank shells. Seamers of this type can use any of the different welding processes and can be quickly adjusted for different size shells of different thicknesses.

A companion piece to the tank seamer is a tank head welding machine. This machine will rotate the tank assembly under two automatic welding heads and will make two cylindrical welds joining the heads to the shell. These can be used for different sizes, lengths, diameters, and material thickness. They are commonly used for making LNG tanks, hot water tanks, expansion tanks, etc.

Another unit is the automatic welding machine for making wide flange beams. There are two types, one that moves along the beam and another type that pulls the three plates making the beam through the machine. In either case two welds are made simultaneously, the assembly is turned over, and the other two welds are made. These machines are becoming more popular as fabricated beams replace hot rolled beams.

Another standardized welding machine is known as the spud welder. This is a machine that rotates an automatic head around a relatively small diameter part such as the spud of a tank. The head rotates the arc around the periphery of the part and makes a fillet weld joining it to a plate. These machines are used in the tank industry and can be used for many small-diameter parts that are welded to flat or even curved plates. They are quickly adjustable for different sizes of spuds and they may clamp the parts as well as make the welds. A typical spud welder is shown by Figure 9-28.